AF334492

The Cancer Degradome

Dylan Edwards • Gunilla Høyer-Hansen
Francesco Blasi • Bonnie F. Sloane
Editors

The Cancer Degradome

Proteases and Cancer Biology

 Springer

Editors
Dylan Edwards
University of East Anglia
School of Biological Sciences
Norwich
United Kingdom NR4 7TJ
Dylan.Edwards@uea.ac.uk

Gunilla Høyer-Hansen
Finsen Laboratory
Copenhagen Biocenter
Ole Maaløes Vej 5
DK–2200 Copenhagen N
Denmark
gunilla@finsenlab.dk

Francesco Blasi
Università Vita Salute San Raffaele
    and IFOM (Fondazione Istituto
    FIRC di Oncologia Molecolare)
Via Adamello 16
20139 Milan
Italy
francesco.blasi@ifom-ieo-campus.it

Bonnie F. Sloane
Wayne State University
540 E. Canfield
Detroit MI 48201
USA
bsloane@med.wayne.edu

ISBN 978-0-387-69056-8   e-ISBN 978-0-387-69057-5

DOI 10.1007/978-0-387-69057-5

Library of Congress Control Number: 2008928134

Preface

Proteolysis is essential for life. From the breakdown of proteins in food for biosynthesis, through to antigen processing in the immune system, the blood clotting cascade, and the hormone-regulated remodelling of female reproductive tissues in adult mammals – proteolysis governs functionality, homeostasis, and fate at the levels of the cell and the entire organism. For the cancer cell, intracellular proteolysis carried out by caspases and the proteasome must be enlisted and controlled to allow it to escape apoptosis. Functioning on the cancer cell surface or in the extracellular milieu, secreted proteases (primarily metalloproteinases, serine proteases, and cathepsins) determine the interactions of cells with their environments. Once considered simply as promoting tumour cell invasion through tissue barriers, proteolysis is now known to be integral to many aspects of cancer biology, including angiogenesis, regulation of the bioavailability of growth factors, cellular adhesion, cytokine/chemokine signalling, inflammatory cell recruitment, and the mobilization of normal cells from their tissue compartments to act as accomplices in metastasis. The last decade has witnessed a revolution in our thinking concerning the role of extracellular proteolysis in cancer biology: this is the primary focus of this book.

The full repertoire of proteases and their inhibitors – collectively called the degradome – has now been revealed from the sequence analysis of several animal genomes. The first section of this book discusses our current perception of the degradome, and the "degradomic" technologies that have been developed for its study. Chapters cover such topics as the bioinformatic analysis of the human degradome, the use of different technology platforms for transcriptomic studies, substrate identification using proteomics and mass spectrometry, and finally the use of activity-based probes to image protease action in cultured cells and whole organisms.

Section II switches focus to deal with particular classes of proteases and inhibitors, discussing new insights into their roles in cancer biology, primarily derived from the study of mouse model systems. A reader looking for comprehensive coverage here will be disappointed, as we felt that there have been many outstanding recent reviews of the basic biology of protease families such as the matrix metalloproteinases (MMPs) and a disintegrin and metalloprotease (ADAMs), and our intention was therefore to highlight other enzymes that have not been covered extensively, such as the transmembrane serine proteases, or new concepts that have emerged. Chapters also discuss model systems that have been employed in angiogenesis and tumour cell invasion. Section III carries the theme of new perspectives of protease function further, dealing in particular with the connections between proteolysis and cell signalling. Chapters discuss invadopodia as membrane regions where the cellular proteolytic and signalling machineries congregate, the role of urokinase plasminogen activator (uPAR) signalling in haematopoietic stem cell mobilization, the connections between MMPs, cytokine signalling and tumour–bone interactions, and the linkage of distinct proteolytic pathways in the "protease web" during tumour metastasis.

In Section IV the use of the degradome as a source of tumour biomarkers is highlighted. There are chapters reviewing the state of play with established markers based on the uPA system, and other valuable indicators such as cysteine cathepsins and TIMP-1. Two chapters cover information from bioinformatic analysis of transcriptomic data. This leads to the final section of the book, in which the potential for targeted cancer therapeutics based on the degradome is evaluated. As well as discussing the problems associated with clinical trials of metalloproteinase inhibitors, chapters in Section V cover the development of novel selective inhibitors based on thorough structural knowledge of specific targets. In addition, exciting new strategies for anti-cancer therapies are discussed that take advantage of tumour-associated proteases to generate cytotoxic payloads from latent pro-drugs, and for improved delivery of drugs to the tumour vasculature.

The five sections share a similar arrangement of subject topics, with proteases or their inhibitors being dealt with in the order of serine proteases, cysteine cathepsins, and metalloproteinases. Sections I–III begin with chapters that the reader may find particularly useful in providing an overview to a particular area.

Following the disappointments of the clinical trials of early synthetic metalloproteinase inhibitors, the cancer protease field is now resurgent, as basic cancer biology and the pharmaceutical industry take on board the new knowledge of the multifaceted roles of proteases. Not only do some proteases antagonize tumour growth, rendering them "anti-targets" that must be spared in the design of novel, more selective agents, but their involvement in the tumour–host interplay identifies entirely new areas for intervention. Also, beyond active site-directed inhibitors, new cancer targets emerge based on knowledge of exosites, substrate cleavages, and protein–protein interactions.

We hope that *The Cancer Degradome: Proteases and Cancer Biology* will convey the prevalent sense of excitement and optimism as protease research enters this new era.

Dylan Edwards
Francesco Blasi
Gunilla Høyer-Hansen
Bonnie F. Sloane

Contents

Section I: The Degradome and Its Analysis

1   **Protease Genomics and the Cancer Degradome** 3
Xose S. Puente, Gonzalo R. Ordóñez, and Carlos López-Otín

2   **The CLIP-CHIP^{TM}: A Focused Oligonucleotide Microarray
Platform for Transcriptome Analysis of the Complete
Human and Murine Cancer Degradomes** 17
Reinhild Kappelhoff, Claire H. Wilson, and Christopher M. Overall

3   **The Hu/Mu ProtIn Chip: A Custom Dual-Species
Oligonucleotide Microarray for Profiling Degradome
Gene Expression in Tumors and Their Microenvironment** 37
Donald R. Schwartz, Kamiar Moin, Ekkehard Weber, and Bonnie F. Sloane

4   **Quantitative Real-Time PCR Analysis of Degradome Gene Expression** 49
Caroline J. Pennington, Robert K. Nuttall, Clara Sampieri-Ramirez,
Matthew Wallard, Simon Pilgrim, and Dylan R. Edwards

5   **Identification of Protease Substrates by Mass
Spectrometry Approaches-1** 67
Mari Enoksson, Wenhong Zhu, and Guy S. Salvesen

6   **Identification of Protease Substrates
by Mass Spectrometry Approaches-2** 83
Anna Prudova, Ulrich auf dem Keller, and Christopher M. Overall

7   **Activity-Based Imaging and Biochemical Profiling Tools
for Analysis of the Cancer Degradome** 101
Vincent Dive, Margot G. Paulick, J. Oliver McIntyre,
Lynn M. Matrisian, and Matthew Bogyo

8 Images of Cleavage: Tumor Proteases in Action 137
Kamiar Moin, Mansoureh Sameni, Christopher Jedeszko,
Quanwen Li, Mary B. Olive, Raymond R. Mattingly, and Bonnie F. Sloane

Section II: Insights into Protease Function

**9 Proteolytic Pathways: Intersecting Cascades
in Cancer Development** 157
Nesrine I. Affara and Lisa M. Coussens

**10 Physiological Functions of Plasminogen Activation:
Effects of Gene Deficiencies in Humans and Mice** 183
Thomas H. Bugge

**11 The Plasminogen Activation System in Tissue
Remodeling and Cancer Invasion** 203
Kasper Almholt, Anna Juncker-Jensen, Kirsty Anne Green,
Helene Solberg, Leif Røge Lund, and John Rømer

**12 The Urokinase Plasminogen Activator Receptor
as a Target for Cancer Therapy** 223
Silvia D'Alessio and Francesco Blasi

**13 The Endocytic Collagen Receptor, uPARAP/Endo180,
in Cancer Invasion and Tissue Remodeling** 245
Thore Hillig, Lars H. Engelholm, and Niels Behrendt

**14 Physiological and Pathological Functions of Type II
Transmembrane Serine Proteases: Lessons from
Transgenic Mouse Models and Human Disease-Associated
Mutations** 259
Karin List and Thomas H. Bugge

**15 Roles of Cysteine Proteases in Tumor Progression:
Analysis of Cysteine Cathepsin Knockout Mice
in Cancer Models** 281
Thomas Reinheckel, Vasilena Gocheva, Christoph Peters,
and Johanna A. Joyce

**16 *In Vitro* and *In Vivo* Models of Angiogenesis
to Dissect MMP Functions** 305
Sarah Berndt, Françoise Bruyère, Maud Jost, and Agnès Noël

17   The Surface Transplantation Model to Study the Tumor–Host Interface 327
Maud Jost, Silvia Vosseler, Silvia Blacher, Norbert E. Fusenig,
Margareta M. Mueller, and Agnès Noël

18   Unravelling the Roles of Proteinases in Cell Migration In Vitro and In Vivo 343
Jelena Gavrilovic and Xanthe Scott

19   New Insights into MMP Function in Adipogenesis 361
Kumari L. Andarawewa and Marie-Christine Rio

20   TIMPs: Extracellular Modifiers in Cancer Development 373
Aditya Murthy, William Cruz-Munoz, and Rama Khokha

Section III: The Interface Between Proteolysis and Cell Signalling

21   Invadopodia: Interface for Invasion 403
Susette C. Mueller, Vira V. Artym, and Thomas Kelly

22   uPAR and Proteases in Mobilization of Hematopoietic Stem Cells 433
Pia Ragno and Francesco Blasi

23   The Urokinase Receptor and Integrins Constitute a Cell Migration Signalosome 451
Bernard Degryse

24   Measuring uPAR Dynamics in Live Cells 475
Moreno Zamai, Gabriele Malengo, and Valeria R. Caiolfa

25   Janus-Faced Effects of Broad-Spectrum and Specific MMP Inhibition on Metastasis 495
Charlotte Kopitz and Achim Krüger

26   Cytokine Substrates: MMP Regulation of Inflammatory Signaling Molecules 519
Jennifer H. Cox and Christopher M. Overall

27   Matrix Metalloproteinases as Key Regulators of Tumor–Bone Interaction 541
Conor C. Lynch and Lynn M. Matrisian

**Section IV: The Degradome as Source of Cancer Diagnostic
and Prognostic Markers**

28 **The Plasminogen Activation System as a Source
of Prognostic Markers in Cancer** 569
Ib Jarle Christensen, Helle Pappot, and Gunilla Høyer-Hansen

29 **Cysteine Cathepsins and Cystatins as Cancer Biomarkers** 587
Tamara T. Lah, Nataša Obermajer, María Beatriz Durán Alonso,
and Janko Kos

30 **Novel Degradome Markers in Breast Cancer** 627
Caroline J. Pennington, Simon Pilgrim, Paul N. Span,
Fred C. Sweep, and Dylan R. Edwards

31 **Meta-Analysis of Gene Expression Microarray Data:
Degradome Genes in Healthy and Cancer Tissues** 645
Kristiina Iljin, Sami Kilpinen, Johanna Ivaska, and Olli Kallioniemi

32 **Degradome Gene Polymorphisms** 663
Ross Laxton and Shu Ye

33 **TIMP-1 as a Prognostic Marker in Colorectal Cancer** 679
Camilla Frederiksen, Anne Fog Lomholt, Hans Jørgen Nielsen,
and Nils Brünner

Section V: Novel Therapeutic Strategies

34 **Structure and Inhibition of the Urokinase-Type
Plasminogen Activator Receptor** 699
Benedikte Jacobsen, Magnus Kjaergaard, Henrik Gårdsvoll,
and Michael Ploug

35 **Engineered Antagonists of uPA and PAI-1** 721
M. Patrizia Stoppelli, Lisbeth M. Andersen, Giuseppina Votta,
and Peter A. Andreasen

36 **MMP Inhibitor Clinical Trials – The Past, Present, and Future** 759
Barbara Fingleton

37 **Tailoring TIMPs for Selective Metalloproteinase Inhibition** 787
Hideaki Nagase and Gillian Murphy

**38 Third-Generation MMP Inhibitors: Recent Advances
in the Development of Highly Selective Inhibitors** 811
Athanasios Yiotakis and Vincent Dive

39 Protease-Activated Delivery and Imaging Systems 827
Gregg B. Fields

**40 Development of Tumour-Selective and
Endoprotease-Activated Anticancer Therapeutics** 853
Jason H. Gill and Paul M. Loadman

**41 Targeting Degradome Genes via Engineered
Viral Vectors** 877
Risto Ala-aho, Andrew H. Baker, and Veli-Matti Kähäri

Appendix 895

Index 899

Contributors

Nesrine I. Affara
Department of Pathology, University of California, San Francisco, CA 94143, USA

Risto Ala-aho
Department of Dermatology and MediCity Research Laboratory, University of Turku, FI-20521, Turku, Finland

Kasper Almholt
Finsen Laboratory, Rigshospitalet, Copenhagen Biocenter, Ole Maaløes Vej 5, DK-2200 Copenhagen, Denmark

María Beatriz Durán Alonso
Department of Genetic Toxicology and Cancer Biology, National Institute of Biology, Večna pot 111, 1000 Ljubljana, Slovenia

Kumari L. Andarawewa
Institut de Génétique et de Biologie Moléculaire et Cellulaire (IGBMC), CNRS UMR 7104, INSERM U596, ULP, BP 163, 67404 Illkirch Cedex, C.U. de Strasbourg, France

Lisbeth M. Andersen
Department of Molecular Biology, University of Aarhus, 10C Gustav Wied's Vej, 8000 Aarhus C, Denmark

Peter A. Andreasen
Department of Molecular Biology, University of Aarhus, 10C Gustav Wied's Vej, 8000 Aarhus C, Denmark

Vira V. Artym
Department of Oncology, Lombardi Comprehensive Cancer Center, Georgetown University Medical School, Washington, DC 20057, USA
and
Laboratory of Cell and Developmental Biology, National Institute of Dental and Craniofacial Research, National Institutes of Health, Bethesda, MD 20892, USA

Andrew H. Baker
British Heart Foundation Glasgow Cardiovascular Research Centre, University of Glasgow, Glasgow G12 8TA, UK

Niels Behrendt
Finsen Laboratory, Rigshospitalet, Copenhagen Biocenter, Ole Maaløes Vej 5, DK-2200 Copenhagen, Denmark

Sarah Berndt
Laboratory of Tumor and Development Biology, University of Liège, Sart-Tilman B23, B-4000 Liège, Belgium
and
Centre de Recherche en Cancérologie Expérimentale (CRCE), Groupe Interdisciplinaire de Génoprotéomique Appliquée-Recherche (GIGA-Research), University of Liège, Sart Tilman B23, B-4000 Liège, Belgium

Silvia Blacher
Laboratory of Tumor and Development Biology, Centre de Recherche en Cancérologie Expérimentale (CRCE), Groupe Interdisciplinaire de Génoprotéomique Appliquée (GIGA-R), University of Liège, Tour de Pathologie (B23), B-4000 Liège, Belgium

Francesco Blasi
Università Vita Salute San Raffaele and IFOM (Fondazione Istituto FIRC di Oncologia Molecolare), Via Adamello 16, 20139 Milan, Italy

Matthew Bogyo
Department of Pathology and Department of Microbiology and Immunology, Stanford University, Stanford, CA 94305, USA

Nils Brünner
Section of Biomedicine, Department of Veterinary Pathobiology, Faculty of Life Sciences, University of Copenhagen, DK-1870 Frederiksberg C, Denmark

Françoise Bruyère
Laboratory of Tumor and Development Biology, University of Liège, Sart-Tilman B23, B-4000 Liège, Belgium
and
Centre de Recherche en Cancérologie Expérimentale (CRCE), Groupe Interdisciplinaire de Génoprotéomique Appliquée-Recherche (GIGA-Research), University of Liège, Sart Tilman B23, B-4000 Liège, Belgium

Thomas H. Bugge
Proteases and Tissue Remodeling Unit, Oral and Pharyngeal Cancer Branch, National Institute of Dental and Craniofacial Research, National Institutes of Health, Bethesda, MD, USA

Valeria R. Caiolfa
Department of Molecular Biology and Functional Genomics, San Raffaele Scientific Institute, Milan, Italy
and
Italian Institute of Technology Network Research, Unit of Molecular Neuroscience, San Raffaele Scientific Institute, Milano, Italy

Ib Jarle Christensen
Finsen Laboratory, Rigshospitalet, Copenhagen Biocenter, Ole Maaløes Vej 5, DK-2200 Copenhagen, Denmark

Lisa M. Coussens
Department of Pathology, Cancer Research Institute and Comprehensive Cancer Center, University of California, San Francisco, CA 94143, USA

Jennifer H. Cox
Centre for Blood Research, Departments of Oral Biological and Medical Sciences and Biochemistry and Molecular Biology, University of British Columbia, Vancouver, BC, Canada

William Cruz-Munoz
Department of Medical Biophysics, Ontario Cancer Institute, University Health Network, Toronto, Canada

Silvia D'Alessio
Department of Molecular Biology and Functional Genomics, DIBIT, Università Vita-Salute San Raffaele, Via Olgettina 58, 20132 Milan, Italy

Bernard Degryse
Department of Molecular Biology and Functional Genomics, DIBIT, University Vita-Salute San Raffaele, Via Olgettina 58, 20132 Milan, Italy

Vincent Dive
CEA, Institut de Biologie et des Technologies de Saclay, Service d'Ingénierie Moléculaire des Proéines, 91191 Gif/Yvette Cedex, France

Dylan R. Edwards
School of Biological Sciences, University of East Anglia, Norwich, NR4 7TJ, UK

Lars H. Engelholm
Finsen Laboratory, Rigshospitalet, Copenhagen Biocenter, Ole Maaløes Vej 5, DK-2200 Copenhagen, Denmark

Mari Enoksson
Burnham Institute for Medical Research, La Jolla, CA 92037, USA

Gregg B. Fields
Department of Chemistry and Biochemistry, Florida Atlantic University, Boca
Raton, FL 33431, USA

Barbara Fingleton
Department of Cancer Biology, Vanderbilt University School of Medicine, Nashville, TN 37232, USA

Camilla Frederiksen
Department of Surgical Gastroenterology 435, Copenhagen University Hospital
Hvidovre, Kettegaard Allé 30, DK-2650 Hvidovre, Denmark

Norbert E. Fusenig
Group Tumor and Microenvironment (A101), German Cancer Research Center
(DKFZ), 69120 Heidelberg, Germany

Henrik Gårdsvoll
Finsen Laboratory, Rigshospitalet, Copenhagen Biocenter, Ole Maaløes Vej 5,
DK-2200 Copenhagen, Denmark

Jelena Gavrilovic
Biomedical Research Centre, School of Biological Sciences, University of East
Anglia, Norwich, NR4 7TJ, UK

Jason H. Gill
Institute of Cancer Therapeutics, University of Bradford, West Yorkshire, BD7
1DP, UK

Vasilena Gocheva
Cancer Biology and Genetics Program, Memorial Sloan Kettering Cancer Center,
New York, NY 10021, USA
and
Weill Graduate School of Medical Sciences, Cornell University, New York, NY
10021, USA

Kirsty Anne Green
Finsen Laboratory, Rigshospitalet, Copenhagen Biocenter, Ole Maaløes Vej 5,
DK-2200 Copenhagen, Denmark

Thore Hillig
Finsen Laboratory, Rigshospitalet, Copenhagen Biocenter, Ole Maaløes Vej 5,
DK-2200 Copenhagen, Denmark

Gunilla Høyer-Hansen
Finsen Laboratory, Rigshospitalet, Copenhagen Biocenter, Ole Maaløes Vej 5,
DK-2200 Copenhagen, Denmark

Kristiina Iljin
Medical Biotechnology, VTT technical Research Centre of Finland and University of Turku, Turku, Finland

Johanna Ivaska
Medical Biotechnology, VTT technical Research Centre of Finland and University of Turku, Turku, Finland

Benedikte Jacobsen
Finsen Laboratory, Rigshospitalet, Copenhagen Biocenter, Ole Maaløes Vej 5, DK-2200 Copenhagen, Denmark

Christopher Jedeszko
Department of Pharmacology and Barbara Ann Karmanos Cancer Institute, Wayne State University School of Medicine, Detroit, MI 48201, USA

Maud Jost
Laboratory of Tumor and Development Biology, Centre de Recherche en Cancérologie Expérimentale (CRCE), Groupe Interdisciplinaire de Génoprotéomique Appliquée (GIGA-R), University of Liège, Tour de Pathologie (B23), B-4000 Liège, Belgium

Johanna A. Joyce
Cancer Biology and Genetics Program, Memorial Sloan Kettering Cancer Center, New York, NY 10021, USA

Anna Juncker-Jensen
Finsen Laboratory, Rigshospitalet, Copenhagen Biocenter, Ole Maaløes Vej 5, DK-2200 Copenhagen, Denmark

Veli-Matti Kähäri
Department of Dermatology and MediCity Research Laboratory, University of Turku, FI-20521, Turku, Finland

Olli Kallioniemi
Medical Biotechnology, VTT technical Research Centre of Finland and University of Turku, Turku, Finland

Reinhild Kappelhoff
Centre for Blood Research, Departments of Oral Biological and Medical Sciences, University of British Columbia, Vancouver, BC, V6T 1Z3, Canada

Thomas Kelly
Department of Oncology, Lombardi Comprehensive Cancer Center, Georgetown University Medical School, Washington, DC 20057, USA
and
Department of Pathology, Arkansas Cancer Research Center, University of Arkansas for Medical Sciences, Little Rock, AR 72205, USA

Ulrich auf dem Keller
The UBC Centre for Blood Research, Department of Oral Biological and Medical Sciences, and Biochemistry and Molecular Biology, University of British Columbia, Vancouver, BC, Canada

Rama Khokha
Department of Medical Biophysics, Ontario Cancer Institute, University Health Network, Toronto, Canada

Sami Kilpinen
Medical Biotechnology, VTT Technical Research Centre of Finland and University of Turku, Turku, Finland

Magnus Kjaergaard
Finsen Laboratory, Rigshospitalet, Copenhagen Biocenter, Ole Maaløes Vej 5, DK-2200 Copenhagen, Denmark

Charlotte Kopitz
Klinikum rechts der Isar der Technischen Universität München, Institut für Experimentelle Onkologie und Therapieforschung

Achim Krüger
Klinikum rechts der Isar der Technischen Universität München, Institut für Experimentelle Onkologie und Therapieforschung

Tamara T. Lah
Department of Genetic Toxicology and Cancer Biology, National Institute of Biology, Večna pot 111, 1000 Ljubljana, Slovenia

Ross Laxton
Department of Clinical Pharmacology, William Harvey Research Institute, St. Barts and the London School of Medicine, John Vane Science Centre, London EC1M 6BQ, UK

Quanwen Li
Department of Pharmacology and Barbara Ann Karmanos Cancer Institute, Wayne State University School of Medicine, Detroit, MI 48201, USA

Karin List
Proteases and Tissue Remodeling Unit, Oral and Pharyngeal Cancer Branch, National Institute of Dental and Craniofacial Research, National Institutes of Health, Bethesda, MD, USA

Paul M. Loadman
Institute of Cancer Therapeutics, University of Bradford, West Yorkshire, BD7 1DP, UK

Anne Fog Lomholt
Department of Surgical Gastroenterology 435, Copenhagen University Hospital
Hvidovre, Kettegaard Allé 30, DK-2650 Hvidovre, Denmark

Carlos López-Otín
Departamento de Bioquímica y Biología Molecular, Facultad de Medicina, Instituto Universitario de Oncología, Universidad de Oviedo, 33006-Oviedo, Spain

Leif Røge Lund
Finsen Laboratory, Rigshospitalet, Copenhagen Biocenter, Ole Maaløes Vej 5, DK-2200 Copenhagen, Denmark

Conor C. Lynch
Departments of Orthopaedics and Rehabilitation and Cancer Biology, Vanderbilt University, Nashville, TN, USA

Gabriele Malengo
Department of Molecular Biology and Functional Genomics, San Raffaele Scientific Institute, Milano, Italy
and
Università Vita-Salute San Raffaele, Milano, Italy

Lynn M. Matrisian
Department of Cancer Biology, Vanderbilt University, Nashville, TN 37232, USA

Raymond R. Mattingly
Department of Pharmacology and Barbara Ann Karmanos Cancer Institute, Wayne State University School of Medicine, Detroit, MI 48201, USA

J. Oliver McIntyre
Department of Cancer Biology, Vanderbilt University, Nashville, TN 37232, USA

Kamiar Moin
Department of Pharmacology and Barbara Ann Karmanos Cancer Institute, Wayne State University School of Medicine, Detroit, MI 48201, USA

Margareta M. Mueller
Group Tumor and Microenvironment (A101), German Cancer Research Center (DKFZ), 69120 Heidelberg, Germany

Susette C. Mueller
Department of Oncology, Lombardi Comprehensive Cancer Center, Georgetown University Medical School, Washington, DC 20057, USA

Gillian Murphy
Department of Oncology, Cambridge University, Cancer Research UK Cambridge Research Institute, Cambridge CB2 0RE, UK

Aditya Murthy
Department of Medical Biophysics, Ontario Cancer Institute, University Health Network, Toronto, Canada

Hideaki Nagase
Kennedy Institute of Rheumatology Division, Imperial College London, London W6 8LH, UK

Hans Jørgen Nielsen
Department of Surgical Gastroenterology 435, Copenhagen University Hospital Hvidovre, Kettegaard Allé 30, DK-2650 Hvidovre, Denmark

Agnès Noël
Laboratory of Tumor and Development Biology, Centre de Recherche en Cancérologie Expérimentale (CRCE), Groupe Interdisciplinaire de Génoprotéomique Appliquée (GIGA-R), University of Liège, Tour de Pathologie (B23), B-4000 Liège, Belgium

Robert K. Nuttall
School of Biological Sciences, University of East Anglia, Norwich, NR4 7TJ, UK

Nataša Obermajer
Faculty of Pharmacy, University of Ljubljana, Aškerčeva 6, 1000 Ljubljana, Slovenia

Mary B. Olive
Department of Pharmacology and Barbara Ann Karmanos Cancer Institute, Wayne State University School of Medicine, Detroit, MI 48201, USA

Gonzalo R. Ordóñez
Departamento de Bioquímica y Biología Molecular, Facultad de Medicina, Instituto Universitario de Oncología, Universidad de Oviedo, 33006-Oviedo, Spain

Christopher M. Overall
University of British Columbia, Centre for Blood Research, Departments of Oral Biological and Medical Sciences, Biochemistry and Molecular Biology, Vancouver, BC, V6T 1Z3, Canada

Helle Pappot
Finsen Laboratory, Rigshospitalet, Copenhagen Biocenter, and Department of Oncology, Rigshospitalet, Copenhagen, Denmark

Margot Paulick
Department of Pathology and Microbiology and Immunology, Stanford University, Stanford, CA 4061, USA

Caroline J. Pennington
School of Biological Sciences, University of East Anglia, Norwich, NR4 7TJ, UK

Christoph Peters
Institut für Molekulare Medizin und Zellforschung, Albert-Ludwigs-Universität, D-79104 Freiburg, Germany

Simon Pilgrim
School of Biological Sciences, University of East Anglia, Norwich, NR4 7TJ, UK

Michael Ploug
Finsen Laboratory, Rigshospitalet, Copenhagen Biocenter, Copenhagen, Denmark

Xose S. Puente
Departamento de Bioquímica y Biología Molecular, Facultad de Medicina, Instituto Universitario de Oncología, Universidad de Oviedo, 33006-Oviedo, Spain

Anna Prudova
The UBC Centre for Blood Research, Department of Oral Biological and Medical Sciences, and Biochemistry and Molecular Biology, University of British Columbia, Vancouver, BC, Canada,

Pia Ragno
Università di Salerno, Dipartimento di Chimica, Via Ponte don Melillo, 84084 Salerno, Italy

Thomas Reinheckel
Institut für Molekulare Medizin und Zellforschung, Albert-Ludwigs-Universität, D-79104 Freiburg, Germany

Marie-Christine Rio
Institut de Génétique et de Biologie Moléculaire et Cellulaire (IGBMC), CNRS UMR 7104, INSERM U596, ULP, BP 163, 67404 Illkirch Cedex, C.U. de Strasbourg, France

John Rømer
Finsen Laboratory, Rigshospitalet, Copenhagen Biocenter, Ole Maaløes Vej 5, DK-2200 Copenhagen, Denmark

Guy Salvesen
Burnham Institute for Medical Research, La Jolla, CA 92037, USA

Mansoureh Sameni
Department of Pharmacology and Barbara Ann Karmanos Cancer Institute, Wayne State University School of Medicine, Detroit, MI 48201, USA

Clara Sampieri-Ramirez
School of Biological Sciences, University of East Anglia, Norwich, NR4 7TJ, UK

Donald R. Schwartz
Biodiscovery, LLC 5692 Plymouth Road Ann Arbor, MI 48105, USA

Xanthe Scott
Biomedical Research Centre, School of Biological Sciences, University of East Anglia, Norwich, NR4 7TJ, UK

Bonnie F. Sloane
Department of Pharmacology and Barbara Ann Karmanos Cancer Institute, Wayne State University School of Medicine, Detroit, MI 48201, USA

Helene Solberg
Finsen Laboratory, Rigshospitalet, Copenhagen Biocenter, Ole Maaløes Vej 5, DK-2200 Copenhagen, Denmark

Paul N. Span
Department of Chemical Endocrinology, Radboud University Nijmegen Medical Centre, The Netherlands

M. Patrizia Stoppelli
Institute of Genetics and Biophysics, National Research Council, Via Castellino, 111-80131 Naples, Italy

Fred C. Sweep
Department of Chemical Endocrinology, Radboud University Nijmegen Medical Centre, The Netherlands

Silvia Vosseler
Group Tumor and Microenvironment (A101), German Cancer Research Center (DKFZ), 69120 Heidelberg, Germany

Giuseppina Votta
Institute of Genetics and Biophysics, National Research Council, Via Castellino, 111-80131 Naples, Italy

Matthew Wallard
School of Biological Sciences, University of East Anglia, Norwich, NR4 7TJ, UK

Ekkehard Weber
Institute of Physiological Chemistry, Martin Luther University, Halle-Wittenber, 06114 Halle, Germany

Claire H. Wilson
Centre for Blood Research, Department of Biochemistry and Molecular Biology, University of British Columbia, Vancouver, BC, V6T 1Z3, Canada

Shu Ye
Department of Clinical Pharmacology, William Harvey Research Institute, St. Barts and the London School of Medicine, John Vane Science Centre, London EC1M 6BQ, UK

Athanasios Yiotakis
Department of Chemistry, Laboratory of Organic Chemistry, University of Athens, Panepistimiopolis Zografou 15771, Athens, Greece

Moreno Zamai
Department of Molecular Biology and Functional Genomics, San Raffaele Scientific Institute, Milano, Italy
and
Italian Institute of Technology Network Research, Unit of Molecular Neuroscience, San Raffaele Scientific Institute, Milano, Italy

Wenhong Zhu
Burnham Institute for Medical Research, La Jolla, CA 92037, USA

Section I
The Degradome and Its Analysis

Chapter 1
Protease Genomics and the Cancer Degradome

Xose S. Puente, Gonzalo R. Ordóñez, and Carlos López-Otín

Abstract Proteases comprise a large group of enzymes involved in multiple physiological and pathological processes, which has made necessary the introduction of global concepts for their study. Thus, the human degradome has been defined as the complete set of proteolytic genes encoded by the human genome. Likewise, the term cancer degradome defines the set of protease genes expressed by a tumour at a specific time. Detailed genomic analyses have revealed that the human degradome is composed of 569 protease-coding genes, whereas mouse and rat degradomes are even more complex, containing 649 and 634 genes, respectively. The precise knowledge of these differences is essential to understand the utility and limitations of these animal models to investigate human diseases, including cancer. In this regard, recent studies with genetically modified mice have shown that proteases contribute to all stages of tumour progression and not only to the later stages as was originally proposed. These studies have also revealed the existence of proteolytic enzymes with tumour-suppressive functions. Accordingly, any attempt to understand the biological and pathological relevance of proteases in cancer must take into account the large structural and functional diversity of proteolytic systems operating in all stages of the disease. Hopefully, the novel information derived from protease genomics may finally lead to the validation of some of these enzymes as important components of future strategies for cancer treatment.

Introduction

Proteases constitute a group of enzymes with the ability to hydrolyze peptide bonds. The irreversibility of this type of reaction makes it suitable for multiple cellular processes, which has contributed to a widespread use of this mechanism in different

C. López-Otín
Departamento de Bioquímica y Biología Molecular, Facultad de Medicina, Universidad de Oviedo, 33006 Oviedo-Spain, e-mail: clo@uniovi.es

D. Edwards et al. (eds.), *The Cancer Degradome.*
© Springer Science + Business Media, LLC 2008

biological contexts, including development, apoptosis, homeostasis, reproduction or host defense (López-Otín and Overall 2002). Since their initial discovery, this group of enzymes has attracted the interest of numerous researchers because of their participation in important physiological processes such as food digestion and blood coagulation. This fact has contributed to clarify important features of this type of enzymes, including the biochemical mechanisms implicated in their catalysis, the structural determinants which define their substrate specificity and the different mechanisms by which their activity is regulated, either by specific activation through limited proteolysis of an inactive precursor or by the action of endogenous protease inhibitors (Rawlings et al. 2004). As novel proteases were identified and additional features discovered, the interest on proteolytic enzymes has grown accordingly. In fact, over the last two decades, proteases have acquired great biomedical interest due to the identification of numerous human pathologies in which proteolytic enzymes are implicated. These protease-associated diseases include inflammatory conditions, cardiovascular alterations, neurodegenerative disorders and cancer (Coussens et al. 2000, Esler and Wolfe 2001, Mohammed et al. 2004, Overall and López-Otín 2002, Puente et al. 2003). In most of these cases, the diseases are linked to an increased proteolytic activity, resulting in enhanced protein degradation and finally leading to tissue damage and destruction. That is the case for inflammatory diseases such as rheumatoid arthritis, in which an excessive protease activity results in cartilage degradation and impaired joint function, or cancer, in which proteases acting at the leading edge are responsible for basal membrane degradation, facilitating the invasion of tumour cells and the further development of metastasis (Zucker et al. 2003). These studies together with the introduction of improved cloning technologies have resulted in the identification of numerous novel human proteases and their association with specific pathologies, which has led to the consideration of these enzymes as promising targets to treat different human diseases (Turk 2006). In this chapter, we will discuss our current knowledge on human proteolytic enzymes and the utility of comparative genomic analysis to understand their evolutionary history and to evaluate experimental data on proteases obtained in animal models. Finally, we will specifically discuss the relevance for cancer of this genomic analysis of proteolytic systems.

The Human Degradome

The importance of proteolysis for life is underscored by the fact that all living organisms contain proteases which are required for normal development or growth (Barrett et al. 2004). Although proteases perform the same catalytic reaction, the hydrolysis of a peptide bond, this activity has evolved independently several times leading to the emergence of numerous enzymes with different mechanisms capable of performing this type of reaction. As a result, proteolytic enzymes can be classified in six different classes according to their catalytic mechanism, including aspartic-, cysteine-, serine-, threonine-, and metalloproteases, as well as the recently identified fungi-specific class of glutamic-peptidases (Fujinaga et al. 2004,

Rawlings et al. 2004). The large number of identified proteases and their importance in human biology and pathology has made necessary the use of novel concepts for the global study of proteolysis. Thus, we have introduced the term degradome to define the complete set of protease genes present in one organism or the repertoire of proteases expressed by a certain tissue (López-Otín and Overall 2002). Likewise, the term cancer degradome has been rapidly coined to define the set of protease genes expressed by a tumour at a specific time (Overall and López-Otín 2002). Although our current understanding of the role of proteases in tumour development is still limited, it is generally accepted that a detailed knowledge of the proteases expressed by a tumour at a certain stage will be extremely useful for early detection and prognosis evaluation of the disease as well as for designing specific treatments based on the degradome of the tumour.

The availability of the human genome sequence opened the possibility to characterize the complete repertoire of human protease genes. To this aim, we first performed a bioinformatic analysis of the human genome to classify all previously known protease-coding genes and to identify novel genes encoding proteins with sequence similarity to previously known proteases from human or other organisms. This allowed us to determine that the human degradome is composed of 569 protease and protease-related genes (López-Otín et al. 2004, Puente et al. 2003). Taking into account that the human genome is estimated to contain less than 25,000 genes (Collins et al. 2004, Hubbard et al. 2007), the analysis of the human degradome indicates that proteases represent more than 2% of the total genes in the human genome, underscoring the importance of proteolysis in human biology.

Human proteases can be divided into five different catalytic classes, with metalloproteases, serine and cysteine proteases being the most abundant ones (194, 176 and 150 genes in the human genome, respectively), while aspartic and threonine peptidases are composed of a limited number of members (21 and 28, respectively) (Fig. 1.1). Nevertheless, it is worthwhile mentioning that among the 569 human proteases, 92 have lost key residues necessary for their proteolytic activity and have been classified as non-protease homologues (Puente et al. 2003). Despite the lack of proteolytic activity, these inactive proteases have acquired different biological properties, and some of them might regulate the activation of other proteases or their access to substrates or inhibitors (Boatright et al. 2004). Although the function for most of these catalytically inactive proteases is not fully understood to date, many of them show a high degree of conservation between human and other mammals, suggesting that they appeared before the mammalian expansion and have been conserved through evolution probably because of their relevance in diverse biological functions. Another interesting characteristic of proteolytic enzymes is the presence in most of them of one or several auxiliary domains. These domains lack proteolytic activity and in most cases can be also found in other types of proteins. The presence of these ancillary regions confers novel biological functions to proteolytic enzymes, facilitating their interaction with specific substrates, activators or inhibitors or their localization on specific cellular compartments (Overall 2002).

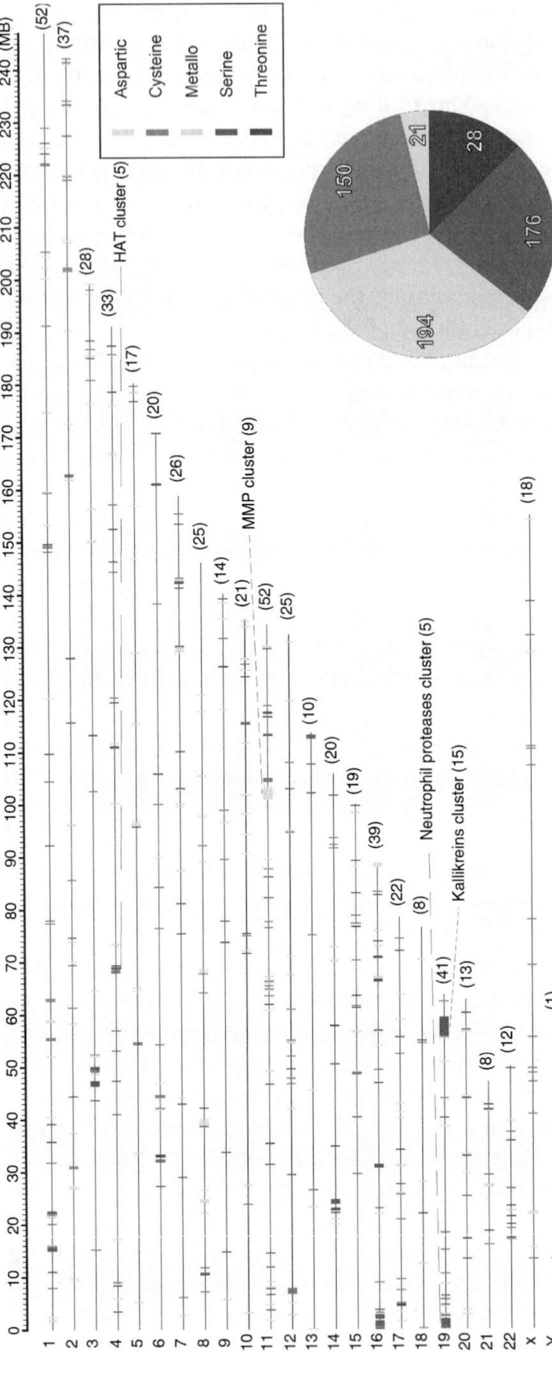

Fig. 1.1 Genomic view of the human degradome. Distribution of protease genes along individual chromosomes showing genes as boxes coloured according to the catalytic class to which they belong. The location of specific protease gene clusters in the human genome is indicated, as well as the number of genes present in each cluster. The number of proteases in each human chromosome is shown on the right side. The circle represents the distribution of human protease genes per catalytic class. (*See also* Color Insert I)

The completion of the human genome sequence has been a starting point to define the complexity of the human degradome. In this regard, the number of proteases and protease homologues currently annotated in the degradome database (*http://www.uniovi.es/degradome*) must be viewed as a current estimate of human protease-coding genes and not as a definitive number of human proteases. Accordingly, the number of human proteases has grown slightly during the last few years as novel structural designs and catalytic mechanisms have been unveiled and the corresponding human orthologs identified (Diaz-Perales et al. 2005). However, it must be taken into account that about one third of all human proteins cannot be classified into any of the protein families currently present in the protein family (Pfam) database (Finn et al. 2006). Therefore, it is expected that the experimental analysis of these orphan proteins could result in the identification of as yet unknown novel catalytic mechanisms which might contribute to expand the dimensions of the human degradome.

Tools for Degradome Research

The growing interest in proteolytic enzymes during the last decade was mainly due to the observation that proteolytic activity was associated with the progression of numerous human diseases, and especially cancer. The ability to determine the set of protease genes expressed by a tumour, or cancer degradome, can be extremely useful to understand the invasive potential of the tumour and to decide personalized treatments based on the use of specific protease inhibitors. In this regard, the definition of the human protease repertoire has opened the possibility to understand the complexity of the human degradome and has allowed the design of novel tools to study the implication of proteases in physiological and pathological processes. Thus, the knowledge of the coding sequences for all human proteases and inhibitors has been first used to develop a cDNA microarray, the CLIP-CHIP, for the detection of all proteases and protease inhibitors in human and mouse samples (Overall et al. 2004). Similar approaches based on oligonucleotide microarrays have also been used by different groups to analyze the expression of proteases and their inhibitors in malignant tumours (Schwartz et al. 2007). More recently, low-density arrays based on specific TaqMan probes have been developed to allow the detection and quantification of even low-expressing protease genes (G.R.O, X.S.P. and C.L.O., unpublished). The recent availability of these discovery tools opens the possibility to analyze in more detail the complexity of biological or pathological processes in terms of proteases, and to understand the molecular mechanisms underlying different human diseases, including cancer.

Increased Complexity of Rodent Degradomes

Although many proteases have been identified due to their expression in different human physiological or pathological conditions, the contribution of a certain protease to processes such as inflammation or tumour growth cannot be inferred from its expression pattern, making necessary the introduction of other approaches to address this question. In this regard, different animal models including mouse, rat, macaque and chicken have proved useful to understand the molecular mechanisms underlying several human diseases. In fact, rat is widely used to study pathologies such as cardiovascular or neurodegenerative diseases, as well as to carry out pharmacological studies of specific drugs. However, the ability to easily manipulate the mouse genome to either mutate or overexpress specific genes has made this species one of the most valuable tools to understand the molecular mechanisms underlying certain human diseases such as cancer (Rosenthal and Brown 2007). In any case, the broad use of animal models to study these processes and to test novel protease inhibitors implies the need to fully define the complexity of their proteolytic systems.

The recent availability of the mouse and rat genome sequences (Gibbs et al. 2004, Waterston et al. 2002) has provided an excellent opportunity to characterize their degradomes and to gain insights into the evolution of mammalian proteases. Surprisingly, despite their smaller genomes, rodent degradomes are more complex than the human degradome, with 649 genes in mouse and 634 in rat, compared to the 569 proteases present in the human degradome (López-Otín and Matrisian 2007, Puente and López-Otín 2004). These evolutionary differences between human and rodent degradomes can be explained by two different mechanisms. The first one proposes that the increased number of proteases in rodents could be due to an expansion of protease-coding genes in their genomes after the rodent and human lineages diverged about 75 million years ago. Alternatively, the reduced number of proteases in the human degradome could be a consequence of the loss or inactivation of protease-coding genes in this lineage. Comparison of the degradomes of these mammalian species has shown that both mechanisms have been acting during evolution yielding the current differences in the complexity of mammalian degradomes.

In relation to the first mechanism, we have reported that the increased complexity of rodent degradomes is mainly due to the expansion of specific families of protease genes, most of them implicated in reproduction and host defense (Puente et al. 2005b). Several examples illustrate the existence of major differences in the functioning of proteolytic systems associated with these important physiological processes. Thus, the comparison of human and rodent proteases has revealed the presence of genes encoding placental cathepsins, testins and testases in mouse and rat, while no human orthologs could be identified for this group of proteases implicated in placental development and fertilization (Deussing et al. 2002, Puente and López-Otín 2004, Puente et al. 2003). However, the increased complexity of rodent degradomes is also due to the expansion of certain protease families which are also present in humans. This is the case for the mast cell protease subfamily of trypsin-like serine proteases, which are implicated in host defense functions. In humans, this group of proteases is composed of just four genes—cathepsin G, chymase and granzymes B and H—which are clustered in a small region of chromosome 14q11. Interestingly, a closer look at the syntenic regions in mouse chromosome 14C1 and rat chromosome 15p13 has shown that this cluster of proteolytic genes has undergone a large expansion, and now contains up to 17 protease genes in mouse and 28 in the rat (Puente and López-Otín 2004). Similarly, the genes encoding kallikreins are also located in a cluster with 26 genes in mouse, 23 in rat and only 15 in human. As reflected by these examples, expansion of proteolytic genes has contributed to the increased complexity of rodent degra-domes. Remarkably, these differences correspond almost exclusively to genes implicated in reproductive or immunological functions, indicating that these pro-cesses have been major forces acting during mammalian evolution. The precise knowledge of these differential genes will be extremely useful when interpreting experimental data obtained using animal models.

The occurrence of rodent-specific protease subfamilies constitutes an extreme case of differences between human and rodent degradomes. Despite the existence of numerous examples that illustrate the importance of gene expansion events during the evolution of rodent degradomes, we have been unable to identify similar expansions in the human genome, with the single exception of a recent duplication involving the gene encoding MMP23, which has originated two almost identical copies of this gene (MMP23A and MMP23B) (Gururajan et al. 1998, Puente et al. 2005a). Based on these data, it appears that gene expansion could be sufficient to explain the larger number of genes present in rodent degradomes when compared to that of humans. However, the loss of protease-coding genes in the human genome has also contributed to the observed differences with rodents. In fact, if for each mouse and rat protease gene absent in humans we analyze the corresponding region in the human genome, we will be able to identify a human sequence with high similarity to the murine counterpart in about 30 cases. However, a detailed analysis of this sequence will rapidly reveal the presence of premature stop codons, frame-shifts or partial gene deletions which have contributed to the inactivation of this protease gene in the human genome. Therefore, pseudogenization has also been an important mechanism contributing to the increased degradome complexity of rodents when compared to humans. Interestingly, a detailed comparative analysis has revealed that most of the pseudogenized proteases which are still functional in rodents are involved in reproductive processes or in immunological functions, reinforcing the importance of these processes during mammalian evolution.

Complexity of the Protease Inhibitor Repertoire

The overall picture emerging from this comparison of human and rodent degra-domes suggests that the larger number of rodent proteases would result in an increased proteolytic activity in rodent tissues. Therefore, it is tempting to speculate that additional mechanisms could have evolved to compensate this increase in proteolysis. As an initial approach to address this question, it should be of interest to investigate whether the genes encoding protease inhibitors are different between these species, as this group of proteins is responsible for the inhibition of specific proteases under physiological conditions. In this regard, determination of the protease inhibitor complement in the genomes of human, mouse and rat has shown that changes in this group of genes might compensate the increased proteolytic potential of rodent tissues. Thus, the repertoire of protease inhibitor genes present in the human genome consists of more than 156 members, while mouse and rat show a higher complexity, with 199 and 183 members, respectively (Puente and López-Otín 2004).

Similar to the case of protease-coding genes, the increased complexity of protease inhibitor genes in rodents is mainly due to the expansion of gene clusters in these species. In fact, a detailed genomic analysis has revealed that a series of protease inhibitor genes expanded in the rodent genomes belong to groups which specifically inhibit some of the protease families which were also expanded in these species. An interesting case is that of a group of serine protease inhibitors of the serine proteinase inhibitor B (SERPINB) family. This group of inhibitors is located in human chromosome 6p25 and is composed of three different genes (SERPINB-1, -6 and -9), while the syntenic regions in mouse chromosome 13A4 or rat 17p12 have undergone a gene expansion process resulting in the presence of eight functional genes encoding SERPINB inhibitors in rat and fifteen in mouse. Similarly, the cystatin gene family, encoding a group of protease inhibitors with high specificity for cathepsins, has also been expanded in rodents. Together, these data suggest that the expansion of protease inhibitor genes in the rodent genomes might constitute a general mechanism to compensate the increased proteolytic activity in rodent tissues which might result from the expansion of protease-coding genes.

Applications of the Comparative Analysis of Mammalian Degradomes

One of the main conclusions that can be raised from the genomic comparison between the human degradome and those of mouse and rat is the increased complexity of the protease complement in these species. Taking into account that these rodents are the most widely used animal models to investigate human diseases, the characterization of their degradomes constitutes a valuable resource to evaluate experimental data obtained with these animals. In this regard, the presence in rodents of large protease families with high sequence identity and similar substrate specificity among their members might generate compensatory mechanisms which complicate the analysis of animals deficient in specific protease genes. Moreover, this increased protease complexity should be taken into consideration when studying the efficacy of novel protease inhibitors. In fact, it is possible that the existence of other family members not affected by the inhibitor might be able to compensate the biological function performed by the target protease, leading to discouraging results with compounds which otherwise could be useful in humans or if experimented in other animal models. The understanding of the different complexity of human and mouse degradomes will be helpful to anticipate this kind of result and to design alternative strategies to investigate the role of proteases in specific physiological or pathological processes.

The above data reflect the utility of degradome comparison between species to identify differential genes which might contribute to differences in physiological processes. The current sequencing of numerous mammalian genomes is extending these findings and will contribute to identify the molecular mechanisms responsible for some of the physiological differences between organisms. In this sense, the availability of the chimpanzee genome sequence (Mikkelsen et al. 2005) has opened the possibility to characterize its degradome and to define small changes in protease-coding genes which might explain differences between these closely related species. The chimpanzee degradome is virtually identical to the human degradome, with 567 protease genes and more than 99.1% identity at the amino acid level between orthologs. However, there are seven protease genes which are differential between human and chimpanzee, most of them being implicated in immunological functions (Puente et al. 2005a, Saleh et al. 2004). On the contrary, more than 75 chimpanzee genes encode proteases which are identical to their human orthologs. These genes can be classified in two different groups, proteases implicated in housekeeping processes, such as the proteasome components, and proteases that participate in neurological processes, including the three secretases implicated in Alzheimer disease or the UCHL1 deubiquitinase which is mutated in patients with Parkinson disease (Esler and Wolfe 2001, Leroy et al. 1998). Other proteases implicated in cancer, such as the deubiquitinating enzyme CYLD, which is mutated in patients with cylindromatosis (Bignell et al. 2000), are also identical between human and chimpanzee. The different susceptibility of humans and chimpanzees to disorders such as Parkinson or Alzheimer disease, and the low incidence of cancer in non-human primates (Beniashvili 1989, Varki 2000), reinforces the hypothesis that changes in gene regulation might be responsible for the different susceptibility to these pathologies in both species (Enard et al. 2002, King and Wilson 1975). Future analysis aimed at identifying the mechanisms involved in the regulation of proteolytic gene expression might shed insights into the molecular basis of these differences.

In summary, the study of mammalian genomes has helped to understand the complexity of proteolytic enzymes and the importance of proteolysis for human biology. Current efforts, aimed at identifying the substrate specificity and biological function for most protease genes, together with the development of novel protease inhibitors with increased specificity and reduced side effects, will greatly benefit the treatment of numerous human pathologies, including cancer.

Protease Genomics and Its Application to Cancer Degradome Research

The classical association of proteases with cancer derives from the proposal that these enzymes could be responsible for the degradation of extracellular matrix components facilitating the release and dissemination of tumour cells and, finally, the generation of metastasis. Accordingly, multiple studies have examined the potential value of these enzymes as targets for cancer therapy. Unfortunately, most clinical trials with protease inhibitors have yielded negative results which has made necessary to re-evaluate the role of proteases in cancer (*see* Chap. 36 by Fingleton, this volume). Recent molecular and cell biology studies as well as loss-of-function and gain-of-function experiments in animal models have led to the somewhat unexpected conclusion that proteolytic enzymes contribute to all stages of tumour progression and not only to the later stages as was originally proposed (Borgono and Diamandis 2004, Egeblad and Werb 2002, Folgueras et al. 2004, Mohamed and Sloane 2006). Thus, proteases act as signalling molecules involved in the regulation of several cellular processes essential for cancer biology. These protease-regulated processes include cell proliferation and adhesion, migration, differentiation, angiogenesis, senescence, apoptosis, autophagy and host defense evasion. Nevertheless, it has been widely assumed that, similar to the previously described pro-metastatic action of these enzymes, the participation of proteases in all these additional cellular processes also favours the tumour instead of the host and facilitates tumour progression. However, the clinical observations showing the acceleration of tumour growth in patients treated with particular broad-spectrum metalloproteinase inhibitors were a clear indication that some proteases might play anti-tumour roles (Coussens et al. 2002). Further studies based on the generation of loss-of-function animal models have provided definitive evidence on the existence of proteases with anti-tumour properties (Balbin et al. 2003, McCawley et al. 2004, Overall and Kleifeld 2006). Interestingly, very recent work has shown that proteases with tumour-suppressive properties are not rare exceptions to the widely assumed rule that protease up-regulation in cancer is synonymous with tumour progression and poor clinical outcome. Thus, it has been reported that proteases of all major catalytic classes may act as tumour suppressors (López-Otín and Matrisian 2007). This growing category of proteases with tumour-defying functions includes several MMPs (matrix metalloproteinases), ADAM-TSs (disintegrin-metalloproteinases with thrombospondin domains), cathepsins, caspases, DUBs (deubiquitinating enzymes) and kallikreins. The molecular mechanisms used by proteases to exert their anti-tumour properties are quite diverse and influence all stages of cancer progression. Thus, tumour-suppressive proteases may negatively regulate cell growth and survival of tumour cells, inhibit angiogenesis, stimulate apoptosis or modulate the inflammatory responses elicited by cancer cells.

The identification of anti-target proteases that favour the host instead of the tumour has made necessary the design of novel approaches to identify the relevant proteases which must be targeted in each individual cancer patient (Acuff et al. 2006, Overall et al. 2004). This aspect is of special interest in the case of proteases belonging to large families identified in the genomic analysis of the human degradome, which are composed of multiple members with structural relationship but playing opposite functions in cancer. As discussed above, recently developed devices such as those called CLIP-CHIP and Hu/Mu-ProtIn have provided the first experimental approaches to facilitate the profiling of tumour proteases. This global analysis of the cancer degradome has already yielded important new information including the identification of the protective role for MMP-12 in lung cancer (Acuff et al. 2006). Likewise, these analyses have pointed to the existence of a number of proteases previously unsuspected to be related to cancer, which may be of future clinical interest as molecular markers for the diagnosis or prognosis of malignant tumours.

In summary, the genomic and functional analysis of proteolytic systems associated with cancer has revealed the large and growing complexity of this field. Over the last few years, our view has evolved from the classical consideration of proteases as non-specific and late-acting prometastatic enzymes, to the recognition of their key roles in early stages of cancer and finally to the identification of their dual functions as pro- or anti-tumourigenic enzymes. Accordingly, any attempt to understand the biological and pathological relevance of proteases in cancer must take into account the large structural and functional diversity of proteolytic systems operating in all stages of the disease. There are still multiples challenges ahead before translating these recent findings into clinical applications but hopefully, the novel information derived from protease genomics may finally lead to the validation of some of these enzymes as important components of the future strategies for cancer treatment.

Acknowledgements The work in our laboratory was supported by grants from Ministerio de Educación y Ciencia-Spain, Fundación M. Botín, Fundación Lilly, Fundación La Caixa and European Union (Cancer Degradome-FP6). The Instituto Universitario de Oncología is supported by Obra Social Cajastur-Asturias.

References

Acuff H. B., Sinnamon M., Fingleton B., et al. (2006). Analysis of host- and tumor-derived proteinases using a custom dual species microarray reveals a protective role for stromal matrix metalloproteinase-12 in non-small cell lung cancer. Cancer Res 66:7968–7975.

Balbin M., Fueyo A., Tester A. M., et al. (2003). Loss of collagenase-2 confers increased skin tumor susceptibility to male mice. Nat Genet 35:252–257.

Barrett A. J., Rawlings N. D., Woessner J. F. (2004). Handbook of proteolytic enzymes. 2nd ed. Amsterdam; Boston: Elsvier Academic Press.

Beniashvili D. S. (1989). An overview of the world literature on spontaneous tumors in nonhuman primates. J Med Primatol 18:423–437.

Bignell G. R., Warren W., Seal S., et al. (2000). Identification of the familial cylindromatosis tumour-suppressor gene. Nat Genet 25:160–165.

Boatright K. M., Deis C., Denault J. B., et al. (2004). Activation of caspases-8 and -10 by FLIP(L). Biochem J 382:651–657.

Borgono C. A., Diamandis E. P. (2004). The emerging roles of human tissue kallikreins in cancer. Nat Rev Cancer 4:876–890.

Collins F. S., Lander E. S., Rogers J., et al. (2004). Finishing the euchromatic sequence of the human genome. Nature 431:931–945.

Coussens L. M., Fingleton B., Matrisian L. M. (2002). Matrix metalloproteinase inhibitors and cancer: Trials and tribulations. Science 295:2387–2392.

Coussens L. M., Tinkle C. L., Hanahan D., et al. (2000). MMP-9 supplied by bone marrow-derived cells contributes to skin carcinogenesis. Cell 103:481–490.

Deussing J., Kouadio M., Rehman S., et al. (2002). Identification and characterization of a dense cluster of placenta-specific cysteine peptidase genes and related genes on mouse chromosome 13. Genomics 79:225–240.

Diaz-Perales A., Quesada V., Peinado J. R., et al. (2005). Identification and characterization of human archaemetzincin-1 and -2, two novel members of a family of metalloproteases widely distributed in Archaea. J Biol Chem 280:30367–30375.

Egeblad M., Werb Z. (2002). New functions for the matrix metalloproteinases in cancer progression. Nat Rev Cancer 2:161–174.

Enard W., Khaitovich P., Klose J., et al. (2002). Intra- and interspecific variation in primate gene expression patterns. Science 296:340–343.

Esler W. P., Wolfe M. S. (2001). A portrait of Alzheimer secretases—New features and familiar faces. Science 293:1449–1454.

Finn R. D., Mistry J., Schuster-Bockler B., et al. (2006). Pfam: Clans, web tools and services. Nucleic Acids Res 34:D247–251.

Folgueras A. R., Pendas A. M., Sanchez L. M., et al. (2004). Matrix metalloproteinases in cancer: From new functions to improved inhibition strategies. Int J Dev Biol 48:411–424.

Fujinaga M., Cherney M. M., Oyama H., et al. (2004). The molecular structure and catalytic mechanism of a novel carboxyl peptidase from Scytalidium lignicolum. Proc Natl Acad Sci U.S.A. 101:3364–3369.

Gibbs R. A., Weinstock G. M., Metzker M. L., et al. (2004). Genome sequence of the Brown Norway rat yields insights into mammalian evolution. Nature 428:493–521.

Gururajan R., Lahti J. M., Grenet J., et al. (1998). Duplication of a genomic region containing the Cdc2L1–2 and MMP21–22 genes on human chromosome 1p36.3 and their linkage to D1Z2. Genome Res 8:929–939.

Hubbard T. J., Aken B. L., Beal K., et al. (2007). Ensembl (2007). Nucleic Acids Res 35:D610–617.

King M. C., Wilson A. C. (1975). Evolution at two levels in humans and chimpanzees. Science 188:107–116.

Leroy E., Boyer R., Auburger G., et al. (1998). The ubiquitin pathway in Parkinson's disease. Nature 395:451–452.

López-Otín C., Matrisian L. M. (2007). Tumour microenvironment: Emerging roles of proteases in tumour suppression. Nat Rev Cancer 7:800–808.

López-Otín C., Overall C. M. (2002). Protease degradomics: A new challenge for proteomics. Nat Rev Mol Cell Biol 3:509–519.

McCawley L. J., Crawford H. C., King L. E., Jr., et al. (2004). A protective role for matrix metalloproteinase-3 in squamous cell carcinoma. Cancer Res 64:6965–6972.

Mikkelsen T. S., Hillier L. W., Eichler E. E., et al. (2005). Initial sequencing of the chimpanzee genome and comparison with the human genome. Nature 437:69–87.

Mohamed M. M., Sloane B. F. (2006). Cysteine cathepsins: Multifunctional enzymes in cancer. Nat Rev Cancer 6:764–775.

Mohammed F. F., Smookler D. S., Taylor S. E., et al. (2004). Abnormal TNF activity in Timp3-/- mice leads to chronic hepatic inflammation and failure of liver regeneration. Nat Genet 36:969–977.

Overall C. M. (2002). Molecular determinants of metalloproteinase substrate specificity: Matrix metalloproteinase substrate binding domains, modules, and exosites. Mol Biotechnol 22:51–86.

Overall C. M., Kleifeld O (2006). Tumour microenvironment—opinion: Validating matrix metalloproteinases as drug targets and anti-targets for cancer therapy. Nat Rev Cancer 6:227–239.

Overall C. M., López-Otín C. (2002). Strategies for MMP inhibition in cancer: Innovations for the post-trial era. Nat Rev Cancer 2:657–672.

Overall C. M., Tam E. M., Kappelhoff R., et al. (2004). Protease degradomics: Mass spectrometry discovery of protease substrates and the CLIP-CHIP, a dedicated DNA microarray of all human proteases and inhibitors. Biol Chem 385:493–504.

Puente X. S., López-Otín C. (2004). A genomic analysis of rat proteases and protease inhibitors. Genome Res 14:609–622.

Puente X. S., Gutiérrez-Fernández A., Ordóñez G. R., et al. (2005a). Comparative genomic analysis of human and chimpanzee proteases. Genomics 86:638–647.

Puente X. S., Sanchez L. M., Gutiérrez-Fernández A., et al. (2005b). A genomic view of the complexity of mammalian proteolytic systems. Biochem Soc Trans 33:331–334.

Puente X. S., Sánchez L. M., Overall C. M., et al. (2003). Human and mouse proteases: A comparative genomic approach. Nat Rev Genet 4:544–558.

Rawlings N. D., Tolle D. P., Barrett A. J. (2004). MEROPS: The peptidase database. Nucleic Acids Res 32 Database issue:D160–164.

Rosenthal N., Brown S. (2007). The mouse ascending: Perspectives for human-disease models. Nat Cell Biol 9:993–999.

Saleh M., Vaillancourt J. P., Graham R. K., et al. (2004). Differential modulation of endotoxin responsiveness by human caspase-12 polymorphisms. Nature 429:75–79.

Schwartz D. R., Moin K., Yao B., et al. (2007). Hu/Mu ProtIn oligonucleotide microarray: Dual-species array for profiling protease and protease inhibitor gene expression in tumors and their microenvironment. Mol Cancer Res 5:443–454.

Turk B (2006). Targeting proteases: Successes, failures and future prospects. Nat Rev Drug Discov 5:785–799.

Varki A (2000). A chimpanzee genome project is a biomedical imperative. Genome Res 10:1065–1070.

Waterston R. H., Lindblad-Toh K., Birney E., et al. (2002). Initial sequencing and comparative analysis of the mouse genome. Nature 420:520–562.

Zucker S., Pei D., Cao J., et al. (2003). Membrane type-matrix metalloproteinases (MT-MMP). Curr Top Dev Biol 54:1–74.

Chapter 2
The CLIP-CHIP^{TM}: A Focused Oligonucleotide Microarray Platform for Transcriptome Analysis of the Complete Human and Murine Cancer Degradomes

Reinhild Kappelhoff, Claire H. Wilson, and Christopher M. Overall

Abstract By modifying the secretome, cancer disturbs normal tissue homeostasis leading to pathological changes, often driven by perturbed proteolysis. Besides the traditional dogma of tumor promotion and metastasis through degradation of the extracellular matrix, extracellular proteases are now recognized to play more important roles as signaling molecules that modulate the web of cytokines and growth factors that orchestrate extracellular homeostasis and cell proliferation, adhesion, migration, differentiation, apoptosis, and evasion from the immune system. Although proteases have long been regarded as promising drug targets in cancer, increasingly it is being recognized that many proteases have beneficial roles in mitigating the detrimental effects of cancer, classifying these proteases as drug anti-targets. Discrimination of protease drug targets and anti-targets in cancer is therefore critical for understanding oncogenesis and for drug development programs. Complicating target and anti-target identification, is the fact that proteases do not act in isolation, but interact in pathways, circuits and cascades with other proteases to form the protease web. The CLIP-CHIP^{TM} is the only microarray platform that focuses on all proteases, non-proteolytic homologues, and protease inhibitors in the human and mouse genomes making it an ideal platform for an unbiased transcriptomic analysis of the cancer degradome – the complete repertoire of proteases and inhibitors expressed in cancer, reactive

R. Kappelhoff
University of British Columbia, Centre for Blood Research, Department of Oral Biological and Medical Sciences, #4.420-2350 Health Sciences Mall, Vancouver, BC, V6T 1Z3, Canada, e-mail: reinhild@interchange.ubc.ca

C. M. Overall
University of British Columbia, Centre for Blood Research, Departments of Oral Biological and Medical Sciences; Biochemistry and Molecular Biology, #4.420-2350. Health Sciences Mall, Vancouver, BC, V6T 1Z3, Canada, e-mail: chris.overall@ubc.ca

D. Edwards et al. (eds.), *The Cancer Degradome.*
© Springer Science + Business Media, LLC 2008

stromal cells, and normal tissue – and the changes wrought in the protease web by cancer. By analysis of cancer-type-specific expression profiles of proteases and inhibitors, the CLIP-CHIP oligonucleotide-based microarray assists in understanding how proteases modulate the tumor and the enveloping secretome.

Introduction: The Cancer Degradome

The information content of the secretome is one of the most important regulatory influences on a cell. Since proteolysis represents one of the last opportunities that a cell can modulate the composition and function of the secretome, proteolysis arguably represents the most important of the posttranslational modifications of protein. Perturbed signal pathways, both extracellular and intracellular, can result in pathology. In cancer, altered signaling can impart selective advantage to the tumor, resulting in loss of regulated cell growth and tissue localization constraints, leading to further oncogenic changes that promote metastasis. Extracellular proteolytic processing of signal molecules in the secretome changes their activity – including activation (Tester et al. 2007), release of inhibitory binding proteins (Dean et al. 2007), loss of agonist function (McQuibban et al. 2001, Dean et al. 2007), switching activity to antagonism (McQuibban et al. 2000), or switching receptor specificity (Vergote et al. 2006) shedding from the cell surface (Dean and Overall 2007) – and so profoundly changes signal pathways and hence cell function. Intracellular proteases play critical roles in the initiation of apoptosis and in the turnover of cytosolic and nuclear proteins, including the regulation of protein ubiquitination. By dynamically altering the levels of ubiquitin-conjugated protein, the deubiquitinating proteases regulate levels of second messenger molecules, transcription factors, and of tumor suppressors such as p53 (Li et al. 2002). Hence, at both the extracellular and intracellular levels, proteases are key components of signaling pathways and in both compartments proteolytic changes can lead to oncogenesis – similarly, oncogenesis can result in altered proteolytic potential which might manifest as increased oncogenesis and metastasis.

Proteolytic processing of signaling molecules, even of only 2–4 amino acids, can profoundly change their biological activity. Elevated expression of 48 genes was found to be associated with breast cancer metastases to the lung and of those validated three signature genes were expressed only in the most virulent of the lung metastases: matrix metalloproteinase (MMP)-2, SPARC, and IL13RA2 (Minn et al. 2005a). In breast carcinoma metastases to the bone MMP-1, connective tissue growth factor (CTGF) and CXCR-4 were key molecules (Minn et al. 2005b). CXCL-12 (SDF-1) binds the chemokine receptor CXCR-4, which is one of the bone metastasis signature genes, and is cleaved and inactivated by MMP-2 and MMP-1 (McQuibban et al. 2001). CTGF is a substrate of MMP-1 and MMP-2 and mobilizes the vascular endothelial growth factor (VEGF) (Dean et al. 2007), providing an alternate explanation for the recent results from the Massague group who suggested that MMP-2 was involved in remodeling of the vascular architecture in aggressive breast cancer metastases to the lung (Gupta et al. 2007). Therefore, simple annotation of normal or transformed cell transcripts or nuclear, cytosolic and secreted proteomes is not sufficient to decipher their information content and pathologic potential. The functional annotation of processed proteins in the cancer proteome is therefore a priority along with the identification of the proteases that mediate these important proteolytic events. That is why cancer degradomics is such a critically important area of research. Degradomics embraces all systems biology studies of proteases. It includes the annotation of the proteases and inhibitors present in a cell or tissue at a particular time (the protease degradome), their substrates (the substrate degradome), and their interactomes – hence the proteolytic potential of the cancer cell or reactive tissue stroma.

Protease Targets and Anti-targets

A long-held assumption has been that proteases that are highly overexpressed in cancer tissues are drug targets. This is wrong. In addition to the destructive role of proteases in cancer and oncogenesis, other proteases have host protective functions. These include mitigating the effects of oncogenic changes, such as by growth suppression and antiangiogenic activities, as well as exerting important roles in host defense. Hence, increases in activity of drug anti-targets are associated with protective host responses. Since these may be expressed by the reactive stroma or by infiltrating inflammatory and immune cells then anti-target proteases are often present in the secretome at the tumor–stroma interface. So, the mere annotation of protease levels by transcript analysis of cancer is therefore insufficient to associate proteases with oncogenesis. For example, MMP-8 is expressed mainly in neutrophils and is associated with inflammatory conditions. However, the absence of MMP-8 expression in the Mmp8$^{-/-}$ mouse showed a strong increase in skin tumors (Balbin et al. 2003). Although overexpression in cancer of some kallikreins can lead to a poor clinical outcome, increased levels of other kallikreins lead to a favorable prognosis (Borgono and Diamandis 2004). So, proteases can both promote tumor progression and metastasis, but a growing number, to date around 30, contribute to tumor suppression (López-Otín and Matrisian 2007), making these drug anti-targets (Overall and Kleifeld 2006). Hence, it is important to annotate both target and anti-target proteases in cancer to design rational drug therapies.

The Protease Web

Complicating the analysis of proteases in cancer as targets and anti-targets is the growing realization that proteases do not operate alone, but interconnect in signaling or proteolytic pathways (Overall and Kleifeld 2006). Proteases form regulatory circuits where information can feed back, and through amplification cascades with a unified flow of information. In so doing, proteases form a dynamic and flexible web of interactions – the protease web – that is embedded in every proteome. Further complexity of this system occurs in that proteases also interact with cofactors, interactor proteins, receptors, inhibitors, substrates, and their cleavage products to modify proteolytic function and biological outcomes. Cancer and cancer therapy disturbs the protease web resulting in altered biological behavior that is not immediately predictable from analysis of protease levels alone. Therefore, although identification of all components in the protease web is critically important in the understanding of the biological role of proteases in the cancer degradome (Overall and Kleifeld 2006), it is as important to understand at the systems level the interactions and substrates of proteases.

The CLIP-CHIP

The protease degradome comprises 1.7% of the human and mouse genomes, the second largest group after the ubiquitin-ligase family and larger than the kinase family. According to their mechanism of catalysis, human and mouse proteases are divided into the well-established five classes of aspartic, cysteine, metallo, serine, and threonine proteases (Fig. 2.1a) (Barrett et al. 1998), which are further sub-

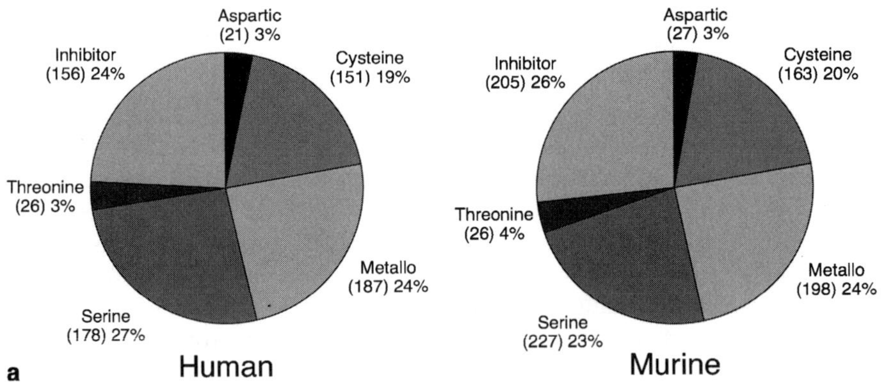

Fig. 2.1a The human and murine degradome. Class distribution of human and murine proteases and inhibitors in the degradome. Serine and metalloproteases are the most highly represented proteases, followed by cysteine proteases. In contrast, there are only a few aspartic and threonine proteases likely reflecting their highly specialized roles

divided into 69 families according to the MEROPS the peptidase database criteria (Rawlings et al., 2008, www.merops.ac.uk). Besides active proteases, there are numbers of protease-like proteins, which are predicted to be inactive in that they have a change in a residue crucial for proteolytic activity and are called non-proteolytic homologues. These molecules are suggested to have important roles as inhibitory or regulatory molecules by titrating inhibitors from the cell medium and thereby increasing the proteolytic activity and by having the ability to bind substrates through the inactive catalytic or exosite domains to act as a dominant negative molecule (Lopez-Otin and Overall 2002). Protease pseudogenes have been found in both human and mouse genomes and have an active counterpart in other species. Pseudogenes are derived by duplication, retrotransposition, frameshifts or by integrated stop codons. The humane degradome is composed of 565 proteases with 156 inhibitor genes. The mouse degradome is more complex with 641 pro-teases and 205 inhibitors annotated to date (Figs. 2.1a and 2.1b). This increased complexity is mainly derived from the expansion of several protease families, including placental cathepsins, testases, glandular kallikreins, and hematopoietic serine proteases, which are involved in immunological or reproductive functions. To compensate for this increase in proteolytic activity, the cysteine and serine protease inhibitors have also expanded in the mouse (Puente et al. 2003).

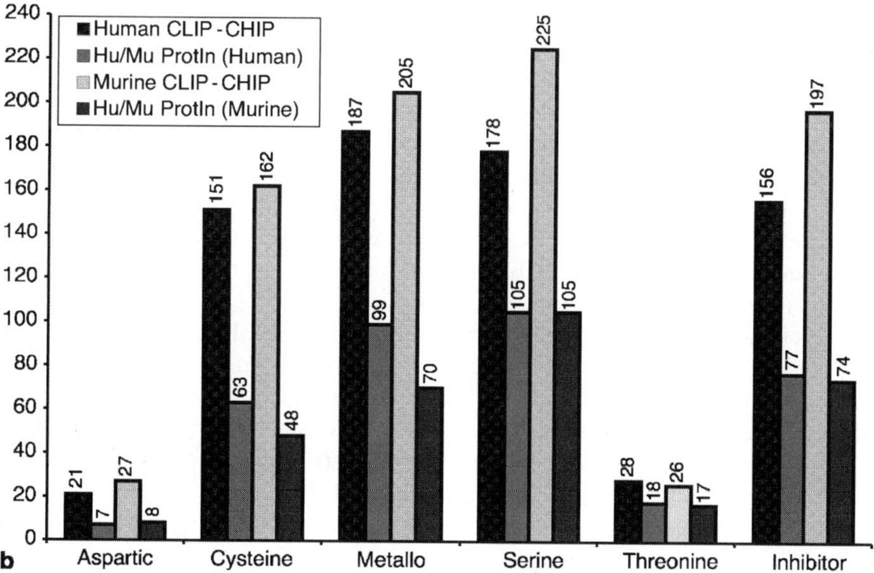

Fig. 2.1b Comparison of protease and inhibitor distribution between the CLIP-CHIP and the Hu/Mu ProtIn Affymetrix Chip. The CLIP-CHIP contains oligonucleotides for 565 human and 646 murine proteases, and 156 human and 197 murine inhibitor gene transcripts; the Hu/Mu ProtIn Affymetrix Chip contains 292 human and 254 murine proteases and 77 human and 74 murine inhibitors (Acuff et al. 2006, Supplementary data)

In cancer research, mice have been the primary models for both understanding the nature of the disease and also as diagnostic and therapeutic tools for tumor cancer treatment (Egeblad and Werb 2002). Because of the differences between the human and mouse degradomes, animal models experiments have to be carefully designed and the results carefully interpreted. But even with this drawback, animal models continue to be important for studying diseases. Therefore, to annotate the human and murine degradomes the CLIP-CHIP microarray was designed. As will be described in this chapter, every human and murine protease, inactive homologue, and inhibitor can be profiled at the transcript level by hybridization against 70-mer oligonucleotides specific for the each of these genes using the CLIP-CHIP.

Transciptomics and the Advent of Microarrays

Microarray technology has significantly changed the way we can measure and observe gene expression at the mRNA level within a given biological sample of interest, allowing the expression of tens or hundreds of thousands of genes to be monitored at a single time point or relative to another sample within a single experiment. Microarrays are thus the ultimate tool for transciptomics, allowing for system-wide analysis of the transcriptome. The transcriptome is defined as being the complete set of messenger RNA molecules that are produced from the genome of a given cell, tissue, or whole organism at any given time. In contrast to the "fixed" genome, the transcriptome can dynamically change in response to environmental and pathological stimuli, thus reflecting the active expression of genes at a particular point in time. Before the advent of microarrays, researchers were limited to using more traditional methods for the analysis of gene expression such as Northern blotting, quantitative real-time polymerase chain reaction (qRT-PCR), and ribonuclease protection assays (RPA). Such techniques limited the researchers to investigating mRNA level expression for only single or a limited number of genes at a time, eliminating the feasibility of performing a transcriptome-wide analysis.

One of the biggest limitations of microarrays is that mRNA level expression does not directly relate to protein abundance. Although proteins are the ultimate products of genes, measuring mRNA level expression is a good starting point for functional gene characterization and is currently a considerably cheaper technology then measuring direct protein levels via mass spectrometry resources or by using ELISA (Enzyme-linked immunosorbent assay) technology.

In contrast with the commonly used whole-genome arrays, several companies currently offer specialized microarrays [e.g., Oligo GEArray® (SuperArray Bioscience Division, MD); DualChip® (Eppendorf, Hamburg, Germany); Amplichip® (Roche Basel, Switzerland); SurePrint Microarray Kits (Agilent, Santa Clara, CA)] or they offer ready-to-print oligonucleotide sets [e.g., Operon® Array-Ready Oligo Sets™ (Operon, Huntsville, AL)] with a smaller gene coverage specifically focused on research topics like apoptosis, biomarkers, cancer, cell cycle, common diseases, cytokine and inflammatory response, extracellular matrix and adhesion molecules, neuroscience, signal transduction, stem cell and development, toxicology and drug metabolism, or tumor metastasis. Therefore, more specific research questions can be addressed while reducing the amount of raw data generated that can be vast when whole-genome arrays are used. The development of focused microarrays has been an important contribution for clinical diagnostic purposes. In 2004, the Ampli-Chip® Cytochrome P450 Genotyping Test manufactured by Roche (Switzerland) for use with the Affymetrix GeneChip (Affymetrix, Santa Clara, CA) became the first microarray genetic-test approved by the U.S. Food and Drug Administration (FDA). Providing comprehensive genotyping of the CYP2D6 and CYP2C19 genes from genomic DNA, the AmpliChip^{TM} CYP450 Test helps determine how a patient may metabolize and therefore respond to a variety of prescription-based drugs, hence allowing for physicians to individualize drug administration and dosage. The AmpliChip Leukemia and AmpliChip p53 are currently in development (www. roche.com).

In February 2007, Agendia's MammaPrint® breast cancer prognosis test (Amsterdam, The Netherlands) became the worlds first In Vitro Diagnostic Multi-variate Index Assay (IVDMIA) to receive U.S. FDA approval for marketing as a prognostic test for the development of distant metastasis in lymph node-negative patients and determines the likelihood of breast cancer returning within 5–10 years after a woman's initial cancer. The MammaPrint®, a 60-mer oligonucleotide-based microarray (manufactured by Agilent, Santa Clara, CA), measures the activity of 70 genes, thereby giving a prognosis expression signature that provides information about the likelihood of tumor recurrence using a specific formula or algorithm to produce a score that determines whether the patient is deemed at low risk or high risk of the cancer spreading to another site (van't Veer et al. 2002, Weigelt et al. 2005, Glas et al. 2006).

Generic Features of Microarray Technology

Microarrays are a type of ligand assay based on the same principles as immunoassays, and Northern and Southern Blots. Regardless of whether the microarrays are manufactured commercially or in-house, cDNA or oligonucleotide based, robotically spotted or in situ synthesized, they all share a number of generic features in regards to their underlying technology such as the probe or "spot," the target or sample probe and the solid-phase medium of the array platform, for example, glass or plastic slides similar to a microscope slide and silica chips. The probe or "spot" typically refers to the single stranded polynucleotide of known or unknown sequence that is fixed onto the array at a specific location. Each individual spot on the grid of an array generally represents an individual gene, thus serving as an experimental assay for the relative level of expression for that given gene. In general, there are two basic approaches for fixing probes onto the array: unique oligonucleotide probes can be individually synthesized nucleotide by nucleotide directly onto the array by in situ synthesis such as by using photolithography (Lockhart et al.

1996). Agilent uses ink jet printing with standard phosphoramidite chemistry in solution. Combimatrix uses electrochemical acid generation in specified locations on a silicon matrix. Presynthesized DNA, cDNA, oligonucleotides, or PCR products are robotically spotted onto the array, where they become directly covalently bound to the array surface or are cross-linked via UV-irradiation. When oligonucleotide probes are used they are generally referred to as being either short, 24–30 bases in length, or long, 60–70 bases in length, while the presynthesized PCR and cDNA products are typically hundreds to thousands of base pairs in length (Yauk et al. 2004). Probes of unknown sequence may also be fixed to the array if a custom unsequenced cDNA library is used. The term "target" or "sample probe" commonly refers to the polynucleotide from the biological sample of interest that hybridizes to a fixed complementary probe sequence present on the array. Typically, the target or sample probe is amplified RNA (aRNA) or cDNA synthesized from total RNA or mRNA extracted from the sample of interest. For detection purposes, this probe is typically synthesized using fluorescently or biotinylated nucleotides or chemically labeled with fluorophores such as Cyanine 3 (Cy3) and Cyanine 5 (Cy5) dyes postsynthesis.

Presently, the two most commonly used techniques for microarray analyses are one-color experiments using in situ synthesized short oligonucleotide arrays such as the Affymetrix GeneChip and competitive two-color hybridization experiments using robotically spotted cDNA or long oligonucleotide arrays or the Agilent, Combimatrix, and Nimblegen arrays. In brief, the two-color array platform allows for the hybridization of two differently labeled samples (e.g., disease vs. nondiseased) to a single array while one-color/channel arrays allow for hybridization of only one labeled sample per array therefore eliminating the ability for direct comparisons to be made. Regardless of platform the underlying basis of microarray technology is complementary base pairing between the array-fixed probes and the target or sample probe. This allows for the determination of the relative levels of mRNA expression of target sequences with the sample via measurement of the quantity of labeled target that binds to each immobilized spot, also termed a feature, of DNA (Colebatch et al. 2002).

Transcriptomic Analysis of the Cancer Degradome with the Two-Color CLIP-CHIP Microarray Platform

Transcriptome analysis of the cancer degradome is an important first step to understand the biology of cancer and is an excellent starting point for identifying which protease targets and anti-targets are expressed. Moreover, identifying the expression levels of all proteases and inhibitors renders possible the determination of the active pathways of the protease web within specific cancer types and hence protease-specific disease biomarkers. Transcriptome-level analysis can lead to the detection of slight subtle changes in the underlying gene expression patterns of the protease web between patients and different tumor types, resulting in the identification of both diagnostic and prognostic markers of disease. Identifying subtle differences can help identify protease pathways warranting further investigation ultimately leading to the development of novel therapeutics and individualized molecular medicine with treatment being tailored specifically for the individual patient. Microarrays such as the highly specialized CLIP-CHIP provide an effective way to investigate the cancer degradome at the transcriptome level.

The Two-Color CLIP-CHIP

The CLIP-CHIP is a dedicated and complete two-color oligonucleotide microarray of all human and mouse protease, non-proteolytic homologues and inhibitor sequences (Overall et al. 2004). Every gene transcript is analyzed by hybridizations to unique 70-mer oligonucleotides, which are robotically spotted onto a glass matrix and cross-linked via UV-irradiation. This format allows for great flexibility in design upgrades. As newly identified proteases enter the databases, unique oligonucleotides can be designed and spotted onto the CLIP-CHIP at the next printing. The CLIP-CHIP platform also gives the additional benefit that oligonucleotides for specifically relevant biological questions, for example, for breast, lung, or prostate cancer-specific genes, can be added to the chip as a separate focused array. In this way, specific correlations in gene expression may be determined for proteases and other cancer-specific genes. A number of oligonucleotides on the CLIP-CHIP, representing synthetic sequences that are nonhomologous with any human or murine sequence (negative controls) and several housekeeping genes (positive controls) have also been printed onto the array, which help in analysis and normalization of the microarray data. Oligonucleotides coding for the three genes gpt (Xanthine guanine phosphoryltransferase), pac (Puromycin N-acetyltransferase), and neo (Neomycin phosphotransferase) commonly used in the development of stable cell lines and knockout mouse models have also been added to the microarray as a positive control for tissue culture or knockout model experiments. An oligonucleotide was designed from the cDNA sequence of the green fluorescence protein (GFP) and printed on the CLIP-CHIP to aid with slide orientation and gridding.

571 human oligonucleotides were obtained from the Human Genome Oligo Set Version 2 and 740 murine oligonucleotides were obtained from the Murine Genome Oligo Set Version 2, 3, and 4 (Operon, Huntsville AL, www.operon.com). Oligonucleotides for genes not represented in the Human Oligo Set (144) or Murine Oligo Set (74) were synthesized after extensive BLAST (basic local alignment search tool) analyses to identify unique sequences. The CLIP-CHIP is printed in two replicate subarrays within one slide; therefore, there are three printed versions of the CLIP-CHIP available: a human CLIP-CHIP with two human subarrays; a murine CLIP-CHIP with two murine subarrays; and a hybrid CLIP-CHIP which contains one human and one murine subarray that is still under validation to

determine cross-species hybridization between the oligonucleotides. The CLIP-CHIP platform is designed in such a way that it is easy to produce customized microarray chips with additional genes, for additional information in a specific study.

Comparison of Technologies for Transcriptomic Analysis of the Cancer Degradome

Within this volume, two other technologies developed for the transcriptomic analysis of the cancer degradome are presented: the Hu/Mu ProtIn microarray chip Schwatz, Chapter 3 and qRT-PCR assay (Pennington, Chapter 4). All three transcriptome platforms have their separate advantages and disadvantages making all three approaches complementary. No one approach will provide comprehensive and qualitative information and a combination of all three platforms is recommended.

The Hu/Mu ProtIn microarray chip is a two-species one-channel Affymetrix-based microarray. It is a focused array, containing selected protease, inhibitor, and interactor oligonucleotides from both human and mouse which have been reported to be important in cancer and has particular use in the determination of cellular origin of proteases and their inhibitors in xenograft models of human cancer (Acuff et al. 2006). Short 25-mer oligonucleotides have been designed and validated to distinguish between proteases and inhibitors expressed by the tumor (human cells) and the host stromal tumor microenvironment (mouse cells) (Schwartz et al. 2007). Being a focused array, the Hu/Mu ProtIn chip contains a selected set of proteases (292 human/254 mouse) and protease inhibitors (77 human/74 mouse). In contrast, the CLIP-CHIP contains the complete repertoire of all proteases (565 human/621 mouse) and protease inhibitors (156 human/205 mouse), thus giving the CLIP-CHIP the advantage of profiling the expression of the full spectrum of the degradome in an unbiased manner (Fig. 2.1b).

Manufactured by Affymetrix the short oligonucleotide probes of the Hu/Mu ProtIn chip are in situ synthesized directly onto the array surface and are an industry standard. While this mode of probe attachment readily produces high quality, reproducible arrays it is not as easily customizable for individual projects as the robotically spotted CLIP-CHIP array where changes or additions to oligonucleotide probes can simply be added in the next round of printing. An integral component of the Hu/Mu ProtIn chip is probe redundancy with each gene being represented by one probe set consisting of 11 probe pairs, each covering a different 25-mer region of the target mRNA. Each of these probe pairs consists of a complementary perfect match (PM) oligonucleotide and a mismatch (MM) oligonucleotide in which the 13th nucleotide in the sequence is changed to its complement thereby functioning as a nonspecific hybridization control and reducing the likelihood of cross-hybridization. In contrast the CLIP-CHIP contains a single 70-mer oligonucleotide probe per gene transcript. Whereas longer probes are much more sensitive and therefore, able to detect subtle changes in gene expression especially for poorly expressed genes. However, they can be more susceptible to specificity issues. The probability for cross-hybridization (both intra-and interspecies) is higher with a 70-mer probe than with a smaller probe. In contrast, a 70-mer is far more sensitive than a smaller probe. Such probes are also more tolerant of sequence mismatching and thus are more suitable for the analysis of highly polymorphic regions as commonly found within cancer. However, species similarity can render species to species cross-hybridization unavoidable for a limited number of genes, despite high stringency oligonucleotide design criteria.

The Hu/Mu ProtIn chip and CLIP-CHIP are based on two different microarray platforms. Being a one-color array the Hu/Mu ProtIn chip allows for the hybridization of only one sample per chip. In contrast the two-color CLIP-CHIP microarray allows for the competitive hybridization of two alternatively labeled samples (e.g., disease vs. non-disease) to the one chip. Competitive hybridization allows for direct comparisons to be made in each array and reduces the number of slides required per experiment and hence the overall cost of the microarray experiment to be performed. While it is probable that the Affymetrix chip may have a higher sequence specificity, its sensitivity to detect low-abundance transcripts is inherently lower then 70-mer oligonucleotide platforms such as the CLIP-CHIP. Further, a typical microarray analysis with the CLIP-CHIP only requires 100 ng of total RNA and 2 g of amplified RNA from a control and test sample labeled and hybridized to the same array (Kappelhoff and Overall 2007). As it is typical for Affymetrix chips, the Hu/ Mu ProtIn chip reported results using 5 g total RNA as 15 g biotinylated cRNA from each sample are needed for the hybridization to individual microarray chips (Acuff et al. 2006). Hence, it is possible to analyze laser capture microdissected tissue samples from tumor versus stroma from the same biopsy by the CLIP-CHIP.

The other approach presented in this volume for transcriptomic analysis of the cancer degradome utilizes one of the most established methods for analysis of gene expression, qRT-PCR. The Edwards lab has developed a qRT-PCR assay that allows for the specific analysis of over 100 human and mouse genes including the MMP, ADAM, ADAMTS, and TIMP gene families and numerous proteases, inhibitor, growth factor, receptor, and regulatory proteins in human- and mouse-derived cells and tissues (Nuttall et al. 2003, Jones et al. 2006). Inherently, the data generated are accurate and quantitative. The huge advantage of the qRT-PCR assay is the quantitative and reliable detection of less than 100 copies of RNA in a 5 ng pool of total RNA, which is equivalent to less than 1 copy per cell (Nuttall et al. 2003). High sensitivity of the qRT-PCR assay has been demonstrated in mice by Pedersen et al. (2005) who demonstrate the assay to be an excellent tool for quantification of small amounts of RNA, without the need for amplification, from laser capture microdissected cancer cells isolated from stromal cells, thus allowing for the quantitative determination of several mRNAs expressed by a very small and specific population of cells. Such an application and sensitivity has at present not been demonstrated for the Hu/Mu ProtIn or CLIP-CHIP arrays. However, with amplification or double amplification of total RNA to obtain higher aRNA amounts it is theoretically possible to perform this, but with the risk that double amplification does not bring an accurate reflection of the true RNA composition.

Out of the three systems presented in this volume, the CLIP-CHIP is the only complete technology capable of analyzing the entire degradome within one experiment for both human and mouse samples. While the Hu/Mu ProtIn chip is particularly useful for applications such as xenograft experiments due to its extensive validation, it still does not cover the entire degradome of either mouse or human. What will be of interest in the future is the use of the Hu/Mu ProtIn chip as a tool for validating results obtained by CLIP-CHIP analysis or vice versa. Although the two chips are two separate microarray platforms strong correlation between results obtained from two-color long-oligonucleotide arrays and one-color short oligonucleotide Affymetrix arrays has been demonstrated (Barczak et al. 2003, Petersen et al. 2005). In our analysis of CLIP-CHIP data generated from breast carcinoma, we have found a high correlation with Affymetrix GeneChip data. The qRT-PCR is highly sensitive; however, its application for full degradome analysis is limited. Nevertheless, qRT-PCR is a quantitative, well established, and very informative tool for follow-up validation experiments from results obtained from the CLIP-CHIP or the Hu/Mu ProtIn microarray chip. In the future, the most suitable application for the qRT-PCR assay will be within a clinical setting where it can be used for fast diagnostic and prognostic purposes, detecting cancer-specific protease gene signatures in patients that have been previously discovered by application of the CLIP-CHIP and/or Hu/Mu ProtIn chip. Utilizing a combination of technologies can help overcome the inherent biases of each approach. While searching for biomarkers within the cancer degradome, it would be ideal if the patterns identified could be detected by the two different microarray platforms providing wide-scale validation then quantified using the qRT-PCR assay.

Scanning, Image Processing, and Microarray Data Analysis of the CLIP-CHIP

A typical CLIP-CHIP array experiment begins with the extraction of total RNA from a specific biological sample of interest followed by amplification of the RNA and fluorescent labeling with Cy3 or Cy5 for detection during the scanning process. Before hybridization to the array, the labeled samples are fragmented and heat denatured to gain single stranded polynucleotides. Following hybridization, the array is washed under stringent conditions to remove nonhybridized sample (Kappelhoff and Overall 2007). Scanning of the CLIP-CHIP is performed for measurement and acquisition of the fluorescent signal present for each spot on the array. This information is stored within an image that is analyzed for the extraction of foreground and background intensity values that are used in subsequent analysis.

Microarray scanners compatible for scanning of the CLIP-CHIP utilize an optical system, similar to that of a confocal laser microscopy system, where a separate laser is used as a source of excitation light for each fluorescent dye and a photomultiplier tube (PMT) is used for detection of the emitted photons. Scanning produces a digital record containing the fluorescence intensity for every pixel at each grid location on the array where the intensity is proportional to the number of sample probes hybridized to the spotted probe (Cheung et al. 1999). Depending on the array platform in use, one typically has little control over selection of scanning equipment; however, as the CLIP-CHIP is of the common two-color array format utilizing Cy3 and Cy5 dyes, it is compatible for use with a number of different scanners provided that they are equipped with 532 and 635nm excitation lasers.

Typically for most microarrays, including the CLIP-CHIP, a 16-bit grey scale image containing the raw data of the experiment is produced for each frequency, for example, Cy3 and Cy5 channels. The images from these two channels can be combined into a single false colored red-yellow-green image to provide an estimate of array quality, for example, spatial effects, before further processing.

CLIP-CHIP Image Analysis

Once the raw data is acquired, the image needs to be analyzed and intensity data extracted. Image analysis and the resulting acquisition of data is an important aspect of microarray experiments and can potentially have a large impact on subsequent data analysis. Currently, there is a wide range of both commercial and freeware image analysis software available that are readily compatible with the two-color array format of the CLIP-CHIP. At present we are using the commercial ImaGene (BioDiscovery, CA); however, other software programs such as Spot (CSIRO Mathematical and Information Sciences, AUS), ScanAlyze (Michael Eisen's lab; Lawrence Berkely National Lab), and TIGR Spotfinder (The Institute of Genomic Research, MA) are also suitable.

There are three fundamental processes that the image analysis software should perform: Gridding, segmentation, and intensity extraction or data acquisition (Yang et al. 2001). Gridding the array involves determining the location of each spot on the array. Although this can be done automatically or semiautomatically, the user should preferably always check the gridding process. To aid in this process, the CLIP-CHIP has been printed with a number of GFP controls that act as "landing lights" when a Cy3-labeled anti-GFP oligonucleotide is spiked into the hybridization solution.

For spotted arrays, segmentation involves the classification of pixels as being foreground or background for both the Cy3 and Cy5 channels. There are a number of segmentation methods available and the type used is dependent on the image analysis software selected (Yang et al. 2001). Data extraction by determining the spot intensity requires computation of the average pixel value of a spot and subtraction of background intensity approximated using a suitable method to provide a corrected foreground fluorescent intensity. Correction for background intensity is necessary as it is likely that not all of the measured spot intensity comes from the fluorescent label. After background correction, the raw microarray data undergo a logarithmic transformation to the base 2 and the log-differential expression ratio, $M = \log_2 Cy5/Cy3$, and the log-intensity (overall brightness), $A = 1/2\log_2 Cy5Cy3$, is calculated for each spot using appropriate data analysis software. Performing logarithmic transformation helps to minimize some of the systematic variation and by converting the intensity ratios into differences between the two channels aids in the identification of differentially expressed genes by ensuring that everything is on the same scale.

Analysis of CLIP-CHIP Microarray Data

After image analysis and logarithmic transformation, the raw CLIP-CHIP data needs to be normalized to correct for systematic variation before any further downstream analysis is performed. Downstream analysis of the CLIP-CHIP data involves determining if a protease, protease inhibitor, or inactive homologue is present or absent within a given tissue (tissue profiling), identifying genes that are differentially expressed between two given samples such as diseased versus non-diseased tissue, or treated versus nontreated tissue on a single-gene basis or performing multiple gene analysis where clusters of genes are analyzed to determine common functionality, pattern identification, gene–gene interaction, and gene regulatory networks within the protease web. The overall success in performing downstream analysis and identifying differentially expressed genes is largely dependent on the suitability of the chosen experimental design which ultimately governs what analysis approach to apply to the data (Yang and Speed 2002, Churchill 2002, Maindonald et al. 2003).

Accurately determining differential expression of a gene requires the selection and calculation of a suitable statistic for gene ranking followed by selection of an appropriate cutoff point whereby genes having a rank value above the cutoff are considered to be differentially expressed and those having a value below are considered not to be. When using the CLIP-CHIP in tissue profiling to identify which degradome members are expressed then a suitable cutoff point where one can have confidence as to the absence or presence of a degradome member from the sample needs to be determined for the spot intensity data only. A number of housekeeping genes have been spotted onto the CLIP-CHIP, which can aid in the clustering of present degradome members as being either a low, medium, or high intensity-expressed gene within the given tissue. For tissue profiling, a reference sample can be cohybridized to the array or the same sample labeled with both Cy3 and Cy5 can be hybridized to the same array as a self–self hybridization experiment to control for experimental variation. After profiling, it will then be possible to look for differential expression of degradome members across different tissue types via the reference sample or by applying a normalization method that allows for single channel analysis.

Within any microarray experiment, not only those performed with the CLIP-CHIP, there are many possible sources of variation throughout the experimental process that can contribute to poor quality or noise within the microarray data. Normalization of the array data is thus required to correct for the effects of experimental or systematic variation while not altering any variation arising from the biological samples themselves (Smyth et al. 2003). For two-color microarrays, a major source of variation is that arising from dye bias that results from the different labeling efficiencies and scanning properties of the Cy3 and Cy5 dyes. Thus, normalization methods applied to the CLIP-CHIP data need to minimize this bias by balancing the fluorescence intensities of the two channels. However, intensity-dependent normalization may not be the only type of normalization required. Yang et al. (2002) address three main forms of normalization being: within slide normalization, paired-slide normalizing for dye-swap experiments, and between-slide normalization. During the process of robotic spotting systematic variation can arise within individual arrays and between arrays due to variations between print-tips of the spotter leading to inconsistencies occurring with location, size and shape of the spots. For CLIP-CHIP array data it is thus recommended that print-tip intensity dependent LOWESS (locally weighted scatterplot smoothing) normalization (Yang et al. 2002) be used as the default method for normalization. It is also possible to perform single channel analysis of CLIP-CHIP array data by utilizing the single channel normalization method for two-color array data as proposed by Yang and Thorne (2003). This normalization approach separately treats the Cy3 and Cy5 data, removing systematic intensity bias that is not due to real gene expression and ultimately allowing for comparison of absolute intensities between separate arrays to be made for which no direct comparisons have been made or allowed for during experimental design, that is, self–self hybridization experiments, errors arising in data from one of the channels making it unusable.

The final stage in the analysis of CLIP-CHIP array data requires bioinformatics analysis of the final list of differentially expressed genes in the degradome. Bioinformatic analysis is used to identify characteristics of the gene and determine the functions and pathways that each gene is involved in within the protease web and ultimately its interaction and role within the cancer degradome. Such analysis helps to determine genes that will be of most interest to follow up in further experiments after confirmation of their differential expression by some of the more traditional methods for measuring gene expression such as Northern blots, qRT-PCR, in situ hybridization, or RPA assays. Further biological studies may involve altering gene function with targeted mutations, antisense technology, or protein inhibition with the ultimate goal to be aiding in the understanding of the biological question at

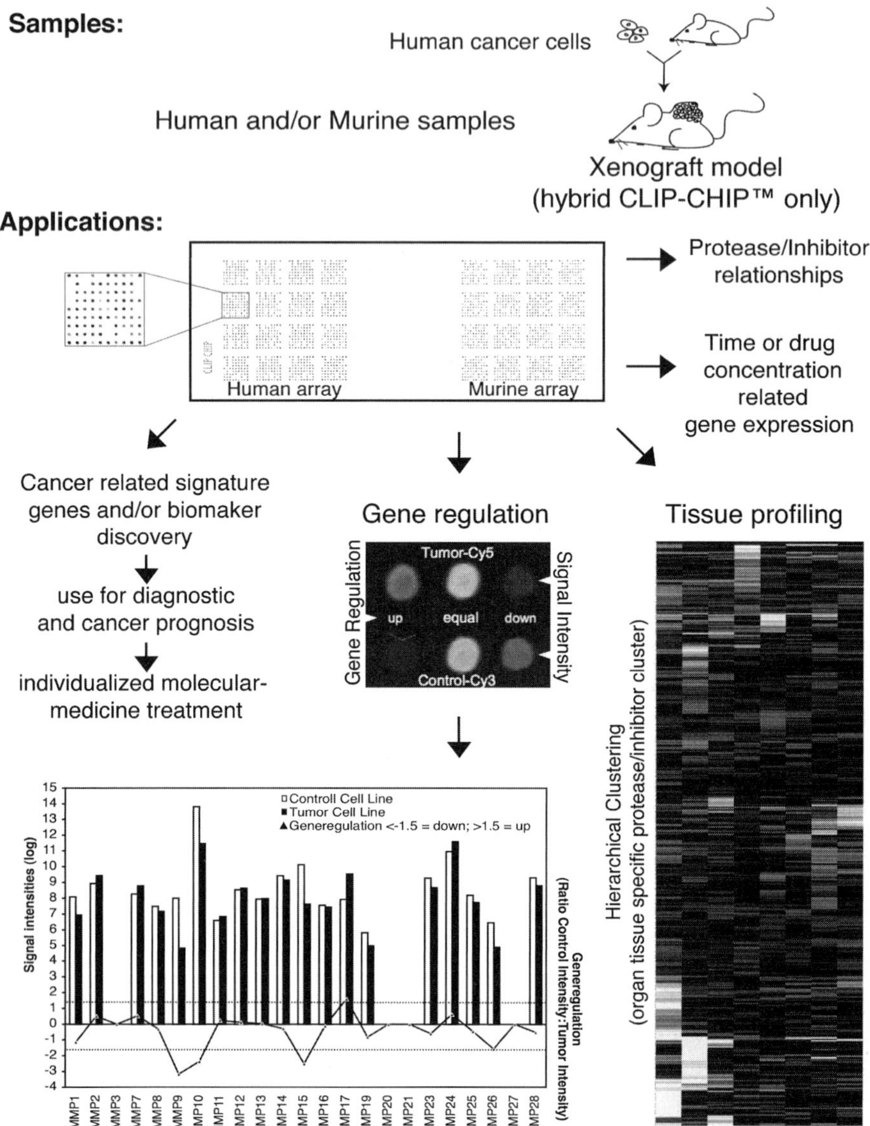

Fig. 2.2 Current and future application of the CLIP-CHIP. The CLIP-CHIP has many uses: it can be used for tissue expression profiling and hierarchical clustering, as shown here in a quick overview in the protease and inhibitor profile of eight human tissues; for gene regulation in tissues, organs or cells, in order to identify up- and downregulated genes. Signal intensities measured in both control and tumor cell lines might show either equal intensities or if up- or downregulated, a higher or lower signal in comparison to the control signal. As an example, equally expressed or differentially regulated MMPs are shown in a human lung cancer cell line A549 versus a control lung cell line. Upregulated genes have a ratio higher than 1.5, downregulated genes have a ratio lower than 1.5. Samples for CLIP-CHIP analysis can either be from human or murine origin, from tissues, organs, tumors, or tissue culture cells, or because of the ability to analyze

hand. In this way the underlying role of the protease web within cancer and the ranking of identified genes and gene products that might be developed for target validation can be more readily made.

Current and Future Applications and Directions of the CLIP-CHIP in Cancer Degradomics

Multiple studies have already demonstrated that microarray-based gene expression profiling enables accurate tumor classification and showed its potential for deciphering tumor diagnosis, prognosis, and therapy. Current applications for the human and murine CLIP-CHIP are the analysis of the complete repertoire of proteases, non-proteolytic homologues, and inhibitor gene transcripts in all human and murine tissues and pathological samples. It can be used for identifying differentially expressed genes or used for gene expression profiling in tissues to find the expressed genes for clustering according to gene ontology (Fig. 2.2). The murine CLIP-CHIP can help to unravel which proteases and their inhibitors are expressed in different murine models of cancer, at different stages of carcinogenesis, such as the transition from benign to malignant tumors, and before and after treatment with chemotherapeutic or radiotherapeutic agents. The human CLIP-CHIP can be used in the same way with human-derived cancer cell lines and patient samples. By identifying the cancer signature genes, this might lead to setting guidelines for their collective use as a signature profile of cancer biomarkers. The human/mouse hybrid CLIP-CHIP is still in its validation phase, but like the Hu/Mu ProtIn chip it should become an important tool for xenograft experiments. However, both the Hu/Mu ProtIn chip and the CLIP-CHIP will not be able to provide relative levels of expression for human and murine transcripts of homologous genes in xenograft studies. These experiments will simply show the presence or absence of expression of a particular protease and its homologues since each primer set for the two species is different and so intrinsically hybridizes with different efficiencies. Therefore, the levels of transcript obtained for one species are not comparable with levels of the transcript from the homologous gene in the other species determined using different primers. Nonetheless, this is still useful information. Finally, the CLIP-CHIP is being made readily available on request for collaborative studies on the cancer degradome.

Fig. 2.2 (Continued) down to 100 ng of RNA in a sample, from laser capture microdissection of cancerous tissues. With the CLIP-CHIP, protease and inhibitor relationships can be identified from which it might be possible to determine the proteolytic potential of a cell or tissue. Time or drug concentration-dependent gene expression can also be monitored and cancer-related signature protease and inhibitor and potential biomarkers can be identified using this microarray platform. Key genes and their products might be useful for diagnostic or prognostic purposes in the clinical environment, leading to individualized molecular medicine in the future. The hybrid human/ murine CLIP-CHIP allows for xenograft experiments to be analyzed where human tumors or tumor cells are implanted in murine mouse models. Such an analysis shows protease and inhibitor expression in the tumor and tumor secretome (human origin) and microenvironmental stroma and stromal secretome (mouse origin), so aiding in identification of protease targets and anti-targets

References

Acuff H.B., Sinnamon M., Fingleton B., et al. (2006). Analysis of host- and tumor-derived proteinases using a custom dual species microarray reveals a protective role for stromal matrix metalloproteinase-12 in non-small cell lung cancer. Cancer Res. 6616:7968–7975.

Balbin M., Fueyo A., Tester A.M., et al. (2003). Loss of collagenase-2 confers increased skin tumor susceptibility to male mice. Nat. Genet. 35(3):252–257.

Barczak A., Rodriguez M.W., Hanspers K., et al. (2003). Spotted long oligonucleotide arrays for human gene expression analysis. Genome Res. 13:1775–1785.

Barrett A.J., Rawlings N.D., and Woessner J.F. (1998). Handbook of proteolytic enzymes. San Diego: Academic Press.

Borgono C.A., and Diamandis E.P. (2004). The emerging roles of human tissue kallikreins in cancer. Nat. Rev. Cancer. 4(11):876–890.

Cheung V.G., Morley M., Aguilar F., et al. (1999). Making and reading microarrays. Nat. Genet. 21:15–19.

Churchill G.A. (2002). Fundamentals of experimental design for cDNA microarrays. Nat. Genet. 32(Suppl):490–495.

Colebatch G., Trevaskis B., and Udvardi M. (2002). Functional genomics: Tools of the trade. New Phytol. 153:27–36.

Dean R.A., and Overall C.M. (2007). Proteomics discovery of metalloproteinase substrates in the cellular context by iTRAQ labeling reveals a diverse MMP-2 substrate degradome. Mol. Cell. Proteomics 6(4):611–623.

Dean R.A., Butler G.S., Hamma-Kourbali Y., et al. (2007). Identification of candidate angiogenic inhibitors processed by MMP-2 in cell based proteomic screens: Disruption of VEGF/HARP (Pleiotrophin) and VEGF/CTGF angiogenic inhibitory complexes by MMP-2 proteolysis. Mol. Cell. Biol. 27(24):8454–65.

Egeblad M., and Werb Z. (2002). New functions for the matrix metalloproteinases in cancer progression. Nat. Rev. Cancer. 2(3):161–174.

Glas A.M., Floore A., Delahaye L.J., et al. (2006). Converting a breast cancer microarray signature into a high-throughput diagnostic test. BMC Genomics 7:278.

Gupta G.P., Nguyen D.X., Chiang A.C., et al. (2007). Mediators of vascular remodelling co-opted for sequential steps in lung metastasis. Nature 446(7137):765–770.

Jones G.C., Corps A.N., Pennington C.J., et al. (2006). Expression profiling of metalloproteinases and tissue inhibitors of metalloproteinases in normal and degenerate human achilles tendon. Arthritis Rheum. 54(3):832–842.

Kappelhoff R., and Overall C.M. (2007). The CLIP-CHIP™ oligonucleotide microarray: Dedicated array for analysis of all protease, nonproteolytic homolog, and inhibitor gene transcripts in human and mouse. Curr. Protoc. Protein Sci. 49:21.19.1–21.19.16.

Li M., Chen D., Shiloh A., et al. (2002). Deubiquitination of p53 by HAUSP is an important pathway for p53 stabilization. Nature 416(6881):648–653.

Lockhart D.J., Dong H., Byrne M.C., et al. (1996). Expression monitoring by hybridization to high-density oligonucleotide arrays. Nat. Biotechnol. 14:1675–1680.

López-Otín C., and Matrisian L.M. (2007). Emerging roles of proteases in tumour suppression. Nat. Rev. Cancer. 7(10):800–808.

López-Otín C., and Overall C.M. (2002). Protease degradomics: A new challenge for proteomics. Nat. Rev. Mol. Cell Biol. 3:509–519.

Maindonald J.H., Pittelkow Y.E., and Wilson S.R. (2003). Some considerations for the design of microarray experiments. Sci. Stat. 40:367–390.

McQuibban G.A., Gong J.H., Tam E.M., et al. (2000). Inflammation dampened by gelatinase A cleavage of monocyte chemoattractant protein-3. Science 289(5482):1202–1206.

McQuibban G.A., Butler G.S., Gong J.H., et al. (2001). Matrix metalloproteinase activity inactivates the CXC chemokine stromal cell-derived factor-1. J. Biol. Chem. 276(47):43503–43508.

Minn A.J., Gupta G.P., Siegel P.M., et al. (2005a). Genes that mediate breast cancer metastasis to lung. Nature 436(7050):518–524.

Minn A.J., Kang Y., Serganova I., et al. (2005b). Distinct organ-specific metastatic potential of individual breast cancer cells and primary tumors. J. Clin. Invest. 115(1):44–55.

Nuttall R.K., Pennington C.J., Taplin J., et al. (2003). Elevated membrane-type matrix metalloproteinases in gliomas revealed by profiling proteases and inhibitors in human cancer cells. Mol. Cancer Res. 1(5):333–345.

Overall C.M., and Kleifeld O. (2006). Validating matrix metalloproteinases as drug targets and anti-targets for cancer therapy. Nat. Rev. Cancer. 6:227–239.

Overall C.M., Tam E.M., Kappelhoff R., et al. (2004). Protease degradomics: Mass spectrometry discovery of protease substrates and the CLIP-CHIP$^{\circledR}$, a dedicated DNA microarray of all human proteases and inhibitors. Biol. Chem. 385(6):493–504.

Pedersen T.X., Pennington C.J., Almholt K., et al. (2005). Extracellular protease mRNAs are predominantly expressed in the stromal areas of microdissected mouse breast carcinomas. Carcinogenesis 26(7):1233–1240.

Petersen D., Chandramouli G.V.R., Geoghegan J., et al. (2005). Three microarray platforms: An analysis of their concordance in profiling gene expression. BMC Genomics 6:63.

Puente X.S., Sánchez L.M., Overall C.M., et al. (2003). Human and mouse proteases: a comparative genomic approach. Nat. Rev. Genet. 4:544–558.

Rawlings N.D., Morton F.R., Kok C Y., Kong J. & Barrett, A.J. (2008). MEROPS: the peptidase database. Nucleic Acids Res. 36: D320–325.

Schwartz D.R., Moin K., Yao B., et al. (2007). Hu/Mu ProtIn oligonucleotide microarray: Dual-species array for profiling protease and protease inhibitor gene expression in tumors and their microenvironment. Mol. Cancer Res. 5(5):443–454.

Smyth G.K., Yang Y.H., and Speed T. (2003). Statistical issues in cDNA microarray data analysis. Methods Mol. Biol. 224:111–136.

Tester A.M., Cox J.H., Connor A.R., et al. (2007). LPS responsiveness and neutrophil chemotaxis in vivo require PMN MMP-8 activity. PLoS ONE 2(3):e312.

Vergote D., Butler G.S., Ooms M., Cox J.H., Silva C., Hollenberg M.D., Jhamandas J.H., Overall C.M., Power C. (2006). Proteolytic processing of SDF-1α reveals a change in receptor specificity mediating HIV-associated neurodegeneration. Proc. Natl. Acad Sci U S A. 103(50): 19182–7.

Van't Veer L.J., Dai H., van de Vijver M.J., et al. (2002). Gene expression profiling predicts clinical outcome of breast cancer. Nature 415(6871):530–536.

Weigelt B., Hu Z., He X., et al. (2005). Molecular portraits and 70-gene prognosis signature are preserved throughout the metastatic process of breast cancer. Cancer Res. 65(20):9155–9158.

Yang Y.H., and Speed T. (2002). Design issues for cDNA microarray experiments. Nat. Rev. 3:579–5888.

Yang Y.H., and Thorne N.P. (2003). Normalization for two-color cDNA microarray data. In: D.R. Goldstein (ed.), Science and Statistics: A Festschrift for Terry Speed, IMS Lecture Notes— Monograph Series, Volume 40, pp. 403–418.

Yang Y.H., Buckley M.J., and Speed T.P. (2001). Analysis of cDNA microarray images. Brief. Bioinform. 2(4):341–349.

Yang Y.H., Dudoit S., Luu P., et al. (2002). Normalization for cDNA microarray data: A robust composite method addressing single and multiple slide systematic variation. Nucleic Acids Res. 30:e15.

Yauk C.L., Berndt M.L., Williams A., et al. (2004). Comprehensive comparison of six microarray technologies. Nucleic Acids Res. 32:1–7.

Chapter 3
The Hu/Mu ProtIn Chip: A Custom Dual-Species Oligonucleotide Microarray for Profiling Degradome Gene Expression in Tumors and Their Microenvironment

Donald R. Schwartz, Kamiar Moin, Ekkehard Weber, and Bonnie F. Sloane

Abstract Proteases and protease inhibitors of the cancer degradome derive from tumor-associated cells as well as tumor cells. Xenograft studies in which human tumor cells are implanted in mice have revealed the contribution of tumor-associated cells to progression and metastasis of the human tumors. The roles of individual host (mice) proteases or protease inhibitors can be defined by implanting the human tumors in mice deficient in the protease or protease inhibitors. Such studies are time consuming and costly. Therefore, to identify proteases and protease inhibitors of interest, we partnered with Affymetrix to develop a dual-species oliogonucleotide microarray chip, the Hu/Mu ProtIn chip, that would discriminate between human and murine proteases and protease inhibitors in the same sample. The Hu/Mu ProtIn chip can be used to determine the expression of human or murine proteases, protease inhibitors, and protease interactors in single-species specimens or in dual-species specimens. Here,we describe our rationale for the design of this chip, its characteristics and its validation. In addition, we present examples of its successful use by ourselves and others to identify host proteases that play functional roles in tumor growth, angiogenesis, and metastasis. The Hu/Mu ProtIn chip helps us to identify interactions among proteases in the tumor and its surrounding microenvironment, interactions that can then be tested for their contribution to tumor progression.

Introduction

More than one protease class has been implicated in the progression of human tumors, as is evidenced in the content of this volume on the cancer degradome. Although the preclinical data implicating matrix metalloproteinases (MMPs) in malignant progression were particularly compelling, the clinical trials on MMP inhibitors (MMPIs) did not fulfill the promise of MMPs as therapeutic targets in cancer. There are several possible explanations for this apparent "disconnect"

Dr. Kamiar Moin

Department of Pharmacology, Wayne State University 540 E, Canfield Detroit, MI 48201 USA

kmoin@med.wayne.edu

D. Edwards et al. (eds.), *The Cancer Degradome.*

© Springer Science + Business Media, LLC 2008

between the preclinical and clinical data. Unfortunately, the failures of MMPIs in clinical trials have resulted in allegations that MMPs and, by extrapolation, other proteases are not appropriate therapeutic targets in cancer. Is this the case or might the failure of the MMPI trials reflect problems in clinical trial design for cytostatic agents (*see* Egeblad and Werb 2002, Chau et al. 2003, Coussens et al. 2002 for discussion)? Clinical trials without surrogate endpoints to monitor and confirm the efficacy of the therapeutic strategies being tested, as is true of the MMPI trials, should not be viewed as definitive (McIntyre and Matrisian 2003, Li and Anderson 2003). Among the critical questions is, whether the MMPIs actually reached and reduced the activity of their target MMPs *in vivo*. Imaging probes, discussed in Chaps. 7 and 8, that are being developed to assess protease activity *in vivo* could provide a surrogate endpoint for clinical trials of agents that modulate proteolytic pathways. The MMPI trials, however, did not include surrogate endpoints to assess reductions in protease activity so it is not known whether MMPs were actually inhibited in the test patients (Chau et al. 2003, McIntyre and Matrisian 2003, Li and Anderson 2003). There are other concerns about the MMPI trials. Coussens et al. (2002) cite data revealing that the tumors studied in the clinical trials did not necessarily express the particular proteases targeted by the MMPIs. None of the clinical trials with BAY 12-9566 or other MMPIs included patients with breast cancers although this is a cancer for which there is strong preclinical evidence that MMPs impact progression. For example, Sledge and colleagues (2003) have demonstrated efficacy for BAY 12-9566 in an orthotopic human breast model. Furthermore, not all MMPs should be inhibited, depending on the type of tumor; this is clearly the case for MMP-8 (collagenase-2), an MMP expressed by inflammatory neutrophils, which plays a protective role in skin cancer (Balbin et al. 2003), and for MMP-12, an MMP expressed by macrophages, which plays a protective role in non-small cell lung cancer (Acuff et al. 2006). More detailed information on possible protective effects of MMPs and other proteases is included in two recent reviews (Lopez-Otin and Matrisian 2007, Martin and Matrisian 2007). Broadspectrum MMPIs might have had unanticipated side effects, indicating how essential it is "to define precisely the tumor degradome" (Balbin et al. 2003) before using MMPIs or other protease inhibitors for cancer therapy. In fact, unanticipated side effects occurred in the clinical trials leading to limitations in the amount of MMPIs that the patients could take and in some patients necessitating MMPI-free holidays (Coussens et al. 2002). How endogenous protease inhibitors factor into the equation is also of relevance. As just one example, tissue inhibitor of metalloproteinase 1 (TIMP-1) has been shown to promote carcinogenesis of squamous cell carcinoma of the skin, exerting "differential regulation on tissues in a stage-dependent manner" (Rhee et al. 2004). We are still far from having a thorough understanding of the roles of proteases in cancer: this includes understanding the roles of MMPs, the roles of other proteases, that those roles are dynamic and may change during the course of malignant progression, whether the proteases playing critical roles in malignant progression come from tumor, stromal, or inflammatory cells, whether the critical proteases are affected by interactions of the tumor with its microenvironment, and so on. In short, we do not yet know which protease(s) is the most appropriate target for antiprotease therapies or when antiprotease therapies might prove most effective.

The initial working hypothesis for those studying proteases in cancer was that invasive processes (local and during metastatic spread) required degradation of extracellular matrices by proteases. The roles of proteases in cancer are now known to be much broader, as discussed in several chapters in Sects. 2 and 3 of this volume. Furthermore, the proteases themselves derive not only from tumor cells, but also from the multiple cell types present in the tumor stroma, for example, fibroblasts, macrophages, mast cells, neutrophils, and endothelial cells. A factor in understanding how proteases function in malignant progression is that a purified protease capable of cleaving a protein substrate *in vitro* may not be the protease or the only protease responsible for degradation of that substrate *in vivo*. Transgenic mice deficient in specific proteases have not only helped elucidate the *in vivo* functions of proteases but have also confirmed a high degree of redundancy. This along with the large number of proteases in the human genome (Rawlings et al. 2007, Puente et al. 2003, Chap. 1), the interplay of proteases with their endogenous inhibitors and activators and the complexity of their biological roles suggest that we should take advantage of array technologies to identify candidate proteases associated with tumor initiation and progression for further study.

The Hu/Mu ProtIn Chip

We have designed, through collaborative efforts of members of the Protease Consortium (Giranda and Matrisian 1999) and a Department of Defense (DOD) Protease Breast Cancer Center of Excellence, the Hu/Mu ProtIn chip, in large part because so many so-called "tumor" proteases originate from tumor-associated cells rather than the tumor cells (e.g., *see* Bamba et al. 2003, Hulkower et al. 2000). Our goal was to develop, in partnership with Affymetrix, a dual-species microarray chip that would allow us to distinguish changes in expression of proteases in human tumor cell xenografts from those in the host mouse cells surrounding and infiltrating into the xenograft. We chose the Affymetrix platform because (1) it is commercially available and thus readily accessible to the research community; (2) this platform is an industry standard and is one that is routinely available in genomic cores at most institutions; (3) genomic cores and bioinformatics facilities have extensive experience with this platform and thus the requisite data analysis capabilities; (4) it uses 25-mer oligonucleotide probes that are highly specific (Relogio et al. 2002) and thus exceptionally suitable for a focused array; and (5) Affymetrix subjects their arrays to rigorous quality control. On the basis of these characteristics, we felt that partnering with Affymetrix would provide us with an array platform that is suitable for use in multiple laboratories, that is, a platform that is both reliable and reproducible. In fact, the Microarray Quality Control Consortium (MAQC Consortium; Shi et al. 2006) found that among the microarray platforms tested ($n = 7$, 6 commercial and 1 spotted array), the Affymetrix platform showed outstanding repeatability of expression signal within the four test sites, lowest signal variation between test sites and the highest concordance with TaqMan validation data (*see* Chap. 4 for further discussion of TaqMan technology). A secondary goal was to have a single microarray platform that would allow us to assess expression of protease-relevant genes in human tissues and cell lines, mouse tissues and cell lines, and in human tumors implanted in mice. We therefore selected proteases, protease inhibitors, and genes that we have designated as protease interactors for inclusion on the chip. The protease interactors include receptors, binding partners, activators, and transcription factors that have been shown to coordinately upregulate expression of protease genes, for example, Ets1. The choice of genes for the chip was not intended to be comprehensive, but was based on the interests of protease consortium and center of excellence members.

The Hu/Mu ProtIn chip contains 516 and 456 custom probe sets, each probe set consisting of 16 probe pairs of Affymetrix's typical Perfect Match and Mismatch design. These custom probe sets survey, respectively, 431 human and 387 mouse genes of interest with some genes surveyed by more than one probe set. The probe sets on the Hu/Mu ProtIn chip differ from those on either the Affymetrix human or mouse genome chips. For a complete list of the custom probe sets for proteases, inhibitors and protease interactors for both mouse and man and for Affymetrix's control probe sets, see Supplementary Table S1 (Schwartz et al. 2007). For the Hu/Mu ProtIn chip, Affymetrix designed a probe set of 16 unique oligonucleotide (25-mer) probes for each gene transcript and blasted these against both the human and mouse genomes to determine the potential for cross-hybridization. We have published an extensive characterization and validation of the Hu/Mu ProtIn chip (Schwartz et al. 2007), including as supplemental data a file that lists all of the probes that might cross-hybridize. As this publication includes detailed information on the methodologies we used, including data processing and computational methods, we will not repeat this information here.

We assessed the selectivity and specificity of the custom probe sets against universal reference RNAs from human and mouse and established that the probe design is sufficiently stringent to prevent detection of transcripts from the other species (Fig. 3.1a; Schwartz et al. 2007). Since transcript signatures of carcinomas are likely to differ from that of universal reference RNAs, we also assessed the selectivity and specificity of the custom probe sets against human and murine mammary carcinomas (Fig. 3.1b; Schwartz et al. 2007) and established that the probe design is sufficiently stringent to prevent detection of transcripts from the other species in tumor tissue as well as in the universal reference RNAs. Nonetheless, those studies do not demonstrate whether the Hu/Mu ProtIn chip is able to detect and differentiate human and murine transcripts in a sample containing both species of transcripts. Therefore, we used the Hu/Mu ProtIn array to analyze two xenograft models in which human tumor cells (A549 human non-small cell lung carcinoma and MDA-MB-231 human breast carcinoma) were implanted orthotopically in immunodeficient mice. We profiled (1) tissue from the orthotopic sites for implantation of the human lung and breast carcinoma cells, that is, normal murine lung and mammary fat pads, respectively (Fig. 3.2a and b, left two bars); (2) the two

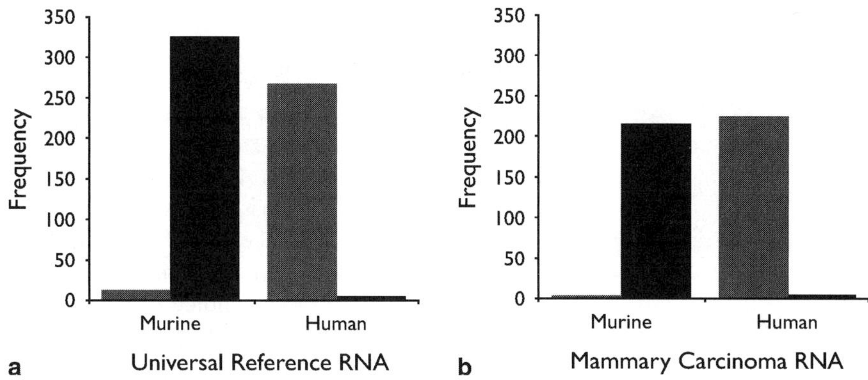

Fig. 3.1 Species identical probe response to universal reference total RNA and mammary carcinoma total RNA. In **a**, three independent replicates of XpressRef (SuperArray Bioscience) universal reference total RNA from mouse or human tissues were profiled with the Hu/Mu ProtIn microarray. The presence of a transcript was determined by the detection algorithm in MAS5 software (Affymetrix), which assigns P as a measure of the detection call confidence. The number of probe sets, giving present call P values smaller than 0.05, for all three replicates are indicated. In the presence of only mouse transcripts, there were extremely few present detection calls by human probe sets (gray bars), and vice versa (black bars), despite the high sequence similarity among protease and protease inhibitor genes and their potential for cross-hybridization. In **b**, total R derived from spontaneous mammary carcinomas from MMTV-PyMT$^+$ (FVB/n) ($n = 8$) and MMTV-PyMT$^+$/uPARAP$^{-/-}$ (FVB/n) ($n = 2$) transgenic mice or human breast ductal carcinoma biopsies ($n = 18$) from women with stage II or III disease was profiled. The number of probe sets, giving present call P values smaller than 0.05, for $\geq 80\%$ are indicated. Transcripts derived from mammary carcinoma were rarely detected by nonspecies identical probes (human, gray bars, mouse, black columns). Frequency represents the number of probe sets that satisfy the selection criterion. Adapted from Schwartz et al. (2007)

human carcinoma lines grown in monolayer culture *in vitro* (Fig. 3.2a and b, middle two bars; and (3) the human lung and breast carcinoma xenografts (Fig. 3.2a and b, right two bars). We confirmed that the probe design is sufficiently stringent to prevent detection of transcripts from the other species in tumor cell lines as well as in the orthotopic sites for implantation of those cell lines, as was the case for human and murine universal reference RNAs and human and murine mammary carcinoma tissues (Fig. 3.1). Both human and murine transcripts are detected in orthotopic xenografts of human lung carcinoma cell lines in murine lung or human breast carcinoma cell lines in murine mammary fat pads. There is a higher frequency of detection of murine transcripts in lung tumor xenografts than in mammary tumor xenografts. This may reflect removal of the entire lung in the former and dissection of the tumor-containing mammary fat pad from surrounding tissues in the latter as well as differences in cellularity between lung and fat pad. Our analyses demonstrate that probes on the Hu/Mu ProtIn array, in response to these biologically diverse RNAs, retain their intended species-identical fidelity, that is, the mouse probes detect mouse transcripts in the mouse orthotopic sites and the human probes detect human transcripts in the human cell lines. In contrast, in

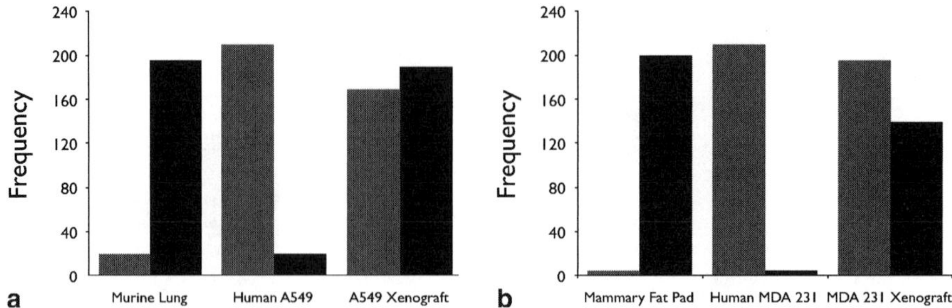

Fig. 3.2 Frequency of transcript detection by mouse and human probe sets in samples derived from normal mouse lung and mammary fat pads (orthotopic sites), cultured human cell lines, and orthotopic xenografts. In a, total RNA derived from normal mouse lung ($n = 3$), cultures of human A549 lung adenocarcinoma cells ($n = 2$) and orthotopically implanted xenografts of A549 cells ($n = 3$) was profiled. In b, total RNA derived from normal mouse mammary fat pads ($n = 4$), MDA-MB-231 breast carcinoma cells ($n = 3$), and orthotopically implanted xenografts of MDA-MB-231 cells ($n = 3$) was profiled. Number of probe sets, giving present call P values smaller than 0.05, for all replicates in each group is indicated. Nonspecies identical probes (gray bars, human and black bars, mouse) rarely detect transcripts in mouse lung, mouse mammary fat pads, A549 cells, and MDA-MB-231 cells. Importantly, transcripts from A549 (**a**) and MDA-MB-231 (**b**) orthotopic xenografts were detected by both mouse (black columns) and human probes (gray columns), suggesting that both mouse (host) and human (tumor) transcripts present in the xenograft can be detected simultaneously by the Hu/Mu ProtIn microarray. Frequency represents the number of probe sets that satisfy the selection criterion. Adapted from Schwartz et al. (2007)

the orthotopic xenografts, both human and mouse transcripts are detected. Furthermore, these data indicate that the Hu/Mu ProtIn array can detect mouse transcripts in the presence of human RNA and thus should be a useful tool for detecting host gene expression in xenograft models of human disease (Schwartz et al. 2007), despite the interspecies homology of proteases (Puente et al. 2003) and protease inhibitors (e.g., Abrahamson et al. 2003, Nagase and Brew 2003).

The most compelling proof of the utility of the Hu/Mu ProtIn chip is its successful use in the identification of host proteases. This includes host proteases that are upregulated in non-small cell lung carcinoma orthotopic xenografts: MMP-12 and -13 and the cysteine cathepsin, cathepsin K (Acuff et al. 2006). All three were shown by microarray analysis to be upregulated in human lung tumors as well as in the xenograft model. Levels of MMP-12 protein are increased in stromal cells, including macrophages, in both the xenografts and the human lung tumors. Further studies on the functional role of MMP-12 in lung tumors, using MMP-12-deficient mice, found increases in tumor size, tumor angiogenesis, and decreases in levels of the angiogenesis inhibitor angiostatin. These findings are consistent with stromal MMP-12 serving a protective function in these tumors.

J.A. Joyce and colleagues (personal communication) are successfully using the Hu/Mu ProtIn chip to profile metastasis-related changes in proteases and protease inhibitors, in this case changes that define site-specific metastases from human mammary tumor xenografts. We have used the Hu/Mu ProtIn chip to profile expression of proteases and protease inhibitors in orthotopic xenografts of

Table 3.1 Hu/Mu ProtIn expression profiling

	IDC	Xenografts	
		DCIS	231[a]
MMP 11	↑	—	—
CTSV	↑	—	↑
CTSH	↑	↑	—
CTSS	⌃	—	
CSTA	↓	↓	—
Serpin a3g	—		⌃
Serpin e1		⌃	⌃
Serpin e2		⌃	⌃

[a] MDA-MB-231 Arrowheads indicate mouse genes

premalignant MCF10 (DCIS) and malignant (MDA-MB-231) human mammary cell lines (Table 3.1), human invasive ductal carcinomas (IDC) of stage I and II (Table 3.1) and 3D culture and coculture models of human cell lines recapitulating changes in mammary architecture during the transition from normal to malignant (Mullins, Sameni, Jedeszko and Sloane, unpublished data). In the xenografts, murine cathepsin S and serpins a3g, e1, and e2 are upregulated and human cathepsins H (CTSH) and V/L2 (CTSV) and cystatin/stefin A (CSTA) are upregulated and downregulated, respectively (Table 3.1). The increase in cathepsin S may reflect angiogenesis induced by the xenograft or macrophage infiltration into the xenograft; both endothelial cells (Shi et al. 2003) and macrophages (Shi et al. 1992) are rich in cathepsin S. In human IDC, we confirmed upregulation of cathepsins H and V and downregulation of cystatin A and, in addition, identified upregulation of MMP-11/stromelysin 3 (Table 3.1). Since MMP-11 is a fibroblast enzyme (Basset et al. 1990), we would not expect to see an increase in human MMP-11 in the xenografts. One might, however, have expected to see an increase in murine MMP-11, but this was not observed. We were pleased that we identified upregulation of MMP-11 and cathepsin V in IDC as these two genes comprise the invasion signature of the recurrence score model, a 21-gene prediction algorithm used in the Oncotype DX clinical test to direct therapeutic decisions for women with early stage breast cancer (Paik et al. 2004). This test has been confirmed independently to accurately predict prognosis of both ER$^+$ and ER$^-$ breast cancers (Fan et al. 2006). We confirmed the increase in expression of both MMP-11 and cathepsin V by quantitative real-time-polymerase chain reaction (qRT-PCR), using primers for MMP-11 designed by Pennington (*see* Chap. 4, this volume). High levels of expression of cathepsin V protein were confirmed by immunohistochemistry (Fig. 3.3). Interestingly, cathepsin V is primarily localized to the nucleus of tumor cells in IDC, a finding consistent with the recent report that cathepsin V, yet not its close homologue cathepsin L, interacts with DNA (Ong et al. 2007). Sinnamon et al. (2008) have recently used the Hu/Mu ProtIn chip in an all murine system, that is, for analyses of proteases and protease inhibitors in intestinal adenomas of Min mice. In this model for the early stages of intestinal tumorigenesis, they identified a number of mast cell proteases that are upregulated and established a protective role for the

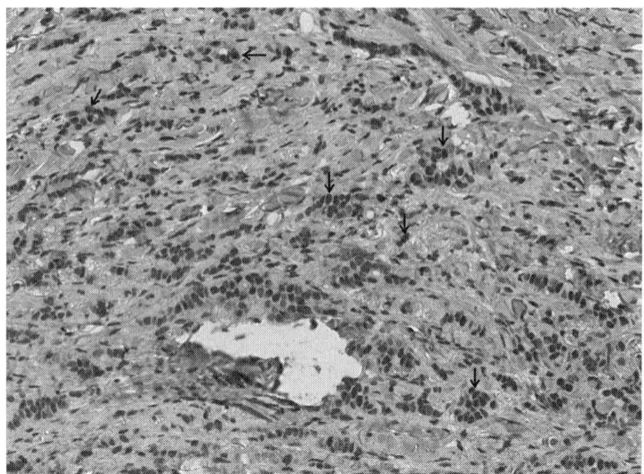

Fig. 3.3 Cathepsin V protein in invasive human ductal carcinoma. Paraffin-embedded sections were stained for cathepsin V with a monoclonal antibody against human cathepsin V. Expression of cathepsin V protein is indicated by dark brown staining (arrows). Magnification, 40×. (*See also* Color Insert I)

mast cells that is mediated in part by mast cell protease/chymase induction of eosinophil migration. In all of the cited studies, hybridizations were performed with 5 μg total RNA; this amount of RNA was readily available from these samples, but it should be emphasized that, as with other Affymetrix platforms, for example, Affymetrix Hu 133A, one can use much less RNA, including RNA amplified from laser capture microdissected tissue (Yang et al. 2005). The cited studies on dual-species xenografts and single-species models illustrate the capabilities of the Hu/Mu ProtIn chip and the xenograft studies validate the ability of the Hu/Mu ProtIn chip to discriminate between human and mouse proteases and protease inhibitors in the same sample.

Use of Hu/Mu ProtIn Chip in the Context of Other Technologies

We designed the Hu/Mu ProtIn chip with specificity as our most important criterion. The small degree of cross-hybridization between murine and human probes (Figs. 3.1 and 3.2; Schwartz et al. 2007) and the ability to identify increases in host (murine) proteases and protease inhibitors in human tumor xenografts (Acuff et al. 2006, Sinnamon et al. 2008) (Table 3.1) indicates that we have achieved specificity. In a dual-species array such as the Hu/Mu ProtIn array when proteases and protease inhibitors of the two species are known to be highly homologous (Abrahamson et al. 2003, Nagase and Brew 2003, Puente et al. 2003), the use of shorter 25-mer probes is desirable as longer probes increase the probability for cross-hybridization. On the contrary, if sensitivity is the most important criterion then longer probes such as the 70-mer probes used in the CLIP-CHIP array, described in Chap. 2, are desirable. In a comparison of 25- and 60-mer probes, the 25-mer probes were found to be 20-fold more specific and the 60-mer probes found to be threefold more sensitive (Relogio et al. 2002). This suggests that the Hu/Mu ProtIn and CLIP-CHIP arrays are complementary and that, depending on the experiment, one may want to use one or the other or the two arrays in tandem. Either array is able to analyze total RNA or amplified RNA so that either can be used to analyze laser capture microdissected tissue samples although analyses with either array of amplified RNA may be subject to errors introduced by the double amplification.

If one is interested only in proteases, protease inhibitors, and protease interactors, then using the Hu/Mu ProtIn chip is warranted. As compared to whole genome chips, the smaller number of genes on the Hu/Mu ProtIn chip makes the analyses less time consuming. Furthermore, one can analyze both human and murine proteases, protease inhibitors, and protease interactors on a single chip, reducing cost. If one, however, is interested in the interactions of proteases with other pathways, then whole genome chips should be used.

Caveats

The recent publication from the pilot phase of the Encyclopedia of DNA Elements (ENCODE Project Consortium; Birney et al. 2007) suggests that all presently available microarrays, including the Hu/Mu ProtIn chip and the CLIP-CHIP, may have to be redesigned. The ENCODE analysis of only 1% of the human genome revealed that it is "pervasively transcribed." Many novel nonprotein-coding transcripts were identified, with many of these overlapping protein-coding transcripts. In short, the ENCODE data have revealed that the complexity of RNA transcripts cannot be accounted for by known genes (Henikoff 2007), and suggests that our analyses of the degradome should concentrate on expression at the protein level and in particular levels of functional proteins, rather than levels of transcripts.

Acknowledgments This work was supported by a Department of Defense Breast Cancer Center of Excellence (DAMD17-02-1-0693).

References

Abrahamson M, Alvarez-Fernandez M, Nathanson C M (2003) Cystatins. Biochem Soc Symp 70: 179–199.
Acuff H B, Sinnamon M, Fingleton B, et al. (2006) Analysis of host- and tumor-derived proteinases using a custom dual species microarray reveals a protective role for stromal matris metalloproteinase-12 in non-small cell lung cancer. Cancer Res 66: 7968–7975.
Balbin M, Fueyo A, Tester A M, et al. (2003) Loss of collagenase-2 confers increased skin tumor susceptibility to male mice. Nat Genet 35: 252–257.

Bamba S, Andoh A, Yasui H, et al. (2003) Matrix metalloproteinase-3 secretion from human colonic subepithelial myofibroblasts: Role of interleukin-17. J Gastroenterol 38: 548–554.

Basset D, Bellocq J P, Wolf C, et al. (1990) A novel metalloproteinase gene specifically expressed in stromal cells of breast carcinomas. Nature 348: 699–704.

Birney E, Stamatoyannopoulos J A, Dutta A, et al. [ENCODE Project Consortium] (2007) Identification and an analysis of functional elements in 1% of the human genome by the ENCODE pilot project. Nature 447: 799–816.

Chau I, Rigg A, Cunningham D (2003) Matrix metalloproteinase inhibitors—an emphasis on gastrointestinal malignancies. Crit Rev Oncol Hematol 45: 151–176.

Coussens L M, Fingleton B, Matrisian L M (2002) Matrix metalloproteinase inhibitors and cancer: Trials and tribulations. Science 295: 2387–2392.

Egeblad M, Werb Z (2002) New functions for the matrix metalloproteinases in cancer progression. Nat Rev 2: 161–174.

Fan C, Oh D S, Wessels L, et al. (2006) Concordance among gene-expression-based predictors for breast cancer. N Engl J Med 355: 615–617.

Giranda V L, Matrisian L M (1999) The protease consortium: An alliance to advance the understanding of proteolytic enzymes as therapeutic targets for cancer. Mol Carcinog 26: 139–142.

Henikoff S (2007) ENCODE and our very busy genome. Nat Genet 39: 817–818.

Hulkower K I, Butler C C, Linebaugh B E, et al. (2000) Fluorescent microplate assay for cancer cell-associated cathepsin B. Eur J Biochem 267: 1–7.

Li W P, Anderson C J (2003) Imaging matrix metalloproteinase expression in tumors. Q J Nucl Med 47: 201–208.

Lopez-Otin C, Matrisian L (2007) Emerging roles of proteases in tumour suppression. Nat Rev Cancer 7: 800–808.

Martin M D, Matrisian L (2007) The other side of MMPs: Protective roles in tumor progression. Cancer Met Rev 267: 717–724.

McIntyre J O, Matrisian L M (2003) Molecular imaging of proteolytic activity in cancer. J Cell Biochem 90: 1087–1097.

Nagase H, Brew K (2003) Designing TIMP (tissue inhibitor of metalloproteinases) variants that are selective metalloproteinase inhibitors. Biochem Soc Symp 70: 201–212.

Nozaki S, Sissons S, Chien D S, et al. (2003) Activity of biphenyl matrix metalloproteinase inhibitor BAY 12-9566 in a human breast cancer orthotopic model. Clin Exp Metastasis 20: 407–412.

Ong P C, McGowan S, Pearce M C, et al. (2007) DNA accelerates the inhibtion of huma cathepsin V by serpins. J Biol Chem, 282: 36980–36986.

Paik S, Shak S, Tang G, et al. (2004) A multigene assay to predict recurrence of tamoxifen-treated, node-negative breast cancer. N Engl J Med 351: 2817–2826.

Puente X S, Sanchez L M, Overall C M, et al. (2003) Human and mouse proteases: A comparative genomic approach. Nat Rev Genetics 4: 544–558.

Rawlings N, Morton F R, Kok C Y, et al. (2008) MEROPS, the peptidase database. Nucleic Acids Res 36: D320–D325.

Relogio A, Schwager C, Richter A, et al. (2002) Optimization of oligonucleotide-based DNA microarrays. Nucleic Acids Res 30: e51.

Rhee J-S, Diaz R, Korets L, et al. (2004) TIMP-1 alters susceptibility to carcinogenesis. Cancer Res 64: 952–961.

Schwartz D R, Moin K, Yao B, et al. (2007) Hu/Mu ProtIn oligonucleotide microarray: Dual species array for profiling protease and protease inhibitor gene expression in tumors and their microenvironment. Mol Cancer Res 5: 443–454.

Shi G P, Munger J S, Mear J P, et al. (1992) Molecular cloning and expression of human alveolar macrophage cathepsin S, an elastinolytic cysteine protease. J Biol Chem 267: 7258–7262.

Shi G P, Sukhova G K, Kuzuya M, et al. (2003) Deficiency of the cysteine protease cathepsin S impairs microvessel growth. Circ Res 92: 492–500.

Shi L, Reid L H, Jones W D, et al. [MAQC Consortium] (2006) The microarray quality control (MAQC) project shows inter- and intraplatform reproducibility of gene expression measurements. Nat Biotechnol 24: 1151–1161.

Sinnamon M J, Carter K J, Sims L P, et al. (2008) A protective role for mast cells in intestinal tumorigenesis. Carcinogenesis 29: 880–886.

Yang F, Foekens J A, Yu J, et al. (2005) Laser microdissection and microarray analysis of breast tumors reveal ER-alpha related genes and pathways. Oncogene 25: 1413–1419.

Chapter 4
Quantitative Real-Time PCR Analysis of Degradome Gene Expression

Caroline J. Pennington, Robert K. Nuttall, Clara Sampieri-Ramirez, Matthew Wallard, Simon Pilgrim, and Dylan R. Edwards

Abstract Dissection of the contribution of proteases and inhibitors in the complex molecular events involved in cancer initiation, growth, and spread requires as a starting point detailed knowledge of the degradome genes that are expressed and dysregulated in cancer. This information identifies candidate genes for functional investigations and also reveals potential markers of disease progression and severity. Quantitative real-time polymerase chain reaction (qRT-PCR) analysis provides optimal sensitivity and specificity for analysis of RNA from human tumors and nonneoplastic tissues. In this chapter, we outline basic qRT-PCR technologies, and approaches for normalization and analysis of expression data. Degradation of RNA is a major problem for microarray analyses, but we demonstrate that TaqMan® qRT-PCR is a remarkably robust technique that can provide reliable information on archival specimens that would not be appropriate for other transcriptomic analyses. We also highlight the utility of low-density TaqMan arrays for degradome expression analysis.

Introduction

Quantitative real-time polymerase chain reaction (qRT-PCR) is one of the principal platform technologies of the genomic age that overcomes some of the main challenges associated with characterization and accurate quantification of protease expression in tissues and cell lines (Bustin et al. 2005). Unlike earlier methods of quantifying gene expression, such as Northern blotting, which requires 5–30 μg of RNA, qRT-PCR can accurately detect as little as 100 copies of target sequence in a 5 ng pool of reverse-transcribed complementary DNA (cDNA), equivalent to about 1 copy per cell (Nuttall et al. 2003). The technique can, therefore, be adapted when clinical tissue or cell samples are limited. As qRT-PCR simultaneously detects and quantifies the presence

C.J. Pennington
School of Biological Sciences, University of East Anglia, Norwich, NR4 7TJ UK,
e-mail: c.j.pennington@uea.ac.uk

D. Edwards et al. (eds.), *The Cancer Degradome.*
© Springer Science + Business Media, LLC 2008

of a specific region of DNA during the early, efficient phase of PCR, results obtained are more accurate and reproducible than those obtained from semiquantitative PCR techniques. The relative ease of methodology has resulted in high throughput profiling studies of large gene families in extensive collections of samples (Wall and Edwards 2002, Morimoto et al. 2004, Porter et al. 2004, Overbergh et al. 2003).

Technologies

qRT-PCR follows the general pattern of polymerase chain reaction, that is, the exponential amplification of target DNA, but with the added advantage of quantification after each round of amplification; this is the "real-time" aspect of the process. The accumulation of data at each cycle of the PCR greatly increases the sensitivity of the reaction and negates the need for post-PCR image processing involved in semiquantitative end-point strategies such as competitive PCR and "primer-dropping" PCR (Wall and Edwards 2002). Quantification of up to 384 samples can be determined in the time it takes to run a 40-cycle PCR, which, depending on the instrument, can be as little as 35 min.

Quantification Strategies

Four quantification strategies are currently available for qRT-PCR amplification: SYBR® green, Taqman® probes (Applied Biosystems, Foster City, CA), and Molecular Beacons and Scorpions® (DxS Ltd., Manchester, UK). Each of these chemistries relies on the detection of a fluorescent signal. SYBR uses a fluorescent molecule in the reaction mix that emits little fluorescence when in solution but a strong signal when incorporated in double-strand DNA (dsDNA) during primer extension. Taqman probes, Molecular Beacons, and Scorpions depend on Förster resonance energy transfer (FRET), in which a fluorogenic dye molecule and a quencher are coupled on a gene-specific oligonucleotide. In both cases, the increase in fluorescence signal is proportional to the amount of product produced during each PCR cycle. Individual samples are quantified relative to each other by determining the cycle at which the signal rises above background fluorescence, termed the cycle threshold or C_t. The lower the C_t, the earlier the signal is detectable above the threshold, the more target is present. The ability to quantify the amount of template in a sample remains accurate over a wide (at least 6 log) dynamic range (Nuttall et al. 2003).

SYBR

SYBR intercalating dyes are perhaps the simplest and certainly the cheapest method of quantifying gene expression using qRT-PCR. Nonlabeled target-specific primers are generated following qRT-PCR primer design protocols (see below), and these are combined with a PCR reaction mix that contains SYBR dye and a passive reference dye. The passive reference dye provides an internal reference to which the SYBR green signal can be normalized during data analysis, which is necessary to correct for fluorescent fluctuations caused by changes in concentration or volume. However, despite the apparent cost-effectiveness of using SYBR reactions one major disadvantage is the ability of SYBR dye to bind to any dsDNA including unspecific products and primer dimers, thus overestimating the amount of target in a sample. The presence of nonspecific amplification can be recognized by detecting amplification products in a "no template" control and by performing a post-run melting curve analysis to display dissociation curves for each target gene. If nonspecific products are amplified it will be necessary to carry out reaction optimization or redesign primers to overcome the problem. The need to run melt curves and conduct PCR optimization adds to the complexity of the analysis increasing the time and potential cost of the reaction. Another drawback of SYBR is that because multiple dye molecules bind to each product, longer amplicons will incorporate more dye molecules, resulting in a higher signal. More efficient reactions will also bind more SYBR molecules than would less-efficient reactions. Both these problems can be overcome by standardizing amplicon size and optimizing PCR efficiencies.

Probe-Based Chemistry

A major advantage of probe-based detection systems over SYBR is the added specificity of using a third gene-specific oligonucleotide or probe in the reaction. In this case, a dual-labeled fluorescent probe is positioned between the forward and reverse primers. These oligonucleotides are combined in a reaction mix that unlike the SYBR does not contain additional signaling dyes. A signal will only be generated if the probe itself hybridizes to its complementary target and fluoresces. Probes can be labeled with dyes of different wavelength emission spectra, so multiplexing qRT-PCR reactions is possible. It should be noted, however, that lack of detection of nonspecific products does not mean that the reaction is completely specific. Undetectable amplification of nontarget products will affect the efficiency of a PCR reaction and, consequently, the relative fold differences between individual samples. Assessing the efficiency of the reaction by analyzing a standard curve should indicate if this is a problem but generally the risk of amplifying and detecting additional products will be significantly reduced compared to SYBR reactions. An additional drawback of using probe-based technology in qRT-PCR is the cost of the individual probes that are required for each target sequence.

Three different probe chemistries are commonly available: Taqman probes, Molecular Beacons, and Scorpion probes. During PCR, a Taqman fluorogenic probe, consisting of an oligonucleotide labeled with a reporter and a quencher dye in close proximity, anneals specifically to the target sequence between the forward and reverse primers. When the probe is cleaved by the 5′ nuclease activity of the DNA polymerase during primer extension, the reporter dye is separated from the quencher dye, FRET no longer occurs, and a sequence-specific signal is generated. With each subsequent cycle, additional reporter dye molecules are cleaved from their respective probes and the increasing fluorescence intensity is monitored. Minor groove binding (MGB) Taqman probes are a modification of Taqman probes. Because of the minor groove binding moiety, probes can be shorter than other Taqman probes; this is an advantage when designing probes for single nucleotide polymorphism (SNP) and allelic discrimination analysis or at GC-rich or other sequence regions for which larger probes are unsuitable. MGB probes have added target specificity but are more expensive than non-MGB Taqman probes.

Molecular Beacons, like Taqman probes, rely on FRET for detection and quantification of target sequences. Unlike Taqman probes, Molecular Beacons form stem-loop structures when not hybridized to target DNA with a reporter dye on one arm and quencher in close proximity on the other arm. When annealed to its complementary target strand, a conformational transition occurs, the stem structure opens, and the entire length of the probe anneals to the target. The reporter and quencher are spread apart and florescence is emitted. Unlike Taqman probes, Molecular Beacons are not hydrolyzed and can be denatured from the target to reform the stem-loop configuration and can then be reused in the next cycle of PCR.

Scorpion probes are described as unimolecular in that the probe is linked, via a nonamplifiable linker to one of the gene-specific primers. Following annealing and extension of the primer, the stem-loop of the attached probe disassociates, stretches out, and anneals to the target, again separating the reporter and quencher dyes and preventing FRET occurring.

The higher initial costs of probe-based assays is partially offset since careful design can result in little requirement of reaction optimization, but there is no doubt that probe-based assays are expensive. One possible compromise between high specificity and cost considerations is the recent introduction of the Roche Universal Probe Library (Roche Applied Science, Burgess Hill, UK). This is a library of short, 8–9 nucleotide (nt) probes which offer transcriptome-wide coverage for specific organisms. Locked nucleic acid (LNA) chemistry means these probes are highly specific for 8–9 nt complementary strands. Added specificity is achieved by designing 5′ and 3′ flanking primers. A full set of 165 Universal Probe Library probes will effectively provide a probe for every gene in a number of specified organisms (including human, mouse, rat). Although initially expensive the Universal Probe Library allows a researcher to design and buy only relatively cheap primers to match an existing "in-house" probe. This makes preliminary studies into potential target genes a much more financially feasible option.

Methodological Aspects of TaqMan qRT-PCR

Primer Design

As with standard end-point PCR, primer design is vital to the success of qRT-PCR. Several design programs are available, for example, Primer Express (Applied Biosystems, Warrington, UK), Primer 3 (Rozen and Skaletsky 2000), or HUSAR (DKFZ,

Heidelberg, Germany). All should allow the strict criteria for qRT-PCR amplicon design to be selected; for example, Taqman probes should have

- not more than 2 Gs or Cs in the last five bases.
- a gauss/call second content of 30–80%.
- an amplicon size range of 50–150 bp.
- a maximum amplicon melting temperature of 85°C.
- a primer length of 9–40 bp.
- a primer melting temperature of 58–60°C with a difference of less than 2°C between primers.
- a probe melting temperature that is 10°C higher than that of the primers.
- more call second than gauss in the probes sequence.
- no gauss on the 5′ end of the probe.
- no self-binding complementarity.

It is important that at least one primer, but preferably the probe, crosses an exon junction. The primers thus created would amplify cDNA but not genomic DNA. This is important since DNase treatment of RNA is rarely 100% efficient and genomic DNA amplification will clearly lead to erroneous results. Primer and probe sets should be checked for sequence specificity by BLAST analysis and by sequencing of the PCR products.

Roche provides an online design center for use with the Universal Probe Library to find the best primer set to flank one of the universal probes. A sequence or accession number can be input and a list of potential primers and corresponding probes are returned in order of suitability. Likewise, PrimerBank provides predesigned and validated primer sequences for SYBR reactions via their Web site (Wang and Seed 2003).

Designing primers and probes for a gene of interest is relatively quick and allows the targeting of specific splice variants that may be of interest and offers more control of the regions targeted. However, time can be saved by purchasing ready-made primer/probe sets or SYBR primer pairs from companies specializing in their manufacture, for example, Qiagen (Crawley, UK) and Applied Biosystems (Warrington, UK) both have a genome-wide stock of fully validated primers and probes. It should be noted, however, that "off-the-shelf" primer/probes are more expensive than custom-designed sets. Whichever option is chosen for primer design, a useful validation of downstream analysis is to use more than one primer/probe set. Nolan et al. (2006a) suggest a 5′:3′ assay for glyceraldehyde 3-phosphate dehydrogenase (GAPDH) in which RNA integrity can be assessed. This assay uses primer sets designed for the 5′, mid, and 3′ region of GAPDH mRNA. Since these authors use oligo(dT) priming for reverse transcription, the ratio of amplicon in this assay indicates whether the reverse transcription reaction represents full-length cDNAs and thus intact starting RNA. This assay has the added benefit of providing a validation that a given gene is correctly amplified and quantified by a chosen primer set. Although this is an expensive option, if broad scale profiling is to be undertaken, it is a useful tool for genes that will be focused on more specifically. This is also an option for validating gene expression of rare transcripts where the sensitivity of the reaction may be borderline.

RNA Quality and Integrity

One of the prime benefits of using qRT-PCR to quantify gene expression is the sensitivity of the technique when using very small amounts of precious clinical specimens or cells, laser-captured tissue, or even formalin-fixed paraffin-embedded archived material. Each of these strategies imposes challenges on the extraction of high quality, intact RNA that is free of genomic DNA, nucleases, and other PCR contaminants. It is easy to argue that the quality of RNA is perhaps the most important determinant of the reproducibility and biological relevance of qRT-PCR data, and so it is important to recognize and remove samples that have degraded to such an extent that their use could result in a misinterpretation of results. Conversely, the meaningful statistical analysis of many tissue-based studies depends on the number of samples included. It is, therefore, equally important to maximize sample numbers and not reject samples that have some degradation but are still of adequate quality for qPCR.

Since most of the actual RNA isolation procedure takes place in a strong denaturant that renders RNases inactive, it is typically before isolation when RNA integrity is most at risk. At this stage, the quality of RNA can be maximized by careful and prompt treatment of starting material; for example, samples should be snap frozen or stored in buffers such as RNAlater® (Ambion, Warrington, UK) immediately following collection and then stored correctly before RNA extraction (Schoor et al. 2003). During tissue homogenization for RNA isolation, it is vital that the denaturant be in contact with the cellular contents from the very moment that the cells are disrupted. This can be problematic when tissues are difficult to break down (e.g., bone) or when samples are numerous, making rapid processing difficult, consequently some optimization of technique and a pilot quality check may be necessary before large numbers of extractions are undertaken. Additionally, where possible, tissue sample collection and extraction protocols should be highly standardized since different protocols will have implications on the type of RNA, the amount, and quality of RNA extracted.

Generally, column-based kits for RNA isolation, for example, SV total RNA isolation kit (Promega Southampton, UK) or High Pure RNA isolation kits (Roche Burgess Hill, UK), are preferable to the more traditional phenol/chloroform extraction methods as these greatly reduce the potential for contamination of extracted RNA with salts, alcohols, and proteins, although it should be noted that small RNAs can remain in silica-based columns during the extraction procedure, and so kits specifically designed for the extraction of microRNAs, for example, mirVana™ (Ambion, Warrington, UK), should be used. The quality and quantity of RNA is initially assessed by measuring UV absorbance in a spectrophotometer. Nucleic acids absorb light at a wavelength of 260 nm, whereas organic compounds like phenol/trizol/rnazol, sugars, and alcohols absorb light at 230 nm and proteins at 280 nm (Sambrook et al. 2001). 260/280 and 260/230 ratios greater than 1.8 are generally indicative of good quality, uncontaminated RNA. We use the NanoDrop® ND-1000 (Nanodrop Technologies, Wilmington, DE), which is highly accurate and requires the use of just 1 µl of RNA that can be used without dilution or the need for cuvettes.

A 260/280 ratio of less than 1.8 can indicate protein contamination or possibly overdried RNA pellets that are not fully in solution. Freezing overnight at $-80°C$ or briefly heating the RNA to $65°C$ can help rectify the latter problem. Contamination of RNA with residual extraction substances is more of a problem and can have wide ranging consequences for qRT-PCR results. A significant reduction in the sensitivity and kinetics of PCR assays is caused by inhibitory components; also different reactions may not be inhibited to the same degree and the effects can be compounded in absolute quantification where an external calibration curve is used to calculate the number of transcripts in the test samples. Various methods can be used to assess the presence of PCR inhibitors in samples with low 260:280/230 ratios. The PCR efficiency can be tested by the standard curve method, in which a single sample is serially diluted and the slope of the curve calculated. This is probably adequate if samples are not limited. Alternatively, the SPUD assay (Nolan et al. 2006b) can be used in which a potentially contaminated RNA sample is spiked with an uncontaminated high-quality control RNA that yields a defined C_t in an uninhibited reaction. PCR inhibitors will result in a higher than expected C_t value when the control RNA is amplified in the presence of contaminants in the test RNA.

A major problem that can be encountered when handling RNA for transcriptomic studies is RNA degradation. Degraded RNA can reduce the yield of reverse-transcribed cDNA and subsequent qPCR resulting in inaccurate representations of gene expression: this is particularly acute if oligo(dT) is used for RT priming. Traditionally, RNA quality has been evaluated by observation of ethidium-bromide stained bands of ribosomal RNA (rRNA) on nondenaturing agarose gels; this is still a good and cheap way of visualizing RNA quality and quantity although alone this would not be adequate for qRT-PCR. More recently, the introduction of the Agilent™ 2100 Bioanalyser (Agilent Technologies UK Ltd., Stockport, UK) and the BioRad Experion microfluidic capillary electrophoresis (BioRad, Hemel Hempstead, UK) systems have provided a more informative, qualitative, and quantitative assessment of RNA by calculating more precisely total RNA concentration based on the 28S:18S rRNA ratio. Additionally, an RNA integrity number (RIN) can be calculated and used to plot samples on a scale of quality and suitability for qRT-PCR (Schroeder et al. 2006).

However, our experience has been that some degradation of RNA for TaqMan qRT-PCR can be tolerated, allowing analysis of samples that would be unsuitable for microarray studies. The rationale for this is that amplicons for qRT-PCR are typically short (70–150 bp), and with random hexamer priming of RT reactions (see below), even partially degraded transcripts are successfully and reproducibly quantified. It is, however, vitally important to recognize and remove samples that have degraded to such an extent that their use could result in a misinterpretation of results. As an example, RNA samples from a bank of 169 archived urothelial carcinomas (UCCs) (Wallard et al. 2006) were assessed by Agilent 2100 Bioanalyser, and found to fall into approximately three equal groups representing "Poor," "Intermediate," and "Good" quality based on 28S:18S rRNA ratios (Fig. 4.1a). All samples were analyzed by TaqMan qRT-PCR for 18S rRNA and 30 protease genes. Although Agilent traces suggested that RNA samples classified as "Poor" quality

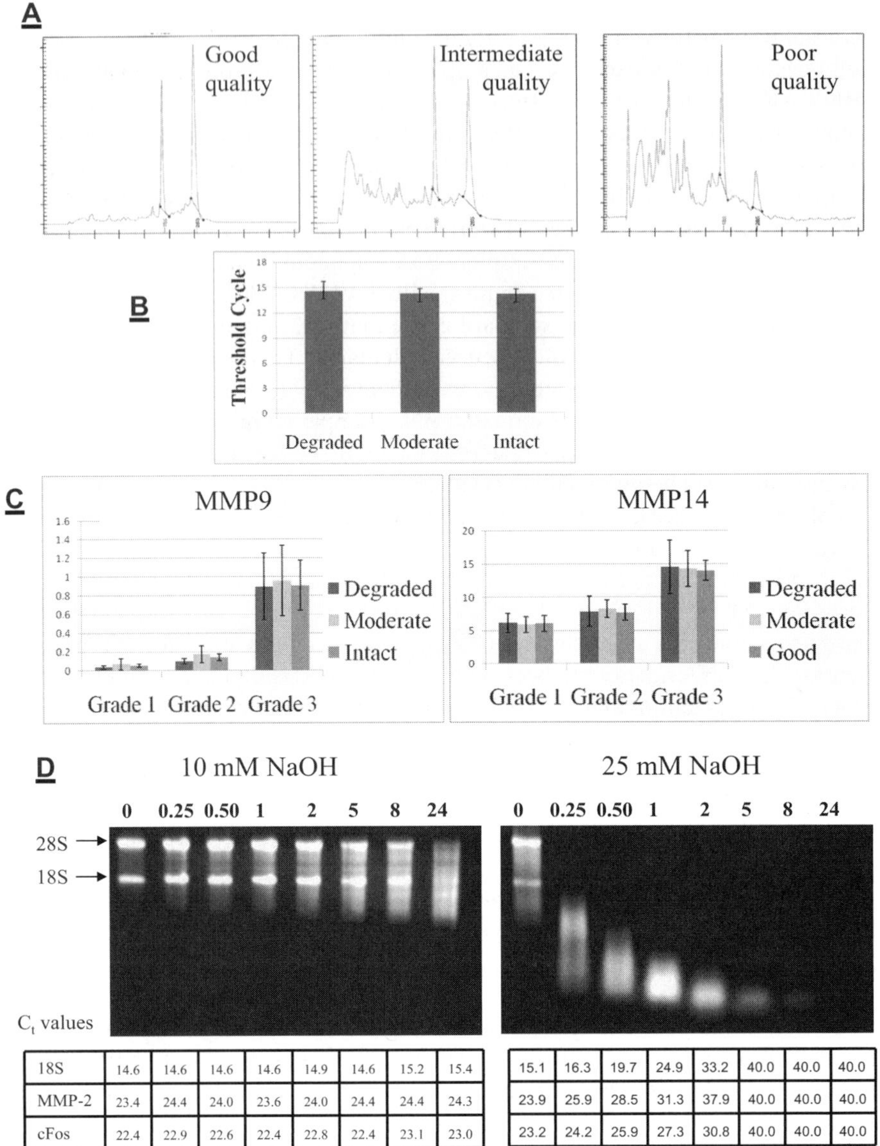

18S	14.6	14.6	14.6	14.6	14.9	14.6	15.2	15.4
MMP-2	23.4	24.4	24.0	23.6	24.0	24.4	24.4	24.3
cFos	22.4	22.9	22.6	22.4	22.8	22.4	23.1	23.0

18S	15.1	16.3	19.7	24.9	33.2	40.0	40.0	40.0
MMP-2	23.9	25.9	28.5	31.3	37.9	40.0	40.0	40.0
cFos	23.2	24.2	25.9	27.3	30.8	40.0	40.0	40.0

Fig. 4.1 TaqMan® quantitative real-time polymerase chain reaction (*qRT-PCR*) allows quantitative analysis of partially degraded *RNAs*. (**a**) Agilent™*RNA* traces: Representative traces obtained for samples from 169 urothelial carcinoma (*UCC*) samples representing intact (*left*), intermediate quality (*middle*) and poor quality *RNA* (*right*). (**b**) Comparison of threshold cycles (C_t) with Agilent total *RNA* analysis. The graph shows the median C_t value ± one standard deviation. (**c**) Relative expression of matrix metalloproteinase (*MMP*) transcripts in relation increasing tumor grades, grouped by Agilent total *RNA* trace quality. Data show the median expression ± one standard deviation for *MMP-9* (*left*) and *MMP-14* (*right*). (**d**) Agarose gel analysis of *RNA* degraded over time on incubation with 10 mM or 25 mM of NaOH and corresponding C_t values for 18S, *MMP-2*, and cFos are shown in the columns underneath

had undergone degradation, the majority of the 18S C_t values for such samples did not differ significantly from those for samples showing an intermediate or good trace (Fig. 4.1**a** and **b**). Conversely, the few samples with high C_t values for 18S *always* corresponded to poor quality or low yield Agilent traces. Exclusion of these 18 out of the initial 169 samples based on high 18S C_t (operationally we exclude any samples that are more than 1 C_t different from the median 18S C_t of the sample set) demonstrated that the remaining 151 samples showed similar patterns of expression of all degradome genes analyzed, regardless of their Agilent quality level. Figure 4.1**c** shows quantification of matrix metalloproteinase (MMP)-9 and MMP-14 in specimens of increasing histological grade, demonstrating that the patterns of expression in relation to tumor grade are evident, regardless of the RNA quality as assessed by Agilent. Agilent analysis would have excluded approximately two-thirds of the specimens, thus reducing the power of the study. These data show that small amplicon size – which is a necessary criterion of the qRT-PCR system – allows the use of moderately degraded samples that might be rejected if inclusion criteria relied on Agilent trace alone.

We extended this analysis by evaluating RNA that had been subjected to alkali degradation following isolation. Figure 4.1**d** shows that 18S and 28S bands were present on agarose gels following up to 24 h incubation with 10 mM NaOH, but these bands were no longer present after 15 min of incubation with 25 mM NaOH. C_t values for 18S rRNA remained stable, within 1 C_t of the control, at all time points during incubation with 10 mM NaOH. After 15 min of incubation with 25 mM NaOH, C_t values had increased by more than 1 C_t and continued to rise until there was no amplification of 18S 5 h after incubation with 25 mM NaOH ($C_t = 40$). Analysis of mRNA expression for MMP-2, a relatively stable mRNA with a long half-life (Overall et al. 1991), showed a similar pattern with C_t values rising by 2 as RNA degradation became detectable by raised 18S C_t and on agarose gels. Expression of cFos mRNA, an immediate early gene with a short half-life, also remained stable alongside TaqMan 18S values and amplification diminished in parallel with 18S as the RNA degraded. This again demonstrates that some RNA degradation is tolerated by qRT-PCR as long as amplicons are kept short and expression is normalized or analyzed alongside an endogenous control gene, a conclusion also reached by (Hamalainen et al. 2001).

Reverse Transcription

Once RNA has been extracted, checked for quality, and accurately quantified, it is necessary to reverse transcribe the RNA to generate cDNA. This can be performed as a one-tube, single-combined reverse transcription and PCR method or as a two-tube method with an initial reverse transcription followed by multiple qRT-PCRs using aliquots of cDNA. This section will focus solely on the more commonly used latter method.

Surprisingly, the relatively small stage of reverse transcribing RNA into cDNA can be an important contributor to variability and lack of reproducibility observed in qRT-PCR (Stahlberg et al. 2004a, 2004b; Bustin et al. 2005). A major source of that variability can be introduced by poor pipetting practice. Also, interassay variability can be introduced because the efficiency of reverse transcription reactions depends somewhat on the relative abundance of transcripts (Karrer et al. 1995, Curry et al. 2002, Bustin and Nolan 2004). Rare transcript templates will not reverse transcribe as efficiently as more abundant transcripts and may be disproportionately affected by background nucleic acid contamination. This problem can be exacerbated by the reverse transcription priming strategy. Three priming strategies are in common use; oligo(dT), random hexamers, and gene-specific priming (GSP). The effectiveness of each of these may vary depending on the concentration of the target genes, the quality and configuration of RNA, and the number of genes the researcher may wish to profile.

Oligo(dT) primers used in reverse transcription target the poly A tail of mRNA and are used in ~40% of reported assays using qRT-PCR (Bustin et al. 2005). Since rRNA will represent 75–80% of total RNA extracted and this abundance can affect the reverse transcription efficiency of rare mRNA species, priming specifically to target mRNA has some advantage. However, oligo(dT) priming is less efficient at generating cDNA from RNAs with significant secondary structures that block elongation of the cDNA strand. Oligo(dT)-primed cDNA will not include targets without a poly A tail, for example, histones, viral RNAs, or rRNAs that may be required as endogenous controls or quality control genes (see later). As discussed in the previous section, oligo(dT) priming is also only possible for good quality, intact RNA since cDNA synthesis of fragmented RNA will fail to reach the qRT-PCR amplicon site if this is located toward the 5′ end of a long mRNA, resulting in false negatives at qRT-PCR.

An alternative to oligo(dT) priming used in ~30% of qRT-PCR assays is priming by random hexamers (Bustin et al. 2005). These are 6 bp oligonucleotides of a varying sequence that prime at multiple origins along all the RNAs in a sample. The drawback of using this priming strategy is that rare mRNA targets may not be primed proportionately due to competition for priming sites by the more abundant rRNA molecules. This may have implications for the accuracy of qRT-PCR quantification. Random priming has also been reported to overestimate mRNA copy numbers compared to a 22 base gene-specific primer (Zhang and Byrne 1999).

Gene-specific priming uses a unique antisense primer or the reverse primer of the subsequent qRT-PCR to target and reverse transcribe only a gene of interest. Because of this specificity, this strategy has been considered sensitive when analyzing rare transcripts. However, if a broader profiling of gene expression is required this strategy becomes both time consuming and financially prohibitive since a reverse transcription reaction would be necessary for every gene to be analyzed.

More recently, the use of random pentadecamer (15-mer) priming has been reported to be 40% more efficient than using random hexamers (Stangegaard et al. 2006). Also, Abgene (Epsom, UK) recommend a 3:1 mix of random hexamers to oligo(dT). We have recently compared qRT-PCR results of rare (MMP-8) and more abundant (MMP-1) protease genes and 18S rRNA primed with either oligo (dT), random hexamers, GSPs for each target, a random pentadecamer or a 3:1 mix of random hexamers:oligo(dT). Gene-specific primers did not increase the sensitivity of the reverse transcription reaction for the rare MMP-8 transcript. There was no significant difference in quantification of targets using any of the priming strategies except that random hexamers allowed for the use of rRNA in qRT-PCR and was cheaper and quicker than using GSPs for each target. For ease and flexibility, random hexamers are probably the most versatile solution to reverse transcription priming, and are the basis of all of our published work.

A third source of variation at the reverse transcription stage of qRT-PCR can be introduced by varying reverse transcriptase enzymes used in reverse transcription reactions. This is particularly relevant if samples are reverse transcribed in separate laboratories and gene expression data are compared. Reverse transcriptases vary in efficiency and it is prudent to standardize protocols between laboratories or indeed experiments within the same laboratory that will be compared at a later date. Finally, it is important to include negative controls in any reverse transcription reaction. In this case, this should include a sample that has been through the reverse transcription reaction except that reverse transcriptase has been omitted and a reverse transcription in which no template RNA has been added. These controls will detect genomic DNA contamination of RNA and contamination of nucleic acids by other reagents or pipettes, tubes, and so on used during the preparation of the reaction. These controls should be included in subsequent qRT-PCR reactions.

Validation and Normalization

Ideally, an internal control used to normalize between samples should be constitutively expressed in all cell types at similar levels to the target gene and should remain constant, independent of disease status or experimental conditions. Historically, "housekeeping" genes, including GAPDH and β-actin, have been used as internal references, but their use has been largely discredited as expression of these genes can alter with varying cell culture conditions, hypoxia, in malignancy, and following treatment with tumor promoters (Hamalainen et al. 2001, Zhong and Simons 1999, Bhatia et al. 1994, Goldsworthy et al. 1993). In contrast, the expression of rRNA has been found to be relatively stable and has become a commonly used endogenous control in qPCR assays (Zhong and Simons 1999, Schmittgen and Zakrajsek 2000, Bhatia et al. 1994) Since rRNA and mRNA are generated by distinct polymerases, their levels are less likely to vary under conditions which would affect the expression of mRNA (Paule and White 2000). These features suggest that rRNA may be an appropriate gene for intersample standardization. However, concern has been expressed regarding possible imbalances between rRNA and mRNA fractions of different samples (Solanas et al. 2001), and the general unsuitability of rRNA as a normalizer for genes expressed at very different levels. Despite these hesitations, we believe that 18S rRNA is the most reliable method of assessing the quality of samples used in qRT-PCR. This is based on the fact that rRNA constitutes 75–80% of total RNA, so if the same amount of RNA of equal quality is used for the reverse transcription and if the same amount of cDNA is used in qRT-PCR, amplification should be identical in all samples. This is operationally equivalent, therefore, to loading equal amounts of total RNA, as determined spectrophotometrically, on a gel for Northern blotting. Based on the analysis shown in Fig. 4.1 on the effects of RNA degradation on 18S rRNA C_t values and our experience, accumulated from analysis of thousands of clinical samples from diverse tissue origins, exclusion of samples that show more than 1 C_t variation from the median 18S C_t value is essential: when such samples are not removed these can lead to a distortion of expression profiles. We believe this level of quality control is central to all qPCR analysis and that 18S rRNA amplification efficiency serves as a quality control and possibly a potential normalizing agent.

The consensus regarding subsequent normalization of qRT-PCR data has been reached following years of publications advocating the use of one or another reference gene followed by contradictions and further propositions. The recent introduction of GeNorm^{TM} (Vandesompele et al. 2002) provides an Excel-based program for determining the most stable reference genes from a panel of potential endogenous control genes. Vandesompele et al. show that the common practice of using a single normalizing gene can lead to erroneous normalization. They suggest that an ideal, universal control gene probably does not exist and that normalizing to more than one endogenous control gene may be a more robust strategy. To determine the most appropriate normalizing genes, GeNorm analyzes a panel of 6 or 12 genes (ACTB, B2M, GAPD, HMBS, HPRT1, RPL13A, RPL32, RPS18, SDHA, TBP, UBC, YWHAZ) to find the optimal genes to use as endogenous controls in each unique experimental system. Primer Design (Southampton, UK) provides a panel of primers for use with SYBR primer or fluorescently labeled probe sets which can be used in conjunction with GeNorm, although other potential normalizes can be added to this panel if appropriate. GeNorm has become the gold standard for determining the number of identity of the most stable normalizing genes for qRT-PCR analysis and overcomes many of the uncertainties that existed before its introduction.

Standard Curve Versus $\Delta/\Delta\ C_t$ Post-Run Analysis

qRT-PCR data can be analyzed using an absolute or relative standard curves method (Nuttall et al. 2004) or a comparative C_t method. Full details of both methods and examples are found in User Bulletin #2 produced by Applied Biosystems (Warrington, UK). It is ideal to include standard curves on every plate so that unknown samples that may fall within the less-sensitive region of amplification are recognized (as detailed earlier). Standard curves also show samples have been pipetted accurately, that the probe and primers are amplifying product in a meaningful way, and that the PCR is efficient (no contaminants etc.). The slope of the standard curve should be close to -3.2, which shows the PCR is 100% efficient. A slope with a value varying from this will indicate that each PCR cycle does not represent a doubling of product. The comparative method of data analysis is commonly used, but the strict rules regarding the efficiency of compared reaction and validation assays are not always adhered to. In brief, the absolute values of the slope of the log input amount versus the ΔC_t should be less than 0.1. See User Bulletin #2 for further clarification.

Degradome Expression in Cancer Cell Lines and Human Tumors

TaqMan qRT-PCR analyses of the MMPs and tissue inhibitor of metalloproteinases (TIMPs) have been reported in human cancer cell lines and gliomas (Nuttall et al. 2003), for the ADAMTS subfamily in breast cancers (Porter et al. 2004, 2006), for broader collections of serine and metalloproteinase and inhibitors in prostate cancer (Riddick et al. 2005), urothelial carcinoma (Wallard et al. 2006), and in comparisons of peripheral blood mononuclear cells and microglia (Nuttall et al. 2007). The developmental profiles of murine MMP and TIMP genes have also been presented (Young et al. 2002, Nuttall et al. 2004). One of the major issues in display of TaqMan data is the comparison of expression for different genes across the same sample set, since numerical values generated by the standard curve method discussed above are specific for particular primer–probe combinations, making inter-gene comparisons difficult. This can be overcome by the $\Delta/\Delta C_t$ method by reporting C_t differences between the genes of interest and 18S rRNA (*see*, for example, Jones et al. 2006). However, using the standard curve method and calibrating against synthetic RNA templates for a collection of MMP and TIMP genes (Young et al. 2002, Nuttall et al. 2003), we found that raw C_t values could be used for approximate intergene comparisons because no two genes showed more than a 5 C_t difference for equivalent absolute levels. Consequently, we developed a "tile-pattern" display that groups genes by 5 C_t ranges of C_t values, represented as very high expression ($C_t \leq 25$), high expression ($C_t = 26$–30), moderate ($C_t = 31$–35), low ($C_t = 36$–39), and not detected ($C_t = 40$). This has proved to be a valuable way to display comparative expression across degradome gene families, as shown for the analysis of MMPs and TIMPs in a panel of human mammary tissues as shown in Fig. 4.2. Another useful display is the box-and-whisker plot of $\Delta/\Delta C_t$ differences relative to 18S rRNA, though it needs to be emphasized that unless amplification efficiency is the same for each gene probe, comparison of levels of expression between different genes is only an approxima-tion. We have also used heatmap displays of C_t data, which is useful in conjunction with hierarchical cluster analysis (Eisen et al. 1998).

We have recently developed a 384-gene TaqMan low-density array (TLDA) that includes the entire human metalloproteinase and serine proteinase families, along with their inhibitors and additional control genes. This TLDA has been used to analyze expression in a small collection of breast cancers and normal mammary tissue, revealing excellent fidelity and sensitivity compared to conventional Taq-Man qRT-PCR. In addition to the confirmation of several MMP genes that are dysregulated (including MMP-1, -3, -10, -11, -12; ADAMTS-14), this analysis revealed for the first time several other genes whose expression is elevated in tumors compared to normal breast (*see* Chap. 30 by Pennington et al., this volume). These genes are being evaluated further using in silico transcriptomic data mining of resources such as Oncomine and In Silico Transcriptomics (*see* Chap. 31 by Iljin et al., this volume).

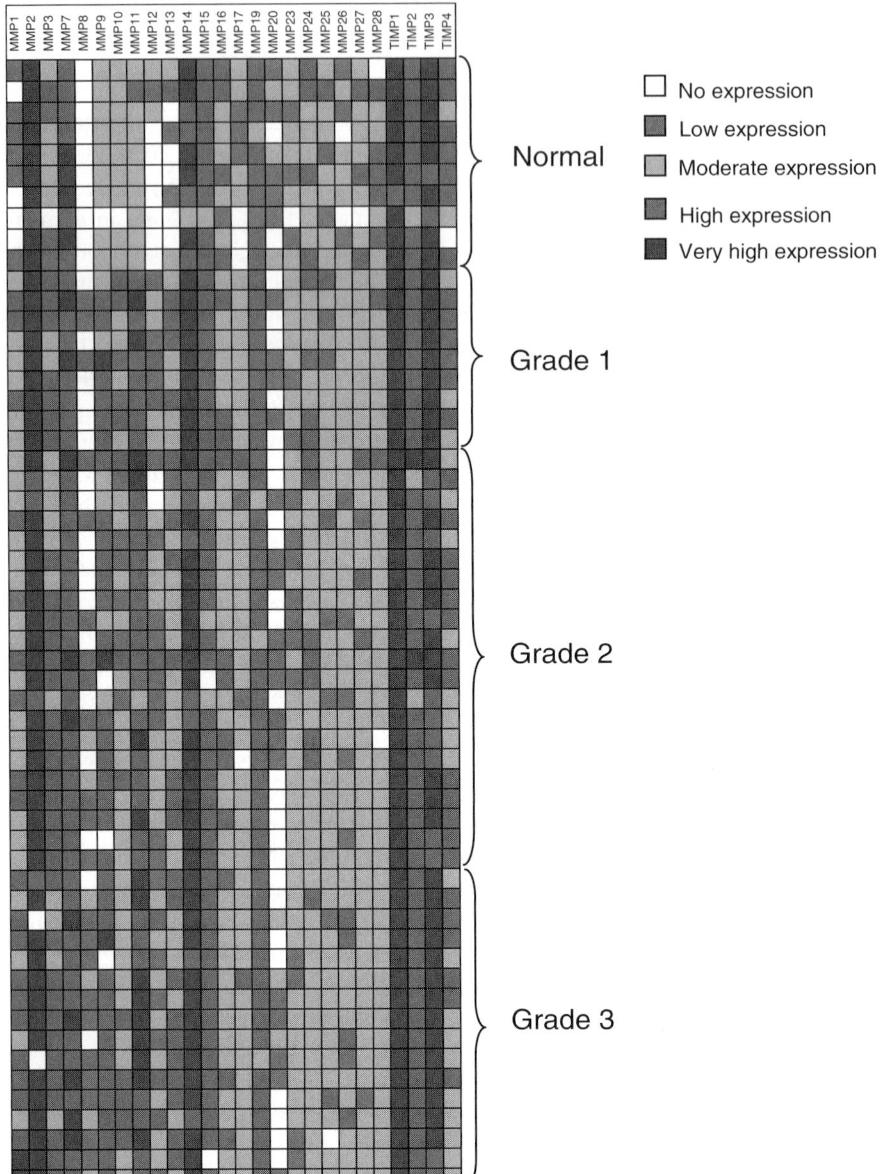

Fig. 4.2 Comparative TaqMan® expression profiles of matrix metalloproteinase (*MMP*) and tissue inhibitor of metalloproteinase (*TIMP*) gene families in normal and malignant human mammary tissues. Data are displayed as the expression levels of the indicated genes in human mammary tissue samples, which have been reported in detail elsewhere (Porter et al. 2004). The tissues are grouped as normal and histopathological grades 1, 2, and 3, respectively. Expression levels (raw C_t levels) are shown in the associated key. (*See also* Color Insert I)

Conclusions and Perspectives

qRT-PCR continues to have value in the analysis of the cancer degradome both as a tool for validation of microarray findings and for primary data generation, the latter, in particular, through the development of low-density arrays that provide comprehensive coverage of gene families. Cost is certainly a factor, but the sensitivity provided by TaqMan and the robustness of the technique, which can tolerate levels of RNA degradation that exclude samples from use on array platforms, make it applicable to analysis of archival material and laser-captured tissue specimens, without a requirement for RNA amplification (Pedersen et al. 2005). The universal probe library strategy also increases its versatility. The specificity of the technique makes it ideal for single nucleotide polymorphism analysis or quantification of alternatively spliced mRNA variants, which are beginning to be studied in detail for degradome genes. We have also applied the technique for parallel quantification of heterogeneous nuclear RNA (hnRNA) and mature mRNA in the same total RNA preparations, the hnRNA providing a surrogate of gene transcription that was shown to correlate well with transcription rates determined in nuclear run-on assays (C. Pennington, unpublished data).

Perhaps one of the major areas where TaqMan analysis will become increasingly valuable is in clinical diagnostics using expression signatures that identify patients at high- and low risk of recurrence, and in prediction of response to therapy (*see* also Chap. 30 by Pennington et al., this volume). The sensitivity, reliability, and speed of TaqMan make it attractive for analysis of small gene subsets, such as the Oncotype 21-gene and the 70-gene Amsterdam "MammaPrint®" signatures (Sotiriou and Piccart 2007). As will be discussed in Chap. 30, it is possible that these signatures could be refined further, potentially by including genes such as MMP-8 and ADAMTS-15 that have been shown to have prognostic value in breast cancer (Porter et al. 2006, Gutierrez-Fernandez et al., submitted), but which likely are expressed at levels that are too low for reliable quantification by microarray analysis.

References

Bhatia P., Taylor W. R., et al. (1994). Comparison of glyceraldehyde-3-phosphate dehydrogenase and 28S-ribosomal RNA gene expression as RNA loading controls for northern blot analysis of cell lines of varying malignant potential. Anal Biochem 216(1): 223–6.

Bustin S. A., and Nolan T. (2004). Pitfalls of quantitative real-time reverse-transcription polymerase chain reaction. J Biomol Tech 15(3): 155–66.

Bustin S. A., Benes V., et al. (2005). Quantitative real-time RT-PCR—a perspective. J Mol Endocrinol 34(3): 597–601.

Curry J., McHale et al. (2002). Low efficiency of the Moloney murine leukemia virus reverse transcriptase during reverse transcription of rare t(8;21) fusion gene transcripts. Biotechniques 32(4): 768, 770, 772, 754–5.

Eisen M. B., Spellman P. T., et al. (1998). Cluster analysis and display of genome-wide expression patterns. Proc Natl Acad Sci USA 95(25): 14863–8.

Goldsworthy S. M., Goldsworthy T. L., et al. (1993). Variation in expression of genes used for normalization of Northern blots after induction of cell proliferation. Cell Prolif 26(6): 511–8.

Hamalainen H. K., Tubman J. C. et al. (2001). Identification and validation of endogenous reference genes for expression profiling of T helper cell differentiation by quantitative real-time RT-PCR. Anal Biochem 299(1): 63–70.

Jones G. C., Corps A. N., et al. (2006). Expression profiling of metalloproteinases and tissue inhibitors of metalloproteinases in normal and degenerate human achilles tendon. Arthritis Rheum 54(3): 832–42.

Karrer E. E., Lincoln J. E., et al. (1995). In situ isolation of mRNA from individual plant cells: creation of cell-specific cDNA libraries. Proc Natl Acad Sci USA 92(9): 3814–8.

Morimoto A. M., Tan N., et al. (2004). Gene expression profiling of human colon xenograft tumors following treatment with SU11248, a multitargeted tyrosine kinase inhibitor. Oncogene 23(8): 1618–26.

Nolan T., Hands R. E., et al. (2006a). Quantification of mRNA using real-time RT-PCR. Nat Protoc 1(3): 1559–82.

Nolan T., Hands R. E., et al. (2006b). SPUD: a quantitative PCR assay for the detection of inhibitors in nucleic acid preparations. Anal Biochem 351(2): 308–10.

Nuttall R. K., Pennington C. J., et al. (2003). Elevated membrane-type matrix metalloproteinases in gliomas revealed by profiling proteases and inhibitors in human cancer cells. Mol Cancer Res 1(5): 333–45.

Nuttall R. K., Sampieri C. L., et al. (2004). Expression analysis of the entire MMP and TIMP gene families during mouse tissue development. FEBS Letters 563: 129–134.

Nuttall R. K., Silva C., et al. (2007). Metalloproteinases are enriched in microglia compared with leukocytes and they regulate cytokine levels in activated microglia. Glia 55(5): 516–26.

Overall C. M., Wrana J. L., et al. (1991). Transcriptional and post-transcriptional regulation of 72-kDa gelatinase/type IV collagenase by transforming growth factor-beta 1 in human fibroblasts. Comparisons with collagenase and tissue inhibitor of matrix metalloproteinase gene expression. J Biol Chem 266(21): 14064–71.

Overbergh L., Giulietti A., et al. (2003). The use of real-time reverse transcriptase PCR for the quantification of cytokine gene expression. J Biomol Tech 14(1): 33–43.

Paule M. R., and White R. J. (2000). Survey and summary: transcription by RNA polymerases I and III. Nucleic Acids Res 28(6): 1283–98.

Pedersen T. X., Pennington C. J., et al. (2005). Extracellular protease mRNAs are predominantly expressed in the stromal areas of microdissected mouse breast carcinomas. Carcinogenesis 26(7): 1233–40.

Porter S., Scott S. D., et al. (2004). Dysregulated expression of adamalysin-thrombospondin genes in human breast carcinoma. Clin Cancer Res 10(7): 2429–40.

Porter S., Span P. N., et al. (2006). ADAMTS8 and ADAMTS15 expression predicts survival in human breast carcinoma. Int J Cancer 118(5): 1241–7.

Riddick A. C. P., Shukla C. J., et al. (2005). Identification of degradome components associated with prostate cancer prostate cancer progression by expression analysis of human prostatic tissues. Br. J. Cancer 92: 2171–2180.

Rozen S., and Skaletsky H. (2000). Primer3 on the WWW for general users and for biologist programmers. Methods Mol Biol 132: 365–86.

Sambrook J., MacCullum P., Russell D. (2001). Molecular Cloning: A Laboratory Manual. Cold Spring Harbor Laboratory Press.

Schmittgen T. D., and Zakrajsek B. A. (2000). Effect of experimental treatment on housekeeping gene expression: validation by real-time, quantitative RT-PCR. J Biochem Biophys Methods 46(1–2): 69–81.

Schoor O., Weinschenk T., et al. (2003). Moderate degradation does not preclude microarray analysis of small amounts of RNA. Biotechniques 35(6): 1192–6, 1198–201.

Schroeder A., Mueller O., et al. (2006). The RIN: an RNA integrity number for assigning integrity values to RNA measurements. BMC Mol Biol 7: 3.

Solanas M., Moral R., et al. (2001). Unsuitability of using ribosomal RNA as loading control for Northern blot analyses related to the imbalance between messenger and ribosomal RNA content in rat mammary tumors. Anal Biochem 288(1): 99–102.

Sotiriou C., and Piccart M. J. (2007). Taking gene-expression profiling to the clinic: when will molecular signatures become relevant to patient care? Nat Rev Cancer 7(7): 545–53.

Stahlberg A., Hakansson J., et al. (2004a). Properties of the reverse transcription reaction in mRNA quantification. Clin Chem 50(3): 509–15.

Stahlberg A., Kubista M., et al. (2004b). Comparison of reverse transcriptases in gene expression analysis. Clin Chem 50(9): 1678–80.

Stangegaard M., Dufva I. H., et al. (2006). Reverse transcription using random pentadecamer primers increases yield and quality of resulting cDNA. Biotechniques 40(5): 649–57.

Vandesompele J., De Preter K., et al. (2002). Accurate normalization of real-time quantitative RT-PCR data by geometric averaging of multiple internal control genes. Genome Biol 3(7): RESEARCH0034.

Wall S. J., and Edwards D. R. (2002). Quantitative reverse transcription-polymerase chain reaction (RT-PCR): a comparison of primer-dropping, competitive, and real-time RT-PCRs. Anal Biochem 300(2): 269–73.

Wallard M. J., Pennington C. J., et al. (2006). Comprehensive profiling and localisation of the matrix metalloproteinases in urothelial carcinoma. Br J Cancer 94(4): 569–77.

Wang X., and Seed B. (2003). A PCR primer bank for quantitative gene expression analysis. Nucleic Acids Res 31(24): e154.

Young D. A., Phillips B. W., et al. (2002). Identification of an initiator-like element essential for the expression of the tissue inhibitor of metalloproteinases-4 (Timp-4) gene. Biochem J 364 (Pt. 1): 89–99.

Zhang J., and Byrne C. D. (1999). Differential priming of RNA templates during cDNA synthesis markedly affects both accuracy and reproducibility of quantitative competitive reverse-transcriptase PCR. Biochem J 337 (Pt. 2): 231–41.

Zhong H., and Simons J. W. (1999). Direct comparison of GAPDH, beta-actin, cyclophilin, and 28S rRNA as internal standards for quantifying RNA levels under hypoxia. Biochem Biophys Res Commun 259(3): 523–6.

Chapter 5
Identification of Protease Substrates by Mass Spectrometry Approaches-1

Mari Enoksson, Wenhong Zhu, and Guy S. Salvesen

Abstract The identification of protease substrates is a challenging task, especially *in vivo*. Bioinformatics-based prediction of cleavage sites and determination of protease preferences on synthetic substrates are important techniques in predicting natural protease substrates. These techniques tell us what a protease can do, but to enable a clear understanding of what a protease actually does *in vivo*, different approaches are required. The introduction of diverse proteomics-based methods is now leading toward the identification of natural protease substrates. These methods can provide a picture of the general protein cleavage pattern at a given time, that is, the proteolytic signature, in a complex biological sample. While gel-based (two-dimensional polyacrylamide gel electrophoresis and two-dimensional differential gel electrophoresis) methods have been used for a long time, there are some disadvantages with these methods. The introduction of protein- and peptide-labeling strategies, such as isotope-coded affinity tags, isobaric tag for relative and absolute quantification, and NHS-SS-biotin, together with enrichment procedures, such as combined fractional diagonal chromatography, have opened up new possibilities for the identification and relative quantification of protease cleavage *in vivo*, and thus have set the scene for a new direction in the study of proteolytic pathways. We give an overview of the different gel-based and solution-based strategies, provide examples of substrate identification, and also discuss the benefits and the pitfalls of the existing strategies.

Introduction

Analysis of protease substrate specificity has come a long way since the 1970s when the activity of new proteases was tested on the oxidized insulin B chain (Blow and Barrett 1977, Kargel et al. 1980, McKay et al. 1983). Today, a number of chemistry- and biology-based technologies are widely available to define the preferences

Guy S. Salvesen

Burnham Institute for Medical Research, 10901 North Torrey Pined Road, La Jolla, CA 92037, USA

D. Edwards et al. (eds.), *The Cancer Degradome.*

© Springer Science + Business Media, LLC 2008

of proteases for certain sequences. They are, naturally, all carried out on unnatural substrates, but they give invaluable information on the inherent properties of proteases to accept certain side chains of peptide or protein sequences in their active-site clefts. These techniques tell us what a protease *can* do.

Identifying natural protease substrates is more challenging. One has to take into account the activity of a protease, or a group of proteases, in their natural location, and somehow subtract all of the noise to give the essential information. A convenient way of studying the general protein cleavage pattern at a given time, that is, the proteolytic signature, in a complex protein mixture such as a cell lysate or tissue sample, is by using proteomic technology. Proteomics provide the necessary separation, protein identification and, in some cases, also the relative quantification of substrate cleavage. This methodology tells us what a protease *does in vivo*.

We will provide a short overview of the different methods, give some examples of substrate identification, and also discuss the benefits and the pitfalls with the different strategies.

Intrinsic Protease Specificity

Most proteases bind their substrates in an extended manner with the peptide chain running through the protease's catalytic cleft. The intrinsic specificity of a protease is defined by our ability to determine the optimal peptide sequence (motif) that binds into the specificity determinants of the active-site cleft (Fig. 5.1). Motifs have been

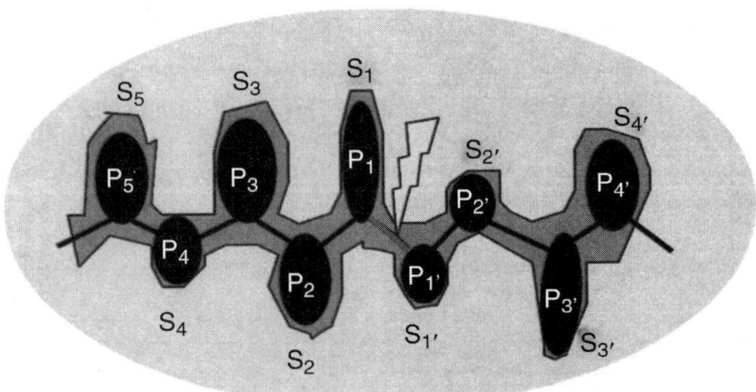

Fig. 5.1 Substrate cleft convention. The convention for naming protease substrate residues came from the work of Schecter and Berger on the size of the active site in papain (Schecter and Berger 1967). Think of the surface of a protease as a cleft with pockets defined by protease side chains. These are known as the specificity (S) pockets. Into these, in an extended conformation, fit the peptide (P) substrate side chains. By convention the scissile bond (yellow bolt) is between P_1 and $P_{1'}$. Residues toward the N-terminus of the substrate are numbered P_1, P_2, P_3, ..., P_n, and toward the C-terminus of the substrate they are named $P_{1'}$, $P_{2'}$, $P_{3'}$ ..., $P_{n'}$. Thus, subsites toward the N-terminus of the scissile bond are "unprimed" but sites toward the C-terminus are "primed." (*See also* Color Insert I)

Table 5.1 A brief survey of methods for determining intrinsic protease specificity

Methods	Description	Selected references
Combinatorial peptide library	Several versions exist: The main objective is to array individual peptides, or pools of peptides, that essentially cover all possible sequences within the peptide (up to four amino acids in length). Peptides are frequently coupled to a fluorescent group that reports cleavage.	(Matthews and Wells 1993, Thornberry et al. 1997, Stennicke et al. 2000, Turk et al. 2001)
Phage/bacterial display	A library of randomized peptides (up to eight residues long) encoded as a fusion protein bound to a support (phage) or as an *Escherichia coli* surface protein bound to a fluorescent reporter. In phage display, protease is added, and released phage are amplified and subjected to several rounds of selection to give a family of putative target sequences. In bacterial display, protease is added and nonfluorescent bacteria are sorted by FACS, amplified, and the surface gene sequenced to determine the target sequence.	(Boulware and Daugherty 2006, Ding et al. 1995, Kridel et al. 2002)
Small pool expression/ mRNA display	In the small pool expression method, a cDNA library is transcribed/translated in vitro, divided into subpools, incubated with the target protease and screened for protease substrates by SDS-PAGE. In mRNA display, a short strand of mRNA is coupled to its encoded protein sequence (80–200 aa residues) and to an affinity tag. Cleaved substrates are enriched after treatment with the protease and several rounds of selection gives a pool of putative target substrates.	(Cryns et al. 1997, Ju et al. 2007, Li et al. 1998, Loeb et al. 2006)

FACS fluorescence-activated cell sorter, *SDS-PAGE* sodium dodecyl sulfate-polyacrylamide gel electrophoresis

suggested for many different proteases, based on information from, for example, synthetic peptide libraries or phage display technology (Deng et al. 2000, Ding et al. 1995, Harris et al. 2000, Nazif and Bogyo 2001, Smith et al. 1995, Stennicke et al. 2000, Thornberry et al. 1997, Turk et al. 2001). These approaches (*see* Table 5.1) have been exploited very successfully to define the intrinsic substrate specificity of a handful of proteases, such as fibrinolytic proteases, matrix metalloproteases (MMPs), caspases, and membrane-associated serine proteases.

Cleavage site specificity predictions have been used as an indirect means for the identification of potential protease substrates containing the consensus cleavage sites. While these rapid methods have accelerated the discovery of intrinsic specificity, they do not provide reliable information on natural protease substrates. On the contrary, this information has sometimes produced substrate predictions that do not occur naturally (Ding et al. 1995). Exposure on the protein surface is usually pivotal for substrate cleavage, and the cleavage site is usually found in regions of the protein that do not take part in the secondary structure, such as in disordered loops. Moreover, the importance of exosites, protein regions that take part in the substrate–protease interaction but are distant from the cleavage site, are often overlooked. Naturally, cell compartmentalization separates protease from substrate as a means of regulating cell signaling. A protein containing a predicted cleavage site may thus often not be a substrate *in vivo*. More important, these methods do not provide information about which substrates are cleaved *in vivo* during a specific signaling event. The only way to determine the natural substrates is by genetic methods (Bugge et al. 1996), or by directly determining the products of proteolysis ex vivo.

Discovery of Natural Substrates

The detection of natural substrates has been made more simple, as well as more high throughput, by the use of proteomic methods. Comparative proteomics involves the identification and quantification of all expressed proteins and assessment of their cellular localization and their posttranslational modification. Proteolysis is the major posttranslational modification that occurs in living organisms, and identification of natural substrates by comparative proteomics thus provides us with information about not only the protease but also the cellular signaling that follows a proteolytic event. We can break down the proteomic methods into gel-based and solution-based methods. Currently, one of the main advantages of solution-based proteomics is that, under the right conditions, direct demonstration of the *in vivo* cleavage sites is possible.

Gel-Based Proteomic Methods

Two-dimensional polyacrylamide gel electrophoresis (2D-PAGE) has been used for decades to identify differential protein expression or posttranslational modifications, including proteolytic processing. First described by O'Farrell in 1975 (O'Farrell 1975), the method evolved after the discovery of immobilized pH gradient (IPG) gel strips and further improved after the introduction of fluorescent dyes, such as SYPRO® Ruby (Lopez et al. 2000). Traditional 2D-PAGE suffers from problems with gel reproducibility. The same sample run on two different gels may look slightly different, thereby making correct protein detection and identification more difficult. This can be circumvented by using 2D-DIGE (two-dimensional differential gel electrophoresis) (Tonge et al. 2001). Protein samples are labeled with fluorescent cyanine dyes (Cy2, Cy3, and Cy5), which have the same gel migration capacity but different fluorescence excitation and emission. Protein samples are labeled with one of the different dyes and samples are subsequently mixed and run on the same gel. The same protein from two different samples will comigrate to the same spot within the gel, but can be differentially visualized by altering excitation and emission filters. A protein that is equally abundant in the two different samples will be detected as a yellow spot in the gel, while proteins detected in only one sample will be either green or red depending on which label was used. The target proteins can then be identified after in-gel digestion and mass spectrometry (MS/MS). The 2D-DIGE system makes the relative quantification of protein abundance within the same gel possible, and available software and hardware platforms are more sensitive than the naked eye, thus allowing detection of more proteins. It is, however, still advisable to run several gels to confirm the reproducibility of the differential protein pattern.

2D-PAGE and 2D-DIGE have been successfully used to identify several protease substrates (Bech-Serra et al. 2006, Bredemeyer et al. 2004, Hwang et al. 2004, Taylor et al. 2007, Zhou et al. 2004). These methods give sufficient evidence for substrate identification, but the substrate hunt is limited to proteins of fairly high abundance, largely due to the limited gel-loading capacity. For example, Gygi and coworkers (Gygi et al. 2000) showed that the 2D-PAGE/MS approach failed to identify proteins that occur in low-copy number (i.e., codon bias values of <0.1) in the yeast genome. Unfortunately, these undetected proteins most likely include many important proteins such as signaling proteins and transcription factors that may be important substrates for proteases. Moreover, although a substrate is cleaved in sufficient quantity for detection by 2D-PAGE or 2D-DIGE, protein identification is not always possible (Bredemeyer et al. 2004). 2D-PAGE has good resolving power, but gel-spot recognition and quantitation can be time consuming and not very accurate, and 2D-PAGE as a separation tool has several drawbacks, such as a limited dynamic range, molecular mass range, and pI range (Gygi et al. 2000). Most importantly, several proteins can be covered in one protein spot, making correct protein ID problematic, and hydrophobic proteins and high molecular weight proteins are underrepresented in the 2D gel system. Some of these problems can be circumvented by sample fractionation and/or substrate enrichment. The 2D resolution can be changed by the use of bigger gels, altering pI, or molecular weight separation, and thus improve the detection of less-abundant proteins and decrease the risk of incorrect protein identification.

Altogether, without prior sample fractionation or enrichment, the applications of these gel-based approaches are limited to profiling of highly expressed proteins. In addition, these methods do not offer information about the protease cleavage site, which makes it more difficult to pinpoint the protease responsible for a protein cleavage.

Solution-Based Proteomic Methods Using Protein or Peptide Labeling

In light of the drawbacks of 2D-PAGE, various solution-based proteomic methods have been developed recently. Different labeling methods have been used for detection of proteolysis substrates, depending on the sample and the availability of reagent and MS/MS instrumentation. These methods rely on the same main principles as 2D-PAGE – protein separation, identification, and relative quantification. Protein quantification by MS/MS is greatly aided by the availability of small tags that have the same chemical composition and properties, but differ in mass. The relative abundance of the differentially labeled peptides in MS/MS reflects the difference in the protein levels between the labeled biological samples. Importantly, several of these labeling strategies provide information of the protease cleavage site.

ICAT Labeling

The introduction of isotope-coded affinity tags (ICATs) had a great impact on the field of comparative proteomics (Gygi et al. 1999, Tao and Aebersold 2003). The basic principle is to mix differentially labeled samples, digest the proteins, and analyze the peptides by liquid chromatography (LC)-MS/MS. Labeling with ICAT and its variants occurs at the free thiol group of Cys residues (which are convenient nucleophiles for specific modifications) through an alkylation reaction using iodo-acetyl- (Gygi et al. 1999, Zhou et al. 2002) or maleimidyl- (Qiu et al. 2002) containing labeling reagents. One sample is labeled with an isotopically light (^{1}H or ^{12}C) reagent and the second sample with the corresponding isotopically heavy (^{2}H or ^{13}C) reagent. The ^{13}C ICAT reagent is preferred by most laboratories since this variant overcomes the separation of the light and heavy forms of labeled peptides in the LC step that otherwise can occur with ^{2}H labels.

ICAT and its variants in combination with 1D- or 2D-LC-MS/MS have been successfully used for protein expression profiling in many fields of biology (Gygi et al. 2002, Tao and Aebersold 2003). ICAT labeling has also been used to identify putative natural substrates of MMPs in cell culture supernatants (Tam et al. 2004). The advantage of ICAT labeling is that it significantly reduces the complexity of the peptide mixture and increases the dynamic range of MS analysis. However, Cys is a relatively rare amino acid. The limited number of Cys residues reduces both protein coverage and proteome coverage, thereby reducing confidence in protein identification. Moreover, the use of ICAT reagent fails for quantification of Cys-free proteins. In fact, 8% of yeast proteins and 11% of human proteins do not contain Cys residues (Wang et al. 2002). Generally, this method has limited use in quantitative analysis of proteolysis or other posttranslational methods and it does not provide cleavage site information.

Amine-Reactive Reagents

Because of the limitation with the Cys-reactive labels, several strategies using other types of labels have been proposed. Most of these methods rely on N-terminal labeling of the proteins. A proteolytic event generates a new N-terminal, which if specifically labeled, provides both substrate identification and cleavage site location. The reactivity of protein N-terminals and the ε-amines on Lys side chains are very similar, but specific N-terminal labeling can be achieved by conversion of Lys to homoarginine by *o*-methylisourea (Beardsley et al. 2000, Kimmel 1967, Warwood et al. 2006). Since the exact location of the natural cleavage site can now be determined, bioinformatics analysis [CutDB (Igarashi et al. 2007), PoPS (Boyd et al. 2005), or Casbah (Luthi and Martin 2007) for example] can be used to suggest or even identify the culprit protease. The majority of the methods described below thus rely on selective analysis of protein N-terminals and/or on the incorporation of a heavy or light isotopic mass tag to detect the protease substrates based on the differential ratio of labeled peptides.

iTRAQ Labeling

A very powerful amine-reactive label that has been used for comparative proteomics is iTRAQ (isobaric tagging for relative and absolute quantitation). The iTRAQ reagent consists of a reporter group, a balance group, and an amino-reactive group. The latter can react with all N-termini, for example, Lys side chains and protein/peptide N-termini. The iTRAQ reagent has been used for measurement of differential protein expression (Ross et al. 2004). Peptides of the same sequence labeled with iTRAQ reagents are identical in mass in single MS mode, but show strong low-mass MS/MS signature ions (114–117 m/z) due to the presence of the reporter group in the labeling reagent. Relative quantification of the labeled peptides is done by analyzing the ratio of the signature ion (114:115:116:117) in the MS/MS spectra, allowing for the quantification of four different samples simultaneously. The latest iTRAQ labeling reagents are able to quantify up to eight protein samples.

iTRAQ has been used for detection of proteolytic substrates using two different approaches. The iTRAQ mass tags were used to label all amine-reactive groups (N-termini and Lys side chains) in proteins from conditioned culture medium of MMP-2 activated and inactivated cells (Dean and Overall 2007). This iTRAQ labeling strategy is similar to ICAT labeling (above), but a comparison between the two labeling reagents showed that the iTRAQ strategy is more powerful in finding protease substrates. In fact, the total number of identified proteins by iTRAQ was ninefold more than that by ICAT and the multiple peptide identified and quantified from the same protein increased the identification confidence.

Because iTRAQ is an amine-specific reagent, it can be used to label N-termini generated during proteolysis. But first, side chain Lys residues must be blocked, and a recent approach reveals that protein guanidination by treatment with *o*-methylisourea serves this purpose (Enoksson et al. 2007). Samples are differentially labeled by iTRAQ reagents, mixed and digested together, thus ensuring a reproducible protein digestion pattern. The peptides are separated by LC, and the labeled peptides detected by MS/MS. The labeled peptides are subsequently mapped into the protein database. While an uncleaved protein is labeled only at the original protein N-terminal [unless this is blocked by, for example, acetylation, which is the case for most human cytosolic proteins (Polevoda and Sherman 2003)], a cleaved protein will show labeling at an internal fragment, at the protease cleavage site. This procedure was evaluated in a mixture of cleaved recombinant proteins (Enoksson et al. 2007). The sensitivity of the method was further demonstrated by detection of caspase-3 substrates in activated cell-free apoptosis. There are several advantages using this iTRAQ method. Since labeling only occurs at the N-termini, the complexity of the sample analysis is greatly decreased. The iTRAQ ratios are higher using this procedure since activation of a protease will show an all-or-nothing effect if the target protease is completely inactive in the control sample. Relative ratios can thus only be demonstrated if several time points are analyzed. The power of this iTRAQ procedure is that it is focused on the new N-terminals formed after proteolytic cleavage and it reveals not only the protease substrate but also directly demonstrates the location of cleavage. A drawback with the iTRAQ reagent is that the results are greatly influenced by MS/MS apparatus availability. Not all mass spectrometers have the accuracy and sensitivity in the low-mass region that is necessary for the optimal sample analysis by iTRAQ-labeled samples. Also, it does not offer an enrichment of labeled peptides over tryptic peptides before the LC-MS/MS. This can, however, be achieved if this labeling strategy is combined with a positive (enrichment of labeled peptides) or negative (removal of tryptic-unlabeled peptides) selection strategy.

Other Potential N-Terminal-Labeling Strategies

Some other types of isotopic amine-reactive labeling reagents such as ICPL (isotope-coded protein label) (Schmidt et al. 2005) and dimethyl labeling (Ji et al. 2005, Ji and Li 2005) can, in principle, be adapted to quantitative protease substrate proteomics. In an improved variant of the dimethylation method, the N-termini were specifically labeled by blocking the Lys side chains by guanidination (Ji et al. 2005). In addition, IVICAT (in vacuo isotope-coded alkylation technique), which specifically labels the N-terminal without prior blocking of Lys side chains, was introduced recently (Simons et al. 2006).

These labeling strategies can potentially be used for the identification of protease substrates and can also provide cleavage site identification. However, a major drawback is the lack of enrichment of the labeled peptide, which has to be achieved either by running the samples on a gel, cut slices, and do in-gel digestion on the fractions and/or by LC separation. Also, in these labeling strategies, as well as other negative selection methods mentioned below, the digestion occurs before the labeling step, which may induce differences in peptide occurrence (Picotti et al. 2007).

Negative Selection of Tryptic Peptides

In a negative selection procedure, the labeled peptides are enriched from the sample simply by extracting the tryptic peptides from the sample after digestion. Here, we discuss two such examples in which the N-termini are acetylated, the proteins digested and the tryptic peptides retained by different methods.

Biotinylation of Tryptic Peptides

In one strategy, the internal tryptic peptides are reacted with biotin after acetylation of N-termini (McDonald et al. 2005). The peptide mixtures are passed over the immobilized streptavidin column, which will retain the internal biotinylated peptides on the column while the flow-through is expected to contain all original N-terminal peptides. This negative selection method simplifies the complexity of peptide mixture and thus enhances the identification of N-terminal peptides, whether native N-terminal peptides or newly generated by protease cleavage. This isolation method has been used to analyze the soluble proteins of mouse skeletal muscle and mouse liver (McDonald et al. 2005). N-terminal processing such as removal of initiator methionine and signal peptide was detected. One could imagine that a negative selection procedure could be used in concert with an amine-reactive isotopic label, such as iTRAQ, to further enhance the detection of labeled peptides and thus strengthen the analysis. Options other than biotin could also be used, such as amine-reactive resin.

COFRADIC – Separation of TNBS-Treated Tryptic Peptides

A different negative selection approach using solution-based proteomics named combined fractional diagonal chromatography (COFRADIC) was described in 2003 (Gevaert et al. 2003). This negative selection procedure is based on the introduction of hydrophobicity to the internal tryptic peptides after digestion of the proteins. In brief, proteins are first denatured, reduced, and alkylated. Free amino groups are acetylated using sulfo-*N*-hydroxysuccinimide acetate and the acetylated proteins are digested. The peptide mixtures are separated by a reverse-phase chromatography into smaller fractions, and each fraction is treated with 2,4,6-trinitrobenzenesulfonic acid (TNBS). The TNBS treatment blocks all the free amines of the internal peptides while the acetylated N-terminal peptides remain unchanged. The hydrophobic TNBS-blocked internal peptides shift on the LC column and can thus be separated from the unaltered N-terminal peptides. The sorted N-terminal peptides are collected in a number of secondary separation fractions and identified by further LC-MS/MS analysis and database searching. By repeating the procedure without the acetylation step, *in vivo* acetylated proteins can also be identified.

The analysis of a cytosolic and membrane skeleton fraction of human thrombocytes was used for validation of this method (Gevaert et al. 2003). The identified N-terminal peptides included signal peptides, peptides with initiator methionine, peptides with the initiator methionine removed, peptides with a small number of N-terminal residues removed, and internal peptides. Since the overall efficiency of the TNBS reaction is estimated about 98%, this N-terminal selection procedure can greatly decrease the complexity of peptides digested from protein mixtures, thereby improving the detection of N-terminal peptides resulting from proteolytic events. However, the identification of small number of internally cleaved and reacetylated peptides may be due to the incomplete reaction of abundant peptides with TNBS.

More recently, isotopic labeling using $^{16}O/^{18}O$ was introduced into the COFRADIC method to identify substrates in apoptosis signaling and their exact cleavage sites *in vivo* (Van Damme et al. 2005). ^{18}O-labeling is achieved by introduction of ^{18}O from $H_2^{18}O$ into all newly produced carboxyl groups during the tryptic hydrolysis step (Yao et al. 2001). By using this isotope-labeling COFRADIC method, more than 1,800 proteins were identified from Fas-induced apoptosis of Jurkat cells. A total of 93 cleavage sites were located in 71 proteins. Most identified cleavage was found at caspase consensus sites. The same method was also used to detect possible HtrA2/Omi cleavage sites (Vande Walle et al. 2007). Jurkat cell lysates were incubated with wild-type or inactive HtrA2/Omi and 50 cleavage sites (in 15 potential substrates) were identified. Since this study was performed *in vitro*, it is hard to judge which of these substrates are in fact cleaved *in vivo* during apoptosis signaling. However, this study shows an example of how a small subset of protease substrates can serve as a base for prediction of the protease cleavage site motif, as compared, for example, to peptide libraries.

The COFRADIC approach is overall quite successful in isolating N-terminal peptides from the internal peptides. The limitation of this approach comes from the size of the sorted N-terminal peptides, which depends on the distance between the N-terminus and the first arginine. The average length of the detected N-terminal peptides was found between 12 and 13 amino acid residues. Very short or very long N-terminal peptides are likely missed during the LC fraction collection. Protein digestion from enzymes other than trypsin should improve this limitation. As mentioned above, incomplete chemistry in two of the steps (TNBS labeling and isotope incorporation) may decrease the power of this method. Also, since the samples are mixed after digestion, differential digestion efficiency will have a substantial impact on reproducibility of the analysis. Also, in these labeling strategies, as well as the other negative selection methods mentioned above, the digestion occurs before the labeling step (Picotti et al. 2007). This can, however, be avoided by introducing the isotopic mass difference by stable isotope labeling by amino acids in cell culture (SILAC). In brief, cells are grown in media where an essential amino acid has been replaced by an isotopic variant (Ong et al. 2002, Ong and Mann 2007). The isotopic difference will, however, then only be detected in peptides containing that specific amino acid residue.

Despite these few limitations, the COFRADIC method coupled with SILAC or $^{16}O/^{18}O$ peptide labeling is currently the only reported MS-based method that can fulfill the two distinct demands in analysis of proteolysis *in vivo*: identification of protease substrates quantitatively, which normally requires isotope labeling of proteins or peptides, and identification of exact cleavage sites within processed proteins, which normally requires N-terminal peptide detection from complex mixtures.

Positive Selection – Enrichment of Specifically Labeled N-Terminal Peptides

During positive selection, the labeled peptide is coupled to an affinity tag that can be used for selection and enrichment of the labeled peptides. A proteomic method based on a combination of N-terminal-specific labeling and positive selection of labeled peptides and LC-MS/MS detection was introduced recently (Timmer et al. 2007). Protein N-termini were labeled with NHS-SS-biotin, a commercially available (Pierce Biotechnology), relatively cheap cleavable amine-reactive label with an affinity (biotin) tag. Similar as in the iTRAQ approach mentioned above (Enoksson et al. 2007), the proteins are specifically labeled at the protein N-termini by first blocking the Lys side chains by guanidination. The proteins are labeled at the N-termini by NHS-SS-biotin and the excess label is quenched and removed by buffer exchange. After digestion, the labeled N-terminal peptides are selected and enriched by streptavidin, which binds strongly to the biotin tag. The labeled peptides are eluted with dithiothreitol (DTT), analyzed by LC-MS/MS, and subsequently mapped into the protein database. The constitutive protolytic events in *Escherichia coli*, yeast, mouse tissue, human cell lysate, and human blood samples were used for validation of this method (Timmer et al. 2007). Methionine removal by methionine aminopeptidases, signal peptide removal by signal peptidases, and mitochondrial transit peptide removal by mitochondrial peptidases were detected using this method. Previously reported and predicted cleavage sites were detected and used as validation for the method. Moreover, some cases of previous misannotations of cleavage sites were found. However, many of the peptides could not be assigned to a protease because the particular cleavage event has not been previously reported or predicted. This problem could be avoided by the development of a similar affinity tag with a heavy and light variant, thus allowing a specific ex vivo comparison between, for example, cells with activated protease signaling and cells with inactivated protease signaling. It is also possible to combine this method with the Cy3/Cy5 dyes that previously have been used in the 2D-DIGE experiments. By only labeling the protein N-terminals, specific protease substrates will light up in a 2D-DIGE system.

Label-Free Analysis

Identification of proteolytic substrates or products by MS can be accomplished using the different chemical labeling approaches mentioned above. Introduction by isotope mass difference can also be achieved by SILAC, *in vivo* incorporation of isotope-labeled amino acids in cell culture. In fact, several putative or natural protease substrates have been detected by SILAC in combination with 2D-electrophoresis and/or MS/MS (Neher et al. 2006, Pinto et al. 2007, Schmidt et al. 2007, Thiede et al. 2006). In addition, Gupta et al. (2007) demonstrated that at least some conservative proteolytic events can be identified by label-free analysis, but this method requires massive numbers of peptide-proteomic data. A total of 14.5 million MS/MS spectra were collected from the MS analysis of trypsin digests of *Shewanella oneidensis*. The large dataset was searched using InsPect (Tanner et al. 2005) and MS-alignment (Tsur et al. 2005) against a six-frame translation of the entire genome. About 28,000 unique peptides and 2,000 proteins were identified. From this dataset, 366 N-terminal peptides revealed the cleavage of N-terminal methionine in 218 proteins. Also, 117 putative signal peptides predicted by MS/MS analysis, 94 of which match predictions made by SignalIP and/or by Prediction of Signal Peptides (PrediSi). This method requires extensive MS/MS as well as dataset analysis, and at this point can only be recommended for laboratories with the adequate equipment and expertise.

Summary

There is an array of different MS-based methods available to find protease substrates and/or detecting protease pathways. The different proteomic methods are all very powerful in separating the starting material and allow for the detection of protease substrates that cannot be detected by other methods. Depending on the setup of the experiment, one can use the data output for protease specificity motif studies, detecting natural protease substrates, or even mapping out protease pathways. The choice of method depends on the type of sample, the available instrumentation, and the type of data that is desired. For example, while 2D-PAGE or 2D-DIGE provides information about the relative difference in cleavage of different substrates, they do not provide cleavage site information and are restricted to proteins of medium or high abundance. Thus, a solution-based method may be preferred since this gives cleavage site information and thus help identify the culprit protease. However, these methods will not tell us to what extent different substrates are cleaved, and every substrate cannot be identified – detection requires that the N-terminal peptide can be analyzed in the MS step. Using different digestion techniques (trypsin, Glu-C, etc.) will help to overcome the latter problem. In the solution-based methods, there are also several different labeling strategies to choose from (*see* Table 5.2). The options for isotope incorporation are diverse; *in vivo* corporation of isotopic amino acid residues (SILAC) introduces isotopes in the starting material, but is limited to peptides that contain the residue of choice, and cannot be used for human samples. Of the different isotopic labeling strategies, one that occurs before mixing and digesting samples is preferred since that assures that the samples have the same tryptic peptides (Picotti et al. 2007). Also, an isotopic label that contains the $^{12}C/^{13}C$ isotopic couple instead of $^{1}H/^{2}H$ is usually preferred since this variant is not separated in the LC step.

While each of the different methods mentioned in this chapter have proved to help identify protease substrates, a combination of the different strategies could increase the efficiency and the sensitivity of the strategies even further. Ideally, a

Table 5.2 Summary of the key characteristics of current mass spectrometry (MS)-based techniques for identification and quantification of protease substrates

Methods	Advantages	Disadvantages
Positive selection	Enrichment, cleavage site determined	Not quantitative
ICAT	Quantitative	No enrichment, cleavage site not determined
iTRAQ	Quantitative, reproducible, cleavage site determined	No enrichment
COFRADIC	Quantitative, enrichment, cleavage site determined	Reproducibility

COFRADIC combined fractional diagonal chromatography, *ICAT* isotope-coded affinity tag, *iTRAQ* isobaric tag for relative and absolute quantification

method that gives us separation as well as enrichment of N-terminal peptides originating from the protease substrates and that carries an isotopic tag would provide the required separation, identification, and relative quantification of protease substrates.

References

Beardsley R.L., Karty J.A., Reilly J.P. (2000). Enhancing the intensities of lysine-terminated tryptic peptide ions in matrix-assisted laser desorption/ionization mass spectrometry. Rapid Commun Mass Spectrom. 14:2147–53.

Bech-Serra J.J., Santiago-Josefat B., Esselens C., et al. (2006). Proteomic identification of desmoglein 2 and activated leukocyte cell adhesion molecule as substrates of ADAM17 and ADAM10 by difference gel electrophoresis. Mol Cell Biol. 26:5086–95.

Blow A.M., Barrett A.J. (1977). Action of human cathepsin G on the oxidized B chain of insulin. Biochem J. 161:17–9.

Boulware K.T., Daugherty P.S. (2006). Protease specificity determination by using cellular libraries of peptide substrates (CLiPS). Proc Natl Acad Sci U S A. 103:7583–8.

Boyd S.E., Pike R.N., Rudy G.B., et al. (2005). PoPS: A computational tool for modeling and predicting protease specificity. J Bioinform Comput Biol. 3:551–85.

Bredemeyer A.J., Lewis R.M., Malone J.P., et al. (2004). A proteomic approach for the discovery of protease substrates. Proc Natl Acad Sci U S A. 101:11785–90.

Bugge T.H., Kombrinck K.W., Flick M.J., et al. (1996). Loss of fibrinogen rescues mice from the pleiotropic effects of plasminogen deficiency. Cell. 87:709–19.

Cryns V.L., Byun Y., Rana A., et al. (1997). Specific proteolysis of the kinase protein kinase C-related kinase 2 by caspase-3 during apoptosis. Identification by a novel, small pool expression cloning strategy. J Biol Chem. 272:29449–53.

Dean R.A., Overall C.M. (2007). Proteomics discovery of metalloproteinase substrates in the cellular context by iTRAQ labeling reveals a diverse MMP-2 substrate degradome. Mol Cell Proteomics. 6:611–23.

Deng S.J., Bickett D.M., Mitchell J.L., et al. (2000). Substrate specificity of human collagenase 3 assessed using a phage-displayed peptide library. J Biol Chem. 275:31422–7.

Ding L., Coombs G.S., Strandberg L., et al. (1995). Origins of the specificity of tissue-type plasminogen activator. Proc Natl Acad Sci U S A. 92:7627–31.

Enoksson M., Li J., Ivancic M.M., et al. (2007). Identification of proteolytic cleavage sites by quantitative proteomics. J Proteome Res. 6:2850–8.

Gevaert K., Goethals M., Martens L., et al. (2003). Exploring proteomes and analyzing protein processing by mass spectrometric identification of sorted N-terminal peptides. Nat Biotechnol. 21:566–9.

Gupta N., Tanner S., Jaitly N., et al. (2007). Whole proteome analysis of post-translational modifications: Applications of mass-spectrometry for proteogenomic annotation. Genome Res. 17:1362–77.

Gygi S.P., Rist B., Gerber S.A., et al. (1999). Quantitative analysis of complex protein mixtures using isotope-coded affinity tags. Nat Biotechnol. 17:994–9.

Gygi S.P., Corthals G.L., Zhang Y., et al. (2000). Evaluation of two-dimensional gel electrophoresis-based proteome analysis technology. Proc Natl Acad Sci U S A. 97:9390–5.

Gygi S.P., Rist B., Griffin T.J., et al. (2002). Proteome analysis of low-abundance proteins using multidimensional chromatography and isotope-coded affinity tags. J Proteome Res. 1:47–54.

Harris J.L., Backes B.J., Leonetti F., et al. (2000). Rapid and general profiling of protease specificity by using combinatorial fluorogenic substrate libraries. Proc Natl Acad Sci U S A. 97:7754–9.

Hwang I.K., Park S.M., Kim S.Y., et al. (2004). A proteomic approach to identify substrates of matrix metalloproteinase-14 in human plasma. Biochim Biophys Acta. 1702:79–87.

Igarashi Y., Eroshkin A., Gramatikova S., et al. (2007). CutDB: A proteolytic event database. Nucleic Acids Res. 35:D546–9.

Ji C., Li L. (2005). Quantitative proteome analysis using differential stable isotopic labeling and microbore LC-MALDI MS and MS/MS. J Proteome Res. 4:734–42.

Ji C., Guo N., Li L. (2005). Differential dimethyl labeling of N-termini of peptides after guanidination for proteome analysis. J Proteome Res. 4:2099–108.

Ji C., Li L., Gebre M., et al. (2005). Identification and quantification of differentially expressed proteins in E-cadherin deficient SCC9 cells and SCC9 transfectants expressing E-cadherin by dimethyl isotope labeling, LC-MALDI MS and MS/MS. J Proteome Res. 4:1419–26.

Ju W., Valencia C.A., Pang H., et al. (2007). Proteome-wide identification of family member-specific natural substrate repertoire of caspases. Proc Natl Acad Sci U S A. 104:14294–9.

Kargel H.J., Dettmer R., Etzold G., et al. (1980). Action of cathepsin L on the oxidized B-chain of bovine insulin. FEBS Lett. 114:257–60.

Kimmel J.R. (1967). Guanidination of proteins. Meth. Enzymol. 11:584–9.

Kridel S.J., Sawai H., Ratnikov B.I., et al. (2002). A unique substrate binding mode discriminates membrane type-1 matrix metalloproteinase from other matrix metalloproteinases. J Biol Chem. 277:23788–93.

Li H., Zhu H., Xu C.J., et al. (1998). Cleavage of BID by caspase 8 mediates the mitochondrial damage in the Fas pathway of apoptosis. Cell. 94:491–501.

Loeb C.R., Harris J.L., Craik C.S. (2006). Granzyme B proteolyzes receptors important to proliferation and survival, tipping the balance toward apoptosis. J Biol Chem. 281:28326–35.

Lopez M.F., Berggren K., Chernokalskaya E., et al. (2000). A comparison of silver stain and SYPRO Ruby Protein Gel Stain with respect to protein detection in two-dimensional gels and identification by peptide mass profiling. Electrophoresis. 21:3673–83.

Luthi A.U., Martin S.J. (2007). The CASBAH: A searchable database of caspase substrates. Cell Death Differ. 14:641–50.

Matthews D.J., Wells J.A. (1993). Substrate phage: Selection of protease substrates by monovalent phage display. Science. 260:1113–7.

McDonald L., Robertson D.H., Hurst J.L., et al. (2005). Positional proteomics: Selective recovery and analysis of N-terminal proteolytic peptides. Nat Methods. 2:955–7.

McKay M.J., Offermann M.K., Barrett A.J., et al. (1983). Action of human liver cathepsin B on the oxidized insulin B chain. Biochem J. 213:467–71.

Nazif T., Bogyo M. (2001). Global analysis of proteasomal substrate specificity using positional-scanning libraries of covalent inhibitors. Proc Natl Acad Sci U S A. 98:2967–72.

Neher S.B., Villen J., Oakes E.C., et al. (2006). Proteomic profiling of ClpXP substrates after DNA damage reveals extensive instability within SOS regulon. Mol Cell. 22:193–204.

O'Farrell P.H. (1975). High resolution two-dimensional electrophoresis of proteins. J Biol Chem. 250:4007–21.

Ong S.E., Mann M. (2007). Stable isotope labeling by amino acids in cell culture for quantitative proteomics. Methods Mol Biol. 359:37–52.

Ong S.E., Blagoev B., Kratchmarova I., et al. (2002). Stable isotope labeling by amino acids in cell culture, SILAC, as a simple and accurate approach to expression proteomics. Mol Cell Proteomics. 1:376–86.

Picotti P., Aebersold R., Domon B. (2007). The implications of proteolytic background for shotgun proteomics. Mol Cell Proteomics. 6:1589–98.

Pinto A.F., Ma L., Dragulev B., et al. (2007). Use of SILAC for exploring sheddase and matrix degradation of fibroblasts in culture by the PIII SVMP atrolysin A: Identification of two novel substrates with functional relevance. Arch Biochem Biophys. 465:11–5.

Polevoda B., Sherman F. (2003). N-terminal acetyltransferases and sequence requirements for N-terminal acetylation of eukaryotic proteins. J Mol Biol. 325:595–622.

Qiu Y., Sousa E.A., Hewick R.M., et al. (2002). Acid-labile isotope-coded extractants: A class of reagents for quantitative mass spectrometric analysis of complex protein mixtures. Anal Chem. 74:4969–79.

Ross P.L., Huang Y.N., Marchese J.N., et al. (2004). Multiplexed protein quantitation in Saccharomyces cerevisiae using amine-reactive isobaric tagging reagents. Mol Cell Proteomics. 3:1154–69.

Schecter I., Berger M. (1967). On the size of the active site in proteases. Biochem. Biophys. Res. Commun. 27:157–162.

Schmidt A., Kellermann J., Lottspeich F. (2005). A novel strategy for quantitative proteomics using isotope-coded protein labels. Proteomics. 5:4–15.

Schmidt F., Hustoft H.K., Strozynski M., et al. (2007). Quantitative proteome analysis of cisplatin-induced apoptotic Jurkat T cells by stable isotope labeling with amino acids in cell culture, SDS-PAGE, and LC-MALDI-TOF/TOF MS. Electrophoresis 28:4359–68.

Simons B.L., Wang G., Shen R.F., et al. (2006). In vacuo isotope coded alkylation technique (IVICAT); an N-terminal stable isotopic label for quantitative liquid chromatography/mass spectrometry proteomics. Rapid Commun Mass Spectrom. 20:2463–77.

Smith M., Shi L., Navre M. (1995). Rapid identification of highly active and selective substrates for stromelysin and matrilysin using bacteriophage peptide display libraries. J Biol Chem. 270:6440–9.

Stennicke H.R., Renatus M., Meldal M., et al. (2000). Internally quenched fluorescent peptide substrates disclose the subsite preferences of human caspases 1, 3, 6, 7 and 8. Biochem J. 350:563–8.

Tam E.M., Morrison C.J., Wu Y.I., et al. (2004). Membrane protease proteomics: Isotope-coded affinity tag MS identification of undescribed MT1-matrix metalloproteinase substrates. Proc Natl Acad Sci U S A. 101:6917–22.

Tanner S., Shu H., Frank A., et al. (2005). InsPecT: Identification of posttranslationally modified peptides from tandem mass spectra. Anal Chem. 77:4626–39.

Tao W.A., Aebersold R. (2003). Advances in quantitative proteomics via stable isotope tagging and mass spectrometry. Curr Opin Biotechnol. 14:110–8.

Taylor R.C., Brumatti G., Ito S., et al. (2007). Establishing a blueprint for CED-3-dependent killing through identification of multiple substrates for this protease. J Biol Chem. 282:15011–21.

Thiede B., Kretschmer A., Rudel T. (2006). Quantitative proteome analysis of CD95 (Fas/Apo-1)-induced apoptosis by stable isotope labeling with amino acids in cell culture, 2-DE and MALDI-MS. Proteomics. 6:614–22.

Thornberry N.A., Rano T.A., Peterson E.P., et al. (1997). A combinatorial approach defines specificities of members of the caspase family and granzyme B. Functional relationships established for key mediators of apoptosis. J Biol Chem. 272:17907–11.

Timmer J.C., Enoksson M., Wildfang E., et al. (2007). Profiling constitutive proteolytic events *in vivo*. Biochem J. 407:41–8.

Tonge R., Shaw J., Middleton B., et al. (2001). Validation and development of fluorescence two-dimensional differential gel electrophoresis proteomics technology. Proteomics. 1:377–96.

Tsur D., Tanner S., Zandi E., et al. (2005). Identification of post-translational modifications by blind search of mass spectra. Nat Biotechnol. 23:1562–7.

Turk B.E., Huang L.L., Piro E.T., et al. (2001). Determination of protease cleavage site motifs using mixture-based oriented peptide libraries. Nat Biotechnol. 19:661–7.

Van Damme P., Martens L., Van Damme J., et al. (2005). Caspase-specific and nonspecific *in vivo* protein processing during Fas-induced apoptosis. Nat Methods. 2:771–7.

Vande Walle L., Van Damme P., Lamkanfi M., et al. (2007). Proteome-wide identification of HtrA2/Omi Substrates. J Proteome Res. 6:1006–15.

Wang S., Zhang X., Regnier F.E. (2002). Quantitative proteomics strategy involving the selection of peptides containing both cysteine and histidine from tryptic digests of cell lysates. J Chromatogr A. 949:153–62.

Warwood S., Mohammed S., Cristea I.M., et al. (2006). Guanidination chemistry for qualitative and quantitative proteomics. Rapid Commun Mass Spectrom. 20:3245–56.

Yao X., Freas A., Ramirez J., et al. (2001). Proteolytic 18O labeling for comparative proteomics: Model studies with two serotypes of adenovirus. Anal Chem. 73:2836–42.

Zhou H., Ranish J.A., Watts J.D., et al. (2002). Quantitative proteome analysis by solid-phase isotope tagging and mass spectrometry. Nat Biotechnol. 20:512–5.

Zhou X.W., Blackman M.J., Howell S.A., et al. (2004). Proteomic analysis of cleavage events reveals a dynamic two-step mechanism for proteolysis of a key parasite adhesive complex. Mol Cell Proteomics. 3:565–76.

Chapter 6
Identification of Protease Substrates by Mass Spectrometry Approaches-2

Anna Prudova, Ulrich auf dem Keller, and Christopher M. Overall

Abstract Proteolysis is a major posttranslational modification of proteins with critical functional consequences to the protein, cell, and organism. The most effective way to monitor proteolytic events is to analyze the proteins directly. This chapter summarizes advantages and limitations of different mass spectrometry-based approaches for detection of proteolysis products. In general, liquid chromatography separation-based proteomics approaches are superior to 2D gel-based techniques and, in turn, quantitative proteomics have a significant advantage over label-free methods. Isotopic labeling of samples helps to identify substrates but fails to detect the exact cleavage site. Techniques that enrich for peptides containing the N-terminus of each protein provide a more relevant context for protease substrate discovery – they focus on the analysis of the neo-N-termini resulting from proteolysis. These techniques identify not only the substrates but also the prime side of the cleavage sites with a potential to extract further information of the protease sequence site specificity, thus setting the gold standard for the future of the degradomics field.

Introduction

Having identified the components of the protease degradome gives the researcher a good picture of the proteolytic potential of the system. The ultimate information about the function of the identified degradome components and the resulting effects on the biological system in question, however, can be evaluated only by defining the direct action of the proteases, that is, the proteolytic modification of their substrates. This requires first, the identification of potential substrates in a given active

C.M. Overall
Centre for Blood Research, University of British Columbia, 4.401 Life Sciences Centre, 2350 Health Sciences Mall, Vancouver, BC V6T 1Z3, Canada, e-mail: chris.overall@ubc.ca

D. Edwards et al. (eds.), *The Cancer Degradome.*

degradome and second, the specific cleavage sites (auf dem Keller et al. 2007). In view of the overwhelming complexity of the protease web (Overall and Kleifeld 2006), exhaustive identification of the protease substrate repertoire with the corresponding cleavage sites can be a very daunting task.

Since proteolysis directly acts on proteins themselves as a posttranslational modification, the most obvious way to monitor this event is to analyze the proteins directly. Until recently, this was done by serial *in vitro* incubations of mostly recombinant substrate candidates with the protease under study and subsequent analysis by sodium dodecyl sulfate-polyacrylamide gel electrophoresis (SDS-PAGE). Thereby, only one substrate candidate could be analyzed at a time under conditions far from *in vivo* physiology. For identification, the cleavage site fragments had to be isolated and subjected to time consuming and expensive chemical amino acid sequencing (Niall 1973). Even if the substrate under study was efficiently cleaved under *in vitro* conditions, it did not mean that the same would have happened in a complex biological system *in vivo*. Simply the fact that it *can* cleave does not mean that it *does* cleave under physiological conditions (Tam et al. 2004). Furthermore, substrate candidates had first to be identified in a separate experiment.

The development of mass spectrometry (MS) technology for the analysis of large biomolecules, particularly proteins, fundamentally changed the way proteins and their modifications are analyzed (Fenn et al. 1989). Now, the protein band of interest could be digested in the gel with a specific protease, such as trypsin, followed by protein identification based on its peptide mass fingerprint and database data analysis (Fenyo 2000). The invention of tandem mass spectrometry (MS/MS) for peptide analysis enabled further fragmentation of the protein-derived peptides, leading to the possibility to unambiguously identify proteins in more and more complex mixtures from their sequences (Wilm et al. 1996). In this chapter, we will summarize recent developments in the application of MS-based techniques for protease substrate discovery.

Two-Dimensional Gel Electrophoresis

The combination of MS with two-dimensional polyacrylamide gel electrophoresis (2D-PAGE) facilitated the simultaneous identification of hundreds of proteins in a complex biological mixture (Shevchenko et al. 1996). The high resolving power of 2D-PAGE and the development of various staining procedures to visualize these protein "spots" made it a popular method of choice for identifying protein abundance changes between two proteome samples.

Hwang et al. (2004) were the first to employ 2D-PAGE in combination with MS for protease substrate discovery. The authors incubated human plasma proteins with matrix metalloproteinase (MMP)-14 and compared the treated and untreated protein mixtures by 2D-PAGE. Subsequently, protein spots which disappeared or, in contrast, appeared in the MMP-14-treated sample due to the proteolytic cleavage were analyzed by peptide mass fingerprinting. This allowed the simultaneous identification of six known and nine new MMP-14 substrates in a complex biological mixture. In a similar study, new substrates for caspase-3 were identified in human breast cancer cell lines by incubating the lysates of caspase-3-deficient MCF-7 cells with the recombinant protease (Lee et al. 2004b). Another study implemented the 2D-PAGE approach to characterize substrates for the intracellular serine protease-1 from *Bacillus subtilis* (Lee et al. 2004a). More recently, Major et al. (2006) used 2D-PAGE to assess the substrate range of the yeast mitochondrial matrix protease Pim1 *in vivo* using wild-type and Pim1delta strains. In addition to peptide mass fingerprinting, the authors extended the coverage of identified proteins by using MS/MS (Major et al. 2006).

One major limitation of conventional 2D-PAGE analyses is the reliability of protein identifications and the relatively high threshold for their quantification. Indeed, the two samples (with and without protease activity) have to be first electrophoresed in a very reproducible manner on two separate gels and then altered spots can be quantified by a densitometric image analysis. Thereby, the detection of only slight changes (which, however, can result in a strong biological phenotype) is still a major challenge. This drawback was recently offset by the introduction of two dimensional difference gel electrophoresis (2D-DIGE), which involves labeling of samples with different fluorescent dyes (Cy3, Cy5) with subsequent analysis of two conditions on a single gel. This technique in combination with MS/MS was successfully used to identify new granzyme A and B substrates by incubating murine cell lysates with the corresponding recombinant proteases (Bredemeyer et al. 2004). A more recent study employed the 2D-DIGE technique to identify ADAMTS-1 substrates in a cell-based screen (Canals et al. 2006).

Two-dimensional gel electrophoresis techniques, however, remain limited in their sensitivity, making it very difficult to identify biologically relevant low-abundant proteins in complex proteomes. Furthermore, very large and small molecular weight proteins evade detection by this method due to their electrophoretic migration behavior on PAGE gels. Difficulties also exist with highly hydrophobic proteins, such as all membrane proteins, and those with extreme pI values. Finally, 2D-PAGE resolution is often insufficient, as shown by the MS-based detection of up to six proteins in one gel spot when analyzing yeast cell extract proteins (Gygi et al. 2000). For many proteases, substrates are cleaved by less than 10 residues Overall and Blobel 2007), often resulting in significant alteration of biological activity. For such substrates, 2D-PAGE lacks the resolution to detect these subtle but important changes to a protein substrate.

Shotgun Proteomics

While 2D-PAGE is still widely used as a reliable, robust, and inexpensive method, researchers tried to enhance the number of identified proteins in a complex proteome by exploring alternative methods for the separation of proteins and their tryptic peptides before MS analysis. This led to the development of so-called "shotgun" proteomics, a solution-based approach which involves fractionation of trypsin-generated peptides in two dimensions of liquid chromatography (LC) before MS/MS analysis, and thereby termed 2D-LC-MS/MS (Washburn et al. 2001). The 2D-LC most commonly involves a combination of strong cation exchange (SCX) and reverse-phase C_{18} chromatography. The first dimension LC can be performed either off-line on a conventional high-performance liquid chromatography (HPLC) system or in-line with the nano LC-MS/MS system—a method also known as multidimensional protein identification technology (MudPIT) (Wolters et al. 2001). Combining this approach with prior protein fractionation further enhances the proteome coverage (Chen et al. 2006). This technique was recently used to obtain comprehensive proteome maps of different eukaryotic samples, including mammalian tissues (Kislinger et al. 2006; Brunner et al. 2007).

ICATs Quantitative Proteomics

The above-described powerful solution-based techniques are aimed for a maximal number of proteins to be unambiguously identified in a complex biological sample. However, initially they lacked an easy possibility to also quantify the identified proteins, a prerequisite for the detection of proteolytic events. This problem was solved by the introduction of stable isotopic tags, with the most widely used being isotope-coded affinity tags (ICATs) (Gygi et al. 1999). ICATs comprise a trifunctional structure: (1) a cysteine-reactive group allowing for covalent binding to reduced cysteine residues of peptides; (2) a linker region with nine carbon atoms which can be synthetized either with "light" ($^{13}C_0$) or with "heavy" ($^{13}C_9$) isotopes; and (3) a cleavable biotin moiety as a handle to isolate ICAT-labeled peptides from the mixture. In this approach, protein samples to be compared are trypsin digested and the peptide mixtures are subsequently reacted with either the light or the heavy ICAT label. Hereby, the labels are incorporated into all cysteine-containing peptides. Afterward, both samples are combined and the labeled peptides positively selected via an avidin affinity column. Upon reductive elution, the peptides are subjected to 2D-LC-MS/MS analysis. Thereby, each peptide is represented in MS1 mode by a pair of peaks with a mass difference of 9 Da corresponding to the heavy and light ICAT labels. The areas of these peaks are integrated and used to determine the relative abundance of the peptide in both samples to be compared.

The first to employ ICAT labeling as a proteomic method for substrate discovery were Tam et al. (2004). Here the authors analyzed the substrate degradome of MT1-MMP in a cell-based system, where the protease and its substrates were present in the relevant context of a complete proteolytic pathway, including cofactors, binding proteins, inhibitors, and other modifying agents. Thereby, numerous novel bioactive substrates including connective tissue growth factor (CTGF), secreted leukocyte protease inhibitor (SLPI), and tumor necrosis factor α (TNFα) as well as the death receptor-6 were identified. These findings underlined the important functions of MMPs as signaling proteases and pioneered the use of quantitative proteomics for protease substrate discovery in a complex biological system under physiological conditions. Consistently, ICAT-based quantitative proteomics also revealed that the inhibition of MT1-MMP overexpressed in MDA-MB-231 cells using small molecule inhibitors, resulted in decreased shedding of cell-surface proteins with concomitant increase in the uncleaved protein levels on the plasma membrane (Butler and Overall 2007). The myriad of substrates so identified using MMP inhibitors underscore the complex problem of using MMPs as targets for disease intervention. Inhibition of proteolysis of many protease substrates may lead to a loss of significant biological functions which cannot be "buffered" by robust compensatory pathways and thus result in drug side effects.

In a more recent and also cell-based study, ICAT labeling was used to identify novel bioactive proteins as MMP-2 substrates and mechanistically dissect the angiogenic function of this MMP (Dean et al. 2007). Here, vascular endothelial growth factor (VEGF)-binding proteins (connective tissue growth factor, CTGF and heparin affin regulatory peptide, HARP) were found to be substrates of MMP-2, the cleavage of which mobilized VEGF and its angiogenic function.

iTRAQ Quantitative Proteomics

While ICAT labeling was successfully used to identify novel protease substrates, the shortcoming of this method is that only cysteine-containing peptides can be analyzed, thus limiting proteome coverage to ~93% of proteins. To overcome this limitation, a new generation of labels for quantitative proteomics that is named iTRAQ (isobaric tag for relative and absolute quantification) (Choe et al. 2005) can be used. The iTRAQ consists of a group that is reactive toward primary amino groups, the linker region, and a reporter group that gives rise to a highly diagnostic low-mass reporter ion upon fragmentation of a tagged peptide in MS/MS mode. The currently available set of iTRAQ reagents contains four different isobaric variants which have the same total mass but are different in the corresponding masses of their linker and reporter regions. Thus, when four different samples containing equal amounts of the same peptide are labeled with the four variants and then combined at a one-to-one ratio, the MS mode will show a single peak representing this peptide. However, upon fragmentation, four different reporter ion peaks will appear in the 114–117 m/z spectra region, with the areas of these peaks corresponding to the peptide amounts in each of the original four samples. Therefore, labeling of a peptide N-terminus and/or lysine residues allows peptide sequence information to be obtained together with its relative quantification in MS/MS mode without doubling the spectra complexity in the MS mode (as observed with any other nonisobaric labeling techniques, such as ICAT, SILAC, acetylation, or reductive dimethylation).

The iTRAQ-based labeling in application to protease substrate discovery was first employed by Dean et al. (2007) (Dean and Overall 2007) using MMP-2 as a model secreted protease. In this study, the authors examined the conditioned medium from *Mmp-2–/–* murine fibroblasts transfected with active MMP-2 or its inactive mutant form (so as to present a "naive" proteome that had not been exposed to the protease). The secreted and shed cell-surface proteins in the serum-free medium were collected and then denatured, alkylated, and digested with trypsin. Following peptide labeling with two different iTRAQ reagents, the samples were mixed in one-to-one ratio and analyzed by 2D-LC-MS/MS. Comparison of the relative abundances of each peptide identified in the two samples (derived from the corresponding iTRAQ ratios for each peptide) identified proteins that were degraded (and therefore represented by low iTRAQ ratios) and the proteins that were shed from the cell surface into the medium (and therefore presented by high iTRAQ ratios). The peptides that belonged to the proteins (or their regions) that were unaffected by MMP-2 cleavages exhibited iTRAQ ratios of ~1. In further experiments with four different iTRAQ labels, different time points were analyzed in a multiplex approach. On the basis of iTRAQ ratios for known previously reported substrates of MMP-2 observed in their analyte mixture, the authors established an iTRAQ ratio cutoff of fourfold. Thus, proteins with ratios less than 0.25 or greater than 4 were considered potential substrates, with some of them tested and confirmed in *in vitro* cleavage assays using purified proteins and recombinant MMP-2. In total, the analysis yielded known substrates (thus validating this approach) and also identified many previously undescribed substrates. In addition, mapping of the peptides with altered iTRAQ ratios compared to those peptides showing no change to the corresponding protein sequences, together with the results of *in vitro* cleavage assays, allowed identification of the region and even the exact site of the cleavage event in some proteins.

To summarize, the above-described solution-based quantitative proteomics approaches have a significant advantage over label-free methods in as much they identify substrates and may highlight a specific region of the molecule which is cleaved based on the differences in label ratios between specific peptides. However, the success of this strategy largely depends on the completeness of sequence coverage of the protein in question. In reality, the exact peptide(s) showing altered label ratios often remain undetected due to the complexity of a proteome mixture in general and its wide dynamic range, resulting in undersampling of lower-abundance proteins. Therefore, identifying potential substrates in a complex proteome still heavily resembles looking for a proverbial needle in a haystack.

ICAT-based substrate discovery methods somewhat address this issue since the proteomic mixture is simplified as it is being enriched for cysteine-containing peptides via label-dependent affinity pullout. As a result, the mixture contains fewer peptides, thus improving statistical chances for the identification of lower-abundance proteins. However, the same chemical bias excludes from analysis all the proteins without any cysteine (7% of all proteins) and limits the coverage of proteins containing only one cysteine residue in their sequence (35% of proteins), thus limiting the utility of this strategy for substrate discovery (Dean and Overall 2007). Consistent with this notion, comparison of iTRAQ and ICAT-based strategies within the same cellular context resulted in identification of higher numbers of total identified proteins (9-fold), known substrates (8-fold), protease inhibitors (4-fold), and proteases (31-fold) in iTRAQ-labeled samples (Dean and Overall 2007). Therefore, while an enrichment of proteomic samples for cysteine-containing peptides by ICAT may offer an advantage in analysis of cysteine-rich potential substrates (e.g., many extracellular matrix proteins and cytokines), it is not beneficial for a system-wide unbiased substrate discovery. On the contrary, enrichment for N-terminal portion(s) of each protein provides a far superior context for protease substrate discovery, as every act of proteolytic processing results in a neo N-terminus representing the prime side of the cleavage site. Thus, by analyzing the neo N-terminal peptide resulting from the proteolytic cleavage one can identify not only the candidate substrate but also the exact cleavage site with a potential to extract further information of the protease sequence site specificity (Overall and Dean 2006). The methods for selective N-termini recovery, their MS and data analysis, and the application to protease substrate identification will be discussed in the next section in the chronological order in which each method was first reported.

N-terminal Enrichment Methods for Protease Substrate Discovery

There have been a number of strategies reported that aim to selectively isolate protein N-terminal peptides for gel-free proteomic characterization of proteolysis. These include (1) differential N-terminal labeling to modify peptide hydrophobicity before diagonal chromatography (Gevaert et al. 2003); (2) N-terminal acetylation, then trypsin digestion, biotinylation, and affinity pullout of the internal peptides (McDonald et al. 2005); and (3) N-terminal-specific protein biotinylation with consequent affinity enrichment (Timmer et al. 2007).

The very first study using N-terminal enrichment to examine proteolytic processing of proteins, utilized an elegantly designed combined fractional diagonal chromatography (COFRADIC) approach (Gevaert et al. 2003). In this strategy, proteins are first acetylated with acetic anhydride at their N-terminal and lysine amino groups, digested with trypsin, and separated by reversed-phase HPLC. Internal peptides in each of the resulting 12 fractions are then chemically modified by 2,4,6-trinitrobenzenesulfonic acid (TNBS) at their N-termini to form very hydrophobic trinitrophenyl (TNP) derivatives, followed by a second reversed-phase fractionation. Because of their higher hydrophobicity, TNP-peptides are bound stronger by the column, elute in later fractions, and can be discarded. The acetylated peptides representing the original N-terminal portion of each protein are eluted in earlier fractions, collected, and then subjected to MS analysis. This secondary reversed-phase fractionation is performed for each of the 12 primary fractions, resulting in a total of 96 fractions and hence in 96 LC-MS/MS analyses per average experiment. The COFRADIC technique was first tested using human thrombocytes-derived proteomes, where 264 proteins were identified from 305 different peptides, with one peptide per protein, on average. About 10% of identified peptides were contaminating internal peptides that start with proline and pyroglutamate residues, and therefore have low or no reactivity toward 2,4, 6-trinitrobenzenesulfonic acid. In addition, another 74 internal tryptic peptides

Table 6.1 Summary of current techniques for protease substrate discovery by enrichment of N-terminal peptides

Technique	COFRADIC	McDonald et al.	Timmer et al.	Enoksson et al.
Major steps	1. Chemical acetylation of protein N-termini and lysines 2. First round RP LC separation 3. Trypsin digest with concomitant C-terminal isotopic labeling of generated peptides 4. TNBS labeling of neo-N-termini of tryptic peptides 5. Second round of RP LC separation; internal peptides are discarded 6. LC-MS/MS	1. Chemical acetylation of protein N-termini and lysines 2. Trypsin digest 3. Biotin coupling and pullout of neo-peptides resulting from trypsin digest or their pullout using NHS-activated Sepharose 4. LC-MS/MS	1. Lysine gyanidinylation 2. Biotin coupling of protein N-termini 3. Trypsin digest 4. Avidin affinity enrichment of biotinylated peptides 5. LC-MS/MS	1. Lysine gyanidinylation 2. iTRAQ labeling of protein N-termini 3. Trypsin digest 4. MALDI MS/MS
Method of N-terminal selection/ enrichment	Negative selection of all N-terminal peptides (including *in vivo* modified, e.g. acetylated, protein N-termini)	Negative selection of all N-terminal peptides (including *in vivo* modified, e.g. acetylated, protein N-termini)	Positive selection of peptides with unblocked amino-termini; *in vivo* modified peptides (e.g. acetylated, myristoylated etc.) are excluded from the analyte mix	Virtual enrichment for protein N-termini during the first "survey MS/MS", followed by their sequencing and quantification in the second MS/MS

Quantification	Quantification in MS mode	Not reported; quantification in MS mode is possible, if stable-isotope labeled acetic anhydrate is used for the acetylation	Not reported; quantification in MS mode is possible, if stable-isotope labeled biotin is used	Quantification in MS/MS mode
Comments for consideration	– Powerful, but labor- and equipment-intense technique – Does not distinguish between chemically and *in vivo* acetylated protein N-termini in a single experiment	– Fast and robust – Has a potential to distinguish between chemically and *in vivo* acetylated protein N-termini in a single experiment if isotope-labeled acetic anhydrate is used	– Allows analysis of only unblocked protein N-termini and not *in vivo* modified proteins	– Limited proteome coverage due to ion suppression – Has a potential for multiplex analysis using 4-plex iTRAQ reagents

were identified, suggesting incomplete TNP-modification, regardless of the first N-terminal amino acid. Based on the theoretical *in silico* digestion of the human proteome yielding 17.5 internal peptides per N-terminal peptide, the authors' results demonstrate a significant N-terminal enrichment. To evaluate the true efficacy of this enrichment technique, it should also be experimentally shown how many N-terminal peptides are omitted from the analysis due to incomplete acetylation, overlapping elution with internal TNP-modified peptides, or sample loss during multiple handling steps, that is, chemical modification and cleanup steps as well as the multiple LC analyses.

Since the COFRADIC technique is based on the MS analysis of the initially chemically acetylated and therefore retained peptides, it does not allow for distinguishing these from the protein N-termini which are retained due to their acetylation *in vivo*. While *in vivo* N-terminal acetylation is absent in prokaryotes, it is estimated to occur in up to 80% of eukaryotic proteins (Polevoda and Sherman 2003). To differentiate between *in vivo* and *in vitro* acetylation, the authors performed two sets of COFRADIC experiments for the same sample, with and without the first acetylation step in the workflow. Omitting the acetylation reaction, results in the retention and analysis of the N-terminal peptides of *in vivo* N-terminally blocked proteins. However, if these N-terminal peptides contain lysine residues, their side-chain amino groups will be TNP-modified and the peptides excluded from the analysis during the secondary reversed-phase separation. According to the authors, this chemical bias results in the loss of approximately half of all *in vivo* acetylated proteins.

Among the identified protein N-termini, the authors observed the following posttranslational proteolytic modifications: (1) removal of the initiator methionine; (2) propeptide or signal peptide removal; and (3) internal cleavages. For example, the study uncovered a previously undescribed truncated form of actin starting at amino acid 29. In some instances, a proteolytic processing predicted by homology with other proteins has been verified and corrected. For example, dihydroorotate dehydrogenase was found to start at residue 28 rather than predicted position 11. While the study demonstrates the utility of the N-terminal enrichment approach to describe proteolytic processes in a biological sample, it does not, however, allow for strict quantification of these events.

To address this issue, the original COFRADIC technique was slightly modified to introduce the quantitative differential aspect (Van Damme et al. 2005). To incorporate the label, the samples are digested with trypsin in the presence of water with ^{18}O isotope. Thus, trypsin-catalyzed incorporation of two ^{18}O atoms at the C-terminus of the newly cleaved-off peptide results in a 4 Da mass difference compared to the same peptide created in the presence of light ^{16}O water (Staes et al. 2004). In this approach, two samples representing protease(s) treated and untreated proteomes are differentially labeled during the digest, then mixed in a one-to-one ratio (total peptide amount), COFRADIC-sorted, and MS analyzed. The N-terminal peptides equally present in both samples will be represented by a 4 Da-different doublet in the first dimension of MS analysis (MS1). Given their equal representation, the area under the corresponding peaks should be the same with the ratio of ~1.

However, if the parent N-terminus is cleaved by a protease, then its area will decrease, resulting in a ratio between protease treated/untreated samples less than 1. In addition, a neo N-terminus resulting from proteolytic cleavage will be generated and represented by a singlet in the MS1 lacking its 4 Da different counterpart due to its presence only in the protease-treated sample. The ratio of such peptides will be greater than 1.

The quantitative COFRADIC approach was applied to describe apoptosis-induced proteolytic events in anti-Fas antibody-treated versus untreated human Jurkat T lymphocytes (Van Damme et al. 2005). In addition to characterizing apoptosis-independent N-terminal processing (i.e., the baseline proteolytic activity of initiator methionine removal, signal peptide trimming, etc.), the analysis identified 93 apoptosis-induced cleavage sites in 71 proteins among 1,834 proteins detected in total. Consistent with the previously well-studied experimental model, most observed cleavages were found to be at the caspase consensus sites. The few cleavages showing other than aspartate P1 specificities represent either noncanonical caspase cleavages, additional protease classes activated during apoptosis, or false positives. To validate the findings, a few caspase-specific cleavage sites were investigated in *in vitro* cleavage assays, where recombinant caspases were used to cleave synthetic peptides harboring the identified candidate cleavage sites. Also, processing of four canonical caspase substrates was successfully detected in the activated Jurkat cell lysates by immunoblotting with the corresponding specific antibodies. However, processing of proteins with caspase unspecific cleavages was not tested by Western blotting, and therefore the possibility that such peptides are due to false-positive identifications was not addressed.

More recently, the same group used the quantitative COFRADIC approach to identify the substrates of an apoptosis-activated mitochondrial serine protease high-temperature requirement protein A2 (HtrA2/Omi) (Vande Walle et al. 2007). Recombinant wild-type HtrA2/Omi or its catalytically inactive S306A mutant was incubated with Jurkat T cell lysates which were then differentially labeled and N-terminally sorted. The analysis yielded 1,162 total protein identifications (from 1,964 peptides) and determined 50 cleavage sites in 15 proteins, represented mostly by cytoskeletal proteins. Several cleavage events were validated by specific immunoblotting in treated Jurkat T cells and by *in vitro* cleavage assays with recombinant HtrA2/Omi and *in vitro* translated susbstrates. Analysis of the 50 detected cleavage sites indicated a HtrA2/Omi preference to cleave after an aliphatic residue at P4 with the four positions C-terminal to the cleavage site being most commonly occupied by small or hydrophobic residues.

To summarize, COFRADIC is a powerful approach that allows for a significant sample simplification and N-terminal peptide enrichment, and therefore enables effective identification and quantification of proteolytic processing in a biological sample. However, the experimental design does not allow the analysis of *in vivo* modified (i.e., acetylated) and unblocked protein N-termini in a single experiment, thus somewhat limiting its use for characterization of N-terminal posttranslational modifications in eukaryotes. In addition, the above-discussed sequence bias of the labeling efficacy and the absence of a clear cutoff between N-terminal and internal

TNP-modified peptides during chromatographic separation limit the number of peptides being analyzed and reduce the proteome coverage. With 2 chemical labeling steps, 2 rounds of HPLC separation, and 96 LC-MS/MS runs per average experiment, this technique is rather time-, equipment-, and labor-intense and so far has not been adopted by a broad scientific community.

A different N-terminal enrichment approach designed by the Beynon laboratory also utilizes protein acetylation as the first step, followed by tryptic digestion (McDonald et al. 2005). Newly formed unblocked internal tryptic peptides are then coupled with *N*-hydroxysuccinimide (NHS), ester-derivative of biotin, retained by immobilized streptavidin and discarded. The remaining mixture consisting of protein N-terminal peptide(s) (naturally blocked or chemically acetylated) is then analyzed to yield information on the proteolytic processes in the sample. To distinguish between and quantify *in vivo* and chemically acetylated proteins, the authors suggest using stable isotope-labeled (^{3}H) acetic anhydrate that would result in a 3 Da mass shift in MS1. By analogy with COFRADIC, labeling with ^{18}O at the C-terminus during trypsin digest (Van Damme et al. 2005) or with acetic anhydrate at the N-terminus has the potential to be utilized for comparative quantification of protease activity between two samples. This technique was tested on the soluble protein fraction of mouse skeletal muscle and a more complex mixture of soluble proteins from mouse liver (McDonald et al. 2005). As a proof of concept, qualitative comparison of matrix assisted laser desorption/ionization–time of flight (MALDI–TOF) spectra of an unfractionated peptide mixture with or without the N-terminal enrichment step indicated a significant spectra simplification, and enabled assignment of the highest intensity signals to true N-terminal peptides in the enriched samples. In contrast, without N-terminal selection the complexity of the sample prevented identification of any N-terminal peptides. LC-MS/MS analysis of N-terminally enriched mouse liver peptides yielded information on the N-terminal processing, such as removal of initiator methionine, loss of a signal peptide or propeptide, either known previously or inferred.

In order to yield more suitable sets of analytes and to increase the N-terminal sequence coverage of any given proteome, the authors suggest performing two parallel digests with proteases of different specificities. To test this hypothesis, *in silico* digest of 8,000 mouse liver proteins with trypsin and/or endopeptidase GluC was filtered to remove the peptides smaller than 500 Da and larger than 5,000 Da which are not suitable for MS analysis. The analysis of the remaining peptides indicated that tryptic or GluC digests alone would result in a respective 50% and 60% unambiguous proteome coverage (using the mass spectrometer with 20 ppm accuracy). The coverage reaches 80% when the sample is digested by the two proteases in parallel. The value can be improved even further when using higher accuracy MS (almost 90% coverage with two digests and 1 ppm instrument accuracy), further underlining the feasibility of proteome characterization via protein identification by a single peptide.

This time-efficient protocol was further improved to decrease the number of steps and therefore to increase sample recovery. The biotinylation step is now excluded and replaced by a direct coupling and removal of internal peptides via a commercially available amino-reactive immobilized reagent, NHS-activated Sepharose (McDonald and Beynon 2006). Thus, in the final protocol the internal peptides can be removed directly after the digest, with the flow-through being analyzed without further treatments. The protocol was tested using LC-MS/MS with soluble proteins from *Escherichia coli* and identified ~300 proteins by their N-termini with relatively few internal peptides. As the authors suggested, the proteome coverage might be further increased by employing additional fractionation steps before LC-MS/MS. In principle, this approach could be used to identify substrates of a specific protease, but this has yet to be reported.

Yet another interesting approach for N-terminal enrichment has been demonstrated by Timmer et al. (2007). Here, the proteins are first denatured, reduced, and alkylated and then the amino groups of lysine side chains are protected by lysine-specific guanidinylation. In the next step, the N-termini of proteins are selectively labeled with NHS-biotin. Following tryptic digest, biotin-labeled N-terminal peptides are positively selected by immobilized streptavidin. The captured peptides are then reductively cleaved off from the column using dithiothreitol (DTT) and analyzed by LC-MS/MS. This approach was tested on *E. coli*, yeast, mouse and human cell lines, and serum proteomes to profile constitutive proteolytic events in these samples. To increase the confidence and coverage, the samples were digested with both GluC and trypsin and run three times using dynamic exclusion criteria. Consistent with previous reports, multiple runs yielded a 50–70% overlap between the same sample analyzed by MS several times. The coverage of the proteome ranges from ~350 peptides in serum to ~500 peptides in *E. coli*, yeast, and mouse tissues and ~1,000 peptides in 293A human embryonic kidney cell line. These values are comparable to the number of peptides identified by McDonald and colleagues in *E. coli* (McDonald and Beynon 2006), but are lower than the ones reported using the COFRADIC technique on a similar cell-line model (Van Damme et al. 2005). This can be due, at least partially, to a higher degree of sample fractionation before LC-MS/MS analysis in COFRADIC or might be inherent to the technique. However, in contrast to COFRADIC where the majority of the identified peptides represented N-terminal peptides (Van Damme et al. 2005) (Gevaert et al. 2003), Timmer et al. (2007) reported that many of the identified peptides belong to internal sequences and can not be ascribed to any known proteolytic modifications (e.g., initiator methionine removal, propeptide removal). With the exception of *E. coli*, where such unascribed peptides constitute less then 50% of total peptides identified, the rest of the samples exhibit a broader range—from 70% of unascribed peptides in yeast and mouse tissues to 80–90% in human cell lines and serum. A possible explanation for such a high percentage of proteolysis in the samples could be a high general protease activity induced by cell disruption that was not completely inhibited before sample denaturation. Using this technique, the authors observed and characterized methionine aminopeptidase activity and removal of signal peptides as well as N-terminal trimming of proteins in serum samples.

In contrast to COFRADIC and McDonald et al., N-terminal enrichment strategies where naturally acetylated N-termini are automatically included in the analyte mix, the present protocol results in retention of only the N-termini of unblocked proteins.

Thus, such selection excludes from the analysis up to 80% of total natural N-termini in eukaryotic samples. In contrast, retaining naturally modified (i.e., acetylated) N-termini helps to curb sample loss and has an additional advantage of higher confidence protein identifications, as it is then based on a positionally anchored original N-terminal peptide (McDonald et al. 2005).

Following from the design of the present method, the degree of N-terminal enrichment (in terms of contamination with internal peptides) of the final analyte largely depends on the efficacy/completeness of protein lysine residue guanidinylation in the beginning of the protocol and on the absence of side reactions during biotin coupling. As noted by the authors, incomplete and/or side reactions will lead to biotin coupling to lysine and/or serine, threonine, or histidine residues and will result in contamination of the final analyte with internal peptides, and therefore must be strictly controlled for. It should be noted that such spectra pollution will decrease true N-termini coverage and further complicate data analysis leading to a higher rate of false-positive identifications.

While the current protocol by Timmer et al. (2007) does not readily allow for quantification of proteolytic events in the analyzed sample, the authors propose that it can be modified to incorporate stable isotope-labeled biotin for N-terminal labeling, or to include C-terminal trypsin-dependent ^{18}O exchange as seen in COFRADIC. Both such potential modifications would result in a mass shift in MS1 and enable relative quantification of the same peptide in two samples.

A different strategy for coping with overwhelming sample complexity has been offered by Enoksson et al. (2007). In this N-terminal pseudo-enrichment approach, the sample is not treated to physically remove all internal peptides, but is left complex, and then filtering for N-terminal peptides is applied at the sample analysis stage on the MALDI–TOF/TOF. Briefly, the proteins are first lysine-specific guanidinylated to block their reactive side-chain amino groups and then labeled with the iTRAQ reagent at the protein N-terminus. Following mixing of two samples (that have been treated or not with a protease) at one-to-one ratio, the sample is digested by trypsin, chromatographically separated and analyzed on a MALDI–TOF/TOF instrument.

This strategy takes advantage of the fact that in contrast to LC-coupled mass spectrometers with electrospray ionization, the MALDI instruments allow for multiple scans/analysis of the same peptide(s) in the sample. Thus, during the first low-energy scan the sample is surveyed for the presence of peptides with the iTRAQ reporter ion in the spectra to form a data-dependent inclusion list for the second scan. The first low-energy scan results in low sample consumption, and its limited fragmentation is not sufficient for peptide sequencing but is suitable for indicating diagnostic iTRAQ reporter ions. Therefore, in the second higher-energy scan only the previously selected peptides with iTRAQ tag will be fragmented for high confidence identification and quantification. In this workflow, the iTRAQ-bearing peptides represent original N-termini of the proteins as well as protease-generated neo-N-termini, with iTRAQ ratios allowing discrimination between the two. Thus, the original N-termini equally present in both samples will have iTRAQ ratio of ~1, while protease-dependent neo N-termini will have a singleton or greater than 1 iTRAQ signal ratio.

This approach was first tested on a mixture of seven purified *E. coli* proteins containing putative caspase cleavage sites, which were treated with wild-type caspase-3 or its catalytically inactive mutant C285A. A total of 12 cleavage sites in 6 proteins were identified in the MS analysis compared to 5 cleavage fragments identified by SDS-PAGE and 8–10 indicated by Western blotting analysis. Further, the method was tested on a cell-free apoptosis model using HEK293 hypotonic extracts, where 20 different cleavage sites were identified, mostly previously undescribed but with canonical caspase cleavage site specificity. Cleavage of one identified substrate, actin, was further confirmed by Western blotting.

As discussed in previous sections of this chapter, iTRAQ labeling does not result in a mass shift in MS1 and therefore does not lead to doubling of the sample complexity. A very serious limitation of this technique is in the fact that identification of iTRAQ-bearing peptides in the first "survey MS/MS" will be limited by ion suppression due to the physical presence of many internal peptides, that is, many N-terminal peptides simply will not be ionized and detected under such conditions. This notion is supported by observations of McDonald et al. made under very similar conditions using a MALDI–TOF instrument to analyze fractionated mouse liver proteins (McDonald et al. 2005). Thus, McDonald et al. reported that high sample complexity prevented N-termini detection when internal peptides are physically present in the sample. Consistent with ion suppression being a limiting factor, Enoksson et al. (2007) detected only 20 cleavage sites compared to 93 cleavages identified by the COFRADIC method in a similar cell-based apoptosis model (Van Damme et al. 2005). However, it should be noted that the use of a different cell line might be a contributing factor as well. Also, in contrast to all the other above-described techniques, the latter method can only be used with MALDI mass spectrometers. Thus, the virtual N-terminal enrichment technique of Enoksson et al. is a suitable method for detection and quantification of protease cleavage sites in defined protein sets of test substrates or in less complex proteomes.

As a future direction, another technique for substrate discovery is in development by the authors' laboratory termed "terminal amine isotope labeling of substrates", TAILS (Kleifeld et al., manuscript in preparation). In this approach, the sample is enriched for N-terminal peptides of each protein, thus allowing for neo-N-termini resulting from proteolysis to be identified with higher probability (sample complexity reduction) and confidence (positional information). Specifically, the proteomes of two samples containing active and inactive protease (control) are first reduced, alkylated, and labeled with amino-reactive isotope-containing reagents (such as formaldehyde or iTRAQ). Such labeling selectively modifies lysine residues and protein N-termini. Following trypsin digestion, the newly created and therefore unblocked internal peptides are selectively removed by an amine scavenging polymer or beads. The remaining N-terminome fraction is then analyzed by MS/MS. For example, when the iTRAQ reagents 114 and 115 are used to differentially label samples containing active and inactive protease, respectively, this results in protease-cleaved neo-N-termini identified as spectra with singletons (i.e., containing only one isotopic signature, 114). These are distinct from noncleaved peptides, which exhibit both isotopic signatures, 114 and 115, at 1:1 ratio. Therefore, MS/MS sequencing identifies protease substrates and defines the sequence of the cleavage site, with iTRAQ labeling allowing for an estimate of how much a particular substrate is processed.

To summarize, this approach utilizes the power of multiplex isotope labeling and negative selection of internal peptides by use of a highly soluble, highly derivatized polymer. Thus, it enriches for protease cleavage neo-peptides and natural N-termini (acetylated and nonacetylated), allowing determination of both protease substrates and their cleavage sites, as well as annotation of N-terminal posttranslational proteome processing in a single experiment.

Conclusions

A number of reported techniques that are different in their labeling, enrichment and quantification strategies all aim at proteome simplification and N-terminal enrichment in order to enable more efficient protease substrate identifications. When selecting a suitable MS technique for determining the substrate(s) of a particular protease, one may choose to consult the following checklist: (1) proteome coverage achieved by the method; in general, the higher it is the better is the chance for finding the substrate(s); (2) quantification aspect; quantification always strengthens qualitative findings and allows for subtraction of basal proteolysis. MS/MS-based quantification methods, such as those based on iTRAQ, offer some advantages, including a possibility for the simultaneous analysis of up to eight samples in one experiment; (3) reagent availability and level of expertise required to perform the protocol and to analyze the data; (4) instrumentation, time, labor, and cost efficiency.

Acknowledgments Christopher M. Overall was supported by a Canada Research Chair in Metalloproteinase Proteomics and Systems Biology, by research grants from the National Cancer Institute of Canada (with funds raised by the Canadian Cancer Association) and from the Canadian Breast Cancer Research Alliance Special Program Grant on Metastasis, and by a Centre Grant from the Michael Smith Research Foundation. Ulrich auf dem Keller was supported by the Deutsche Forschungsgemeinschaft Germany. Anna Prudova was supported by a postdoctoral fellowship from the Centre for Blood Research, UBC.

References

auf dem Keller, U., Doucet, A., et al. 2007. "Protease research in the era of systems biology." Biol Chem 388: 1159–62.

Bredemeyer, A. J., Lewis, R. M., et al. 2004. "A proteomic approach for the discovery of protease substrates." Proc Natl Acad Sci USA 101: 11785–90.

Brunner, E., Ahrens, C. H., et al. 2007. "A high-quality catalog of the Drosophila melanogaster proteome." Nat Biotechnol 25: 576–83.

Butler, G. S. and Overall, C. M. 2007. "Proteomic validation of protease drug targets: Pharmacoproteomics of matrix metalloproteinase inhibitor drugs using isotope-coded affinity tag labelling and tandem mass spectrometry." Curr Pharm Des 13: 263–70.

Canals, F., Colome, N., et al. 2006. "Identification of substrates of the extracellular protease ADAMTS1 by DIGE proteomic analysis." Proteomics 6(Suppl. 1): S28–35.

Chen, E. I., Hewel, J., et al. 2006. "Large scale protein profiling by combination of protein fractionation and multidimensional protein identification technology (MudPIT)." Mol Cell Proteomics 5: 53–6.

Choe, L. H., Aggarwal, K., et al. 2005. "A comparison of the consistency of proteome quantitation using two-dimensional electrophoresis and shotgun isobaric tagging in Escherichia coli cells." Electrophoresis 26: 2437–49.

Dean, R. A. and Overall, C. M. 2007. "Proteomics discovery of metalloproteinase substrates in the cellular context by iTRAQ labeling reveals a diverse MMP-2 substrate degradome." Mol Cell Proteomics 6: 611–23.

Dean, R. A., Butler, G. S., et al. 2007. "Identification of candidate angiogenic inhibitors processed by matrix metalloproteinase 2 (MMP-2) in cell-based proteomic screens: Disruption of vascular endothelial growth factor (VEGF)/heparin affin regulatory peptide (pleiotrophin) and VEGF/Connective tissue growth factor angiogenic inhibitory complexes by MMP-2 proteolysis." Mol Cell Biol 27: 8454–65.

Enoksson, M., Li, J., et al. 2007. "Identification of proteolytic cleavage sites by quantitative proteomics." J Proteome Res 6: 2850–8.

Fenn, J. B., Mann, M., et al. 1989. "Electrospray ionization for mass spectrometry of large biomolecules." Science 246: 64–71.

Fenyo, D. 2000. "Identifying the proteome: Software tools." Curr Opin Biotechnol 11: 391–5.

Gevaert, K., Goethals, M., et al. 2003. "Exploring proteomes and analyzing protein processing by mass spectrometric identification of sorted N-terminal peptides." Nat Biotechnol 21: 566–9.

Gygi, S. P., Rist, B., et al. 1999. "Quantitative analysis of complex protein mixtures using isotope-coded affinity tags." Nat Biotechnol 17: 994–9.

Gygi, S. P., Corthals, G. L., et al. 2000. "Evaluation of two-dimensional gel electrophoresis-based proteome analysis technology." Proc Natl Acad Sci USA 97: 9390–5.

Hwang, I. K., Park, S. M., et al. 2004. "A proteomic approach to identify substrates of matrix metalloproteinase-14 in human plasma." Biochim Biophys Acta 1702: 79–87.

Kislinger, T., Cox, B., et al. 2006. "Global survey of organ and organelle protein expression in mouse: Combined proteomic and transcriptomic profiling." Cell 125: 173–86.

Lee, A. Y., Goo Park, S., et al. 2004a. "Identification of the degradome of Isp-1, a major intracellular serine protease of Bacillus subtilis, by two-dimensional gel electrophoresis and matrix-assisted laser desorption/ionization-time of flight analysis." Proteomics 4: 3437–45.

Lee, A. Y., Park, B. C., et al. 2004b. "Identification of caspase-3 degradome by two-dimensional gel electrophoresis and matrix-assisted laser desorption/ionization-time of flight analysis." Proteomics 4: 3429–36.

Major, T., von Janowsky, B., et al. 2006. "Proteomic analysis of mitochondrial protein turnover: Identification of novel substrate proteins of the matrix protease pim1." Mol Cell Biol 26: 762–76.

McDonald, L. and Beynon, R. J. 2006. "Positional proteomics: Preparation of amino-terminal peptides as a strategy for proteome simplification and characterization." Nat Protoc 1: 1790–8.

McDonald, L., Robertson, D. H., et al. 2005. "Positional proteomics: Selective recovery and analysis of N-terminal proteolytic peptides." Nat Methods 2: 955–7.

Niall, H. D. 1973. "Automated Edman degradation: The protein sequenator." Methods Enzymol 27: 942–1010.

Overall, C. M. and Blobel, C. P. 2007. "In search of partners: linking extracellular proteases to substrates." Nat Rev Mol Cell Biol 8: 245–57.

Overall, C. M. and Dean, R. A. 2006. "Degradomics: Systems biology of the protease web. Pleiotropic roles of MMPs in cancer." Cancer Metastasis Rev 25: 69–75.

Overall, C. M. and Kleifeld, O. 2006. Validating matrix metalloproteinases as drug targets and anti-targets for cancer therapy." Nat Rev Cancer 6: 227–39.

Polevoda, B. and Sherman, F. 2003. "N-terminal acetyltransferases and sequence requirements for N-terminal acetylation of eukaryotic proteins." J Mol Biol 325: 595–622.

Shevchenko, A., Jensen, O. N., et al. 1996. "Linking genome and proteome by mass spectrometry: Large-scale identification of yeast proteins from two dimensional gels." Proc Natl Acad Sci USA 93: 14440–5.

Staes, A., Demol, H., et al. 2004. "Global differential non-gel proteomics by quantitative and stable labeling of tryptic peptides with oxygen-18." J Proteome Res 3: 786–91.

Tam, E. M., Morrison, C. J., et al. 2004. "Membrane protease proteomics: Isotope-coded affinity tag MS identification of undescribed MT1-matrix metalloproteinase substrates." Proc Natl Acad Sci USA 101: 6917–22.

Timmer, J. C., Enoksson, M., et al. 2007. "Profiling constitutive proteolytic events in vivo." Biochem J 407: 41–8.

Van Damme, P., Martens, L., et al. 2005. "Caspase-specific and nonspecific in vivo protein processing during Fas-induced apoptosis." Nat Methods 2: 771–7.

Vande Walle, L., Van Damme, P., et al. 2007. "Proteome-wide identification of HtrA2/Omi substrates." J Proteome Res 6: 1006–15.

Washburn, M. P., Wolters, D., et al. 2001. "Large-scale analysis of the yeast proteome by multidimensional protein identification technology." Nat Biotechnol 19: 242–7.

Wilm, M., Shevchenko, A., et al. 1996. "Femtomole sequencing of proteins from polyacrylamide gels by nano-electrospray mass spectrometry." Nature 379: 466–9.

Wolters, D. A., Washburn, M. P., et al. 2001. "An automated multidimensional protein identification technology for shotgun proteomics." Anal Chem 73: 5683–90.

Chapter 7
Activity-Based Imaging and Biochemical Profiling Tools for Analysis of the Cancer Degradome

Vincent Dive, Margot G. Paulick, J. Oliver McIntyre, Lynn M. Matrisian, and Matthew Bogyo

Abstract Proteases represent one of the largest and most well-characterized families of enzymes in the human genome. Furthermore, there are many human health conditions associated with alterations in protease activity and function, most notably cancer. Frequently, associations between specific proteases and a given disease are correlative, and there is a need to determine the significance of direct causal relationships. Unfortunately, our understanding of protease function in the context of complex proteolytic cascades involved in human biology remains in its infancy. This gap in our knowledge has to do with the high degree of complexity both at the level of expression and also at the level of posttranslational regulation of proteases involved in the pathways that regulate human physiology or disease pathology. Thus, in order to begin to decipher complex regulatory networks and to assign function to the more than 500 proteases in the human genome, tools will need to be generated that allow direct assessment of protease activity in the context of complex biological systems. In this chapter, we will focus on the recent advances in the development and application of protease probes that can be used to directly monitor levels of active proteases in biological environments ranging from whole cell lysates to whole organisms. This chapter focuses on new developments in the field of small molecule activity-based probes and the development of protease sensors based on substrates. The goal of this chapter is to give the reader an update on our ability to spy on proteases at the level of their enzymatic activity.

M. Bogyo
Dept. of Pathology Stanford University 300 Pasteur Dr. Stanford, CA 94305-5324,
e-mail: mbogyo@stanford.edu

D. Edwards et al. (eds.), *The Cancer Degradome.*
© Springer Science + Business Media, LLC 2008

Introduction: Why Monitor Protease Activity?

Proteases, like most other classes of enzymes, are rarely regulated at the level of transcription and translation. Instead, virtually all proteases are synthesized as inactive zymogens that are subsequently activated by any of a number of stimuli that can range from change of pH to direct proteolytic processing by other proteases. To further complicate the regulatory process, many classes of endogenous, protein-based inhibitors exist that serve to regulate the activity of proteases even after processing of an inactive zymogen has occurred. Thus, simple measurement of location and abundance of a protease is not sufficient to be able to define its functional roles in a given disease process. In response to this challenge, significant efforts have been made to develop tools that allow direct monitoring of protease activity in the context of their native biological environment. One of the most common ways to monitor activity of a protease is to develop substrates whose processing by a protease can be easily monitored. Thus, substrate turnover can be measured and the overall binding specificity and enzyme efficiency can be determined. Simple fluorogenic substrates have been designed for use with purified proteases, and modifications of this design have resulted in substrate-based probes that have been used to visualize proteolytic activity in living cells and tissues. The specificity of these reagents is often controlled both by the localization of the probes and by the use of extended peptide sequences that show high selectivity for a given protease target. However, these reporters often lack the required selectivity for use in complex biological samples containing hundreds of proteases. Thus, in order for reagents to be useful for profiling protease activities in complex samples they must have a high degree of selectivity or allow direct identification of the target proteases. The use of small molecule inhibitors with exquisite selectivity for specific proteases has resulted in significant advances in the development and application of activity-based reagents that make use of these highly selective reactive functional groups to limit the complexity of protease that are targeted by a single probe. Furthermore, the permanent nature of activity-based probe (ABP) labeling allows identification of targets using biochemical methods. While it remains difficult to make either ABPs or substrate-based imaging agents with absolute specificity for a given protease, these reagents can often be highly effective tools for monitoring the regulation of a defined and relatively small subset of target proteases. As we outline below, these tools have led to a number of important discoveries and have greatly increased our understanding of protease function in disease states such as cancer.

Small Molecule Activity-Based Probes for Proteases

The field of activity-based proteomics is a relatively new discipline that makes use of small molecules, termed ABPs, to tag and monitor distinct sets of proteins within a complex proteome. These activity-dependent labels facilitate analysis of system-wide changes at the level of enzyme activity rather than simple protein

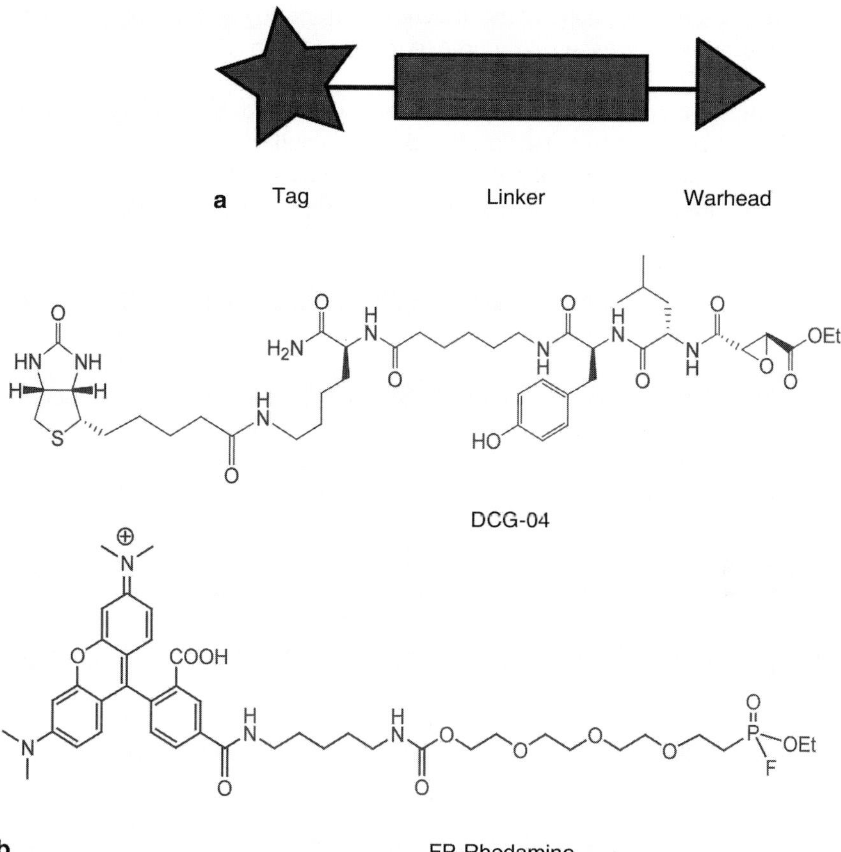

Fig. 7.1 General and specific structures of activity-based probes (*ABPs*). (**a**) General structure of an *ABP*. (**b**) DCG-04 is an *ABP* that targets cysteine proteases and contains a biotin tag (red), a dipeptide-containing linker (blue), and an epoxide as a warhead (green). Fluorophosphonate (*FP*)-rhodamine is an *ABP* that targets the serine hydrolase superfamily of enzymes and contains a rhodamine fluorophore (red), a poly (ethylene glycol) linker (blue), and a fluorophosphonate as a warhead (green). (*See also* Color Insert I)

abundance. In their most basic form, ABPs consist of three distinct functional elements (Fig. 7.1): a reactive group for covalent attachment to the enzyme, a linker region that can modulate reactivity and specificity of the reactive group, and a tag for identification and purification of modified enzymes.

Perhaps the greatest challenge in the design of a chemical probe is the selection of a reactive group that provides the necessary covalent modification of a target protein. It is often difficult to select effective reactive functional groups as they must be both reactive toward a specific residue on a protein and inert toward other reactive species within the cell or cell extract. In general, the reactive groups of most of the successfully designed chemical probes have been based on the chemistries of covalent, mechanism-based inhibitors of various enzyme families. Protease inhibitors provide a rich source of reactive groups that have been designed based on subtle differences in reaction mechanisms for the major protease families (for an extensive review, *see* Powers et al. 2002). However, only the serine, threonine, and cysteine protease families utilize a catalytic mechanism that allows direct covalent modification of the primary active-site nucleophile. For these two protease families, the majority of probes have been designed based on the prior work of medicinal and natural product chemists that have pioneered the use of suicide inhibitors (for reviews, *see* Powers et al. 2002, Jeffery and Bogyo 2003, Speers and Cravatt 2004, Evans and Cravatt 2006, Sadaghiani et al. 2007b). For the metalloprotease family, the lack of a direct acyl-enzyme intermediate in the catalytic hydrolysis reaction has required the use of alternate strategies to direct covalent modification of target proteases. All of the current examples of metalloprotease ABPs make use of a pharmacophore that binds with high specificity and affinity to the catalytic metal in the active site. This tight binding element is attached to a photocrosslinker that can be used to secure the probe in place through the formation of a nonspecific, light-activated covalent bond. Because of the limitations in developing metalloprotease ABPs, there has been a significant effort in developing substrate-based imaging probes for metalloproteases, as discussed below.

In ABPs, the linker region of a chemical probe connects the reactive group to the tag used for identification and/or purification. The linker region can serve multiple purposes. Its primary function is to provide enough space between the reactive group and the tag to prevent steric hindrance that could block access of the reactive group or accessibility of the tag for the purpose of purification. The linker can also incorporate specificity elements used to target the probe to a desired enzyme or family of enzymes. These specificity elements normally take the form of a peptide or a peptide-like structure, particularly for the ABPs used to target proteases.

The purpose of the tag on a chemical probe is to allow quick and simple identification and purification of probe-modified proteins. The most commonly used tags are biotin, fluorescent, and radioactive tags (for review *see* Sadaghiani et al. 2007b). Biotin facilitates detection by simple Western blot approaches using a reporter avidin molecule in place of the standard secondary antibody. Furthermore, biotin allows direct isolation of labeled targets through affinity chromatography. Fluorescent and radioactive tags can be visualized by direct scanning of gels with a fluorescence or phosphorimager scanner and typically have a greater dynamic range than do streptavidin–biotin detection methods. Fluorescent tags also have the added advantage of allowing direct microscopic imaging of proteases that have been modified by an ABP. Specific application of fluorescent ABPs for imaging applications are discussed later in this chapter.

The last 10 years have seen a significant increase in the number and types of ABPs that have been designed by chemists. While the field of activity-based protein profiling has continued to expand in the scope of enzymatic targets that can be profiled, the greatest progress has been made with probes that target proteases (for a relatively complete list, *see* Table 7.1). Profiling of the active forms of cysteine, serine, and threonine proteases in complex proteomes has been successfully

Table 7.1 List of commonly used protease activity-based probes (ABPs)

	Papain family	Probe name	References
Cysteine proteases	Biotin tags		
	Cathepsins/ Calpains	DCG-04	(Greenbaum et al. 2000)
	Cathepsin B	NS-196	(Schaschke et al. 2000)
	Legumain	Asp-AOMK	(Kato et al. 2005)
	Fluorescent tags		
	Cathespins/ Calpains	BODIPY-DCG-04	(Greenbaum et al. 2002)
	Cathepsins	GB-111, GB-123, GB-137	(Blum et al. 2005, 2007)
	Cathepsin B	SV5	(Verhelst and Bogyo 2005)
	Legumain	Fam-XPD-AOMK	(Sexton et al. 2007b)
	Radiolabels		
	Cathespins/ Calpains	Cbz-Try-Ala-N2	Enzyme Systems Products
		DCG-04	(Greenbaum et al. 2000)
		JPM-565	(Shi et al. 1992)
		JPM-OEt	(Bogyo et al. 2000)
		LHVS-PhOH	(Bogyo et al. 2000)
	Cathepsin B	MB-074	(Bogyo et al. 2000)
	Caspases		
	Biotin tags		
	General caspase	Biotin-X-VAD(OMe)-fmk	Calbiochem
		Z-VK-X-(biotin)-D (Ome)-fmk	Calbiochem
		KMB-01	(Berger et al. 2006)
	Caspase-3	b-EvaD-epoxide	(Sexton et al. 2007a)
	Caspase-8	bAB06, bAB13	(Berger et al. 2006)
	Caspase-9	bAB19	(Berger et al. 2006)
		bAB28	(Berger et al. 2006)
	Fluorescent tags		
	General caspase	SR-VAD-fmk	Immunochemistry Technologies, LLC
		FAM-VAD-fmk	
	Deubiquitinating enzymes		
	HA tags	HA-Ub-VS	(Borodovsky et al. 2002)
		HA-Ub-Cl	(Borodovsky et al. 2002
		HA-Ub-Br2	(Borodovsky et al. 2002)
		HA-Ub-Br3	(Borodovsky et al. 2002)
		HA-Ub-VME	(Borodovsky et al. 2002)
		HA-Ub-VSPh	(Borodovsky et al. 2002)
		HA-Ub-VCN	(Borodovsky et al. 2002)
	Arg-Gingipain		
	Biotin tags	BiRK	(Mikolajczyk et al. 2003)

(continued)

Table 7.1 (continued)

	Papain family	Probe name	References
Serine proteases	Radiolabels		
	All hydrolyases	^{3}H-DFP	PerkinElmer
	Biotin tags		
	All hydrolyases	FP-Biotin	(Liu, Patricelli and Cravatt 1999)
		FP-Peg-Biotin	(Kidd, Liu and Cravatt 2001)
	Chymotrypsin-like	Biotin-AAF-cmk	Enzyme Systems Products
	Trypsin-like	Bio-PK-DPP, Bio-NK-DPP	(Pan et al. 2006)
	Granzyme B	Bio-x-IEPD^P-(OPh)$_2$	(Mahrus and Craik 2005)
	Granzyme A	Bio-x-IGN(AmPhg)^P-(OPh)$_2$	(Mahrus and Craik 2005)
	Fluorescent tags		(Patricelli et al. 2001)
	All hydrolyases		(Liu et al. 1999)
			(Patricelli et al. 2001)
Threonine proteases	*The proteasome*		
	Radiolabels	NP-LLL-VS	(Bogyo et al. 1997)
	Biotin tags	Epoxomicin biotin	(Meng et al. 1999)
		AdaLys(Bio)AhX$_3$L$_3$-VS	(Kessler et al. 2001)
Metalloproteases	*General probes*		
	Biotin tags		
	Hydroxamates	TFMPD-K(Bio)-GGX-NHOH	(Chan et al. 2004)
	Fluorescent tags	HxBP-Rh	(Saghatelian et al. 2004)
	Hydroxamates	TFMPD-K(Cy3)-GGX-NHOH	(Chan et al. 2004)
	Radiolabel		
	Phosphinic peptides	Compound 1	(David et al. 2007)

achieved through the development of ABPs that exploit the presence of conserved active-site nucleophiles in these protease families. These probes selectively interact with the enzyme active site, free of propeptide or inhibitor, and then form specific mechanism-based covalent bonds between the active-site nucleophile and the reactive warhead group. After modification by the probe, target proteases can be isolated, biochemically monitored, or imagined by virtue of the tagging group used on the probe. For metalloproteases, the lack of corresponding conserved nucleophiles in the active site has required the use of a photochemical group which, upon light excitation, forms a stable linkage with amino acid residues of enzyme active site. Several ABPs have been successfully developed for zinc proteases, leading in some cases to the discovery of elevated zinc protease activities in pathological samples (Chan et al. 2004, Saghatelian et al. 2004). However, up to now, the ABPs developed for matrix metalloproteases (MMPs) have failed to detect endogenous active forms of these proteases in samples known to overexpress MMPs (cancer cells or tumors). Possible reasons for this failure as well as future directions to overcome the actual limitations are discussed below.

Substrate-Based Imaging Agents

The principle of using substrate-based imaging agents as a means to detect proteolytic activity in vivo has been demonstrated by a number of investigators. The Weissleder group has made inroads into the application of near infrared (NIR) probes for proteolytic activity as a means to measure protease activity and subsequent inhibition (Weissleder 2002). NIR fluorophores have been attached to a linear copolymer either directly (Weissleder et al. 1999) or via a peptide containing a proteolytic cleavage site (Bremer et al. 2001). These probes function on the principle of fluorescence/Förster resonance energy transfer (FRET): close proximity of the fluorophores results in resonance transfer that quenches the fluorescent signal, and the fluorescence intensity increases after cleavage of the peptide linker abrogating the quenching of the fluorophore. These probes have been injected into tumor-bearing mice and have applications relevant to many diseases, including cancer (McIntyre and Matrisian 2003). For example, Cy 5.5-based reagents containing a peptide cleavable by cathepsin B have been used as a method to distinguish well-differentiated and undifferentiated breast cancers (Bremer et al. 2002) and the early detection of intestinal adenomas (Marten et al. 2002) in preclinical models.

More than three decades ago, reagents to measure intermolecular distances by FRET (Förster 1948) between donor (D) and acceptor (A) chromophores (Stryer and Haugland 1967) were incorporated into substrates designed to measure hydrolase activity based on increased fluorescence attendant on loss of FRET (Latt et al. 1972). In these kinds of FRET-based reagents, FRET reduces the apparent lifetime of the excited state of the chromophore resulting in a reduction in the amplitude of fluorescence, particularly of the D chromophore. The efficient quenching of EDANS by DABCYL, introduced in a novel fluorogenic substrate for assaying retroviral proteases by FRET (Matayoshi et al. 1990), has been used in a plethora of FRET substrates with specificity for various proteases afforded by the sequence of the peptide linking the D fluorophore (EDANS) and the A quencher (DABCYL), for example, Calbiochem, http://www.emdbiosciences.com; Sigma-Aldrich, http://www.sigmaaldrich.com. In such FRET-based protease substrates, the intrinsic fluorescence of a specific chromophore incorporated into the substrate is quenched by resonance energy transfer to an acceptor; FRET is optimized by using fluorophores with relatively long fluorescence lifetimes and acceptors with resonance absorption bands that overlap the fluorescence emissions (Haugland et al. 1969, Matayoshi et al. 1990). In the design of FRET-substrates for proteases, though, optimal FRET quenching is afforded by attachment of donor and acceptor chromophores in proximity on either side of the cleavable peptide bond; interference of the chromophores with peptide recognition and cleavage must be considered. For example, a FRET-peptide substrate for ADAMTS13 with fluorophore donor located at P_7, the seventh amino acid from the scissile bond, and acceptor (quencher) at P_5', that is, separated by 11 amino acids, serves as an efficient and selective substrate for the protease, while a substrate with the donor located at P_4 to reduce the D–A separation was cleaved less efficiently, attributed to proximity of the modified residue to the cleavage site (Kokame et al. 2005). The design of these kinds of fluorogenic peptide substrates is particularly pertinent in the development of fluorescent triple-helical peptide collagen-like substrates (Lauer-Fields et al. 2003). In such reagents, the conformation of the substrate, known to be critical for platelet–collagen interactions (Smethurst et al. 2007), has a significant effect on protease affinity and specificity (Lauer-Fields et al. 2007b) and has resulted in the development of a new class of selective MMP inhibitors based on the structure of the triple-helical transition state (Lauer-Fields et al. 2007a).

FRET-based proteases probes have also been developed based on fluorescence homotransfer (Erijman and Weber 1993), in which the fluorescence is attenuated by resonance energy transfer to an adjacent identical chromophore. For example, the fluorescence of multiply labeled fluorescein (FL)-Dipyrromethene boron Difuoride (BODIPY) albumin is 98% quenched due to homotransfer, providing a reagent that can be used to monitor proteolytic cleavage of albumin (Reis et al. 1998). Likewise, collagen and gelatin heavily labeled with FL, referred to as dye-quenched (DQ)-collagen and DQ-gelatin, respectively, are minimally fluorescent with fluorescence being manifest following proteolytic cleavage; these kinds of reagents have been used for in vitro imaging of proteolysis by human breast cancer cells (Sameni et al. 2000). The principle of homotransfer self-quenching has been used in a number of polymer-based NIR fluorescent (NIRF) protease probes developed by Weissleder's group (Weissleder et al. 1999, Bremer et al. 2001, Weissleder 2002). A similar strategy has been used recently to measure the degradation of a NIRF-labeled polyglutamic acid by cysteine proteases (Melancon et al. 2007).

In addition to the classical fluorescent chromophores, FRET has also been demonstrated between variants of green fluorescent protein (GFP) (Heim and Tsien 1996, Zhang et al. 2002) as well as with a number of new classes of fluorophores such as nanocrystals and nanoparticles (Sapsford et al. 2006). FRET within a fusion protein of blue- and green-fluorescent proteins (BFPs–GFPs) was abrogated by proteolytic cleavage of the linker between the two domains (Heim and Tsien 1996). The development of GFPs with different spectral properties (Piston et al. 1999) has opened the possibility for generating reagents for in situ assay of various proteases. For example, a modified GFP–*Discosoma* sp. Red (DsRed) sensor with a linker cleavable by the 2A-protease encoded by enterovirus 71 (EV71) has been developed to detect EV71 virus infection manifest by increased lifetime of the GFP donor following proteolytic cleavage of the substrate sensor (Ghukasyan et al. 2007). Another recently developed GFP-based sensor uses a nonfluorescent yellow-FP as acceptor to measure FRET in living cells (Ganesan et al. 2006); in such resonance energy-accepting chromoprotein (REACh) reagents, the donor GFP is quenched by FRET to a dark nonfluorescent acceptor. Analogous to the DABCYL quenching of EDANS (Matayoshi et al. 1990) and the more recently described "DQ" reagents such as those with "black hole quencher" (Zheng et al. 2007), the REACh reagents have only a single fluorogenic chromophore (donor fluorescence), facilitating fluorescence lifetime imaging (FLIM). The quenched GFP–REACh reagents have been used to visualize the distribution of specific biological processes, for example, ubiquitination machinery, without the attenuation of signal that accompanies the spectral selection required with reagents that contain more than one fluorophore (Ganesan et al. 2006).

ABPs for Cysteine and Serine Proteases

The development of ABPs for cysteine proteases has been particularly successful mainly due to the availability of a large number of covalently reactive functional groups and to the fact that this catalytic class is divided into relatively small subfamilies with overlapping substrate specificity. Thus, probes with optimal reactive functional groups and selective linker sequences can be used to monitor small sets of related cysteine proteases.

By far the largest family of cysteine proteases is the family of deubiquitinating proteases (DUBs) with a total of more then 60 members in the human genome (Lopez-Otin and Overall 2002). These proteases regulate the removal of ubiquitin from target proteins, thus controlling their rates of degradation by the proteasome. In addition, there are a number of small ubiquitin-like modifiers (SUMOs) that also are attached to protein substrates and eventually are removed by proteases related to the DUBs. Members of this family of cysteine proteases are somewhat unique in that they recognize a folded protein (ubiquitin) as a substrate and therefore only inefficiently process small peptide substrates. Thus, ABPs designed to target this family of enzymes have required the use of the full 76 amino acid ubiquitin chain as the linker region of the probe. A number of useful ABPs for the DUB family have been synthesized by native ligation of a range of small electrophiles to the C-terminus of epitope-tagged ubiquitin (*see* Table 7.1). These probes have been used to identify new protease families and to monitor changes in activities of DUBs in cancer cells as outlined later in this chapter.

Two other significant subfamilies of cysteine proteases that have been studied with ABPs are the primarily lysosomal cysteine proteases of the papain family (clan CA/CB) and the cytosolic caspases (clan CD) involved in the regulation of cell death. A number of different classes of probes for both papain family and caspases have been developed (*see* Table 7.1). The majority of these probes are short, tri- or tetrapeptides that carry reactive functional groups such as epoxides, acyloxymethyl ketones (AOMK), or vinyl sulfones. Since members of both of these protease families play important roles in regulation of cellular processes involved in cancer progression or response to chemotherapy, probes have already found widespread use for imaging and biomarker discovery. These applications are outlined later in this chapter.

Serine proteases, while similar in overall number of total family members to the cysteine protease family, have been somewhat more difficult to study using ABPs. This is primarily because the serine protease family is made of many highly specialized subfamilies that have relatively few or even single members. In addition, serine proteases are part of a larger group of serine hydrolyase enzymes, thus complicating the development of protease-specific probes. Regardless of these limitations, general serine hydrolyase probes have proven to be highly valuable reagents for monitoring this larger family of enzymes (*see* Table 7.1). In particular, a general fluorophosphonate probe has been used to identify a number of significant new cancer biomarkers as outlined later in this chapter. In addition, probes containing peptide-based scaffolds in combination with a less reactive diphenyl phosphonate warhead have proven to be quite selective as ABPs for serine proteases (*see* Table 7.1). By changing the sequence of the primary peptide scaffold, probes of this family have been designed to target diverse enzymes in both trypsin and chymotrypsin family proteases and granzyme family proteases with distinct specificities and functional roles. It is clear that further development of probes for serine proteases will likely yield new tools with direct applications to cancer in the near future.

Probes for Metalloprotease Activity

Small Molecule Activity-Based Probes

Development of ABPs to covalently modify the active site of zinc metalloproteases relies first on the selection of an optimal broad-spectrum or selective synthetic inhibitor and second on the identification of a suitable photolabile group that can be incorporated into the inhibitor structure. The main classes of photolabile groups (or photophores) yielding reliable and reproducible labeling of target proteins are depicted in Fig. 7.2. These photophores fulfill a number of important criteria for photoaffinity labeling experiments: reasonable stability under ambient light, a photochemically generated

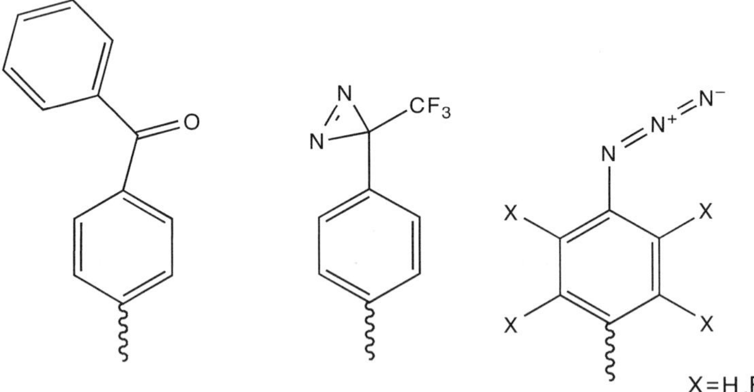

Fig. 7.2 Structures of the three major photophores incorporated in zinc metalloprotease activity-based probes (*ABPs*): benzophenone, diazirine, and phenylazide.

excited state with lifetime shorter than the dissociation of the inhibitor–enzyme complex but long enough to spend sufficient time in close proximity to the target site for covalent linkage, unambiguous photochemistry to provide a single covalent adduct, an activated form that reacts with CH groups as well as nucleophilic X–H bonds, and an activation wavelength longer than the ultraviolet absorption of protein targets (>300 nm) (Fleming 1995, Dorman and Prestwich 2000). These common photophores are hydrophobic groups and thus their incorporation into an inhibitor may compromise inhibitor solubility. In complex proteomes, potentially useful ABPs might be missed because of high nonspecific background labeling. Also, given the bulkiness of these photophores, the site of incorporation into the inhibitor should be chosen carefully in order to preserve the inhibitor affinity for its targets. Commercially available derivatives of amino acids incorporating these photophores can be used during classical peptide synthesis. Introduction of the photophore at a specific site on the inhibitor is also possible; however, this often requires significant synthetic efforts that often make such strategies impractical.

While attachment of a highly reactive photolabile group may yield the desired results for cross-linking, in some cases synthesis of multiple analogues may be required to ensure an efficient labeling of the targeted enzymes. In fact, the chemical composition of the residues surrounding the photoreactive group in the enzyme active site often varies leading to highly variable yields of covalent cross-linking among related protease targets. Efficient labeling is expected to occur when, in the enzyme–ABP complex, the reactive group points toward cavities of the enzyme active site, while weak labeling may arise if the reactive moiety is highly exposed to solvent molecules. Thus, the positioning of the photolabile group in the inhibitor structure requires significant optimization. This was illustrated in a series of probes developed to target MMP active forms. In this study, a benzophenone photolabile group was incorporated either in the P_2' or P_3' position of hydroxamate peptide inhibitors (Fig. 7.3) (Sieber et al. 2006). Whereas a strong labeling of MMP-1 was observed when the benzophenone occupied the inhibitor P_2' position, only weak labeling of MMP-1 was observed when this group was positioned at the P_3' position. The same trends were reported for MMP-9 and MMP-12. These results can be explained by the fact that the S_3' subsite of MMPs is much more solvent exposed than the S_2' subsite. Variation in the cross-linking yield was also found to depend on the identity of the target protein, even within the same protein subfamily. The presence of a very deep cavity (the S_1' subsite) in most MMP active sites (except MMP-1 and -7) was exploited to introduce a photolabile group (an azide group) in the P_1' position of a phosphinic peptide inhibitor of MMPs (Fig. 7.3) (David et al. 2007). In this MMP probe, the photolabile group was incorporated on the distal part of an unnatural side chain, placing the azido group deep inside the S_1' cavity of MMPs, in a position protected from bulk solvent. This probe was shown to selectively label only the active site of MMP-12 with a cross-linking yield of 42%. Determination of the probe efficiency in modifying other MMPs revealed some unexpected results. In fact, this probe was observed to cross-link the active site of eight MMPs (MMP-2, -3, -8, -9, -11, -12, -13, and -14), but with a 40-fold difference in efficacy, a variation that may limit the detection of some MMP active forms in complex proteomes. The

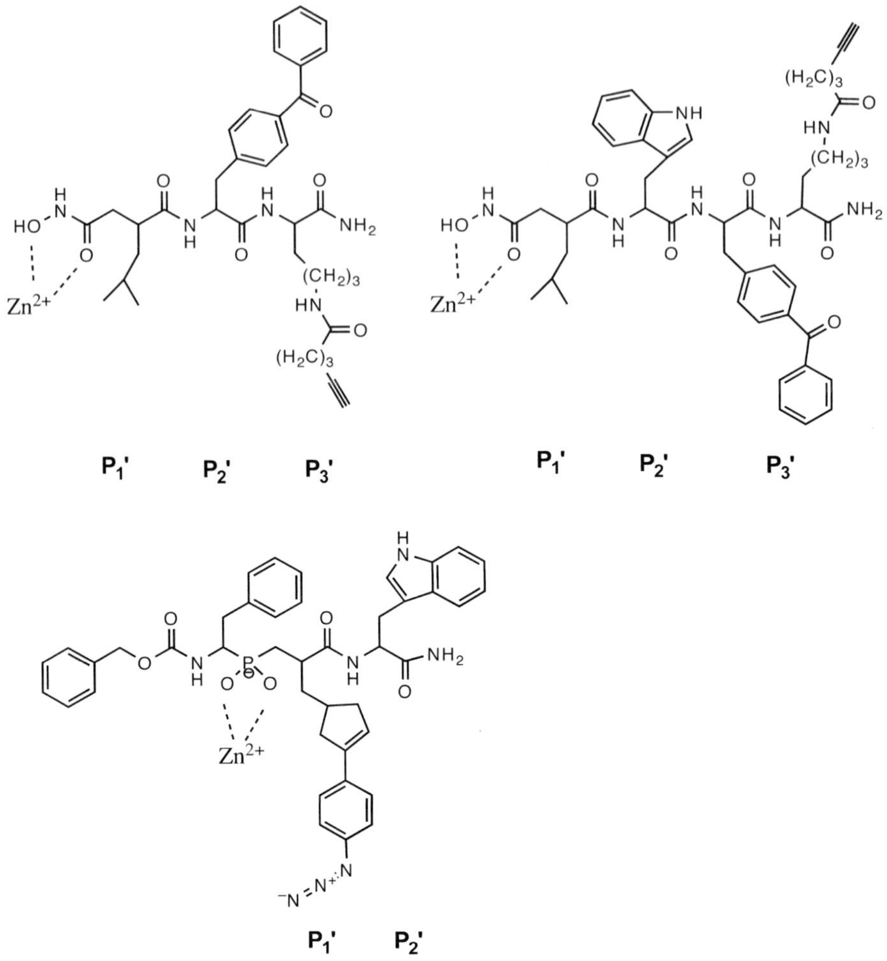

Fig. 7.3 Examples of activity-based probes (*ABPs*) developed to detect matrix metalloproteases (*MMPs*), using hydroxamate or phosphinic peptide templates. The hydroxamate *ABPs* incorporate an alkyne group in the C-terminal position making it possible to add, post-photoactivation, either a fluorescent or a biotin tag using click chemistry

lowest yield of covalent modification with this probe was observed for MMP-3 and MMP-8, suggesting that it may be difficult to detect these two MMPs with this probe. These results highlight one of the main drawbacks of using a photolabile group to cross-link ABPs to their target, namely, that even modest changes in the active sites may result in dramatic variation in cross-linking yields and thus a strong impact on the detection threshold for a given target protease.

Many ABPs developed to target the MMPs have made use of hydroxamate-containing peptides as templates (Chan et al. 2004, Saghatelian et al. 2004, Sieber et al. 2006). Given the avidity of the hydroxamate group for the zinc atom, hydroxamate inhibitors were observed to display only modest selectivity toward zinc metalloproteases (Brown et al. 2004, Cuniasse et al. 2005). Thus, ABPs using this pharmacophore bind many classes of metalloproteases. Ilomastat (GM6001), a highly potent hydroxamate-based MMP inhibitor, was used as a template to derive one of the first generation metalloprotease ABPs (Saghatelian et al. 2004). By incorporating a benzophenone photolabile group in the P_2' position of Ilomastat, a highly potent ABP was developed, allowing efficient labeling of recombinant MMP-2. However, treatment of tumor cells known to overexpress MMPs led to strong labeling of three zinc proteases (NEP, LAP, and DPPIII), but not of any of the MMP family proteases. Labeling of these targets was rationalized by showing that Ilomastat displayed nanomolar potency toward these zinc proteases. It is worth noting that these zinc proteases display low sequence homology with MMPs and possess very different active site topologies. Thus, the use of the relatively nonselective hydroxamate inhibitor scaffold may allow broad coverage of the zinc metalloprotease family; however, it may also be possible that the lack of probe selectivity toward MMPs is responsible for the inability to detect MMPs using this class of probes. Furthermore, the potentially low abundance of MMP active forms may require the generation of extremely selective ABPs to isolate and enrich the MMPs from the proteome to facilitate their detection and identification. The use of lower affinity zinc-binding elements, such as the phosphoryl group, can be exploited to develop phosphinic peptide inhibitors exhibiting high potency and restricted selectivity profile toward MMPs (Devel et al. 2006).

The difficulty in detecting active forms of MMPs can be explained by arguing that these proteases are mostly expressed as inactive zymogen forms and that most of the activated MMP fractions are blocked by endogenous inhibitors (i.e., the tissue inhibitors of metalloproteinases, TIMPs). Thus, active forms of MMPs may exist at levels that are well below the current detection limit of ABP technologies, even when using the exceptional resolution and sensitivity of multidimensional liquid chromatography coupled to mass spectrometry (MS). Analysis of MMP expression by cells or tumor tissues by gelatin zymography allows the detection of extremely low levels of activated forms of MMP-2 and MMP-9 in complex proteomes. In fact, for MMP-9, a threshold of detection of 100 attomoles has been reported (Masure et al. 1991). Given the sensitivity of this method, the detection of MMP-9 and MMP-2 active forms in cell supernatants and tumors tissues extracts is often reported. However, these data should be interpreted with great caution since the "active" forms of MMP-2 and MMP-9 detected at 82 and 62 kDa in gelatin zymography may account for both active forms and their TIMP complexes, since the denaturing conditions of electrophoresis dissociate the MMP–TIMP complexes. Using affinity capture approaches, which allow enrichment of only active forms of MMPs, it has been reported that the conditioned media of HT1080 fibrosarcoma cells, pretreated with concanavalin A to activate pro-MMP-2, mostly contains TIMP-2/MMP-2 complexes and only trace amounts of active MMP-2 (Hesek et al. 2006). Here again, trace amounts of active MMP-2, as detected by gelatin zymography, may represent only a few femtomoles of protein, a quantity poorly

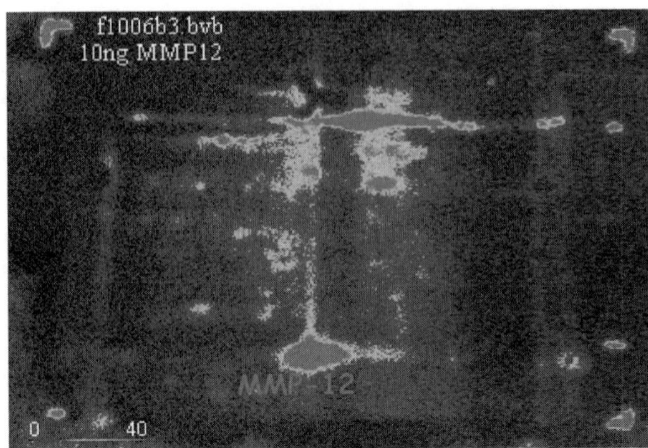

Fig. 7.4 Labeling of mice tumor extract (colon carcinoma) by the phosphinic probe displayed in scheme 2. Before photoactivation, 0.5 pmol of h-matrix metalloprotease (*MMP*)-12 were added to the sample as an internal standard for protein quantification. The sample was analyzed by 2D sodium dodecyl sulfate–polyacrylamide gel electrophoresis (*SDS–PAGE*) followed by detection. (*See also* Color Insert I)

detected even by MS. In confirmation of this finding, labeling of murine tumor extracts with a phosphinic radioactive probe led to similar conclusions. In order to quantify the amounts of MMP active forms labeled by the probe in tumor extracts, a known quantity of the catalytic domain of human MMP-12 was added to the sample, before photoactivation of the probe. The catalytic domain of h-MMP-12 was chosen due to its unique molecular weight and isoelectric point that allow it to be differentiated from murine-derived MMPs. Furthermore, sensitive detection of h-MMP-12 was achieved with the phosphinic probe; as low as 2.5 fmol of h-MMP-12 can be detected (David et al. 2007). As shown in Fig. 7.4 (unpublished results), labeling of exogenous MMP-12 (0.5 pmol) can be observed, confirming that the probe is able to find its targets, even in complex proteomes. Other labeled proteins may represent either specific labeling of endogenous MMP active forms or nonspecific labeling of proteins present in high abundance in the tumor extract. Competitive experiments with excess of MMP inhibitors suggest that several proteins detected in this experiment represent MMP active forms. Based on radioactivity counting, it can be estimated that, as a whole, ~0.5 pmol of endogenous proteins are labeled by the ABP in this sample, with the most intense spots representing about 50 fmol of MMP active forms, a level preventing direct identification of the labeled proteins by 2DE-MS. These experiments again suggest that MMP active forms are present in extremely low amounts, a result in agreement with gelatin zymography data and which may explain the failure to detect MMP active forms in previous reports. These results support the notion that MMPs are mostly present in their zymogen form and in complex with TIMPs, a situation that could be very specific to the MMP family, as compared to other classes of zinc metalloproteases. In this respect, it is worth noting that many zinc metalloproteases identified by the ABP profiling approach are expressed directly as active forms, for which no natural inhibitors have been reported. Included in this family are the three metalloproteases (neprilysin, dipeptidylpeptidase, and leucine aminopeptidase) observed to be covalently labeled by an ABP developed to target MMPs. Using a more sensitive approach to detect labeled proteins, based on a liquid chromatography–MS platform and a series of new ABPs, Sieber et al. (2006) identified several zinc metalloproteases whose activity is not regulated by zymogen activation and natural inhibitors. In a recent report, using sepharose resin functionalized with a hydroxamate-based MMP inhibitor, detection of MMP active forms in tumor extracts was achieved (Hesek et al. 2006). However, the quantity of total protein loaded on this affinity column was not specified, thus it is not possible to estimate the percentage of MMP active forms relative to total protein levels present in these tumor extracts.

While a number of valuable new metalloprotease probes have been reported that are highly useful reagents for monitoring the activities of a number of metalloprotease families (*see* Table 7.1), optimal covalent labeling of active MMPs remains challenging. As discussed above, the yield of cross-linking by photoaffinity probes is likely to be controlled by several factors, which are not easy to explicitly take into consideration when designing a probe. Thus, the development of ABPs that are able to label all MMPs with high efficiency will require more systematic studies evaluating the influence of the photolabile group, as well as its site of incorporation on the probe scaffold. The specificity of the inhibitors selected for developing ABP probes is likely to be critical to allow successful detection of MMP active forms when these forms are present at a much lower abundance compared to other related zinc proteases. Thus, fine-tuning the yield of covalent modification and selectivity of the ABP will require dedicated efforts. Additional factors, such as tissue extraction procedures, stability of the MMP active forms in extraction buffers, storage of the sample before processing, and analysis conditions, may also be critical for detection of proteins expressed in low amounts. Thus, optimization of all these factors may prove to be critical to fill the gap in MMP detection. Finally, since the photoactivation step of the metalloprotease ABPs limits their use to ex vivo experiments, future developments in reactive moieties to incorporate in ABPs will be necessary to generate ABPs that can be used to profile the zinc metalloproteases in vivo.

Substrate-Based Imaging Agents for Metalloprotease Activity

A NIR FRET substrate-based probe containing the peptide sequence GPLGVRGK was developed and used to detect MMP-2 activity in HT1080 human fibrosarcoma xenografts, giving a fluorescence response that could be inhibited by treatment with a synthetic MMP inhibitor (Bremer et al. 2001). In more recent studies, these kinds of polymer-based protease substrates have been used to assess proteases activities in murine arthritis (Izmailova et al. 2007) and in cardiovascular disease (Jaffer et al. 2007). Interestingly, a peptide-based NIRF probe, quenched by heterotransfer to a NIR absorber and designed to detect MMP-7 activity (Pham et al. 2004), appears to provide detection of tumor-associated MMP activity without the use of a polymer delivery vehicle (Wellington Pham, personal communication). For optical imaging, Achilefu and colleagues have prepared a number of NIR optical contrast agents designed to either bind to or be metabolized by tumors and, together with Britton Chance, have demonstrated the feasibility of detecting 2 cm-deep subsurface tumors using a metabolism-enhanced NIR fluorescent contrast agent and NIRF in vivo imaging (Achilefu et al. 2002, Chen et al. 2003, Achilefu 2004). The Tsien group have described a new strategy to use activatable cell-penetrating peptides (ACPPs), consisting of a polyarginine membrane-translocating motif linked via an MMP-cleavable peptide (PLG*LAG) to an appropriate masking polyanionic domain (a cleavable peptide hairpin), to deliver fluorescent labels to within tumor cells both in vitro and in vivo after cleavage by tumor-associated proteases (Jiang et al. 2004). Such ACPPs offer a general strategy toward both imaging and delivery of therapeutics in a variety of diseases in which extracellular proteases have been implicated.

McIntyre and Matrisian have generated dendrimer-based optical proteolytic beacons (PBs) for MMP-7 detection (McIntyre and Matrisian 2003, McIntyre et al. 2004) (Fig. 7.5). Their PBs are built on a dendrimeric polymer core, such as Generations 4 Starburst® Polyamidoamine (PAMAM) (nominal MW, 14,215). The prototype visible-range PB for MMP-7, PBvisM7, consisted of dendrimer coupled to substrate peptide previously labeled with fluorescein (FL) linked to its N-terminus (McIntyre et al. 2004). The peptide sequence is selectively cleaved by MMP-7 as compared to other MMPs (Welch et al. 1995). A caproyl linker (Ahx) is included adjacent to the N-terminal FL so as to diminish the solubility of the FL-RPLA peptide produced by proteolysis. The FL-labeled cleavable peptide serves as the

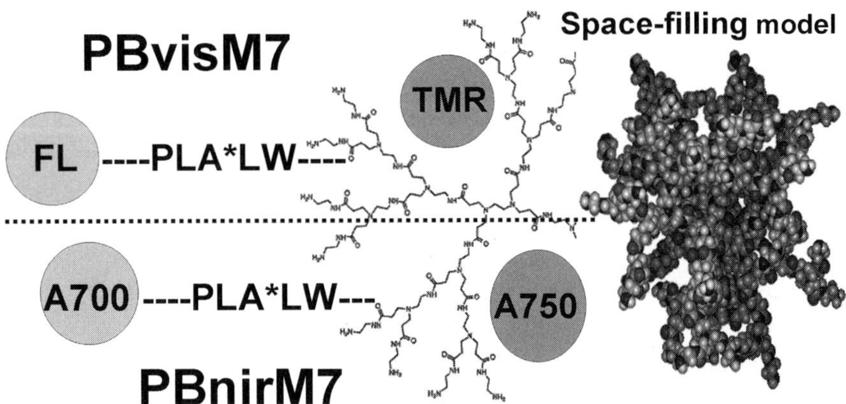

Fig. 7.5 Schematic structure of PBvis and PBnir. In PBvisM7, fluorescein (*FL*), linked at the N-terminus of the matrix metalloprotease (*MMP*)-selective cleavable peptide (cleavage site denotated by *), serves as the optical sensor. The internal reference, tetramethylrhodamine (TMR), is linked directly to the Starburst (PAMAM) dendrimer (generation 2 shown at the left, generation 4 in the space-filling model). For PBnirM7, AlexaFluor700 serves as the sensor and AlexaFluor750 as the reference

optical sensor and the PBs also include tetramethylrhodamine (TMR) linked direct-ly to the dendrimer scaffold; the TMR serves not only to quench the fluorescence of the FL on the protease sensor, but also as an internal reference that is used to track both substrate and product. Treatment of PBvisM7 with MMP-7 results in a significant enhancement in the FL fluorescence with a minimal change in the fluorescence of TMR, with the maximal MMP-7 enhancement in green fluorescent signal ~17-fold (McIntyre et al. 2004). The more recently reported NIR version of the MMP-7-PB, PBnirM7, uses AlexaFluor700 or Cy5.5 as the sensor instead of FL and AlexaFluor750 instead of TMR for the reference (Scherer et al. 2008).

Applications of Activity-Based Probes

Once a small-molecule ABP has been designed and its targets identified by affinity purification, it is possible to use this probe in a number of diverse applications. In one of the most useful applications of an ABP, levels of active proteases can be monitored in a range of samples that differ in stages or types of disease pathology. This application for ABPs allows specific proteases to be identified that may serve as useful biomarkers of that disease. A number of elegant examples of the use of ABPs to identify proteases as cancer biomarkers are outlined below. Proteases that show altered levels during progression toward disease may also represent valid targets for drug development. ABPs also serve as potentially valuable tools to monitor inhibi-tion of target proteases by small molecule drug leads. ABPs allow therapeutic effects of a drug to be linked with inhibition of specific protease targets, thus helping to validate that target for further drug development efforts. Finally, since ABPs form direct covalent bonds with their targets, it is possible to directly visualize the localization of active proteases in whole cells and even in whole animals. The final application of ABPs covered in this chapter is their use for in vitro and in vivo imaging. Note that the applications of ABPs are also pertinent to the use of substrate-based imaging probes, that is, as noninvasive cancer biomarkers, tools for target modulation by small molecule protease inhibitors, and in cancer detection.

Using ABPs to Identify Protease Biomarkers in Cancer

One of the potentially most promising applications for ABPs is for the discovery of new biomarkers of disease. Since members of all of the major classes of proteases have been implicated in some stage of cancer progression, it is likely that proteases will be a rich source for new markers for disease diagnosis and prognosis. Several ABPs have been developed to target enzymes implicated in cancer progression and tumorigenesis, including metalloproteases, cysteine cathepsins, and esterases (Evans and Cravatt 2006, Schmidinger et al. 2006, Fonovic and Bogyo 2007). These ABPs have been used to profile human tumors and tumor cell lines and identify novel enzyme activities for the diagnosis and treatment of cancer (Table 7.2). In a typical experiment, normal and disease proteomes are labeled with an ABP and the proteins are separated and analyzed by gel electrophoresis (Fig. 7.6). Enzymes that differ in their activity levels can then be identified as potentially interesting new biomarkers.

In one example of an application of ABPs to cancer biomarker discovery, FP-rhodamine, an ABP that targets the serine hydrolase superfamily of enzymes, was used to profile the activities of these enzymes in a set of human breast and melanoma cancer cell lines (Jessani et al. 2002). This study confirmed that highly invasive cancer cells from several different tumor types upregulate a distinct set of serine hydrolase activities, including the protease urokinase and a novel integral membrane hydrolase, KIAA1363. Although urokinase was known to be involved in tumor progression, KIAA1363 had never been implicated in cancer and therefore represents a potentially important new cancer biomarker (Jessani et al. 2002). In a related study, a panel of primary human breast cancer tissues was probed with the biotin-labeled version of FP-rhodamine (Jessani et al. 2002). Probe-labeled proteins were enriched using avidin-conjugated beads, digested by trypsin, and subjected to semiquantitative MS analysis. A set of enzymes, including KIAA1363, with elevated activities in the most aggressive tumor tissues, was identified as potential breast cancer biomarkers. Recently, both FP-rhodamine and FP-biotin were used to identify enzymes that are involved in cancer cell intravasation, the process by which tumor cells enter into the vasculature (Madsen et al. 2006). The activity level of the serine protease urokinase-type plasminogen activator (uPA) was substantially elevated in the high intravasating (HT-high/diss) variants of the human fibrosarcoma cell line HT-1080. Inhibition of uPA activity significantly reduced the rate of intravasation and metastasis of HT-high/diss cells, suggesting that active uPA is a key determinant of these processes (Madsen et al. 2006).

Metalloproteases also play key roles in cancer progression events such as angiogenesis and metastasis (Deryugina and Quigley 2006). Several metalloprotease genes are overexpressed in metastatic cancers, and inhibitors of these enzymes reduce tumor angiogenesis in animal models of cancer (Egeblad and Werb 2002). A library of metalloprotease probes based on a peptide hydroxamate scaffold carrying a photocrosslinker was used to profile the activities of metalloproteases in both breast carcinoma and melanoma cell lines (Saghatelian et al. 2004, Sieber et al. 2006). Neprilysin, alanyl aminopeptidase, and ADAM10 activities were found to be elevated in invasive cells. Although neprilysin has historically been considered a negative regulator of tumorigenesis, its high activity in invasive melanoma cells suggests that this enzyme may contribute to cancer progression.

Histone deacetylases (HDACs) are enzymes that remove acetyl groups from lysine residues on histone tails and therefore are important regulators of gene expression. These enzymes have also been implicated in tumor growth and development (Minucci and Pelicci 2006). While not proteases, these enzymes carry out a hydrolysis reaction using catalytic residues that are similar to a metalloprotease. Thus, ABPs have been designed to target HDACs that are similar to ABPs that target metalloproteases. An HDAC-selective ABP has been designed based on a

Table 7.2 List of probes used for proteomic profiling of protease activity in cancer models

ABP	ABP structure	Proteome	Enzyme activity	References
FP-rhodamine		Human breast melanoma cell lines	Serine hydrolases	(Jessani et al. 2004)
		Human ER(+) and ER(−) breast tissue samples		(Jessani et al. 2002)
		HT-high/diss and HT-low/diss human fibrosarcoma cell lines		(Jessani et al. 2005)
		MDA-MB-231 breast cancer cells before and after passage in mice		(Madsen et al. 2006)
FP-biotin		Human ER(+) and ER(−) breast cancer tissue samples	Serine hydrolases	(Jessani et al. 2005)
		HT-high/diss and HT-low/diss human fibrosarcoma cell lines		(Madsen et al. 2006)
HxBP-Rh		Human melanoma cell lines	Metalloproteases	(Saghatelian et al. 2004)
		Human breast cancer cell lines		(Sieber et al. 2006)

(continued)

Table 7.2 (continued)

ABP	ABP structure	Proteome	Enzyme activity	References
SAHA-BPyne	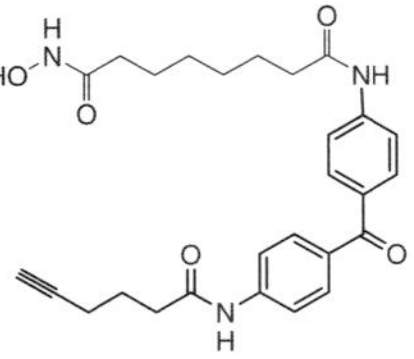	Human melanoma and breast cancer cell lines	Class I/II HDACs	(Salisbury and Cravatt 2007)
HAUb-VME		Cervical carcinoma biopsies tissue samples and cell lines Human tumor cell lines	USPs	(Ovaa et al. 2004) (Rolen et al. 2006)
DCG-04	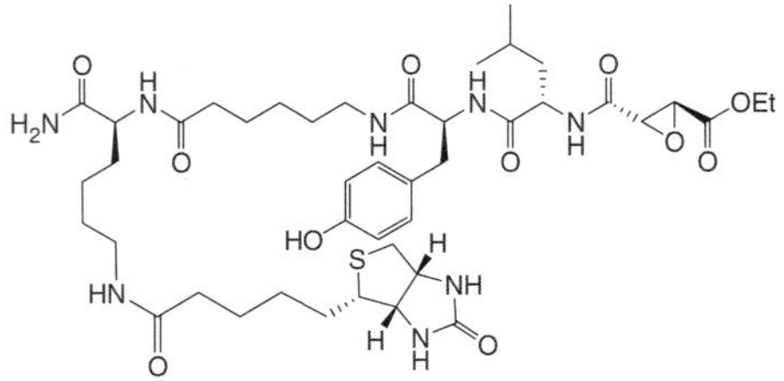	PyMT/ctsb$^{+/+}$, PyMT/ctsb$^{-/-}$, and PyMT/ctsb$^{+/-}$ murine tumor cells	Cathepsins	(Vasiljeva et al. 2006)

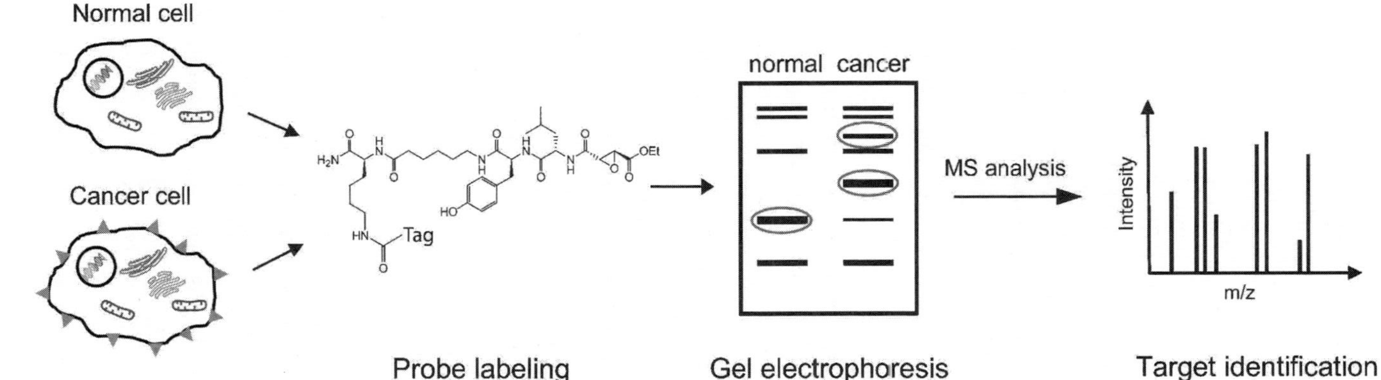

Fig. 7.6 Tumor biomarker discovery using activity-based probes (*ABPs*). Enzymes from cancer cells and normal cells are reacted with a biotin-containing *ABP* and are then separated and analyzed by gel electrophoresis. Probe-labeled enzymes are visualized, and enzymes with altered activities in normal and cancerous cells are identified. These potential tumor biomarkers can then be identified by mass spectrometry (*MS*) analysis

hydroxamic acid zinc-chelating moiety and a photoactivatible benzophenone group (Salisbury and Cravatt 2007). This probe was used to analyze HDAC activity in melanoma and ovarian cancer cell proteomes (Salisbury and Cravatt 2007). Differences in the composition and activity of HDACs were found among cancer cells indicating that the members of the HDAC enzyme family may have a variety of functional roles in cancer.

The ubiquitin-specific proteases (USPs) are a large family of proteolytic enzymes that regulate the production and recycling of ubiquitin and are involved in cell growth and differentiation (Rolen et al. 2006, Ovaa 2007). ABPs containing a reactive electrophile conjugated to the full-length ubiquitin protein have been shown to be highly selective probes of the USPs. A number of these probes were used to identify unique and tumor-specific activities in a variety of human tumor cell lines (Ovaa et al. 2004). One specific USP, UCH-L1, was highly active in numerous malignant tumor cell lines. UCH-L1 activity was also found to be upregulated in normal B cells after in vitro Epstein-Barr virus infection. This increase in activity correlated with a transition from slow to rapid proliferation of the cells, implicating UCH-L1 in this adaptation. USP-specific ABPs have also been used to profile USP activity in human cervical cancer biopsies (Rolen et al. 2006). The activities of two USPs, UCH-L3 and UCH-37, were elevated in tumor tissue when compared to normal tissue. Additionally, the activities of four USPs were upregulated in primary keratinocytes upon infection with human papilloma virus oncogenes, suggesting that the USPs are involved in growth transformation (Rolen et al. 2006).

ABPs have also been applied to functionally characterize enzyme activities in mouse models of cancer. The biotinylated ABP DCG-04 that targets the papain family of cysteine proteases was used to evaluate cysteine cathepsin activity in mammary tumor cells from PyMT;ctsb$^{-/-}$ mice, a mouse mammary cancer model deficient in cathepsin B (Vasiljeva et al. 2006). Although cathepsin B is the most active cysteine cathepsin on the surface of PyMT;ctsb$^{+/+}$ mammary cells, tumor cells lacking this protease (from PyMT;ctsb$^{-/-}$ mice) show an upregulation of active cathepsin X on their cell surfaces. Cathepsin X activity partially compensates for the deficiency of cathepsin B in these tumor cells. Data from these experiments suggest that proteases can dynamically compensate for each other, thus complicating the analysis of data from genetically deficient "knock-out" mice. In a similar study using the DCG-04 probe, the levels of multiple cysteine cathepsins were found to be highly upregulated in tumors that developed in the beta cells of the pancreas of the RIP1-Tag2 mouse, a mouse model of pancreatic cancer. Cathepsin expression was found to be linked to processes such as angiogenesis, and levels of active protease correlated with overall invasiveness of tumors (Joyce et al. 2004).

In an effort to more fully characterize the enzyme activity profiles of xenografted mouse tumors, ABPs such as FP-rhodamine have been used to characterize enzyme activities in MDA-MB-231 breast cancer cells both before and after growth as tumors in the mammary fat pad of immune-deficient mice (Jessani et al. 2004). Many serine hydrolase activities, such as uPA and tissue plasminogen activator (tPA), were highly elevated in the in vivo-derived lines of MDA-MB-231 and correlated with increased tumor growth rates and metastasis upon reintroduction into mice.

ABPs in Enzyme Inhibitor Discovery and Verification

In traditional drug discovery, libraries of small molecules are screened in vitro against purified, often recombinant, protein targets to identify inhibitors. However, in vitro assays provide only limited information regarding the in vivo potency and selectivity of an inhibitor for a related series of enzymes and provide no information about the selectivity pattern that will be observed once the compound is used in vivo. Since ABPs bind to the active sites of their enzyme targets, probes have been used to develop small molecule inhibitor screens that resolve many of the shortcomings that plague standard in vitro inhibitor assays (Greenbaum et al. 2002, Leung et al. 2003, Evans and Cravatt 2006, Fonovic and Bogyo 2007, Sadaghiani et al. 2007a). In an ABP-based screen, whole cells, cell lysates, or even whole organisms are treated with a range of concentrations of a potential inhibitor (Fig. 7.7). Total tissue or cell extracts are then reacted with an ABP and subjected to gel electrophoresis to separate the labeled enzymes. Small molecule inhibitor binding to a target is then measured as a decrease in enzyme labeling by the ABP. The resulting percent competition values can be measured by quantification of labeled proteins and used to generate IC_{50} values of the small molecule for each of the primary targets of the ABP. In contrast to standard inhibitor assays, ABP-based assays can be performed in complex proteome mixtures (including cells and whole organisms) containing multiple related enzymes, thus allowing for the evaluation of both potency and selectivity in a native cellular environment. These assays also eliminate the need for time-consuming expression and purification of drug targets and can be used to identify inhibitors for enzymes that lack known substrates. Finally, when used in vivo, ABP-based drug screens provide information regarding potency, selectivity, and biodistribution of an inhibitor in the context of a whole organism.

In one example of an ABP-based competition study, the potency and selectivity of a series of cysteine protease inhibitors was monitored in rat liver extracts (Greenbaum et al. 2002). This screen identified a small molecule that selectively targeted cathepsin B activity. Since cathepsin B is suspected of facilitating tumor invasion, this compound could potentially be used as a lead target for cancer therapy. ABP-based assays using the serine hydrolase probe FP-rhodamine have also been applied to the discovery of novel, selective inhibitors of KIAA1363, a poorly characterized enzyme with highly elevated activity in invasive cancer cells (Leung et al. 2003, Chiang et al. 2006). This screen yielded valuable lead compounds that facilitated further study of the function of KIAA1363 in the metabolism of lipids. Highly selective inhibitors of the caspases, cysteine proteases involved in apoptosis that are often dysregulated in cancer, have also been identified using an ABP-based competition assay (Berger et al. 2006).

In addition to the discovery of new inhibitors for cancer-related enzymes, ABPs have been applied to the characterization of existing drugs. Proteasome-directed ABPs have been used to evaluate the specificity of bortezomib, a clinically approved proteasome inhibitor for the treatment of multiple myeloma (Altun et al. 2005, Berkers et al. 2005). Myeloma cells were cultured in the presence or

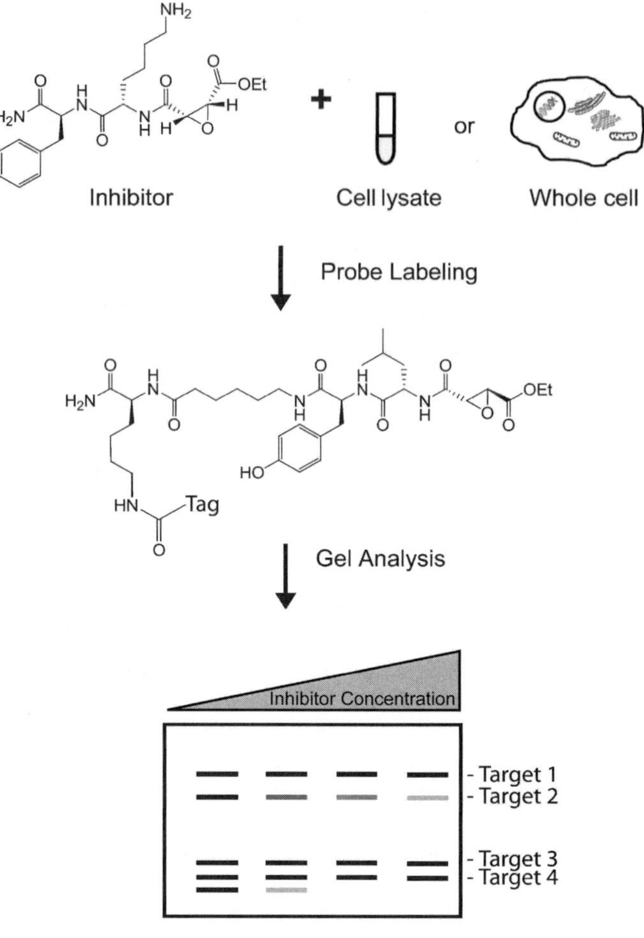

Fig. 7.7 Enzyme inhibitor discovery using an activity-based probe (*ABP*)-based assay. Cell lysates or whole cells are treated with a range of concentrations of an inhibitor. These samples are then reacted with an *ABP* and subjected to gel electrophoresis to separate active enzymes. A decrease in residual activity corresponds to more potent inhibition by the inhibitor. Additionally, the selectivity of the inhibitor for one or multiple enzymes can be determined using this assay

absence of bortezomib, incubated with a cell-permeable, proteasome-specific ABP, lysed, and then analyzed by gel electrophoresis. Results from these experiments revealed that only the activities of the β1/β1i and β5/β5i subunits of the proteasome were inhibited by bortezomib (Altun et al. 2005, Berkers et al. 2005).

ABPs have also been employed for the in vivo evaluation of inhibitors that target enzymes involved in cancer. Kraus and colleagues used a proteasome-directed ABP to identify the active human proteasomal subunits targeted by bortezomib in patients receiving this drug (Kraus et al. 2007). Blood cells obtained from a patient receiving bortezomib monotherapy for multiple myeloma were treated with a proteasome-specific ABP. Bortezomib treatment was shown to reversibly eliminate both $\beta 1$ and $\beta 5$ proteasomal activities and reduce $\beta 2$ proteasomal activity in human blood cells. Since proteasomal subunits in cancer cells are known to have variable activity, the preferences of bortezomib for certain subunits may explain the differences in patient sensitivity to this cancer drug (Kraus et al. 2007). The in vivo specificity and biodistribution of another proteasome inhibitor, MG262, was monitored in murine tissues using a fluorescently labeled proteasome-specific ABP (Verdoes et al. 2006).

In another example of an application of ABPs to monitor the in vivo potency and selectivity of small molecule inhibitors, the cathepsin-specific ABP DCG-04 was used to evaluate inhibition of cysteine cathepsins in the RIP1-TAG2 transgenic mice, a mouse model of pancreatic cancer (Joyce et al. 2004, Sadaghiani et al. 2007b). In these studies, inhibitors were injected into mice, and normal and tumor tissue samples were collected and analyzed for residual cathepsin activity. The inhibition of these proteases by the cathepsin-specific inhibitor JPM-OEt resulted in a reduction in invasion, angiogenesis, and tumor growth (Joyce et al. 2004). Importantly, fluorescently labeled DCG-04 enabled the biochemical identification and monitoring of the cysteine cathepsins during tumorigenesis in these mice. In a follow-up study, a panel of cathepsin inhibitors was evaluated in this same mouse model using radiolabeled DCG-04 (Sadaghiani et al. 2007b). Inhibitors that had been optimized for selectivity and potency against target proteases in crude tissue extracts were tested for overall potency, biodistribution, and selectivity in vivo. From these studies, a set of inhibitors was identified that showed optimal potency and selectivity in tumor tissues and can be used as lead compounds for cancer therapy. These studies demonstrate that ABPs can be a valuable tool for the analysis of drug specificity and pharmacodynamic properties in vivo.

The substrate-based proteolytic probes have also been used as pharmacodynamic markers to demonstrate efficacy of small molecule protease inhibitors. A NIR FRET substrate-based probe detected proteolytic activity in HT1080 human fibrosarcoma xenografts, giving a fluorescence response that could be inhibited by treatment with the synthetic MMP inhibitor prinomastat (Bremer et al. 2001). Using the dendrimer-based proteolytic beacon PBvisM7, treatment of tumor-bearing mice with the broad-spectrum MMP inhibitor BB-94 resulted in a marked reduction in sensor FL fluorescence to ~40% of that before treatment (McIntyre et al. 2004). Interestingly, this effect was observed only in MMP-7-transfected tumors, with no effect on the FL fluorescence detected over the control tumor, suggesting that the selectivity of the peptide sequence in substrate-based proteolytic probes may be a useful tool in assessing the inhibitory profile of small molecule protease inhibitors in vivo.

Imaging Protease Activity in Tumors

One of the major challenges in cancer diagnosis is the early detection of small primary tumors (Weissleder et al. 1999). Since many enzyme activities are upregulated in tumor cells, probes that report on enzymatic activity represent valuable tools for early diagnostic imaging strategies (Weissleder et al. 1999, Mahmood and Weissleder 2003, Sloane et al. 2006). In general, optical imaging techniques are used to image protease activity in vivo. The cost, space, and time involved in optical imaging are less demanding compared to other imaging modalities. Furthermore, the advantage of optical imaging methods include the use of nonionizing low-energy radiation, high sensitivity with the possibility of detecting micron-sized objects, and continuous data acquisition in real time and in an intact environment. Optical imaging in the NIR region between 700 and 900 nm has a low absorption by intrinsic photoactive biomolecules and allows light to penetrate several centimeters into the tissue, a depth that is sufficient to image practically all small animals (Zuzak et al. 2002). Imaging in the NIR region has less tissue autofluorescence, markedly improving the target/background ratio as compared with the visible region of the spectrum (Rudin and Weissleder 2003). The detection sensitivity depends both on selection of the fluoroscent probe and optimization of imaging geometry for detection with a highly sensitive charge-coupled device (CCD) camera. These kinds of optical imaging systems are capable of detecting a small number of photons that are transmitted through living tissues permitting real-time images to be collected within a few seconds. A fast and relatively easy imaging procedure makes this modality attractive for potential clinical use. Fluorescence-mediated tomography (FMT) has recently been shown to three-dimensionally localize and quantify fluorescent probes in deep tissues at high sensitivity (Ntzia-christos et al. 2002).

Current methods for imaging enzymes mainly rely on antibody labeling or on substrates that become fluorescent after enzyme cleavage (Baruch et al. 2004, Sloane et al. 2006). Although antibodies are specific for their enzyme targets, they are not cell permeable and do not give information about enzyme activity. Fluorescent substrates are useful for the activity-based imaging of proteases; however, these compounds often suffer from a lack of specificity, leading to cleavage by multiple classes of proteases (Baruch et al. 2004, Sloane et al. 2006). Furthermore, there is no way to determine which protease is responsible for sub-strate processing in vivo using fluorescent substrate reporters. In contrast, ABPs covalently bind to active enzymes, thus permitting assignment of imaging signals to specific enzymes (Fig. 7.8). In fact, a number of ABPs that target cysteine proteases have been used to image enzyme activity in tumor cells both in vitro and in vivo (Joyce et al. 2004, Blum et al. 2005, 2007).

A fluorescently tagged DCG-04 analogue has been used to image cysteine cathepsin activity during tumorigenesis in RIP1-TAG2 transgenic mice (Joyce et al. 2004). In this study, the ABP was administered systemically by intravenous injection into a mouse. After allowing the probe to circulate for several hours, pancreatic tumor tissue was collected and imaged using fluorescent microscopy. Cathepsin activity was found to be elevated in tumors and at the invasive edges of islet carcinomas. After imaging, tumor tissues were lysed and analyzed by gel electrophoresis, providing an activity profile that could be used to identify and quantify the levels of probe-modified cathepsins that produced the fluorescent signals. Additionally, this ABP was applied to the imaging of cathepsin activity

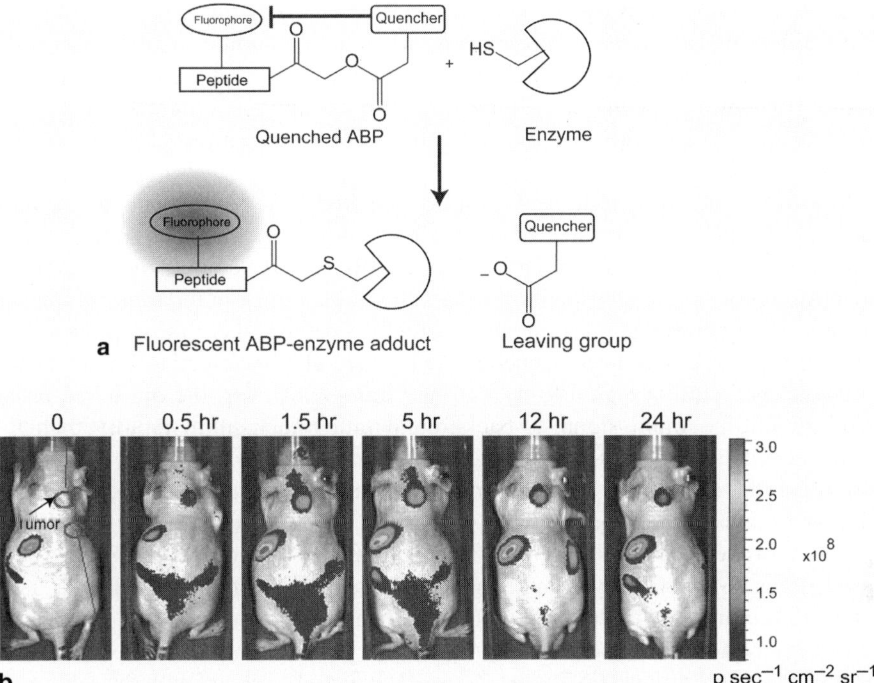

Fig. 7.8 Quenched activity-based probes (*ABPs*) for the noninvasive imaging of tumors in vivo. (**a**) Covalent labeling of a cysteine protease target by a quenched *ABP*. Activity-based labeling of the target enzyme results in the loss of the quenching group and subsequent generation of a fluorescently-labeled enzyme. (**b**) Optical imaging of MDA-MB-231 breast cancer xenograft tumors in nude mice using a quenched cysteine cathepsin-specific *ABP*. The quenched probe was injected intravenously, and fluorescent images of the mice were taken at various time points after injection. Images taken from Blum et al. (2007). (*See also* Color Insert I)

in a mouse model of cervical carcinogenesis (K14-HPV/E$_2$ mice) (Joyce et al. 2004). Cathepsin activity levels were also elevated in cervical tumor tissues, further confirming that cysteine cathepsin activity can serve as a useful cancer biomarker and that cathepsin-specific ABPs have potential value as imaging agents to monitor tumor progression in whole animals.

In a recent advance, a cathepsin-specific ABP that becomes fluorescent only upon binding to its enzyme target has been developed (Blum et al. 2005). Since tagged ABPs used in imaging are constitutively fluorescent, they generate a high nonspecific fluorescent background when used in living cells. The newly designed quenched ABP (qABP) makes use of the acyloxy leaving group found on the acyloxymethyl ketone warhead. By attaching a fluorescent quencher, the probe is rendered nonfluorescent when free in solution. Covalent modification of cysteine cathepsins by this probe liberates the quencher moiety, and the probe becomes fluorescent. This cell-permeable, quenched probe has been used to image cathepsin activity levels in both the murine fibroblast cell line NIH-3T3 and the human MCF-10A breast cancer cell line

(Blum et al. 2005). Cells treated with the qABP showed distinct punctuate fluorescent staining of lysosomal compartments, whereas the unquenched control probe produced bright, nonspecific intracellular fluorescence that required extensive washing to reveal specific target labeling. Importantly, the fluorescent signals could be specifically blocked by pretreatment of cells with a general cysteine protease inhibitor.

In a follow-up study, Blum et al. used a related series of quenched and nonquenched ABPs to noninvasively image cathepsin activity in a xenografted mouse model of breast cancer (Blum et al. 2007). NIR-labeled versions of the cysteine cathepsin probes produced spatially resolvable fluorescence in the tumor tissues of live mice that correlated with the levels of active cathepsins in those tissues (Fig. 7.8). Both quenched and nonquenched ABPs were able to selectively label tumor tissue and had similar signal-to-background ratios; however, the quenched probe achieved its maximum signal-to-background ratio much more rapidly than the nonquenched probe. Ex vivo analysis of tumor tissues from these mice further confirmed that the signals observed in the live animals were due to specific probe labeling of active cathepsins.

The substrate-based PBs selective for MMP-7, PBvisM7, and PBnirM7, have been used to detect MMP-7 activity in xenograft tumors in mice. In these studies, pairs of xenograft tumors were established on the rear flanks of each animal; one tumor with human colorectal tumor cells that express several MMP family members but do not express detectable amounts of endogenous MMP-7, and a second with the same cells transfected with an MMP-7 expression vector (Witty et al. 1994). Imaging was achieved following intravenous injection of a single bolus of either PBvisM7 or PBnirM7. Approximately 2–4 h following PB injection, the reference (R) signal was low in both the control and MMP-7-transfected tumors while the sensor (S) channel showed an approximately tenfold difference between the control and MMP-7-expressing tumors with a comparable difference in sensor/reference (S/R) ratio (Scherer et al. 2008). The second generation PBnirM7 has been used to detect intestinal adenomas in the multiple intestinal neoplasia (Min) mouse model of familial polyposis (Scherer et al. 2008). In those studies, the animals were sacrificed post intravenous administration of PBnirM7 revealing enhanced fluorescence of the MMP-7 sensor associated with a number of adenomas in the intestinal tract examined ex vivo (Fig. 7.9). The S/R ratio was consistently and significantly higher in Min adenomas compared to normal intestinal tissue of mice lacking the Min mutation, and in Min adenomas from MMP-7-null mice. Similar studies by the Weissleder group also demonstrated the detection of adenoma-associated protease activity in the intestines of Min mice (Marten et al. 2002). Taken together, these fluorescence imaging studies in living mice indicate that PB-M7s can be used to detect and selectively image MMP-7 activity in vivo due to the enhanced fluorescence of the sensor in the proteolyzed reagent that results in an increase in S/R ratio. The optical imaging approach using new optical reporters has the potential for highly sensitive, noninvasive, in vivo detection and imaging of tumor-associated proteolytic activity.

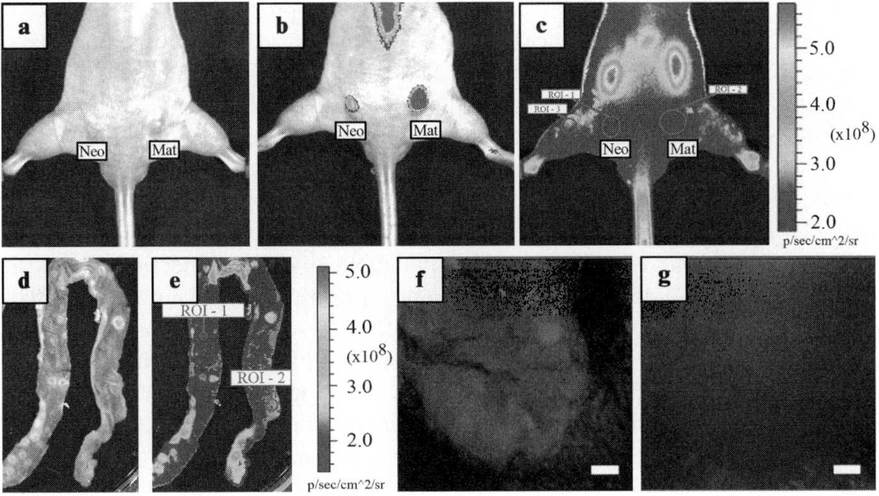

Fig. 7.9 In vivo imaging with substrate-based proteolytic beacon PB-M7NIR. (**a–c**) In vivo imaging of mouse subcutaneous xenograft tumors with PB-M7NIR. Dorsal, caudal view of a nude mouse of 4 weeks following subcutaneous injection of SW480neo (Neo) or MMP-7-expressing SW480mat (Mat) cells. Tumor areas (~57 mm^2 each) are shown in white light (**a**), the Cy5.5 sensor channel (**b**), and the reference channels (**c**) 4 h post retro-orbital intravenous injection of 1.0 nmol PB-M7NIR. Encircled areas represent regions of interest for quantitative assessment. Note accumulation of PB-M7NIR in the kidneys of the mouse as detected with the reference channel (**c**), but selective accumulation of sensor signal in the Mat tumor indicative of proteolytic activity (**b**). Sensor signal on the spine and tail are presumed to be due to low levels of circulating, activated probe that become detectable when they are close to the surface of the mouse. (**d–g**) Ex vivo imaging of PB-M7NIR in APC^{MIN} intestinal adenomas. Explanted mouse intestine from an APC^{MIN} mouse with spontaneous polyps in 60 min. Post-injection of 1 nmol of PB-M7NIR Beacon. (**d**) White light image and (**e**) NIRF image in the Cy5.5 (sensor) channel. (**e–f**). High power images of a single, intact adenoma (10×-objective) from an APC^{Min} mouse (**e**). False-red coloring in the Cy5.5 (sensor) channel (**f**) Cy7 (reference-green) channel. White line = 100 microns. Adapted fr om Scherer et al. (2008). *ROI*, region of interest. (*See also* Color Insert I)

Conclusion and Future Directions

Over the last several years, the field of activity-based proteomics has produced a wealth of new technologies for the direct biological study of enzymes. Protease probes that monitor the activity of numerous diverse enzyme classes have been synthesized, and these probes have been applied to many biologically and pathologically relevant fields. Additionally, a number of new tools, including gel-free screening systems and quenched probes, have been developed that allow rapid identification and visualization of enzyme activity in vitro and in vivo. Both ABPs and substrate-based imaging probes have been applied to the identification and evaluation of potential enzyme inhibitors in the physiologically relevant environments of a complex proteome, cell, or even whole animal. However, challenges in the field of activity-based proteomics still remain to be addressed. In order to identify new probe scaffolds that allow for greater proteomic coverage by ABPs, structurally diverse probe libraries need to be developed. Furthermore, advances in gel-free analysis systems will be required to profile proteins with low activities or abundances and to rapidly identify large numbers of proteins targeted by ABPs. Perhaps the most important challenge facing activity-based proteomics is the need to combine the data from activity-based assays with relevant biological experiments to gain a more complete understanding of enzyme function in cancer and other biological processes and diseases. The continued use of optical imaging of proteolytic activity has exciting potential both for the understanding of cancer and in applications to cancer detection, diagnosis, and treatment. For preclinical studies, extension into the use of multiphoton fluorescence microscopy for intravital imaging of protease activities should facilitate the further delineation of specific roles of proteases in processes critical to tumor progression. New developments in NIR optical tomography research (Nioka and Chance 2005, Ntziachristos et al. 2005) are yielding promising optical approaches for imaging in clinical practice, particularly as a complementary modality for breast cancer detection (Chance et al. 2005, Zhu et al. 2005). The development of new kinds of targeted optical reagents, including those providing for both imaging and therapy (Chen et al. 2005, Zheng et al. 2007), will likely provide new paradigms for the clinician. Noninvasive imaging techniques for proteolytic activity provide an extraordinary opportunity to increase the sensitivity of detecting early-stage tumors and to identify tumors that require particularly aggressive therapy. With time and the rapid advance in technology, we are likely to see a sharp increase in the number and types of applications of protease probes to oncology.

Acknowledgments We thank the members of the Bogyo lab for helpful discussions and manuscript comments. This work was supported by funding from a NIH National Technology Center for Networks and Pathways (NTCNP) grant U54 RR020843 (MB) and R01 CA084360 (LMM).

References

Achilefu S. (2004). Lighting up tumors with receptor-specific optical molecular probes. Technol Cancer Res Treat 3(4): 393–409.

Achilefu S., Jimenez H. N., Dorshow R. B., et al. (2002). Synthesis, in vitro receptor binding, and in vivo evaluation of fluorescein and carbocyanine peptide-based optical contrast agents. J Med Chem 45(10): 2003–2015.

Altun M., Galardy P. J., Shringarpure R., et al. (2005). Effects of PS-341 on the activity and composition of proteasomes in multiple myeloma cells. Cancer Res 65(17): 7896–7901.

Baruch A., Jeffery D. A., and Bogyo M. (2004). Enzyme activity—It's all about image. Trends Cell Biol 14(1): 29–35.

Berger A. B., Witte M. D., Denault J. B., et al. (2006). Identification of early intermediates of caspase activation using selective inhibitors and activity-based probes. Mol Cell 23(4): 509–521.

Berkers C. R., Verdoes M., Lichtman E., et al. (2005). Activity probe for in vivo profiling of the specificity of proteasome inhibitor bortezomib. Nat Methods 2(5): 357–362.

Blum G., Mullins S. R., Keren K., et al. (2005). Dynamic imaging of protease activity with fluorescently quenched activity-based probes. Nat Chem Biol 1(4): 203–209.

Blum G., von Degenfeld G., Merchant M. J., et al. (2007). Noninvasive optical imaging of cysteine protease activity using fluorescently quenched activity-based probes. Nat Chem Biol 3(10): 668–677.

Bogyo M., McMaster J. S., Gaczynska M., et al. (1997). Covalent modification of the active site threonine of proteasomal beta subunits and the Escherichia coli homolog HslV by a new class of inhibitors. Proc Natl Acad Sci U S A 94(13): 6629–6634.

Bogyo M., Verhelst S., Bellingard-Dubouchaud V., et al. (2000). Selective targeting of lysosomal cysteine proteases with radiolabeled electrophilic substrate analogs. Chem Biol 7(1): 27–38.

Borodovsky A., Ovaa H., Kolli N., et al. (2002). Chemistry-based functional proteomics reveals novel members of the deubiquitinating enzyme family. Chem Biol 9(10): 1149–1159.

Bremer C., Tung C. H., Bogdanov A., Jr., et al. (2002). Imaging of differential protease expression in breast cancers for detection of aggressive tumor phenotypes. Radiology 222(3): 814–818.

Bremer C., Tung C. H., and Weissleder R. (2001). In vivo molecular target assessment of matrix metalloproteinase inhibition. Nat Med 7(6): 743–748.

Brown S., Meroueh S. O., Fridman R., et al. (2004). Quest for selectivity in inhibition of matrix metalloproteinases. Curr Top Med Chem 4(12): 1227–1238.

Chan E. W., Chattopadhaya S., Panicker R. C., et al. (2004). Developing photoactive affinity probes for proteomic profiling: Hydroxamate-based probes for metalloproteases. J Am Chem Soc 126(44): 14435–14446.

Chance B., Nioka S., Zhang J., et al. (2005). Breast cancer detection based on incremental biochemical and physiological properties of breast cancers: A six-year, two-site study. Acad Radiol 12(8): 925–933.

Chen Y., Gryshuk A., Achilefu S., et al. (2005). A novel approach to a bifunctional photosensitizer for tumor imaging and phototherapy. Bioconjug Chem 16(5): 1264–1274.

Chen Y., Zheng G., Zhang Z. H., et al. (2003). Metabolism-enhanced tumor localization by fluorescence imaging: In vivo animal studies. Opt Lett 28(21): 2070–2072.

Chiang K. P., Niessen S., Saghatelian A., et al. (2006). An enzyme that regulates ether lipid signaling pathways in cancer annotated by multidimensional profiling. Chem Biol 13(10): 1041–1050.

Cuniasse P., Devel L., Makaritis A., et al. (2005). Future challenges facing the development of specific active-site-directed synthetic inhibitors of MMPs. Biochimie 87(3–4): 393–402.

David A., Steer D., Bregant S., et al. (2007). Cross-linking yield variation of a potent matrix metalloproteinase photoaffinity probe and consequences for functional proteomics. Angew Chem Int Ed Engl 46(18): 3275–3277.

Deryugina E. I., and Quigley J. P. (2006). Matrix metalloproteinases and tumor metastasis. Cancer Metastasis Rev 25(1): 9–34.

Devel L., Rogakos V., David A., et al. (2006). Development of selective inhibitors and substrate of matrix metalloproteinase-12. J Biol Chem 281(16): 11152–11160.

Dorman G., and Prestwich G. D. (2000). Using photolabile ligands in drug discovery and development. Trends Biotechnol 18(2): 64–77.

Egeblad M., and Werb Z. (2002). New functions for the matrix metalloproteinases in cancer progression. Nat Rev Cancer 2(3): 161–174.

Erijman L., and Weber G. (1993). Use of sensitized fluorescence for the study of the exchange of subunits in protein aggregates. Photochem Photobiol 57(3): 411–415.

Evans M. J., and Cravatt B. F. (2006). Mechanism-based profiling of enzyme families. Chem Rev 106(8): 3279–3301.

Fleming S. A. (1995). Chemical reagents in photoaffinity labeling. Tetrahedron 51: 12479–12520.

Fonovic M., and Bogyo M. (2007). Activity based probes for proteases: Applications to biomarker discovery, molecular imaging and drug screening. Curr Pharm Des 13(3): 253–261.

Förster T. (1948). Intermolecular energy migration and fluorescence. Ann. Physik (Leipzig) 2: 55–75.

Ganesan S., Ameer-Beg S. M., Ng T. T., et al. (2006). A dark yellow fluorescent protein (YFP)-based resonance energy-accepting chromoprotein (REACh) for Forster resonance energy transfer with GFP. Proc Natl Acad Sci U S A 103(11): 4089–4094.

Ghukasyan V., Hsu Y. Y., Kung S. H., et al. (2007). Application of fluorescence resonance energy transfer resolved by fluorescence lifetime imaging microscopy for the detection of enterovirus 71 infection in cells. J Biomed Opt 12(2): 024016.

Greenbaum D., Baruch A., Hayrapetian L., et al. (2002). Chemical approaches for functionally probing the proteome. Mol Cell Proteomics 1(1): 60–68.

Greenbaum D., Medzihradszky K. F., Burlingame A., et al. (2000). Epoxide electrophiles as activity-dependent cysteine protease profiling and discovery tools. Chem Biol 7(8): 569–581.

Haugland R. P., Yguerabide J., and Stryer L. (1969). Dependence of the kinetics of singlet-singlet energy transfer on spectral overlap. Proc Natl Acad Sci U S A 63(1): 23–30.

Heim R., and Tsien R. Y. (1996). Engineering green fluorescent protein for improved brightness, longer wavelengths and fluorescence resonance energy transfer. Curr Biol 6(2): 178–182.

Hesek D., Toth M., Meroueh S. O., et al. (2006). Design and characterization of a metalloproteinase inhibitor-tethered resin for the detection of active MMPs in biological samples. Chem Biol 13(4): 379–386.

Izmailova E. S., Paz N., Alencar H., et al. (2007). Use of molecular imaging to quantify response to IKK-2 inhibitor treatment in murine arthritis. Arthritis Rheum 56(1): 117–128.

Jaffer F. A., Libby P., and Weissleder R. (2007). Molecular imaging of cardiovascular disease. Circulation 116(9): 1052–1061.

Jeffery D. A., and Bogyo M. (2003). Chemical proteomics and its application to drug discovery. Curr Opin Biotechnol 14(1): 87–95.

Jessani N., Humphrey M., McDonald W. H., et al. (2004). Carcinoma and stromal enzyme activity profiles associated with breast tumor growth in vivo. Proc Natl Acad Sci U S A 101(38): 13756–13761.

Jessani N., Liu Y., Humphrey M., et al. (2002). Enzyme activity profiles of the secreted and membrane proteome that depict cancer cell invasiveness. Proc Natl Acad Sci U S A 99(16): 10335–10340.

Jessani N., Niessen S., Wei B. Q., et al. (2005). A streamlined platform for high-content functional proteomics of primary human specimens. Nat Methods 2(9): 691–697.

Jiang T., Olson E. S., Nguyen Q. T., et al. (2004). Tumor imaging by means of proteolytic activation of cell-penetrating peptides. Proc Natl Acad Sci U S A 101(51): 17867–17872.

Joyce J. A., Baruch A., Chehade K., et al. (2004). Cathepsin cysteine proteases are effectors of invasive growth and angiogenesis during multistage tumorigenesis. Cancer Cell 5(5): 443–453.

Kato D., Boatright K. M., Berger A. B., et al. (2005). Activity-based probes that target diverse cysteine protease families. Nat Chem Biol 1(1): 33–38.

Kessler B. M., Tortorella D., Altun M., et al. (2001). Extended peptide-based inhibitors efficiently target the proteasome and reveal overlapping specificities of the catalytic beta-subunits. Chem Biol 8(9): 913–929.

Kidd D., Liu Y., and Cravatt B. F. (2001). Profiling serine hydrolase activities in complex proteomes. Biochemistry 40(13): 4005–4015.

Kokame K., Nobe Y., Kokubo Y., et al. (2005). FRETS-VWF73, a first fluorogenic substrate for ADAMTS13 assay. Br J Haematol 129(1): 93–100.

Kraus M., Ruckrich T., Reich M., et al. (2007). Activity patterns of proteasome subunits reflect bortezomib sensitivity of hematologic malignancies and are variable in primary human leukemia cells. Leukemia 21(1): 84–92.

Latt S. A., Auld D. S., and Vallee B. L. (1972). Fluorescende determination of carboxypeptidase A activity based on electronic energy transfer. Anal Biochem 50(1): 56–62.

Lauer-Fields J., Brew K., Whitehead J. K. et al. (2007a). Triple-helical transition state analogues: A new class of selective matrix metalloproteinase inhibitors. J Am Chem Soc 129(34): 10408–10417.

Lauer-Fields J. L., Minond D., Sritharan T., et al. (2007b). Substrate conformation modulates aggrecanase (ADAMTS-4) affinity and sequence specificity. Suggestion of a common topological specificity for functionally diverse proteases. J Biol Chem 282(1): 142–150.

Lauer-Fields J. L., Sritharan T., Stack M. S., et al. (2003). Selective hydrolysis of triple-helical substrates by matrix metalloproteinase-2 and -9. J Biol Chem 278(20): 18140–18145.

Leung D., Hardouin C., Boger D. L., et al. (2003). Discovering potent and selective reversible inhibitors of enzymes in complex proteomes. Nat Biotechnol 21(6): 687–691.

Liu Y., Patricelli M. P., and Cravatt B. F. (1999). Activity-based protein profiling: The serine hydrolases. Proc Natl Acad Sci U S A 96(26): 14694–14699.

Lopez-Otin C., and Overall C. M. (2002). Protease degradomics: A new challenge for proteomics. Nat Rev Mol Cell Biol 3(7): 509–519.

Madsen M. A., Deryugina E. I., Niessen S., et al. (2006). Activity-based protein profiling implicates urokinase activation as a key step in human fibrosarcoma intravasation. J Biol Chem 281(23): 15997–16005.

Mahmood U., and Weissleder R. (2003). Near-infrared optical imaging of proteases in cancer. Mol Cancer Ther 2(5): 489–496.

Mahrus S., and Craik C. S. (2005). Selective chemical functional probes of granzymes A and B reveal granzyme B is a major effector of natural killer cell-mediated lysis of target cells. Chem Biol 12(5): 567–577.

Marten K., Bremer C., Khazaie K., et al. (2002). Detection of dysplastic intestinal adenomas using enzyme-sensing molecular beacons in mice. Gastroenterology 122(2): 406–414.

Masure S., Proost P., Van Damme J., et al. (1991). Purification and identification of 91-kDa neutrophil gelatinase. Release by the activating peptide interleukin-8. Eur J Biochem 198: 391–398.

Matayoshi E. D., Wang G. T., and Krafft G. A. (1990). Novel fluorogenic substrates for assaying retroviral proteases by resonance energy transfer. Science 247(4945): 954–958.

McIntyre J. O., Fingleton B., Wells K. S., et al. (2004). Development of a novel fluorogenic proteolytic beacon for in vivo detection and imaging of tumour-associated matrix metalloproteinase-7 activity. Biochemical J 377(Pt. 3): 617–628.

McIntyre J. O., and Matrisian L. M. (2003). Molecular imaging of proteolytic activity in cancer. J Cell Biochem 90(6): 1087–1097.

Melancon M. P., Wang W., Wang Y., et al. (2007). A novel method for imaging in vivo degradation of poly(L-glutamic acid), a biodegradable drug carrier. Pharm Res 24(6): 1217–1224.

Meng L., Mohan R., Kwok B. H., et al. (1999). Epoxomicin, a potent and selective proteasome inhibitor, exhibits in vivo antiinflammatory activity. Proc Natl Acad Sci U S A 96(18): 10403–10408.

Mikolajczyk J., Boatright K. M., Stennicke H. R., et al. (2003). Sequential autolytic processing activates the zymogen of Arg-gingipain. J Biol Chem 278(12): 10458–10464.

Minucci S., and Pelicci P. G. (2006). Histone deacetylase inhibitors and the promise of epigenetic (and more) treatments for cancer. Nat Rev Cancer 6(1): 38–51.

Nioka S., and Chance B. (2005). NIR spectroscopic detection of breast cancer. Technol. Cancer Res. Treat 4(5): 497–512.

Ntziachristos V., Ripoll J., Wang L. V., et al. (2005). Looking and listening to light: The evolution of whole-body photonic imaging. Nat Biotechnol 23(3): 313–320.

Ntziachristos V., Tung C. H., Bremer C., et al. (2002). Fluorescence molecular tomography resolves protease activity in vivo. Nature Med 8(7): 757–760.

Ovaa H. (2007). Active-site directed probes to report enzymatic action in the ubiquitin proteasome system. Nat Rev Cancer 7(8): 613–620.

Ovaa H., Kessler B. M., Rolen U., et al. (2004). Activity-based ubiquitin-specific protease (USP) profiling of virus-infected and malignant human cells. Proc Natl Acad Sci U S A 101(8): 2253–2258.

Pan Z., Jeffery D. A., Chehade K., et al. (2006). Development of activity-based probes for trypsin-family serine proteases. Bioorg Med Chem Lett 16(11): 2882–2885.

Patricelli M. P., Giang D. K., Stamp L., et al. (2001). Direct visualization of serine hydrolase activities in complex proteomes using fluorescent active site-directed probes. Proteomics 1(9): 1067–1071.

Pham W., Choi Y., Weissleder R., et al. (2004). Developing a peptide-based near-infrared molecular probe for protease sensing. Bioconjug Chem 15(6): 1403–1407.

Piston D. W., Patterson G. H., and Knobel S. M. (1999). Quantitative imaging of the green fluorescent protein (GFP). Methods Cell Biol 58: 31–48.

Powers J. C., Asgian J. L., Ekici O. D., et al. (2002). Irreversible inhibitors of serine, cysteine, and threonine proteases. Chem Rev 102(12): 4639–4750.

Reis R. C., Sorgine M. H., and Coelho-Sampaio T. (1998). A novel methodology for the investigation of intracellular proteolytic processing in intact cells. Eur J Cell Biol 75(2): 192–197.

Rolen U., Kobzeva V., Gasparjan N., et al. (2006). Activity profiling of deubiquitinating enzymes in cervical carcinoma biopsies and cell lines. Mol Carcinog 45(4): 260–269.

Rudin M., and Weissleder R. (2003). Molecular imaging in drug discovery and development. Nat Rev Drug Discov 2(2): 123–131.

Sadaghiani A. M., Verhelst S. H., and Bogyo M. (2007a). Tagging and detection strategies for activity-based proteomics. Curr Opin Chem Biol 11(1): 20–28.

Sadaghiani A. M., Verhelst S. H., Gocheva V., et al. (2007b). Design, synthesis, and evaluation of in vivo potency and selectivity of epoxysuccinyl-based inhibitors of papain-family cysteine proteases. Chem Biol 14(5): 499–511.

Saghatelian A., Jessani N., Joseph A., et al. (2004). Activity-based probes for the proteomic profiling of metalloproteases. Proc Natl Acad Sci U S A 101(27): 10000–10005.

Salisbury C. M., and Cravatt B. F. (2007). Activity-based probes for proteomic profiling of histone deacetylase complexes. Proc Natl Acad Sci U S A 104(4): 1171–1176.

Sameni M., Moin K., and Sloane B. F. (2000). Imaging proteolysis by living human breast cancer cells. Neoplasia 2(6): 496–504.

Sapsford K. E., Berti L., and Medintz I. L. (2006). Materials for fluorescence resonance energy transfer analysis: Beyond traditional donor-acceptor combinations. Angew Chem Int Ed Engl 45(28): 4562–4589.

Schaschke N., Assfalg-Machleidt I., Lassleben T., et al. (2000). Epoxysuccinyl peptide-derived affinity labels for cathepsin B. FEBS Lett 482(1–2): 91–96.

Scherer, R. L., Vansaun, M. N., McIntyre, J., et al. (2008). Optical imaging of Matrix Metalloproteinase 7 activity in vivo using a proteolytic nanobeacon. Mol Imag: in Press.

Schmidinger H., Hermetter A., and Birner-Gruenberger R. (2006). Activity-based proteomics: Enzymatic activity profiling in complex proteomes. Amino Acids 30(4): 333–350.

Sexton K. B., Kato D., Berger A. B., et al. (2007a). Specificity of aza-peptide electrophile activity-based probes of caspases. Cell Death Differ 14(4): 727–732.

Sexton K. B., Witte M. D., Blum G., et al. (2007b). Design of cell-permeable, fluorescent activity-based probes for the lysosomal cysteine protease asparaginyl endopeptidase (AEP)/legumain. Bioorg Med Chem Lett 17(3): 649–653.

Shi G. P., Munger J. S., Meara J. P., et al. (1992). Molecular cloning and expression of human alveolar macrophage cathepsin S, an elastinolytic cysteine protease. J Biol Chem 267(11): 7258–7262.

Sieber S. A., Niessen S., Hoover H. S., et al. (2006). Proteomic profiling of metalloprotease activities with cocktails of active-site probes. Nat Chem Biol 2(5): 274–281.

Sloane B. F., Sameni M., Podgorski I., et al. (2006). Functional imaging of tumor proteolysis. Annu Rev Pharmacol Toxicol 46: 301–315.

Smethurst P. A., Onley D. J., Jarvis G. E., et al. (2007). Structural basis for the platelet-collagen interaction: The smallest motif within collagen that recognizes and activates platelet Glycoprotein VI contains two glycine-proline-hydroxyproline triplets. J Biol Chem 282(2): 1296–1304.

Speers A. E., and Cravatt B. F. (2004). Chemical strategies for activity-based proteomics. Chembiochem 5(1): 41–47.

Stryer L., and Haugland R. P. (1967). Energy transfer: A spectroscopic ruler. Proc Natl Acad Sci U S A 58(2): 719–726.

Vasiljeva O., Papazoglou A., Kruger A., et al. (2006). Tumor cell-derived and macrophage-derived cathepsin B promotes progression and lung metastasis of mammary cancer. Cancer Res 66(10): 5242–5250.

Verdoes M., Florea B. I., Menendez-Benito V., et al. (2006). A fluorescent broad-spectrum proteasome inhibitor for labeling proteasomes in vitro and in vivo. Chem Biol 13(11): 1217–1226.

Verhelst S. H., and Bogyo M. (2005). Solid-phase synthesis of double-headed epoxysuccinyl activity-based probes for selective targeting of papain family cysteine proteases. Chembiochem 6(5): 824–827.

Weissleder R. (2002). Scaling down imaging: Molecular mapping of cancer in mice. Natl Rev Cancer 2(1): 11–18.

Weissleder R., Tung C. H., Mahmood U., et al. (1999). In vivo imaging of tumors with protease-activated near-infrared fluorescent probes. Nat Biotechnol 17(4): 375–378.

Welch A. R., Holman C. M., Browner M. F., et al. (1995). Purification of human matrilysin produced in escherichia coli and characterization using a new optimized fluorogenic peptide substrate. Arch Biochem Biophys 324(1): 59–64.

Witty J. P., McDonnell S., Newell K., et al. (1994). Modulation of matrilysin levels in colon carcinoma cell lines affects tumorigenicity in vivo. Cancer Res 54: 4805–4812.

Zhang J., Campbell R. E., Ting A. Y., et al. (2002). Creating new fluorescent probes for cell biology. Nat Rev Mol Cell Biol 3(12): 906–918.

Zheng G., Chen J., Stefflova K., et al. (2007). Photodynamic molecular beacon as an activatable photosensitizer based on protease-controlled singlet oxygen quenching and activation. Proc Natl Acad Sci U S A 104(21): 8989–8994.

Zhu Q., Cronin E. B., Currier A. A., et al. (2005). Benign versus malignant breast masses: Optical differentiation with US-guided optical imaging reconstruction. Radiology 237(1): 57–66.

Zuzak K. J., Schaeberle M. D., Lewis E. N., et al. (2002). Visible reflectance hyperspectral imaging: Characterization of a noninvasive, in vivo system for determining tissue perfusion. Anal Chem 74(9): 2021–2028.

Chapter 8
Images of Cleavage: Tumor Proteases in Action

Kamiar Moin, Mansoureh Sameni, Christopher Jedeszko, Quanwen Li, Mary B. Olive, Raymond R. Mattingly, and Bonnie F. Sloane

Abstract The roles of proteases in cancer are now known to be much broader than simply degradation of extracellular matrices during tumor invasion and metastasis. Furthermore, proteases from tumor-associated cells (e.g., fibroblasts, inflammatory cells, and endothelial cells) as well as tumor cells are recognized to contribute to proteolytic pathways critical to neoplastic progression. Although increased expression of proteases at the level of transcripts and protein has been observed in many tumors, the functional roles of proteases remain to be determined. Novel techniques for imaging activity of proteases, both *in vitro* and *in vivo*, are available as are selective imaging probes and substrates that allow discrimination of the activity of one class of protease from another or one individual protease from another. In this chapter, we describe *in vitro* models and assays for the functional imaging of proteases and proteolytic pathways. These models and assays can serve as screening platforms for the identification of pathways that are potential therapeutic targets and for further development of technologies and imaging probes for *in vivo* use. Such uses might include diagnosis and patient follow-up during the course of therapies that alter protease activities, perhaps even providing the crucial data needed to alter the course of treatment and/or the therapies used.

Introduction and Historical Background

Elevated expression of proteases can be documented at the transcript and protein levels in many tumors, as discussed in Chaps. 28–30. This elevated expression does not necessarily translate into elevated activity as proteases are synthesized as inactive proenzymes that require activation and there may also be elevated expression of endogenous protease inhibitors. Thus, to assess whether a protease is active

Dr. Kamiar Moin
Department of Pharmacology, Wayne State University, 540 E. Canfield Detroit, MI 48201 USA
kmoin@med.wayne.edu

D. Edwards et al. (eds.), *The Cancer Degradome.*

and has the potential of playing a functional role, we need techniques that will allow us to measure protease activity. Novel techniques and probes for functional imaging of protease activity, both *in vitro* and *in vivo*, are being developed (for review, *see* Moin et al. 2007). This chapter will discuss assays and probes for imaging protease activity *in vitro*, and Chap. 7 by Bogyo and others will discuss assays and probes for imaging protease activity *in vivo*.

The recent discordance between the results of preclinical and clinical trials with matrix metalloprotease inhibitors (MMPIs) has led us to reevaluate what we know and do not know about the functions of proteases in tumors. The MMPI clinical trials did not include functional assays to determine whether the MMPIs actually reached the sites of action of their target MMPs and most importantly reduced the MMP activity at those sites. Thus, those in the cancer protease community are left wondering whether their preclinical studies were inaccurate or perhaps poorly designed, and whether the clinical studies did not use efficacious doses or dosing regimens for the MMPIs. Our hope is that imaging protease activity in 4D (i.e., 3D in time) organotypic coculture models and in preclinical models will improve our understanding of how proteases contribute to neoplastic progression. Our intent is to use functional imaging techniques to delineate the roles played during neoplastic progression by protease classes and ultimately by individual proteases. This is obviously an ambitious goal that will require collaborative efforts among many investigators. Those partners presently in this effort are the Protease Consortium (Giranda and Matrisian 1999), a Department of Defense Breast Cancer Center of Excellence and the Center for Proteolytic Pathways (http://cpp.burnham.org/metadot/index.pl).

There is an extensive body of literature documenting the association of proteases with cancer. Indeed, a search of PubMed for the phrase "proteases and cancer" brings up a list of >43,000 papers, including >4,800 reviews. Nonetheless, the protease community still has neither identified and validated all of the proteases that play causal roles in neoplastic progression nor determined which proteases would be appropriate therapeutic targets in premalignant lesions as compared to end-stage cancers or in any one type of cancer. To help in this regard, we designed in partnership with Affymetrix a custom oligonucleotide microarray, the Hu/Mu ProtIn chip, for use in identifying the proteases that are expressed in human cancer (Schwartz et al. 2007). The Hu/Mu ProtIn chip has on a single-chip oligos for human and mouse proteases, protease inhibitors, and protease interactors. Thus, using a single platform we can identify the proteases expressed in normal, premalignant and malignant human specimens and determine whether similar patterns of expression are found in mouse transgenic and xenograft models, that is, in models required to test causality of the proteases and for preclinical studies of therapeutic agents that target proteolytic pathways and protease imaging probes. Although the Hu/Mu ProtIn chip is proving invaluable in discovering potential protease targets (Schwartz et al. 2007, Sinnamon et al. 2008), identifying transcripts for proteases, protease inhibitors and protease interactors does not tell us whether the proteases are active. Thus, assays are needed that will assess protease activity, assays such as those described in this chapter.

Imaging Activity of Individual Proteases/Protease Classes with Synthetic Substrates

Early attempts to define the protease activity in tumors and other tissues, such as those pioneered by Robert E. Smith, used histochemical techniques (Smith and van Frank 1975, for review *see* Boonacker and Van Noorden 2001). Smith's probes were designed primarily to assess the activities of lysosomal proteases. Cryostat sections of the tumors were subjected to histochemical analyses employing the substrates in solution or later in a cellulose membrane overlay. The latter is a technique detecting fluorescence after isoelectric focusing and was first described in 1985 (Garrett et al. 1985). This is similar to the primary imaging technique that has been used to evaluate tumor activities of serine proteases and MMPs, that is, in situ zymography. The first reported use of in situ zymography to determine protease activity in tumors was for the localization of the urokinase-type plasminogen activator (uPA) and tissue-type plasminogen activator (tPA) in tissue sections of skin carcinomas (Sappino et al. 1991); use of in situ zymography to detect MMPs was not reported until 1999 when it was used to establish that the gelatinolytic activity of human thyroid carcinomas arose from tumor cell nests rather than the stromal cells (Nakamura et al. 1999). By modifying the substrate (amino acid derivatives of 4-methoxy-beta-napthylamine) originally developed by Smith (Smith and van Frank 1975), Van Noorden and colleagues developed a technique to continuously monitor activity of cathepsin B, a cysteine protease, in both cells and tissue sections (Van Noorden et al. 1987). They recorded the accumulation of the final fluorescent product (5-nitrosalicylaldehyde) over time and were able to perform kinetic analyses. Furthermore, they were able to localize cathepsin B activity to specific cell types and concluded that cysteine proteases play a significant role in intracellular collagen degradation by fibroblasts and chondrocytes. Later, Van Noorden in collaboration with Smith, developed a new set of peptide-based fluorogenic substrates with cresyl violet as the reporting group to assess the kinetic parameters of cysteine proteases in living cells and tissues (Van Noorden et al. 1997, for review *see* Boonacker and Van Noorden 2001). Using such a probe, Van Noorden and colleagues were able to demonstrate not only elevated cathepsin B activity in well-differentiated human colon carcinomas but also a change in localization of this activity from apical to basal within the same tissues (Hazen et al. 2000).

Imaging Activity of Proteases with Protein Substrates

Another key development in imaging of proteolytic activity has been the availability of protein substrates for functional imaging of live cells. Our laboratory has led in the development of assay systems utilizing such substrates (for review *see* Sloane et al. 2006). Initially, we grew tumor cells on fluorescein isothiocyanate

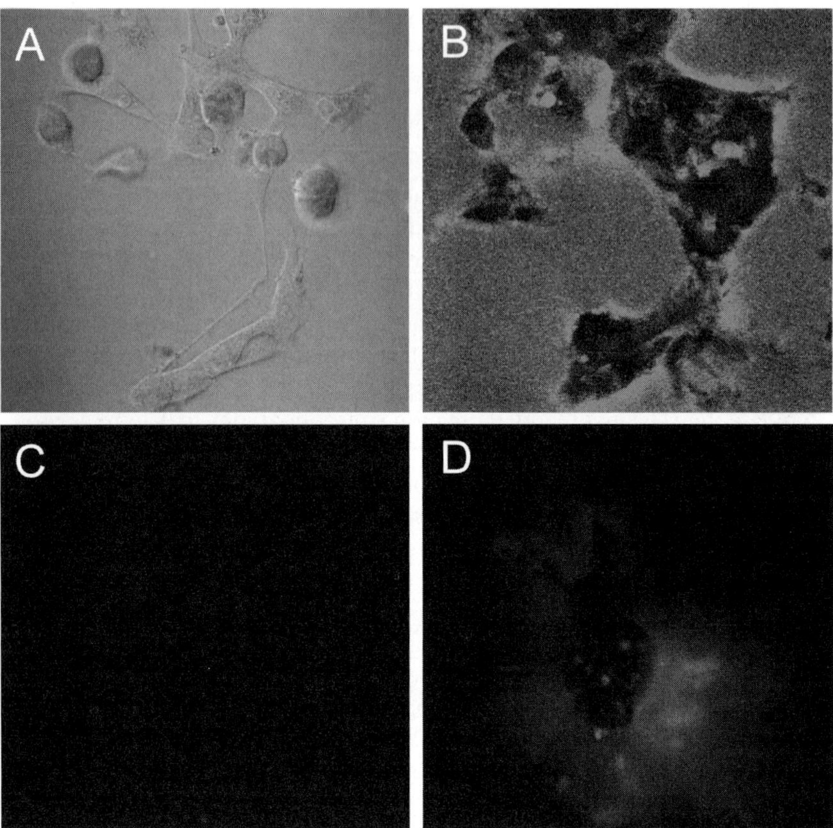

Fig. 8.1 Images of degradation of fluorescent and quenched fluorescent protein substrates by tumor and stromal cells growing on the substrate. In panels **a–b**, BT549 human breast tumor cells (2×10^4) were grown on fluorescein isothiocyanate (*FITC*)-labeled collagen IV matrix for 2 days. In panel **d**, BT549 tumor cells and WS1-2Ti human breast fibroblasts (5:1 ratio) were grown on reconstituted basement membrane (*rBM*) containing dye-quenched (*DQ*)-collagen IV for 2 days. All images were recorded with a Zeiss LSM 510 confocal microscope. (**a**) Differential interference contrast (*DIC*) image showing BT549 cells growing on the collagen matrix. (**b**) Clearing of the fluorescence background indicating proteolysis of the collagen IV substrate by BT549 cells corresponding to the areas where cells are present. (**c**) Background image, without cells, of the matrix containing *DQ*-collagen IV. (**d**) Areas of intense fluorescence representing cleavage products of *DQ*-collagen IV adjacent to and inside cells. Magnification, 40×

(FITC)-labeled extracellular matrix (ECM) protein substrates and imaged the ability of those cells to degrade their underlying ECM (Fig. 8.1a,b, Sloane 1996), as had others (Chen et al. 1985). The advantage of such an assay is that it allows one to image the degradation of large ECM substrates that may be relevant to the ability of tumor cells, or other migrating cells such as endothelial cells, to move through ECM *in vivo*. A limitation is that one is assessing discrete areas in which there is loss of fluorescence in a high background of fluorescently tagged proteins. Furthermore, in studies using FITC-labeled proteins, live cells and matrices are not normally observed in real time but after fixation. This is due, in part, to the need to counterstain for other markers, for example, paxillin, by immunocytochemistry. Therefore, one is observing proteolysis of a matrix that occurred at an earlier time as a result of cells actively migrating on and invading into the matrix. Indeed, when we imaged proteolysis of FITC-labeled laminin by U87 glioblastoma cells, we were initially surprised to find trails in which there was loss of fluorescence that did not coincide with the location of the fixed tumor cells, but rather appeared to reflect migration on this matrix (Sloane 1996). Nonetheless, if one does not require counterstaining, FITC-labeled substrates can be used in real time as was done for the image in Fig. 8.1b. Such images of FITC-labeled substrates are, however, difficult to interpret and quantify due to loss of fluorescence in a high background of fluorescence. This led us to use dye-quenched (DQ) fluorescent protein substrates, the DQ-substrates (Invitrogen, Carlsbad, CA), for imaging proteolysis by live cells in real time.

Imaging Proteolytic Activity of Live Cells

The DQ-substrates are protein substrates (e.g., DQ-collagen IV, DQ-collagen I, DQ-BSA) into which have been incorporated a large quantity of FITC molecules that are situated very closely to each other on the protein backbone. This molecular proximity causes the substrate to be self-quenched due to a FRET (Forester resonance energy transfer) effect, as evidenced by the absence of fluorescence in panel **c** of Fig. 8.1. Cleavage of the protein backbone by a protease results in emission of fluorescent degradation products, that is, fluorescence due to loss of the FRET effect. The gain in fluorescence due to cleavage of DQ-collagen IV by cocultures of tumor cells and fibroblasts is illustrated in panel **d** of Fig. 8.1. We have developed a novel confocal microscopy assay for functional imaging of DQ-substrate degradation by live tumor cells grown in 2D monolayer (Sameni et al. 2001, Ahram et al. 2000), 3D monotypic (Sameni et al. 2001, 2003; Podgorski et al. 2005, Cavallo-Medved et al. 2005), and 3D multicellular or organotypic (Sameni et al. 2003) cultures. We illustrate schematically in Fig. 8.2 the various elements comprising a 3D culture in which tumor cells are grown on a reconstituted basement membrane (rBM) of either Matrigel or Cultrex containing a DQ-substrate and then overlaid with 2% rBM. Under such conditions, tumor cells invade into the underlying matrix and in doing so generate green fluorescent degradation products of the DQ-substrate, that is, pericellular proteolysis. Degradation products are also observed inside the cells and may represent intracellular proteolysis or endocytosis of cleavage products generated outside the cell. Our studies suggest that there is both pericellular and intracellular proteolysis, the extent of each being dependent on the proteolytic pathways used by the cells being imaged (Sameni et al. 2000). Importantly, the proteolysis imaged in these assays can be quantified per cell utilizing appropriate analytical software (*see* image analysis section below, Jedeszko et al., 2008).

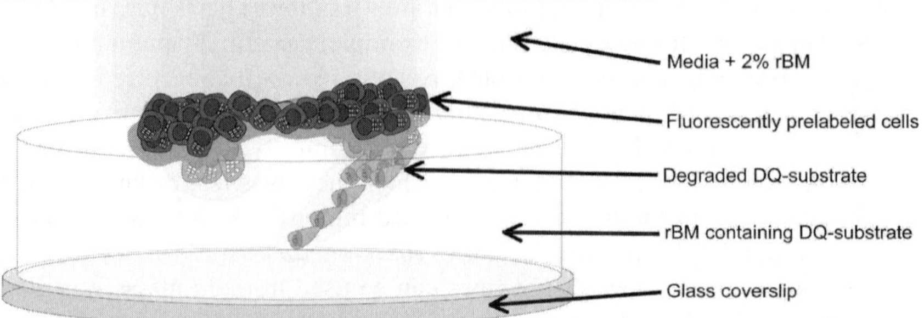

Fig. 8.2 Schematic representation of a 3D reconstituted basement membrane (*rBM*) overlay culture grown on *rBM* containing a dye-quenched (*DQ*)-substrate illustrating intracellular and extracellular proteolysis. Green, red, and dark blue represent cleavage fragments of *DQ*-substrate, cells, and nuclei, respectively. (*See also* Color Insert I)

We have analyzed the ability of live cells to degrade DQ-protein substrates when growing in monolayers on those substrates mixed with a nonfluorescent matrix (gelatin, rBM, collagen I). To date, monolayers of a variety of cells have been shown to degrade DQ-BSA: BT20 and BT549 human breast carcinoma cells (Sameni et al. 2001); DQ-collagen IV: BT20 and BT549 human breast carcinoma cells (Sameni et al. 2001), U937 human macrophages (Sameni et al. 2003), U87 human glioblastoma cells (Sameni et al. 2001), and human endothelial cells (Cavallo-Medved and Sloane, unpublished data); and DQ-collagen I: DU145, PC3, and LNCaP (lymph node carcinoma of the prostate) human prostate carcinoma cells (Podgorski et al. 2005) and human WS-12Ti breast and CCD colon fibroblasts (M. Sameni, J. Dosescu, B.F. Sloane, unpublished data). In early studies, we used DQ-protein substrates mixed with gelatin (Sameni et al. 2001, 2003; Ahram et al. 2000). As a denatured protein, gelatin is of course not representative of a matrix encountered by migrating or invading cells *in vivo*. In addition, we were concerned that DQ-protein substrates embedded in a denatured matrix might be more readily endocytosed. If this was the case, their being degraded intracellularly might not represent a normal pathway for degradation or alternatively enhanced endocytosis might increase the ratio of intracellular to pericellular proteolysis. Therefore, we now routinely use for our studies either DQ-collagen IV mixed with rBM, which itself contains collagen IV, or DQ-collagen I mixed with collagen I.

Recent studies have allayed our concerns about endocytosis as they have shown that internalization of collagen for intracellular degradation occurs normally as a result of uPARAP (urokinase plasminogen activator receptor-associated protein)-mediated endocytosis. uPARAP is a protein associated with the cell surface receptor for uPA (for review, *see* Behrendt 2004) and uPARAP-mediated intracellular degradation represents a major pathway of ECM turnover in murine mammary tumors (Berger et al. 2004). These findings are consistent with our observations of intracellular proteolysis of collagens IV and I by human breast, colon and prostate carcinoma cells, and glioblastoma cells (Sameni et al. 2001, 2003; Podgorski et al.

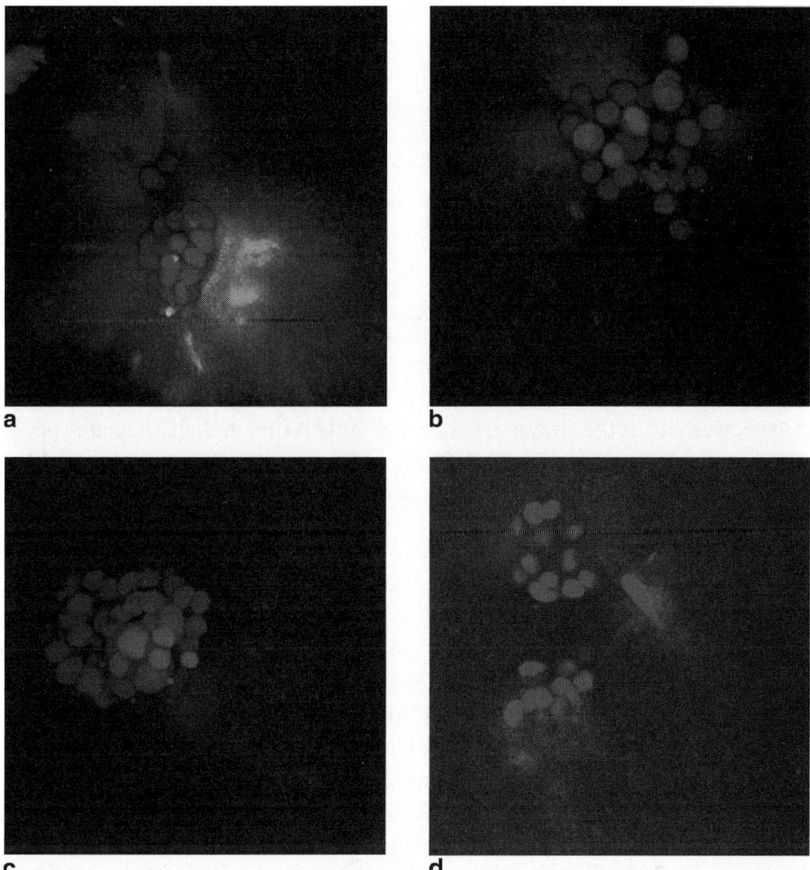

Fig. 8.3 Images of abrogation of proteolysis by broad-spectrum inhibitors of several classes of proteases. BT549 breast tumor cell and WS-12Ti breast fibroblast (5:1 ratio) cocultures were grown on reconstituted basement membrane (*rBM*) containing dye-quenched (*DQ*)-collagen IV with and without inhibitors and imaged with a Zeiss LSM 510 confocal microscope. Red, green, and blue fluorescence depict pre-labeled fibroblasts, cleavage products of *DQ*-collagen IV, and Hoescht-labeled nuclei, respectively. (**a**) Cocultures in the absence of inhibitors (same image as in Fig. 8.1d, but including red and blue channels to illustrate the fibroblasts and nuclei, respectively) showing strong green fluorescence representing proteolysis of *DQ*-collagen IV. Proteolysis, as evidenced by a decrease in intensity of the green fluorescent cleavage products, is substantially reduced in cocultures by 2 μM aprotinin (serine protease inhibitor); (**b**) 10 μM CAO74/CAO74Me cocktail (cell permeable and impermeable, respectively, cysteine protease inhibitors that are selective for cathepsins B and L); (**c**) 25 μM GM6001 [matrix metalloprotease (*MMP*) inhibitor]. (**d**) Magnification, 40×. (*See also* Color Insert I)

2005, Cavallo-Medved et al. 2005) and collagen IV by human macrophages (Sameni et al. 2003).

MMPs, serine, and cysteine proteases all contribute to pericellular tumor proteolysis, as demonstrated by use of broad-spectrum and selective protease inhibitors (Fig. 8.3, Sameni et al. 2000, 2003, Podgorski et al. 2005). Fluorescent degradation products that accumulate intracellularly do so in vesicles that stain for lysosomal markers such as LysoTracker and the cysteine protease cathepsin B. The latter was demonstrated either by immunostaining to localize cathepsin B protein or by histochemical staining to localize cathepsin B activity (Sameni et al. 2000, 2003; Ahram et al. 2000). Inhibition of endocytosis reduces the accumulation of fluorescent degradation products without increasing the amount of degradation products observed outside the cells (Sameni et al. 2000), whereas stimulation of endocytosis by Rac1 increases accumulation of degradation products intracellularly (Ahram et al. 2000). These findings are similar to those observed for uPARAP-mediated endocytosis, as discussed above.

Another approach for imaging protease activity in living cells is to use activity-based probes (ABPs), such as those developed by Bogyo and colleagues (Berger et al. 2004, Bogyo et al. 2000, Kato et al. 2005). These probes are based on inhibitors and target the active site of the enzyme, hence the name ABPs. In fact, they are specific to the mature active enzyme and will not bind to the target if the active site is blocked (Berger et al. 2004, Bogyo et al. 2000, Kato et al. 2005). The Bogyo group has synthesized probes based on the molecular structure of the epoxide-derived E-64, a potent inhibitor of cysteine proteases (Veerhelst and Bogyo 2005), to specifically target these enzymes (Berger et al. 2004, Bogyo et al. 2000, Kato et al. 2005). They have also developed quenched ABPs by including a quencher on the probes that can be cleaved by proteolysis (Blum et al. 2005). Since ABPs bind to the proteases covalently, they can be used to isolate and characterize the target protease biochemically, for example, subsequent to live cell imaging, cell extracts can be prepared and the ABP-tagged proteases analyzed by sodium dodecyl sulfate–polyacrylamide gel electrophoresis (SDS–PAGE) (Blum et al. 2005).

Tsien and colleagues have introduced exciting new fluorescent probes for imaging of protease activity in living tumor cells (Jiang et al. 2004). In this approach, a cationic cell penetrating peptide (CPP) carrying a fluorogenic cargo (e.g., Cy 5) is fused to a polyanionic peptide via a cleavable linker sequence selective for a particular protease. Normally, this complex cannot enter the cell due to hindrance caused by the anionic moiety. When the linker is cleaved, the CPP is released (hence activatable CPP) and can then enter the cell carrying its payload (Jiang et al. 2004). Utilizing this technique, Tsein and coworkers were able to image HT-1080 fibrosarcoma tumors *in vitro* and *in vivo* (Jiang et al. 2004). The same concept can be used to deliver drugs to tumors *in vivo* (Jiang et al. 2004). In fact, one can envision a scenario where drug and reporter are both delivered to the tumor site, thus allowing for monitoring of drug delivery.

Imaging Proteolytic Activity of Live Cell Interactions

In tumors, proteases secreted from tumor cells, fibroblasts, or inflammatory cells have been shown to bind to receptors/binding proteins in specialized regions of the tumor cell surface such as invadopodia and caveolae. These specialized regions alter functional interactions among proteases and protease classes, in many cases enhancing proteolysis (for reviews, *see* Cavallo-Medved and Sloane 2003, Chen and Wang 1999). In this regard, we have found that stromal and inflammatory cells (monocytes, macrophages, and fibroblasts) alone either exhibited no degradation or only modest degradation of DQ-collagen IV, whereas in cocultures there is greatly enhanced degradation (Sameni et al. 2003). Fibroblasts are important enhancers of colon and breast tumor proteolysis of DQ-collagen IV (Sameni et al. 2003), resulting in an increase in proteolysis of as great as 15-fold. Interaction of tumor cells with monocytes results in their activation to macrophages (Mohamed et al. in press), with subsequent threefold increases in proteolysis. On the contrary, the high levels of proteolysis that occur when fibroblasts are cocultured with tumor cells are not further increased by addition of macrophages. The intensity of fluorescent degradation products is highest at sites of fibroblast–tumor cell interactions, consistent with cell–cell contact being required for enhancement of proteolysis by fibroblasts (Sameni et al. 2000). In contrast, our most recent studies of cocultures of tumor cells with myofibroblasts and mammary myoepithelial cells show that whereas the interaction of tumor cells with myofibroblasts enhances proteolysis, addition of myoepithelial cells results in a reduction of proteolysis and reversal of the malignant phenotype (Sameni and Sloane, unpublished data). Thus, tumor proteolysis can be modulated by tumor–stromal–inflammatory interactions as well as by the tumor cells themselves.

Model Systems for Imaging Proteolytic Activity of Tumor Microenvironment

The contribution of the tumor microenvironment to neoplastic progression is now well accepted. This was appreciated by the cancer protease community early as many of the so-called tumor proteases are known to derive from tumor-associated cells such as fibroblasts and inflammatory cells (also see above) rather than the tumor cells themselves (Hewitt and Dano 1996, DeClerck 2000, Elenbaas and Weinberg 2001, Owen et al. 2004, McCawley and Matrisian 2001). There are several reviews that discuss the role of the tumor microenvironment in regulating expression, activation, and localization of tumor proteases and thereby tumor invasion (Hernandez-Barrantes et al. 2002, Quaranta and Giannelli 2003, Schmeichel and Bissell 2003).

Elegant studies by Bissell, Brugge, and coworkers (for reviews *see* Schmeichel and Bissell 2003; Shaw et al. 2004) have established the importance of studying cells in vitro in a 3D context. Therefore, we have analyzed proteolysis of DQ-collagen IV by cells grown as 3D spheroids, aggregates or acini, in the latter case using the protocols established by Brugge and colleagues (Debnath and Brugge 2005, Debnath et al. 2003) for growing MCF-10A human breast epithelial cells in 3D rBM overlay cultures (Mullins and Sloane, unpublished data). As depicted in Fig. 8.4a, MCF-10A acini degrade DQ-collagen IV at the periphery of the acini. In

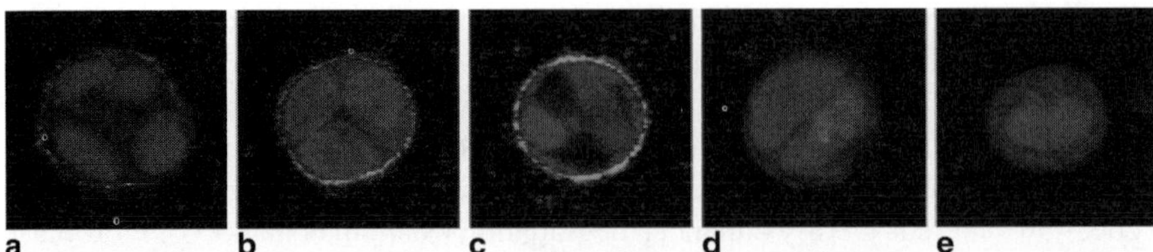

Fig. 8.4 Images of PAK1 (p21-activated kinase 1) regulation of pericellular proteolysis by MCF-10A human breast epithelial acini. MCF-10A cells were infected with control retroviruses that only expressed RFP (red fluorescent protein) (panel **a**), or bicistronic constructs expressing RFP plus wild-type PAK1 (panel **b**), constitutively active PAK1, 423E mutant (panel **c**), or two forms of dominant-negative PAK1 (83, 86L, 299R; panel **c** or 299R mutants; panel **e**) and cultured for 2 days in a 3D reconstituted basement membrane (*rBM*) overlay culture containing dye-quenched (*DQ*)-collagen IV. Nuclei of live cells were stained with Draq5 for 30 min and confocal images were recorded with a Zeiss LSM 510 confocal microscope. Expression of retroviral constructs (red), and nuclei (blue) are shown in equatorial sections of the cells along with the associated cleavage products of *DQ*-collagen IV (green). Expression of activated PAK1 increases pericellular proteolysis, whereas dominant-negative forms of PAK1 block pericellular proteolysis. Magnification, 40×. (*See also* Color Insert I)

general, we observe both pericellular and intracellular proteolysis with spheroids and aggregates of other cell lines examined to date: human carcinoma cells of breast (Sameni et al. 2003), colon (Sameni et al. 2003, Cavallo-Medved et al. 2005), and prostate (Podgorski et al. 2005) origin, human fibroblasts of breast and colon origin (Sameni et al. 2003), and human glioblastoma cells (Sameni et al. 2001). The ratios of pericellular to intracellular proteolysis vary from one cell line to another irrespective of their tissue of origin and presumably reflect the degradome of the particular cell line. For example, BT20 human breast carcinoma cells grown as either monolayers (Sameni et al. 2000) or spheroids (Sameni et al. 2003) exhibit primarily pericellular degradation of DQ-collagen IV. In contrast, another human breast carcinoma cell line, BT549, exhibits only intracellular degradation of DQ-collagen IV in monolayer cultures (Sameni et al. 2003), yet both pericellular and intracellular degradation when grown as spheroids (Sameni et al. 2003). The cohesiveness of spheroids varies from one cell line to another, which might influence the ability to endocytose substrates or degradation products. For example, spheroids formed by the HCT116 human colon carcinoma cell line are cohesive, whereas those formed by a daughter cell line in which the Ki-ras allele has been deleted are amorphous (Sameni et al. 2003). Nonetheless, HCT116 spheroids exhibit both more pericellular and intracellular degradation of DQ-collagen IV than does the daughter cell line (Sameni et al. 2003). Other studies we have conducted in which we examined tubule formation by endothelial cells have revealed that even fibrillar matrices like collagen I are endocytosed (Cavallo-Medved and Sloane, unpublished data), a finding consistent with those showing that uPARAP-mediated endocytosis is associated with intracellular collagen degradation (Curino et al. 2005).

Imaging Modulation of Proteolysis

The individual proteases responsible for pericellular and intracellular proteolysis have not been unequivocally identified (Sameni et al. 2000, 2003; Ahram et al. 2000, Podgorski et al. 2005). This is primarily due to the lack of specificity of substrates available, especially ones for imaging proteolytic activity. Therefore, we have used protease inhibitors in an effort to identify individual proteases or protease classes (Fig. 8.3) as well as some other regulatory proteins that are involved in the pericellular and intracellular proteolysis we are imaging (Figs. 8.4 and 8.5). The intracellular degradation appears to be mediated primarily by lysosomal cysteine cathepsins (Sameni et al. 2000, 2003; Ahram et al. 2000, Podgorski et al. 2005), as

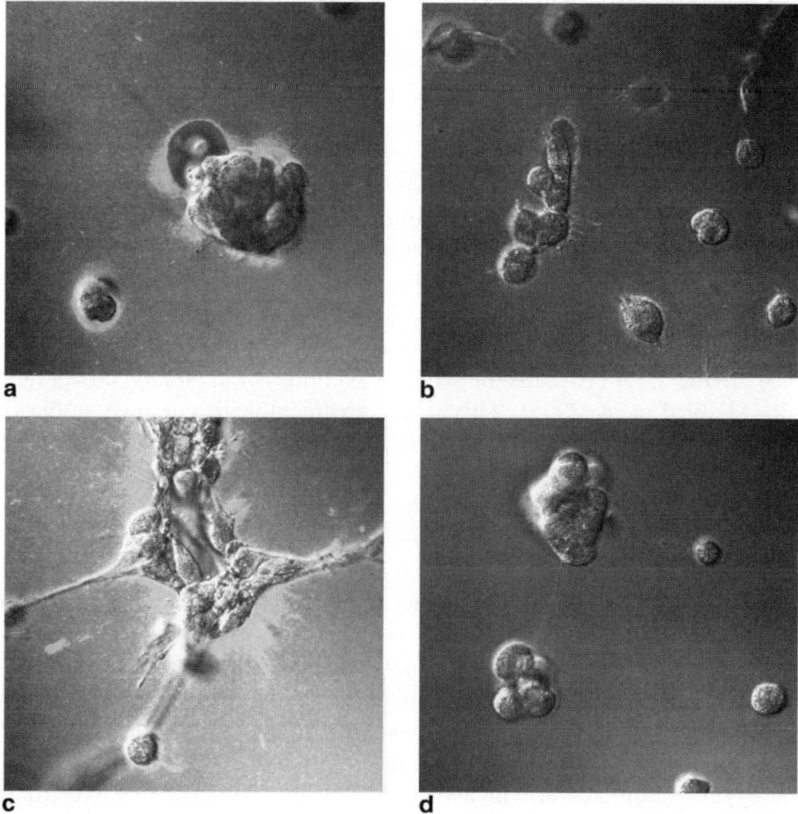

Fig. 8.5 Differential interference contrast (*DIC*) and fluorescent overlay images of abrogation of proteolysis by inhibitors of signaling pathways. BT549 breast tumor cells (2×10^4) were grown on reconstituted basement membrane (*rBM*) containing dye-quenched (*DQ*)-collagen IV alone (**a**), or in the presence of 20 μg/ml anti-β1 integrin blocking antibody, AIIB2 (**b**), 10 μM Rho-kinase inhibitor H1152 (**c**) or 30 μM Rac1 inhibitor, NSC23766 (**d**). All images were recorded with a Zeiss LSM 510 confocal microscope. Magnification, 40×. (*See also* Color Insert I)

in human breast carcinoma cells the selective, cell-permeable cysteine–cathepsin inhibitor, CA074Me (Sloane et al. 2005), reduces intracellular degradation of BSA by 90% and of type IV collagen by 80% (Sameni et al. 2000). Broad-spectrum inhibitors of MMPs and serine and cysteine proteases reduce pericellular degradation of type IV collagen by human prostate carcinoma cells ~95% (Podgorski et al. 2005). Abrogation of either pericellular or intracellular proteolysis seems to depend on the degradome of the tumor cell line as inhibition varied from one line to another, even lines of the same tissue origin, and among the protein substrates tested (Sameni et al. 2000, Podgorski et al. 2005).

Proteolysis may also be modulated by targeting protease-binding partners or membrane domains in which proteases are localized. For example, stably down-regulating the expression of caveolin-1, the structural protein of caveolae, in HCT116 cells reduces their degradation of DQ-collagen IV in parallel with reducing the expression and localization to caveolae of cathepsin B, uPA and their cell-surface receptors, p11/S100A10 (Mai et al. 2000a; Mai et al. 2000b), and uPAR, respectively, and reducing their ability to invade through Matrigel (Cavallo-Medved et al. 2005). By using an ECM protein substrate in these studies rather than a substrate selective for an individual protease or protease class, we were able to identify a potential proteolytic pathway involved in degradation of DQ-collagen IV by live HCT116 cells rather than simply an individual protease(s). Given the recent findings on uPARAP and intracellular collagen degradation, it will be of interest to see whether uPARAP is involved in this pathway (Curino et al. 2005).

Another potential target for modulating protease activity is $\beta1$ integrin. Friedl, Sahai, and colleagues (Mayer et al. 2004, Brockbank et al. 2005) have linked tumor cell invasion to $\beta1$ integrin, a protein, which also exhibits reduced localization to caveolae in the HCT116 cells in which caveolin 1 was stably downregulated (Cavallo-Medved et al. 2005). This is in agreement with our own studies (Fig. 8.5b), demonstrating that a blocking antibody to $\beta1$ integrin decreases pericellular proteolysis. Furthermore, Friedl and Sahai (Friedl and Wolf 2003, Sahai and Marshall 2003) have suggested that in the tumor microenvironment, there are two forms of cellular migration, one that requires proteolysis (cell movement via cell elongation) and another that does not (cell movement via rounded bleb formation). In agreement with Sahai and Marshall, we have found that Rho signaling does not involve elongated cell movement that is dependent on pericellular proteolysis (Sahai and Marshall 2003). In fact, in our hands, inhibition of the Rho pathway promotes cellular elongation and invasion as well as increases pericellular proteolysis (Fig. 8.5c). Wolf and Friedl (2005) have reported that tumor cells treated with a cocktail of protease inhibitors at high concentrations become amoeboid in shape and thereby remain motile and invasive in the absence of an ability to degrade the ECM. In our hands, cocktails of protease inhibitors at those high concentrations resulted in cytotoxicity for most cell lines (Sloane et al. 2006). Wolf and Friedl recently described a battery of sophisticated techniques by which to image "proteolytic tumor cell invasion" through 3D fibrillar collagen I matrices *in vitro*, perhaps indicating a modulation of their earlier views on a dissociation of proteolysis and tumor cell invasion (Brockbank et al. 2005). Indeed, in their most recent work (Wolf and Friedl 2007), they show that tumor cells in their forward movement will reorganize the fibers impeding motility, thereby forming microtracks which are subsequently expanded into wider tracks by ECM remodeling via MT1–MMP or MMP-14. This will be discussed further in Chap. 18 by Scott and Gavrilovic.

Sahai and Marshall (2003) have divided tumor cell motility in 3D rBM into one mode mediated by Rho/ROCK (Rho-associated coiled coil-containing protein kinase) signaling and one mode mediated by proteases. If one mode of motility is blocked, then the tumor cells switch to the other. Blocking both prevents the tumor cells from invading. Therefore, we determined whether fibroblasts migrating on collagen I containing DQ-collagen I degraded the DQ-substrates as we had observed for tumor cells migrating on rBM containing DQ-collagen IV. We observed fluorescent degradation products of DQ-collagen I as a result of WS-12Ti human breast fibroblasts migrating on the collagen I matrix (Sloane et al. 2006). This degradation could be reduced by either a broad-spectrum MMPI or a cell-permeable cysteine protease inhibitor, with the MMPI increasing the amount of fluorescent degradation products observed intracellularly. Once again this may relate to the recent observations on dual pathways for collagen degradation (Behrendt 2004, Curino et al. 2005).

In light of studies by Abba and colleagues showing that PAK1 (p21-activated kinase 1) overexpression occurs during the transition from normal epithelium to ductal carcinoma in situ (DCIS) (Abba et al., 2004), and our own observations that (1) Ras and Rac small GTPases regulate protease activity in concert with changes in cellular morphology and motility (Premzl et al. 2001, Bervar et al. 2003, Cavallo-Medved et al. 2003, Sahai and Marshall 2003, Menard et al, 2005) and (2) proteolysis is decreased by a Rac1 inhibitor (Fig. 8.5d), we hypothesized that PAK1 activation might also alter pericellular proteolysis. Indeed, in MCF-10A acini, expression of activated PAK1 increases pericellular proteolysis, whereas abrogation of PAK1 blocks pericellular proteolysis (Fig. 8.4). These results are consistent with a pathway in which Rac activation of PAK1 regulates pericellular proteolysis during premalignant progression.

Image Analysis: Methods for Data Processing and Quantification of Proteolysis

Analysis of images of proteolysis involves both morphology (e.g., distribution, visualization in 3D) and quantification of the proteolysis. Localization of proteolysis is important to our understanding of its biological or pathological significance. This usually requires some sort of reference labeling. In our laboratory, we often pre-label cells with CellTracker dyes (Invitrogen, Carlsbad, CA) in order to define the cell boundaries for distinguishing between pericellular and intracellular proteolysis. Advanced image analysis software available commercially, such as Volocity (Improvision, Waltham, MA), Iamris (Bitplane, St. Paul, MN), and Metamorph (Molecular Devices, Sunnyvale, CA), includes algorithms for visualization and quantification.

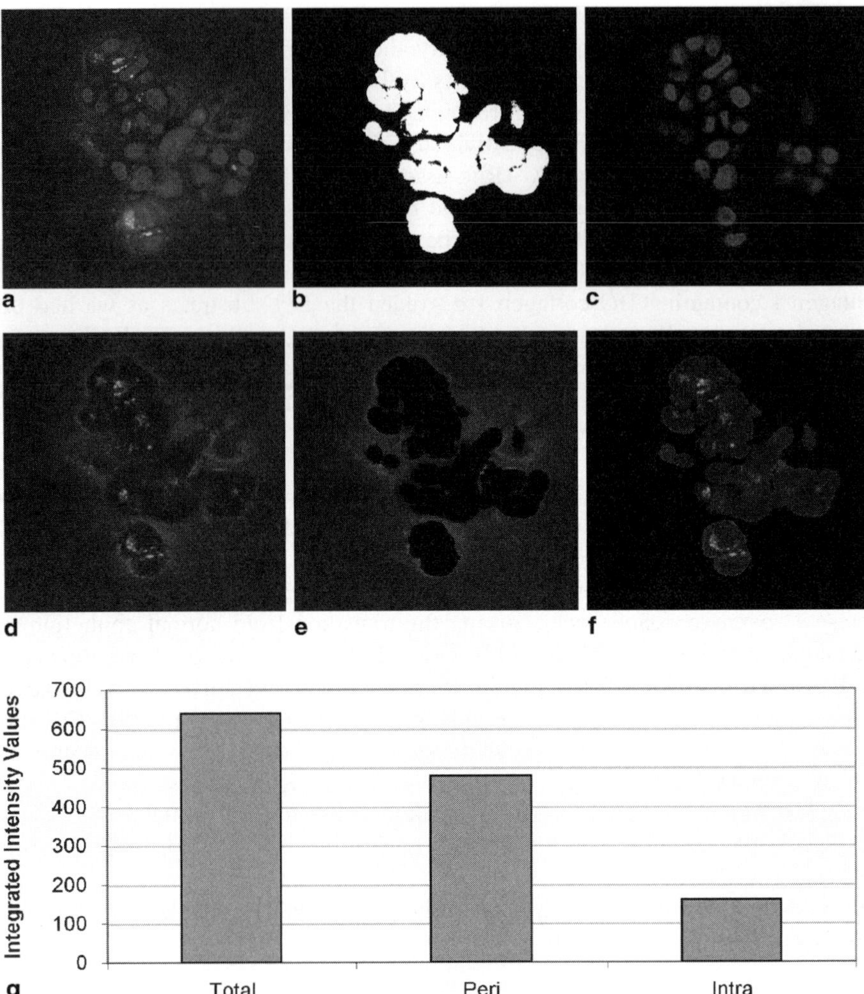

Fig. 8.6 Quantification of intracellular and pericellular degradation of dye-quenched (*DQ*)-collagen IV by MCF10–DCIS (ductal carcinoma in situ) cells. (**a**) Single optical section taken at equatorial plane showing fluorescence from cleaved DQ-collagen IV (green), pre-labeled cells (red) and nuclei (blue). (**b**) Same optical section as (**a**) but a binarized image of the cells (red channel) was used to designate x, y coordinates of intracellular areas within the same optical section. (**c**) Same optical section but binarized image of nuclei (blue channel) was used to create a 3D reconstruction for counting nuclei, thus allowing us to determine the number of cells within the field of view. (**d**) Cleavage products of *DQ*-collagen IV from same optical plane, representing both intracellular and pericellular cleavage products of *DQ*-collagen IV. (**e**) Pericellular cleavage products of *DQ*-collagen IV obtained by masking image from panel **d** with the image from panel **b** to eliminate all signal from cytoplasmic areas. (**f**) Image of intracellular cleavage products of *DQ*-collagen IV obtained by subtracting the image in panel **e** from that in panel **d**. (**g**) Quantification of total proteolysis of *DQ*-collagen IV in all optical sections throughout the volume of the structure was determined as normalized integrated intensity based on cell number. Total fluorescence was separated into intracellular (Intra) and pericellular (Peri) degradation using image arithmetic in Metamorph software (Molecular Devices, Inc.) Magnification, 40×. (*See also* Color Insert I)

We illustrate in Fig. 8.6, the method that we have developed to quantify both intra- and pericellular proteolysis based on fluorescent intensity. Pre-labeling of cells allows us to define the entire volume of the cell and/or structure under study. This information is used to mask the intracellular proteolysis so that only extracellular proteolysis is considered in the initial determination. Once extracellular proteolysis is measured, the value obtained can be subtracted from the total proteolysis to determine the value for intracellular proteolysis (Fig. 8.6, Jedeszko et al. in press). This type of measurement allows us to determine what fraction of proteolysis is due to endocytosis of the substrate as opposed to proteolysis due to secreted and/or cell surface enzymes.

Future Directions

An ability to image the activity of proteolytic pathways, and in some cases the activity of individual proteases, is essential if we are to define the biological and pathobiological roles of these enzymes. Most importantly, these efforts will be critical to evaluate therapeutic strategies that target proteolytic pathways, whether those strategies be ones that target proteases specifically or target upstream signaling pathways, and thereby proteases that are downstream of those signaling pathways. Our ultimate goal is to develop and optimize technologies and imaging probes and substrates that can be used in vitro as a screening tool in the clinic. In this regard, we are developing Raman spectroscopy techniques for microscopic profiling of proteolytic alterations in the tumor microenvironment. Raman spectroscopy has already been shown to be capable of detecting structural changes in peritumoral collagen, which occur as a result of proteolysis (Short et al. 2006). Our technique is based on the principles of Raman spectroscopy that when light in the near-infrared region is allowed to interact with a biological structure, it will cause molecular vibrations that are unique to that entity (Rao et al. 2007). This molecular signature (Raman signature) can be recorded and then compared to that of another structure, for example, tumor tissue as opposed to normal tissue. Eventually, we hope to have technologies and imaging probes that would be useful for diagnosis and for patient follow-up during the course of therapies that alter protease activities, perhaps even providing the crucial data needed to alter the course of treatment and/or the therapies used.

Acknowledgments This work was supported by U.S. Public Health Service Grant CA56586 and a Department of Defense Breast Cancer Center of Excellence (DAMD17-02-1-0693). The Microscopy and Imaging Resources Laboratory is supported by National Institutes of Health Center Grants P30ES06639 and P30CA22453 and a Roadmap Grant U54RR020843.

References

Abba M C, Drake J A, Hawkins K A, et al. (2004) Transcriptomic changes in human breast cancer progression as determined by serial analysis of gene expression. Breast Cancer Res 6: R499–R513.

Ahram M, Sameni M, Qiu R-G, et al. (2000) Rac1-induced endocytosis is associated with intracellular proteolysis during migration through a 3-dimensional matrix. Exptl Cell Res 260:292–303.

Behrendt N (2004) The urokinase receptor (uPAR) and the uPAR-associated protein (uPARAP/ Endo180): Membrane proteins engaged in matrix turnover during tissue remodeling. Biol Chem 385:103–136.

Berger A, Vitorino P, Bogyo M (2004) Activity-based protein profiling: Application to biomarker discovery, *in vivo* imaging and drug discovery. Am J Pharmacogenomics 4:371–381.

Bervar A, Zajc I, Sever N, et al. (2003) Invasiveness of transformed human breast epithelial cell lines is related to cathepsin B and inhibited by cysteine proteinase inhibitors. Biol Chem 384:447–455.

Blum G, Mullins S, Keren K, et al. (2005) Dynamic imaging of protease activity with fluorescently quenched activity-based probes. Nat Chem Biol 1:203–209.

Bogyo M, Verhelst S, Bellingard-Dubouchaud V, et al. (2000) Selective targeting of lysosomal cysteine proteases with radiolabeled electrophilic substrate analogs. Chem Biol 7:27–38.

Boonacker E, Van Noorden C J F (2001) Enzyme cytochemical techniques for metabolic mapping in living cells, with special reference to proteolysis. J. Histochem Cytochem 49:1473–1486.

Brockbank E C, Bridges J, Marshall C J, et al. (2005) Integrin beta1 is required for the invasive behaviour but not proliferation of squamous cell carcinoma cells *in vivo*. Br J Cancer 17:102–112.

Cavallo-Medved D, Sloane B F (2003) Cell surface cathepsin B: Understanding its functional significance. In: Zucker S, Chen W T eds. Cell Surface Proteases. San Diego: Academic Press 313–341.

Cavallo-Medved D, Dosescu J, Linebaugh B E, et al. (2003) Mutant K-ras regulates cathepsin B localization in caveolae of human colorectal carcinoma cells. Neoplasia 5:507–519.

Cavallo-Medved D, Mai J, Dosescu J, et al. (2005) Caveolin-1 mediates expression and localization of cathepsin B, pro-urokinase plasminogen activator and their cell surface receptors in human colorectal carcinoma cells. J Cell Sci 118:1493–1503.

Chen W T, Wang J (1999) Specialized surface protrusions of invasive cells, invadopodia and lamellipodia, have differential MT1-MMP, MMP-2, and TIMP-2 localization. Ann NY Acad Sci 878:361–371.

Chen W T, Chen J M, Parsons S J, et al. (1985) Local degradation of fibronectin at sites of expression of the transforming gene product pp60src. Nature 316:156–158.

Curino A C, Engelholm L H, Yamada S S, et al. (2005) Intracellular collagen degradation mediated by uPARAP/Endo180 is a major pathway of extracellular matrix turnover during malignancy. J Cell Biol 169:977–985.

DeClerck Y A (2000) Interactions between tumour cells and stromal cells and proteolytic modification of the extracellular matrix by metalloproteinases in cancer. Eur J Cancer 36:1258–1268.

Debnath J, Brugge J S (2005) Modelling glandulat epithelial cancers in three-dimensional cultures. Nat Rev Cancer 5:675–688.

Debnath J, Muthuswamy S K, Brugge J S (2003) Morphogenesis and oncogenesis of MCF-10A mammary epithelial acini grown in three-dimensional basement membrane cultures. Methods 30:256–268.

Elenbaas B, Weinberg R A (2001) Heterotypic signaling between epithelial tumor cells and fibroblasts in carcinoma formation. Exp Cell Res 264:169–184.

Friedl P, Wolf K (2003) Proteolytic and non-proteolytic migration of tumour cells and leucocytes. Biochem Soc Symp (70):277–285.

Garrett J R, Kidd A, Kyriacou K, et al. (1985) Use of different derivatives of D-Val-Leu-Arg for studying kallikrein activities in cat submandibular glands and saliva. Histochem J 17:805–818.

Giranda V L, Matrisian L M (1999) The protease consortium: An alliance to advance the understanding of proteolytic enzymes as therapeutic targets for cancer. Mol Carcinog 26:139–142.

Hazen L G M, Bleeker F E, Lauritzen B, et al. (2000) Comparative localization of cathepsin B protein and activity on colorectal cancer. J Histochem Cytochem 48:1421–1430.

Hernandez-Barrantes S, Bernardo M, Toth M, et al. (2002) Regulation of membrane type-matrix metalloproteinases. Semin Cancer Biol 12:131–138.

Hewitt R, Dano K (1996) Stromal cell expression of components of matrix-degrading protease systems in human cancer. Enzyme Protein 49:163–173.

Jedeszko C, Sameni M, Olive M, et al. (2008) Visualizing protease activity in living cells: From 2D to 4D. Current Protocols Cell Biol 39:4.20.1–4.20.15.

Jiang T, Olson E, Nguyen Q, et al. (2004) Tumor imaging by means of proteolytic activation of cell-penetrating peptides. PNAS 101:17867–17872.

Kato D, Boatright K, Berger A, et al. (2005) Activity-based probes that target diverse cysteine protease families. Nat Chem Biol 1:33–38.

Mai J, Finley R, Waisman D M, et al. (2000a) Human procathepsin B interacts with the annexin II tetramer on the surface of tumor cells. J Biol Chem 275:12806–12812.

Mai J, Waisman D M, Sloane B F (2000b) Cell surface complex of cathepsin B/annexin II tetramer in malignant progression. Biophys Biochim Acta 1477:215–230.

Mayer C, Maaser K, Daryab N, et al. (2004) Release of cell fragments by invading melanoma cells. Eur J Cell Biol 83:709–715.

McCawley L J, Matrisian L M (2001) Tumor progression: Defining the soil round the tumor seed. Curr Biol 11:R25–R27.

Menard R E, Jovanovski A P, Mattingly R R (2005) Active p21-activated kinase 1 rescues MCF10A breast epithelial cells from undergoing anoikis. Neoplasia 7:638–645.

Moin K, McIntyre O J, Matrisian L M, et al. (2007) Fluorescent imaging in tumors. In: Shield A, Price P eds. In vivo Imaging of Cancer Therapy. Totowa, NJ, USA· Humana Press 281–302.

Mohamed M M, Cavallo-Medved D, Sloane B F. Human monocytes augment invasiveness and proteolytic activity of inflammatory breast cancer. Biol chem, in press.

Nakamura H, Ueno H, Yamashita K, et al. (1999) Enhanced production and activation of progelatinase A mediated by membrane-type 1 matrix metalloproteinase in human papillary thyroid carcinomas. Cancer Res 59:467–473.

Owen J L, Iragavarapu-Charyulu V, Lopez D M (2004) T cell-derived matrix metalloproteinase-9 in breast cancer: Friend or foe? Breast Dis 20:143–153.

Podgorski I, Linebaugh B E, Sameni M, et al. (2005) Bone microenvironment modulates expression and activity of cathepsin B in prostate cancer. Neoplasia 7:207–223.

Premzl A, Puizdar V, Bergant V Z, et al. (2001) Invasion of ras-transformed breast epithelial cells depends on proteolytic activity of cysteine and aspartic proteinases. Biol Chem 382:853–857.

Quaranta V, Giannelli G (2003) Cancer invasion: Watch your neighbourhood! Tumori 89:343–348.

Rao A R, Hanchanale V, Javle P, et al. (2007) Spectroscopic view of life and work of the Nobel Laureate Sir C.V. Raman. J Endourol 21:8–11.

Sahai E, Marshall C J (2003) Differing modes of tumour cell invasion have distinct requirements for Rho/ROCK signaling and extracellular proteolysis. Nat Cell Biol 5:690–692.

Sameni M, Moin K, Sloane B F (2000) Imaging proteolysis by living human breast cancer cells. Neoplasia 2:496–504.

Sameni M, Dosescu J, Sloane B F (2001) Imaging proteolysis by living human glioma cells. Biol Chem 382:785–788.

Sameni M, Dosescu J, Moin K, et al. (2003) Functional imaging of proteolysis: Stromal and inflammatory cells increase tumor proteolysis. Mol Imaging 2:159–175.

Sappino A P, Belin D, Huarte J, et al. (1991) Differential protease expression by cutaneous squamous and basal cell carcinomas. J Clin Invest 88:1073–1079.

Schmeichel K L, Bissell M J (2003) Modeling tissue-specific signaling and organ function in three dimensions. J Cell Sci 116:2377–2388.

Schwartz D R, Moin K, Yao B, et al. (2007) The Hu/Mu ProtIn oligonucleotide microarray: Application of a custom dual species array for mining protease, protease inhibitor and protease interactor gene expression in tumors and their microenvironment. Mol Cancer Res 5:443–454.

Shaw K R, Wrobel C N, Brugge J S (2004) Use of three-dimensional basement membrane cultures to model oncogene-induced changes in mammary epithelial morphogenesis. J Mammary Gland Biol Neoplasia 9:297–310.

Short M A, Lui H, McLean D, et al. (2006) Changes in nuclei and peritumoral collagen within nodular basal cell carcinomas via confocal micro-Raman spectroscopy. J Biomed Opt 11:34004.

Sinnamon M J, Carter K J, Sims L P, et al. (2008) A protective role for mast cells in intestinal tumorigenesis. Carcinogenesis 29:880–886.

Sloane B F (1996) Suicidal tumor proteases. Nat Biotech 14:826–827.

Sloane B F, Yan S, Podgorski I, et al. (2005) Cathepsin B and tumor proteolysis: Contribution of the tumor microenvironment. Semin Cancer Biol 15:149–157.

Sloane B F, Sameni M, Podgorski I, et al. (2006) Functional imaging of tumor proteolysis. Annu Rev Pharmacol Toxicol 46:301–315.

Smith R E, van Frank R M (1975) The use of amino acid derivatives of 4-methoxy-beta-napthylamine for the assay and subcellular localization of tissue proteinases. Front Biol 43:193–249.

Van Noorden C J, Vogels I M, Everts V, et al. (1987) Localization of cathepsin B in fibroblasts and chondrocytes by continuous monitoring of the formation of final fluorescent reaction product using 5-nitrosalicylaldehyde. Histochem J 19:483–487.

Van Noorden C J F, Boonacker E, Bissell E R, et al. (1997) Ala-Pro-Cresyl Violet, a synthetic fluorogenic substrate for the analysis of kinetic parameters of dipeptityl peptidase IV (CD26) in individual rat hepatocytes. Anal Biochem 252:71–77.

Veerhelst S, Bogyo M (2005) Solid-phase synthesis of double-headed epoxysuccinyl activity-based probes for selective targeting of papian family cysteine proteases. ChemBioChem 6:824–827.

Wolf K, Friedl P (2005) Functional imaging of pericellular proteolysis in cancer cell invasion. Biochimie 87:315–320.

Wolf K, Friedl P (2007) Multistep pericellular proteolysis controls the transition from individual to collective cancer cell invasion. Nat Cell Biol 9:893–904.

Section II
Insights into Protease Function

Chapter 9
Proteolytic Pathways: Intersecting Cascades in Cancer Development

Nesrine I. Affara and Lisa M. Coussens

Abstract Matrix remodeling proteases, including metalloproteinases, serine proteases, and cysteine cathepsins, have emerged as important regulators of cancer development due to the realization that many provide a significant protumor advantage to developing neoplasms through their ability to modulate extracellular matrix metabolism, bioavailability of growth and proangiogenic factors, regulation of bioactive chemokines and cytokines, and processing of cell–cell and cell–matrix adhesion molecules. While some proteases directly regulate these events, others contribute to cancer development by regulating posttranslational activation of other significant protease activities. Thus, understanding the cascade of enzymatic activities contributing to overall proteolysis during carcinogenesis may identify rate-limiting steps or pathways that can be targeted with anticancer therapeutics. This chapter reviews recent insights into the complexity of roles played by extracellular and intracellular proteases that regulate tissue remodeling accompanying cancer development and focuses on the intersecting proteolytic activities that amplify protumor programming of tissues to favor cancer development.

Introduction

It is well established that cancer arises as a consequence of genetic alterations in genes that provide a survival and/or proliferative advantage to mitotically active cells. By studying mouse models of de novo cancer development, it is now clear that genetically altered neoplastic cells co-opt important physiologic host–response processes, that is, extracellular matrix (ECM) remodeling, angiogenesis, activation/ recruitment of innate, and adaptive leukocytes (inflammation), early during cancer development to favor their own survival. While it was initially believed that matrix

L.M. Coussens
Department of Pathology and Comprehensive Cancer Center, University of California, 2340 Sutter Street, N-221, Box 0875, San Francisco, CA 94115, e-mail: coussens@cc.ucsf.edu

D. Edwards et al. (eds.), *The Cancer Degradome.*
© Springer Science + Business Media, LLC 2008

remodeling proteases merely regulated migration and/or invasion of neoplastic cells into ectopic tissues, it is now clear that their more significant contribution has to do with regulating the bioactivity of a diverse array of growth factors, chemokines, soluble and insoluble matrix molecules that regulate activation and/or maintenance of overall tissue homeostasis, as well as inflammatory and angiogenic programs in pathologically damaged tissues, including cancer (Van Kempen et al. 2006). Moreover, it has also become clear that *in vivo*, many critical proteolytic cofactors for cancer development derive from activated stromal cells and thus reinforces the concept that cancer development requires reciprocal interaction between genetically altered neoplastic cells with activated diploid stromal cells and the dynamic microenvironment in which they both live (Bissell and Aggeler 1987, Bissell and Radisky 2001). Recent advances in activity-based profiling of protease function (Blum et al. 2005, 2007; Salomon et al. 2003; Sieber and Cravatt 2006; Sloane et al. 2006) have enabled tracking the distribution and magnitude of proteolytic activities in cells and tissues (Kato et al. 2005). Together with insights gained from examining individual protease gene functions in mouse models of de novo carcinogenesis, have emerged insights into the multitude of enzymatic activities that participate in tissue remodeling associated with cancer development (Egeblad and Werb 2002, Lopez-Otin and Overall 2002). Recently, sequencing of the human genome enabled characterization of the human degradome, which has been found to consist of at least 569 proteases and homologues that belong to various classes, including metalloproteinases, serine proteases, and cysteine cathepsins (Lopez-Otin and Matrisian 2007; Fig. 9.1 and *see* also Chap. 1 by Puente, Ordonez and López-Otín, this volume). For many of these enzymes, their most significant protumor activity may lie in their ability to posttranslationally regulate other proteases initially secreted as either inactive zymogens or sequestered by matrix in nonactive forms (Lopez-Otin and Matrisian 2007). This realization has led to the notion that embedded within tissues are complex, interconnecting protease networks that, depending on the tissue perturbation, selectively engage specific protease amplification circuits (Lopez-Otin and Overall 2002). These coordinated efforts regulate overall tissue homeostasis, response to acute damage, and subsequent tissue repair, as well as contribute to the pathogenesis of chronic disease states such as cancer.

Proteases Implicated in Cancer Development

Requisite for neoplastic, vascular, or inflammatory cell invasion during tumorigenic processes are the remodeling events that occur within the stroma or the ECM and on cell surfaces. ECM-remodeling proteases are universally expressed during tumor progression and metastasis, where in addition to interacting with ECM and cell surface substrates, many in turn regulate activation of other proteases initially secreted as inactive zymogens or sequestered in ECM (Dano et al. 1999, DeClerck et al. 2004, Egeblad and Werb 2002). Several classes of ECM-remodeling proteases have been identified (metallo-, serine, and cysteine), some have emerged as important regulators

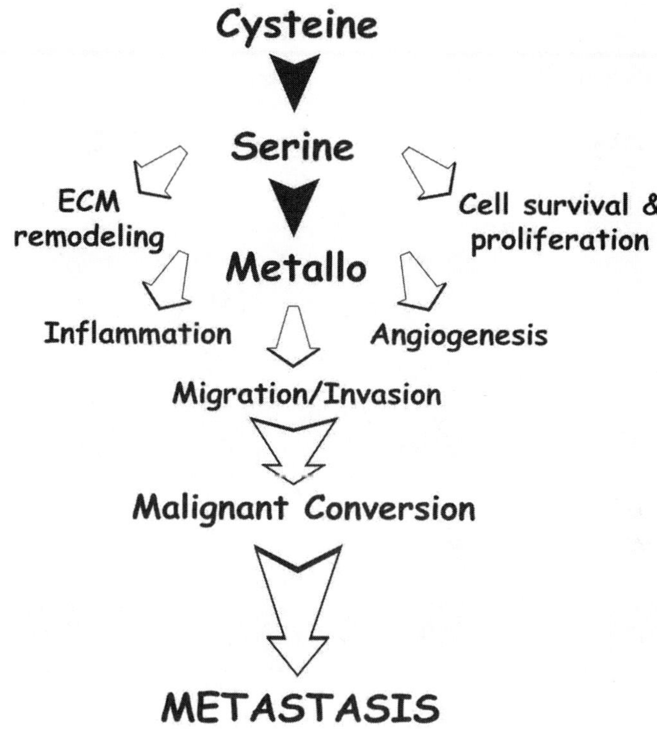

Fig. 9.1 Proteases act as critical cofactors for cancer development. Proteases belonging to several catalytic classes, including metalloproteinases, serine, and cysteine proteases, have emerged as important regulators of tissue remodeling, inflammation, angiogenesis, and acquisition of invasive capabilities, processes that accompany and potentiate cancer development

of tissue remodeling, inflammation, and angiogenesis accompanying cancer development. In epithelial tumors, a majority of ECM-remodeling proteases are made by activated stromal cells, a large percentage of which being infiltrating leukocytes such as mast cells, immature myeloid cells, monocytes, macrophages, granulocytes, and lymphocytes (Van Kempen et al. 2006). *In vivo* assessment of individual protease gene functions have indeed identified some proteases as significant cofactors for cancer development due to their ability to regulate important aspects of neoplastic progression, others are significant in that they set in motion interconnecting protease cascades resulting in amplification of enzymatic activity of "terminal" proteases, such as matrix metalloproteinase (MMP)-9 (Fig. 9.2). While these cascades of activation important for carcinogenesis resemble those regulating coagulation (Hoffman and Monroe 2005) and/or complement (Carroll

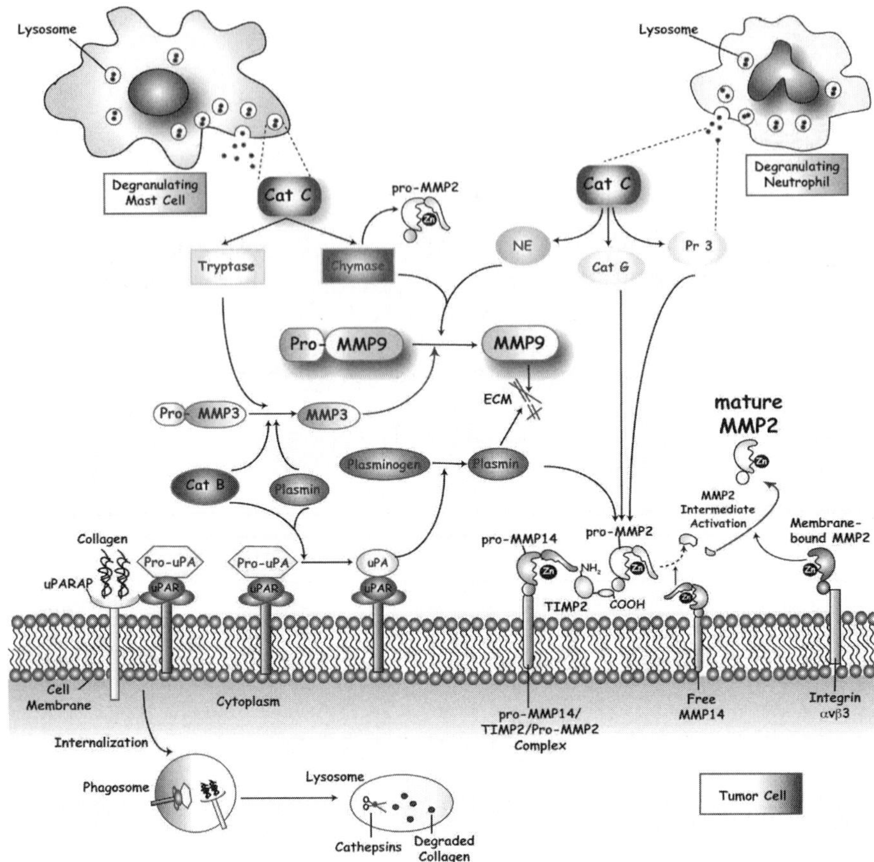

Fig. 9.2 Intersecting protease pathways during neoplastic progression. Perturbation of extracellular matrix (*ECM*) components through proteolysis during tumor progression results from the activity of combined protease pathways belonging to diverse proteolytic enzyme systems, including matrix metalloproteinases (*MMPs*), serine proteases, such as uPA, plasmin, mast cell chymase, mast cell tryptase, neutrophil elastase, and proteinase 3, and cathepsins, such as cathepsin C. Rather than functioning individually, each protease functions as a "signaling molecule" that exerts its effects as part of a proteolytic pathway, where proteases potentially interact and activate other proteases in a cascade-like manner, culminating in amplification of enzymatic activity of "terminal" proteases, such as MMP-9

2004), *in vivo* experimental studies in mouse models have revealed organ- and tumor type-specific regulation of protease bioactivities, as well as involvement of proteases emanating from multiple enzymatic classes, that is, cysteine, serine, and metallo. In the sections that follow, we discuss a significant role for MMP-9 during tumor progression in multiple organ sites, and examine the diversity of proteases whose bioactivity result in amplification of MMP-9-mediated effects in tumor tissue.

Metalloprotease Activation and Function During Cancer Development

MMPs, also known as matrixins, are a family of zinc-dependent endopeptidases that facilitate neoplastic progression not only by degrading structural components of the ECM but also by triggering release of growth and angiogenic factors sequestered by neoplastic tissues and by processing of cell–cell and cell–matrix adhesion molecules (Egeblad and Werb 2002). To date, 23 vertebrate MMPs have been identified and classified into distinct categories based on domain structure and substrate specificity (Egeblad and Werb 2002, Puente et al. 1996). Bioactivity of MMP function is controlled posttranslationally. Secreted MMPs (with the exception of stromelysin-3/MMP-11) remain as inactive zymogens, requiring enzymatic and/or autolytic removal of propeptide domains. Once activated, however, MMPs are further regulated by two major types of endogenous inhibitors: α_2-macroglobulin and tissue inhibitors of metalloporoteinases [TIMPs; reviewed in Egeblad and Werb (2002)]. Bioactivity of TIMPs is further regulated posttranslationally where some are known to be inactivated by serine protease cleavage (Frank et al. 2001).

Serine proteinases, such as plasmin or urokinase-type plasminogen activator (uPA), neutrophil elastase, mast cell chymase, and trypsin, cleave propeptide domains of secreted pro-MMPs and consequently induce autocatalytic activation of MMP-1, -3, and -9 (Egeblad and Werb 2002). Some activated MMPs can further activate other pro-MMPs. For example, MMP-3 activates pro-MMP-1 and pro-MMP-9, whereas pro-MMP-2 is resistant (Egeblad and Werb 2002). Thus, some serine proteinases act as initiators of activation cascades regulating bioactivity of pro-MMPs *in vivo*.

Cell-mediated activation mechanisms have also been identified, most notably represented by a plasma membrane-associated ternary complex formed by pro-MMP-2, TIMP-2, and the transmembrane-spanning MMP/Membrane Type I-MMP [MMP-14/MT1-MMP; Itoh et al. (2001), Wang et al. (2000), Worley et al. (2003)] that is activated intracellularly (Pei and Weiss 1995, Yana and Weiss 2000). Several advantages to having degradative enzymes in a bound state at the cell surface have been proposed. Namely, bound proenzymes may be more readily activated and the bound enzymes generated may be more active than the same enzymes found in the soluble phase. Bound enzymes may be protected from inactivation by inhibitors, binding of an enzyme to a cell surface may provide a means of concentrating components of a multistep pathway, thereby increasing rate of reactions. Immobilizing enzymes on the cell surface or in matrix may provide a means of restricting activity of an enzyme so that substrates only in the vicinity of the cell or adjacent matrix components are degraded. Hence, activation at the cell surface links MMP expression with proteolysis and invasion and may actually provide the most significant control point in MMP activity.

To address the functional roles for MMPs during cancer development, tumor-prone organ-specific transgenic mouse models harboring homozygous null gene deletions in individual MMPs have been utilized. Using a transgenic mouse model of multistage skin carcinogenesis where the early region genes of human papillomavirus type 16 (HPV16) are expressed as transgenes under control of the human keratin 14 (K14) promoter, for example, K14-HPV16 mice (Coussens et al. 1996), genetic elimination of MMP-9 significantly reduced the incidence of carcinomas in K14-HPV16 mice, while in contrast, reconstitution of K14-HPV16/MMP-9^{null} mice with wild type bone marrow-derived cells restored characteristics of neoplastic development and tumor incidence to levels similar to control HPV16 mice (Coussens et al. 2000). More specifically, the cellular source of cells supplying MMP-9 in neoplastic tissues was predominately chronically activated innate immune cells, whose infiltration coincided with development of angiogenic vasculature in premalignant skin (Coussens et al. 2000). Similarly, angiogenesis and tumor development were significantly inhibited during pancreatic islet carcinogenesis (Bergers et al. 2000) and cervical carcinogenesis (Giraudo et al. 2004) when MMP-9 was either genetically deleted or inhibited pharmacologically. Similarly, in each of these distinct tissue microenvironments, infiltrating leukocytes were the predominant sources of MMP-9 (Bergers et al. 2000, Coussens et al. 2000, Giraudo et al. 2004). During development of ovarian carcinomas using a xenograph model (Huang et al. 2002), as well as during skin carcinogenesis (Coussens et al. 2000) and neuroblastoma development (Jodele et al. 2005), reconstitution of MMP-9-deficient mice with MMP-9-proficient bone marrow-derived cells restored cellular programs necessary for development of angiogenic vasculature, tissue remodeling, and overt tumor development, thus implicating leukocyte-derived MMP-9 as a significant cofactor for cancer development.

Proangiogenic roles for leukocyte-derived MMP-9 are now well accepted. During islet carcinogenesis, although vascular endothelial growth factor (VEGF) is constitutively expressed in normal β-cells and at all stages of islet carcinogenesis, it only becomes bioavailable for interaction with its receptor on microvascular endothelial cells following infiltration of leukocytes expressing MMP-9, thereby triggering activation of angiogenic programs (Bergers et al. 2000). During development of experimental neuroblastomas, MMP-9 may regulate bioavailability of VEGF, but also regulates pericyte recruitment to developing angiogenic vessels, thus inducing stabilization of newly formed tumor vasculature (Chantrain et al. 2004, 2006). An additional line of evidence supporting a role for MMP-9 in promoting neovascularization comes from studies reporting the unique ability of MMP-9 to induce release of soluble kit-ligand, that thereby initiates mobilization of hematopoietic stem cells/progenitor cells in bone marrow (Heissig et al. 2002, Jodele et al. 2005).

Moreover, while MMP-9 induces a tissue microenvironment that is permissive not only for primary tumor development but its bioactivities also regulate secondary metastasis formation. Some clues into the later aspects of events regulating metastasis have implicated MMP-9 made by macrophages and alveolar VEGF receptor (VEGFR1)$^{+}$ endothelial cells in microenvironmental remodeling necessary for metastatic cell survival in lung (Hiratsuka et al. 2002). Using a mouse model of experimental metastasis formation, Hiratsuka et al. (2002) reported that following recruitment to sites of primary tumor growth, macrophages circulate to distal organs. Distal organs exhibiting low-level expression of VEGFR1 fail to induce MMP-9 in response to leukocyte presence and are therefore not suitable environments for subsequent metastatic cell growth. In contrast, distal organs that are VEGFR1-positive and contain a population of endothelial cells capable of inducing expression of MMP-9 above that supplied by circulating macrophages are "fertile" sites for productive metastatic growth. While induced expression of the VEGFR1 ligand VEGF-A does not appear to be involved, presence of an active VEGFR1 tyrosine kinase domain is necessary; thus, it seems reasonable that activated MMP-9 releases matrix-sequestered VEGF-A rendering it bioavailable for interaction with its receptors as has been reported by Bergers and colleagues (Bergers et al. 2000), thus, stimulating efficient vascular remodeling and angiogenesis necessary for metastatic cell growth and survival. Studies by Matrisian and colleagues have demonstrated MMP-9 derived from inflammatory cells (possibly neutrophils) present in premetastatic lung facilitates survival/establishment of early metastatic cells, but not growth of metastatic foci (Acuff et al. 2006), while MMP-9 derived from Kupffer cells in liver parenchyma, and not from bone marrow-derived cells, facilitate ability of metastatic colon cancer tumor foci to grow (Gorden et al. 2007). Taken together, these findings indicate that mechanisms by which premetastatic niches enhance metastatic outgrowth are organ and cancer-type specific.

In addition to MMP-9, other MMPs have also emerged as important cofactors in cancer development. For instance, studies using MMP gene knockout mice have indicated a key role of MMP-7 in the development of intestinal adenomas in the multiple intestinal neoplasia (Min) mouse model of intestinal neoplasia (Wilson et al. 1997). Genetic elimination of MMP-11 (stromelysin 3) resulted in a decreased tumor incidence and tumor size in 7,12-dimethylbenzanthracene (DMBA)-induced carcinomas (Masson et al. 1998). In contrast to the role of MMPs in promoting tumor progression, studies using MMP-3 (stromelysin 1)-deficient mice revealed a protective role for MMP-3 during chemically induced squamous cell carcinoma development (McCawley et al. 2004). Similarly, loss of MMP-8 (collagenase 2) enhanced rather than reduced skin tumor susceptibility in MMP-8-deficient male mice (Balbin et al. 2003). In contrast, while MMP-14-deficient mice have not been specifically examined using de novo models of cancer development, the role of MMP-14 as a cancer cofactor is undisputed. MMP-14, alone or in concert with MMP-2, has been found to activate procollagenase-3 (pro-MMP-13) (Knauper et al. 1996), that in turn mediates degradation of ECM components, including proteoglycans as well as type II collagen (Fosang et al. 1996). MMP-14 and MMP-2 have also been shown to release cryptic fragments of laminin-5 γ2 chain domain III, which due to presence of epidermal growth factor (EGF)-like repeats, binds to EGF-receptor on tumor cells, thus activating downstream signaling events that lead to tumor cell motility (Giannelli et al. 1997, Koshikawa et al. 2000). Taken together, the coordinated expression of MMP-14, MMP-2, and MMP-13 induces formation of a cascade of zymogen activation in close proximity to the tumor cell surface, resulting in amplification of proteolysis within pericellular tumor microenvironments.

It is not surprising that MMPs have attracted significant attention as anticancer therapeutic targets. Unfortunately, clinical evaluation of MMP inhibitors revealed no efficacy in patients suffering from the advanced stages of various types of cancer (Coussens et al. 2002). These failed clinical experiments have nonetheless enabled revisiting of the upstream regulatory mechanisms controlling activation of important proteases, like MMP-9, that play a clear and undisputed role in cancer development (Fig. 9.2). Active MMP-9 represents a terminal protease target whose proteolytic activity is amplified by several proteolytic pathways that converge or act in parallel to activate the latent proform of the enzyme, and thus indicating that anti-protease-based therapeutics may achieve better efficacy when targeting a "pathway" as opposed to a single class or single species of enzyme(s).

Serine Protease Regulation of MMP Activity During Cancer Development

Several serine proteases have been implicated as important regulators of cancer development, some of which are known regulators of MMP-9 bioactivity. Some of these include enzymes involved in regulating activation of plasminogen [urokinase-type and tissue-type plasminogen activators, uPA and tPA, respectively; *see* Chap. 10 by Bugge and Chap. 11 by Almholt et al., this volume; Bugge et al. (1998)], as well as serine proteases stored in secretory lysosomes of leukocytes, namely mast cell chymase (Coussens et al. 1999), mast cell tryptase (Coussens et al. 1999), and neutrophil elastase [NE; Starcher et al. (1996)].

Serine proteases are synthesized as inactive zymogens whose activation involves a two-step mechanism (Caughey 2002, Reiling et al. 2003). Following synthesis and passage through the endoplasmic reticulum where signal peptides are removed, the proprotease containing an "activation dipeptide" is generated characterized by presence of two amino-terminal residues that blocks substrate access to the catalytic site cleft, thereby maintaining the latent state and preventing premature protease activation. In addition, serine protease proteolytic activity is tightly regulated by low pH of the environment typical of secretory granules.

Plasminogen Activators

Several mechanistic studies have reported that the uPA/plasmin proteolytic cascade functionally contributes to neoplastic progression, including acquisition of a migratory and invasive phenotype by tumor cells, as well as remodeling of ECM components via activation of a number of MMPs, such as MMP-9 (Ramos-DeSimone et al. 1999). Enzymatic activity of plasmin is tightly regulated by two plasminogen activators, uPA and tPA (tissue-type plasminogen activator). Initiation of plasmin activation occurs following binding of uPA to its receptor uPAR (urokinase receptor), a gycosyl-phosphathidylinosityl (GPI)-anchored cell membrane protein that has been found to localize at discrete focal contacts (Pollanen et al. 1988). Subsequently, binding of uPA to uPAR catalyzes conversion of plasminogen to its active form, plasmin. In contrast, uPAR-bound pro-uPA is also activated by plasmin, which in turn results in a feedback pathway that accelerates plaminogen activation (Ellis et al. 1991; Fig. 9.2). Once active, cell-bound plasmin contributes to ECM remodeling by directly activating pro-MMPs, including pro-MMP-1, -2, -3, and -9 (Hahn-Dantona et al. 1999, He et al. 1989, Monea et al. 2002, Ramos-DeSimone et al. 1999).

Maintenance of this cascade depends upon the balance between uPAR-bound uPA proteolytic activity and endogenous inhibitors, including plasminogen activator inhibitor-1 (PAI-1) (Wun and Reich 1987). PAI-1 not only regulates proteolytic activity of uPA but also induces rapid internalization of the uPA/PAI-1/uPAR complex (Conese and Blasi 1995), thus regulating levels of cell surface-bound uPA as well. Taken together, through focally localized uPA/uPAR proteolytic activity at the cell surface, plasmin is considered a key event in conferring tumor cells with their ability to migrate through fibrinous matrices.

Mast Cell Serine Proteases

Mast cells represent a rich source of serine protease activity (uPA, chymases, and tryptases) stored in secretory granules that is released into the extracellular milieu following activation/degranulation in response to cross-linking of FcεRI, a high affinity receptor for immunoglobulin E (IgE) (Blank and Rivera 2004) as well as by mechanisms including components of the complement system (C3a and C5a) (el-Lati et al. 1994), neuropeptides such as substance P (Karimi et al. 2000), cytokines including stem cell factor (Hogaboam et al. 1998), as well as engagement of the Toll-like receptors (Kulka et al. 2004), that induce mast cell activation independently of IgE. While tryptases are known to cleave substrates at the carboxyl-terminal side of basic amino acids, chymases in contrast exert chymotrypsin-like activity and cleave peptides at the carboxyl-terminal side of aromatic amino acids (Schechter and Berger 1967, Schechter et al. 1986). To date, one human chymase gene belonging to the α-chymase family has been identified, while rodents express α-chymase (murine mast cell protease-5/mMCP-5), as well as several β-chymases [mMCP-1, -2, and -4; (Tchougounova et al. (2003)]. Interestingly, genetic elimination of mMCP4 completely abolishes chymotrypsin-like activities in peritoneal cells and cutaneous tissue (skin) isolated from mMCP-4-deficient mice (Tchougounova et al. 2005), demonstrating that mMCP-4 is the major source of stored chymotrypsin-like activity in these tissues. In contrast, mast cell tryptases in rodents include mast cell protease-6 (mMCP-6) and mMCP-7. While mMCP-7 is released into the circulation following mast cell degranulation, mMCP-6 is released into the vicinity of degranulated mast cells (Ghildyal et al. 1996). Recently, a novel mast cell tryptase, denoted mMCP-11/mastin, was also reported in dogs, pigs, and mice (Wong et al. 2004). In humans, tryptases fall into two major categories, including membrane-anchored tryptases, such as γ-tryptase, and soluble tryptases, such as α-, β-, and δ-tryptases, the latter that has also been termed mMCP-7-like tryptase, given the similarity with mMCP-7 genetic organization. In particular, only β-tryptase plays important roles following mast cell secretion since α- and β-tryptases are secreted in their inactive zymogen forms due to the presence of catalytic domain defects in addition to the presence of propeptide mutations [reviewed in Caughey (2007)].

The functional significance of mast cell-derived chymases and tryptases in neoplastic progression has been recently appreciated. Mast cell chymase has been shown to elicit proinflammatory effects through its ability to mediate recruitment of granulocytes into inflamed tissues (He and Walls 1998). Indeed, injection of human chymase into skin of guinea pigs resulted in a significant increase in neutrophil influx as well as increased vascular permeability (He and Walls 1998). Although many enzymes have been shown to activate pro-MMP-9, mast cell-derived mMCP-4 plays a critical role in activation of MMP-9, given that only the proform of MMP-9 was detected in tissue extracts isolated from mMCP-4-deficient mice (Tchougounova et al. 2005). Moreover, levels of active MMP-2 were significantly lower in the absence of mMCP-4 as compared to tissues from wild-type mice, indicating that mMCP-4 is important but not essential for activation of pro-MMP-2 *in vivo* (Tchougounova et al. 2005). In contrast, α-chymase has also been reported to cleave and inactivate free TIMP-1, in addition to TIMP-1 when bound in a complex with pro-MMP-9, thus enabling conversion of inhibited MMP-9 to active MMP-9 (Frank et al. 2001). Thus, chymase possesses indirect proangiogenic activities via regulating release of sequestered VEGF from the matrix following activation of MMP-9 (Coussens et al. 1999). In addition, by evaluating skin, heart, and lung tissues in mMCP-4 homozygous null mice, a major role of mast cell-derived chymase in regulating turnover of connective tissue components, including thrombin, fibronectin, and collagen, has been revealed (Tchougounova et al. 2005).

In contrast to chymases, mast cell tryptases have been implicated in cancer development due to their ability to act as direct mitogens for stromal fibroblasts (Hartmann et al. 1992, Ruoss et al. 1991) and epithelial cells (Cairns and Walls 1996), as well as their ability to activate MMP-9 indirectly through their ability to initially activate pro-MMP-3 (Gruber et al. 1989; Fig. 9.2). Furthermore, mast cell-derived tryptases have been found to modulate neoplastic microenvironments during skin carcinogenesis by stimulating synthesis of α_1 procollagen mRNA and proliferation of dermal fibroblasts (Coussens et al. 1999). Tryptases also act as potent proinflammatory factors, given that injection of mMCP-6, but not mMCP-7, induced neutrophil infiltration into peritoneal cavities (Huang et al. 1998). Although the mechanism by which mMCP-6 mediates accumulation of neutrophils in inflamed tissues remains to be determined, recent studies indicate that tryptases induce leukocyte recruitment indirectly by stimulating chemokine release, such as IL-8, from endothelial and epithelial cells (Cairns and Walls 1996, Compton et al. 1999). Taken together, these findings indicate that mast cell-derived tryptases may functionally contribute to ECM remodeling and inflammation that accompanies cancer development.

While specific gene knock-outs of murine chymases and/or tryptases have yet to assessed in a de novo tumor model, their role in cancer appears clear; thus, mechanisms regulating their activity in neoplastic tissues has been examined. In contrast to most MMPs that are secreted as zymogens requiring pericellular activation, mast cell proteases are stored in mast cell secretory granules in their mature, enzymatically active forms ready for exocytic release. Upstream intracellular activators of mast cell serine proteases include cathepsin C (Wolters et al. 2001), a cysteine protease particularly abundant in mast cell secretory granules, thus implicating presence of a cascade of proteolytic activations (Fig. 9.1), that may serve as potential therapeutic intervention strategies during tumor progression.

Neutrophil-Derived Serine Proteases

Neutrophil elastase (NE) is a serine protease transcriptionally activated during early myeloid development and subsequently stored in azurophilic granules of neutrophils (Fouret et al. 1989, Zimmer et al. 1992). While the role of NE in pulmonary disorders, including emphysema and fibrosis, has been well documented (Chua and Laurent 2006), much less is known about its role in cancer. Interest in NE during neoplastic processes stems from its ability to directly modulate ECM components or indirectly through initiation of protease cascades that culminate in activation of MMPs, including pro-MMP-9 (Ferry et al. 1997). In addition, NE has been found to regulate inflammatory processes by enhancing neutrophil migration into inflamed tissues (Nakamura et al. 1992). Through its ability to localize to plasma membranes following exocytosis, membrane-bound NE facilitates transendothelial migration of neutrophils (Cepinskas et al. 1999).

Early studies implicated NE as a key effector molecule regulating neutrophils, given its potent ability to mediate intracellular clearance of bacteria (Belaaouaj et al. 1998). More recently, a novel antimicrobial mechanism including secretion of extracellular structures, also termed neutrophil extracellular traps (NET), has been reported (Fuchs et al. 2007). NETs contain chromatin and sequestered neutrophil-derived granule proteases that enable neutrophils to deliver high concentration of proteolytic enzymes, including NE, to mediate bacterial clearance extracellularly (Brinkmann et al. 2004, Fuchs et al. 2007).

Recent clinical reports have correlated elevated NE expression with poor survival rates in patients with primary breast cancer (Akizuki et al. 2007) and nonsmall cell lung cancer (Yamashita et al. 1996), as well as having recently been found to initiate development of acute promyelocytic leukemia by mediating direct catalytic cleavage of PML-RARα, a protein generated by a chromosomal translocation fusing promyelocytic leukemia (PML) and retinoic acid receptor-α (RARα) genes (Lane and Ley 2003).

Experimentally, localization of active NE to the outer surface of the plasma membrane enables neutrophil transmigration *in vivo* (Young et al. 2007). Using *in vivo* intravital microscopy, recent studies revealed that neutrophil adhesion to postcapillary venules and emigration out of vasculature were attenuated in the presence of selective NE inhibitors (Woodman et al. 1993). This role is further supported by studies using a mouse model of acute experimental arthritis that found reduced neutrophil infiltration into subsynovial tissue spaces in NE-deficient mice (Sato et al. 2006), which were also found to exhibit reduced incidence of ultraviolet B- and chemically (Benzopyrene)-induced skin tumors (Starcher et al. 1996). Whether reduced tumor incidence was due to impaired NE-mediated cleavage of ECM substrates, deficient MMP-9 activation, or diminished cleavage of NE substrates such as ECM components or E-selectin on endothelial cells (Nozawa et al. 2000) remains to be established. In addition, genetic elimination of both NE and cathepsin G (a cysteine protease) in mice completely attenuated neutrophil emigration to sites of inflammation in response to zymosan particles in an air-pouch model of inflammation (Sato et al. 2006). However, absence of NE did not alter neutrophil recruitment to inflamed tissues in response to lipopolysaccharide (Hirche et al. 2004). While neutrophils migrated normally to sites of bacterial infection in NE-deficient mice, their ability to initiate intracellular killing of gram-negative bacteria was altered (Young et al. 2004), thus indicating some degree of specificity for recruitment and response regulated by NE.

The ability of NE to differentially regulate neutrophil recruitment and presence in "damaged" tissues, while directly significant for acute inflammatory responses, is also significant in the context of the role of neutrophils in resulting remodeling of matrix in tissues. Neutrophil-derived proteases have been identified as important regulators of insoluble elastin (Baugh and Travis 1976), a structural component of tissues such as blood vessels, skin, and lung, in addition to hydrolysis of other ECM components, including fibronectin (McDonald and Kelley 1980), proteoglycans, and type IV collagen catabolism (Mainardi, Dixit and Kang et al. 1980, Sato et al. 2006, Cepinskas et al. 1999). Some of these ECM molecules are direct substrates for NE, while others are substrates for enzymes whose bioactivity is regulated by NE.

Central to the proposed roles of NE in tumorigenesis is the mechanism of NE activation and identification of potential target substrates. NE is synthesized as an inactive zymogen requiring posttranslational removal of the amino-terminal dipeptide for enzymatic activation (Adkison et al. 2002). Propeptide cleavage occurs prior or during transport of NE to neutrophil azurophil granules through the catalytic activity of cathepsin C (Fig. 9.2). Following activation of neutrophils by various cytokines and chemoattractants, NE is secreted in its catalytically active form. One mechanism by which NE resists inhibition by circulating proteinase inhibitors, including α_1-antitrypsin (Owen et al. 1995), is by localizing to neutrophil cell surface following fusion of primary granules with plasma membranes during exocytosis (Lee and Downey 2001).

NE can indirectly modulate structural components of tumor ECM and facilitate tumor cell migration by activating MMPs. Following binding of pro-MMP-2 to membrane-anchored MT1-MMP/MMP-14, pro-MMP-2 becomes susceptible to activation by NE, as opposed to soluble pro-MMP-2 that is resistant to activation by NE (Shamamian et al. 2001). Furthermore, NE has also been shown to activate pro-MMP-3 (Okada and Nakanishi 1989), as well as pro-MMP-9 (Ferry et al.

1997). In turn, MMP-9 can cleave the NE inhibitor α1-antitrypsin, and thereby indirectly enhance NE enzymatic activity (Liu et al. 2000).

Taken together, these observations extend the role of NE to a modulator of inflammatory responses, given its ability to facilitate neutrophil extravasation into inflamed tissues by directly acting on ECM components such as elastin (Baugh and Travis 1976). Furthermore, NE indirectly induces ECM remodeling by activating MMPs (Ferry et al. 1997), and thus promoting tumor cell migration and invasion. Following excessive neutrophil influx at sites of inflammation, catalytically active NE is rapidly released from azurophil granules, along with terminal proteases including pro-MMP-9 stored in its zymogen form in neutrophil gelatinase granules. In turn, catalytically active NE within tumor microenvironments induces activation of secreted pro-MMP-9 (Fig. 9.2). One way NE may induce amplification of MMP-9-mediated effects is by catalyzing activation of pro-MMP-9 originating from other cell populations, including infiltrated mast cells. Thus, neutrophil-derived NE can significantly contribute to net tumor proteolysis. Moreover, current insights regarding protease degradomics reveal that in addition to targeting distinct matrix substrates, serine proteases exhibit partially overlapping target substrate profiles, such as activation of pro-MMP-9, whose net activity is significantly amplified during pathological processes and following simultaneous activation of various serine protease circuits originating from different cellular compartments, such as mast cells and neutrophils (Fig. 9.1). Alternatively, despite the notion that these protease pathways may be individually redundant, the above observations support the notion that serine proteases profoundly influence neoplastic progression by acting collectively.

Cysteine Cathepsin Proteases in Cancer Development

Cathepsins are lysosomal cysteine proteases that belong to the papain-like superfamily, characterized by presence of an active site cleft in which amino acid residues cysteine and histidine constitute the catalytic ion pair (Turk et al. 2001). Human cathepsins comprise 11 members: cathepsins B, C, H, F, K, L, O, S, V, W, and X/Z. While several family members have been identified as important regulators of cancer development, we focus here on cathepsin C due to the significant role it plays in regulating multiple serine proteases that together regulate important immune-based aspects of cancer development include regulating overall MMP-9 bioactivity.

Cathepsin proteolytic activity is regulated at various levels. All members of the cathepsin family are synthesized as zymogens, sharing a signal peptide and propeptide sequence removed at maturation (Vasiljeva et al. 2007). Interestingly, a residual portion of the propeptide, termed the exclusion domain, remains bound to the catalytic part of active cathepsin C (Dolenc et al. 1995) that contributes to formation and stabilization of the tetrameric structure of the mature enzyme (Cigic et al. 2000). *In vitro* studies indicate that procathepsin C cannot autolytically activate (Dahl et al. 2001), but identity of cathepsin C activators are yet to be determined.

While most cathepsins act as endopeptidases, cathepsin C, also known as dipeptidyl peptidase I (DPPI), represents the only exception by acting as a dipeptidyl aminopeptidase, cleaving two-residue units from the N-terminus of a polypeptide chain (McGuire et al. 1992). Localization of active sites of cathepsin C to the external surface of the protein confers cathepsin C with an advantage of hydrolyzing diverse groups of chymotrypsin-like proteases in their native state, regardless of size (Turk et al. 2001), as opposed to other oligomeric proteases, including tryptases, where active site are located inside of the protein (Hallgren and Pejler 2006).

While cathepsins were initially thought to only mediate terminal intracellular protein degradation within lysosomes, it is now evident that individual cathepsins exert diverse biological functions, including interstitial thrombin and fibronectin metabolism, cytotoxic lymphocyte-mediated apoptotic clearance of virus-infected and tumor cells (Shresta et al. 1998), survival from sepsis (Mallen-St Clair et al. 2004) and experimental arthritis (Adkison et al. 2002). Using cathepsin C homozygous null mice, cathepsin C has been found to activate several serine proteases, thus supporting its role as an important upstream regulator of multiple proteolytic events. In the absence of cathepsin C, cytotoxic T lymphocyte-derived granzymes A and B were found to be inactive and present in their proforms (Pham and Ley 1999). Similarly, cathepsin C has been found to be essential for intracellular activation of neutrophil-derived NE, cathepsin G and proteinase 3 (Adkison et al. 2002). In contrast, although cathepsin C has been shown to activate both human mast cell pro-α-chymase and pro-β-tryptase in vitro (Muramatsu et al. 2000, Sakai et al. 1996), cathepsin C was essential for activation of only mouse mast cell chymase *in vivo*, as opposed to mMCP-6, a mouse tryptase sharing proregion sequence homology with human β-tryptase (Wolters et al. 2001). Indeed, *in vitro* studies indicate that pro-β-tryptase undergoes auto-cleavage, resulting in a two amino acid residue activation dipeptide that is subsequently removed by cathepsin C catalytic activity (Sakai et al. 1996). Nonetheless, these results indicate that cathepsin C may not be essential for activation of all mouse mast cell tryptases (Sheth et al. 2003, Wolters et al. 2001).

Interestingly, cathepsin C-dependent activation cascades induce multiple pathways that may be important for cancer development, namely those culminating in the conversion of pro-MMP-9 and pro-MMP-2 to their mature forms, thus greatly amplifying MMP bioactivity (Fig. 9.1). Indeed, following activation by cathepsin C, mMCP-4, the major source of stored chymotrypsin-like activity in mouse peritoneum and skin (Tchougounova et al. 2003), further activates pro-MMP-2 and pro-MMP-9 (Tchougounova et al. 2005). In addition, neutrophil-derived NE, cathepsin G, and proteinase 3 have also been found to regulate MMP-2 activation (Shamamian et al. 2001). While these proteases are unable to process soluble pro-MMP-2, recent studies indicate that binding of pro-MMP-2 to the membrane-tethered MMP-14 induces conformational change, rendering a cleavage site within pro-MMP-2 prodomain region available for cleavage by neutrophil-derived serine proteases (Shamamian et al. 2001). These observations raise an interesting point – that being redundancy as a common theme among proteolytic cascades.

In addition to the role of cathepsins in maintaining tissue homeostasis and regulating diverse enzymatic activities, recent studies have revealed their involvement as mediators of inflammation. While mice harboring homozygous deletions in the cathepsin C gene exhibit normal neutrophil chemotactic responses to thioglycollate, mice were resistant to experimental acute arthritis and displayed altered neutrophil recruitment in response to zymosan and immune complexes, a defective response rescued by administration of a neutrophil chemoattractant (Adkison et al. 2002). Likewise, using the air pouch model of inflammation, the number of infiltrating neutrophils was significantly attenuated in mice deficient in both NE and cathepsin G ($NE^{-/-} \times CG^{-/-}$; Adkison et al. 2002). Notably, a significant decrease in local levels of chemokines, including tumor necrosis factor (TNF)-α and interleukin-1β (IL-1β), was observed in air pouch microenvironment of cathepsin C-deficient mice as well as $NE^{-/-} \times CG^{-/-}$ mice. Moreover, injection of IL-8, a neutrophil-specific chemokine, restored infiltration of neutrophils into air pouches of cathepsin C-deficient mice (Adkison et al. 2002). These results provide novel insights into the role of neutrophil-derived proteases in inflammation, in addition to their ability to modulate ECM components, by regulating local levels of chemoattractants at sites of inflammation (Adkison et al. 2002).

Using de novo carcinogenesis models in cathepsin-deficient tumor-prone mice has implicated roles for individual cathepsins in distinct tumorigenesis processes. Profiling of differential expression and activity of cathepsins in normal, premalignant, and malignant islets of RIP1-Tag2 (rat insulin promoter (RIP)-simian virus 40 tumor antigen (Tag)) transgenic mice revealed a complex cascaTagde of sequential cathepsin expression correlating with tumor development (Gocheva et al. 2006, Joyce et al. 2004). Taking a genetic approach, Joyce and colleagues eliminated single cathepsin genes and assessed their unique roles to islet carcinogenesis and found that while tumor-associated angiogenesis was significantly reduced in the absence of cathepsin B or S, genetic elimination of cathepsin B or L instead attenuated tumor cell proliferation and decreased tumor volume, while absence of cathepsin C was without consequence (Gocheva et al. 2006). Similar studies by Peters and colleagues found that during mammary carcinogenesis in MMTV-PymT (mouse mammary tumor virus-polyoma middle T antigen) transgenic mice (Guy et al. 1992), absence of cathepsin B emanating from macrophages significantly limited primary tumor development as well as pulmonary metastasis formation (Vasiljeva et al. 2006). Interestingly, whereas cathepsin B was found to be a significant protumor regulator of pancreatic and mammary carcinogenesis, cathepsin B does not appear to be functionally significant during skin carcinogenesis (Junankar and Coussens, unpublished observations). Moreover, whereas absence of cathepsin C during islet carcinogenesis is without consquence, its absence during squamous carcinoma development in K14-HPV16 transgenic mice is profound (Junankar and Coussens, unpublished observations). During murine skin carcinogenesis, like in humans afflicted with loss of mutations in the cathepsin C gene (de Haar et al. 2004, Frezzini et al. 2004, Hewitt et al. 2004, Noack et al. 2004), myeloid cells fail to infiltrate damaged tissue (Adkison et al. 2002, Pham et al.

2004, Pham and Ley 1999); thus, protumor programs (angiogenesis, matrix remodeling) regulated by inflammation, fail to be activated.

In summary, profiling cysteine cathepsin activities in various mouse models of multistage cancer revealed interconnecting protease cascades that are initiated by a common upstream protease activator, cathepsin C, and culminating in amplification of enzymatic activity of "terminal" proteases such as MMP-9 (Fig. 9.2). Importantly, these studies identified individual cathepsins that play differential roles in specific cancers emanating from multiple organ sites, and thus affirming the significance of organ- and tumor type-specific regulation of protease bioactivities.

Functional Role for Amplifying Protease Activities

Collectively, a vast body of literature indicates that lysosomal and pericellular proteases act as cofactors for cancer development. Rather than functioning individually, each protease can be regarded as a "signaling molecule" that exerts its effects as part of a proteolytic pathway, where proteases potentially interact and activate other proteases in a cascade-like manner, thus triggering formation of interconnecting protease circuits that form the so-called "protease web" (Overall and Kleifeld 2006). Notably, these linear protease circuits converge, leading to amplification of net proteolytic activity of enzymes like MMP-9 within tissues and consequently enhancing development of pathological conditions (Fig. 9.2). Alternatively, by using separate protease circuits, the same tissue may achieve comparable net substrate activation under different pathological settings. These observations shed light on the importance of characterizing how a tissue responds under distinct circumstances by differentially activating specific proteolytic pathways. Nonetheless, in an attempt to understand the functional roles of proteases in cancer development, it is necessary to further view proteolysis as a "system" by incorporating all the elements of the protease web, including not only proteases but inhibitors, cofactors, cleaved substrates, and receptors as well.

Perturbations of specific protease circuits may profoundly influence the normal balance of enzymatic activities, affecting the net proteolysis as a whole and leading to development of distinct pathological conditions. As an example, loss-of-function mutations in the cathepsin C gene are associated with the autosomal recessive disease Papillon-Lefevre syndrome (PLS) and Haim-Munk syndrome, characterized by severe periodontal inflammation and skin lesions, as well as increased susceptibility to infections (Hart et al. 2000, Noack et al. 2004). These symptoms may be attributed to a severe reduction in the levels and activities of multiple neutrophil granule-associated serine proteases, including NE, cathepsin G, proteinase 3, and mast cell chymases. Thus, to understand the underlying pathology of PLS, it is important to first define physiological substrates for cathepsin C – that is if we step back and analyze the consequence of cathepsin C activation within the context of a proteolytic pathway, such studies may reveal the mechanisms by which cathepsin C-deficiency mediates disease processes. More importantly, protease biology is a dynamic field, where potential activators and substrates are continuously identified. For example, using a membrane permeable inhibitor of cathepsins, intracellular levels of collagen degradation were reduced in human breast carcinoma cells (Sameni et al. 2000), indicating a novel mechanism by which tumor cells degrade collagen intracellularly within lysosomes, as opposed to the traditional role of collagenolytic MMPs as the ultimate extracellular effectors of ECM catabolism (Mott and Werb 2004). Such mechanisms include identification of new molecules that link lysosomal cathepsin B and cell surface-bound pro-uPA (Sameni et al. 2000), such as uPAR-associated protein (uPARAP; *see* Chap. 13 by Hillig, Engelholm and Behrendt, this volume; Behrendt et al. (2000); Fig. 9.2]. Following formation of a ternary complex with pro-uPA and uPAR, uPARAP is internalized along with bound collagen to endosomal compartments where cathepsins mediate collagen degradation (Curino et al. 2005, Mohamed and Sloane 2006). These studies indicate that strategies aimed at inhibiting proteases should be carefully designed, given the possibility that pharmacological inhibition of MMP pericellular activities may be counteracted with increased cathepsin-dependent intracellular catabolism of ECM components. Taken together, the net proteolytic activity of neoplastic tissues can no longer be understood without taking into consideration the cascade of protease activities and their potential interconnections – that is the flow of information within the protease network as a whole.

Targeting Proteases or Novel Use of Proteases Within the Tumor Microenvironment?

Important challenges for the future include characterization of the "cancer degradome" at the protease and substrate levels. Which proteases are differentially active in specific tumor types? What substrates do they activate and what interconnections and networks can they potentially form? Alternatively, it is clear that disturbance of individual proteases induces alterations in multiple levels of protease cascades. Should future studies focus on combinatorial approaches targeting multiple proteases as opposed to individual proteases? Answering this question is illustrated by recent studies using direct intratumoral injections of interfering RNAs (RNAi) that demonstrated simultaneous down-regulation of cathepsin B and MMP-9 (Lakka et al. 2004) as well as MMP-9 and uPAR (Lakka et al. 2005, Lakka et al. 2003) resulting in regression of preestablished tumors more effectively than targeting either protease alone. Nonetheless, although these approaches are promising, future studies should take into consideration the possibility that alternative pathways may be activated to compensate for complete loss of one enzymatic activity. For instance, following genetic elimination of cathepsin B, tumor cells have been found to induce cathepsin X expression, resulting in a partial compensation for loss of cathepsin B-mediated effects (Vasiljeva et al. 2006).

In contrast, will future strategies take advantage of unique characteristics limited to specific protease pathways? Proteases may be restricted to certain regions of the cell, such as proteases that are tethered to cell surfaces, including uPA, thus directing proteolysis to discrete focal areas versus intracellular proteases within lysosomes, such as cathepsins. Recent studies exploiting unique localization of specific protease pathways to direct activation of nontoxic "prodrugs," such as doxorubicin (Dox), selectively to tumor sites [*see* also Chap. 39 by Fields and Chap. 40 by Gill and Loadman, this volume; Devy et al. (2004)]. Here, incorporation of a tripeptide specifier recognized exclusively by tumor-associated plasmin initiated release of free Dox from its prodrug form selectively in the vicinity of tumor cells, given that the systemic presence of physiological plasmin inhibitors, including α_2-antiplasmin and α_2-macroglobulin, limits Dox activation within the circulation (Devy et al. 2004). Using such approaches, Dox has been shown to exert its cytotoxic and/or cytostatic effects locally while restricting cardiotoxicity, which may limit Dox dosage intake. Similar strategies have been used to design prodrugs that are activated in tumor cells over-expressing selective protease pathways, including cathepsin B. Indeed, Panchal et al. (1996) engineered α-hemolysin, which once activated by tumor cells expressing high levels of cathepsin B, induces selective apoptosis of tumor cells, thus reducing side effects associated with the use of this drug.

Furthermore, recent advances in novel anticancer therapies took advantage of selective expression of the antiapoptotic protein survivin in malignant ovarian cells to specifically activate expression of proapoptotic proteases, such as cytotoxic T lymphocyte-derived granzyme B (Caldas et al. 2006). Driven by the survivin promoter, active granzyme B not only reduced tumor incidence and size of xeno-grafted human ovarian carcinoma in nude mice but also prevented metastatic spread (Caldas et al. 2006). Such approaches illustrate the utility of proteases that are normally employed by immune cells to eliminate tumor cells in designing future therapeutics. Taken together, to advance the understanding of individual proteases and delineate their physiological as well as pathological roles, the complexity by which proteases interact must be taken into account as opposed to merely investigating individual proteases.

Acknowledgments The authors acknowledge all the scientists who made contributions to the areas of research reviewed here that were not cited due to space constraints. The authors acknowledge support from the National Institutes of Health and a Department of Defense Era of Hope Scholar Award to LMC.

References

Acuff, H.B., Carter, K.J., Fingleton, B., Gorden, D.L., and Matrisian, L.M. 2006 Matrix metallo-proteinase-9 from bone marrow-derived cells contributes to survival but not growth of tumor cells in the lung microenvironment. Cancer Res 66, 259–66.

Adkison, A.M., Raptism, S.Z., Kelleym, D.G.M., and Phamm, C.T. 2002 Dipeptidyl peptidase I activates neutrophil-derived serine proteases and regulates the development of acute experimental arthritis. J Clin Invest 109, 363–71.

Akizuki, M., Fukutomi, T., Takasugi, M., Takahashi, S., Sato, T., Harao, M., Mizumoto, T., and Yamashita, J. 2007 Prognostic significance of immunoreactive neutrophil elastase in human breast cancer: Long-term follow-up results in 313 patients. Neoplasia 9, 260–4.

Balbin, M., Fueyo, A., Tester, A.M., Pendas, A.M., Pitiot, A.S., Astudillo, A., Overall, C.M., Shapiro, S.D., and Lopez-Otin, C. 2003 Loss of collagenase-2 confers increased skin tumor susceptibility to male mice. Nat Genet 35, 252–7.

Baugh, R.J., and Travis, J. 1976 Human leukocyte granule elastase: Rapid isolation and characterization. Biochemistry 15, 836–41.

Behrendt, N., Jensen, O.N., Engelholm, L.H., Mortz, E., Mann, M., and Dano, K. 2000 A urokinase receptor-associated protein with specific collagen binding properties. J Biol Chem 275, 1993–2002.

Belaaouaj, A., McCarthy, R., Baumann, M., Gao, Z., Ley, T.J., Abraham, S.N., and Shapiro, S.D. 1998 Mice lacking neutrophil elastase reveal impaired host defense against gram negative bacterial sepsis. Nat Med 4, 615–618.

Bergers, G., Brekken, R., McMahon, G., Vu, T.H., Itoh, T., Tamaki, K., Tanzawa, K., Thorpe, P., Itohara, S., Werb, Z. et al.. 2000 Matrix metalloproteinase-9 triggers the angiogenic switch during carcinogenesis. Nat Cell Biol 2, 737–744.

Bissell, M.J., and Aggeler, J. 1987 Dynamic reciprocity: How do extracellular matrix and hormones direct gene expression? Prog Clin Biol Res 249, 251–262.

Bissell, M.J., and Radisky, D. 2001 Putting tumours in context. Nat Rev Cancer 1, 46–54.

Blank, U., and Rivera, J. 2004 The ins and outs of IgE-dependent mast-cell exocytosis. Trends Immunol 25, 266–73.

Blum, G., Mullins, S.R., Keren, K., Fonovic, M., Jedeszko, C., Rice, M.J., Sloane, B.F., and Bogyo, M. 2005 Dynamic imaging of protease activity with fluorescently quenched activity-based probes. Nat Chem Biol 1, 203–9.

Blum, G., von Degenfeld, G., Merchant, M.J., Blau, H.M., and Bogyo, M. 2007 Noninvasive optical imaging of cysteine protease activity using fluorescently quenched activity-based probes. Nat Chem Biol 3, 668–77.

Brinkmann, V., Reichard, U., Goosmann, C., Fauler, B., Uhlemann, Y., Weiss, D.S., Weinrauch, Y., and Zychlinsky, A. 2004 Neutrophil extracellular traps kill bacteria. Science 303, 1532–1535.

Bugge, T.H., Lund, L.R., Kombrinck, K.K., Nielsen, B.S., Holmback, K., Drew, A.F., Flick, M.J., Witte, D.P., Dano, K., and Degen, J.L. 1998 Reduced metastasis of Polyoma virus middle T antigen-induced mammary cancer in plasminogen-deficient mice. Oncogene 16, 3097–3104.

Cairns, J.A., and Walls, A.F. 1996 Mast cell tryptase is a mitogen for epithelial cells. Stimulation of IL-8 production and intercellular adhesion molecule-1 expression. J Immunol 156, 275–83.

Caldas, H., Jaynes, F.O., Boyer, M.W., Hammond, S., and Altura, R.A. (2006) Survivin and Granzyme B-induced apoptosis, a novel anticancer therapy. Mol. Cancer Ther. 5, 693–703.

Carroll, M.C. 2004 The complement system in regulation of adaptive immunity. Nat Immunol 5, 981–6.

Caughey, G.H. 2002 New developments in the genetics and activation of mast cell proteases. Mol Immunol 38, 1353–7.

Caughey, G.H. 2007 Mast cell tryptases and chymases in inflammation and host defense. Immunol Rev 217, 141–54.

Cepinskas, G., Sandig, M., and Kvietys, P.R. 1999 PAF-induced elastase-dependent neutrophil transendothelial migration is associated with the mobilization of elastase to the neutrophil surface and localization to the migrating front. J Cell Sci 112 (Pt 12), 1937–45

Chantrain, C.F., Shimada, H., Jodele, S., Groshen, S., Ye, W., Shalinsky, D.R., Werb, Z., Coussens, L.M., and DeClerck, Y.A. 2004 Stromal matrix metalloproteinase-9 regulates the vascular architecture in neuroblastoma by promoting pericyte recruitment. Cancer Res 64, 1675–86.

Chantrain, C.F., Henriet, P., Jodele, S., Emonard, H., Feron, O., Courtoy, P.J., DeClerck, Y.A., and Marbaix, E. 2006 Mechanisms of pericyte recruitment in tumour angiogenesis: A new role for metalloproteinases. Eur J Cancer 42, 310–8.

Chua, F., and Laurent, G.J. 2006 Neutrophil elastase: Mediator of extracellular matrix destruction and accumulation. Proc Am Thorac Soc 3, 424–7.

Cigic, B., Dahl, S.W., and Pain, R.H. 2000 The residual pro-part of cathepsin C fulfills the criteria required for an intramolecular chaperone in folding and stabilizing the human proenzyme. Biochemistry 39, 12382–90.

Compton, S.J., Cairns, J.A., Holgate, S.T., and Walls, A.F. 1999 Interaction of human mast cell tryptase with endothelial cells to stimulate inflammatory cell recruitment. Int Arch Allergy Immunol 118, 204–5.

Conese, M., and Blasi, F. 1995 Urokinase/urokinase receptor system: Internalization/degradation of urokinase-serpin complexes: Mechanism and regulation. Biol Chem Hoppe Seyler 376, 143–55.

Coussens, L.M., Hanahan, D., and Arbeit, J.M. 1996 Genetic predisposition and parameters of malignant progression in K14-HPV16 transgenic mice. Am J Path 149, 1899–1917.

Coussens, L.M., Raymond, W.W., Bergers, G., Laig-Webster, M., Behrendtsen, O., Werb, Z., Caughey, G.H., and Hanahan, D. 1999 Inflammatory mast cells up-regulate angiogenesis during squamous epithelial carcinogenesis. Genes Dev 13, 1382–97.

Coussens, L.M., Tinkle, C.L., Hanahan, D., and Werb, Z. 2000 MMP-9 supplied by bone marrow-derived cells contributes to skin carcinogenesis. Cell 103, 481–90.

Coussens, L.M., Fingleton, B., and Matrisian, L.M. 2002 Matrix metalloproteinase inhibitors and cancer: Trials and tribulations. Science 295, 2387–92.

Curino, A.C., Engelholm, L.H., Yamada, S.S., Holmbeck, K., Lund, L.R., Molinolo, A.A., Behrendt, N., Nielsen, B.S., and Bugge, T.H. 2005 Intracellular collagen degradation mediated by uPARAP/Endo180 is a major pathway of extracellular matrix turnover during malignancy. J Cell Biol 169, 977–85.

Dahl, S.W., Halkier, T., Lauritzen, C., Dolenc, I., Pedersen, J., Turk, V., and Turk, B. 2001 Human recombinant pro-dipeptidyl peptidase I (cathepsin C) can be activated by cathepsins L and S but not by autocatalytic processing. Biochemistry 40, 1671–8.

Dano, K., Romer, J., Nielsen, B.S., Bjorn, S., Pyke, C., Rygaard, J., and Lund, L.R. 1999 Cancer invasion and tissue remodeling—cooperation of protease systems and cell types. Apmis 107, 120–7.

DeClerck, Y.A., Mercurio, A.M., Stack, M.S., Chapman, H.A., Zutter, M.M., Muschel, R.J., Raz, A., Matrisian, L.M., Sloane, B.F., Noel, A. et al. 2004 Proteases, extracellular matrix, and cancer: A workshop of the path B study section. Am J Pathol 164, 1131–39.

de Haar, S.F., Jansen, D.C., Schoenmaker, T., De Vree, H., Everts, V., and Beertsen, W. 2004 Loss-of-function mutations in cathepsin C in two families with Papillon-Lefevre syndrome are associated with deficiency of serine proteinases in PMNs. Hum Mutat 23, 524.

Devy, L., de Groot, F.M., Blacher, S., Hajitou, A., Beusker, P.H., Scheeren, H.W., Foidart, J.M., and Noel, A. 2004 Plasmin-activated doxorubicin prodrugs containing a spacer reduce tumor growth and angiogenesis without systemic toxicity. Faseb J 18, 565–7.

Dolenc, I., Turk, B., Pungercic, G., Ritonja, A., and Turk, V. 1995 Oligomeric structure and substrate induced inhibition of human cathepsin C. J Biol Chem 270, 21626–31.

Egeblad, M. and Werb, Z. 2002 New functions for the matrix metalloproteinases in cancer progression. Nat Rev Cancer 2, 161–174.

el-Lati, S.G., Dahinden, C.A., and Church, M.K. 1994 Complement peptides C3a- and C5a-induced mediator release from dissociated human skin mast cells. J Invest Dermatol 102, 803–6.

Ellis, V., Behrendt, N., and Dano, K. 1991 Plasminogen activation by receptor-bound urokinase. A kinetic study with both cell-associated and isolated receptor. J Biol Chem 266, 12752–8.

Ferry, G., Lonchampt, M., Pennel, L., de Nanteuil, G., Canet, E., and Tucker, G.C. 1997 Activation of MMP-9 by neutrophil elastase in an in vivo model of acute lung injury. FEBS Lett 402, 111–5.

Fosang, A.J., Last, K., Knauper, V., Murphy, G., and Neame, P.J. 1996 Degradation of cartilage aggrecan by collagenase-3 (MMP-13). FEBS Lett 380, 17–20.

Fouret, P., du Bois, R.M., Bernaudin, J.F., Takahashi, H., Ferrans, V.J., and Crystal, R.G. 1989 Expression of the neutrophil elastase gene during human bone marrow cell differentiation. J Exp Med 169, 833–45.

Frank, B.T., Rossall, J.C., Caughey, G.H., and Fang, K.C. 2001 Mast cell tissue inhibitor of metalloproteinase-1 is cleaved and inactivated extracellularly by alpha-chymase. J Immunol 166, 2783–92.

Frezzini, C., Leao, J.C., and Porter, S. 2004 Cathepsin C involvement in the aetiology of Papillon-Lefevre syndrome. Int J Paediatr Dent 14, 466–467.

Fuchs, T.A., Abed, U., Goosmann, C., Hurwitz, R., Schulze, I., Wahn, V., Weinrauch, Y., Brinkmann, V., and Zychlinsky, A. 2007 Novel cell death program leads to neutrophil extracellular traps. J Cell Biol 176, 231–41.

Ghildyal, N., Friend, D.S., Stevens, R.L., Austen, K.F., Huang, C., Penrose, J.F., Sali, A., and Gurish, M.F. 1996 Fate of two mast cell tryptases in V3 mastocytosis and normal BALB/c mice undergoing passive systemic anaphylaxis: Prolonged retention of exocytosed mMCP-6 in connective tissues, and rapid accumulation of enzymatically active mMCP-7 in the blood. J Exp Med 184, 1061–73.

Giannelli, G., Falk-Marzillier, J., Schiraldi, O., Stetler-Stevenson, W.G., and Quaranta, V. 1997 Induction of cell migration by matrix metalloprotease-2 cleavage of laminin-5. Science 277, 225–8.

Giraudo, E., Inoue, M., and Hanahan, D. 2004 An amino-bisphosphonate targets MMP-9-expressing macrophages and angiogenesis to impair cervical carcinogenesis. J Clin Invest 114, 623–33.

Gocheva, V., Zeng, W., Ke, D., Klimstra, D., Reinheckel, T., Peters, C., Hanahan, D., and Joyce, J. A. 2006 Distinct roles for cysteine cathepsin genes in multistage tumorigenesis. Genes Dev 20, 543–56.

Gorden, D.L., Fingleton, B., Crawford, H.C., Jansen, D.E., Lepage, M., and Matrisian, L.M. 2007 Resident stromal cell-derived MMP-9 promotes the growth of colorectal metastases in the liver microenvironment. Int J Cancer 121, 495–500.

Gruber, B.L., Marchese, M.J., Suzuki, K., Schwartz, L.B., Okada, Y., Nagase, H., and Ramamurthy, N.S. 1989 Synovial procollagenase activation by human mast cell tryptase dependence upon matrix metalloproteinase 3 activation. J Clin Invest 84, 1657–62.

Guy, C.T., Cardiff, R.D., and Muller, W.J. 1992 Induction of mammary tumors by expression of polyomavirus middle T oncogene: A transgenic mouse model for metastatic disease. Mol Cell Biol 12, 954–61.

Hahn-Dantona, E., Ramos-DeSimone, N., Sipley, J., Nagase, H., French, D.L., and Quigley, J.P. 1999 Activation of proMMP-9 by a plasmin/MMP-3 cascade in a tumor cell model. Regulation by tissue inhibitors of metalloproteinases. Ann N Y Acad Sci 878, 372–87.

Hallgren, J., and Pejler, G. 2006 Biology of mast cell tryptase. An inflammatory mediator. Febs J 273, 1871–95.

Hart, T.C., Hart, P.S., Michalec, M.D., Zhang, Y., Firatli, E., Van Dyke, T.E., Stabholzm, A., Zlotogorski, A., Shapira, L., and Soskolne, W.A. 2000 Haim-Munk syndrome and Papillon-Lefevre syndrome are allelic mutations in cathepsin C. J Med Genet 37, 88–94.

Hartmann, T., Ruoss, S.J., Raymond, W.W., Seuwen, K., and Caughey, G.H. 1992 Human tryptase as a potent, cell-specific mitogen: Role of signaling pathways in synergistic responses. Am J Physiol 262, L528–34.

He, S., and Walls, A.F. 1998 Human mast cell chymase induces the accumulation of neutrophils, eosinophils and other inflammatory cells in vivo. Br J Pharmacol 125, 1491–500.

He, C.S., Wilhelm, S.M., Pentland, A.P., Marmer, B.L., Grant, G.A., Eisen, A.Z., and Goldberg, G.I. 1989 Tissue cooperation in a proteolytic cascade activating human interstitial collagenase. Proc Natl Acad Sci U S A 86, 2632–6.

Heissig, B., Hattori, K., Dias, S., Friedrich, M., Ferris, B., Hackett, N.R., Crystal, R.G., Besmer, P., Lyden, D., Moore, M.A. et al. 2002 Recruitment of stem and progenitor cells from the bone marrow niche requires MMP-9 mediated release of kit-ligand. Cell 109, 625–37.

Hewitt, C., McCormick, D., Linden, G., Turk, D., Stern, I., Wallace, I., Southern, L., Zhang, L., Howard, R., Bullon, P. et al. 2004 The role of cathepsin C in Papillon-Lefevre syndrome, prepubertal periodontitis, and aggressive periodontitis. Hum Mutat 23, 222–28.

Hiratsuka, S., Nakamura, K., Iwai, S., Murakami, M., Itoh, T., Kijima, H., Shipley, J.M., Senior, R.M., and Shibuya, M. 2002 MMP9 induction by vascular endothelial growth factor receptor-1 is involved in lung-specific metastasis. Cancer Cell 2, 289–300.

Hirche, T.O., Atkinson, J.J., Bahr, S., and Belaaouaj, A. 2004 Deficiency in neutrophil elastase does not impair neutrophil recruitment to inflamed sites. Am J Respir Cell Mol Biol 30, 576–84.

Hoffman, M.M., and Monroe, D.M. (2005) Rethinking the coagulation cascade. Curr Hematol Rep 4, 391–6.

Hogaboam, C., Kunkel, S.L., Strieter, R.M., Taub, D.D., Lincoln, P., Standiford, T.J., and Lukacs, N.W. 1998 Novel role of transmembrane SCF for mast cell activation and eotaxin production in mast cell-fibroblast interactions. J Immunol 160, 6166–71.

Huang, C., Friend, D.S., Qiu, W.T., Wong, G.W., Morales, G., Hunt, J., and Stevens, R.L. 1998 Induction of a selective and persistent extravasation of neutrophils into the peritoneal cavity by tryptase mouse mast cell protease 6. J Immunol 160, 1910–9.

Huang, S., Van Arsdall, M., Tedjarat, S., McCarty, M., Wu, W., Langley, R., and Fidler, I.J. 2002 Contributions of stromal metalloproteinase-9 to angiogenesis and growth of human ovarian carcinoma in mice. J. Natl. Cancer Inst. 94, 1134–42.

Itoh, Y., Takamura, A., Ito, N., Maru, Y., Sato, H., Suenaga, N., Aoki, T., and Seiki, M. 2001 Homophilic complex formation of MT1-MMP facilitates proMMP-2 activation on the cell surface and promotes tumor cell invasion. Embo J 20, 4782–93.

Jodele, S., Chantrain, C.F., Blavier, L., Lutzko, C., Crooks, G.M., Shimada, H., Coussens, L.M., and Declerck, Y.A. 2005 The contribution of bone marrow-derived cells to the tumor vasculature in neuroblastoma is matrix metalloproteinase-9 dependent. Cancer Res 65, 3200–8.

Joyce, J.A., Baruch, A., Chehade, K., Meyer-Morse, N., Giraudo, E., Tsai, F.Y., Greenbaum, D.C., Hager, J.H., Bogyo, M., and Hanahan, D. 2004 Cathepsin cysteine proteases are effectors of invasive growth and angiogenesis during multistage tumorigenesis. Cancer Cell 5, 443–53.

Karimi, K., Redegeld, F.A., Blom, R., and Nijkamp, F.P. 2000 Stem cell factor and interleukin-4 increase responsiveness of mast cells to substance P. Exp Hematol 28, 626–34.

Kato, D., Boatright, K.M., Berger, A.B., Nazif, T., Blum, G., Ryan, C., Chehade, K.A., Salvesen, G.S., and Bogyo, M. 2005 Activity-based probes that target diverse cysteine protease families. Nat Chem Biol 1, 33–8.

Knauper, V., Will, H., Lopez-Otin, C., Smith, B., Atkinson, S.J., Stanton, H., Hembry, R.M., and Murphy, G. 1996 Cellular mechanisms for human procollagenase-3 (MMP-13) activation. Evidence that MT1-MMP (MMP-14) and gelatinase a (MMP-2) are able to generate active enzyme. J Biol Chem 271, 17124–31.

Koshikawa, N., Giannelli, G., Cirulli, V., Miyazaki, K., and Quaranta, V. 2000 Role of cell surface metalloprotease MT1-MMP in epithelial cell migration over laminin-5. J Cell Biol 148, 615–24.

Kulka, M., Alexopoulou, L., Flavell, R.A., and Metcalfe, D.D. 2004 Activation of mast cells by double-stranded RNA: Evidence for activation through Toll-like receptor 3. J Allergy Clin Immunol 114, 174–82.

Lakka, S.S., Gondi, C.S., Yanamandra, N., Dinh, D.H., Olivero, W.C., Gujrati, M., and Rao, J.S. 2003 Synergistic down-regulation of urokinase plasminogen activator receptor and matrix metalloproteinase-9 in SNB19 glioblastoma cells efficiently inhibits glioma cell invasion, angiogenesis, and tumor growth. Cancer Res 63, 2454–61.

Lakka, S.S., Gondi, C.S., Yanamandra, N., Olivero, W.C., Dinh, D.H., Gujrati, M., and Rao, J.S. 2004 Inhibition of cathepsin B and MMP-9 gene expression in glioblastoma cell line via RNA interference reduces tumor cell invasion, tumor growth and angiogenesis. Oncogene 23, 4681–9.

Lakka, S.S., Gondi, C.S., Dinh, D.H., Olivero, W.C., Gujrati, M., Rao, V.H., Sioka, C., and Rao, J.S. 2005 Specific interference of urokinase-type plasminogen activator receptor and matrix metalloproteinase-9 gene expression induced by double-stranded RNA results in decreased invasion, tumor growth, and angiogenesis in gliomas. J Biol Chem 280, 21882–92.

Lane, A., and Ley, T.J. 2003 Neutrophil elastase cleaves PML-RARalpha and is important for the development of acute promyelocytic leukemia in mice. Cell 1115, 305–18.

Lee, W.L., and Downey, G.P. 2001 Leukocyte elastase: Physiological functions and role in acute lung injury. Am J Respir Crit Care Med 164, 896–904.

Liu, Z., Zhou, X., Shapiro, S.D., Shipley, J.M., Twining, S.S., Diaz, L.A., Senior, R.M., and Werb, Z. 2000 The serpin alpha1-proteinase inhibitor is a critical substrate for gelatinase B/MMP-9 in vivo. Cell 102, 647–55.

Lopez-Otin, C., and Matrisian, L.M. 2007 Tumour microenvironment: Emerging roles of proteases in tumour suppression. Nat Rev Cancer 7, 800–8.

Lopez-Otin, C., and Overall, C.M. 2002 Protease degradomics: A new challenge for proteomics. Nat Rev Mol Cell Biol 3, 509–19.

Mainardi, C.L., Dixit, S.N., and Kang, A.H. 1980 Degradation of type IV (basement membrane) collagen by a proteinase isolated from human polymorphonuclear leukocyte granules. J Biol Chem 255, 5435–41.

Mallen-St Clair, J., Pham, C.T., Villalta, S.A., Caughey, G.H., and Wolters, P.J. 2004 Mast cell dipeptidyl peptidase I mediates survival from sepsis. J Clin Invest 113, 628–34.

Masson, R., Lefebvre, O., Noel, A., Fahime, M.E., Chenard, M.P., Wendling, C., Kebers, F., LeMeur, M., Dierich, A., Foidart, J.M. et al. 1998 In vivo evidence that the stromelysin-3 metalloproteinase contributes in a paracrine manner to epithelial cell malignancy. J Cell Biol 140, 1535–41.

McCawley, L.J., Crawford, H.C., King, L.E., Jr., Mudgett, J., and Matrisian, L.M. 2004 A protective role for matrix metalloproteinase-3 in squamous cell carcinoma. Cancer Res 64, 6965–72.

McDonald, J.A., and Kelley, D.G. 1980 Degradation of fibronectin by human leukocyte elastase. Release of biologically active fragments. J Biol Chem 255, 8848–58.

McGuire, M.J., Lipsky, P.E., and Thiele, D.L. 1992 Purification and characterization of dipeptidyl peptidase I from human spleen. Arch Biochem Biophys 295, 280–8.

Mohamed, M.M., and Sloane, B.F. 2006 Cysteine cathepsins: Multifunctional enzymes in cancer. Nat Rev Cancer 6, 764–75.

Monea, S., Lehti, K., Keski-Oja, J., and Mignatti, P. 2002 Plasmin activates pro-matrix metalloproteinase-2 with a membrane-type 1 matrix metalloproteinase-dependent mechanism. J Cell Physiol 192, 160–70.

Mott, J.D., and Werb, Z. 2004 Regulation of matrix biology by matrix metalloproteinases. Curr Opin Cell Biol 16, 558–64.

Muramatsu, M., Katada, J., Hayashi, I., and Majima, M. 2000 Chymase as a proangiogenic factor. A possible involvement of chymase-angiotensin-dependent pathway in the hamster sponge angiogenesis model. J Biol Chem 275, 5545–52.

Nakamura, H., Yoshimura, K., McElvaney, N.G., and Crystal, R.G. 1992 Neutrophil elastase in respiratory epithelial lining fluid of individuals with cystic fibrosis induces interleukin-8 gene expression in a human bronchial epithelial cell line. J Clin Invest 89, 1478–84.

Noack, B., Gorgens, H., Hoffmann, T., Fanghanel, J., Kocher, T., Eickholz, P., and Schackert, H. K. 2004 Novel mutations in the cathepsin C gene in patients with pre-pubertal aggressive periodontitis and Papillon-Lefevre syndrome. J Dent Res 83, 368–70.

Nozawa, F., Hirota, M., Okabe, A., Shibata, M., Iwamura, T., Haga, Y., and Ogawa, M. 2000 Elastase activity enhances the adhesion of neutrophil and cancer cells to vascular endothelial cells. J Surg Res 94, 153–8.

Okada, Y., and Nakanishi, I. 1989 Activation of matrix metalloproteinase 3 (stromelysin) and matrix metalloproteinase 2 ('gelatinase') by human neutrophil elastase and cathepsin G. FEBS Lett 249, 353–6.

Overall, C.M., and Kleifeld, O. 2006 Tumour microenvironment—opinion: Validating matrix metalloproteinases as drug targets and anti-targets for cancer therapy. Nat Rev Cancer 6, 227–39.

Owen, C.A., Campbell, M.A., Sannes, P.L., Boukedes, S.S., and Campbell, E.J. 1995 Cell surface-bound elastase and cathepsin G on human neutrophils: A novel, non-oxidative mechanism by which neutrophils focus and preserve catalytic activity of serine proteinases. J Cell Biol 131, 775–89.

Panchal, R.G., Cusack, E., Cheley, S., and Bayley, H. 1996 Tumor protease-activated, pore-forming toxins from a combinatorial library. Nat Biotechnol 14, 852–6.

Pei, D., and Weiss, S.J. 1995 Furin-dependent intracellular activation of the human stromelysin-3 zymogen. Nature 375, 244–7.

Pham, C.T., and Ley, T.J. 1999 Dipeptidyl peptidase I is required for the processing and activation of granzymes A and B in vivo. Proc Natl Acad Sci USA 96, 8627–32.

Pham, C.T., Ivanovich, J.L., Raptis, S.Z., Zehnbauer, B., and Ley, T.J. 2004 Papillon-Lefevre syndrome: Correlating the molecular, cellular, and clinical consequences of cathepsin C/dipeptidyl peptidase I deficiency in humans. J Immunol 173, 7277–81.

Pollanen, J., Hedman, K., Nielsen, L.S., Dano, K., and Vaheri, A. (1988) Ultrastructural localization of plasma membrane-associated urokinase-type plasminogen activator at focal contacts. J Cell Biol 106, 87–95.

Puente, X.S., Pendás, A.M., Llano, E., Velasco, G., and López-Otín, C. 1996 Molecular cloning of a novel membrane-type matrix metalloproteinase from a human breast carcinoma. Cancer Research 56, 944–9.

Ramos-DeSimone, N., Hahn-Dantona, E., Sipley, J., Nagase, H., French, D.L., and Quigley, J.P. 1999 Activation of matrix metalloproteinase-9 (MMP-9) via a converging plasmin/stromely-sin-1 cascade enhances tumor cell invasion. J Biol Chem 274, 13066–76.

Reiling, K.K., Krucinski, J., Miercke, L.J., Raymond, W.W., Caughey, G.H., and Stroud, R.M. 2003 Structure of human pro-chymase: A model for the activating transition of granule-associated proteases. Biochemistry 42, 2616–24.

Ruoss, S.J., Hartmann, T., and Caughey, G.H. 1991 Mast cell tryptase is a mitogen for cultured fibroblasts. J Clin Invest 88, 493–9.

Sakai, K., Ren, S., and Schwartz, L.B. 1996 A novel heparin-dependent processing pathway for human tryptase. Autocatalysis followed by activation with dipeptidyl peptidase I. J Clin Invest 97, 988–95.

Salomon, A.R., Ficarro, S.B., Brill, L.M., Brinker, A., Phung, Q.T., Ericson, C., Sauer, K., Brock, A., Horn, D.M., Schultz, P.G. et al. 2003 Profiling of tyrosine phosphorylation pathways in human cells using mass spectrometry. Proc Natl Acad Sci USA 100, 443–8.

Sameni, M., Moin, K., and Sloane, B.F. 2000 Imaging proteolysis by living human breast cancer cells. Neoplasia 2, 496–504.

Sato, T., Takahashi, S., Mizumoto, T., Harao, M., Akizuki, M., Takasug, M., Fukutomi, T., and Yamashita, J. 2006 Neutrophil elastase and cancer. Surg Oncol 15, 217–22.

Schechter, I., and Berger, A. 1967 On the size of the active site in proteases. I. Papain. Biochem Biophys Res Commun 27, 157–62.

Schechter, N.M., Choi, J.K., Slavin, D.A., Deresienski, D.T., Sayama, S., Dong, G., Lavker, R.M., Proud, D., and Lazarus, G.S. 1986 Identification of a chymotrypsin-like proteinase in human mast cells. J Immunol 137, 962–70.

Shamamian, P., Schwartz, J.D., Pocock, B.J., Monea, S., Whiting, D., Marcus, S.G., and Mignatti, P. 2001 Activation of progelatinase A (MMP-2) by neutrophil elastase, cathepsin G, and proteinase-3: A role for inflammatory cells in tumor invasion and angiogenesis. J Cell Physiol 189, 197–206.

Sheth, P.D., Pedersen, J., Walls, A.F., and McEuen, A.R. 2003 Inhibition of dipeptidyl peptidase I in the human mast cell line HMC-1: Blocked activation of tryptase, but not of the predominant chymotryptic activity. Biochem Pharmacol 66, 2251–62.

Shresta, S., Pham, C.T., Thomas, D.A., Graubert, T.A., and Ley, T.J. 1998 How do cytotoxic lymphocytes kill their targets? Curr Opin Immunol 10, 581–7.

Sieber, S.A., and Cravatt, B.F. 2006 Analytical platforms for activity-based protein profiling–exploiting the versatility of chemistry for functional proteomics. Chem Commun (Camb) 2311–9.

Sloane, B.F., Sameni, M., Podgorski, I., Cavallo-Medved, D., and Moin, K. 2006 Functional imaging of tumor proteolysis. Annu Rev Pharmacol Toxicol 46, 301–15.

Starcher, B., O'Neal, P., Granstein, R.D., and Beissert, S. 1996 Inhibition of neutrophil elastase suppresses the development of skin tumors in hairless mice. J Invest Dermatol 107, 159–63.

Tchougounova, E., Pejler, G., and Abrink, M. 2003 The chymase, mouse mast cell protease 4, constitutes the major chymotrypsin-like activity in peritoneum and ear tissue. A role for mouse mast cell protease 4 in thrombin regulation and fibronectin turnover. J Exp Med 198, 423–31.

Tchougounova, E., Lundequist, A., Fajardo, I., Winberg, J.O., Abrink, M., and Pejler, G. 2005 A key role for mast cell chymase in the activation of pro-matrix metalloprotease-9 and pro-matrix metalloprotease-2. J Biol Chem 280, 9291–6.

Turk, D., Janjic, V., Stern, I., Podobnik, M., Lamba, D., Dahl, S.W., Lauritzen, C., Pedersen, J., Turk, V., and Turk, B. 2001 Structure of human dipeptidyl peptidase I (cathepsin C): Exclusion domain added to an endopeptidase framework creates the machine for activation of granular serine proteases. Embo J 20, 6570–82.

Van Kempen, L.C., de Visser, K.E., and Coussens, L.M. 2006 Inflammation, proteases and cancer. Eur J Cancer 42, 728–34.

Vasiljeva, O., Papazoglou, A., Kruger, A., Brodoefel, H., Korovin, M., Deussing, J., Augustin, N., Nielsen, B.S., Almholt, K., Bogyo, M., et al. 2006 Tumor cell-derived and macrophage-derived cathepsin B promotes progression and lung metastasis of mammary cancer. Cancer Res 66, 5242–50.

Vasiljeva, O., Reinheckel, T., Peters, C., Turk, D., Turk, V., and Turk, B. 2007 Emerging roles of cysteine cathepsins in disease and their potential as drug targets. Curr Pharm Des 13, 387–403.

Wang, Z., Juttermann, R., and Soloway, P.D. 2000 TIMP-2 is required for efficient activation of proMMP-2 in vivo. J Biol Chem 275, 26411–5.

Wilson, C.L., Heppner, K.J., Labosky, P.A., Hogan, B.L., and Matrisian, L.M. 1997 Intestinal tumorigenesis is suppressed in mice lacking the metalloproteinase matrilysin. Proc Natl Acad Sci U S A 94, 1402–7.

Wolters, P.J., Pham, C.T., Muilenburg, D.J., Ley, T.J., and Caughey, G.H. 2001 Dipeptidyl peptidase I is essential for activation of mast cell chymases, but not tryptases, in mice. J Biol Chem 276, 18551–66.

Wong, G.W., Yasuda, S., Morokawa, N., Li, L., and Stevens, R.L. 2004 Mouse chromosome 17A3.3 contains 13 genes that encode functional tryptic-like serine proteases with distinct tissue and cell expression patterns. J Biol Chem 279, 2438–52.

Woodman, R.C., Reinhardt, P.H., Kanwar, S., Johnston, F.L., and Kubes, P. 1993 Effects of human neutrophil elastase (HNE) on neutrophil function in vitro and in inflamed microvessels. Blood 82, 2188–95.

Worley, J.R., Thompkins, P.B., Lee, M.H., Hutton, M., Soloway, P., Edwards, D.R., Murphy, G., and Knauper, V. 2003 Sequence motifs of tissue inhibitor of metalloproteinases 2 (TIMP-2) determining progelatinase A (proMMP-2) binding and activation by membrane-type metalloproteinase 1 (MT1-MMP). Biochem J 372, 799–809.

Wun, T.C., and Reich, E. 1987 An inhibitor of plasminogen activation from human placenta. Purification and characterization. J Biol Chem 262, 3646–53.

Yamashita, J., Tashiro, K., Yoneda, S., Kawahara, K., and Shirakusa, T. 1996 Local increase in polymorphonuclear leukocyte elastase is associated with tumor invasiveness in non-small cell lung cancer. Chest 109, 1328–34.

Yana, I., and Weiss, S.J. 2000 Regulation of membrane type-1 matrix metalloproteinase activation by proprotein convertases. Mol Biol Cell 11, 2387–401.

Young, R.E., Thompson, R.D., Larbi, K.Y., La, M., Roberts, C.E., Shapiro, S.D., Perretti, M., and Nourshargh, S. 2004 Neutrophil elastase (NE)-deficient mice demonstrate a nonredundant role for NE in neutrophil migration, generation of proinflammatory mediators, and phagocytosis in response to zymosan particles in vivo. J Immunol 172, 4493–502.

Young, R.E., Voisin, M.B., Wang, S., Dangerfield, J., and Nourshargh, S. 2007 Role of neutrophil elastase in LTB(4)-induced neutrophil transmigration in vivo assessed with a specific inhibitor and neutrophil elastase deficient mice. Br J Pharmacol.

Zimmer, M., Medcalf, R.L., Fink, T.M., Mattmann, C., Lichter, P., and Jenne, D.E. 1992 Three human elastase-like genes coordinately expressed in the myelomonocyte lineage are organized as a single genetic locus on 19pter. Proc Natl Acad Sci USA 89, 8215–9.

Chapter 10
Physiological Functions of Plasminogen Activation: Effects of Gene Deficiencies in Humans and Mice

Thomas H. Bugge

Abstract Exhaustive analysis of humans and mice with genetic deficiencies in plasminogen, plasminogen activators, plasmin inhibitor, and plasminogen activator inhibitors have yielded fundamental new insights into both the mechanisms of activation and the physiological functions of the plasminogen activation system. At least five different pathways for the activation of plasminogen are operative in vivo, and these five pathways display a remarkable functional redundancy. Plasminogen as well as the components that govern the activation and inhibition of the plasminogen activation system are dispensable for development. However, the cleavage of fibrin and other extracellular substrates by plasmin is critical for postnatal remodeling and repair of multiple epithelial and mesenchymal tissues. As a consequence, life without plasminogen is associated with high morbidity and mortality due to progressive multiorgan damage. Conversely, genetic deficiencies in inhibitors of plasmin and plasminogen activators not only markedly accelerate tissue repair but also result in a lifelong bleeding predisposition due to premature fibrin dissolution.

Introduction

The plasminogen activation system was the first proteolytic system to be implicated in human tumor progression, and for long it has served as a paradigm for extracellular proteolysis in cancer, as well as a prospective target for cancer therapy (Dano et al. 1985, 1999). Plasminogen, its activators, inhibitors, and cellular receptors

T.H. Bugge
Proteases and Tissue Remodeling Unit, Oral and Pharyngeal Cancer Branch, National Institute of Dental and Craniofacial Research, National Institutes of Health, Room 211, 30 Convent Drive, Bethesda, MD 20892, e-mail: thomas.bugge@nih.gov

D. Edwards et al. (eds.), *The Cancer Degradome.* 183

have been exhaustively studied in the context of development, invasion, metastasis, and prognosis of human tumors, as reflected by the publication of more than 4,000 primary research publications and 600 review articles containing the keywords "plasminogen" and "cancer."

This chapter will describe some of the key insights into the physiological functions of plasminogen activation that have been gained over the last two decades from the study of humans and mice deficient in plasminogen, plasmin inhibitor, and plasminogen activator inhibitors. In the chapter, the term "physiological" is broadly defined as processes that can be regarded as evolutionarily beneficial (e.g., development, reproduction, restoration of tissue homeostasis). The many putative proteolytic and nonproteolytic functions of components of the plasminogen activation system that are unrelated to the activation of plasminogen will not be covered [for recent reviews of these exciting topics, *see*, e.g., Blasi and Carmeliet (2002), Loskutoff et al. (1999), Melchor and Strickland (2005), Stefansson and Lawrence (2003), Tsirka (2002), and Yepes and Lawrence (2004)].

The Components of the Plasminogen Activation System – A Short Synopsis

Plasminogen is a modular trypsin-like serine protease zymogen that is converted to the active protease plasmin, by a single endoproteolytic cleavage within the activation site of the serine protease domain. Plasminogen is predominantly synthesized by the liver and is present in a very high concentration (1–2 μm) in plasma and other extravascular fluids (Collen and Lijnen 1986). Lower level extrahepatic synthesis of plasminogen has also been documented (Zhang et al. 2002). Plasminogen is converted to the active protease plasmin by urokinase plasminogen activator (uPA), tissue-type plasminogen activator (tPA), and by an as yet unidentified serine protease (Lund et al. 2006). tPA and uPA are closely related trypsin-like serine proteases that are expressed at many extrahepatic sites, either constitutively or after disruption of homeostasis, leading to local activation of plasminogen (Dano et al. 1985). Following its formation, plasmin is inhibited primarily by α_2-antiplasmin: a fast-acting, serpin-type protease inhibitor that is synthesized by the liver and is present in high concentration in plasma and extravascular fluids (Coughlin 2005, Lijnen and Collen 1985). Three serpin-type inhibitors for uPA and tPA have been described: plasminogen activator inhibitor (PAI-1), PAI-2, and neuroserpin. Of these, PAI-1 appears to be the most critical physiological inhibitor of uPA and tPA, while the functions of PAI-2 and neuroserpin in the inhibition of plasminogen activation at present are unclear (Dougherty et al. 1999, Galliciotti and Sonderegger 2006, Loskutoff 1993, Yepes and Lawrence 2004). After inhibition by their cognate serpins, plasmin and plasminogen activators are internalized for lysosomal degradation by members of the low-density lipoprotein receptor family (Herz and Strickland 2001, Strickland et al. 1995).

The Many Pathways to Plasmin

Five pathways are currently known to lead to the conversion of plasminogen to plasmin during physiological conditions (Fig. 10.1). tPA is a poor activator of plasminogen in solution, but a very potent activator of plasminogen in the context of a fibrin clot. Fibrin strongly promotes tPA-mediated plasminogen activation by serving as a scaffold for the binding of tPA and plasminogen that brings the two

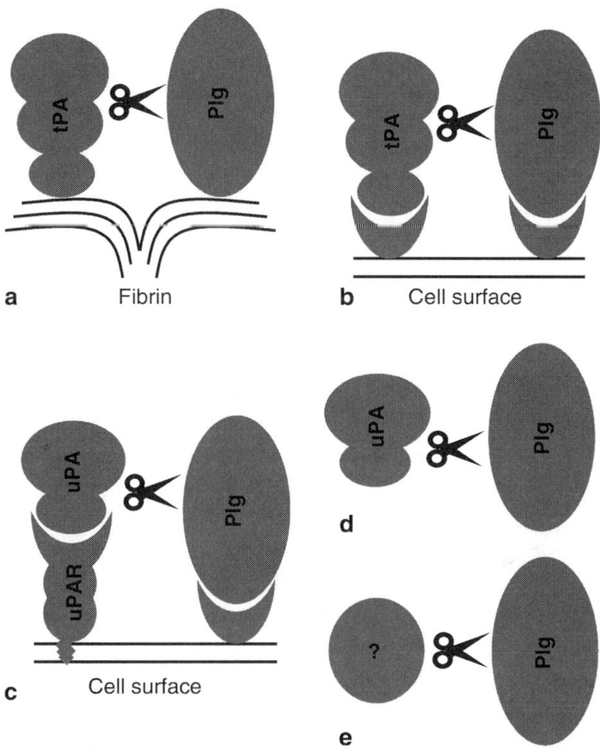

Fig. 10.1 The five known physiological pathways for plasminogen activation. **a** Fibrin-dependent activation of plasminogen (*Plg*) by tissue-type plasminogen activator (*tPA*). Fibrin serves as a scaffold for binding of tPA and Plg and protects newly formed plasmin from inactivation by α_2-antiplasmin. **b** Cell surface-mediated Plg activation by tPA. tPA and Plg bind to specific cell surface receptors that provide a spatially favorable alignment and productive plasmin generation. The cell surface protects plasmin from inactivation by α_2-antiplasmin. **c** urokinase plasminogen activator receptor (*uPAR*)-dependent Plg activation by urokinase plasminogen activator (*uPA*). Pro-uPA binds to uPAR and Plg binds to specific cell surface receptors leading to feedback activation of Plg to plasmin by uPA, and, conversely, pro-uPA to active uPA by plasmin. Cell surface-bound plasmin is protected from inactivation by α_2-antiplasmin. **d** uPAR-independent activation of Plg by uPA. uPA converts Plg to plasmin in a process that does not involve uPAR and may be cell surface dependent or independent. **e** uPA and tPA-independent Plg activation. Plg is proteolytically converted to plasmin by an unknown serine protease that is different from uPA and tPA

molecules in close apposition and by protecting plasmin from inactivation by α_2-antiplasmin (Collen 1980, Collen and Lijnen 2005, Hoylaerts et al. 1982, Thorsen 1992).

A second pathway for plasminogen activation by tPA is cell mediated and involves the simultaneous binding of tPA and plasminogen to the cell surface, which leads to productive plasmin generation. The annexin II-S100A10 heterotetramer recently has emerged as one strong candidate receptor for mediating physiologically relevant tPA-dependent cell surface plasminogen activation, as annexin II-deficient mice develop widespread spontaneous fibrin deposition and cells from annexin II-deficient mice display reduced tPA-mediated plasminogen activation in vitro (Beebe et al. 1989; Felez et al. 1991; Hajjar et al. 1987, 1986; Hajjar et al. 1994; Kim and Hajjar 2002; Ling et al. 2004; Plow et al. 1986).

A principal pathway for plasminogen activation by uPA involves the binding of uPA to a specific cell surface receptor, the urokinase plasminogen activator receptor (uPAR). uPA is synthesized as a single chain proenzyme (pro-uPA) with low intrinsic activity that is efficiently converted to active two-chain uPA by plasmin. Two-chain uPA, in turn, is a potent activator of plasminogen. The concomitant binding of pro-uPA to uPAR, and of plasminogen to as yet not fully characterized cell surface receptors strongly potentiates uPA-mediated plasminogen activation, probably through the formation of ternary complexes that align the two proenzymes in a way that exploits their low intrinsic activity and thereby favors a mutual activation process. The net result of this process is the efficient and localized generation of active uPA and plasmin on the cell surface (Ellis et al. 1991, Ellis and Dano 1993, Ellis et al. 1989, Ronne et al. 1991, Stephens et al. 1989).

Although uPAR appears to be critical for cell-mediated plasminogen activation by uPA in vivo (Liu et al. 2003, Liu et al. 2001), it is not the only physiologically relevant pathway, or even the dominant pathway, for the activation of plasminogen by uPA in mice. Thus, mice deficient in uPAR or mice with combined deficiencies in uPAR and tPA develop a much milder spectrum of phenotypic abnormalities than uPA-deficient mice or mice with combined uPA and tPA deficiencies. Furthermore, neither uPAR-deficient nor uPAR and tPA double-deficient mice display the pronounced defects in tissue repair that are characteristic of uPA-deficient, uPA and tPA double-deficient, or plasminogen-deficient mice (Dewerchin et al. 1996; Bezerra et al. 2001; Bugge et al. 1995a,b; 1996a,b Carmeliet et al. 1998, 1994; Deindl et al. 2003; Kitching et al. 1997; Ploplis et al. 1995; Shanmukhappa et al. 2006; *see* the following section). The dominant uPAR-independent pathway of plasminogen activation by uPA that was defined from these gene inactivation studies may involve as yet uncharacterized cellular receptors for uPA (Longstaff et al. 1999), or, alternatively, be cell-independent.

Recently, it was noted that skin wound healing was more severely impaired in plasminogen-deficient mice than congenic mice with combined uPA and tPA deficiencies. Furthermore, active plasmin and plasmin-α_2-antiplasmin complexes could be readily detected in extracts from uPA and tPA double-deficient wounds. This exciting finding demonstrates the existence of a third physiologically relevant pathway for plasminogen activation in mice (Lund et al. 2006). The protease responsible for the activation of plasminogen in the combined absence of uPA and tPA is currently unknown, but appears to be an ecotin-inhibitable serine protease (Lund et al. 2006).

The various physiological pathways for plasminogen activation display significant functional redundancy in vivo. Thus, mice with single deficiencies in either uPA or tPA do not display the pervasive multiorgan pathology that befalls mice with combined uPA and tPA deficiency (Bugge et al. 1996a, Carmeliet et al. 1994, Drew et al. 1998). Furthermore, the capacity to repair injured tissues is generally much more severely affected in combined uPA and tPA-deficient mice than in mice with single deficiencies in the two plasminogen activators, which often display only modest impairments of tissue repair (Bezerra et al. 2001, Bugge et al. 1996a, Carmeliet et al. 1994, Kitching et al. 1997, Leonardsson et al. 1995, Shanmukhappa et al. 2006). In the central nervous system, however, uPA is expressed at very low levels, and tPA appears to be the dominant physiological plasminogen activator with little or no contribution by uPA under nonpathological conditions (Hoover-Plow et al. 2001, Mataga et al. 2004, Melchor et al. 2003, Mizutani et al. 1996, Nakagami et al. 2000, Oray et al. 2004, Pang et al. 2004, Wu et al. 2000).

Physiological Functions of Plasminogen Activation

Plasminogen deficiency is now established as the principal cause of ligneous conjunctivitis, a rare, multisyndromic, inherited disease that was described as early as 1847 (Bouisson 1847; Schuster et al. 1997, 1999b; Tefs et al. 2006). Critical insights into the physiological roles of plasminogen activation have been gained from the clinical examination of individuals with this disease. As of 2007, about 75 cases of severe or complete plasminogen deficiency in humans have been reported. These deficiencies are caused by homozygosity or compound heterozygosity for a wide assortment of missense, frameshift, splice site, and nonsense mutations that can be found throughout the plasminogen gene. The disease is characterized by the formation of chronic, disfiguring, wood-like (ligneous), fibrin-rich lesions on mucous membranes of multiple body sites. The onset and severity of lesion formation varies considerably from individual to individual. Ligneous lesions generally reappear rapidly after surgical removal due to defective wound healing. Tissues that can be affected include the conjunctiva and cornea of the eye, the gingiva, the ears, the sinuses, the larynx, the vocal cords, the bronchi, the gastrointestinal tract, the female genital tract, and the skin. Congenital occlusive hydrocephalus caused by floating thombi within the cerebrospinal fluid has also been described (Ciftci et al. 2003, Kraft et al. 2000, Ozcelik et al. 2001, Pantanowitz et al. 2004, Schott et al. 1998, Schuster et al. 1999a, Tefs et al. 2006; Table 10.1). Ligneous lesions are often quite debilitating, leading to impaired vision, tooth loss, chronic respiratory infections, infertility, and premature death (Baykul and Bozkurt 2004, Beck et al. 1999, Ciftci et al. 2003, Kraft et al. 2000, Ozcelik et al. 2001, Pantanowitz et al. 2004, Schuster et al. 1999a, Tefs et al. 2006). Interestingly, however, of about 75 patients with documented severe plasminogen deficiency described in the literature, no episodes of venous thrombosis were documented. Histological examination of ligneous lesions often reveals extensive epithelial ulcerations with reactive hyperplasia, surrounding large amorphous, fibrin-rich masses that contain acute and/or chronic inflammatory cell infiltrates composed of neutrophils, T cells, macrophages, B cells, and mast cells. Neovascularization and deposition of plasma proteins such as immunoglobulin and albumin are frequently observed in ligneous lesions, whereas lipid, amyloid, and keratin are generally not detectable (Chambers et al. 1969, Cooper et al. 1979, Eagle et al. 1986, Gunhan et al. 1994, Hidayat and Riddle 1987, Holland et al. 1989, Mingers et al. 1997). Consistent with plasminogen deficiency as the underlying cause, ligneous lesions have been treated very efficiently by systemic or topical administration of plasmin, plasminogen, or even fresh-frozen plasma. This often leads to complete resolution of ligneous lesions (Schott et al. 1998, Tabbara 2004, Watts et al. 2002).

The precise etiology of the ligneous lesions that accompanies plasminogen deficiency in humans has not been definitively established. However, the pronounced accumulation of fibrin and inflammatory cells in ligneous lesions, when combined with data that plasminogen-deficient mice that are also genetically deficient in fibrinogen are completely protected from ligneous lesions, implicates insufficient extravascular fibrinolysis as the principal underlying cause. In this scenario, topical irritation, minor trauma, or infection of plasminogen-deficient mucus membranes triggers an inflammatory response that leads to local fibrin deposition. In the absence of sufficient extravascular fibrin clearance, a "vicious cycle" of fibrin-triggered inflammatory cell recruitment, and inflammatory cell-induced fibrin deposition occurs, causing the characteristic features of ligneous lesions. In summary, the study of the clinical effects of severe plasminogen deficiency in humans suggests a principal role of plasminogen activation in extravascular fibrin surveillance in multiple tissues, in particular those subjected to frequent trauma from environmental exposure.

Plasminogen-deficient mice display a spectrum of phenotypic abnormalities that are very similar to those observed in humans with plasminogen deficiency (Table 10.1). Fetal plasminogen is dispensable for mouse embryonic development, and plasminogen-deficient mice are generally unremarkable at birth (Bugge et al. 1995a, 1996b; Ploplis et al. 1995), although occlusive hydrocephalus has been noted in rare cases (Drew et al. 1998). However, like plasminogen-deficient humans, plasminogen-deficient mice with time develop focal lesions of multiple epithelial tissues, leading to wasting, impaired organ function, and premature death (Bugge et al. 1995a, 1996b; Drew et al. 1998; Ploplis et al. 1995). These lesions can be found in most organ systems and organs in the body, including the gastrointestinal tract (esophagus, squamous and glandular stomach, liver, pancreas, duodenum, colon, rectum), respiratory system (trachea, bronchi, lungs), female genital tract (vagina, uterus of parous females), eye (cornea, conjunctiva), and auditory system (middle ear, tympanic membrane, and external ear; Bugge et al. 1995a, 1996b; Drew et al. 1998; Eriksson et al. 2006; Ploplis et al. 1995). Plasminogen-deficient female mice display substantially diminished ability to nurture their litters due to impaired mammary gland involution and impaired milk secretion, secondary to fibrin accumulation in the alveoli and ducts of the mammary gland (Green et al. 2006, Lund et al. 2000). Although the brain of plasminogen-deficient mice is anatomically normal, the mice display learning deficits that are associated with impaired long-term potentiation and synaptic plasticity (Hoover-Plow et al. 2001, Mataga et al. 2004, Mizutani et al. 1996, Nakagami et al. 2000, Oray et al. 2004, Pang et al. 2004).

The epithelial lesions that develop in plasminogen-deficient mice are histologically very similar to the epithelial lesions observed in plasminogen-deficient humans. They typically exhibit epithelial disruption with focal necrosis of underlying tissue, reactive hyperplasia, and extensive fibrin deposition with profuse inflammatory cell infiltration (Bugge et al. 1995a, 1996b; Drew et al. 1998; Eriksson et al. 2006; Ploplis et al. 1995). Impaired and aberrant tissue repair after chance trauma is likely to be the primary underlying cause of the spontaneous lesions that accumulate in plasminogen-deficient mice. Thus, studies of the kinetics and the overall outcome of the healing of defined injuries of plasminogen-deficient mice that are generated in a wide variety of epithelial and mesenchymal tissues have revealed a generalized and severe impairment of tissue repair (Table 10.1). The delay and aberrant healing documented in plasminogen-deficient mice include intravascular thrombus dissolution (Lijnen et al. 1996), incisional skin wounds (Lund et al. 1999, Romer et al. 1996), tympanic membrane perforation (Li et al. 2006), scrape and excimer-induced corneal wounds (Drew et al. 2000), experimental glomerulonephritis (Kitching et al. 1997), organic solvent-induced liver necrosis (Bezerra et al. 1999, Pohl et al. 2001), antigen-induced arthritis (Busso et al. 1998), skeletal muscle crush injury (Suelves et al. 2002), peripheral nerve injury (Akassoglou et al. 2000, 2002; Siconolfi and Seeds, 2001), neuronal remodeling after kainate-induced seizure (Wu et al. 2000), amyloid deposition in the brain (Melchor et al. 2003), and experimental myocardial infarction (Creemers et al. 2000).

Outside of the central nervous system, the impaired dissolution of fibrin appears to be the principal molecular defect that underlies the spontaneous multiorgan pathology and defective tissue repair associated with plasminogen deficiency in mice (Table 10.1). Thus, the genetic elimination of fibrinogen in plasminogen-deficient mice prevents wasting, normalizes the life span, and also completely prevents the formation of spontaneous lesions in the digestive tract, the respiratory tract, the urogenital tract, the cornea, the conjunctiva, and other tissues and organs (Bugge et al. 1996b, Drew et al. 1998). Likewise, with the notable exception of resolution of carbon tetrachloride-induced liver necrosis (Bezerra et al. 1999, Ng et al. 2001), fibrinogen gene disruption, or fibrinogen depletion have normalized the kinetics and overall outcome of the healing of all experimental tissue injuries where this has been tested. These include skin wound healing (Bugge et al. 1996b), corneal wound healing (Kao et al. 1998), antigen-induced arthritis (Busso et al. 1998), muscle regeneration (Suelves et al. 2002), and peripheral nerve damage (Akassoglou et al. 2000, 2002). Furthermore, heterozygosity for the fibrinogen gene markedly improved the lactational competence of plasminogen-deficient females (Green et al. 2006).

Table 10.1 Physiological effects of congenital plasminogen deficiency in humans and mice

	Fibrinogen dependence[b]
Humans	
Ligneous lesions[a]	
Eyes (cornea, conjunctiva)	
Auditory canal (middle ear)	
Mouth (gingiva)	
Respiratory tract (sinuses, larynx, vocal cords, trachea, bronchi)	
Gastrointestinal tract (stomach)	
Female genital tract (vagina, cervix)	
Occlusive hydrocephalus	
Impaired fertility	
Impaired wound healing	
Increased mortality	
Mice	
Ligneous lesions	
Eyes (cornea, conjunctiva)	Yes
Auditory canal (middle ear, tympanic membrane)	ND
Respiratory tract (trachea, bronchi, lungs)	Yes
Gastrointestinal tract (esophagus, stomach, liver, pancreas, duodenum, colon, rectum)	Yes
Female genital tract (vagina, uterus)	Yes
Occlusive hydrocephalus	ND
Impaired fertility	ND
Impaired lactation	Yes
Wasting	Yes
Impaired learning	No
Increased mortality	Yes
Impaired tissue repair	
Intravascular thrombosis	Yes
Skin (incisional wounds)	Yes
Tympanic membrane (rupture)	ND
Cornea (excimer laser ablation, scrape)	Yes
Kidney (glomerulonephritis)	ND
Joints (antigen-induced arthritis)	Yes
Skeletal muscle (crush injury)	Yes
Heart (myocardial infarction)	ND
Peripheral nervous system (crush injury)	Yes
Liver (necrosis)	No
Central nervous system (kainite-induced seizure, amyloid deposition)	No

[a]Fibrin and inflammatory cell-rich white, yellowish or reddish pseudomembranous, occasionally vascularized lesions named after their wood-like (ligneous) appearance

[b]Phenotype alleviated by fibrinogen deficiency, fibrinogen haploinsufficiency, or fibrinogen depletion

ND Not determined

Compiled from Akassoglou et al. (2000, 2002), Baykul and Bozkurt (2004), Bezerra et al. (1999), Bugge et al. (1995a, 1996b), Busso et al. (1998), Chambers et al. (1969), Ciftci et al. (2003), Cooper et al. (1979), Creemers et al. (2000), Drew et al. (1998), Eagle et al. (1986), Eriksson et al.

Fibrinogen expression is neglectable within the central nervous system under nonpathological conditions, and plasmin exerts its critical functions in brain homeostasis, injury repair, and learning independent of fibrin cleavage. In this regard, a number of candidate plasmin substrates have been identified, including probrain-derived neurotropic factor, laminin, proteoglycans, and amyloid-β. This indicates that plasminogen has multiple proteolytic targets in the brain (Melchor, Pawlak and Strickland 2003, Nakagami et al. 2000, Pang et al. 2004, Wu et al. 2000).

Physiological Functions of Inhibitors of Plasmin and Plasminogen Activators

α_2-Antiplasmin

Congenital autosomal α_2-antiplasmin deficiency is a rare disorder in humans that was first described in 1978 (Koie et al. 1978). Affected individuals are normal at birth, but suffer a moderate to severe lifelong predisposition for spontaneous and trauma-induced bleeding, as well as rebleeding after hemostasis has been achieved (Table 10.2). Episodes reported in these individuals include umbilical cord bleeding, urinary tract bleeding, bleeding gums, bleeding into the chest cavity, joint bleeding, spontaneous and traumatic subcutaneous bleeding, subarachnoid, epidural, and cerebral bleeding, and excessive bleeding after tooth extraction (Harish et al. 2006, Kluft et al. 1979, Kluft et al. 1982, Miles et al. 1982, Yoshinaga et al. 2000, Yoshioka et al. 1982). These bleeding episodes appear to be secondary to accelerated fibrinolysis, and they have been effectively treated with the plasmin inhibitor tranexamic acid, which inhibits the binding of plasminogen to fibrin (Kettle and Mayne 1985, Kluft et al. 1982, Yoshioka et al. 1982). The underlying cause of congenital α_2-antiplasmin deficiency, where determined, was attributed to splice site mutations, frameshift mutations, in frame deletions, and missense mutations within the α_2-antiplasmin gene (Hanss et al. 2003, Holmes et al. 1987, Lind and Thorsen 1999, Miura et al. 1989a, Miura et al. 1989b, Yoshinaga et al. 2000). Interestingly, relatives of affected individuals with heterozygous α_2-antiplasmin deficiency display a mild bleeding tendency (Hanss et al. 2003, Kluft et al. 1982, Leebeek et al. 1988, Miles et al. 1982).

Unlike humans, α_2-antiplasmin deficiency in mice does not appear to be associated with spontaneous bleeding under standard animal housing conditions.

Table 10.2 (Continued) (2006), Green et al. (2006), Gunhan et al. (1994), Hidayat and Riddle (1987), Holland et al. (1989), Hoover-Plow et al. (2001), Kao et al. (1998), Kitching et al. (1997), Kraft et al. (2000), Li et al. (2006), Lijnen et al. (1996), Lund et al. (1999, 2000), Mataga et al. (2004), Melchor et al. (2003), Mingers et al. (1997), Mizutani et al. (1996), Nakagami et al. (2000), Ng et al. (2001), Oray et al. (2004), Ozcelik et al. (2001), Pang et al. (2004), Pantanowitz et al. (2004), Ploplis et al. (1995), Pohl et al. (2001), Romer et al. (1996), Schott et al. (1998), Schuster et al. (1997, 1999b), Siconolfi and Seeds (2001), Suelves et al. (2002), Tabbara (2004), Tefs et al. (2006), Watts et al. (2002), and Wu et al. (2000)

Table 10.2 Physiological effects of a_2-antiplasmin-deficiency[a] in humans and mice

Humans
Spontaneous and trauma-induced bleeding episodes
Umbilical cord
Urinary tract
Gums
Chest cavity
Joints
Subcutaneous
Subarachnoid
Epidural
Cerebral
Tooth extraction socket
Mice[b]
Accelerated tissue repair
Intravascular thrombosis (endotoxin induced) Skin (incisional wounds)
Liver (necrosis)

[a]Congenital α_2-antiplasmin deficiency (heterozygous or homozygous deficiency)
[b]Spontaneous or trauma-induced bleeding episodes have not been reported in α_2-antiplasmin-deficient mice
Compiled from Hanss et al. (2003), Harish et al. (2006), Holmes et al. (1987), Kanno et al. (2006), Kettle and Mayne (1985), Kluft et al. (1979), Kluft et al. (1982), Koie et al. (1978), Leebeek et al. (1988), Lijnen et al. (1999), Lind and Thorsen (1999), Miles et al. (1982), Miura et al. (1989a), Miura et al. (1989b), Okada et al. (2004), Yoshinaga et al. (2000), and Yoshioka et al. (1982)

Furthermore, although lysis of fibrin clots was accelerated in these mice, bleeding times were not increased after tail tip or toe amputation (Lijnen et al. 1999). Interestingly, the increased plasmin activity caused by α_2-antiplasmin deficiency provided increased protection from endotoxin-induced thrombosis (Lijnen et al., 1999), accelerated the regeneration after toxic liver injuries, and enhanced skin wound healing (Kanno et al. 2006, Okada et al. 2004).

PAI-1

Humans with very low or undetectable PAI-1 have been identified (Dieval et al. 1991; Fay et al. 1997, 1992; Lee et al. 1993; Minowa et al. 1999; Schleef et al. 1989; Takahashi et al. 1996). In most cases, the molecular deficiency that underlies this autosomal recessive disorder has not been determined. However, thorough analysis of one large kindred, in which PAI-1 deficiency was frequent, uncovered a frameshift mutation in exon 4 of the PAI-1 gene, which leads to the generation of a null allele. Homozygosity for this null allele was documented in seven people and heterozygosity for the null allele was documented in 19 relatives of these inviduals (Fay et al. 1997, 1992). The collective analysis of PAI-1-deficient humans has revealed a critical role of PAI-1 in hemostasis (Table 10.3). PAI-1-deficient humans display supra-physiological levels of plasminogen activator activity, which causes a lifelong predisposition for spontaneous and trauma-induced bleeding. Reported

Table 10.3 Physiological effects of congenital PAI-1 deficiency in humans and mice

Humans
Spontaneous and trauma-induced bleeding episodes
Joints
Periosteum
Epidural
Tooth extraction socket
Chest cavity
Subcutaneous
Prolonged menstrual bleeding
Mice[a]
Accelerated tissue repair
Intravascular thrombosis (endotoxin induced)
Skin (incisional wounds)
Joints (antigen-induced arthritis)
Lungs (bleomycin-induced fibrosis)

[a]Spontaneous or trauma-induced bleeding episodes have not been reported in PAI-1-deficient mice
Compiled from Carmeliet et al. (1993a, 1993b), Chan et al. (2001), Dieval et al. (1991), Eitzman et al. (1996), Fay et al. (1997, 1992), Kawasaki et al. (2000), Lee et al. (1993), Minowa et al. (1999), Oda et al. (2001), Schleef et al. (1989), Suelves et al. (2005), Takahashi et al. (1996), Van Ness et al. (2002), and Zhu et al. (1999)

cases include recurrent bleeding into knee and elbow joints, subperiosteal bleeding after jaw trauma, epidural bleeding after head trauma, delayed bleeding after inguinal hernia surgery, prolonged bleeding after tooth extraction, and frequent bruising. Excessive menstrual bleeding is also a common predilection of PAI-1-deficient females. Bleeding episodes in PAI-1 deficient humans have been treated effectively by oral administration of α-aminocaproic acid or tranexamic acid (Dieval et al. 1991, Fay et al. 1997, 1992; Lee et al. 1993; Minowa et al. 1999; Schleef et al. 1989; Takahashi et al. 1996). Heterozygous siblings and parents of affected individuals were unremarkable. Given the many critical functions proposed for PAI-1 in cell migration, cell adhesion, angiogenesis, and immunity, it is curious that detailed physical examinations of humans with complete PAI-1-deficiency have failed to uncover any physiological abnormalities besides excessive spontaneous or trauma-induced bleeding.

PAI-1-deficient mice develop and reproduce normally, but present a hyperfibrinolytic state characterized by accelerated lysis of intravascular and ex vivo fibrin clots. However, like α_2-antiplasmin-deficient mice, the physiological consequences of loss of PAI-1 in mice appear to be less severe than in humans. Spontaneous bleeding episodes were not recorded in PAI-1-deficient mice under standard housing conditions, and the mice did not display increased bleeding or rebleeding after partial amputation of the tail or cecum (Carmeliet et al. 1993a, 1993b). However, consistent with the general impairment of tissue repair observed in mice with reduced plasminogen activation (plasminogen-deficient mice, plasminogen activator-deficient mice), the surpraphysiological state of activation of plasminogen that accompanies PAI-1 deficiency appears to improve the time to healing and the overall outcome of a diverse number of tissue injuries in mice. These include intravascular thrombosis (Carmeliet et al. 1993b, Kawasaki et al. 2000, Zhu et al. 1999), obstructive kidney damage (Oda et al. 2001), skeletal muscle injury (Suelves et al. 2005), incisional skin wound healing (Chan et al. 2001), antigen-induced arthritis (Van Ness et al. 2002), and bleomycin-induced lung injury (Eitzman et al. 1996). The collective findings from these studies have made PAI-1 an increasingly attractive drug candidate.

Conclusions

Two decades of exhaustive analysis of humans and mice with genetic deficiencies in plasminogen, plasminogen activators, plasmin and plasminogen activator inhibitors have yielded fundamental new insights into the physiological role of the activation of plasminogen. Plasminogen and the components that govern plasminogen activation are dispensable for development. However, the cleavage of fibrin and other extracellular substrates by plasmin is critical to the postnatal remodeling and repair of multiple epithelial and mesenchymal tissues, and life without plasminogen is associated with high morbidity and mortality. At least five different pathways for the activation of plasminogen are operative in vivo, and the five pathways display a remarkable functional redundancy. Genetic deficiencies in inhibitors of plasmin and plasminogen activators not only cause lifelong bleeding predispositions but also accelerate tissue repair and regeneration.

Acknowledgments The author thanks Dr. Stella Tsirka for advice on the physiology of plasminogen activation in the central nervous system, and Drs. Mary Jo Danton and Silvio Gutkind for critically reading the manuscript. This work was supported by the National Institute of Dental and Craniofacial Research Intramural Program.

References

Akassoglou, K., Kombrinck, K.W., Degen, J.L., Strickland, S. 2000. Tissue plasminogen activator-mediated fibrinolysis protects against axonal degeneration and demyelination after sciatic nerve injury. J Cell Biol 149(5):1157–1166.

Akassoglou, K., Yu, W.M., Akpinar, P., Strickland, S. 2002. Fibrin inhibits peripheral nerve remyelination by regulating Schwann cell differentiation. Neuron 33(6):861–875.

Baykul, T., Bozkurt, Y. 2004. Destructive membranous periodontal disease (ligneous periodontitis): A case report and 3 years follow-up. Br Dent J 197(8):467–468.

Beck, J.M., Preston, A.M., Gyetko, M.R. 1999. Urokinase-type plasminogen activator in inflammatory cell recruitment and host defense against Pneumocystis carinii in mice. Infect Immun 67(2):879–884.

Beebe, D.P., Miles, L.A., Plow, E.F. 1989. A linear amino acid sequence involved in the interaction of t-PA with its endothelial cell receptor. Blood 74(6):2034–2037.

Bezerra, J.A., Bugge, T.H., Melin-Aldana, H., Sabla, G., Kombrinck, K.W., Witte, D.P., Degen, J.L. 1999. Plasminogen deficiency leads to impaired remodeling after a toxic injury to the liver. Proc Natl Acad Sci USA 96(26):15143–15148.

Bezerra, J.A., Currier, A.R., Melin-Aldana, H., Sabla, G., Bugge, T.H., Kombrinck, K.W., Degen, J.L. 2001. Plasminogen activators direct reorganization of the liver lobule after acute injury. Am J Pathol 158(3):921–929.

Blasi, F., Carmeliet, P. 2002. uPAR: A versatile signalling orchestrator. Nat Rev Mol Cell Biol 3 (12):932–943.

Bouisson, M. 1847. Ophthalmie sur-aigue avec formation de pseudomembranes a la surface de la conjunctive. Ann Ocul 17:100–104.

Bugge, T.H., Flick, M.J., Daugherty, C.C., Degen, J.L. 1995a. Plasminogen deficiency causes severe thrombosis but is compatible with development and reproduction. Genes Dev 9(7):794–807.

Bugge, T.H., Suh, T.T., Flick, M.J., Daugherty, C.C., Romer, J., Solberg, H., Ellis, V., Dano, K., Degen, J.L. 1995b. The receptor for urokinase-type plasminogen activator is not essential for mouse development or fertility. J Biol Chem 270(28):16886–16894.

Bugge, T.H., Flick, M.J., Danton, M.J., Daugherty, C.C., Romer, J., Dano, K., Carmeliet, P., Collen, D., Degen, J.L. 1996a. Urokinase-type plasminogen activator is effective in fibrin clearance in the absence of its receptor or tissue-type plasminogen activator. Proc Natl Acad Sci USA 93(12):5899–5904.

Bugge, T.H., Kombrinck, K.W., Flick, M.J., Daugherty, C.C., Danton, M.J., Degen, J.L. 1996b. Loss of fibrinogen rescues mice from the pleiotropic effects of plasminogen deficiency. Cell 87 (4):709–719.

Busso, N., Peclat, V., Van Ness, K., Kolodziesczyk, E., Degen, J., Bugge, T., So, A. 1998. Exacerbation of antigen-induced arthritis in urokinase-deficient mice. J Clin Invest 102 (1):41–50.

Carmeliet, P., Kieckens, L., Schoonjans, L., Ream, B., van Nuffelen, A., Prendergast, G., Cole, M., Bronson, R., Collen, D., Mulligan, R.C. 1993a. Plasminogen activator inhibitor-1 gene-deficient mice. I. Generation by homologous recombination and characterization. J Clin Invest 92 (6):2746–2755.

Carmeliet, P., Stassen, J.M., Schoonjans, L., Ream, B., van den Oord, J.J., De Mol, M., Mulligan, R.C., Collen, D. 1993b. Plasminogen activator inhibitor-1 gene-deficient mice. II. Effects on hemostasis, thrombosis, and thrombolysis. J Clin Invest 92(6):2756–2760.

Carmeliet, P., Schoonjans, L., Kieckens, L., Ream, B., Degen, J., Bronson, R., De Vos, R., van den Oord, J.J., Collen, D., Mulligan, R.C. 1994. Physiological consequences of loss of plasminogen activator gene function in mice. Nature 368(6470):419–424.

Carmeliet, P., Moons, L., Dewerchin, M., Rosenberg, S., Herbert, J.M., Lupu, F., Collen, D. 1998. Receptor-independent role of urokinase-type plasminogen activator in pericellular plasmin and matrix metalloproteinase proteolysis during vascular wound healing in mice. J Cell Biol 140 (1):233–245.

Chambers, J.D., Blodi, F.C., Golden, B., McKee, A.P. 1969. Ligneous conjunctivitis. Trans Am Acad Ophthalmol Otolaryngol 73(5):996–1004.

Chan, J.C., Duszczyszyn, D.A., Castellino, F.J., Ploplis, V.A. 2001. Accelerated skin wound healing in plasminogen activator inhibitor-1-deficient mice. Am J Pathol 159(5):1681–1688.

Ciftci, E., Ince, E., Akar, N., Dogru, U., Tefs, K., Schuster, V. 2003. Ligneous conjunctivitis, hydrocephalus, hydrocele, and pulmonary involvement in a child with homozygous type I plasminogen deficiency. Eur J Pediatr 162(7–8):462–465.

Collen, D. 1980. On the regulation and control of fibrinolysis. Edward Kowalski memorial lecture. Thromb Haemost 43(2):77–89.

Collen, D., Lijnen, H.R. 1986. The fibrinolytic system in man. Crit Rev Oncol Hematol 4(3):249–301.

Collen, D., Lijnen, H.R. 2005. Thrombolytic agents. Thromb Haemost 93(4):627–630.

Cooper, T.J., Kazdan, J.J., Cutz, E. 1979. Lingeous conjunctivitis with tracheal obstruction. A case report, with light and electron microscopy findings. Can J Ophthalmol 14(1):57–62.

Coughlin, P.B. 2005. Antiplasmin: The forgotten serpin? Febs J 272(19):4852–4857.

Creemers, E., Cleutjens, J., Smits, J., Heymans, S., Moons, L., Collen, D., Daemen, M., Carmeliet, P. 2000. Disruption of the plasminogen gene in mice abolishes wound healing after myocardial infarction. Am J Pathol 156(6):1865–1873.

Dano, K., Andreasen, P.A., Grondahl-Hansen, J., Kristensen, P., Nielsen, L.S., Skriver, L. 1985. Plasminogen activators, tissue degradation, and cancer. Adv Cancer Res 44:139–266.

Dano, K., Romer, J., Nielsen, B.S., Bjorn, S., Pyke, C., Rygaard, J., Lund, L.R. 1999. Cancer invasion and tissue remodeling—Cooperation of protease systems and cell types. Apmis 107(1):120–127.

Deindl, E., Ziegelhoffer, T., Kanse, S.M., Fernandez, B., Neubauer, E., Carmeliet, P., Preissner, K. T., Schaper, W. 2003. Receptor-independent role of the urokinase-type plasminogen activator during arteriogenesis. Faseb J 17(9):1174–1176.

Dewerchin, M., Nuffelen, A.V., Wallays, G., Bouche, A., Moons, L., Carmeliet, P., Mulligan, R. C., Collen, D. 1996. Generation and characterization of urokinase receptor-deficient mice. J Clin Invest 97(3):870–878.

Dieval, J., Nguyen, G., Gross, S., Delobel, J. Kruithof, E.K. 1991. A lifelong bleeding disorder associated with a deficiency of plasminogen activator inhibitor type 1. Blood 77(3):528–532.

Dougherty, K.M., Pearson, J.M., Yang, A.Y., Westrick, R.J., Baker, M.S., Ginsburg, D. 1999. The plasminogen activator inhibitor-2 gene is not required for normal murine development or survival. Proc Natl Acad Sci USA 96(2):686–691.

Drew, A.F., Kaufman, A.H., Kombrinck, K.W., Danton, M.J., Daugherty, C.C., Degen, J.L., Bugge, T.H. 1998. Ligneous conjunctivitis in plasminogen-deficient mice. Blood 91 (5):1616–1624.

Drew, A.F., Schiman, H.L., Kombrinck, K.W., Bugge, T.H., Degen, J.L., Kaufman, A.H. 2000. Persistent corneal haze after excimer laser photokeratectomy in plasminogen-deficient mice. Invest Ophthalmol Vis Sci 41(1):67–72.

Eagle, R.C. Jr., Brooks, J.S., Katowitz, J.A., Weinberg, J.C., Perry, H.D. 1986. Fibrin as a major constituent of ligneous conjunctivitis. Am J Ophthalmol 101(4):493–494.

Eitzman, D.T., McCoy, R.D., Zheng, X., Fay, W.P., Shen, T., Ginsburg, D., Simon, R.H. 1996. Bleomycin-induced pulmonary fibrosis in transgenic mice that either lack or overexpress the murine plasminogen activator inhibitor-1 gene. J Clin Invest 97(1):232–237.

Ellis, V., Dano, K. 1993. Potentiation of plasminogen activation by an anti-urokinase monoclonal antibody due to ternary complex formation. A mechanistic model for receptor-mediated plasminogen activation. J Biol Chem 268(7):4806–4813.

Ellis, V., Scully, M.F., Kakkar, V. 1989. Plasminogen activation initiated by single-chain uroki-nase-type plasminogen activator. Potentiation by U937 monocytes. J Biol Chem 264(4):2185–2188.

Ellis, V., Behrendt, N., Dano, K. 1991. Plasminogen activation by receptor-bound urokinase. A kinetic study with both cell-associated and isolated receptor. J Biol Chem 266(19):12752–12758.

Eriksson, P.O., Li, J., Ny, T., Hellstrom, S. 2006. Spontaneous development of otitis media in plasminogen-deficient mice. Int J Med Microbiol 296(7):501–509.

Fay, W.P., Shapiro, A.D., Shih, J.L., Schleef, R.R., Ginsburg, D. 1992. Brief report: Complete deficiency of plasminogen-activator inhibitor type 1 due to a frame-shift mutation. N Engl J Med 327(24):1729–1733.

Fay, W.P., Parker, A.C., Condrey, L.R., Shapiro, A.D. 1997. Human plasminogen activator inhibitor-1 (PAI-1) deficiency: Characterization of a large kindred with a null mutation in the PAI-1 gene. Blood 90(1):204–208.

Felez, J., Chanquia, C.J., Levin, E.G., Miles, L.A., Plow, E.F. 1991. Binding of tissue plasminogen activator to human monocytes and monocytoid cells. Blood 78(9):2318–2327.

Galliciotti, G., Sonderegger, P. 2006. Neuroserpin. Front Biosci 11:33–45.

Green, K.A., Nielsen, B.S., Castellino, F.J., Romer, J., Lund, L.R. 2006. Lack of plasminogen leads to milk stasis and premature mammary gland involution during lactation. Dev Biol 299 (1):164–175.

Gunhan, O., Celasun, B., Perrini, F., Covani, U., Perrini, N., Ozdemir, A., Bostanci, H., Finci, R. 1994. Generalized gingival enlargement due to accumulation of amyloid-like material. J Oral Pathol Med 23(9):423–428.

Hajjar, K.A., Harpel, P.C., Jaffe, E.A., Nachman, R.L. 1986. Binding of plasminogen to cultured human endothelial cells. J Biol Chem 261(25):11656–11662.

Hajjar, K.A., Hamel, N.M., Harpel, P.C., Nachman, R.L. 1987. Binding of tissue plasminogen activator to cultured human endothelial cells. J Clin Invest 80(6):1712–1719.

Hajjar, K.A., Jacovina, A.T., Chacko, J. 1994. An endothelial cell receptor for plasminogen/tissue plasminogen activator. I. Identity with annexin II. J Biol Chem 269(33):21191–21197.

Hanss, M.M., Farcis, M., Ffrench, P.O., de Mazancourt, P., Dechavanne, M. 2003. A splicing donor site point mutation in intron 6 of the plasmin inhibitor (alpha2 antiplasmin) gene with heterozygous deficiency and a bleeding tendency. Blood Coagul Fibrinolysis 14(1):107–111.

Harish, V.C., Zhangm, L., Huff, J.D., Lawson, H., Owen, J. 2006. Isolated antiplasmin deficiency presenting as a spontaneous bleeding disorder in a 63-year-old man. Blood Coagulat Fibrinol 17(8):673–675.

Herz, J., Strickland, D.K. 2001. LRP: A multifunctional scavenger and signaling receptor. J Clin Invest 108(6):779–784.

Hidayat, A.A., Riddle, P.J. 1987. Ligneous conjunctivitis. A clinicopathologic study of 17 cases. Ophthalmology 94(8):949–959.

Holland, E.J., Chan, C.C., Kuwabara, T., Palestine, A.G., Rowsey, J.J., Nussenblatt, R.B. 1989. Immunohistologic findings and results of treatment with cyclosporine in ligneous conjunctivitis. Am J Ophthalmol 107(2):160–166.

Holmes, W.E., Lijnen, H.R., Nelles, L., Kluft, C., Nieuwenhuis, H.K., Rijken, D.C., Collen, D. 1987. Alpha 2-antiplasmin Enschede: Alanine insertion and abolition of plasmin inhibitory activity. Science 238(4824):209–211.

Hoover-Plow, J., Skomorovska-Prokvolit, O., Welsh, S. 2001. Selective behaviors altered in plasminogen-deficient mice are reconstituted with intracerebroventricular injection of plasminogen. Brain Res 898(2):256–264.

Hoylaerts, M., Rijken, D.C., Lijnen, H.R., Collen, D. 1982. Kinetics of the activation of plasminogen by human tissue plasminogen activator. Role of fibrin. J Biol Chem 257(6):2912–2919.

Kanno, Y., Hirade, K., Ishisaki, A., Nakajima, K., Suga, H., Into, T., Matsushita, K., Okada, K., Matsuo, O., Matsuno, H. 2006. Lack of alpha2-antiplasmin improves cutaneous wound healing via over-released vascular endothelial growth factor-induced angiogenesis in wound lesions. J Thromb Haemost 4(7):1602–1610.

Kao, W.W., Kao, C.W., Kaufman, A.H., Kombrinck, K.W., Converse, R.L., Good, W.V. Bugge, T.H., Degen, J.L. 1998. Healing of corneal epithelial defects in plasminogen- and fibrinogen-deficient mice. Invest Ophthalmol Vis Sci 39(3):502–508.

Kawasaki, T., Dewerchin, M., Lijnen, H.R., Vermylen, J., Hoylaerts, M.F. 2000. Vascular release of plasminogen activator inhibitor-1 impairs fibrinolysis during acute arterial thrombosis in mice. Blood 96(1):153–160.

Kettle, P., Mayne, E.E. 1985. A bleeding disorder due to deficiency of alpha 2-antiplasmin. J Clin Pathol 38(4):428–429.

Kim, J., Hajjar, K.A. 2002. Annexin II: A plasminogen-plasminogen activator co-receptor. Front Biosci 7:d341–348.

Kitching, A.R., Holdsworth, S.R., Ploplis, V.A., Plow, E.F., Collen, D., Carmeliet, P., Tipping, P.G. 1997. Plasminogen and plasminogen activators protect against renal injury in crescentic glomerulonephritis. J Exp Med 185(5):963–968.

Kluft, C., Vellenga, E., Brommer, E.J. 1979. Homozygous alpha 2-antiplasmin deficiency. Lancet 2(8135):206.

Kluft, C., Vellenga, E., Brommer, E.J., Wijngaards, G. 1982. A familial hemorrhagic diathesis in a Dutch family: An inherited deficiency of alpha 2-antiplasmin. Blood 59(6):1169–1180.

Koie, K., Kamiya, T., Ogata, K., Takamatsu, J. 1978. Alpha2-plasmin-inhibitor deficiency (Miyasato disease). Lancet 2(8104–5):1334–1336.

Kraft, J., Lieb, W., Zeitler, P., Schuster, V. 2000. Ligneous conjunctivitis in a girl with severe type I plasminogen deficiency. Graefes Arch Clin Exp Ophthalmol 238(9):797–800.

Lee, M.H., Vosburgh, E., Anderson, K., McDonagh, J. 1993. Deficiency of plasma plasminogen activator inhibitor 1 results in hyperfibrinolytic bleeding. Blood 81(9):2357–2362.

Leebeek, F.W., Stibbe, J., Knot, E.A., Kluft, C., Gomes, M.J., Beudeker, M. 1988. Mild haemostatic problems associated with congenital heterozygous alpha 2-antiplasmin deficiency. Thromb Haemost 59(1):96–100.

Leonardsson, G., Peng, X.R., Liu, K., Nordstrom, L., Carmeliet, P., Mulligan, R., Collen, D., Ny, T. 1995. Ovulation efficiency is reduced in mice that lack plasminogen activator gene function: Functional redundancy among physiological plasminogen activators. Proc Natl Acad Sci USA 92(26):12446–12450.

Li, J., Eriksson, P.O., Hansson, A., Hellstrom, S., Ny, T. 2006. Plasmin/plasminogen is essential for the healing of tympanic membrane perforations. Thromb Haemost 96(4):512–519.

Lijnen, H.R., Collen, D. 1985. Protease inhibitors of human plasma. Alpha-2-antiplasmin. J Med 16(1–3):225–284.

Lijnen, H.R., Carmelie, P., Bouche, A., Moons, L., Ploplis, V.A., Plow, E.F., Collen, D. 1996. Restoration of thrombolytic potential in plasminogen-deficient mice by bolus administration of plasminogen. Blood 88(3):870–876.

Lijnen, H.R., Okada, K., Matsuo, O., Collen, D., Dewerchin, M. 1999. Alpha2-antiplasmin gene deficiency in mice is associated with enhanced fibrinolytic potential without overt bleeding. Blood 93(7):2274–2281.

Lind, B., Thorsen, S. 1999. A novel missense mutation in the human plasmin inhibitor (alpha2-antiplasmin) gene associated with a bleeding tendency. Br J Haematol 107(2):317–322.

Ling, Q., Jacovina, A.T., Deora, A., Febbraio, M., Simantov, R., Silverstein, R.L., Hempstead, B., Mark, W.H., Hajjar, K.A. 2004. Annexin II regulates fibrin homeostasis and neoangiogenesis in vivo. J Clin Invest 113(1):38–48.

Liu, S., Bugge, T.H., Leppla, S.H. 2001. Targeting of tumor cells by cell surface urokinase plasminogen activator-dependent anthrax toxin. J Biol Chem 276(21):17976–17984.

Liu, S., Aaronson, H., Mitola, D.J., Leppla, S.H., Bugge, T.H. 2003. Potent antitumor activity of a urokinase-activated engineered anthrax toxin. Proc Natl Acad Sci USA 100(2):657–662.

Longstaff, C., Merton, R.E., Fabregas, P., Felez, J. 1999. Characterization of cell-associated plasminogen activation catalyzed by urokinase-type plasminogen activator, but independent of urokinase receptor (uPAR, CD87). Blood 93(11):3839–3846.

Loskutoff, D.J. 1993. A slice of PAI. J Clin Invest 92(6):2563.

Loskutoff, D.J., Curriden, S.A. Hu, G., Deng, G. 1999. Regulation of cell adhesion by PAI-1. Apmis 107(1):54–61.

Lund, L.R., Romer, J., Bugge, T.H., Nielsen, B.S., Frandsen, T.L., Degen, J.L., Stephens, R.W., Dano, K. 1999. Functional overlap between two classes of matrix-degrading proteases in wound healing. Embo J 18(17):4645–4656.

Lund, L.R., Bjorn, S.F., Sternlicht, M.D., Nielsen, B.S., Solberg, H., Usher, P.A., Osterby, R., Christensen, I.J., Stephens, R.W., Bugge, T.H., Dano, K., Werb, Z. 2000. Lactational competence and involution of the mouse mammary gland require plasminogen. Development 127 (20):4481–4492.

Lund, L.R., Green, K.A., Stoop, A.A., Ploug, M., Almholt, K., Lilla, J., Nielsen, B.S., Christensen, I.J., Craik, C.S., Werb, Z., Dano, K., Romer, J. 2006. Plasminogen activation independent of uPA and tPA maintains wound healing in gene-deficient mice. Embo J 25(12):2686–2697.

Mataga, N., Mizuguchi, Y., Hensch, T.K. 2004. Experience-dependent pruning of dendritic spines in visual cortex by tissue plasminogen activator. Neuron 44(6):1031–1041.

Melchor, J.P., Strickland, S. 2005. Tissue plasminogen activator in central nervous system physiology and pathology. Thromb Haemost 93(4):655–660.

Melchor, J.P., Pawlak, R., Strickland, S. 2003. The tissue plasminogen activator-plasminogen proteolytic cascade accelerates amyloid-beta (Abeta) degradation and inhibits Abeta-induced neurodegeneration. J Neurosci 23(26):8867–8871.

Miles, L.A., Plow, E.F., Donnelly, K.J., Hougie, C., Griffin, J.H. 1982. A bleeding disorder due to deficiency of alpha 2-antiplasmin. Blood 59(6):1246–1251.

Mingers, A.M., Heimburger, N., Zeitler, P., Kreth, H.W., Schuster, V. 1997. Homozygous type I plasminogen deficiency. Semin Thromb Hemost 23(3):259–269.

Minowa, H., Takahashi, Y., Tanaka, T., Naganuma, K., Ida, S., Maki, I., Yoshioka, A. 1999. Four cases of bleeding diathesis in children due to congenital plasminogen activator inhibitor-1 deficiency. Haemostasis 29(5):286–291.

Miura, O., Hirosawa, S., Kato, A., Aoki, N. 1989a. Molecular basis for congenital deficiency of alpha 2-plasmin inhibitor. A frameshift mutation leading to elongation of the deduced amino acid sequence. J Clin Invest 83(5):1598–1604.

Miura, O., Sugahara, Y., Aoki, N. 1989b. Hereditary alpha 2-plasmin inhibitor deficiency caused by a transport-deficient mutation (alpha 2-PI-Okinawa). Deletion of Glu137 by a trinucleotide deletion blocks intracellular transport. J Biol Chem 264(30):18213–18219.

Mizutani, A., Saito, H., Matsuki, N. 1996. Possible involvement of plasmin in long-term potentiation of rat hippocampal slices. Brain Res 739(1–2):276–281.

Nakagami, Y., Abe, K., Nishiyama, N., Matsuki, N. 2000. Laminin degradation by plasmin regulates long-term potentiation. J Neurosci 20(5):2003–2010.

Ng, V.L., Sabla, G.E., Melin-Aldana, H., Kelley-Loughnane, N., Degen, J.L., Bezerra, J.A. 2001. Plasminogen deficiency results in poor clearance of non-fibrin matrix and persistent activation of hepatic stellate cells after an acute injury. J Hepatol 35(6):781–789.

Oda, T., Jung, Y.O., Kim, H.S., Cai, X., Lopez-Guisa, J.M., Ikeda, Y., Eddy, A.A. 2001. PAI-1 deficiency attenuates the fibrogenic response to ureteral obstruction. Kidney Int 60(2):587–596.

Okada, K., Ueshima, S., Imano, M., Kataoka, K., Matsuo, O. 2004. The regulation of liver regeneration by the plasmin/alpha 2-antiplasmin system. J Hepatol 40(1):110–116.

Oray, S., Majewska, A., Sur, M. 2004. Dendritic spine dynamics are regulated by monocular deprivation and extracellular matrix degradation. Neuron 44(6):1021–1030.

Ozcelik, U., Akcoren, Z., Anadol, D., Kiper, N., Orhon, M., Gocmen, A., Irkec, M., Schuster, V. 2001. Pulmonary involvement in a child with ligneous conjunctivitis and homozygous type I plasminogen deficiency. Pediatr Pulmonol 32(2):179–183.

Pang, P.T., Teng, H.K., Zaitsev, E., Woo, N.T., Sakata, K., Zhen, S., Teng, K.K., Yung, W.H., Hempstead, B.L., Lu, B. 2004. Cleavage of proBDNF by tPA/plasmin is essential for long-term hippocampal plasticity. Science 306(5695):487–491.

Pantanowitz, L., Bauer, K., Tefs, K., Schuster, V., Balogh, K., Pilch, B.Z., Adcock, D., Cirovic, C., Koche, O. 2004. Ligneous (pseudomembranous) inflammation involving the female genital tract associated with type-1 plasminogen deficiency. Int J Gynecol Pathol 23(3):292–295.

Ploplis, V.A., Carmeliet, P., Vazirzadeh, S., Van Vlaenderen, I., Moons, L., Plow, E.F., Collen, D. 1995. Effects of disruption of the plasminogen gene on thrombosis, growth, and health in mice. Circulation 92(9):2585–2593.

Plow, E.F., Freaney, D.E., Plescia, J., Miles, L.A. 1986. The plasminogen system and cell surfaces: Evidence for plasminogen and urokinase receptors on the same cell type. J Cell Biol 103(6 Pt 1):2411–2420.

Pohl, J.F., Melin-Aldana, H., Sabla, G., Degen, J.L., Bezerra, J.A. 2001. Plasminogen deficiency leads to impaired lobular reorganization and matrix accumulation after chronic liver injury. Am J Pathol 159(6):2179–2186.

Romer, J., Bugge, T.H., Pyke, C., Lund, L.R., Flick, M.J., Degen, J.L., Dano, K. 1996. Impaired wound healing in mice with a disrupted plasminogen gene. Nat Med 2(3):287–292.

Ronne, E., Behrendt, N., Ellis, V., Ploug, M., Dano, K., Hoyer-Hansen, G. 1991. Cell-induced potentiation of the plasminogen activation system is abolished by a monoclonal antibody that recognizes the NH2-terminal domain of the urokinase receptor. FEBS Lett 288(1–2):233–236.

Schleef, R.R., Higgins, D.L., Pillemer, E., Levitt, L.J. 1989. Bleeding diathesis due to decreased functional activity of type 1 plasminogen activator inhibitor. J Clin Invest 83(5):1747–1752.

Schott, D., Dempfle, C.E., Beck, P., Liermann, A., Mohr-Pennert, A., Goldner, M., Mehlem, P., Azuma, H., Schuster, V., Mingers, A.M., Schwarz, H.P., Kramer, M.D. 1998. Therapy with a purified plasminogen concentrate in an infant with ligneous conjunctivitis and homozygous plasminogen deficiency. N Engl J Med 339(23):1679–1686.

Schuster, V., Mingers, A.M., Seidenspinner, S., Nussgens, Z., Pukrop, T., Kreth, H.W. 1997. Homozygous mutations in the plasminogen gene of two unrelated girls with ligneous conjunctivitis. Blood 90(3):958–966.

Schuster, V., Seidenspinner, S., Muller, C., Rempen, A. 1999a. Prenatal diagnosis in a family with severe type I plasminogen deficiency, ligneous conjunctivitis and congenital hydrocephalus. Prenat Diagn 19(5):483–487.

Schuster, V., Seidenspinner, S., Zeitler, P., Escher, C., Pleyer, U., Bernauer, W., Stiehm, E.R., Isenberg, S., Seregard, S., Olsson, T., Mingers, A.M., Schambeck, C., Kreth, H.W. 1999b. Compound-heterozygous mutations in the plasminogen gene predispose to the development of ligneous conjunctivitis. Blood 93(10):3457–3466.

Shanmukhappa, K., Sabla, G.E., Degen, J.L., Bezerra, J.A. 2006. Urokinase-type plasminogen activator supports liver repair independent of its cellular receptor. BMC Gastroenterol 6:40.

Siconolfi, L.B., Seeds, N.W. 2001. Mice lacking tPA, uPA, or plasminogen genes showed delayed functional recovery after sciatic nerve crush. J Neurosci 21(12):4348–4355.

Stefansson, S., Lawrence, D.A. 2003. Old dogs and new tricks: Proteases, inhibitors, and cell migration. Sci STKE 2003(189):pe24.

Stephens, R.W., Pollanen, J., Tapiovaara, H., Leung, K.C., Sim, P.S., Salonen, E.M., Ronne, E., Behrendt, N., Dano, K., Vaheri, A. 1989. Activation of pro-urokinase and plasminogen on human sarcoma cells: A proteolytic system with surface-bound reactants. J Cell Biol 108 (5):1987–1995.

Strickland, D.K., Kounnas, M.Z., Argraves, W.S. 1995. LDL receptor-related protein: A multi-ligand receptor for lipoprotein and proteinase catabolism. Faseb J 9(10):890–898.

Suelves, M., Lopez-Alemany, R., Lluis, F., Aniorte, G., Serrano, E., Parra, M., Carmeliet, P., Munoz-Canoves, P. 2002. Plasmin activity is required for myogenesis in vitro and skeletal muscle regeneration in vivo. Blood 99(8):2835–2844.

Suelves, M., Vidal, B., Ruiz, V., Baeza-Raja, B., Diaz-Ramos, A., Cuartas, I., Lluis, F., Parra, M., Jardi, M., Lopez-Alemany, R., Serrano, A.L., Munoz-Canoves, P. 2005. The plasminogen activation system in skeletal muscle regeneration: Antagonistic roles of urokinase-type plas-minogen activator (uPA) and its inhibitor (PAI-1). Front Biosci 10:2978–2985.

Tabbara, K.F. 2004. Prevention of ligneous conjunctivitis by topical and subconjunctival fresh frozen plasma. Am J Ophthalmol 138(2):299–300.

Takahashi, Y., Tanaka, T., Minowa, H., Ookubo, Y., Sugimoto, M., Nakajima, M., Miyauchi, Y., Yoshioka, A. 1996. Hereditary partial deficiency of plasminogen activator inhibitor-1 asso-ciated with a lifelong bleeding tendency. Int J Hematol 64(1):61–68.

Tefs, K., Gueorguieva, M., Klammt, J., Allen, C.M., Aktas, D., Anlar, F.Y., Aydogdu, S.D., Brown, D., Ciftci, E., Contarini, P., Dempfle, C.E., Dostalek, M., Eisert, S., Gokbuget, A., Gunhan, O., Hidayat, A.A., Hugle, B., Isikoglu, M., Irkec, M., Joss ,S.K., Klebe, S., Kneppo, C., Kurtulus, I., Mehta, R.P., Ornek, K., Schneppenheim, R., Seregard, S. Sweeney, E., Turtschi, S., Veres, G., Zeitler, P., Ziegler, M., Schuster, V. 2006. Molecular and clinical spectrum of type I plasminogen deficiency: A series of 50 patients. Blood 108(9):3021–3026.

Thorsen, S. 1992. The mechanism of plasminogen activation and the variability of the fibrin effector during tissue-type plasminogen activator-mediated fibrinolysis. Ann N Y Acad Sci 667:52–63.

Tsirka, S.E. 2002. Tissue plasminogen activator as a modulator of neuronal survival and function. Biochem Soc Trans 30(2):222–225.

Van Ness, K., Chobaz-Peclat, V., Castellucci, M., So, A., Busso, N. 2002. Plasminogen activator inhibitor type-1 deficiency attenuates murine antigen-induced arthritis. Rheumatology (Oxford) 41(2):136–141.

Watts, P., Suresh, P., Mezer, E., Ells, A., Albisett, I.M., Bajzar, L., Marzinotto, V., Andrew M., Massicotle, P., Rootman, D. 2002. Effective treatment of ligneous conjunctivitis with topical plasminogen. Am J Ophthalmol 133(4):451–455.

Wu, Y.P., Siao, C.J., Lu, W., Sung, T.C., Frohman, M.A., Milev, P., Bugge, T.H., Degen, J.L., Levine, J.M., Margolis, R.U., Tsirka, S.E. 2000. The tissue plasminogen activator (tPA)/plasmin extracellular proteolytic system regulates seizure-induced hippocampal mossy fiber outgrowth through a proteoglycan substrate. J Cell Biol 148(6):1295–1304.

Yepes, M., Lawrence, D.A. 2004. Tissue-type plasminogen activator and neuroserpin: A well-balanced act in the nervous system? Trends Cardiovasc Med 14(5):173–180.

Yoshinaga, H., Hirosawa, S., Chung, D.H., Miyasaka, N., Aoki, N., Favier, R. 2000. A novel point mutation of the splicing donor site in the intron 2 of the plasmin inhibitor gene. Thromb Haemost 84(2):307–311.

Yoshioka, A., Kamitsuji, H., Takase, T., Iida, Y., Tsukada, S., Mikami, S., Fukui, H. 1982. Congenital deficiency of alpha 2-plasmin inhibitor in three sisters. Haemostasis 11(3):176–184.

Zhang, L., Seiffert, D., Fowler, B.J., Jenkins, G.R., Thinnes, T.C., Loskutoff, D.J., Parmer, R.J., Miles, L.A. 2002. Plasminogen has a broad extrahepatic distribution. Thromb Haemost 87 (3):493–501.

Zhu, Y., Carmeliet, P., Fay, W.P. 1999. Plasminogen activator inhibitor-1 is a major determinant of arterial thrombolysis resistance. Circulation 99(23):3050–3055.

Chapter 11
The Plasminogen Activation System in Tissue Remodeling and Cancer Invasion

Kasper Almholt, Anna Juncker-Jensen, Kirsty Anne Green, Helene Solberg, Leif Røge Lund, and John Rømer

Abstract Plasminogen (Plg) is perhaps the most abundant and widely distributed protease zymogen in the organism. The local activation of Plg by the highly specific Plg activators is involved in a number of processes that require proteolytic restructuring of the extracellular milieu. We review the experimental evidence that implicates Plg activation in four very different tissue-remodeling events. These are wound healing, embryo implantation, mammary gland involution, and cancer metastasis. In all cases, the urokinase-type plasminogen activator (uPA) is produced by specialized cell populations within the remodeling tissue. In several of these processes, we have identified a functional overlap between the Plg activation system and one or more proteases of the matrix metalloprotease (MMP) family. We anticipate that similar functional redundancies are widespread in physiological as well as pathological proteolytic processes, and it therefore represents a challenge to identify the critical proteases for therapeutic targeting.

Introduction

Ontogenesis of any higher organism is an unbroken chain of coordinated tissue-remodeling events that execute the development of an organism from the fertilized egg to its adult form. In the adult organism, however, tissue-remodeling events only occur in certain tissues or in response to injury and are an exception to the general state of tissue homeostasis. Even so, tissue remodeling is strictly required in the adult organism for a number of crucial functions. In this chapter, we will focus on the role of extracellular proteases in selected tissue-remodeling events, with particular

K. Almholt
Novo Nordisk A/S, Novo Nordisk Park, DK-2760, Måløv, Denmark, e-mail: KAhl@novonordisk.com

D. Edwards et al. (eds.), *The Cancer Degradome.*
© Springer Science + Business Media, LLC 2008

emphasis on the serine proteases that govern the activation of plasminogen (Plg). We will look at the Plg activation system in three examples of normal postdevelopmental tissue remodeling. These are skin wound healing, embryo implantation in the uterine wall, and mammary gland remodeling during pregnancy cycling. Finally, we will focus on the Plg activation system in malignant tissue remodeling, with the mouse mammary tumor virus (MMTV)-polyomavirus middle T antigen (PymT) transgenic breast cancer model as the case study.

The extracellular proteases have been extensively examined in physiological tissue-remodeling events, often using gene-targeted mice. The goal of these studies has been to uncover the role of the proteases during normal tissue remodeling in its own right. A second purpose has always been to unravel the role of the proteases in cancer, based on the assumption that these physiological processes can be used as meaningful surrogate models for cancer invasion. In a classic comparison, cancer was described as "wounds that do not heal" (Dvorak 1986), which can be broadened to state that cancer invasion resembles normal tissue remodeling gone out of control (Johnsen et al. 1998, Rømer 2003, Danø et al. 2005). In any nonneoplastic tissue-remodeling event, one normal tissue is replaced by another in a controlled fashion, while cancer invasion is characterized by an uncontrolled substitution of normal tissue with expanding neoplastic tissue. The lack of control during cancer invasion is naturally a crucial difference, while other features such as protease expression patterns can be very similar when comparing, for example, skin wound healing and skin cancer or comparing mammary gland cycling and breast cancer (Johnsen et al. 1998, Rømer 2003, Green and Lund 2005).

Physiological tissue-remodeling events often have the advantages of greater reproducibility and ease of quantification compared to cancer invasion and metastasis; advantages that have increased the rate of discovery. An important insight obtained from physiological tissue-remodeling events is the high degree of functional overlap among the extracellular proteases. It will be crucial to identify these built-in redundancies and compensatory mechanisms, if cancer invasion and metastasis are to be successfully combatted.

Extracellular Proteolysis

Extracellular proteases first received attention due to their collective ability to degrade any component of the extracellular matrix (ECM), which would allow them to pave the way for invasion and metastasis. In addition to their role as molecular scissors, proteases are now known to convey extracellular signals including the release and/or direct modification of signaling molecules, regulation of cell–cell and cell–ECM contacts, and generation of bioactive ECM fragments. Several extracellular protease families are involved in physiological tissue-remodeling processes and cancer invasion (for detailed reviews, *see* Andreasen et al. 2000, López-Otín and Overall 2002, Danø et al. 2005, Page-McCaw et al. 2007). The serine proteases and related molecules of the Plg activation cascade have been extensively studied in this regard. Plg is synthesized in the liver, circulates in blood at a concentration of ~2 μM, and is found at a somewhat lower concentration in all interstitial fluids. The conversion of Plg to plasmin is catalyzed by two well-studied activators, urokinase-type Plg activator (uPA) and tissue-type Plg activator (tPA), and there is good evidence that at least one additional enzyme can catalyze this activation in vivo, namely, plasma kallikrein [pKal; Selvarajan et al. (2001), Lund et al. (2006)]. Both uPA and tPA are also synthesized as precursors. The catalytically inactive uPA precursor, pro-uPA, binds to a cell surface receptor, the urokinase receptor (uPAR), and this interaction accelerates the conversion of pro-uPA to uPA in a positive feedback mechanism between (pro-) uPA and Plg/plasmin (Ploug 2003). The tPA precursor, which has some intrinsic activity, and tPA itself, bind strongly to fibrin in blood clots and appear to be particularly well situated to achieve thrombolysis after intravascular clotting (Collen and Lijnen 2005). Plg/plasmin also binds to cell surfaces and to fibrin in blood clots and is in both cases protected from inactivation by the specific inhibitor α_2-antiplasmin (Ploug 2003, Collen and Lijnen 2005). Whether the Plg activators have a physiological role in addition to the activation of Plg to plasmin is not yet definitively established, but there is evidence that uPAR interacts strongly with connective tissue components such as vitronectin with or without bound uPA (Wei et al. 1994, Høyer-Hansen et al. 1997, Ploug 2003), and it may be involved in intracellular signaling processes (Fazioli et al. 1997, Blasi and Carmeliet 2002). Two specific Plg activator inhibitors (PAIs) exist, PAI-1 and PAI-2 [for reviews *see* Andreasen et al. (1997), Andreasen et al. (2000), Dellas and Loskutoff (2005)]. PAI-1 in particular interacts on the cell surface with uPA bound to uPAR and may affect cell surface interactions with integrins and matrix components (Stefansson and Lawrence 1996, Loskutoff et al. 1999, Czekay et al. 2003). Plasmin has a wide range of substrate proteins, including the proenzyme forms of uPA and tPA, fibrin in blood clots, and many other structural ECM proteins. It also participates in the activation of a number of proproteins, including pro-MMPs, cytokines, growth factors, and other potential messengers (Andreasen et al. 2000).

In addition to plasmin and its activators, there are a number of other extracellular protease families. We will introduce the matrix metalloproteinase (MMP) family here because there is considerable evidence of a functional overlap between the Plg activation system and the MMPs. Currently, 23 human MMPs have been identified (22 in mice that do not have MMP-26). An updated list of human and mouse proteases is available from the López-Otín lab at http://web.uniovi.es/degradome/. MMPs are zinc-dependent proteases with varying specificities for ECM proteins (Folgueras et al. 2004, Nagase et al. 2006, Page-McCaw et al. 2007). A few MMPs, including the three classical collagenases (MMP-1, -8, and -13), are able to degrade native collagen. Two MMPs, the so-called gelatinases (MMP-2 and -9), are particularly active against denatured collagen. The two typical stromelysins (MMP-3 and -10) have a broad specificity and at least MMP-3 is particularly active against proteoglycans. The matrilysins (MMP-7 and -26) have the simplest domain structure in the MMP family and a broad substrate specificity. All of these enzymes are among the 16 MMPs, which are secreted as soluble proteins. The remaining seven

MMPs are membrane-bound enzymes (MMP-14, -15, -16, -17, -23, -24, and -25), which have varying specificities that in the case of MMP-14 includes native collagen (Page-McCaw et al. 2007). The seven membrane-bound MMPs, and the soluble MMP-11, -21, and -28 all have a furin cleavage site and may be activated by furin-like proprotein convertases during posttranslational modification. The 13 other soluble MMPs are activated extracellularly, either by plasmin, other MMPs, or other extracellular proteases (Folgueras et al. 2004, Nagase et al. 2006, Page-McCaw et al. 2007). The MMP system is counterbalanced by a group of four endogenous inhibitors, the tissue inhibitor of metalloproteases (TIMP-1, -2, -3, and -4). All four TIMPs are capable of inhibiting most of the MMPs, and even certain metalloproteases outside the MMP family, but the TIMPs are also involved in the *activation* of some MMPs (Nagase et al. 2006).

Plasminogen Activation in Tissue Remodeling and Cancer Invasion

Skin Wound Healing

Tissue remodeling in response to injury has been thoroughly investigated in skin wound models [for detailed reviews *see* Martin (1997), Singer and Clark (1999), Werner and Grose (2003)]. Skin responds to physical insults, such as freezing, burning or a surgical procedure by initiating a tissue-remodeling response appropriate for the particular insult. The model we use in our laboratory is the full thickness incision on the dorsal skin of the mouse (Rømer et al. 1996, Lund et al. 1999). A cut of 20 mm in length is made under general anesthesia with a sharp blade. The wound is left without stitching or dressing and the mice are caged individually. The edges of the wound initially separate by 4–6 mm in the center. Hemostasis occurs spontaneously and a desiccated crust is established within 1–2 days of wounding. Healing is judged complete when the wound area is completely reepithelialized and all scab has dropped off. In wild-type mice, the healing time is relatively constant and very reproducible, from 16 to 19 days dependent on the background strain of mice (Rømer et al. 1996; Lund et al. 1999, 2006). Within several hours of wounding, the cells at the cut edge commence migration across the provisional matrix toward the center of the wound. It is generally believed that with this type of stratified squamous epithelium, there is a rolling process, whereby cells break free from the basal lamina and intercellular linkages, possibly via expression of extracellular proteases, to adopt a motile phenotype with ruffled edges (Clark 1996). Coincident with this change in behavior the mRNA levels of several extracellular proteases are greatly increased. The keratinocytes of the regenerating epidermis express uPA (Fig. 11.1a), uPAR, PAI-1, MMP-3, -9, -10, and -13 (Rømer et al. 1991, 1994, 1996; Madlener et al. 1998; Lund et al. 1999). The epidermis in the wound field is thickened and consists of a broad proliferative zone and a

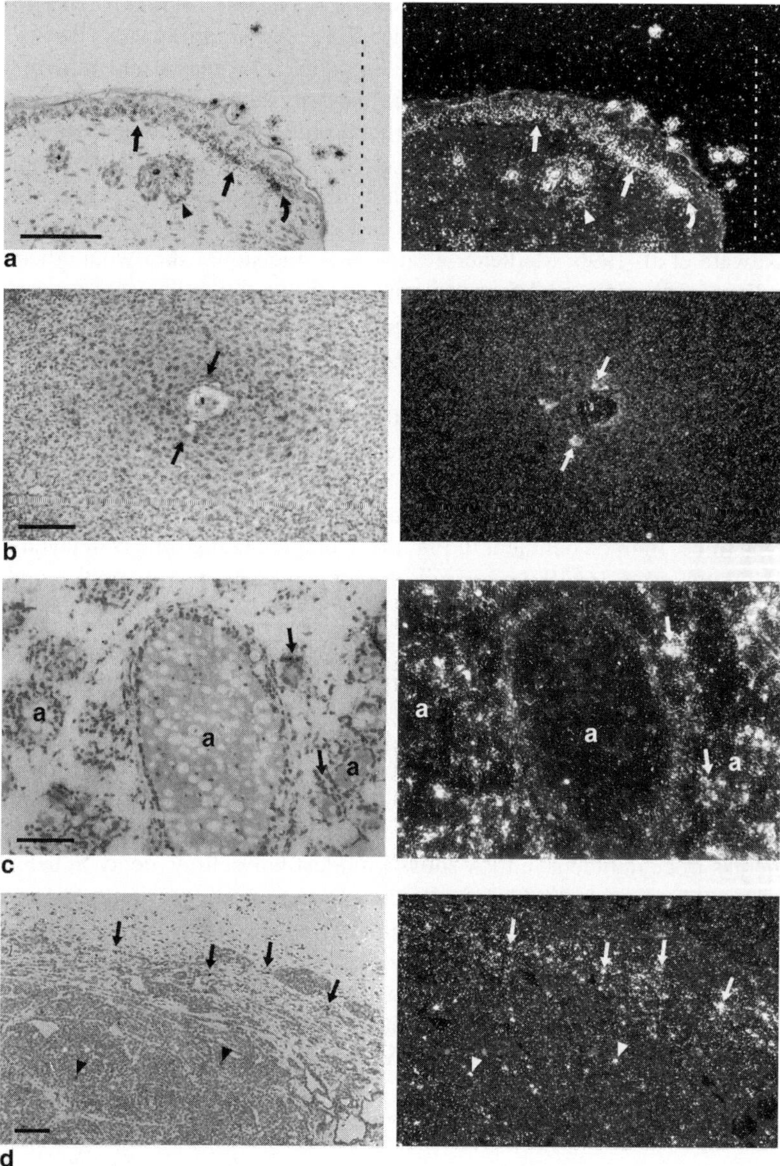

Fig. 11.1 Expression of urokinase-type Plg activator (*uPA*) mRNA in three normal tissue-remodeling events and in cancer invasion. Mouse uPA mRNA was detected by in situ hybridization with S-35-labeled RNA probes. In the left column are bright field images; in the right column are the corresponding dark field images. (**a**) Full thickness skin wound 24 h after the incision. uPA mRNA is present in the basal and suprabasal keratinocytes in the zone of proliferative and migrating keratinocytes (straight arrows) extending to the tip of the keratinocyte wedge (curved arrow) adjacent to the wound edge (dashed line) (Rømer et al. 1991, 1996). Note that keratinocytes of nearby hair follicles also express uPA mRNA (arrowhead) and can contribute to the regenerating epidermis. (**b**) Section through embryo at day 6.5 of gestation. uPA mRNA is present in

migrating zone just at the invasive front. Expression of uPAR and MMP-9 is confined to this very narrow zone of leading edge keratinocytes (Rømer et al. 1994, Lund et al. 1999). Stromal cells, including fibroblasts and macrophages, in the adjacent uninjured dermis and/or in the provisional matrix and granulation tissue that is formed beneath the wound crust produce uPA, PAI-1, MMP-2, -3, -11, -12, -13, -14, and TIMP-1 (Rømer et al. 1991, 1994, 1996; Madlener et al. 1998; Lund et al. 1999). MMP-8 is also abundant in skin wounds (Gutiérrez-Fernández et al. 2007) and PAI-2 and TIMP-2 and -3 are constitutively expressed in mouse skin (Kawata et al. 1996, Madlener et al. 1998). Incisional skin wound healing is an excellent surrogate model for cancer invasion due to the invasive properties of the migrating keratinocytes and due to the similarities in protease expression between wound healing and squamous cell skin cancer (Johnsen et al. 1998, Rømer 2003).

The tightly regulated expression of the proteases in migrating keratinocytes suggests that these enzymes are important for the movement of the epidermal layer beneath the wound scab. In mice with protease genes removed, we have studied differences in wound histology and the rate of wound healing compared to wild-type mice. In mice deficient in Plg, the average healing time is delayed to 43–55 days compared to 16–19 days in wild-type mice (Rømer et al. 1996; Lund et al. 1999, 2006, and unpublished data). In particular, the movement of the keratinocytes is impaired, likely due to accumulation of fibrin observed in front of and underneath the leading edge keratinocytes (Rømer et al. 1996, Lund et al. 1999). The prolonged healing time of skin wounds in Plg-deficient mice is thus largely corrected by simultaneous fibrin deficiency (Bugge et al. 1996b). In contrast to the situation in Plg-deficient mice, when the *activators* of Plg are lacking, there is far less effect on the healing rate (Lund et al. 2006). We find that mice deficient in tPA show no discernable difference from wild-type mice, neither in healing time nor in histology, whereas mice deficient in uPA show a modest but definite delay in healing, of ~2–3 days (Lund et al. 2006). Despite the modest effects of single Plg activator deficiencies, wound healing is severely impaired by a combination of uPA and tPA deficiency (Bugge et al. 1996a). The average healing time in uPA-tPA double-deficient mice is delayed ~2 weeks to about 31 days (Lund et al. 2006). This shows that although lack of tPA does not in itself impair the healing process, tPA is capable of substituting for uPA when uPA is absent. The role of tPA during normal wound healing remains to be elucidated. The difference in healing times between Plg-deficient mice (43–55 days) and mice deficient in both of the two classical Plg

---

Fig. 11.1 (Continued) invasive trophoblast giant cells in the uterine stroma surrounding the embryo (arrows) (Sappino et al. 1989, Teesalu et al. 1996). (**c**) Involuting mammary gland 5 days after removing the pups at the peak of lactation. uPA mRNA is abundant in fibroblast-like cells and macrophages (arrows) in between regressing alveoli (**a**) and is also seen in some intraluminal cells [apoptotic cells and macrophages; Lund et al. (1996)]. (**d**) Mammary cancer from a 13-week-old MMTV-PymT-transgenic mouse. uPA mRNA is present in fibroblast-like cells and macrophages mainly at the invasive front of the cancer (arrows) but also in the interior (arrowheads), as well as in intraluminal macrophages (Almholt et al. 2005, unpublished data). Bars, 100 μm

activators uPA and tPA (31 days) has led us to identify a third Plg activator, pKal (Lund et al. 2006). The use of a pKal-specific inhibitor demonstrated that this protease can account for the discrepant healing times and thus plays a physiological role in wound healing, at least when uPA and tPA are absent (Lund et al. 2006). It is of interest that mice double-deficient in uPAR and tPA do not suffer any healing delay (Bugge et al. 1996a). This finding indicates that epidermal regeneration is not dependent on the cell surface anchorage of pro-uPA to uPAR, with the consequent acceleration of its activation to uPA. Although PAI-1 is upregulated during wound healing and PAI-2 is constitutively expressed in the epidermis, deficiency for PAI-1 or PAI-2 or the combination of both does not affect the rate of wound healing in an excisional skin wound model (Dougherty et al. 1999).

Ultimately, healing occurs, even in Plg-deficient mice. On the basis of the colocalization of uPA, uPAR, PAI-1, MMP-3, -9, -10, and -13 in leading-edge keratinocytes (Rømer et al. 1991, 1994, 1996; Madlener et al. 1998; Lund et al. 1999), we therefore proposed that there might be a functional overlap between the two proteolytic systems. We tested the role of the MMPs in skin wound healing by the use of the broad spectrum MMP inhibitor galardin (Lund et al. 1999). When wild-type mice are treated with galardin (100 mg kg^{-1} day^{-1} i.p.), the healing process is significantly delayed to an average healing time of 22–35 days dependent on the background strain of mice (Lund et al. 1999, 2006, unpublished data). This outcome suggests that the contribution of the galardin-sensitive MMPs to wound healing is comparable to the contribution of plasmin. Immunohistochemical analysis reveals an accumulation of fibrin in front of and underneath the migrating keratinocytes in galardin-treated wild-type mice, similar to what is seen in mock-treated Plg-deficient mice (Lund et al. 1999). This suggests that the galardin-induced delay in healing is at least partially a result of impaired fibrinolysis and thus, in a qualitative sense, similar to the fibrin-dependent delay observed in Plg-deficient mice (Bugge et al. 1996b, Rømer et al. 1996, Lund et al. 1999). In a quantitative sense though, the contribution of the MMPs may be somewhat less than that of plasmin, considering that the average healing time is 43–55 days in Plg-deficient mice (Rømer et al. 1996; Lund et al. 1999, 2006, unpublished data). Consistent with a critical convergence point of the Plg activation and MMP protease systems, we found that skin wound healing is *completely prevented* during galardin treatment of Plg-deficient mice (Lund et al. 1999).

It must be noted, however, that the MMPs are not the only metalloproteinases present in wound tissue and galardin may have inhibitory effects on other metalloproteinases such as the ADAMs (a disintegrin and metalloprotease domain). Because of the diverse expression patterns of the MMPs during skin wound healing, it is difficult to dissect out the roles of the individual MMPs in the healing process, even though there are a number of knockouts available (Page-McCaw et al. 2007). So far, deficiencies of five MMPs have been analyzed (MMP-3, -8, -9, -13, and -14), including 3 that are expressed in the leading-edge keratinocytes (MMP-3, -9, and -13), and separately they appear to have modest effects on healing time (Bullard et al. 1999, Mohan et al. 2002, Mirastschijski et al. 2004, Hartenstein et al. 2006, Gutiérrez-Fernández et al. 2007). Owing to the fact that the galardin-sensitive

MMPs are the major collagenolytic proteases in the organism, it is of relevance that Col1a-1 mutant mice, which have a mutation in the helical domain of the collagen $\alpha_1(I)$ chain at the site where all collagenases degrade the molecule (Wu et al. 1994, Liu et al. 1995), display a significantly slower healing than wild-type mice (Beare et al. 2003, Lapiere 2003). The phenotype of the Col1a-1 mutant strongly suggests the involvement of collagenolytic MMPs in wound healing in addition to the fibrinolytic MMPs implicated by galardin treatment.

Wound healing analyses have become a major driving force in the extracellular protease field. The functional redundancies within the Plg activation system and between the Plg and MMP protease families are best described in this in vivo model. We foresee that there will be functionally overlapping proteases within the MMP family as well. To explore these, it may be of great value to inhibit individual MMPs in combination with Plg deficiency, since the presence of Plg during skin wound healing is likely to mask the effect of the absence of single MMPs.

Embryo Implantation

Rodents and humans alike form a hemochorial placenta in which extraembryonic trophoblast cells invade through the maternal uterine epithelium and into the underlying uterine stroma to make direct contact with the maternal blood supply (Abrahamsohn and Zorn 1993, Cross et al. 1994, Rossant and Cross 2001). The invasive nature of hemochorial placentation mimics cancer invasion to the extent that normal trophoblast cells have been called pseudomalignant (Kirby 1965, Strickland and Richards 1992). The invasive process starts at day 4.5 of gestation in mice and is initially accomplished by cells derived from the outer epithelial layer of the blastocyst, the so-called primary trophoblast giant cells. From day 6.5, these cells are joined by a new lineage of cells, the secondary trophoblast giant cells. Normal placentation depends on the invasion of these cell types into the uterine stroma. In response to embryonic attachment to and invasion of the uterine epithelia, uterine stromal cells in the immediate vicinity of the implantation site begin to proliferate and differentiate, forming a dense cellular matrix. This process is called decidualization. Although the purpose of the decidua is not fully understood, one of its functions is probably to impede the invasion of trophoblast giant cells and thus protect the mother from the invasive nature of the embryo [for reviews, *see* Cross et al. (1994), Rinkenberger et al. (1997), Rinkenberger and Werb (2000)]. This protection is accomplished at least in part by forming a physical barrier and by producing protease inhibitors, such as PAI-1 and TIMP-1, -2, and -3 that are produced by the decidual cells (Reponen et al. 1995, Alexander et al. 1996a, Teesalu et al. 1996, Das et al. 1997). These protease inhibitors form one half of a finely tuned balance between activated proteases and protease inhibitors, enabling a controlled degradation of ECM by the trophoblast giant cells at this early stage. The expression of uPA (Sappino et al. 1989, Teesalu et al. 1996; Fig. 11.1b) and uPAR (Teesalu et al. 1996, Solberg et al. 2003) and several MMPs, including MMP-1A, -2, -9, -11, and -14 (Reponen et al. 1995, Alexander et al. 1996a, Das et al. 1997,

Balbín et al. 2001, Solberg et al. 2003) in trophoblast giant cells and/or decidual cells at this stage suggest a functional role of these proteases in the invasive phase of implantation. The generation of gene-deficient mice has, however, demonstrated that ablation of single components of the Plg activation system, including uPA, uPAR, tPA, PAI-1, and PAI-2, does not affect the fertility of homozygous knockout animals (Carmeliet et al. 1993, 1994; Bugge et al. 1995b; Dewerchin et al. 1996; Dougherty et al. 1999). Even in Plg-deficient mice (Bugge et al. 1995a, Ploplis et al. 1995), embryo implantation is unaffected and trophoblast invasion and decidualization proceed normally (Solberg et al. 2003). Plg-deficient mice thus give birth to litters of normal size, but the pups rarely survive until weaning as a consequence of severely impaired lactation [Lund et al. 2000, Green et al. 2006; *see* discussion below]. When Plg deficiency is combined with MMP inhibition using galardin treatment (150 mg kg^{-1} day^{-1} i.p.), trophoblast invasion and decidualization are markedly reduced resulting in frequent embryonic lethality. The number of resorption sites in the galardin-treated Plg-deficient postpartum uteri was 2–3 times the number of full-term pups (Solberg et al. 2003). Galardin treatment alone did not affect trophoblast invasion or decidualization, and there was only a modest number of resorption sites in the uteri. Consistent with this, none of the MMP-null mice, except the runted MMP-14-deficient mice, have been demonstrated to be infertile (Folgueras et al. 2004, Page-McCaw et al. 2007). Taken together, these data indicate a functional overlap in the invasive phase of implantation between Plg/ plasmin and one or more MMPs, while each proteolytic system on its own is dispensable for this process.

Postlactational Mammary Gland Involution

Extracellular proteases are generally absent in resting mammary glands from adult mice but are present as the gland undergoes the pregnancy cycle. Following lactation, the mammary gland is prepared for future pregnancies through an involution phase. The levels of a number of extracellular proteolytic components are particularly upregulated during this phase. These include uPA, tPA, MMP-2, -3, -9, -11, and their corresponding inhibitors PAI-1 and TIMP-1 (Lefebvre et al. 1992; Talhouk et al. 1992; Li et al. 1994; Lund et al. 1996, 2000; Alexander et al. 2001). Postlactational involution can to some extent be regarded as a surrogate model for breast cancer invasion due to the extensive tissue remodeling and extracellular proteolysis taking place at this stage of the pregnancy cycle and the similar protease expression pattern during involution and in breast cancers (Johnsen et al. 1998, Green and Lund 2005). In involuting mammary glands and breast cancers alike, at least uPA, MMP-2, -3, and -11 are all expressed in connective tissue cells, and MMP-9 in macrophages (Green and Lund 2005, Almholt et al. 2007).

Involution can be artificially synchronized by removal of suckling pups at the peak of lactation, typically at day 10 after birth. The ensuing involution consists of a primary reversible phase of apoptosis, a secondary irreversible phase of tissue remodeling, and a tertiary phase of adipogenesis or biosynthesis. The first, reversible stage of involution is prematurely activated in Plg-deficient mice most likely due to fibrin occlusion of the ducts leading to increased milk stasis during lactation (Green et al. 2006). As a result of milk stasis and early onset of involution, Plg-deficient mice are severely lactation-impaired. In certain mouse background strains, they are essentially unable to support a litter. However, if they succeed in establishing and maintaining the lactating state, the second, protease-dependent and irreversible stage of involution becomes *delayed* (Lund et al. 2000). The induction of Plg activator expression in unchallenged involuting mice coincides temporally with the compromised second stage involution phenotype in Plg-deficient mice. Plg activation is primarily increased through an upregulation of the mRNA level of uPA (Lund et al. 1996). In early involution, a few scattered fibroblast-like cells or macrophages express uPA. Later in involution, uPA is abundant in these stromal cells surrounding the regressing alveoli and is also seen in some intraalveolar cells [apoptotic cells and macrophages; Lund et al. (1996); Fig. 11.1c]. The third, adipogenic phase of involution is also impaired in Plg-deficient mice (Selvarajan et al. 2001). The adipocyte differentiation phenotype in Plg-deficient mice was instrumental in the identification of a third Plg activator, pKal, which has a physiological role in adipogenesis that could not be accounted for by uPA or tPA (Selvarajan et al. 2001).

Despite the delay in the second and third stages, involution is eventually accomplished in Plg-deficient mice (Lund et al. 2000), probably due to the compensating action of other proteases. When Plg-deficient mice are challenged with the MMP inhibitor galardin during involution ($100 \text{ mg kg}^{-1} \text{ day}^{-1}$ i.p.), the glands accumulate fibrin to an extent that greatly exceeds what could be detected in untreated Plg-deficient mice or in galardin-treated wild-type mice (Lund et al. 2000, unpublished data), although fibrin does accumulate in the mammary glands of Plg-deficient mice (Green et al. 2006). This is consistent with a functional overlap of Plg/plasmin and the MMPs in fibrinolysis during involution. A number of MMPs are expressed during mammary gland involution, including MMP-2, -3, -9, and -11 (Lefebvre et al. 1992; Talhouk et al. 1992; Li et al. 1994; Lund et al. 1996, 2000; Alexander et al. 2001). TIMP-1 mRNA is present during pregnancy and further elevated during early involution, declining in the second and third phases of involution (Talhouk et al. 1992, Alexander et al. 2001). Implantation of TIMP-1 slow release pellets into involuting mammary glands delays alveolar regression (Talhouk et al. 1992), indicative of a regulatory role of TIMP-1 in the early phase and a general proinvolutory role of the MMPs. Somewhat at odds with this observation, ubiquitous TIMP-1 overexpression through a β-actin-TIMP-1 transgene does not delay alveolar apoptosis in the early stages (Alexander et al. 2001), but this could be a question of differing TIMP-1 concentrations in the two approaches. Consistent with the involution delay induced by TIMP-1 implants, all stages of involution are *accelerated* in TIMP-3-deficient mice, and the first phase of involution is no longer reversible in these mice (Fata et al. 2001). These results suggest a key role for one or more MMPs in mammary involution, although TIMP-3 does have the ability to inhibit several metalloproteases of the ADAM and ADAMTS (ADAM with a thrombospondin type-1 motif) families (Nagase et al. 2006).

To date, only a few functional studies have started to dissect the contribution of the individual MMPs. MMP-3 was originally identified in myoepithelial cells during involution but has since been shown to be expressed in the mammary stroma (Dickson and Warburton 1992, Li et al. 1994, Lund et al. 1996). Overexpression of MMP-3 in luminal epithelium leads to premature involution during late pregnancy (Sympson et al. 1994, Alexander et al. 1996b), consistent with a proinvolutory role of the MMPs. However, MMP-3 deficiency has no effect on mammary apoptosis in early phase involution (Alexander et al. 2001), suggesting that the proinvolutory effect of MMP-3 overexpression may be an effect of overexpressing a stromal protein in luminal cells. In fact, MMP-3 deficiency *accelerates* adipocyte maturation during the final stage of involution and the same is observed when TIMP-1 is ubiquitously overexpressed (Alexander et al. 2001). All in all, MMP-3 is probably a poor candidate for a direct functional overlap with the Plg activation system, bearing in mind that Plg deficiency causes premature onset of involution followed by delayed alveolar regression and adipogeneis (Lund et al. 2000, Selvarajan et al. 2001, Green et al. 2006).

In conclusion, the data demonstrate that Plg activation and MMP activity once again coincide in a physiological tissue-remodeling process. Although the two proteolytic systems may be able to compensate for each other to some extent, it is likely that neither of them is entirely dispensable as evidenced by the compromised involution phenotypes observed by perturbing the Plg or MMP systems individually.

MMTV-PymT Breast Cancer Metastasis

We expect that the functional overlap between the Plg activation and MMP protease systems is not restricted to normal physiological processes, such as skin wound healing and embryo implantation, but also plays a role during cancer invasion. We have used the MMTV-PymT transgenic breast cancer model (Guy, Cardiff and Muller 1992) for direct studies of extracellular proteolysis in cancer invasion. In this model, the PymT is induced specifically in the mammary epithelium by the MMTV promoter. Tumors develop spontaneously in virgin females at an approximate age of 1.5 months, depending on the genetic background (Guy et al. 1992; Almholt et al. 2003, 2005). The tumors metastasize primarily not only to the lungs (Guy et al. 1992) but also to the regional lymph nodes (Almholt et al. 2005). The expression of several extracellular proteases has been analyzed in the MMTV-PymT transgenic breast cancer model (Almholt et al. 2007). One study used laser microdissection to isolate cancer cells and stromal cells and concluded that stromal tissue adjacent to cancer cells expresses higher levels of uPA, PAI-1, and MMP-2, -3, -11, -13, and -14 than the cancer tissue (Pedersen et al. 2005). The predominantly stromal expression patterns of all these components in the MMTV-PymT model have been confirmed by in situ hybridization (Bugge et al. 1998, Almholt et al. 2003, 2005; Pedersen et al. 2005; Szabova et al. 2005). The uPA mRNA is present not only in macrophages and fibroblast-like cells in the primary tumor stroma, mainly at the invasive front (Almholt et al. 2005; Fig. 11.1d), but also in a population of intraluminal macrophages (unpublished data). In situ hybridization was used in another study on MMTV-PymT tumors to determine the expression pattern of four of the membrane-type MMPs: MMP-14, -15, -16, and -17 (Szabova et al. 2005). MMP-14 and -16 were detected in the stroma, whereas MMP-15 was the only protease found mainly in the epithelium. MMP-17 expression was not observed in the MMTV-PymT tumors. The expression patterns of Plg activation and MMP components in the MMTV-PymT model have been directly compared to human breast cancer (Almholt et al. 2007). The proteases expressed in the MMTV-PymT model generally mirror those identified in human breast cancer both in terms of expression and localization. In the mouse model as well as in invasive ductal breast cancer, the expression of uPA, uPAR, PAI-1, MMP-2, -3, -11, -13, and -14, is concentrated in the stroma of the tumors (Almholt et al. 2007). These findings suggest that the MMTV-PymT model is a useful tool for the study of protease involvement in breast cancer invasion and metastasis. In the MMTV-PymT model, tumors are formed and grow equally fast in the genetic absence or presence of Plg (Bugge et al. 1998), uPA (Almholt et al. 2005), PAI-1 (Almholt et al. 2003), or uPAR (unpublished data), indicating that these proteolytic compents are not rate limiting for tumor onset or growth. Adequate vascularization is prerequisite to tumor growth. When new vessels are formed by sprouting from existing vasculature, endothelial cells penetrate a newly deposited fibrin-rich matrix and use it as a temporary scaffold for vessel structure (Dvorak et al. 1995). Accordingly, the fibrinolytic capacity of the Plg activation system is required for vascularization in certain experimental conditions (Bajou et al. 1998, 2001; Gutierrez et al. 2000), and PAI-1 may play a dual role as a protease inhibitor and cell migration mediator (Andreasen et al. 2000). Nevertheless, absence of PAI-1 is of no consequence for the vascular density of MMTV-PymT mammary cancers (Almholt et al. 2003). It is likely that functional redundancy with other proteolytic components compensates for the absence of a single component in the intact tumor. It is also possible that vascularization of spontaneous tumors can be partially accomplished by endothelial precursor cells in a process that is independent of proteases or depends on a different subset of proteases (Jodele et al. 2006).

In contrast, the Plg activation system seems to play a unique role in metastasis in the MMTV-PymT model. uPA (Almholt et al. 2005), PAI-1 (Almholt et al. 2003), and uPAR (unpublished data) are all present in the lung metastasis stroma. In the context of unaltered primary tumor size, metastasis is significantly reduced in both Plg-deficient (Bugge et al. 1998) and uPA-deficient mice (Almholt et al. 2005), although the phenotype in Plg-deficient mice is mouse strain dependent (unpublished data). Absence of the specific uPA inhibitor PAI-1 has a slight metastasis-*promoting* effect, but the difference is not significant, perhaps due to the presence of alternative PAIs (Almholt et al. 2003). Deficiency of uPAR does not on its own significantly affect metastasis in the inbred FVB strain (unpublished data).

Less is known about the impact of MMPs on the MMTV-PymT tumors (Almholt et al. 2007). We have treated MMTV-PymT-transgenic mice with the MMP inhibitor galardin (slow release implants) from age 6 weeks and until sacrifice.

In the galardin-treated mice, the average lung metastasis volume is reduced by ~99%, in the context of primary tumors that are only reduced to half the size of placebo-treated tumors (unpublished data), suggesting that one or more MMPs play a vital role in metastasis in this model. Metastasis incidence is also reduced in MMTV-PymT mice that overexpress TIMP-1 through a liver-specific transgene, but the reduction is less pronounced than in galardin-treated mice and the reduction is absent with a mammary gland-specific TIMP-1 transgene (Yamazaki et al. 2004). Galardin is a broad spectrum MMP inhibitor and may inhibit certain ADAMs as well. The endogenous inhibitor TIMP1 also inhibits most MMPs, although it is a poor inhibitor of MMP-14, -16, -19, and -24 (Nagase et al. 2006). It will take a considerable effort to define which MMPs are responsible for the metastasis reduction observed with these inhibitors.

In summary, inhibition of either the Plg activation system or the MMPs reduces spontaneous metastasis. The effect on primary tumor growth was less pronounced in both cases, even with one of the most aggressively growing transgenic tumors available, suggesting that the Plg activation or MMP systems may not be rate limiting for spontaneous tumor growth as such. It will be interesting to combine inhibition of the Plg activation and MMP systems for a possible additive/synergistic reduction of spontaneous metastasis.

Conclusions and Future Perspectives

Limited proteolysis in the extracellular milieu is vital during ontogenesis and tissue-remodeling processes in the adult organism. Therefore, it was a surprise when genetic analyses demonstrated that practically all protease-deficient mice are viable, and in most cases are without obvious phenotypes. A likely explanation for the apparent normalcy of protease-deficient mice is a built-in redundancy among the many extracellular proteases. Even distantly related proteases often have several common substrates in vitro, which may translate into an ability for adaptive compensation in vivo. In the foreseeable future, the majority of genes for the most intensely investigated protease networks, including the Plg activation and MMP systems, will have been "knocked out." This will ultimately provide a phenotypic baseline for the mouse "degradome."

In the meantime, the patchwork of overlapping protease functions can be partially mapped by inhibiting several proteases simultaneously. The breeding setup required to obtain a sufficient amount of double, triple, or even quadruple gene-deficient mice will be quite elaborate and costly if the relevant control mice are to be generated in a stringent manner. To circumvent this problem, progress can perhaps be made by combining gene-deficient mice with monospecific protease inhibitors, such as monoclonal antibodies against mouse proteases (Pass et al. 2007), or with protease inhibitors of intermediate specificity that target, for example, a subgroup of MMPs with similar catalytic sites. The development of specific chemical inhibitors is, however, inevitably driven by a focus on human proteases, and these often do not perform well as mouse protease inhibitors.

A complementary approach to unravel the complexity and redundancy of so many proteases may be to identify critical convergence points for a number of different protease activities. The proteolysis-resistant Col1a-1 mutant demonstrates the value of interrupting a cellular pathway at such a convergence point, in much the same way, that analysis of the lesser number of growth factor *receptors*, compared to growth factors themselves, has enabled progress to be made in that field (Werner and Grose 2003). We have focused on the Plg activation and MMP systems in three normal tissue-remodeling processes and in an example of cancer invasion. In skin wound healing and in embryo implantation, we have established a functional overlap of the Plg activation system with a group of galardin-sensitive metalloproteases. It remains to be determined what the critical common substrates for this functional redundancy are. Preliminary data suggest that fibrin is a crucial convergence point since fibrin accumulates in skin wounds and in involuting mammary glands when the Plg activation and MMP systems are blunted. It will be of great value to analyze these tissue-remodeling processes in the absence of fibrin. Unfortunately, the spontaneous bleeding phenotype of fibrinogen-deficient mice precludes certain studies since these mice do not survive a pregnancy (Suh et al. 1995).

We have emphasized that extracellular proteolysis in the context of cancer invasion and metastasis often has a striking similarity to physiological tissue-remodeling processes. Insight into normal tissue processes has, therefore, often contributed to a better understanding of cancer. We anticipate that a symbiotic relationship between these research endeavors will continue in the coming years.

Acknowledgments   The authors thank Mr. John Post for assistance in creating the artwork. This work was supported by the European Commission Grant LSHC-CT-2003-503297, the European Union research training grant for mammary gland biology RTN-2002-00246, the Danish Cancer Society, the Copenhagen Hospital Corporation, and the "Grosserer Alfred Nielsen og hustrus" foundation.

References

Abrahamsohn P. A., and Zorn T. M. T. 1993. Implantation and decidualization in rodents. J. Exp. Zool. 266:603–628.

Alexander C. M., Hansell E. J., Behrendtsen O., Flannery M. L., Kishnani N. S., Hawkes S. P., and Werb Z. 1996a. Expression and function of matrix metalloproteinases and their inhibitors at the maternal-embryonic boundary during mouse embryo implantation. Development 122:1723–1736.

Alexander C. M., Howard E. W., Bissell M. J., and Werb Z. 1996b. Rescue of mammary epithelial cell apoptosis and entactin degradation by a tissue inhibitor of metalloproteinases-1 transgene. J. Cell Biol. 135:1669–1677.

Alexander C. M., Selvarajan S., Mudgett J., and Werb Z. 2001. Stromelysin-1 regulates adipogenesis during mammary gland involution. J. Cell Biol. 152:693–703.

Almholt K., Nielsen B. S., Frandsen T. L., Brünner N., Danø K., and Johnsen M. 2003. Metastasis of transgenic breast cancer in plasminogen activator inhibitor-1 gene-deficient mice. Oncogene 22:4389–4397.

Almholt K., Lund L. R., Rygaard J., Nielsen B. S., Danø K., Rømer J., and Johnsen M. 2005. Reduced metastasis of transgenic mammary cancer in urokinase-deficient mice. Int. J. Cancer 113:525–532.

Almholt K., Green K. A., Juncker-Jensen A., Nielsen B. S., Lund L. R., and Rømer J. 2007. Extracellular proteolysis in transgenic mouse models of breast cancer. J. Mammary Gland Biol. Neoplasia 12:83–97.

Andreasen P. A., Kjøller L., Christensen L., and Duffy M. J. 1997. The urokinase-type plasminogen activator system in cancer metastasis: A review. Int. J. Cancer 72:1–22.

Andreasen P. A., Egelund R., and Petersen H. H. 2000. The plasminogen activation system in tumor growth, invasion, and metastasis. Cell. Mol. Life Sci. 57:25–40.

Bajou K., Noël A., Gerard R. D., Masson V., Brünner N., Holst-Hansen C., Skobe M., Fusenig N. E., Carmeliet P., Collen D., and Foidart J. M. 1998. Absence of host plasminogen activator inhibitor 1 prevents cancer invasion and vascularization. Nat. Med. 4:923–928.

Bajou K., Masson V., Gerard R. D., Schmitt P. M., Albert V., Praus M., Lund L. R., Frandsen T. L., Brünner N., Danø K., Fusenig N. E., Weidle U., Carmeliet G., Loskutoff D., Collen D., Carmeliet P., Foidart J. M., and Noël A. 2001. The plasminogen activator inhibitor PAI-1 controls in vivo tumor vascularization by interaction with proteases, not vitronectin: Implications for antiangiogenic strategies. J. Cell Biol. 152:777–784.

Balbín M., Fueyo A., Knäuper V., López J. M., Álvarez J., Sánchez L. M., Quesada V., Bordallo J., Murphy G., and López-Otín C. 2001 Identification and enzymatic characterization of two diverging murine counterparts of human interstitial collagenase (MMP-1) expressed at sites of embryo implantation. J. Biol. Chem. 276:10253–10262.

Beare A. H., O'Kane S., Krane S. M., and Ferguson M. W. 2003. Severely impaired wound healing in the collagenase-resistant mouse. J. Invest. Dermatol. 120:153–163.

Blasi F., and Carmeliet P. 2002. uPAR: A versatile signalling orchestrator. Nat. Rev. Mol. Cell Biol. 3:932–943.

Bugge T. H., Flick M. J., Daugherty C. C., and Degen J. L. 1995a. Plasminogen deficiency causes severe thrombosis but is compatible with development and reproduction. Genes Dev. 9:794–807.

Bugge T. H., Suh T. T., Flick M. J., Daugherty C. C., Rømer J., Solberg H., Ellis V., Danø K., and Degen J. L. 1995b. The receptor for urokinase-type plasminogen activator is not essential for mouse development or fertility. J. Biol. Chem. 270:16886–16894.

Bugge T. H., Flick M. J., Danton M. J., Daugherty C. C., Rømer J., Danø K., Carmeliet P., Collen D., and Degen J. L. 1996a. Urokinase-type plasminogen activator is effective in fibrin clearance in the absence of its receptor or tissue-type plasminogen activator. Proc. Natl. Acad. Sci. USA 93:5899–5904.

Bugge T. H., Kombrinck K. W., Flick M. J., Daugherty C. C., Danton M. J., and Degen J. L. 1996b. Loss of fibrinogen rescues mice from the pleiotropic effects of plasminogen deficiency. Cell 87:709–719.

Bugge T. H., Lund L. R., Kombrinck K. K., Nielsen B. S., Holmbäck K., Drew A. F., Flick M. J., Witte D. P., Danø K., and Degen J. L. 1998. Reduced metastasis of Polyoma virus middle T antigen-induced mammary cancer in plasminogen-deficient mice. Oncogene 16:3097–3104.

Bullard K. M., Lund L., Mudgett J. S., Mellin T. N., Hunt T. K., Murphy B., Ronan J., Werb Z., and Banda M. J. 1999. Impaired wound contraction in stromelysin-1-deficient mice. Ann. Surg. 230:260–265.

Carmeliet P., Kieckens L., Schoonjans L., Ream B., van Nuffelen A., Prendergast G., Cole M., Bronson R., Collen D., and Mulligan R. C. 1993. Plasminogen activator inhibitor-1 gene-deficient mice. I. Generation by homologous recombination and characterization. J. Clin. Invest. 92:2746–2755.

Carmeliet P., Schoonjans L., Kieckens L., Ream B., Degen J., Bronson R., De Vos R., van den Oord J. J., Collen D., and Mulligan R. C. 1994. Physiological consequences of loss of plasminogen activator gene function in mice. Nature 368:419–424.

Clark R. A. F. (ed) 1996. The Molecular and Cellular Biology of Wound Repair. 2nd edn. New York: NY Plenum Press.

Collen D., and Lijnen H. R. 2005. Thrombolytic agents. Thromb. Haemost. 93:627–630.

Cross J. C., Werb Z., and Fisher S. J. 1994. Implantation and the placenta: Key pieces of the development puzzle. Science 266:1508–1518.

Czekay R. P., Aertgeerts K., Curriden S. A., and Loskutoff D. J. 2003. Plasminogen activator inhibitor-1 detaches cells from extracellular matrices by inactivating integrins. J. Cell Biol. 160:781–791.

Danø K., Behrendt N., Høyer-Hansen G., Johnsen M., Lund L. R., Ploug M., and Rømer J. 2005. Plasminogen activation and cancer. Thromb. Haemost. 93:676–681.

Das S. K., Yano S., Wang J., Edwards D. R., Nagase H., and Dey S. K. 1997. Expression of matrix metalloproteinases and tissue inhibitors of metalloproteinases in the mouse uterus during the peri-implantation period. Dev. Genet. 21:44–54.

Dellas C., and Loskutoff D. J. 2005. Historical analysis of PAI-1 from its discovery to its potential role in cell motility and disease. Thromb. Haemost. 93:631–640.

Dewerchin M., Nuffelen A. V., Wallays G., Bouche A., Moons L., Carmeliet P., Mulligan R. C., and Collen D. 1996. Generation and characterization of urokinase receptor-deficient mice. J. Clin. Invest. 97:870–878.

Dickson S. R., and Warburton M. J. 1992. Enhanced synthesis of gelatinase and stromelysin by myoepithelial cells during involution of the rat mammary gland. J. Histochem. Cytochem. 40:697–703.

Dougherty K. M., Pearson J. M., Yang A. Y., Westrick R. J., Baker M. S., and Ginsburg D. 1999. The plasminogen activator inhibitor-2 gene is not required for normal murine development or survival. Proc. Natl. Acad. Sci. USA 96:686–691.

Dvorak H. F. 1986. Tumors: Wounds that do not heal. Similarities between tumor stroma generation and wound healing. N. Engl. J. Med. 315:1650–1659.

Dvorak H. F., Brown L. F., Detmar M., and Dvorak A. M. 1995. Vascular permeability factor/vascular endothelial growth factor, microvascular hyperpermeability, and angiogenesis. Am. J. Pathol. 146:1029–1039.

Fata J. E., Leco K. J., Voura E. B., Yu H. Y., Waterhouse P., Murphy G., Moorehead R. A., and Khokha R. 2001. Accelerated apoptosis in the *Timp-3*-deficient mammary gland. J. Clin. Invest. 108:831–841.

Fazioli F., Resnati M., Sidenius N., Higashimoto Y., Appella E., and Blasi F. 1997. A urokinase-sensitive region of the human urokinase receptor is responsible for its chemotactic activity. EMBO J. 16:7279–7286.

Folgueras A. R., Pendás A. M., Sánchez L. M., and López-Otín C. 2004. Matrix metalloproteinases in cancer: From new functions to improved inhibition strategies. Int. J. Dev. Biol. 48:411–424.

Green K. A., and Lund L. R. 2005. ECM degrading proteases and tissue remodelling in the mammary gland. BioEssays 27:894–903.

Green K. A., Nielsen B. S., Castellino F. J., Rømer J., and Lund L. R. 2006. Lack of plasminogen leads to milk stasis and premature mammary gland involution during lactation. Dev. Biol. 299:164–175.

Gutierrez L. S., Schulman A., Brito-Robinson T., Noria F., Ploplis V. A., and Castellino F. J. 2000. Tumor development is retarded in mice lacking the gene for urokinase-type plasminogen activator or its inhibitor, plasminogen activator inhibitor-1. Cancer Res. 60:5839–5847.

Gutiérrez-Fernández A., Inada M., Balbín M., Fueyo A., Pitiot A. S., Astudillo A., Hirose K., Hirata M., Shapiro S. D., Noël A., Werb Z., Krane S. M., López-Otín C., and Puente X. S. 2007. Increased inflammation delays wound healing in mice deficient in collagenase-2 (MMP-8). FASEB J. 21:2580–2591.

Guy C. T., Cardiff R. D., and Muller W. J. 1992. Induction of mammary tumors by expression of polyomavirus middle T oncogene: A transgenic mouse model for metastatic disease. Mol. Cell. Biol. 12:954–961.

Hartenstein B., Dittrich B. T., Stickens D., Heyer B., Vu T. H., Teurich S., Schorpp-Kistner M., Werb Z., and Angel P. 2006. Epidermal development and wound healing in matrix metalloproteinase 13-deficient mice. J. Invest. Dermatol. 126:486–496.

Høyer-Hansen G., Behrendt N., Ploug M., Danø K., and Preissner K. T. 1997. The intact urokinase receptor is required for efficient vitronectin binding: Receptor cleavage prevents ligand interaction. FEBS Lett. 420:79–85.

Jodele S., Blavier L., Yoon J. M., and DeClerck Y. A. 2006. Modifying the soil to affect the seed: role of stromal-derived matrix metalloproteinases in cancer progression. Cancer Metastasis Rev. 25:35–43.

Johnsen M., Lund L. R., Rømer J., Almholt K., and Danø K. 1998. Cancer invasion and tissue remodeling: Common themes in proteolytic matrix degradation. Curr. Opin. Cell Biol. 10:667–671.

Kawata Y., Mimuro J., Kaneko M., Shimada K., and Sakata Y. 1996. Expression of plasminogen activator inhibitor 2 in the adult and embryonic mouse tissues. Thromb. Haemost. 76:569–576.

Kirby D. R. S. 1965. The 'invasiveness' of the trophoblast. In The Early Conceptus, Normal and Abnormal, ed. W. W. Park, pp. 68–74. Edinburgh: University of St. Andrews Press.

Lapiere C. M. 2003. Collagenase and impaired wound healing. J. Invest. Dermatol. 120:12–13.

Lefebvre O., Wolf C., Limacher J. M., Hutin P., Wendling C., LeMeur M., Basset P., and Rio M.-C. 1992. The breast cancer-associated stromelysin-3 gene is expressed during mouse mammary gland apoptosis. J. Cell Biol. 119:997–1002.

Li F., Strange R., Friis R. R., Djonov V., Altermatt H. J., Saurer S., Niemann H., and Andres A. C. 1994. Expression of stromelysin-1 and TIMP-1 in the involuting mammary gland and in early invasive tumors of the mouse. Int. J. Cancer 59:560–568.

Liu X., Wu H., Byrne M., Jeffrey J., Krane S., and Jaenisch R. 1995. A targeted mutation at the known collagenase cleavage site in mouse type I collagen impairs tissue remodeling. J. Cell Biol. 130:227–237.

López-Otín C., and Overall C. M. 2002. Protease degradomics: A new challenge for proteomics. Nat. Rev. Mol. Cell Biol. 3:509–519.

Loskutoff D. J., Curriden S. A., Hu G., and Deng G. 1999. Regulation of cell adhesion by PAI-1. APMIS 107:54–61.

Lund L. R., Rømer J., Thomasset N., Solberg H., Pyke C., Bissell M. J., Danø K., and Werb Z. 1996. Two distinct phases of apoptosis in mammary gland involution: Proteinase-independent and -dependent pathways. Development 122:181–193.

Lund L. R., Rømer J., Bugge T. H., Nielsen B. S., Frandsen T. L., Degen J. L., Stephens R. W., and Danø K. 1999. Functional overlap between two classes of matrix-degrading proteases in wound healing. EMBO J. 18:4645–4656.

Lund L. R., Bjørn S. F., Sternlicht M. D., Nielsen B. S., Solberg H., Usher P. A., Østerby R., Christensen I. J., Stephens R. W., Bugge T. H., Danø K., and Werb Z. 2000. Lactational competence and involution of the mouse mammary gland require plasminogen. Development 127:4481–4492.

Lund L. R., Green K. A., Stoop A. A., Ploug M., Almholt K., Lilla J., Nielsen B. S., Christensen I. J., Craik C. S., Werb Z., Danø K., and Rømer J. 2006. Plasminogen activation independent of uPA and tPA maintains wound healing in gene-deficient mice. EMBO J. 25:2686–2697.

Madlener M., Parks W. C., and Werner S. 1998. Matrix metalloproteinases (MMPs) and their physiological inhibitors (TIMPs) are differentially expressed during excisional skin wound repair. Exp. Cell Res. 242:201–210.

Martin P. 1997. Wound healing—Aiming for perfect skin regeneration. Science 276:75–81.

Mirastschijski U., Zhou Z., Rollman O., Tryggvason K., and \gren M. S. 2004. Wound healing in membrane-type-1 matrix metalloproteinase-deficient mice. J. Invest. Dermatol. 123:600–602.

Mohan R., Chintala S. K., Jung J. C., Villar W. V., McCabe F., Russo L. A., Lee Y., McCarthy B. E., Wollenberg K. R., Jester J. V., Wang M., Welgus H. G., Shipley J. M., Senior R. M., and Fini M. E. 2002. Matrix metalloproteinase gelatinase B (MMP-9) coordinates and effects epithelial regeneration. J. Biol. Chem. 277:2065–2072.

Nagase H., Visse R., and Murphy G. 2006. Structure and function of matrix metalloproteinases and TIMPs. Cardiovasc. Res. 69:562–573.

Page-McCaw A., Ewald A. J., and Werb Z. 2007. Matrix metalloproteinases and the regulation of tissue remodelling. Nat. Rev. Mol. Cell Biol. 8:221–233.

Pass J., Jögi A., Lund I. K., Rønø B., Rasch M. G., Gårdsvoll H., Lund L. R., Ploug M., Rømer J., Danø K., and Høyer-Hansen G. 2007. Murine monoclonal antibodies against murine uPA receptor produced in gene-deficient mice: Inhibitory effects on receptor-mediated uPA activity *in vitro* and *in vivo*. Thromb. Haemost. 97:1013–1022.

Pedersen T. X., Pennington C. J., Almholt K., Christensen I. J., Nielsen B. S., Edwards D. R., Rømer J., Danø K., and Johnsen M. 2005. Protease mRNAs are predominantly expressed in the stromal areas of microdissected mouse breast carcinomas. Carcinogenesis 26:1233–1240.

Ploplis V. A., Carmeliet P., Vazirzadeh S., Van Vlaenderen I., Moons L., Plow E. F., and Collen D. 1995. Effects of disruption of the plasminogen gene on thrombosis, growth, and health in mice. Circulation 92:2585–2593.

Ploug M. 2003. Structure-function relationships in the interaction between the urokinase-type plasminogen activator and its receptor. Curr. Pharm. Des. 9:1499–1528.

Reponen P., Leivo I., Sahlberg C., Apte S. S., Olsen B. R., Thesleff I., and Tryggvason K. 1995. 92-kDa type IV collagenase and TIMP-3, but not 72-kDa type IV collagenase or TIMP-1 or TIMP-2, are highly expressed during mouse embryo implantation. Dev. Dyn. 202:388–396.

Rinkenberger J., and Werb Z. 2000. The labyrinthine placenta. Nat. Genet. 25:248–250.

Rinkenberger J. L., Cross J. C., and Werb Z. 1997. Molecular genetics of implantation in the mouse. Dev. Genet. 21:6–20.

Rossant J., and Cross J. C. 2001. Placental development: Lessons from mouse mutants. Nat. Rev. Genet. 2:538–548.

Rømer J. 2003. Skin cancer and wound healing. Tissue-specific similarities in extracellular proteolysis. APMIS Suppl. 107:5–36.

Rømer J., Lund L. R., Eriksen J., Ralfkiær E., Zeheb R., Gelehrter T. D., Danø K., and Kristensen P. 1991. Differential expression of urokinase-type plasminogen activator and its type-1 inhibitor during healing of mouse skin wounds. J. Invest. Dermatol. 97:803–811.

Rømer J., Lund L. R., Eriksen J., Pyke C., Kristensen P., and Danø K. 1994. The receptor for urokinase-type plasminogen activator is expressed by keratinocytes at the leading edge during re-epithelialization of mouse skin wounds. J. Invest. Dermatol. 102:519–522.

Rømer J., Bugge T. H., Pyke C., Lund L. R., Flick M. J., Degen J. L., and Danø K. 1996. Impaired wound healing in mice with a disrupted plasminogen gene. Nat. Med. 2:287–292.

Sappino A. P., Huarte J., Belin D., and Vassalli J. D. 1989. Plasminogen activators in tissue remodeling and invasion: mRNA localization in mouse ovaries and implanting embryos. J. Cell Biol. 109:2471–2479.

Selvarajan S., Lund L. R., Takeuchi T., Craik C. S., and Werb Z. 2001. A plasma kallikrein-dependent plasminogen cascade required for adipocyte differentiation. Nat. Cell Biol. 3:267–275.

Singer A. J., and Clark R. A. 1999. Cutaneous wound healing. N. Engl. J. Med. 341:738–746.

Solberg H., Rinkenberger J., Danø K., Werb Z., and Lund L. R. 2003. A functional overlap of plasminogen and MMPs regulates vascularization during placental development. Development 130:4439–4450.

Stefansson S., and Lawrence D. A. 1996. The serpin PAI-1 inhibits cell migration by blocking integrin alpha V beta 3 binding to vitronectin. Nature 383:441–443.

Strickland S., and Richards W. G. 1992. Invasion of the trophoblasts. Cell 71:355–357.

Suh T. T., Holmbäck K., Jensen N. J., Daugherty C. C., Small K., Simon D. I., Potter S. S., and Degen J. L. 1995. Resolution of spontaneous bleeding events but failure of pregnancy in fibrinogen-deficient mice. Genes Dev. 9:2020–2033.

Sympson C. J., Talhouk R. S., Alexander C. M., Chin J. R., Clift S. M., Bissell M. J., and Werb Z. 1994. Targeted expression of stromelysin-1 in mammary gland provides evidence for a role of proteinases in branching morphogenesis and the requirement for an intact basement membrane for tissue-specific gene expression. J. Cell Biol. 125:681–693.

Szabova L., Yamada S. S., Birkedal-Hansen H., and Holmbeck K. 2005. Expression pattern of four membrane-type matrix metalloproteinases in the normal and diseased mouse mammary gland. J. Cell. Physiol. 205:123–132.

Talhouk R. S., Bissell M. J., and Werb Z. 1992. Coordinated expression of extracellular matrix-degrading proteinases and their inhibitors regulates mammary epithelial function during involution. J. Cell Biol. 118:1271–1282.

Teesalu T., Blasi F., and Talarico D. 1996. Embryo implantation in mouse: Fetomaternal coordination in the pattern of expression of uPA, uPAR, PAI-1 and alpha 2MR/LRP genes. Mech. Dev. 56:103–116.

Wei Y., Waltz D. A., Rao N., Drummond R. J., Rosenberg S., and Chapman H. A. 1994. Identification of the urokinase receptor as an adhesion receptor for vitronectin. J. Biol. Chem. 269:32380–32388.

Werner S., and Grose R. 2003. Regulation of wound healing by growth factors and cytokines. Physiol. Rev. 83:835–870.

Wu H., Liu X., and Jaenisch R. 1994. Double replacement: Strategy for efficient introduction of subtle mutations into the murine Col1a-1 gene by homologous recombination in embryonic stem cells. Proc. Natl. Acad. Sci. USA 91:2819–2823.

Yamazaki M., Akahane T., Buck T., Yoshiji H., Gomez D. E., Schoeffner D. J., Okajima E., Harris S. R., Bunce O. R., Thorgeirsson S. S., and Thorgeirsson U. P. 2004. Long-term exposure to elevated levels of circulating TIMP 1 but not mammary TIMP-1 suppresses growth of mammary carcinomas in transgenic mice. Carcinogenesis 25:1735–1746.

Chapter 12
The Urokinase Plasminogen Activator Receptor as a Target for Cancer Therapy

Silvia D'Alessio and Francesco Blasi

Abstract Proteolytic processes are necessary for normal physiological functions in the body, including normal blood vessel maintenance, clot formation and dissolution, bone remodeling, and ovulation. The same enzyme system for the above roles is also used by the cancer cells for their growth and spread. These enzymes are produced by the tumor cells or cells surrounding them and can degrade the basement membrane and extracellular matrix (ECM) which consist of several components including collagens, glycoproteins, proteoglycans, and glycosaminoglycans. A major protease system responsible for ECM degradation is the plasminogen activation system, which generates the potent serine protease plasmin. The subject of this chapter, the urokinase plasminogen activator (uPA) receptor, plays an impressive range of distinct but overlapping functions in the process of cancer invasion and metastasis. Indeed, overexpression of this molecule is strongly correlated with poor prognosis in a variety of malignant tumors. Impairment of uPAR function, or inhibition of its expression, impedes the metastatic potential of many tumors. Several approaches have been employed to target uPAR with the aim of disrupting its ligand-independent action or interaction with uPA, Vn, or integrins, including the more recent antisense technology. This chapter also discusses the *in vivo* and *in vitro* use of antisense approaches and other similar techniques for downregulating uPAR as a potential therapy for cancer.

Introduction

The major cause of death in patients with malignant solid tumors such as carcinomas is the ability of cancer cells to invade surrounding tissues and form distant metastases. The spread of cancer cells from the primary site to a distant location is

S. D'Alessio and F. Blasi

Department of Molecular Biology and Functional Geneomics, DIBIT, Università Vita-Salute San Raffaele, Via Olgettina 58, 20132 Milan, Italy, e-mail: dalessio.silvia@hsr.it, blasi.francesco@hsr.it

D. Edwards et al. (eds.), *The Cancer Degradome.*

© Springer Science + Business Media, LLC 2008

known to follow a sequence that requires their detachment from the primary site, migration through the local stroma, invasion into and then extravasation from the vascular tree, before finally migrating toward, adhering to and proliferating at a distant site to form a metastatic tumor.

A major determinant for the invasive and metastatic potential of tumor cells is their ability to proteolytically degrade extracellular matrix (ECM) and the basement membrane surrounding the primary tumor, which facilitates local invasion and intravasation, leading to distant dissemination of the disease (Muehlenweg et al. 2001, Romer et al. 2004). Although several extracellular protease systems have been implicated in the tissue degradation and remodeling that often accompanies cancer invasion, several studies show that the urokinase plasminogen activator (uPA) system is central to these processes, as reviewed previously (Blasi and Carmeliet 2002, Mazar 2001, Mazzieri and Blasi 2005, Romer et al. 2004, Wang et al. 2001). Components of the uPA system thus represent promising candidates for targeted cancer therapies.

The uPA/uPAR System

The uPA/uPAR system is involved in a variety of cell functions, including extra-cellular proteolysis, adhesion, proliferation, chemotaxis, neutrophil priming for oxidant production and cytokine release, all of which variously contribute to the development, implantation, angiogenesis, inflammation, and metastasis of tumors (Ge and Elghetany 2003).

The uPA/uPAR system consists mainly of the serine protease uPA, its cell membrane-associated receptor (uPAR), a substrate (plasminogen), and plasmino-gen activator inhibitors (PAI-1 and PAI-2) (Mazar et al. 1999, Wang 2001). uPA is produced and secreted as single-chain polypeptide—a zymogen known as pro-uPA—that lacks plasminogen-activating activity. Upon binding of pro-uPA to uPAR, it is cleaved by various proteases into an active two-chain uPA molecule. This active uPA enzyme then converts the zymogen plasminogen to the active serine protease plasmin, which is involved in the degradation of the ECM and basement membranes by either direct proteolytic digestion or by activation of other zymogen proteases, such as pro-metalloproteases and pro-collagenases, thereby promoting tumor migration (Mazzieri and Blasi 2005, Wang 2001). Binding of uPA to uPAR provides an inducible, transient, and localized cell-surface proteolytic activity (Mazzieri and Blasi 2005, Wang 2001).

Studies have suggested that among the uPA/uPAR system components, uPAR might have a more crucial role in the metastasis process (Wang 2001). Many of the activities of uPA, including its activation by plasmin, are dependent on its binding to uPAR (Mazar et al. 1999). However, uPAR knock-out mice do not show any evidence of deficient fibrinolysis, even when in combination with the $tPA^{-/-}$ mutation, which suggests that pro-uPA activation takes place naturally also in the absence of uPAR, at least under nonstimulated conditions.

uPAR

The human uPAR cDNA encodes a 335 amino acid polypeptide, which during the cell surface sorting is posttranslationally modified in several ways to generate the mature receptor. An amino-terminal signal peptide and a carboxyterminal GPI-anchor peptide are removed during cell surface sorting and processing for GPI anchoring. Finally, the protein is extensively glycosylated. The mature uPAR protein consists of 3 homologous cysteine-rich repeats of about 90 amino acids each. Three-dimensional structure of uPAR and the details of the uPA binding site are discussed in Chapter 34 by Jacobson et al.

uPAR Cleavage and Shedding

Cell surface uPAR has been shown to undergo two major types of covalent modifications which alter the function of the receptor. The first type of uPAR modification is either a proteolytic cleavage close to the GPI anchor or a hydrolysis of the GPI-anchor by a phospholipase (Pedersen et al. 1993, Sier et al. 1998). This cleavage releases the entire receptor from the cell surface (suPAR), with concomitant functional changes, but does apparently not alter the ligand-binding properties of the receptor notably. We refer to this process as uPAR shedding. The second type is a proteolytic cleavage in the linker region connecting DI and DII and results in the release of the D1 fragment from the rest of the receptor. As described below, this cleavage changes the biochemical properties of uPAR completely. We will refer to this phenomenon as uPAR cleavage.

Recent advances in the study of uPAR shedding and cleavage supports the possibility that these processes are important in the malignant process of tumor invasion and metastasis. The GPI-anchoring of uPAR renders the protein prone to release from the cell surface and soluble forms of uPAR can indeed be found both *in vitro* and *in vivo*. The mechanism of uPAR shedding may be catalyzed by cellular phosphatidylinositol-specific phospholipase D (PIPL-D), which is able to release GPI-anchored proteins, including uPAR, from the cell surface (Metz et al. 1994, Wilhelm et al. 1999). Also plasmin appears to shed uPAR, although the mechanism is still not clear (Beaufort et al. 2004). Shedding of uPAR releases the receptor from the cell surface but does not apparently alter the affinity for its two major ligands, uPA and Vn (Ronne et al. 1994, Wei et al. 1994). Shedding occurs both *in vitro* and *in vivo*: suPAR is present in the conditioned medium from a variety of cultured cells (Chavakis et al. 1998, Holst-Hansen et al. 1999, Lau and Kim 1994, Ploug et al. 1991, Sidenius et al. 2000), as well as in biological fluids such as plasma, urine, cerebrospinal fluid, ascites, and ovarian cyst fluid (Garcia-Monco et al. 2002, Mustjoki et al. 2000a, Pedersen et al. 1993, Sier et al., 1999, Stephens et al. 1997, Wahlberg et al. 1998). The uPAR linker region connecting DI and DII is prone to hydrolysis by a variety of different proteases. This region may be cleaved by trypsin, chymotrypsin and, physiologically more important, uPA, plasmin, neutrophil elastase, as well as by a number of different matrix metalloproteinases (MMPs) (Andolfo et al. 2002, Behrendt et al. 1991, Hoyer-Hansen et al. 1997a, Koolwijk et al. 2001, Ploug et al. 1994). While the uPA-catalyzed cleavage of suPAR *in vitro* is independent of the uPA/uPAR interaction (Hoyer-Hansen et al., 1992), cleavage of cell-surface uPAR is accelerated through a mechanism which requires the binding of uPA to uPAR (Hoyer-Hansen et al. 1997b). The reason for this acceleration of uPAR cleavage is not clear, but evidence has been presented that GPI-anchored and soluble uPAR may have different conformations (Andolfo et al. 2002, Hoyer-Hansen et al. 2001) with the GPI-anchored form being more susceptible to cleavage (Andolfo et al. 2002, Hoyer-Hansen et al. 2001). Besides uPA, also MMPs are capable of cleaving uPAR in cell culture (Koolwijk et al. 2001) and *in vitro* (Andolfo et al. 2002). The protease(s) responsible for uPAR cleavage *in vivo* has not been determined. Cleaved uPAR [DIIDIII] has been found on the surface of many cells including endothelial cells, lymphocytes, and in several different cancer cell lines (Holst-Hansen et al. 1999, Hoyer-Hansen et al. 1992, Ragno et al. 1998, Sidenius et al. 2000, Solberg et al. 1994). Cleaved uPAR has also been identified in detergent extracts of human tumors xenografted in nude mice and in the Lewis Lung tumor in mouse (Holst-Hansen et al. 1999, Solberg et al. 1994). Cleavage of uPAR releases D1 to the surroundings and this fragment can indeed be found in the conditioned medium from cells with cleaved uPAR on the surface (Koolwijk et al. 2001, Mustjoki, Sidenius and Vaheri 2000b, Sidenius et al. 2000). Soluble forms of the DIIDIII fragment are also found in culture medium of cancer cells (Mustjoki et al. 2000b, Sidenius et al. 2000), in the fluid from human malignant ovarian cysts (Wahlberg et al. 1998), in urine from healthy individuals and cancer patients (Mustjoki et al. 2000a, Sidenius et al. 2000, Sier et al. 1999), and in blood from patients with acute myeloid leukemia (Mustjoki et al. 2000a).

uPA/uPAR Interactions and Signal Transduction

The important role of uPAR in tumor cell adhesion, migration, invasion, and proliferation makes this receptor an attractive drug target in cancer treatment; however, this is complicated by the extent of the published uPAR "interactome." For this reason, the most important question becomes which of the many molecular interactions are really essential to mediate uPAR function. Recently, the crystal structures of uPAR in complex with a peptide antagonist (Llinas et al. 2005) and with the N-terminal fragment of uPA (Barinka et al. 2006) were presented, providing the first rational basis toward understanding how uPAR may organize its multiple molecular interactions.

The second well-characterized ligand for uPAR is Vn, a glycoprotein produced in the liver and present at high concentrations in plasma (Preissner 1989). The uPAR binding site in Vn has been mapped to the amino-terminal somatomedin B domain of Vn (Deng et al. 1996, Okumura et al. 2002), a region which also contains the binding sites for PAI-1 and for the integrin Vn receptor ($\alpha V\beta 3$) (Hoyer-Hansen et al. 1997a, Sidenius and Blasi 2000). Several antibodies against D1 inhibit the interaction between uPAR and Vn (Hoyer-Hansen et al. 1997a, Kanse et al. 1996, Sidenius and Blasi 2000). The binding to Vn is connected to the occupancy of uPAR by uPA (Hoyer-Hansen et al. 1997a, Sidenius and Blasi 2000, Waltz and Chapman 1994, Wei et al. 1994) and is, at least *in vitro*, controlled by uPAR dimerization (Sidenius et al. 2002). The interaction between uPA, Vn, and uPAR is profoundly altered by receptor cleavage. The released D1 fragment has a >1,000-fold reduced affinity for uPA as compared to the intact receptor and the DIIDIII fragment has no measurable affinity at all (Ploug et al. 1994). Also the uPAR interaction with Vn is lost as none of the generated fragments binds with measurable affinity (Hoyer-Hansen et al. 1997a, Sidenius and Blasi 2000). Sidenius and colleagues (2007) have also demonstrated that a direct Vn interaction is both necessary and sufficient to initiate uPAR-induced changes in cell morphology, migration, and signaling independently of other direct lateral protein–protein interactions. Their data suggest that the single interaction between uPAR and Vn may be responsible for many of the proteolysis-independent biological effects initiated by uPAR (Madsen et al. 2007).

Most of the cellular responses modulated by the uPA/uPAR system, including migration, cellular adhesion, differentiation, and proliferation (Blasi and Carmeliet 2002) require transmembrane signaling, which cannot be mediated directly by a GPI-anchored protein such as uPAR. For this reason, besides the well-established interactions with uPA and Vn, uPAR has been reported to entertain direct contacts with a variety of extracellular proteins and membrane receptors, such as integrins (Chapman and Wei 2001, Ossowski and Aguirre-Ghiso 2000), epidermal growth factor (EGF) receptor (Liu et al. 2002), high molecular weight kininogen (Colman et al. 1997), caveolin, and the G-protein-coupled receptor FPRL1 (Resnati et al. 2002). As a result, uPAR activates intracellular signaling molecules such as tyrosine- and serine-protein kinases, Src, focal adhesion kinase (FAK), and extracellular-signal-regulated kinase (ERK)/mitogen-activated protein kinase (MAPK).

The interaction of uPAR with integrins is supported by co-immunoprecipitation experiments and by the effect of uPAR-binding peptides isolated from phages libraries (Aguirre-Ghiso et al. 1999, Bohuslav et al. 1995, Carriero et al. 1999, Tarui et al. 2001, Wei et al. 1996). Although uPAR can interact with many integrins, it appear to have the highest affinity for the fibronectin receptors $\alpha3\beta1$ and $\alpha5\beta1$ (*see* Chapter 23 by Degryse). Ligand-induced signaling necessary for normal $\beta1$ integrin function requires caveolin and is indeed regulated by uPAR. Caveolin and uPAR may operate within adhesion sites to organize kinase-rich lipid domains in proximity to integrins, promoting efficient signal transduction (Wei et al. 1999). Furthermore, disruption of uPAR-integrins association by uPAR-binding peptides broadly impairs integrin function, suggesting a novel strategy for regulation of integrins in the settings of inflammation and tumor progression (Simon et al. 2000). The best characterized uPAR-dependent signaling pathway is the one described by Aguirre-Ghiso and colleagues. They propose that even cancer cells with multiple mutations may use the uPAR surface receptor and ECM components to regulate signaling pathways that control cell cycle progression and/or arrest (Aguirre-Ghiso et al. 2003). They describe a uPAR-dependent mechanism by which the majority of tumor cells modulate the activity ratio between the proliferation inducer ERK (Hoshino et al. 1999) and the negative growth regulator p38 (Chen et al. 2000). On the basis of the study of ten different cell lines, their results show how uPAR and $\alpha5\beta1$ activate the EGFR in a EGF-independent but FAK-dependent manner (Liu et al. 2002) and generate high ERK and low p38 activity necessary for the *in vivo* growth of cancer cells. A positive loop is activated in which high ERK activity increases uPAR and uPA expression (Aguirre-Ghiso et al. 2001, Lengyel et al. 1995, 1997, 1996) and the high uPAR level maintains high ERK activity by activating $\alpha5\beta1$ (Aguirre-Ghiso et al. 1999, Liu et al. 2002). The cancer cell proliferation loop can be interrupted by a reduction of uPAR level by cleavage of its domain 1, important for the uPAR/$\alpha5\beta1$ interaction and activation (Aguirre-Ghiso et al. 1999, Liu et al. 2002, Montuori et al. 2002) or by loss of uPA and/or FN.

uPAR in Cell Motility

uPAR plays a role in the migration of a variety of cell types, and evidence is accumulating that uPAR-dependent migration is mediated through integrins (*see* Chapter 23, Degryse). In addition to $\beta1$- and $\beta2$-integrin involvement in the uPAR-dependent adhesion and migration of leukocytes (Aguirre-Ghiso et al. 1999, Gyetko et al. 1995, 1994; Liu et al. 2002, Montuori et al. 2002, Sitrin et al. 1996, Wong et al. 1996), the interaction of uPAR with integrins has also been demonstrated on tumor cells. Xue et al. have demonstrated the interaction of uPAR on HT1080 cells with various α and β integrins including $\beta1$ and $\beta3$ and αv, $\alpha3$, $\alpha5$, and $\alpha6$ (Xue et al. 1997). Migration, but not adhesion on Vn, of FG cells, which express av $\beta5$, was uPA–uPAR dependent. However, migration of several melanoma cell lines, which express only $\alpha v\beta3$ occurred independently of uPAR (Yebra et al. 1996). The adhesive and pro-migratory effects of uPAR, as well as the identity of the integrin adapter, depend on the cell type and the ECM component in question. The interactions of uPAR in cell migration and invasion may change as matrix barriers are remodeled or as the cell migrates through areas of different matrix composition. Several signaling pathways have been implicated in uPAR-mediated cell migration *in vitro*. uPAR-dependent signaling via the JAK/STAT pathway may be involved in the migration of vascular smooth muscle cells (Dumler et al. 1998). A second, uPAR-dependent signaling pathway involving Src-like protein tyrosine kinases has also been described in these cells, although the functional relevance of this second pathway is not yet understood. The JAK/STAT pathway was also activated by clustering the uPA–uPAR complex using a monoclonal antibody in the human kidney epithelial cell line TCL-598, resulting in the migration of this cell line (Koshelnick et al. 1997, Nguyen et al. 1998). In MCF-7 breast cancer cells, uPAR occupancy resulted in cell migration, which occurred through the activation of ERK1/ERK2. An inhibitor targeting MAPK kinase, a member of the JAK family of kinases, suppressed ligand-induced uPAR-dependent activation of ERK1/ERK2 in these cells (Nguyen et al. 1998). The MAPK pathway is activated in cytokine-mediated signaling and has been implicated as a major signal-transducing pathway in angiogenesis (stimulated by vascular endotheliel growth factor (VEGF) and basic fibroblast growth factor (bFGF) (D'Angelo et al. 1995, Jones et al. 1998).

Chymotrypsin-cleaved suPAR is a potent chemoattractant for several different cell lines (Fazioli et al. 1997, Resnati et al. 1996). The chemotactic response and kinetics of p56/59hck phosphorylation induced by proteolytically inactive uPA derivatives and that of cleaved suPAR are similar, suggesting that the same signaling pathway may be activated by these molecules (Resnati et al. 1996). Inhibitors of tyrosine kinases and heterotrimeric G proteins block the chemotactic response and the induction of phosphorylation of p56/59hck. The fact that pertussis toxin inhibits chemotaxis and the phosphorylation of p56/59hck suggests that heterotrimeric G proteins are involved and that they act upstream of the tyrosine kinase in the signaling pathway. The fact that cleaved soluble uPAR and peptides containing the uPAR chemotactic epitope are strong chemokine-like molecules strongly suggests the existence of one or more membrane "adapter" molecule(s) capable of transmitting the chemotactic signal over the membrane (Fazioli et al. 1997, Resnati et al. 1996). Indeed, it was shown that the DIIDIII fragment generated by chymotrypsin cleavage of suPAR interacts with, and signals through, the FPRL1/LXA4R chemokine receptor (Resnati et al. 2002). Both GPI-anchored and soluble forms of cleaved uPAR have been observed on different cell types and in diverse biological fluids. The strong chemotactic properties of suPAR fragments, together with the fact that similar fragments are found at high concentrations in cancer, suggests that the chemotactic activities of these fragments may play a role in the process of tumor invasion, maybe as an autocrine or paracrine signal for tumor cell motility.

uPAR and Cell Proliferation

In addition to regulating cell migration, uPAR also regulates cell proliferation. Work by Ossowski and co-workers has described a mechanism by which the uPAR/ integrin interaction may not only affect tumor growth and invasion through its regulation of extracellular proteolysis and integrin-dependent cell migration but also by directly promoting tumor cell proliferation (Liu et al. 2002). The model cell system used by these researchers is the HEp3 cell line, which is highly malignant, grows rapidly *in vivo* on the chicken chorioallantoic membrane, expresses uPAR, and has a high level of active ERK (Aguirre-Ghiso et al. 1999). Downregulation of uPAR expression in these cells, by antisense technology or prolonged culture *in vitro*, results in strong reduction in the level of ERK activation and causes concomitant tumor dormancy. ERK activation by uPAR in HEp3 cells is dependent upon the interaction between uPAR and $\alpha 5 \beta 1$-integrin and is maximal when both of these receptors are engaged by their respective ligands (uPA and Fn). Because ERK is a downstream effector molecule of both integrins and the EGF receptor (EGFR), which is expressed in HEp3 cells, Ossowski and colleagues proceeded to analyze the possible role of EGFR in uPAR-mediated ERK activation and tumor growth. Indeed, they succeeded in demonstrating that in HEp3 cells the EGFR, independently of EGF, mediates the uPA/uPAR/$\alpha5\beta1$/Fn-induced tumor growth pathway. Interestingly, the uPAR-mediated growth promoting signaling pathway did not require high levels of EGFR expression, distinguishing it from the fibronectin-dependent integrin-mediated EGFR-activation previously described (Moro et al. 1998). uPAR-dependent EGFR-activation requires high levels of intact uPAR and is paralleled by a physical association between the EGFR and $\alpha5\beta1$ in a FAK-dependent manner. These data present, to our knowledge, the first example of how carcinoma cells can utilize a normal expression level of a EGFR and a high expression level of uPAR to activate a growth factor-independent mitogenic pathway.

uPAR and Cancer

uPAR is expressed across a variety of tumor cell lines and tissues, including colon, breast, ovary, lung, kidney, liver, stomach, bladder, endometrium, and bone (Ge and Elghetany 2003, Mazzieri and Blasi 2005, Wang 2001). uPAR expression is not confined to the tumor cells themselves: several tumor-associated cell types, including macrophages, mast cells, endothelial cells, NK cells, and fibroblasts, are all capable of uPAR expression in various tumor types (Mazar et al. 1999, Mazar 2001, Sidenius and Blasi 2003). Indeed, the involvement of stromal cells in the generation of extracellular proteolysis argues that cancer invasion is the result of an interaction between cancer cells and stromal cells. It is not only the cancer cells that invade but also a mixed cell population. Tha cancer cells are the initiators and probably the organizers, but each cell type contributes in a distinct way to the overall process.

Several experimental evidences support the importance of the uPA/uPAR system in cancer with respect to its ability to modulate cell migration and cell adhesion and therefore determine the invasive and metastatic properties of tumor cells both *in vitro* and *in vivo* (for review, *see* Sidenius and Blasi 2003). uPAR levels have been strongly correlated with metastatic potential and advanced disease, which has been demonstrated in tumor samples obtained from patients with colon and breast cancer (Ge and Elghetany 2003, Dano et al. 2005). For example, uPAR is overexpressed in malignant breast cancer tissues but not in normal and benign breast tumors. uPAR has been found to be particularly abundant at the leading edge of tumors, that is, in those areas where tumor cells invade normal tissue (Lindberg et al. 2006, Skriver et al. 1984, Yamamoto et al. 1994). Tumor angiogenesis, a necessary event in tumor progression to sustain tumor growth and metastasis dissemination, is also modulated by the uPA/uPAR system (Carmeliet and Jain 2000, Jain and Carmeliet 2001). It requires a finely regulated cell proliferation, differentiation, and migration. After activation, endothelial cells express increased amount of uPA and uPAR at their leading migratory front to modulate ECM degradation, redeposition and cell adhesion (Blasi and Carmeliet 2002). The increased levels of uPAR expression typically associated with tumor tissue, its relative absence from normal, quiescent tissue, and its central role in regulating angiogenesis and tumor progression suggest that uPAR represents an attractive target for cancer therapy. Most experimental strategies have been focused on reducing pericellular uPA-mediated plasminogen activation, a goal which (in theory) may be obtained by a variety of approaches. First, by reducing the expression of uPA and/or uPAR expression. Second, by interfering with the uPA/uPAR interaction. Third, by a direct inhibition of uPA activity. Alternative approaches to interfere with the uPA-system includes interference with the uPAR/Vn and uPAR/integrin interactions. Direct approaches aimed at directly killing the tumor cells using toxins targeting one or more of the components of the uPA-system have also been developed and hold great promise.

There are several reasons why pharmacological targeting of uPA/uPAR may be attempted without major side effects. First, animal models suggest that the uPA-system is not essential for fertility or survival under physiological conditions. Second, the thrombosis risk of nocturnal hemoglobinuria patients does not appear to depend on the lack of uPAR (Bessler et al. 2002).

Antisense Therapeutic Strategies for Downregulation of uPAR **In Vivo**

Kook and co-workers were the first to evaluate the effect of antisense inhibition of uPAR on invasion and metastasis in human squamous cell carcinoma. Using a vector that is capable of expressing an antisense uPAR transcript, Kook et al. (1994) demonstrated that in highly malignant human squamous carcinoma cells downregulation of uPAR reduced the invasive potential. Furthermore, the tumors that developed from antisense clones were less invasive and nontumorigenic in chick embryos 7 days after injection and less invasive when injected in nude mice. This was the seminal *in vivo* demonstration that tumor growth, invasion, and metastasis could be inhibited by an antisense approach through Watson–Crick base-specific complementarity.

Downregulation of uPAR can also be obtained by the classic antisense oligodeoxynucleotide (asODN) technology, which consists of the injection of antisense DNA strands complementary to uPAR mRNA, or the antisense RNA technology, based on transfection with a vector capable of expressing the antisense transcript complementary to uPAR mRNA. Margheri et al. (2005) and D'Alessio et al. (2004) both investigated the antimetastatic and/or antitumor potential of the same 18mer phosphorothioate asODN in two different experimental models. Using a rodent model of bone metastasis, Margheri et al. (2005) injected malignant human prostate carcinoma cells into the heart (left ventricle) of CD1 nude mice. The animals were then subjected to daily intraperitoneal injections of asODNs and analyzed at 28 days after the heart injection or at the first signs of serious distress. Treatment with the asODN resulted in complete inhibition of bone metastases in 80% of the mice as well as complete inhibition of lymphonode and lung metastases.

D'Alessio et al. (2004) injected human melanoma cells with high metastatic potential into the hind leg muscles of CD1 male nude mice. Four days after cell implant, when a mean tumor mass of 350 mg was evident, the mice were treated intravenously with the asODN for five consecutive days. A second and third cycle of treatment was administered at 2-day intervals, and the mice were sacrificed and analyzed 25 days after tumor implantation. The asODN treatment resulted in 45% reduction of primary tumor mass and 78% reduction of lung metastases. Thus, these two studies show the ability of uPAR downregulation to reduce tumor growth in two types of cancers, which was effective both when administration was intraperitoneal and by the more clinically relevant intravenous way.

Similarly, Nozaki et al. (2005) showed that injection of highly malignant human oral squamous carcinoma cells pretreated with an 18mer phosphorothiate uPAR asODN into the chorioallantoic membrane vein of 10-day-old chick embryos yielded 86% inhibition of liver metastasis. They also showed that orthotopic implantation of cells pretreated with the asODN into the submucosa of the oral floor of 6-week-old female BALB/c immunocompromised mice inhibited the invasive capacity of these cells.

Research groups investigating antisense RNA technology for downregulation of uPAR *in vivo* have employed both plasmids (Dass et al., 2005, Go et al. 1997, Kook et al. 1994, Wang et al. 2001) and adenovirus (Gondi et al. 2004a, Lakka et al. 2001, Mohan et al. 1999) constructs for this purpose. Go et al. (1997) produced their plasmid construct by cloning the same 300 bp fragment as (Kook et al. 1994), corresponding to the 5′ end of uPAR in an antisense orientation. Human glioma cells stably transfected with the antisense construct were injected intracerebrally into 7-week-old female athymic nude mice. Animals were sacrificed and analyzed 1, 2, 3, and 4 weeks postinjection. Stable transfectants failed to form tumors and were negative for uPAR expression.

Wang et al. (2001) constructed plasmid expression vectors containing either a 585 bp 3′ uPAR cDNA fragment or a 498 bp 5′ uPAR cDNA fragment in the antisense orientation. Human colon cancer cell lines were transfected with each of the antisense clones and injected into the dilated lateral tail veins of 3–4-week-old athymic mice. At 9–12 weeks postinjection, the mice were sacrificed and examined. Metastasis of cells to the lungs was observed in 63% to 78% of mice injected with the parental colon cancer or control cells. By contrast, only 19% and 9% pulmonary metastases were observed in mice injected with the 3′ and 5′ antisense clones, respectively.

Downregulation of uPAR levels by an antisense strategy using adenovirus constructs resulted also in inhibition of growth and invasion of human glioblastoma (Gondi et al. 2004b, Mohan et al. 1999) and human lung cancer cells (Lakka et al. 2001). These three studies used the same adenovirus construct for the two types of cancers, highlighting its cross-cancer potency.

Mohan et al. (1999) also performed tumor regression experiments by injecting human glioblastoma cells subcutaneously into nude mice and then injecting the mice every other day with the antisense-containing adenovirus construct when the tumor size had reached 4–5 mm (after 8–10 days). They showed that injection of the antisense construct into pre-established tumors in nude mice caused regression of those tumors. Mohan et al. showed complete inhibition of tumor formation after intracranial injections of glioma cells transfected with the antisense construct, Gondi et al. (2004a) reported a reduction in intracranial tumor growth of 60% or greater.

The discovery of RNA interference (RNAi) has provided new opportunities for cancer therapy. Small interfering RNAs (siRNAs) are more potent inhibitors of gene expression compared with ribozymes and deoxyribozymes (Beale et al. 2003). However, this evaluation was only performed *in vitro* using human squamous carcinoma cells A431 and EGFR as targets, and the findings might not be directly relevant to the *in vivo* efficiency of these constructs. Surprisingly, not much has been accomplished to compare different gene downregulation systems, and more studies should be devoted to this examination.

Nevertheless, investigators have already used an shRNA-based RNAi plasmid system for the downregulation of uPAR in prostate cancer (Pulukuri et al. 2005) and glioblastoma cells (Gondi et al. 2004a, Lakka et al. 2005). These researchers have all used a plasmid construct expressing the same small hairpin RNA (shRNAs), which they also refer to as siRNAs, targeted to uPAR. Importantly, a clear majority of *in vivo* studies looking at the downregulation of uPAR have used sequences antisense to the 5' region of the uPAR mRNA.

For the above studies, human glioblastoma cells were intracranially injected into athymic male and female nude mice. Eight to ten days after tumor growth, sustained-release infusion of 150 µg of the shRNA-expressing plasmid construct was performed into the brain of each animal. The mice were sacrificed and analyzed at the end of the 5-week follow-up period or when the control mice started showing symptoms. Gondi et al. (2004b) reported a 65% regression of pre-established intracranial tumor growth. These findings were supported by Lakka et al. (2005) who reported 70% inhibition of pre-established intracranial tumor growth.

uPAR Downregulation may Affect Cellular Function and Signal Transduction

Researchers that have used either antisense or siRNA technologies for the successful *in vivo* downregulation of uPAR in various cancers have concurrently tested these same technologies in *in vitro* biological assays. Evaluation of the results of these *in vitro* assays reveals that downregulation of uPAR has led, in most cases, to inhibition of invasion (D'Alessio et al. 2004, Dass et al. 2005, Lakka et al. 2005, 2001; Margheri et al. 2005, Mohan et al. 1999), migration (Dass et al. 2005, Lakka et al. 2001), adhesion (Dass et al. 2005), and proliferation (D'Alessio et al. 2004, Lakka et al. 2005, Margheri et al. 2005). In addition, reduced uPAR levels lead to inhibition of tumor-induced angiogenesis (Lakka et al. 2005) and ECM degradation (Nozaki et al. 2005, Wang et al. 2001).

As stated earlier, some of the biological functions of uPAR, such as proliferation, are facilitated by the regulation of several different signaling molecules. In an attempt to understand and/or elucidate the involvement of uPAR in downstream signaling pathways, studies have investigated the effect of uPAR downregulation on components of the relevant signaling pathways. D'Alessio et al. (2004) reported that melanoma cells exhibited a strong decrease in ERK1/2 activation when an 18mer asODN was used to downregulate uPAR. Using this same asODN for the downregulation of uPAR in prostate cancer cells, Margheri et al. (2005) reported a strong decrease of FAK/JNK/Jun phosphorylation (thereby causing a decrease in the activation of the FAK/JNK/Jun pathway). At the same time, the synthesis of cyclins A, B, D1, and D3 was inhibited, and these prostate cancer cells accumulated in the G2 phase of the cell cycle. The downregulation of uPAR by a plasmid construct expressing shRNA for uPAR resulted in significantly reduced levels of the phosphorylated forms of MAPK, ERK, and AKT signaling pathway molecules (Lakka et al. 2005).

However, the majority of studies applying uPAR downregulation for cancer *in vivo* failed to look at which signaling pathways are perturbed as a result. In any case, different laboratories choose to elucidate effects on different pathways and, although there is an abundance of literature looking at individual pathways *in vitro*, it is difficult to compare results from separate studies because various parameters, including cell line, passage number, minor technical differences, the antisense sequence, the concentration of constructs, the time-points evaluated, and the way the data are reported, often prevent such comparisons.

Inhibition of the uPAR/Vn Interaction

Tumor cells often express reduced levels of adhesion receptors and also often fail to deposit ECM around themselves. The fact that uPAR is upregulated in many tumor cells suggests that the cells may use this alternative adhesion pathway as a response to the altered expression of normal cell adhesion proteins. In glioblastomas and in hepatocellular carcinomas, both Vn and uPAR are present at relatively high levels (De Petro et al. 1998, Gladson and Cheresh 1991, Gladson et al. 1995, Kondoh et al. 1999) and since the interaction between uPAR and Vn induces cytoskeleton rearrangements and increases cell motility (Kjoller and Hall 2001), this interaction may contribute to the highly malignant phenotype of these tumors. Therapeutical antitumor approaches aimed at blocking the uPAR/Vn interaction therefore seems warranted. Development of inhibitors of the uPAR/Vn interaction may possibly start from the uPAR-binding somatomedin B domain of Vn, which is a natural and potent uPAR/Vn-interaction antagonist. As it was shown that uPAR-binding to Vn involves dimerization of uPAR (Sidenius et al. 2002), inhibitors of uPAR-dimerization might also become useful.

Inhibition of the uPAR/Integrin Interaction

As described above, uPAR/integrin interactions affects several cellular properties including cell adhesion, migration, and proliferation, which may potentially be important in the malignant process of tumor invasion and metastasis. Specific peptide-based inhibitors of the uPAR/integrin interaction have been identified (Wei et al. 1996), and direct evidence supporting a functional role of the uPAR/integrin interaction in tumor progression has come from an *in vivo* bone xenograft model (van der Pluijm et al. 2001). In this study, stably transfected MDA-MB-231 cells that express a peptide which blocks the uPAR/integrin interaction (peptide 25, Wei et al. 1996) showed a significant reduction in tumor progression in bone. Also the continuous systemic administration of peptide 25 resulted in significantly reduced MDA-MB-231 tumor progression when compared to scrambled control peptide. Along the same lines, it will be important to identify uPAR peptides preventing the formation of or dissociating already formed uPAR/integrin complexes (*see* Chapter 23, Degryse).

Combination of uPAR Downregulation with Gene Modulation of Other Molecular Targets

Downregulation of more than one component involved in tumor invasion and metastasis might possibly have a synergistic or additive effect in preventing tumor dissemination. Lakka et al. (2003) reported that intracranial injection of human glioma cells infected with an adenovirus bicistronic construct capable of simultaneously expressing antisense uPAR and matrix metalloproteinase-9 (MMP-9) antisense, showed decreased invasiveness and tumorigenicity in mice. Subcutaneous injections of the bicistronic construct into established tumors caused tumor regression. MMP-9 is involved in the metastasis of various types of cancers, although its inhibition has not led to significant improvements in clinical trials (Klein et al. 2004). Thus, it is hoped that a dual targeted approach, combining MMP-9 and uPAR downregulation, will lead to better efficacy *in vivo*.

Lakka et al. also used the bicistronic plasmid construct targeting both uPAR and MMP-9 simultaneously (Lakka et al. 2005) with total regression of pre-established intracerebral tumor growth in mice. Similarly, Gondi et al. (2004b) showed that RNAi of uPAR and cathepsin B reduced glioma cell invasion and angiogenesis in *in vivo* models. Furthermore, intratumoral injections of these plasmid vectors expressing shRNA for uPAR and cathepsin B resulted in the regression of pre-established intracranial tumors (Sloane et al. 2005).

Injection into SCID mice of an adenovirus construct capable of simultaneously expressing uPAR and MMP-9 antisense has been reported to cause the regression of subcutaneous H1299 tumors (Rao et al. 2005). In addition, lung metastasis was inhibited with A549 cells (Rao et al. 2005). Furthermore, uPAR downregulation might be coupled to overexpression of tumor-inhibiting genes, such as tumor-suppressor genes, for enhanced therapeutic effect. Such an approach was tested by Adachi et al. (2002) by combining uPAR downregulation with p16 tumor suppressor overexpression, to demonstrate a dramatic inhibition of orthotopic and ectopic glioma growth *in vivo*. However, this is the only study that combines downregulation of uPAR with overexpression of growth-inhibiting genes, and more work needs to be done, although this seems to be a promising and efficacious option: another feasible option would be to overexpress one of the plasminogen activator inhibitors. In most of the studies that combine uPAR downregulation with downregulation of a second important pro-cancer target, the authors fail to report whether the effect was additive or synergistic.

uPAR and Apoptosis

Oncogenic cell transformation is currently viewed as a multistep process in which a series of genetic lesions change cellular physiology leading to the acquisition of new capabilities, such as an enhanced ability to proliferate, migrate, and escape apoptotic cell death (Hanahan and Weinberg 2000). Apoptosis can be viewed as a safe-lock mechanism that could prevent the establishment of a fully transformed phenotype. For instance, it is currently accepted that uncontrolled proliferation could by itself prime the transforming cell to apoptotic cell death (Hood and Cheresh 2002, Pelengaris et al. 2002). This is why the acquired capabilities of resistance to apoptotic cell death and tissue invasion are considered to be obligate steps in tumor progression. Recent findings indicate that a decreased uPAR expression may promote apoptosis. This is the case of SNB19 glioblastoma cells expressing antisense uPAR constructs that are less invasive than parental cells when injected *in vivo* and undergo loss of mitochondrial transmembrane potential, release of cytochrome c, caspase-9 activation, and apoptosis (Yanamandra et al. 2000). Furthermore, glioma cells bearing a reduced uPAR number are more susceptible to tumor necrosis factor-α-related apoptosis than parental cells (Krishnamoorthy et al. 2001). Alfano and colleagues (Alfano et al. 2006) provide a causal link between uPAR signaling and protection from programmed cell death. They show that ligand engagement of uPAR counteracts the pro-apoptotic effect triggered by UV light, cisplatin, and forced detachment from the culture dish. Furthermore, they demonstrate that the expression level of uPAR positively correlates with resistance to anoikis in embryonic kidney epithelial (HEK-293) cell lines. They also show that the uPA/uPAR interaction results in a marked upregulation of the anti-apoptotic factor Bcl-xL, which is required for the uPA-dependent anti-apoptotic activity. In agreement with these observations, targeting the uPAR with inhibitory peptides leads to a reduction of glioma tumor size in mice through inhibition of cell proliferation and increased tumor cell apoptosis (Bu et al. 2004).

Similarly, Besch and co-workers suggest a new function of uPAR acting as a survival factor for melanoma by downregulating p53. They show that uPAR inhibition results in massive cell death via apoptosis. Apoptosis was mediated by p53 and occurred independently of ERK or FAK signaling (Besch et al. 2006). In the emerging picture, the uPA/uPAR system has the ability to support the malignant phenotype through several mechanisms: first, by virtue of its matrix-degrading ability that favors tumor dissemination; second, by stimulating cell motility;

third, by eliciting cell proliferation; and fourth, by protecting cells from apoptosis, thus enhancing tumor survival.

The Next Generation

Delivery of uPAR downregulation constructs, whether plasmid vectors, adenoviral vectors, or synthetic strands, remains to be tested appropriately. With the current state of cancer gene therapy, delivery is a major stumbling block, and various carriers such as liposomes, polymers, and microparticles (Dass et al. 2002) are being evaluated to address this issue. The closest to clinical relevancy in terms of delivery (administration) achieved by the studies listed above was the use of mini-osmotic pumps delivering downregulating agents directly into the brain (Gondi et al. 2004a, Lakka et al. 2005). Anyway, the issue of side-effects of uPAR downregulation on normal tissue and organs were not looked at, even at the cell culture level. Surely, a system as central as uPA–uPAR, which has various physiological functions in the body besides being pro-tumorigenic, needs to be properly respected and monitored.

Little is known about tumor growth and dissemination in mice with targeted distruption of the uPAR gene. uPAR deficiency does not compromise the embryonic development and viability of uPAR$^{-/-}$ mice (Bugge et al. 1995): homozygous uPAR-deficient mice do not display major growth and fertility problems, do not show histological abnormalities in tissues, and do not differ from wild-type mice for spontaneous lysis of experimental pulmonary plasma clot (Dewerchin et al. 1996). This is similar to what is also noted in uPA deficient mice (Carmeliet et al. 1994). Thus, the apparent lack of toxicity from inhibiting this proteolytic system makes it an ideal candidate for targeting as a cancer therapeutic.

References

Adachi, Y., Chandrasekar, N., Kin, Y., et al. (2002). Suppression of glioma invasion and growth by adenovirus-mediated delivery of a bicistronic construct containing antisense uPAR and sense p16 gene sequences. Oncogene 21, 87–95.

Aguirre-Ghiso, J. A., Kovalski, K., and Ossowski, L. (1999). Tumor Dormancy Induced by Downregulation of Urokinase Receptor in Human Carcinoma Involves Integrin and MAPK Signaling, vol. 147, pp. 89–104.

Aguirre-Ghiso, J. A., Liu, D., Mignatti, A., et al. (2001). Urokinase receptor and fibronectin regulate the ERK(MAPK) to p38(MAPK) activity ratios that determine carcinoma cell proliferation or dormancy *in vivo*. Mol Biol Cell 12, 863–79.

Aguirre-Ghiso, J. A., Estrada, Y., Liu, D., et al. (2003). ERK(MAPK) activity as a determinant of tumor growth and dormancy; regulation by p38(SAPK). Cancer Res 63, 1684–95.

Alfano, D., Iaccarino, I., and Stoppelli, M. P. (2006). Urokinase signaling through its receptor protects against anoikis by increasing BCL-xL expression levels. J Biol Chem 281, 17758–67.

Andolfo, A., English, W. R., Resnati, M., et al. (2002). Metalloproteases cleave the urokinase-type plasminogen activator receptor in the D1-D2 linker region and expose epitopes not present in the intact soluble receptor. Thromb Haemost 88, 298–306.

Barinka, C., Parry, G., Callahan, J., et al. (2006). Structural basis of interaction between urokinase-type plasminogen activator and its receptor. J Mol Biol 363, 482–95.

Beale, G., Hollins, A. J., Benboubetra, M., et al. (2003). Gene silencing nucleic acids designed by scanning arrays: Anti-EGFR activity of siRNA, ribozyme and DNA enzymes targeting a single hybridization-accessible region using the same delivery system. J Drug Target 11, 449–56.

Behrendt, N., Ploug, M., Patthy, L., et al. (1991). The ligand-binding domain of the cell surface receptor for urokinase-type plasminogen activator. J Biol Chem 266, 7842–7.

Besch, R., Berking, C., Kammerbauer, C., et al. (2006). Inhibition of urokinase-type plasminogen activator receptor induces apoptosis in melanoma cells by activation of p53. Cell Death Differ 14, 818–29.

Bessler, M., Rosti, V., Peng, Y., et al. (2002). Glycosylphosphatidylinositol-linked proteins are required for maintenance of a normal peripheral lymphoid compartment but not for lymphocyte development. Eur J Immunol 32, 2607–16.

Blasi, F., and Carmeliet, P. (2002). uPAR: A versatile signalling orchestrator. Nat Rev Mol Cell Biol 3, 932–43.

Bohuslav, J., Horejsi, V., Hansmann, C., et al. (1995). Urokinase plasminogen activator receptor, beta 2-integrins, and Src-kinases within a single receptor complex of human monocytes. J Exp Med 181, 1381–90.

Bu, X., Khankaldyyan, V., Gonzales-Gomez, I., et al. (2004). Species-specific urokinase receptor ligands reduce glioma growth and increase survival primarily by an antiangiogenesis mechanism. Lab Invest 84, 667–78.

Bugge, T. H., Suh, T. T., Flick, M. J., et al. (1995). The receptor for urokinase-type plasminogen activator is not essential for mouse development or fertility. J Biol Chem 270, 16886–94.

Carmeliet, P., and Jain, R. K. (2000). Angiogenesis in cancer and other diseases. Nature 407, 249–57.

Carmeliet, P., Schoonjans, L., Kieckens, L., et al. (1994). Physiological consequences of loss of plasminogen activator gene function in mice. Nature 368, 419–24.

Carriero, M. V., Del Vecchio, S., Capozzoli, M., et al. (1999). Urokinase receptor interacts with alpha(v)beta5 vitronectin receptor, promoting urokinase-dependent cell migration in breast cancer. Cancer Res 59, 5307–14.

Chapman, H. A., and Wei, Y. (2001). Protease crosstalk with integrins: The urokinase receptor paradigm. Thromb Haemost 86, 124–9.

Chavakis, T., Kanse, S. M., Yutzy, B., et al. (1998). Vitronectin concentrates proteolytic activity on the cell surface and extracellular matrix by trapping soluble urokinase receptor-urokinase complexes. Blood 91, 2305–12.

Chen, G., Hitomi, M., Han, J., et al. (2000). The p38 pathway provides negative feedback for Ras proliferative signaling. J Biol Chem 275, 38973–80.

Colman, R. W., Pixley, R. A., Najamunnisa, S., et al. (1997). Binding of high molecular weight kininogen to human endothelial cells is mediated via a site within domains 2 and 3 of the urokinase receptor. J Clin Invest 100, 1481–7.

D'Alessio, S., Margheri, F., Pucci, M., et al. (2004). Antisense oligodeoxynucleotides for urokinase-plasminogen activator receptor have anti-invasive and anti-proliferative effects *in vitro* and inhibit spontaneous metastases of human melanoma in mice. Int J Cancer 110, 125–33.

D'Angelo, G., Struman, I., Martial, J., et al. (1995). Activation of mitogen-activated protein kinases by vascular endothelial growth factor and basic fibroblast growth factor in capillary endothelial cells is inhibited by the antiangiogenic factor 16-kDa N-terminal fragment of prolactin. Proc Natl Acad Sci U S A 92, 6374–8.

Dano, K., Behrendt, N. Hoyer-Hansen, G. et al. (2005). Plasminogen Activation and cancer. Thromb. Hemort. 93, 676–81.

Dass, C. R., Walker, T. L., and Burton, M. A. (2002). Liposomes containing cationic dimethyl dioctadecyl ammonium bromide: Formulation, quality control, and lipofection efficiency. Drug Deliv 9, 11–8.

Dass, C. R., Nadesapillai, A. P., Robin, D., et al. (2005). Downregulation of uPAR confirms link in growth and metastasis of osteosarcoma. Clin Exp Metastasis 22, 643–52.

Deng, G., Curriden, S. A., Wang, S., et al. (1996). Is plasminogen activator inhibitor-1 the molecular switch that governs urokinase receptor-mediated cell adhesion and release? J Cell Biol 134, 1563–71.

De Petro, G., Tavian, D., Copeta, A., et al. (1998). Expression of urokinase-type plasminogen activator (u-PA), u-PA receptor, and tissue-type PA messenger RNAs in human hepatocellular carcinoma. Cancer Res 58, 2234–9.

Dewerchin, M., Nuffelen, A. V., Wallays, G., et al. (1996). Generation and characterization of urokinase receptor-deficient mice. J Clin Invest 97, 870–8.

Dumler, I., Weis, A., Mayboroda, O. A., et al. (1998). The Jak/Stat pathway and urokinase receptor signaling in human aortic vascular smooth muscle cells. J Biol Chem 273, 315–21.

Fazioli, F., Resnati, M., Sidenius, N., et al. (1997). A urokinase-sensitive region of the human urokinase receptor is responsible for its chemotactic activity. Embo J 16, 7279–86.

Garcia-Monco, J. C., Coleman, J. L., and Benach, J. L. (2002). Soluble urokinase receptor (uPAR, CD 87) is present in serum and cerebrospinal fluid in patients with neurologic diseases. J Neuroimmunol 129, 216–23.

Ge, Y., and Elghetany, M. T. (2003). Urokinase plasminogen activator receptor (CD87): Something old, something new. Lab Hematol 9, 67–71.

Gladson, C. L., and Cheresh, D. A. (1991). Glioblastoma expression of vitronectin and the alpha v beta 3 integrin. Adhesion mechanism for transformed glial cells. J Clin Invest 88, 1924–32.

Gladson, C. L., Pijuan-Thompson, V., Olman, M. A., et al. (1995). Up-regulation of urokinase and urokinase receptor genes in malignant astrocytoma. Am J Pathol 146, 1150–60.

Go, Y., Chintala, S. K., Mohanam, S., et al. (1997). Inhibition of *in vivo* tumorigenicity and invasiveness of a human glioblastoma cell line transfected with antisense uPAR vectors. Clin Exp Metastasis 15, 440–6.

Gondi, C. S., Lakka, S. S., Dinh, D. H., et al. (2004a). RNAi-mediated inhibition of cathepsin B and uPAR leads to decreased cell invasion, angiogenesis and tumor growth in gliomas. Oncogene 23, 8486–96.

Gondi, C. S., Lakka, S. S., Yanamandra, N., et al. (2004b). Adenovirus-mediated expression of antisense urokinase plasminogen activator receptor and antisense cathepsin B inhibits tumor growth, invasion, and angiogenesis in gliomas. Cancer Res 64, 4069–77.

Gyetko, M. R., Todd, R. F., 3rd., Wilkinson, C. C., et al. (1994). The urokinase receptor is required for human monocyte chemotaxis *in vitro*. J Clin Invest 93, 1380–7.

Gyetko, M. R., Sitrin, R. G., Fuller, J. A., et al. (1995). Function of the urokinase receptor (CD87) in neutrophil chemotaxis. J Leukoc Biol 58, 533–8.

Hanahan, D., and Weinberg, R. A. (2000). The hallmarks of cancer. Cell 100, 57–70.

Holst-Hansen, C., Hamers, M. J., Johannessen, B. E., et al. (1999). Soluble urokinase receptor released from human carcinoma cells: A plasma parameter for xenograft tumour studies. Br J Cancer 81, 203–11.

Hood, J. D., and Cheresh, D. A. (2002). Role of integrins in cell invasion and migration. Nat Rev Cancer 2, 91–100.

Hoshino, R., Chatani, Y., Yamori, T., et al. (1999). Constitutive activation of the 41-/43-kDa mitogen-activated protein kinase signaling pathway in human tumors. Oncogene 18, 813–22.

Hoyer-Hansen, G., Ronne, E., Solberg, H., et al. (1992). Urokinase plasminogen activator cleaves its cell surface receptor releasing the ligand-binding domain. J Biol Chem 267, 18224–9.

Hoyer-Hansen, G., Behrendt, N., Ploug, M., et al. (1997a). The intact urokinase receptor is required for efficient vitronectin binding: Receptor cleavage prevents ligand interaction. FEBS Lett 420, 79–85.

Hoyer-Hansen, G., Ploug, M., Behrendt, N., et al. (1997b). Cell-surface acceleration of urokinase-catalyzed receptor cleavage. Eur J Biochem 243, 21–6.

Hoyer-Hansen, G., Pessara, U., Holm, A., et al. (2001). Urokinase-catal cleavage of the urokinase receptor requires an intact glycolipid anchor. Biochem J 358, 673–9.

Jain, R. K., and Carmeliet, P. F. (2001). Vessels of death or life. Sci Am 285, 38–45.

Jones, M. K., Sarfeh, I. J., and Tarnawski, A. S. (1998). Induction of *in vitro* angiogenesis in the endothelial-derived cell line, EA hy926, by ethanol is mediated through PKC and MAPK. Biochem Biophys Res Commun 249, 118–23.

Kanse, S. M., Kost, C., Wilhelm, O. G., et al. (1996). The urokinase receptor is a major vitronectin-binding protein on endothelial cells. Exp Cell Res 224, 344–53.

Kjoller, L., and Hall, A. (2001). Rac mediates cytoskeletal rearrangements and increased cell motility induced by urokinase-type plasminogen activator receptor binding to vitronectin. J Cell Biol 152, 1145–57.

Klein, G., Vellenga, E., Fraaije, M. W., et al. (2004). The possible role of matrix metalloproteinase (MMP)-2 and MMP-9 in cancer, e.g. acute leukemia. Crit Rev Oncol Hematol 50, 87–100.

Kondoh, N., Wakatsuki, T., Ryo, A., et al. (1999). Identification and characterization of genes associated with human hepatocellular carcinogenesis. Cancer Res 59, 4990–6.

Kook, Y. H., Adamski, J., Zelent, A., et al. (1994). The effect of antisense inhibition of urokinase receptor in human squamous cell carcinoma on malignancy. Embo J 13, 3983–91.

Koolwijk, P., Sidenius, N., Peters, E., et al. (2001). Proteolysis of the urokinase-type plasminogen activator receptor by metalloproteinase-12: Implication for angiogenesis in fibrin matrices. Blood 97, 3123–31.

Koshelnick, Y., Ehart, M., Hufnagl, P., et al. (1997). Urokinase receptor is associated with the components of the JAK1/STAT1 signaling pathway and leads to activation of this pathway upon receptor clustering in the human kidney epithelial tumor cell line TCL-598. J Biol Chem 272, 28563–7.

Krishnamoorthy, B., Darnay, B., Aggarwal, B., et al. (2001). Glioma cells deficient in urokinase plasminogen activator receptor expression are susceptible to tumor necrosis factor-alpha-related apoptosis-inducing ligand-induced apoptosis. Clin Cancer Res 7, 4195–201.

Lakka, S. S., Rajagopal, R., Rajan, M. K., et al. (2001). Adenovirus-mediated antisense urokinase-type plasminogen activator receptor gene transfer reduces tumor cell invasion and metastasis in non-small cell lung cancer cell lines. Clin Cancer Res 7, 1087–93.

Lakka, S. S., Gondi, C. S., Yanamandra, N., et al. (2003). Synergistic down-regulation of urokinase plasminogen activator receptor and matrix metalloproteinase-9 in SNB19 glioblastoma cells efficiently inhibits glioma cell invasion, angiogenesis, and tumor growth. Cancer Res 63, 2454–61.

Lakka, S. S., Gondi, C. S., Dinh, D. H., et al. (2005). Specific interference of urokinase-type plasminogen activator receptor and matrix metalloproteinase-9 gene expression induced by double-stranded RNA results in decreased invasion, tumor growth, and angiogenesis in gliomas. J Biol Chem 280, 21882–92.

Lau, H. K., and Kim, M. (1994). Soluble urokinase receptor from fibrosarcoma HT-1080 cells. Blood Coagul Fibrinolysis 5, 473–8.

Lengyel, E., Stepp, E., Gum, R., et al. (1995). Involvement of a mitogen-activated protein kinase signaling pathway in the regulation of urokinase promoter activity by c-Ha-ras. J Biol Chem 270, 23007–12.

Lengyel, E., Wang, H., Stepp, E., et al. (1996). Requirement of an upstream AP-1 motif for the constitutive and phorbol ester-inducible expression of the urokinase-type plasminogen activator receptor gene. J Biol Chem 271, 23176–84.

Lengyel, E., Wang, H., Gum, R., et al. (1997). Elevated urokinase-type plasminogen activator receptor expression in a colon cancer cell line is due to a constitutively activated extracellular signal-regulated kinase-1-dependent signaling cascade. Oncogene 14, 2563–73.

Lindberg, P., Larsson, A., and Nielsen, B. S. (2006). Expression of plasminogen activator inhibitor-1, urokinase receptor and laminin gamma-2 chain is an early coordinated event in incipient oral squamous cell carcinoma. Int J Cancer 118, 2948–56.

Liu, D., Aguirre-Ghiso, J., Estrada, Y., et al. (2002). EGFR is a transducer of the urokinase receptor initiated signal that is required for *in vivo* growth of a human carcinoma. Cancer Cell 1, 445–57.

Llinas, P., Le Du, M. H., Gardsvoll, H., et al. (2005). Crystal structure of the human urokinase plasminogen activator receptor bound to an antagonist peptide. Embo J 24, 1655–63.

Madsen, C. D., Ferraris, G. M., Andolfo, A., et al. (2007). uPAR-induced cell adhesion and migration: Vitronectin provides the key. J Cell Biol 177, 927–39.

Margheri, F., D'Alessio, S., Serrati, S., et al. (2005). Effects of blocking urokinase receptor signaling by antisense oligonucleotides in a mouse model of experimental prostate cancer bone metastases. Gene Ther 12, 702–14.

Mazar, A. P. (2001). The urokinase plasminogen activator receptor (uPAR) as a target for the diagnosis and therapy of cancer. Anticancer Drugs 12, 387–400.

Mazar, A., Henkin, J., and Goldfarb, R. (1999). The urokinase plasminogen activator system in cancer: Implications for tumor angiogenesis and metastasis. Angiogenesis 3, 15–32.

Mazzieri, R., and Blasi, F. (2005). The urokinase receptor and the regulation of cell proliferation. Thromb Haemost 93, 641–6.

Mazzieri, R., D'Alessio, S., Kenmoe, R. K., et al. (2006). An uncleavable uPAR mutant allows dissection of signaling pathways in uPA-dependent cell migration. Mol Biol Cell 17, 367–78.

Metz, C. N., Brunner, G., Choi-Muira, N. H., et al. (1994). Release of GPI-anchored membrane proteins by a cell-associated GPI-specific phospholipase D. Embo J 13, 1741–51.

Mohan, P. M., Chintala, S. K., Mohanam, S., et al. (1999). Adenovirus-mediated delivery of antisense gene to urokinase-type plasminogen activator receptor suppresses glioma invasion and tumor growth. Cancer Res 59, 3369–73.

Montuori, N., Carriero, M. V., Salzano, S., et al. (2002). The cleavage of the urokinase receptor regulates its multiple functions. J Biol Chem 277, 46932–9.

Moro, L., Venturino, M., Bozzo, C., et al. (1998). Integrins induce activation of EGF receptor: Role in MAP kinase induction and adhesion-dependent cell survival. Embo J 17, 6622–32.

Muehlenweg, B., Sperl, S., Magdolen, V., et al. (2001). Interference with the urokinase plasminogen activator system: A promising therapy concept for solid tumours. Expert Opin Biol Ther 1, 683–91.

Mustjoki, S., Sidenius, N., Sier, C. F., et al. (2000a). Soluble urokinase receptor levels correlate with number of circulating tumor cells in acute myeloid leukemia and decrease rapidly during chemotherapy. Cancer Res 60, 7126–32.

Mustjoki, S., Sidenius, N., and Vaheri, A. (2000b). Enhanced release of soluble urokinase receptor by endothelial cells in contact with peripheral blood cells. FEBS Lett 486, 237–42.

Nguyen, D. H. D., Hussaini, I. M., and Gonias, S. L. (1998). Binding of Urokinase-type Plasminogen Activator to Its Receptor in MCF-7 Cells Activates Extracellular Signal-regulated Kinase 1 and 2 Which Is Required for Increased Cellular Motility 273, 8502–7.

Nozaki, S., Endo, Y., Nakahara, H., et al. (2005). Inhibition of invasion and metastasis in oral cancer by targeting urokinase-type plasminogen activator receptor. Oral Oncol 41, 971–7.

Okumura, Y., Kamikubo, Y., Curriden, S. A., et al. (2002). Kinetic analysis of the interaction between vitronectin and the urokinase receptor. J Biol Chem 277, 9395–404.

Ossowski, L., and Aguirre-Ghiso, J. A. (2000). Urokinase receptor and integrin partnership: Coordination of signaling for cell adhesion, migration and growth. Curr Opin Cell Biol 12, 613–20.

Pedersen, N., Schmitt, M., Ronne, E., et al. (1993). A ligand-free, soluble urokinase receptor is present in the ascitic fluid from patients with ovarian cancer. J Clin Invest 92, 2160–7.

Pelengaris, S., Khan, M., and Evan, G. (2002). c-MYC: More than just a matter of life and death. Nat Rev Cancer 2, 764–76.

Ploug, M., Ronne, E., Behrendt, N., et al. (1991). Cellular receptor for urokinase plasminogen activator. Carboxyl-terminal processing and membrane anchoring by glycosyl-phosphatidylinositol. J Biol Chem 266, 1926–33.

Ploug, M., Ellis, V., and Dano, K. (1994). Ligand interaction between urokinase-type plasminogen activator and its receptor probed with 8-anilino-1-naphthalenesulfonate. Evidence for a hydrophobic binding site exposed only on the intact receptor. Biochemistry 33, 8991–7.

Preissner, K. T. (1989). The role of vitronectin as multifunctional regulator in the hemostatic and immune systems. Blut 59, 419–31.

Pulukuri, S. M., Gondi, C. S., Lakka, S. S., et al. (2005). RNA interference-directed knockdown of urokinase plasminogen activator and urokinase plasminogen activator receptor inhibits prostate cancer cell invasion, survival, and tumorigenicity *in vivo*. J Biol Chem 280, 36529–40.

Ragno, P., Montuori, N., Covelli, B., et al. (1998). Differential expression of a truncated form of the urokinase-type plasminogen-activator receptor in normal and tumor thyroid cells. Cancer Res 58, 1315–9.

Rao, J. S., Gondi, C., Chetty, C., et al. (2005). Inhibition of invasion, angiogenesis, tumor growth, and metastasis by adenovirus-mediated transfer of antisense uPAR and MMP-9 in non-small cell lung cancer cells. Mol Cancer Ther 4, 1399–408.

Resnati, M., Guttinger, M., Valcamonica, S., et al. (1996). Proteolytic cleavage of the urokinase receptor substitutes for the agonist-induced chemotactic effect. Embo J 15, 1572–82.

Resnati, M., Pallavicini, I., Wang, J. M., et al. (2002). The fibrinolytic receptor for urokinase activates the G protein-coupled chemotactic receptor FPRL1/LXA4R. Proc Natl Acad Sci U S A 99, 1359–64.

Romer, J., Nielsen, B. S., and Ploug, M. (2004). The urokinase receptor as a potential target in cancer therapy. Curr Pharm Des 10, 2359–76.

Ronne, E., Behrendt, N., Ploug, M., et al. (1994). Quantitation of the receptor for urokinase plasminogen activator by enzyme-linked immunosorbent assay. J Immunol Methods 167, 91–101.

Sidenius, N., and Blasi, F. (2000). Domain 1 of the urokinase receptor (uPAR) is required for uPAR-mediated cell binding to vitronectin. FEBS Lett 470, 40–6.

Sidenius, N., and Blasi, F. (2003). The urokinase plasminogen activator system in cancer: Recent advances and implication for prognosis and therapy. Cancer Metastasis Rev 22, 205–22.

Sidenius, N., Sier, C. F., and Blasi, F. (2000). Shedding and cleavage of the urokinase receptor (uPAR): Identification and characterisation of uPAR fragments *in vitro* and *in vivo*. FEBS Lett 475, 52–6.

Sidenius, N., Andolfo, A., Fesce, R., et al. (2002). Urokinase regulates vitronectin binding by controlling urokinase receptor oligomerization. J Biol Chem 277, 27982–90.

Sier, C. F., Stephens, R., Bizik, J., et al. (1998). The level of urokinase-type plasminogen activator receptor is increased in serum of ovarian cancer patients. Cancer Res 58, 1843–9.

Sier, C. F., Sidenius, N., Mariani, A., et al. (1999). Presence of urokinase-type plasminogen activator receptor in urine of cancer patients and its possible clinical relevance. Lab Invest 79, 717–22.

Simon, D. I., Wei, Y., Zhang, L., et al. (2000). Identification of a urokinase receptor-integrin interaction site. Promiscuous regulator of integrin function. J Biol Chem 275, 10228–34.

Sitrin, R. G., Todd, R. F., 3rd, Albrecht, E., et al. (1996). The urokinase receptor (CD87) facilitates CD11b/CD18-mediated adhesion of human monocytes. J Clin Invest 97, 1942–51.

Skriver, L., Larsson, L. I., Kielberg, V., et al. (1984). Immunocytochemical localization of urokinase-type plasminogen activator in Lewis lung carcinoma. J Cell Biol 99, 752–7.

Sloane, B. F., Yan, S., Podgorski, I., et al. (2005). Cathepsin B and tumor proteolysis: Contribution of the tumor microenvironment. Semin Cancer Biol 15, 149–57.

Solberg, H., Romer, J., Brunner, N., et al. (1994). A cleaved form of the receptor for urokinase-type plasminogen activator in invasive transplanted human and murine tumors. Int J Cancer 58, 877–81.

Stephens, R. W., Pedersen, A. N., Nielsen, H. J., et al. (1997). ELISA determination of soluble urokinase receptor in blood from healthy donors and cancer patients. Clin Chem 43, 1868–76.

Tarui, T., Mazar, A. P., Cines, D. B., et al. (2001). Urokinase-type plasminogen activator receptor (CD87) is a ligand for integrins and mediates cell-cell interaction. J Biol Chem 276, 3983–90.

van der Pluijm, G., Sijmons, B., Vloedgraven, H., et al. (2001). Urokinase-receptor/integrin complexes are functionally involved in adhesion and progression of human breast cancer *in vivo*. Am J Pathol 159, 971–82.

Wahlberg, K., Hoyer-Hansen, G., and Casslen, B. (1998). Soluble receptor for urokinase plasminogen activator in both full-length and a cleaved form is present in high concentration in cystic fluid from ovarian cancer. Cancer Res 58, 3294–8.

Waltz, D. A., and Chapman, H. A. (1994). Reversible cellular adhesion to vitronectin linked to urokinase receptor occupancy. J Biol Chem 269, 14746–50.

Wang, Y. (2001). The role and regulation of urokinase-type plasminogen activator receptor gene expression in cancer invasion and metastasis. Med Res Rev 21, 146–70.

Wang, Y., Liang, X., Wu, S., et al. (2001). Inhibition of colon cancer metastasis by a 3'-end antisense urokinase receptor mRNA in a nude mouse model. Int J Cancer 92, 257–62.

Wei, Y., Waltz, D. A., Rao, N., et al. (1994). Identification of the urokinase receptor as an adhesion receptor for vitronectin. J Biol Chem 269, 32380–8.

Wei, Y., Lukashev, M., Simon, D. I., et al. (1996). Regulation of integrin function by the urokinase receptor. Science 273, 1551–5.

Wei, Y., Yang, X., Liu, Q., et al. (1999). A role for caveolin and the urokinase receptor in integrin-mediated adhesion and signaling. J Cell Biol 144, 1285–94.

Wilhelm, O. G., Wilhelm, S., Escott, G. M., et al. (1999). Cellular glycosylphosphatidylinositol-specific phospholipase D regulates urokinase receptor shedding and cell surface expression. J Cell Physiol 180, 225–35.

Wong, W. S., Simon, D. I., Rosoff, P. M., et al. (1996). Mechanisms of pertussis toxin-induced myelomonocytic cell adhesion: Role of Mac-1(CD11b/CD18) and urokinase receptor (CD87). Immunology 88, 90–7.

Xue, W., Mizukami, I., Todd, R. F., 3rd, et al. (1997). Urokinase-type plasminogen activator receptors associate with beta1 and beta3 integrins of fibrosarcoma cells: Dependence on extracellular matrix components. Cancer Res 57, 1682–9.

Yamamoto, M., Sawaya, R., Mohanam, S., et al. (1994). Expression and localization of urokinase-type plasminogen activator in human astrocytomas *in vivo*. Cancer Res 54, 3656–61.

Yanamandra, N., Konduri, S. D., Mohanam, S., et al. (2000). Downregulation of urokinase-type plasminogen activator receptor (uPAR) induces caspase-mediated cell death in human glioblastoma cells. Clin Exp Metastasis 18, 611–5.

Yebra, M., Parry, G. C., Stromblad, S., et al. (1996). Requirement of receptor-bound urokinase-type plasminogen activator for integrin alphavbeta5-directed cell migration. J Biol Chem 271, 29393–9.

Chapter 13
The Endocytic Collagen Receptor, uPARAP/ Endo180, in Cancer Invasion and Tissue Remodeling

Thore Hillig, Lars H. Engelholm, and Niels Behrendt

Abstract uPARAP/Endo180 is a constitutively recycling endocytosis receptor of 180 kDa. It is a type-1 membrane protein and includes an N-terminal Cys-rich domain followed by a fibronectin type-II domain, eight C-type lectin-like domains, a transmembrane segment, and a small cytoplasmic domain. The receptor binds and internalizes collagen, which is then directed to lysosomal degradation. The internalization efficiency increases when the collagen is in a gelatin-like state and, in line with this notion, the uptake of defined ¼ and ¾ collagen fragments is more efficient than the internalization of intact collagen. Thus, uPARAP/Endo180 most likely has a preferential role in the clearance of precleaved collagen, occurring after the initial attack of a collagenolytic MMP. Mesenchymal cell types such as fibroblasts, osteoblasts, some endothelial cells, and some macrophages express uPARAP/Endo180, with a strong expression in areas with dominant collagen turnover, such as developing bone. PyMT mice, which develop genetically induced, invasive mammary tumors, have reduced tumor growth and increased tumor collagen content when uPARAP/Endo180 is absent due to gene inactivation. Cancer cells have not been found to express uPARAP/Endo180 but some of the same cell types that express the receptor in healthy tissue show a strong increase in expression when they take part in the stroma that surrounds the cancer islets in some invasive cancers. Thus, in some situations involving invasive growth, collagen clearance by uPARAP/Endo180 is likely to take active part in the outgrowth and escape of cancer cells from a confined tissue compartment.

After the submission of this manuscript, it has been reported that uPARAP/ Endo180 is expressed in basal-like breast cancer cells (Wienke, D. et al. (2007) Cancer Res. 67:10230–10240).

N. Behrendt
Finsen Laboratory section 3735, Copenhagen Biocenter, Rigshospitalet, Ole Maaløes Vej 5, DK-2200 Copenhagen N, Denmark, e-mail: Niels.behrendt@finsenlab.dk

D. Edwards et al. (eds.), *The Cancer Degradome.*
© Springer Science + Business Media, LLC 2008

Identification of uPARAP/Endo180

At the protein level the endocytic receptor uPARAP/Endo180 was identified by two independent groups. In one study, the protein was identified in a trimolecular complex by the use of cross-linking experiments with prourokinase and the urokinase-type plasminogen activator receptor (uPAR), on the surface of U937 cells (Behrendt et al. 1993). This observation indicated that the protein was situated in close proximity to uPAR on the surface of U937 cells in order for the cross-linking to take place, thus giving rise to the name uPAR-associated protein (uPARAP). A tryptic digest of the cross-linked complex was subjected to mass spectrometry-based analysis, and the protein was sequenced and cloned and proved to be a novel member of the macrophage mannose receptor family (Behrendt et al. 2000). This protein family consists of four members: the macrophage mannose receptor (MMR), the secretory phospholipase A2 receptor (PLA2R), the receptor DEC-205, and uPARAP (Behrendt 2004).

In an independent work, an unknown protein of molecular weight 180 kDa was observed by use of monoclonal antibodies obtained after immunization of mice with intact, or membrane fractions of, human fibroblasts (Isacke et al. 1990), and this protein was subsequently cloned and shown to be identical with uPARAP (Sheikh et al. 2000). The antibody in question was reactive with a cell-surface protein occurring on several cultured cell types including fibroblasts, macrophages, and endothelial cells. Moreover, the protein showed evidence of endocytosis with internalization into endosomes via clathrin-coated pits and recycling to the plasma membrane; thus the name Endo180 was chosen (Sheikh et al. 2000). In the following, the designation uPARAP/Endo180 will be used.

It should be noted that the uPARAP/Endo180 encoding cDNA was identified already in 1996 in a third, independent work (Wu et al. 1996), although no characterization of the protein was performed at that point.

Protein Structure

Based on sequence alignment with the macrophage mannose receptor, the domain structure of uPARAP/Endo180 could be deduced. From the extracellular amino terminus, uPARAP/Endo180 consists of a Cys-rich domain followed by a fibronectin type-II (FN-II) domain, eight C-type carbohydrate recognition domains (CRDs), a transmembrane segment, and a cytoplasmic domain (Behrendt et al. 2000, Sheikh et al. 2000). The crystal structure of uPARAP/Endo180 has not yet been determined but three-dimensional structures have been determined for two domain types in the presumably similar MMR. These are the MMR Cys-rich domain (Liu et al. 2000) and the MMR CRD4 (Feinberg et al. 2000).

Cys-Rich Domain

The Cys-rich domain of the MMR is a globular structure with a β-trefoil as the structural basis, where three similar β-sheets, each composed of four strands, are organized in a near symmetric manner (Liu et al. 2000). The MMR Cys-rich domain binds sulfated sugars in a loop of six amino acids between β-strand 11 and 12, but this loop is absent from the Cys-rich domains of uPARAP/Endo180 and the other members of the MMR protein family. Thus, the Cys-rich domain of uPARAP/Endo180 probably does not bind sulfated sugars (Liu et al. 2000, East and Isacke 2002).

FN-II Domain of uPARAP/Endo180

The FN-II domain structure has not been determined for any member of the MMR family, but the sequence similarity in this region is high between the MMR family members and there is also a large degree of homology with FN II domains of proteins outside the MMR family (East and Isacke 2002, Wienke et al. 2003). Because of this large degree of homology, it seems reasonable to assume that the known structure of, for example, the third FN-II domain of MMP-2, a domain important for the collagen binding ability of this collagenolytic enzyme, has some similarity with the FN-II domain structure of uPARAP/Endo180 (Behrendt 2004). The former FN-II domain is a compact structure with two double-stranded, antiparallel β-sheets oriented at a right angle to each other (Briknarova et al. 2001). A large region of the exposed surface is a hydrophobic region that binds to peptides mimicking gelatin (Briknarova et al. 2001). FN-II domains are known to take part in collagen binding in several proteins (Banyai et al. 1990) and this prompted a competition experiment with various collagens in the above- mentioned cross-linking setup with pro-uPA/uPAR and uPARAP/Endo180. This experiment demonstrated a robust inhibition of the pro-uPA–uPARAP/Endo180 cross-linking with collagen type V and a moderate inhibitory effect of collagen types I and IV (Behrendt et al. 2000), thus providing the first indication for a collagen-binding function of uPARAP/Endo180. An open question relates to the notion that, in various other collagen-binding proteins, two or more FN-II domains may be required for an efficient binding to collagen (Banyai et al. 1994, Pickford et al. 1997) whereas only one FN-II domain is present in uPARAP/Endo180. On the contrary, an NMR study on a structure including two FN-II domains of MMP-2 revealed little interaction between these domains (Briknarova et al. 2001) and, thus, it is uncertain whether there is indeed a requirement for more than one FN-II domain for collagen binding in the latter protein. Even if the single FN-II domain of uPARAP/Endo180 is not sufficient for efficient binding to collagen, several mechanisms could contribute to compensate for this. Thus, a clustering of several uPARAP/Endo180 molecules is not an unlikely event, as the protein is already concentrated in areas of clathrin-coated pits, and since the collagen itself could aid in the clustering process as one weak collagen binding could be stabilized with the binding of more uPARAP/Endo180 molecules.

Furthermore, as support for the ability of proteins with a single FN-II domain to bind collagen, studies on the closely related MMR (Napper et al. 2006. Martinez-Pomares et al. 2006) show binding to collagen via the FN-II domain independently of CRDs.

CRDs

In several proteins, domains with sequence homology with the CRDs of uPARAP/Endo180 are involved in carbohydrate binding. This is exemplified by one or two CRDs in the homologue MMR. Just like the latter protein, uPARAP/Endo180 includes eight putative CRDs. However, in uPARAP/Endo180 only CRD2 has retained the structural elements for chelating critical Ca^{++} ions involved in the carbohydrate-binding mechanism of this domain type (Sheikh et al. 2000, East and Isacke 2002). In accordance with this notion, probably only CRD2 has carbohydrate-binding activity (East et al. 2002). The structure of a typical sugar-binding CRD is stabilized by two disulfide bonds and has an organized core with two α-helices and two β-sheets, of which one has two strands and the other has three (East and Isacke 2002), and with a loop extending from the largest β-sheet. The actual binding to sugars is dependent on two Ca^{++} ions in a hydrophobic fold, where two coordination sites exist. One Ca^{++} ion is directly involved in sugar binding (the principal site) and the other functions as a structural support of the hydrophobic part of the CRD (East and Isacke 2002).

A series of affinity columns containing different immobilized sugars were tested for binding of recombinant, soluble uPARAP/Endo180. Whereas no binding was found to mannose, fucose, or galactose, uPARAP/Endo180 could be bound to and eluted from a column containing *N*-acetylglucosamine-agarose (East et al. 2002). So far, however, no biological ligand has been found for the lectin function of uPARAP/Endo180.

Cytoplasmic Domain

The cytoplasmic domain of uPARAP/Endo180 contains two motifs that were initially both considered candidates for governing the endocytic function: a tyrosine-based endocytosis motif (Tyr1452) and a di-hydrophobic (Leu1468-Val1469) endocytosis motif with a negatively charged residue (Glu1464) being positioned 4 amino acids prior to the motif (Sheikh et al. 2000, Howard and Isacke 2002). In MHC class II proteins, a di-hydrophobic motif, similar to that found in uPARAP/Endo180, has been shown to target internalized vesicles to early endosomes (Pond et al. 1995). A mutagenesis experiment revealed that substitution of the tyrosine motif does not interfere with internalization (Howard and Isacke 2002). In contrast, mutagenesis of the di-hydrophobic motif to Ala-Ala led to a strong decrease in internalization and mutation of the -4 glutamate also led to a reduction (Howard and Isacke 2002). The mechanism behind uPARAP/Endo180 receptor recycling from early endosomes back to the plasma membrane is still unclear, as a di-aromatic motif important for receptor recycling of MMR and PLA2R is not present in uPARAP/Endo180 (East and Isacke 2002). uPARAP/Endo180 is a constitutively active internalization receptor; however, it has been suggested that some regulation of the internalization process takes place through phosphorylation of serine residues adjacent to the endocytosis motif. Thus, a low level of serine phosphorylation was observed in immunoprecipitated uPARAP/Endo180 protein from human fibroblasts, with a strong increase in phosphorylation being found in phorbol ester-treated cells (Sheikh et al. 2000).

Interdomain Organization

Analysis of the spatial domain organization of uPARAP/Endo180 by single particle electron microscopy with recombinant uPARAP/Endo180 variants has led to a model in which the Cys-rich domain folds back to contact CRD2 (Rivera-Calzada et al. 2003). A later study supports this notion and furthermore suggests the possibility of a pH-dependent conformational change with a more open conformation of the backfold at acidic pH (Boskovic et al. 2006). It is tempting to speculate that this type of conformational change has a function in the release of bound, internalized collagen in a more acidic endosomal compartment (Boskovic et al. 2006), where uPARAP/Endo180 enters during the constitutive recycling of the receptor (Sheikh et al. 2000).

uPARAP/Endo180-Deficient Mice

Mice with genetic uPARAP/Endo180 deficiency were generated using two strategies, leading to very similar genetic constructs. In both cases, a complete functional deficiency was obtained in terms of collagen interactions although, as detailed below, certain targeted cells in culture may express low levels of truncated uPARAP/Endo180 antigen. In one study, the targeting strategy included a construct leading to deletion of the entire region containing exons 2–6. An RNA transcript of the targeted uPARAP/Endo180 DNA was seen expressed in fibroblasts from newborn uPARAP/Endo180 targeted mice, but no protein expression could be detected at that point in these animals (Engelholm et al. 2003). However, later work has suggested that low levels of a truncated protein with a size in accordance with the stated deletion of exons may indeed be present in some cultured cells from these mice (D.H. Madsen, the Finsen Laboratory, unpublished observation). In another study, exons 2–5 and part of exon 6 were deleted. In the targeted mouse embryos, low levels of a truncated protein of molecular weight 140 kD could be identified (East et al. 2003). Thus, these truncated proteins would be lacking the Cys-rich domain, the FN-II domain, and the first CRD. In the study by East et al., it was positively demonstrated that the protein product retained the sugar-binding properties previously reported to originate from CRD2, whereas the collagen-binding function was lost (East et al. 2003). In both studies, the mice deficient in uPARAP/Endo180 were viable with no obvious aberrant phenotype and were able to reproduce (East et al. 2003, Engelholm et al. 2003). However, a recent detailed study on bone development, performed with the mice generated by Engelholm et al., has revealed a small retardation in the growth of long bones and cranial closure (Wagenaar-Miller et al. 2007).

Collagen Internalization

As detailed above, a binding to collagen was noted already in the early characterization of uPARAP/Endo180 (Behrendt et al. 2000). The ability of the receptor to internalize collagen was subsequently demonstrated for collagens I, IV, and V in a study using radioactively labeled collagens (Engelholm et al. 2003). Newborn mouse fibroblasts from wild-type and uPARAP/Endo180-deficient animals were incubated with 125I-labeled collagens for determination of collagen internalization. Strikingly, whereas wild-type fibroblasts displayed an efficient uptake of collagen, the uPARAP/Endo180-deficient fibroblasts were completely unable to perform this process, thus identifying uPARAP/Endo180 as a crucial collagen internalization receptor (Engelholm et al. 2003).

In another study, the ability of uPARAP/Endo180 to internalize collagen was likewise demonstrated with fibroblasts from uPARAP/Endo180 +/+ and −/− mice, but utilizing an assay where cells were incubated with fluorescence-labeled collagen or gelatin (East et al. 2003). Fibroblasts from the uPARAP/Endo180-targeted mice were unable to bind collagen type IV and gelatin, and also unable to internalize these proteins (East et al. 2003). Binding to and internalization of collagen by uPARAP/Endo180 was also demonstrated in a study using both soluble, recombinant purified uPARAP/Endo180 and cell-based experiments (Wienke et al. 2003). The receptor bound to the different collagens tested (collagen types I, II, IV, and gelatin), and in a competition experiment, collagen IV and gelatin could mutually compete the binding of each other. Furthermore, binding to collagens was unaffected in engineered uPARAP/Endo180 proteins without CRDs 5–8, and also maintained for all tested collagen types (except collagen IV) in a protein without CRDs 2–8 (Wienke et al. 2003). Cells transfected with uPARAP/Endo180 (MCF-7 and T47D breast carcinoma cells) and cells expressing endogenous uPARAP/Endo180 (MG-63 osteosarcoma cells) were shown to bind to collagen IV and gelatin, and this binding was blocked by uPARAP/Endo180 siRNA treatment (Wienke et al. 2003). Furthermore, mutation of the intracellular dihydrophobic internalization motif resulted in unaltered collagen binding, but absence of collagen internalization (Wienke et al. 2003). The subcellular fate of the internalized collagen was investigated in fibroblasts from newborn mice (Kjoller et al. 2004). Treatment with E64D, an inhibitor of lysosomal cysteine proteases, resulted in a strong accumulation of vesicles containing collagen, giving an indication that collagen is directed to lysosomes. This observation was confirmed by costaining with the lysosomal marker Lamp-1 (*see* Fig. 13.1; Kjoller et al. 2004). In mouse tissue explants, the uptake and intracellular degradation of collagen was seen in uPARAP/Endo180-expressing chondrocytes and osteoblasts of developing bone (Wagenaar-Miller et al. 2007). A recent study with mouse fibroblasts showed that the efficiency of uPARAP/Endo180 to internalize collagen was increased when the collagen was precleaved into defined large fragments by a collagenase (Madsen et al. 2007). The improved efficiency was a result of the cleaved collagen attaining a gelatin-like state at physiological temperature. Thus, heat-denatured intact collagen was likewise internalized more efficiently, whereas heat-denatured collagen fragments were internalized to the same degree as nonheated fragments. The importance of this observation was studied by analyzing culture media from uPARAP/Endo180-containing and uPARAP/Endo180-deficient fibroblasts grown on a collagen matrix. It was evident that there was an accumulation of collagen fragments in the media from uPARAP/Endo180-deficient cells but not in the media from uPARAP/Endo180-positive cells. This observation provided a strong indication of a sequential mechanism of collagen degradation by fibroblasts, with initial cleavage by collagenolytic metalloproteases and subsequent clearance of collagen fragments through a uPARAP/Endo180-dependent internalization route (Madsen et al. 2007). The whole process is then finalized by intracellular degradation by lysosomal proteases.

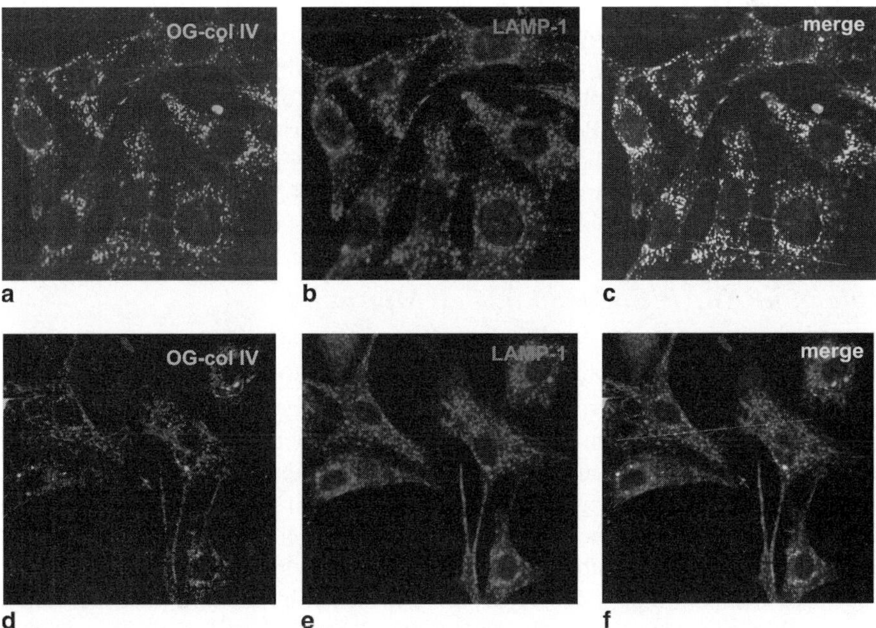

Fig. 13.1 Internalized collagen is directed to lysosomal degradation. Wild-type (**a–c**) and *uPARAP/Endo*180-deficient (**d–f**) fibroblasts were treated with an inhibitor of lysosomal cysteine proteases, E64d. Cells were then incubated with Oregon green-labeled collagen IV for 16 h at 37°C. Lysosomes were labeled with red fluorescence, using a LAMP-1-specific antibody. A strong lysosomal accumulation of OG-collagen is seen in wild-type fibroblasts, whereas cells deficient for uPARAP/Endo180 do not internalize collagen. (Reproduced from Kjoller et al. 2004 with permission from the publisher) (*See also* Color Insert I)

A recent work has demonstrated that the related MMR is also a collagen clearance receptor. This receptor is strongly expressed by liver sinusoidal endothelial cells and was found to be responsible for approximately half of the liver retention of collagen fragments from plasma (Malovic et al. 2007). The remaining half was also cleared by livers from MMR-deficient mice. The latter observation was not accounted for, but could be a result of other MMR family members also being expressed in the liver.

uPA-System Interaction

On certain types of cultured cells, for example U937 cells, uPARAP/Endo180 is located in close proximity to uPAR on the cell surface [shown in cross-linking experiments (Behrendt et al. 2000)]. The physical intimacy between the proteins led to speculation regarding a cooperative pathway of matrix degradation, initially by envisioning uPARAP/Endo180 as a receptor that brought the plasminogen activation event and downstream collagenolytic enzymes in close proximity to matrix substrates (Behrendt et al. 2000). Later, as the internalization properties of the protein were discovered, this hypothesis was refined to represent the intracellular component of a matrix-degrading machinery (Engelholm et al. 2003). It is unclear if such a cooperative complex exists in vivo in areas of tissue remodeling. Many cell types that are uPARAP/Endo180 positive probably express little or no uPAR but in some cases, such as the osteoblasts of developing bone, a strong expression of both receptors is indeed noted (Engelholm et al. 2001).

Role of uPARAP/Endo180 in Cell Migration

The generation of mice deficient for uPARAP/Endo180 enabled experiments with murine fibroblasts to determine a possible role of the receptor in motility and migration. Fibroblasts from uPARAP/Endo180-deficient mice were found to be less motile than those from uPARAP/Endo180-expressing littermate wild-type control cells, on both matrigel and rat tendon collagen matrices (East et al. 2003, Engelholm et al. 2003). The mechanistic background for this phenomenon is still unresolved, but could possibly include a stronger adhesion provided by uPARAP/Endo180, thus providing a firmer grip on the matrix by the migrating fibroblasts and resulting in more efficient movement. Also, a more indirect effect of uPARAP/Endo180 on migration is possible as a putative interplay with integrins could result in signaling events, giving rise to increased motility. In another set of studies, the directional migration against a uPA gradient was found to be inhibited in a uPARAP/Endo180 si-RNA-treated breast cancer cell line (Sturge et al. 2003). The directional sensing was maintained in cells expressing a uPARAP/Endo180 variant which was mutated in the di-hydrophobic internalization motif (Sturge et al. 2003), whereas no sense of direction was observed after treatment of the cells with antibodies directed against two of the four outermost domains of uPARAP/Endo180 (CRD2 and the Cys-rich domain) (Sturge et al. 2003). The results were interpreted as evidence for a specific uPARAP/Endo180-mediated regulation of uPA-uPAR-induced cellular orientation. As the orientation effect was independent of internalization, it is possible that a better adhesion in uPARAP/Endo180-expressing cells contributed to the effects observed. This also might be consistent with the above-mentioned results with blocking antibodies directed against domains close to the FN-II domain responsible for collagen binding. A more recent study by the same investigators, examining the involvement of uPARAP/Endo180 in migration, indicated that the pathway including Rho, Rho-kinase (ROCK), and myosin light chain 2 (MLC2) phosphorylation was necessary for rear cell deadhesion during cell migration, and that this pathway was dependent on the uPARAP/Endo180-containing endosome compartment (Sturge, Wienke, and Isacke 2006).

Histological Localization

uPARAP/Endo180 in Healthy Tissue

The expression of uPARAP/Endo180 has been observed in several types of mesenchymal cells such as fibroblasts, osteoblasts, certain endothelial cells, and some macrophages in various organs, including developing bone, placenta, bladder, skin, kidney, spleen, liver, and lung (Isacke et al. 1990. Wu et al. 1996, Sheikh et al. 2000, Engelholm et al. 2001, Mousavi et al. 2005, Honardoust et al. 2006). Tissue undergoing remodeling of its collagen matrix often shows expression of uPARAP/Endo180 in some cell populations. In gingival wounds, a strong upregulation was noted in several cell types including myofibroblasts, macrophages, and endothelial cells (Honardoust et al. 2006). Furthermore, uPARAP/Endo180 is strongly expressed in the developing bones of mouse embryos and newborn mice (Wu et al. 1996, Engelholm et al. 2001, Howard et al. 2004, Wagenaar-Miller et al. 2007). In this case, the protein is expressed by osteoblasts and chondrocytes (Engelholm et al. 2001, Wagenaar-Miller et al. 2007). By screening for uPARAP/Endo180 mRNA expression in several tissues, high levels were found in lung and kidney, whereas the liver and brain were found to be almost negative (Wu et al. 1996), although with expression of an alternatively spliced, truncated RNA in fetal liver. In apparent contradiction to a lack of expression in the liver, hepatic stellate cells in culture have been shown to internalize collagen in a uPARAP/Endo180-dependent manner (Mousavi et al. 2005). As stellate cells comprise only 10% of the cells in the liver, an expression in these cells may have been difficult to detect in the study by Wu et al. Another region reported to be without uPARAP/Endo180 is the epidermis, whereas dermal macrophages have been shown to be uPARAP/Endo180 positive (Sheikh et al. 2000).

uPARAP/Endo180 in Cancer

In a recent study, uPARAP/Endo180 deficiency was combined with the MMTV-PyMT mouse breast tumor model in which mice develop spontaneous malignant breast tumors. Mice deficient for uPARAP/Endo180 showed a significantly delayed primary tumor growth and strikingly, the tumors had a very high content of undegraded collagen compared to the tumors of the littermate uPARAP/Endo180-positive controls (Curino et al. 2005). Moreover, no intracellular collagen was found in explanted fibroblast-like cells from the tumors of uPARAP/Endo180-deficient mice, in contrast to what was seen in the uPARAP/Endo180-sufficient cells. The expression of uPARAP/Endo180 in the tumors from wild-type PyMT mice mimicked that found in the human disease (see further below), with expression in fibroblast-like cells surrounding the mammary ducts and no expression in tumor cells (Nielsen et al. 2002, Curino et al. 2005). It was concluded that collagen clearance by intracellular degradation via the uPARAP/Endo180-dependent pathway was involved in invasive tumor growth in this system.

A study of uPARAP/Endo180 protein and mRNA expression (Nielsen et al. 2002) showed only very little expression in the normal human breast, with detectable staining only in a few fibroblast-like or myoepithelial cells. In benign breast lesions, uPARAP/Endo180 staining was strongly positive in intralobular fibroblasts and, depending on the type of lesion, also in myoepithelium and tumor-associated fibroblasts. In ductal carcinoma in situ (DCIS; *see* Fig. 13.2), uPARAP/Endo180 was observed in both myoepithelial cells and some tumor-associated fibroblasts, with more uPARAP/Endo180-positive fibroblasts around some of the DCIS foci. Invasive carcinomas of both ductal and lobular type also showed uPARAP/Endo180 staining of fibroblast-like cells, whereas the cancer cells were uPARAP/Endo180 negative in all cases. The pattern of staining by immunohistochemistry was indistinguishable from the pattern obtained by in situ hybridization (Nielsen et al. 2002). Similarly, in a study of 112 human squamous cell carcinomas of the head and neck, uPARAP/Endo180 showed increased expression especially in fibroblast-like cells from the tumor stroma (Sulek et al. 2007). Whereas no case of expression of uPARAP/Endo180 in cancer cells has been reported so far*, the total expression pattern is nevertheless in complete accordance with a function in cancer. In all cases studied, uPARAP/Endo180 is expressed in fibroblast-like cells in close proximity to the tumor and in some cases as part of the enveloping stroma barrier (Nielsen et al. 2002). It is indeed plausible that uPARAP/Endo180 promotes the mechanism of invasion and escape of tumor cells in the local tissue compartment, based on the ability of the receptor to assist collagen degradation and clearance. It is noteworthy that, in a study of colorectal cancer, searching for tumor endothelial markers by expressed sequence tags, uPARAP/Endo180 was identified as one of the most upregulated genes in endothelial cells adjacent to the cancer compared to endothelial cells from normal colorectal tissue

---

* Please refer to note added in proof on the first page.

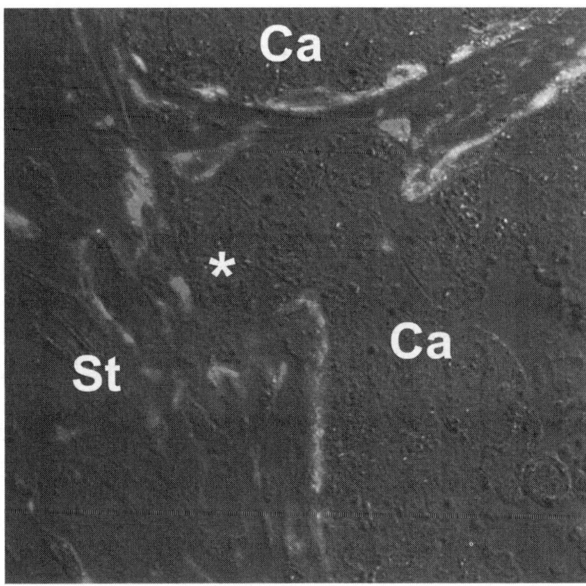

Fig. 13.2 Expression of *uPARAP/Endo*180 at a tumor–stroma interface. Double immunostaining of uPARAP/Endo180 (rabbit antibody coupled to Cy3; red) and cytokeratin 17 [myoepithelial marker, monoclonal mouse antibody coupled to fluorescein isothiocyanate (*FITC*); green] in human breast ductal carcinoma in situ with focal invasion (indicated by star). uPARAP/ Endo180 is expressed in fibroblasts and some myoepithelial cells surrounding the carcinoma cells (*Ca*) and in fibroblasts located in the stroma (*St*). For material and methods, see Nielsen et al. (2007). (Boye Schnack Nielsen, the Finsen Laboratory, Copenhagen, unpublished work; figure kindly made available by B.S. Nielsen) (*See also* Color Insert I)

(St Croix et al. 2000). It was not completely clear, however, if fibroblast-like cells may have contributed to the cell populations analyzed in this study.

Biological Role of uPARAP/Endo180 and Concluding Remarks

Even though the knowledge on the function of uPARAP/Endo180 in development and in different tissues is still very incomplete, the local expression patterns and analyses of the biochemical properties of the protein do provide strong suggestions as to some roles of the receptor in vivo. As noted above, the uPARAP/Endo180 mRNA is expressed in developing bone from newborn mice and mouse embryos (Wu et al. 1996, Engelholm et al. 2001), and the cellular origin of the in situ signal is most likely the osteoblast-osteocyte and possibly also osteoclasts and endothelial cells (Engelholm et al. 2001, Wagenaar-Miller et al. 2007). Moreover, uPARAP/ Endo180 expression has been observed in cartilage-forming sites in young mice (Howard et al. 2004). All of these observations are compatible with the hypothesis that uPARAP/Endo180 has a role in collagen turnover in developing tissue. This is consistent with the notion that although mice deficient in uPARAP/Endo180 appear normal and are fertile (East et al. 2003, Engelholm et al. 2003), a close examination of the developing bone did reveal a small delay in the time-course of bone growth in the uPARAP/Endo180-deficient, newborn mouse (Wagenaar-Miller et al. 2007). In the same study, however, this became even much more clear when uPARAP/Endo180 deficiency was combined with deficiency for MT1-MMP, a membrane-bound protease involved in collagen degradation which is expressed in areas of collagen turnover in developing bone, overlapping with those of uPARAP/Endo180 expression (Wagenaar-Miller et al. 2007). Mice deficient for MT1-MMP alone were small and affected in bone and connective tissues, with arthritis and osteopenia and reduced survival, and with ~30% of mice dying before weaning and the remaining before 90 days of age (Holmbeck et al. 1999, Wagenaar-Miller et al. 2007), but in the study on combined deficiency, these conditions were severely worsened. Although MT1-MMP; uPARAP/Endo180 double-deficient mice were born in the expected mendelian ratio, they were uniformly prone to early postnatal death, with all mice dying within the first 21 days of age. The development of bone was heavily retarded, including the calvarial bone, and the cranial closure was very incomplete in the double-deficient mice (Wagenaar-Miller et al. 2007).

Altogether the above observations give strong indications pointing to uPARAP/Endo180 being physiologically relevant in several remodeling events involving collagen degradation. Importantly, extracellular matrix degradation/remodeling is a crucial step in cancer invasion (Hanahan and Weinberg 2000), and therefore the above-mentioned role of uPARAP/Endo180 in the mouse tumor model (Curino et al. 2005) is most likely related to the same basic functions as those relevant to the healthy remodeling processes. The likely importance of uPARAP/Endo180 in invasion makes it a candidate therapeutic target in cancer treatment, aiming to prevent the spread of malignant disease.

Acknowledgments We thank Dr. Boye Schnack Nielsen for generously providing the photograph (immunolocalization of uPARAP/Endo180 in DCIS) used in Fig. 13.2. This work was supported by EU contract LSHC-CT-2003-503297.

References

Banyai, L., Trexler, M., Koncz, S., Gyenes, M., Sipos, G., and Patthy, L. 1990. The collagen-binding site of type-II units of bovine seminal fluid protein PDC-109 and fibronectin. Biochem. J. 193:801–806.

Banyai, L., Tordai, H., and Patthy, L. 1994. The gelatin-binding site of human 72 kDa type IV collagenase (gelatinase A). Eur. J. Biochem. 298 (Pt 2):403–407.

Behrendt, N. 2004. The urokinase receptor (uPAR) and the uPAR-associated protein (uPARAP/Endo180): membrane proteins engaged in matrix turnover during tissue remodeling. FEBS Lett. 385:103–136.

Behrendt, N., Ronne, E., and Dano, K. 1993. A novel, specific pro-urokinase complex on monocyte-like cells, detected by transglutaminase-catalyzed cross-linking. Biol. Chem. 336:394–396.

Behrendt, N., Jensen, O. N., Engelholm, L. H., Mortz, E., Mann, M., and Dano, K. 2000. A urokinase receptor-associated protein with specific collagen binding properties. J. Biol. Chem. 275:1993–2002.

Boskovic, J., Arnold, J. N., Stilion, R., Gordon, S., Sim, R. B., Rivera-Calzada, A., Wienke, D., Isacke, C. M., Martinez-Pomares, L., and Llorca, O. 2006. Structural model for the mannose receptor family uncovered by electron microscopy of Endo180 and the mannose receptor. J. Biol. Chem. 281:8780–8787.

Briknarova, K., Gehrmann, M., Banyai, L., Tordai, H., Patthy, L., and Llinas, M. 2001. Gelatin-binding region of human matrix metalloproteinase-2: solution structure, dynamics, and function of the COL-23 two-domain construct. J. Biol. Chem. 276:27613–27621.

Curino, A. C., Engelholm, L. H., Yamada, S. S., Holmbeck, K., Lund, L. R., Molinolo, A. A., Behrendt, N., Nielsen, B. S., and Bugge, T. H. 2005. Intracellular collagen degradation mediated by uPARAP/Endo180 is a major pathway of extracellular matrix turnover during malignancy. J. Cell Biol. 169:977–985.

East, L. and Isacke, C. M. 2002. The mannose receptor family. Biochim. Biophys. Acta 1572:364–386.

East, L., Rushton, S., Taylor, M. E., and Isacke, C. M. 2002. Characterization of sugar binding by the mannose receptor family member, Endo180. J. Biol. Chem. 277:50469–50475.

East, L., McCarthy, A., Wienke, D., Sturge, J., Ashworth, A., and Isacke, C. M. 2003. A targeted deletion in the endocytic receptor gene Endo180 results in a defect in collagen uptake. EMBO Rep. 4:710–716.

Engelholm, L. H., Nielsen, B. S., Netzel-Arnett, S., Solberg, H., Chen, X. D., Lopez Garcia, J. M., Lopez-Otin, C., Young, M. F., Birkedal-Hansen, H., Dano, K., Lund, L. R., Behrendt, N., and Bugge, T. H. 2001. The urokinase plasminogen activator receptor-associated protein/endo180 is coexpressed with its interaction partners urokinase plasminogen activator receptor and matrix metalloprotease-13 during osteogenesis. Lab. Invest. 81:1403–1414.

Engelholm, L. H., List, K., Netzel-Arnett, S., Cukierman, E., Mitola, D. J., Aaronson, H., Kjoller, L., Larsen, J. K., Yamada, K. M., Strickland, D. K., Holmbeck, K., Dano, K., Birkedal-Hansen, H., Behrendt, N., and Bugge, T. H. 2003. uPARAP/Endo180 is essential for cellular uptake of collagen and promotes fibroblast collagen adhesion. J. Cell Biol. 160:1009–1015.

Feinberg, H., Park-Snyder, S., Kolatkar, A. R., Heise, C. T., Taylor, M. E., and Weis, W. I. 2000. Structure of a C-type carbohydrate recognition domain from the macrophage mannose receptor. J. Biol. Chem. 275:21539–21548.

Hanahan, D. and Weinberg, R. A. 2000. The hallmarks of cancer. Cell 100:57–70.

Holmbeck, K., Bianco, P., Caterina, J., Yamada, S., Kromer, M., Kuznetsov, S. A., Mankani, M., Robey, P. G., Poole, A. R., Pidoux, I., Ward, J. M., and Birkedal-Hansen, H. 1999. MT1-MMP-deficient mice develop dwarfism, osteopenia, arthritis, and connective tissue disease due to inadequate collagen turnover. Cell 99:81–92.

Honardoust, H. A., Jiang, G., Koivisto, L., Wienke, D., Isacke, C. M., Larjava, H., and Hakkinen, L. 2006. Expression of Endo180 is spatially and temporally regulated during wound healing. Histopathology 49:634–648.

Howard, M. J. and Isacke, C. M. 2002. The C-type lectin receptor Endo180 displays internalization and recycling properties distinct from other members of the mannose receptor family. J. Biol. Chem. 277:32320–32331.

Howard, M. J., Chambers, M. G., Mason, R. M., and Isacke, C. M. 2004. Distribution of Endo180 receptor and ligand in developing articular cartilage. Osteoarthritis Cartilage 12:74–82.

Isacke, C. M., van der Geer, P., Hunter, T., and Trowbridge, I. S. 1990. p180, a novel recycling transmembrane glycoprotein with restricted cell type expression. Mol. Cell Biol. 10:2606–2618.

Kjoller, L., Engelholm, L. H., Hoyer-Hansen, M., Dano, K., Bugge, T. H., and Behrendt, N. 2004. uPARAP/endo180 directs lysosomal delivery and degradation of collagen IV. Exp. Cell Res. 293:106–116.

Liu, Y., Chirino, A. J., Misulovin, Z., Leteux, C., Feizi, T., Nussenzweig, M. C., and Bjorkman, P. J. 2000. Crystal structure of the cysteine-rich domain of mannose receptor complexed with a sulfated carbohydrate ligand. J. Exp. Med. 191:1105–1116.

Madsen, D. H., Engelholm, L. H., Ingvarsen, S., Hillig, T., Wagenaar-Miller, R. A., Kjoller, L., Gardsvoll, H., Hoyer-Hansen, G., Holmbeck, K., Bugge, T. H., and Behrendt, N. 2007. Extracellular collagenases and the endocytic receptor, uPARAP/Endo180, cooperate in fibroblast-mediated collagen degradation. J. Biol. Chem. 282:27037–27045.

Malovic, I., Sorensen, K. K., Elvevold, K. H., Nedredal, G. I., Paulsen, S., Erofeev, A. V., Smedsrod, B. H., and McCourt, P. A. 2007. The mannose receptor on murine liver sinusoidal endothelial cells is the main denatured collagen clearance receptor. Hepatology 45:1454–1461.

Martinez-Pomares, L., Wienke, D., Stillion, R., McKenzie, E. J., Arnold, J. N., Harris, J., McGreal, E., Sim, R. B., Isacke, C. M., and Gordon, S. 2006. Carbohydrate-independent recognition of collagens by the macrophage mannose receptor. Eur. J. Immunol. 36:1074–1082.

Mousavi, S. A., Sato, M., Sporstol, M., Smedsrod, B., Berg, T., Kojima, N., and Senoo, H. 2005. Uptake of denatured collagen into hepatic stellate cells: evidence for the involvement of urokinase plasminogen activator receptor-associated protein/Endo180. Biochem. J. 387:39–46.

Napper, C. E., Drickamer, K., and Taylor, M. E. 2006. Collagen binding by the mannose receptor mediated through the fibronectin type II domain. Biochem. J. 395:579–586.

Nielsen, B. S., Rank, F., Engelholm, L. H., Holm, A., Dano, K., and Behrendt, N. 2002. Urokinase receptor-associated protein (uPARAP) is expressed in connection with malignant as well as benign lesions of the human breast and occurs in specific populations of stromal cells. Int. J. Cancer 98:656–664.

Nielsen, B. S., Rank, F., Illemann, M., Lund, L. R., and Dano, K. 2007. Stromal cells associated with early invasive foci in human mammary ductal carcinoma in situ coexpress urokinase and urokinase receptor. Int. J. Cancer 120:2086–2095.

Pickford, A. R., Potts, J. R., Bright, J. R., Phan, I., and Campbell, I. D. 1997. Solution structure of a type 2 module from fibronectin: implications for the structure and function of the gelatin-binding domain. Structure 5:359–370.

Pond, L., Kuhn, L. A., Teyton, L., Schutze, M. P., Tainer, J. A., Jackson, M. R., and Peterson, P. A. 1995. A role for acidic residues in di-leucine motif-based targeting to the endocytic pathway. J. Biol. Chem. 270:19989–19997.

Rivera-Calzada, A., Robertson, D., MacFadyen, J. R., Boskovic, J., Isacke, C. M., and Llorca, O. 2003. Three-dimensional interplay among the ligand-binding domains of the urokinase-plasminogen-activator-receptor-associated protein, Endo180. EMBO Rep. 4:807–812.

Sheikh, H., Yarwood, H., Ashworth, A., and Isacke, C. M. 2000. Endo180, an endocytic recycling glycoprotein related to the macrophage mannose receptor is expressed on fibroblasts, endothelial cells and macrophages and functions as a lectin receptor. J. Cell Sci. 113 (Pt 6):1021–1032.

St Croix, B., Rago, C., Velculescu, V., Traverso, G., Romans, K. E., Montgomery, E., Lal, A., Riggins, G. J., Lengauer, C., Vogelstein, B., and Kinzler, K. W. 2000. Genes expressed in human tumor endothelium. Science 289:1197–1202.

Sturge, J., Wienke, D., East, L., Jones, G. E., and Isacke, C. M. 2003. GPI-anchored uPAR requires Endo180 for rapid directional sensing during chemotaxis. J. Cell Biol. 162:789–794.

Sturge, J., Wienke, D., and Isacke, C. M. 2006. Endosomes generate localized Rho-ROCK-MLC2-based contractile signals via Endo180 to promote adhesion disassembly. J. Cell Biol. 175:337–347.

Sulek, J., Wagenaar-Miller, R. A., Shireman, J., Molinolo, A., Madsen, D. H., Engelholm, L. H., Behrendt, N., and Bugge, T. H. 2007. Increased expression of the collagen internalization receptor uPARAP/Endo180 in the stroma of head and neck cancer. J. Histochem. Cytochem. 55:347–353.

Wagenaar-Miller, R. A., Engelholm, L. H., Gavard, J., Yamada, S. S., Gutkind, J. S., Behrendt, N., Bugge, T. H., and Holmbeck, K. 2007. Complementary roles of intracellular and pericellular collagen degradation pathways in vivo. Mol. Cell Biol. 27:6309–6322.

Wienke, D., MacFadyen, J. R., and Isacke, C. M. 2003. Identification and characterization of the endocytic transmembrane glycoprotein Endo180 as a novel collagen receptor. Mol. Biol. Cell 14:3592–3604.

Wu, K., Yuan, J., and Lasky, L. A. 1996. Characterization of a novel member of the macrophage mannose receptor type C lectin family. J. Biol. Chem. 271:21323–21330.

Chapter 14
Physiological and Pathological Functions of Type II Transmembrane Serine Proteases: Lessons from Transgenic Mouse Models and Human Disease-Associated Mutations

Karin List and Thomas H. Bugge

Abstract Recent advances in the mouse and human genome projects have resulted in the identification of a surprisingly large number of genes encoding new proteolytic enzymes which include members of the type II transmembrane serine protease (TTSP) family. In the past few years, the TTSP family has undergone a rapid transformation from a group of largely unexplored proteins into an established family of cell surface–associated proteases playing important roles in development, homeostasis, and pathogenesis. This chapter summarizes current knowledge about the biochemical properties and the biological functions of the TTSP family members in mammalian physiology and disease.

Introduction

Within the past 50 years, extensive knowledge about the role of proteolysis in health and disease has been gained, including insights into the functions of specific proteases belonging to the families of plasminogen activators, matrix metalloproteinases, proteins containing A Disintegrin and A Metalloprotease domain (ADAMs), and cathepsins. The completion of the mouse and human genome sequence projects has facilitated an explosion in the discovery of novel candidate

K. List
Department of Pharmacology and Barbara Ann Karmanos Cancer Institute, Wayne State University School of Medicine, Detroit, MI, e-mail: klist@med.wayne.edu

T.H. Bugge
Proteases and Tissue Remodeling Unit, Oral and Pharyngeal Cancer Branch, National Institute of Dental and Craniofacial Research, National Institutes of Health, 30 Convent Drive, Bethesda, MD 20892, e-mail: tbugge@mail.nih.gov

D. Edwards et al. (eds.), *The Cancer Degradome.*

protease genes, suggesting that an exceedingly greater repertoire of proteases are engaged in pericellular proteolysis than previously anticipated. The uncovering of an unexpectedly large family of type II transmembrane serine proteases (TTSPs) presents the challenge to determine the functions of an, as yet, largely unexplored protease family. In recent years, several studies have emerged, including the identification of disease-associated human TTSP mutations and the characterization of knock-out/transgenic mouse models, describing critical and diverse functions for TTSPs in pathogenesis and development. An overview of these findings will be presented in this chapter.

The Members of the TTSP Family

All members of the TTSP family are characterized by the presence of an N-terminal signal anchor and a C-terminal serine protease domain, separated by a stem region containing an array of protein domains that varies widely between individual TTSPs (Fig. 14.1). The TTSP family currently encompasses 20 members in both humans and mice and they are divided into subfamilies based on the composition of the domains in the stem region, the phylogenetic relationship of the serine protease domain, and the chromosomal location of their genes (Hooper et al. 2001, Netzel-Arnett et al. 2003, Szabo et al. 2003, Szabo and Bugge 2007).

The Hepsin/TMPRSS Subfamily

Enteropeptidase (originally termed enterokinase) was the first discovered member of the family, when Pavlov demonstrated more than 100 years ago that a protease that is present in the intestine is essential for the activation of pancreatic proenzymes. Later, trypsinogen was identified as the target for enteropeptidase in the duodenal lumen and almost a century after Pavlov's groundbreaking studies the human and mouse enteropeptidase cDNAs were cloned (Kitamoto et al. 1994, Yuan et al. 1998). Enteropeptidase is a large TTSP (1,019 amino acids) with a complex domain structure including a SEA (*s*ea urchin sperm protein, *e*nteropeptidase, *a*grin) domain, two CUB (*c*omplement C1r/C1s, *u*rchin embryonic growth factor, *b*one morphogenetic protein 1) domains, two LDLA (*l*ow-*d*ensity *l*ipoprotein receptor type-*A* repeat) domains, a group A scavenger receptor domain, a MAM (meprin, AS antigen, receptor protein phosphatose μ) domain and a serine protease domain. Enteropeptidase expression is restricted to the small intestine where it converts trypsinogen into trypsin, which not only hydrolyses peptide bonds of dietary proteins but also activates a number of pancreatic zymogens including chymotrypsinogen and proelastase. Due to the role of enteropeptidase as an initiator of the trypsinogen intestinal zymogen cascade, patients with the rare genetic disorder primary/congenital enteropeptidase deficiency suffer from impaired dietary protein digestion, causing severe intestinal malabsorption and growth failure (Holzinger et al. 2002, Rutgeert et al. 1972, Rutgeert and Eggermon 1976).

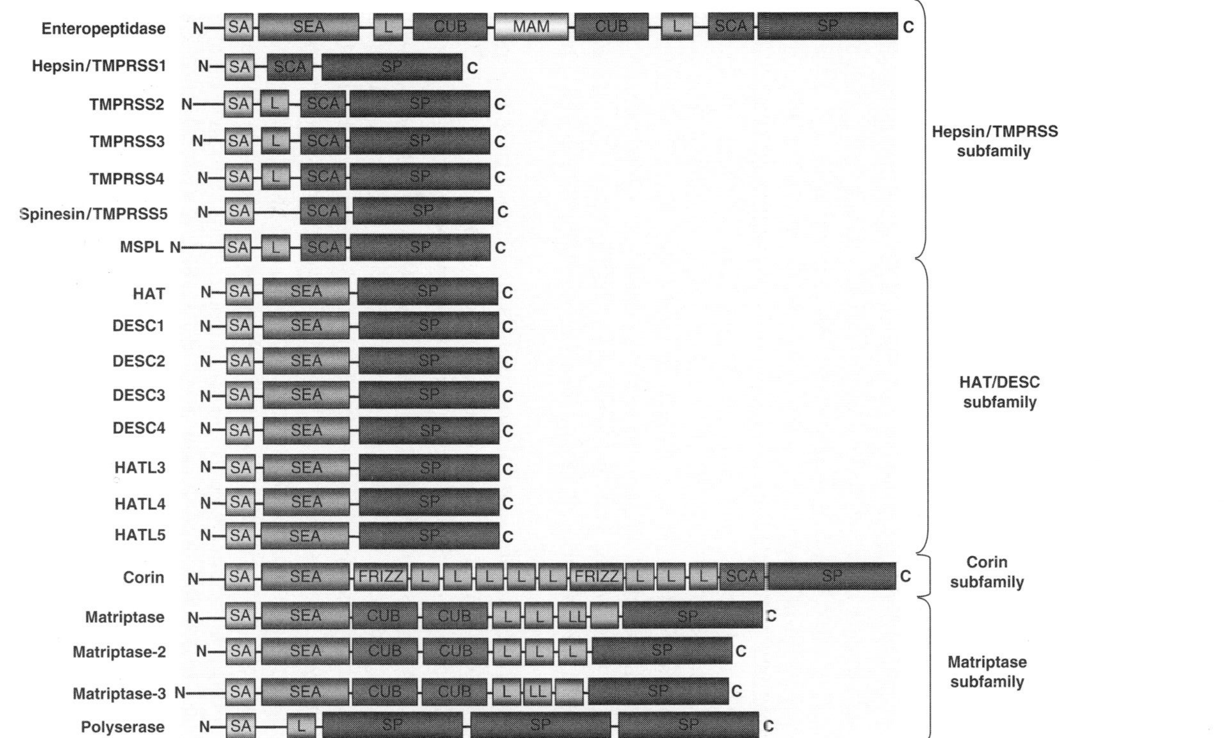

Fig. 14.1 Schematic representation of type II transmembrane serine protease (TTSP) family members, which are divided into four subfamilies based on the composition of the domains in the stem region, the phylogenetic relationship of the serine protease domain, and the chromosomal location of their genes. SA, signal anchor; SEA, sea urchin sperm protein, enteropeptidase, agrin domain; CUB, complement, urchin embryonic growth factor, bone morphogenetic protein 1 domain, L, low-density lipoprotein receptor type-A repeat domain; SCA, group A scavenger receptor domain; MAM, meprin, A5 antigen, receptor protein phosphatase μ domain; FRIZZ, frizzled domain; SP, serine protease domain

The symptoms can be alleviated with pancreatic enzyme replacement, which includes trypsin required to activate the patient's own pancreatic proenzymes (Moroz et al. 2001). Secondary/acquired enteropeptidase deficiency has been described in pancreatic disease, such as cystic fibrosis and other types of exocrine pancreatic insufficiency, and in mucosal diseases such as chronic intractable diarrhea, where the degree of enzyme deficiency correlates with the severity of mucosal injury (Grant 1986, Lebenthal et al. 1976).

Hepsin/TMPRSS1 (Transmembrane Protease, Serine 1) was the first TTSP to be cloned and got its name because of its high expression in hepatocytes and hepatoma cells (Leytus et al. 1988, Tsuji et al. 1991). It has a simple domain structure with a single group A scavenger receptor domain in the stem region. Structural studies of the extracellular part of hepsin showed that the group A scavenger receptor domain and the serine protease domain are associated with each other through a single disulfide bond and an extensive network of noncovalent interactions, and that the protease domain contains a large, hydrophobic $S1'$ pocket that is similar to the $S1'$ pockets of other trypsin-like serine proteases (Somoza et al. 2003). Despite its name, hepsin is fairly widely expressed including the kidney, pancreas, lung, thyroid, pituitary gland, testis, and prostate, and is also detectable in human endothelial cells (Aimes et al. 2003, Tsuji et al. 1991). Interestingly, hepsin can be activated autocatalytically on the cell surface, and is an *in vitro* activator of factor VII, factor IX, factor XII, microsomal glutathione *S*-transferase, pro-urokinase, and pro-hepatocyte growth factor (pro-HGF) (Herter et al. 2005, Kazama et al. 1995, Kirchhofer et al. 2005, Kunii et al. 2006, Moran et al. 2006, Qiu et al. 2007).

Since hepsin activates several coagulation factors *in vitro*, it is considered a candidate player in coagulation activation and fibrin deposition (Herter et al. 2005, Kazama et al. 1995). In addition, hepsin has been proposed to function in controlling cell growth, based on findings that anti-hepsin antibodies and hepsin-specific antisense oligonucleotides arrest growth of hepatoma cells, and that highly proliferative cells in various tissues in the developing mouse embryo display high hepsin expression levels (Torres-Rosado et al. 1993). However, in two independent studies of hepsin-deficient mice, no effects on hemostasis, cell growth and embryo-genesis, or liver function and regeneration were observed (Table 14.1) (Wu et al. 1998, Yu et al. 2000). The serum concentration of the bone-derived form of alkaline phosphatase is elevated in hepsin-deficient mice, the significance of which is still unclear (Wu et al. 1998, Yu et al. 2000).

Importantly, a recent study demonstrated an essential role for hepsin in auditory function, since hepsin-deficient mice exhibit profound hearing loss caused by abnormal cochleae development including tectorial membrane formation and defects in the compaction of spiral ganglion neurons (Guipponi et al. 2007). Interestingly, the hepsin-deficient mice display altered thyroid hormone metabolism with a reduction of serum T4 levels. This finding may offer important clues to the molecular mechanism underlying hepsin-dependent cochlear development, since thyroid hormones have been shown to be involved in the development of the auditory function (Crofton 2004, Guipponi et al. 2007).

Table 14.1 TTSP genetic mouse models and/or identified human mutations

TTSP	Mouse model	Mouse phenotype	Human mutation	References[a]
Corin	Knock-out	Hypertensive with further blood pressure increase on high-salt diet and during pregnancy Essential for activation of pro-ANP	Corin gene allele is associated with an enhanced cardiac hypotrophic response to hypertension	1,2
Enteropeptidase			Congenital enteropeptidase deficiency causing chronic diarrhea, edema, growth retardation. Compound heterozygosity for nonsense and frameshift mutations associated with duodenopancreatic reflux disease	3, 4, 5, 6
Hepsin	Knock-out	Hearing loss and reduced free thyroxine levels Increased serum concentrations of bone-derived alkaline phosphatase	Variants in the hepsin gene associated with prostate cancer in men of European origin	7, 8, 9, 10
	Transgenic	Disorganization of the basement membrane and promotion of primary prostate cancer progression and metastasis		11
Matriptase	Knock-out	Perinatal lethality, defective epidermal differentiation, Impaired epidermal and oral barrier function, Hypomorphic hair follicles, ichthyosis, decreased thymocyte survival	Missense mutation associated with the skin and hair disorder autosomal recessive ichthyosis with hypotrichosis (ARIH)	12, 13, 14
	Hypomorphic	Hyperproliferative and retention ichthyosis with impaired desquamation, hypotrichosis and tooth defects		15
	Transgenic	Spontaneous epidermal squamous cell carcinoma and increased carcinogen-induced tumor formation		16
TMPRSS2	Knock-out	No effects on development, overall survival or fertility	Fusion of TMPRSS2 to ERG sequences correlates to poor survival in human prostate cancer	17, 18
TMPRSS3			Congenital and childhood onset autosomal recessive deafness associated with TMPRSS3 mutation	19, 20, 21

[a]References 1. Chang et al. (2005), 2. Dries et al. (2005), 3. Rutgeert et al. (1972), 4. Holzinger et al. (2002), 5. Grant et al. (1986), 6. Rutgeert et al. (1976), 7. Guipponl et al. (2007), 8. Wu et al. (1998), 9. Pal et al. (2006), 10. Yu et al. (2000), 11. Klezovitch et al. (2004), 12. Basel-Vanagaite et al. (2007), 13. List et al. (2002), 14. List et al. (2003), 15. List et al. (2007b), 16. List et al. (2005), 17. Kim et al. (2006), 18. Wilson, (2005), 19. Elbracht et al. (2007), 20. Guipponi et al. (2002), 21. Wattenhofer et al. (2002).

Hepsin has been implicated in tumorigenesis, based on expression studies of human cancer where the protease is highly upregulated. This is particularly well documented in prostate cancer (Landers et al. 2005, Magee et al. 2001, Riddick et al. 2005, Stephan et al. 2004, Wu and Parry 2007), and also reported in renal cell carcinoma and in ovarian carcinomas (Betsunoh et al. 2007, Tanimoto et al. 1997).

Genetic analysis has demonstrated an association of five single nucleotide polymorphisms (SNPs) in the hepsin gene with susceptibility to prostate cancer among men of European descent and an association of one of the SNPs with Gleason score (Pal et al. 2006). The effect of dysregulated hepsin expression in prostate carcinogenesis was studied experimentally in a transgenic mouse model, where it was demonstrated that hepsin overexpression in prostate epithelium causes disorganization of the basement membrane, promotion of primary prostate cancer progression, and metastasis to liver, lung, and bone (Klezovitch et al. 2004). Hepsin may be directly involved in degradation of extracellular matrix proteins and/or activation or other proteases or growth factors, but the molecular target(s) for hepsin in carcinogenesis has yet to be identified.

Mosaic serine protease (MSP) was cloned from human lung and two different cDNAs were identified, encoding a long form (MSPL), which contains a transmembrane domain, and a short form (MSPS) of the protease that lacks the transmembrane domain. Both cDNAs are predominantly expressed in lung, placenta, pancreas, and prostate (Kim et al. 2001).

Epitheliasin/TMPRSS2 was first identified using exon trapping techniques and was later found to exhibit increased expression upon exposure to androgens in a study of hormone-regulated transcripts in prostate cancer (Lin et al., 1999a, Paoloni-Giacobino et al. 1997). TMPRSS2 is expressed mainly in the epithelium of the prostate, kidney, gastrointestinal, urogenital, and respiratory tracts (Jacquinet et al. 2000, Vaarala et al. 2001a, 2001b). TMRSS2 can undergo autocatalytic cleavage resulting in the secretion of the protease domain into the media by cultured prostate cancer cells and into the serum of prostate tumor-bearing mice (Afar et al., 2001). TMPRSS2 is a candidate activator of influenza viruses via cleavage of viral hemagglutinin and may play a role in respiratory tract viral infections (Bottcher et al. 2006). The *Protease Activated Receptor 2* (PAR-2) is another putative substrate for TMPRSS2 (Wilson et al. 2005).

An impressive amount of literature (more than 30 published articles in the last 3 years) has documented recurrent chromosomal rearrangements creating fusions between the TMPRSS2 gene and E26 transformation-specific (ETS) transcription factor family genes in prostate cancer. It has been demonstrated in various studies that 36–78% of prostate cancers from prostate-specific antigen-screened surgical cohorts harbor TMPRSS2-ETS (ERG, ETV1, or ETV4) fusion genes and that ~90% of samples with ERG outlier expression contain TMPRSS2-ETS fusions (Tomlins et al. 2007). Interestingly, the role of TMPRSS2 in this context is proteolysis-independent, since the majority of the fusion transcripts contain no coding TMPRSS2 sequences (Rubin and Chinnaiyan 2006). However, TMPRSS2-ETS fusions place the ERG/ETV genes under the regulation of the androgen-inducible

TMPRSS2 promoter, thus causing overexpression of active transcription factors in the prostate epithelium, an event that has been demonstrated to confer neoplastic phenotypes in benign cultured prostate cells and in mouse prostate (Tomlins et al. 2007). Considering the high incidence of prostate cancer and the high frequency of this gene fusion, the TMPRSS2-ETS gene fusion is among the most common genetic aberrations described so far in human malignancies. These alterations may prove to be important for diagnosis, prognosis, treatment assessment, and development of novel therapeutic approaches in prostate cancer.

TMPRSS2-mediated proteolysis may also directly play a role in cancer since expression of the protease is significantly upregulated in prostate cancer, but further evidence is still needed to support this hypothesis (Lin et al. 1999a).

The physiological functions of TMPRSS2 are not clear at this time because TMPRSS2 deficiency does not influence survival or fertility, or affect the development of major organs, including the prostate (Kim et al. 2006).

TMPRSS3 was identified as a TTSP that is expressed in pancreatic cancer whereas expression was undetectable in normal pancreas and in chronic pancreatitis. TMPRSS3 is also present in gastric, colorectal, and ampullary cancer, and a transcript variant of the protease is overexpressed in ovarian carcinomas (Iacobuzio-Donahue et al. 2003, Sawasaki et al. 2004, Wallrapp et al. 2000). The TMPRSS3 expression profile in normal tissues is quite broad and includes the gastrointestinal and urogenital tracts, and inner ear (Guipponi et al. 2007, Wallrapp et al. 2000). The epithelial amiloride-sensitive sodium channel (ENaC), which is expressed in many sodium-reabsorbing tissues, is a potential substrate of TMPRSS3 (Guipponi et al. 2002).

TMPRSS3 has been linked to both congenital and childhood onset autosomal recessive deafness in multiple studies that identified mutations in the TMPRSS3 gene (Elbracht et al. 2007, Guipponi et al. 2002, Masmoudi et al. 2001, Wattenhofer et al. 2002, Wattenhofer et al. 2005). It was demonstrated that TMPRSS3 mutant proteins that cause deafness, including mutations in the LDLA domain, group A scavenger receptor domain, and serine protease domain, are inactive or have reduced proteolytic activity, indicating that TMPRSS3-mediated proteolysis is required for inner ear development (Guipponi et al. 2002, Wattenhofer et al. 2002). It has been proposed that proteolytic cleavage of ENaC may be functionally important, since it is co-expressed with TMPRSS3 in the epithelium of the inner ear and is efficiently activated by the protease *in vitro* (Guipponi et al. 2002).

Like several other TTSPs, TMPRSS4 has been reported to be dysregulated in cancer. Thus, the protease is upregulated in thyroid neoplasms and pancreatic cancer (Dawelbait et al. 2007, Kebebew et al. 2005, Kebebew et al. 2006). It has been proposed that tissue factor pathway inhibitor 2 (TFPI2), which belongs to the family of small Kunitz-type inhibitors, is an inhibitor of TMPRSS4 (Dawelbait et al. 2007). TMPRSS4 is capable of ENaC activation *in vitro*, but the physiological functions of the protease have yet to be established (Planes and Caughey 2007).

Spinesin/TMPRSS5 expression is strictly confined to the central nervous system where it is predominantly expressed in neurons, in their axons, at the synapses of motoneurons in the spinal cord, and in some oligodendrocytes (Watanabe et al. 2004,

Yamaguchi et al. 2002). Like hepsin, it has a simple domain composition containing a single group A-scavenger receptor-like domain in the stem region. Spinesin/TMPRSS5 needs proteolytic cleavage to be activated, and enteropeptidase is an activator *in vitro* (Yamaguchi et al. 2002). The elucidation of the physiological role for Spinesin/TMPRSS5 and the identification of relevant substrates in the central nervous system still await further investigation.

The HAT/DESC Subfamily

This family is the least studied of the TTSP subfamilies and knowledge about some of the members is just starting to be acquired. The highly homologous genes of the Human Airway Tryptase (HAT)/DESC subfamily are located in a single chromosomal cluster and encode proteases that all have the same domain composition with a single SEA domain in the stem region (Hobson et al. 2004, Netzel-Arnett et al. 2003, Szabo and Bugge 2007, Szabo et al. 2003).

As the name indicates, HAT was isolated from airway secretions, and immunohistochemical studies demonstrated that the enzyme is located on the cells of the submucosal serous glands of the bronchi and trachea (Yamaoka et al. 1998, Yasuoka et al. 1997). Several functions of HAT have been documented *in vitro* including stimulation of interleukin-8 synthesis in bronchial epithelial cells, stimulation of human bronchial fibroblast proliferation in a PAR-2-dependent pathway, and proteolytic activation of influenza virus hemagglutinin (Bottcher et al. 2006, Matsushima et al. 2006, Miki et al. 2003).

DESC1 (Differentially Expressed in Squamous Cell Carcinoma 1) was identified as a protease whose expression was downregulated in head and neck cancer as compared to matched normal tissue (Lang and Schuller 2001, Sedghizadeh et al. 2006). DESC1 is expressed in a variety of tissues including skin, salivary gland, prostate, epididymis, and testis (Lang and Schuller 2001, Hobson et al. 2004). Biochemical studies have confirmed that DESC1 is an active protease with gelatinolytic and caseinolytic activities, and that it can form stable inhibitory complexes with both plasminogen activator inhibitor-1 and protein C inhibitor (Hobson et al. 2004). The crystal structure of the DESC1 proteinase domain in complex with benzamidine was recently resolved and DESC1 exhibits a typical trypsin-like serine proteinase fold with a thrombin-like S1 pocket, a urokinase-type plasminogen activator-type S2 pocket, and an open hydrophobic S3/S4 cavity (Kyrieleis et al. 2007). DESC4 was identified in oral epithelium and is also expressed in the brain, heart, liver, salivary glands, the respiratory epithelium of the nasal cavity, and the tear gland ducts of the eyes (Behrens et al. 2004).

The HAT-like (HATL) 3–5 genes and DESC2-3 genes are predicted to encode functional proteases based on bioinformatic analysis (Hobson et al. 2004, Szabo and Bugge 2007), but a biochemical characterization of these proteases has not as yet been performed and the physiological function of the members of the family still awaits investigation.

The Matriptase Subfamily

Matriptase (also known as MT-SP1, TAGD-15, and epithin) was identified as an ovarian, prostate, and breast carcinoma-associated protein, and has since been documented to be overexpressed in a wide variety of human carcinomas, and to be causally involved in cutaneous carcinogenesis (Kim et al. 1999, Lin et al. 1999c, 1997; List et al. 2005, List, Bugge and Szabo 2006a, Takeuchi et al. 2000, Uhland 2006). Furthermore, several studies using cell culture systems or xenograft models have suggested that matriptase activity may directly contribute to carcinogenesis since inhibiting the protease in established tumor cell lines from prostate, colon, and ovarian cancer using small interfering RNAs, antisense matriptase oligodeoxyribonucleotides, hammerhead ribozyme transgenes, or synthetic active-site matriptase inhibitors impairs cancer cell invasive properties and tumor progression (Forbs et al. 2005, Galkin et al. 2004, Satomi et al. 2001, Sanders et al. 2006).

Matriptase is synthesized as an inactive, single chain zymogen and autoactivation of matriptase can be induced in cultured cells by lysophospholipids, androgens, and the polyanionic compound suramin (Benaud et al. 2002, Kiyomiya et al. 2006). The inhibition of activated matriptase by HGF activator inhibitor (HAI)-1 was first documented by the identification of matriptase/HAI-1 complexes in human milk and in a number of cancer cell lines (Lin et al. 1999b). Several candidate downstream targets for matriptase have been described in cell-free systems and in cell culture studies including PAR-2, pro-HGF, pro-urokinase-type plasminogen activator (pro-uPA), stromelysin, SIMA135/TRASK/CDCP1, and insulin-like growth factor binding protein-related protein-1 (Ahmed et al. 2006, Bhatt et al., 2005, Lee et al. 2000, Takeuchi et al. 2000).

As mentioned above, matriptase is overexpressed in a remarkable variety of human carcinomas, and also serves as a prognostic marker in several types of cancer (List et al. 2006a, Uhland 2006). It has been hypothesized that an imbalance in the matriptase:HAI-1 ratio, which has been reported in several different cancers, may be indicative of an insufficient inhibition of matriptase and that this cancer-associated dysregulation of matriptase-mediated proteolysis is causally involved in tumor progression (Oberst et al. 2002, Vogel et al. 2006, Zeng et al. 2005). In support of this hypothesis, mice with transgene-induced matriptase dysregulation in the epidermis develop hyperplasia, dysplasia, and spontaneous malignant conversion into invasive squamous cell carcinoma, and display increased susceptibility to chemical carcinogens (List et al. 2005). The functional relevance of HAI-1 inhibition of matriptase was confirmed by the finding that the oncogenic effects of dysregulated matriptase expression were completely reversed by the simultaneous epidermal transgenic expression of HAI-1 (List et al. 2005). The molecular mechanism underlying matriptase-induced carcinogenesis is not known, however, several different modes of action can be envisioned. Matriptase activity may directly affect the cell microenvironment by processing of extracellular components and cell-matrix adhesion proteins. Matriptase might also act through activation or inactivation of downstream target molecules such as protease zymogens, growth factors, growth factor receptors, and chemokines.

Under normal physiological conditions, matriptase is expressed constitutively at low levels in most epithelia, suggestive of a general role in epithelial morphogenesis and homeostasis (List et al. 2006b, 2007a; Oberst et al. 2003). Studies of matriptase-deficient mice have revealed an essential role of the membrane protease in oral epithelium, epidermis, hair follicles, and thymic epithelium (List et al. 2002, 2003). Matriptase is widely expressed during development, and can be detected as early as embryonic day (E) 10 in the epithelial lining of several tissues of the embryo proper and at day E8.5 in chorionic trophoblasts of the mouse placenta (Fan et al. 2007, Szabo et al. 2007). Matriptase-deficient mice develop to term, showing that the membrane protease does not have critical nonredundant functions in the development of either extra-embryonic or embryonic tissues (List et al. 2002). However, strict regulation of matriptase proteolytic activity is required for placental development and embryo survival, since embryos with a null mutation in the HAI-1 gene die at E10 due to disruption of the chorionic membranes and complete loss of undifferentiated chorionic trophoblasts in the placenta (Fan et al. 2007, Szabo et al. 2007, Tanaka et al. 2005). The finding that concomitant ablation of matriptase expression in HAI-1-deficient embryos allows for normal placentation and development to term verified that the defects in HAI-1-deficient mice are due to dysregulated matriptase activity (Szabo et al. 2007).

Matriptase is indispensable for postnatal life because it has essential nonredundant functions in epidermal differentiation, hair follicle development, and thymocyte survival (List et al. 2002, 2003). Matriptase-deficient mice succumb shortly after birth due to severe defects in the cornified layer of the epidermis and oral epithelium, which leads to the loss of a proper water barrier function and fatal dehydration. When skin from newborn matriptase-deficient mice was grafted onto adult athymic nude mice, it developed an ichthyosis-like phenotype with thick, and dry scales, associated with an almost complete absence of erupted pelage hairs (List et al. 2003). The lack of matriptase perturbs several key processes in the formation of the cornified layer during terminal differentiation, including lipid matrix formation, cornified envelope morphogenesis, and shedding of corneocytes. On the molecular level, matriptase deficiency impairs proteolytic processing of a major epidermal polyprotein, profilaggrin, which results in the loss of filaggrin monomer units and the NH_2-terminal filaggrin S-100 regulatory protein (List et al. 2002, 2003). Interestingly, these findings could be reproduced in a human organotypic skin model using specific siRNA-mediated inhibition of matriptase expression (Mildner et al. 2006).

The molecular pathways through which epidermal proteases, including matriptase, exert their functions are still not yet fully understood. However, increasing evidence suggests that terminal epidermal differentiation is regulated by a complex cascade of serine proteases that undergo sequential activation during the maturation and shedding of the cornified layer. Among the proteases that have been implicated in epidermal differentiation and function are prostasin, furin, calpain, the profilaggrin processing endopeptidase1, and the tissue kallikreins stratum corneum tryptic enzyme (KLK5) and stratum corneum chymotryptic enzyme (KLK7) (Brattsand et al. 2005, Caubet et al. 2004, Egelrud et al. 2005, Kim and Bae 1998, Leyvraz et al. 2005, Pearton et al. 2001, Resing et al. 1993a, Resing et al. 1993b, Zeeuwen

2004). Since matriptase is an auto-activating protease, it may play a key role as an activator of a zymogen-cascade, and a recent study indeed demonstrated that matriptase is the physiological activator of prostasin, a GPI-anchored serine protease that is required for epidermal differentiation, epidermal barrier function and postnatal survival (Leyvraz et al. 2005 Netzel-Arnett et al. 2006).

Recently, the identification of a point mutation in the human matriptase gene was reported in patients suffering from the rare skin and hair disorder Autosomal Recessive Ichthyosis with Hypotrichosis (ARIH) (Basel-Vanagaite et al. 2007). The patients present with dry, scaly skin, and sparse, fragile hair. The epidermis displays hyperkeratosis and at the ultrastructural level an impaired degradation of corneodesmosomes in the cornifed layer was observed, indicative of reduced shedding of corneocytes. Importantly, the ARIH-associated mutation causes a Gly→Arg substitution in residue 827 (G827R) of the matriptase serine protease domain (Basel-Vanagaite et al. 2007). The G827 is a highly conserved residue that is located at the entrance to the primary specificity pocket and the mutation in this residue is analogous to the mutation of Gly216 in tryptase-a, protein C, and trypsin which in all cases causes a reduced catalytic activity (Craik et al. 1985, Marchetti et al. 1993, Selwood et al. 2002). Thus, it is plausible that the symptoms in ARIH patients are caused by reduced matriptase activity, substantiated by recent findings that recombinant G827R matriptase has highly reduced proteolytic activity, and that matriptase hypomorphic mice with very low matriptase mRNA levels phenocopy the key features of ARIH (K. List, T. Bugge, unpublished).

Matriptase-2 and matriptase-3 were named on the basis of their structural similarity to matriptase, but at the present time, the physiological functions of these two proteases are not known. Matriptase-2 is mainly expressed in liver, kidney, and uterus, and expression can be detected during development as early as E7.5 with a peak in expression at day E13.5 (Hooper et al. 2003, Velasco et al. 2002). Matriptase-2 was reported to degrade protein substrates including fibronectin, fibrinogen, type I collagen, and to cleave pro-uPA *in vitro* (Velasco et al. 2002). In a recent study, it was proposed that matriptase-2 overexpression in cultured breast cancer cells inhibits growth and invasion and that high expression of the protease in tumors correlates with favorable prognosis for breast cancer patients (Parr et al. 2007). Furthermore, an SNP in the matriptase-2 gene has been reported to be associated with increased breast cancer risk in a Finnish population (Hartikainen et al. 2006).

Matriptase-3 displays a relatively restricted, species-conserved expression pattern with the highest levels in the brain, skin, reproductive, and oropharyngeal tissues (Szabo et al. 2005). Activated matriptase-3 can form stable inhibitor complexes with an array of serpins, including plasminogen activator inhibitor-1, protein C inhibitor, α1-proteinase inhibitor, α2-antiplasmin, and antithrombin III (Szabo et al. 2005).

Polyserase-1 is a polyprotease that contains three tandem protease domains in a single translation product and undergoes a series of posttranslational processing events to generate three distinct and independent serine protease domains called serase-1, -2, and -3 of which serase-1 and -2 are proteolytically active (Cal et al. 2003). Several splice variants of polyserase-1 have been identified containing one, two, or three serase domains adjacent to the stem region of the protein, and the serase-1B variant contains an extra SEA domain while it lacks the serase-2 and -3 domains. The serase-1B variant activates pro-uPA and the proteolytic activation is negatively regulated by glycosaminoglycans *in vitro* (Okumura et al. 2006). Polyserase-1 is detected in different human tissues, including kidney, liver, lung, and brain, and in a variety of tumor cell lines (Cal et al. 2003).

The Corin Subfamily

Corin was cloned from heart and is primarily expressed in atrial and ventricular cardiomyocytes (Hooper et al. 2000, Yan et al. 1999, 2000). As it is the case for many of the members of the TTSP family, activation cleavage is required to convert the corin zymogen to an active enzyme (Knappe et al. 2003). It was recently demonstrated that N-linked oligosaccharide glycosylation of corin is critical for this process, but the mechanism responsible for the zymogen activation is unknown (Liao et al. 2007). Corin activates the pro-form of atrial natriuretic peptide (ANP), which is a cardiac hormone essential for regulation of water—sodium balance and blood pressure. The physiological significance of corin-mediated pro-ANP activation was later confirmed in studies of corin-deficient mice, demonstrating that corin plays a critical role in cardiovascular homeostasis (Chan et al. 2005, Yan et al. 2000). Corin-deficient mice develop spontaneous hypertension, and display elevated systolic, diastolic, and mean arterial blood pressure as compared to that of control mice. The hypertensive phenotype in ANP-deficient mice is exacerbated when the animals are fed a high-salt diet and during pregnancy (Chan et al. 2005). Also, several lines of experimental evidence point to pro-ANP as the critical substrate for corin: (1) corin is an efficient activator of pro-ANP *in vitro*, (2) corin-deficient mice have elevated levels of pro-ANP but no detectable ANP, (3) Infusion of active recombinant soluble corin transiently restores pro-ANP conversion, resulting in the release of circulating biologically active ANP, and (4) ANP-deficient mice display a phenotype very similar to that of corin-deficient mice. In addition, it was recently reported that the human corin I555/P568 gene allele is associated with an enhanced cardiac hypertrophic response to hypertension in patients with African ancestry, who are not treated with antihypertensive drugs (Chan et al. 2005, Rame et al. 2007, Yan et al. 1999). These findings may be of importance in efforts to design new strategies for treatment of patients with cardiovascular diseases.

Corin may also contribute to the disruption of the water and sodium homeostasis that is frequently observed in patients with lung cancer since it is able to process pro-ANP in small cell lung cancer cells *in vitro* (Wu and Wu, 2003).

Conclusions and Perspectives

Cell surface proteolysis is a complex process involving many different proteases, receptors, and inhibitors. Considerable biochemical and genetic evidence has been generated to include members of the TTSP family as important players in physiological and pathological cell surface proteolysis. Studies of mouse models and knowledge from human disease genomics have demonstrated that the TTSPs have extraordinarily diverse functions in physiological processes, ranging from auditory development, skin maturation, cardiovascular homeostasis, and digestive function. Furthermore, data is accumulating suggesting that several of the TTSPs have functional roles in cancer development and progression, thus presenting exciting future diagnostic, prognostic, and therapeutic opportunities. However, it is clear that, much more information is still needed about the activities, inhibitors, natural substrates, and regulation of these proteases. This era of functional genomics has the potential for the integrated use of bioinformatic analyses, disease association research, knockout and transgenic technologies, and cellular and molecular biology to gain further insights into TTSP functions.

Acknowledgments   The authors thank Drs. Mary Jo Danton, Roman Szabo, and Silvio Gutkind for critically reading the manuscript. This work was supported by the National Institute of Dental and Craniofacial Research Intramural Research Program.

References

Afar, D.E.H., Vivanco, I., Hubert, R.S., Kuo, J., Chen, E., Saffran, D.C., Raitano, A.B., and Jakobovits, A. 2001. Catalytic cleavage of the androgen regulated TMPRSS2 protease results in its secretion by prostate and prostate cancer epithelia. Cancer Research 61(4):1686–1692.

Ahmed, S., Jin, X., Yagi, M., Yasuda, C., Sato, Y., Higashi, S., Lin, C.Y., Dickson, R. B., and Miyazaki, K. 2006. Identification of membrane-bound serine proteinase matriptase as processing enzyme of insulin-like growth factor binding protein-related protein-1 (IGFBP-rP1/ angiomodulin/mac25). The FEBS Journal 273(3):615–627.

Aimes, R.T., Zijlstra, A., Hooper, J.D., Ogbourne, S.M., Sit, M.L., Fuchs, S., Gotley, D.C., Quigley, J.P., and Antalis, T.M. 2003. Endothelial cell serine proteases expressed during vascular morphogenesis and angiogenesis. Thrombosis and Haemostasis 89(3):561–572.

Basel-Vanagaite, L., Attia, R., Ishida-Yamamoto, A., Rainshtein, L., Ben Amitai, D., Lurie, R., Pasmanik-Chor, M., Indelman, M., Zvulunov, A., Saban, S., Magal, N., Sprecher, E., and Shohat, M. 2007. Autosomal recessive ichthyosis with hypotrichosis caused by a mutation in ST14, encoding type II transmembrane serine protease matriptase. American Journal of Human Genetics 80(3):467–477.

Behrens, M., Bufe, B., Schmale, H., and Meyerhof, W. 2004. Molecular cloning and characterisation of DESC4, a new transmembrane serine protease. Cellular and Molecular Life Sciences 61 (22):2866–2877.

Benaud, C., Oberst, M., Hobson, J.P., Spiegel, S., Dickson, R.B., and Lin, C.Y. 2002. Sphingosine 1-phosphate, present in serum-derived lipoproteins, activates matriptase. The Journal of Biological Chemistry 277(12):10539–10546.

Betsunoh, H., Mukai, S., Akiyama, Y., Fukushima, T., Minamiguchi, N., Hasui, Y., Osada, Y., and Kataoka, H. 2007. Clinical relevance of hepsin and hepatocyte growth factor activator inhibitor type 2 expression in renal cell carcinoma. Cancer Science 98(4):491–498.

Bhatt, A.S., Erdjument-Bromage, H., Tempst, P., Craik, C.S., and Moasser, M.M. 2005. Adhesion signaling by a novel mitotic substrate of src kinases. Oncogene 24(34):5333–5343.

Bottcher, E., Matrosovich, T., Beyerle, M., Klenk, H.D., Garten, W., and Matrosovich, M. 2006. Proteolytic activation of influenza viruses by serine proteases TMPRSS2 and HAT from human airway epithelium. Journal of Virology 80(19):9896–9898.

Brattsand, M., Stefansson, K., Lundh, C., Haasum, Y., and Egelrud, T. 2005. A proteolytic cascade of kallikreins in the stratum corneum. The Journal of Investigative Dermatology 124(1):198–203.

Cal, S., Quesada, V., Garabaya, C., and Lopez-Otin, C. 2003. Polyserase-I, a human polyprotease with the ability to generate independent serine protease domains from a single translation product. Proceedings of the National Academy of Sciences of the United States of America 100 (16):9185–9190.

Caubet, C., Jonca, N., Brattsand, M., Guerrin, M., Bernard, D., Schmidt, R., Egelrud, T., Simon, M., and Serre, G. 2004. Degradation of corneodesmosome proteins by two serine proteases of the kallikrein family, SCTE/KLK5/hK5 and SCCE/KLK7/hK7. Journal of Investigative Dermatology 122(5):1235–1244.

Chan, J.C.Y., Knudson, O., Wu, F.Y., Morser, J., Dole, W.P., and Wu, Q.Y. 2005. Hypertension in mice lacking the proatrial natriuretic peptide convertase corin. Proceedings of the National Academy of Sciences of the United States of America 102(3):785–790.

Craik, C.S., Largman, C., Fletcher, T., Roczniak, S., Barr, P.J., Fletterick, R., and Rutter, W.J. 1985. Redesigning trypsin: alteration of substrate specificity. Science 228(4697):291–297.

Crofton, K.M. 2004. Developmental disruption of thyroid hormone: correlations with hearing dysfunction in rats. Risk Analysis 24(6):1665–1671.

Dawelbait, G., Winter, C., Zhang, Y., Pilarsky, C., Grutzmann, R., Heinrich, J.C., and Schroeder, M. 2007. Structural templates predict novel protein interactions and targets from pancreas tumour gene expression data. Bioinformatics 23(13):i115–i124.

Dries, D.L., Victor, R.G., Rame, J.E., Cooper, R.S, Wu, X., Zhu, X., Leonard, D., Ho, S.I., Wu, Q., Post, W., Drazner, M.H. 2005. Corin gene minor allele defined by 2 missense mutations is common in blacks and associated with high blood pressure and hypertension. Circulation 112(16):2403–2410.

Egelrud, T., Brattsand, M., Kreutzmann, P., Walden, M., Vitzithum, K., Marx, U.C., Forssmann, W.G., and Magert, H.J. 2005. hK5 and hK7, two serine proteinases abundant in human skin, are inhibited by LEKTI domain 6. The British Journal of Dermatology 153(6):1200–1203.

Elbracht, M., Senderek, J., Eggermann, T., Thurmer, C., Park, J., Westhofen, M., and Zerres, K. 2007. Autosomal recessive postlingual hearing loss (DFNB8): compound heterozygosity for two novel TMPRSS3 mutations in German siblings. Journal of Medical Genetics 44(6):e81.

Fan, B., Brennan, J., Grant, D., Peale, F., Rangell, L., and Kirchhofer, D. 2007. Hepatocyte growth factor activator inhibitor-1 (HAI-1) is essential for the integrity of basement membranes in the developing placental labyrinth. Developmental Biology 303(1):222–230.

Forbs, D., Thiel, S., Stella, M.C., Sturzebecher, A., Schweinitz, A., Steinmetzer, T., Sturzebecher, J., and Uhland, K. 2005. In vitro inhibition of matriptase prevents invasive growth of cell lines of prostate and colon carcinoma. International Journal of Oncology 27(4):1061–1070.

Galkin, A.V., Mullen, L., Fox, W.D., Brown, J., Duncan, D., Moreno, O., Madison, E.L., and Agus, D.B. 2004. CVS-3983, a selective matriptase inhibitor, suppresses the growth of androgen independent prostate tumor xenografts. The Prostate 61(3):228.

Grant, D. 1986. Acute necrotising pancreatitis—a role for enterokinase. International Journal of Pancreatology 1(3–4):167–183.

Guipponi, M., Vuagniaux, G., Wattenhofer, M., Shibuya, K., Vazquez, M., Dougherty, L., Scamuffa, N., Guida, E., Okui, M., Rossier, C., Hancock, M., Buchet, K., Reymond, A., Hummler, E., Marzella, P.L., Kudoh, J., Shimizu, N., Scott, H.S., Antonarakis, S.E., and Rossier, B.C. 2002. The transmembrane serine protease (TMPRSS3) mutated in deafness DFNB8/10 activates the epithelial sodium channel (ENaC) in vitro. Human Molecular Genetics 11(23):2829–2836.

Guipponi, M., Tan, J., Cannon, P.Z., Donley, L., Crewther, P., Clarke, M., Wu, Q., Shepherd, R.K., and Scott, H.S. 2007. Mice deficient for the type II transmembrane serine protease, TMPRSS1/ hepsin, exhibit profound hearing loss. The American Journal of Pathology.

Hartikainen, J.M., Tuhkanen, H., Kataja, V., Eskelinen, M., Uusitupa, M., Kosma, V.M., and Mannermaa, A. 2006. Refinement of the 22q12-q13 breast cancer-associated region: evidence of TMPRSS6 as a candidate gene in an eastern Finnish population. Clinical Cancer Research 12(5):1454–1462.

Herter, S., Piper, D.E., Aaron, W., Gabriele, T., Cutler, G., Cao, P., Bhatt, A.S., Choe, Y., Craik, C.S., Walker, N., Meininger, D., Hoey, T., and Austin, R.J. 2005. Hepatocyte growth factor is a preferred *in vitro* substrate for human hepsin, a membrane-anchored serine protease implicated in prostate and ovarian cancers. Biochemical Journal 390:125–136.

Hobson, J.P., Netzel-Arnett, S., Szabo, R., Rehault, S.M., Church, F.C., Strickland, D.K., Lawrence, D.A., Antalis, T.M., and Bugge, T.H. 2004. Mouse DESC1 is located within a cluster of seven DESC1-like genes and encodes a type II transmembrane serine protease that forms serpin inhibitory complexes. Journal of Biological Chemistry 279(45):46981–46994.

Holzinger, A., Maier, E.M., Buck, C., Mayerhofer, P.U., Kappler, M., Haworth, J.C., Moroz, S.P., Hadorn, H.B., Sadler, J.E., and Roscher, A.A. 2002. Mutations in the proenteropeptidase gene are the molecular cause of congenital enteropeptidase deficiency. American Journal of Human Genetics 70(1):20–25.

Hooper, J.D., Scarman, A.L., Clarke, B.E., Normyle, J.F., and Antalis, T.M. 2000. Localization of the mosaic transmembrane serine protease corin to heart myocytes. European Journal of Biochemistry 267(23):6931–6937.

Hooper, J.D., Clements, J.A., Quigley, J.P., and Antalis, T.M. 2001. Type II transmembrane serine proteases—Insights into an emerging class of cell surface proteolytic enzymes. Journal of Biological Chemistry 276(2):857–860.

Hooper, J.D., Campagnolo, L., Goodarzi, G., Truong, T.N., Stuhlmann, H., and Quigley, J.P. 2003. Mouse matriptase-2: identification, characterization and comparative mRNA expression analysis with mouse hepsin in adult and embryonic tissues. Biochemical Journal 373:689–702.

Iacobuzio-Donahue, C.A., Maitra, A., Olsen, M., Lowe, A.W., Van Heek, N.T., Rosty, C., Walter, K., Sato, N., Parker, A., Ashfaq, R., Jaffee, E., Ryu, B., Jones, J., Eshleman, J.R., Yeo, C.J., Cameron, J.L., Kern, S.E., Hruban, R.H., Brown, P.O., and Goggins, M. 2003. Exploration of global gene expression patterns in pancreatic adenocarcinoma using cDNA microarrays. American Journal of Pathology 162(4):1151–1162.

Jacquinet, E., Rao, N.V., Rao, G.V., and Hoidal, J.R. 2000. Cloning, genomic organization, chromosomal assignment and expression of a novel mosaic serine proteinase: epitheliasin. FEBS Letters 468(1):93–100.

Kazama, Y., Hamamoto, T., Foster, D.C., and Kisiel, W. 1995. Hepsin, a putative membrane-associated serine-protease, activates human factor-Vii and initiates a pathway of blood-coagulation on the cell-surface leading to thrombin formation. Journal of Biological Chemistry 270(1):66–72.

Kebebew, E., Peng, M., Reiff, E., Duh, Q.Y., Clark, O.H., and McMillan, A. 2005. ECM1 and TMPRSS4 are diagnostic markers of malignant thyroid neoplasms and improve the accuracy of fine needle aspiration biopsy. Annals of Surgery 242(3):353–363.

Kebebew, E., Peng, M., Reiff, E., and McMillan, A. 2006. Diagnostic and extent of disease multigene assay for malignant thyroid neoplasms. Cancer 106(12):2592–2597.

Kim, S.Y., and Bae, C.D. 1998. Calpain inhibitors reduce the cornified cell envelope formation by inhibiting proteolytic processing of transglutaminase 1. Experimental & Molecular Medicine 30(4):257–262.

Kim, M.G., Chen, C., Lyu, M.S., Cho, E.G., Park, D., Kozak, C., and Schwartz, R.H. 1999. Cloning and chromosomal mapping of a gene isolated from thymic stromal cells encoding a new mouse type II membrane serine protease, epithin, containing four LDL receptor modules and two CUB domains. Immunogenetics 49(5):420–428.

Kim, D.R., Sharmin, S., Inoue, M., and Kido, H. 2001. Cloning and expression of novel mosaic serine proteases with and without a transmembrane domain from human lung. Biochimica Et Biophysica Acta-Gene Structure and Expression 1518(1–2):204–209.

Kim, T.S., Heinlein, C., Hackman, R.C., and Nelson, P.S. 2006. Phenotypic analysis of mice lacking the Tmprss2-encoded protease. Molecular and Cellular Biology 26(3):965–975.

Kirchhofer, D., Peek, M., Lipari, M.T., Billeci, K., Fan, B., and Moran, P. 2005. Hepsin activates pro-hepatocyte growth factor and is inhibited by hepatocyte growth factor activator inhibitor-1B (HAI-1B) and HAI-2. FEBS Letters 579(9):1945–1950.

Kitamoto, Y., Yuan, X., Wu, Q.Y., Mccourt, D.W., and Sadler, J.E. 1994. Enterokinase, the initiator of intestinal digestion, is a mosaic protease composed of a distinctive assortment of domains. Proceedings of the National Academy of Sciences of the United States of America 91 (16):7588–7592.

Kiyomiya, K., Lee, M.S., Tseng, I.C., Zuo, H., Barndt, R.J., Johnson, M.D., Dickson, R.B., and Lin, C.Y. 2006. Matriptase activation and shedding with HAI-1 is induced by steroid sex hormones in human prostate cancer cells, but not in breast cancer cells. American Journal of Physiology. Cell Physiology 291(1):C40–49.

Klezovitch, O., Chevillet, J., Mirosevich, J., Roberts, R.L., Matusik, R.J., and Vasioukhin, V. 2004. Hepsin promotes prostate cancer progression and metastasis. Cancer Cell 6(2):185–195.

Knappe, S., Wu, F., Masikat, M.R., Morser, J., and Wu, Q. 2003. Functional analysis of the transmembrane domain and activation cleavage of human corin: design and characterization of a soluble corin. Journal of Biological Chemistry 278(52):52363–52370.

Kunii, D., Shimoji, M., Nakama, S., Ikebe, M., Hachiman, T., Sato, I., Tamaki, A., Yamazaki, K., and Aniya, Y. 2006. Purification of liver serine protease which activates microsomal glutathione S-transferase: possible involvement of hepsin. Biological . . . Pharmaceutical Bulletin 29 (5):868–874.

Kyrieleis, O.J., Huber, R., Ong, E., Oehler, R., Hunter, M., Madison, E.L., and Jacob, U. 2007. Crystal structure of the catalytic domain of DESC1, a new member of the type II transmembrane serine proteinase family. FEBS Journal 274(8):2148–2160.

Landers, K.A., Burger, M.J., Tebay, M.A., Purdie, D.M., Scells, B., Samaratunga, H., Lavin, M.F., and Gardiner, R.A. 2005. Use of multiple biomarkers for a molecular diagnosis of prostate cancer. International Journal of Cancer 114(6):950–956.

Lang, J.C., and Schuller, D.E. 2001. Differential expression of a novel serine protease homologue in squamous cell carcinoma of the head and neck. British Journal of Cancer 84(2):237–243.

Lebenthal, E., Antonowicz, I., and Shwachman, H. 1976. Enterokinase and trypsin activities in pancreatic insufficiency and diseases of the small intestine. Gastroenterology 70(4):508–512.

Lee, S.L., Dickson, R.B., and Lin, C.Y. 2000. Activation of hepatocyte growth factor and urokinase/plasminogen activator by matriptase, an epithelial membrane serine protease. Journal of Biological Chemistry 275(47):36720–36725.

Leytus, S.P., Loeb, K.R., Hagen, F.S., Kurachi, K., and Davie, E.W. 1988. A Novel trypsin-like serine protease (Hepsin) with a putative transmembrane domain expressed by human-liver and hepatoma-cells. Biochemistry 27(3):1067–1074.

Leyvraz, C., Charles, R.P., Rubera, I., Guitard, M., Rotman, S., Breiden, B., Sandhoff, K., and Hummler, E. 2005. The epidermal barrier function is dependent on the serine protease CAP1/Prss8. Journal of Cell Biology 170(3):487–496.

Liao, X., Wang, W., Chen, S., and Wu, Q. 2007. Role of glycosylation in corin zymogen activation. Journal of Biological Chemistry 282(38):27728–27735.

Lin, C.Y., Wang, J.K., Torri, J., Dou, L., Sang, Q.X.A., and Dickson, R.B. 1997. Characterization of a novel, membrane-bound, 80-kDa matrix-degrading protease from human breast cancer cells—Monoclonal antibody production, isolation, and localization. Journal of Biological Chemistry 272(14):9147–9152.

Lin, B.Y., Ferguson, C., White, J.T., Wang, S.Y., Vessella, R., True, L.D., Hood, L., and Nelson, P.S. 1999a. Prostate-localized and androgen-regulated expression of the membrane-bound serine protease TMPRSS2. Cancer Research 59(17):4180–4184.

Lin, C.Y., Anders, J., Johnson, M., and Dickson, R.B. 1999b. Purification and characterization of a complex containing matriptase and a Kunitz-type serine protease inhibitor from human milk. Journal of Biological Chemistry 274(26):18237–18242.

Lin, C.Y., Anders, J., Johnson, M., Sang, Q.A., and Dickson, R.B. 1999c. Molecular cloning of cDNA for matriptase, a matrix-degrading serine protease with trypsin-like activity. Journal of Biological Chemistry 274(26):18231–18236.

List, K., Haudenschild, C.C., Szabo, R., Chen, W.J., Wahl, S.M., Swaim, W., Engelholm, L.H., Behrendt, N., and Bugge, T.H. 2002. Matriptase/MT-SP1 is required for postnatal survival, epidermal barrier function, hair follicle development, and thymic homeostasis. Oncogene 21 (23):3765–3779.

List, K., Szabo, R., Wertz, P.W., Segre, J., Haudenschild, C.C., Kim, S.Y., and Bugge, T.H. 2003. Loss of proteolytically processed filaggrin caused by epidermal deletion of matriptase/MT-SP1. Journal of Cell Biology 163(4):901–910.

List, K., Szabo, R., Molinolo, A., Sriuranpong, V., Redeye, V., Murdock, T., Burke, B., Nielsen, B. S., Gutkind, J.S., and Bugge, T.H. 2005. Deregulated matriptase causes ras-independent multistage carcinogenesis and promotes ras-mediated malignant transformation. Genes ... Development 19(16):1934–1950.

List, K., Bugge, T.H., and Szabo, R. 2006a. Matriptase: potent proteolysis on the cell surface. Molecular Medicine 12(1–3):1–7.

List, K., Szabo, R., Molinolo, A., Nielsen, B.S., and Bugge, T.H. 2006b. Delineation of matriptase protein expression by enzymatic gene trapping suggests diverging roles in barrier function, hair formation, and squamous cell carcinogenesis. American Journal of Pathology 168(5):1513–1525.

List, K., Hobson, J.P., Molinolo, A., and Bugge, T.H. 2007. Co-localization of the channel activating protease prostasin/(CAP1/PRSS8) with its candidate activator, matriptase. Journal of Cell Physiology 213(1):237–245.

List, K., Currie, B., Scharschmidt, T.C., Szabo, R., Shireman, J., Molinolo, A., Cravatt, B.F., Segre, J. and Bugge, T.H. 2007b. Autosomal Ichthyosis with Hypotrichosis Syndrome displays low matriptase proteolytic activity and is phenocopied in ST14 hypomorphic mice. Journal of Biological Chemistry 282(20):36714–36723.

Magee, J.A., Araki, T., Patil, S., Ehrig, T., True, L., Humphrey, P.A., Catalona, W.J., Watson, M. A., and Milbrandt, J. 2001. Expression profiling reveals hepsin overexpression in prostate cancer. Cancer Research 61(15):5692–5696.

Marchetti, G., Patracchini, P., Gemmati, D., Castaman, G., Rodeghiero, F., Wacey, A., Cooper, D. N., Tuddenham, E.G., and Bernardi, F. 1993. Symptomatic type II protein C deficiency caused by a missense mutation (Gly 381→Ser) in the substrate-binding pocket. British Journal of Haematology 84(2):285–289.

Masmoudi, S., Antonarakis, S.E., Schwede, T., Ghorbel, A.M., Gratri, M., Pappasavas, M.P., Drira, M., Elgaied-Boutila, A., Wattenhofer, M., Rossier, C., Scott, H.S., Ayadi, H., and Guipponi, M. 2001. Novel missense mutations of TMPRSS3 in two consanguineous Tunisian families with non-syndromic autosomal recessive deafness. Human Mutation 18(2):101–108.

Matsushima, R., Takahashi, A., Nakaya, Y., Maezawa, H., Miki, M., Nakamura, Y., Ohgushi, F., and Yasuoka, S. 2006. Human airway trypsin-like protease stimulates human bronchial fibroblast proliferation in a protease-activated receptor-2-dependent pathway. American Journal of Physiology-Lung Cellular and Molecular Physiology 290(2):L385–L395.

Miki, M., Nakamura, Y., Takahashi, A., Nakaya, Y., Eguchi, H., Masegi, T., Yoneda, K., Yasuoka, S., and Sone, S. 2003. Effect of human airway trypsin-like protease on intracellular free Ca2+ concentration in human bronchial epithelial cells. Journal of Medical Investigation 50(1–2): 95–107.

Mildner, M., Ballaun, C., Stichenwirth, M., Bauer, R., Gmeiner, R., Buchberger, M., Mlitz, V., and Tschachler, E. 2006. Gene silencing in a human organotypic skin model. Biochemical and Biophysical Research Communications 348(1):76–82.

Moran, P., Li, W., Fan, B., Vij, R., Eigenbrot, C., and Kirchhofer, D. 2006. Pro-urokinase-type plasminogen activator is a substrate for hepsin. Journal of Biological Chemistry 281(41): 30439–30446.

Moroz, S.P., Hadorn, B., Rossi, T.M., and Haworth, J.C. 2001. Celiac disease in a patient with a congenital deficiency of intestinal enteropeptidase. American Journal of Gastroenterology 96 (7):2251–2254.

Netzel-Arnett, S., Hooper, J.D., Szabo, R., Madison, E.L., Quigley, J.P., Bugge, T.H., and Antalis, A.M. 2003. Membrane anchored serine proteases: a rapidly expanding group of cell surface proteolytic enzymes with potential roles in cancer. Cancer and Metastasis Reviews 22(2–3): 237–258.

Netzel-Arnett, S., Currie, B.M., Szabo, R., Lin, C.Y., Chen, L.M., Chai, K.X., Antalis, T.M., Bugge, T.H., and List, K. 2006. Evidence for a matriptase-prostasin proteolytic cascade regulating terminal epidermal differentiation. Journal of Biological Chemistry 281 (44):32941–32945.

Oberst, M.D., Johnson, M.D., Dickson, R.B., Lin, C.Y., Singh, B., Stewart, M., Williams, A., al-Nafussi, A., Smyth, J.F., Gabra, H., and Sellar, G.C. 2002. Expression of the serine protease matriptase and its inhibitor HAI-1 in epithelial ovarian cancer: correlation with clinical outcome and tumor clinicopathological parameters. Clinical Cancer Research 8(4):1101–1107.

Oberst, M.D., Singh, B., Ozdemirli, M., Dickson, R.B., Johnson, M.D., and Lin, C.Y. 2003. Characterization of matriptase expression in normal human tissues. Journal of Histochemistry and Cytochemistry 51(8):1017–1025.

Okumura, Y., Hayama, M., Takahashi, E., Fujiuchi, M., Shimabukuro, A., Yano, M., and Kido, H. 2006. Serase-1B, a new splice variant of polyserase-1/TMPRSS9, activates urokinase-type plasminogen activator and the proteolytic activation is negatively regulated by glycosaminoglycans. Biochemical Journal 400:551–561.

Pal, P., Xi, H., Kaushal, R., Sun, G., Jin, C.H., Jin, L., Suarez, B.K., Catalona, W.J., and Deka, R. 2006. Variants in the HEPSIN gene are associated with prostate cancer in men of European origin. Human Genetics 120(2):187–192.

Paoloni-Giacobino, A., Chen, H., Peitsch, M.C., Rossier, C., and Antonarakis, S.E. 1997. Cloning of the TMPRSS2 gene, which encodes a novel serine protease with transmembrane, LDLRA, and SRCR domains and maps to 21q22.3. Genomics 44(3):309–320.

Parr, C., Sanders, A.J., Davies, G., Martin, T., Lane, J., Mason, M.D., Mansel, R.E., and Jiang, W. G. 2007. Matriptase-2 inhibits breast tumor growth and invasion and correlates with favorable prognosis for breast cancer patients. Clinical Cancer Research 13(12):3568–3576.

Pearton, D.J., Nirunsuksiri, W., Rehemtulla, A., Lewis, S.P., Presland, R.B., and Dale, B.A. 2001. Proprotein convertase expression and localization in epidermis: evidence for multiple roles and substrates. Experimental Dermatology 10(3):193–203.

Planes, C., and Caughey, G.H. 2007. Regulation of the epithelial Na+ channel by peptidases. Current Topics in Developmental Biology 78:23–46.

Qiu, D., Owen, K., Gray, K., Bass, R., and Ellis, V. 2007. Roles and regulation of membrane-associated serine proteases. Biochemical Society Transactions 35(Pt 3):583–587.

Rame, J.E., Drazner, M.H., Post, W., Peshock, R., Lima, J., Cooper, R.S., and Dries, D.L. 2007. Corin I555(P568) allele is associated with enhanced cardiac hypertrophic response to increased systemic afterload. Hypertension 49(4):857–864.

Resing, K.A., al-Alawi, N., Blomquist, C., Fleckman, P., and Dale, B.A. 1993a. Independent regulation of two cytoplasmic processing stages of the intermediate filament-associated protein filaggrin and role of Ca2+ in the second stage. Journal of Biological Chemistry 268 (33):25139–25145.

Resing, K.A., Johnson, R.S., and Walsh, K.A. 1993b. Characterization of protease processing sites during conversion of rat profilaggrin to filaggrin. Biochemistry 32(38):10036–10045.

Riddick, A.C.P., Shukla, C.J., Pennington, C.J., Bass, R., Nuttall, R.K., Hogan, A., Sethia, K.K., Ellis, V., Collins, A.T., Maitland, N.J., Ball, R.Y., and Edwards, D.R. 2005. Identification of degradome components associated with prostate cancer progression by expression analysis of human prostatic tissues. British Journal of Cancer 92(12):2171–2180.

Rubin, M.A., and Chinnaiyan, A.M. 2006. Bioinformatics approach leads to the discovery of the TMPRSS2:ETS gene fusion in prostate cancer. Laboratory Investigation 86(11):1099–1102.

Rutgeert, L., and Eggermon, E. 1976. Human enterokinase. Tijdschr Gastroenterol 19(4):231–246.

Rutgeerts, L., Tytgat, G., and Eggermont, E. 1972. Human Enterokinase. Tijdschr Gastroenterol. 15(6):379–384.

Sanders, A.J., Parr, C., Davies, G., Martin, T.A., Lane, J., Mason, M.D., and Jiang, W.G. 2006. Genetic reduction of matriptase-1 expression is associated with a reduction in the aggressive phenotype of prostate cancer cells *in vitro* and *in vivo*. Journal of Experimental Therapeutics & Oncology 6(1):39–48.

Satomi, S., Yamasaki, Y., Tsuzuki, S., Hitomi, Y., Iwanaga, T., and Fushiki, T. 2001. A role for membrane-type serine protease (MT-SP1) in intestinal epithelial turnover. Biochemical and Biophysical Research Communications 287(4):995–1002.

Sawasaki, T., Shigemasa, K., Gu, L.J., Beard, J.B., and O'Brien, T.J. 2004. The transmembrane protease serine (TMPRSS3/TADG-12) D variant: a potential candidate for diagnosis and therapeutic intervention in ovarian cancer. Tumor Biology 25(3):141–148.

Sedghizadeh, P.P., Mallery, S.R., Thompson, S.J., Kresty, L., Beck, F.M., Parkinson, E.K., Biancamano, J., and Lang, J.C. 2006. Expression of the serine protease Desc1 correlates directly with normal keratinocyte differentiation and inversely with head and neck squamous cell carcinoma progression. Head and Neck-Journal for the Sciences and Specialties of the Head and Neck 28(5):432–440.

Selwood, T., Wang, Z.-M., McCaslin, D.R., and Schechter, N.M. 2002. Diverse stability and catalytic properties of human tryptase alpha and beta isoforms are mediated by residue differences at the S1 pocket. Biochemistry 41(10):3329–3340.

Somoza, J.R., Ho, J.D., Luong, C., Ghate, M., Sprengeler, P.A., Mortara, K., Shrader, W.D., Sperandio, D., Chan, H., McGrath, M.E., and Katz, B.A. 2003. The structure of the extracellular region of human hepsin reveals a serine protease domain and a novel scavenger receptor cysteine-rich (SRCR) domain. Structure 11(9):1123–1131.

Stephan, C., Yousef, G.M., Scorilas, A., Jung, K., Jung, M., Kristiansen, G., Hauptmann, S., Kishi, T., Nakamura, T., Loening, S.A., and Diamandis, E.P. 2004. Hepsin is highly over expressed in and a new candidate for a prognostic indicator in prostate cancer. Journal of Urology 171(1):187–191.

Szabo, R., and Bugge, T.H. 2007. Type II transmembrane serine proteases in development and disease. The International Journal of Biochemistry and Cell Biology, Int J Biochem Cell Biol. 40(6-7): 1297–1316.

Szabo, R., Wu, Q.Y., Dickson, R.B., Netzel-Arnett, S., Antalis, T.M., and Bugge, T.H. 2003. Type II transmembrane serine proteases. Thrombosis and Haemostasis 90(2):185–193.

Szabo, R., Netzel-Arnett, S., Hobson, J.P., Antalis, T.M., and Bugge, T.H. 2005. Matriptase-3 is a novel phylogenetically preserved membrane-anchored serine protease with broad serpin reactivity. The Biochemical Journal 390(Pt 1):231–242.

Szabo, R., Molinolo, A., List, K., Bugge, T.H. 2007. Matriptase inhibition by hepatocyte growth factor activator inhibitor-1 is essential for placental development. Oncogene 26(11):1546–1556.

Takeuchi, T., Harris, J.L., Huang, W., Yan, K.W., Coughlin, S.R., Craik, C.S. 2000. Cellular localization of membrane-type serine protease 1 and identification of protease-activated receptor-2 and single-chain urokinase-type plasminogen activator as substrates. Journal of Biological Chemistry 275(34):26333–26342.

Tanaka, H., Nagaike, K., Takeda, N., Itoh, H., Kohama, K., Fukushima, T., Miyata, S., Uchiyama, S., Uchinokura, S., Shimomura, T., Miyazawa, K., Kitamura, N., Yamada, G., and Kataoka, H. 2005. Hepatocyte growth factor activator inhibitor type 1 (HAI-1) is required for branching morphogenesis in the chorioallantoic placenta. Molecular and Cellular Biology 25(13):5687–5698.

Tanimoto, H., Yan, Y., Clarke, J., Korourian, S., Shigemasa, K., Parmley, T.H., Parham, G.P., and OBrien, T.J. 1997. Hepsin, a cell surface serine protease identified in hepatoma cells, is overexpressed in ovarian cancer. Cancer Research 57(14):2884–2887.

Tomlins, S.A., Laxman, B., Dhanasekaran, S.M., Helgeson, B.E., Cao, X., Morris, D.S., Menon, A., Jing, X., Cao, Q., Han, B., Yu, J., Wang, L., Montie, J.E., Rubin, M.A., Pienta, K.J., Roulston, D., Shah, R.B., Varambally, S., Mehra, R., and Chinnaiyan, A.M. 2007. Distinct classes of chromosomal rearrangements create oncogenic ETS gene fusions in prostate cancer. Nature 448(7153):595–599.

Torres-Rosado, A., O'Shea, K.S., Tsuji, A., Chou, S.H., and Kurachi, K. 1993. Hepsin, a putative cell-surface serine protease, is required for mammalian cell growth. Proceedings of the National Academy of Sciences of the United States of America 90(15):7181–7185.

Tsuji, A., Torresrosado, A., Arai, T., Lebeau, M.M., Lemons, R.S., Chou, S.H., and Kurachi, K. 1991. Hepsin, a cell membrane-associated protease—characterization, tissue distribution, and gene localization. Journal of Biological Chemistry 266(25):16948–16953.

Uhland, K. 2006. Matriptase and its putative role in cancer. Cellular and Molecular Life Sciences 63(24):2968–2978.

Vaarala, M.H., Porvari, K., Kyllonen, A., Lukkarinen, O., and Vihko, P. 2001a. The TMPRSS2 gene encoding transmembrane serine protease is overexpressed in a majority of prostate cancer patients: detection of mutated TMPRSS2 form in a case of aggressive disease. International Journal of Cancer 94(5):705–710.

Vaarala, M.H., Porvari, K.S., Kellokumpu, S., Kyllonen, A.P., and Vihko, P.T. 2001b. Expression of transmembrane serine protease TMPRSS2 in mouse and human tissues. Journal of Pathology 193(1):134–140.

Velasco, G., Cal, S., Quesada, V., Sanchez, L.M., and Lopez-Otin, C. 2002. Matriptase-2, a membrane-bound mosaic serine proteinase predominantly expressed in human liver and showing degrading activity against extracellular matrix proteins. Journal of Biological Chemistry 277(40):37637–37646.

Vogel, L.K., Saebo, M., Skjelbred, C.F., Abell, K., Pedersen, E.D.K., Vogel, U., and Kure, E.H. 2006. The ratio of Matriptase/HAI-1 mRNA is higher in colorectal cancer adenomas and carcinomas than corresponding tissue from control individuals. BMC Cancer 6.

Wallrapp, C., Hahnel, S., Muller-Pillasch, F., Burghardt, B., Iwamura, T., Ruthenburger, M., Lerch, M.M., Adler, G., and Gress, T.M. 2000. A novel transmembrane serine protease (TMPRSS3) overexpressed in pancreatic cancer. Cancer Research 60(10):2602–2606.

Watanabe, Y., Okui, A., Mitsui, S., Kawarabuki, K., Yamaguchi, T., Uemura, H., and Yamaguchi, N. 2004. Molecular cloning and tissue-specific expression analysis of mouse spinesin, a type II transmembrane serine protease 5. Biochemical and Biophysical Research Communications 324(1):333–340.

Wattenhofer, M., Di Iorio, M.V., Rabionet, R., Dougherty, L., Pampanos, A., Schwede, T., Montserrat-Sentis, B., Arbones, M.L., Iliades, T., Pasquadibisceglie, A., D'Amelio, M., Alwan, S., Rossier, C., Dahl, H.H.M., Petersen, M.B., Estivill, X., Gasparini, P., Scott, H.S., and Antonarakis, S.E. 2002. Mutations in the TMPRSS3 gene are a rare cause of childhood nonsyndromic deafness in Caucasian patients. Journal of Molecular Medicine 80(2):124–131.

Wattenhofer, M., Sahin-Calapoglu, N., Andreasen, D., Kalay, E., Caylan, R., Braillard, B., Fowler-Jaeger, N., Reymond, A., Rossier, B.C., Karaguzel, A., and Antonarakis, S.E. 2005. A novel TMPRSS3 missense mutation in a DFNB8/10 family prevents proteolytic activation of the protein. Human Genetics 117(6):528–535.

Wilson, S., Greer, B., Hooper, J., Zijlstra, A., Walker, B., Quigley, J., and Hawthorne, S. 2005. The membrane-anchored serine protease, TMPRSS2, activates PAR-2 in prostate cancer cells. Biochemical Journal 388(Pt 3):967–972.

Wu, Q., Parry, G. 2007. Hepsin and prostate cancer. Front Bioscience 12:5052–5059.

Wu, F., Wu, Q. 2003. Corin-mediated processing of pro-atrial natriuretic peptide in human small cell lung cancer cells. Cancer Research 63(23):8318–8322.

Wu, Q.Y., Yu, D.Y., Post, J., Halks-Miller, M., Sadler, J.E., and Morser, J. 1998. Generation and characterization of mice deficient in hepsin, a hepatic transmembrane serine protease. Journal of Clinical Investigation 101(2):321–326.

Yamaguchi, N., Okui, A., Yamada, T., Nakazato, H., and Mitsui, S. 2002. Spinesin/TMPRSS5, a novel transmembrane serine protease, cloned from human spinal cord. Journal of Biological Chemistry 277(9):6806–6812.

Yamaoka, K., Masuda, K., Ogawa, H., Takagi, K., Umemoto, N., and Yasuoka, S. 1998. Cloning and characterization of the cDNA for human airway trypsin-like protease. Journal of Biological Chemistry 273(19):11895–11901.

Yan, W., Sheng, N., Seto, M., Morser, J., and Wu, Q.Y. 1999. Corin, a mosaic transmembrane serine protease encoded by a novel cDNA from human heart. Journal of Biological Chemistry 274(21):14926–14935.

Yan, W., Wu, F.Y., Morser, J., and Wu, Q.Y. 2000. Corin, a transmembrane cardiac serine protease, acts as a pro-atrial natriuretic peptide-converting enzyme. Proceedings of the National Academy of Sciences of the United States of America 97(15):8525–8529.

Yasuoka, S., Ohnishi, T., Kawano, S., Tsuchihashi, S., Ogawara, M., Masuda, K., Yamaoka, K., Takahashi, M., and Sano, T. 1997. Purification, characterization, and localization of a novel trypsin-like protease found in the human airway. American Journal of Respiratory Cell and Molecular Biology 16(3):300–308.

Yu, I.S., Chen, H.J., Lee, Y.S.E., Huang, P.H., Lin, S.R., Tsai, T.W., and Lin, S.W. 2000. Mice deficient in hepsin, a serine protease, exhibit normal embryogenesis and unchanged hepatocyte regeneration ability. Thrombosis and Haemostasis 84(5):865–870.

Yuan, X., Zheng, X.L., Lu, D.S., Rubin, D.C., Pung, C.Y.M., and Sadler, J.E. 1998. Structure of murine enterokinase (enteropeptidase) and expression in small intestine during development. American Journal of Physiology-Gastrointestinal and Liver Physiology 37(2):G342–G349.

Zeeuwen, P.L. 2004. Epidermal differentiation: the role of proteases and their inhibitors. European Journal of Cell Biology 83(11–12):761–773.

Zeng, L., Cao, J., and Zhang, X. 2005. Expression of serine protease SNC19/matriptase and its inhibitor hepatocyte growth factor activator inhibitor type 1 in normal and malignant tissues of gastrointestinal tract. World Journal of Gastroenterology 11(39):6202–6207.

Chapter 15
Roles of Cysteine Proteases in Tumor Progression: Analysis of Cysteine Cathepsin Knockout Mice in Cancer Models

Thomas Reinheckel*, Vasilena Gocheva*, Christoph Peters, and Johanna A. Joyce

Abstract Cysteine cathepsins are a family of proteases that are frequently upregulated in various human cancers, including breast, prostate, lung, and brain. Indeed, elevated expression and/or activity of certain cysteine cathepsins correlates with increased malignancy and poor patient prognosis. In normal cells, cysteine cathepsins are typically localized in lysosomes and other intracellular compartments, and are involved in protein degradation and processing. However, in certain diseases such as cancer, cysteine cathepsins are translocated from their intracellular compartments to the cell surface and can be secreted into the extracellular milieu. Pharmacological studies and in vitro experiments have suggested general roles for the cysteine cathepsin family in distinct tumorigenic processes such as angiogenesis, proliferation, apoptosis, and invasion. Understanding which individual cathepsins are the key mediators, what their substrates are, and how they may be promoting these complex roles in cancer are important questions to address. Here, we discuss recent results that begin to answer some of these questions, illustrating in particular the lessons learned from studying several mouse models of multistage carcinogenesis, which have identified distinct, tissue-specific roles for individual cysteine cathepsins in tumor progression.

T. Reinheckel, C. Peters
Institut für Molekulare Medizin und Zellforschung, Albert-Ludwigs-Universität, D-79104 Freiburg, Germany

V. Gocheva, J.A. Joyce
Cancer Biology and Genetics Program, Memorial Sloan Kettering Cancer Center, New York, NY 10021, USA

* These authors made an equal contribution

D. Edwards et al. (eds.), *The Cancer Degradome.*

Introduction

Proteolytic degradation is critically important during tumor development to facilitate angiogenesis, invasion, and metastasis, biological processes known to require localized matrix remodeling. A broader role for proteolysis in other aspects of tumor development, including promotion of cell proliferation and resistance to apoptosis, is increasingly recognized, emphasizing the importance of identifying the enzymes involved and understanding their mechanism of action. Aspartic, cysteine, serine, threonine, and matrix metalloprotease (MMP) families have been evaluated in multiple cancer types, with the major focus having been on the MMPs. Recently, however, cathepsins of the cysteine protease family have received increased attention for their proposed roles in angiogenesis, apoptosis, cell proliferation, and tumor invasion (Mohamed and Sloane 2006, Gocheva and Joyce 2007). In this chapter, we discuss recent analyses of the effects of cysteine cathepsin deletion in several mouse models of multistage carcinogenesis, which have uncovered distinct, tissue-specific roles for individual cysteine cathepsins in tumorigenesis.

Cysteine cathepsins are a group of enzymes that belong to the papain family of cysteine proteases. There are 11 members in humans and 18 in mice (Table 15.1), all of which share a conserved catalytic site formed by cysteine, histidine, and asparagine residues (Lecaille et al. 2002). Despite similarities in sequence and fold, cysteine cathepsin family members differ among each other in specificity. While most of them are endopeptidases, some possess exopeptidase activity. For example, cathepsin X is a carboxypeptidase, while cathepsin C can cleave dipeptides from the N-terminus of its protein substrates. Cathepsins B and H, however, have both endo- and exopeptidase activity. This difference in specificity among family members could influence their individual ability to cleave certain substrates, while ensuring efficient and complete degradation of proteins by their collective action in the lysosomes. Therefore, it was initially believed that the major function of cysteine cathepsins is nonspecific, terminal protein proteolysis. However, we now know from the studies of cathepsin knockout mice that these enzymes play important, nonredundant roles in many other physiological and pathological processes (discussed in detail in the next section).

Cysteine cathepsins are synthesized as inactive precursors containing an amino-terminal signal peptide that mediates their transport across the endoplasmic reticulum membrane (Turk et al. 2000). All cysteine cathepsins possess N-glycosylation sites, which are used to target the enzymes to the lysosomal compartment via the mannose 6-phosphate receptor pathway. Activation is fully achieved once the enzymes reach the acidic environment of the late endosomes or lysosomes, where the propeptide that blocks access of substrates to the catalytic cleft of the proenzyme is removed (Coulombe et al. 1996, Cygler et al. 1996, Sivaraman et al. 1999). Endoproteolytic removal of the propeptide is accomplished by either autocatalytic processing (in the case of the endopeptidases) or limited proteolysis by other proteases. This process is triggered by low pH and is enhanced by the presence of glycosaminoglycans (GAGs) (Wiederanders and Kirschke 1989). GAGs support

Table 15.1 Nomenclature and expression patterns of cysteine cathepsins in human and mouse

Protease (alternative names)	Human (11)	Mouse (18)	Expression pattern
Cathepsin B	CTSB	CtsB	Ubiquitous
Cathepsin C (J, Dipeptidyl peptidase I)	CTSC	CtsC	Ubiquitous
Cathepsin F	CTSF	CtsF	Ubiquitous
Cathepsin H	CTSH	CtsH	Ubiquitous
Cathepsin K (O, O2)	CTSK	CtsK	Osteoclasts, lung-epithelium, thyroid gland
Cathepsin L	CTSL	–	Ubiquitous
Cathepsin L2 (V)	CTSL2	CtsL	Thymus, testis, cornea, epidermis, macrophages
Cathepsin O	CTSO	CtsO	Ubiquitous
Cathepsin S	CTSS	CtsS	Lymphatic tissues, antigen-presenting cells (APC), muscle
Cathepsin W (Lymphopain)	CTSW	CtsW	Natural killer (NK) cells, cytotoxic T lymphocytes
Cathepsin X (Z, P, Y)	CTSZ	CtsZ	Ubiquitous
Cathepsin J	–	CtsJ	Placenta
Cathepsin M	–	CtsM	Placenta
Cathepsin Q	–	CtsQ	Placenta
Cathepsin R	–	CtsR	Placenta
Cathepsin 1	–	Cts1	Placenta
Cathepsin 2	–	Cts2	Placenta
Cathepsin 3	–	Cts3	Placenta
Cathepsin 6	–	Cts6	Placenta

autoactivation and activity of cysteine cathepsins at neutral pH, which is highly relevant for the activation of secreted procathepsins (Vasiljeva et al. 2005). Similarly, cathepsin K requires binding to chondroitin sulfate for optimal collagenolytic activity (Lecaille et al. 2002).

Not all cysteine cathepsins are exclusively present in the lysosomes. For example, cathepsin W is retained in the endoplasmic reticulum (Ondr and Pham 2004), while an isoform of cathepsin L was found in the nucleus, where it can process the CDP/Cux transcription factor (Goulet et al. 2004). Moreover, in certain pathological conditions, cysteine cathepsins, which are considered to be intracellular proteases, have been detected at the cell surface (Mai et al. 2000) and cathepsin activity can also be detected in the extracellular milieu (Mort et al. 1985). In fact, early descriptions of cathepsin L referred to this protease as "major excreted protein (MEP)" of malignantly transformed fibroblasts (Mason et al. 1987). Cysteine cathepsin secretion from several cancer cell lines has also been reported, and will be reviewed in Chap. 29 (Lah et al.). Although the precise mechanism of cysteine cathepsin secretion is an area of active investigation, it has been shown that at the cell surface, cathepsins are associated with various binding partners such as integrins or annexin II in discrete regions in the plasma membrane such as caveolae

(Mai et al. 2000, Lechner et al. 2006). Procathepsin B can interact with the annexin II heterotetramer, a protein involved in plasminogen activation. The varied localization of cysteine cathepsins in cancer cells will undoubtedly influence the range of substrates available, and therefore the mechanisms by which they contribute to a wide range of biological processes, including key steps in the progression of primary and metastatic tumors.

In addition to differences in their subcellular localization, cysteine cathepsin family members also exhibit a diverse tissue distribution. The majority of cathepsins are ubiquitously expressed in most normal tissues, suggesting housekeeping functions (Lecaille et al. 2002) (Table 15.1). In contrast, the expression of other members, in particular cathepsin K, S, L2, and W, is restricted to specific cell or tissue types (Table 15.1) where they carry out specialized roles. Cathepsin K is primarily localized to osteoclasts (Drake et al. 1996), while cathepsin S is present mostly in lymphatic tissues and immune cell types such as antigen-presenting cells (Shi et al. 1994). Cathepsin L2, which is the orthologue of mouse cathepsin L, is specifically expressed in the thymus, testis, and corneal epithelium (Adachi et al. 1998; Santamaria et al. 1998; Yasuda et al. 2004; Cheng et al. 2006). Cathepsin W also shows a restricted expression pattern in cytotoxic lymphocytes and natural killer cells (Brown et al. 1998). Finally, there are eight cysteine cathepsins unique to the mouse with placental-specific expression (Deussing et al. 2002). Cell-type-specific cysteine cathepsin expression likely contributes to some unique cathepsin functions, which are discussed in the next section.

Given the destructive potential of the cysteine cathepsins, several safeguarding mechanisms exist to protect normal cells from uncontrolled proteolysis: they are synthesized as inactive precursors, compartmentalized within lysosomes, and are targets of endogenous inhibitors, which can bind any inappropriately activated cathepsins. Several families of endogenous inhibitors exist including stefins, cystatins, and kininogens (Turk et al. 2001). Stefins (A and B) act mostly intracellularly, while the kininogens and cystatins (C, D, E/M, F, S, SN, and SA) are extracellular proteins. Cystatin C is the most potent inhibitor of cathepsins B, H, L, and S, and displays broad tissue distribution (Kopitar-Jerala 2006). Other inhibitors include thyropins and the general inhibitor α2-macroglobulin (Turk et al. 2000). Maintaining the appropriate balance between cysteine cathepsins and their inhibitors is essential given the important roles these proteases can play in numerous pathological conditions when misregulated, particularly in cancer development.

Phenotypes of Cysteine Cathepsin Knockout Mice

In recent years, knockout mice have been generated for nearly every member of the cysteine cathepsin family (Saftig et al. 1998, Pham and Ley 1999, Halangk et al. 2000, Roth et al. 2000, Ondr and Pham 2004, Tang et al. 2006, Reinheckel et al. unpublished data). Phenotypic analysis of these mice has revealed specific and nonredundant roles for individual cysteine cathepsins. While protein degradation in the lysosomes is not the exclusive property of any one cathepsin, as there is no widespread defect in that process in any of the cysteine cathepsin null animals, other physiological functions such as epidermal homeostasis, normal cardiac and brain function, antigen presentation, and apoptosis are affected. For example, *cathepsin S* null ($CtsS^{-/-}$) mice have defects in MHC class II-associated antigen processing and presentation, indicating an important role for this protease in mediating humoral immunity (Shi et al. 1999). Cathepsin L was shown to be involved in skin morphogenesis, hair follicle morphogenesis and cycling, and heart function since its ablation causes epidermal hyperplasia and periodic hair loss (Roth et al. 2000), and $CtsL^{-/-}$ mice develop cardiomyopathy with age (Stypmann et al. 2002; Petermann et al. 2006). Cathepsin C is a processing enzyme involved in the activation of several serine proteases such as granzymes A and B, and $CtsC^{-/-}$ mice have corresponding defects in cytotoxic T cell-induced apoptosis (Pham and Ley 1999). *Cathepsin K* null ($CtsK^{-/-}$) mice have impaired resorption of the bone matrix and develop osteopetrosis, which fits well with the restricted expression of cathepsin K in osteoclasts (Saftig et al. 1998). The generation of *cathepsin F* null ($CtsF^{-/-}$) mice was recently reported (Tang et al. 2006) and proposed as a model of late-onset neurological disease. $CtsF^{-/-}$ mice develop progressive defects in motor coordination and hind leg weakness, and the authors infer this phenotype may result from the accumulation of lipofuscin in CNS neurons (Tang et al. 2006).

On the contrary, mice mutant for other family members such as *cathepsins B, W, H*, and *X* appear normal (Ondr and Pham 2004, Reinheckel et al. unpublished data). However, further detailed analyses may be required to identify phenotypic consequences of their ablation, if indeed there are any. As an example, the phenotype of *cathepsin B* null ($CtsB^{-/-}$) mice is very subtle and these animals can be distinguished from their littermates only when subjected to pathological stress, such as experimental pancreatitis and liver injury, to which they are resistant (Halangk et al. 2000, Guicciardi et al. 2001).

Cysteine Cathepsins in Cancer

Cysteine cathepsin upregulation has been reported in many human tumors, including breast (Poole et al. 1978, Castiglioni et al. 1994, Bervar et al. 2003), lung (Werle et al. 1999; Fujise et al. 2000), brain (Rempel et al. 1994, Strojnik et al. 1999), gastrointestinal (Watanabe et al. 1989, Liu et al. 1998, Ebert et al. 2005), prostate (Sinha et al. 1995, Nagler et al. 2004) cancers and melanoma (Kos et al. 1997, Frohlich et al. 2001). The increased expression of certain cysteine cathepsins, such as cathepsin B and cathepsin L, has frequently been positively correlated with a poor prognosis for patients with a variety of malignancies (Saad et al. 1998, Frohlich et al. 2001, Harbeck et al. 2001), although contrary results have also been reported (Nasu et al. 2001). Moreover, increased antigen levels of cysteine cathepsins in the serum have been reported for several types of cancer (Thomssen et al. 1995, Lah et al. 2000, Schweiger et al. 2000, Miyake et al. 2004), indicating their potential use as prognostic biomarkers. A detailed discussion of cysteine cathepsins in human cancer can be found in Chap. 29 (Lah et al.).

However, while clinical association studies emphasize the importance of understanding the roles of these proteases in cancer, they do not provide causal insights into the in vivo functions of cysteine cathepsins in tumor initiation and progression. Thus, we set out to investigate the roles of a subset of cathepsins that had been implicated in key tumorigenic processes (Joyce et al. 2004), by breeding cathepsin knockout mice with mouse models of human pancreatic islet cell, breast, or skin carcinogenesis (Fig. 15.1). Tumor progression and metastasis in the cathepsin deficient, cancer-prone mice were subsequently analyzed and will be discussed in detail below. The hallmarks of malignancy in the three mouse models of human cancer are as follows:

RIP1-Tag2 Model of Pancreatic Islet Cell Carcinogenesis

Carcinomas of the endocrine islets in the pancreas are induced in transgenic mice by expression of the simian virus type-40 large and small T-antigens (Tag) under the control of the rat insulin promoter (RIP) (Hanahan 1985). In the RIP1-Tag2 (RT2) mice, approximately 50% of the 400 pancreatic islets of newborn mice develop hyperplasia during the first month of life. Subsequently, 45–50 of these islets induce neovascularization, in a process that is termed "angiogenic switching" (Hanahan and Folkman 1996), in which the preneoplastic lesions develop an independent blood supply, a step that is essential for subsequent tumor development. Approximately 10 islet cell tumors arise per mouse by 12–14 weeks of age, which can be classified as encapsulated, microinvasive, and invasive carcinomas that are, however, rarely metastatic (Lopez and Hanahan 2002) (Fig. 15.1).

MMTV-PyMT Model of Mammary Epithelial Carcinogenesis

Breast cancer in this model is induced by transgenic expression of the Polyoma virus middle-T-oncogene (PyMT) in the ductal epithelium of the mammary gland (Guy et al. 1992). The specificity of oncogene expression is determined by the long terminal repeat (LTR) promoter of the mouse mammary tumor virus (MMTV). This viral promoter is costimulated by ovarian steroid hormones and therefore tumorigenesis is initiated in female mice during puberty at ~4–5 weeks of age. The first pathological alterations, atypical hyperplasias and adenosis, are still benign. These lesions progress within a few weeks to premalignant ductal carcinoma in situ (DCIS) and at later stages to malignant invasive ductal carcinomas (IDC) (Fig. 15.1). At ~14 weeks of age, all female MMTV-PyMT mice exhibit large and mostly poorly differentiated IDCs in each of their ten mammary glands. At this stage, all mice have developed multiple metastases in the lungs. Metastasis also occurs in the axillary and cervical lymph nodes; however, the penetrance of lymph node metastases is not 100%, as for lung metastases (Almholt et al. 2005).

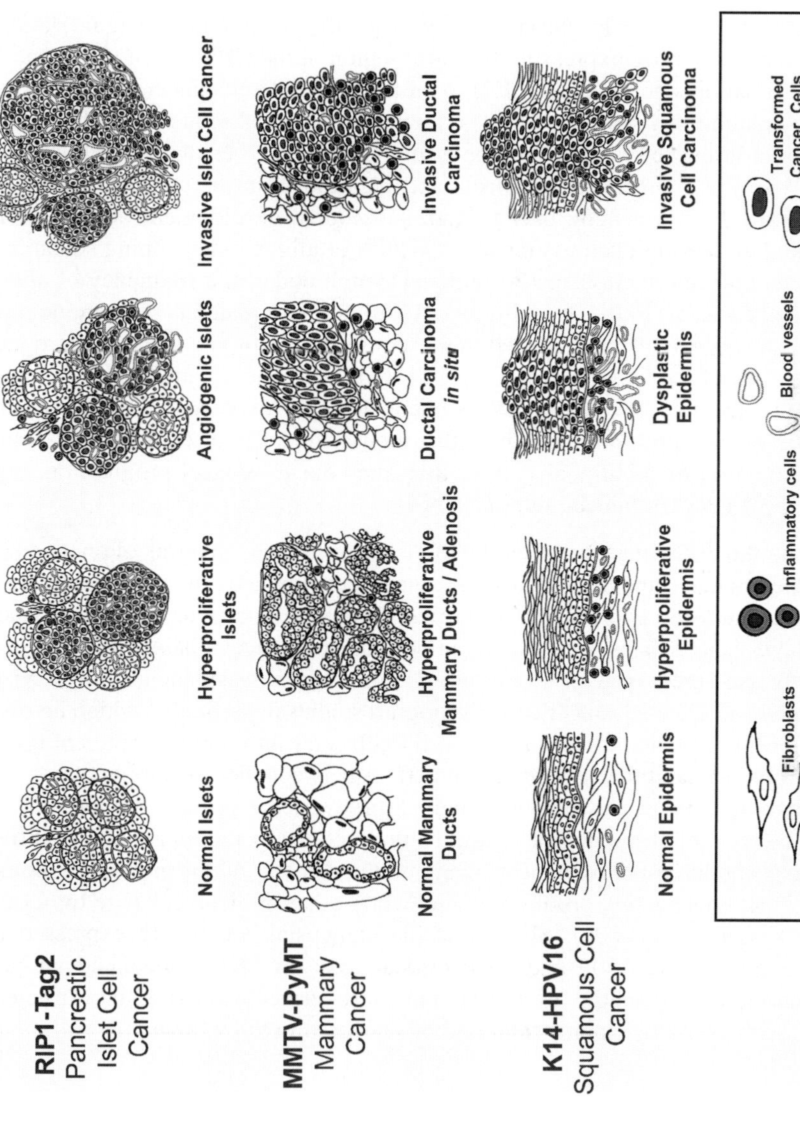

Fig. 15.1 Transgenic mouse models of cancer used to study the effects of cathepsin ablation on tumorigenesis

K14-HPV16 Model of Human Skin and Cervical Cancer

Human papillomaviruses (HPV) cause proliferative lesions in the skin and squamous mucosa, which are mostly hyperplasias, papillomas, and condylomas. However, epithelial lesions of some HPV subtypes, with HPV16 among them, often progress to high-grade dysplasias and occasionally to carcinomas (zur Hausen 1987). Transgenic mice expressing the early region of the HPV16 genome (including the critical oncogenes E6 and E7) in basal epithelial cells under control of the human cytokeratin 14 promoter (K14) undergo the classical sequence of malignant progression during the first 12 months of life (Fig. 15.1). Female as well as male K14-HPV16 mice develop hyperplasia and papillomas in the skin (Arbeit et al. 1994). In the FVB/n genetic background, subsequent development of dysplasias with abundant neoangiogenesis results in ~50% of all mice developing squamous cell carcinomas that metastasize to regional lymph nodes at a frequency of about 20% (Coussens et al. 1996). In addition to the epidermal squamous cell carcinomas, cervical cancers can also be induced in these mice by costimulation with estrogen. Notably, most of the tumor-bearing mice develop only one squamous cell carcinoma despite the fact that the HPV16 oncogenes are expressed throughout the epidermis of the entire animal. Thus, this mouse model is of considerable value for investigation of additional factors that alter the incidence, progression, and metastasis of HPV16-induced tumors.

Despite the differences between the three cancer models in terms of oncogenes, tissue specificities, and timescales of cancer development, they all undergo a sequential, multistep evolution from normal (although oncogene-positive) tissues to invasive cancer. This is a highly important conceptual distinction from the frequently used models involving orthotopic or ectopic transplantation of ex vivo maintained and selected cancer cells. Xenograft studies in particular, which involve injection or transplantation of human cancer cell lines into immunodeficient mice, are unlikely to recapitulate essential interactions between the developing tumor and its surrounding tissue microenvironment. Moreover, the primary tumor models studied by us exhibit the complete and natural actions of the innate and adaptive immune system in response to the growing tumor. Aspects of this immune response can be proangiogenic and protumorigenic, while other immune cells are tumoricidal (de Visser et al. 2006). Since cysteine cathepsins are highly expressed in immune cells (Table 15.1) and have critical roles in MHCII-mediated antigen presentation, it is essential to investigate these proteases in immunosufficient models, such as in the experimental animals covered in this chapter.

Analysis of Cysteine Cathepsin Deficiency in Cancer-Prone Mice

In order to determine the involvement of cysteine cathepsins in cancer, several labs have crossed cathepsin knockout mice to different models of tumorigenesis, which have revealed distinct roles for these proteases in regulating cancer pathogenesis and invasion. In this section, we will describe the phenotypic analysis for each of these crosses, and in the following section, we will discuss the mechanistic insights into cysteine cathepsin functions in vivo that have been revealed by these genetic experiments.

Cathepsin Deficiencies in the RIP1-Tag2 Pancreatic Cancer Model

The most thorough genetic analysis of the involvement of cathepsins in tumorigenesis to date has been completed in the RIP1-Tag2 (RT2) model. Increased expression of six members of the cysteine cathepsin family (cathepsins B, C, H, L, S, and X) was reported in this model over the course of tumor progression (Joyce et al. 2004). The collective importance of the cathepsin family was demonstrated using a broad-spectrum cysteine cathepsin inhibitor, which significantly perturbed all stages of tumor development. Subsequently, mice null for *cathepsins B, C, L, S* (Gocheva et al. 2006), *H*, and *X* (Gocheva and Joyce unpublished data) were crossed with RT2 mice in order to determine the individual functions, if any, of these cysteine cathepsins in pancreatic islet cell cancer. Indeed, it was found that RT2 mice null for *cathepsin B, L*, or *S* had a significant reduction in tumor burden and general perturbations throughout the course of tumor development (Table 15.2). This detailed analysis also revealed some process-specific functions of individual cathepsin family members.

Specifically, we showed that deletion of *cathepsin B* or *S* reduced angiogenic switching and caused defects in the tumor vasculature, including decreased microvessel density (MVD), indicating that these proteases are not only important for neovascularization but are also essential for the continuous maintenance of tumor angiogenesis. An independent study reported a similar decrease in tumor burden in $CtsS^{-/-}$ RT2 mice and also suggested a functional role for this enzyme in angiogenesis (Wang et al. 2006).

Moreover, ablation of *cathepsins B, L*, or *S* caused a pronounced increase in tumor cell apoptosis, which correlated well with the significantly decreased tumor burden in these mice (Gocheva et al. 2006) (Table 15.2). Another process that was affected was tumor cell proliferation, which was significantly decreased in $CtsB^{-/-}$ and $CtsL^{-/-}$ RT2 mice, again closely correlating with the substantial reduction in their tumor burdens (Table 15.2). Surprisingly, proliferation was not affected in the $CtsS^{-/-}$ RT2 animals, indicating that the nearly 50% decrease in tumor burden is due to a combination of increased cell death and defects in angiogenesis (Gocheva et al. 2006).

Critically, deletion of any one of these three cathepsins (B, L, or S) caused a significant reduction in the invasiveness of the resulting tumors, with a marked decrease in the percentage of microinvasive as well as frankly invasive lesions. Thus, in these three cathepsin knockouts, not only was there a significant reduction in pancreatic tumor burden, but also the tumors that did arise were mostly benign, encapsulated lesions (Gocheva et al. 2006). These results demonstrated that

Table 15.2 Summary of the effects of cathepsin ablation in the different mouse models of human cancer analyzed

	$CtsB^{-/-}$		$CtsL^{-/-}$		$CtsS^{-/-}$	$CtsC^{-/-}$
	RIP1-Tag2[a]	MMTV-PyMT[b]	RIP1-Tag2[a]	K14-HPV16[c]	RIP1-Tag2[a,d]	RIP1-Tag2[a]
Development of premalignant lesions	Reduced	Reduced	Reduced	Enhanced	Reduced	No change
Tumor volume	Reduced	Reduced	Reduced	Accelerated development	Reduced	No change
Metastatic tumor burden	N/A[e]	Reduced	N/A[e]	Enhanced	N/A[e]	N/A[e]
Proliferation	Reduced	Reduced	Reduced	Enhanced	No change	No change
Apoptosis	Enhanced	No change	Enhanced	No change	Enhanced	No change
MVD[f]/angiogenic switch	Reduced	No change	No change	No change	Reduced	No change
Tumor grade/invasion	Reduced	Reduced	Reduced	Increased	Reduced	Reduced

The data summarized in this table is compiled from the following references:

[a] Gocheva et al. (2006)

[b] Vasiljeva et al. (2006)

[c] Lohmuller and Reinheckel, manuscript in preparation

[d] Wang et al. (2006)

[e] Tumors in the RIP1-Tag2 model very rarely metastasize

[f] Microvessel density (MVD)

mutating any one of the cysteine cathepsin proteases B, L, or S interferes with the progression of benign lesions to invasive carcinomas, indicating that each enzyme plays an important and nonredundant role in the process of tumor invasion.

Interestingly, although cathepsin C was identified as one of the six cysteine cathepsins upregulated during islet tumor progression (Joyce et al. 2004), when RT2 mice are null for *cathepsin C*, there was no effect on any of the parameters analyzed (Gocheva et al. 2006) (Table 15.2). This result illustrates the importance of thorough functional validation for candidates arising from microarray profiling of cancers, as there will naturally be genes whose expression is altered as a result of changes in the transcriptional program, but which do not functionally contribute to cancer development.

Cathepsin B Deficiency in the MMTV-PyMT Mammary Cancer Model

The involvement of cathepsin B in promoting tumorigenesis has also been established in the Polyoma middle-T-oncogene (PyMT) mouse model of mammary cancer, indicating that cathepsin B plays important protumorigenic roles in cancers of different tissue types and different cells of origin. In this model, it was determined that deleting one or both alleles of *CtsB* delayed the onset of mammary tumor formation and caused a reduction in tumor growth rates, with the resulting lesions being smaller in size (Vasiljeva et al. 2006). This impaired tumor progression is caused by reduced proliferation rates and diminished invasive potential of $CtsB^{-/-}$ cancer cells, while apoptotic rates did not differ between the $CtsB^{+/+}$ and $CtsB^{-/-}$ tumors (Table 15.2).

An important characteristic of the PyMT model is that the tumors spontaneously metastasize to the lung, thus allowing the authors to investigate whether cathepsin B plays a role in metastasis, in addition to its involvement in primary tumor growth. In fact, ablation of one or both copies of *CtsB* caused a reduction in the volume of pulmonary metastases to 45% and to 65%, respectively. The authors extended their analysis of the role of cathepsin B in the process of tumor cell seeding by using a lung colonization experimental assay, where LacZ-tagged primary PyMT cells were intravenously injected into wild-type female mice. Injection of $CtsB^{+/-}$ PyMT or $CtsB^{-/-}$ PyMT cells resulted in a significant decrease in both the number and the size of the lung colonies compared to wild-type PyMT cells, indicating the importance of cathepsin B in the formation and growth of distant metastases (Vasiljeva et al. 2006).

Cathepsin L Deficiency in the K14-HPV16 Skin Cancer Model

While the aforementioned studies demonstrated protumorigenic functions for a subset of cysteine cathepsins, with their deletion causing significant defects in tumor development, this seems to depend on the particular type of cancer and mouse model investigated. A recent study assessed the functional contribution of cathepsin L to epidermal carcinogenesis by crossing $CtsL^{-/-}$ mice with transgenic K14-HPV16 mice (Lohmuller and Reinheckel manuscript in preparation). Surprisingly, ablation of $CtsL$ actually enhanced tumor progression, with an earlier onset of formation of palpable skin tumors (Table 15.2). Furthermore, these epithelial tumors were more malignant and there was increased metastasis to the axillary lymph nodes compared to wild-type K14-HPV16 mice. While angiogenesis and inflammation during neoplastic progression were unaffected, there was a significant increase in proliferation of the epidermis in $CtsL^{-/-}$ K14-HPV16 mice compared to $CtsL^{+/+}$ K14-HPV16 controls. Although these findings may seem surprising as they challenge the large body of evidence supporting procancerous roles for cysteine cathepsins, including cathepsin L, they may actually be related to the fact that cathepsin L plays a major role in the maintenance of epidermal homeostasis (Roth et al. 2000). In fact, $CtsL^{-/-}$ mice without any oncogene expression also develop epidermal hyperproliferation that can be normalized by keratinocyte-specific reexpression of the protease, indicating a cell-type-specific role of cathepsin L in the regulation of epidermal proliferation (Reinheckel et al. 2005).

Therefore, detailed and comprehensive analyses are necessary in order to fully understand the roles that individual cysteine cathepsin family members, such as cathepsin L, play in multiple types of cancer, as the involvement of certain cathepsins may vary between the different models analyzed. For example, it could be very interesting to examine the effects of cathepsin L deletion on cervical carcinogenesis in the K14-HPV16 mice and ask whether the tumor-suppressing effects of this cysteine cathepsin are specific to the skin, or are observed in other epithelial tissues. Similarly, are there unique protective functions for cysteine cathepsins other than cathepsin L in preventing transformation of the skin epithelium? Analysis of K14-HPV16 mice mutant in other cathepsin family members will allow us to determine whether there are examples where these proteases are tumor-suppressive, or whether this is a unique function of cathepsin L. Certainly the continued investigation of cysteine cathepsin function in these and other transgenic and "knock-in" models of cancer is essential to address the general importance of cysteine cathepsins in cancer promotion, a critical question that should be answered in preclinical models before advancing to the clinic with cysteine cathepsin inhibitors.

Mechanistic Insights into Cysteine Cathepsin Function from Knockout Mice

The studies described above in the cathepsin-deficient mice have uncovered some process- and tissue-specific roles for cysteine cathepsin family members in a variety of mouse models of cancer to date, and the subsequent investigation of the biological mechanisms underlying these processes in vitro has provided novel insights into cysteine cathepsin functions in tumorigenesis. In some cases, the investigation of cathepsin deficiencies in vivo has confirmed candidates suggested from the previous biochemical analysis of cathepsin substrates, and in other cases the knockout mice have enabled the identification of completely new substrates, as will be discussed below. Angiogenesis and invasion are two of the stages in tumor development that have long been known to require proteoloytic activity. As discussed in the previous section, analysis of several cathepsin knockouts in different tumor models has confirmed important roles for cysteine cathepsin proteases in each of these processes.

Angiogenesis

The process of new blood vessel formation, termed angiogenesis or neovascularization, is critical for the growth of a tumor beyond 1–2 mm^3 in size (Hanahan and Folkman 1996, Carmeliet and Jain 2000). Angiogenesis is typically initiated by the release of proangiogenic growth factors, such as vascular endothelial growth factor (VEGF) from the tumor. Endothelial cells migrate into the growing lesion and divide and differentiate to form new blood vessels, which may be additionally supported by pericytes and smooth muscle cells. Induction of proteolysis is necessary for the controlled degradation of the extracellular matrix (ECM) and vascular basement membrane (BM), a critical step in vessel sprouting. The outcome of whether angiogenesis is activated ultimately depends on the balance between proangiogenic (VEGF, FGFs, proteases, etc.) and antiangiogenic (endostatin, tumstatin, thrombospondin, etc.) factors.

To date, cathepsin B and S have been shown to be involved in tumor angiogenesis in the RT2 model (Table 15.2). One mechanism by which cathepsin B and S could promote blood vessel formation is through proteolysis of BM constituent proteins, such as laminin, collagen IV, and fibronectin, which cathepsin B has been shown to cleave in vitro (Buck et al. 1992). Moreover, $CtsS^{-/-}$ endothelial cells show a decreased ability to cleave elastin and collagen IV, leading to an impaired ability of these cells to invade through Matrigel or collagen I gel membranes (Shi et al. 2003). Therefore, $CtsS^{-/-}$ mice have defects in microvessel formation under different conditions, including wound healing and tumor angiogenesis.

A more direct mechanism by which cathepsins might influence the formation of new blood vessels is by inactivating angiogenic inhibitors or by processing proangiogenic factors. For example, protein levels of three antiangiogenic factors, tumstatin, arresten, and canstatin, are increased in the $CtsS^{-/-}$ RT2 tumors compared to the $CtsS^{+/+}$ RT2 controls (Wang et al. 2006). Biochemical data confirmed that cathepsin S degrades arresten and canstatin in vitro, in addition to generating proangiogenic γ2 fragments from laminin-5, a common basement membrane component (Wang et al. 2006).

Similarly another family member, cathepsin L, can generate endostatin, a well-known endogenous inhibitor of angiogenesis, from collagen XVIII in vitro (Felbor et al. 2000), thereby suggesting an active role for this protease in regulating angiogenesis. However, another report has suggested that after its rapid generation, cathepsin L can then degrade endostatin very efficiently in vitro (Ferreras et al. 2000). Cathepsin B was also found to cleave endostatin, though much less efficiently than cathepsin L, and might also participate in its degradation (Ferreras et al. 2000). Thus, the precise role of cathepsins in influencing the balance between the generation and destruction of this antiangiogenic factor, and others, in the complex tumor microenvironment still remains to be elucidated, and the analysis of the different cysteine cathepsin null tumor models discussed here now represents an ideal system to address this question.

Cathepsin L was also shown to be required for neovascularization after ischemia by enhancing the ability of endothelial progenitor cells to invade ischemic tissues and incorporate into the newly forming vessels (Urbich et al. 2005). Whether cathepsin L plays a similar role in the neovascularization of tumors is still unknown, as the contribution of endothelial progenitor cells to tumor angiogenesis is still an area of active debate and may be very mouse model specific (*see* Nolan et al. 2007 for discussion). While we did not detect angiogenic defects in the few, small tumors that develop in the $CtsL^{-/-}$ RT2 mice (Gocheva et al. 2006) (Table 15.2), we know that islet cell tumors do not become dependent on new blood vessel formation until they require a significant expansion in size (Bell-McGuinn et al. 2007). Since the $CtsL^{-/-}$ RT2 tumors were up to 90% smaller on average than the control RT2 tumors, it is possible that the critical switch in which angiogenesis is activated had not yet been triggered, resulting in unchanged microvessel density.

Tumor Invasion

Cysteine cathepsins may contribute not only to the invasion and migration of endothelial cells during angiogenesis, but they could also promote the dissemination of cancer cells into the surrounding tissue. There have been several possible mechanisms proposed for how cathepsins could be involved in the process of invasion and metastasis including degradation of the ECM, activation of other proteases, or cleavage of cell–cell adhesion factors (Gocheva and Joyce 2007).

Indeed, using the individual cathepsin null/RT2 tumors described above, we identified the cell-adhesion protein, E-cadherin, as a novel substrate for the proinvasive cathepsins B, L, and S (Gocheva et al. 2006). E-cadherin is the principal component of adherens junctions that maintain cell–cell adhesion, thereby limiting cell migration (Cavallaro and Christofori 2004). In fact, downregulation of E-cadherin through mechanisms such as gene silencing, point mutations, or posttranslational modifications is a common hallmark of invasive cancers (Cavallaro and Christofori 2004). In RT2 tumors, E-cadherin protein expression is typically lost as lesions become progressively more invasive, particularly at the tumor margin (Perl et al. 1998). However, in RT2 tumors null for *cathepsin B, L*, or *S*, the levels of E-cadherin protein were maintained, in stark contrast to the control RT2 tumors or the $CtsC^{-/-}$ RT2 tumors (Gocheva et al. 2006). It was reasoned that posttranslational cleavage of E-cadherin was perturbed in these cathepsin null RT2 mice, and subsequent biochemical analysis confirmed that E-cadherin is cleaved in its extracellular domain by cathepsins B, L, and S, but not cathepsin C. In other systems, cleavage of the extracellular portion of E-cadherin has been shown to abrogate its adhesive functions and promote tumor cell invasion (Cavallaro and Christofori 2004), and ongoing experiments should further clarify the exact mechanism of action for cathepsin-mediated E-cadherin cleavage and tumor cell invasion (Gocheva and Joyce unpublished data).

Additional mechanisms by which cysteine cathepsins promote invasion may involve direct degradation of components of the ECM, thus creating space for the invading cells to migrate into. For example, cathepsins B and L were shown to cleave components of the BM/ECM including laminin (Lah et al. 1989), type-IV collagen (Buck et al. 1992, Guinec et al. 1993), fibronectin (Buck et al. 1992, Ishidoh and Kominami 1995), and tenascin-C (Mai et al. 2002), leading to limited proteolysis of the surrounding matrix. During cancer development, cysteine cathepsins are often translocated to the cell surface where they can be associated with proteins such as annexin II, or secreted, as discussed in the Introduction. Once at the plasma membrane or the extracellular space, cysteine cathepsins have improved access to other cell-surface proteases or to their ECM substrates, although some studies have demonstrated that degradation of ECM components can also occur intracellularly (Sameni et al. 2000, 2001). Finally, cysteine cathepsins also participate in the activation of other proteases by initiating a proteolytic cascade leading to the cleavage of multiple downstream targets, which could collectively enhance the process of tumor invasion. In vitro experiments have shown that cathepsins cleave and activate MMP-1 and MMP-3 (Eeckhout and Vaes 1977), as well as convert the precursor form of urokinase-type plasminogen activator (uPA) to the active enzyme, which in turn catalyzes the cleavage of plasminogen into plasmin (Goretzki et al. 1992, Guo et al. 2002). Plasmin is a broad-spectrum serine protease that can also directly degrade ECM components and activate other MMPs (Rao 2003).

Thus, the invasion of cancer cells is likely facilitated by a combination of these proteolytic mechanisms, including dissociation of cell surface adhesion complexes and degradation of the surrounding ECM to allow local invasion and ultimately tumor cell metastasis. The generation of tumor-bearing mice deficient in individual cysteine cathepsins represents the ideal experimental model to now determine the relative contributions of these different pathways in different tumor microenvironments under physiologically relevant conditions.

Tumor Cell Proliferation and Apoptosis

Tumor growth is largely influenced by the balance between tumor cell proliferation and apoptosis, and previous in vitro data had suggested that certain cysteine cathepsins may have both proproliferative and antiapoptotic roles that could significantly contribute to increased tumor growth and malignancy. In fact, reduced tumor cell proliferation rates were found in $CtsB^{-/-}$ RT2, $CtsL^{-/-}$ RT2, and $CtsB^{-/-}$ PyMT mice (Table 15.2), suggesting that these proteases might normally be activating growth factors or cleaving key proteins implicated in cell cycle progression. Indeed, cathepsin B was previously shown to proteolytically cleave and process the insulin-like growth factor-1 (IGF-1) in endosomes following IGF-1-induced IGF-1 receptor internalization, resulting in appropriate downstream cellular signaling (Authier et al. 2005).

In addition, cathepsin L might also play a more direct role in proliferation through cleavage and activation of the transcription factor CDP/Cux-1, thus promoting cell cycle progression and cellular proliferation (Goulet et al. 2004). However, cathepsin L's proproliferative role might be strictly tissue specific as $CtsL^{-/-}$ mice exhibit epidermal hyperplasia (Roth et al. 2000) and $CtsL^{-/-}$ HPV16 tumors have enhanced proliferation indexes (Reinheckel et al. unpublished data) as discussed above. In the skin, cathepsin L was shown to increase the recycling of growth factors in keratinocytes resulting in sustained keratinocyte proliferation due to the increased availability of growth factors (Reinheckel et al. 2005), again pointing toward the more specialized progrowth functions of this protease in the skin epithelium.

The involvement of cathepsins in tumor cell proliferation has also been confirmed indirectly by Wang and colleagues, who analyzed the phenotype of *cystatin C* null RT2 mice (Wang et al. 2006). Cystatin C inhibits cathepsins B, L, S, and H and its absence resulted in a general increase in cathepsin activity. It was found that cell proliferation was significantly increased in *cystatin C$^{-/-}$* RT2 tumors, again indicating that at least some of the cathepsins repressed by this endogenous inhibitor have progrowth functions.

In terms of the effects of cysteine cathepsins on apoptosis, it was shown that ablating *cathepsins B, L,* or *S* in the RT2 model resulted in a significant increase in programmed cell death (Gocheva et al. 2006) (Table 15.2). These results indicate that these enzymes could have antiapoptotic roles during tumor development, although it might be context and stimulus dependent since other studies have reported that under certain conditions cathepsins can actually promote cell death. In fact, cathepsin B acts as an executioner protease in TNF-alpha-induced tumor cell death in a fibrosarcoma cell line (Foghsgaard et al. 2001) and was able to trigger cytochrome *c* release from the mitochrondria in TNF-alpha-mediated hepatocyte apoptosis (Guicciardi et al. 2000), potentially explaining why $CtsB^{-/-}$ mice are resistant to TNF-alpha-induced liver injury (Guicciardi et al. 2000). However, additional cathepsin cleavage substrates need to be identified before the precise mechanism of action of these proteases in the process of cell death is fully understood, which could also help resolve whether the local cellular environment controls whether cysteine cathepsins are ultimately pro- or antiapoptotic.

Functional Compensation by Cathepsin Family Members in Cysteine Cathepsin Knockout Mice

One concern that is often raised in the generation and analysis of knockout mice for entire families, as for the cysteine cathepsins discussed here, is that functional compensation by other family members may occur to take over the role of the deleted gene, which could potentially obscure the mutant phenotype. Given this consideration, we have analyzed the total expression and activity of the cysteine cathepsin family in the different cathepsin null cancer-prone mice discussed above, to determine whether there is compensatory upregulation. Analysis of cysteine cathepsin expression and activity in the different cathepsin null RT2 tumors did not reveal any changes when compared to wild-type RT2 tumors (Gocheva et al. unpublished data). This was also the case in an independent analysis of $CtsS^{-/-}$ RT2 tumor cell extracts by Wang et al. (2006). However, it was determined that ablation of *cathepsin B* in PyMT tumor cells results in increased levels of cathepsin X on the cell surface, suggesting that there might be some compensatory mechan isms within the family (Vasiljeva et al. 2006). Surprisingly, there was no significant difference in cathepsin X mRNA or protein levels between wild-type PyMT and $CtsB^{-/-}$ PyMT animals, indicating that altered intracellular trafficking and redistri bution of the enzyme is responsible for the increased levels on the cell membrane. These changes in cathepsin X levels seem to be functionally relevant since inhibition of cathepsin X by a neutralizing antibody caused an even further reduction in invasion of $CtsB^{-/-}$ PyMT cells in culture, indicating that there might be some redundancy in the action of certain proteases in tumorigenic processes.

The generation of $CtsB^{-/-}$; $CtsL^{-/-}$ double-knockout mice has also indicated that these two cathepsins play redundant roles in the maintenance of the central nervous system. Most double-mutant mice die around 12 days of age from severe brain atrophy associated with selective neuronal loss in the cerebellar Purkinje and granule cell layers (Felbor et al. 2002), while the individual $CtsB^{-/-}$ and $CtsL^{-/-}$ mice are viable with normal life expectancy. If the $CtsB^{-/-}$; $CtsL^{-/-}$ mice were carefully nursed, the authors could extend their lifespan to an endpoint of 50 days, and analysis at that stage indicated further neuronal loss and marked atrophy in the cerebral cortex (Felbor et al. 2002). This result suggests that there is some functional overlap between certain cysteine cathepsins in performing specific physiological roles, and demonstrates the combined importance of cathepsins B and L in maintaining normal brain function. This functional redundancy was confirmed by rescue of the neuronal defects in $CtsB^{-/-}$; $CtsL^{-/-}$ mice following restoration of cathepsin L by transgenic expression (Sevenich et al. 2006). Double mutants of other cathepsin genes have been generated: $CtsK^{-/-}$; $CtsL^{-/-}$ mice (Friedrichs et al. 2003), $CtsF^{-/-}$; $CtsL^{-/-}$ mice (Tang et al. 2006), $CtsF^{-/-}$; $CtsS^{-/-}$ mice (Tang et al. 2006), $CtsL^{-/-}$; $CtsS^{-/-}$ mice (Mallen-St Clair et al. 2006), and $CtsB^{-/-}$; $CtsS^{-/-}$ mice (Garfall et al. unpublished data), which are all viable, without any obvious additional phenotype compared to the individual knockouts, at least in the analyses that have been performed to date. Thus, whether compensatory mechanisms are activated could depend on several parameters, including the cell type, tissue type, and the physiological or pathological process involved, as well as the normal function of the particular cysteine cathepsin that is deleted.

Stromal Contributions of Cysteine Cathepsins to Tumor Development

In addition to an increase in the expression of proteolytic enzymes by tumor cells over the course of cancer progression, coopted stromal cells within the tumor microenvironment can also produce proteases. In fact, in many cancers, the stromal contribution of matrix-degrading enzymes is significantly greater than that from the transformed cells (Liotta and Kohn 2001, Jedeszko and Sloane 2004). For example, in the RT2 model, flow cytometry was used to sort whole tumors into the three major constituent cell types: endothelial, tumor, and innate immune cells (Joyce et al. 2004). When the purified cell lysates were profiled for cysteine cathepsin expression and activity, it was determined that while tumor cells preferentially expressed cathepsin L, the majority of cathepsin expression and activity was derived from the infiltrating innate immune cells (Joyce et al. 2004), namely, tumor-associated macrophages (TAMs). Indeed in certain cancers, TAMs appear to promote tumorigenesis (reviewed in Gadea and Joyce 2006); one possible mechanism could involve supplying tumor-promoting matrix-degrading enzymes, such as cathepsins B and S. In fact, cathepsin B is also expressed by macrophages in the PyMT model, where host-derived cathepsin B was shown to be more critical than tumor-derived cathepsin B in promoting lung metastasis (Vasiljeva et al. 2006). However, whether cysteine cathepsin expression from the host cells alone is generally sufficient to promote tumorigenesis in these mouse models, and more importantly in human cancer, remains an area of active research.

Conclusions and Future Directions

To date, our investigation of mouse models of human cancers with deficiencies in individual cysteine cathepsins have revealed highly context-specific functions for these proteases. In most instances, deletion of cysteine cathepsins results in reduced tumor growth and metastasis; however, the example of the protective role of cathepsin L in K14-HPV16-induced skin cancers calls for detailed investigation of additional cysteine cathepsin family members in a variety of mouse tumor models. This is even more important with regard to the discussion on the use of specific versus broad-spectrum cysteine cathepsin inhibitors as a therapeutic approach for treating different malignancies including cancer (Palermo and Joyce 2008). A second conclusion from our studies is the concept that the function of cysteine cathepsins in tumorigenesis is not restricted to the cancer cell itself.

Rather, cysteine cathepsins appear to be highly instrumental for the activation of the tumor stroma, in which tumor-associated inflammatory cells, such as macrophages, contribute cathepsin activity with tumorigenic and prometastatic effects. A major goal in the next several years will be to identify the cell types within the microenvironment that supply the relevant cathepsins and to determine the mechanisms by which cysteine cathepsins exert their effects on the tumor through the identification of novel substrates. Finally, it will be essential to further study the interactions between cysteine cathepsin family members and their inhibitors, as well as their interplay with other proteases, in order to identify the key proteolytic pathways critical for tumor progression and metastasis.

Acknowledgements   The authors thank Eva Schill-Wendt for preparation of the illustrations in Fig. 15.1. The work of TR and CP was supported by a grant of the European Union Framework Programme 6, Project LSHC-CT-2003-503297, CancerDegradome, by grant of the Deutsche Krebshilfe (Re106977), and grant 23-7532.22-33-11/1 of the Ministerium fuer Wissenschaft und Kunst, Baden Wuerttemberg. Research in JAJ's laboratory was supported in part by the following: NIH R01 CA125162, NIH U54 CA125518, Sidney Kimmel Foundation for Cancer Research, V Foundation for Cancer Research, Rita Allen Foundation, Emerald Foundation, and the Geoffrey Beene Foundation. VG is the recipient of a Frank L. Horsfall Fellowship.

Reference

Adachi, W., Kawamoto, S., Ohno, I., Nishida, K., Kinoshita, S., Matsubara, K., and Okubo, K. 1998. Isolation and characterization of human cathepsin V: a major proteinase in corneal epithelium. Invest Ophthalmol Vis Sci **39**(10): 1789–1796.

Almholt, K., Lund, L.R., Rygaard, J., Nielsen, B.S., Dano, K., Romer, J., and Johnsen, M. 2005. Reduced metastasis of transgenic mammary cancer in urokinase-deficient mice. Int J Cancer **113**(4): 525–532.

Arbeit, J.M., Munger, K., Howley, P.M., and Hanahan, D. 1994. Progressive squamous epithelial neoplasia in K14-human papillomavirus type 16 transgenic mice. J Virol **68**(7): 4358–4368.

Authier, F., Kouach, M., and Briand, G. 2005. Endosomal proteolysis of insulin-like growth factor-I at its C-terminal D-domain by cathepsin B. FEBS Lett **579**(20): 4309–4316.

Bell-McGuinn, K.M., Garfall, A.L., Bogyo, M., Hanahan, D., and Joyce, J.A. 2007. Inhibition of cysteine cathepsin protease activity enhances chemotherapy regimens by decreasing tumor growth and invasiveness in a mouse model of multistage cancer. Cancer Res **67**(15): 7378–7385.

Bervar, A., Zajc, I., Sever, N., Katunuma, N., Sloane, B.F., and Lah, T.T. 2003. Invasiveness of transformed human breast epithelial cell lines is related to cathepsin B and inhibited by cysteine proteinase inhibitors. Biol Chem **384**(3): 447–455.

Brown, J., Matutes, E., Singleton, A., Price, C., Molgaard, H., Buttle, D., and Enver, T. 1998. Lymphopain, a cytotoxic T and natural killer cell-associated cysteine proteinase. Leukemia **12** (11): 1771–1781.

Buck, M.R., Karustis, D.G., Day, N.A., Honn, K.V., and Sloane, B.F. 1992. Degradation of extracellular-matrix proteins by human cathepsin B from normal and tumour tissues. Biochem J **282**(Pt 1): 273–278.

Carmeliet, P., and Jain, R.K. 2000. Angiogenesis in cancer and other diseases. Nature **407**(6801): 249–257.

Castiglioni, T., Merino, M.J., Elsner, B., Lah, T.T., Sloane, B.F., and Emmert-Buck, M.R. 1994. Immunohistochemical analysis of cathepsins D, B, and L in human breast cancer. Hum Pathol **25**(9): 857–862.

Cavallaro, U., and Christofori, G. 2004. Cell adhesion and signalling by cadherins and Ig-CAMs in cancer. Nat Rev Cancer **4**(2): 118–132.

Cheng, T., Hitomi, K., van Vlijmen-Willems, I.M., de Jongh, G.J., Yamamoto, K., Nishi, K., Watts, C., Reinheckel, T., Schalkwijk, J., and Zeeuwen, P.L. 2006. Cystatin M/E is a high affinity inhibitor of cathepsin V and cathepsin L by a reactive site that is distinct from the legumain-binding site. A novel clue for the role of cystatin M/E in epidermal cornification. J Biol Chem **281**(23): 15893–15899.

Coulombe, R., Grochulski, P., Sivaraman, J., Menard, R., Mort, J.S., and Cygler, M. 1996. Structure of human procathepsin L reveals the molecular basis of inhibition by the prosegment. Embo J **15**(20): 5492–5503.

Coussens, L.M., Hanahan, D., and Arbeit, J.M. 1996. Genetic predisposition and parameters of malignant progression in K14-HPV16 transgenic mice. Am J Pathol **149**(6): 1899–1917.

Cygler, M., Sivaraman, J., Grochulski, P., Coulombe, R., Storer, A.C., and Mort, J.S. 1996. Structure of rat procathepsin B: model for inhibition of cysteine protease activity by the proregion. Structure **4**(4): 405–416.

de Visser, K.E., Eichten, A., and Coussens, L.M. 2006. Paradoxical roles of the immune system during cancer development. Nat Rev Cancer **6**(1): 24–37.

Deussing, J., Kouadio, M., Rehman, S., Werber, I., Schwinde, A., and Peters, C. 2002. Identification and characterization of a dense cluster of placenta-specific cysteine peptidase genes and related genes on mouse chromosome 13. Genomics **79**(2): 225–240.

Drake, F.H., Dodds, R.A., James, I.E., Connor, J.R., Debouck, C., Richardson, S., Lee-Rykaczewski, E., Coleman, L., Rieman, D., Barthlow, R., Hastings, G., and Gowen, M. 1996. Cathepsin K, but not cathepsins B, L, or S, is abundantly expressed in human osteoclasts. J Biol Chem **271**(21): 12511–12516.

Ebert, M.P., Kruger, S., Fogeron, M.L., Lamer, S., Chen, J., Pross, M., Schulz, H.U., Lage, H., Heim, S., Roessner, A., Malfertheiner, P., and Rocken, C. 2005. Overexpression of cathepsin B in gastric cancer identified by proteome analysis. Proteomics **5**(6): 1693–1704.

Eeckhout, Y., and Vaes, G. 1977. Further studies on the activation of procollagenase, the latent precursor of bone collagenase. Effects of lysosomal cathepsin B, plasmin and kallikrein, and spontaneous activation. Biochem J **166**(1): 21–31.

Felbor, U., Dreier, L., Bryant, R.A., Ploegh, H.L., Olsen, B.R., and Mothes, W. 2000. Secreted cathepsin L generates endostatin from collagen XVIII. Embo J **19**(6): 1187–1194.

Felbor, U., Kessler, B., Mothes, W., Goebel, H.H., Ploegh, H.L., Bronson, R.T., and Olsen, B.R. 2002. Neuronal loss and brain atrophy in mice lacking cathepsins B and L. Proc Natl Acad Sci U S A **99**(12): 7883–7888.

Ferreras, M., Felbor, U., Lenhard, T., Olsen, B.R., and Delaisse, J. 2000. Generation and degradation of human endostatin proteins by various proteinases. FEBS Lett **486**(3): 247–251.

Foghsgaard, L., Wissing, D., Mauch, D., Lademann, U., Bastholm, L., Boes, M., Elling, F., Leist, M., and Jaattela, M. 2001. Cathepsin B acts as a dominant execution protease in tumor cell apoptosis induced by tumor necrosis factor. J Cell Biol **153**(5): 999–1010.

Friedrichs, B., Tepel, C., Reinheckel, T., Deussing J., von Figura, K., Herzog, V., Peter, C., Saftig, P., Brix, K., 2003. Thyroid functions of mouse cathepsins B, K, and L. J Clin Invest. **111**(11): 1733–1745.

Frohlich, E., Schlagenhauff, B., Mohrle, M., Weber, E., Klessen, C., and Rassner, G. 2001. Activity, expression, and transcription rate of the cathepsins B, D, H, and L in cutaneous malignant melanoma. Cancer **91**(5): 972–982.

Fujise, N., Nanashim, A., Taniguchi, Y., Matsuo, S., Hatano, K., Matsumoto, Y., Tagawa, Y., and Ayabe, H. 2000. Prognostic impact of cathepsin B and matrix metalloproteinase-9 in pulmonary adenocarcinomas by immunohistochemical study. Lung Cancer **27**(1): 19–26.

Gadea, B.B., and Joyce, J.A. 2006. Tumour-host interactions: implications for developing anti-cancer therapies. Expert Rev Mol Med **8**(30): 1–32.

Gocheva, V., and Joyce, J.A. 2007. Cysteine cathepsins and the cutting edge of cancer invasion. Cell Cycle **6**(1): 60–64.

Gocheva, V., Zeng, W., Ke, D., Klimstra, D., Reinheckel, T., Peters, C., Hanahan, D., and Joyce, J. A. 2006. Distinct roles for cysteine cathepsin genes in multistage tumorigenesis. Genes Dev **20** (5): 543–556.

Goretzki, L., Schmitt, M., Mann, K., Calvete, J., Chucholowski, N., Kramer, M., Gunzler, W.A., Janicke, F., and Graeff, H. 1992. Effective activation of the proenzyme form of the urokinase-type plasminogen activator (pro-uPA) by the cysteine protease cathepsin L. FEBS Lett **297**(1–2): 112–118.

Goulet, B., Baruch, A., Moon, N.S., Poirier, M., Sansregret, L.L., Erickson, A., Bogyo, M., and Nepveu, A. 2004. A cathepsin L isoform that is devoid of a signal peptide localizes to the nucleus in S phase and processes the CDP/Cux transcription factor. Mol Cell **14**(2): 207–219.

Guicciardi, M.E., Deussing, J., Miyoshi, H., Bronk, S.F., Svingen, P.A., Peters, C., Kaufmann, S. H., and Gores, G.J. 2000. Cathepsin B contributes to TNF-alpha-mediated hepatocyte apoptosis by promoting mitochondrial release of cytochrome c. J Clin Invest **106**(9): 1127–1137.

Guicciardi, M.E., Miyoshi, H., Bronk, S.F., and Gores, G.J. 2001. Cathepsin B knockout mice are resistant to tumor necrosis factor-alpha-mediated hepatocyte apoptosis and liver injury: implications for therapeutic applications. Am J Pathol **159**(6): 2045–2054.

Guinec, N., Dalet-Fumeron, V., and Pagano, M. 1993. "In vitro" study of basement membrane degradation by the cysteine proteinases, cathepsins B, B-like and L. Digestion of collagen IV, laminin, fibronectin, and release of gelatinase activities from basement membrane fibronectin. Biol Chem Hoppe Seyler **374**(12): 1135–1146.

Guo, M., Mathieu, P.A., Linebaugh, B., Sloane, B.F., and Reiners, J.J., Jr. 2002. Phorbol ester activation of a proteolytic cascade capable of activating latent transforming growth factor-betaL a process initiated by the exocytosis of cathepsin B. J Biol Chem **277**(17): 14829–14837.

Guy, C.T., Cardiff, R.D., and Muller, W.J. 1992. Induction of mammary tumors by expression of polyomavirus middle T oncogene: a transgenic mouse model for metastatic disease. Mol Cell Biol **12**(3): 954–961.

Halangk, W., Lerch, M.M., Brandt-Nedelev, B., Roth, W., Ruthenbuerger, M., Reinheckel, T., Domschke, W., Lippert, H., Peters, C., and Deussing, J. 2000. Role of cathepsin B in intracellular trypsinogen activation and the onset of acute pancreatitis. J Clin Invest **106**(6): 773–781.

Hanahan, D. 1985. Heritable formation of pancreatic beta-cell tumours in transgenic mice expressing recombinant insulin/simian virus 40 oncogenes. Nature **315**(6015): 115–122.

Hanahan, D., and Folkman, J. 1996. Patterns and emerging mechanisms of the angiogenic switch during tumorigenesis. Cell **86**(3): 353–364.

Harbeck, N., Alt, U., Berger, U., Kruger, A., Thomssen, C., Janicke, F., Hofler, H., Kates, R.E., and Schmitt, M. 2001. Prognostic impact of proteolytic factors (urokinase-type plasminogen activator, plasminogen activator inhibitor 1, and cathepsins B, D, and L) in primary breast cancer reflects effects of adjuvant systemic therapy. Clin Cancer Res **7**(9): 2757–2764.

Ishidoh, K., and Kominami, E. 1995. Procathepsin L degrades extracellular matrix proteins in the presence of glycosaminoglycans in vitro. Biochem Biophys Res Commun **217**(2): 624–631.

Jedeszko, C., and Sloane, B.F. 2004. Cysteine cathepsins in human cancer. Biol Chem **385**(11): 1017–1027.

Joyce, J.A., Baruch, A., Chehade, K., Meyer-Morse, N., Giraudo, E., Tsai, F.Y., Greenbaum, D.C., Hager, J.H., Bogyo, M., and Hanahan, D. 2004. Cathepsin cysteine proteases are effectors of invasive growth and angiogenesis during multistage tumorigenesis. Cancer Cell **5**(5): 443–453.

Kopitar-Jerala, N. 2006. The role of cystatins in cells of the immune system. FEBS Lett **580**(27): 6295–6301.

Kos, J., Stabuc, B., Schweiger, A., Krasovec, M., Cimerman, N., Kopitar-Jerala, N., and Vrhovec, I. 1997. Cathepsins B, H, and L and their inhibitors stefin A and cystatin C in sera of melanoma patients. Clin Cancer Res **3**(10): 1815–1822.

Lah, T.T., Buck, M.R., Honn, K.V., Crissman, J.D., Rao, N.C., Liotta, L.A., and Sloane, B.F. 1989. Degradation of laminin by human tumor cathepsin B. Clin Exp Metastasis 7(4): 461–468.

Lah, T.T., Cercek, M., Blejec, A., Kos, J., Gorodetsky, E., Somers, R., and Daskal, I. 2000. Cathepsin B, a prognostic indicator in lymph node-negative breast carcinoma patients: comparison with cathepsin D, cathepsin L, and other clinical indicators. Clin Cancer Res 6(2): 578–584.

Lecaille, F., Kaleta, J., and Bromme, D. 2002. Human and parasitic papain-like cysteine proteases: their role in physiology and pathology and recent developments in inhibitor design. Chem Rev 102(12): 4459–4488.

Lechner, A.M., Assfalg-Machleidt, I., Zahler, S., Stoeckelhuber, M., Machleidt, W., Jochum, M., and Nagler, D.K. 2006. RGD-dependent binding of procathepsin X to integrin alphavbeta3 mediates cell-adhesive properties. J Biol Chem 281(51): 39588–39597.

Liotta, L.A., and Kohn, E.C. 2001. The microenvironment of the tumour-host interface. Nature 411(6835): 375–379.

Liu, Y., Xiao, S., Shi, Y., Wang, L., Ren, W., and Sloane, B.F. 1998. Cathepsin B on invasion and metastasis of gastric carcinoma. Chin Med J (Engl) 111(9): 784–788.

Lopez, T., and Hanahan, D. 2002. Elevated levels of IGF-1 receptor convey invasive and metastatic capability in a mouse model of pancreatic islet tumorigenesis. Cancer Cell 1(4): 339–353.

Mai, J., Finley, R.L., Jr., Waisman, D.M., and Sloane, B.F. 2000. Human procathepsin B interacts with the annexin II tetramer on the surface of tumor cells. J Biol Chem 275(17): 12806–12812.

Mai, J., Sameni, M., Mikkelsen, T., and Sloane, B.F. 2002. Degradation of extracellular matrix protein tenascin-C by cathepsin B: an interaction involved in the progression of gliomas. Biol Chem 383(9): 1407–1413.

Mallen-St Clair, J., Shi, G.P., Sutherland, R.E., Chapman, H.A., Caughey, G.H., and Wolters, P.J. 2006. Cathepsins L and S are not required for activation of dipeptidyl peptidase I (cathepsin C) in mice. Biol Chem 387(8): 1143–1146.

Mason, R.W., Gal, S., and Gottesman, M.M. 1987. The identification of the major excreted protein (MEP) from a transformed mouse fibroblast cell line as a catalytically active precursor form of cathepsin L. Biochem J 248(2): 449–454.

Miyake, H., Hara, I., and Eto, H. 2004. Serum level of cathepsin B and its density in men with prostate cancer as novel markers of disease progression. Anticancer Res 24(4): 2573–2577.

Mohamed, M.M., and Sloane, B.F. 2006. Cysteine cathepsins: multifunctional enzymes in cancer. Nat Rev Cancer 6(10): 764–775.

Mort, J.S., Recklies, A.D., and Poole, A.R. 1985. Release of cathepsin B precursors from human and murine tumours. Prog Clin Biol Res 180: 243–245.

Nagler, D.K., Kruger, S., Kellner, A., Ziomek, E., Menard, R., Buhtz, P., Krams, M., Roessner, A., and Kellner, U. 2004. Up-regulation of cathepsin X in prostate cancer and prostatic intraepithelial neoplasia. Prostate 60(2): 109–119.

Nasu, K., Kai, K., Fujisawa, K., Takai, N., Nishida, Y., and Miyakawa, I. 2001. Expression of cathepsin L in normal endometrium and endometrial cancer. Eur J Obstet Gynecol Reprod Biol 99(1): 102–105.

Nolan, D.J., Ciarrocchi, A., Mellick, A.S., Jaggi, J.S., Bambino, K., Gupta, S., Heikamp, E., McDevitt, M.R., Scheinberg, D.A., Benezra, R., and Mittal, V. 2007. Bone marrow-derived endothelial progenitor cells are a major determinant of nascent tumor neovascularization. Genes Dev 21(12): 1546–1558.

Ondr, J.K., and Pham, C.T. 2004. Characterization of murine cathepsin W and its role in cell-mediated cytotoxicity. J Biol Chem 279 (26): 27525–27533.

Palermo, C., and Joyce, J.A. 2008. Cysteine cathepsin proteases as pharmacological targets in cancer. Trends Pharmacol Sci 29(1):22–28.

Perl, A.K., Wilgenbus, P., Dahl, U., Semb, H., and Christofori, G. 1998. A causal role for E-cadherin in the transition from adenoma to carcinoma. Nature 392(6672): 190–193.

Petermann, I., Mayer, C., Stypmann, J., Biniossek, M.L., Tobin, D.J., Engelen, M.A., Dandekar, T., Grune, T., Schild, L., Peters, C., and Reinheckel, T. 2006. Lysosomal, cytoskeletal, and metabolic alterations in cardiomyopathy of cathepsin L knockout mice. Faseb J **20**(8): 1266–1268.

Pham, C.T., and Ley, T.J. 1999. Dipeptidyl peptidase I is required for the processing and activation of granzymes A and B in vivo. Proc Natl Acad Sci U S A **96**(15): 8627–8632.

Poole, A.R., Tiltman, K.J., Recklies, A.D., and Stoker, T.A. 1978. Differences in secretion of the proteinase cathepsin B at the edges of human breast carcinomas and fibroadenomas. Nature **273**(5663): 545–547.

Rao, J.S. 2003. Molecular mechanisms of glioma invasiveness: the role of proteases. Nat Rev Cancer **3**(7): 489–501.

Reinheckel, T., Hagemann, S., Dollwet-Mack, S., Martinez, E., Lohmuller, T., Zlatkovic, G., Tobin, D.J., Maas-Szabowski, N., and Peters, C. 2005. The lysosomal cysteine protease cathepsin L regulates keratinocyte proliferation by control of growth factor recycling. J Cell Sci **118**(Pt 15): 3387–3395.

Rempel, S.A., Rosenblum, M.L., Mikkelsen, T., Yan, P.S., Ellis, K.D., Golembieski, W.A., Sameni, M., Rozhin, J., Ziegler, G., and Sloane, B.F. 1994. Cathepsin B expression and localization in glioma progression and invasion. Cancer Res **54**(23): 6027–6031.

Roth, W., Deussing, J., Botchkarev, V.A., Pauly-Evers, M., Saftig, P., Hafner, A., Schmidt, P., Schmahl, W., Scherer, J., Anton-Lamprecht, I., Von Figura, K., Paus, R., and Peters, C. 2000. Cathepsin L deficiency as molecular defect of furless: hyperproliferation of keratinocytes and pertubation of hair follicle cycling. Faseb J **14**(13): 2075–2086.

Saad, Z., Bramwell, V.H., Wilson, S.M., O'Malley, F.P., Jeacock, J., and Chambers, A.F. 1998. Expression of genes that contribute to proliferative and metastatic ability in breast cancer resected during various menstrual phases. Lancet **351**(9110): 1170–1173.

Saftig, P., Hunziker, E., Wehmeyer, O., Jones, S., Boyde, A., Rommerskirch, W., Moritz, J.D., Schu, P., and von Figura, K. 1998. Impaired osteoclastic bone resorption leads to osteopetrosis in cathepsin-K-deficient mice. Proc Natl Acad Sci U S A **95**(23): 13453–13458.

Sameni, M., Moin, K., and Sloane, B.F. 2000. Imaging proteolysis by living human breast cancer cells. Neoplasia **2**(6): 496–504.

Sameni, M., Dosescu, J., and Sloane, B.F. 2001. Imaging proteolysis by living human glioma cells. Biol Chem **382**(5): 785–788.

Santamaria, I., Velasco, G., Cazorla, M., Fueyo, A., Campo, E., and Lopez-Otin, C. 1998. Cathepsin L2, a novel human cysteine proteinase produced by breast and colorectal carcinomas. Cancer Res **58**(8): 1624–1630.

Schweiger, A., Staib, A., Werle, B., Krasovec, M., Lah, T.T., Ebert, W., Turk, V., and Kos, J. 2000. Cysteine proteinase cathepsin H in tumours and sera of lung cancer patients: relation to prognosis and cigarette smoking. Br J Cancer **82**(4): 782–788.

Sevenich, L., Pennacchio, L.A., Peters, C., and Reinheckel, T. 2006. Human cathepsin L rescues the neurodegeneration and lethality in cathepsin B/L double-deficient mice. Biol Chem **387**(7): 885–891.

Shi, G.P., Webb, A.C., Foster, K.E., Knoll, J.H., Lemere, C.A., Munger, J.S., and Chapman, H.A. 1994. Human cathepsin S: chromosomal localization, gene structure, and tissue distribution. J Biol Chem **269**(15): 11530–11536.

Shi, G.P., Villadangos, J.A., Dranoff, G., Small, C., Gu, L., Haley, K.J., Riese, R., Ploegh, H.L., and Chapman, H.A. 1999. Cathepsin S required for normal MHC class II peptide loading and germinal center development. Immunity **10**(2): 197–206.

Shi, G.P., Sukhova, G.K., Kuzuya, M., Ye, Q., Du, J., Zhang, Y., Pan, J.H., Lu, M.L., Cheng, X. W., Iguchi, A., Perrey, S., Lee, A.M., Chapman, H.A., and Libby, P. 2003. Deficiency of the cysteine protease cathepsin S impairs microvessel growth. Circ Res **92**(5): 493–500.

Sinha, A.A., Gleason, D.F., Staley, N.A., Wilson, M.J., Sameni, M., and Sloane, B.F. 1995. Cathepsin B in angiogenesis of human prostate: an immunohistochemical and immunoelectron microscopic analysis. Anat Rec **241**(3): 353–362.

Sivaraman, J., Lalumiere, M., Menard, R., and Cygler, M. 1999. Crystal structure of wild-type human procathepsin K. Protein Sci **8**(2): 283–290.

Strojnik, T., Kos, J., Zidanik, B., Golouh, R., and Lah, T. 1999. Cathepsin B immunohistochemical staining in tumor and endothelial cells is a new prognostic factor for survival in patients with brain tumors. Clin Cancer Res **5**(3): 559–567.

Stypmann, J., Glaser, K., Roth, W., Tobin, D.J., Petermann, I., Matthias, R., Monnig, G., Haverkamp, W., Breithardt, G., Schmahl, W., Peters, C., and Reinheckel, T. 2002. Dilated cardiomyopathy in mice deficient for the lysosomal cysteine peptidase cathepsin L. Proc Natl Acad Sci USA **99**(9): 6234–6239.

Tang, C.H., Lee, J.W., Galvez, M.G., Robillard, L., Mole, S.E., and Chapman, H.A. 2006. Murine cathepsin f deficiency causes neuronal lipofuscinosis and late-onset neurological disease. Mol Cell Biol **26**(6): 2309–2316.

Thomssen, C., Schmitt, M., Goretzki, L., Oppelt, P., Pache, L., Dettmar, P., Janicke, F., and Graeff, H. 1995. Prognostic value of the cysteine proteases cathepsins B and cathepsin L in human breast cancer. Clin Cancer Res **1**(7): 741–746.

Turk, B., Turk, D., and Turk, V. 2000. Lysosomal cysteine proteases: more than scavengers. Biochim Biophys Acta **1477**(1–2): 98–111.

Turk, V., Turk, B., and Turk, D. 2001. Lysosomal cysteine proteases: facts and opportunities. Embo J **20**(17): 4629–4633.

Urbich, C., Heeschen, C., Aicher, A., Sasaki, K., Bruhl, T., Farhadi, M.R., Vajkoczy, P., Hofmann, W.K., Peters, C., Pennacchio, L.A., Abolmaali, N.D., Chavakis, E., Reinheckel, T., Zeiher, A. M., and Dimmeler, S. 2005. Cathepsin L is required for endothelial progenitor cell-induced neovascularization. Nat Med **11**(2): 206–213.

Vasiljeva, O., Dolinar, M., Pungercar, J.R., Turk, V., and Turk, B. 2005. Recombinant human procathepsin S is capable of autocatalytic processing at neutral pH in the presence of glycosaminoglycans. FEBS Lett **579**(5): 1285–1290.

Vasiljeva, O., Papazoglou, A., Kruger, A., Brodoefel, H., Korovin, M., Deussing, J., Augustin, N., Nielsen, B.S., Almholt, K., Bogyo, M., Peters, C., and Reinheckel, T. 2006. Tumor cell-derived and macrophage-derived cathepsin B promotes progression and lung metastasis of mammary cancer. Cancer Res **66**(10): 5242–5250.

Wang, B., Sun, J., Kitamoto, S., Yang, M., Grubb, A., Chapman, H.A., Kalluri, R., and Shi, G.P. 2006. Cathepsin S controls angiogenesis and tumor growth via matrix-derived angiogenic factors. J Biol Chem **281**(9): 6020–6029.

Watanabe, M., Higashi, T., Watanabe, A., Osawa, T., Sato, Y., Kimura, Y., Tominaga, S., Hashimoto, N., Yoshida, Y., Morimoto, S., and et al. 1989. Cathepsin B and L activities in gastric cancer tissue: correlation with histological findings. Biochem Med Metab Biol **42**(1): 21–29.

Werle, B., Lotterle, H., Schanzenbacher, U., Lah, T.T., Kalman, E., Kayser, K., Bulzebruck, H., Schirren, J., Krasovec, M., Kos, J., and Spiess, E. 1999. Immunochemical analysis of cathepsin B in lung tumours: an independent prognostic factor for squamous cell carcinoma patients. Br J Cancer **81**(3): 510–519.

Wiederanders, B., and Kirschke, H. 1989. The processing of a cathepsin L precursor in vitro. Arch Biochem Biophys **272**(2): 516–521.

Yasuda, Y., Li, Z., Greenbaum, D., Bogyo, M., Weber, E., and Bromme, D. 2004. Cathepsin V, a novel and potent elastolytic activity expressed in activated macrophages. J Biol Chem **279**(35): 36761–36770.

zur Hausen, H. 1987. Papillomaviruses in human cancer. Cancer **59**(10): 1692–1696.

Chapter 16
In Vitro and *In Vivo* Models of Angiogenesis to Dissect MMP Functions

Sarah Berndt, Françoise Bruyère, Maud Jost, and Agnès Noël

Abstract Angiogenesis and lymphangiogenesis, the formation of new blood and lymphatic vessels from preexisting ones, are important processes associated with cancer growth and metastatic dissemination. It has become clear that matrix metalloproteinases contribute more to angiogenesis than by just degrading matrix components. They are capable to process a large array of extracellular and cell-surface proteins, and they contribute both in the onset and in the maintenance of angiogenesis. Their implication during lymphangiogenesis is expected, but not yet documented. This chapter describes *in vitro* and *in vivo* models which have proven suitability for investigating each step of (lymph)angiogenic processes. Their rationale and limitation is discussed and emerging functions of matrix metalloproteinases are reviewed.

Introduction

Angiogenesis, the formation of new blood vessels from a preexisting vascular network, is associated with normal developmental processes, physiological tissue remodeling, and a wide range of pathologies, such as tumor development, metastasis, inflammation, and ocular illness (Carmeliet 2003). In tumors, angiogenesis is reinforced by vasculogenesis, the recruitment and functional incorporation of bone marrow-derived cells into the newly forming vessels (Carmeliet 2003). Both angiogenesis and vasculogenesis contribute to tumor growth by providing nutrients and oxygen, as well as to the formation of metastases by offering a route for dissemination. In addition, cancer cells can hijack the lymphatic vasculature which is amplified in tumor through lymphangiogenesis (Adams and Alitalo 2007).

A. Noël
Laboratory of Tumor and Development Biology, University of Liège, Tour de Pathologie (B23), Sart-Tilman, B-4000 Liège, e-mail: agnes.noel@ulg.ac.be

D. Edwards et al. (eds.), *The Cancer Degradome.*
© Springer Science + Business Media, LLC 2008

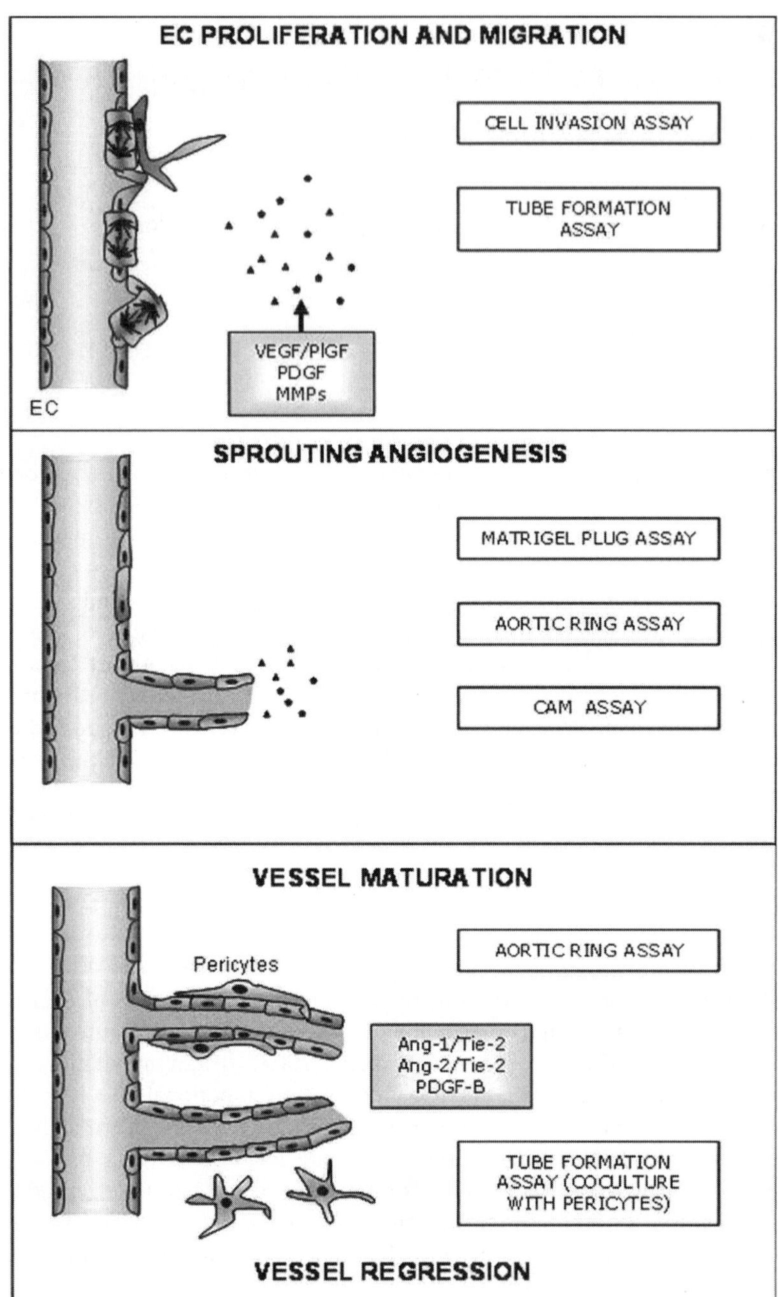

Fig. 16.1 Schematic representation of the different steps of angiogenesis and relevance of *in vitro* and *in vivo* models to study these events. Endothelial cells (*EC*) proliferation and migration is initiated by angiogenic factors [such as vascular endothelial growth factor (*VEGF*), placental-like growth factor (*PlGF*)] and by matrix metalloproteinases (*MMPs*). Relevant models to study the

During the angiogenic growth, some endothelial cells (EC) within the capillary vessel wall are activated for sprouting (Fig. 16.1). Sprouting is controlled by a balance between proangiogenic signals such as vascular endothelial growth factor (VEGF) family members (VEGF, placental-like growth factor, PlGF) and elements that promote quiescence such as the presence of covering pericytes. Expansion of endothelial sprouts requires the induction of proliferation, motile and invasive activities, as well as the modulation of cell–cell interactions and local matrix degradation (Fig. 16.1). Further steps are then required to convert endothelial sprouts into functional and blood-carrying vessels. Strong adhesive interactions and EC–EC junctional contacts need to be established and blood flow requires the formation of a vascular lumen. Vessel stabilization is dependent on the recruitment of perivascular covering cells (pericytes and/or smooth muscle cells). Although proteases were initially viewed as simple regulators of matrix destruction, they are now recognized as active players in the different steps of the angiogenic process. The different proteolytic systems involved comprise serine proteases (Noel et al. 2004), cathepsins, and metalloproteinases, as metalloproteinases (MMPs) and related enzymes (a disintegrin and metalloprotease or ADAM, ADAM with thrombospondin-like domain or ADAMTS) (Handsley and Edwards 2005, Noel et al. 2007). The majority of MMPs are secreted and some are membrane anchored (MT-MMP-1, -2, -3, -5) or bound with a glycosyl phosphatidyl inositol link to the cell surface (MT4-MMP and MT6-MMP) (Hernandez-Barrantes et al. 2002). Their enzymatic activities are regulated by a class of natural inhibitors named TIMP-1 to -4 for tissue inhibitor of metalloproteinases (Brew et al. 2000). A number of studies including gene deletions in mice have pinpointed the role of MMP-2, MMP-9, and MT1-MMP (MMP-14) in the onset of angiogenesis in tumors and in development and in bone formation (Itoh et al. 1998, Holmbeck et al. 1999, Bergers et al. 2000, Zhou et al. 2000, Masson et al. 2005). Well-coordinated extracellular and pericellular proteolytic activities control the extracellular matrix (ECM) remodeling and modulate the bioavailability and the activity of regulatory proteins such as growth factors, growth factor-binding proteins, cytokines, chemokines, membrane receptors and cell adhesion molecules (Overall and Dean 2006, van Hinsbergh et al. 2006, Cauwe et al. 2007, Hu et al. 2007, Noel et al. 2007). The major mechanisms of action of MMPs in angiogenesis and vasculogenesis are summarized in Table 16.1.

In the present chapter, we focus on MMP-related proteolytic activities involved in angiogenesis. We discuss the most relevant models of angiogenesis which have proven valuable for unravelling the multiple roles of MMPs in this complex biological process. For a general description of MMPs, the reader is referred to other chapters of the present volume, as well as to reviews published previously

---

Fig. 16.1 (Continued) onset of angiogenesis (activation of EC) are the cell invasion and the tube formation assays. Vessel sprouting can be mimicked in the Matrigel plug assay, the aortic ring and the chick chorioallantoic membrane (*CAM*) assays. Vessel maturation relying on perivascular cell recruitment and vessel coverage can be studied in the aortic ring assay and the tube assay in coculture with pericytes or smooth muscle cells

Table 16.1 Implication of MMPs in different steps of angiogenesis

Steps of angiogenesis	Mechanisms	Representative examples
Angiogenic switch	Production, activation of angiogenic factors	MT1-MMP enhances VEGF gene expression in tumor cells (Deryugina et al. 2002, Sounni et al. 2004).
	or enhancement of their bioavailability	MMP-9 mobilizes VEGF sequestered in ECM (Bergers et al. 2000).
		VEGF is released after proteolytic cleavage of CTGF engaged in CTGF/VEGF complex (Hashimoto et al. 2002).
		The cleavage of VEGF 165 by MMP-3 or MMP-9 results in the generation of a smaller molecule with properties similar to VEGF 121 (Lee et al. 2005).
Cell migration and pericellular proteolysis	Degradation of matrix components	MT1-MMP acts as a fibrinolysin and facilitates capillary outgrowth (Hiraoka et al. 1998, Hotary et al. 2002).
	Cell–cell and cell–matrix interactions	MT1-MMP colocalizes with $\beta 1$ integrin in cell–cell contact, and is associated with $\alpha v\beta 3$ integrins in migrating EC (Galvez et al. 2002).
		MMP-9 interacts with the cell adhesion molecule CD44, and CD44 cleavage by MT1-MMP promotes cell migration (Mori et al. 2002).
	Pericellular proteolysis	Internalization via caveolae is involved in MT1-MMP-mediated migration of EC. (Galvez et al. 2004).
		The localization of MMP-2 on the cell membrane is associated with $\alpha v\beta 3$ integrin which aids in focusing the proteolytic activity pericellularly (Brooks et al. 1998, Silletti et al. 2001).
		MT1-MMP can process several membrane proteins (integrin subunit, t-transglutaminase, syndecan-1) Deryugina et al. 2002; Belkin et al. 2000, Endo et al. 2003.
Angiogenesis inhibition	Generation of angiogenic inhibitors by the cleavage of matrix components	MMP-9 cleaves type-IV collagen and generates Tumstatin (Hamano and Kalluri 2005).
		The cleavage of collagen type-VIII by MMP-12 generates endostatin which inhibits VEGF-induced EC migration and promotes EC apoptosis (Dixelius et al. 2000, Rehn et al. 2001).
Vessel maturation/ stabilization	Mobilization of pericytes to cover EC	MMP-9 plays a crucial role in the recruitment of bone marrow-derived cells and vessel coverage by pericytes (Jodele et al. 2005).

Table 16.1 (continued)

Steps of angiogenesis	Mechanisms	Representative examples
		Pericyte-derived TIMP-2 inhibits MT1-MMP dependent activation of EC (Anand-Apte et al. 1997, Lafleur et al. 2001).
Vasculogenesis	Recruitment of hematopoetic/ endothelial precursor cells (EPCs) from the bone marrow	MMP-9 recruits EPCs from the vascular niche to the proliferation compartment in the bone marrow via the release of soluble c-kit ligand (Heissig et al. 2002).

(Egeblad and Werb 2002, Overall and Lopez-Otin 2002, Handsley and Edwards 2005, Overall and Kleifeld 2006). General descriptions of *in vitro* and *in vivo* models of angiogenesis are also available (Desbaillets et al. 2000, Auerbach et al. 2003, Norrby 2006, Wartenberg et al. 2006, Noel et al. 2007b). The models of angiogenesis used for MMP investigation and presented here include cell migration, endothelial tube formation, and aortic ring assays (for the *in vitro* assays) and chick chorioallantoic membrane (CAM), Matrigel plug assay, and zebrafish (for the *in vivo* assays). We will critically present their advantages, limitations, and interests in evaluating MMP implication in the angiogenic process. Since lymphangiogenesis (i.e., the formation of new lymphatic vessels) is emerging as an important process contributing to metastatic dissemination, a brief section will focus on *in vitro* and *in vivo* models of lymphangiogenesis.

Relevance of Models of Angiogenesis

It is obvious that no single model is able to elucidate the entire process of angiogenesis. Because of the complexity of the cellular and molecular mechanisms underlying the angiogenic reaction, *in vivo* studies are more informative and more relevant than *in vitro* investigations. However, *in vivo* assays are time-consuming, expensive, and the implication of inflammatory reactions in these systems renders complex the interpretation of the cellular and molecular mechanisms. *In vitro* studies allowing defined experimental conditions are therefore a necessary complement to the *in vivo* experiments. An "ideal" model of angiogenesis should fulfill several requirements. It should (1) be easy, rapid to use, reproducible, and reliable; (2) identify which EC function is affected by the experimental condition (cell proliferation, migration, invasion, survival); (3) provide a quantitative measure of the vasculature and its complexity (number, length and surface of vessels, number of branchings); and (4) give information on the functional characteristic of the new vasculature (permeability, blood flow) and its level of maturation/stabilization (coverage with perivascular cells, regression in the absence of angiogenic stimuli). Any response seen *in vitro* should be confirmed *in vivo*. Unfortunately, no single assay can fulfill all these criteria and a panel of complementary models is required to address this issue.

Cell Invasion Assay

Both macrovascular (human umbilical vascular endothelial cells) and microvascular endothelial cells (HDMEC or human dermal microvascular endothelial cells, BAEC or bovine aortic endothelial cells, PAEC or porcine aortic endothelial cells, HBMEC or human brain microvascular endothelial cells) are used in vitro (McLaughlin et al. 2006, Albini and Benelli 2007). There is also a batch of commercial microvascular EC isolated from various organs (aorta, coronary artery, dermis, lung, bladder, pulmonary artery, saphenous vein, lymphatic origin).

Among the tests that have been used for evaluating the migrative properties of a specific cell population in response to several factors is the chemoinvasion chambers (so-called Boyden chambers) (Albini and Benelli 2007). Cells are seeded on the top of a cell-permeable filter coated with a matrix component (collagen, fibronectin, laminin) or a reconstituted matrix such as Matrigel [a crude extract of Engelbreth-Holm-Swarm (EHS) tumors mainly composed of laminin (Norrby 2006)]. Chemoattractant is added in the culture medium below the filter to promote cell migration. Measurements are carried out by counting cells that have migrated on the lower side of the filter. This assay is largely used to study the migrative properties of endothelial cells transfected or not transfected with the MMP of interest. Natural or synthetic inhibitors of MMPs such as galardin (Roeb et al. 2005), depsipeptide (Klisovic et al. 2005), marimastat (Wagner et al. 1998), IP6 (inositol hexaphosphate) (Tantivejkul et al. 2003) and short-chain fatty acids (Emenaker and Basson 1998) have shown an antimigrative effect in this model. Recently, MT1-MMP expressed by several head and neck squamous cell carcinoma cell lines has been shown to be required for the processing and the release of semaphorin 4D into its soluble form from these cells, thereby inducing endothelial cell chemotaxis *in vitro* (Basile et al. 2007). Such *in vitro* system has the following advantages: (i) defined experimental conditions can be achieved, (ii) the EC population is relatively uniform, (iii) the function of individual genes or protein can be addressed, and (iv) the quantification is easy. However, this model reflects only the migrative properties of EC and/or chemotactic response of EC to specific attractants.

Tube Formation Assay

A reliable test to investigate EC morphogenesis is based on the ability of endothelial cells to form three-dimensional (3D) structures (tube formation) on an appropriate ECM environment (Madri et al. 1988). *In vitro* EC organization into tube-like structures, also called capillaries, has been studied for decades on 2D-coated plates or on 3D gels (Davis et al. 2002). In the initial assay established by Montesano et al. (1992), EC are seeded as a monolayer onto the surface of collagen or fibrin gels, and some EC invade the matrix to form tube structures. In more recent assays, EC are suspended as single cell in 3D-matrix (Davis et al. 2002). Alternatively, HUVEC cell monolayer can be seeded between two layers of collagen (Deroanne et al. 2001). In fact, the most widely used matrix is Matrigel. Although tube formation assay on Matrigel has gained a proeminent place in the angiogenesis field, it is worth noting that some cultured cells of nonendothelial origin such as fibroblasts may also respond to Matrigel by forming tube-like structures (Noel et al. 1991). One critical concern when using Matrigel is to standardize the protein concentration that may lead to discrepancy in the results generated. Furthermore, a strong word of caution is that these tube formation assays, by creating a *de novo* vascular-like network from isolated EC or EC monolayer, do not mimic the sprouting process of angiogenesis but rather mimic vasculogenesis (Davis et al. 2002). In addition, one crucial limitation is the lack of a standardized quantification method, measurements being often made manually.

The advantages of these "vasculogenic assays" are to offer the possibility to investigate the mechanisms underlying EC morphogenesis, lumen formation, and tube stabilization or regression. In this context, matrix-integrin-cytoskeletal signaling appears as a major pathway (Davis et al. 2002). EC tubulogenesis is sensitive to TIMP-2 and TIMP-4, but not to TIMP-1 (Lafleur et al. 2002, Davis and Saunders 2006). Inhibition of MMP-9 reduced tube formation (Jadhav et al. 2004). MT1-MMP can act as a fibrinolysin and promote capillary formation in a fibrin gel (Hiraoka et al. 1998). Overexpression of MT1-, MT2-, or MT3-MMP, but not MT4-MMP, enhances the fibrin-invasive activity of EC (Hotary et al. 2002, Plaisier et al. 2004). MT1-MMP also colocalizes with NO synthase in migratory endothelial cells, and thus appears to be a key molecular effector of NO during EC migration (Genis et al. 2007). While some MMPs such as MT-MMPs can stimulate tube formation (Jeong et al. 1999, Davis and Saunders 2006), others regulate tube regression (e.g., MMP-1, MMP-10, and MMP-13) in 3D collagen matrices (Davis et al. 2001, Bayless and Davis 2003). Interestingly, when pericytes are added to EC during the regression phase, they strongly inhibit MMP-1 and MMP-10-dependent regression (Saunders et al. 2006). In this model, EC-derived TIMP-2 and pericyte-derived TIMP-3 are responsible, in concert, for tube regression (Saunders et al. 2006). Although TIMP-1 does not affect tube formation, it strongly inhibits tube regression (Davis and Saunders 2006). Altogether, these observations led to the concept that distinct MMPs primarily act as promorphogenic (i.e., tube formation) or proregression agents (Davis et al. 2002, Davis and Senger 2005, Handsley and Edwards 2005, Saunders et al. 2005).

Aortic Ring Assay

Since angiogenesis involves not only EC but also perivascular cells, an *ex vivo* vascular tissue culture method has been developed (Nicosia and Ottinetti 1990). When aortic fragments isolated either from rat or from mice are cultured in a type-I collagen gel, they spontaneously give rise to a microvascular network within 7–9 days. Microvessels originated mostly from the two wounded edges of aortic fragments, with only a few growing out from the intimal zone (Villaschi and Nicosia 1993). Growth factors or inhibitors can be added onto the medium in order to evaluate their pro- or antiangiogenic impact. Previously, quantification used to be done manually by blinded observers. Nowadays, quantification is often performed by computer-assisted methods (Masson 2002). Main parameters measured are the length, the number of vessels, and their branchings. Perivascular cells which do not associate with the forming vascular network and their distribution around the aortic explant can also be quantified (Blacher et al. 2001).

This model has gained broad acceptance (Masson 2002) since it bridges the gap between *in vitro* and *in vivo* models. The aortic ring assay mimics the sprouting of EC from a preexisting vessel and takes into account the importance of perivascular cells (Zhu and Nicosia 2002, Li et al. 2005). In addition, in this system, EC are not preselected by passaging and thus are not in a proliferative state (Auerbach et al. 2003). Other advantages are (i) the possibility of generating many assays per animal, (ii) the lack of inflammatory complications to unravel molecular mechanisms of angiogenesis, and (iii) the unique opportunity to exploit the recent generation of MMP-deficient mice. However, one point of caution must be paid to the variability of the angiogenic responses between different mice strains and aging (Burbridge et al. 2002, Zhu et al. 2003).

In this model, MMP expression levels increased gradually during the angiogenic growth phase and remained high when vessels regressed and collagen is lysed around the aortic rings. The profile of MMP expression is modulated by both matrix composition and exogenous addition of growth factors. For example, while MMP-2 and MMP-3 are present in large amount in fibrin cultures, MMP-11 and MT1-MMP are more highly expressed during vessel formation in collagen gels. The angiogenic bFGF (basic fibroblast growth factor) upregulates the expression of MMP-2, MMP-3, MMP-9, MMP-10, MMP-11, and MMP-13 (Burbridge et al. 2002). Synthetic MMP inhibitors such as Ro-28-2653 (Maquoi et al. 2004), batimastat, and marimastat (Zhu et al. 2000, Burbridge et al. 2002) block the formation of microvessels when added in the culture medium at the beginning of the experiment (Zhu et al. 2000). However, batimastat and marimastat stabilized the microvessels and prevented vascular regression after the angiogenic growth phase. MMPs are thus implicated in the microvascular outgrowth phase, in the regression process as well as in the degradation of the neovasculature in latter stages (Zhu et al. 2000).

The aortic ring assay has been recently applied to different MMP-deficient mice (Masson et al. 2002). MMP-11 and MMP-19 are not required for EC spreading out from the aortic rings (Masson et al. 2002, Pendas et al. 2004). Similarly, a single or combined lack of MMP-2 and MMP-9 does not impair the *in vitro* capillary outgrowth from aortic rings (Masson et al. 2005). In sharp contrast, aortic explants isolated from MT1-MMP-null mice display defective capillary sprouting in collagen gels compared with wild-type counterparts. However, wild-type and MT1-MMP-null explants display comparable neovessel outgrowth when embedded in a 3D-gel of cross-linked fibrin, revealing matrix-dependent effect (Chun et al. 2004). MT1-MMP may contribute to the angiogenic process through different mechanisms including at least ECM remodeling and processing of cell-surface molecules

(Sounni et al. 2003, Handsley and Edwards 2005, van Hinsbergh et al. 2006). Indirect effects of MT1-MMP on angiogenesis can also rely on the enhancement of VEGF gene expression by tumor cells (Deryugina et al. 2002, Sounni et al. 2002). Indeed, conditioned media of MT1-MMP overexpressing MCF-7 clones upregulates microvessel outgrowth from aortic rings and this effect can be abrogated by blocking VEGF (Sounni et al. 2002). Therefore, among different individual MMPs investigated in the aortic ring assay, MT1-MMP appears as a key regulator of angiogenic sprouting.

The CAM Assay

The CAM assay was set up by Folkman et al. in 1974 (Auerbach et al. 1974, Ausprunk et al. 1974). First used by embryologists, it has been transposed for the study of tumor angiogenesis and the screening of anti- or proangiogenic factors. CAM are highly vascularized membranes whose EC display morphological characteristics of immature and undifferentiated cells with a high mitotic rate until day 10 of development (Ausprunk et al. 1975). The angiogenic process can be divided into three phases. In the early phase (Day 5–7), the majority of the angiogenic process is achieved by sprouting. The intermediate phase (Day 8–12) is characterized by an intussusception growth process that replaces the sprouts: intussusception involves the formation of transluminal pillars that expand and modify vessel form and function (Patan et al. 1992, Schlatter et al. 1997). By Day 12 or 13, the chorioallantois encircles the entire shell membrane and its expansion is complete (Ausprunk et al. 1974). The CAM assay is carried out *in ovo* by placing growth factors directly onto the CAM through an opening in the eggshell. Test molecules are prepared in carriers (such as slow-release polymer pellets, gelatin sponges, or air-dried on plastic discs). The quantification of angiogenesis is made after 3–4 days of engraftment. Recently, Blacher et al. (2005) have reported an accurate method for assessment of microvascular parameters.

Advantages of the CAM are that it is technically very simple and inexpensive and thus suitable for large scale screening within a short response period (2–3 days). A major limitation of this assay is the standardization of the method used to apply the compound to be tested. In addition, attention should be paid not to misinterpret the de novo angiogenic process since molecules of interest are placed onto preexisting vessels that could appear artifactually to be increased following contraction of the membrane (Ribatti and Vacca 1999) and thus lead to difficulty in discrimination between new capillaries and already existing ones. As the immune system of the CAM is not fully developed, this model also allows the study of tumor-induced angiogenesis by tumor engraftment and subsequent metastasis to chick organs (Auerbach et al. 1976, Ausprunk et al. 1975, Gordon and Quigley 1986, Hagedorn et al. 2005, Zijlstra et al. 2006).

The CAM has been helpful to investigate tumor-derived MMPs. Application of this model to MMP-tumor-secreted studies had been extensively reported (Baum et al. 2007, Schneiderhan et al. 2007). A model of 3D collagen engraftment on the CAM has been developed to analyze spatial and temporal associations *in vivo* between inflammatory cell-derived MMPs and the angiogenesis induced by tumor cells. The onset of angiogenesis is critically dependent on the stromal collagenase MMP13 (chMMP-13), supplied mainly by a hematopoietic lineage (monocytes-macrophages). Initiation of HT1080 cell-induced angiogenesis onto the CAM is dependent on an initial influx of MMP9-containing heterophils (avian counterparts of mammalian neutrophils) followed by an accumulation of chMMP-9 protein in the collagen engraftment and the later arrival of monocytes/macrophages (Zijlstra et al. 2004). Accordingly, disruption of this inflammatory cell influx by anti-inflammmatory drugs significantly reduced angiogenesis. This indicates a possible role of inflammatory cells in the CAM angiogenic process (Zijlstra et al. 2006).

The CAM is also suitable for studying intravasation, a critical step of the metastatic process. MMP-9 expression in human cell lines including HT-1080 cells correlates with the ability of human cells to intravasate and a synthetic inhibitor (marimastat) inhibits significantly tumor cell intravasation and metastasis (Kim et al. 1998). However, MMP-9 downregulation by siRNA in HT-1080 cells showed an unexpected two- to threefold increase in levels of intravasation and metastasis, while intravasation was sensitive to a broad-range MMP inhibitor (Deryugina et al. 2005).

The Matrigel Plug Assay

Matrigel supplemented with either cells or angiogenic molecules (bFGF, VEGF) is injected subcutaneously into mice and allowed to solidify where it forms a plug (Akhtar et al. 2002). This plug can be removed after 7–21 days from the animal and examined histologically to determine the blood vessel infiltration. Plugs can also be quantified for their hemoglobin contents (Passaniti et al. 1992) or fluorescein measurements of plasma volume can be assessed using FITC-dextran (Johns et al. 1996). The Matrigel plug assay has been modified to permit a clear delineation of the neovascularization zone. In this sponge/Matrigel plug assay, Matrigel is first injected alone into the mouse followed by an insertion of a tissue fragment or a sponge into the plug. Measurements of new vessels are then achieved by FITC-dextran injection (Akhtar et al. 2002).

Although this *in vivo* model does not require any surgical procedure and is easy to administer, it suffers from several drawbacks. First, the histological quantification on sections is quite tedious (Passaniti et al. 1992). Second, it is somewhat an artificial model because Matrigel is a reconstituted matrix, not chemically defined and which contains a large variety of growth factors that can influence results. In more recent studies, growth factor-depleted Matrigel is used (Norrby 2006). Finally, it is subject to considerable variability because of the difficulty to obtain similar 3D plugs, even though total Matrigel volume is kept constant (Auerbach et al. 2000). To overcome this problem, a modification can be introduced in the assay in mice and rats using subcutaneous chambers that allow constant 3D form and volume of the Matrigel plug which increases reproducibility (Kragh et al. 2003, Ley et al. 2004). To minimize the amount of Matrigel used, angioreactors have been set up, which consist of semiclosed silicone cylinders that are implanted subcutaneously into nude mice (Guedez et al. 2003). The Matrigel plug assay is viewed as a valuable *in vivo* model for the rapid screening of potential pro- and antiangiogenic agents. In this assay, the oral administration of an MMP inhibitor (BAY12-9566) inhibits FGF-induced angiogenesis. Injection of MMP-9 antisense in mice also decreases EC migration and Matrigel vascularization (London et al. 2003). Surprisingly, angiogenesis in Matrigel plugs is increased rather than decreased in MMP-$19^{-/-}$ mice (Jost et al. 2006). This observation further supports the emerging opposite effects of various MMPs during the process of angiogenesis, some being proangiogenic agents (e.g., MMP-9) and other acting as negative regulators (e.g., MMP-19) of angiogenesis. Chantrain et al. (2004) have adapted the Matrigel plug assay by incorporating tumor cells into the matrix. A significant inhibition of angiogenesis is then observed in immunodeficient RAG1/MMP-9 double-deficient mice orthotopically implanted with a mixture of neuroblastoma cells and Matrigel. In this system, stromal-derived MMP-9 contributes to angiogenesis by promoting blood vessel morphogenesis and pericyte recruitment (Chantrain et al. 2004). Altogether, these data in accordance with previous ones (Coussens et al. 2000) have underlined the key contribution of MMP-9 in angiogenesis and in the mobilization of bone marrow-derived cells.

Zebrafish

In 1999, zebrafish was depicted as a whole animal model for screening drugs that affect the angiogenic process (Serbedzija et al. 1999). These tropical freshwater fish have a short generation time ($\sim$3 months) and can be housed in large numbers and in a small space. A striking organ similarity is observed between zebrafish and mammals at the anatomical, physiological and molecular levels despite their phylogenetic lineage differences (more than 400 million years) (Ny et al. 2006a). The optical transparency of embryos makes them easy to study for diverse developmental processes, from gastrulation to organogenesis. Small test molecules are directly added to the water and diffuse into the embryos. Anti- and proangiogenic molecules already tested in mammals have been shown to exert similar effects in the zebrafish (Norrby 2006). One major advantage is the use of fluorescent labels that can stain a single cell population (e.g., endothelial cells). This assay is useful for embryonic and organogenic angiogenesis. A very large catalog of genetic tools is now available to act on the zebrafish genome. A strong way to understand molecular events in angiogenesis or vasculogenesis is based on the morpholino (MO) knock-down technology which permits reverse genetic analysis of gene function (Ober et al. 2004, Chen et al. 2005, Kajimura et al. 2006).

This model has recently enlightened the evolutionary pattern of the metzincin family. In the zebrafish genome, 83 metzincin genes have been identified. Further phylogenetic analyses reveal that the expansion of the metzincin gene superfamily in vertebrates has occurred predominantly by the simple duplication of preexisting genes rather than the appearance and subsequent expansion of new metzincin subtypes. Evolution of the related TIMP gene family identifies four zebrafish TIMP genes (Huxley-Jones et al. 2007). A study conducted on zebrafish MMP-9 and its developmental expression pattern suggests that zMMP-9 serves as a useful marker of mature myeloid cells (Yoong et al. 2007). The role of MMP-2 during embryogenesis is assessed by an *in situ* analysis showing zMMP2 expression at one-cell stage until 72 h stage of development (Zhang et al. 2003b). Injection of zMMP2 antisense MO oligonucleotides into the embryo resulted in a truncated axis, indicating that this MMP plays an important role in zebrafish embryogenesis (Zhang et al. 2003). In contrast, knockout studies indicate that MMP-2 does not play a key role in mouse embryogenesis (Itoh et al. 1998). Concerning the MT-MMPs, two isoforms isolated from the zebrafish are structurally similar to MT1-MMP (named zebrafish MT-MMP alpha and beta). These two metalloproteinases are expressed through at least the first 72 h of development and this expression is triggered at the cell surface (Zhang et al. 2003c). In addition, TIMP-2 appears to be required for the normal development of zebrafish embryos (Zhang et al. 2003a). By employing flurorescent MMP substrates, an *in vivo* model of zymography has been developed. MMP activity is primarily depicted in ECM-rich structures predicted to undergo active remodeling, such as the pericordal sheath and somite boundaries (Crawford and Pilgrim 2005).

Lymphangiogenesis Models

Only a few lymphatic culture systems have been developed. Initial attempts used 2D cultures of human dermal lymphatic cells isolated either by immunopurification with fluorescence-activated cell sorting (FACS) or by magnetic beads (Davison et al. 1980, Kriehuber et al. 2001). Lymphatic EC can be isolated from the thoracic duct of different species (rat, mouse, dog, cow) by enzymatic digestion (Gnepp and Chandler 1985, Pepper et al. 1994, Tan 1998, Mizuno et al. 2003). They can also be generated by culturing cells induced by the intraperitoneal injection of incomplete Freund's adjuvant into mice (Gnepp and Chandler 1985, Tan 1998, Pepper et al. 1994, Mizuno et al. 2003). Isolated EC have been immortalized with human telomerase reverse transcriptase (hTERT-HDLEC) (Nisato et al. 2004). Furthermore, lymphatic EC differentiation can be induced in embryoid bodies (Liersch et al. 2006). These culture systems suffer from limitations that include (1) the limited number of cells that can be obtained by isolating nontransformed cells, (2) the nonphysiological features of immortalized cells, and (3) the putative dedifferentiation of cells in 2D cultures (Tammela et al. 2005). In addition, none of these culture systems adequately represent the 3D growth of lymphatic microvessels with a lumen. The adaptation of the aortic ring assay to the lymphatic ring assay using the thoracic duct issued from rat led to disappointing results, generating two types of tube-like structures

(LLC or lymphatic-like channels and HLC or hematic-like channels) (Nicosia 1987). The recent setting up of 3D-lymphatic ring cultures from mouse thoracic duct overcomes the main obstacle of the 2D systems and offers the possibility to exploit the panel of MMP-deficient mice recently generated (Bruyère et al. 2008).

Several *in vivo* models of lymphangiogenesis have been developed and mainly consist in the induction of tumor-associated lymphangiogenesis by VEGF-C over-expression in tumor cells (Skobe et al. 2001, Pepper and Skobe 2003) or in transgenic mice (Mandriota et al. 2001). Furthermore, lymphatic endothelial benign tumors (lymphangioma) are induced by intraperitoneal injection of incomplete Freund's adjuvant (Mancardi et al. 1999, Nakamura et al. 2004). This lymphatic cell hyperplasia is formed by VEGFR-3 and podoplanin-positive lesions growing on the surface of the diaphragm and liver, with leukocyte infiltration (Mancardi et al. 1999, Nakamura et al. 2004). Lymphedema are also induced by the excision of a circumferential band of skin in mouse tail (Rutkowski et al. 2006) or microsurgical ablation of tail lymph vessels (Tabibiazar et al. 2006). The corneal assay permits to study lymphangiogenesis by implanting growth factors containing pellets into corneal micropockets (Cao et al. 2004). In contrast to zebrafish, Xenopus develops a lymphatic system, and therefore, the Xenopus tadpole appears as a new genetic model to investigate lymphangiogenesis (Ny et al. 2006). One of the main limitations of these *in vivo* models to identify key regulators of lymphangiogenesis is the important implication of the inflammatory reaction, which does not allow discriminating between a direct effect on lymphatic endothelial cells and an indirect effect through a modulation of inflammation.

Comments and Conclusions

No single model can mimic the entire angiogenic process. Major differences are seen between species (including also animal strains, gender), the specific environments (organ, tissue), the stage of development and age (embryonic versus adult including age-related differences) and the mode of administration of molecules of interest. Undoubtedly, there is a hierarchy of complexity between the different models (Fig. 16.1). The simplest *in vitro* ones focus on one single cell type and address specific EC function (e.g., EC proliferation, migration, chemotaxis). More complex systems (aortic ring assay and *in vivo* models) take into account several cell properties and several cell types (EC, pericytes, fibroblasts-like cells, smooth muscle cells). In these systems, the different events occurring during the angiogenic process are regulated sequentially and spatially, and thus better mimic the *in vivo* situation. Since quantification analysis may lead to some discrepancies when it is done manually, (semi)automatic computer-assisted analysis is required to allow a rapid, objective evaluation of pro- and antiangiogenic molecules/genes of interest (Blacher et al. 2001, 2005).

In the 1990s, clinical trials with MMP inhibitors were based on the concept that MMPs are mainly produced by cancer cells and contribute to tumor progression by degrading matrix components. However, it is now widely accepted that different cell types such as EC, fibroblasts, inflammatory cells and adipocytes are the main source of MMPs. Initially viewed as major regulators of tissue destruction or remodeling, MMPs were expected to regulate angiogenesis by controlling EC migration. Recent studies underlined their key contribution in all steps of angiogenesis including the activation of EC and the promotion of their sprouting, migration, survival, differentiation, coverage by perivascular cells recruited from adjacent tissue or from the bone marrow and even by mediating the regression of tube-like structures in the absence of continuous angiogenic stimuli. MMPs are recognized as modulators of a large panel of molecules which generate new biologically active fragments from the matrix, cell surface-associated proteins and soluble factors (Overall and Lopez-Otin 2002, Handsley and Edwards 2005, Overall and Dean 2006, van Hinsbergh et al. 2006). It became apparent that MMPs have multiple functions, sometime opposite ones. Therefore, a better understanding of their mechanisms of action at different steps of the angiogenic, vasculogenic and lymphangiogenic processes is urgently needed. Instead of directly targeting MMPs, it is possible that substrates and products of MMPs will be preferred targets for treating angiogenesis-related disease. The success of such applications depends on knowledge of how proteases are acting in different contexts and require a panel of complementary models of angiogenesis.

References

Adams, R.H., and Alitalo, K. 2007. Molecular regulation of angiogenesis and lymphangiogenesis. Nat. Rev. Mol. Cell Biol. 8:464–478.

Akhtar, N., Dickerson, E.B., and Auerbach, R. 2002. The sponge/Matrigel angiogenesis assay. Angiogenesis 5:75–80.

Albini, A., and Benelli, R. 2007. The chemoinvasion assay: a method to assess tumor and endothelial cell invasion and its modulation. Nat. Protoc. 2:504–511.

Anand-Apte, B., Pepper, M.S., Voest, E., Montesano, R., Olsen, B., Murphy, G., Apte, S.S., and Zetter, B. 1997. Inhibition of angiogenesis by tissue inhibitor of metalloproteinase-3. Invest. Ophthalmol. Vis. Sci. 38:817–823.

Auerbach, R., Kubai, L., Knighton, D., and Folkman, J. 1974. A simple procedure for the long-term cultivation of chicken embryos. Dev. Biol. 41:391–394.

Auerbach, R., Kubai, L., and Sidky, Y. 1976. Angiogenesis induction by tumors, embryonic tissues, and lymphocytes. Cancer Res. 36:3435–3440.

Auerbach, R., Akhtar, R., Lewis, R.L., and Shinners, B.L. 2000. Angiogenesis assays: problems and pitfalls. Cancer Metastasis Rev. 19:167–172.

Auerbach, R., Lewis, R., Shinners, B., Kubai, L., and Akhtar, N. 2003. Angiogenesis assays: a critical overview. Clin. Chem. 49:32–40.

Ausprunk, D.H., Knighton, D.R., and Folkman, J. 1974. Differentiation of vascular endothelium in the chick chorioallantois: a structural and autoradiographic study. Dev. Biol. 38:237–248.

Ausprunk, D.H., Knighton, D.R., and Folkman, J. 1975. Vascularization of normal and neoplastic tissues grafted to the chick chorioallantois. Role of host and preexisting graft blood vessels. Am. J. Pathol. 79:597–628.

Basile, J.R., Holmbeck, K., Bugge, T.H., and Gutkind, J.S. 2007. MT1-MMP controls tumor-induced angiogenesis through the release of semaphorin 4D. J Biol. Chem. 282:6899–6905.

Baum, O., Hlushchuk, R., Forster, A., Greiner, R., Clezardin, P., Zhao, Y., Djonov, V., and Gruber, G. 2007. Increased invasive potential and up-regulation of MMP-2 in MDA-MB-231 breast cancer cells expressing the beta3 integrin subunit. Int. J. Oncol. 30:325–332.

Bayless, K.J., and Davis, G.E. 2003. Sphingosine-1-phosphate markedly induces matrix metalloproteinase and integrin-dependent human endothelial cell invasion and lumen formation in three-dimensional collagen and fibrin matrices. Biochem. Biophys. Res. Commun. 312:903–913.

Belkin, A.M., Akimov, S.S., Zaritskaya, L.S., Ratkinov, B.I., Deryugina, E.I., and Strongn, A.Y. 2001. Matrix-dependent proteolysis of surface transgutaminase by membrane-type metalloproteinase regulates cancer cell adhesion and locomotion. J. Biol. Chem. 276:18415–22.

Bergers, G., Brekken, R., McMahon, G., Vu, T.H., Itoh, T., Tamaki, K., Tanzawa, K., Thorpe, P., Itohara, S., Werb, Z., and Hanahan, D. 2000. Matrix metalloproteinase-9 triggers the angiogenic switch during carcinogenesis. Nat. Cell Biol. 2:737–744.

Blacher, S., Devy, L., Burbridge, M.F., Roland, G., Tucker, G., Noel, A., and Foidart, J.M. 2001. Improved quantification of angiogenesis in the rat aortic ring assay. Angiogenesis 4:133–142.

Blacher, S., Devy, L., Hlushchuck, R., Larger, E., Lamandé, N., Burri, P., Corvol, P., Djonov, V., Foidart, J.M., and Noel, A. 2005. Quantification of angiogenesis in the chicken chorioallantoic membrane (CAM). Image Anal Stereol. 24:169–180.

Brew, K., Dinakarpandian, D., and Nagase, H. 2000. Tissue inhibitors of metalloproteinases: evolution, structure and function. Biochim. Biophys. Acta 1477:267–283.

Brooks, P.C., Silletti, S., von Schalscha, T.L., Friedlander, M., and Cheresh, D.A. 1998. Disruption of angiogenesis by PEX, a noncatalytic metalloproteinase fragment with integrin binding activity. Cell 92:391–400.

Bruyere, F., Melen-Lamalle, L., Blagier, S., Roland, G., Thiry, M., Meons, L., Frankenne, F., Carmeliet, P., Alitalo, K., Libert, C., Sveetian, J.P., Foidaet, J.M. and Noel, A. 2008. Modeling-lymphangiogenesis in a three-dimensional culture system. Nature Methods 5:431–7.

Burbridge, M.F., Coge, F., Galizzi, J.P., Boutin, J.A., West, D.C., and Tucker, G.C. 2002. The role of the matrix metalloproteinases during in vitro vessel formation. Angiogenesis 5:215–226.

Carmeliet, P. 2003. Angiogenesis in health and disease. Nat. Med. 9:653–660.

Cao, R., Bjorndhal, M.A., Religa, P., Clasper, S., Garvin, S., Galter, D., Meister, B., Ikomi, F., Tritsaris, K., Dissing, S., Ohhashi, T., Jackson, D.G., and Cao, Y. 2004. PDGF-BB induces intratumoral lymphangiogenesis and promotes lymphatic metastasis. Cancer Cell 6:333–45.

Cauwe, B., Steen, P.E., and Opdenakker, G. 2007. The biochemical, biological, and pathological kaleidoscope of cell surface substrates processed by matrix metalloproteinases. Crit. Rev. Biochem. Mol. Biol. 42:113–185.

Chantrain, C.F., Shimada, H., Jodele, S., Groshen, S., Ye, W., Shalinsky, D.R., Werb, Z., Coussens, L.M., and Declerck, Y.A. 2004. Stromal matrix metalloproteinase-9 regulates the vascular architecture in neuroblastoma by promoting pericyte recruitment. Cancer Res. 64:1675–1686.

Chen, E., Stringer, S.E., Rusch, M.A., Selleck, S.B., and Ekker, S.C. 2005. A unique role for 6-O sulfation modification in zebrafish vascular development. Dev. Biol. 284:364–376.

Chun, T.H., Sabeh, F., Ota, I., Murphy, H., McDonagh, K.T., Holmbeck, K., Birkedal-Hansen, H., Allen, E.D., and Weiss, S.J. 2004. MT1-MMP-dependent neovessel formation within the confines of the three-dimensional extracellular matrix. J. Cell Biol. 167:757–767.

Coussens, L.M., Tinkle, C.L., Hanahan, D., and Werb, Z. 2000. MMP-9 supplied by bone marrow-derived cells contributes to skin carcinogenesis. Cell 103:481–490.

Crawford, B.D., and Pilgrim, D.B. 2005. Ontogeny and regulation of matrix metalloproteinase activity in the zebrafish embryo by in vitro and in vivo zymography. Dev. Biol. 286:405–414.

Davis, G.E., and Senger, D.R. 2005. Endothelial extracellular matrix: biosynthesis, remodeling, and functions during vascular morphogenesis and neovessel stabilization. Circ. Res. 97:1093–1107.

Davis, G.E., and Saunders, W.B. 2006. Molecular balance of capillary tube formation versus regression in wound repair: role of matrix metalloproteinases and their inhibitors. J. Investig. Dermatol. Symp. Proc. 11:44–56.

Davis, G.E., Pintar Allen, K.A., Salazar, R., and Maxwell, S.A. 2001. Matrix metalloproteinase-1 and -9 activation by plasmin regulates a novel endothelial cell-mediated mechanism of collagen gel contraction and capillary tube regression in three-dimensional collagen matrices. J. Cell Sci. 114:917–930.

Davis, G.E., Bayless, K.J., and Mavila, A. 2002. Molecular basis of endothelial cell morphogenesis in three-dimensional extracellular matrices. Anat. Rec. 268:252–275.

Davison, P.M., Bensch, K., and Karasek, M.A. 1980. Isolation and growth of endothelial cells from the microvessels of the newborn human foreskin in cell culture. J. Invest. Dermatol. 75:316–321.

Deroanne, C.F., Lapiere, C.M., and Nusgens, B.V. 2001. In vitro tubulogenesis of endothelial cells by relaxation of the coupling extracellular matrix-cytoskeleton. Cardiovasc. Res. 49:647–658.

Deryugina, E.I., Soroceanu, L., and Strongin, A.Y. 2002. Up-regulation of vascular endothelial growth factor by membrane-type 1 matrix metalloproteinase stimulates human glioma xenograft growth and angiogenesis. Cancer Res. 62:580–588.

Deryugina, E.I., Ratnikov, B.I., Postnova, T.I., Rozanov, D.V., and Strongiv, A.Y. 2002. Processing of integrin alpha (α) subunit by membrane type a metalloproteinase stimulates migration of breast carcinoma cells on vitroneqin and enhances tyrosine phosphorylation of focal adhesion kinase. J. Biol. Chem. 277:9749–56.

Deryugina, E.I., Zijlstra, A., Partridge, J.J., Kupriyanova, T.A., Madsen, M.A., Papagiannakopoulos, T., and Quigley, J.P. 2005. Unexpected effect of matrix metalloproteinase down-regulation on vascular intravasation and metastasis of human fibrosarcoma cells selected in vivo for high rates of dissemination. Cancer Res. 65:10959–10969.

Desbaillets, I., Ziegler, U., Groscurth, P., and Gassmann, M. 2000. Embryoid bodies: an in vitro model of mouse embryogenesis. Exp. Physiol. 85:645–651.

Dixelius, J., Larsson, H., Sasaki, T., Holmqvist, K., Lu, L., Engstrom, A., Timpl, R., Welsh, M., and Claesson-Welsh, L. 2000. Endostatin-induced tyrosine kinase signaling through the Shb adaptor protein regulates endothelial cell apoptosis. Blood 95:3403–3411.

Egeblad, M., and Werb, Z. 2002. New functions for the matrix metalloproteinases in cancer progression. Nat. Rev. Cancer 2:161–174.

Emenaker, N.J., and Basson, M.D. 1998. Short chain fatty acids inhibit human (SW1116) colon cancer cell invasion by reducing urokinase plasminogen activator activity and stimulating TIMP-1 and TIMP-2 activities, rather than via MMP modulation. J. Surg. Res. 76:41–46.

Endo, K., Takino, T., Miyamori, H., Kinsen, H., Yoshizaki, T., Furukawa, M., and Sato, H. 2003. Cleavage of syndecan-1 by membrane type matrix metalloproteinase-1 stimulates cell migration. J. Biol. Chem. 278:40764–70.

Galvez, B.G., Matias-Roman, S., Yanez-Mo, M., Sanchez-Madrid, F., and Arroyo, A.G. 2002. ECM regulates MT1-MMP localization with beta1 or alphavbeta3 integrins at distinct cell compartments modulating its internalization and activity on human endothelial cells. J Cell Biol. 159:509–521.

Galvez, B.G., Matias-Roman, S., Yanez-Mo, M., Vicente-Manzanares, M., Sanchez-Madrid, F., and Arroyo, A.G. 2004. Caveolae are a novel pathway for membrane-type 1 matrix metalloproteinase traffic in human endothelial cells. Mol. Biol. Cell 15:678–687.

Genis, L., Gonzalo, P., Tutor, A.S., Galvez, B.G., Martinez-Ruiz, A., Zaragoza, C., Lamas, S., Tryggvason, K., Apte, S.S., and Arroyo, A.G. 2007. Functional interplay between endothelial nitric oxide synthase and membrane type 1-matrix metalloproteinase in migrating endothelial cells. Blood 110:2916–2923.

Gnepp, D.R., and Chandler, W. 1985. Tissue culture of human and canine thoracic duct endothelium. In Vitro Cell Dev. Biol. 21:200–206.

Gordon, J.R., and Quigley, J.P. 1986. Early spontaneous metastasis in the human epidermoid carcinoma HEp3/chick embryo model: contribution of incidental colonization. Int. J Cancer 38:437–444.

Guedez, L., Rivera, A.M., Salloum, R., Miller, M.L., Diegmueller, J.J., Bungay, P.M., and Stetler-Stevenson, W.G. 2003. Quantitative assessment of angiogenic responses by the directed in vivo angiogenesis assay. Am. J Pathol. 162:1431–1439.

Hagedorn, M., Javerzat, S., Gilges, D., Meyre, A., de Lafarge, B., Eichmann, A., and Bikfalvi, A. 2005. Accessing key steps of human tumor progression in vivo by using an avian embryo model. Proc. Natl. Acad. Sci. USA 102:1643–1648.

Hamano, Y., and Kalluri, R. 2005. Tumstatin, the NC1 domain of alpha3 chain of type IV collagen, is an endogenous inhibitor of pathological angiogenesis and suppresses tumor growth. Biochem. Biophys. Res. Commun. 333:292–298.

Handsley, M.M., and Edwards, D.R. 2005. Metalloproteinases and their inhibitors in tumor angiogenesis. Int. J. Cancer 115:849–860.

Hashimoto, G., Inoki, I., Fujii, Y., Aoki, T., Ikeda, E., and Okada, Y. 2002. Matrix metalloproteinases cleave connective tissue growth factor and reactivate angiogenic activity of vascular endothelial growth factor 165. J. Biol. Chem. 277:36288–36295.

Heissig, B., Hattori, K., Dias, S., Friedrich, M., Ferris, B., Hackett, N.R., Crystal, R.G., Besmer, P., Lyden, D., Moore, M.A., Werb, Z., and Rafii, S. 2002. Recruitment of stem and progenitor cells from the bone marrow niche requires MMP-9 mediated release of kit-ligand. Cell 109:625–637.

Hernandez-Barrantes, S., Bernardo, M., Toth, M., and Fridman, R. 2002. Regulation of membrane type-matrix metalloproteinases. Semin. Cancer Biol. 12:131–138.

Hiraoka, N., Allen, E., Apel, I.J., Gyetko, and Weiss, S.J. 1998. Matrix metalloproteinases regulate neovascularization by acting as pericellular fibrinolysins. Cell 95:365–377.

Holmbeck, K., Bianco, P., Caterina, J., Yamada, S., Kromer, M., Kuznetsov, S.A., Mankani, M., Robey, P.G., Poole, A.R., Pidoux, I., Ward, J.M., and Birkedal-Hansen, H. 1999. MT1-MMP-deficient mice develop dwarfism, osteopenia, arthritis, and connective tissue disease due to inadequate collagen turnover. Cell 99:81–92.

Hotary, K.B., Yana, I., Sabeh, F., Li, X.Y., Holmbeck, K., Birkedal-Hansen, H., Allen, E.D., Hiraoka, N., and Weiss, S.J. 2002. Matrix metalloproteinases (MMPs) regulate fibrin-invasive activity via MT1-MMP-dependent and -independent processes. J. Exp. Med. 195:295–308.

Hu, J., Van den Steen, P.E., Sang, Q.X., and Opdenakker, G. 2007. Matrix metalloproteinase inhibitors as therapy for inflammatory and vascular diseases. Nat. Rev. Drug Discov. 6:480–498.

Huxley-Jones, J., Clarke, T.K., Beck, C., Toubaris, G., Robertson, D.L., and Boot-Handford, R.P. 2007. The evolution of the vertebrate metzincins; insights from Ciona intestinalis and Danio rerio. BMC. Evol. Biol. 7:63.

Itoh, T., Tanioka, M., Yoshida, H., Yoshioka, T., Nishimoto, H., and Itohara, S. 1998. Reduced angiogenesis and tumor progression in gelatinase A-deficient mice. Cancer Res. 58:1048–1051.

Jadhav, U., Chigurupati, S., Lakka, S.S., and Mohanam, S. 2004. Inhibition of matrix metalloproteinase-9 reduces in vitro invasion and angiogenesis in human microvascular endothelial cells. Int. J. Oncol. 25:1407–1414.

Jeong, J.W., Cha, H.J., Yu, D.Y., Seiki, M., and Kim, K.W. 1999. Induction of membrane-type matrix metalloproteinase-1 stimulates angiogenic activities of bovine aortic endothelial cells. Angiogenesis 3:167–174.

Jodele, S., Chantrain, C.F., Blavier, L., Lutzko, C., Crooks, G.M., Shimada, H., Coussens, L.M., and Declerck, Y.A. 2005. The contribution of bone marrow-derived cells to the tumor vasculature in neuroblastoma is matrix metalloproteinase-9 dependent. Cancer Res. 65:3200–3208.

Johns, A., Freay, A.D., Fraser, W., Korach, K.S., and Rubanyi, G.M. 1996. Disruption of estrogen receptor gene prevents 17 beta estradiol-induced angiogenesis in transgenic mice. Endocrinology 137:4511–4513.

Jost, M., Folgueras, A.R., Frerart, F., Pendas, A.M., Blacher, S., Houard, X., Berndt, S., Munaut, C., Cataldo, D., Alvarez, J., Melen-Lamalle, L., Foidart, J.M., Lopez-Otin, C., and Noel, A. 2006. Earlier onset of tumoral angiogenesis in matrix metalloproteinase-19-deficient mice. Cancer Res. 66:5234–5241.

Kajimura, S., Aida, K., and Duan, C. 2006. Understanding hypoxia-induced gene expression in early development: in vitro and in vivo analysis of hypoxia-inducible factor 1-regulated zebra fish insulin-like growth factor binding protein 1 gene expression. Mol. Cell Biol. 26:1142–1155.

Kim, J., Yu, W., Kovalski, K., and Ossowski, L. 1998. Requirement for specific proteases in cancer cell intravasation as revealed by a novel semiquantitative PCR-based assay. Cell 94:353–362.

Klisovic, D.D., Klisovic, M.I., Effron, D., Liu, S., Marcucci, G., and Katz, S.E. 2005. Depsipeptide inhibits migration of primary and metastatic uveal melanoma cell lines in vitro: a potential strategy for uveal melanoma. Melanoma Res. 15:147–153.

Kragh, M., Hjarnaa, P.J., Bramm, E., Kristjansen, P.E., Rygaard, J., and Binderup, L. 2003. In vivo chamber angiogenesis assay: an optimized Matrigel plug assay for fast assessment of anti-angiogenic activity. Int. J Oncol. 22:305–311.

Kriehuber, E., Breiteneder-Geleff, S., Groeger, M., Soleiman, A., Schoppmann, S.F., Stingl, G., Kerjaschki, D., and Maurer, D. 2001. Isolation and characterization of dermal lymphatic and blood endothelial cells reveal stable and functionally specialized cell lineages. J. Exp. Med. 194:797–808.

Lafleur, M.A., Forsyth, P.A., Atkinson, S.J., Murphy, G., and Edwards, D.R. 2001. Perivascular cells regulate endothelial membrane type-1 matrix metalloproteinase activity. Biochem. Biophys. Res. Commun. 282:463–473.

Lafleur, M.A., Handsley, M.M., Knauper, V., Murphy, G., and Edwards, D.R. 2002. Endothelial tubulogenesis within fibrin gels specifically requires the activity of membrane-type-matrix metalloproteinases (MT-MMPs). J Cell Sci. 115:3427–3438.

Lee, S., Jilani, S.M., Nikolova, G.V., Carpizo, D., and Iruela-Arispe, M.L. 2005. Processing of VEGF-A by matrix metalloproteinases regulates bioavailability and vascular patterning in tumors. J Cell Biol. 169:681–691.

Ley, C.D., Olsen, M.W., Lund, E.L., and Kristjansen, P.E. 2004. Angiogenic synergy of bFGF and VEGF is antagonized by Angiopoietin-2 in a modified in vivo Matrigel assay. Microvasc. Res. 68:161–168.

Li, X., Tjwa, M., Moons, L., Fons, P., Noel, A., Ny, A., Zhou, J.M., Lennartsson, J., Li, H., Luttun, A., Ponten, A., Devy, L., Bouche, A., Oh, H., Manderveld, A., Blacher, S., Communi, D., Savi, P., Bono, F., Dewerchin, M., Foidart, J.M., Autiero, M., Herbert, J.M., Collen, D., Heldin, C. H., Eriksson, U., and Carmeliet, P. 2005. Revascularization of ischemic tissues by PDGF-CC via effects on endothelial cells and their progenitors. J Clin. Invest 115:118–127.

Liersch, R., Nay, F., Lu, L., and Detmar, M. 2006. Induction of lymphatic endothelial cell differentiation in embryoid bodies. Blood 107:1214–1216.

London, C.A., Sekhon, H.S., Arora, V., Stein, D., Iversen, P.L., and Devi, G.R. 2003. A novel antisense inhibitor of MMP-9 attenuates angiogenesis, human prostate cancer cell invasion and tumorigenicity. Cancer Gene Ther. 10:823–832.

Madri, J.A., Pratt, B.M., and Yannariello-Brown, J. 1988. Matrix-driven cell size change modulates aortic endothelial cell proliferation and sheet migration. Am. J. Pathol. 132:18–27.

Mancardi, S., Stanta, G., Dusetti, N., Bestagno, M., Jussila, L., Zweyer, M., Lunazzi, G., Dumont, D., Alitalo, K., and Burrone, O.R. 1999. Lymphatic endothelial tumors induced by intraperitoneal injection of incomplete Freund's adjuvant. Exp Cell Res. 246:368–375.

Mandriota, S.J., Jussila, L., Jeltsch, M., Compagni, A., Baetens, D., Prevo, R., Banerji, S., Huarte, J., Montesano, R., Jackson, D.G., Orci, L., Alitalo, K., Christofori, G., and Pepper, M.S. 2001. Vascular endothelial growth factor-C-mediated lymphangiogenesis promotes tumour metastasis. EMBO J 20:672–682.

Maquoi, E., Sounni, N.E., Devy, L., Olivier, F., Frankenne, F., Krell, H.W., Grams, F., Foidart, J. M., and Noel, A. 2004. Anti-invasive, antitumoral, and antiangiogenic efficacy of a pyrimidine-2, 4, 6-trione derivative, an orally active and selective matrix metalloproteinases inhibitor. Clin. Cancer Res. 10:4038–4047.

Masson, V., Devy, L., Grignet-Debrus, C., Berndt, S., Bajou, K., Blacher, S., and Noel, A. 2002. Mouse Aortic Ring Assay: A New Approach of the Molecular Genetics of Angiogenesis. Biological Procedure Online 4:24–31.

Masson, V., de la Ballina, L.R., Munaut, C., Wielockx, B., Jost, M., Maillard, C., Blacher, S., Bajou, K., Itoh, T., Itohara, S., Werb, Z., Libert, C., Foidart, J.M., and Noel, A. 2005. Contribution of host MMP-2 and MMP-9 to promote tumor vascularization and invasion of malignant keratinocytes. FASEB J. 19:234–236.

McLaughlin, N., Annabi, B., Sik, K.K., Bahary, J.P., Moumdjian, R., and Beliveau, R. 2006. The response to brain tumor-derived growth factors is altered in radioresistant human brain endothelial cells. Cancer Biol. Ther. 5:1539–1545.

Mizuno, R., Yokoyama, Y., Ono, N., Ikomi, F., and Ohhashi, T. 2003. Establishment of rat lymphatic endothelial cell line. Microcirculation. 10:127–131.

Montesano, R., Pepper, M.S., Vassalli, J.D., and Orci, L. 1992. Modulation of angiogenesis in vitro. EXS 61:129–136.

Mori, H., Tomari, T., Koshikawa, N., Kajita, M., Itoh, Y., Sato, H., Tojo, H., Yana, I., and Seiki, M. 2002. CD44 directs membrane-type 1 matrix metalloproteinase to lamellipodia by associating with its hemopexin-like domain. EMBO J 21:3949–3959.

Nakamura, E.S., Koizumi, K., Kobayashi, M., and Saiki, I. 2004. Inhibition of lymphangiogenesis-related properties of murine lymphatic endothelial cells and lymph node metastasis of lung cancer by the matrix metalloproteinase inhibitor MMI270. Cancer Sci. 95:25–31.

Nicosia, R.F. 1987. Angiogenesis and the formation of lymphaticlike channels in cultures of thoracic duct. In Vitro Cell Dev. Biol. 23:167–174.

Nicosia, R.F., and Ottinetti, A. 1990. Growth of microvessels in serum-free matrix culture of rat aorta. A quantitative assay of angiogenesis in vitro. Lab. Invest. 63:115–122.

Nisato, R.E., Harrison, J.A., Buser, R., Orci, L., Rinsch, C., Montesano, R., Dupraz, P., and Pepper, M.S. 2004. Generation and characterization of telomerase-transfected human lymphatic endothelial cells with an extended life span. Am. J Pathol. 165:11–24.

Noel, A.C., Calle, A., Emonard, H.P., Nusgens, B.V., Simar, L., Foidart, J., Lapiere, J., and Foidart, J.M. 1991. Invasion of reconstituted basement membrane matrix is not correlated to the malignant metastatic cell phenotype. Cancer Res. 51:405–414.

Noel, A., Maillard, C., Rocks, N., Jost, M., Chabottaux, V., Sounni, N.E., Maquoi, E., Cataldo, D., and Foidart, J.M. 2004. Membrane associated proteases and their inhibitors in tumour angiogenesis. J. Clin. Pathol. 57:577–584.

Noel, A., Jost, M., and Maquoi, E. 2007. Matrix metalloproteinases at cancer tumor-host interface. Semin. Cell Dev. Biol.

Norrby, K. 2006. In vivo models of angiogenesis. J. Cell Mol. Med. 10:588–612.

Ny, A., Autiero, M., and Carmeliet, P. 2006. Zebrafish and Xenopus tadpoles: small animal models to study angiogenesis and lymphangiogenesis. Exp Cell Res. 312:684–693.

Ober, E.A., Olofsson, B.O., Makinen, T., Jin, S.W., Shoji, W., Koh, G.Y., Alitalo, K., and Stainier, D.Y. 2004. Vegfc is required for vascular development and endoderm morphogenesis in zebrafish. EMBO Rep. 5:78–84.

Overall, C.M., and Dean, R.A. 2006. Degradomics: systems biology of the protease web. Pleiotropic roles of MMPs in cancer. Cancer Metastasis Rev. 25:69–75.

Overall, C.M., and Kleifeld, O. 2006. Tumour microenvironment – opinion: validating matrix metalloproteinases as drug targets and anti-targets for cancer therapy. Nat. Rev. Cancer 6:227–239.

Overall, C.M., and Lopez-Otin, C. 2002. Strategies for MMP inhibition in cancer: innovations for the post-trial era. Nat. Rev. Cancer 2:657–672.

Passaniti, A., Taylor, R.M., Pili, R., Guo, Y., Long, P.V., Haney, J.A., Pauly, R.R., Grant, D.S., and Martin, G.R. 1992. A simple, quantitative method for assessing angiogenesis and anti-angiogenic agents using reconstituted basement membrane, heparin, and fibroblast growth factor. Lab. Invest. 67:519–528.

Patan, S., Alvarez, M.J., Schittny, J.C., and Burri, P.H. 1992. Intussusceptive microvascular growth: a common alternative to capillary sprouting. Arch. Histol. Cytol. 55(Suppl):65–75.

Pendas, A.M., Folgueras, A.R., Llano, E., Caterina, J., Frerard, F., Rodriguez, F., Astudillo, A., Noel, A., Birkedal-Hansen, H., and Lopez-Otin, C. 2004. Diet-induced obesity and reduced skin cancer susceptibility in matrix metalloproteinase 19-deficient mice. Mol. Cell Biol. 24:5304–5313.

Pepper, M.S., and Skobe, M. 2003. Lymphatic endothelium: morphological, molecular and functional properties. J. Cell Biol. 163:209–213.

Pepper, M.S., Wasi, S., Ferrara, N., Orci, L., and Montesano, R. 1994. In vitro angiogenic and proteolytic properties of bovine lymphatic endothelial cells. Exp. Cell Res. 210:298–305.

Plaisier, M., Kapiteijn, K., Koolwijk, P., Fijten, C., Hanemaaijer, R., Grimbergen, J.M., Mulder-Stapel, A., Quax, P.H., Helmerhorst, F.M., and van Hinsbergh, V.W. 2004. Involvement of membrane-type matrix metalloproteinases (MT-MMPs) in capillary tube formation by human endometrial microvascular endothelial cells: role of MT3-MMP. J. Clin. Endocrinol. Metab. 89:5828–5836.

Rehn, M., Veikkola, T., Kukk-Valdre, E., Nakamura, H., Ilmonen, M., Lombardo, C., Pihlaja-niemi, T., Alitalo, K., and Vuori, K. 2001. Interaction of endostatin with integrins implicated in angiogenesis. Proc. Natl. Acad. Sci. USA 98:1024–1029.

Ribatti, D., and Vacca, A. 1999. Models for studying angiogenesis in vivo. Int. J Biol. Markers 14:207–213.

Roeb, E., Bosserhoff, A.K., Hamacher, S., Jansen, B., Dahmen, J., Wagner, S., and Matern, S. 2005. Enhanced migration of tissue inhibitor of metalloproteinase overexpressing hepatoma cells is attributed to gelatinases: relevance to intracellular signaling pathways. World J. Gastroenterol. 11:1096–1104.

Rutkowski, J.M., Boardman, K.C., and Swartz, M.A. 2006. Characterization of lymphangiogenesis in a model of adult skin regeneration. Am. J. Physiol. Heart Circ. Physiol. 291:H1402–H1410.

Saunders, W.B., Bayless, K.J., and Davis, G.E. 2005. MMP-1 activation by serine proteases and MMP-10 induces human capillary tubular network collapse and regression in 3D collagen matrices. J. Cell Sci. 118:2325–2340.

Saunders, W.B., Bohnsack, B.L., Faske, J.B., Anthis, N.J., Bayless, K.J., Hirschi, K.K., and Davis, G.E. 2006. Coregulation of vascular tube stabilization by endothelial cell TIMP-2 and pericyte TIMP-3. J. Cell Biol. 175:179–191.

Schlatter, P., Konig, M.F., Karlsson, L.M., and Burri, P.H. 1997. Quantitative study of intussusceptive capillary growth in the chorioallantoic membrane (CAM) of the chicken embryo. Microvasc. Res. 54:65–73.

Schneiderhan, W., Diaz, F., Fundel, M., Zhou, S., Siech, M., Hasel, C., Moller, P., Gschwend, J.E., Seufferlein, T., Gress, T., Adler, G., and Bachem, M.G. 2007. Pancreatic stellate cells are an important source of MMP-2 in human pancreatic cancer and accelerate tumor progression in a murine xenograft model and CAM assay. J. Cell Sci. 120:512–519.

Serbedzija, G.N., Flynn, E., and Willett, C.E. 1999. Zebrafish angiogenesis: a new model for drug screening. Angiogenesis 3:353–359.

Silletti, S., Kessler, T., Goldberg, J., Boger, D.L., and Cheresh, D.A. 2001. Disruption of matrix metalloproteinase 2 binding to integrin alpha vbeta 3 by an organic molecule inhibits angiogenesis and tumor growth in vivo. Proc. Natl. Acad. Sci. USA 98:119–124.

Skobe, M., Hawighorst, T., Jackson, D.G., Prevo, R., Janes, L., Velasco, P., Riccardi, L., Alitalo, K., Claffey, K., and Detmar, M. 2001. Induction of tumor lymphangiogenesis by VEGF-C promotes breast cancer metastasis. Nat. Med. 7:192–198.

Sounni, N.E., Devy, L., Hajitou, A., Frankenne, F., Munaut, C., Gilles, C., Deroanne, C., Thompson, E.W., Foidart, J.M., and Noel, A. 2002. MT1-MMP expression promotes tumor growth and angiogenesis through an up-regulation of vascular endothelial growth factor expression. FASEB J. 16:555–564.

Sounni, N.E., Janssen, M., Foidart, J.M., and Noel, A. 2003. Membrane type-1 matrix metalloproteinase and TIMP-2 in tumor angiogenesis. Matrix Biol. 22:55–61.

Sounni, N.E., Roghi, C., Chabottaux, V., Janssen, M., Munaut, C., Maquoi, E., Galvez, B.G., Gilles, C., Frankenne, F., Murphy, G., Foidart, J.M., and Noel, A. 2004. Up-regulation of vascular endothelial growth factor-A by active membrane-type 1 matrix metalloproteinase through activation of Src-tyrosine kinases. J Biol. Chem. 279:13564–13574.

Tabibiazar, R., Cheung, L., Han, L.J., Swanson, J., Beilhack, A., An, A., Dadras, S.S., Rockson, N., Joshi, S., Wagner, R., and Rockson, S.G. 2006. Inflammatory manifestations of experimental lymphatic insufficiency. PLoS. Med. 3:e254.

Tammela, T., Petrova, T.V., and Alitalo, K. 2005. Molecular lymphangiogenesis: new players. Trends Cell Biol. 15:434–441.

Tan, Y. 1998. Basic fibroblast growth factor-mediated lymphangiogenesis of lymphatic endothelial cells isolated from dog thoracic ducts: effects of heparin. Jpn. J Physiol 48:133–141.

Tantivejkul, K., Vucenik, I., and Shamsuddin, A.M. 2003. Inositol hexaphosphate (IP6) inhibits key events of cancer metastasis: I. in vitro studies of adhesion, migration and invasion of MDA-MB 231 human breast cancer cells. Anticancer Res. 23:3671–3679.

van Hinsbergh, V.W., Engelse, M.A., and Quax, P.H. 2006. Pericellular proteases in angiogenesis and vasculogenesis. Arterioscler. Thromb. Vasc. Biol. 26:716–728.

Villaschi, S., and Nicosia, R.F. 1993. Angiogenic role of endogenous basic fibroblast growth factor released by rat aorta after injury. Am. J. Pathol. 143:181–190.

Wagner, S., Stegen, C., Bouterfa, H., Huettner, C., Kerkau, S., Roggendorf, W., Roosen, K., and Tonn, J.C. 1998. Expression of matrix metalloproteinases in human glioma cell lines in the presence of IL-10. J. Neurooncol. 40:113–122.

Wartenberg, M., Donmez, F., Budde, P., and Sauer, H. 2006. Embryonic stem cells: a novel tool for the study of antiangiogenesis and tumor-induced angiogenesis. Handb. Exp. Pharmacol. 53–71.

Yoong, S., O'Connell, B., Soanes, A., Crowhurst, M.O., Lieschke, G.J., and Ward, A.C. 2007. Characterization of the zebrafish matrix metalloproteinase 9 gene and its developmental expression pattern. Gene Expr. Patterns 7:39–46.

Zhang, J., Bai, S., Tanase, C., Nagase, H., and Sarras, M.P. Jr. 2003a. The expression of tissue inhibitor of metalloproteinase 2 (TIMP-2) is required for normal development of zebrafish embryos. Dev. Genes Evol. 213:382–389.

Zhang, J., Bai, S., Zhang, X., Nagase, H., and Sarras, M.P. Jr. 2003b. The expression of gelatinase A (MMP 2) is required for normal development of zebrafish embryos. Dev. Genes Evol. 213:456–463.

Zhang, J., Bai, S., Zhang, X., Nagase, H., and Sarras, M.P. Jr. 2003c. The expression of novel membrane-type matrix metalloproteinase isoforms is required for normal development of zebrafish embryos. Matrix Biol. 22:279–293.

Zhou, Z., Apte, S.S., Soininen, R., Cao, R., Baaklini, G.Y., Rauser, R.W., Wang, J., Cao, Y., and Tryggvason, K. 2000. Impaired endochondral ossification and angiogenesis in mice deficient in membrane-type matrix metalloproteinase I. Proc. Natl. Acad. Sci. USA 97:4052–4057.

Zhu, W.H., and Nicosia, R.F. 2002. The thin prep rat aortic ring assay: a modified method for the characterization of angiogenesis in whole mounts. Angiogenesis 5:81–86.

Zhu, W.H., Guo, X., Villaschi, S., and Francesco, N.R. 2000. Regulation of vascular growth and regression by matrix metalloproteinases in the rat aorta model of angiogenesis. Lab. Invest. 80:545–555.

Zhu, W.H., Iurlaro, M., MacIntyre, A., Fogel, E., and Nicosia, R.F. 2003. The mouse aorta model: influence of genetic background and aging on bFGF- and VEGF-induced angiogenic sprouting. Angiogenesis 6:193–199.

Zijlstra, A., Aimes, R.T., Zhu, D., Regazzoni, K., Kupriyanova, T., Seandel, M., Deryugina, E.I., and Quigley, J.P. 2004. Collagenolysis-dependent angiogenesis mediated by matrix metalloproteinase-13 (collagenase-3). J. Biol. Chem. 279:27633–27645.

Zijlstra, A., Seandel, M., Kupriyanova, T.A., Partridge, J.J., Madsen, M.A., Hahn-Dantona, E.A., Quigley, J.P., and Deryugina, E.I. 2006. Proangiogenic role of neutrophil-like inflammatory heterophils during neovascularization induced by growth factors and human tumor cells. Blood 107:317–327.

Chapter 17
The Surface Transplantation Model to Study the Tumor–Host Interface

Maud Jost, Silvia Vosseler, Silvia Blacher, Norbert E. Fusenig, Margareta M. Mueller, and Agnès Noël

Abstract The tumor is a complex system comprising neoplastic genetically altered cells and a tumor stroma composed of remodeled extracellular matrix, newly formed vessels, and infiltrating host cells. The development of a cancer is a progressive multistep process in which neoplastic cells progress to malignancy by activating their microenvironment and by responding to the tumor-supporting cues of the surrounding tissue. Because of the recently recognized importance of a permissive stroma for tumor development and invasion, the host compartment is now viewed as an interesting new target for tumor therapy. Among positive regulators contributing to the elaboration of this permissive stroma are growth factors, cytokines/chemokines, proteases, and their inhibitors. The present review summarizes what we learned during the last decade on the contribution of these factors at the tumor–host interface by exploiting a useful in vivo surface transplantation model of skin carcinomas.

Introduction

Tumors are not only composed of neoplastic cells, but they are also heterogeneous, structurally complex, and result from an evolving crosstalk between tumor cells and different host cell types. Genetic alterations in tumor cells are essential for tumor progression but not sufficient to generate malignant tumors. Indeed, the stromal

M.M. Mueller

Group Tumor- and Microenvironment (A101), German Cancer Research Center (DKFZ), 69120 Heidelberg, Germany, e-mail: ma.mueller@dkfz-heidelberg.de

A. Noël

Laboratory of Tumor and Development Biology, Centre de Recherche en Cancérologie Expérimentale (CRCE), Groupe Interdisciplinaire de Génoprotéomique Appliqué (GIGA-R), University of Liège, Tour de pathologie (B23), B-4000 Liège, Belgium, e-mail: agnes.noel@ulg.ac.be

D. Edwards et al. (eds.), *The Cancer Degradome*.

environment is required to create a permissive soil for the invasion of the seed, the genetically altered tumor cells. Since the "seed and soil" hypothesis proposed by Paget in 1889 (Paget 1889), the importance of tumor–stroma interactions have been documented (Hanahan and Weinberg 2000, Stetler-Stevenson and Braylan 2001, Fidler 2003, Carmeliet 2005, Noel et al. 2008). Indeed, the stroma of malignant tumors resembles the granulation tissue of a healing wound (Dvorak 1986, Mueller and Fusenig 2002) and alterations of the stromal environment (Zigrino et al. 2005) include enhanced vascularization through angiogenesis and vasculogenesis (Carmeliet 2005, Li et al. 2006), modified extracellular matrix (ECM) composition, recruitment of fibroblastic cells (Kalluri and Zeisberg 2006) and inflammatory cells, and unbalanced protease activities (Folgueras et al. 2004, Zigrino et al. 2005). Consequently, the evolution of tumor xenografts in mice is known to depend on the presence of several host cells including fibroblasts (Noel et al. 1994, Noel et al. 1998, Zhang et al. 2006), adipocytes (Kuperwasser et al. 2004, Andarawewa et al. 2005) immune as well as inflammatory cells (Coussens and Werb 2001), and endothelial cells (Skobe et al. 1997, Vosseler et al. 2005).

In addition to inflammation, the acquisition of an angiogenic phenotype is viewed as a prerequisite for tumor progression. Neovascularization is crucial for sustained tumor growth since it allows oxygenation and nutrient perfusion of the tumor as well as removal of waste products (Carmeliet and Jain 2000). Additionally, vascular endothelial cells can stimulate tumor growth in a paracrine manner by inducing tumor proliferation and invasion. Finally, increased angiogenesis coincides with increased tumor cell entry into the blood circulation and thus facilitates metastasis. Several lines of evidence indicate that induction of angiogenesis precedes the formation of malignant tumors, suggesting that angiogenesis may be rate limiting not only for tumor expansion but also for the onset of malignancy (Skobe and Fusenig 1998, Mueller and Fusenig 2004). As a consequence of the recognition of the essential role of the tumor–stroma interface for tumor progression, the tumor microenvironment has emerged as a new putative target for tumor therapy. Extensive research during more than 30 years led to the entry of anti-VEGF (vascular endothelial growth factor) therapeutics in clinical practice for the therapy of cancers (Duda et al. 2007). Bevacizumab (Avastin®, Genentech Inc.), an anti-VEGF antibody, was the first antiangiogenic compound approved in 2004 by the US Food and Drug Administration (Carmeliet 2005, Ferrara and Kerbel 2005). However, despite the remarkable rapid clinical development of anti-VEGF agents, a growing body of preclinical evidence suggests that other angiogenic pathways are as important in disease progression and might explain the resistance appearing after anti-VEGF therapy (Duda et al. 2007).

In this context, it is essential to keep in mind that the modifications of stromal features are controlled by tumor cells themselves, depending on their degree of aggressiveness and invasiveness. Tumor cells can regulate the elaboration of a permissive stromal environment via the aberrant expression of angiogenic factors (VEGF, placental-like growth factor or PlGF, platelet-derived growth factor or PDGF), proteases (matrix metalloproteinases or MMPs, serine proteases, cathepsins), and chemotactic proteins (stromal cell-derived factor 1 or SDF1α, macrophage chemotactic protein-1 or MCP1) (Bergers and Benjamin 2003).

These upregulated factors disrupt normal tissue homeostasis and act in a paracrine manner to induce stromal reactions such as angiogenic and inflammatory responses (Coussens and Werb 2001, Balkwill and Coussens 2004, Mueller and Fusenig 2004). The recruited host cells are important producers of growth factors, cytokines, chemokines, and proteases, all essential for ECM remodeling, cell migration, and angiogenesis (Benelli et al. 2001, Balkwill and Coussens 2004). Proteases contribute to the remodeling of ECM, promoting host cell migration (inflammation and angiogenesis) and tumor cell invasion. Proteases act not only by disrupting physiological barriers such as basement membranes but importantly also by releasing growth and chemotactic factors from the ECM and demasking cryptic domains of matrix components (Kalluri 2003). In addition, they are key regulators of the shedding, activation, and/or degradation of cell surface molecules including adhesion molecules, mediators of apoptosis, receptors of chemokines/cytokines, and intercellular junction proteins (Cauwe et al. 2007).

The role of proteases in the regulation of angiogenesis and tumor progression made them initially very desirable as therapeutic targets. However, the failure of clinical trials with broad-spectrum MMP inhibitors in cancer (Coussens et al. 2002) (*see* also Chap. 36) made very clear that the role of proteases during tumor growth and progression as well as stromal activation and angiogenesis is much more complex than initially expected (Matrisian and Lopez-Otin 2007). Therefore, the development of new therapies requires an in-depth understanding of the complex interactions established between host and tumor. The functional role of the stroma is difficult to delineate in classical in vivo models of spontaneous or transplanted tumors, due to the intermingled close association of tumor and stroma elements. Therefore, the analysis of tumor–stroma interactions and their role in tumor development requires experimental in vivo systems reflecting different tumor stages. To fulfill those requirements, an in vivo model has been set up to study tumor–stroma interactions: the surface transplantation model (Fusenig et al. 1983).

Surface Tranplants of Squamous Cell Carcinoma of the Skin

The surface transplantation model, which was initially developed to study the interactions of normal epithelial and stromal cells and their impact on growth and differentiation (Fusenig 1992), allows the complete reconstitution of a skin epithelium under the influence of the connective tissue environment, without direct contact between epithelial and stromal cells (schematically shown in Fig. 17.1a). In this model, keratinocytes (of mouse or human origin) are precultured on a 2–3 mm thick type-I collagen gel mounted between two concentric Teflon rings. When a confluent monolayer has formed, the culture is covered by a silicone transplantation chamber and transplanted in toto onto the back muscle fascia of mice where it is held in place by fixing it by wound clips with the surrounding mouse skin. Although separated from the host stroma by the collagen matrix, the grafted cells rapidly develop into highly proliferative stratified epithelia (Fig. 17.1b, e)

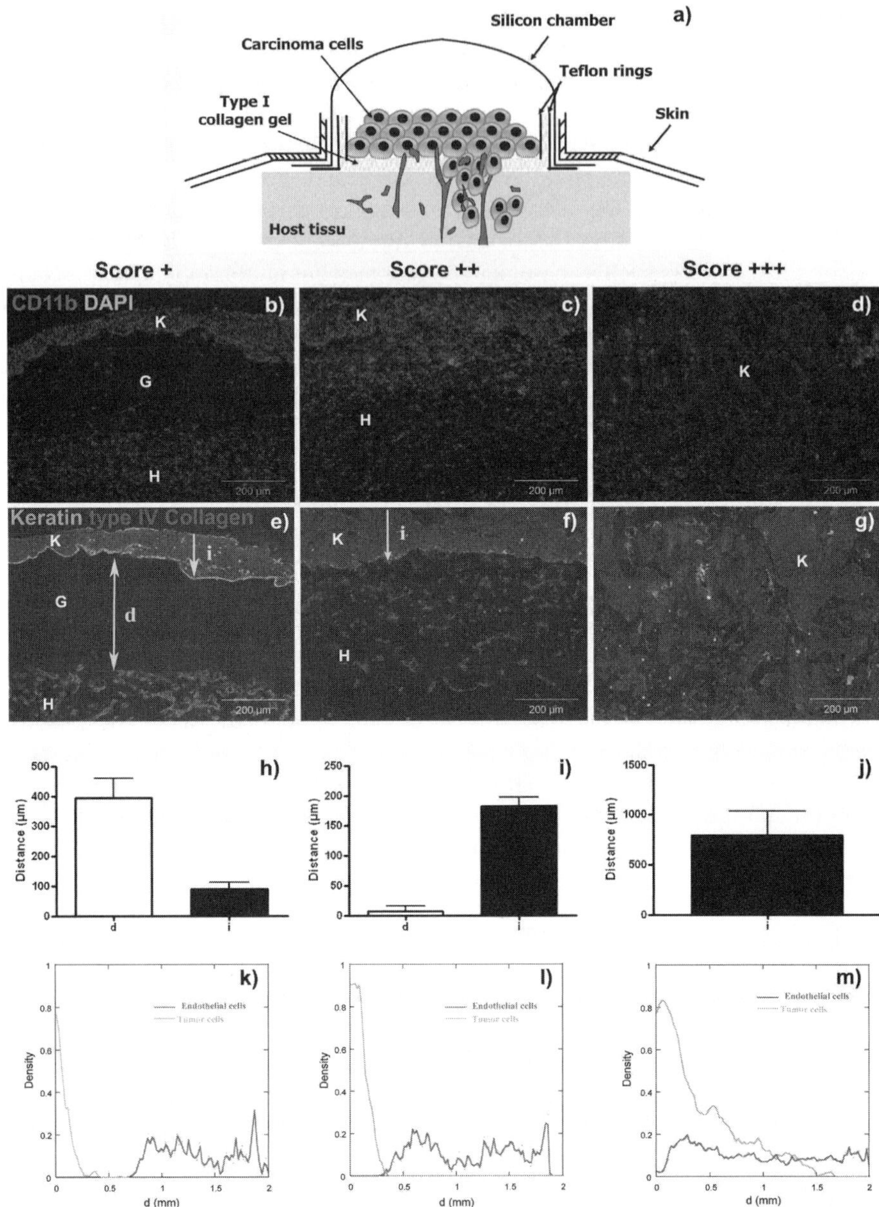

Fig. 17.1 Representation of skin carcinoma cell invasion in the surface transplantation model. **a** Schematic representation of the model: keratinocytes are cultured on the type-I collagen gel, mounted in concentric teflon rings and covered with a hat-shaped silicone chamber. **b–g** Different stages of tumor invasion (score $+$, $++$, $+++$) are observed after 1 (**b, e**), 2 (**c, f**), or 3 (**d, g**) weeks of transplantation. After 1 week of transplantation, tumor cells proliferate to form a multilayered epithelium, and few inflammatory cells infiltrate the collagen gel (**b**). The collagen gel is

within 1–3 weeks. The collagen gel is gradually replaced by a highly vascularized granulation tissue, which eventually gets in close contact to the epithelium (Fig. 17.1c, d, f, g). Therefore, this model recapitulates different steps of skin carcinoma progression, mimics the microenvironment of a developing skin carcinoma, and allows a kinetic analysis of tumor–stroma interactions during tumor development and angiogenic switch (Mueller and Fusenig 2004).

The stromal compartment of malignant tumors is important and continuous interaction between tumor and stromal cells is prerequisite for carcinoma development and progression. One important finding of studies using the transplantation chamber assay is that although tumor cells rapidly proliferate forming multilayered epithelia on top of the gel, invasive growth, the hallmark of malignancy, does not manifest until the vascularized granulation tissue has replaced the gel and approached the tumor cells. Thus, a close association and interaction between tumor and stromal tissue is obviously needed for tumor invasion (Skobe et al. 1997; Mueller and Fusenig 2002). This sequential course of stromal activation and tumor invasion indicates that rapid interactions between tumor and host cells occur on transplantation, resulting first in activation of stromal tissue. However, such an early sequence of events became apparent only by using the matrix-inserted transplantation assay, in which a collagen gel is interposed between tumor cells and stromal compartments. This particular transplantation model displays several crucial advantages. The collagen gel provides an appropriate substratum for tumor cell attachment and serves as a temporal "barrier" preventing immediate contact between grafted tumor cells and host cells. However, it allows a dialogue between these cells via diffusible factors (growth factors, angiogenic factors, cytokines/chemokines) allowing to characterize the kinetics of the different stromal responses in depth (infiltration of inflammatory cells, fibroblastic cells, and angiogenesis). Indeed, the differential tumorigenic potential of cells is even more evident in this surface transplant system in which benign clones form slightly dysplastic keratinizing epithelia, while malignant cells develop into invasive carcinoma (Boukamp et al. 1990, Breitkreutz et al. 1991). Interestingly, the onset strength of the stromal reaction clearly correlates with the stage of malignancy, being later and weakest in transplants of benign and earliest and strongest in those of metastatic cells.

Fig. 17.1 (Continued) progressively replaced by a granulation tissue at 2 weeks (**c**), leading to a vascularized tumor at 3 weeks (**d**). **b–d**: Immunostaining of CD11b, a marker of granulocytes and monocytes/macrophages (K: tumor cells; G: collagen gel; H: host tissue). **e–g** Immunostaining of keratinocytes (green) and vessels (red). **e** A score of + is attributed when blood vessels remain below the collagen gel or particularly the matrix. **f** When blood vessels get into close contact with malignant epithelial layer, a score of ++ is assigned. **g** The score is +++ when tumor and blood vessels are intermingled. **h–j** Tumor invasion can be quantified by manual measurements of the distance "*i*," between the top of tumor cell layer and the deepest front of tumor spread (yellow arrows) (**e, f**). In this system, vascularization is estimated by measuring the distance "*d*" separating tumor cells from the front blood vessel migration (**e**). Quantification by computerized image analysis consists in determining the distribution of tumor/endothelial cell density as a function of the distance to the upper boundary of tumor (**k–m**). (*See also* Color Insert I)

Methods to Quantify Tumor Cell Invasion

Different methods are used to quantify tumor cell invasion and the extent of the angiogenic response (Fig. 17.1). The more simple appreciation of tumor development and progression is a semiquantitative scoring of cell invasion (Fig. 17.1). Score + to +++ is assigned according to the infiltration of blood vessels into the collagen gels toward tumor cells (Bajou et al. 1998, 2001, Jost et al. 2006) (Fig. 17.1b–g). A more objective method of quantification relies on the manual measurements of (1) tumor invasion by determining the distance "i" between the top of the tumor cell layer and the deepest front of tumor spread (Fig. 17.1d, e) and (2) angiogenesis by estimating the distance "d" separating the tumor cells from the front of blood vessel migration (Fig. 17.1d) (Bajou et al. 2004). More recently, an original image analysis algorithm for computerized processing has been set up (Jost et al. 2006, Blacher et al. 2008). This method determines the tumor/endothelial cell density as a function of the distance to the upper boundary of the tumor layer (Fig. 17.1). It provides more information relating to the morphology of the studied structures, and can precisely estimate the intermingling between tumor cells and blood vessels (Jost et al. 2006). Another method to quantify tumor cell invasion and angiogenesis is the morphometric analysis using analySIS software (Olympus). Using this tool, tumor transplants immunostained for endothelial cell and tumor cell markers were photographed and divided into two major compartments, 300 μm below and 500 μm within the tumor, respectively. The CD31-stained areas were calculated by analySIS software for these two compartments, leading to a quantitative estimation for tumor cell invasion (within tumor) and for mean vessel density (below tumor) (Vosseler et al. 2005, Obermueller et al. 2004). Interestingly, in addition to estimating the malignant features of different tumor cells, the surface model offers the possibility to determine the key molecular determinants of both tumor and host compartments.

Proteases and Inhibitors as Key Molecular Determinants of the Host Compartment

Different classes of proteases have been implicated during different stages of cancer progression. The principal classes of proteases involved are the MMPs, serine proteases, and cathepsins (van Hinsbergh et al. 2006). The functional association between MMPs and the plasminogen activator (PA)/plasmin systems, in particular the role of plasmin as a pro-MMP activator, has generated substantial attention in the context of both physiological and pathological tissue remodeling (Folgueras et al. 2004, Lee and Huang 2005, van Hinsbergh et al. 2006). Members of the MMP family and components of the PA system are coexpressed during development, tissue remodeling, tissue repair, but also in multiple diseases such as tumor invasion and metastasis (Lijnen 2001). These proteases control cell proliferation, migration, and invasion by remodeling the ECM and releasing growth factors sequestered in the matrix. Furthermore, by cleaving extracellular components and shedding cell surface molecules, the proteases have been implicated in the activation and bioavailability of cytokines/chemokines, growth factor receptors, and integrins (Rakic et al. 2003, Egeblad and Werb 2002, Noel et al. 2004, Overall and Kleifeld 2006). Although it was initially believed that high production of proteases (MMPs and PA/plasmin system) came from neoplastic cells themselves, host stromal cells are now recognized as essential producers of proteases (Noel et al. 2008, Egeblad and Werb 2002).

The PA–plasmin system is a pericellular proteolytic system with pleiotropic functions in physiological and pathological tissue remodeling (Rakic et al. 2003, Durand et al. 2004, Noel et al. 2004, Binder et al. 2007) (*see* also Chaps. 10 and 11). It is a complex system of serine proteases, protease inhibitors, and protease receptors that governs the conversion of the abundant protease zymogen, plasminogen (Plg), into active plasmin. Activation of Plg appears to be strictly associated with the cell surface via the binding to specific receptors, as well as with other surfaces that present kinetically favorable circumstances for Plg activation, such as the fibrin thrombus (Myohanen and Vaheri 2004). Surface-generated plasmin is relatively protected from its primary physiological inhibitor α2-antiplasmin. Cell surface Plg activation by the two PAs, urokinase-type PA (uPA) and tissue-type PA (tPA), is regulated by two physiological inhibitors, Plg activator inhibitor-1 and -2 (PAI-1 and PAI-2), each forming a 1:1 complex with uPA and tPA. As an inhibitor of proteases, PAI-1 was initially viewed as an antiangiogenic and antitumoral factor. However, unexpected and novel results were obtained when the surface transplantation model using mouse malignant keratinocytes was applied to PAI-1-deficient mice. Indeed, the grafted mouse skin carcinoma cells failed to invade the stroma of PAI-1-deficient mice (Table 17.1) (Bajou et al. 1998, Bajou et al. 2001). These results were opposite to the initial hypothesis that a deficiency of protease inhibitor would enhance tumor growth and invasion. In these deficient mice, tumor cells can induce granulation tissue formation beneath the collagen gel and angiogenesis, but new blood vessels cannot reach the tumor layer. These results have been confirmed in other experimental models (Gutierrez et al. 2000, Devy et al. 2002, McMahon et al. 2001). The fact that the PAI-1 production by tumor cells cannot circumvent host cell deficiency, even at high concentration by transfecting malignant keratinocytes with PAI-1 cDNA (Bajou et al. 2004), clearly emphasizes the stroma tissue as the most important source of PAI-1 (Bajou et al. 1998, Maillard et al. 2005). Additionally, the local variation of PAI-1 concentration is very important for tumor development. Indeed, a dose-dependent proangiogenic effect of this inhibitor has been demonstrated in the SCC model (Bajou et al. 2004) and in another model of pathological angiogenesis, the choroidal neovascularization assay (Lambert et al. 2003a). The proangiogenic effect of PAI-1 in the surface transplantation model relies on its capacity to interact with uPA (Bajou et al. 2001). However, lack of uPA, tPA, or uPA receptor, as well as combined deficiencies of uPA and tPA, did not affect tumor angiogenesis, whereas lack of Plg reduced it (Table 17.1). Overall, these data indicate that plasmin proteolysis, even though

Table 17.1 Effects of serine protease gene deletion on tumor angiogenesis and invasion

KO mice	MMP cellular sources		In vivo transplantations scoring		MMP effect on tumor	References
	Host cells	Epithelial tumor cells	Tumor invasion	Angiogenesis		
Plg	−	−	+	+	Positive regulator	Bajou et al. 2001
uPA	+	+	+++	+++	/	Bajou et al. 2001
tPA	+	+	+++	+++	/	Bajou et al. 2001
uPA/tPA	+	+	+++	+++	/	Bajou et al. 2001
uPAR	+	+	+++	+++	/	Bajou et al. 2001
PAI-1 (C57Bl/6)	+	+	0	0	Positive regulator	Bajou et al. 1998; Bajou et al. 2001
PAI-1 (Rag-1 KO) (nu/nu)	+	+	0	0	Positive regulator	Maillard et al. 2005

Components of PA/plasmin system can be expressed by epithelial tumor cells (keratinocytes) or healthy stromal cells. Three weeks after tumor transplantation, angiogenesis and tumor invasion are determined by scoring + to +++ as described in Fig. 17.1. All corresponding wild-type (WT) mice presented an invasive and angiogenic score (+++).

essential, must be tightly controlled during tumor angiogenesis and other enzymes may, at least in part, contribute to the angiogenic phenotype (Bajou et al. 2001).

The MMPs constitute an additional large family of structurally related matrix-degrading proteases that have pivotal roles in development, tissue remodeling, and cancer. MMPs share a number of common structural and functional features (Lopez-Otin and Overall, 2002, Folgueras et al. 2004; Greenlee et al. 2007). All MMPs have essential zinc and calcium ions, are synthesized as zymogens, and are inhibited by endogenous inhibitors, such as α2-macroglobuline and specific MMP inhibitors or TIMPs (tissue inhibitors of metalloproteinases) that reversibly inhibit proteases in a 1:1 enzyme–inhibitor complex (Sternlicht and Werb 2001). MMPs have multiple domains that control their secretion, specificity, and substrate binding. Their function is tightly regulated at the level of gene expression, zymogen activation, enzyme activity, and cell-surface localization (Greenlee et al. 2007). The MMP protease family includes soluble enzymes secreted into the extracellular milieu and others associated with the cell surface (MT-MMPs, membrane-type metalloproteinases) (for review: Egeblad and Werb 2002, Folgueras et al. 2004). MMPs target a large diversity of substrates including growth factor receptors, cell adhesion molecules, chemokines, cytokines, apoptotic ligands, and pro-MMPs (Egeblad and Werb 2002, Overall and Kleifeld 2006, Cauwe et al. 2007). The characterization of new substrates as well as the generation of genetically modified animal models of gain or loss of MMP function has demonstrated the relevance of MMP activities in cancer development and progression (Folgueras et al. 2004). The gelatinase subgroup of MMPs, represented by MMP-2 (gelatinase A) and MMP-9 (gelatinase B), was the first class of MMP to be described as protumoral (Kleiner and Stetler-Stevenson, 1999, Duffy et al. 2000, Tester et al. 2004). Thereafter, their protumoral role has been extended to other MMPs (Pendas et al. 2004, Rio 2005, Overall and Kleifeld 2006). In this context, it is worth noting the key contribution of membrane-type MMP in cancer cell invasion. MT1-, MT2-, and MT3-MMP appear as a key triad for cancer cell invasion through the basement membrane (Hotary et al. 2002), while MT4-MMP is involved in metastatic dissemination of breast carcinomas (Chabottaux et al. 2006). Surprisingly, some MMPs display a protective function toward cancer progression (Matrisian and Lopez-Otin 2007). In fact, MMP-8-deficient mice challenged with carcinogens showed a markedly increased susceptibility to tumorigenesis in comparison to corresponding wild-type (WT) mice (Balbin et al. 2003). This study was the first report of an MMP having a protective role in cancer progression, validating MMP-8 as an antitarget in cancer therapy (Overall and Kleifeld 2006).

Key Determinants of the Tumor Compartment

In vivo, MMPs, in particular gelatinases and collagenases, have been found to be differentially regulated in premalignant and malignant skin SCC and breast carcinoma tumor cells and in their adjacent stroma (Borchers et al. 1997, Airola and Fusenig 2001, Werb et al. 1999). Yet, MMP expression data obtained in monolayer cultures of skin SCC cells did not identify a significant difference between benign and malignant tumor cells (Ala-Aho et al. 2000, Bachmeier et al. 2000) (Meade-Tollin et al. 1998). Secretion of a number of proteases such as proMMP-2, proMMP-9, and MMP-13 as well as very low levels of MMP-3 was already observed in monocultures of human immortalized nontumorigenic HaCaT cells (Papakonstantinou et al. 2005) and expression of MMP-3 and to a lesser extend MMP-9 increased with progression to benign (HaCaT-ras A-5) and even more so to enhanced malignant (HaCaT-ras II-4RT) tumor cells (Bachmeier and Nerlich 2002). However, when immortal and tumorigenic HaCaT cells were cultured in an in vivo-like environment on a collagen type-1 gel containing normal human dermal fibroblasts, a profound influence of the microenvironment, that is of the ECM and of stroma fibroblasts, on MMP expression became apparent. In these cocultures, immortal nontumorigenic HaCaT cells, benign (A-5) cells, and enhanced malignant (A-5RT3) cells exhibited a striking difference in their MMP expression pattern. In this tissue environment, MMP-1 mRNA and protein were strongly upregulated in malignant A-5RT3 cells, only weakly expressed in benign A-5 cells, and almost absent in immortalized HaCaT cells (Airola and Fusenig 2001). This enhanced expression of MMP-1 was further confirmed in two other malignant HaCaT-ras clones as well as in 2/2 primary squamous cell carcinoma lines. Finally, in vivo, malignant A-5RT3 tumors expressed MMP-1 mRNA consistently, preferentially at the tumor border. In contrast, MMP-1 expression was absent in the transplants of A-5 cells and HaCaT cells (Airola and Fusenig 2001).

This prominent difference in MMP-expression dependent on ECM and tissue organization suggests a very strong influence of the microenvironment with its stromal cells on the regulation of MMPs. The time course of malignant tumor growth that begins with an early onset of stromal activation, the rapid penetration of vessels and perivascular cells through the collagen gel toward the tumor cells, and their eventual infiltration into the malignant tumor tissue highlights another striking difference between benign and malignant transplants that lies in the differential dynamics of angiogenesis induction. Angiogenesis induction was transient in benign yet persistent in malignant tumors and found to be controlled by the regulation of the VEGF receptors 1 and 2. VEGFR-1 (vascular endothelial growth factor receptor) and -2 were downregulated in the stroma of benign, but continuously expressed in the malignant transplants. In contrast, VEGF-A expression persisted in both types of tumor cell transplants independently of the kinetics of angiogenesis (Skobe et al. 1997).

These observations gave a clear indication for the essential inductive or permissive role of the stroma for tumor invasion. In line with this essential role, VEGFR-2 blockade caused vessel regression and normalization as well as stromal maturation that ultimately resulted in a reversion from a highly malignant and invasive to a noninvasive tumor phenotype. Vessel regression was followed by downregulation of expression of both VEGFR-2 and VEGFR-1 on endothelial cells and increased association of α-smooth muscle actin-positive cells with small vessels indicating their normalization that was further supported by a regular ultrastructure. The phenotypic regression of an invasive carcinoma to a well-demarcated dysplastic squamous epithelium was accentuated by the establishment of a clearly structured epithelial basement membrane and the accumulation of collagen bundles in the stabilized connective tissue. This normalization of the tumor stroma border coincided with downregulated expression of the stromal MMP-9 and -13, which supposedly resulted in attenuated turnover of ECM components, permitting their structural organization (Fig. 17.2) (Vosseler et al. 2005). Thus, analysis of tumor–stroma interaction of skin SCCs in the matrix-inserted surface transplantation model provided abundant evidence for an essential role of the tumor stroma in regulating tumor malignancy as well as the expression of progression-associated MMPs. In particular, (1) a clear association of MMP expression in tumor and stromal cells with tumor progression was observed only in the tissue context of either an in vitro organotypic model or the transplantation model and (2) a stromal normalization achieved by VEGFR-2 blockade in highly malignant tumors induced a phenotypic reversion to a premalignant dysplasia that was in part mediated by a downregulation of stromal MMPs.

MMP-Deficient Mice as Models to Investigate the Stromal Contribution to Tumor Progression

To further elucidate the role of stromal MMPs in tumor progression and angiogenesis, the surface transplantation model has been recently applied to different MMP-deficient mice. The single deficiency of MMP-3, -8, and -11 or the combined

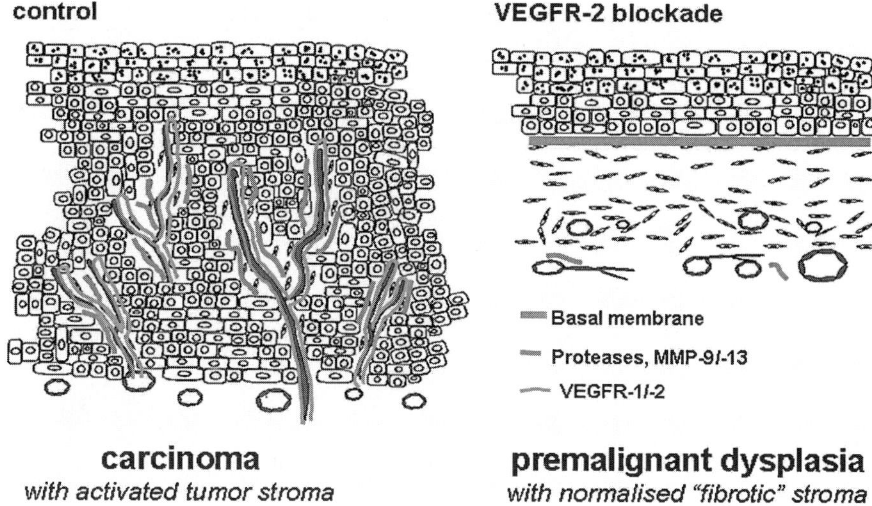

Fig. 17.2 Downregulation of stromal matrix metalloproteinase (*MMP*) expression by blockade of angiogenesis. Inhibiting angiogenesis in malignant transplants by the vascular endothelial growth factor receptor-2 (*VEGFR*-2) blocking antibody DC101 completely abrogates tumor vascularization and stromal MMP expression, and as a consequence, tumor invasion (Vosseler et al. 2005). (*See also* Color Insert I)

MMP-3/-9 deficiency did not affect tumor invasion and angiogenesis (Table 17.2) (Masson et al. 2005). Similarly, the absence of MMP-2 or MMP-9 in host tissue did not impair tumor progression (Table 17.2) (Masson et al. 2005). In sharp contrast, both tumor invasion and vascularization were impaired by the combined deficiency of MMP-2 and -9 (Masson et al. 2005). These results indicate that the concomitant stromal production of MMP-2 and -9 is required for tumor invasion and angiogenesis. Of particular importance is the necessity of specific interactions occurring between tumor cells and mesenchymal cells producing MMP-2, as well as inflammatory cells secreting MMP-9 (Masson et al. 2005). A synergistic contribution of MMP-2 and -9 in pathological angiogenesis has also been demonstrated in choroidal neoangiogenesis induced by laser burn (Lambert et al. 2003b). Interestingly and in contrast to most MMP deficiencies in mice described so far, the angiogenic response was accelerated and tumor invasion increased in MMP-19-deficient mice in comparison to WT mice (Jost et al. 2006). Indeed, endothelial cell recruitment was significantly increased 2 weeks after the transplantation, leading to an acceleration of tumor vascularization. As tumor vascularization precedes malignant invasion, this acceleration induced an early tumor invasion in MMP-19-deficient mice in comparison to corresponding WT mice, 21 days after transplantation. These data support the recent discovery of some MMPs as protective molecules toward cancer progression (Matrisian and Lopez-Otin 2007).

Table 17.2 Effects of MMP gene deletion on tumor angiogenesis and invasion

KO mice	MMP cellular sources		In vivo transplantations scoring		MMP effect on tumor	References
	Host cells	Epithelial tumor cells	Tumor invasion	Angiogenesis		
MMP-2	+	+	+++	+++	/	Masson et al. 2005
MMP-3	+	+	+++	+++	/	Masson et al. 2005
MMP-8	+	/	+++	+++	/	Unpublished data
MMP-9	+	/	+++	+++	/	Masson et al. 2005
MMP-11	+	/	+++	+++	/	Unpublished data
MMP-19	+	/	++++	++++	Negative regulator	Jost et al. 2006
MMP-2/-9	−	−	0	0	Positive regulator	Masson et al. 2005
MMP-3/-9	−	−	+++	+++	/	Masson et al. 2005

MMPs can be expressed by epithelial tumor cells (keratinocytes) or healthy stromal cells. Three weeks after tumor transplantation, angiogenesis and tumor invasion are determined by scoring + to +++ as described in the text and in Fig. 17.1. All corresponding wild-type (WT) mice presented an invasive and angiogenic score (+++).

Conclusion

Taken together all these data highlight several major aspects in protease expression in cancer. (1) Protease expression in tumor and stromal cells seems to contribute to tumor malignancy. Yet expression of tumor-derived proteases is clearly regulated by the in vivo tissue context, that is the tumor microenvironment. (2) Different proteases produced by stromal cells are the most important regulators of cancer development and progression. Yet, one has to keep in mind that the contribution of different stromal proteases has to be carefully evaluated especially in lieu of their potential usefulness as therapeutic targets since they sometimes act in an opposite manner. The matrix-inserted surface transplantation assay is therefore a highly valuable tool to identify the target (positive regulators) and anti-target (negative regulators) nature of different MMPs.

References

Airola, K., and Fusenig, N.E. 2001. Differential stromal regulation of MMP-1 expression in benign and malignant keratinocytes. Journal of Investigative Dermatology 116:85–92.
Ala-Aho, R., Johansson, N., Grenman, R., Fusenig, N.E., Lopez-Otin, C., and Kahari, V.M. 2000. Inhibition of collagenase-3 (MMP-13) expression in transformed human keratinocytes by

interferon-gamma is associated with activation of extracellular signal-regulated kinase-1,2 and STAT1. Oncogene 19:248–257.

Andarawewa, K.L., Motrescu, E.R., Chenard, M.P., Gansmuller, A., Stoll, I., Tomasetto, C., and Rio, M.C. 2005. Stromelysin-3 is a potent negative regulator of adipogenesis participating to cancer cell-adipocyte interaction/crosstalk at the tumor invasive front. Cancer Research 65:10862–10871.

Bachmeier, B.E., and Nerlich, A.G. 2002. Immunohistochemical pattern of cytokeratins and MMPs in human keratinocyte cell lines of different biological behaviour. International Journal of Oncology 20:495–499.

Bachmeier, B.E., Boukamp, P., Lichtinghagen, R., Fusenig, N.E., and Fink, E. 2000. Matrix metalloproteinases-2,-3,-7,-9 and -10, but not MMP-11, are differentially expressed in normal, benign tumorigenic and malignant human keratinocyte cell lines. Biological Chemistry 381:497–507.

Bajou, K., Noel, A., Gerard, R.D., Masson, V., Brunner, N., Holst-Hansen, C., Skobe, M., Fusenig, N.E., Carmeliet, P., Collen, D., and Foidart, J.M. 1998. Absence of host plasminogen activator inhibitor 1 prevents cancer invasion and vascularization. Nature Medicine 4:923–928.

Bajou, K., Masson, V., Gerard, R.D., Schmitt, P.M., Albert, V., Praus, M., Lund, L.R., Frandsen, T.L., Brunner, N., Dano, K., Fusenig, N.E., Weidle, U., Carmeliet, G., Loskutoff, D., Collen, D., Carmeliet, P., Foidart, J.M., and Noel, A. 2001. The plasminogen activator inhibitor PAI-1 controls in vivo tumor vascularization by interaction with proteases, not vitronectin. Implications for antiangiogenic strategies. The Journal of Cell Biology 152:777–784.

Bajou, K., Maillard, C., Jost, M., Lijnen, R.H., Gils, A., Declerck, P., Carmeliet, P., Foidart, J.M., and Noel, A. 2004. Host-derived plasminogen activator inhibitor-1 (PAI-1) concentration is critical for in vivo tumoral angiogenesis and growth. Oncogene 23:6986–6990.

Balbin, M., Fueyo, A., Tester, A.M., Pendas, A.M., Pitiot, A.S., Astudillo, A., Overall, C.M., Shapiro, S.D., and Lopez-Otin, C. 2003. Loss of collagenase-2 confers increased skin tumor susceptibility to male mice. Nature Genetics 35:252–257.

Balkwill, F., and Coussens, L.M. 2004. Cancer: an inflammatory link. Nature 431:405–406.

Benelli, R., Morini, M., Carrozzino, F., Ferrari, N., Minghelli, S., Santi, L., Cassatella, M., Noonan, D.M., and Albini, A. 2001. Neutrophils as a key cellular target for angiostatin: implications for regulation of angiogenesis and inflammation. FASEB Journal 15:267–269.

Bergers, G., and Benjamin, L.E. 2003. Tumorigenesis and the angiogenic switch. Nature reviews Cancer 3:401–410.

Binder, B.R., Mihaly, J., and Prager, G.W. 2007. uPAR-uPA-PAI-1 interactions and signaling: avascular biologist's view. Thrombosis and Haemostasis 97:336–342.

Blacher, S., Jost, M., Melen, L., Lund, L.R., Romer, J., Foidart, J.M. and Noel, A. 2008. Quantification of in vivo tumor invasion and vascularization by computerized image analysis. Microvascular Research 75(2):169–178.

Borchers, A.H., Steinbauer, H., Schafer, B.S., Kramer, M., Bowden, G.T., and Fusenig, N.E. 1997. Fibroblast-directed expression and localization of 92-kDa type IV collagenase along the tumor–stroma interface in an in vitro three-dimensional model of human squamous cell carcinoma. Molecular Carcinogenesis 19:258–266.

Boukamp, P., Stanbridge, E.J., Foo, D.Y., Cerutti, P.A., and Fusenig, N.E. 1990. C-Ha-Ras oncogene expression in immortalized human keratinocytes (HaCaT) alters growth-potential in vivo but lacks correlation with malignancy. Cancer Research 50:2840–2847.

Breitkreutz, D., Boukamp, P., Ryle, C.M., Stark, H.J., Roop, D.R., and Fusenig, N.E. 1991. Epidermal morphogenesis and keratin expression in C-Ha-Ras-transfected tumorigenic clones of the human HaCaT cell-line. Cancer Research 51:4402–4409.

Carmeliet, P. 2005. Angiogenesis in life, disease and medicine. Nature 438:932–936.

Carmeliet, P., and Jain, R.K. 2000. Angiogenesis in cancer and other diseases. Nature 407:249 257.

Cauwe, B., Van den Steen, P.E., and Opdenakker, G. 2007. The biochemical, biological, and pathological kaleidoscope of cell surface substrates processed by matrix metalloproteinases. Critical Reviews in Biochemistry and Molecular Biology 42:113–185.

Chabottaux, V., Sounni, N.E., Pennington, C.J., English, W.R., van den Brule, F., Blacher, S., Gilles, C., Munaut, C., Maquoi, E., Lopez-Otin, C., Murphy, G., Edwards, D.R., Foidart, J.M., and Noel, A. 2006. Membrane-type 4 matrix metalloproteinase promotes breast cancer growth and metastases. Cancer Research 66:5165–5172.

Coussens, L.M., and Werb, Z. 2001. Inflammatory cells and cancer: think different! Journal of Experimental Medicine 193:F23–F26.

Coussens, L.M., Fingleton, B., and Matrisian, L.M. 2002. Matrix metalloproteinase inhibitors and cancer: trials and tribulations. Science 295:2387–2392.

Devy, L., Blacher, S., Grignet-Debrus, C., Bajou, K., Masson, R., Gerard, R.D., Gils, A., Carmeliet, G., Carmeliet, P., Declerck, P.J., Noel, A., and Foidart, J.M. 2002. The pro- or antiangiogenic effect of plasminogen activator inhibitor 1 is dose dependent. FASEB Journal 16:147–154.

Duda, D.G., Batchelor, T.T., Willet, C.G., and Jain, R.K. 2007. VEGF-targeted cancer therapy strategies: current progress, hurdles and future prospects. Trends in Molecular Medicine 13:223–230.

Duffy, M.J., Maguire, T.M., Hill, A., McDermott, E., and O'Higgins, N. 2000. Metalloproteinases: role in breast carcinogenesis, invasion and metastasis. Breast Cancer Research 2:252–257.

Durand, M.K., Bodker, J.S., Christensen, A., Dupont, D.M., Hansen, M., Jensen, J.K., Kjelgaard, S., Mathiasen, L., Pedersen, K.E., Keldal, S., Wind, T., and Andreasen, P.A. 2004. Plasminogen activator inhibitor-1 and tumour growth, invasion, and metastasis. Thrombosis and Haemostasis 91:438–449.

Dvorak, H.F. 1986. Tumors: wounds that do not heal. Similarities between tumor stroma generation and wound healing. The New England Journal of Medicine 315:1650–1659.

Egeblad, M., and Werb, Z. 2002. New functions for the matrix metalloproteinases in cancer progression. Nature Reviews Cancer 2:161–174.

Ferrara, N., and Kerbel, R.S. 2005. Angiogenesis as a therapeutic target. Nature 438:967–974.

Fidler, I.J. 2003. Timeline – The pathogenesis of cancer metastasis: the 'seed and soil' hypothesis revisited. Nature Reviews Cancer 3:453–458.

Folgueras, A.R., Pendas, A.M., Sanchez, M., Lopez-Otin, C. 2004. Matrix metalloproteinases in cancer: from new functions to improved inhibition strategies. The International Journal of Developmental Biology 48:411–424.

Fusenig, N.E. 1992. Cell interaction and epithelial differentiation. In R.I. Freshney (ed.) Culture of epithelial cells: New york: Wiley-Liss Inc; 25–27.

Fusenig, N.E., Breitkreutz, D., Dzarlieva, R.T., Boukamp, P., Bohnert, A., and Tilgen, W. 1983. Growth and differentiation characteristics of transformed keratinocytes from mouse and human skin in vitro and in vivo. Journal of Investigative Dermatology 81(1 Suppl):168s–175s.

Greenlee, K.J., Werb, Z., and Kheradmand, F. 2007. Matrix metalloproteinases in lung: multiple, multifarious, and multifaceted. Physiological Reviews 87:69–98.

Gutierrez, L.S., Schulman, A., Brito-Robinson, T., Noria, F., Ploplis, V.A., and Castellino, F.J. 2000. Tumor development is retarded in mice lacking the gene for urokinase-type plasminogen activator or its inhibitor, plasminogen activator inhibitor-1. Cancer Research 60:5839–5847.

Hanahan, D., and Weinberg, R.A. 2000. The hallmarks of cancer. Cell 100:57–70.

Hotary, K.B., Yana, I., Sabeh, F., Li, X.Y., Holmbeck, K., Birkedal-Hansen, H., Allen, E.D., Hiraoka, N., and Weiss, S.J. 2002. Matrix metalloproteinases (MMPs) regulate fibrin-invasive activity via MT1-MMP-dependent and -independent processes. The Journal of Experimental Medicine 195:295–308.

Jost, M., Folgueras, A.R., Frerart, F., Pendas, A.M., Blacher, S., Houard, X., Berndt, S., Munaut, C., Cataldo, D., Alvarez, J., Melen-Lamalle, L., Foidart, J.M., Lopez-Otin, C., and Noel, A. 2006. Earlier onset of tumoral angiogenesis in matrix metalloproteinase-19-deficient mice. Cancer Research 66:5234–5241.

Kalluri, R. 2003. Basement membranes: structure, assembly and role in tumour angiogenesis. Nature Reviews Cancer 3:422–433.

Kalluri, R., and Zeisberg, M. 2006. Fibroblasts in cancer. Nature Reviews Cancer 6:392–401.

Kleiner, D.E., and Stetler-Stevenson, W.G. 1999. Matrix metalloproteinases and metastasis. Cancer Chemotherapy and Pharmacology 43:S42–S51.

Kuperwasser, C., Chavarria, T., Wu, M., Magrane, G., Gray, J.W., Carey, L., Richardson, A., and Weinberg, R.A. 2004. Reconstruction of functionally normal and malignant human breast tissues in mice. PNAS 101:4966–4971.

Lambert, V., Munaut, C., Carmeliet, P., Gerard, R.D., Declerck, P., Gils, A., Claes, C., Foidart, J.M., Noel, A., and Rakic, J.M. 2003a. Dose-dependent modulation of choroidal neovascularization by plasminogen activator inhibitor type 1: implications for clinical trials. Investigative Ophthalmology & Visual Science 44:2791–2797.

Lambert, V., Wielockx, B., Munaut, C., Galopin, C., Jost, M., Itoh, T., Werb, Z., Baker, A., Libert, C., Krell, H.W., Foidart, J.M., Noel, A., and Rakic, J.M. 2003b. MMP-2 and MMP-9 synergize in promoting choroidal neovascularization. FASEB Journal 17:2290–2292.

Lee, C., and Huang, T. 2005. Plasminogen activator inhibitor-1: the expression, biological functions, and effects on tumorigenesis and tumor cell adhesion and migration. Journal of Cancer Molecules 1:25–36.

Li, B., Sharpe, E.E., Maupin, A.B., Teleron, A.A., Pyle, A.L., Carmeliet, P., and Young, P.P. 2006. VEGF and PlGF promote adult vasculogenesis by enhancing EPC recruitment and vessel formation at the site of tumor neovascularization. FASEB Journal 20:1495–1497.

Lijnen, H.R. 2001. Plasmin and matrix metalloproteinases in vascular remodeling. Thrombosis and Haemostasis 86:324–333.

Lopez-Otin, C., and Overall, C.M. 2002. Protease degradomics: a new challenge for proteomics. Nature Reviews Molecular Cell Biology 3:509–519.

Maillard, C., Jost, M., Romer, M.U., Brunner, N., Houard, X., Lejeune, A., Munaut, C., Bajou, K., Melen, L., Dano, K., Carmeliet, P., Fusenig, N.E., Foidart, J.M., and Noel, A. 2005. Host plasminogen activator inhibitor-1 promotes human skin carcinoma progression in a stage-dependent manner. Neoplasia 7:57–66.

Masson, V., de la Ballina, L.R., Munaut, C., Wielockx, B., Jost, M., Maillard, C., Blacher, S., Bajou, K., Itoh, T., Itohara, S., Werb, Z., Libert, C., Foidart, J.M., and Noel, A. 2005. Contribution of host MMP-2 and MMP-9 to promote tumor vascularization and invasion of malignant keratinocytes. FASEB Journal 19:234–236.

Matrisian, L., and Lopez-Otin, C. 2007. Emerging roles of proteases in tumour suppression. Nature Reviews Cancer 7:800–808.

McMahon, G.A., Petitclerc, E., Stefansson, S., Smith, E., Wong, M.K., Westrick, R.J., Ginsburg, D., Brooks, P.C., and Lawrence, D.A. 2001. Plasminogen activator inhibitor-1 regulates tumor growth and angiogenesis. The Journal of Biological Chemistry 276:33964–33968.

Meade-Tollin, L.C., Boukamp, P., Fusenig, N.E., Bowen, C.P.R., Tsang, T.C., and Bowden, G.T. 1998. Differential expression of matrix metalloproteinases in activated c-ras(Ha)-transfected immortalized human keratinocytes. British Journal of Cancer 77:724–730.

Mueller, M.M., and Fusenig, N.E. 2002. Tumor–stroma interactions directing phenotype and progression of epithelial skin tumor cells. Differentiation 70:486–497.

Mueller, M.M., and Fusenig, N.E. 2004. Friends or foes – Bipolar effects of the tumour stroma in cancer. Nature Reviews Cancer 4:839–849.

Myohanen, H., and Vaheri, A. 2004. Regulation and interactions in the activation of cell-associated plasminogen. Cellular and Molecular Life Sciences 61:2840–2858.

Noel, A., Emonard, H., Polette, M., Birembaut, P., and Foidart, J.M. 1994. Role of matrix, fibroblasts and type IV collagenases in tumor progression and invasion. Pathology, Research and Practice 190:934–941.

Noel, A., Hajitou, A., L'Hoir, C., Maquoi, E., Baramova, E., Lewalle, J.M., Remacle, A., Kebers, F., Brown, P., Calberg-Bacq, C.M., and Foidart, J.M. 1998. Inhibition of stromal matrix metalloproteases: effects on breast-tumor promotion by fibroblasts. International Journal of Cancer 76:267–273.

Noel, A., Maillard, C., Rocks, N., Jost, M., Chabottaux, V., Sounni, N.E., Maquoi, E., Cataldo, D., and Foidart, J.M. 2004. Membrane associated proteases and their inhibitors in tumour angiogenesis. Journal of Clinical Pathology 57:577–584.

Noel, A., Jost, M., and Maquoi, E. 2008. Matrix metalloproteinases at cancer tumor–host interface. Seminars in Cell & Developmental Biology 19:52–60.

Obermueller, E., Vosseler, S., Fusenig, N.E., and Mueller, M.M. 2004. Cooperative autocrine and paracrine functions of granulocyte colony-stimulating factor and granulocyte-macrophage colony-stimulating factor in the progression of skin carcinoma cells. Cancer Research 64:7801–7812.

Overall, C.M., and Kleifeld, O. 2006. Tumour microenvironment – opinion: validating matrix metalloproteinases as drug targets and anti-targets for cancer therapy. Nature Reviews Cancer 6:227–239.

Paget, S. 1889. The distribution of secondary growths in cancer of the breast. Lancet 1:571–573.

Papakonstantinou, E., Aletras, A.J., Glass, E., Tsogas, P., Dionyssopoulos, A., Adjaye, J., Fimmel, S., Gouvousis, P., Herwig, R., Lehrach, H., Zouboulis, C.C., and Karakiulakis, G. 2005. Matrix metalloproteinases of epithelial origin in facial sebum of patients with acne and their regulation by isotretinoin. Journal of Investigative Dermatology 125:673–684.

Pendas, A.M., Folgueras, A.R., Llano, E., Caterina, J., Frerard, F., Rodriguez, F., Astudillo, A., Noel, A., Birkedal-Hansen, H., and Lopez-Otin, C. 2004. Diet-induced obesity and reduced skin cancer susceptibility in matrix metalloproteinase 19-deficient mice. Molecular and Cellular Biology 24:5304–5313.

Rakic, J.M., Maillard, C., Jost, M., Bajou, K., Masson, V., Devy, L., Lambert, V., Foidart, J.M., and Noel, A. 2003. Role of plasminogen activator-plasmin system in tumor angiogenesis. Cellular and Molecular Life Sciences 60:463–473.

Rio, M.C. 2005. From a unique cell to metastasis is a long way to go: clues to stromelysin-3 participation. Biochimie 87:299–306.

Skobe, M., and Fusenig, N.E. 1998. Tumorigenic conversion of immortal human keratinocytes through stromal cell activation. PNAS 95:1050–1055.

Skobe, M., Rockwell, P., Goldstein, N., Vosseler, S., and Fusenig, N.E. 1997. Halting angiogenesis suppresses carcinoma cell invasion. Nature Medicine 3:1222–1227.

Sternlicht, M.D., and Werb, Z. 2001. How matrix metalloproteinases regulate cell behavior. Annual Review of Cell and Developmental Biology 17:463–516.

Stetler-Stevenson, M., and Braylan, R.C. 2001. Flow cytometric analysis of lymphomas and lymphoproliferative disorders. Seminars in Hematology 38:111–123.

Tester, A.M., Waltham, M., Oh, S.J., Bae, S.N., Bills, M.M., Walker, E.C., Kern, F.G., Stetler-Stevenson, W.G., Lippman, M.E., and Thompson, E.W. 2004. Pro-matrix metalloproteinase-2 transfection increases orthotopic primary growth and experimental metastasis of MDA-MB-231 human breast cancer cells in nude mice. Cancer Research 64:652–658.

van Hinsbergh, V.W., Engelse, M.A., and Quax, P.H. 2006. Pericellular proteases in angiogenesis and vasculogenesis. Arteriosclerosis, Thrombosis, and Vascular Biology 26:716–728.

Vosseler, S., Mirancea, N., Bohlen, P., Mueller, M.M., and Fusenig, N.E. 2005. Angiogenesis inhibition by vascular endothelial growth factor receptor-2 blockade reduces stromal matrix metalloproteinase expression, normalizes stromal tissue, and reverts epithelial tumor phenotype in surface heterotransplants. Cancer Research 65:1294–1305.

Werb, Z., Vu, T.H., Rinkenberger, J.L., and Coussens, L.M. 1999. Matrix-degrading proteases and angiogenesis during development and tumor formation. APMIS 107:11–18.

Zhang, W.Y., Matrisian, L.M., Holmbeck, K., Vick, C.C., and Rosenthal, E.L. 2006. Fibroblast-derived MT1-MMP promotes tumor progression in vitro and in vivo. BMC Cancer 6:1–9.

Zigrino, P., Loffek, S., and Mauch, C. 2005. Tumor–stroma interactions: their role in the control of tumor cell invasion. Biochimie 87:321–328.

Chapter 18
Unravelling the Roles of Proteinases in Cell Migration In Vitro and In Vivo

Jelena Gavrilovic and Xanthe Scott

Abstract Metalloproteinases have been implicated in cell migration in many in vitro model systems which involving tumour cell or leukocyte migration. Here the similarities between the migration mechanisms of amoebae, leukocytes and tumour cells are discussed with a particular focus on recent studies of metalloproteinase dependence in tumour cell migration in three dimensional matrices. Some novel in vivo model systems where metalloproteinases or serine proteinases have been explored are discussed. Finally certain matricryptic sites exposed following metalloproteinase remodelling of the extracellular matrix are also considered.

Introduction

Metalloproteinase involvement in cell migration has been an area of research interest for several decades. Cellular interactions with the extracellular matrix have been investigated with molecular approaches and sophisticated microscopy allowing visualisation of a range of cell types in two and three dimensions (2D and 3D) in vitro as well as more recently in vivo. Many reviews have covered the mechanisms underlying cell migration, including the roles of the cytoskeleton in driving the cell forward as well as signalling networks regulating this process. Here, approaches to the study of cell migration will be reviewed, with an emphasis on recent studies regarding the roles of metalloproteinases and some serine proteinases in cell migration in 2D and 3D as well as the roles of some matricryptic sites in extracellular matrix-driven migration.

J. Gavrilovic
Biomedical Research Centre, School of Biological Sciences, University of East Anglia, Norwich, NR4 7TJ, UK, e-mail: j.gavrilovic@uea.ac.uk

D. Edwards et al. (eds.), *The Cancer Degradome.*
© Springer Science + Business Media, LLC 2008

Overview of Cell Migration

Leukocyte Cell Migration and Some Comparisons with Amoeba

The amoeba *Dictyostelium discoideum* has provided much knowledge regarding chemokinetic migration and cytoskeletal organisation. Over several decades the similarities between lymphoid cell migration with that of *Dictyostelium* have been identified (reviewed in Bagorda et al. 2006). Chemotactic migration is essential to *Dictyostelium's* life cycle: free-living amoebae detect bacteria (the food source) through chemotactic migration towards folic acid, a bacterial metabolism product. In periods of stress such as low food availability, amoebae become responsive to cAMP (cyclic adenosine monophosphate), which is secreted by neighbouring amoebae and to which the cells respond chemotactically (Bagorda et al. 2006). Cell migration is a critical part of the inflammatory response to injury when leukocytes are recruited to the site of injury. During an inflammatory response the endothelium becomes 'activated' where endothelial cells produce various pro-inflammatory molecules that include cytokines, chemokines and cell adhesion molecules (CAMs), all of which play a role in the recruitment of leukocytes and/ or platelets to the damaged tissue (reviewed in Libby et al. 2002). In normal physiology the damaged area of, for example, wounded skin is cleared of dead cells and tissue debris by phagocytes and repair begins by cell proliferation and the laying down of new matrix. Under normal conditions inflammation subsides but, for reasons not yet clear, in diseases such as cancer and atherosclerosis the tissue becomes chronically inflamed.

Leukocyte—Endothelial Cell Interactions

The recruitment of monocytes to the endothelium and subsequent transendothelial migration involves multiple steps and requires tight regulation. Initial tethering of monocytes is followed by rolling along the endothelial surface and finally firm, sustained adhesion (reviewed in Worthylake and Burridge 2001). Monocyte binding to the endothelium elicits signals that facilitate the migration of monocytes through the endothelial monolayer. Initial tethering and rolling are primarily mediated by members of the selectin family, whereas firm adhesion involves members of the integrin family and their ligands. However, it is becoming increasingly apparent that there are a series of overlapping roles between different classes of CAMs with evidence of integrins also participating in initial tethering events (reviewed in Steeber et al. 2005). Integrins and their ligands play an essential role in the firm adhesion and subsequent migration of leukocytes.

Selectins are a family of transmembrane glycoproteins. Leukocyte (L-) selectin is constitutively expressed, whereas endothelial (E-) selectin is induced by inflammatory cytokines (reviewed in Blankenberg et al. 2003). Platelet (P-) selectin is primarily expressed in platelets but is also present in endothelial cells. Initial tethering of monocytes to the endothelium can be mediated by selectins and their carbohydrate-based ligands. Whilst selectins have not been described in *Dictyostelium*, this organism does express cell surface cadherin-like molecule DdCAD-1 (reviewed in Bowers-Morrow et al. 2004), which may be important in amoeba cell–cell interactions. Aggregation in *Dictyostelium* is mediated by the lectins discoidins I and II, which interestingly both contain an RGD site (Gabius et al. 1985).

Inflammation is an important feature of tumour progression and features described initially for cellular transmigration in, for example, atherosclerosis are also relevant in cancer. Thus, tumour cell extravasation is influenced by shear flow (Dong et al. 2005) just as the development of atherosclerotic lesions occurs primarily in regions where blood flow is disturbed (reviewed in Libby et al. 2002). Under conditions of laminar shear stress, endothelial cells are ellipsoid and orientated in the direction of flow whereas in regions of disturbed flow endothelial cells lose their uniform orientation and become more polygonal in shape (reviewed in Lusis 2000). In addition, disturbed flow can augment the expression of leukocyte adhesion molecules (reviewed in Libby et al. 2002).

In the context of cancer, cytokines and chemokines are secreted by tumour cells which attract leukocytes including neutrophils and macrophages (reviewed in Coussens and Werb 2002), which in turn produce a wide range of mediators. Cytokine production by tumour cells and associated macrophages also impacts on the development of angiogenesis and thus on tumour invasion. A number of cytokines produced by tumour cells, including anti-inflammatory IL10 (reviewed in Tedgui and Mallat 2006), reduce the T cell response to tumours (Coussens and Werb 2002). Cytokines have a very wide range of activity related to inflammation, including the induction of other cytokines, chemokines, CAMs and proteases, and therefore make a huge contribution to the chronic inflammatory state.

Infection has been associated with tumour development in a number of cancers (reviewed in Coussens and Werb 2002). In general, at sites of infection neutrophils respond to fMLP (formyl-Met-Leu-Phe) released by bacteria which signals through a G Protein-coupled receptor (GPCR). Chemokines (chemotactic cytokines) such as monocyte chemoattractant protein (MCP)-1 and fractalkine are small proteins that direct the migration of circulating leukocytes to sites of inflammation or injury, again signalling through GPCR (reviewed in Charo and Taubman 2004). In addition to chemotactic roles, chemokines can regulate the adhesion of leukocytes to the endothelium by modulating integrin-mediated adhesion (reviewed in Worthylake and Burridge 2001). *Dictyostelium* chemotaxis towards both folic acid and cAMP is mediated through GPCRs (Bagorda et al. 2006), indicating some of the close parallels between neutrophil and *Dictyostelium* migration.

Integrins in Leukocyte Transendothelial Cell Migration

In order for tumour cells or leukocytes to migrate across the endothelial monolayer during extravasation, regulation of CAMs is required. In inflammation, following initial tethering of leukocytes to the endothelium via selectins, which also mediate rolling of cells over the endothelial surface, sustained adhesion to the endothelium is induced through the binding of integrins $\alpha4\beta1$ to VCAM-1 and $\alpha_{L/M}\beta2$ to ICAM-1 (reviewed in Worthylake and Burridge 2001). Cell migration requires cycles of adhesion and detachment in order for a cell to move forward, which suggests crosstalk between different classes of integrins. Tumour cells may also adhere to the endothelium through selectins but do not seem to roll along the endothelial surface (reviewed in Miles et al. 2007). Very similar mechanisms are used by tumour cells to attach firmly to the endothelium, particularly $\alpha4\beta1$ integrin–VCAM-1 interactions as well as $\alpha6\beta1$ binding to as yet to be defined ligands, but presumably including laminin(s), depending on the integrin profile of particular tumour cells (reviewed in Miles et al. 2007).

The integrin $\alpha v\beta3$ is up-regulated on the lumen of the activated endothelium (Hoshiga et al. 1995). Expressed in monocytes and certain tumour cells, this integrin pair can bind to several ECM proteins including vitronectin and fibronectin (Wayner et al. 1991). In addition, integrin $\alpha v\beta3$ binds platelet endothelial (PE) CAM-1, a member of the immunoglobulin family that is expressed in monocytes and at the intercellular junctions of endothelial cells (Piali et al. 1995) and these molecules have been implicated in monocyte transendothelial migration. In monocytic cell lines engagement of $\alpha v\beta3$ integrin decreases $\beta2$ integrin binding to ICAM-1 and promotes transendothelial cell migration (Weerasinghe et al. 1998), indicating that $\alpha v\beta3$ integrin may play a key role in integrin crosstalk. In tumour cell extravasation, this hierarchy of integrin ligation has not been explored although crosstalk between different integrin partners has been reported during cell migration over ECM components (Galvez et al. 2002).

Until very recently, *Dictyostelium* were thought to lack integrins but Cornillon et al. (2006) have now identified an adhesion molecule sibA (similar to integrin beta) which has certain β-integrin features. Mutants in sibA demonstrate a smaller contact area with the underlying substrate than wild-type cells (which could be thought of as a defect in cell spreading) and numbers of cells adhering to the substrate were significantly reduced under a flow of medium. At the molecular level, sibA has a von Willebrand A (VWA) domain highly homologous to β integrin VWA domain, a transmembrane domain and conserved NPxY motifs in the cytoplasmic domain. SibA is expressed at the cell surface and can interact with talin, a cytoskeletal protein important in cell adhesion (Cornillon et al. 2006). A disintegrin containing protein AmpA, possibly related to ADAM or ADAMTS proteinases but lacking a metalloprotease domain, is critical to cell migration in *Dictyostelium* and it has been postulated that this protein may bind to DdCad-1 (Varney et al. 2002). It is also interesting to speculate whether AmpA could be a ligand for the integrin-like sibA.

Roles of Proteinases in Migration of Leukocytes and Tumour Cells Through Endothelial Cell Barriers

Interaction of leukocytes with the endothelium and sub-endothelial ECM results in the up-regulation of metalloproteinases, including MMP2 (Romanic and Madri 1994) and more recently MT1-MMP (Matias-Roman et al. 2005). In fact in the latter study, MT1-MMP was shown to be required during human monocyte migration through filters coated with fibronectin or the extracellular portion of ICAM-1 or VCAM-1 (made as Fc fusions) in a chemotactic assay. In the same study, monocyte transmigration, induced by the chemoattractant MCP-1 through TNFα-activated endothelial cells in vitro, was also dependent on MT1-MMP, demonstrated by antibody inhibition studies (Matias-Roman et al. 2005). However, transmigration of monocytes through resting endothelial cells (i.e. not treated with TNFα) was independent of MT1-MMP. Of relevance here is the observation that TIMP1 inhibits monocyte migration across a resting endothelial layer also implying a role for MMPs but independent of MT-MMPs (Bar-Or et al. 2003). Matias-Roman et al. (2005) also further demonstrated that MT1-MMP co-localised with profilin, a marker of the leading edge of migrating cells, in monocytes cultured on TNFα-activated endothelial cells. Arroya's group also showed that monocytes migrating over fibronectin, VCAM-1 or ICAM-1 displayed clustered MT1-MMP, whereas cells on BSA had diffuse MT1-MMP staining.

More recently, MT1-MMP has been shown to shed ICAM-1 and to play a key role in transendothelial cell migration (Sithu et al. 2007). In this study, endothelial cells over-expressing ICAM-1 promoted transendothelial migration in a TIMP2- and TIMP3-dependent manner. ICAM-1 was shown to bind to MT1-MMP, co-localising in endothelial cell surface ruffles. This distribution of MT1-MMP was disrupted when ICAM-1 cytoplasmic tail mutants were expressed and since trans-endothelial cell migration was markedly reduced in these mutants but partially rescued by MT1-MMP, a complex role for MT1-MMP is suggested in this process. These results are in keeping with those of Matias-Roman et al. (2005) since MT1-MMP was shown to cluster when the cells were migrating over an ICAM-1 substrate. The study of leukocyte extravasation in MT1-MMP-null mice would be of interest, if these mice could tolerate such investigation. Leukocytes derived from MT1-MMP-null mice implanted into wild-type mice would answer the question whether MT1-MMP is required on the leukocyte side of the equation. In models of shear stress, ICAM-1 is shed by MMP9 (Sultan et al. 2004) and by ADAM17 in response to PMA stimulation (Tsakadze et al. 2006), indicating that both cellular and stimulus context are critical to specific proteinase involvement.

Just as leukocytes exit the vasculature at sites of inflammation or injury, tumour cell extravasation involves interaction with the endothelium at metastatic locations as well as sub-endothelial matrix. Tumour cells use many of the signals described above for leukocytes including selectins and chemokines to orchestrate invasion and metastasis (reviewed in Coussens and Werb 2002). In the next section, recent studies exploring roles of metalloproteinases in both tumour cells and leukocytes in in vitro and in vivo models will be reviewed, highlighting recent studies where different models have yielded intriguing data.

Leukocyte and Tumour Cell Migration in 2D and 3D

Many studies have been performed which have shed light on metalloprotease involvement in cell migration including MMPs and ADAMs in cancer (VanSaun and Matrisian 2006, Arribas et al. 2006) and MMPs in vascular remodelling (Newby 2005). Studies in 2D, where visualisation of cell migration has been possible using either time-lapse videomicroscopy for assessing random cell migration or the Dunn chemotaxis chamber, were exploited to great effect by Anne Ridley and colleagues in the study of Rho GTPases in cell migration (Ridley et al. 2003 and references therein). Over the last 10 years, increasingly 3D studies and, more recently, in vivo studies have been performed, revealing many important insights into cell migration including potential roles for metalloproteinases. Several recent reviews by Ken Yamada and colleagues raised important issues regarding the study of cell migration in 2D including the as yet unresolved question of whether cells migrating in 3D have a lamellipodium which is seen extensively in 2D (reviewed in Evan-Ram and Yamada 2005). Other studies suggest that cells migrating in a 3D matrix do have invadapodia (*see* Chapter 10 by Mueller et al. this volume). In addition, it has been observed that microglia (brain macrophages) exhibit highly motile filapodia sampling their environment (Nimmerjahn et al. 2005) although Even-Ram and Yamada (2005) have speculated that such activity may relate to the unavoidable damage incurred when preparing such specimens. In vivo models clearly will provide vital information regarding the migration of cells within their social context but these studies are still complemented by those in vitro where dissection of mechanisms is more tractable.

A halfway house between 2D systems and in vivo migration are the various 3D matrices which have been developed starting from predominantly type-I collagen gels through tumour extracts (e.g. Matrigel where laminin-1 predominates) and fibrin gels. Each of these systems have yielded interesting data and enabled hypotheses to be tested. Most relevant to this chapter are the data which have emerged from the groups of Steve Weiss and Peter Friedl.

Early studies on leukocyte cell migration in 3D collagen gels suggested that collagen degradation was not required in this context (Schor et al. 1983). This area was taken up by Friedl and colleagues using sophisticated microscopic techniques to explore tumour-cell and T-cell migration in detail (Wolf et al. 2003a, b). These authors have used quenched-fluorescent collagen to track degradation of collagen gels during cell migration. Friedl and colleagues have suggested that when proteinases are inhibited, T cells and tumour cells adopt an amoeboid-like migration and can still penetrate collagen gels. Similarly, Sahai and Marshall (2003) demonstrated that tumour cells migrating in a 3D collagen gel in the presence of a cocktail of proteinase inhibitors adopt a rounded morphology and require Rho signalling.

More recently, Friedl and colleagues have explored mechanisms of migration of tumour cells in greater detail using collagen gels where most of the collagen is unlabelled with a small component of quenched fluorescent collagen (Wolf et al. 2007). The authors suggest that MT1-MMP is involved in 'collective cell migration' of tumour cells. HT1080 cells over-expressing MT1-MMP expressed this MMP at the leading edge but an area of proteolysis is detected, with a collagen-cleavage site antibody, just behind the leading edge of the cell (Wolf et al. 2007). In this chapter, the authors show significant inhibition of cell migration in a 3D matrix with the MP inhibitor BB2516 (marimastat). The authors then go on to observe that when cultured in a cocktail of proteinase inhibitors (inhibiting all classes of proteinases) HT1080 cells can continue to migrate, deforming the collagen matrix and migrating as single cells. They comment that twofold enhanced migration speed is observed in proteinase competent cells. Overall the Friedl group argues that 'collective-cell migration' requires collagenolysis but single-cell migration is not impaired in the absence of MT1-MMP activity (*see* Fig. 18.1a and b for summary). These authors have the view that it is not necessary for epithelial cells to undergo EMT in order to become invasive, but this remains controversial and probably tumour dependent. The single-cell migration observed is reminiscent of earlier studies in 3D-collagen gels from a number of groups, including that of Jean Paul Thiery where in the absence of proteinase inhibitors (though in the presence of serum and thus in the presence of some inhibitors) cell migration in chains of cells was observed following EMT (e.g. Tucker et al. 1990).

The question arises as to how do the Friedl group's recent studies fit in with studies by those groups who have reported critical roles for MMPs in, for example, tumour cell migration (reviewed in Sounni and Noel 2005)? Notable amongst these studies are those of the Weiss group. In a series of papers, Weiss and co-workers have elegantly demonstrated that there is a requirement of three members of the MT-MMP family in the migration of tumour cells through collagen gels, chick chorioallantoic membrane (CAM) or intact peritoneal basement membrane barriers (Hotary et al. 2000, Sabeh et al. 2004, Hotary et al. 2006). Hotary et al. (2000) investigated collagen gel invasion by MDCK cells in response to HGF and showed that transfection of these cells with MT1-, MT2- or MT3-MMPs accelerated invasion into collagen gels (with variations observed between these enzymes). It is relevant to note that invasion by MDCK cells in response to HGF is observed following 12 days of culture and the accelerated invasion is observed after 3 days of culture. Studies with cells isolated from MT1-MMP-/- mice showed that these cells display no collagen invasion over a 6-day period, and crucially this phenotype is rescued by transfection with MT1-MMP (Sabeh et al. 2004, Fig. 18.1C and D). In addition, these authors report that MT1-MMP-/- fibroblasts become trapped in the collagen-rich dermis of 4-week-old mice (Sabeh et al. 2004). The recent work (Hotary et al. 2006) indicates that proteolytic degradation of basement membrane ECM components is essential for the invasion over 8 days seen with several different tumour cell lines of carcinoma origin (summarised in Fig. 18.2B). The Friedl group study migration within a type-I collagen gel is observed over a 24 h period, and it is possible that the differences reported may reflect the

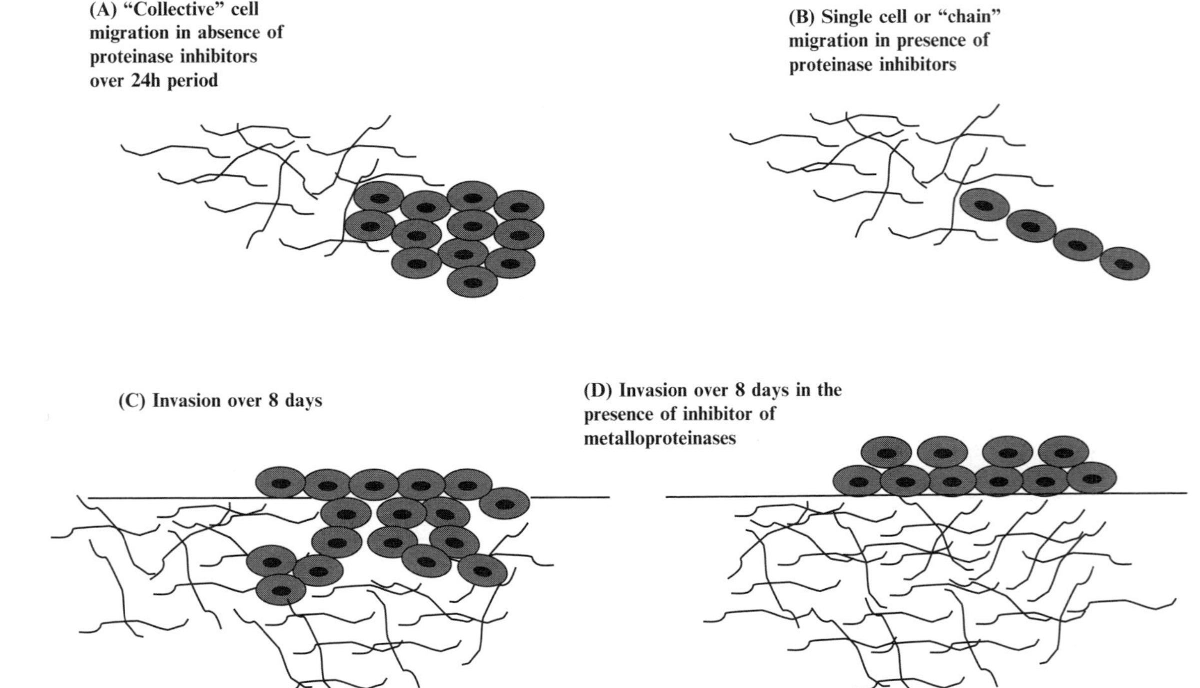

Fig. 18.1 Short- and long-term invasion of collagen gels by tumour cells. **a** Tumour cells invading through collagen gels up to 24 h seem to move as a group. **b** When proteinases are inhibited in these short-term assays, cells migrate as single cells. **c** Tumour cells invading collagen gels over 6 days in the absence of proteinase inhibitors. **d** Tumour cell invasion is completely blocked by metalloproteinase inhibitors

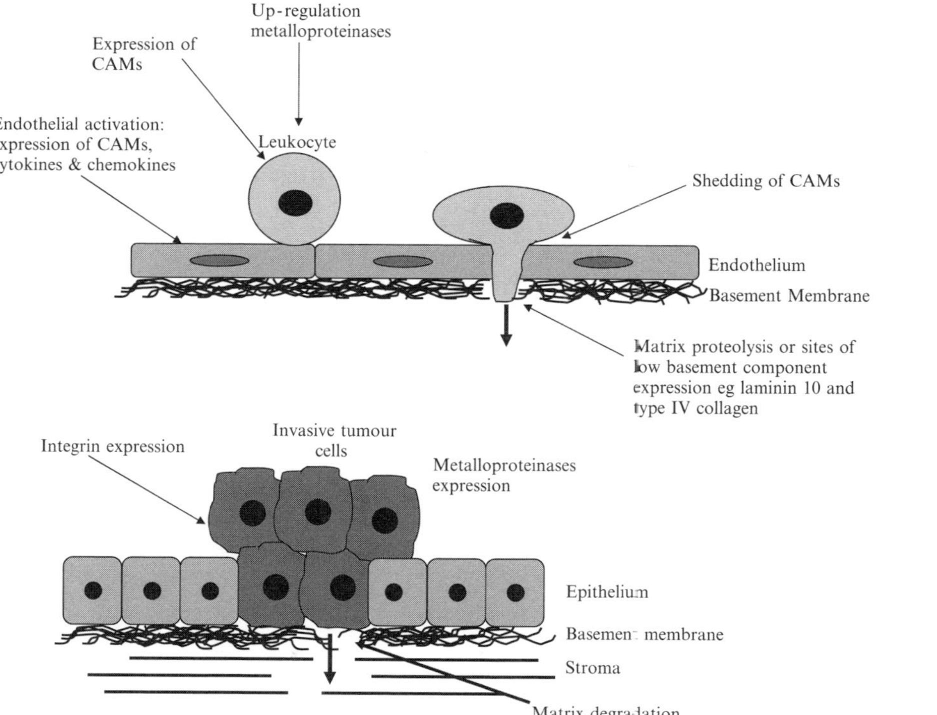

Fig. 18.2 Models of cell migration during normal or pathological events. **a** Activated endothelium attracts leukocytes to regions of infection or injury. Leukocytes attach to endothelial cells and migrate through the monolayer. Matrix proteolysis by the up-regulation of metalloproteinase expression or low expression of matrix proteins allows leukocytes to migrate through the basement membrane barrier. **b** Tumour invasion across intact basement membrane barriers over long time periods (12 days) in vitro. Expression of metalloproteinases enables invasive cells to migrate through an intact basement membrane and in vivo would allow migration to regions distant from the site of origin (*CAM*, cell adhesion molecule)

time-points studied. The data from Hotary et al. (2006) would suggest that the single-cell migration observed by Friedl and colleagues in cells treated with proteinase inhibitors and migrating in reconstructed 3D type-I collagen matrices is not sufficient to allow migration through an intact basement membrane. Friedl's group has largely used the highly motile HT1080 cell line of fibrosarcoma origin in their studies, although in their most recent study they have performed experiments with MDA-MB-231 cells, again revealing proteolytic degradation behind the leading edge, although the localisation of MT1-MMP was not determined in these cells (Wolf et al. 2007). As Mueller et al. (Chap. 21 this volume) comment, the temporal generation of the ¾ collagen neoepitope is unknown and it is not inconceivable that this may take some time to be revealed as the cell moves forward.

The recent data reviewed above suggest that short-term invasion assays in 3D collagen matrices are not dependent on MMP function but invasion of intact basement membranes and longer-term invasion of collagen gels requires certain MT-MMPs.

Cell Migration Studies In Vivo

Several studies in genetically manipulated mice have suggested roles of proteinases in cell migration. For example, TIMP1-null mice exposed to bleomycin develop an enhanced neutrophilia in the lung (Kim et al. 2005), which is suggestive of increased cell migration due to proteolytic action of a TIMP1-sensitive proteinase (ADAM10 or MMPs excluding some MT-MMP family members). Possible sites of action include MMP7 degradation of syndecan-1 (Li et al. 2002) or E-cadherin (McGuire et al. 2003) or cleavage of basement membrane components. The role of MMP7 is particularly supported by the fact that MMP7-null mice are protected from bleomycin lung injury with few cells penetrating the alveolar epithelial cell layer to enter the lung space (Li et al. 2002). In this case, in wild-type mice generation of a gradient of the chemokine KC is set up by MMP7 cleavage of syndecan-1. A number of recent studies in MMP-deficient mice have demonstrated crucial roles for several MMPs in cheomokine (and cytokine) processing which impact on neutrophil cell migration (reviewed in Van Lint and Libert 2007).

Intravital miscroscopy has begun to revolutionise our knowledge of cell migration in vivo, particularly with respect to leukocytes and tumour cells. Early intravital microscopy tumour studies by Ann Chambers and co-workers showed that the MP inhibitor batimastat suppressed B16 melanoma liver metastases-associated angiogenesis although tumour extravasation was not altered (Wylie et al. 1999). The pioneering work of Sussan Nourshargh and colleagues has allowed the mechanisms underlying leukocyte migration in vivo to be elucidated, for example during migration through the basement membrane encountered by leukocyte and tumour cells alike, once the endothelial barrier has been crossed. Recently, this group has established that neutrophils transmigrate through venules and then through the pericyte basement membrane by locating sites of lower expression of key components of the basement membrane (Wang et al. 2006). Immunolocalization studies reveal that whilst expression of the proteoglycan perlecan was uniform, identifiable 'exit points' had severely decreased immunostaining for type-IV collagen and laminin-10 (summarised in Fig. 18.2A). Of relevance to these data, Van Agtmael et al. (2005) have shown that in the normal retina the basement membrane has a non-uniform thickness, indicating that there may be sites of differential expression of ECM components in this tissue as well. Nourshargh's group has also shown that neutrophil elastase and other serine proteinases play an important role in migration through the venule BM (Young et al. 2007). Whilst NE-null mice do not show a defect in cytokine-induced transmigration, the serine proteinase inhibitor aprotinin blocks transmigration, indicating that additional serine proteinases are involved. Any role for metalloproteinases in generation of sites of lower basement membrane component expression has yet to be explored.

Thus far few studies have addressed the potential roles of MPs in transendothelial cell migration in vivo. Ann Ager and colleagues demonstrated that MPs have some roles in transendothelial migration of lymphocytes by pre-incubating lymphocytes with a hydroxamate MP inhibitor and thus inhibiting ADAMs and MMPs (Faveeuw et al. 2001). These authors demonstrated that lymphocytes pre-treated with the hydroxamate inhibitor had higher levels of surface L-selectin and accumulated within the endothelial lining of high endothelial venules. Venturi et al. (2003) have uncovered a role for L-selectin shedding in neutrophil entry into the peritoneum in a mouse model of inflammation. L-selectin shedding, prevented by replacement of the membrane proximal cleavage site with the equivalent sequence from the E-selectin molecule, resulted in a significantly increased leukocyte migration into the peritoneum, although rolling was not altered, as visualised by intravital microscopy (Venturi et al. 2003). In studies of EAE, the transmembrane receptor dystroglycan, expressed in the brain parenchymal basement membrane, is cleaved by MMPs 2 and 9 (but not MMPs 1, 3, 7 and 8) and as a result macrophages penetrate the blood–brain barrier in this inflammatory condition (Agrawal et al. 2006). Mice null for MMPs 2 or 9 are resistant to EAE and macrophages fail to infiltrate.

Zebrafish is a very attractive model system for in vivo analysis of cell migration due to its transparency, the availability of genetic mutants and the accessibility of pharmacological inhibitors. Very recently, Philippe Herbomel's group has observed neutrophil and macrophage migration in vivo in zebrafish and uncovered the remarkable fact that whilst primitive neutrophils are rapidly attracted to sites of injury or infection they barely phagocytose bacteria, whereas macrophages attracted simultaneously phagocytose bacteria in great quantities (Le Guyader et al. 2007). Zebrafish have also been used to study tumour cell migration, and a recent study with intraperitoneal injection of several human tumour cell lines and murine B16 melanoma shows that HT1080 cells adopt an amoeboid-like migration and are highly invasive whereas B16 cells show a mesenchymal type of migration (Stoletov et al. 2007). MDA-435 cells showed mixed amoeboid and mesenchymal morphology and were poorly invasive but MDA-435 carcinoma cells over-expressing RhoC (associated with human metastasis) became highly invasive and exhibited amoeboid-like cell migration in the zebrafish (Stoletov et al. 2007). Co-injection experiments with parent MDA-435 cells (which do not invade in the zebrafish) revealed that the Rho-C expressing MDA cells did not modify the migration rate of control MDA-435 cells. The authors conclude that this suggests that the alteration in invasion of Rho-C is intrinsic to the cell line and that proteolytic mechanisms are not implicated in the enhanced migration of RhoC-expressing cells. Whilst the data seem compatible with the idea that path clearing is not occurring (which could enhance cell migration of control cells) perhaps a further experiment with co-injection of proteinase inhibitors is warranted. Another recent study indicates that chronic inflammation, with hallmarks of human diseases such as psoriasis, is induced in a zebrafish with a mutation in HAI-1 (hepatocyte growth factor activator inhibitor-1), an inhibitor of the serine proteinase matriptase (Mathias et al. 2007). Intravital microscopy of this mutant crossed with zebrafish with GFP-tagged neutrophils reveals that these cells display periods of random migration with a loss of polarity and adoption of a rounded morphology whilst pausing as well as periods of persistent migration. The authors observed that a COX-2 inhibitor blocked neutrophil migration and induced a similar rounded morphology, suggesting that inflammatory mediators whose production involves COX-2 are active in vivo. Zebrafish is thus emerging as an excellent model for the study of cell migration in vivo. Embryonic development in *Xenopus laevis* is another very interesting model for in vivo analysis of metalloproteinases in cell migration. MMP7 has been localised to migrating macrophages in *Xenopus* (Harrison et al. 2004), and recent studies indicate that in vivo macrophage migration inhibited morpholino knockdown of XMMP7, 9 and 18 (Matt Tomlinson and Grant Wheeler, personal communication).

Matricryptic Sites in Cell Migration

Matricryptic sites in matrix components (termed matricryptins by Davis et al. 2000) are sites which become exposed largely through proteolytic cleavage and which have novel biological activities. Recent reviews cover some of the major matricryptic sites (Bellon et al. 2004, Tran et al. 2005), and so here brief consideration will be given to those which relate most closely to the cell migration studies described above.

Type-IV Collagen

A number of cryptic anti-angiogenic factors have been described in type-IV collagen generated, for example, by MMP cleavage of tumour basement membrane sources (reviewed in Mundel and Kalluri 2007). In some cases, these fragments have been reported to block tumour cell migration as well through effects on

MT1-MMP (reviewed in Pasco et al. 2005), although other studies reveal pro-migratory cryptic sites within type-IV collagen which bind to $\alpha v \beta 3$ integrins (Xu et al. 2001). It would be of interest to determine the integrins involved in migration through intact basement membrane in the type of experiment described in Hotary et al. (2006) as well as the nature of any type-IV collagen fragments generated in this model system.

Laminin 5

Degradation of the $\gamma 2$ chain of rat laminin 5 has been reported to expose a cryptic site resulting in enhanced migration of tumour cells (Koshikawa et al. 2000). However, other studies have shown that human $\gamma 2$ and $\alpha 3$ chains of laminin 5 are degraded by BMP-1 and not by MT1-MMP or MMP2 (Amano et al. 2000). Recent studies indicate that MMP7 cleaves the $\beta 3$ chain of laminin 5 generating a 90 kDa fragment (Remy et al. 2006). This group demonstrates that MMP-7-degraded laminin 5 promotes migration of the colon carcinoma cell line HT29 and that MMP7 and laminin 5 co-localise in cells on the outer borders of cellular colonies. Signalling events generated by this cleavage event remain to be determined but this model opens new avenues for investigation.

Proteoglycans

Proteoglycans are inhibitory to axon regeneration in both the peripheral nervous system and the central nervous system (CNS; reviewed in Busch and Silver 2007). Work from several laboratories has demonstrated that treatment of CNS with chondroitinase ABC (ChABC), which cleaves GAG chains, results in enhanced neural repair (Barritt et al. 2006). Recent studies indicate that perineuronal nets in rat brain contain several proteoglycans including aggrecan, brevican, neurocan and phosphocan (Deepa et al. 2006). Whilst the mechanisms by which ChABC exerts its effects remain unclear (reviewed in Crespo et al. 2007), exposure of pro-migratory sites in proteoglycans or other ECM components remains a possibility. Previous work indicated that MMPs could also promote regeneration in the PNS again by cleaving proteoglycans, though in this case one would presume through cleavage of the protein core (Krekoski et al. 2002). Larsen et al. (2003) demonstrated that MMP9 enhances remyelination by degradation of another inhibitory proteoglycan NG2. Nerves and blood vessels grow into normally avascular inter-vertebral discs, and ChABC treatment of intervertebral discs results in endothelial cell migration and proliferation (Johnson et al. 2005). It is still unclear whether proteoglycan degradation exposes cryptic sites or simply removes a barrier to migration. It is of relevance to note that exencephaly occurs in the brains of perlecan-null mice, attributed to the removal of an intact basement membrane barrier (Costell et al. 1999).

Type-I Collagen

In these recent investigations of cell migration in 3D (described above), the potential roles of matricryptic sites in the ECM have yet to be explored. Studies of cell migration in 2D have revealed that denaturation of type-I collagen (Davis 1992) or degradation of type-I collagen by MMP-13 into classical ¾ and ¼ fragments (Messent et al. 1998) leads to the exposure of an RGD site which can then become available for binding through $\alpha v \beta 3$ integrins. Exposure of vascular smooth cells to purified ¾ collagen fragments results in markedly enhanced migration in response to PDGF-BB, visualised by time-lapse videomicroscopy (Stringa et al. 2000). This is of interest since $\alpha v \beta 3$ integrin has some role in pathological vascular smooth muscle cell migration (reviewed in Newby 2005).

Conclusion

Much remains to be uncovered regarding proteolytic mechanisms involved in cell migration in 3D as well as in vivo. The use of complex in vitro model systems should result in the generation of exciting hypotheses to be tested in vivo, where organisms such as zebrafish should allow both genetic and microscopic analysis. The new era of intravital imaging using quantum dots with their advantages of photostability, tunability to narrow emission spectra and potential as drug delivery vehicles (Stroh et al. 2005) as well as second harmonic imaging of collagen will undoubtedly extend dramatically our understanding of cell migration in vivo. In combination with genetic approaches, these advanced microscopic techniques will allow the interplay of cells with neighbouring cells and with their surrounding matrix to be determined at a level which may allow more astute in vivo discrimination between, for example, proteinase inhibitors with subtly different biochemical specificities.

Acknowledgements Work in the authors' laboratory was supported by Diabetes UK, the Norwich and Norfolk Diabetes Trust and the British Heart Foundation.

References

Agrawal S., Anderson P., Durbeej M. et al. (2006). Dystroglycan is selectively cleaved at the parenchymal basement membrane at sites of leukocyte extravasation in experimental autoimmune encephalomyelitis. J. Exp. Med. 203: 1007–1019.

Amano S., Scott I.C., Takahara K. et al. (2000). Bone morphogenetic protein 1 is an extracellular processing enzyme of the laminin 5 gamma 2 chain. J. Biol. Chem. 275: 22728–22735.

Arribas J., Bech-Serra J.J., Santiago-Josefat B. (2006). ADAMs, cell migration and cancer. Cancer Metastasis Rev. 25: 57–68.

Bagorda A., Mihaylov V.A., and Parent C.A. (2006). Chemotaxis: moving forward and holding on to the past. Thromb. Haemost. 95: 12–21.

Bar-Or A., Nuttall R.K., Duddy M. et al. (2003). Analyses of all matrix metalloproteinase members in leukocytes emphasize monocytes as major inflammatory mediators in multiple sclerosis. Brain 126: 2738–2749.

Barritt A.W., Davies M., Marchand F. et al. (2006). Chondroitinase ABC promotes sprouting of intact and injured spinal systems after spinal cord injury. J. Neurosci. 26: 10856–10867.

Bellon G., Martiny L., and Robinet A. (2004). Matrix metalloproteinases and matrikines in angiogenesis. Crit. Rev. Oncol. Hematol. 49: 203–220.

Blankenberg S., Barbaux S., and Tiret L. (2003). Adhesion molecules and atherosclerosis. Atherosclerosis 170: 191–203.

Bowers-Morrow V.M., Ali S.O., and Williams K.L. (2004) Comparison of molecular mechanisms mediating cell contact phenomena in model developmental systems: an exploration of universality. Biol. Rev. Camb. Philos. Soc. 79: 611–642.

Busch S.A. and Silver J. (2007). The role of extracellular matrix in CNS regeneration. Curr. Opin. Neurobiol. 17: 120–127.

Charo I.F. and Taubman M.B. (2004). Chemokines in the pathogenesis of vascular disease. Circ. Res. 95: 858–866.

Cornillon S., Gebbie L., Benghezal M. et al. (2006). An adhesion molecule in free-living Dictyostelium amoebae with integrin beta features. EMBO Rep. 7: 617–621.

Costell M., Gustafsson E., Aszodi A. et al. (1999). Perlecan maintains the integrity of cartilage and some basement membranes. J. Cell Biol. 147: 1109–1122.

Coussens L.M. and Werb Z. (2002). Inflammation and cancer. Nature 420: 860–867.

Crespo D., Asher R.A., Lin R. et al. (2007). How does chondroitinase promote functional recovery in the damaged CNS? Exp. Neurol. 206: 159–171.

Davis G.E. (1992). Affinity of integrins for damaged extracellular matrix: $\alpha_v\beta_3$ binds to denatured collagen type I through RGD sites. Biochem. Biophys. Res. Commun. 182: 1025–1031.

Davis G.E., Bayless K.J., Davis M.J. et al. (2000). Regulation of tissue injury responses by the exposure of matricryptic sites within extracellular matrix molecules. Am. J. Pathol. 156: 1489–1498.

Deepa S.S., Carulli D., Galtrey C. et al. (2006). Composition of perineuronal net extracellular matrix in rat brain: a different disaccharide composition for the net-associated proteoglycans. J. Biol. Chem. 281: 17789–17800.

Dong C., Slattery M., and Liang S. (2005). Micromechanics of tumor cell adhesion and migration under dynamic flow conditions. Front Biosci. 10: 379–384.

Even-Ram S. and Yamada K.M. (2005). Cell migration in 3D matrix. Curr. Opin. Cell Biol. 17: 524–532.

Faveeuw C., Preece G., and Ager A. (2001). Transendothelial migration of lymphocytes across high endothelial venules into lymph nodes is affected by metalloproteinases. Blood 98: 688–695.

Gabius H.J., Springer W.R., and Barondes S.H. (1985). Receptor for the cell binding site of discoidin I. Cell 42: 449–456.

Galvez B.G., Matias-Roman S., Yanez-Mo M. et al. (2002). ECM regulates MT1-MMP localization with beta1 or alphavbeta3 integrins at distinct cell compartments modulating its internalization and activity on human endothelial cells. J. Cell Biol. 159: 509–521.

Harrison M., Abu-Elmagd M., Grocott T. et al. (2004). Matrix metalloproteinase genes in Xenopus development. Dev. Dyn. 231: 214–220.

Hoshiga M., Alpers C.E., Smith L.L. et al. (1995). $\alpha_v\beta_3$ Integrin expression in normal and atherosclerotic artery. Circ. Res. 77: 1129–1135.

Hotary K., Allen E., Punturieri A. et al. (2000). Regulation of cell invasion and morphogenesis in a three-dimensional type I collagen matrix by membrane-type matrix metalloproteinases 1, 2, and 3 1. J. Cell Biol. 149: 1309–1323.

Hotary K., Li X.Y., Allen E. et al. (2006). A cancer cell metalloprotease triad regulates the basement membrane transmigration program. Genes Dev. 20: 2673–2686.

Johnson W.E., Caterson B., Eisenstein S.M. et al. (2005). Human intervertebral disc aggrecan inhibits endothelial cell adhesion and cell migration in vitro. Spine 30: 1139–1147.

Kim K.H., Burkhart K., Chen P. et al. (2005). Tissue inhibitor of metalloproteinase-1 deficiency amplifies acute lung injury in bleomycin-exposed mice. Am. J. Respir. Cell Mol. Biol. 33: 271–279.

Koshikawa N., Giannelli G., Cirulli V. et al. (2000). Role of cell surface metalloprotease MT1-MMP in epithelial cell migration over laminin-5. J. Cell Biol. 148: 615–624.

Krekoski C.A., Neubauer D., Graham J.B. et al. (2002). Metalloproteinase-dependent predegeneration in vitro enhances axonal regeneration within acellular peripheral nerve grafts. J. Neurosci. 22: 10408–10415.

Larsen P.H., Wells J.E., Stallcup W.B. et al. (2003). Matrix metalloproteinase-9 facilitates remyelination in part by processing the inhibitory NG2 proteoglycan. J. Neurosci. 23: 11127–11135.

Le Guyader D., Redd M.J., Colucci-Guyon E. et al. (2008). Origins and unconventional behavior of neutrophils in developing zebrafish. Blood 111: 132–141.

Li Q., Park P.W., Wilson C.L. et al. (2002). Matrilysin shedding of syndecan-1 regulates chemokine mobilization and transepithelial efflux of neutrophils in acute lung injury. Cell 111: 635–646.

Libby P., Ridker P.M., and Maseri A. (2002). Inflammation and atherosclerosis. Circulation 105: 1135–1143.

Lusis A.J. (2000). Atherosclerosis. Nature 407: 233–241.

Mathias J.R., Dodd M.E., Walters K.B. et al. (2007). Live imaging of chronic inflammation caused by mutation of zebrafish Hai1. J. Cell Sci. 120: 3372–3383.

Matias-Roman S., Galvez G., Genis L. et al. (2005). Membrane type 1-matrix metalloproteinase is involved in migration of human monocytes and is regulated through their interaction with fibronectin or endothelium. Blood 105: 3956–3964.

McGuire J.K., Li Q., and Parks W.C. (2003). Matrilysin (matrix metalloproteinase-7) mediates E-cadherin ectodomain shedding in injured lung epithelium. Am. J. Pathol. 162: 1831–1843.

Messent A.J., Tuckwell D.S., Knäuper V. et al. (1998). Effects of collagenase-cleavage of type I collagen on $\alpha 2\beta 1$ integrin-mediated cell adhesion. J. Cell Sci. 111: 1127–1135.

Miles F.L., Pruitt F.L., van Golen K.L. et al. (2008). Stepping out of the flow: capillary extravasation in cancer metastasis. Clin. Exp. Metastasis 25: 305–324.

Newby A.C. (2005). Dual role of matrix metalloproteinases (matrixins) in intimal thickening and atherosclerotic plaque rupture. Physiol. Rev. 85: 1–31.

Nimmerjahn A., Kirchhoff F. and Helmchen F. (2005). Resting microglial cells are highly dynamic surveillants of brain parenchyma in vivo. Science 308: 1314–1318.

Pasco S., Brassart B., Ramont L. et al. (2005). Control of melanoma cell invasion by type IV collagen. Cancer Detect. Prev. 29: 260–266.

Piali L., Hammel P., Uherek C. et al. (1995). CD31/PECAM-1 is a ligand for $\alpha_v\beta_3$ integrin involved in adhesion of leukocytes to endothelium. J. Cell Biol. 130: 451–460.

Remy L., Trespeuch C., Bachy S. et al. (2006). Matrilysin 1 influences colon carcinoma cell migration by cleavage of the laminin-5 beta3 chain. Cancer Res. 66: 11228–11237.

Ridley A.J., Schwartz M.A., Burridge K. et al. (2003). Cell migration: integrating signals from front to back. Science 302: 1704–1709.

Romanic A.M. and Madri J.A. (1994). The induction of 72-kD gelatinase in T cells upon adhesion to endothelial cells is VCAM-1 dependent. J. Cell Biol. 125: 1165–1178.

Sabeh F., Ota I., Holmbeck K. et al. (2004). Tumor cell traffic through the extracellular matrix is controlled by the membrane-anchored collagenase MT1-MMP. J. Cell Biol. 167: 769–781.

Sahai E. and Marshall C.J. (2003). Differing modes of tumour cell invasion have distinct requirements for Rho/ROCK signalling and extracellular proteolysis. Nat. Cell Biol. 5: 711–719.

Schor S.L., Allen T.D. and Winn B. (1983). Lymphocyte migration into three-dimensional collagen matrices: a quantitative study. J. Cell Biol. 96: 1089–1096.

Sithu S.D., English W.R., Olson P. et al. (2007). Membrane-type 1-matrix metalloproteinase regulates intracellular adhesion molecule-1 (ICAM-1)-mediated monocyte transmigration. J. Biol. Chem. 282: 25010–25019.

Sounni N.E. and Noel A. (2005). Membrane type-matrix metalloproteinases and tumor progression. Biochimie 87: 329–342.

Steeber D.A., Venturi G.M., and Tedder T.F. (2005). A new twist to the leukocyte adhesion cascade: intimate cooperation is key. Trends Immunol. 26: 9–12.

Stoletov K., Montel V., Lester R.D. et al. (2007). High-resolution imaging of the dynamic tumor cell vascular interface in transparent zebrafish. Proc. Natl. Acad. Sci. USA 104: 17406–17411.

Stringa E., Knauper V., Murphy G. et al. (2000). Collagen degradation and platelet-derived growth factor stimulate the migration of vascular smooth muscle cells. J. Cell Sci. 113: 2055–2064.

Stroh M., Zimmer J.P., Duda D.G. et al. (2005). Quantum dots spectrally distinguish multiple species within the tumor milieu in vivo. Nat. Med. 11: 678–682.

Sultan S., Gosling M., Nagase H. et al. (2004). Shear stress-induced shedding of soluble intercellular adhesion molecule-1 from saphenous vein endothelium. FEBS Lett. 564: 161–165.

Tedgui A. and Mallat Z. (2006). Cytokines in atherosclerosis: pathogenic and regulatory pathways. Physiol. Rev. 86: 515–581.

Tran K.T., Lamb P., and Deng J.S. (2005). Matrikines and matricryptins: implications for cutaneous cancers and skin repair. J. Dermatol. Sci. 40: 11–20.

Tsakadze N.L., Sithu S.D., Sen U. et al. (2006). Tumor necrosis factor-alpha-converting enzyme (TACE/ADAM-17) mediates the ectodomain cleavage of intercellular adhesion molecule-1 (ICAM-1). J. Biol. Chem. 281: 3157–3164.

Tucker G.C., Boyer B., Gavrilovic J. et al. (1990). Collagen-mediated dispersion of NBT-II rat bladder carcinoma cells. Cancer Res. 50: 129–137.

Van Agtmael T., Schlotzer-Schrehardt U., McKie L. et al. (2005). Dominant mutations of Col4a1 result in basement membrane defects which lead to anterior segment dysgenesis and glomerulopathy. Hum. Mol. Genet. 14: 3161–3168.

Van Lint P., and Libert C. (2007). Chemokine and cytokine processing by matrix metalloproteinases and its effect on leukocyte migration and inflammation. J. Leukoc. Biol. 82: 1375–1381.

VanSaun M.N., and Matrisian L.M. (2006). Matrix metalloproteinases and cellular motility in development and disease. Birth Defects Res. C. Embryo. Today 78: 69–79.

Varney T.R., Ho H., Petty C. et al. (2002). A novel disintegrin domain protein affects early cell type specification and pattern formation in Dictyostelium. Development 129: 2381–2389.

Venturi G.M., Tu L., Kadono T. et al. (2003). Leukocyte migration is regulated by L-selectin endoproteolytic release. Immunity 19: 713–724.

Wang S., Voisin M.B., Larbi K.Y. et al. (2006). Venular basement membranes contain specific matrix protein low expression regions that act as exit points for emigrating neutrophils. J. Exp. Med. 203: 1519–1532.

Wayner E.A., Orlando R.A., and Cheresh D.A. (1991). Integrins $\alpha_v\beta_3$ and $\alpha_v\beta_5$ contribute to cell attachment to vitronectin but differentially distribute on the cell surface. J. Cell Biol. 113(4): 919–929.

Weerasinghe D., McHugh K.P., Ross F.P. et al. (1998). A role for the $\alpha_v\beta_3$ integrin in the transmigration of monocytes. J. Cell Biol. 142: 595–607.

Wolf K., Mazo I., Leung H. et al. (2003a). Compensation mechanism in tumor cell migration: mesenchymal-amoeboid transition after blocking of pericellular proteolysis. J.Cell Biol. 160: 267–277.

Wolf K., Muller R., Borgmann S. et al. (2003b). Amoeboid shape change and contact guidance: T-lymphocyte crawling through fibrillar collagen is independent of matrix remodeling by MMPs and other proteases. Blood 102: 3262–3269.

Wolf K., Wu Y.I., Liu Y. et al. (2007). Multi-step pericellular proteolysis controls the transition from individual to collective cancer cell invasion. Nat. Cell Biol. 9: 893–904.

Worthylake R.A. and Burridge K. (2001). Leukocyte transendothelial migration: orchestrating the underlying molecular machinery. Curr. Opin. Cell Biol. 13: 569–577.

Wylie S., MacDonald I.C., Varghese H.J. et al. (1999). The matrix metalloproteinase inhibitor batimastat inhibits angiogenesis in liver metastases of B16F1 melanoma cells. Clin. Exp. Metastasis 17: 111–117.

Xu J., Rodriguez D., Petitclerc E. et al. (2001). Proteolytic exposure of a cryptic site within collagen type IV is required for angiogenesis and tumor growth in vivo. J. Cell Biol. 154: 1069–1079.

Young R.E., Voisin M.B., Wang S. et al. (2007). Role of neutrophil elastase in LTB4-induced neutrophil transmigration in vivo assessed with a specific inhibitor and neutrophil elastase deficient mice. Br. J. Pharmacol. 151: 628–637.

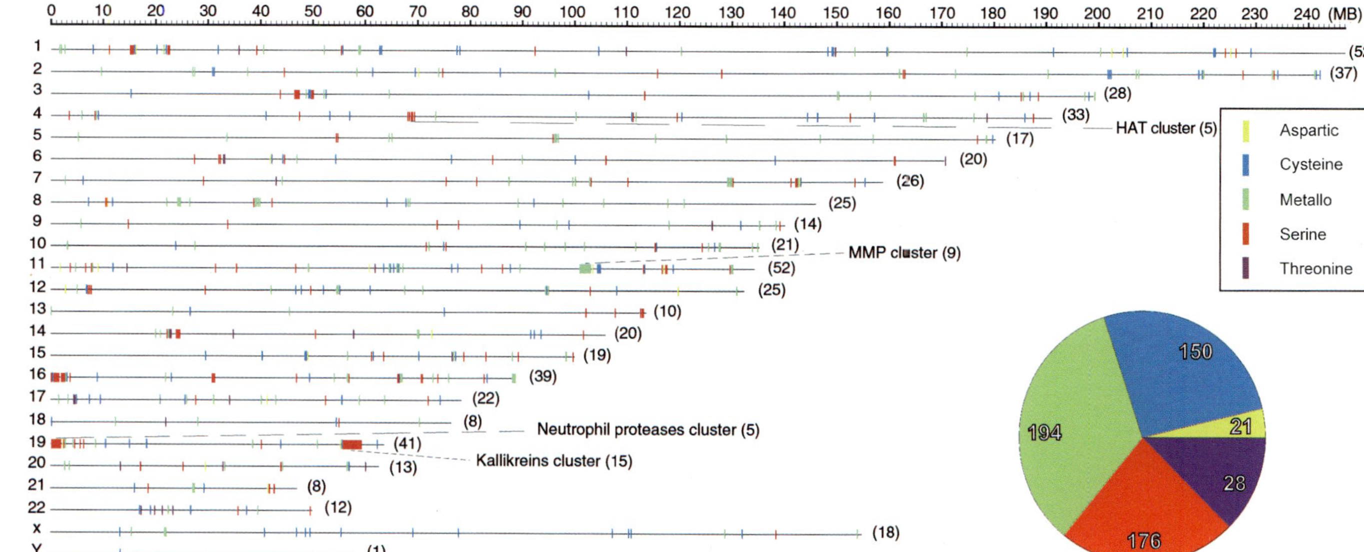

Fig. 1.1 Hierarchical clustering of mean degradome gene expression profiles for18 malignant and 23 healthy tissues. The number of samples for each tissue is marked in the parentheses. Values of heatmap are scaled genewise to mean 0 and standard deviation 1. Degradome genes ($n = 493$) consist of 145 metalloproteases (presented by the green color above the image), 103 cysteine proteases (red), 101 protease inhibitors (blue), 112 serine proteases (light blue), 18 threonine proteases (yellow), and 13 aspartic proteases (cyan). Few significant degradome gene clusters are marked by the dark green dashed rectangles. The leftmost one indicates blood-related degradome genes, the middle cluster contains mainly gastrointestinal genes, and the rightmost indicates nervous system-specific degradome genes

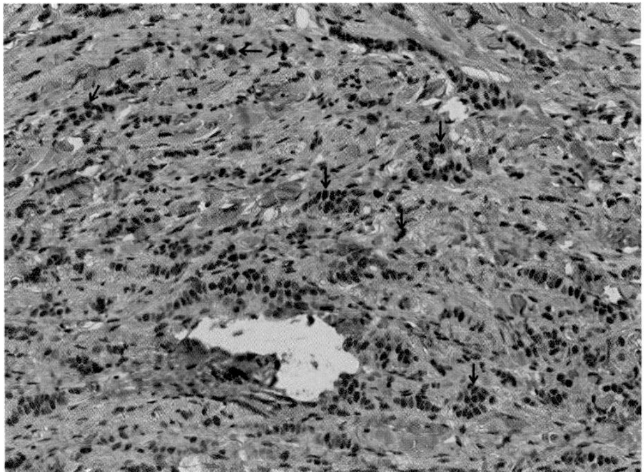

Fig. 3.3 Cathepsin V protein in invasive human ductal carcinoma. Paraffin-embedded sections were stained for cathepsin V with a monoclonal antibody against human cathepsin V (a kind gift from Dr. E. Weber, Martin Luther University, Halle, Germany). Expression of cathepsin V protein is indicated by dark brown staining (arrows). Magnification, 40×.

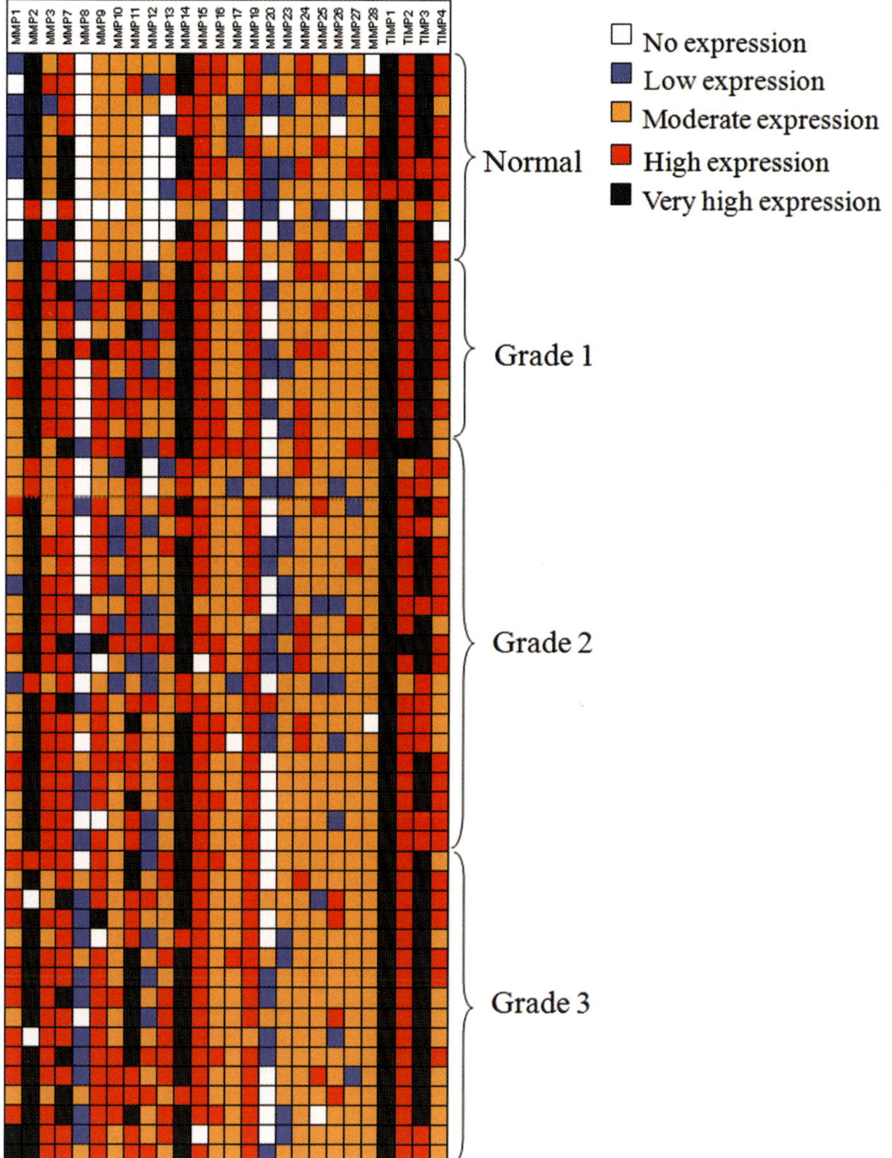

Fig. 4.2 Comparative TaqMan[®] expression profiles of matrix metalloproteinase (*MMP*) and tissue inhibitor of metalloproteinase (*TIMP*) gene families in normal and malignant human mammary tissues. Data are displayed as the expression levels of the indicated genes in human mammary tissue samples, which have been reported in detail elsewhere (Porter et al. 2004). The tissues are grouped as normal and histopathological grades 1, 2, and 3, respectively. Expression levels (raw C_t levels) are shown in the associated key.

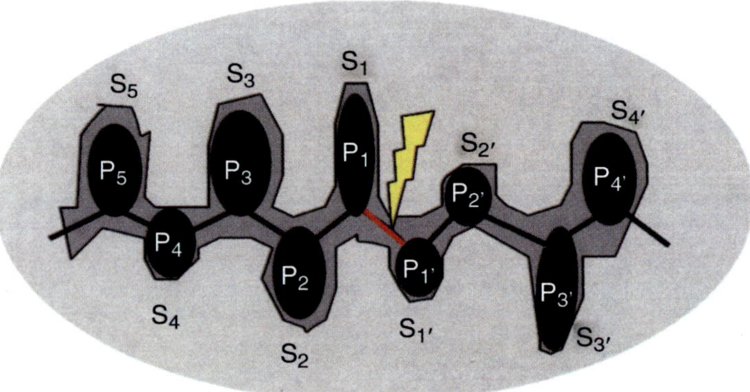

Fig. 5.1 Substrate cleft convention. The convention for naming protease substrate residues came from the work of Schecter and Berger on the size of the active site in papain (Schecter and Berger 1967). Think of the surface of a protease as a cleft with pockets defined by protease side chains. These are known as the specificity (S) pockets. Into these, in an extended conformation, fit the peptide (P) substrate side chains. By convention the scissile bond (yellow bolt) is between P_1 and $P_{1'}$. Residues toward the N-terminus of the substrate are numbered P_1, P_2, P_3, . . . , P_n, and toward the C-terminus of the substrate they are named $P_{1'}$, $P_{2'}$, $P_{3'}$ . . . , $P_{n'}$. Thus, subsites toward the N-terminus of the scissile bond are "unprimed" but sites toward the C-terminus are "primed."

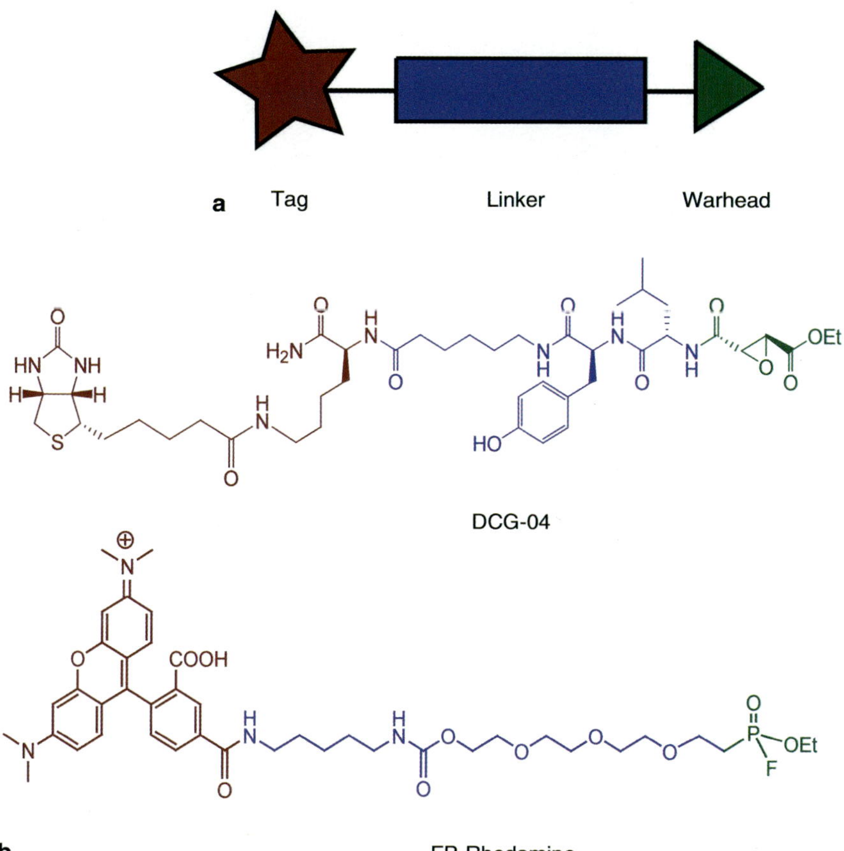

Fig. 7.1 General and specific structures of activity-based probes (*ABPs*). (**a**) General structure of an *ABP*. (**b**) DCG-04 is an *ABP* that targets cysteine proteases and contains a biotin tag (red), a dipeptide-containing linker (blue), and an epoxide as a warhead (green). Fluorescent protein (*FP*)-rhodamine is an *ABP* that targets the serine hydrolase superfamily of enzymes and contains a rhodamine fluorophore (red), a poly (ethylene glycol) linker (blue), and a fluorophosphonate as a warhead (green).

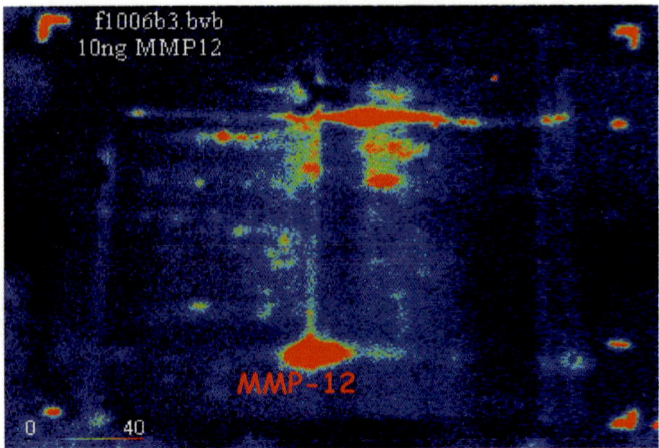

Fig. 7.4 Labeling of mice tumor extract (colon carcinoma) by the phosphinic probe displayed in scheme 2. Before photoactivation, 0.5 pmol of h-matrix metalloprotease (*MMP*)-12 were added to the sample as an internal standard for protein quantification. The sample was analyzed by 2D sodium dodecyl sulfate–polyacrylamide gel electrophoresis (*SDS–PAGE*) followed by detection.

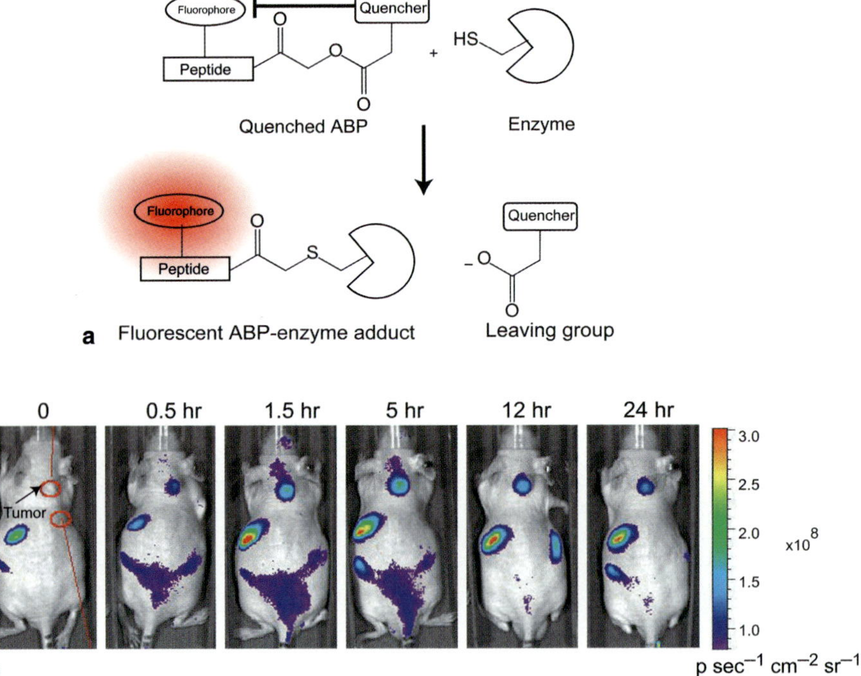

Fig. 7.8 Quenched activity-based probes (*ABPs*) for the noninvasive imaging of tumors in vivo. (**a**) Covalent labeling of a cysteine protease target by a quenched *ABP* (GB137). Activity-based labeling of the target enzyme results in the loss of the quenching group and subsequent generation of a fluorescently labeled enzyme. (**b**) Optical imaging of MDA-MB-231 breast cancer xenograft tumors in nude mice using a quenched cysteine cathepsin-specific *ABP*. The quenched probe was injected intravenously and fluorescent images of the mice were taken at various time points after injection. Images taken from Blum et al. (2007).

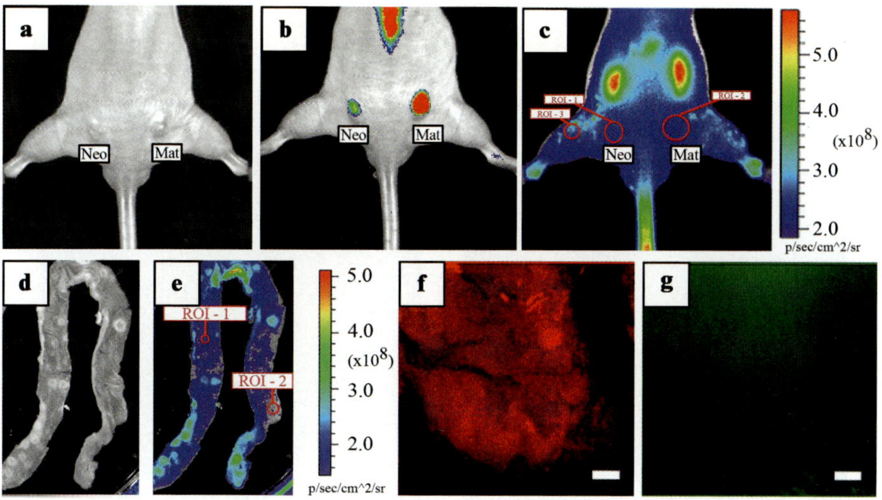

Fig. 7.9 In vivo imaging with substrate-based proteolytic beacon PB-M7NIR. (**a–c**) In vivo imaging of mouse subcutaneous xenograft tumors with PB-M7NIR. Dorsal, caudal view of a nude mouse of 4 weeks following subcutaneous injection of SW480neo (Neo) or MMP-7-expressing SW480mat (Mat) cells. Tumor areas (~57 mm^2 each) are shown in white light (**a**), the Cy5.5 sensor channel (**b**), and the reference channels (**c**) 4 h post retro-orbital intravenous injection of 1.0 nmol PB-M7NIR. Encircled areas represent regions of interest for quantitative assessment. Note accumulation of PB-M7NIR in the kidneys of the mouse as detected with the reference channel (**c**), but selective accumulation of sensor signal in the Mat tumor indicative of proteolytic activity (**b**). Sensor signal on the spine and tail are presumed to be due to low levels of circulating, activated probe that become detectable when they are close to the surface of the mouse. (**d–g**) Ex vivo imaging of PB-M7NIR in APC^{MIN} intestinal adenomas. Explanted mouse intestine from an APC^{MIN} mouse with spontaneous polyps in 60 min. Post-injection of 1 nmol of PB-M7NIR Beacon. (**d**) White light image and (**e**) NIRF image in the Cy5.5 (sensor) channel. (**e–f**). High power images of a single, intact adenoma (10×-objective) from an APC^{Min} mouse (**e**). False-red coloring in the Cy5.5 (sensor) channel (**f**) Cy7 (reference-green) channel. White line = 100 microns. Adapted fr om Scherer et al. (2008). *ROI*, region of interest.

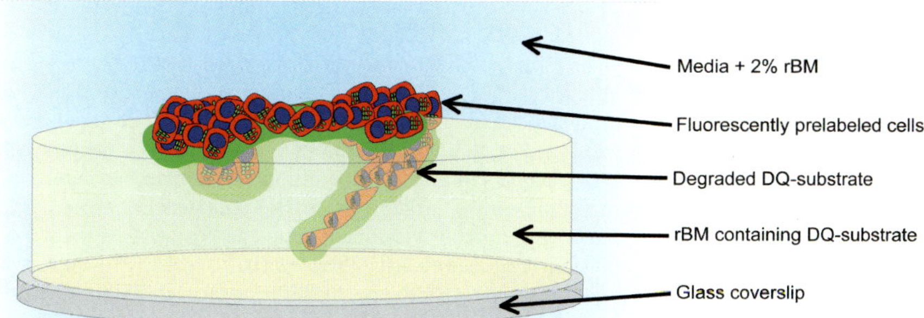

Fig. 8.2 Schematic representation of a 3D reconstituted basement membrane (*rBM*) overlay culture grown on *rBM* containing a dye-quenched (*DQ*)-substrate illustrating intracellular and extracellular proteolysis. Green, red, and dark blue represent cleavage fragments of *DQ*-substrate, cells, and nuclei, respectively.

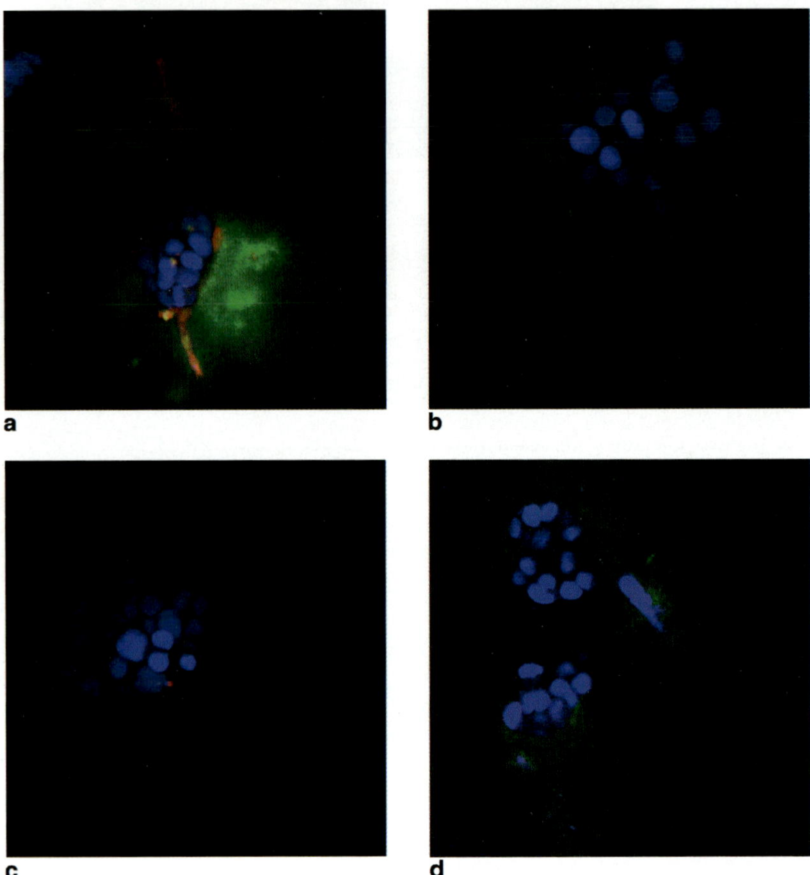

Fig. 8.3 Images of abrogation of proteolysis by broad-spectrum inhibitors of several classes of proteases. BT549 breast tumor cell and WS-12Ti breast fibroblast (5:1 ratio) cocultures were grown on reconstituted basement membrane (*rBM*) containing dye-quenched (*DQ*)-collagen IV with and without inhibitors and imaged with a Zeiss LSM 510 confocal microscope. Red, green, and blue fluorescence depict pre-labeled fibroblasts, cleavage products of *DQ*-collagen IV, and Hoescht-labeled nuclei, respectively. (**a**) Cocultures in the absence of inhibitors (same image as in Fig. 8.1d, but including red and blue channels to illustrate the fibroblasts and nuclei, respectively) showing strong green fluorescence representing proteolysis of *DQ*-collagen IV. Proteolysis, as evidenced by a decrease in intensity of the green fluorescent cleavage products, is substantially reduced in cocultures by 2 μM aprotinin (serine protease inhibitor); (**b**) 10 μM CAO74/CAO74Me cocktail (cell permeable and impermeable, respectively, cysteine protease inhibitors that are selective for cathepsins B and L); (**c**) 25 μM GM6001 [matrix metalloprotease (*MMP*) inhibitor]. (**d**) Magnification, 40×.

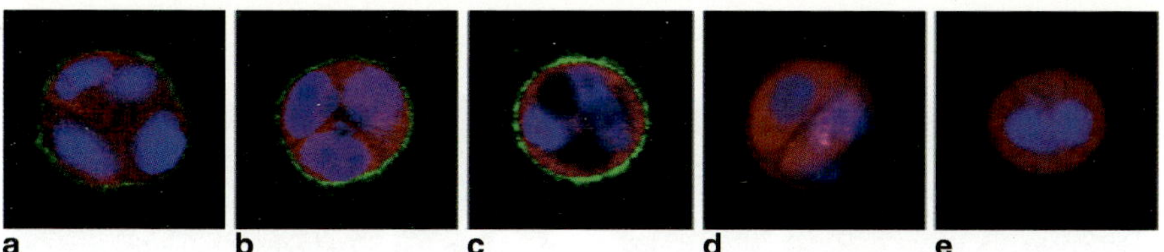

Fig. 8.4 Images of PAK1 (p21-activated kinase 1) regulation of pericellular proteolysis by MCF-10A human breast epithelial acini. MCF-10A cells were infected with control retroviruses that only expressed RFP (panel **a**), or bicistronic constructs expressing RFP plus wild-type PAK1 (panel **b**), constitutively active PAK1, 423E mutant (panel **c**), or two forms of dominant-negative PAK1 (83, 86L, 299R; panel **c** or 299R mutants; panel **e**) and cultured for 2 days in a 3D reconstituted basement membrane (*rBM*) overlay culture containing dye-quenched (*DQ*)-collagen IV. Nuclei of live cells were stained with Draq5 for 30 min and confocal images were recorded with a Zeiss LSM 510 confocal microscope. Expression of retroviral constructs (red), and nuclei (blue) are shown in equatorial sections of the cells along with the associated cleavage products of *DQ*-collagen IV (green). Expression of activated PAK1 increases pericellular proteolysis, whereas dominant-negative forms of PAK1 block pericellular proteolysis. Magnification, 40×.

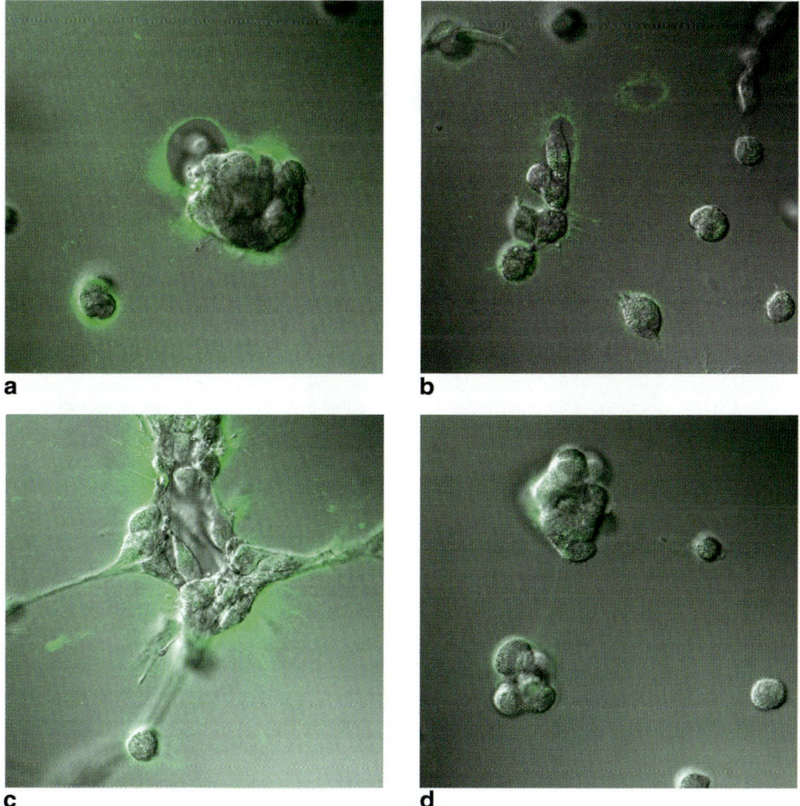

Fig. 8.5 Differential interference contrast (*DIC*) and fluorescent overlay images of abrogation of proteolysis by inhibitors of signaling pathways. BT549 breast tumor cells (2×10^4) were grown on reconstituted basement membrane (*rBM*) containing dye-quenched (*DQ*)-collagen IV alone (**a**), or in the presence of 20 μg/ml anti-β1 integrin blocking antibody, AIIB2 (**b**), 10 μM Rho-kinase inhibitor H1152 (**c**) or 30 μM Rac1 inhibitor, NSC23766 (**d**). All images were recorded with a Zeiss LSM 510 confocal microscope. Magnification, 40×.

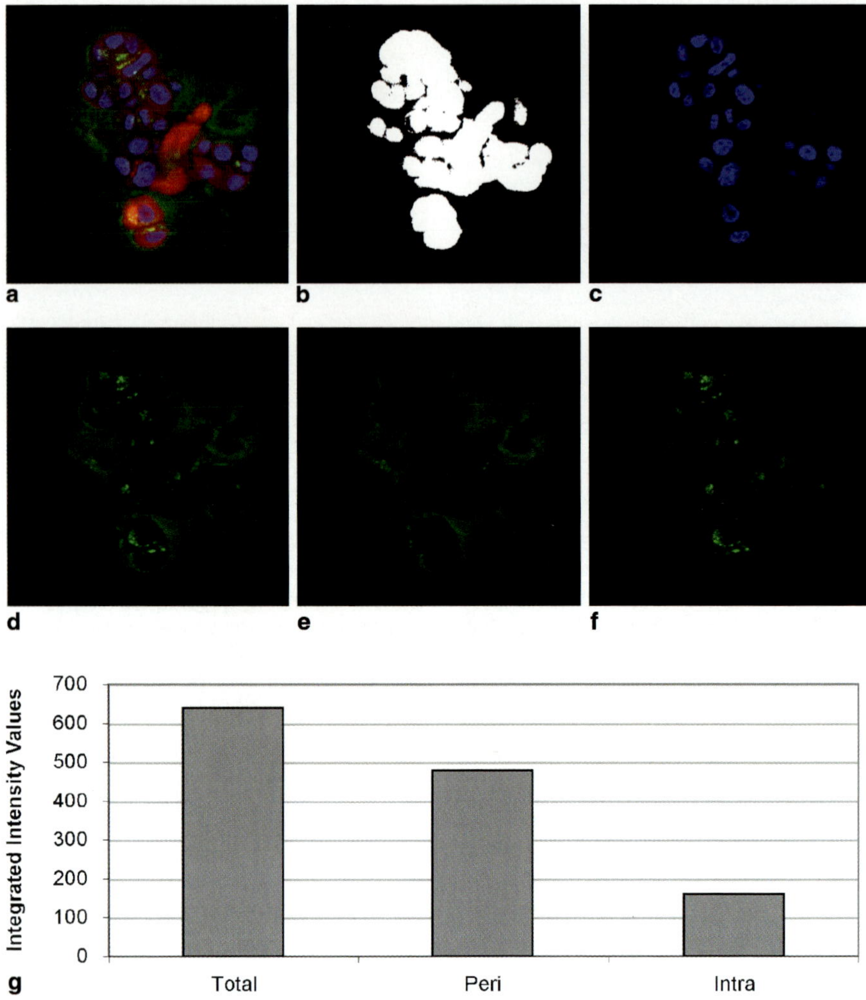

Fig. 8.6 Quantification of intracellular and pericellular degradation of dye-quenched (*DQ*)-collagen IV by MCF10–DCIS (ductal carcinoma in situ) cells. (**a**) Single optical section taken at equatorial plane showing fluorescence from cleaved DQ-collagen IV (green), pre-labeled cells (red) and nuclei (blue). (**b**) Same optical section as (**a**) but a binarized image of the cells (red channel) was used to designate x, y coordinates of intracellular areas within the same optical section. (**c**) Same optical section but binarized image of nuclei (blue channel) was used to create a 3D reconstruction for counting nuclei, thus allowing us to determine the number of cells within the field of view. (**d**) Cleavage products of *DQ*-collagen IV from same optical plane, representing both intracellular and pericellular cleavage products of *DQ*-collagen IV. (**e**) Pericellular cleavage products of *DQ*-collagen IV obtained by masking image from panel **d** with the image from panel **b** to eliminate all signal from cytoplasmic areas. (**f**) Image of intracellular cleavage products of *DQ*-collagen IV obtained by subtracting the image in panel **e** from that in panel **d**. (**g**) Quantification of total proteolysis of *DQ*-collagen IV in all optical sections throughout the volume of the structure was determined as normalized integrated intensity based on cell number. Total fluorescence was separated into intracellular (Intra) and pericellular (Peri) degradation using image arithmetic in Metamorph software (Molecular Devices, Inc.) Magnification, 40×.

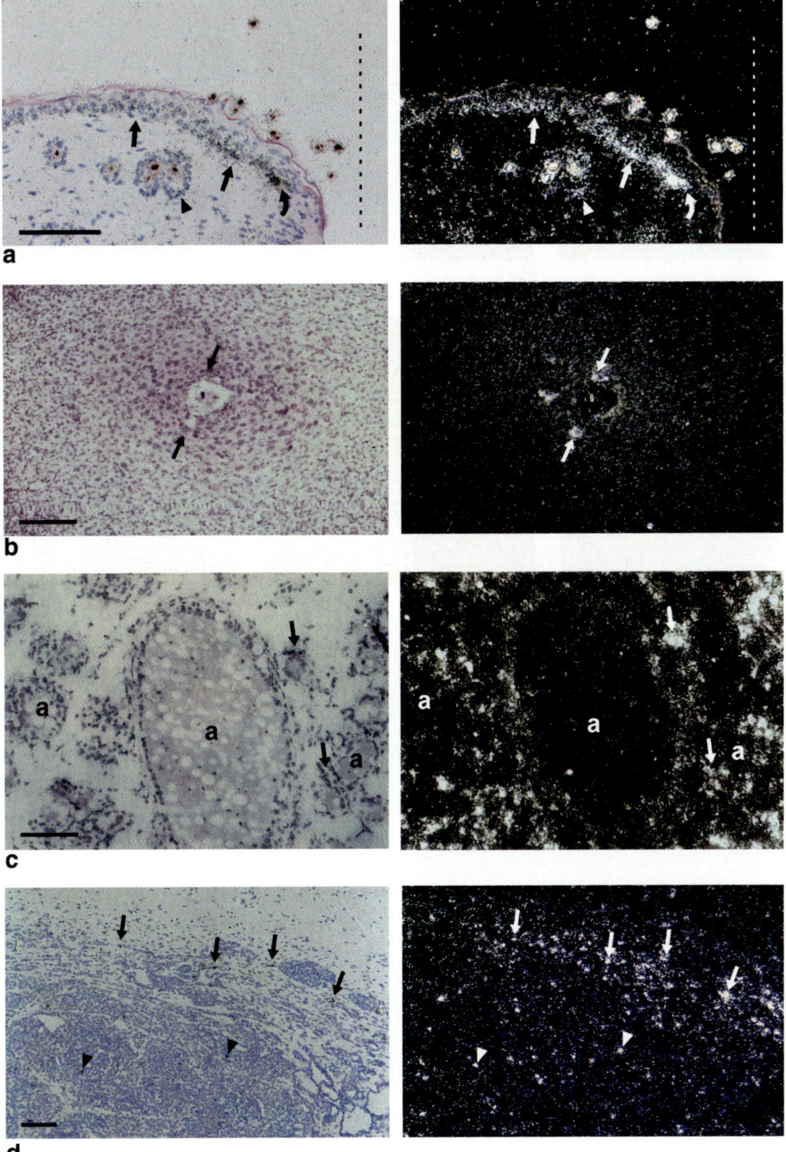

Fig. 11.1 Expression of urokinase-type Plg activator (*uPA*) mRNA in three normal tissue-remodeling events and in cancer invasion. Mouse uPA mRNA was detected by in situ hybridization with S-35-labeled RNA probes. In the left column are bright field images; in the right column are the corresponding dark field images. (**a**) Full thickness skin wound 24 h after the incision. uPA mRNA is present in the basal and suprabasal keratinocytes in the zone of proliferative and migrating keratinocytes (straight arrows) extending to the tip of the keratinocyte wedge (curved arrow) adjacent to the wound edge (dashed line) (Rømer et al. 1991, 1996). Note that keratinocytes of nearby hair follicles also express uPA mRNA (arrowhead) and can contribute to the

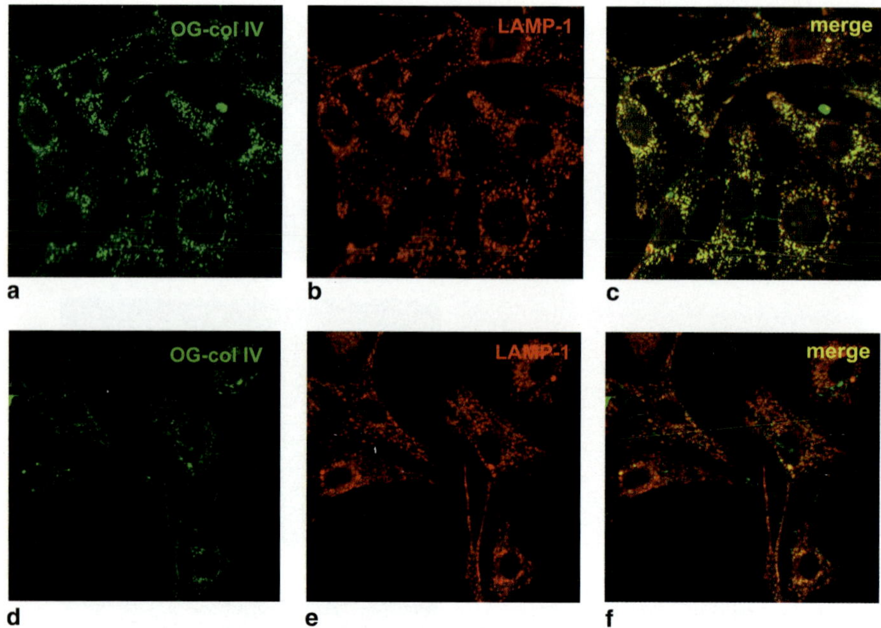

Fig. 13.1 Internalized collagen is directed to lysosomal degradation. Wild-type (**a–c**) and uroki-nase-type plasminogen activator receptor associated protein (*uPARAP/Endo*180)-deficient (**d–f**) fibroblasts were treated with an inhibitor of lysosomal cysteine proteases, E64d. Cells were then incubated with Oregon green-labeled collagen IV for 16 h at 37°C. Lysosomes were labeled with red fluorescence, using a LAMP-1-specific antibody. A strong lysosomal accumulation of OG-collagen is seen in wild-type fibroblasts, whereas cells deficient for uPARAP/Endo180 do not internalize collagen. (Reproduced from Kjoller et al. 2004 with permission from the publisher).)

regenerating epidermis. (**b**) Section through embryo at day 6.5 of gestation. uPA mRNA is present in invasive trophoblast giant cells in the uterine stroma surrounding the embryo (arrows) (Sappino et al. 1989, Teesalu et al. 1996). (**c**) Involuting mammary gland 5 days after removing the pups at the peak of lactation. uPA mRNA is abundant in fibroblast-like cells and macrophages (arrows) in between regressing alveoli (**a**) and is also seen in some intraluminal cells [apoptotic cells and macrophages; Lund et al. (1996)]. (**d**) Mammary cancer from a 13-week-old MMTV-PymT-transgenic mouse. uPA mRNA is present in fibroblast-like cells and macrophages mainly at the invasive front of the cancer (arrows) but also in the interior (arrowheads), as well as in intraluminal macrophages (Almholt et al. 2005, unpublished data). Bars, 100 μm

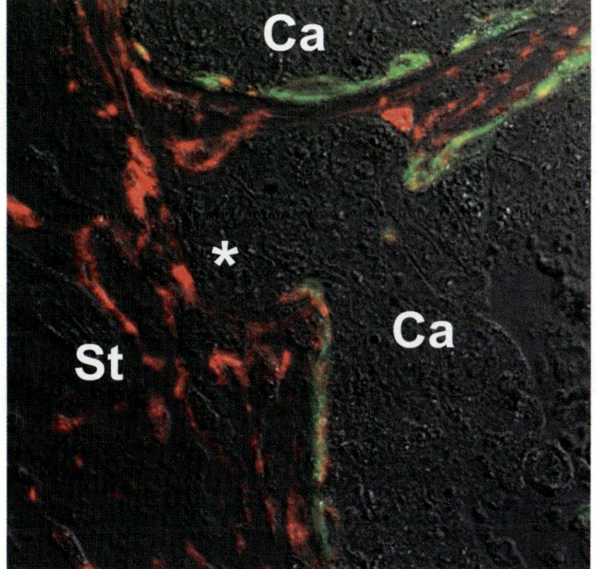

Fig. 13.2 Expression of urokinase-type plasminogen activator receptor associated protein (*uPARAP/Endo*180) at a tumor–stroma interface. Double immunostaining of uPARAP/Endo180 (rabbit antibody coupled to Cy3; red) and cytokeratin 17 [myoepithelial marker, monoclonal mouse antibody coupled to fluorescein isothiocyanate (*FITC*); green] in human breast ductal carcinoma in situ with focal invasion (indicated by star). uPARAP/Endo180 is expressed in fibroblasts and some myoepithelial cells surrounding the carcinoma cells (*Ca*) and in fibroblasts located in the stroma (*St*). For material and methods, see Nielsen et al. (2007). (Boye Schnack-Nielsen, the Finsen Laboratory, Copenhagen, unpublished work; figure kindly made available by B.S. Nielsen).

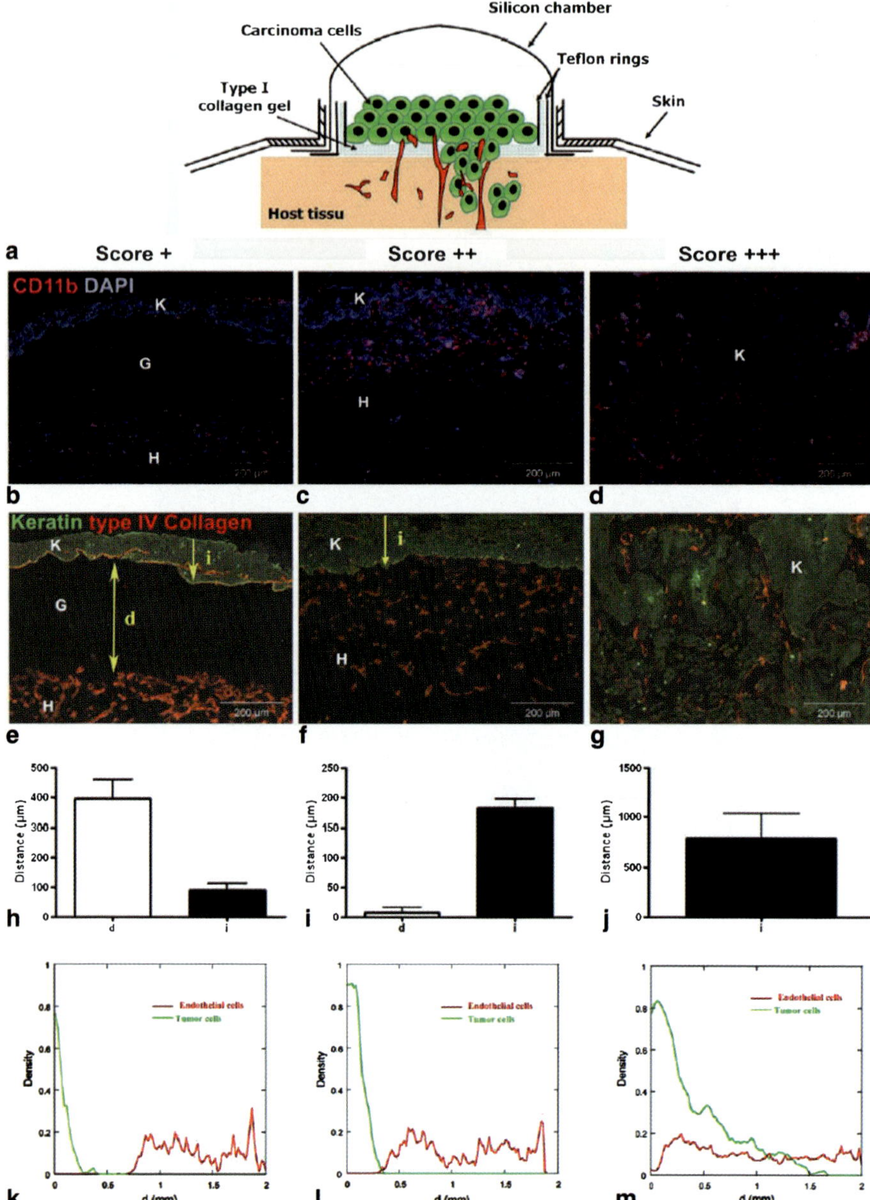

Fig. 17.1 Representation of skin carcinoma cell invasion in the surface transplantation model. **a** Schematic representation of the model: keratinocytes are cultured on the type-I collagen gel, mounted in concentric teflon rings and covered with a hat-shaped silicone chamber. **b–g** Different stages of tumor invasion (score +, ++, +++) are observed after 1 (**b, e**), 2 (**c, f**), or 3 (**d, g**) weeks of transplantation. After 1 week of transplantation, tumor cells proliferate to form a multilayered epithelium, and few inflammatory cells infiltrate the collagen gel (**b**). The collagen gel is

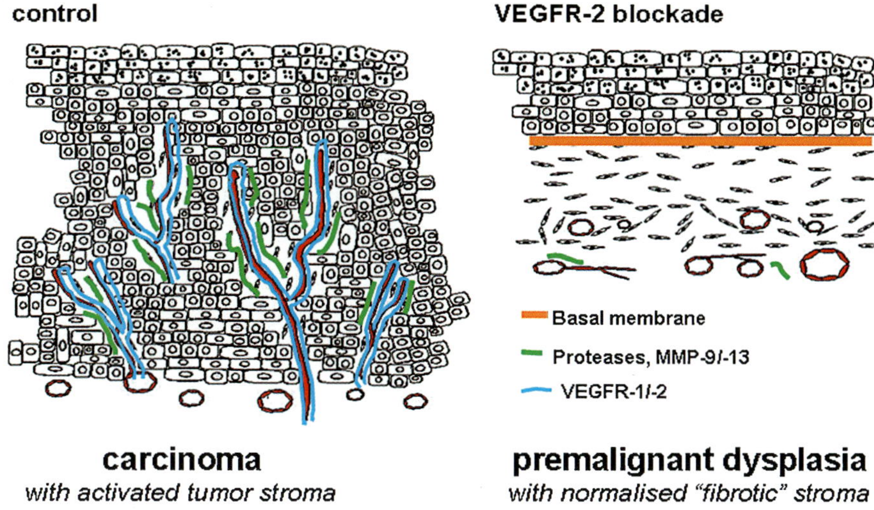

Fig. 17.2 Downregulation of stromal matrix metalloproteinase (*MMP*) expression by blockade of angiogenesis. Inhibiting angiogenesis in malignant transplants by the vascular endothelial growth factor receptor-2 (*VEGFR*-2) blocking antibody DC101 completely abrogates tumor vascularization and stromal MMP expression, as a consequence, tumor invasion (Vosseler et al. 2005).

progressively replaced by a granulation tissue at 2 weeks (**c**), leading to a vascularized tumor at 3 weeks (**d**). **b–d**: Immunostaining of CD11b, a marker of granulocytes and monocytes/macrophages. **e–g** Immunostaining of keratinocytes (green) and vessels (red). **e** A score of + is attributed when blood vessels remain below the collagen gel or particularly the matrix. **f** When blood vessels get into close contact with malignant epithelial layer, a score of ++ is assigned. **g** The score is +++ when tumor and blood vessels are intermingled. **h–j** Tumor invasion can be quantified by manual measurements of the distance "*i*," between the top of tumor cell layer and the deepest front of tumor spread (yellow arrows) (**e, f**). In this system, vascularization is estimated by measuring the distance "*d*" separating tumor cells from the front blood vessel migration (**e**). Quantification by computerized image analysis consists in determining the distribution of tumor/endothelial cell density as a function of the distance to the upper boundary of tumor (**k–m**).

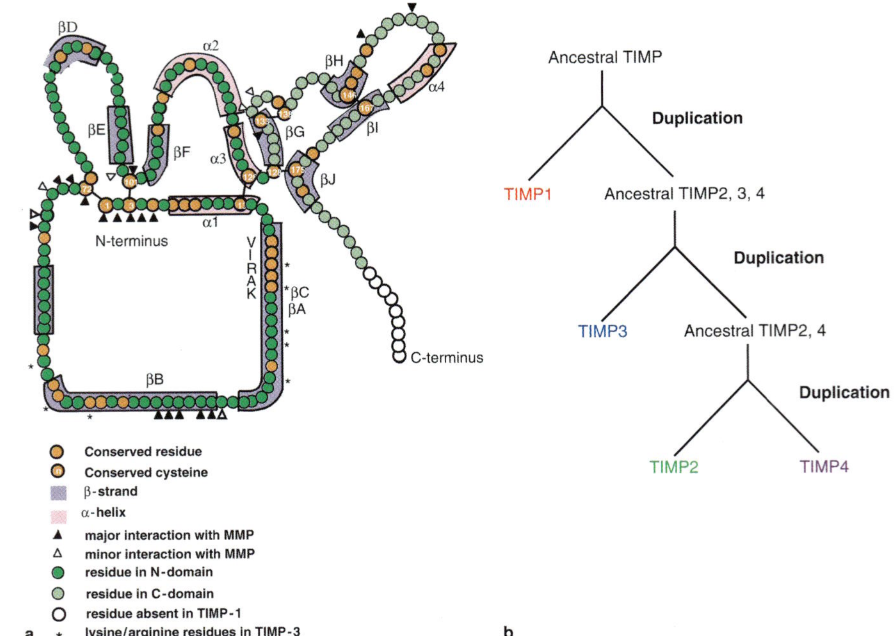

Fig. 20.1 Proposed structure of the prototypical tissue inhibitor of metalloproteinase (*TIMP*). The two domains (N-terminal and C-terminal) are shaded in green, and major differences between TIMP1 and TIMP3 architecture are highlighted. The conserved VIRAK sequence in TIMP is required for MMP inhibition. **b** Possible evolutionary pathway in the generation of the four mammalian TIMPs. Brew et al. (2000) suggest that an ancestral TIMP underwent a duplication even to generate two paralogous TIMPs, thereby creating TIMP1. A second duplication event created TIMP3 and the ancestor to TIMP2 and 4. A final duplication resulted in the generation of TIMP2 and 4. TIMP1 most highly resembles the ancestral TIMP as it has accumulated the fewest mutations across species

Chapter 19
New Insights into MMP Function in Adipogenesis

Kumari L. Andarawewa and Marie-Christine Rio

Abstract Matrix metalloproteinases (MMPs) mediate homeostasis of the extracellular environment. This equilibrium is modified and/or altered during normal (adipocyte differentiation, mammary gland involution, wound healing) or pathological (obesity, cancer) biological processes. While the role of MMPs is well known during tissue involution, wound healing, and cancer, their implication during adipogenesis has just begun to unfold. Although not designed to be comprehensive, this chapter provides *in vitro* and *in vivo* evidence that matrix degradation is essential for adipogenesis, and that the proteolytic activity of MMPs is critical for adipose tissue development. Moreover, obesity is currently a sign of poor prognosis in various human carcinomas. In this context, the involvement of MMP11 in adipocyte–cancer cell interaction/cross talk during the early invasive steps of carcinomas provides evidence that the MMP system participates in this process, and highlights a new link between obesity and cancer. How adipocytes and MMPs might cooperate to favor tumor progression, and notably the possible role of adipokines and adipose tissue angiogenesis, will be discussed.

Introduction

Carcinoma is deleterious for the patients due to the systemic dissemination of the disease and metastases. The prerequisite to this process is the local colonization of adjacent connective tissues by epithelial cancer cells at the site of the primary tumor. This event leads to illegitimate connective cell–cancer cell interaction/cross talk. Connective cells include fibroblasts, adipocytes, inflammatory cells, and endothelial cells. In this context, most of the reported data have emphasized fibroblast and inflammatory cell participation, notably via enzymatic mechanisms

M.-C. Rio
Institut de Génétique et de Biologie Moléculaire et Cellulaire (IGBMC), CNRS UMR 7104, INSERM U596, ULP, BP 163, 67404 Illkirch Cedex, C.U. de Strasbourg, France,
e-mail: rio@igbmc.u-strasbg.fr

D. Edwards et al. (eds.), *The Cancer Degradome.*
© Springer Science + Business Media, LLC 2008

that include the plasmin and matrix metalloproteinase (MMP) proteolytic systems, leading to extracellular matrix (ECM) remodeling and activation of latent factors. Surprisingly, very little attention has been given to adipocytes, although it is obvious that early local tumor invasion occurs in the immediate proximity of preadipocytes and/or fully differentiated adipocytes in numerous organs.

In this chapter, we will examine critical aspects of MMPs in the biology of fat cells of the "adipose organ" and highlight their involvement in adipogenesis. Their function during tumor invasion will be presented with an emphasis on MMP11 (previously named stromelysin-3), a bad prognosis factor in various human carcinomas (Basset et al. 1990, 1997; Rio 2005). Finally, we will discuss future directions for investigation that are essential for a more complete understanding of MMP function in adipogenesis in normal and pathological conditions.

Definition of "The Adipose Organ"

The formation of adipose tissue is a complex process requiring the commitment of mesodermal stem cells to a preadipocyte lineage and the conversion of preadipocytes into mature adipocytes (Smas and Sul 1995, Gregoire 2001). This differentiation switch activates a specific program of gene expression, notably increased expression of two adipogenic markers, peroxisome proliferator-activated receptor (PPAR; Fajas et al. 2001) and adipocyte protein (aP2; Tontonoz et al. 1994). Thus, preadipocytes are "adipose precursor cells" which are believed to be present throughout life. Preadipocyte proliferation and differentiation is defined by many intrinsic and extrinsic factors. Differentiation is characterized by a change in morphology from the elongated fibroblast-like form to the round unilocular appearance of mature lipid-filled fat cells. Adipose tissue is present in all mammals. The combination of all adipose tissue depots in a given organism has been referred to as "the adipose organ." It is the only organ in the body that can markedly change in mass during adult life (Hausman et al. 2001). The basic components of an adipose depot are mature adipocytes, stromal-vascular (SV) cells, blood vessels, lymph nodes, and nerves. A population of mature adipocytes includes fat cells of variable size. Size heterogeneity is lost in obesity, and adipocyte hypertrophy (increase in cell size) often precedes adipocyte hyperplasia (increase in cell number) in this pathology. Adipocyte hypertrophy results from excess triglyceride accumulation in existing adipocytes. Hyperplasia, referred to as adipogenesis, results from the recruitment of new adipocytes from precursor cells in the adipose tissue, and involves the proliferation and differentiation of these preadipocytes. Finally, there are substantial regional differences in the growth and cellularity of adipose tissues (Fain et al. 2004, Dusserre et al. 2000), which may depend on the nature of the ECM in which preadipocytes proliferate and differentiate (Nakajima et al. 1998).

Mature lipid-filled adipocytes are terminally differentiated cells and are generally considered incapable of division, but this point is a matter of debate. Several studies have shown that adipocyte dedifferentiation may occur after either acute injury or in a malignant context (Meng et al. 2001, Andrade et al. 1998). Indeed, treatment of adipocyte-differentiated cells with tumor necrosis factor-alpha (TNF-α) (Souza et al. 2003) or matrix metalloproteinase 11 (MMP11; Andarawewa et al. 2005) leads to lipid depletion and reversal of adipocyte phenotype accompanied by a decrease in PPARg expression. These data indicate that adipocytes exhibit a plasticity depending on the microenvironment.

The adipocyte is a unique cell since it is surrounded by two membranes, a plasma membrane and a basement membrane. Differentiation is associated with an increase in the secretion of basement membrane components such as laminin, proteoglycans, and type-IV and -VI collagens. Once cells are committed preadipocytes (Bernlohr et al. 1984), the original ECM is remodeled, leading to mature lipid-filled adipocytes through proteinase action (Nakajima et al. 1998, Lilla et al. 2002). Thus, adipogenesis is characterized by the conversion of a fibronectin-rich matrix to a basement membrane (Gregoire 2001, Selvarajan et al. 2001).

Evidence that MMPs Play a Role in Adipogenesis

Several *in vitro* and *in vivo* data strongly support that MMPs are involved in the regulation of adipocyte differentiation, in addition to ECM protein cleavage.

MMP Expression/Function During In Vitro Adipocyte Differentiation

The first indication that MMPs might function in adipocyte metabolism came from the mouse cell culture systems of 3T3L1 adipocyte precursors. 3T3L1 cells are committed preadipocytes when they reach confluence. Treatment of confluent cultures with a differentiation-inducing mix (DM) leads to the expression of adipogenic proteins (i.e., PPARg and aP2) and the accumulation of lipids (Bernlohr et al. 1984). MMP2, MMP3, MMP9, MMP13, and MMP14, and two of their natural inhibitors—the tissue inhibitors of MMPs (TIMPs), TIMP2 and TIMP4—are highly expressed during the 3T3L1 adipocyte differentiation process. On the contrary, MMP1, MMP7, MMP11 and MMP19, and TIMP1 and TIMP3 are not expressed or decreased during adipogenesis (Alexander et al. 2001, Croissandeau et al. 2002, Chavey et al. 2003). Collectively, these data suggest that MMPs participate either positively or negatively in adipocyte metabolism.

Mouse embryonic fibroblasts (MEFs) also have the capacity to differentiate into adipocytes in response to DM (Lazar 2002). Similar to the 3T3L1 cells, MMP2 is overexpressed and MMP11 is reduced during this process (Andarawewa et al. 2005). This cell model has been used to study the function of MMP11 on adipogenesis. The efficacy of wild-type or MMP11-deficient MEF primary cultures to differentiate into adipocytes has been compared. MMP11-deficient MEFs showed higher number and size of intracellular lipid droplets, and higher PPARg and aP2 mRNA levels compared with wild type. Thus, the potential of MMP11-deficient

MEFs to differentiate into adipocytes is markedly increased compared with MMP11-positive MEFs, indicating that MMP11 is a negative regulator of adipogenesis. Indeed, an enzymatically active mouse recombinant MMP11 protein reverts adipocyte differentiation in 8-day DM-differentiated MMP11-deficient MEFs. Thus, at least one MMP, MMP11, is a potent negative regulator of adipogenesis and has the ability to dedifferentiate adipocytes. Lastly, a separate study showed that adipocyte maturation proceeds only in combination with MMP14-dependent remodeling of 3D type-1 collagen scaffolding, indicating that MMP14 is a positive regulator of adipose tissue development and function (Chun et al. 2006).

Role of MMPs in Adipose Homeostasis: Lessons from MMP-Deficient Mice

Human adipocytes produce MMP2 and MMP9 (Bouloumie et al. 2001). Several nonmalignant biological processes alter adipose homeostasis, most notably in obesity and postweaning mammary gland involution. Several studies have investigated the expression/role of MMPs in these processes using either wild-type or MMP-deficient mice.

High-fat diet (HFD) treatment of wild-type mice results in increased levels of MMP2, MMP3, MMP11, MMP12, MMP13, MMP14, MMP19, and TIMP1, and decreased levels of MMP7, MMP9, MMP16, MMP24, and TIMP3, and TIMP4 in adipose tissues compared with mice fed a normal-fat diet (NFD) (Maquoi et al. 2002, Chavey et al. 2003). Moreover, HFD-fed MMP3-deficient (Maquoi et al. 2003), MMP11-deficient (Lijnen et al. 2002), and MMP19-deficient (Pendas et al. 2004) mice develop more adipose tissue.

They show a higher body weight, increased adiposity, and adipocyte hypertrophy compared with control littermates. In contrast, TIMP1-deficient mice have lower adipose tissue weight (Lijnen et al. 2003). These data indicate that MMP3, MMP11, and MMP19 are involved in a negative regulatory process that controls fat mass homeostasis *in vivo*. Finally, the effect of MMPs on adipogenesis during obesity has also been confirmed using MMP inhibitors (Christiaens and Lijnen 2006). Thus, obesity is associated with profound changes in the MMP/TIMP balance, supporting their role in the control of matrix remodeling events during this disease.

During mammary gland involution, secretory epithelial cells die and are replaced by adipocytes. The relationship between adipocyte metabolism and MMP function/activity was investigated during this process, and it was shown that MMP2, MMP3, MMP9, and MMP11 are involved (Lefebvre et al. 1992) (Green and Lund 2005). Moreover, MMP3-deficient mice accumulate more fat in their mammary fat pad (Alexander et al. 2001). TIMP1-overexpressing mice exhibit a similar phenotype (Alexander et al. 2001), whereas TIMP3-deficient mice show accelerated adipose reconstitution in their mammary glands (Fata et al. 2001). These results indicate that MMPs determine the rate of adipocyte differentiation during involutive mammary gland remodeling. In this context, it has been proposed that alteration in the fatty stroma in involuting breast tissue can promote metastasis (Watson 2006).

Adipocytes and MMPs in Carcinomas: Two Sides of the Same Coin?

Adipocyte function during cancer invasive processes has been neglected until recently. This is mainly because adipocytes disappear rapidly via the desmoplastic response of connective tissues during early invasive steps, leading to tumors devoid of adipocytes. Nevertheless, the adipocyte is an excellent candidate to play a role in influencing tumor behavior through heterotypic signaling processes and might prove to be critical for tumor survival, growth, and metastasis (Iyengar et al. 2003, Manabe et al. 2003).

Adiposis and Cancer: Epidemiological Data

Obesity is a worldwide problem which impacts on the risk and prognosis of some of the more common forms of cancer. This association is poorly understood, but solid epidemiological data support the role of fat mass/distribution in the development of risk factors, morbidity, and mortality (Hausman et al. 2001, Hursting et al. 2003). The International Agency for Research on Cancer (IARC) has concluded that there is sufficient evidence of a cancer-preventive effect for avoidance of weight gain in several cancers (Vainio et al. 2002). A similar conclusion was drawn from a Shangai Breast Cancer Study for breast cancer (Malin et al. 2005). More recently, it has been proposed that adiposity might rather be related to the risk of dying from cancer than to cancer incidence (Wright et al. 2007). Indeed, large amounts of adipose tissues are closely associated with poor prognoses in breast cancer of obese postmenopausal women (Manabe et al. 2003). A large American prospective analysis of the weight—cancer relationship shows that excess body mass index is the cause of ∼14% of all cancer deaths in men and 20% in women aged 50 years or older (Calle et al. 2003). Whether obesity operates directly, via true biologic effects of adiposity, or indirectly remains largely unknown.

Furthermore, it has been demonstrated that MMPs are involved at several steps of cancer development, and represent markers of poor prognosis (Jodele et al. 2006, VanSaun and Matrisian 2006, Overall and Kleifeld 2006, Chapter by Pennington et al. this volume). Several explanations have been proposed for this pejorative effect. One striking point is that most MMPs are paracrine factors which are expressed and secreted by connective tissue cells, and which act on ECM and cancer cells, notably at the early time of invasion. It is, therefore, tempting to speculate that the role of adipocytes in cancer might be mediated, at least in part, via MMPs, when pioneer invading cancer cells encounter neighboring adipocytes.

MMP11 Plays a Role in Adipocyte–Cancer Cell Interaction/Cross Talk

This hypothesis has been tested using MMP11, well known to promote tumor development in *in vivo* mouse tumor experimental models (Masson et al. 1998, Andarawewa et al. 2003, Deng et al. 2005). Although the MMP11 substrate(s) remains unknown, it has been shown that the promoting effect of MMP11 on tumorigenesis is dependent on its catalytic function (Noel et al. 2000). Moreover, MMP11 does not favor cancer cell proliferation, invasive properties, or tumor angiogenesis, but cancer cell survival (Boulay et al. 2001, Wu et al. 2001).

The impact of MMP11 has been investigated on forced cancer cell–adipocyte interactions/cross talk, 4 days after subcutaneous injection of C26 syngeneic cancer cells in wild-type or MMP11-deficient mice. This mimics the reaction of connective tissue to early local cancer cell invasion, in the presence of adipocytes. Indeed, the desmoplastic reaction is achieved very rapidly in these experiments and the tumor stroma, containing very few or totally devoid of adipocytes, is fully constituted at day 6 (Boulay et al. 2001). In MMP11-deficient conditions, the plasma and basement membranes of adipocytes are altered, allowing the passage of lipids from adipocytes to cancer cells and the ECM, ultimately leading to the progressive decay of adipocytes and cancer cell death. These results indicate that MMP11 functions in host adipocytes to favor the survival of invading cancer cells (Andarawewa et al. 2005).

Invasive Cancer Cells Induce MMP11 Expression in Neighboring Adipocytes

Numerous clinical data have shown that high levels of MMP11 are associated with aggressiveness of various human carcinomas and poor patient clinical outcome (reviewed in Basset et al. 1997). MMP11 involvement in adipocyte–cancer cell interaction/cross talk in human carcinomas has been investigated at the invasive front of breast carcinomas. This area is devoid of constituted stroma, and therefore exhibits a high ratio of adipocytes to fibroblasts. In contrast to normal adipose tissue devoid of MMP11, MMP11 is expressed by adipocytes located in the proximity of invading cancer cells. Thus, cancer cells induce MMP11 expression in proximal adipocytes. Interestingly, a reduction in adipocyte size occurs simultaneously and the number of peritumoral fibroblast-like cells increases. The origin of peritumoral fibroblasts, which provide structural and biochemical support for cancer cells (Tlsty 2001), remains debated. Interestingly, it has been proposed that in addition to increased proliferation of fibroblasts, dedifferentiation of preexisting adipocytes in the adjacent adipose tissue and/or prevention of differentiation of preadipocytes to mature adipocytes may also explain the extremely high fibroblast:adipocyte ratio observed in the stroma surrounding cancer cells (Hennighausen and Robinson 2001, Meng et al. 2001).

Together, these data indicate that the negative regulatory function of MMP11 on adipogenesis and dedifferentiation is aberrantly restored during the early steps of local tumor invasion. MMP11 participates, therefore, in the accumulation of fibroblast-like cells that might in reality be preadipocytes or dedifferentiated adipocytes, since these cells are morphologically undistinguishable, to the detriment of adipocytes. Indeed, at later timepoints of invasive carcinomas where the constituted stroma no longer contains adipocytes, MMP11 is restricted to a particular subpopulation of fibroblast-like cells located in the immediate vicinity of cancer cells and that are not myofibroblasts (Andarawewa et al. 2005). Collectively, these data support an essential role for adipocytes during the first steps of the tumor desmoplastic response, and constitute the first evidence that MMPs are implicated in such a phenomenon.

Potential Mechanisms

The biological and molecular processes underlying the function of adipocytes in cancer remain largely unknown. The implication that this function might be mediated via MMPs suggests two emerging directions involving adipokines and/or angiogenesis.

Via Regulation of the MMP System by Adipokines?

Besides their energy-storing function, adipocytes are also active endocrine cells that produce various biologically active polypeptides, the adipokines (Maeda et al. 1997). Leptin and adiponectin are secreted to the circulation (as hormones) by the adipocytes. Other adipokines are paracrine factors, most of them not released by adipocytes but by nonfat cells of the adipose tissue (Fain et al. 2004). Leptin is in direct proportion to the amount of adipose tissue, and is therefore positively correlated with obesity, whereas adiponectin is inversely correlated with obesity (Fischer-Posovszky et al. 2007).

There is increasing evidence for a role of adipokines (such as adiponectin and leptin) secreted by peritumoral adipose tissues in several cancers (Schaffler et al. 2007). High circulating levels of leptin are associated with increased prostate cancer risk and increased aggressiveness, while adiponectin levels decrease in prostate cancer and are inversely correlated with grade of disease (Mistry et al. 2007). Leptin may also have a promoting effect on carcinogenesis and metastasis of breast cancer (Kaur and Zhang 2005, Miyoshi et al. 2006), but there are conflicting results (Vona-Davis and Rose 2007). Conversely, adiponectin inhibits peritoneal metastasis development of gastric cancer (Ishikawa et al. 2007). Thus, leptin appears to be a positive factor for tumor development and aggressiveness, while adiponectin protects against cancers.

Several *in vitro* studies have shown that leptin and adiponectin regulate the expression and/or activity of some MMPs in various cells. Leptin enhances MMP2,

MMP9, TIMP1, and TIMP2 in HUVEC cells (Park et al. 2001), and MMP2 and MMP9 in cytotrophoblastic cells (Castellucci et al. 2000). On the contrary, leptin represses MMP1 expression in hepatic stellate cells (Cao et al. 2007), and MMP2 in glomerular mesangial cells (Lee et al. 2005). Finally, adiponectin increases TIMP1 in human macrophages (Kumada et al. 2004). From these data, it is tempting to speculate that adipocytes act in carcinomas via adipokine-mediated regulation of MMP expression/function.

Via Adipose Tissue Angiogenesis?

A relationship between adipogenesis and angiogenesis has been reported (Rose et al. 2004, Hausman and Richardson 2004). First, the same stem cells have been shown to give rise to vascular cells and adipocytes (Zangani et al. 1999). Second, adipocytes need a blood supply for survival and it has been postulated that each adipocyte is in close proximity to a blood capillary. In obesity, neovascularization of the expanding adipose tissue is critical for maintaining proper function. Human microvascular endothelial cells can promote proliferation of neighboring preadipocytes via unidentified paracrine signaling (Hutley et al. 2001). Park et al. (2001) have proposed that leptin acts as a functional link between adipocytes and the vasculature. Indeed, leptin may directly induce angiogenesis, or indirectly by potentiating vascular endothelial growth factor (VEGF) secretion and uptake. VEGF is expressed and secreted by adipocytes, and is critical for maintaining local vascularity and adipose tissue accretion, as well as in the expansion and retraction of the adipose tissue mass (Hausman and Richardson 2004). Interestingly, there is also a relationship between VEGF and the MMP system, and MMPs are well known to be involved in angiogenesis (Arroyo et al. 2007). In contrast, adiponectin is a negative regulator of angiogenesis, and it has been postulated that adiponectin antitumor effects might result via its antiangiogenic activities (Ishikawa et al. 2007).

The angiogenic activity of adipokines is likely to provide new insights into the relationship between obesity and cancer. It is tempting to speculate that adipose tissue angiogenesis, which is greatly MMP-dependent, might be aberrantly restored during carcinogenesis, thereby favoring tumor progression to metastases.

Conclusion

The studies summarized in this chapter reveal that MMPs participate in adipogenesis, some acting positively and others as negative factors. Moreover, studies on MMP11 provide new insights into the function of adipose tissue in tumor progression. One issue that requires clarification is the precise role that other MMPs might play in adipocyte–cancer cell interaction/cross talk. This chapter strongly supports the concept that the loss of adipose tissue homeostasis and the related alterations of MMP production/activation are major contributors to the aggressive behavior of human cancers. Thus, delineation of the role of the tumor microenvironment in carcinogenesis should include adipocytes in the future. From a clinical viewpoint, it is important to know how adipocytes would serve to promote tumor progression. The adipokine-mediated regulation of the MMP system is likely to provide additional insight into the relationship between fat/obesity and cancer. In particular, understanding the effects of adipokines produced by the adipose tissue microenvironment on tumor-mediated angiogenesis would be fundamental. The overall role of the adipose tissue vascular bed, which is expanded as a result of obesity, on tumor proliferation and invasion should be investigated. These studies should aid in identifying adipocyte-related molecules as possible intervention points for innovative anticancer strategies.

Acknowledgment We thank Susan Chan for helpful discussion. This work was supported by funds from the Institut National de la Santé et de la Recherche Médicale, the Centre National de la Recherche Scientifique, the Hôpital Universitaire de Strasbourg, the Association pour la Recherche sur le Cancer, the Ligue Nationale Contre le Cancer and the Comités du Haut-Rhin et du Bas-Rhin, the Fondation de France, and the European Commission (FP5 QLK3-CT-2002-02136, FP6 LSHC-CT-2003-503297). K.L.A. was a recipient of an EU fellowship.

References

Alexander, C.M., Selvarajan, S., Mudgett, J., Werb, Z. 2001. Stromelysin-1 regulates adipogenesis during mammary gland involution. J Cell Biol. 152:693–703.

Andarawewa, K.L., Boulay, A., Masson, R., Mathelin, C., Stoll, I., Tomasetto, C., Chenard, M.P., Gintz, M., Bellocq, J.P., Rio, M.C. 2003. Dual stromelysin-3 function during natural mouse mammary tumor virus-ras tumor progression. Cancer Res. 63:5844–9.

Andarawewa, K.L., Motrescu, E.R., Chenard, M.P., Gansmuller, A., Stoll, I., Tomasetto, C., Rio, M.C. 2005. Stromelysin-3 is a potent negative regulator of adipogenesis participating to cancer cell-adipocyte interaction/crosstalk at the tumor invasive front. Cancer Res. 65:10862–71.

Andrade, Z.A., de-Oliveira-Filho, J., Fernandes, A.L. 1998. Interrelationship between adipocytes and fibroblasts during acute damage to the subcutaneous adipose tissue of rats: an ultrastructural study. Braz J Med Biol Res. 31:659–64.

Arroyo, A.G., Genis, L., Gonzalo, P., Matias-Roman, S., Pollan, A., Galvez, B.G. 2007. Matrix metalloproteinases: new routes to the use of MT1-MMP as a therapeutic target in angiogenesis-related disease. Curr Pharm Des. 13:1787–1802.

Basset, P., Bellocq, J.P., Wolf, C., Stoll, I., Hutin, P., Limacher, J.M., Podhajcer, O.L., Chenard, M.P., Rio, M.C., Chambon, P. 1990. A novel metalloproteinase gene specifically expressed in stromal cells of breast carcinomas. Nature. 348:699–704.

Basset, P., Bellocq, J.P., Lefebvre, O., Noel, A., Chenard, M.P., Wolf, C., Anglard, P., Rio, M.C. 1997. Stromelysin-3: a paradigm for stroma-derived factors implicated in carcinoma progression. Crit Rev Oncol Hematol. 26:43–53.

Bernlohr, D.A., Angus, C.W., Lane, M.D., Bolanowski, M.A., Kelly, T.J., Jr. 1984. Expression of specific mRNAs during adipose differentiation: identification of an mRNA encoding a homologue of myelin P2 protein. Proc Natl Acad Sci U S A. 81:5468–72.

Boulay, A., Masson, R., Chenard, M.P., El Fahime, M., Cassard, L., Bellocq, J.P., Sautes-Fridman, C., Basset, P., Rio, M.C. 2001. High cancer cell death in syngeneic tumors developed in host mice deficient for the stromelysin-3 matrix metalloproteinase. Cancer Res. 61:2189–93.

Bouloumie, A., Sengenes, C., Portolan, G., Galitzky, J., Lafontan, M. 2001. Adipocyte produces matrix metalloproteinases 2 and 9: involvement in adipose differentiation. Diabetes. 50:2080–6.

Calle, E.E., Rodriguez, C., Walker-Thurmond, K., Thun, M.J. 2003. Overweight, obesity, and mortality from cancer in a prospectively studied cohort of U.S. adults. N Engl J Med. 348:1625–38.

Cao, Q., Mak, K.M., Lieber, C.S. 2007. Leptin represses matrix metalloproteinase-1 gene expression in LX2 human hepatic stellate cells. J Hepatol. 46:124–33.

Castellucci, M., De Matteis, R., Meisser, A., Cancello, R., Monsurro, V., Islami, D., Sarzani, R., Marzioni, D., Cinti, S., Bischof, P. 2000. Leptin modulates extracellular matrix molecules and metalloproteinases: possible implications for trophoblast invasion. Mol Hum Reprod. 6:951–8.

Chavey, C., Mari, B., Monthouel, M.N., Bonnafous, S., Anglard, P., Van Obberghen, E., Tartare-Deckert, S. 2003. Matrix metalloproteinases are differentially expressed in adipose tissue during obesity and modulate adipocyte differentiation. J Biol Chem. 278:11888–96.

Christiaens, V., Lijnen, H.R. 2006. Role of the fibrinolytic and matrix metalloproteinase systems in development of adipose tissue. Arch Physiol Biochem. 112:254–9.

Chun, T.H., Hotary, K.B., Sabeh, F., Saltiel, A.R., Allen, E.D., Weiss, S.J. 2006. A pericellular collagenase directs the 3-dimensional development of white adipose tissue. Cell. 125:577–91.

Croissandeau, G., Chretien, M., Mbikay, M. 2002. Involvement of matrix metalloproteinases in the adipose conversion of 3T3-L1 preadipocytes. Biochem J. 364:739–46.

Deng, H., Guo, R.F., Li, W.M., Zhao, M., Lu, Y.Y. 2005. Matrix metalloproteinase 11 depletion inhibits cell proliferation in gastric cancer cells. Biochem Biophys Res Commun. 326:274–81.

Dusserre, E., Moulin, P., Vidal, H. 2000. Differences in mRNA expression of the proteins secreted by the adipocytes in human subcutaneous and visceral adipose tissues. Biochim Biophys Acta. 1500:88–96.

Fain, J.N., Madan, A.K., Hiler, M.L., Cheema, P., Bahouth, S.W. 2004. Comparison of the release of adipokines by adipose tissue, adipose tissue matrix, and adipocytes from visceral and subcutaneous abdominal adipose tissues of obese humans. Endocrinology. 145:2273–82.

Fajas, L., Debril, M.B., Auwerx, J. 2001. Peroxisome proliferator-activated receptor-gamma: from adipogenesis to carcinogenesis. J Mol Endocrinol. 27:1–9.

Fata, J.E., Leco, K.J., Voura, E.B., Yu, H.Y., Waterhouse, P., Murphy, G., Moorehead, R.A., Khokha, R. 2001. Accelerated apoptosis in the Timp-3-deficient mammary gland. J Clin Invest. 108:831–41.

Fischer-Posovszky, P., Wabitsch, M., Hochberg, Z. 2007. Endocrinology of adipose tissue - an update. Horm Metab Res. 39:314–21.

Green, K.A., Lund, L.R. 2005. ECM degrading proteases and tissue remodelling in the mammary gland. Bioessays. 27:894–903.

Gregoire, F.M. 2001. Adipocyte differentiation: from fibroblast to endocrine cell. Exp Biol Med (Maywood). 226:997–1002.

Hausman, D.B., DiGirolamo, M., Bartness, T.J., Hausman, G.J., Martin, R.J. 2001. The biology of white adipocyte proliferation. Obes Rev. 2:239–54.

Hausman, G.J., Richardson, R.L. 2004. Adipose tissue angiogenesis. J Anim Sci. 82:925–34.

Hennighausen, L., Robinson, G.W. 2001. Signaling pathways in mammary gland development. Dev Cell. 1:467–75.

Hursting, S.D., Lavigne, J.A., Berrigan, D., Perkins, S.N., Barrett, J.C. 2003. Calorie restriction, aging, and cancer prevention: mechanisms of action and applicability to humans. Annu Rev Med. 54:131–52.

Hutley, L.J., Herington, A.C., Shurety, W., Cheung, C., Vesey, D.A., Cameron, D.P., Prins, J.B. 2001. Human adipose tissue endothelial cells promote preadipocyte proliferation. Am J Physiol Endocrinol Metab. 281:E1037–44.

Ishikawa, M., Kitayama, J., Yamauchi, T., Kadowaki, T., Maki, T., Miyato, H., Yamashita, H., Nagawa, H. 2007. Adiponectin inhibits the growth and peritoneal metastasis of gastric cancer through its specific membrane receptors AdipoR1 and AdipoR2. Cancer Sci. 98(7):1120–7.

Iyengar, P., Combs, T.P., Shah, S.J., Gouon-Evans, V., Pollard, J.W., Albanese, C., Flanagan, L., Tenniswood, M.P., Guha, C., Lisanti, M.P., Pestell, R.G., Scherer, P.E. 2003. Adipocyte-secreted factors synergistically promote mammary tumorigenesis through induction of anti-apoptotic transcriptional programs and proto-oncogene stabilization. Oncogene. 22:6408–23.

Jodele, S., Blavier, L., Yoon, J.M., DeClerck, Y.A. 2006. Modifying the soil to affect the seed: role of stromal-derived matrix metalloproteinases in cancer progression. Cancer Metastasis Rev. 25:35–43.

Kaur, T., and Zhang, Z.F. 2005. Obesity, breast cancer and the role of adipocytokines. Asian Pac J Cancer Prev. 6:547–52.

Kumada, M., Kihara, S., Ouchi, N., Kobayashi, H., Okamoto, Y., Ohashi, K., Maeda, K., Nagaretani, H., Kishida, K., Maeda, N., Nagasawa, A., Funahashi, T., Matsuzawa, Y. 2004. Adiponectin specifically increased tissue inhibitor of metalloproteinase-1 through interleukin-10 expression in human macrophages. Circulation. 109:2046–9.

Lazar, M.A. 2002. Becoming fat. Genes Dev. 16:1–5.

Lee, M.P., Madani, S., Sekula, D., Sweeney, G. 2005. Leptin increases expression and activity of matrix metalloproteinase-2 and does not alter collagen production in rat glomerular mesangial cells. Endocr Res. 31:27–37.

Lefebvre, O., Wolf, C., Limacher, J.M., Hutin, P., Wendling, C., LeMeur, M., Basset, P., Rio, M.C. 1992. The breast cancer-associated stromelysin-3 gene is expressed during mouse mammary gland apoptosis. J Cell Biol. 119:997–1002.

Lijnen, H.R., Van, H.B., Frederix, L., Rio, M.C., Collen, D. 2002. Adipocyte hypertrophy in stromelysin-3 deficient mice with nutritionally induced obesity. Thromb Haemost. 87:530–5.

Lijnen, H.R., Demeulemeester, D., Van Hoef, B., Collen, D., Maquoi, E. 2003. Deficiency of tissue inhibitor of matrix metalloproteinase-1 (TIMP-1) impairs nutritionally induced obesity in mice. Thromb Haemost. 89:249–55.

Lilla, J., Stickens, D., Werb, Z. 2002. Metalloproteases and adipogenesis: a weighty subject. Am J Pathol. 160:1551–4.

Maeda, K., Okubo, K., Shimomura, I., Mizuno, K., Matsuzawa, Y., Matsubara, K. 1997. Analysis of an expression profile of genes in the human adipose tissue. Gene. 190:227–35.

Malin, A., Matthews, C.E., Shu, X.O., Cai, H., Dai, Q., Jin, F., Gao, Y.T., Zheng, W. 2005. Energy balance and breast cancer risk. Cancer Epidemiol Biomarkers Prev. 14:1496–1501.

Manabe, Y., Toda, S., Miyazaki, K., Sugihara, H. 2003. Mature adipocytes, but not preadipocytes, promote the growth of breast carcinoma cells in collagen gel matrix culture through cancer-stromal cell interactions. J Pathol. 201:221–8.

Maquoi, E., Munaut, C., Colige, A., Collen, D., Lijnen, H.R. 2002. Modulation of adipose tissue expression of murine matrix metalloproteinases and their tissue inhibitors with obesity. Diabetes. 51:1093–1101.

Maquoi, E., Demeulemeester, D., Voros, G., Collen, D., Lijnen, H.R. 2003. Enhanced nutritionally induced adipose tissue development in mice with stromelysin-1 gene inactivation. Thromb Haemost. 89:696–704.

Masson, R., Lefebvre, O., Noel, A., Fahime, M.E., Chenard, M.P., Wendling, C., Kebers, F., LeMeur, M., Dierich, A., Foidart, J.M., Basset, P., Rio, M.C. 1998. In vivo evidence that the stromelysin-3 metalloproteinase contributes in a paracrine manner to epithelial cell malignancy. J Cell Biol. 140:1535–41.

Meng, L., Zhou, J., Sasano, H., Suzuki, T., Zeitoun, K.M., Bulun, S.E. 2001. Tumor necrosis factor alpha and interleukin 11 secreted by malignant breast epithelial cells inhibit adipocyte differentiation by selectively down-regulating CCAAT/enhancer binding protein alpha and peroxisome proliferator-activated receptor gamma: mechanism of desmoplastic reaction. Cancer Res. 61:2250–5.

Mistry, T., Digby, J.E., Desai, K.M., Randeva, H.S. 2007. Obesity and prostate cancer: a role for adipokines. Eur Urol. 52:46–53.

Miyoshi, Y., Funahashi, T., Tanaka, S., Taguchi, T., Tamaki, Y., Shimomura, I., Noguchi, S. 2006. High expression of leptin receptor mRNA in breast cancer tissue predicts poor prognosis for patients with high, but not low, serum leptin levels. Int J Cancer. 118:1414–9.

Nakajima, I., Yamaguchi, T., Ozutsumi, K., Aso, H. 1998. Adipose tissue extracellular matrix: newly organized by adipocytes during differentiation. Differentiation. 63:193–200.

Noel, A., Boulay, A., Kebers, F., Kannan, R., Hajitou, A., Calberg-Bacq, C.M., Basset, P., Rio, M. C., Foidart, J.M. 2000. Demonstration *in vivo* that stromelysin-3 functions through its proteolytic activity. Oncogene. 19:1605–12.

Overall, C.M., Kleifeld, O. 2006. Tumour microenvironment–opinion: validating matrix metalloproteinases as drug targets and anti-targets for cancer therapy. Nat Rev Cancer. 6:227–39.

Park, H.Y., Kwon, H.M., Lim, H.J., Hong, B.K., Lee, J.Y., Park, B.E., Jang, Y., Cho, S.Y., Kim, H. S. 2001. Potential role of leptin in angiogenesis: leptin induces endothelial cell proliferation and expression of matrix metalloproteinases *in vivo* and *in vitro*. Exp Mol Med. 33:95–102.

Pendas, A.M., Folgueras, A.R., Llano, E., Caterina, J., Frerard, F., Rodriguez, F., Astudillo, A., Noel, A., Birkedal-Hansen, H., Lopez-Otin, C. 2004. Diet-induced obesity and reduced skin cancer susceptibility in matrix metalloproteinase 19-deficient mice. Mol Cell Biol. 24:5304–13.

Rio, M.C. 2005. From a unique cell to metastasis is a long way to go: clues to stromelysin-3 participation. Biochimie. 87:299–306.

Rose, D.P., Komninou, D.P., Stephenson, G.D. 2004. Obesity, adipocytokines, and insulin resistance in breast cancer. Obes Rev. 5:153–65.

Schaffler, A., Scholmerich, J., Buechler, C. 2007. Mechanisms of disease: adipokines and breast cancer–endocrine and paracrine mechanisms that connect adiposity and breast cancer. Nat Clin Pract Endocrinol Metab. 3:345–54.

Selvarajan, S., Lund, L.R., Takeuchi, T., Craik, C.S., Werb, Z. 2001. A plasma kallikrein-dependent plasminogen cascade required for adipocyte differentiation. Nat Cell Biol. 3:267–75.

Smas, C.M., Sul, H.S. 1995. Control of adipocyte differentiation. Biochem J. 309 (Pt. 3):697–710.

Souza, S.C., Palmer, H.J., Kang, Y.H., Yamamoto, M.T., Muliro, K.V., Paulson, K.E., Greenberg, A.S. 2003. TNF-alpha induction of lipolysis is mediated through activation of the extracellular signal related kinase pathway in 3T3-L1 adipocytes. J Cell Biochem. 89:1077–86.

Tlsty, T.D. 2001. Stromal cells can contribute oncogenic signals. Semin Cancer Biol. 11:97–104.

Tontonoz, P., Hu, E., Graves, R.A., Budavari, A.I., Spiegelman, B.M. 1994. mPPAR gamma 2: tissue-specific regulator of an adipocyte enhancer. Genes Dev. 8:1224–34.

Vainio, H., Kaaks, R., Bianchini, F. 2002. Weight control and physical activity in cancer prevention: international evaluation of the evidence. Eur J Cancer Prev. 11 Suppl 2:S94–S100.

VanSaun, M.N., Matrisian, L.M. 2006. Matrix metalloproteinases and cellular motility in development and disease. Birth Defects Res C Embryo Today. 78:69–79.

Vona-Davis, L., Rose, D.P. 2007. Adipokines as endocrine, paracrine, and autocrine factors in breast cancer risk and progression. Endocr Relat Cancer. 14:189–206.

Watson, C.J. 2006. Post-lactational mammary gland regression: molecular basis and implications for breast cancer. Expert Rev Mol Med. 8:1–15.

Wright, M.E., Chang, S.C., Schatzkin, A., Albanes, D., Kipnis, V., Mouw, T., Hurwitz, P., Hollenbeck, A., Leitzmann, M.F. 2007. Prospective study of adiposity and weight change in relation to prostate cancer incidence and mortality. Cancer. 109:675–84.

Wu, E., Mari, B.P., Wang, F., Anderson, I.C., Sunday, M.E., Shipp, M.A. 2001. Stromelysin-3 suppresses tumor cell apoptosis in a murine model. J Cell Biochem. 82:549–55.

Zangani, D., Darcy, K.M., Masso-Welch, P.A., Bellamy, E.S., Desole, M.S., Ip, M.M. 1999. Multiple differentiation pathways of rat mammary stromal cells *in vitro*: acquisition of a fibroblast, adipocyte or endothelial phenotype is dependent on hormonal and extracellular matrix stimulation. Differentiation. 64:91–101.

Chapter 20
TIMPs: Extracellular Modifiers in Cancer Development

Aditya Murthy, William Cruz-Munoz, and Rama Khokha

Abstract The tissue microenvironment impacts health and disease by providing a dynamic media where cells interact with the extracellular matrix scaffold, and a sink for critical ligands that dictate cell function. These dynamics become altered in cancer development through dysregulated proteolysis that is in part controlled by tissue inhibitors of metalloproteinases (TIMPs). This chapter reviews TIMP evolution and structure, the metalloproteinase targets, and the substrate complexity arising from metalloproteinase function. Individually, TIMPs inhibit proteolysis in a temporally and spatially distinct manner, resulting in a variety of phenotypes arising from specific TIMP deficiencies. It also discusses the consequences of altered TIMP gene expression on cell proliferation, apoptosis, angiogenesis, invasion, and metastasis in cell culture systems and mouse models. Although TIMPs inhibit angiogenesis, invasion, and metastasis, their effects on cell proliferation and apoptosis are tissue specific and context dependent. These studies highlight the importance of TIMPs as cancer modifier genes.

Introduction

The tissue inhibitor of metalloproteinase (TIMP) family has evolved to regulate tissue homeostasis through its ability to operate at the stromal–cellular interface. It controls remodeling of the extracellular matrix (ECM) as well as the cell surface by inhibiting the activity of several classes of metalloproteinases. TIMPs are ancient proteins found in invertebrates and vertebrates including nematodes, insects, fish, and mammals. However, TIMPs have not been reported in plants despite the known

R. Khokha
Department of Medical Biophysics, Ontario Cancer Institute, University Health Network, Toronto, Canada, e-mail: rkhokha@uhnresearch.ca

D. Edwards et al. (eds.), *The Cancer Degradome.*
© Springer Science + Business Media, LLC 2008

existence of metalloproteinases in this kingdom (Maidment et al. 1999). The mammalian genome contains four distinct TIMP proteins, each with different yet overlapping metalloproteinase inhibitory profiles (Murphy et al. 2003). The mammalian TIMP is a two-domain protein consisting of N- and C-termini and six disulphide linkages, and can exist in glycosylated and unglycosylated forms. Well-known targets of TIMPs include the enzymes from matrix metalloproteinases (MMPs), a disintegrin and metalloproteinases (ADAMs), and ADAM with thrombospondin motif (ADAM-TS) classes (*see also* Chap. 37). During metalloproteinase inhibition, a single netrin domain forms a wedge-like structure to interact with the active site of the enzyme forming a 1:1 stoichiometric complex that sterically inhibits metalloproteinase activity (Brew et al. 2000) (Fig. 20.1a).

TIMP was independently identified as a collagenase inhibitor (Cawston et al. 1981) and for its erythroid potentiating activity (Docherty et al. 1985), and additionally murine Timp1 was identified as a cell cycle-responsive gene (Edwards et al. 1986). It was then discovered that antisense ribonucleic acid (RNA)-mediated downregulation of TIMP1 conferred oncogenic properties on immortal but non-transformed murine fibroblasts showing a central role for TIMP in tumorigenesis (Khokha et al. 1989). The systems initially used to examine TIMPs included reproductive biology (Brenner et al. 1989, Waterhouse et al. 1993), wound healing (reviewed in Parks et al. 2004), tumorigenesis, and metastasis (Khokha et al. 1989, 1992). The cloning of all four human and murine Timp genes (Docherty et al. 1985, Gasson et al. 1985, Stetler-Stevenson et al. 1989, Apte et al. 1994, Greene et al. 1996, Leco et al. 1997), followed by analysis of their regulatory elements and genetic expression studies (Edwards et al. 1992, Leco et al. 1992, 1994; Wick et al. 1995, Dean et al. 2000), has revealed a more complex role in orchestrating tissue homeostasis. Cellular and matrix turnover are both affected by TIMPs, and each TIMP distinctly influences cell function. Clinical studies covering diverse human cancers also document the tremendous heterogeneity of TIMP expression and their correlation with disease stage and prognosis. Furthermore, the understanding of TIMP function has coevolved with that of the expanding field of metalloproteinase biology, and together they have highlighted complementary concepts essential for the health and survival of the organism.

The Evolution and Structure of TIMPs

The significant overlap in TIMP inhibition of metalloproteinases has likely arisen due to the events leading to the generation of each TIMP, and phylogenetic analysis shows that the Timp genes were created by multiple duplication events (Huxley-Jones et al. 2007) (Fig. 20.1b). Comparative genomic analyses of Timp evolution between invertebrate and vertebrate genomes provide insight into these events. The vertebrate Timp family arose from an earlier whole genome duplication before vertebrate and invertebrate divergence (Yu et al. 2003). Furthermore, whole genome duplications are probably responsible for the formation of the four

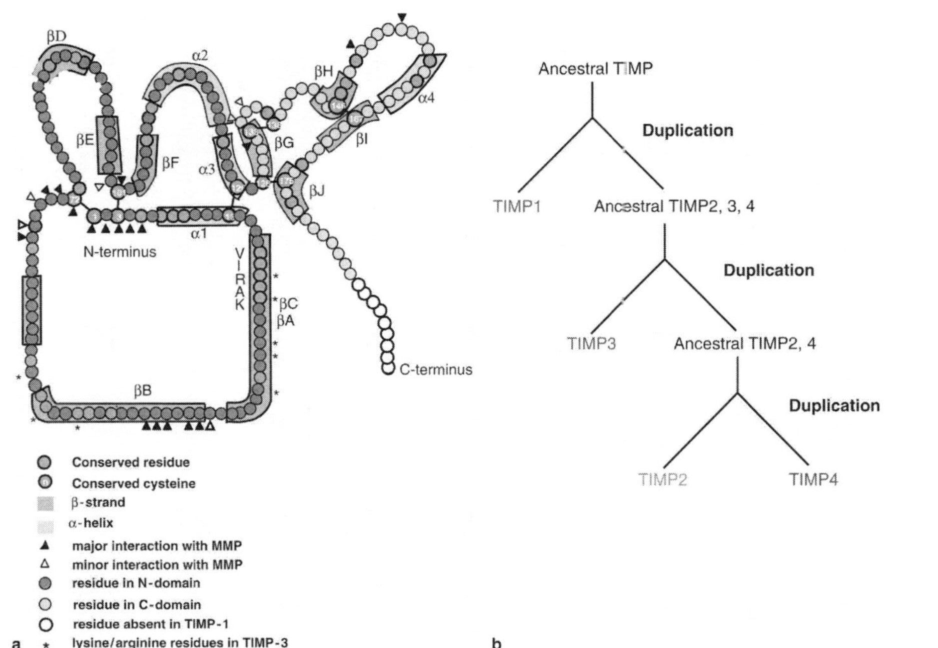

Fig. 20.1 a Proposed structure of the prototypical tissue inhibitor of metalloproteinase (*TIMP*). The two domains (N-terminal and C-terminal) are shaded in green, and major differences between TIMP1 and TIMP3 architecture are highlighted. The conserved VIRAK sequence in TIMP is required for MMP inhibition. **b** Possible evolutionary pathway in the generation of the four mammalian TIMPs. Brew et al. (2000) suggest that an ancestral TIMP underwent a duplication even to generate two paralogous TIMPs, thereby creating TIMP1. A second duplication event created TIMP3 and the ancestor to TIMP2 and 4. A final duplication resulted in the generation of TIMP2 and 4. TIMP1 most highly resembles the ancestral TIMP as it has accumulated the fewest mutations across species

mammalian Timp genes, before the divergence of tetrapods and teleosts. Timp1, Timp3, and Timp4 genes are located within introns of three Synapsin genes, suggesting that the Syn-Timp locus was duplicated at least three times in *Mus musculus* and *Homo sapiens*. Timp2 on the contrary is not located within a Synapsin intron, and could have arisen via duplication of Timp alone or degeneration of the Syn locus after duplication. While *Ciona intestinalis* has a single Timp orthologous to the four human Timps, *Danio rerio* has four Timp2 genes (Timp2a, b,c,d) orthologous to the human Timp2 (Huxley-Jones et al. 2007). This indicates that duplication events occurred in the Timp2 locus in *D. rerio* after the tetrapod/teleost divergence. Among the four mammalian Timp genes, Timp1 is considered to most closely resemble the ancestral Timp, as it demonstrates the lowest rate of evolutionary change (Brew et al. 2000). Figure 20.1b illustrates the sequence of gene duplications leading to the creation of each Timp gene. Timp3 is the second most ancient while Timp2 and Timp4 arose from the final duplication event and are the newest members of the family.

Invertebrate and vertebrate TIMPs differ significantly in both structure and function. The two *Caenorhabditis elegans* TIMPs are single-domain proteins consisting of only the N-terminal region of mammalian TIMP, whereas the single *Drosophila melanogaster* TIMP has the two-domain structure, and functionally resembles the mammalian TIMP3. It is able to inhibit the activity of *D. melanogaster* ADAM17 along with MMP inhibition (Pohar et al. 1999, Wei et al. 2003) and associates strongly with the ECM via interactions with hyaluronic acid, both properties unique to the mammalian TIMP3. Until recently, the N-terminal end of TIMP3 was thought to be solely responsible for the TIMP3:ECM binding. However, elegant work by Lee et al. (2007) has revealed that in fact both N- and C-termini are involved in ECM binding. Both termini utilize basic amino acids such as lysine and arginine in this interaction, as exhaustive mutations to specific amino acids at either terminus, together with domain swapping to generate ECM-adhering TIMP1 that clearly identify the six amino acids (N-terminus: Lys-26, 27, 30, and 76; C-terminus: Lys-165 and Arg-163) required for ECM binding. Figure 20.1a represents a schematic of the prototypical two-domain TIMP, with the MMP-inhibitory and ECM-binding residues indicated. As TIMP1 and TIMP3 are structurally similar, the figure also illustrates the residues of TIMP3 that are absent in TIMP1. The crystal structure of a full length TIMP has yet to be solved, as only the N-termini in isolation or complexed with MMP have been obtained thus far (Fernandez-Catalan et al. 1998, Morgunova et al. 2002, Iyer et al. 2007).

Targets of TIMPs and Phenotypes of TIMP Deficiency

As the number of published substrates processed by each MMP, ADAM, and ADAMTS steadily increases, the biological importance of each TIMP in regulating tissue homeostasis comes to light. Table 20.1 lists the inhibitory capability of individual TIMPs on some common metalloproteinases (MMP2, MMP7, MMP9,

Table 20.1 Murine phenotypes arising from individual Tissue Inhibitor of Metalloproteinase (*TIMP*) knockouts

	Phenotypes	Reference
Timp1$^{-/-}$	Reduced luminal obliteration/increased reepithelialization after tracheal transplantation	Chen et al. 2006
	Enhanced acute lung injury after bleomycin exposure	Kim et al. 2005
	Increased HGF activity in regenerating livers	Mohammed et al. 2005
	Altered LV geometry and cardiac function	Roten et al. 2000
	Exacerbated LV remodeling after myocardial infarction	Creemers et al. 2003, Ikonomidis et al. 2005
	Enhanced estrogen-induced uterine edema	Nothnick et al. 2004
	Decreased serum total testosterone levels	Nothnick et al. 1998
	Reduced serum progesterone levels during corpus luteum development	Nothnick 2003
	Decreased adipose tissue development during nutritionally induced obesity	Lijnen et al. 2003
Timp2$^{-/-}$	Increased nerve branching and acetylcholine receptor expression	Jaworski et al. 2006
	Weakened muscle and reduced fast twitch muscle mass	Lluri et al. 2006
	Deficits in preattentional sensorimotor gating	Jaworski et al. 2005
	Required for efficient pro-MMP-2 activation both in vivo and in vitro	Caterina et al. 2000, Wang et al. 2000
Timp3$^{-/-}$	Enhanced metastatic dissemination to multiple organs	Cruz-Munoz et al. 2006b
	Increased susceptibility to LPS-induced mortality	Smookler et al. 2006
	Enhanced tumor angiogenesis in response to FGF-2	Cruz-Munoz et al. 2006a
	LV dilation and dilated cardiomyopathy following aortic banding	Kassiri and Khokha 2006
	Increased pulmonary compliance following LPS challenge	Martin et al. 2005
	Increased inflammatory response to intra-articular antigen injection and TNF-alpha	Mahmoodi et al. 2005
	Spontaneous LV dilatation, cardiomyocyte hypertrophy, and contractile dysfution	Fedak et al. 2004
	Impaired bronchiole branching morphogenesis	Gill et al. 2003
	Chronic hepatic inflammation and failure of liver regeneration	Mohammed et al. 2004
	Spontaneous air space enlargement and impaired lung fuction in aged mice	Leco et al. 2001
	Accelerated apoptosis during mammary gland involution	Fata et al. 2001
Timp3$^{+/-}$	Acceleration of type 2 diabetes when combined with insulin receptor heterozygosity	Federici et al. 2005

MT1-MMP, ADAM10, ADAM12, ADAM7, ADAM33, and ADAMTS4), as well as the substrates processed by each of these enzymes. Given the substrate repertoire that represents ECM and cell-surface molecules, the complexity is evident in TIMP regulation of multiple signaling pathways. At the stromal–cellular interface, TIMPs can elicit a paracrine response by ligand processing or modulate cell autonomous function. The immune response is a classical example where such proteolytic processing alters immune cell activation, migration, function of clearing antigen, and finally resolution of inflammation as reviewed in Murphy et al. (2008). Extracellular proteolytic cascades also trigger "start" or "stop" signals for proliferation to guide cell division. The process of liver regeneration, where ~70% of the liver is surgically removed to initiate compensatory hepatocyte proliferation, offers a powerful in vivo system to study the contribution of metalloproteinases and TIMPs in cell division. Factors important for liver regeneration, as demonstrated by genetic mutant models, are often direct or indirect target of metalloproteinases, and therefore regulated by TIMPs as reviewed in Mohammed and Khokha (2005).

Although there is significant overlap in the repertoire of metalloproteinases inhibited by each TIMP, the expression and localization patterns of these inhibitors limit the ability of an individual TIMP to comprehensively regulate MMP activity in vivo. Murine expression analyses indicate that TIMP1 is highly expressed in the muscle, lung, and bone, TIMP2 is ubiquitously expressed, TIMP3 is enriched in the heart, kidney, lung, and thymus, and TIMP4 in heart, brain, and muscle (Leco et al. 1994, 1997; Fata et al. 1999, Nuttall et al. 2004). Reproductive organs are enriched in most TIMPs and demonstrate cell-type specificity within each tissue. For instance, TIMP2 is present in stromal cells, and TIMP3 and 4 in epithelial cells of the developing mouse mammary gland (Fata et al. 1999, Nuttall et al. 2004). Therefore, despite having common targets, each TIMP can regulate unique cellular processes by inhibiting an ADAM, ADAMTS, or MMP in a specific tissue compartment (Chirco et al. 2006). It is important to note that in addition to inhibiting metalloproteinase function, TIMPs have been shown to operate via MMP-independent mechanisms. Below, we discuss phenotypes arising in Timp-deficient mice and attempt to describe the many consequences of regulating metalloproteinase function, connecting ECM remodeling, intracellular signaling, and pathology.

TIMP1

TIMP1 was identified as having erythroid potentiating activity (EPA) owing to its ability to augment red blood cell colony formation (Gasson et al. 1985). TIMP1-deficient mice display mild phenotypes when challenged in several models including elevated cardiac ECM breakdown (Roten et al. 2000, Creemers et al. 2003, Ikonomidis et al. 2005), enhanced hepatocyte proliferation (Mohammed et al. 2005), and altered metabolic control of obesity (Lijnen et al. 2003) (Table 20.1). In accord with its high expression in reproductive tissue, male TIMP1-deficient mice have modestly lower testosterone levels. Importantly, each TIMP exhibits a unique pattern of expression during sexual maturation in both females and males, suggesting their specific roles at puberty (Nothnick et al. 1998). The well-studied MMPs and ADAMs inhibited by TIMP1 include MMP1, MMP9, ADAM10, and ADAMTS4 (Table 20.2), while TIMP1 is a poor inhibitor of membrane-type matrix

Table 20.2 Profiling Tissue Inhibitor of Metalloproteinase (*TIMP*) target Matrix Metalloproteinases (*MMP*s) and A Disintegrin and Metalloproteinases (*ADAMs*), and their downstream substrates

	MMP2	MMP7	MMP9	MT1-MMP	ADAM10	ADAM12	ADAM17	ADAM33	ADAMTS4
TIMP1	•		•		•				•
TIMP2	•	•	•	•					•
TIMP3	•	•	•	•	•	•	•	•	•
TIMP4		•						•	•
Target	**MMP2**	**MMP7**	**MMP9**	**MT1-MMP**	**ADAM10**	**ADAM12**	**ADAM17**	**ADAM33**	**ADAMTS4**
ECM	Collagen I, IV, VII Decorin Fibrillin Galectin-3 Gelatin Plasminogen β Dystroglycan	β4 Integrin Decorin E-Cadherin Fibronectin I Laminin Entactin Syndecan 1	β Dystroglycan Collagen IV, V, XI Elastin Fibrillin MBP TGFβ1	α5 Integrin Collagen I, II, III Fibronectin Gelatin Laminin Progelatinase A ProMMP2 Syndecan 1 Testican 1	Aggrecan I Collagen IV MBP	α2 Actinin α9β1 Integrin Aggrecan I	L-Selectin		α9 Integrin Aggrecan I Fibronectin
Cell Surface	Pro-HGF FGF1 FGFR1	FasL Fas	CCL7 CCL11 CXCL12 ICAM1 IGF2 IGFBP3 IL8 RECK Pro-HGF	CCL7 CXCL12	βAPP Delta-Like1 CD23 CD40 NGFR	HB-EGF IGFBP3 IGFBP5	CD30 c-Met Fractalkine GHBP IL1R II IL6R IL15RA MUC1 MxL1 Notch TGFα TNFα TNFR1, 2	βAPP peptide CD23 Kit-L1 peptide Insulin β chain TRANCE	MT4-MMP

• indicates inhibition by the respective TIMP.

metalloproteinases (MT-MMPs). The MMP-inhibitory function of TIMP1 correlates well with the phenotypes of heart tissue remodeling and ventricle function (Roten et al. 2000, Creemers et al. 2003, Ikonomidis et al. 2005). TIMP1 has also been shown to regulate cell proliferation by inhibiting the release of hepatocyte growth factor (HGF) by MMP2 and MMP9. HGF is an important growth factor required for liver regeneration (Lindroos et al. 1991, Schmidt et al. 1995), and hepatocytes of Timp1$^{-/-}$ mice show an accelerated entry into the cell cycle. Specifically, Timp1$^{-/-}$ mice exhibit elevated HGF signaling culminating in accelerated hepatocyte cell division. In this model, the MMP-inhibitory function of TIMP1 is important in regulating cell proliferation (Kim et al. 2000, Mohammed et al. 2005).

TIMP2

The well-established biological function of the TIMP family is to inhibit activated metalloproteinases. While this is role is performed by all TIMPs, TIMP2 paradoxically has a central function in MMP2 activation at the cell surface. It acts as an adaptor for MMP2 by allowing the formation of a trimolecular complex involving MT1-MMP/TIMP2/Pro-MMP2, where pro-MMP-2 is activated in a two-step process (Caterina et al. 2000, Wang et al. 2000, English et al. 2006). Thus, TIMPs are capable of regulating metalloproteinase activity via multiple mechanisms (Toth et al. 2000, Wang et al. 2000). Despite its previously mentioned ubiquitous expression pattern, only neurological phenotypes have been reported in TIMP2 null mice (Jaworski et al. 2005, Jaworski et al. 2006). Lluri et al. (Lluri et al. 2006) have shown that TIMP2 is expressed at neuromuscular junctions and colocalizes with ß1 integrin. Interestingly, ß1 integrin expression is decreased in TIMP2-deficient muscle, suggesting a role for this cell adhesion molecule in maintaining muscle fiber integrity. The molecular mechanism explaining the decrease in ß1 integrin expression in Timp2$^{-/-}$ tissue has not been investigated, but one can speculate that the resulting loss of ECM stability and enhanced metalloproteinase activity indirectly contributes to integrin proteolysis (Seo et al. 2003) (Table 20.2).

TIMP3

TIMP3 is the only TIMP genetically linked to a human disease. Individuals harboring mutations in the C-terminal of TIMP3 suffer from a macular degenerative disease termed Sorsby's fundus dystrophy (SFD) (Weber et al. 1994a, b). Interestingly, the MMP-inhibitory property of the mutant TIMP3 is maintained in patients with SFD, indicating that the mechanism underlying the disease is MMP independent. In fact, the elevated production and resulting accumulation of mutant TIMP3 in the Bruch's membrane of patients with SFD is causal to macular degeneration (Langton et al. 2000).

Our group has investigated the physiological role of TIMP3 in multiple tissues by exposing Timp3$^{-/-}$ mice to specific stimuli (Table 20.1). Two of the earliest phenotypes identified were that of air space enlargement in the lung and accelerated involution of the mammary gland of TIMP3 knockout mice (Fata et al. 2001, Leco et al. 2001). Study by Fata et al. (2001) shows that the loss of TIMP3 is conducive to accelerated apoptosis during mammary involution, in part owing to greater matrix proteolysis; the molecular mechanisms contributing to cell death still remain to be understood. TIMP3 is a potent inhibitor of many MMP, ADAM, and ADAMTS enzymes (Table 20.2), and the observations of pulmonary air space enlargement (Leco et al. 2001), dilated cardiomyopathy (Fedak et al. 2004), and cartilage degradation (Sahebjam et al. 2007) in Timp3$^{-/-}$ mice indicate that enhanced metalloproteinase activity leads to compromized ECM homeostasis as a function of aging. Later studies by Mohammed et al. (2004), Smookler et al. (2006), and Mahmoodi et al. (2005) investigated the role of TIMP3 in regulating inflammation and proliferation dependence on the tumor necrosis factor (TNF)-signaling pathway, establishing TIMP3 as an important negative regulator of TNF bioavailability owing to its unique ability to inhibit ADAM17. The lack of Timp3 results in increased circulating levels of TNF as well as its two receptors, TNFR1 and TNFR2. Federici et al. (2005) identified a novel role for TIMP3 in providing protection from type-2 diabetes via modulation of TNF shedding by transarterial chemoembolization (TACE). Here, elevated vascular inflammation accompanied by insensitivity to insulin signals caused the development of glucose intolerance and hyperglycemia in Timp3±;InsR± mice at 6 months of age (Table 20.1). Using a heart disease model of pressure overload, Kassiri et al. (2005) dissected the dysregulation of MMP and ADAM activities in Timp3$^{-/-}$ mice and found that increased TNF transcriptionally upregulated several specific MMPs (MMP2, MT1-MMP, MMP13) while exerting no effect on others (MMP7, MMP9). Intriguingly, the combination of Timp3 and TNF deletion led to a greater neutrophil influx and production of MMP8 in cardiac tissue. Here, TIMP3 molecularly linked ECM turnover with that of cytokine activity when cardiac tissue homeostasis was perturbed.

TIMP4

The last member of the TIMP family, TIMP4, has yet to be investigated through the gene targeting approach. TIMP4 protein is present in cardiomyocytes and smooth muscle cells, with lower levels in the brain and muscle (Koskivirta et al. 2006). Its colocalization with inflammatory cells such as macrophages and CD3^{+} T cells suggests a role in inflammatory cardiac pathologies such as atherosclerotic lesions where it localizes within necrotic regions. As TIMP4 exhibits the most restricted expression pattern of all the four TIMPs in mice, its effects may be limited to the target organs despite its property of being a secreted protein. Future work will no doubt reveal new roles of this inhibitor as a regulator of processes influencing cardiac and inflammatory homeostasis.

Summary

None of the Timps are essential during mouse development as individual knockouts of Timps do not exhibit in utero lethality, although *D. melanogaster* Timp mutant phenocopies integrin mutants displaying inflated wings and premature lethality (Godenschwege et al. 2000). Overall, we see regulation of three important systems from the in vivo analysis of TIMP function: ECM remodeling, cytokine and growth factor bioavailability, and inflammatory cell function, critical in maintaining tissue homeostasis. It is evident that deletion of a single Timp does not result in a complete loss of regulation of these processes raising the possibility of functional compensation by other three Timps. However, when tested through measurement of RNA expression in several tissues, we have not observed alteration in the expression of TIMPs 1, 2, and 4 in Timp3$^{-/-}$ mice. Interestingly, TIMP3 demonstrates the ability to simultaneously affect all three systems. These systems are intricately connected in vivo as ECM cleavage releases not only the structural constraints but also ECM-bound ligands, and TIMP regulation of receptor shedding influences ligand:receptor kinetics. Thus, TIMPs alter the amplitude of a signaling stimulus as well as the triggers that serve to recruit infiltrating cells. The tissue- and stimulus-specific requirement of each TIMP makes them important in maintaining tissue homeostasis.

TIMPs in Cancer

When it comes to understanding the role played by TIMPs in tumorigenesis, we have at our disposal literature from a vast number of clinical studies and also following TIMP manipulation in experimental systems. Generally, there is a lack of consensus on the significance of TIMP expression patterns in human cancers and patient outcome, and this complexity will not be discussed in this chapter. Below we summarize how different TIMPs affect processes fundamental to cancer development (apoptosis, proliferation, angiogenesis, invasion), as revealed through in vitro studies followed by studies of cancer development in genetic models.

Cell Proliferation

TIMP1 and TIMP2 were initially identified as important factors in erythropoeisis due to their erythroid potentiating activity (Stetler-Stevenson et al. 1992, Murate et al. 1993). Surprisingly, an antisense RNA-mediated downregulation of TIMP1 transformed murine 3T3 fibroblasts into tumorigenic cells (Khokha et al. 1989). Current research investigating the roles played by TIMPs in regulating proliferation of both normal and malignant cells demonstrate that the cellular context within which each TIMP is expressed may dictate its effect on proliferation (Hayakawa et al. 1990, Baker et al. 1998, Fata et al. 1999, Celiker et al. 2001, Hoegy et al. 2001, van der Laan et al. 2003). Table 20.3 shows that each TIMP can exert opposing effects on proliferation, either enhancing or inhibiting the process depending on the cell type involved. Additionally, the mechanisms by which these effects are propagated differ between each study, involving pathways such as extracellular signal-regulated kinases (ERK) (Petitfrere et al. 2000), epidermal growth factor receptor (EGFR) (Hoegy et al. 2001), vascular endothelial growth factor (VEGF) (Seo et al. 2003), hepatocyte growth factor (HGF) (Mohammed et al. 2005), fibroblast growth factors (FGF), nuclear factor-kappa B (NFκB) (Lizarraga et al. 2004), and cyclin D1. However, most of these responses have not been studied in depth to identify the direct connection between these signaling effectors and the ligands under TIMP regulation. It is also important to note that not all of the mentioned pathways are investigated in each system, and doing so would reveal common signals influenced by each TIMP regardless of the cell type.

Apoptosis

As shown in Table 20.3, TIMPs are able to trigger apoptosis in a variety of cell types including fibroblasts, endothelial, epithelial, and hematopoietic cells. Resistance to apoptosis is an important early trait in cellular transformation. While the intracellular signaling pathways involved in apoptosis take centre stage, the direct link between TIMPs and cell death is not well defined. This is partly because of their target repertoire, many of which act as triggers of apoptosis. Hepatic stellate cells overexpressing TIMP1 exhibit an antiapoptotic phenotype in vitro, and Murphy et al. (2002) demonstrate that the antiapoptotic effect is dependent on its MMP inhibitory function, as a specific loss-of-function mutation at the MMP-inhibitory region enhances stellate cells susceptibility to various apoptotic stimuli. Studies on a variety of cell lines such as the erythroleukemia cell line UT-7, endothelial cells, and breast epithelial cells by independent groups show that TIMP1 inhibits apoptosis via a PI(3)K-dependent manner and modulation of the Bcl family of proteins (Lambert et al. 2003, Boulday et al. 2004, Liu et al. 2005). In contrast to TIMP1, adenoviral TIMP3 overexpression leads to enhanced apoptosis in several cell types (Baker et al. 1998). Given the specificity of TIMP3 for inhibiting ADAM17, and ADAM17-mediated shedding of TNF and its receptors, TIMP3 overexpression in vitro results in stabilization of the death receptors Fas and TNFR1, thereby sensitizing cells to receptor-mediated apoptosis via a caspase-dependent mechanism (Smith et al. 1997, Bond et al. 2002, Ahonen et al. 2003). Consistent with these roles of TIMP3, clinical studies have demonstrated that epigenetic silencing of TIMP3 via methylation occurs in several types of human

Table 20.3 TIMP effects on apoptosis, angiogenesis, proliferation and invasion in vitro

	Method	Apoptosis	Reference
TIMP-1	rTIMP1	Reduce caspase 3 activity, enhance Bcl-2 expression in hepatic stellate cells (HSC)	Murphy et al. 2002
	rTIMP1	Increase activity of PI-3K/AKT, JAK2 tyrosine, and Bad phosphorylation, maintained Bcl-XL expression	Lambert et al. 2003
	rTIMP1	Inhibit TNF-induced apoptosis, activation of PI-3K/AKT in endothelial cells	Boulday et al. 2004
	rTIMP1	Protect MCF10 cells from TRAIL-induced cell death, caspase (3,8,9) activity, FAK and PI-3K activation	Liu et al. 2005
TIMP-2	rTIMP2	Increase apoptosis in activated T lymphocytes and Tsup or Jurkat lymphoma cell lines	Lim et al. 1999
	TIMP2 vector	Increase apoptosis in HCC tumor	Tran et al. 2003
TIMP3	Ad-TIMP3	Induction of caspase 8/9 activation and cleavage of PPAR and FAK, mitochondrial acitivation	Bond et al. 2002
	rTIMP3, Ad-TIMP3	Stabilization of death receptors (TNF-R, FAS, TRAIL-RI), caspase 8 activation	Ahonen et al. 2003
	Ad-TIMP3	Reverses antiapopotic effect of TNF-α on Fas-induced apoptosis, inhibits NF-κB activation	Drynda et al. 2005
	AdTIMP3	Increased apoptosis of rat aortic smooth muscle cells	Baker et al. 1998
TIMP4	rTIMP4	Decrease apoptosis in MDA-MB-435-derived tumors, increase expression of Bcl-2 and Bcl-XL	Jiang et al. 2001
	Purified TIMP4	Induce apoptosis in transformed cardiac fibroblast but not in normal fibroblastst	Tummalapalli et al. 2001

	Method	Angiogenesis	Reference
TIMP1	rTIMP1	Inhibit FGF-2-induced neovasucularization, decrease HMVEC migration tumor angiogenesis	Johnson et al. 1994
	retrovirus-TIMP1	Decrease EC migration and angiogenesis in Burkitt's lymphoma	Guedez et al. 2001
	rTIMP1	Decrease HDMEC migration, increased levels of VE cadhering, dephosphorylation of FAK, and paxillin and PTEN expression	Akahane et al. 2004
TIMP2	TIMP2 vector	Decrease invasiveness/migration of EC and tumor angiogenesis	Valente et al. 1998
	rTIMP2	Decrease SHP1-integrin association and increase phosphatase activity against FGF-R1 and VEGF-R2	Seo et al. 2003
	retrovirus-TIMP2	Inhibit angiogenesis, MPK-1 phosphatase upregulation, inactivation of MAPK pathways	Feldman et al. 2004
TIMP3	rTIMP3	Inhibit SP1-induced EC invasiveness in fibrin and collagen	Bayless and Davis 2003

Table 20.3 (continued)

	Method	Angiogenesis	Reference
	rTIMP3	Compete for binding to VEGFR2	Qi et al. 2003
	Ad-TIMP3	Inhibit VEGF-induced tubulogenesis of HEMVEC	Plaisier et al. 2004
	KO	Enhance angiogenesis in tumor and in response o FGF-2	Cruz-Munoz et al. 2006a
TIMP-4	rTIMP4	Decrease HUVEC and HDMEC tubulogenesis in fibrin	Lafleur et al. 2002
	rTIMP4	Inhibit migration of EC	Fernandez and Moses 2006

	Method	Cell proliferation	Reference
TIMP1	purified TIMP1	Stimulate erythroid burst-forming units	Hayakawa et al. 1990
	rTIMP1	Decrease proliferation of mammary ductal epithelial cells	Fata et al. 1999
	rTIMP1	Increase proliferation of MDA-MB-435 and activation of ERK and p38 pathways	Porter et al. 2004
TIMP-2	rTIMP2	Suppression of TYK growth factor-induced proliferation, disrupt EGFR phosphorylation/Grb-2 association	Hoegy et al. 2001
	rTIMP2	Inhibit HMVEC proliferation in response to FGF/VEGF, decrease SHP1-integrin association	Seo et al. 2003
	rTIMP2	Increase proliferation of A549 lung epithelial, cyclin D1 upregulation, NF-κB activation, IkBβ decrease	Lizarraga et al. 2004
TIMP-3	AdTIMP3	Increase proliferation of cardiac fibroblasts	Lovelock et al. 2005
	AdTIMP3	Decrease proliferation of RA-synovial fibroblasts	van der Laan et al. 2003
TIMP4	AdTIMP4	Increase proliferation of cardiac fibroblasts	Lovelock et al. 2005
	rTIMP4	Decrease proliferation of G401 Wilm's tumor cells	Celiker et al. 2001

	Method	Invasion	Reference
TIMP1	Inducible TIMP1	Decreased invasive potential in B16F10 melanoma	Khokha et al. 1992
	AdTIMP1	Reduced SK-Mel5 and A2058 melanoma invasiveness	Ahonen et al. 1998
	βCat-induced TIMP1	Decreased invasiveness of fribromatosis cells	Kong et al. 2004
TIMP2	AdTIMP2	Reduced invasion and attachment by SK-Mel5 and A2058 melanoma	Ahonen et al. 1998
	AdTIMP2	AdTIMP2 expression decreases invasion of PANC-1	Rigg and Lemoine 2001
	rTIMP2	Inhibit migration and invasion of MCF10A	Ahn et al. 2004

Table 20.3 (continued)

	Method	Apoptosis	Reference
	rTIMP2	Reduced invasion by MDA-MB-435 and MDA-MB-231 breast carcinoma cells	Lee et al. 2005
TIMP3	TIMP3 vector	Reduced invasion by SK-Mel5 and A2058 melanoma	Ahonen et al. 1998
	AdTIMP3	Decreased invasiveness of leiomyosarcoma cells	Castagnino et al. 1998
	AdTIMP3	Decreased invasiveness of smooth muscle cells	Baker et al. 1998
	AdTIMP3	AdTIMP3 reduced invasion by Hela and HT1080	Baker et al. 1999
TIMP4	rTIMP4	Reduced invasive potential of MDA-MB-435	Wang et al. 1997

cancers such as melanoma, breast, pancreatic, prostate, colon, and cervical, suggesting its role as a tumor suppressor (House et al. 2003, Karan et al. 2003, van der Velden et al. 2003, Han et al. 2004, Widschwendter et al. 2004, Lui et al. 2005, Riddick et al. 2005, Kim et al. 2006, Bai et al. 2007). Compared to TIMP1 and TIMP3, less is known of the effects of TIMP2 and TIMP 4 on apoptosis. Current literature suggests that the MMP inhibitory function of TIMP2 protects macrophages from apoptosis in an in vitro overexpression model (Johnson et al. 2006) while enhancing apoptosis in T lymphocytes (Lim et al. 1999).

The in vitro studies that demonstrate elevated TIMP3 expression results in enhanced apoptosis in normal and transformed cells (Ahonen et al. 1998, 2003; Baker et al. 1999, Bond et al. 2002) are in conflict with the in vivo study where a lack of TIMP3 leads to accelerated mammary epithelial apoptosis during mammary involution. The level of TIMP overexpression achieved in these in vitro systems is well above physiological thresholds. Regardless of the manner of TIMP3 manipulation, the cell death response remains the same, suggesting that the balance between metalloproteinases and their inhibitors, rather than each component per se, may dictate cell death.

Angiogenesis

Table 20.3 summarizes the mode of TIMP manipulation and their effects on the steps involved during angiogenesis, and these studies consistently show that elevated TIMP levels inhibit angiogenesis. Addition of recombinant TIMP1 results in enhanced cell adhesion caused by increased expression of adhesion molecules VE-cadherin and PECAM-1 on endothelial cells. Additionally, the phosphorylation of intracellular focal adhesion kinase (FAK) required for focal adhesion is reversed in this model, further inhibiting migration in an MMP-independent manner (Akahane et al. 2004). However, consistent with the antiapoptotic role of TIMP1, initial proliferation of tumorigenic cells is enhanced in models overexpressing TIMP1 as demonstrated in a Burkitt's lymphoma cell line (Guedez et al. 2001).

TIMP2 and TIMP3 act on similar intracellular pathways in cancer cell lines when overexpressed, inhibiting the function of FGF and VEGF networks (Seo et al. 2003, Feldman et al. 2004). Both TIMP2 and TIMP3 appear to antagonize FGF and VEGF interaction with their receptors and ablate downstream signaling, a possible mechanism for inhibiting angiogenesis (Spurbeck et al. 2002, Qi et al. 2003). Additionally, the MMP and ADAM inhibitory capabilities of both TIMP2 and TIMP3 have been implicated in regulating endothelial cell tube morphogenesis and stabilization of vascular networks in vitro owing to the inhibition of MMP1, MMP10, MT1-MMP, and ADAM10 (Saunders et al. 2006). Compartmentalization of TIMP function is evident here as TIMP3 is supplied by pericytes and TIMP2 by endothelial cells, both required for blood vessel stabilization. The role of TIMP3 as a competitive inhibitor of VEGF to VEGFR binding is one of the earliest examples of MMP-independent functions of TIMPs.

Cell Contact and Motility

The previously discussed cellular processes of apoptosis and proliferation indicate that TIMP effects are dependent on the target cell type tested, and the mode or level of TIMP manipulation. As a consequence, it is difficult to generalize their effects in these basic processes. On the contrary, as with angiogenesis, it is evident that enhancing TIMP activity in vitro results in a decreased invasive and metastatic capacity of many normal and malignant cells (Ahonen et al. 1998, Baker et al. 1999, Engers et al. 2001, Ahn et al. 2004, Kong et al. 2004). TIMP-modulated cell lines show altered cell density and reduced cell adhesion owing to FAK redistribution and lowered cadherin expression in fibroblasts (Ho et al. 2001). Beta catenin signaling is important in modulating cell adhesion, and cells lacking Timp3 display increased signaling and altered target gene expression, specifically elevated MMP7. The functional consequence of altered beta catenin signaling on cell adhesion varies between epithelial and mesenchymal cell types. Notably, TIMP3 inhibition of this pathway seems to be an important feature of mammary gland morphogenesis in vivo (Hojilla et al. 2007).

A diverse set of mechanisms is involved in cell motility and tumor cell invasion. Cell motility depends on both cytoskeletal changes and altered intercellular contact. Rho and Rac GTPases are important early signals that control the actin cytoskeleton and several groups have reviewed their role in cell movement in both normal and malignant cells (Fryer and Field 2005, Cancelas et al. 2006, Moldovan et al. 2006, Rose et al. 2007). Additionally, Rac-mediated expression of cadherins enhances cell–cell adhesion and antagonizes cell motility (Yamada and Nelson 2007). Engers et al. (2001) demonstrate that Rac induces the expression of TIMP1 and TIMP2, which, in turn, inhibit the invasive capacity of human renal cell carcinomas. These findings suggest that TIMP1 and TIMP2 are induced by GTPases and inhibit metalloproteinase activity in a very localized microenvironment. Over-expression of TIMP1 and TIMP2 is known to inhibit invasion of B16F10 melanoma cell lines in vitro (Khokha et al. 1992, Ahn et al. 2004).

Membrane-type metalloproteinases such as MT1-MMP are critical promoters of invasion in three-dimensional (3D) collagen gels (*see also* Chap. 18). Comparative analysis of multiple MMPs with MT1-MMP has shown that MT1-MMP activity is required for motility of several cell types (Hotary et al. 2003, 2006; Sabeh et al. 2004, Zhai et al. 2005). This metalloproteinase is a potent collagenase, and, as mentioned earlier, provides the anchoring mechanism for MMP2 at the cell surface in addition to mediating its activation. TIMPs display variable efficiencies at inhibiting MT1-MMP. For instance, TIMP1 is unable to inhibit this metalloproteinase, while TIMP2 is the critical adaptor between MT1-MMP and MMP2 at the cell surface (Barbolina and Stack 2007). At higher concentrations, TIMP2 is an effective inhibitor of MT1-MMP. Comparative analysis with mouse embryonic fibroblasts deficient in individual TIMP, in a culture system designed to study MT1-MMP activity and pro-MMP-2 activation at the cell surface, demonstrates that TIMP2 and TIMP3 are key modulators of this metalloproteinase (English et al. 2006). How this complex may operate in vivo, on individual and combined Timp2 and Timp3 deficiencies, to regulate cell motility and tumor cell invasion is an important question that remains to be answered.

In Vivo Models of Cancer Biology

Analysis of TIMP function in vitro provides insight into the biochemical processes that drive MMP inhibition, downstream signaling, and the consequences to cellular function in a cell autonomous or monoculture system. However, the tissue microenvironment is an important biological niche that harbors TIMPs, regulating the processes discussed above in a paracrine or systemic manner. This is best investigated using in vivo models, where TIMP levels have been modulated. Genetic models allow us to investigate TIMP function in a system that is physiologically relevant in terms of tissue structure, function, pathology, and molecular biology.

Transgenic Timp expression or Timp deletion has been used to map effects on cell proliferation and cell death during physiological phenomena (mammary morphogenesis, liver regeneration; Table 20.1), and on malignant processes (neoplasia, metastasis; Table 20.4). Timp1 knockdown or loss of Timp3 in the mammary gland causes accelerated ductal morphogenesis indicating their natural inhibitory function during pubertal mammary development. In contrast, elevating TIMP levels through implanted slow-release pellets show that while TIMP1, TIMP3, and TIMP4 inhibit ductal morphogenesis, TIMP2 accelerates this growth. Consistently, hepatic Timp1 overexpression affects liver regeneration by delaying hepatocyte entry into the cell cycle, whereas Timp1 knockdown accelerates this process. In a model of liver fibrosis, transgenic expression of TIMP1 has an antiapoptotic effect on hepatic

Table 20.4 Cancer phenotypes of Tissue Inhibitor of Metalloproteinase (*TIMP1*) and TIMP3 modulation

	Method	Phenotype	Reference
TIMP1	Transgenic TIMP1	Inhibition of initiation and growth of TAg-induced hepatocellular carcinoma	Martin et al. 1996, 1999b
	Transgenic TIMP1	Inhibited growth of intradermally injected lymphoma cells	Kruger et al. 1997
	Antisense TIMP1	Elevated tumor metastasis to brain and liver	Kruger et al. 1998
	Albumin-TIMP1	Inhibition of polyoma Middle T-induced mammary tumorigenesis	Yamazaki et al. 2004
	MMTV-LRP-TIMP1	No effect on polyoma Middle T-induced mammary tumorigenesis	Yamazaki et al. 2004
	Adenoviral TIMP1	Reduced macrometastases but elevated micrometastases to liver	Kopitz et al. 2007
TIMP3	TIMP3 KO	TIMP3 loss in host enhances tumor growth and angiogenesis	Cruz-Munoz et al. 2006a
	TIMP3 KO	Enhanced metastasis by melanoma and lymphoma cells	Cruz-Munoz et al. 2006b

stellate cells in vivo. Timp3$^{-/-}$ hepatocytes have a more rapid entry into the cell cycle; however, this is followed by enhanced cell death and a failure of liver regeneration. Timp3$^{-/-}$ mice also exhibit accelerated apoptosis during postlactation mammary gland involution. These studies provide insights into the effects of individual Timps on cell proliferation and cell death in a tissue context. However, since the above in vivo systems have often utilized readouts at the cellular or morphological levels, how many steps removed is the extracellular TIMP from the intracellular pathways remains to be defined.

To date, studies on TIMP1 transgenic mice have yielded significant findings on its role in tumor initiation, growth, and invasion. The viral oncogene Simian Virus 40/T antigen (SV40/TAg) induces heritable hepatocellular carcinoma when overexpressed in mice (Martin et al. 1996). Transgenic mice that overexpress hepatic TIMP1 display inhibited initiation and growth of TAg-induced hepatocellular carcinoma (Martin et al. 1996, 1999b). This counters some in vitro findings that TIMP1 has antiapoptotic properties when overexpressed (Lambert et al. 2003, Boulday et al. 2004, Liu et al. 2005). The molecular mechanism by which this inhibition occurs is through the insulin growth factor (IGF)-signaling pathway, an important mitogenic network whose significance is well established in both normal and malignant cells (Miller and Yee 2005, Karamouzis and Papavassiliou 2006, Kurmasheva and Houghton 2006, Renehan et al. 2006). Normally, circulating IGF is sequestered by IGF-binding proteins (IGFBPs), preventing excessive signaling via the type-I IGF receptor which could lead to a hyperproliferative and antiapoptotic phenotype. Proteolytic cleavage of IGFBPs occurs in order to release IGF, and metalloproteinases have been implicated in this process (Coppock et al. 2004, Hemers et al. 2005, Mitsui et al. 2006, Miyamoto et al. 2007). Martin et al. (1999a) demonstrate that TIMP1 regulates IGF bioavailability in vivo by inhibiting its release from IGFBPs, attenuating downstream mitogenic MAP kinase signaling through ERK-1 and -2. The tumor-suppressive role of transgenic TIMP1 extends beyond TAg-induced hepatocellular carcinoma, as Kruger et al. demonstrate that Timp1-transgenic mice have inhibited growth of intradermally injected lymphoma cells. Knocking down Timp1 via antisense RNA exhibits the opposite effect of accelerating tumor growth. They also show that metastastic colonization is elevated in mice exhibiting lower levels of TIMP1 when compared to normal mice in models of experimental metastases to liver and brain (Kruger et al. 1997, 1998). Furthermore, adenovirus-mediated TIMP1 overexpression efficiently protects the liver from T cell lymphoma and colon carcinoma experimental metastases (Elezkurtaj et al. 2004). However, in a recent study, Kopitz et al. report that nude mice infected with adenoviral vectors expressing TIMP1 exhibit an interesting pattern of decreased macrometastasis but significantly elevated micrometastasis to the liver, in experimental metastasis assays performed with lymphoma and fibrosarcoma cells. Here, micrometastases are defined as colonies measuring <0.2 mm in size. These seemingly paradoxical findings are consistent with the proliferative and antiapoptotic roles of TIMP1 (Table 20.3) that may constitute a signal conducive to metastasis. The authors suggest that HGF signaling is elevated in these mice due to the presence of increased cMet at the cell surface (Kopitz et al. 2007). Finally, in a breast cancer model, systemic elevation of TIMP1 inhibited mammary polyoma middle T-induced tumorigenesis, while mammary-specific TIMP1 overexpression showed no effect (Yamazaki et al. 2004).

As with TIMP1, TIMP3 has been studied in models of tumor growth where either the host or the tumor exhibits TIMP3 deficiency. These studies demonstrate the importance of understanding host versus tumor characteristics in tumorigenesis. Interestingly, a loss of TIMP3 in the tumor does not affect its growth or angiogenic potential as observed in a model of injected ES-cell teratomas in nude mice (Cruz-Munoz et al. 2006a). Similarly, host expression of TIMP3 impacts metastasis in models of injected lymphoma and melanoma cells; Timp3-knockout mice exhibit enhanced metastasis of these cell lines to the kidney and liver (lymphoma) or bone and lung (melanoma) (Cruz-Munoz et al. 2006b). The molecular pathways that lead to TIMP3 effects on tumor initiation, growth, and metastasis still remain to be elucidated. The interaction of tumor cells with host stroma (or host cells with tumor stroma) provides a tissue context, an ECM, and additional sources of MMPs, ADAMs, and growth factors which influence all stages of tumorigenesis.

Conclusions

The extracellular localization of TIMPs provides a "sink" for regulating the final step of metalloproteinase activity. Evolutionarily, the four TIMPs have evolved to counter the activity of the majority of the metalloproteinases, suggesting that they operate upstream of the specific protease-ligand processing unit. Given the large metalloproteinase repertoire in the mammalian genome (reviewed in Puente et al.

2003) and only four Timp genes, inhibition of multiple metalloproteinases is achieved by a single TIMP. This apparent lack of specificity seems to be advantageous for tissue homeostasis, especially as the metalloproteinases are ideally placed for rapid responses to stimuli. These responses simultaneously impact the structural ECM barrier along with the cell surface landscape, which then distills down to intracellular signaling transducers able to dictate cell fate and behavior. The examples of cytokine signaling, cell division, death, contact, and motility discussed throughout the chapter all utilize the metalloproteinase capacity to rapidly propagate or shut down a signal. TIMPs act as "brakes" to inhibit metalloproteinase activity, thus moderating the final cell response. This does not imply that TIMPs' function is always relegated to inhibiting cellular processes. In fact they regulate inhibitors of cell signaling, for instance, by preventing the shedding of receptors by metalloproteinases, thereby allowing downstream signaling to proceed. But why does the functional outcome vary with cell type on manipulation of TIMP levels? There exists a tremendous diversity in the metalloproteinase substrate repertoire dependent on the cell type, adding yet another level of complexity to the impact of TIMP. Despite the consistency in biochemical processing (i.e., metalloproteinase shedding of a substrate, and TIMP inhibition of the active enzyme), functional outcome is dictated by the substrate composition of the target cell and its surrounding stroma.

The maturation of systems biology approaches now allows us to identify genome- and proteome-wide changes in response to transforming signals. In parallel, it is possible to observe system-wide consequences of TIMP modulation. Independent of TIMPs, a major focus in cancer biology has been the study of oncogenes and tumor suppressors. It is important to note that not all TIMP targets impact oncogenesis, and not all signals modulating the cellular processes of proliferation, apoptosis, and motility are critical to this disease. There is a need to identify

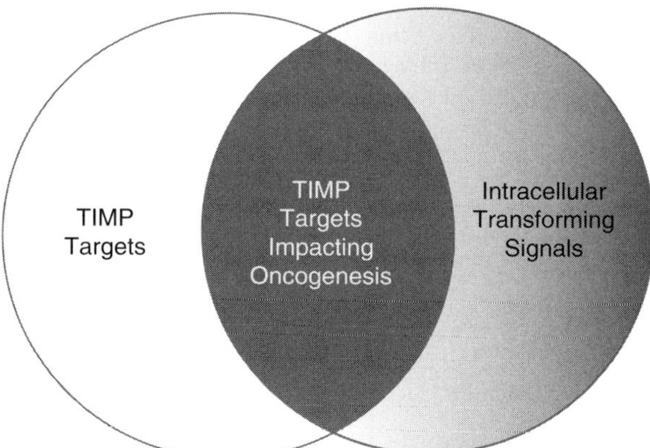

Fig. 20.2 Tissue inhibitor of metalloproteinase (*TIMP*) targets influencing transformation and ultimately tumorigenesis. The challenge is to segregate TIMP targets that are insignificant in cancer biology from those which modulate oncogenic signals

which of these factors are downstream of TIMP function, and additionally if TIMP expression and function is in turn dictated by them. Comparing these transformation and TIMP modulation models will help isolate the TIMP targets that are physiologically relevant contributors to cancer development (Fig. 20.2).

References

Ahn, S.M., Jeong, S.J., Kim, Y.S., et al. 2004. Retroviral delivery of TIMP-2 inhibits H-ras-induced migration and invasion in MCF10A human breast epithelial cells. Cancer Lett 207(1): 49–57.

Ahonen, M., Baker, A.H. and Kahari, V.M. 1998. Adenovirus-mediated gene delivery of tissue inhibitor of metalloproteinases-3 inhibits invasion and induces apoptosis in melanoma cells. Cancer Res 58(11): 2310–2315.

Ahonen, M., Poukkula, M., Baker, A.H., et al. 2003. Tissue inhibitor of metalloproteinases-3 induces apoptosis in melanoma cells by stabilization of death receptors. Oncogene 22(14): 2121–2134.

Akahane, T., Akahane, M., Shah, A., et al. 2004. TIMP-1 inhibits microvascular endothelial cell migration by MMP-dependent and MMP-independent mechanisms. Exp Cell Res 301 (2): 158–167.

Apte, S.S., Mattei, M.G. and Olsen, B.R. 1994. Cloning of the cDNA encoding human tissue inhibitor of metalloproteinases-3 (TIMP-3) and mapping of the TIMP3 gene to chromosome 22. Genomics 19(1): 86–90.

Bai, Y.X., Yi, J.L., Li, J.F., et al. 2007. Clinicopathologic significance of BAG1 and TIMP3 expression in colon carcinoma. World J Gastroenterol 13(28): 3883–3885.

Baker, A.H., Zaltsman, A.B., George, S.J., et al. 1998. Divergent effects of tissue inhibitor of metalloproteinase-1, -2, or -3 overexpression on rat vascular smooth muscle cell invasion, proliferation, and death in vitro. TIMP-3 promotes apoptosis. J Clin Invest 101(6): 1478–1487.

Baker, A.H., George, S.J., Zaltsman, A.B., et al. 1999. Inhibition of invasion and induction of apoptotic cell death of cancer cell lines by overexpression of TIMP-3. Br J Cancer 79(9–10): 1347–1355.

Barbolina, M.V. and Stack, M.S. 2008. Membrane type 1-matrix metalloproteinase: substrate diversity in pericellular proteolysis. Semin Cell Dev Biol 19(1):24–33.

Bayless, K.J. and Davis, G.E. 2003. Sphingosine-1-phosphate markedly induces matrix metalloproteinase and integrin-dependent human endothelial cell invasion and lumen formation in three-dimensional collagen and fibrin matrices. Biochem Biophys Res Commun 312(4): 903–913.

Bond, M., Murphy, G., Bennett, M.R., et al. 2002. Tissue inhibitor of metalloproteinase-3 induces a Fas-associated death domain-dependent type II apoptotic pathway. J Biol Chem 277(16): 13787–13795.

Boulday, G., Fitau, J., Coupel, S., et al. 2004. Exogenous tissue inhibitor of metalloproteinase-1 promotes endothelial cell survival through activation of the phosphatidylinositol 3-kinase/Akt pathway. Ann N Y Acad Sci 1030: 28–36.

Brenner, C.A., Adler, R.R., Rappolee, D.A., et al. 1989. Genes for extracellular-matrix-degrading metalloproteinases and their inhibitor, TIMP, are expressed during early mammalian development. Genes Dev 3(6): 848–859.

Brew, K., Dinakarpandian, D. and Nagase, H. 2000. Tissue inhibitors of metalloproteinases: evolution, structure and function. Biochim Biophys Acta 1477(1–2): 267–283.

Cancelas, J.A., Jansen, M. and Williams, D.A. 2006. The role of chemokine activation of Rac GTPases in hematopoietic stem cell marrow homing, retention, and peripheral mobilization. Exp Hematol 34(8): 976–985.

Castagnino, P., Soriano, J.V., Montesano, R., et al. 1998. Induction of tissue inhibitor of metalloproteinases-3 is a delayed early cellular response to hepatocyte growth factor. Oncogene 17(4): 481–492.

Caterina, J.J., Yamada, S., Caterina, N.C., et al. 2000. Inactivating mutation of the mouse tissue inhibitor of metalloproteinases-2(Timp-2) gene alters proMMP-2 activation. J Biol Chem 275 (34): 26416–26422.

Cawston, T.E., Galloway, W.A., Mercer, E., et al. 1981. Purification of rabbit bone inhibitor of collagenase. Biochem J 195(1): 159–165.

Celiker, M.Y., Wang, M., Atsidaftos, E., et al. 2001. Inhibition of Wilms' tumor growth by intramuscular administration of tissue inhibitor of metalloproteinases-4 plasmid DNA. Oncogene 20(32): 4337–4343.

Chen, P., Farivar, A.S., Mulligan, M.S., et al. 2006. Tissue inhibitor of metalloproteinase-1 deficiency abrogates obliterative airway disease after heterotopic tracheal transplantation. Am J Respir Cell Mol Biol 34(4): 464–472.

Chirco, R., Liu, X.W., Jung, K.K., et al. 2006. Novel functions of TIMPs in cell signaling. Cancer Metastasis Rev 25(1): 99–113.

Coppock, H.A., White, A., Aplin, J.D., et al. 2004. Matrix metalloprotease-3 and -9 proteolyze insulin-like growth factor-binding protein-1. Biol Reprod 71(2): 438–443.

Creemers, E.E., Davis, J.N., Parkhurst, A.M., et al. 2003. Deficiency of TIMP-1 exacerbates LV remodeling after myocardial infarction in mice. Am J Physiol Heart Circ Physiol 284(1): H364–H371.

Cruz-Munoz, W., Kim, I. and Khokha, R. 2006a. TIMP-3 deficiency in the host, but not in the tumor, enhances tumor growth and angiogenesis. Oncogene 25(4): 650–655.

Cruz-Munoz, W., Sanchez, O.H., Di Grappa, M., et al. 2006b. Enhanced metastatic dissemination to multiple organs by melanoma and lymphoma cells in timp-3$^{-/-}$ mice. Oncogene 25(49): 6489–6496.

Dean, G., Young, D.A., Edwards, D.R., et al. 2000. The human tissue inhibitor of metalloproteinases (TIMP)-1 gene contains repressive elements within the promoter and intron 1. J Biol Chem 275(42): 32664–32671.

Docherty, A.J., Lyons, A., Smith, B.J., et al. 1985. Sequence of human tissue inhibitor of metalloproteinases and its identity to erythroid-potentiating activity. Nature 318(6041): 66–69.

Drynda, A., Quax, P.H., Neumann, M., et al. 2005. Gene transfer of tissue inhibitor of metalloproteinases-3 reverses the inhibitory effects of TNF-alpha on Fas-induced apoptosis in rheumatoid arthritis synovial fibroblasts. J Immunol 174(10): 6524–6531.

Edwards, D.R., Waterhouse, P., Holman, M.L., et al. 1986. A growth-responsive gene (16C8) in normal mouse fibroblasts homologous to a human collagenase inhibitor with erythroid-potentiating activity: evidence for inducible and constitutive transcripts. Nucleic Acids Res 14(22): 8863–8878.

Edwards, D.R., Rocheleau, H., Sharma, R.R., et al. 1992. Involvement of AP1 and PEA3 binding sites in the regulation of murine tissue inhibitor of metalloproteinases-1 (TIMP-1) transcription. Biochim Biophys Acta 1171(1): 41–55.

Elezkurtaj, S., Kopitz, C., Baker, A.H., et al. 2004. Adenovirus-mediated overexpression of tissue inhibitor of metalloproteinases-1 in the liver: efficient protection against T-cell lymphoma and colon carcinoma metastasis. J Gene Med 6(11): 1228–1237.

Engers, R., Springer, E., Michiels, F., et al. 2001. Rac affects invasion of human renal cell carcinomas by up-regulating tissue inhibitor of metalloproteinases (TIMP)-1 and TIMP-2 expression. J Biol Chem 276(45): 41889–41897.

English, J.L., Kassiri, Z., Koskivirta, I., et al. 2006. Individual Timp deficiencies differentially impact pro-MMP-2 activation. J Biol Chem 281(15): 10337–10346.

Fata, J.E., Leco, K.J., Moorehead, R.A., et al. 1999. Timp-1 is important for epithelial proliferation and branching morphogenesis during mouse mammary development. Dev Biol 211(2): 238–254.

Fata, J.E., Leco, K.J., Voura, E.B., et al. 2001. Accelerated apoptosis in the Timp-3-deficient mammary gland. J Clin Invest 108(6): 831–841.

Fedak, P.W., Smookler, D.S., Kassiri, Z., et al. 2004. TIMP-3 deficiency leads to dilated cardiomyopathy. Circulation 110(16): 2401–2409.

Federici, M., Hribal, M.L., Menghini, R., et al. 2005. Timp3 deficiency in insulin receptor-haploinsufficient mice promotes diabetes and vascular inflammation via increased TNF-alpha. J Clin Invest 115(12): 3494–3505.

Feldman, A.L., Stetler-Stevenson, W.G., Costouros, N.G., et al. 2004. Modulation of tumor-host interactions, angiogenesis, and tumor growth by tissue inhibitor of metalloproteinase 2 via a novel mechanism. Cancer Res 64(13): 4481–4486.

Fernandez, C.A. and Moses, M.A. 2006. Modulation of angiogenesis by tissue inhibitor of metalloproteinase-4. Biochem Biophys Res Commun 345(1): 523–529.

Fernandez-Catalan, C., Bode, W., Huber, R., et al. 1998. Crystal structure of the complex formed by the membrane type 1-matrix metalloproteinase with the tissue inhibitor of metalloproteinases-2, the soluble progelatinase A receptor. Embo J 17(17): 5238–5248.

Fryer, B.H. and Field, J. 2005. Rho, Rac, Pak and angiogenesis: old roles and newly identified responsibilities in endothelial cells. Cancer Lett 229(1): 13–23.

Gasson, J.C., Golde, D.W., Kaufman, S.E., et al. 1985. Molecular characterization and expression of the gene encoding human erythroid-potentiating activity. Nature 315(6022): 768–771.

Gill, S.E., Pape, M.C., Khokha, R., et al. 2003. A null mutation for tissue inhibitor of metalloproteinases-3 (Timp-3) impairs murine bronchiole branching morphogenesis. Dev Biol 261(2): 313–323.

Godenschwege, T.A., Pohar, N., Buchner, S., et al. 2000. Inflated wings, tissue autolysis and early death in tissue inhibitor of metalloproteinases mutants of Drosophila. Eur J Cell Biol 79(7): 495–501.

Greene, J., Wang, M., Liu, Y.E., et al. 1996. Molecular cloning and characterization of human tissue inhibitor of metalloproteinase 4. J Biol Chem 271(48): 30375–30380.

Guedez, L., McMarlin, A.J., Kingma, D.W., et al. 2001. Tissue inhibitor of metalloproteinase-1 alters the tumorigenicity of Burkitt's lymphoma via divergent effects on tumor growth and angiogenesis. Am J Pathol 158(4): 1207–1215.

Han, X., Zhang, H., Jia, M., et al. 2004. Expression of TIMP-3 gene by construction of a eukaryotic cell expression vector and its role in reduction of metastasis in a human breast cancer cell line. Cell Mol Immunol 1(4): 308–310.

Hayakawa, T., Yamashita, K., Kishi, J., et al. 1990. Tissue inhibitor of metalloproteinases from human bone marrow stromal cell line KM 102 has erythroid-potentiating activity, suggesting its possibly bifunctional role in the hematopoietic microenvironment. FEBS Lett 268(1): 125–128.

Hemers, E., Duval, C., McCaig, C., et al. 2005. Insulin-like growth factor binding protein-5 is a target of matrix metalloproteinase-7: implications for epithelial-mesenchymal signaling. Cancer Res 65(16): 7363–7369.

Ho, A.T., Voura, E.B., Soloway, P.D., et al. 2001. MMP inhibitors augment fibroblast adhesion through stabilization of focal adhesion contacts and up-regulation of cadherin function. J Biol Chem 276(43): 40215–40224.

Hoegy, S.E., Oh, H.R., Corcoran, M.L., et al. 2001. Tissue inhibitor of metalloproteinases-2 (TIMP-2) suppresses TKR-growth factor signaling independent of metalloproteinase inhibition. J Biol Chem 276(5): 3203–3214.

Hojilla, C.V., Kim, I., Kassiri, Z., et al. 2007. Metalloproteinase axes increase beta-catenin signaling in primary mouse mammary epithelial cells lacking TIMP3. J Cell Sci 120(Pt 6): 1050–1060.

Hotary, K.B., Allen, E.D., Brooks, P.C., et al. 2003. Membrane type I matrix metalloproteinase usurps tumor growth control imposed by the three-dimensional extracellular matrix. Cell 114 (1): 33–45.

Hotary, K., Li, X.Y., Allen, E., et al. 2006. A cancer cell metalloprotease triad regulates the basement membrane transmigration program. Genes Dev 20(19): 2673–2686.

House, M.G., Herman, J.G., Guo, M.Z., et al. 2003. Aberrant hypermethylation of tumor suppressor genes in pancreatic endocrine neoplasms. Ann Surg 238(3): 423–431; discussion 431–432.

Huxley-Jones, J., Clarke, T.K., Beck, C., et al. 2007. The evolution of the vertebrate metzincins; insights from Ciona intestinalis and Danio rerio. BMC Evol Biol 7: 63.

Ikonomidis, J.S., Hendrick, J.W., Parkhurst, A.M., et al. 2005. Accelerated LV remodeling after myocardial infarction in TIMP-1-deficient mice: effects of exogenous MMP inhibition. Am J Physiol Heart Circ Physiol 288(1): H149–H158.

Iyer, S., Wei, S., Brew, K., et al. 2007. Crystal structure of the catalytic domain of matrix metalloproteinase-1 in complex with the inhibitory domain of tissue inhibitor of metalloproteinase-1. J Biol Chem 282(1): 364–371.

Jaworski, D.M., Boone, J., Caterina, J., et al. 2005. Prepulse inhibition and fear-potentiated startle are altered in tissue inhibitor of metalloproteinase-2 (TIMP-2) knockout mice. Brain Res 1051 (1–2): 81–89.

Jaworski, D.M., Soloway, P., Caterina, J., et al. 2006. Tissue inhibitor of metalloproteinase-2 (TIMP-2)-deficient mice display motor deficits. J Neurobiol 66(1): 82–94.

Jiang, Y., Wang, M., Celiker, M.Y., et al. 2001. Stimulation of mammary tumorigenesis by systemic tissue inhibitor of matrix metalloproteinase 4 gene delivery. Cancer Res 61(6): 2365–2370.

Johnson, J.L., Baker, A.H., Oka, K., et al. 2006. Suppression of atherosclerotic plaque progression and instability by tissue inhibitor of metalloproteinase-2: involvement of macrophage migration and apoptosis. Circulation 113(20): 2435–2444.

Johnson, M.D., Kim, H.R., Chesler, L., et al. 1994. Inhibition of angiogenesis by tissue inhibitor of metalloproteinase. J Cell Physiol 160(1): 194–202.

Karamouzis, M.V. and Papavassiliou, A.G. 2006. The IGF-1 network in lung carcinoma therapeutics. Trends Mol Med 12(12): 595–602.

Karan, D., Lin, F.C., Bryan, M., et al. 2003. Expression of ADAMs (a disintegrin and metalloproteases) and TIMP-3 (tissue inhibitor of metalloproteinase-3) in human prostatic adenocarcinomas. Int J Oncol 23(5): 1365–1371.

Kassiri, Z. and Khokha, R. 2005. Myocardial extra-cellular matrix and its regulation by metalloproteinases and their inhibitors. Thromb Haemost 93(2): 212–219.

Kassiri, Z., Oudit, G.Y., Sanchez, O., et al. 2005. Combination of tumor necrosis factor-alpha ablation and matrix metalloproteinase inhibition prevents heart failure after pressure overload in tissue inhibitor of metalloproteinase-3 knock-out mice. Circ Res 97(4): 380–390.

Khokha, R., Waterhouse, P., Yagel, S., et al. 1989. Antisense RNA-induced reduction in murine TIMP levels confers oncogenicity on Swiss 3T3 cells. Science 243(4893): 947–950.

Khokha, R., Zimmer, M.J., Graham, C.H., et al. 1992. Suppression of invasion by inducible expression of tissue inhibitor of metalloproteinase-1 (TIMP-1) in B16-F10 melanoma cells. J Natl Cancer Inst 84(13): 1017–1022.

Kim, K.H., Burkhart, K., Chen, P., et al. 2005. Tissue inhibitor of metalloproteinase-1 deficiency amplifies acute lung injury in bleomycin-exposed mice. Am J Respir Cell Mol Biol 33(3): 271–279.

Kim, T.H., Mars, W.M., Stolz, D.B., et al. 2000. Expression and activation of pro-MMP-2 and pro-MMP-9 during rat liver regeneration. Hepatology 31(1): 75–82.

Kim, Y.II., Petko, Z., Dzicciatkowski, S., et al. 2006. CpG island methylation of genes accumulates during the adenoma progression step of the multistep pathogenesis of colorectal cancer. Genes Chromosomes Cancer 45(8): 781–789.

Kong, Y., Poon, R., Nadesan, P., et al. 2004. Matrix metalloproteinase activity modulates tumor size, cell motility, and cell invasiveness in murine aggressive fibromatosis. Cancer Res 64(16): 5795–5803.

Kopitz, C., Gerg, M., Bandapalli, O.R., et al. 2007. Tissue inhibitor of metalloproteinases-1 promotes liver metastasis by induction of hepatocyte growth factor signaling. Cancer Res 67 (18): 8615–8623.

Koskivirta, I., Rahkonen, O., Mayranpaa, M., et al. 2006. Tissue inhibitor of metalloproteinases 4 (TIMP4) is involved in inflammatory processes of human cardiovascular pathology. Histochem Cell Biol 126(3): 335–342.

Kruger, A., Fata, J.E. and Khokha, R. 1997. Altered tumor growth and metastasis of a T-cell lymphoma in Timp-1 transgenic mice. Blood 90(5): 1993–2000.

Kruger, A., Sanchez-Sweatman, O.H., Martin, D.C., et al. 1998. Host TIMP-1 overexpression confers resistance to experimental brain metastasis of a fibrosarcoma cell line. Oncogene 16 (18): 2419–2423.

Kurmasheva, R.T. and Houghton, P.J. 2006. IGF-I mediated survival pathways in normal and malignant cells. Biochim Biophys Acta 1766(1): 1–22.

Lafleur, M.A., Handsley, M.M., Knauper, V., et al. 2002. Endothelial tubulogenesis within fibrin gels specifically requires the activity of membrane-type-matrix metalloproteinases (MT-MMPs). J Cell Sci 115(Pt 17): 3427–3438.

Lambert, E., Boudot, C., Kadri, Z., et al. 2003. Tissue inhibitor of metalloproteinases-1 signalling pathway leading to erythroid cell survival. Biochem J 372(Pt 3): 767–774.

Langton, K.P., McKie, N., Curtis, A., et al. 2000. A novel tissue inhibitor of metalloproteinases-3 mutation reveals a common molecular phenotype in Sorsby's fundus dystrophy. J Biol Chem 275(35): 27027–27031.

Leco, K.J., Hayden, L.J., Sharma, R.R., et al. 1992. Differential regulation of TIMP-1 and TIMP-2 mRNA expression in normal and Ha-ras-transformed murine fibroblasts. Gene 117(2): 209–217.

Leco, K.J., Khokha, R., Pavloff, N., et al. 1994. Tissue inhibitor of metalloproteinases-3 (TIMP-3) is an extracellular matrix-associated protein with a distinctive pattern of expression in mouse cells and tissues. J Biol Chem 269(12): 9352–9360.

Leco, K.J., Apte, S.S., Taniguchi, G.T., et al. 1997. Murine tissue inhibitor of metalloproteinases-4 (Timp-4): cDNA isolation and expression in adult mouse tissues. FEBS Lett 401(2–3): 213–217.

Leco, K.J., Waterhouse, P., Sanchez, O.H., et al. 2001. Spontaneous air space enlargement in the lungs of mice lacking tissue inhibitor of metalloproteinases-3 (TIMP-3). J Clin Invest 108(6): 817–829.

Lee, M.H., Atkinson, S. and Murphy, G. 2007. Identification of the extracellular matrix (ECM) binding motifs of tissue inhibitor of metalloproteinases (TIMP)-3 and effective transfer to TIMP-1. J Biol Chem 282(9): 6887–6898.

Lee, Y.K., So, I.S., Lee, S.C., et al. 2005. Suppression of distant pulmonary metastasis of MDA-MB 435 human breast carcinoma established in mammary fat pads of nude mice by retroviral-mediated TIMP-2 gene transfer. J Gene Med 7(2): 145–157.

Lijnen, H.R., Demeulemeester, D., Van Hoef, B., et al. 2003. Deficiency of tissue inhibitor of matrix metalloproteinase-1 (TIMP-1) impairs nutritionally induced obesity in mice. Thromb Haemost 89(2): 249–255.

Lim, M.S., Guedez, L., Stetler-Stevenson, W.G., et al. 1999. Tissue inhibitor of metalloproteinase-2 induces apoptosis in human T lymphocytes. Ann N Y Acad Sci 878: 522–523.

Lindroos, P.M., Zarnegar, R. and Michalopoulos, G.K. 1991. Hepatocyte growth factor (hepatopoietin A) rapidly increases in plasma before DNA synthesis and liver regeneration stimulated by partial hepatectomy and carbon tetrachloride administration. Hepatology 13(4): 743–750.

Liu, X.W., Taube, M.E., Jung, K.K., et al. 2005. Tissue inhibitor of metalloproteinase-1 protects human breast epithelial cells from extrinsic cell death: a potential oncogenic activity of tissue inhibitor of metalloproteinase-1. Cancer Res 65(3): 898–906.

Lizarraga, F., Maldonado, V. and Melendez-Zajgla, J. 2004. Tissue inhibitor of metalloproteinases-2 growth-stimulatory activity is mediated by nuclear factor-kappa B in A549 lung epithelial cells. Int J Biochem Cell Biol 36(8): 1655–1663.

Lluri, G., Langlois, G.D., McClellan, B., et al. 2006. Tissue inhibitor of metalloproteinase-2 (TIMP-2) regulates neuromuscular junction development via a beta1 integrin-mediated mechanism. J Neurobiol 66(12): 1365–1377.

Lovelock, J.D., Baker, A.H., Gao, F., et al. 2005. Heterogeneous effects of tissue inhibitors of matrix metalloproteinases on cardiac fibroblasts. Am J Physiol Heart Circ Physiol 288(2): H461–H468.

Lui, E.L., Loo, W.T., Zhu, L., et al. 2005. DNA hypermethylation of TIMP3 gene in invasive breast ductal carcinoma. Biomed Pharmacother 59(Suppl 2): S363–S365.

Mahmoodi, M., Sahebjam, S., Smookler, D., et al. 2005. Lack of tissue inhibitor of metalloproteinases-3 results in an enhanced inflammatory response in antigen-induced arthritis. Am J Pathol 166(6): 1733–1740.

Maidment, J.M., Moore, D., Murphy, G.P., et al. 1999. Matrix metalloproteinase homologues from Arabidopsis thaliana. Expression and activity. J Biol Chem 274(49): 34706–34710.

Martin, D.C., Ruther, U., Sanchez-Sweatman, O.H., et al. 1996. Inhibition of SV40 T antigen-induced hepatocellular carcinoma in TIMP-1 transgenic mice. Oncogene 13(3): 569–576.

Martin, D.C., Fowlkes, J.L., Babic, B., et al. 1999a. Insulin-like growth factor II signaling in neoplastic proliferation is blocked by transgenic expression of the metalloproteinase inhibitor TIMP-1. J Cell Biol 146(4): 881–892.

Martin, D.C., Sanchez-Sweatman, O.H., Ho, A.T., et al. 1999b. Transgenic TIMP-1 inhibits simian virus 40 T antigen-induced hepatocarcinogenesis by impairment of hepatocellular proliferation and tumor angiogenesis. Lab Invest 79(2): 225–234.

Martin, E.L., McCaig, L.A., Moyer, B.Z., et al. 2005. Differential response of TIMP-3 null mice to the lung insults of sepsis, mechanical ventilation, and hyperoxia. Am J Physiol Lung Cell Mol Physiol 289(2): L244–L251.

Miller, B.S. and Yee, D. 2005. Type I insulin-like growth factor receptor as a therapeutic target in cancer. Cancer Res 65(22): 10123–10127.

Mitsui, Y., Mochizuki, S., Kodama, T., et al. 2006. ADAM28 is overexpressed in human breast carcinomas: implications for carcinoma cell proliferation through cleavage of insulin-like growth factor binding protein-3. Cancer Res 66(20): 9913–9920.

Miyamoto, S., Nakamura, M., Yano, K., et al. 2007. Matrix metalloproteinase-7 triggers the matricrine action of insulin-like growth factor-II via proteinase activity on insulin-like growth factor binding protein 2 in the extracellular matrix. Cancer Sci 98(5): 685–691.

Mohammed, F.F. and Khokha, R. 2005. Thinking outside the cell: proteases regulate hepatocyte division. Trends Cell Biol 15(10): 555–563.

Mohammed, F.F., Smookler, D.S., Taylor, S.E., et al. 2004. Abnormal TNF activity in Timp3$^{-/-}$ mice leads to chronic hepatic inflammation and failure of liver regeneration. Nat Genet 36(9): 969–977.

Mohammed, F.F., Pennington, C.J., Kassiri, Z., et al. 2005. Metalloproteinase inhibitor TIMP-1 affects hepatocyte cell cycle via HGF activation in murine liver regeneration. Hepatology 41(4): 857–867.

Moldovan, L., Mythreye, K., Goldschmidt-Clermont, P.J., et al. 2006. Reactive oxygen species in vascular endothelial cell motility. Roles of NAD(P)H oxidase and Rac1. Cardiovasc Res 71(2): 236–246.

Morgunova, E., Tuuttila, A., Bergmann, U., et al. 2002. Structural insight into the complex formation of latent matrix metalloproteinase 2 with tissue inhibitor of metalloproteinase 2. Proc Natl Acad Sci USA 99(11): 7414–7419.

Murate, T., Yamashita, K., Ohashi, H., et al. 1993. Erythroid potentiating activity of tissue inhibitor of metalloproteinases on the differentiation of erythropoietin-responsive mouse erythroleukemia cell line, ELM-I-1–3, is closely related to its cell growth potentiating activity. Exp Hematol 21(1): 169–176.

Murphy, F.R., Issa, R., Zhou, X., et al. 2002. Inhibition of apoptosis of activated hepatic stellate cells by tissue inhibitor of metalloproteinase-1 is mediated via effects on matrix metalloproteinase inhibition: implications for reversibility of liver fibrosis. J Biol Chem 277 (13): 11069–11076.

Murphy, G., Knauper, V., Lee, M.H., et al. 2003. Role of TIMPs (tissue inhibitors of metalloproteinases) in pericellular proteolysis: the specificity is in the detail. Biochem Soc Symp (70): 65–80.

Murphy, G., Murthy, A., Khokha, R. 2008. Clipping shedding and RIPping keep immunity on cue. Trends Immunol 29(2):75–82.

Nothnick, W.B. 2003. Tissue inhibitor of metalloproteinase-1 (TIMP-1) deficient mice display reduced serum progesterone levels during corpus luteum development. Endocrinology 144(1): 5–8.

Nothnick, W.B., Soloway, P.D. and Curry, T.E., Jr. 1998. Pattern of messenger ribonucleic acid expression of tissue inhibitors of metalloproteinases (TIMPs) during testicular maturation in male mice lacking a functional TIMP-1 gene. Biol Reprod 59(2): 364–370.

Nothnick, W.B., Zhang, X. and Zhou, H.E. 2004. Steroidal regulation of uterine edema and tissue inhibitors of metalloproteinase (TIMP)-3 messenger RNA expression is altered in TIMP-1-deficient mice. Biol Reprod 70(2): 500–508.

Nuttall, R.K., Sampieri, C.L., Pennington, C.J., et al. 2004. Expression analysis of the entire MMP and TIMP gene families during mouse tissue development. FEBS Lett 563(1–3): 129–134.

Parks, W.C., Wilson, C.L. and Lopez-Boado, Y.S. 2004. Matrix metalloproteinases as modulators of inflammation and innate immunity. Nat Rev Immunol 4(8): 617–629.

Petitfrere, E., Kadri, Z., Boudot, C., et al. 2000. Involvement of the p38 mitogen-activated protein kinase pathway in tissue inhibitor of metalloproteinases-1-induced erythroid differentiation. FEBS Lett 485(2–3): 117–121.

Plaisier, M., Kapiteijn, K., Koolwijk, P., et al. 2004. Involvement of membrane-type matrix metalloproteinases (MT-MMPs) in capillary tube formation by human endometrial microvascular endothelial cells: role of MT3-MMP. J Clin Endocrinol Metab 89(11): 5828–5836.

Pohar, N., Godenschwege, T.A. and Buchner, E. 1999. Invertebrate tissue inhibitor of metalloproteinase: structure and nested gene organization within the synapsin locus is conserved from Drosophila to human. Genomics 57(2): 293–296.

Porter, J.F., Shen, S. and Denhardt, D.T. 2004. Tissue inhibitor of metalloproteinase-1 stimulates proliferation of human cancer cells by inhibiting a metalloproteinase. Br J Cancer 90(2): 463–470.

Puente, X.S., Sanchez, L.M., Overall, C.M., et al. 2003. Human and mouse proteases: a comparative genomic approach. Nat Rev Genet 4(7): 544–558.

Qi, J.H., Ebrahem, Q., Moore, N., et al. 2003. A novel function for tissue inhibitor of metalloproteinases-3 (TIMP3): inhibition of angiogenesis by blockage of VEGF binding to VEGF receptor-2. Nat Med 9(4): 407–415.

Renehan, A.G., Frystyk, J. and Flyvbjerg, A. 2006. Obesity and cancer risk: the role of the insulin-IGF axis. Trends Endocrinol Metab 17(8): 328–336.

Riddick, A.C., Shukla, C.J., Pennington, C.J., et al. 2005. Identification of degradome components associated with prostate cancer progression by expression analysis of human prostatic tissues. Br J Cancer 92(12): 2171–2180.

Rigg, A.S. and Lemoine, N.R. 2001. Adenoviral delivery of TIMP1 or TIMP2 can modify the invasive behavior of pancreatic cancer and can have a significant antitumor effect in vivo. Cancer Gene Ther 8(11): 869–878.

Rose, D.M., Alon, R. and Ginsberg, M.H. 2007. Integrin modulation and signaling in leukocyte adhesion and migration. Immunol Rev 218: 126–134.

Roten, L., Nemoto, S., Simsic, J., et al. 2000. Effects of gene deletion of the tissue inhibitor of the matrix metalloproteinase-type 1 (TIMP-1) on left ventricular geometry and function in mice. J Mol Cell Cardiol 32(1): 109–120.

Sabeh, F., Ota, I., Holmbeck, K., et al. 2004. Tumor cell traffic through the extracellular matrix is controlled by the membrane-anchored collagenase MT1-MMP. J Cell Biol 167(4): 769–781.

Sahebjam, S., Khokha, R. and Mort, J.S. 2007. Increased collagen and aggrecan degradation with age in the joints of Timp3 ($^{-/-}$) mice. Arthritis Rheum 56(3): 905–909.

Saunders, W.B., Bohnsack, B.L., Faske, J.B., et al. 2006. Coregulation of vascular tube stabilization by endothelial cell TIMP-2 and pericyte TIMP-3. J Cell Biol 175(1): 179–191.

Schmidt, C., Bladt, F., Goedecke, S., et al. 1995. Scatter factor/hepatocyte growth factor is essential for liver development. Nature 373(6516): 699–702.

Seo, D.W., Li, H., Guedez, L., et al. 2003. TIMP-2 mediated inhibition of angiogenesis: an MMP-independent mechanism. Cell 114(2): 171–180.

Smith, M.R., Kung, H., Durum, S.K., et al. 1997. TIMP-3 induces cell death by stabilizing TNF-alpha receptors on the surface of human colon carcinoma cells. Cytokine 9(10): 770–780.

Smookler, D.S., Mohammed, F.F., Kassiri, Z., et al. 2006. Tissue inhibitor of metalloproteinase 3 regulates TNF-dependent systemic inflammation. J Immunol 176(2): 721–725.

Spurbeck, W.W., Ng, C.Y., Strom, T.S., et al. 2002. Enforced expression of tissue inhibitor of matrix metalloproteinase-3 affects functional capillary morphogenesis and inhibits tumor growth in a murine tumor model. Blood 100(9): 3361–3368.

Stetler-Stevenson, W.G., Krutzsch, H.C. and Liotta, L.A. 1989. Tissue inhibitor of metalloproteinase (TIMP-2). A new member of the metalloproteinase inhibitor family. J Biol Chem 264(29): 17374–17378.

Stetler-Stevenson, W.G., Bersch, N. and Golde, D.W. 1992. Tissue inhibitor of metalloproteinase-2 (TIMP-2) has erythroid-potentiating activity. FEBS Lett 296(2): 231–234.

Toth, M., Bernardo, M.M., Gervasi, D.C., et al. 2000. Tissue inhibitor of metalloproteinase (TIMP)-2 acts synergistically with synthetic matrix metalloproteinase (MMP) inhibitors but not with TIMP-4 to enhance the (Membrane type 1)-MMP-dependent activation of pro-MMP-2. J Biol Chem 275(52): 41415–41423.

Tran, P.L., Vigneron, J.P., Pericat, D., et al. 2003. Gene theraphy for hepatocellular carcinoma using non-viral vectors composed of bis guanidinium-tren-cholesterol and plasmids encoding the tissue inhibitors of metalloproteinases TIMP-2 and TIMP-3. Cancer Gene Ther 10(6): 435–444.

Tummalapalli, C.M., Heath, B.J. and Tyagi, S.C. 2001. Tissue inhibitor of metalloproteinase-4 instigates apoptosis in transformed cardiac fibroblasts. J Cell Biochem, 80(4): 512–521.

Valente, P., Fassina, G., Melchiori, A., et al. 1998. TIMP-2 over-expression reduces invasion and angiogenesis and protects B16F10 melanoma cells from apoptosis. Int J Cancer 75(2): 246–253.

van der Laan, W.H., Quax, P.H., Seemayer, C.A., et al. 2003. Cartilage degradation and invasion by rheumatoid synovial fibroblasts is inhibited by gene transfer of TIMP-1 and TIMP-3. Gene Ther 10(3): 234–242.

van der Velden, P.A., Zuidervaart, W., Hurks, M.H., et al. 2003. Expression profiling reveals that methylation of TIMP3 is involved in uveal melanoma development. Int J Cancer 106(4): 472–479.

Wang, M., Liu, Y.E., Greene, J., et al. 1997. Inhibition of tumor growth and metastasis of human breast cancer cells transfected with tissue inhibitor of metalloproteinase 4. Oncogene 14(23): 2767–2774.

Wang, Z., Juttermann, R. and Soloway, P.D. 2000. TIMP-2 is required for efficient activation of proMMP-2 in vivo. J Biol Chem 275(34): 26411–26415.

Waterhouse, P., Denhardt, D.T. and Khokha, R. 1993. Temporal expression of tissue inhibitors of metalloproteinases in mouse reproductive tissues during gestation. Mol Reprod Dev 35(3): 219–226.

Weber, B.H., Vogt, G., Pruett, R.C., et al. 1994a. Mutations in the tissue inhibitor of metalloproteinases-3 (TIMP3) in patients with Sorsby's fundus dystrophy. Nat Genet 8(4): 352–356.

Weber, B.H., Vogt, G., Wolz, W., et al. 1994b. Sorsby's fundus dystrophy is genetically linked to chromosome 22q13-qter. Nat Genet 7(2): 158–161.

Wei, S., Xie, Z., Filenova, E., et al. 2003. Drosophila TIMP is a potent inhibitor of MMPs and TACE: similarities in structure and function to TIMP-3. Biochemistry 42(42): 12200–12207.

Wick, M., Haronen, R., Mumberg, D., et al. 1995. Structure of the human TIMP-3 gene and its cell cycle-regulated promoter. Biochem J 311(Pt 2): 549–554.

Widschwendter, A., Muller, H.M., Fiegl, H., et al. 2004. DNA methylation in serum and tumors of cervical cancer patients. Clin Cancer Res 10(2): 565–571.

Yamada, S. and Nelson, W.J. 2007. Localized zones of Rho and Rac activities drive initiation and expansion of epithelial cell-cell adhesion. J Cell Biol 178(3): 517–527.

Yamazaki, M., Akahane, T., Buck, T., et al. 2004. Long-term exposure to elevated levels of circulating TIMP-1 but not mammary TIMP-1 suppresses growth of mammary carcinomas in transgenic mice. Carcinogenesis 25(9): 1735–1746.

Yu, W.P., Brenner, S. and Venkatesh, B. 2003. Duplication, degeneration and subfunctionalization of the nested synapsin-Timp genes in Fugu. Trends Genet 19(4): 180–183.

Zhai, Y., Hotary, K.B., Nan, B., et al. 2005. Expression of membrane type 1 matrix metalloproteinase is associated with cervical carcinoma progression and invasion. Cancer Res 65(15): 6543–6550.

Section III
The Interface Between Proteolysis and Cell Signalling

Chapter 21
Invadopodia: Interface for Invasion

Susette C. Mueller, Vira V. Artym, and Thomas Kelly

Abstract The term invadopodia is essentially a functional definition linking membrane protrusions with proteolysis of the matrix at focused points beneath cells cultured on two-dimensional ECM matrices. Their molecular definition awaits more complete characterization of the interactions of molecular components that organize into morphologically distinct structures recognized by characteristic actin/cortactin cores which subsequently attain a proteolytic functionality that facilitates cellular invasion. Because of their similarity to podosomes and because there are progressive stages in the formation of invadopodia, confusion exists regarding the precise definition of invadopodia. In this chapter, we discuss sources of confusion prevalent in the literature to date, delineate characteristic features of invadopodia, review molecular components required for invadopodia-mediated matrix degradation, and discuss the assembly of functional invadopodia and the relationship of the invadopodia membrane to the actin/cortactin core. Finally, we discuss what is known about the functional consequences of invadopodia-mediated matrix degradation and the role of proteases associated with invadopodia in cellular interaction with other cells and the matrix, and their contribution to the tumor microenvironment.

Introduction

Early studies of adhesion, migration, and invasion in Rous sarcoma-transformed cells revealed that the v-Src tyrosine kinase is transforming and produces a highly invasive cellular phenotype (Martin 2004, Thomas and Brugge 1997). In particular, the transformed cellular adhesions in fibroblasts that resulted from v-Src expression were termed invadopodia, rosettes, or podosomes and they displayed distinctive actin-associated morphology. Tumor cells were subsequently found to produce

S.C. Mueller
Department of Oncology, Lombardi Comprehensive Cancer Center, Georgetown University Medical School, Washington, District of Columbia 20057-1469, USA,
e-mail: mueller@georgetown.edu

D. Edwards et al. (eds.), *The Cancer Degradome.*
© Springer Science + Business Media, LLC 2008

invadopodia, and they possess at least some of the characteristics described for transformed fibroblasts (Ayala et al. 2006, Chen 1989, 1990; David-Pfeuty and Singer 1980, Gavazzi et al. 1989, Kelly et al. 1994, Wang et al. 1984, Weaver, 2006, Yamaguchi et al. 2006). Invadopodia were originally named and described based on the ability of invadopodial membranes to penetrate the extracellular matrix (ECM) and their association with sites of localized degradation in fluorescent matrices (Chen 1989). In summary, invadopodia adhesion sites in tumor cells are recognized by dot-like aggregates of actin and cortactin, and their membranes penetrate the matrix in the form of filopodia-like extensions assisted by membrane-associated proteolytic enzymes. Localized matrix-degradation, podosomes, and invadopodia have been reviewed recently (Ayala et al. 2006, Calle et al. 2004, Chen et al. 1994, Chen 1996, Linder 2007, McNiven et al. 2004, Weaver 2006, Yamaguchi et al. 2006, Yamaguchi and Condeelis 2007).

Invadopodia or podosomes have been reported in fibroblasts, epithelial and mesenchymal tumor cells, and also in endothelial cells, smooth muscle cells, and hematopoietic cells, the latter primarily of the monocyte lineage. Podosomes and invadopodia all appear to be Src dependent (Martin 2004). However, confusion exists as to the distinction between podosomes and invadopodia (for review, *see* Ayala et al. 2006, for example). There is no single antibody that uniquely and exclusively identifies invadopodia or podosomes. A list of proteins, compiled based on a variety of data, details possible constituents of podosomes, invadopodia, and circular dorsal ruffles (Buccione et al. 2004), but the criteria for the identification of invadopodia and the types of experiments used to support the conclusions concerning them have not been uniformly agreed upon from study to study.

Confusion between these structures arises from a variety of other factors as well. While they have overlapping constituents and morphological features, there appear to be distinct differences from cell to cell with regard to number and size, association of internal and external membranes with the actin core, and molecular constituents such as integrin heterodimers (compare epithelial cell podosomes from normal cells to macrophages to tumor cell invadopodia, e.g., Artym et al. 2006, Clark et al. 2007, Evans et al. 2003, Spinardi et al. 2004 and in reviews mentioned above). Another major difference relates to cellular response to ECM. Adhesions, including invadopodia, do not attain full functionality unless the cell receives proper signals from the ECM (e.g., Cukierman et al. 2001, Mueller et al. 1999), and deregulation of integrins can promote invasiveness via invadopodia (Nakahara et al. 1996, 1998). For example, the ECM-degrading enzyme fibroblast activation protein-α (FAP or seprase) directly interacts with α3β1 integrin at invadopodia, but only when cells are cultured on collagen and not plastic (Mueller et al. 1999). Thus, one cannot assume that podosomes or invadopodia in cells cultured on plastic are the same morphologically or functionally as those observed when cells are cultured on 2D or 3D ECM proteins.

Other sources of confusion arise from the use of different cell types with different signal transduction backgrounds combined with inadequate criteria for invadopodia identification. Actin core assembly might proceed, and proteolytic function coupling might lag behind or not be present when cells are stimulated with a given growth factor or pharmaceutical agent. For example, phorbol ester treatment of A7r5 smooth muscle cells elicits podosomes (actin core formation) but not matrix degradation; in contrast, the same phorbol ester elicited both actin core formation and matrix degradation in primary vascular smooth muscle cells (VSMC) (Furmaniak-Kazmierczak et al. 2007). Although the primary role of podosomes is thought to be the mediation of adhesion and migration in hematopoietic cells, their role in localized matrix degradation, similar to invadopodia, has recently been established using tests for localized matrix degradation in vitro (Table 4.1). And, an invasive function for podosomes in leukocytes has recently been proposed (Carman et al. 2007).

We will focus in this chapter on the role of invadopodia in localized ECM degradation emphasizing what is known about the stages of their formation and their role during tumor invasion and metastasis.

Molecular Components of Invadopodia

Molecular components of invadopodia that have been studied in conjunction with localization of MT1-MMP protease, or with sites of localized degradation in fluorescent matrices, are listed in Table 4.1.

Invadopodia components, in general, fall into four classes of molecules: (1) proteins involved with actin polymerization, (2) proteins involved with integrin-mediated adhesion, (3) signaling proteins regulating actin polymerization and membrane remodeling (kinases and Ras-related GTPases), and (4) ECM-degrading proteases (Artym et al. 2006, Ayala et al. 2006, Buccione et al. 2004, Gimona and Buccione 2006, Linder 2007, Weaver 2006 Yamaguchi and Condeelis 2007). Many of these components play important roles in lamellipodia formation, cell spreading, and dorsal ruffling in response to growth factor, and some are found in focal adhesions. Many more components have been implicated in podosome formation and stability including microtubule-related HDAC activity and KIF1C proteins, for example, but their involvement in protease delivery or invadopodia-mediated matrix degradation are not known (Linder 2007). Rather than reviewing all the potential components of invadopodia, since these are covered well in the numerous reviews mentioned above, we will focus on reviewing papers in which the protease activity of invadopodia has been considered including the impact of signaling molecules on the functional outcome of invadopodia formation.

Identification of Invadopodia

To provide a framework for understanding the cellular signaling required for the formation of functional invadopodia, it is necessary to discuss the steps in their genesis and to further define criteria for their identification.

Membranes of Invadopodia

Invadopodia are proteolytically active membrane protrusions that adhere to and interact with the ECM. The membrane protrusions closely resemble filopodia in size and they are closely associated with actin/cortactin cores (Fig. 21.1). Electron microscopy of invadopodia from MDA-MB-231 breast cancer, LOX and A375 melanoma, and Rous sarcoma virus transformed chicken embryonic fibroblasts cells invading into thick matrices indicates that invadopodia are of approximately the same dimension as filopodia, as thin as about one to several hundred nanometers in diameter but many microns in length, thus distinguishing them from other protrusions, such as larger cellular protrusions, pseudopodia, or lamellipodia (Fig. 21.1; Bowden et al. 1999, 2006; Chen 1989, Coopman et al. 1996, Gimona and Buccione 2006, Kelly et al. 1994, Monsky et al. 1994, Mueller et al. 1992, Mueller and Chen 1991). Other immunofluorescence studies have illustrated invadopodial extensions of slightly wider dimensions, on the order of microns (e.g., Bourguignon et al. 1998, Hauck et al. 2002, Hiratsuka et al. 2006). An interesting feature high-lighted in studies using the N-WASP biosensor to detect localization of N-WASP activity at sites of degradation in MTLn3 cells cultured on fluorescent matrix was that the penetration was progressive, with the hole in the matrix broadening and lengthening over time [*see* Fig. 21.1 (Lorenz et al. 2004)]. N-WASP activity is high in early stages of matrix degradation and the holes start out at something less than 2 μm in diameter, but progressively N-WASP activity decreases as the hole is deepened and widened.

Correlative light and electron microscopy of A375MM melanoma cells was used by Baldassarre et al. (2003) to describe an "extracellular matrix-degrading structure" (EDS) that included invaginations of up to 8 μm width associated with gelatin fragments (*see* Fig. 21.1). Surface protrusions associated with them were several hundred nanometers to several micrometers in diameter. In breast cancer cells, invadopodial actin/cortactin cores were ∼0.5–1 μm in diameter, similar in size to the holes in the matrix formed under them on 2D gelatin films which start out less than 1 μm in diameter and become wider (Artym et al. 2006, unpublished data). These were detected on films that were just under 50 nm thick in contrast to the much thicker matrices evidenced in vertical confocal sections in the N-WASP study discussed above (Lorenz et al. 2004). In the latter study, the use of time lapse and confocal sectioning was sufficient to differentiate early steps from later steps in cellular invasion. However, Z-depth resolution in detecting invadopodia-mediated matrix degradation becomes an issue, as the sensitivity of detection is decreased with increasing thickness of the matrix, and is inherently of a lower resolution than in the *X–Y* axis. Thus, in the absence of time lapse imaging, and on matrices of variable thickness, early steps in formation of invadopodia could be potentially confused with later steps.

As cells invade into a thick layer of matrix, what becomes of the initial invadopodium relative to the leading edge of the cell moving into the matrix? Martins et al. have reported that the formation of filopodia occurs over the entire

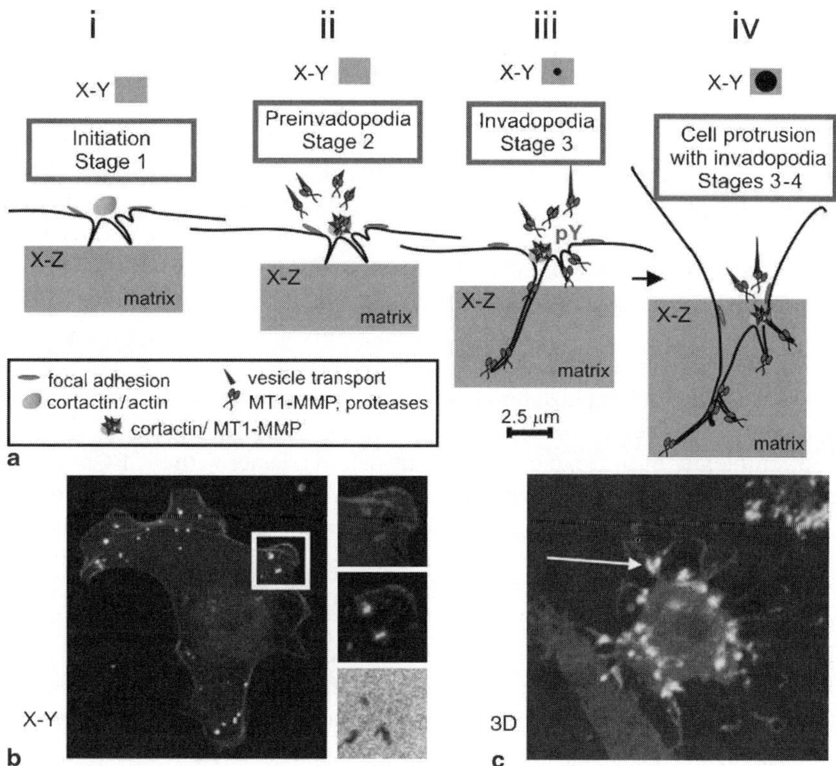

Fig. 21.1 Stages of invadopodia in 2D and 3D culture settings. **a** *i–iv* illustrate the initial formation of invadopodia beginning with cortactin recruitment (**i**), followed by recruitment of proteases such as MT1-MMP (**ii**), extension of invadopodia membrane into matrix facilitated by membrane-associated proteases and (**iii**), the enlargement of the site of degradation and invasion as the cell extends a pseudopodia/protrusion containing invadopodia and adhesion sites. **b** MDA-MB-231/c-Src(Y527F) breast cancer cells were immunostained using anticortactin 4F11 mAb (*green*) and counterstained with phalloidin (*red*) and imaged using confocal microscopy. The gelatin matrix (*grey*) contained sites of degradation and colocalizing F-actin and cortactin (*contained in insets*). **c** MDA-MB-231/c-Src(Y527F) cells were cultured on a thicker gelatin matrix whose autofluorescence is visualized together with cortactin in the red channel, DAPI in the blue channel, and antiphosphotyrosine 4G10 mAb staining in the green channel of this confocal image. Invadopodia are seen projecting into the matrix around the cell and aggregates containing colocalized cortactin and phosphotyrosine are visualized as yellow dots (*arrow*) (*See also* Color Insert I)

cell surface when endothelial cells are first introduced into a three-dimensional (3D) collagen gel culture (Martins and Kolega 2006). However, the focus of their formation increasingly localizes to the so-called peripheral zones of pseudopodial branches. Intermediate pseudopodia, the source of the pseudopodial branches, were described to be 2–5 μm in width. But, the relationship of the filopodia that formed in response to the matrix in this study and invadopodia is not clear as no markers for invadopodia such as cortactin, phosphotyrosine (pY), Tks5/FISH, or MT1-MMP were examined (*see* Table 21.1). On the contrary, this study highlights that there is a hierarchy of protrusion formation with fine filaments associated with the ends of branched pseudopodia. Thus, some of the confusion in the literature is at least partially due to questions of whether invadopodia (less than 0.5 µm) or pseudopodial branches (2–5 µm) were examined. There is a dearth of high-resolution studies utilizing a number of invadopodial and focal adhesion markers to identify pseudopodia, which might contain invadopodia together with focal or matrix adhesions. An interesting question with regard to the role of focal adhesion kinase (FAK) in invadopodia is its colocalization with Src and integrins in protrusions of v-Src transformed fibroblasts (Hauck et al. 2002): Do these structures contain cortactin and MT1-MMP aggregates in the same sites? This appears to be an unresolved question and conflicts with the situation in breast cancer cells where FAK localizes to focal adhesions but not phosphotyrosine-rich aggregates resembling invadopodia complexes (Bowden et al. 2006).

The complexity of the leading edge of the cell in contact with the ECM has also been illustrated by the ECM-degradation structure (EDS) or degrading structures described by Baldassarre et al. (2003) using electron microscopy. Their images of invading cells in 3D collagen gels reveal fine protrusions similar in size to invadopodia (Wolf et al. 2007 and references therein, Burgstaller and Gimona 2005, Furmaniak-Kazmierczak et al. 2007). The data can be interpreted collectively to suggest that invadopodia form in association with larger cellular protrusions and at the leading edge of the cell (Fig. 21.1a, *iv*). When cells are cultured on extremely thin matrixes, the insertion of the invadopodia into the matrix is thwarted by glass and perhaps, consequently, protrusion size remains limited.

On 2D substrates, invadopodia are often formed under the cell, and the resolution at the light microscope level, even using a confocal microscope, is not sufficient to visualize the localization of the membranes associated with actin/cortactin cores. Simultaneous imaging of invadopodial membranes together with the adhesive, actin-rich core is typically not reported, and is technically difficult as the membranes of invadopodia have no unique markers. However, a number of transmembrane or membrane-bound proteins have been localized to actin/cortactin cores associated in the same study with sites of localized degradation of the fluorescent matrix. These include ADAM 12 in Src-transformed NIH 3T3 cells (Abram et al. 2003), CD44 in Hela cells (Vikesaa et al. 2006), FAP/seprase (Mueller et al. 1999), uPAR (Furmaniak-Kazmierczak et al. 2007), $\alpha3\beta1$ integrin (Mueller et al. 1999), $\beta1$ integrin (Mueller and Chen 1991), and MT1-MMP (Artym et al. 2006, Furmaniak-Kazmierczak et al. 2007). In 2D studies, these all appear dot-like; however, the ultrastructure of the membranes at these sites is not known. Thus, the membrane could be an invadopodia protrusion, or, alternatively, these membrane proteins might be concentrated within vesicles associated with the invadopodia. In v-Src transformed chicken embryo fibroblasts, immunoelectron microscopy reveals the association of $\beta1$ integrins with the membrane extensions and vacuoles, again highlighting the complexity of the invadopodial structure in cells invading into 3D matrices or thick 2D substrata (Mueller and Chen 1991).

Table 21.1 Stages of invadopodia formation and function and their molecular components

Stage[a]	Component[b]	Evidence for role in invadopodia[c]	Citation
I–III	Actin	Dynamic actin is associated with invadopodia.	Baldassarre et al. 2006, Yamaguchi and Condeelis 2007
I–III	ADAM 12	Colocalized with invadopodia and sites of localized degradation in breast tumor cells and v-Src transformed fibroblasts; associated with Tsk5/FISH.	Abram et al. 2003, Courtneidge et al. 2005, Huovila et al. 2005, Seals et al. 2005
I–II	AFAP-110	See PKCα entry	Gatesman et al. 2004
None	AP-2	No evidence that AP-2 was involved with invadopodia in glioblastoma cells. siRNA targeting AP-2 blocked endocytosis but not phagocytosis.	Chuang et al. 2004
I–III	Arf6/Erk	Arf6 expression was correlated in breast cancer cell lines to invasion and gelatin degradation which could be blocked by siRNA targeting Arf6. Arf6-eGFP, but not Arf1-eGFP, localized at the degraded holes. HGF stimulated invadopodia formation in LOX and paxillin, Arf6, and holes colocalized.	D'Souza-Schorey and Chavrier 2006, Hashimoto et al. 2004, Hoover et al. 2005, Tague et al. 2004
I–III	Arp2/3	siRNA of components of Arp2/3 complex members blocked formation of invadopodia and localized matrix degradation. Components localized to invadopodia actin cores.	Yamaguchi et al. 2005
I–III	ASAP1/AMAP1	Required for cortactin/actin cores and invadopodia-mediated matrix degradation in breast cancer cells. Recruited cortactin and paxillin downstream of Arf6.	Hashimoto et al. 2006, Onodera et al. 2005
I–III	CD44/IMP1 and IMP3	In the Met-1 tumor line, CD44v3,8–10 associated with MMP-9 in invadopodial protrusions. In Hela cells, IMPs stabilized CD44 mRNA as shown by siRNA knockdown of IMP1 and 3. Knockdown blocked colocalization of actin cores and CD44 as well as matrix degradation.	Bourguignon et al. 1998, Vikesaa et al. 2006
I–III	Cdc42, RhoA	In breast cancer cells, Cdc42 is required for invadopodia and localized matrix degradation: demonstrated by siRNA knockdown. In melanoma cells, dominant negative (d.n.) Rho did	Furmaniak-Kazmierczak et al. 2007, Hai et al. 2002, Lener et al. 2006, Linder, 2007, Moreau et al. 2003, 2006, Nakahara et al.

(continued)

Table 21.1 (continued)

Stage[a]	Component[b]	Evidence for role in invadopodia[c]	Citation
		not inhibit invadopodia degradation activity, but d.n. Rac, Cdc42, or frabin, a Cdc42 effector, did inhibit. Of the latter group, activated forms increased degradation. Cdc42 promoted podosome formation in primary aortic endothelial cells (actin-dots).	2003, Tatin et al. 2006, Varon et al. 2006, Webb et al. 2005, Yamaguchi et al. 2005
I–III	Cin85	Cin85 was required together with Cbl for invadopodia-mediated matrix degradation downstream of AMAP1 (ASAP1). Cin85 colocalized at holes and with AMAP1 on gelatin films.	Nam et al. 2007
I–III	Cofilin	siRNA reduced the population of immobile, long-lived invadopodia (actin cores) and reduced the number of cells with localized degradation.	Yamaguchi et al. 2005
I–III	Cortactin	In osteoclasts, siRNA targeting did not inhibit cell differentiation on bone, but prevented formation of sealing rings and bone resorption; wild type and SH3 deleted, but not 3Y mutant cortactin reconstituted podosome formation after siRNA knockdown. In breast, head, and neck squamous carcinoma: siRNA eliminated actin cores and localized gelatin degradation; colocalization with MT1-MMP and holes was demonstrated. Cortactin was required for protease secretion and surface membrane levels of MT1-MMP. Original description of stages of invadopodia formation based on time course experiments was proposed. In A7r5 smooth muscle cells and NIH 3T3 cells, active Src and PMA induced podosomes that required cortactin; 3Y mutant was translocated to the cell surface and endogenous pY cortactin was still present at podosomes. Cortactin is required for invadopodia and matrix degradation in fibroblasts; 3Y mutant of cortactin failed to rescue cortactin siRNA depletion.	Artym et al. 2006, Clark et al. 2007, Tatin et al. 2006, Tehrani et al. 2006, Varon et al. 2006, Webb et al. 2007, Zhou et al. 2006

Table 21.1 (continued)

Stage[a]	Component[b]	Evidence for role in invadopodia[c]	Citation
I–III	Dynamin	In human melanoma cells, dynamin mutants (GTPase inactive or deltaPRD domain) blocked localized matrix degradation. Actin, cortactin, and pY colocalized with dynamin at holes.	Baldassarre et al. 2003, Buccione et al. 2004, McNiven et al. 2004
I–III	EGF	Increased invadopodia activity in breast cancer cells with elevated EGFR.	Yamaguchi et al. 2005
I–III	ERK	Erk localized to actin punctae; U0126 inhibited but Src(Y527F) induced matrix degradation in VSMC. In LOX melanoma cells, phospho-Erk localized to sites of degradation. See also Arf6/Erk.	Furmaniak-Kazmierczak et al. 2007, Tague et al. 2004
II–IV	FAP/seprase	FAP/seprase, a serine-type protease, interacted with integrins and associated with invadopodia in fibroblast, endothelial, and melanoma cells. In fibroblasts, DPPIV colocalized with FAP/seprase in invadopodia of cells embedded in 3D collagen. FAP/seprase has not been localized to actin/cortactin cores.	Artym et al. 2002, Chen and Kelly 2003, Ghersi et al. 2002, 2006, Goodman et al. 2003, Kelly, 2005, Monsky et al. 1994, Mueller et al. 1999
I–II	Glycerol-phospho-inositols	These compounds blocked localized matrix degradation in breast and melanoma cancer cells.	Buccione et al. 2005
None	Grb2	Grb2 siRNA had no effect on breast cancer cell invadopodia formation and GFP-Grb2 did not localize to invadopodia.	Yamaguchi et al. 2005
I–III	HGF	Stimulated invadopodia formation in LOX melanoma cells determined by localized degradation.	Tague et al. 2004
II–III	High pY	Colocalization of pY and cortactin or phosphocortactin in Src-activated cells identified invadopodia at sites of matrix degradation.	Bowden et al. 2006, Furmaniak-Kazmierczak et al. 2007
II–IV	MMP-2	Exogenous MMP-2 and anti-MMP-2 staining localized to sites of degradation in v-Src transformed fibroblasts. Anti-MMP-2 staining was not detected in invadopodia of VSMC, although uPAR, MMP-9, and MT1-MMP were.	Furmaniak-Kazmierczak et al. 2007, Galvez et al. 2002, Monsky et al. 1993
II–IV	MMP-9	Localization of anti-MMP-9 at Src (Y527F) induced invadopodia of VSMC.	Furmaniak-Kazmierczak et al. 2007

(continued)

Table 21.1 (continued)

Stage[a]	Component[b]	Evidence for role in invadopodia[c]	Citation
II–IV	MT1-MMP	Transfection of CHO or MCF-7 tumor cells with MT1-MMP was sufficient to induce invadopodia-mediated matrix degradation. MT1-MMP siRNA and GM6001 inhibited TGFα-induced invadopodia in primary aortic endothelial cells. MT1-MMP was required for invasion into collagen by a number of cell types. Also required for collagen degradation and phagocytosis by human gingival fibroblasts. High resolution confocal imaging colocalized MT1-MMP, F-actin, and sites of degradation in 2D and 3D cultures. Many other 3D studies have imaged integrin, MT1-MMP, and F-actin on the cell surface of invading cells. In 2D, it was present with cortactin prior to formation of sites of degradation and lingered after cortactin has disengaged in invadopodia in breast tumor cells; was required for localized degradation in melanoma cells; colocalized with pY-cortactin at sites of matrix degradation in VSMC.	Artym et al. 2006, d'Ortho et al. 1998, Furmaniak-Kazmierczak et al. 2007, Hotary et al. 2003, Itoh and Seiki 2006, Lee et al. 2006a, Nakahara et al. 1997, Sabeh et al. 2004, Strongin, 2006, Taniwaki et al. 2007, Varon et al. 2006, Wolf et al. 2007, Wolf and Friedl 2005, Yana et al. 2007
I–III	Nck1	GFP-Nck1 was localized at invadopodia of breast cancer cells. siRNA targeting Nck1 blocked actin core formation and matrix degradation and rescue with wild-type Nck1 partially restored invadopodia.	Yamaguchi et al. 2005
I–III	N-WASP	siRNA-mediated knockdown demonstrated that N-WASP was required for invadopodia formation and localized degradation in mammary carcinoma cells. Mutants of N-WASP that were unable to activate the N-WASP effector Arp2/3 complex blocked podosome formation in v-Src-transformed 3Y1 rat fibroblasts and fibronectin degradation (i.e., invadopodia).	Chen, 1989, Lorenz et al. 2004, Mizutani et al. 2002, Yamaguchi et al. 2005

Table 21.1 (continued)

Stage[a]	Component[b]	Evidence for role in invadopodia[c]	Citation
I–III	p190RhoGAP	Colocalized with actin at invadopodia and signaling through beta 1 integrin in melanoma cells resulted in p190RhoGAP tyrosine phosphorylation. Microinjected anti-p190RhoGAP antibodies inhibited localized matrix degradation.	Nakahara et al. 1998
None	P38 MAPk	The MAPk inhibitor SB203580 failed to block Src(Y527F)-induced invadopodia in VSMC.	Furmaniak-Kazmierczak et al. 2007
III–IV	p61Hck	In macrophages, p61 Hck, Src-family member tyrosine kinase, localized in lysosomes and induced podosomes when constitutively activated. In NIH3T3 cells, it induced rosettes of podosome with localized gelatin degrading activity.	Cougoule et al. 2005
I–III	PAK	Activated or open conformation of kinase dead PAK induced invadopodia and localized matrix degradation in VSMC. PAK localized to holes of degraded gelatin.	Furmaniak-Kazmierczak et al. 2007
I–III	Paxillin	Localized together with cortactin to sites of matrix degradation in breast cancer and melanoma cells.	Bowden et al. 1999, Tague et al. 2004
II–IV	PI3K	PI3K inhibitors blocked matrix degradation in melanoma cells and primary aortic endothelial cells. Wortmannin blocked hole formation and also blocked colocalization of MMP-9 with actin cores.	Nakahara et al. 2003, Redondo-Munoz et al. 2006, Varon et al. 2006
I–III	PKC	PMA-induced podosomes were similar to Src-induced podosomes in HUVEC cells and were associated with matrix degradation. Src-induced podosomes, that is, invadopodia, required MT1-MMP, PKC, Src, and Cdc42. Immunofluorescence colocalization and a time course demonstrated transient induction of podosomes after PMA treatment. The phorbol ester PDBu induced podosomes in VSMC and A7r5 cells, but not degradation in the latter.	Burgstaller and Gimona 2005, Furmaniak-Kazmierczak et al. 2007, Hai et al. 2002, Lener et al. 2006, Webb et al. 2005, Zhou et al. 2006

(continued)

Table 21.1 (continued)

Stage[a]	Component[b]	Evidence for role in invadopodia[c]	Citation
I–II	PKCα	Constitutively activated PKCα or PMA activated c-Src and podosome formation in an AFAP-110- dependent manner in ovarian carcinoma cells. In Src(Y527F)-activated fibroblasts, AFAP-110 was required for podosome formation. Degradation was not examined.	Gatesman et al. 2004
I–III	PKD1 (PKCμ)	A complex of PKD1, paxillin, and cortactin was formed in invasive breast cancer cell lines and its role in invadopodia was suggested by colocalization of its components with invadopodia at sites of degradation.	Bowden et al. 1999
II–IV	Rab8	In MDA-MB-231 breast cancer cells, MT1-MMP and Rab 8 colocalized at the cell surface. Effects of Rab8 mutants on collagen gel invasion and thick gelatin films was noted: Rab 8 inhibited and its dominant negative mutants stimulated delivery of MT1-MMP to the cell surface.	Bravo-Cordero et al. 2007
I–III	Rac	D.n. Rac1 blocked Src(Y527F)-induced invadopodia in VSMC. Rac1 is required for peripheral cortactin localization in fibroblasts. However, d.n. Rac1 had no influence on PMA-induced podosomes in HUVEC, whereas d. n. Cdc42 and d.n. RhoA blocked their formation (number of cells with podosomes).	Furmaniak-Kazmierczak et al. 2007, Tatin et al. 2006, Weed et al. 1998
I–III	Rac1/ synaptojanin	Number of glioblastoma cells degrading the matrix was reduced by siRNA targeting Rac1 or synaptojanin 2 (downstream of Rac1). Synaptojanin 2 localized to sites of degradation.	Chuang et al. 2004
I–III	Rho	In NIH 3T3 cells transfected with activated Src(Y527F), activated Rho colocalized with F-actin, cortactin, and Fish and it was necessary for podosome structure and localized degradation. However, d.n. Rho had no effect on localized degradation in melanoma cells.	Berdeaux et al. 2004, Nakahara et al. 2003

Table 21.1 (continued)

Stage[a]	Component[b]	Evidence for role in invadopodia[c]	Citation
I–III	Src	Src is required for invadopodia formation and degrading activity in breast cancer cells and fibroblasts. Constitutively, active Src(Y527F) promoted podosomes and invadopodia formation in multiple cell types.	Artym et al. 2006, Bowden et al. 2006, Chen et al. 1985, Chen, 1989, Okamura and Resh 1995, Osiak et al. 2005, Varon et al. 2006
I–III	SSeCKs	Prevented podosome formation measured by colocalization of Tks5/Fish and actin in NIH3T3 cells expressing temperature-sensitive v-Src. Matrigel invasion was inhibited.	Gelman and Gao 2006
III–IV	TGFβ, SnoN	In breast cancer cells, absence of SnoN (shRNA) promoted TGFβ induction of holes formation by invadopodia; and degradation was inhibited by GM6001. TGFβ alone increased holes formation.	Zhu et al. 2007
I–IV	TGFβ/SMAD	Primary aortic endothelial cells form rosettes of invadopodia with associated MT1-MMP and requiring Src, PI3Kinase, and RhoA GTPase signaling. Collagen invasion was correlated with invadopodia and an extensive survey of colocalizing podosome components was performed.	Varon et al. 2006
I–III	Tks5/FISH	PX domain was required for Tks5/FISH plus Src to activate localized degradation in fibroblasts and breast cancer cells. Expression of deleted PX domain Tks5 plus Src repressed actin core formation. Matrix degradation by invadopodia was diminished by siRNA targeting Tks5.	Abram et al. 2003, Courtneidge et al. 2005, Seals et al. 2005
III–IV	TNFα, VEGF	Subconfluent HUVEC form invadopodia with associated MT1-MMP localized at sites of degradation in response to cytokines. This required Src and RhoGTPase signaling.	Osiak et al. 2005
II–IV	uPAR	uPAR colocalized with actin cores and sites of localized degradation in VSMC.	Furmaniak-Kazmierczak et al. 2007
II–III	WAVE1	siRNA blocking WAVE1 did not affect actin initiation sites but blocked degradation by invadopodia in mammary carcinoma cells.	Yamaguchi et al. 2005

(continued)

Table 21.1 (continued)

Stage[a]	Component[b]	Evidence for role in invadopodia[c]	Citation
I–III	WIP	siRNA targeting WIP blocked actin core of invadopodia. WIP interaction with N-WASP but not with cortactin was required for actin core formation. Localized to invadopodia.	Yamaguchi et al. 2005
I–III	α3β1, α6β1, integrin β1	In melanoma cells, activation of α6β1 integrin by laminin peptides induced association of FAP/ seprase with integrin, and increased invadopodia-mediated degradation. Integrin, FAP/ seprase, and uPA were associated at the cell surface of melanoma cells.	Artym et al. 2002, Mueller et al. 1999, Mueller and Chen 1991, Nakahara et al. 1996, 1998
II–IV	α4β1 integrin	α4β1 integrin and CXCR4 upregulate MMP-9. Integrin signaling regulated its localization and activity in podosomes of B-cell chronic lymphocytic leukemia and HUVEC cells.	Redondo-Munoz et al. 2006
	αvβ3 and β1 integrins	Associated with MT1-MMP in ovarian carcinoma cells and endothelial cells.	Deryugina et al. 2001, Ellerbroek et al. 2001, Galvez et al. 2002
	αvβ3 integrin	Associated with MT1-MMP and MMP-2 at the cell surface.	Brooks et al. 1996, Galvez et al. 2002

[a] Stage refers to Stages I–IV defined in Artym et al. (2006) and is inferred by comparison to cortactin/F-actin and MT1-MMP colocalization, and the degree of matrix degradation at the invadopodia:

Stage I: Initiation (cortactin/actin present)
Stage II: Preinvadopodia (cortactin/actin + MT1-MMP)
Stage III: Mature invadopodia (cortactin/actin + MT1-MMP + foci of matrix degradation)
Stage IV: Late invadopodia (MT1-MMP + degraded matrix).

[b] Invadopodia-related molecule.
[c] Evidence of colocalization with ECM-degrading proteases such as MT1-MMP or localized degradation in a fluorescent gelatin cross-linked matrix.

Proteolytic Activity of Invadopodia

Invadopodia are operationally defined as possessing membrane-associated proteolytic activity. Protease activity is localized to discrete sites that can be detected by small holes formed in a fluorescent matrix beneath the cell using microscopic imaging in the presence versus the absence of protease inhibitors or siRNA knockdown of proteases (Artym et al. 2006, Bowden et al. 2001, Chen et al. 1984, Kelly et al. 1994, 1998). The rationale for this assay is that membrane association of proteases limits proteolysis to sites of membrane contact. In the assay for localized degradation by invadopodia, small holes are visualized by high resolution microscopic imaging since the holes are initially about 0.5 μm in size. The gelatin layer is either directly labeled with fluorescence or fluorescent fibronectin is bound to the unlabelled gelatin via its gelatin-binding domain. Assays are conducted in the presence of serum because, for the most part, invadopodia fail to degrade the matrix in the absence of serum factors (Mueller et al. unpublished data). Distinctive actin cores are formed in serum-free conditions in the presence but not in the absence of EGF (Yamaguchi et al. 2005). Other ECM components can be coupled to the matrix covalently and are similarly degraded by invadopodia (Kelly et al. 1994).

The term invadopodial complex has been used to define a subset of actin/cortactin-containing aggregates at the membrane that contain cortactin colocalizing with high levels of tyrosine-phosphorylated proteins (Bowden et al. 2006). This particular combination singularly identifies invadopodia that are active in matrix proteolysis since using the fluorescent-gelatin degradation assay, within individual cells, aggregates of cortactin/phosphotyrosine colocalize at sites of localized degradation (dark holes in the fluorescent layer), and comparison of cell lines reveals a strong correlation between numbers of aggregates of cortactin/phosphotyrosine and numbers of holes formed beneath cells. In addition to cortactin, others have reported a variety of invadopodia-related proteins such as Tsk5/FISH that colocalize with the center of the degradation hole and the actin core (Table 21.1).

MT1-MMP is one marker of invadopodia, particularly when it is colocalized with cortactin. It is required for invadopodia-mediated matrix degradation of cross-linked gelatin films (Artym et al. 2006), Table 21.1), it degrades collagen, it is required for invasion through collagen matrices, and it is a major enzyme involved in remodeling ECM during development and pathology (Itoh and Seiki 2006). TIMP-2, but not TIMP-1, and GM6001, or siRNA knockdown of MT1-MMP block invadopodia mediated degradation of 2D fluorescent gelatin films as well as collagen invasion in 3D systems (Artym et al. 2006, d'Ortho et al. 1998, Hotary et al. 2000, 2002, 2006; Sabeh et al. 2004, Varon et al. 2006, Wolf et al. 2007). The presence of MT1-MMP aggregates may not always be associated with proteolysis (Wolf et al. 2007), and thus its arrival at invadopodia might precede maturation and activation. However, there is also uncertainty as to when matrix proteolysis truly begins since visualization of the formation of a hole or the appearance of dequenched, proteolyzed DQ-collagen has a finite sensitivity. And, MT1-MMP is not the only enzyme that localizes to invadopodia and participates in invasion; others are present, for example, uPAR, FAP/seprase, MMP-2, MMP-9, and ADAM-12 [Table 21.1 (Linder 2007)]. The target substrates for these additional proteases and their function during invasion vis-a-vis invadopodia are not fully appreciated.

Actin/Cortactin Cores

The simple presence of cortactin/actin aggregates or "cores" at the membrane does not necessarily indicate proteolytic activity and thus, they do not identify mature invadopodia (Artym et al. 2006, Bowden et al. 2006). In the absence of proteolysis, visualization of a cortactin/actin core might occur before the invadopodia has matured and acquired MT1-MMP and proteolytic activity [Fig. 21.1 (Artym et al. 2006)]. Alternatively, cortactin/actin aggregates might represent other structures unrelated to invadopodia or with an alternate function such as endocytic vesicles (Cosen-Binker and Kapus 2006, Kaksonen et al. 2000).

Podosomes of monocytic cells and v-Src transformed fibroblasts are often defined by F-actin cores surrounded by a ring of vinculin (Buccione et al. 2004, Linder, 2007). Similarly, the actin cores of invadopodia often have a distinctive appearance as do the resulting pattern of holes formed in the fluorescent matrix (Bowden et al. 1999, 2006, Linder, 2007, Weaver 2006). In the case of v-Src transformed chicken embryo fibroblasts, these podosomes may be collections of invadopodia with matrix-degrading capability on matrix (Chen, 1989). The ring shape that contains focal adhesion proteins vinculin, α-actinin, paxillin, and tensin serves to distinguish podosomes from focal adhesions which have a linear, elliptical shape. The presence of the vinculin ring has been reported in podosomes of transformed fibroblasts, smooth muscle cells, endothelial cells, and activated blood cells such as macrophages (Linder and Aepfelbacher 2003). Vinculin associated with podosomes can take several forms, a simple ring around each individual podosome (Marchisio et al. 1987) or associated with clusters of podosomes to produce a ribbon-like pattern of vinculin staining (Schliwa et al. 1984).

However, vinculin rings appear to be absent in breast cancer cells forming invadopodia (Artym et al. 2006) and eosinophils (Johansson et al. 2004). Since vinculin is associated with adhesions, the presence or absence of vinculin surrounding actin cores might relate to the diversity of integrin-mediated adhesions in different cell types or on different matrices. For example, a potential collaboration between $\alpha5\beta1$ integrin and $\alpha3\beta1$ integrin in melanoma cells was suggested by the close proximity of their staining patterns, $\alpha3\beta1$ integrin colocalized with the hole at the site of degradation and $\alpha5\beta1$ integrin in areas surrounding the site of degradation (Mueller et al. 1999). The $\alpha5\beta1$ integrin adhesions were suggested to be focal adhesion-like.

Ultimately, high resolution imaging and related techniques must be used to sort out the molecular associations and structures that comprise the invadopodia and closely associated adhesive sites at the leading edge of invasive cells; this is a prerequisite to determining the mechanisms of cellular invasion. However, a ring of vinculin, although it might be diagnostic for podosomes in some cell types, is not a useful identifier for tumor cell invadopodia.

Stages of Invadopodia Formation and Acquisition of Proteolytic activity

The formation of invadopodia can be elucidated using time lapse imaging and multicolor immunofluorescence to identify the actin core components or proteases together with the functional proteolytic activity of mature invadopodia (Artym et al.

2006, Lorenz et al. 2004). Thus, de novo formation of invadopodia can be studied using confocal microscopy of samples fixed at increasing times after initially seeding cells on a fluorescent matrix (gelatin). Pairs of invadopodial components including cortactin, pY, Tks5/FISH, and MT1-MMP together with phalloidin staining to detect F-actin and fluorescent matrices to detect localized degradation can be used to determine the relative prevalence of each with time after seeding cells. Invadopodia begin forming within 30 min after seeding cells on the fluorescent matrix, depending on the cell type. The preponderance of cortactin/actin cores, cortactin/MT1-MMP, or cortactin/MT1-MMP/holes with time after seeding cells was used to infer the existence of various stages of invadopodia formation (Artym et al. 2006).

Using the more accurate live cell, time lapse imaging approach, Artym et al. (2006) found that the time span between the arrival of cortactin at invadopodia initiation sites and the arrival of MT1-MMP and detection of the hole forming in the matrix is on the order of minutes. Relative to cortactin, an actin-binding protein, MT1-MMP, an ECM-degrading enzyme, and detection of localized degradation of the fluorescent matrix, four stages of invadopodia formation were characterized: initiation, preinvadopodia, mature invadopodia, and late invadopodia (Artym et al. 2006). Initiation (stage I in this model) is marked by the appearance of cortactin aggregates at the membrane. Formation of preinvadopodia is indicated by the appearance of MT1-MMP aggregates colocalizing with the cortactin aggregates (stage II). Maturation of invadopodia begins with the appearance of a hole in the fluorescent matrix just under the aggregates of cortactin and MT1-MMP (stage III), and full maturation of invadopodia is signaled by the disappearance of cortactin from the site leaving the MT1-MMP aggregate behind (stage IV). Figure 21.2 illustrates these stages in a time line of invadopodia formation.

Studies which have connected the presence and signaling pathways of invadopodia or podosomes to the activity of localized degradation of a gelatin matrix are listed in Table 21.1. The original description of these stages relied on the knowledge that at a given cortactin/actin core appearing at the cell membrane, a hole or site of localized degradation would or would not eventually form underneath it. Thus, each stage has a specific criterion for membership including the eventual formation of a hole, and molecular membership of invadopodia components. Invadopodia proteins were hypothetically assigned to stages based on whether cortactin/actin, MT1-MMP, or proteolytic degradation were simultaneously detected in the study (Table 21.1).

Invadopodia dynamics can be followed by measuring the time that elapses between the appearances of each stage (Artym et al. 2006). The delays between cortactin accumulation, MT1-MMP accumulation, and initiation of matrix degradation were different for migrating versus nonmigrating breast cancer cells suggesting cross talk during cell migration and membrane-associated matrix degradation (Artym et al. 2006). Invadopodia are also dynamic from a molecular point of view. Baldassarre et al. (2006) reported actin comets associated with sites of degradation, and perhaps corresponding to this activity, other investigators have shown highly dynamic exchange of actin at podosomes and invadopodia (Destaing et al. 2003).

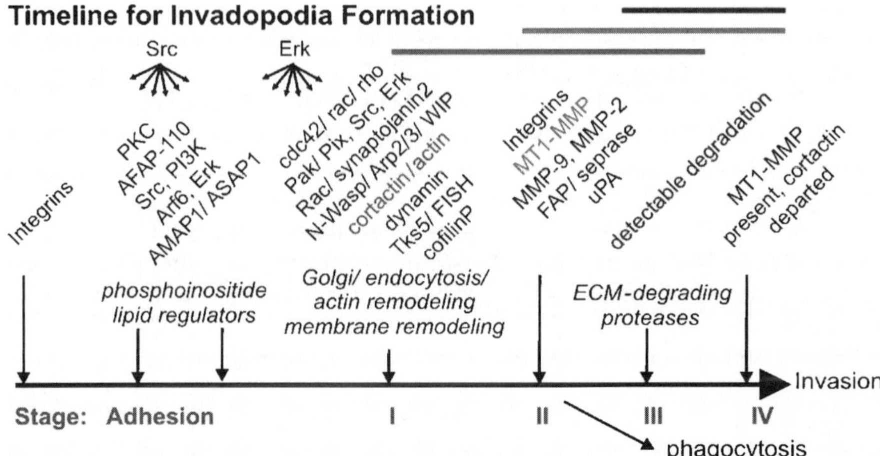

Fig. 21.2 Time line for invadopodia formation. The stages of invadopodia are presented along a time line and prominent invadopodia-associated proteins are hypothetically assigned to positions along the time line according to studies colocalizing them with cortactin/actin, MT1-MMP, or gelatin degradation. In color, the template for the time line is illustrated by cortactin/actin (*red*), MT1-MMP (*green*), and detectable degradation (*blue*) (*See also* Color Insert I)

Cortactin was associated with retrograde actin flow in NIH 3T3 cells and vesicles appear to be propelled by comets of actin and cortactin (Kaksonen et al. 2000). Yamaguchi et al. have characterized invadopodia visualized using GFP-actin. Actin cores appeared to be motile over the membrane and their motility was inversely related to the intensity, that is, actin content (Yamaguchi et al. 2005). They found that the longer lived cores tended to stay in one spot and were even left behind as cells migrated away.

In summary, invadopodia, like podosomes, are highly dynamic as demonstrated using time lapse imaging (*see* Artym et al. 2006, Ayala et al. 2006, Baldassarre et al. 2003, Evans et al. 2003, Linder, 2007, Linder and Aepfelbacher 2003, Yamaguchi et al. 2005 for discussion and review).

Relating Functional Components to Stages of Invadopodia

In reviewing published studies, those invadopodia/podosomes that were described by characterization of the presence or absence of actin/cortactin cores before and after drug treatment, mutant protein expression, or siRNA knockdown, but containing no investigation of localized matrix degradation, could tentatively be termed Stage I invadopodia (cortactin/actin localization, no MT1-MMP, no hole formed). However, if the study correlated the presence or absence of cortactin/actin with MT1-MMP, a tentative assignment to Stages II–III might also be made. Similarly, if presence or absence of holes was simultaneously determined, then the drug, siRNA, or mutant protein effect could be attributed to effects on Stages I–IV.

Finally, if cortactin/actin, MT1-MMP, and holes were simultaneously investigated, then the outcome of each experiment could be understood more fully in terms of the time course of invadopodia formation and maturation that was characterized in breast cancer cell lines. Table 21.1 represents the results of a review conducted in this fashion. The molecular classes of invadopodia components have been integrated into the time course of invadopodia formation based on the molecular identification and functional studies associated with each report in the literature (Figs. 21.1 and 21.2, Table 21.1). For comparison some other molecules for which studies suggest a similar assignment are depicted in Fig. 21.2.

Invadopodia can have long- or short-term life spans, from minutes to hours (Yamaguchi et al. 2005). Since only GFP-actin was monitored, we presume these invadopodia to be members of a stage between I and III (Fig. 21.2). This was determined by comparison with the Src-activated MDA-MB-231 cells in which these stages were originally defined. In the latter case, presence of cortactin aggregates was tightly correlated with presence of F-actin aggregates suggesting their interdependency (Artym et al. 2006). Yamaguchi et al. (2005) suggest that invadopodia arise and operate in three stages: initiation, searching, and maturation. Time lapse imaging indicates that the short-lived invadopodia may be a precursor to the long-lived, immobile ones and that these two groups could be, respectively, thought of as a "searching" and "maturation" phase. The difference between the two relates to changes in strength of adhesion to the matrix. One possibility is that searching and maturation are equivalent to Stages I–III (Stage IV lacks cortactin/ phalloidin-bound F-actin), where the difference in adhesion relates to the arrival of MT1-MMP and other associated proteins. Exploring further the dynamics of the EGF-induced invadopodia, they found that the lifetimes of the invadopodia were affected by cofilin knockdown (Yamaguchi et al. 2005) with the consequence that the longer lived population of invadopodia was lost. Consequently, degradation was reduced, although not eliminated. This suggests that, depending on the cell type, invadopodia lifetimes may reflect differences in prominence of signaling pathways, particularly cofilin-signaling pathways. And, it suggests that long-lived invadopodia might be equivalent to Stage III and contributing most heavily to matrix degradation, whereas short-lived invadopodia are equivalent to Stages I and II.

Invadopodia and Their Relationship with Pseudopodia and Cellular Protrusions in 3D

Returning to the discussion of invadopodia as forming on the tips of pseudopodia or pseudopodial branches, examination of cortactin and MT1-MMP localization together with protease activity in 3D cultures is the next logical step to relating 2D studies to in vivo formation and function of invadopodia. In thicker matrices with high rigidity, such as cross-linked gelatin layers, it appears that the cell migrates to "follow" the initial invadopodia membrane that has penetrated the matrix in advance, engulfing partially degraded matrix (Fig. 21.1, compare a, *iii* and a, *iv*) (Baldassarre et al. 2003,

Coopman et al. 1996, Lorenz et al. 2004, Mueller and Chen 1991). However, cortactin and MT1-MMP have not been colocalized in these settings.

However, in studies of tumor cells invading through a 3D collagen lattice, MT1-MMP appears punctate over much of the cell surface, suggesting that formation of pseudopodia and their branch formation was minimal (Wolf et al. 2007). However, under these conditions, cells formed linear tracks of advancement, with cell–to–cell interactions occurring along their lengths. Evidence of protease-mediated collagen degradation was obtained for a region of the leading cell subtending the leading edge and thus probably associated with only a subset of the MT1-MMP punctae (Wolf et al. 2007). It is interesting to compare this 3D study with that of the 2D study of Artym et al. (2006) since hypothetically, the MT1-MMP punctae at the leading edge of the cell in the collagen lattice might represent the newest sites of invadopodia formation, that is, Stage II (Fig. 21.2a, *ii*), and those more rearward in the cell might represent invadopodia that have acquired proteolytic activity, that is, Stage III (Fig. 21.2a, *iii*), and then later maintained proteolytic activity but lost cortactin, that is, Stage IV (Fig. 21.2a, *iv*). Confirmation of this could be obtained by staining both cortactin and MT1-MMP.

Functional Consequence of Invadopodia-Mediated Degradation

Proteases Mediate Matrix Degradation as well as Stimulate Cross-Talk Between Cells and the Extracellular Environment

Matrix degradation is perhaps the best known function of invadopodia. Indeed they are distinguished from focal adhesions by their matrix-degrading capability and their more transient existence (Chen et al. 1984, Chen 1989). Invadopodia are capable of degrading a wide range of ECM molecules including fibronectin, laminin, type-I collagen, as well as glutaraldehyde cross-linked gelatin (Kelly et al. 1994, Mueller and Chen 1991). It is evident that increased activities of a variety of proteases must occur at invadopodia. Early work indicated that a 170 kDa serine protease now known as seprase or fibroblast activation protein-α (FAP) and MMP-2 were associated with invadopodia (Aoyama and Chen 1990, Emonard et al. 1992, Monsky et al. 1993). There are now at least eight proteases known to localize to invadopodia (Chen et al. 2003, Ghersi et al. 2003, Linder 2007, Weaver 2006). Importantly, concentration to invadopodia is critical for the heightened proteolytic activity. Integrins are key players in ligating proteases to invadopodia. For example, FAP association with $\alpha_3\beta_1$ integrin that is stimulated by collagen is well documented (Mueller et al. 1999) as is the increased surface localization and concentration to invadopodia of FAP when $\alpha_6\beta_1$ integrin is ligated with laminin peptides (Nakahara et al. 1996). In addition to binding $\alpha_3\beta_1$ integrin, FAP also ligates other proteases to invadopodia through its ability to form larger complexes with the related serine protease dipeptidyl peptidase IV (DPPIV) (Ghersi et al. 2002, 2006). The close association of FAP with urokinase-type plasminogen activator receptor (uPAR) in the plasma membrane of aggressive melanoma cells is also mediated by β_1 integrins (Artym et al. 2002). Indeed, $\alpha_3\beta_1$ integrin that is known to bind to FAP (Mueller et al. 1999) also binds to uPAR (Zhang et al. 2003). Thus, FAP, DPPIV, and uPAR are organized in invadopodia through binding interactions that hinge around $\alpha_3\beta_1$ integrin and FAP.

MT1-MMP forms functionally important complexes with MMP-2 at invadopodia (Nakahara et al. 1997) that likely cooperate with FAP–DPPIV–uPAR–integrin complexes to efficiently degrade ECM (Kelly 2005). The active form of MMP-2 was the first MMP shown to localize to invadopodia (Emonard et al. 1992, Monsky et al. 1993). More recently, MMP-9 has been identified in podosomes of B-chronic lymphocytic leukemia cells and to have an important role in transmigration through endothelial cell barriers (Redondo-Munoz et al. 2006). MMP-mediated proteolysis is critical for the matrix degrading and invasive functions of invadopodia (Kelly et al. 1998). The molecule that binds MT1-MMP to invadopodia is not known, but the linkage involves the transmembrane domain of MT1-MMP (Nakahara et al. 1997) and the intact invadopodia cytoskeleton that is formed only when cortactin is present (Artym et al. 2006). Evidence has suggested a possible interaction between $\alpha_v\beta_3$ integrin and the MT1-MMP, MMP-2 complexes (Brooks et al. 1996, Deryugina et al. 2001). Not only are integrins involved in localizing MMPs, but they are also often involved in upregulating their expression (e.g., Lochter et al. 1999).

Together FAP/seprase, DPPIV, uPA, MT1-MMP, MMP-2, and MMP-9 account for six of the eight proteases known to reside in invadopodia. The other two are a disintegrin and metalloprotease-12 (ADAM-12) that has been found in the podosomes of Src-transformed cells (Abram et al. 2003) and a distinct type of metalloproteinase called invadolysin initially discovered in *Drosophila melanogaster* mutants that produced aberrant mitotic spindles (McHugh et al. 2004). The roles of these enzymes in invadopodia are not known but they may contribute to other functions of invadopodia such as modification of chemokines and cytokines.

These eight proteases that localize to invadopodia may not reflect the entire armament of invadopodial proteases because there may be others that have not been investigated. However, between them they can degrade most ECM molecules. Moreover, uPAR presumably concentrates uPA to invadopodia where it can activate plasmin in extracellular fluids. Plasmin degrades multiple ECM substrates and activates MMPs that degrade a broad spectrum of matrix molecules. Thus, invadopodia are equipped with all the proteases needed to break down matrix barriers and promote cell invasion. But, eroding matrix barriers is only one function of the proteases at invadopodia. They also participate in the ongoing "conversation" between motile cells, the ECM, and the cells present in the microenvironment.

Invadopodial Proteases and the Tumor Microenvironment

The proteolytic activity of DPPIV is known to have roles in blood glucose regulation, leukocyte migration, and angiogenesis through cleavage of glucagon-like peptide and glucose-dependent insulinotropic polypeptide; CXCL12 and CCL22 chemokines; and neuropeptide-Y (Chen et al. 2003). In this way, invadopodial proteases can serve as cellular traffic directors modulating the response of cells surrounding invading cells that express invadopodia.

FAP appears to have multiple roles in tumor—host interactions. For example, overexpression of FAP on tumor cells stimulates angiogenesis and rapid tumor growth in an animal model of human breast cancer (Huang et al. 2004). Moreover, FAP was identified in a homology cloning approach as important for capillary morphogenesis and angiogenesis of microvascular endothelial cells (Aimes et al. 2003). FAP–DPPIV complexes are found on microvascular endothelial cells and their action is required for endothelial cell invasion of collagen (Ghersi et al. 2006). Thus, FAP apparently has a direct role in communications between tumor cells and the extracellular environment to stimulate the angiogenesis that sustains rapid tumor growth. The pro-angiogenic function of FAP may be enhanced by MMP-9. MMP-9 has been implicated in proangiogenic functions through release of matrix-bound vascular endothelial growth factor (Bergers et al. 2000, Vu et al. 1998). However, MMP-9 also has antiangiogenic functions through the production of tumstatin, an antiangiogenic fragment derived from MMP-9 cleavage of type-IV collagen (Hamano et al. 2003). More evidence for potential cooperation between FAP and MMP-9 comes from the recent finding that FAP and MMP-9 produce distinct but comparable low-molecular weight fragments of denatured type-I collagen (Christiansen et al. 2007). Moreover, new work has shown that recombinant and soluble FAP activity can be increased up to sevenfold by truncation of FAP by EDTA-sensitive proteases (Chen et al. 2006). Indeed, this elevated activity of FAP is seen in tumors of ovarian cancer (Chen et al. 2006) and breast cancer (Kelly 1999). While the identities of the EDTA-sensitive proteases that mediate the hyperactivation of FAP are unknown, it is tempting to speculate that they are MMPs such as MMP-9. Although recombinant soluble FAP was used to detect the phenomenon of hyperactivation of FAP, there is now evidence that a soluble form of FAP occurs naturally in blood (Lee et al. 2006b).

McKee's group has identified a soluble form of FAP in the plasma as the α_2-antiplasmin cleaving enzyme (APCE) (Lee et al. 2006b). FAP cleaves the N-terminal 12 amino acids from α_2-antiplasmin, rendering it 13 times more capable of penetrating fibrin aggregates (Lee et al. 2006b). α_2-antiplasmin inhibits plasmin, the major fibrinolytic enzyme in serum, making it possible that overexpression of FAP observed in melanoma, breast, and many epithelial cancers could directly contribute to stabilizing fibrin, making fibrin microclots and microemboli more likely in these cancer patients.

In summary, invadopodial proteases are complexed together on the surface of the plasma membrane. These complexes are both adhesive and lytic. They are sites of cell signaling because integrins are included in the complexes but they are also sites where cells proteolytically modify the matrix. These modifications not only facilitate migration of cells through matrix barriers, they are also avenues for cell–cell communication. Through the actions of invadopodial proteases, growth and angiogenic factors are released from the matrix, cytokine functions are altered, and bioactive fragments of matrix molecules are produced. These evoke biological responses from surrounding cells. In the case of malignant tumors the result of these responses is the angiogenesis that fuels rapid tumor growth. Thus, the invadopodial proteases, particularly FAP and DPPIV, are appealing targets for therapeutic interventions (Kelly 2005). Future work will determine the roles of proteolytic activities at invadopodia and those functions mediated by complex formation and subsequent signaling.

Various reports suggest that invadopodia-like protrusions might also be involved in a variety of processes, for example, diapedesis (Carman et al. 2007), eosinophils adhering to VCAM-1 via podosomes (Johansson et al. 2004), anchor cell invasion through basement membrane in *Caenorhabditis* elegans (Sherwood et al. 2005), and migration of border cells during Drosophila development (Fulga and Rorth 2002). In one of these cases, degradation of the cell–cell adhesion molecule VCAM-1 required the podosome-associated disintegrin metalloprotease ADAM-8 (Johansson et al. 2004). Thus, invadopodia might serve other functions in addition to ECM degradation and might participate in a wide array of developmental and normal physiologic processes in addition to tumor cell invasion. Furthermore, invadopodia might possess more than passing resemblance to growth cone filopodia, for example, and opportunities for direct comparison could shed light on the variety of functions that these most interesting cellular structures mediate.

Conclusions and Future Perspective

Moving forward, clarification of the relationship between podosomes and invadopodia will require careful comparisons between cell types, between cells on defined matrices, and using multiple imaging approaches including time lapse imaging. Previously, studies of podosomes were conducted on glass substrata, whereas studies of invadopodia were performed on matrix-coated substrata, by necessity, in order to detect localized matrix degradation. The ultimate challenge, in the future, will be to detect and study the role of invadopodia in living tissues and to identify their association and participation in pathological cell invasion in vivo. The study of podosomes and invadopodia using advanced imaging techniques in model systems, and within 3D settings in vitro and in vivo is just taking off and is facilitated by the advent of new high-resolution imaging capabilities including multiphoton imaging, multispectral imaging, fluorescence detection of protease activity in vivo, second harmonic visualization of collagen fibers, antibodies to detect cleaved collagen, fluorescence resonance energy transfer (FRET), fluorescence recovery after photobleaching (FRAP), and the development of siRNA-knockdown techniques to catalogue required molecular components, structures, molecular interactions, and functional consequences of invadopodia formation [(Artym et al. 2006, Destaing et al. 2003, Evans et al. 2003, Lorenz et al. 2004, Wolf et al. 2007) and references therein].

Thus, major questions in invadopodia research include determining the functional relationship between podosomes and invadopodia, understanding structural and functional variations from cell type to cell type; identifying exact placement of each invadopodia-related molecule in the time course of invadopodia and then determining their functional roles and molecular interdependencies. Ultimately, the goal is to determine the in vivo relevance of invadopodia and podosomes for disease progression, and to discover how they might be therapeutically targeted to control pathological cell invasion during tumor formation and metastasis.

References

Abram C.L., Seals D.F., Pass I., et al. (2003). The adaptor protein fish associates with members of the ADAMs family and localizes to podosomes of Src-transformed cells. J. Biol. Chem. 278, 16844–16851.

Aimes R.T., Zijlstra A., Hooper J.D., et al. (2003). Endothelial cell serine proteases expressed during vascular morphogenesis and angiogenesis. Thromb. Haemost. 89, 561–572.

Aoyama A. and Chen W.T. (1990). A 170-kDa membrane-bound protease is associated with the expression of invasiveness by human malignant melanoma cells. Proc. Natl. Acad. Sci. USA. 87, 8296–8300.

Artym V.V., Kindzelskii A.L., Chen W.T., et al. (2002). Molecular proximity of seprase and the urokinase-type plasminogen activator receptor on malignant melanoma cell membranes: dependence on beta1 integrins and the cytoskeleton. Carcinogenesis 23, 1593–1601.

Artym V.V., Zhang Y., Seillier-Moiseiwitsch F., et al. (2006). Dynamic interactions of cortactin and membrane type 1 matrix metalloproteinase at invadopodia: defining the stages of invadopodia formation and function. Cancer Res. 66, 3034–3043.

Ayala I., Baldassarre M., Caldieri G., et al. (2006). Invadopodia: a guided tour. Eur. J. Cell Biol. 85, 159–164.

Baldassarre M., Ayala I., Beznoussenko G., et al. (2006). Actin dynamics at sites of extracellular matrix degradation. Eur. J. Cell Biol. 85, 1217–1231.

Baldassarre M., Pompeo A., Beznoussenko G., et al. (2003). Dynamin participates in focal extracellular matrix degradation by invasive cells. Mol. Biol. Cell. 14, 1074–1084.

Berdeaux R.L., Diaz B., Kim L., et al. (2004). Active Rho is localized to podosomes induced by oncogenic Src and is required for their assembly and function. J. Cell Biol. 166, 317–323.

Bergers G., Brekken R., McMahon G., et al. (2000). Matrix metalloproteinase-9 triggers the angiogenic switch during carcinogenesis. Nat. Cell Biol. 2, 737–744.

Bourguignon L.Y., Gunja-Smith Z., Iida N., et al. (1998). CD44v(3,8–10) is involved in cytoskeleton-mediated tumor cell migration and matrix metalloproteinase (MMP-9) association in metastatic breast cancer cells. J. Cell. Physiol. 176, 206–215.

Bowden E.T., Barth M., Thomas D., et al. (1999). An invasion-related complex of cortactin, paxillin and PKCmu associates with invadopodia at sites of extracellular matrix degradation. Oncogene 18, 4440–4449.

Bowden E.T., Coopman P.J., and Mueller S.C. (2001). Invadopodia: unique methods for measurement of extracellular matrix degradation in vitro. Methods Cell Biol. 63, 613–627.

Bowden E.T., Onikoyi E., Slack R., et al. (2006). Co-localization of cortactin and phosphotyrosine identifies active invadopodia in human breast cancer cells. Exp. Cell Res. 312, 1240–1253.

Bravo-Cordero J.J., Marrero-Diaz R., Megias D., et al. (2007). MT1-MMP proinvasive activity is regulated by a novel Rab8-dependent exocytic pathway. EMBO J. 26, 1499–1510.

Brooks P.C., Strömblad S., Sanders L.C., et al. (1996). Localization of matrix metalloproteinase MMP-2 to the surface of invasive cells by interaction with integrin avb3. Cell 85, 683–693.

Buccione R., Baldassarre M., Trapani V., et al. (2005). Glycerophosphoinositols inhibit the ability of tumour cells to invade the extracellular matrix. Eur. J. Cancer 41, 470–476.

Buccione R., Orth J.D., and McNiven M.A. (2004). Foot and mouth: podosomes, invadopodia and circular dorsal ruffles. Nat. Rev. Mol. Cell Biol. 5, 647–657.

Burgstaller G. and Gimona M. (2005). Podosome-mediated matrix resorption and cell motility in vascular smooth muscle cells. Am. J. Physiol. Heart Circ. Physiol. 288, H3001–H3005.

Calle Y., Chou H.C., Thrasher A.J., et al. (2004). Wiskott-Aldrich syndrome protein and the cytoskeletal dynamics of dendritic cells. J. Pathol. 204, 460–469.

Carman C.V., Sage P.T., Sciuto T.E., et al. (2007). Transcellular diapedesis is initiated by invasive podosomes. Immunity 26, 784–797.

Chen D., Kennedy A., Wang J.Y., et al. (2006). Activation of EDTA-resistant gelatinases in malignant human tumors. Cancer Res. 66, 9977–9985.

Chen W.T. (1990). Transmembrane interactions at cell adhesion and invasion sites. Cell Differ. Dev. 32, 329–335.

Chen W.T. and Kelly T. (2003). Seprase complexes in cellular invasiveness. Cancer Metastasis Rev. 22, 259–269.

Chen W.T., Kelly T., and Ghersi G. (2003). DPPIV, seprase, and related serine peptidases in multiple cellular functions. Curr. Top. Dev. Biol. 54, 207–232.

Chen W.-T. (1989). Proteolytic activity of specialized surface protrusions formed at rosette contact sites of transformed cells. J. Exp. Zool. 251, 167–185.

Chen W.-T. (1996). Proteases associated with invadopodia, and their role in degradation of extracellular matrix. Enzyme Protein 49, 59–71.

Chen W.-T., Chen J.M., Parsons S.J., et al. (1985). Local degradation of fibronectin at sites of expression of the transforming gene product pp60src. Nature 316, 156–158.

Chen W.-T., Lee C.C., Goldstein L., et al. (1994). Membrane proteases as potential diagnostic and therapeutic targets for breast malignancy. Breast Cancer Res. Treat. 31, 217–226.

Chen W.-T., Olden K., Bernard B.A., et al. (1984). Expression of transformation-associated protease(s) that degrade fibronectin at cell contact sites. J. Cell Biol. 98, 1546–1555.

Christiansen V.J., Jackson K.W., Lee K.N., et al. (2007). Effect of fibroblast activation protein and alpha2-antiplasmin cleaving enzyme on collagen types I, III, and IV. Arch. Biochem. Biophys. 457, 177–186.

Chuang Y.Y., Tran N.L., Rusk N., et al. (2004). Role of synaptojanin 2 in glioma cell migration and invasion. Cancer Res. 64, 8271–8275.

Clark E.S., Whigham A.S., Yarbrough W.G., et al. (2007). Cortactin is an essential regulator of matrix metalloproteinase secretion and extracellular matrix degradation in invadopodia. Cancer Res. 67, 4227–4235.

Coopman P.J., Thomas D.M., Gehlsen K.R., et al. (1996). Integrin a3b1 participates in the phagocytosis of extracellular matrix molecules by human breast cancer cells. Mol. Biol. Cell 7, 1789–1804.

Cosen-Binker L.I. and Kapus A. (2006). Cortactin: the gray eminence of the cytoskeleton. Physiology (Bethesda.) 21, 352–361.

Cougoule C., Carreno S., Castandet J., et al. (2005). Activation of the lysosome-associated p61Hck isoform triggers the biogenesis of podosomes. Traffic 6, 682–694.

Courtneidge S.A., Azucena E.F., Jr., Pass I., et al. (2005). The Src substrate Tks5, podosomes (Invadopodia), and cancer cell invasion. Cold Spring Harb. Symp. Quant. Biol. 70, 1–6.

Cukierman E., Pankov R., Stevens D.R., et al. (2001). Taking cell-matrix adhesions to the third dimension. Science 294, 1708–1712.

d'Ortho M.P., Stanton H., Butler M., et al. (1998). MT1-MMP on the cell surface causes focal degradation of gelatin films. FEBS Lett. 421, 159–164.

D'Souza-Schorey C. and Chavrier P. (2006). ARF proteins: roles in membrane traffic and beyond. Nat. Rev. Mol. Cell Biol. 7, 347–358.

David-Pfeuty T. and Singer S.J. (1980). Altered distributions of the cytoskeletal proteins vinculin and alpha-actinin in cultured fibroblasts transformed by Rous sarcoma virus. Proc. Natl. Acad. Sci. USA. 77, 6687–6691.

Deryugina E.I., Ratnikov B., Monosov E., et al. (2001). MT1-MMP initiates activation of pro-MMP-2 and integrin alphavbeta3 promotes maturation of MMP-2 in breast carcinoma cells. Exp. Cell Res. 263, 209–223.

Destaing O., Saltel F., Geminard J.C., et al. (2003). Podosomes display actin turnover and dynamic self-organization in osteoclasts expressing actin-green fluorescent protein. Mol. Biol. Cell 14, 407–416.

Ellerbroek S.M., Wu Y.I., Overall C.M., et al. (2001). Functional interplay between type I collagen and cell surface matrix metalloproteinase activity. J. Biol. Chem. 276, 24833–24842.

Emonard H.P., Remacle A.G., Noël A.C., et al. (1992). Tumor cell surface-associated binding site for the Mr 72,000 type IV collagenase. Cancer Res. 52, 5845–5848.

Evans J.G., Correia I., Krasavina O., et al. (2003). Macrophage podosomes assemble at the leading lamella by growth and fragmentation. J. Cell Biol. 161, 697–705.

Fulga T.A. and Rorth P. (2002). Invasive cell migration is initiated by guided growth of long cellular extensions. Nat. Cell Biol. 4, 715–719.

Furmaniak-Kazmierczak E., Crawley S.W., Carter R.L., et al. (2007). Formation of extracellular matrix-digesting invadopodia by primary aortic smooth muscle cells. Circ. Res. 100, 1328–1336.

Galvez G., Matias-Roman S., Yanez-Mo M., et al. (2002). ECM regulates MT1-MMP localization with beta1 or alphavbeta3 integrins at distinct cell compartments modulating its internalization and activity on human endothelial cells. J. Cell Biol. 159, 509–521.

Gatesman A., Walker V.G., Baisden J.M., et al. (2004). Protein kinase Calpha activates c-Src and induces podosome formation via AFAP-110. Mol. Cell. Biol. 24, 7578–7597.

Gavazzi I., Nermut M.V., and Marchisio P.C. (1989). Ultrastructure and gold-immunolabelling of cell-substratum adhesions (podosomes) in RSV-transformed BHK cells. J. Cell Sci. 94, 85–99.

Gelman I.H. and Gao L. (2006). SSeCKS/Gravin/AKAP12 metastasis suppressor inhibits podosome formation via RhoA- and Cdc42-dependent pathways. Mol. Cancer Res. 4, 151–158.

Ghersi G., Dong H., Goldstein L.A., et al. (2002). Regulation of fibroblast migration on collagenous matrix by a cell surface peptidase complex. J. Biol. Chem. 277, 29231–29241.

Ghersi G., Dong H., Goldstein L.A., et al. (2003). Seprase-DPPIV association and prolyl peptidase and gelatinase activities of the protease complex. Adv. Exp. Med. Biol. 524, 87–94.

Ghersi G., Zhao Q., Salamone M., et al. (2006). The protease complex consisting of dipeptidyl peptidase IV and seprase plays a role in the migration and invasion of human endothelial cells in collagenous matrices. Cancer Res. 66, 4652–4661.

Gimona M. and Buccione R. (2006). Adhesions that mediate invasion. Int. J. Biochem. Cell Biol. 38, 1875–1892.

Goodman J.D., Rozypal T.L., and Kelly T. (2003). Seprase, a membrane-bound protease, alleviates the serum growth requirement of human breast cancer cells. Clin. Exp. Metastasis 20, 459–470.

Hai C.M., Hahne P., Harrington E.O., et al. (2002). Conventional protein kinase C mediates phorbol-dibutyrate-induced cytoskeletal remodeling in A7r5 smooth muscle cells. Exp. Cell Res. 280, 64–74.

Hamano Y., Zeisberg M., Sugimoto H., et al. (2003). Physiological levels of tumstatin, a fragment of collagen IV alpha3 chain, are generated by MMP-9 proteolysis and suppress angiogenesis via alphaV beta3 integrin. Cancer Cell 3, 589–601.

Hashimoto S., Hirose M., Hashimoto A., et al. (2006). Targeting AMAP1 and cortactin binding bearing an atypical src homology 3/proline interface for prevention of breast cancer invasion and metastasis. Proc. Natl. Acad. Sci. USA 103, 7036–7041.

Hashimoto S., Onodera Y., Hashimoto A., et al. (2004). Requirement for Arf6 in breast cancer invasive activities. Proc. Natl. Acad. Sci. USA 101, 6647–6652.

Hauck C.R., Hsia D.A., Ilic D., et al. (2002). v-Src SH3-enhanced interaction with focal adhesion kinase at beta 1 integrin-containing invadopodia promotes cell invasion. J. Biol. Chem. 277, 12487–12490.

Hiratsuka S., Watanabe A., Aburatani H., et al. (2006). Tumour-mediated upregulation of chemoattractants and recruitment of myeloid cells predetermines lung metastasis. Nat. Cell Biol. 8, 1369–1375.

Hoover H., Muralidharan-Chari V., Tague S., et al. (2005). Investigating the role of ADP-ribosylation factor 6 in tumor cell invasion and extracellular signal-regulated kinase activation. Methods Enzymol. 404, 134–147.

Hotary K., Allen E., Punturieri A., et al. (2000). Regulation of cell invasion and morphogenesis in a three-dimensional type I collagen matrix by membrane-type matrix metalloproteinases 1, 2, and 3. J. Cell Biol. 149, 1309–1323.

Hotary K., Li X.Y., Allen E., et al. (2006). A cancer cell metalloprotease triad regulates the basement membrane transmigration program. Genes Dev. 20, 2673–2686.

Hotary K.B., Allen E.D., Brooks P.C., et al. (2003). Membrane type I matrix metalloproteinase usurps tumor growth control imposed by the three-dimensional extracellular matrix. Cell 114, 33–45.

Hotary K.B., Yana I., Sabeh F., et al. (2002). Matrix metalloproteinases (MMPs) regulate fibrin-invasive activity via MT1-MMP-dependent and -independent processes. J. Exp. Med. 195, 295–308.

Huang Y., Wang S., and Kelly T. (2004). Seprase promotes rapid tumor growth and increased microvessel density in a mouse model of human breast cancer. Cancer Res. 64, 2712–2716.

Huovila A.P., Turner A.J., Pelto-Huikko M., et al. (2005). Shedding light on ADAM metalloproteinases. Trends Biochem. Sci. 30, 413–422.

Itoh Y. and Seiki M. (2006). MT1-MMP: a potent modifier of pericellular microenvironment. J. Cell Physiol. 206, 1–8.

Johansson M.W., Lye M.H., Barthel S.R., et al. (2004). Eosinophils adhere to vascular cell adhesion molecule-1 via podosomes. Am. J. Respir. Cell Mol. Biol. 31, 413–422.

Kaksonen M., Peng H.B., and Rauvala H. (2000). Association of cortactin with dynamic actin in lamellipodia and on endosomal vesicles. J. Cell Sci. 113, 4421–4426.

Kelly T. (1999). Evaluation of seprase activity. Clin. Exp. Metastasis 17, 57–62.

Kelly T. (2005). Fibroblast activation protein-alpha and dipeptidyl peptidase IV (CD26): cell-surface proteases that activate cell signaling and are potential targets for cancer therapy. Drug Resist. Updat. 8, 51–58.

Kelly T., Mueller S.C., Yeh Y., et al. (1994). Invadopodia promote proteolysis of a wide variety of extracellular matrix proteins. J. Cell Physiol. 158, 299–308.

Kelly T., Yan Y., Osborne R.L., et al. (1998). Proteolysis of extracellular matrix by invadopodia facilitates human breast cancer cell invasion and is mediated by matrix metalloproteinases. Clin. Exp. Metastasis 16, 501–512.

Lee H., Overall C.M., McCulloch C.A., et al. (2006a). A critical role for the membrane-type 1 matrix metalloproteinase in collagen phagocytosis. Mol. Biol. Cell. 17, 4812–4826.

Lee K.N., Jackson K.W., Christiansen V.J., et al. (2006b). Antiplasmin-cleaving enzyme is a soluble form of fibroblast activation protein. Blood 107, 1397–1404.

Lener T., Burgstaller G., Crimald L., et al. (2006). Matrix-degrading podosomes in smooth muscle cells. Eur. J. Cell Biol. 85, 183–189.

Linder S. (2007). The matrix corroded: podosomes and invadopodia in extracellular matrix degradation. Trends Cell Biol. 17, 107–117.

Linder S. and Aepfelbacher M. (2003). Podosomes: adhesion hot-spots of invasive cells. Trends Cell Biol. 13, 376–385.

Lochter A., Navre M., Werb Z., et al. (1999). alpha1 and alpha2 integrins mediate invasive activity of mouse mammary carcinoma cells through regulation of stromelysin-1 expression. Mol. Biol. Cell 10, 271–282.

Lorenz M., Yamaguchi H., Wang Y., et al. (2004). Imaging sites of N-wasp activity in lamellipodia and invadopodia of carcinoma cells. Curr. Biol. 14, 697–703.

Marchisio P.C., Cirillo D., Tete A., et al. (1987). Rous sarcoma virus-tranformed fibroblasts and cells of monocytic origin display a peculiar dot-like organization of cytoskeletal proteins invovled in microfilament-membrane interactions. Exp. Cell Res. 169, 202–214.

Martin G.S. (2004). The road to Src. Oncogene 23, 7910–7917.

Martins G.G. and Kolega J. (2006). Endothelial cell protrusion and migration in three-dimensional collagen matrices. Cell Motil. Cytoskeleton 63, 101–115.

McHugh B., Krause S.A., Yu B., et al. (2004). Invadolysin: a novel, conserved metalloprotease links mitotic structural rearrangements with cell migration. J. Cell Biol. 167, 673–686.

McNiven M.A., Baldassarre M., and Buccione R. (2004). The role of dynamin in the assembly and function of podosomes and invadopodia. Front Biosci. 9, 1944–1953.

Mizutani K., Miki H., He H., et al. (2002). Essential role of neural Wiskott-Aldrich syndrome protein in podosome formation and degradation of extracellular matrix in src-transformed fibroblasts. Cancer Res. 62, 669–674.

Monsky W.L., Kelly T., Lin C.-Y., et al. (1993). Binding and localization of M_r 72,000 matrix metalloproteinase at cell surface invadopodia. Cancer Res. 53, 3159–3164.

Monsky W.L., Lin C.-Y., Aoyama A., et al. (1994). A potential marker protease of invasiveness, seprase, is localized on invadopodia of human malignant melanoma cells. Cancer Res. 54, 5702–5710.

Moreau V., Tatin F., Varon C., et al. (2003). Actin can reorganize into podosomes in aortic endothelial cells, a process controlled by Cdc42 and RhoA. Mol. Cell Biol. 23, 6809–6822.

Moreau V., Tatin F., Varon C., et al. (2006). Cdc42-driven podosome formation in endothelial cells. Eur. J. Cell Biol. 85, 319–325.

Mueller S.C. and Chen W.-T. (1991). Cellular invasion into matrix beads: localization of beta 1 integrins and fibronectin to the invadopodia. J. Cell Sci. 99, 213–225.

Mueller S.C., Yeh Y., and Chen W.-T. (1992). Tyrosine phosphorylation of membrane proteins mediates cellular invasion by transformed cells. J. Cell Biol. 119, 1309–1325.

Mueller S.C., Ghersi G., Akiyama S.K., et al. (1999). A novel protease-docking function of integrin at invadopodia. J. Biol. Chem. 274, 24947–24952.

Nakahara H., Nomizu M., Akiyama S.K., et al. (1996). A mechanism for regulation of melanoma invasion. Ligation of alpha6beta1 integrin by laminin G peptides. J. Biol. Chem. 271, 27221–27224.

Nakahara H., Howard L., Thompson E.W., et al. (1997). Transmembrane/cytoplasmic domain-mediated membrane type 1-matrix metalloprotease docking to invadopodia is required for cell invasion. Proc. Natl. Acad. Sci. U. S. A. 94, 7959–7964.

Nakahara H., Mueller S.C., Nomizu M., et al. (1998). Activation of b1 integrin signaling stimulates tyrosine phosphorylation of $p190^{RhoGAP}$ and membrane-protrusive activities at invadopodia. J. Biol. Chem. 273, 9–12.

Nakahara H., Otani T., Sasaki T., et al. (2003). Involvement of Cdc42 and Rac small G proteins in invadopodia formation of RPMI7951 cells. Genes Cells 8, 1019–1027.

Nam J.M., Onodera Y., Mazaki Y., et al. (2007). CIN85, a Cbl-interacting protein, is a component of AMAP1-mediated breast cancer invasion machinery. EMBO J. 26, 647–656.

Okamura H. and Resh M.D. (1995). p80/85 cortactin associates with the Src SH2 domain and colocalizes with v-Src in transformed cells. J. Biol. Chem. 270, 26613–26618.

Onodera Y., Hashimoto S., Hashimoto A., et al. (2005). Expression of AMAP1, an ArfGAP, provides novel targets to inhibit breast cancer invasive activities. EMBO J. 24, 963–973.

Osiak A.E., Zenner G., and Linder S. (2005). Subconfluent endothelial cells form podosomes downstream of cytokine and RhoGTPase signaling. Exp. Cell Res. 307, 342–353.

Redondo-Munoz J., Escobar-Diaz E., Samaniego R., et al. (2006). MMP-9 in B-cell chronic lymphocytic leukemia is up-regulated by alpha4beta1 integrin or CXCR4 engagement via distinct signaling pathways, localizes to podosomes, and is involved in cell invasion and migration. Blood 108, 3143–3151.

Sabeh F., Ota I., Holmbeck K., et al. (2004). Tumor cell traffic through the extracellular matrix is controlled by the membrane-anchored collagenase MT1-MMP. J. Cell Biol. 167, 769–781.

Schliwa M., Nakamura T., Porter K.R., et al. (1984). A tumor promoter induces rapid and coordinated reorganization of actin and vinculin in cultured cells. J. Cell Biol. 99, 1045–1059.

Seals D.F., Azucena E.F., Pass I., et al. (2005). The adaptor protein Tks5/Fish is required for podosome formation and function, and for the protease-driven invasion of cancer cells. Cancer Cell 7, 155–165.

Sherwood D.R., Butler J.A., Kramer J.M., et al. (2005). FOS-1 promotes basement-membrane removal during anchor-cell invasion in C. elegans. Cell 121, 951–962.

Spinardi L., Rietdorf J., Nitsch L., et al. (2004). A dynamic podosome-like structure of epithelial cells. Exp. Cell Res. 295, 360–374.

Strongin A.Y. (2006). Mislocalization and unconventional functions of cellular MMPs in cancer. Cancer Metastasis Rev. 25, 87–98.

Tague S.E., Muralidharan V., and D'Souza-Schorey C. (2004). ADP-ribosylation factor 6 regulates tumor cell invasion through the activation of the MEK/ERK signaling pathway. Proc. Natl. Acad. Sci. USA 101, 9671–9676.

Taniwaki K., Fukamachi H., Komori K., et al. (2007). Stroma-derived matrix metalloproteinase (MMP)-2 promotes membrane type 1-MMP-dependent tumor growth in mice. Cancer Res. 67, 4311–4319.

Tatin F., Varon C., Genot E., et al. (2006). A signalling cascade involving PKC, Src and Cdc42 regulates podosome assembly in cultured endothelial cells in response to phorbol ester. J. Cell Sci. 119, 769–781.

Tehrani S., Faccio R., Chandrasekar I., et al. (2006). Cortactin has an essential and specific role in osteoclast actin assembly. Mol. Biol. Cell. 17(7), 2882–2895.

Thomas S.M. and Brugge J.S. (1997). Cellular functions regulated by Src family kinases. Annu. Rev. Cell Dev. Biol. 13, 513–609.

Varon C., Tatin F., Moreau V., et al. (2006). Transforming growth factor beta induces rosettes of podosomes in primary aortic endothelial cells. Mol. Cell Biol. 26, 3582–3594.

Vikesaa J., Hansen T.V., Jonson L., et al. (2006). RNA-binding IMPs promote cell adhesion and invadopodia formation. EMBO J. 25, 1456–1468.

Vu T.H., Shipley J.M., Bergers G., et al. (1998). MMP-9/gelatinase B is a key regulator of growth plate angiogenesis and apoptosis of hypertrophic chondrocytes. Cell 93, 411–422.

Wang E., Yin H.L., Krueger J.G., et al. (1984). Unphosphorylated gelsolin is localized in regions of cell-substratum contact or attachment in Rous sarcoma virus-transformed rat cells. J. Cell Biol. 98, 761–771.

Weaver A.M. (2006). Invadopodia: specialized cell structures for cancer invasion. Clin. Exp. Metastasis 23, 97–105.

Webb B.A., Eves R., Crawley S.W., et al. (2005). PAK1 induces podosome formation in A7r5 vascular smooth muscle cells in a PAK-interacting exchange factor-dependent manner. Am. J. Physiol Cell Physiol. 289, C898–C907.

Webb B.A., Jia L., Eves R., et al. (2007). Dissecting the functional domain requirements of cortactin in invadopodia formation. Eur. J. Cell Biol. 86, 189–206.

Weed S.A., Du Y., and Parsons J.T. (1998). Translocation of cortactin to the cell periphery is mediated by the small GTPase Rac1. J. Cell Sci. 111, 2433–2443.

Wolf K. and Friedl P. (2005). Functional imaging of pericellular proteolysis in cancer cell invasion. Biochimie 87, 315–320.

Wolf K., Wu Y.I., Liu Y., et al. (2007). Multi-step pericellular proteolysis controls the transition from individual to collective cancer cell invasion. Nat. Cell Biol. 9(8), 893–904.

Yamaguchi H. and Condeelis J. (2007). Regulation of the actin cytoskeleton in cancer cell migration and invasion. Biochim. Biophys. Acta. 1773, 642–652.

Yamaguchi H., Lorenz M., Kempiak S., et al. (2005). Molecular mechanisms of invadopodium formation: the role of the N-WASP-Arp2/3 complex pathway and cofilin. J. Cell Biol. 168, 441–452.

Yamaguchi H., Pixley F., and Condeelis J. (2006). Invadopodia and podosomes in tumor invasion. Eur. J. Cell Biol. 85, 213–218.

Yana I., Sagara H., Takaki S., et al. (2007). Crosstalk between neovessels and mural cells directs the site-specific expression of MT1-MMP to endothelial tip cells. J. Cell Sci. 120, 1607–1614.

Zhang F., Tom C.C., Kugler M.C., et al. (2003). Distinct ligand binding sites in integrin alpha3-beta1 regulate matrix adhesion and cell-cell contact. J. Cell Biol. 163, 177–188.

Zhou S., Webb B.A., Eves R., et al. (2006). Effects of tyrosine phosphorylation of cortactin on podosome formation in A7r5 vascular smooth muscle cells. Am. J. Physiol. Cell Physiol. 290, C463–C471.

Zhu Q., Krakowski A.R., Dunham E.E., et al. (2007). Dual role of SnoN in mammalian tumorigenesis. Mol. Cell Biol. 27, 324–339.

Chapter 22
uPAR and Proteases in Mobilization of Hematopoietic Stem Cells

Pia Ragno and Francesco Blasi

Abstract The specific cell-surface receptor for the urokinase-type plasminogen activator (uPA) was identified in 1985; since then, a large body of evidence showed that urokinase-type plasminogen activator receptor (uPAR) was not only the receptor for a proteolytic enzyme but also a multifunctional signaling molecule capable of various interactions and activities. In fact, uPAR is implicated in several biological events, such as cell adhesion, migration, proliferation, and survival. Cell-surface uPAR can be cleaved by neutrophil proteases, matrix metalloproteases, plasmin, and uPA itself. Both full-length and cleaved uPAR can be shed from the cell surface, generating full-length and cleaved forms of soluble uPAR (suPAR and c-suPAR). suPAR retains most of uPAR activities which are lost by c-suPAR. However, c-suPAR becomes a potent chemoattractant for cells expressing receptors of the FPR (fMLP: formyl-methionine-leucine-proline) family.

Very recently, uPAR involvement has been reported in the trafficking of human and mouse hematopoietic stem cells (HSCs) from and to bone marrow (BM). In humans, c-suPAR levels are strongly increased during HSC mobilization induced by the granulocyte-colony-stimulating factor (G-CSF). Increased c-suPAR could contribute to HSC migration into the circulation both directly, by inducing HSC migration, and indirectly, by inactivating CXCR4, a key receptor in HSC retention in BM. In vivo, a specific c-suPAR-derived peptide induces mobilization of mouse $CD34^+$ HSCs to the same extent as G-CSF. In mice, uPAR is expressed on HSCs and seems to act as a retention signal in BM, likely by interacting with the VLA-4 integrin. uPAR shedding from the cell surface induces HSC release from BM and mobilization into the circulation. Proteases, which are largely produced in BM during G-CSF-induced mobilization, and, in particular, plasmin are strongly involved in these events and play an important role in HSC trafficking.

P. Ragno
Università di Salerno, Dipartimento di Chimica, Via Ponte don Melillo, 84084 Salerno, Italy,
e-mail: ragno@unina.it

D. Edwards et al. (eds.), *The Cancer Degradome.*

Introduction

This chapter will focus on the interactions of the urokinase receptor (uPAR) and will highlight the recently discovered role of uPAR in mobilization of human and mouse hematopoietic stem cells (HSCs). HSC mobilization may also require neutrophil proteases and plasmin; thus, a section of the chapter will review their implication and potential functions in HSC trafficking from and to the bone marrow (BM). This topic has a potentially direct practical application as HSC transplantation is an important therapeutic tool in a series of human malignancies.

Structure of uPAR

uPAR is a GPI-anchored protein composed of three consecutive three-finger domains organized in "three quarters of a circle" shape, which generates both a deep internal cavity where an inhibitory peptide or the growth factor domain of urokinase-type plasminogen activator (uPA) can bind, and a large external surface. The receptor-binding module of uPA engages the uPAR central cavity, leaving the external receptor surface accessible for other protein interactions (Llinas et al. 2005, Huang et al. 2005, Barinka et al. 2006). Alanine scanning mutagenesis has identified both the uPA-binding epitope on the internal cavity and the vitronectin (VN)-binding epitope on the external surface (Gardsvoll et al. 2006, Madsen et al. 2007). The structure of uPAR and the uPA and VN-binding epitopes will be described in Chap. 34 by Jacobson et al.

Ligands and Interactors

uPA and VN

The original discovery of uPAR as a receptor for uPA occurred over 20 years ago almost at the same time in the laboratories of J.D. Vassalli and F. Blasi (Vassalli et al. 1985, Stoppelli et al. 1985). This finding brought much excitement in the field since it was the first cell-surface receptor identified for a protease. The fact that uPA was a plasminogen activator suggested that the binding to the cell surface would induce a cell surface–associated proteolytic cascade (Blasi et al. 1987), a suggestion that has been since confirmed (*see*, for example, Stephens et al. 1989). However, the physiological role of uPAR in plasminogen activation turned out not to be essential, since uPAR-knockout (Ko) mice did not show any sign of fibrin deposition (Dewerchin et al. 1996). Moreover, mice doubly deficient in uPAR and tPA (Bugge et al. 1996) did not show the signs of fibrinolytic deficiency that were observed in the uPA-tPA double Ko mice. Since both uPA and tPA are required to achieve optimal fibrinolytic activity (Carmeliet et al. 1994), this result also showed that uPA was probably able to bind other cell-surface proteins and hence be proteolytically active also in the absence of uPAR. Even though the role of uPAR in fibrinolysis has not been further investigated, subsequent findings showed that the uPAR Ko mice did have several pathologic phenotypes, but these were more likely caused by cell signaling defects (see below) (Blasi and Carmeliet 2002). In particular, these mice showed deficiencies in neutrophils and macrophages migration *in vivo* (May et al. 1998, Gyetko et al. 2000), acceleration of kidney fibrosis (Zhang et al. 2003a, 2003b), increased bone density and osteoblasts' osteogenic potential, decreased osteoclasts formation, and altered cytoskeletal reorganization in mature osteoclasts (Furlan et al. 2007). Moreover, these mice also display deficient mobilization of HSCs (Tjwa, M. unpublished) and are protected from skin carcinogenesis (D'Alessio, S. unpublished). Several data show that uPAR is important also in human HSCs mobilization (Selleri et al. 2005, 2006) (see below).

About 10 years after uPAR discovery, Chapman and coworkers identified a second uPAR ligand, the extracellular matrix and serum protein, VN (Wei et al. 1994). The affinities of uPA and VN for uPAR are very different, as VN does not bind uPAR in the absence of uPA (Gardsvoll and Ploug 2007) and the binding of uPA and VN is not mutually exclusive. VN is also a ligand for the alphaV-beta3 and alphaV-beta5 integrins, and it was realized that the binding site for uPAR, present in the somatomedin B domain (Deng et al. 1996), is quite close to the RGD, the recognition sequence for most integrins, and overlaps at least in part with the binding site for PAI-1. This opens the possibility that in fact VN acts as an adaptor for the interaction of uPAR and integrins and suggests that PAI-1 can disrupt the binding of VN to uPAR and hence modulate uPAR-dependent adhesion.

Integrins

The presence of uPAR allows cells to directly adhere onto VN, spread, and migrate. The addition of uPA stimulates both adhesion and migration (Wei et al. 1996). This adhesion is not RGD-dependent as it does not involve integrins. However, several proofs have been provided that uPAR can interact with integrins. Bohuslav et al. (1995) reported that they could immunoprecipitate integrins using an anti-uPAR antibody. Since then, several confirmatory reports have been published, although coimmunoprecipitation of uPAR with integrins has not been shown with all cells. In some cell lines, uPAR coprecipitates with integrins, but in others it is necessary to treat the cells with uPA (Wei et al. 1996, Aguirre Ghiso et al. 1999, Mazzieri et al. 2006). Using purified proteins, coimmunoprecipitation of iodinated suPAR with alpha3-beta1 or alpha5-beta1 was shown, but only in the presence of uPA (Degryse et al. 2005). Evidence in favor of a direct interaction was obtained using specific peptides. Chapman and colleagues isolated a peptide from a phage-display library because it bound suPAR but did not interfere with the binding of uPA to uPAR. This peptide prevented many of the functions of uPAR, like cell adhesion onto VN and migration (Wei et al. 1996). It was then noticed that this peptide resembled a sequence present (with various degrees of conservation) in the w4 repeat of the beta-propeller domain of various alpha integrins. Although these peptides are active at rather high concentrations, interestingly they inhibited several integrins functions, that is, cell adhesion and cell migration as well as uPAR–integrins coimmunoprecipitation (Wei et al. 1996, 2005, Mazzieri et al. 2006). Moreover, exogenously supplied suPAR in fact inhibited integrin functions, like the internalization of fibrinogen by the macrophage integrin alpha2_M-beta3 (Simon et al. 2000). In other cells, suPAR activates integrin-dependent ERK phosphorylation (Aguirre Ghiso et al. 1999). All these data suggest an interaction of uPAR and integrins.

On the uPAR side, potential-binding sites have been identified in the second and the third domains (Degryse et al. 2005, Chaurasia et al. 2006). Importantly, single-site mutants of uPAR have been identified that are unable to bind either alpha3-beta1 or alpha5-beta1 integrins or both (Wei et al. 2007).

Despite the wealth of indications of a direct interaction between uPAR and integrins, the possibility that this interaction is mediated by other proteins, like VN or uPA, cannot be discarded. Indeed, in addition to the binding of the RGD of VN to integrins, direct evidence has also been obtained for the binding of uPA, in particular the kringle domain or the region connecting the kringle to the protease domain, to the integrins (Pluskota et al. 2003, Tarui et al. 2006, Franco et al. 2006).

G Protein-Coupled Receptors

A peptide from the DI-DII linker region of uPAR, or a fragment of uPAR (DIIDIII) containing this peptide at the N-terminus, was shown to induce chemotaxis at very low concentrations (Fazioli et al. 1997, Nguyen et al. 2000). These peptides appear to activate the FPRL1 (fMLP receptor-like protein 1) receptor, which was known to respond at high concentrations of the bacterial chemoattractant fMLP (Formyl-methionine-leucine-proline) (Resnati et al. 2002). Pretreatment with this peptide or with DIIDIII desensitized cells not only from the action of these peptides but also from the stimulus of other chemototactic proteins like MCP-1 (monocyte chemoattracting protein-1) and RANTES (Furlan et al. 2004). Direct binding studies showed that the DIIDIII, but not the full-length protein, had a relatively weak affinity for FPRL1 (Resnati et al. 2002). It was subsequently shown that cells expressing not FPRL1, but its homologs FPR, which is activated at low concentrations of fMLP, or FPRL-2, would also respond to the same peptides in chemotaxis (Montuori et al. 2002, de Paulis et al. 2004).

Tyrosine Kinase Receptors

Liu et al. (2002) identified the EGF-R as a possible functional interactor of uPAR, by showing that overexpression of uPAR allowed coimmunoprecipitation of the two receptors, and activated the EGF-R tyrosine autophosphorylation even in the absence of EGF. This resulted in activation of the signaling pathway and of proliferation. All these data in cell culture receive a strong support by the finding that the EGF responsiveness of the EGF-R in uPAR Ko mouse keratinocytes is essentially lost, even though the protein and its ligand are still present, demonstrating a requirement for uPAR of the EGF-R activation (D'Alessio and Blasi, to be published). These results are in good agreement with previous data from Gonias' laboratory, showing that EGFR selectively cooperates with uPAR to mediate mitogenesis (Jo et al. 2005, 2007).

Another tyrosine kinase receptor that has been linked to uPAR is the platelet-derived growth factor receptor (PDGFR). Indeed, LRP1B, a member of lipoprotein receptors, modulates the migration of smooth muscle cells (SMCs) by increasing the degradation of membrane uPAR and PDGFR-beta (Tanaga et al. 2004). Moreover, uPAR activation by uPA induces its association with PDGFR-beta, PDGF-independent PDGFR-beta phosphorylation of cytoplasmic tyrosine kinase domains, and receptor dimerization. This induces the uPA-dependent downstream signaling that regulates vascular SMC migration and proliferation (Kiyan et al. 2005).

uPAR-dependent Signaling Pathways

Induction of migration through uPA, DIIDIII, or the DI-DII chemotactic peptide induces a variety of signaling pathways. Indeed, inhibitors of tyrosine kinases and of the ERK1/2 kinases completely block uPA or DIIDIII-induced chemotaxis (Resnati et al. 1996, Degryse et al. 1999, Nguyen et al. 2000). The requirement for other important signaling molecules like Ras, Raf, Src (Degryse et al. 1999, Nguyen et al. 2000), Fak, and Rac (Kjoller and Hall 2001) in uPAR-dependent migration or cell motility has been demonstrated. The uPA induction of migration has been shown to also activate other signaling pathways, including the PI3-kinase, Tyk2, and the Jak/Stat pathway (Koshelnick et al. 1997, Dumler et al. 1999, Kusch et al. 2000).

Nowadays, it is clear that migration mediated by uPAR can be obtained by activating different signaling pathways. However, no overall picture has yet emerged that reveals the basis of the choice of the induced pathway (Blasi and Carmeliet, 2002).

In many cancer cell lines, uPAR abundance is directly related to the growth of the cells. In particular, Ossowski's work has outlined that in Hep3 epidermoid carcinoma cells, the downregulation of uPAR expression leads to a decreased growth rate *in vivo* in the chorion-allantoic membrane (Kook et al. 1994). Likewise, Hep3 cells can undergo a period of latency in which they are unable to grow *in vivo*; this phenotype is reversed by the reexpression of uPAR (Yu et al. 1997). The growth-inducing property of uPAR in cancer cells was explained when the uPAR–integrins interaction was investigated. Indeed, in these cells the growth rate *in vivo* is affected by the interference of uPAR with the alpha5-beta1 (fibronectin receptor) activity. In fact, uPAR and fibronectin lead to the induction of Erk1/2 and inhibition of p38MAP-kinase activation and this unbalance results in the increase in growth rate *in vivo* (Aguirre-Ghiso et al. 2001). As indicated above, uPAR overexpression in Hep3 cells and its interaction with alpha5-beta1 affects the activity of the EGF-R, resulting in constitutive activation also in the absence of EGF or other ligands (Liu et al. 2002).

In conclusion, despite the wealth of evidence that shows a direct involvement of uPAR in cancer cell growth, these studies do not outline a precise mechanism. Cancer cell lines suffer from the lack of characterization of the oncogenic mutations present. In fact, uPAR involvement in cell growth and cancer is reinforced by the comparison of the growth properties of mouse embryo fibroblasts (MEFs) isolated from uPAR wild-type (WT) or Ko mice, a system genetically different only at the level of the uPAR gene. The latter, in fact, have a faster growth rate which can be reversed by the reintroduction of uPAR. The faster growth rate is maintained also when MEFs are transformed by oncogenes like RasV12 and E1a, indicating that uPAR controls cell growth even in the presence of activated oncogenes. In fact, uPAR Ko MEFs are also transformed more efficiently from these oncogenes and induce larger and faster-growing tumors when inoculated in nude mice (Mazzieri et al. 2007). The effect of uPAR deletion in cell growth of otherwise WT cells is cell-type specific. While osteoblasts behave like MEFs (Furlan et al. 2007), keratinocytes on the contrary are unable to grow in culture and are deficient in skin wound healing (D'Alessio and Blasi, to be published). The nature of this discrepancy is still unclear. In any case, overall, all these data show that uPAR expression is important in regulating cell growth not only in culture but also *in vivo*.

uPAR and Proteases in Hematopoietic Stem Cell Mobilization

Hematopoietic Stem Cells

HSCs are clonogenic cells capable of self-renewal and multilineage differentiation; they must be distinguished by hematopoietic progenitor cells (HPCs) which are oligo-lineage cells, incapable of self-renewal and with an absolute lower differentiation potential. The expression of the CD34 antigen and lineage negativity are commonly used in clinics to identify human HSCs/HPCs. CD34$^+$ cells, sorted through the fluorescence-activated cell sorter, contain cell populations capable of both short-term and long-term reconstitution of myeloablated transplant recipients. Under steady-state conditions, the majority of CD34$^+$ HSCs reside in the BM

(BM-HSCs), while circulating HSCs (PB-HSCs) amount to only 0.06% (Fruehauf et al. 1995). BM and PB CD34$^+$ cells are functionally different and, correspondingly, show a differential gene expression (Fruehauf et al. 1995). Physiologically, HSCs recirculate in the bloodstream and repopulate distinct anatomical districts (Wright et al. 2001). HSC capability to circulate from blood to BM (homing) and from BM to blood (mobilization) has been conserved in different species, but its biological significance remains unknown. However, this physiological and peculiar capability can explain the success of BM transplantation. In 1951 two independent groups reported that injection of cells from BM or from spleen (where also hematopoiesis occurs in adult mice) can protect lethally irradiated animals (Jacobson et al. 1951) and a few years later it became evident that the rescue effect was due to cellular factors (Ford et al. 1956). The existence of HSCs or HPCs was then demonstrated when specific assays became available and the formation of colonies (containing granulocyte/macrophage and erythroid cells) in the spleen of irradiated mice, following injection of BM cells, could be measured (Till et al. 1964).

Actually, the number of multipotent or more restricted hematopoietic progenitors can be counted by *in vitro* HSC culture on irradiated BM stroma and subsequent enumeration of colony-forming cells (CFCs) in methylcellulose in the presence of specific growth factors.

Hematopoietic Stem Cell Mobilization

The story of HSC mobilization started with the finding that HSCs were present in peripheral blood (PB) of different animals, including humans, during steady-state homeostasis (Korbling and Fliedner 1994). Increased HSC levels were subsequently detected in patients affected by myeloproliferative disorders (Hibbin et al. 1984), and some authors, at the end of 1970s, reported increased circulating HSC levels in the blood of patients following treatments with cyclophosphamide and other drugs. Technical improvements allowing large-scale harvesting (the continuous-flow leukopheresis) and studies showing faster BM repopulation in patients transplanted with autologous mobilized PB-HSCs led to the wide use of PB-HSCs in transplantation therapy (Korbling and Fliedner 1994). Currently, mobilized PB-HSCs represent the major source of stem cells for autologous stem cell transplantion and are also increasingly used in allogeneic HSC-transplantation, because of the relative ease of collection, the higher yield of stem cells, and the shorter time to engraftment (Fruehauf and Seggewiss 2003). At the beginning, mobilization protocols were based on chemotherapy alone; following the discovery of human G-CSF (Molineux et al. 1990), clinicians started to use G-CSF, which has now become the standard mobilizing agent, even though various other cytokines are able to induce mobilization, including GM-CSF, interleukin-8 (IL-8), IL-3, and SCF (Fruehauf and Seggewiss 2003). The majority of HSCs resides in the BM, which provides them with a suitable microenvironment, regulating their proliferation/survival and differentiation. At steady-state, proliferating and differentiating HSCs are located within specialized niches, whereas differentiated cells and a very low number of HSCs leave the BM and migrate into the circulation. HSCs are retained in the BM mainly by their adhesive interactions with the cellular and matrix components of the stroma and by interaction with specific, diffusible, BM chemokines.

$CD34^+$ cells express a wide variety of adhesion molecules, including integrins, selectins, CD44 family, and immunoglobulin superfamily members. The ligands of these adhesion molecules are expressed in the BM stroma, both on cells and in the matrix. Interestingly, the same molecules which play a role in HSC homing to BM are often implicated in the pharmacological mobilization of these cells into the circulation. Indeed, mobilization is believed to require functional changes in the adhesion profile of BM-HSCs. Among the many relevant well-characterized adhesion molecules is the very late antigen (VLA)-4, integrin receptor for fibronectin (FN) and for the vascular cell adhesion molecule-1 (VCAM-1). In experimental models, VLA-4 (alpha4-beta1)- and VLA-5 (alpha5-beta1 integrin)-blocking antibodies prevent engraftment of NOD/SCID mice by human $CD34^+$ cells and inhibition of VLA-4 blocks homing to BM by murine HSCs (Papayannopoulou et al. 1995), even though targeted deletion of these integrins fails to block HSC–BM localization (Wagers et al. 2002). Downregulation or inactivation of these integrins thus seems crucial for HSC mobilization, as expected, since they mediate HSC interactions with BM stroma.

Another important antigen present on both human and mouse HSCs is c-Kit, a tyrosine-kinase receptor. The long-term repopulating (LTR) activity of BM HSC correlates with c-Kit expression. Before chemotherapy, LTR cells in the BM are present within the c-Kit^{hi} population. After chemotherapy, mobilized LTR cells are found in the c-Kit$^-$ subpopulation. Thus, the reduced expression of c-Kit on HSCs correlates with the exit of HSCs from BM (Randall and Weissman 1997).

Various recent reports identify other key molecules regulating HSC trafficking from and to BM stroma, which likely also regulate integrin-mediated functions. These are the CXC receptor 4 (CXCR4) and its ligand, the stromal-derived factor 1 (SDF1). Murine and human HSC express CXCR4 and efficiently migrate toward SDF1, which is largely produced by BM endothelium (Kollet et al. 2001, Wright et al. 2002). The observation that the engraftment of human HSCs in NOD/SCID mice was strongly reduced by CXCR4-blocking antibodies (Peled et al. 1999) suggested a role for this receptor in HSC BM localization, which has been confirmed by subsequent results. Indeed, administration of a single dose of a pharmacologic inhibitor of CXCR4, AMD3100, induces leukocytosis, mobilization of $CD34^+$ cells, and an increase in circulating CFCs in humans to the same extent as 5 days of treatment with G-CSF; the same effect was also observed in mice (Liles et al. 2003, Broxmeyer et al. 2005). Besides directly attracting HSCs to the BM stroma, SDF1 is also effective in inducing proliferation and differentiation of HSCs and may enhance the activity of integrins involved in BM retention of HSCs, such as VLA-4 and LFA-1 (Peled et al. 2000).

Mobilization of HSCs is thus strongly dependent on the disruption of BM retention forces. In this context, a role for proteases capable of cleaving molecules that retain HSCs in BM might be hypothesized.

Proteases in HSC Mobilization

Total body irradiation and/or chemotherapy induce, per se, HSC mobilization; these agents determine tissue damage that, in turn, leads to a dramatic increase in the levels of released chemokines, cytokines, and proteolytic enzymes in many organs, as part of the regeneration and repair process. G-CSF-dependent HSC mobilization correlates with an increase in the number of leukocytes and of polymorphonuclear (PMN) cells. The importance of PMN cells was studied in mice made neutropenic by administration of antineutrophil antibodies. The IL-8-induced mobilization was reduced significantly during the neutropenic phase, reappeared with the presence of peripheral PMN cells, and was increased proportionally during the neutrophilic phase. In neutropenic mice, the IL-8-induced mobilization was restored by the infusion of purified PMN cells but not by infusion of mononuclear cells (Pruijt et al. 2002).

Neutrophils represent a source of proteases which could contribute to HSC mobilization since, on activation, they release a large amount of proteases from their specific and azurophilic granules (matrix metalloprotease-9, lactoferrin and elastase, cathepsin G, proteinase 3, respectively). In fact, mobilization by either cyclophosphamide or G-CSF transforms the BM into a highly proteolytic environment (Lévesque et al. 2002). Administration of either G-CSF or cyclophosphamide results in the accumulation of granulocytic precursors and release of active neutrophil elastase and cathepsin G, that directly cleave VCAM-1 *in vitro*, which contribute to HSC anchoring to BM stroma. These events correlate with the kinetics of HPC mobilization into the peripheral blood (Lévesque et al. 2001).

Also matrix metalloproteinase-9 (MMP-9) increases in BM and PB during mobilization of progenitor cells by G-CSF (Carstanjen et al. 2002, Carion et al. 2003) and a significant relationship between the levels of circulating HPC, both at steady state and after mobilization, and those of secreted MMP-9 was also reported (Carion et al. 2003). Inhibitory antibodies against MMP-9 prevented IL-8-induced mobilization (Pruijt et al. 1999).

The increased levels of different proteases during HSC mobilization suggested that they could play a role in HSC release from BM. Indeed, proteases can specifically cleave molecules implicated in HSC retention in BM. BM extracellular fluids isolated from G-CSF-mobilized mice contain serine proteases capable of cleaving c-Kit into discrete fragments, thus indicating that the direct proteolytic cleavage of c-Kit by neutrophil and macrophage proteases may be responsible, at least in part, for the downregulation of c-KIT expression on mobilized hematopoietic progenitors *in vivo* (Lévesque et al. 2003a). Neutrophil proteases can also affect the CXCR4/SDF1 axis. Leukocyte elastase is able to *in vitro* cleave the N-terminus of both CXCR4 and its ligand, thus inhibiting their binding (Valenzuela-Fernández et al. 2002). CXCR4 and SDF1 cleavage in the BM, during G-CSF treatment, has also been demonstrated (Lévesque et al. 2003b).

SDF1 can be also cleaved by CD26/dipeptidylpeptidase IV (DPPIV), a membrane-bound extracellular peptidase that is expressed by a subpopulation of CD34[+]

hemopoietic cells isolated from cord blood. Cleaved SDF1 is unable to induce $CD34^+$ cell migration and inhibits uncleaved SDF1 activity (Christopherson et al. 2002). In vivo, the effect of G-CSF in $CD26^{-/-}$ mice was significantly lower than in WT mice (Christopherson et al. 2003); in agreement with these results, inhibition or deletion of CD26 greatly increased the efficiency of transplantation due to a better engraftment (Christopherson et al. 2004).

The recent observation that administration of Serpin-1, inhibitor of serine-proteases, inhibits G-CSF-induced HSC/HPC mobilization (van Pel et al. 2006) is in total agreement with previous results and strongly supports the hypothesis of a crucial role for protease activity, in particular serine proteases, in cytokine-induced HSC/HPC mobilization. By contrast, HPC mobilization by G-CSF was normal in MMP-9-deficient mice, elastase- and cathepsin G-deficient mice, or mice lacking dipeptidyl peptidase I (Levesque et al. 2004). Surprisingly, also the combined inhibition of these proteases had no significant effect on HPC mobilization. G-CSF induced anyhow a significant decrease in SDF1 expression in the BM of elastase- and cathepsin G-deficient mice, which suggests the involvement of other, unexplored proteases, which could substitute or act independently of previously identified neutrophil proteases.

A potential effect of the uPA has been explored only in elapsed and anticoagulated cord blood, in which uPA increased the yield of progenitor cells during red cell depletion (Lee et al. 2002). Recent reports clearly indicate a strong involvement of the uPAR and of plasmin in HSC mobilization.

uPAR and Plasminogen Activation in HSC Mobilization

G-CSF-induced HSC mobilization is a multistep process which includes HSC detachment from the BM microenvironment, motility and subsequent migration, and intravasation. Indeed, uPAR is strongly involved in migration and adhesion of normal and malignant cells, even independently of the proteolytic activity of its ligand (Blasi and Carmeliet 2002). During human HSC mobilization, uPAR expression increased significantly on peripheral blood mononuclear cells (PBMNCs), in particular on $CD33^+$ myeloid precursors and on $CD14^+$ monocytic cells released from BM into the circulation. By contrast, $CD34^+$ cells and T and B lymphocytes were uPAR-negative. Upregulation of cell-surface uPAR in $CD33^+$ and $CD14^+$ monocytic cells coincides with the increase in the soluble form of the receptor (suPAR) in the serum, thus suggesting its release from the monocytic cell surface. Both the previously described intact and cleaved forms (c-suPAR) of shed suPAR were observed, even if PBMNCs from G-CSF-treated donors expressed mainly the full-length form of uPAR; this observation suggested that the cleavage of the receptor may occur after its release from the cell surface (Selleri et al. 2005). Interestingly, uPAR can be a substrate of neutrophil proteases, MMP-9, cathepsin G and elastase, as well as of plasmin, which all cleave uPAR upstream of the sequence endowed with chemotactic properties (Andolfo et al. 2002, Beaufort et al.

2004). Thus, protease cleavage might generate potentially chemotactic forms of suPAR (c-suPAR) present in the serum of G-CSF-treated donors. Chemotactic c-suPAR is a potent chemoattractant for BM-HSCs, since they express the high-affinity receptor for fMLP (FPR). c-suPAR is also able to interfere with the CXCR4/SDF1 axis, the key step in HSC mobilization. Indeed, *in vitro* SDF-1 dependent migration of $CD34^+$ BM-HSCs is inhibited by c-suPAR or a c-suPAR-derived peptide including the chemotactic sequence, corresponding to aa 84–95 ($uPAR_{84-95}$). Interestingly, the opposite effect is observed with full-length suPAR, which stimulates human HSC migration toward SDF1, which suggests a role for full-length suPAR in the homing process, rather than in mobilization (Wysoczynski et al. 2005).

Altogether, these data suggest that during G-CSF-induced HSC mobilization, uPAR expression is first upregulated on $CD33^+$ and $CD14^+$ cells and is then cleaved, thus leading to increased uPAR shedding in the serum. suPAR could be rapidly cleaved both in the serum and in the BM. In the first case, c-suPAR might chemoattract BM HSCs, inducing their migration into the circulation through a positive gradient. In the second case, c-suPAR might inactivate CXCR4 by heterologous desensitization and promote HSC release from BM (Fig. 22.1).

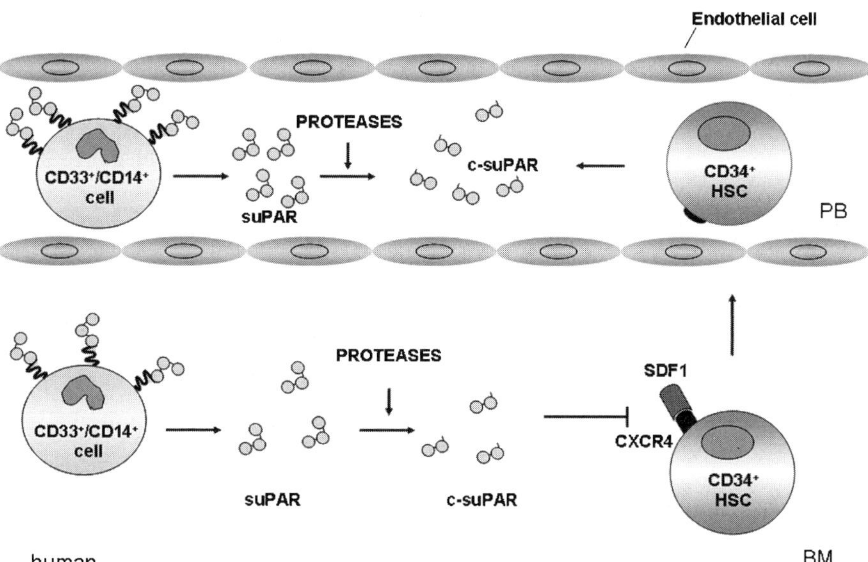

Fig. 22.1 uPAR involvement in human HSC mobilization. G-CSF administration upregulates uPAR expression on $CD33^+$ myeloid and $CD14^+$ monocytic cells, thus leading to increased uPAR shedding. Soluble uPAR (*suPAR*) could be cleaved by proteases both in bone marrow (*BM*) and in peripheral blood (*PB*), thus generating the chemotactically active form of suPAR (*c-suPAR*). PB c-suPAR could chemoattract BM hematopoietic stem cells (*HSCs*) into the circulation through a positive gradient and/or it could inactivate BM-HSC CXCR4 by heterologous desensitization, thus promoting HSC release from BM

The finding of increased levels of both full-length and cleaved forms of uPAR in allogenic HSC donors following G-CSF treatment has been subsequently confirmed (Fietz et al. 2006). Moreover, the chemotactic human c-suPAR peptide has been shown to possess mobilizing activity also *in vivo* in the mice (Selleri et al. 2006). Indeed, administration of human $uPAR_{84-95}$ induces migration of mouse $CD34^+$ HSCs/HPCs into the circulation to an extent similar to that observed in G-CSF. $uPAR_{84-95}$-mobilized leukocytes are strongly enriched in PMN cells as previously observed with other mobilizing agents.

The availability of specific mutants makes the mice an excellent tool to investigate the role of proteases and uPAR in HSC mobilization. Carmeliet et al. have carried out a careful study utilizing mice deleted for the uPA, tPA, uPAR, and plasminogen (Plg) genes. This study has clearly established that uPAR and Plg are essential for both 5-fluorouracil (5-FU) and G-CSF-induced HSC mobilization (Tjwa et al. to be published). The mouse system has given somewhat different indications than the human studies, which possibly underscores the species difference. However, the main differences may be due to the different types of available experimental approaches in the two species and may be reconciled in a single picture.

In humans, peripheral $CD34^+$ cells did not express uPAR; in mouse, uPAR marks BM cells which are in close contact with osteoblasts as well as a subset of HPCs. At steady state, uPAR Ko mice are partially depleted of HPCs in the BM which show a decreased cell cycle quiescence and chemoprotection. uPAR Ko mice are impaired in HPC mobilization, homing, and short-term engraftment. Thus, uPAR must be an engraftment/retention signal for HPCs. In WT mice, in response to 5-FU myeloablation or G-CSF stimulation, the membrane-anchored uPAR retention signal on HPCs is inactivated by plasmin via proteolytic cleavage into a soluble uPAR cleavage product (suPAR), which stimulates mobilization. In agreement with these findings, the mobilization of HPCs is impaired in mice lacking either uPAR or Plg. The retention effect seems to depend on the interaction in the BM of the membrane-anchored uPAR of HPCs with the alpha4-beta1 integrin, which is important in the retention of HSC in the osteoblastic cells niche. This interaction is functionally interrupted when plasmin cleaves membrane-anchored uPAR or on addition of exogenous soluble forms of uPAR, suPAR, and soluble DIIDIII (Fig. 22.2). Indeed, these fragments stimulate mobilization and partially restore the mobilization activity and the chemoresistance in uPAR Ko mice. Plasmin, therefore, appears to convert uPAR from a membrane-anchored retention into a soluble mobilization signal. Loss of uPAR also impairs long-term engraftment and multilineage repopulation of primary and secondary myeloablated recipients. These findings implicate uPAR and plasmin as novel regulators of the maintenance, homing, engraftment, retention, and mobilization of HPCs (Tjwa, M. to be published).

In light of these results, the chemotactic hypothesis based on the results of human studies (Selleri et al. 2005, 2006) might represent one of the at least two important events in HSC mobilization in which uPAR is involved: retention in the BM and mobilization into the peripheral blood.

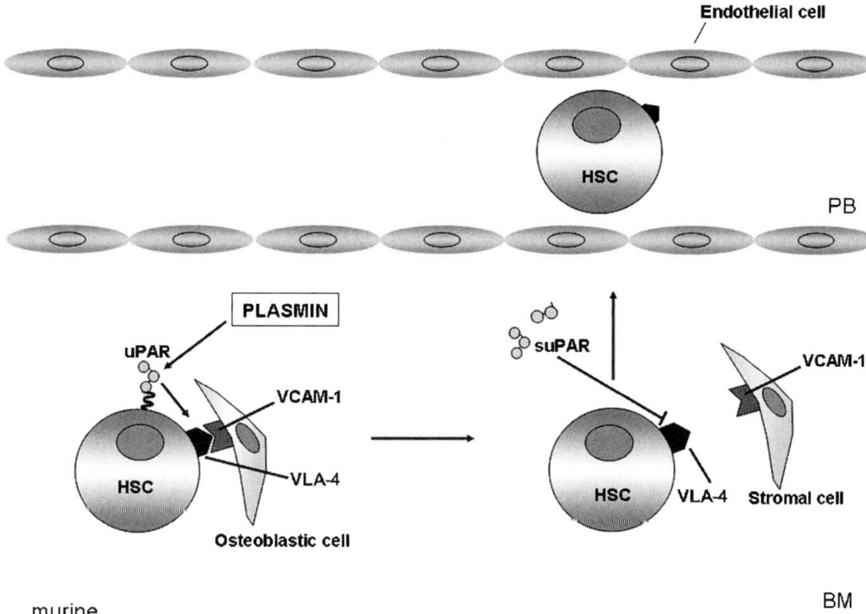

Fig. 22.2 Urokinase-type plasminogen activator receptor (*uPAR*) involvement in mouse HSC retention in bone marrow (*BM*) and mobilization. In mice, uPAR is expressed on hematopoietic stem cells (*HSC*) in BM in close contact with cells of the osteoblastic niche through the interaction with the alpha4-beta1 integrin (*VLA*-4). This interaction is functionally interrupted when plasmin cleaves membrane-anchored uPAR or on addition of exogenous soluble forms of uPAR, soluble uPAR (*suPAR*), and soluble DIIDIII. Thus, plasmin may convert uPAR from a membrane-anchored retention into a soluble mobilization signal. (Tjwa, to be published)

Conclusions

These results open the way to further studies in which important questions on uPAR function in HSC mobilization are better defined and the mechanisms involved outlined in detail. Outstanding questions are the role of uPAR in HSC self-renewal, the molecular nature of the retention signal and uPAR's role, the regulation of uPAR's cleavage by plasmin, and the molecules interacting with uPAR, suPAR, and c-suPAR. One very important question is, in any case, the definition of the molecular state of the HSC in uPAR Ko mice. Preliminary experiments indicate that the gene expression profile of sorted HSC is profoundly different in WT and uPAR Ko mice (Eden, O & Balsi to be published). Understanding the basis of these differences will lead to a much better understanding of HSC physiology and of the uPAR molecular plasticity. This will represent a priority, albeit very difficult, challenge in the next few years.

References

Aguirre Ghiso J.A., Kovalski K., and Ossowski L. (1999). Tumor dormancy induced by down-regulation of urokinase receptor in human carcinoma involves integrin and MAPK signaling. J. Cell Biol. 147:89–104.

Aguirre-Ghiso J.A., Liu D., Mignatti A., et al. (2001). Urokinase receptor and fibronectin regulate the ERK(MAPK) to p38(MAPK) activity ratios that determine carcinoma cell proliferation or dormancy *in vivo*. Mol. Biol. Cell. 12:863–879.

Andolfo A., English W.R., Resnati M., et al. (2002). Metalloproteases cleave the urokinase-type plasminogen activator receptor in the D1-D2 linker region and expose epitopes not present in the intact soluble receptor. Thromb. Haemost. 88:298–306.

Barinka C., Parry G., Callahan J., et al. (2006). Structural basis of interaction between urokinase-type plasminogen activator and its receptor. J. Mol. Biol. 363:482–495.

Beaufort N., Leduc D., Rousselle J.C., et al. (2004). Plasmin cleaves the juxtamembrane domain and releases truncated species of the urokinase receptor (CD87) from human bronchial epithelial cells. FEBS Lett. 574:89–94.

Blasi F. and Carmeliet P. (2002). uPAR: a versatile signalling orchestrator. Nat. Rev. Mol. Cell Biol. 3:932–943.

Blasi F., Vassalli J.D., and Danø K. (1987). Urokinase-type plasminogen activator: proenzyme, receptor and inhibitors. J. Cell Biol. 104:801–804.

Bohuslav J., Horejsi V., Hansmann C., et al. (1995). Urokinase plasminogen activator receptor, beta 2-integrins, and Src-kinases within a single receptor complex of human monocytes. J. Exp. Med. 181:1381–1390.

Broxmeyer H.E., Orschell C.M., et al. (2005). Rapid mobilization of murine and human hematopoietic stem and progenitor cells with AMD3100, a CXCR4 antagonist. J. Exp. Med. 201:1307–1318.

Bugge T.H., Flick M.J., Danton M.J., et al. (1996). Urokinase-type plasminogen activator is effective in fibrin clearance in the absence of its receptor or tissue-type plasminogen activator. Proc. Natl. Acad. Sci. USA 93:5899–5904.

Carion A., Benboubker L., Hérault O., et al. (2003). Stromal-derived factor 1 and matrix metalloproteinase 9 levels in bone marrow and peripheral blood of patients mobilized by granulocyte colony-stimulating factor and chemotherapy. Relationship with mobilizing capacity of haematopoietic progenitor cells. Br. J. Haematol. 122:918–926.

Carmeliet P., Schoonjans L., Kieckens L., et al. (1994). Physiological consequences of loss of plasminogen activator gene function in mice. Nature. 368:419–424.

Carstanjen D., Ulbricht N., Iacone A., et al. (2002). Matrix metalloproteinase-9 (gelatinase B) is elevated during mobilization of peripheral blood progenitor cells by G-CSF. Transfusion. 42:588–596.

Chaurasia P., Aguirre-Ghiso J.A., et al. (2006). A region in urokinase plasminogen receptor domain III controlling a functional association with alpha5beta1 integrin and tumor growth. J. Biol. Chem. 281:14852–14863.

Christopherson K.W., Hangoc G., and Broxmeyer H.E. (2002). Cell surface peptidase CD26/dipeptidylpeptidase IV regulates CXCL12/stromal cell-derived factor-1 alpha-mediated chemotaxis of human cord blood CD34$^+$ progenitor cells. J. Immunol. 169:7000–7008.

Christopherson K.W., Cooper S., Hangoc G., et al. (2003). CD26 is essential for normal G-CSF-induced progenitor cell mobilization as determined by CD26$^{-/-}$ mice. Exp. Hematol. 31:1126–1134.

Christopherson K.W., Hangoc G., Mantel C.R., et al. (2004). Modulation of hematopoietic stem cell homing and engraftment by CD26. Science. 305:1000–1003.

Degryse B., Resnati M., Rabbani S.A., et al. (1999). Src-dependence and pertussis-toxin sensitivity of urokinase receptor-dependent chemotaxis and cytoskeleton reorganization in rat smooth muscle cells. Blood. 94:649–662.

Degryse B., Resnati M., Czekay R.P., et al. (2005). Domain 2 of the urokinase receptor contains an integrin-interacting epitope with intrinsic signaling activity: generation of a new integrin inhibitor. J. Biol. Chem. 280:24792–24803.

Deng G., Curriden S.A., Wang S., et al. (1996). Is plasminogen activator inhibitor-1 the molecular switch that governs urokinase receptor-mediated cell adhesion and release? J. Cell Biol. 134:1563–1571.

de Paulis A., Montuosi N., Prevete N., et al. (2004). Urokinase induces basophil chemotaxis through a urokinase receptor epitope that is an endogenous ligand for formyl peptide receptor-like 1 and -like 2. J. Immunol. 173:5739–5748.

Dewerchin M., Nuffelen A.V., Wallays G., et al. (1996). Generation and characterization of urokinase receptor-deficient mice. J. Clin. Invest. 97:870–878.

Dumler I., Kopmann A., Wagner K., et al. (1999). Urokinase induces activation and formation of Stat4 and Stat1-Stat2 complexes in human vascular smooth muscle cells. J. Biol. Chem. 274:24059–24056.

Fazioli F., Resnati M., Sidenius N., et al. (1997). The urokinase-sensitive region of the urokinase receptor is responsible for its potent chemotactic activity. EMBO J. 16:7279–7286.

Fietz T., Hattori K., Thiel E., et al. (2006). Increased soluble urokinase plasminogen activator receptor (suPAR) serum levels after granulocyte colony-stimulating factor treatment do not predict successful progenitor cell mobilization *in vivo*. Blood. 107:3408–3409.

Ford C.E., Hamerton J.L., Barnes D.W., et al. (1956). Cytological identification of radiation-chimaeras. Nature. 177:452–454.

Franco P., Vocca I., Carriero M.V., et al. (2006). Activation of urokinase receptor by a novel interaction between the connecting peptide region of urokinase and alpha v beta 5 integrin. J. Cell Sci. 119:3424–3434.

Fruehauf S. and Seggewiss R. (2003). It's moving day: factors affecting peripheral blood stem cell mobilization and strategies for improvement. Br. J. Haematol. 122:360–375.

Fruehauf S., Haas R., Conradt C., et al. (1995). Peripheral blood progenitor cell (PBPC) counts during steady-state hematopoiesis allow to estimate the yield of mobilized PBPC after filgrastim (R-metHuG-CSF)-supported cytotoxic chemotherapy. Blood. 85:2619–2626.

Furlan F., Orlando S., Laudanna C., et al. (2004). The soluble D2D3(88–274) fragment of the urokinase receptor inhibits monocyte chemotaxis and integrin-dependent cell adhesion. J. Cell Sci. 117:2909–2916.

Furlan F., Galbiati C., Jorgensen N.R., et al. (2007). Urokinase plasminogen activator receptor affects bone homeostasis by regulating osteoblast and osteoclast function. J. Bone Miner. Res. 22:1387–1396.

Gardsvoll H. and Ploug M. (2007). Mapping of the vitronectin-binding site on the urokinase receptor: involvement of a coherent receptor interface consisting of residues from both domain I and the flanking interdomain linker region. J. Biol. Chem. 282:13561–13572.

Gardsvoll H., Gilquin B., Le Du M.H., et al. (2006). Characterization of the functional epitope on the urokinase receptor. Complete alanine scanning mutagenesis supplemented by chemical cross-linking. J. Biol. Chem. 281:19260–19272.

Gyetko M.R., Sud S., Kendall T., et al. (2000). Urokinase receptor-deficient mice have impaired neutrophil recruitment in response to pulmonary Pseudomonas aeruginosa infection. J. Immunol. 165:1513–1519.

Hibbin J.A., Njoku O.S., Matutes E., et al. (1984). Myeloid progenitor cells in the circulation of patients with myelofibrosis and other myeloproliferative disorders. Br. J. Haematol. 57:495–503.

Huang M., Mazar A.P., Parry G., et al. (2005). Crystallization of soluble urokinase receptor (suPAR) in complex with urokinase amino-terminal fragment (1–143). Acta Crystallogr. D Biol. Crystallogr. 61:697–700.

Jacobson L.O., Simmons E.L., Marks E.K., et al. (1951). Recovery from radiation injury. Science. 113:510–511.

Jo M., Thomas K.S., Marozkina N., et al. (2005). Dynamic assembly of the urokinase-type plasminogen activator signaling receptor complex determines the mitogenic activity of urokinase-type plasminogen activator. J. Biol. Chem. 280:17449–17457.

Jo M., Thomas K.S., Takimoto S., et al. (2007). Urokinase receptor primes cells to proliferate in response to epidermal growth factor. Oncogene. 26:2585–2594.

Kiyan J., Kiyan R., Haller H., et al. (2005). Urokinase-induced signaling in human vascular smooth muscle cells is mediated by PDGFR-beta. EMBO J. 24:1787–1797.

Kjoller L. and Hall A. (2001). Rac mediates cytoskeletal rearrangements and increased cell motility induced by urokinase-type plasminogen activator receptor binding to vitronectin. J. Cell Biol. 152:1145–1157.

Kollet O., Spiegel A., Peled A., et al. (2001). Rapid and efficient homing of human CD34(+)CD38 (−/low)CXCR4(+) stem and progenitor cells to the bone marrow and spleen of NOD/SCID and NOD/SCID/B2m(null) mice. Blood. 97:3283–3291.

Kook Y.H., Adamski J., Zelent A., et al. (1994). The effect of antisense inhibition of urokinase receptor in human squamous cell carcinoma on malignancy. EMBO J. 13:3983–3991.

Korbling M. and Fliedner T.M. (1994). History of blood stem cell transplants. Blood stem cell transplants. In: Peripheral blood stem cell autographts, eds. Gale R.P., Juttner C.A., Henon P. pp. 1994–1999. NewYork: Cambridge University Press.

Koshelnick Y., Ehart M., Hufnagl P., et al. (1997). Urokinase receptor is associated with the components of the JAK1/STAT1 signaling pathway and leads to activation of this pathway upon receptor clustering in the human kidney epithelial tumor cell line TCL-598. J. Biol. Chem. 272:28563–28567.

Kusch A., Tkachuk S., Haller H., et al. (2000). Urokinase stimulates human vascular smooth muscle cell migration via a phosphatidylinositol 3-kinase-Tyk2 interaction. J. Biol. Chem. 275:39466–39473.

Lee Y.H., Han J.Y., Seo S.Y., et al. (2002). Urokinase could increase the yield of progenitor cells during red cell depletion in elapsed and anticoagulated cord blood. Am. J. Hematol. 71:336–339.

Lévesque J.P., Takamatsu Y., Nilsson S.K., et al. (2001). Vascular cell adhesion molecule-1 (CD106) is cleaved by neutrophil proteases in the bone marrow following hematopoietic progenitor cell mobilization by granulocyte colony-stimulating factor. Blood. 98:1289–1297.

Lévesque J.P., Hendy J., Takamatsu Y., et al. (2002). Mobilization by either cyclophosphamide or granulocyte colony-stimulating factor transforms the bone marrow into a highly proteolytic environment. Exp. Hematol. 30:440–449.

Lévesque J.P., Hendy J., Winkler I.G., et al. (2003a). Granulocyte colony-stimulating factor induces the release in the bone marrow of proteases that cleave c-KIT receptor (CD117) from the surface of hematopoietic progenitor cells. Exp. Hematol. 31:109–117.

Lévesque J.P., Hendy J., Takamatsu Y., et al. (2003b). Disruption of the CXCR4/CXCL12 chemotactic interaction during hematopoietic stem cell mobilization induced by GCSF or cyclophosphamide. J. Clin. Invest. 111:187–196.

Levesque J.P., Liu F., Simmons P.J., et al. (2004). Characterization of hematopoietic progenitor mobilization in protease-deficient mice. Blood. 104:65–72.

Liles W.C., Broxmeyer H.E., Rodge E., et al. (2003). Mobilization of hematopoietic progenitor cells in healthy volunteers by AMD3100, a CXCR4 antagonist. Blood. 102:2728–2730.

Liu D., Aguirre-Ghiso J., Estrada Y., et al. (2002). EGFR is a transducer of the urokinase receptor initiated signal that is required for *in vivo* growth of a human carcinoma. Cancer Cell. 1:445–457.

Llinas P., Le Du M.H., Gardsvoll H., et al. (2005). Crystal structure of the human urokinase plasminogen activator receptor bound to an antagonist peptide. EMBO J. 24:1655–1663.

Madsen C.D., Ferraris G.M., Andolfo A., et al. (2007). uPAR-induced cell adhesion and migration: vitronectin provides the key. J. Cell Biol. 177:927–939.

May A.E., Kanse S.M., Lund L.R., et al. (1998). Urokinase receptor (CD87) regulates leukocyte recruitment via beta 2 integrins *in vivo*. J. Exp. Med. 188:1029–1037.

Mazzieri R., D'Alessio S., Kamgang-Kemhoe R., et al. (2006). An uncleavable uPAR mutant allows dissection of signaling pathways in uPA-dependent cell migration. Mol. Biol. Cell. 17:367–378.

Mazzieri R., Furlan F., D'Alessio S., et al. (2007). A direct link between expression of urokinase plasminogen activator receptor (uPAR), growth rate and oncogenic transformation in mouse embryonic fibroblasts. Oncogene. 26:725–732.

Molineux G., Pojda Z., Hampson I.N., et al. (1990). Transplantation potential of peripheral blood stem cells induced by granulocyte colony-stimulating factor. Blood. 76:2153–2158.

Montuori N., Carriero M.V., Salzano S., et al. (2002). The cleavage of the urokinase receptor regulates its multiple functions. J. Biol. Chem. 277:46932–46939.

Nguyen D.H., Webb D.J., Catling A.D., et al. (2000). Urokinase-type plasminogen activator stimulates the Ras/Extracellular signal-regulated kinase (ERK) signaling pathway and MCF-7 cell migration by a mechanism that requires focal adhesion kinase, Src, and Shc. Rapid dissociation of GRB2/Sps-Shc complex is associated with the transient phosphorylation of ERK in urokinase-treated cells. J. Biol. Chem. 275:19382–19388.

Papayannopoulou T., Craddock C., Nakamoto B., et al. (1995). The VLA4/VCAM-1 adhesion pathway defines contrasting mechanisms of lodgement of transplanted murine hemopoietic progenitors between bone marrow and spleen. Proc. Natl. Acad. Sci. USA 92:9647–9651.

Peled A., Petit I., Kollet O., et al. (1999). Dependence of human stem cell engraftment and repopulation of NOD/SCID mice on CXCR4. Science. 283:845–848.

Peled A., Kollet O., Ponomaryov T., et al. (2000). The chemokine SDF-1 activates the integrins LFA-1, VLA-4, and VLA-5 on immature human CD34(+) cells: role in transendothelial/stromal migration and engraftment of NOD/SCID mice. Blood. 95:3289–3296.

Pluskota E., Soloviev D.A., and Plow E.F. (2003). Convergence of the adhesive and fibrinolytic systems: recognition of urokinase by integrin alpha Mbeta 2 as well as by the urokinase receptor regulates cell adhesion and migration. Blood. 101:1582–1590.

Pruijt J.F., Fibbe W.E., Laterveer L., et al. (1999). Prevention of interleukin-8-induced mobilization of hematopoietic progenitor cells in rhesus monkeys by inhibitory antibodies against the metalloproteinase gelatinase B (MMP-9). Proc. Natl. Acad. Sci. USA 96:10863–10868.

Pruijt J.F., Verzaal P., van Os R., et al. (2002). Neutrophils are indispensable for hematopoietic stem cell mobilization induced by interleukin-8 in mice. Proc. Natl. Acad. Sci. USA 99:6228–6233.

Randall T.D. and Weissman I.L. (1997). Phenotypic and functional changes induced at the clonal level in hematopoietic stem cells after 5-fluorouracil treatment. Blood. 89:3596–3606.

Resnati M., Guttinger M., Valcamonica S., et al. (1996). Proteolytic cleavage of the urokinase receptor substitutes for the agonist-induced chemotactic effect. EMBO J. 15:1572–1582.

Resnati M., Pallavicini I., Wang J.M., et al. (2002). The fibrinolytic receptor for urokinase activates the G protein-coupled chemotactic receptor FPRL1/LXA4R. Proc. Natl. Acad. Sci. USA 99:1359–1364.

Selleri C., Montuori N., Ricci P., et al. (2005). Involvement of the urokinase-type plasminogen activator receptor in hematopoietic stem cell mobilization. Blood. 105:2198–2205.

Selleri C., Montuori N., Ricci P., et al. (2006). In vivo activity of the cleaved form of soluble urokinase receptor: a new hematopoietic stem/progenitor cell mobilizer. Cancer Res. 66:10885–10890.

Simon D.I., Wei Y., Zhang L., et al. (2000). Identification of a urokinase receptor-integrin interaction site. Promiscuous regulator of integrin function. J. Biol. Chem. 275:10228–10234.

Stephens R.W., Pollanen J., Tapiovaara H., et al. (1989). Activation of pro-urokinase and plasminogen on human sarcoma cells: a proteolytic system with surface-bound reactants. J. Cell Biol. 108:1987–1995.

Stoppelli M.P., Corti A., Soffientini A., et al. (1985). Differentiation-enhanced binding of the amino terminal fragment of human urokinase plasminogen activator to a specific receptor on U937 monocytes. Proc. Natl. Acad. Sci. USA. 82:4939–4943.

Tanaga K., Bujo H., Zhu Y., et al. (2004). LRP1B attenuates the migration of smooth muscle cells by reducing membrane localization of urokinase and PDGF receptors. Arterioscler. Thromb. Vasc. Biol. 24:1422–1428.

Tarui T., Akakura N., Majumdar M., et al. (2006). Direct interaction of the kringle domain of urokinase-type plasminogen activator (uPA) and integrin alpha v beta 3 induces signal transduction and enhances plasminogen activation. Thromb. Haemost. 95:524–534.

Till J.E., McCulloch E.A., and Siminovitch L. (1964). A stochastic model of stem cell proliferation, based on the growth of spleen colony-forming cells. Proc. Natl. Acad. Sci. USA 51:29–5136.

Valenzuela-Fernández A., Planchenault T., et al. (2002). Leukocyte elastase negatively regulates stromal cell-derived factor-1 (SDF-1)/CXCR4 binding and functions by amino-terminal processing of SDF-1 and CXCR4. J. Biol. Chem. 277:15677–15689.

van Pel M., van Os R., Velders G.A., et al. (2006). Serpina1 is a potent inhibitor of IL-8-induced hematopoietic stem cell mobilization. Proc. Natl. Acad. Sci. USA 103:1469–1474.

Vassalli J.-D., Baccino D., and Belin D. (1985). A cellular binding site for the Mr 55,000 form of the human plasminogen activator urokinase. J. Cell Biol. 100:88–92.

Wagers A.J., Allsopp R.C., and Weissman I.L. (2002). Changes in integrin expression are associated with altered homing properties of Lin(-/lo)Thy1.1(lo)Sca-1(+)c-kit(+) hematopoietic stem cells following mobilization by cyclophosphamide/granulocyte colony-stimulating factor. Exp. Hematol. 30:176–185.

Wei Y., Waltz D.A., Rao N., et al. (1994). Identification of the urokinase receptor as an adhesion receptor for vitronectin. J Biol Chem. 269:32380–32388.

Wei Y., Lukashev M., Simon D.I., et al. (1996). Regulation of integrin function by the urokinase receptor. Science. 273:1551–1555.

Wei Y., Czekay R.P., Robillard L., et al. (2005). Regulation of alpha5beta1 integrin conformation and function by urokinase receptor binding. J. Cell Biol. 168:501–511.

Wei Y., Tang C.H., Kim Y., et al. (2007). Urokinase receptors are required for alpha 5 beta 1 integrin-mediated signaling in tumor cells. J. Biol. Chem. 282:3929–3939.

Wright D.E., Wagers A.J., Gulati A.P., et al. (2001). Physiological migration of hematopoietic stem and progenitor cells. Science. 294:1933–1936.

Wright D.E., Bowman E.P., Wagers A.J., et al. (2002). Hematopoietic stem cells are uniquely selective in their migratory response to chemokines. J. Exp. Med. 195:1145–1154.

Wysoczynski M., Reca R., Ratajczak J., et al. (2005). Incorporation of CXCR4 into membrane lipid rafts primes homing-related responses of hematopoietic stem/progenitor cells to an SDF-1 gradient. Blood. 105:40–48.

Yu W., Kim J., and Ossowski L. (1997). Reduction in surface urokinase receptor forces malignant cells into a protracted state of dormancy. J. Cell Biol. 137:767–777.

Zhang G., Kim H., Cai X., et al. (2003a). Urokinase receptor modulates cellular and angiogenic responses in obstructive nephropathy. J. Am. Soc. Nephrol. 14:1234–1253.

Zhang G., Kim H., Cai X., et al. (2003b). Urokinase receptor deficiency accelerates renal fibrosis in obstructive nephropathy. J. Am. Soc. Nephrol. 14:1254–7121.

Chapter 23
The Urokinase Receptor and Integrins Constitute a Cell Migration Signalosome

Bernard Degryse

Abstract Initially identified as a permissive receptor involved in the regulation of pericellular proteolysis, the receptor of urokinase, urokinase-type plasminogen activator receptor (uPAR), is also a signaling receptor capable of regulating tissue remodeling, cell adhesion, differentiation, proliferation, and migration. Therefore, the uPA/uPAR system can permit a tumor cell to modify its environment or to move across it. uPAR exerts these effects by interacting laterally with other membrane receptors such as seven-transmembrane domain receptors, tyrosine kinase receptors, and integrins. This latter family of receptors mediates bidirectional signaling and is connected to the cell cytoskeleton. Beside their well-documented roles in cell adhesion and migration, integrins promote cell survival and resistance to genotoxic injury, crucial properties that allow a tumor cell to adapt to new environments and survive in hostile conditions. Therefore, combining the uPA/uPAR system to the integrin system provides an ultimate advantage to the tumoral cell. Moreover, thanks to the large array of ligands and membrane-bound partners, formation of the uPAR–integrin complex represents the cornerstone that consents to build up larger signaling complexes and to adjust their compositions in order to satisfy the various cellular/tumoral requirements as the tumor cells grow, invade, or disseminate. Therefore, the uPAR–integrin complexes constitute convenient adaptable signaling complexes, some kind of "chameleon signalosome." This chapter will discuss these issues, describing the influence of uPAR on integrin activity (and conversely) and signaling, and summarize our understanding of the molecular basis of uPAR–integrin interactions. In addition, new challenging data that may revolutionize the classical view of uPAR–integrin interaction will be also discussed.

B. Degryse
Dept. of Molecular Biology and Functional Genomics, DIBIT, University Vita-Salute
San Raffaele, Via Olgettina 58, 20132 Milan, Italy, e-mail: degryse.bernard@hsr.it

D. Edwards et al. (eds.), *The Cancer Degradome.*

Introduction

The most original and important characteristic of urokinase-type plasminogen activator receptor (uPAR) is its glycosyl-phosphatidyl-inositol (GPI) anchor which implies that uPAR is entirely located on the outer side of the plasma membrane. This fact is quite challenging as uPAR is nevertheless a signaling receptor capable of regulating gene expression, cell proliferation, adhesion, and migration. Most probably, uPAR achieves these functions by interacting laterally with other molecules. So far, various uPAR partners have been identified, including seven-transmembrane domain receptors such as FPRL1; endocytic receptors such as LRP (LDL receptor-related protein), very low-density lipoprotein receptor (VLDL-R), the mannose 6-phosphate/IGF-II receptor (CD222, CIMPR), or uPAR-associated protein (uPARAP) (Endo180); caveolin, the gp130 cytokine receptor; and tyrosine kinase receptors such as the EGF receptor (EGF-R), the PDGF receptor (PDGF-R), or the IGF-1 receptor (IGF-1-R) (for reviews, see Blasi and Carmeliet 2002, Degryse 2003, Ragno 2006). Integrins represent another large family of uPAR partners, and this chapter is particularly dedicated to the description of the interactions between uPAR and integrins.

Introducing the Urokinase Receptor

uPAR is a 283-residue single-chain protein constituted of three homologous domains with domain I located at its N-terminus. uPAR is bound to the cell surface by its GPI anchor added posttranslationally at the C-terminus of domain III (for reviews on uPAR, see Blasi and Carmeliet 2002, Degryse 2003, Ragno 2006).

uPAR is primarily known as the receptor of urokinase (uPA) that increases the rate of activation of pro-uPA into uPA, thereby enhancing the activation of plasminogen into active plasmin (Ellis et al. 1989). In addition, uPAR can localize uPA activity at discrete points on the cell surface. Thus, uPAR was first identified as a major player in the regulation of pericellular proteolysis, a process of particular importance in the context of cell migration (Blasi and Carmeliet 2002). Degradation of the extracellular matrix (ECM) and of basement membrane proteins, intravasation, and extravasation represent key steps in tumor invasion and/or metastatic dissemination. Moreover, the catalytic-dependent effects of uPA and thus uPAR can also depend on uPA ability to activate growth factors such as basic fibroblast growth factor (bFGF) (Odekon et al. 1992), protransforming growth factor-β (pro-TGF-β) (Odekon et al. 1994), and prohepatocyte growth factor (pro-HGF) (Naldini et al. 1992, 1995).

However, uPAR is also an effective signaling receptor mediating proliferative and migrating signals. Therefore, uPAR has not only permissive (as regulator of pericellular proteolysis) but also inducing functions (as a signaling receptor).

A major advance in our understanding of uPAR function was provided by the X-ray crystal structure that has been published recently (Huang et al. 2005, Llinas

Table 23.1 The Soluble Ligands of uPAR

Ligands	Classification	Functions
Pro-uPA	Zymogen	Inactive precursor of uPA promotes cell adhesion, migration, and proliferation
High molecular weight form urokinase (uPA)	Serine protease	Proteolytic enzyme promotes cell adhesion, migration, and proliferation
uPA–PAI-1 complex	Complex of uPA and its physiological inhibitor PAI-1	No proteolytic activity. Inhibits uPA-dependent cell migration. Induces uPAR and integrin internalization
Vitronectin (VN)	Extracellular matrix (ECM) and plasma protein	Promotes cell adhesion and migration. Cofactor of PAI-1
Two-chain high molecular weight kininogen (HKa)	Plasma protein	Proinflammatory. Inhibits VN-dependent adhesion and migration
Streptococcal surface deshydrogenase (SDH)	Anchor-less microbial surface protein from *Streptococcus pyogenes*	Glycolytic enzyme. Promotes bacterial adherence to host cells[a]

[a] According to Jin et al. (2005).

et al. 2005, Barinka et al. 2006, Huai et al. 2006). uPAR has the form of a "croissant" forming a pocket in which uPA is bound. Thus, this structure leaves the whole external surface of uPAR available for other interactions, in perfect agreement with the numerous reported soluble (Table 23.1) and membrane-bound ligands of uPAR (Table 23.2). The extracellular ligands are interconnected by complex relationships, and might regulate every step of the cell migration cycle by conveying "Stop" and "Go" signals to the cell (Fig. 23.1).

Interestingly, uPAR itself can be listed among the membrane-bound ligands of uPAR. Dimerization of uPAR plays a key role in the binding of vitronectin (VN) and in the (re)distribution of uPAR into the lipid rafts (Sidenius et al. 2002, Cunningham et al. 2003). Being a GPI-anchored protein, uPAR has more mobility onto the plasma membrane and has been shown to relocalize on particular place of the cell surface (Resnati et al. 1996, Degryse et al. 1999) where it can be found into large signaling complexes (Bohuslav et al. 1995). In addition, uPAR clustering initiates uPAR-dependent signaling (Sitrin et al. 2000). Lipids rafts and caveolae are considered as signaling platforms where uPAR (or other membrane receptors) can be conveniently located to establish interactions with membrane-bound ligands, and thus downstream signaling molecules such as c-Src and FAK (focal adhesion kinase) (for a review, see Simons and Toomre 2000). The presence of uPAR in these microdomains appears connected to signaling as uPAR was reported to associate and stabilize caveolin–β_1 integrin complex (Chapman et al. 1999, Wei et al. 1999). In addition, uPA-induced chemotaxis is inhibited both by anti-uPAR and anti-$\alpha v\beta 3$ antibodies (Degryse et al. 1999). The reverse relationship has also been proven true; both anti-uPAR and anti-$\alpha v\beta 3$ antibodies efficiently block VN-induced cell migration (Degryse et al. 2001a).

Table 23.2 The Membrane-bound Partners of uPAR

Partner	Type	Functions
uPAR	GPI-anchored protein	Plasminogen activation Pericellular proteolysis Cell adhesion, migration, and proliferation
FPRL1, FPR, FPRL2	Seven-transmembrane domain receptors	Cell migration
$\alpha M\beta 2$, $\alpha L\beta 2$, $\alpha X\beta 2$, $\alpha 3\beta 1$, $\alpha 4\beta 1$, $\alpha 5\beta 1$, $\alpha 6\beta 1$, $\alpha 9\beta 1$, $\alpha v\beta 3$, $\alpha v\beta 5$, $\alpha v\beta 6$	Integrins	Cell adhesion and migration
EGF-R, PDGF-R, IGF-1-R	Receptor protein tyrosine kinases	Cell proliferation and migration
LRP, LRP1B, VLDL-R, mannose 6-phosphate-R, uPARAP	Endocytic receptors	uPAR internalization (LRP: uPAR and integrin internalization)
Caveolin	Scaffolding/structural protein	Signaling
Gp130	Cytokine receptor	Cell migration
gC1qR	Complement receptor	Complement activation
L-selectin	Adhesion receptor	Cell adhesion
Seprase	Serine protease	Pericellular proteolysis

The difference between the two uPAR functions, permissive versus signaling receptor, appears to be correlated with the oligomerization state, that is, monomeric uPAR versus oligomeric (dimeric) uPAR, respectively. In both situations, the binding of uPA to uPAR is the decisive step that triggers the increase in pericellular proteolysis and induces a conformational change that results in the exposition of the SRSRY chemotactic epitope located in the linker region between domains I and II of uPAR, which may promote lateral interactions with other membrane receptors such as FPRL1 and integrins. The SRSRY sequence is responsible for the migratory properties of uPAR, implicating it as a cell surface chemokine (Blasi 1999, Fazioli et al. 1997, Degryse et al. 1999) or MACKINE (membrane-anchored chemokine-like proteins) (Degryse 2003).

In addition, various forms of uPAR have been described. Degradation of the GPI anchor generates soluble uPAR (suPAR). Both membrane-bound uPAR and suPAR can be cleaved in the linker region between domains I and II by a variety of proteases including uPA giving DI fragment and DIIDIII-uPAR (Ploug and Ellis 1994). Interestingly, DIIDIII-uPAR is chemotactic reproducing the effects of uPA, thereby indicating that DIIDIII-uPAR actually binds to FPRL1 (Fazioli et al. 1997). Moreover, high levels of soluble forms of uPAR are markers of cancers and correlate with poor prognosis (Stephens et al. 1999, Brünner et al. 1999, Sier et al. 1999).

The presence of an intact receptor and its uPA-binding capacities are very important for the internalization of uPAR. However, the internalization of uPAR is not promoted by uPA but rather by its physiological inhibitor, the plasminogen activator inhibitor-1 (PAI-1), which induces the formation of PAI-1–uPA complex

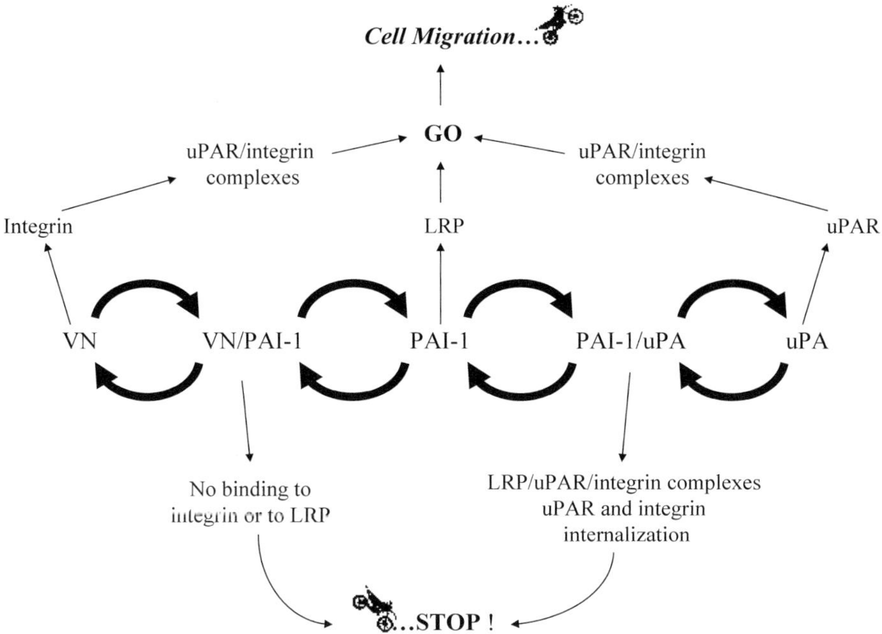

Fig. 23.1 VN, PAI-1, and uPA, and their respective receptor integrins, LRP and uPAR, constitute a cell migration signalosome or "chameleon" signalosome. This model explains the control of cell migration by VN, PAI-1, and uPA that correlates with their different forms. "Go" signals leading to cell migration are generated by the binding of VN, PAI-1, and uPA to their respective motogenic receptor (i.e., integrins, LRP, and uPAR). At the leading edge of the motile cell, "Go" signals induce cellular responses such as extension of protrusions in the direction of migration, cell polarization, focal adhesion assembly, and motility. "Stop" signals result from the formation of VN/PAI-1 and PAI-1–uPA complexes that inhibit integrin-, LRP-, and uPAR-dependent cell migration (Stefansson and Lawrence 1996, Degryse et al. 2001b, Kamikubo et al. submitted). Thus, "Stop" signals can control the disassembly of adhesion sites at the rear of the migrating cell. Interestingly, VN, PAI-1, and uPA (each protein under both free and complex forms) can control the fate of protrusions at the leading edge that the motile cell uses to probe the ECM according to the "sticky fingers" mechanism recently described (Galbraith et al. 2007). Therefore, VN, PAI-1, and uPA are capable of regulating all steps of the migration cycle that are required sequentially or even simultaneously at various times and locations of the migrating cell. In addition, the uPAR–integrin complexes can recruit their numerous extracellular ligands and membrane-bound partners to build up larger signaling complexes in order to meet the cellular requirements and to adjust their compositions according to the environment of the cells as the cells move and/or grow. Therefore, the uPAR–integrin complexes constitute some kind of "chameleon signalosome," that is, migrating/adapting signaling complexes.

and the subsequent internalization of uPAR (Nykjaer et al. 1992, 1994; Conese et al. 1995). Internalization requires the formation of contacts not only between PAI-1, uPA, the LDL receptor-related protein (LRP or LRP-1), and uPAR but also directly between LRP and uPAR (Nykjaer et al. 1992, 1994; Conese et al. 1995, Czekay et al. 2001). Once internalized, the PAI-1–uPA complex is degraded while uPAR is recycled back to the cell surface (Nykjaer et al. 1997). Regeneration of free uPAR is thought to maintain a functioning uPAR system on the cell membrane, permitting new interactions with uPA or VN and thus a tight control of pericellular proteolysis and signaling. The impact of the internalization of uPAR on the cell machinery is real. Blocking this process results in increased expression of uPAR on the cell surface and higher uPA production, and subsequently enhances plasminogen activation and cell motility (Weaver et al. 1997, Webb et al. 1999, 2000). Conversely, in the absence of uPAR regeneration on the cell membrane, cells stop migrating (Degryse et al. 2001b). PAI-1–uPA complexes inhibit uPA-induced cell migration by promoting LRP-dependent internalization of uPAR and integrins (Conese et al. 1995, Degryse et al. 2001b, Czekay et al. 2003).

In contrast, uPA, uPA/PAI-1, and LRP do not bind to DIIDIII-uPAR, and consequently this shorter form of uPAR is not efficiently internalized (Hoyer-Hansen et al. 1992, Nykjaer et al. 1998, Ragno et al. 1998). The mannose 6-phosphate/insulin-like growth factor-II receptor (cation-independent mannose 6-phosphate receptor, CIMPR) can bind and internalize both uPAR and DIIDIII-uPAR in an uPA-independent manner (Nykjaer et al. 1998). Again, this internalization receptor negatively regulates uPAR functions (Leksa et al. 2002). Tumor cells have been reported to bear more copies of DIIDIII-uPAR than of full-length uPAR on their cell surface (Ragno et al. 1998). This shorter form of uPAR cannot bind to uPA or VN. In contrast, the two-chain kinin-free high molecular weight kininogen (HKa) is capable of binding domains II and III of both DIIDIII-uPAR and full-length uPAR in a Zn^{2+}-dependent manner (Colman et al. 1997, Chavakis et al. 2000). HKa and VN are competitive-binding partners, providing uPAR with interesting adhesive and antiadhesive properties.

The Integrins, a Large Family of uPAR Partners

The fact that uPAR was capable of mediating different signals to the cell challenged the paradigm that a signaling receptor requires a cytoplasmic domain to transduce a signal. Quite logically, it was hypothesized that these signaling capacities were due to lateral interactions between uPAR and other membrane receptors (Resnati et al. 1996, Fazioli et al. 1997). This hypothesis was later fully supported by the identification of FPRL1 as a transducer of uPAR (Resnati et al. 2002). Further investigations revealed a large array of uPAR partners such as the integrins (for reviews, see Blasi and Carmeliet 2002, Degryse 2003, Ragno 2006). Presently, the relationships between uPAR and integrins are probably the most studied.

Integrins are type-I transmembrane heterodimers consisting of one α and one β subunit. In mammals, 18 α-subunits and 8 β-subunits have been identified generating 24 heterodimers (for reviews, see Ffrench-Constant and Colognato 2004, Ginsberg et al. 2005, Kinashi 2005). Each subunit is constituted of a large extracellular domain, a short single transmembrane domain, and a small cytoplasmic domain. This intracellular domain can be directly connected to downstream signaling molecules such as FAK and c-Src, and linked to the cell cytoskeleton.

Integrins are well known for their role in the regulation of cell adhesion and migration. Their major ligands are ECM proteins. Integrins also have numerous membrane-bound ligands that can finely modulate their activity (Porter and Hogg 1998, Kinashi 2005). However, integrins are capable of bidirectional signaling which represents a major difference between uPAR- and integrin-dependent signaling. Inside-out signaling regulates the extracellular-binding activity of integrins, whereas outside-in signaling induced by binding of ligands such as ECM proteins generates signals that are transmitted into the cell.

Integrin activity is accurately regulated via several mechanisms that control integrin conformation and clustering, thereby adjusting affinity and avidity. Integrins exhibit folded, intermediate, and fully opened conformations characterized by increasing affinity that reflects their state of activation. High-affinity binding of ligands is regulated by these conformational changes, a process referred to as integrin activation that can be achieved by inside-out signaling initiated, for instance, by chemokines (Kinashi 2005). On the other hand, integrin ligation by their ligands, which kicks off outside-in signaling, induces also conformational changes and integrin clustering. Last, the numerous integrin partners such as integrin-associated protein (IAP), tetraspanins, and uPAR can adapt integrin affinity, avidity, and signaling in order to precisely meet cell requirements (Porter and Hogg 1998, Kinashi 2005).

uPAR interacts with almost half of the members of the integrin family including $\alpha M\beta 2$ (Mac-1, CR3), $\alpha L\beta 2$ (LFA-1), $\alpha X\beta 2$ (CR4), $\alpha 3\beta 1$, $\alpha 4\beta 1$, $\alpha 5\beta 1$, $\alpha 6\beta 1$, $\alpha 9\beta 1$, $\alpha v\beta 3$, $\alpha v\beta 5$, and $\alpha v\beta 6$. However, this list is certainly longer as uPAR has been reported to interact with integrins from the $\beta 1$, $\beta 2$, $\beta 3$, and $\beta 5$ subfamilies. Most of these are cis-interactions within the same cell, but transinteractions have also been reported, suggesting that uPAR–integrin association also plays a role in cell–cell contacts and signaling (Tarui et al. 2001a). Experimental evidences of uPAR–integrin interactions were mainly based on colocalization (Pöllänen et al. 1988, Reinartz et al. 1995, Xue et al. 1997, Ghosh et al. 2000, Wei et al. 2001, Bass et al. 2005), cocapping (Xue et al. 1994, Bohuslav et al. 1995), fluorescence resonance energy transfer (FRET) (Xue et al. 1997, Kindzelskii et al. 1997, Xia et al. 2002), coimmunoprecipitation (Bohuslav et al. 1995, Xue et al. 1997, Ghosh et al. 2000, Wei et al. 2001), and in vitro pull-down assay using purified proteins (Degryse et al. 2005). Colocalization, cocapping, and coimmunoprecipitation do not really demonstrate a direct interaction in vivo. However, FRET is a widely used method to monitor protein–protein interactions in living cells, and does support direct uPAR–integrin interactions. FRET is a process in which energy is transferred from the donor fluorophore to the acceptor, and is usually detectable up to 10 nm. In fact, Xue et al. (1997) stated a distance of 7 nm between uPAR and $\beta 1$ or $\beta 3$ integrins.

Functional assays were also used and brought a more vivid view of the wide influence of uPAR on integrin activity. Perhaps the best example is given by the $\beta 2$ integrins $\alpha M\beta 2$, and $\alpha L\beta 2$, which are inactive in uPAR knockout mice preventing neutrophils recruitment, due to altered leukocyte adhesion to the endothelial wall (May et al. 1998, Simon et al. 2000). Similarly, targeting uPAR with anti-uPAR antibodies blocked VN/$\alpha v\beta 3$-dependent migration of rat smooth muscle cells (SMC) (Degryse et al. 1999). Another impressive example is the forced change in ligand binding observed in uPAR-transfected HEK 293 cells where uPAR inhibited β1 integrin-dependent cell adhesion to fibronectin (FN) but promoted β1-dependent adhesion to VN (Simon et al. 1996, Wei et al. 1996). This uPAR-enforced switch in ligand preference may provide fast, easily regulated adaptative properties which might be of great benefit for migrating cells such as inflammatory, invasive, and metastatic cells as they have to adapt to the different encountered environments.

uPAR can exert these effects using several mechanisms that can eventually be combined in different manners. uPAR can modulate integrin function by modifying affinity and avidity, or regulate the spatial distribution of integrins. Furthermore, uPAR controls the number of integrins present on the cell surface by acting on integrin expression and internalization. In other words, uPAR is a real integrin manager controlling how integrins work, their number, and their location.

uPAR behaves as a modulator of integrin function. Thus, in this role uPAR is not very different from other integrin-associated molecules such as IAP or tetraspanins that alter, inhibit, or stimulate integrin functions (Porter and Hogg 1998, Kinashi 2005). Furthermore, since knocking out uPAR expression has no lethal effect in mice, it may be hypothesized that at least some integrins can use other partners as a substitute. However, it is surprising that despite the high number of studies on uPAR and integrins published so far, none have reported the effects of uPAR on integrin affinity. A recent report has shown that the presence or absence of uPAR has clear effects on integrin conformation and adhesion of tumor cells (Wei et al. 2005). In addition, uPAR stimulates integrin-dependent adhesion and migration (Sitrin et al. 1996, Yebra et al. 1996, May et al. 1998, 2000; Degryse et al. 2005, Wei et al. 2005); thus it is quite tempting to assume that uPAR effectively modulates integrin affinity but the final proof is still lacking. Nevertheless, uPAR promotes integrins clustering (this step comes after the conformational changes of integrins) (Myohanen et al. 1993, Wei et al. 1996, Degryse et al. 1999, Gellert et al. 2004), providing evidences that uPAR positively regulates integrin avidity.

In fact, uPAR also contributes to the regulation of the spatial distribution of the integrins on the cell surface (Ghosh et al. 2000). In leukocytes, uPA induces the colocalization of β2 integrins and uPAR. Similarly, pro-uPA induces uPAR and αvβ3 colocalization into the membrane ruffles of migrating SMC (Degryse et al. 1999). There is also a correlation between the level of uPAR expression and the formation of focal adhesion (Chintala et al. 1997, Kjøller and Hall 2001, Abu-Ali et al. 2005). Moreover, uPAR was shown to redistribute integrins in lipid rafts where uPAR promotes the formation of caveolin–β₁ integrin complexes by associating and stabilizing these complexes (Stahl and Mueller 1995, Chapman et al. 1999, Wei et al. 1999, Schwab et al. 2001). VN, the ligand of both uPAR and integrins, induces the clustering of uPAR and αvβ3 (Ciambrone and McKeown-Longo 1992, Xue et al. 1997, Stepanova et al. 2002).

Most of the reports of the literature mention a positive regulation of integrin activity by uPAR but some negative effects were also reported. In addition to the examples cited above, β2, α5β1, and αvβ5 integrins requires uPAR occupancy by uPA in order to function correctly (Simon et al. 1996, Yebra et al. 1996, Chavakis et al. 1999, Silvestri et al. 2002, Margheri et al. 2006). In fact, it appears that uPA is a key player in inducing uPAR–integrin interactions, suggesting that these interactions are dependent on uPAR conformation (Simon et al. 1996, Yebra et al. 1996, 1999; Carriero et al. 1999, Wei et al. 2001, Degryse et al. 2005). The fact that suPAR binds poorly to integrins supports this hypothesis (Degryse et al. 2005). However, in light of a few recent reports showing that uPA actually binds to integrins, the situation might be very different (Pluskota et al. 2003, 2004; Demetriou et al. 2004, Kwak et al. 2005, Franco et al. 2006, Tarui et al. 2006). Indeed, uPA may bridge uPAR and integrins together, permitting their interactions. uPA can bind to uPAR via its growth factor domain and simultaneously to integrin through its kringle domain or even through its catalytic domain (Pluskota et al. 2003, 2004; Demetriou et al. 2004, Kwak et al. 2005, Franco et al. 2006, Tarui et al. 2006, Pawar et al. 2007). Indeed, a single uPA was shown to bind to both uPAR and $\alpha M\beta2$ (Pluskota et al. 2003). In addition, both kringle and proteolytic domain of uPAR bind to the I-domain of $\alpha M\beta2$, inducing cell adhesion and migration, and boosting plasminogen activation and fibrinolysis (Pluskota et al. 2003, 2004). These reports are in line with previous observation showing that uPA negatively regulates $\alpha M\beta2$ (Mac-1) activity (Sitrin et al. 1996, Simon et al. 1996). uPA may block $\alpha M\beta2$ binding to fibrinogen by steric hindrance of the I-domain of the integrin. Furthermore, the uPA kringle binds to $\alpha v\beta3$, $\alpha4\beta1$, and $\alpha9\beta1$ integrins (Kwak et al. 2005, Tarui et al. 2006). Ligation of $\alpha v\beta3$ has functional consequences such as expression of cytokines, cell adhesion (neutrophils, CHO cells), migration (chemotaxis and haptotaxis) of murine SMC and CHO cells, and plasminogen activation, suggesting that kringle binding actually regulates integrin activity (Kwak et al. 2005, Tarui et al. 2006). In murine SMC, the signaling pathway activated by the kringle domain includes $G_{i/o}$ protein, PI-3 kinase, ERK and p38 MAP kinases, and the EGF-R (Roztocil et al. 2007). However, the isolated kringle domain does not exactly mimic the effects of uPA. In contrast to uPA, the kringle inhibits angiogenesis and tumor growth (Kim et al. 2003a, b, 2007). Angiostatin (K1-4) and other kringle fragments K1-3 and K1-5 of plasminogen exert similar inhibitory effects through binding to $\alpha v\beta3$ (Tarui et al. 2001b). By removing the β-propeller domain, uPA (acting as a protease) can also regulate the activity of at least $\alpha6$ integrins (Demetriou et al. 2004). However, a recent report suggests that the cleavage of integrin $\alpha6$ may be connected to invasion (Pawar et al. 2007). First, the cleaved form of $\alpha6$ is observed only on invasive human prostate cancer tissue. Second, prostate tumoral cells overexpressing wild-type $\alpha6$ exhibit a threefold increase in migration on laminin when compared to their cellular counterpart expressing a noncleavable form of $\alpha6$ (Pawar et al. 2007). Recently, a synthetic peptide derived from the linker region of uPA has been shown to bind to $\alpha v\beta5$ and promotes cytoskeleton reorganization and cell migration (Franco et al. 2006). Taken together, these studies suggest that uPA binding to integrins may play a crucial role by regulating integrin activity. However, uPA binding to integrins may promote effects that are different from those induced by its binding to uPAR. Moreover, it remains to be clearly determined whether uPA brings uPAR and integrins together, permitting their interactions or, alternatively, binds to a preformed uPAR–integrin complex.

The influence of the shorter forms of uPAR on integrin activity is not well described. It has been demonstrated that DIIDIII-uPAR cannot bind to uPA or VN and does not bind to integrins too (Ragno et al. 1998, Montuori et al. 1999, 2002). However, DIIDIII-uPAR may still exert an effect on integrin activity, as shown by the longer DIIDIII-uPAR$_{88-274}$ (this form still possesses the SRSRY chemotactic sequence and binds to FPRL1), which inhibits integrin-dependent adhesion by blocking chemokine-induced inside-out signaling (Furlan et al. 2004).

Both uPAR and integrins can crossregulate uPAR–integrin interactions by controlling the expression of the partner. The level of expression of uPAR (and/or its ligand uPA) is under the control of integrins, and the opposite relationship has also been proven to be true (Bianchi et al. 1996, Wang et al. 1998, Ghosh et al. 2000, Adachi et al. 2001, Hapke et al. 2001a, b, Besta et al. 2002). For instance, the expression of uPAR and αvβ3 is strictly correlated (Nip et al. 1995, Adachi et al. 2001, Khatib et al. 2001). This observation agrees that uPAR and integrins are closely involved in the regulation of cell migration. Both uPAR- and αvβ3-dependent cell migration is totally inhibited by anti-uPAR and anti-αvβ3 antibodies (Degryse et al. 1999, 2001a). Thus, there is a clear correlation between the level of uPAR expression and the activation state of the integrins as the formation and disruption of the uPAR–integrin complex corresponds to an activation or deactivation of the integrin (Chintala et al. 1997, Aguirre Ghiso et al. 1999, Simon et al. 2000). According to the literature, higher levels of uPAR expression lead to integrin activation and lower levels to integrin deactivation. For example, expression of uPAR correlates with activation of various integrins such as β2-integrins αMβ2, αLβ2 (May et al. 2002), and α5β1 (thereby promoting tumor growth) (Aguirre-Ghiso et al. 2001). Again, uPA may play a key role. uPAR occupancy by uPA increased αvβ5 expression and activity (Silvestri et al. 2002). Conversely, downregulating uPAR expression with antisense oligonucleotides decreases αMβ2 adhesive functions, that is, integrin activity (Sitrin et al. 1996). Similarly, the downregulation of uPAR reduces αvβ3 expression and subsequent signaling and cell migration (Adachi et al. 2001, Gondi et al. 2006). This is also true for α5β1, a reduced uPAR expression leads to decreased avidity and tumor dormancy (Aguirre Ghiso et al. 1999).

uPAR internalization can also regulate the activity and the number of integrins present on the cell surface (Czekay et al. 2003). This process is initiated by PAI-1 and is LRP dependent. PAI-1 induces integrins' deactivation by promoting uPA–uPAR–integrins complexes' internalization. PAI-1 detaches cells from VN, FN, and collagen 1, suggesting a mechanism valid for various integrins (Czekay et al. 2003). Therefore, uPAR internalization appears to be a fast and convenient way to regulate integrin activity and underline the importance of the uPAR–integrin interactions.

Interestingly, integrins also have a concrete influence on uPAR activity, localization, and expression as exemplified in various reports. αMβ2 and α4β1 activate uPAR (Wong et al. 1996, May et al. 2000). The integrin ligands FN, VN, and laminin induce the interaction of uPAR (particularly at focal adhesion) with β1, β3, α3, α5, α6, and αv integrins (Xue et al. 1997). Similarly, αMβ2 or α3β1 aggregation redirects uPAR to integrin clusters (Xue et al. 1994, Ghosh et al. 2000, 2006). Partial or complete absence of integrins CD11 (αL, αM, αX, αD)/CD18 (β2) results in decreased capping of uPAR (Kindzelskii et al. 1994). Expression of uPAR is β1 and β2 dependent in T lymphocytes (Bianchi et al. 1996). Last, IL-1-induced upregulation of α6β1 and uPAR correlates with increased migration while down-regulation with anti-α6, β1, and uPAR antibodies reduces signaling, cell proliferation, adhesion, and migration of pancreatic cancer cells (Sawai et al. 2006). Furthermore, integrins have also been shown to exploit uPAR to regulate the activity of other integrins; for instance, α4β1 (VLA-4) uses uPAR as mediator for the activation of β2 integrins (May et al. 2000).

Molecular Basis of uPAR–Integrin Interactions

The mechanism of uPAR–integrin interactions is unknown. Since uPA positively or negatively regulates most uPAR–integrin interactions, domain I of uPAR (which is essential for uPA binding) is directly or indirectly very important for the formation of uPAR–integrin complexes (Myohanen et al. 1993, Montuori et al. 2002). Furthermore, DIIDIII-uPAR neither binds uPA nor associates with integrins (Ragno et al. 1998, Montuori et al. 1999, 2002). However, recent advances have revealed the important role played by domains II and III in uPAR–integrin interactions (Degryse et al. 2005, Chaurasia et al. 2006, Wei et al. 2007).

The first identified sequence of uPAR involved in uPAR–integrin interactions is located in domain II (Degryse et al. 2005). This sequence baptized D2A consists of the residues $_{130}$IQEGEEGRPKDDR$_{142}$ of human uPAR. D2A-derived synthetic peptide binds to αvβ3 and α5β1 integrins and induces integrin-, not uPAR-, dependent signaling, thereby stimulating cell migration. The minimum chemotactic sequence was also reported and is composed of the four residues GEEG. Introducing mutations (changing the two glutamic acids into two alanines) in this sequence generated D2A-Ala and GAAG peptides and abolished the chemotactic activity of both D2A and GEEG. Very interestingly, D2A-Ala and GAAG were shown to be very potent inhibitors of integrin-dependent signaling and chemotaxis. Furthermore, subtle differences between the mechanism of action of D2A and D2A-Ala were also reported. The migration-promoting effect of D2A is uPAR-dependent, whereas the inhibitory action of D2A-Ala is uPAR-independent. In addition, our most recent data showed that D2A also has a mitogenic activity, and thus represents the first identified mitogenic sequence of uPAR (Eden et al. in preparation). On the basis of these data, together with our observation that cells expressing DIDII-uPAR are less sensitive (but not insensitive) to low doses of VN, we proposed the existence of other sites of interaction between uPAR and integrins most probably in domain III (Degryse et al. 2005).

Two recent papers confirmed this latter hypothesis (Chaurasia et al. 2006, Wei et al. 2007). Synthetic peptide $_{240}$GCATASMCQ$_{248}$ derived from the sequence of domain III of human uPAR disrupts suPAR–α5β1 integrin complex (Chaurasia et al. 2006). Furthermore, single amino acid mutants of uPAR S245A or H249A, and D262A failed to associate with α5β1 and α3β1, respectively, suggesting that this part of uPAR is involved in uPAR–integrin interaction (Chaurasia et al. 2006, Wei et al. 2007). However, peptide 240–248 does not promote cell signaling but rather reduced integrin-dependent ERK activation (Chaurasia et al. 2006). Therefore, taken together, data from these three papers indicates one integrin-binding site in domain II and one additional binding site in domain III of uPAR (Degryse et al. 2005, Chaurasia et al. 2006, Wei et al. 2007). However, only the D2A sequence from domain II has intrinsic signaling, chemotactic, and mitogenic activities (Degryse et al. 2005).

Sequences involved in uPAR–integrin interactions have also been identified both on α and β subunit of integrins. A uPAR-binding sequence spanning residues 424–440 is present within the αM subunit of Mac-1, representing a non-I-domain-binding site located in the β propeller (Simon et al. 2000). M25, the derived synthetic peptide, $_{424}$PRYQHIGLVAMFRQNTG$_{440}$, abolishes uPAR-αMβ2 and uPAR–β1–integrin complexes (Simon et al. 2000). M25 corresponds to peptide 25, STYHHLSLGYMYTLN, which is capable to bind to uPAR and was previously identified from a phage display library (Wei et al. 1996). Surprisingly, M25 peptide did not block ligand binding to αM but nonetheless inhibited adhesion of leukocytes to fibrinogen, VN, and cytokine-stimulated endothelial cells (Simon et al. 2000). Moreover, M25 also blocked uPAR–β1 integrin interactions, thereby inhibiting integrin-dependent migration of SMC on FN and collagen (Simon et al. 2000). The M25 sequence is relatively conserved among the α subunits associating with the β1 chain, and comparable data were obtained with α325, a synthetic peptide homologous to M25 but derived from α3β1 (Wei et al. 2001). Disruption of uPAR-α3β1 interaction by peptide α325 prevents uPAR-α3β1-dependent uPA production and cell invasion (Ghosh et al. 2006) as well as cell migration and EGF-R activation (Mazzieri et al. 2006).

In the β1 chain of α5β1, two sequences located close to the β-propeller of the α5 subunit have been shown to bind to uPAR (Wei et al. 2005). These sequences are also close to the RGD-binding site of α5β1. The two derived synthetic peptides, β1P1 $_{224}$NLDSPEGGF$_{232}$ and β1P2 $_{262}$FHFAGDGKL$_{270}$, disrupt uPAR–α5β1 complexes, thereby modifying integrin conformation, and exert radical effects on cell adhesion. Indeed, uPAR-bound integrin binds to FN both in an RGD- and β1 peptide-dependent manner while free α5β1 binds to FN in RGD-dependent manner only (Wei et al. 2005). These data indicate a correlation between uPAR–α5β1 complexes, integrin conformation, and integrin activity but also show that disrupting uPAR–α5β1 complexes with β1P1 and β1P2 peptides does not completely abolish integrin activity. Therefore, α and β subunits may have different functions in the uPAR–integrin connection.

The identification of these binding sites on both uPAR and integrins demonstrates direct uPAR–integrin interactions as suggested by diverse methods including coimmunoprecipitation and FRET (Myohanen et al. 1993, Xue et al. 1994, 1997; Wei et al. 1996, Kindzelskii et al. 1997). Further studies are, however, required to understand which mechanism(s) uPAR uses to regulate integrin activity and, conversely, how integrins modulate uPAR activity. On the basis of the data reported so far in the literature, it is possible to propose four different models that can explain the uPAR–integrin relationship (Fig. 23.2).

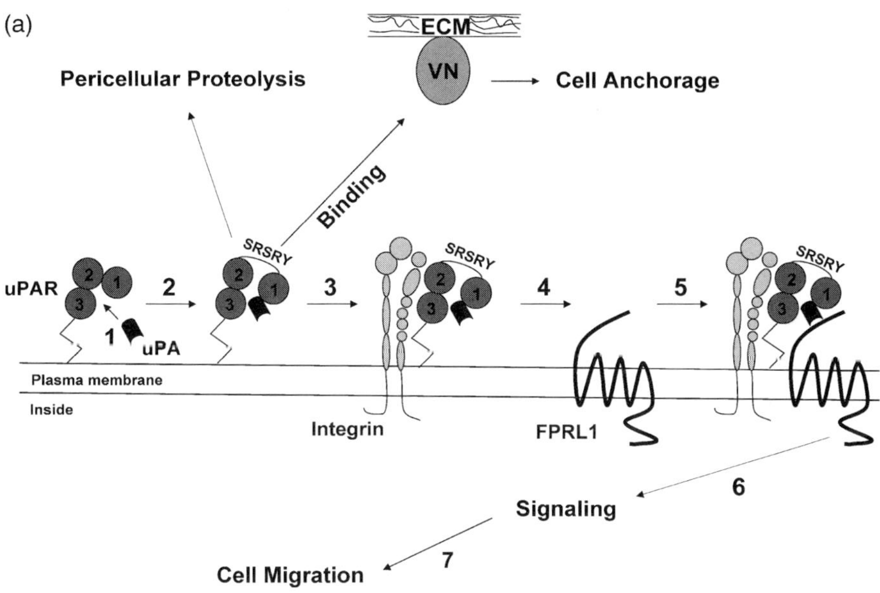

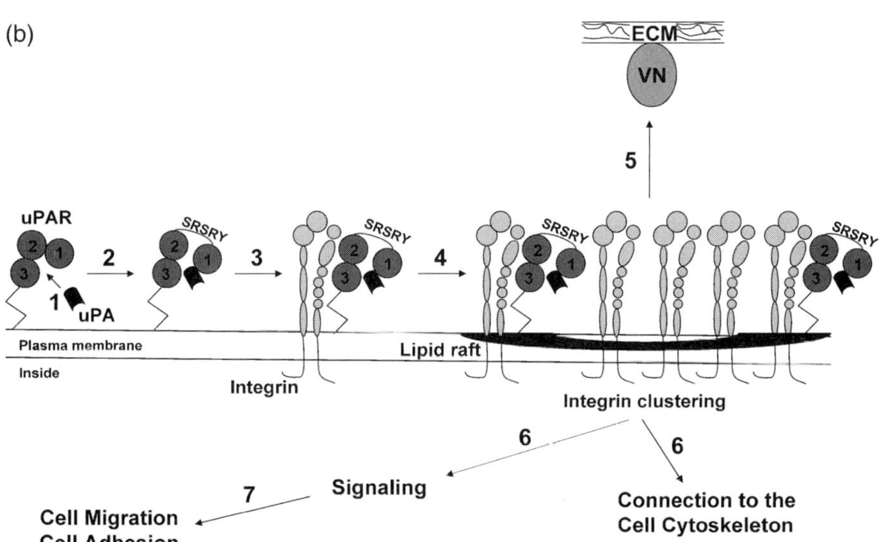

                                                                 B. Degryse

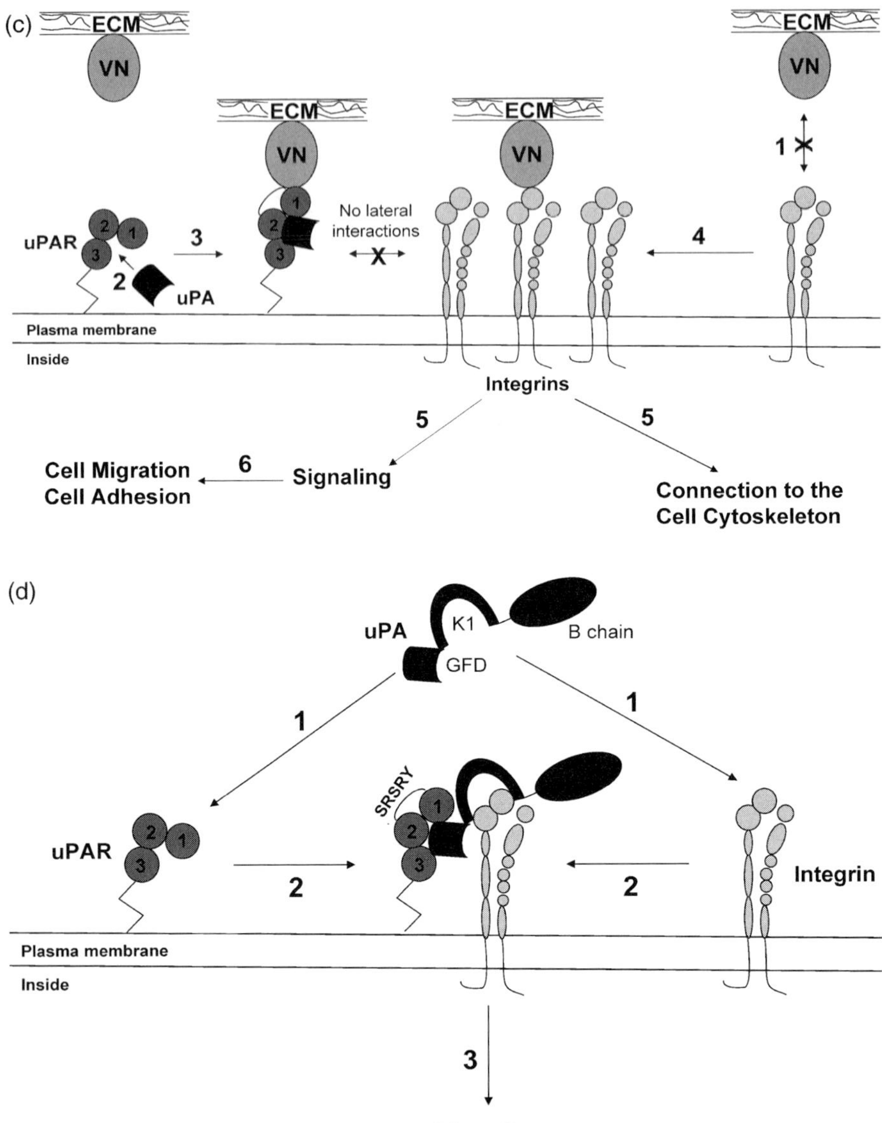

Fig. 23.2 Four models are proposed to illustrate uPAR–integrin interactions. (**a**) Classical model. The binding of uPA to uPAR (1) induces a conformational change of uPAR leading to the exposition of the chemotactic sequence SRSRY located in the linker region of uPAR (2). uPA binding to uPAR also controls pericellular proteolysis and high-affinity binding of uPAR to VN. This conformational change of uPAR promotes lateral interactions with integrins and the formation of uPAR–integrin signaling complex (3). For a simplified representation, this complex is shown as a single integrin and a single uPAR. However, larger complexes including other signaling proteins such as FAK or c-Src are more likely to exist. The conformational change

uPAR- and Integrin-Dependent Signaling, an Intricate Network

Numerous connections between uPAR and integrins already exist in the extracellular and plasma membrane compartments. Therefore, discriminating uPAR- from integrin-dependent intracellular signaling pathways is a tough nut to crack.

However, the very first problem that awaits the researcher (who is no squirrel) is to properly define what is a uPAR-dependent signaling pathway because uPAR, which has no intracellular domain, uses a large panel of transducers to mediate its signals to the cell. In these conditions, one may rather refer to these pathways according to the name of the transducer, for instance, as FPRL-1- or EGF-R-dependent pathway (Resnati et al. 2002, Liu et al. 2002). Thus, in our present case, because uPAR also uses integrins as transducers, the same signaling pathways can be equally referred to as uPAR- and integrin-dependent pathways. On the integrin side, the situation is also confusing; for example, $\beta2$ integrins do not function in the absence of uPAR, consequently, the signaling pathways controlled by these integrins can be similarly considered uPAR- and integrin-dependent pathways.

Beside these quite philosophical considerations, more concrete points further justify why it is so difficult to distinguish uPAR- from integrin-dependent signaling pathways. First, uPAR and integrins share common ligands, uPA and VN (Pluskota et al. 2003, 2004; Demetriou et al. 2004, Kwak et al. 2005, Franco et al. 2006, Tarui et al. 2006). Second, by forming complexes, uPAR and integrins create complicated direct and indirect connections that impact both uPAR and integrin activities. Third, uPAR and integrins have no intrinsic signaling kinase activity and thus have to rely on intracellular kinases in order to mediate signals to the cell. These kinases may be identical or interconnected. Last, diverse downstream signaling pathways can converge on the same cellular targets and generate similar effects. For example,

---

Fig. 23.2 (Continued) of uPAR has turned it into a ligand of seven-transmembrane domain receptors such as FPRL1 (4). The binding of uPAR to FPRL1 (5) induces cell signaling (6), resulting in the stimulation of cell migration (7). (**b**) Recruiting officer model. uPA binding to uPAR (1) induces a conformational change that results in the exposition of the chemotactic epitope SRSRY (2). This conformational change of uPAR promotes lateral interactions with integrins (3). Acting as a "recruiting officer, " uPAR brings integrins into lipid rafts, thereby generating integrin cluster (4). Fully activated and aggregated integrins can bind to VN (5), mediate signals to cell, (6) and be connected to the cell cytoskeleton (7) Integrin-dependent signaling stimulates cell adhesion and spreading or cell migration (7). (**c**) No lateral interactions model. Probably standing too far away from the ECM, integrins cannot bind to ECM-bound VN (1). The binding of uPA to uPAR (2) induces a change of conformation of uPAR that permits high-affinity binding to VN, thereby promoting cell anchorage to the ECM (3). Being close enough, integrin can now bind to VN and form clusters (4), resulting in induction of cellular signaling (5) and connection to the cell cytoskeleton (5). Integrin-induced signaling leads to cell adhesion and spreading, changes in cell morphology and eventually to cell migration (6). (**d**) uPA bridge model. One single molecule of uPA can simultaneously bind to both uPAR and integrin (1). uPA binds to uPAR through its growth factor domain (GFD) and to integrin via its kringle domain (K1) as shown here. Binding of uPA to integrin through its catalytic chain (B chain) has also been reported. uPA binding brings uPAR and integrin together (2), resulting in the formation of uPAR–integrin complex and the generation of cellular signaling (3).

different chemoattractants acting through various membrane receptors induce cell cytoskeleton reorganization by controlling similar small GTP-binding proteins.

As a result of all these issues, one might ask whether real differences between uPAR- and integrin-dependent signaling pathways exist. Numerous reports have already stated that uPAR- and integrin-controlled pathways share common signaling molecules, and this is not surprising because uPAR and integrins form signaling complexes on the cell surface. uPAR binding to integrins potentiates integrin-dependent signaling pathways (Simon et al. 1996, Chavakis et al. 1999, Yebra et al. 1999). For example, uPAR regulates in this way FAK, the MAP kinase pathway (Ras, MEK, and ERK), and MLCK (myosin light chain kinase) (Simon et al. 1996, Chavakis et al. 1999, Nguyen et al. 1999, Yebra et al. 1999). The c-Src dependence of chemotaxis induced by uPA/uPAR seems as well to rely on uPAR–integrin complexes and on the interaction of c-Src with integrins (Degryse et al. 1999, Wei et al. 1999). In addition, caveolin, Jaks (Janus kinases), and Stats (signal transducer and activator of transcription) appear to also belong to both uPAR- and integrin-dependent pathways (Koshelnick et al. 1997, Wei et al. 1999, Dumler et al. 1998, 1999a, b; Degryse et al. 2005).

In view of the just mentioned data, it may seem quite unexpected that subtle differences between uPAR- and integrin-dependent signaling pathways have been observed. In fact, in rat SMC (RSMC), uPA/uPAR and VN/αvβ3 stimulate different signaling pathways (Degryse et al. 2001a). Some VN-treated RSMC even exhibited a particular morphology (as shown by actin and microtubules organization) with two opposite extremities looking like the leading edge of migrating RSMC (Degryse et al. 2001a). A morphological effect also observed on murine fibroblasts and dependent on Rac (Kjøller and Hall 2001). Other reports showed that VN binding to uPAR provided cell anchorage but failed to induce downstream cellular signaling and cell spreading (Stahl and Mueller 1997, Sidenius and Blasi 2000). In leukocytes, phosphoinositide hydrolysis and Ca^{2+} mobilization is uPAR- but not β2-dependent (Sitrin et al. 1999). Only uPAR aggregation induced the activation of phospholipase C, the generation of inositol triphosphate (Ins-1,4,5)P_3, and the subsequent release of intracellular calcium (Sitrin et al. 1999). The activation of uPAR- or integrin-dependent signaling pathways may rely on the cleavage of uPAR in the linker region between domains I and II (Mazzieri et al. 2006). As expected, wild-type uPAR associates with FPRL1 and mediates uPA-induced cell migration, thus activating the classical uPAR-dependent pathway (Resnati et al. 2002), a process which also requires the interaction with α3β1 integrin and results in uPAR cleavage and activation of the MAP kinase pathway (Mazzieri et al. 2006). On the other hand, an uncleavable mutant of uPAR (hcr-uPAR) also interacts with α3β1 but on addition of uPA, the uPAR–α3β1 complex associates with the EGF-R, leading to the transactivation of the EGF-R. In that case, no interactions between uPAR and FPRL1 were observed (Mazzieri et al. 2006). Transactivation of the EGF-R or other receptor tyrosine kinase has been extensively documented, and results from integrin-dependent signaling (for reviews, see Giancotti and Tarone 2003, Ffrench-Constant and Colognato 2004, Cabodi et al. 2004). Even more surprising is the paper stating that uPAR binding to VN is sufficient to promote signaling, cell migration, and changes in cell morphology without the need for lateral interactions with integrins (Madsen et al. 2007). The proposed mechanism is that uPAR ligation by VN promotes cell attachment, allowing the cells to come close enough to the VN-treated surface to permit the binding of integrins. Previous observations support this idea. uPAR binding to VN was shown to induce changes in cell morphology (similar to those reported by Madsen et al. 2007), and increased cell motility through a p130Cas/Rac-dependent signaling pathway (Kjøller and Hall 2001). In addition, VN-promoted recruitment of lymphocytes, i.e., cell motility, in liver tumors is uPAR dependent and $\alpha v \beta 3$ independent (Edwards et al. 2006). In conclusion, all these data reinforce the idea that uPAR is a fully functional signaling membrane receptor that can mediate various motogenic, mitogenic, and adhesive signals to the cells through the regulation of specific signaling pathways. In addition, the most recent hypothesis that uPAR does not require lateral interactions with transducer(s) is extremely challenging.

New Anti-Cancer Strategy, Future Therapies?

The plasminogen activator (uPA/uPAR) system has critical functions in tumor development, invasion, and metastasis. This system has been involved in cell growth, angiogenesis, cell adhesion and migration, pericellular proteolysis, plasminogen and growth factors activation, tissue remodeling, invasion, and metastasis. On the other hand, the integrin system is formed by a large family of membrane receptors that convey bidirectional signaling and are connected to the cell cytoskeleton. Integrins are also indispensable partners of growth factor and cytokine receptors and (as just described above) of uPAR. The integrin system exerts crucial functions in embryonic development and in adult tissue homeostasis, inflammation, cell differentiation, adhesion, migration, cell cycle progression, angiogenesis, and tumor metastasis. In particular, by mediating cell attachment integrins promote cell survival and confer resistance to genotoxic injury, that is, adhesion-mediated radioresistance/drug resistance of tumor cells. Considering the vital roles of these two systems it is no surprise that they are widely misused in disease states such as cancers. Combining the uPA/uPAR system, which allows the cell to modify its environment or to move across it, with the integrin system, which permits the same cell to adapt to its new environment and survive, confers a grand advantage to the tumor cells. This combination is clearly reflected at the molecular levels by the formation of uPAR–integrin complexes. In addition, uPAR and integrins have numerous extracellular ligands and associate to a large array of membrane-bound partners; thus the uPAR–integrin complexes constitute the cornerstone that permits to build up large signaling complexes and to adjust their compositions according to the environment of the cells and to the cellular requirements as the cells grow and/or move. Therefore, the uPAR–integrin complexes constitute some kind of "chameleon signalosome," i.e., adaptable signaling complexes (Fig. 23.1).

uPAR and integrins have already being considered as therapeutic targets (Mazar 2001, Hehlgans et al. 2007). However, targeting uPAR–integrin complexes may represent a better anticancer strategy providing the great advantage to affect various receptors and signaling pathways. This strategy will leave less possibility to cancer cells to counteract and escape by using other receptors/signaling pathways as substitute. The most recent developed molecules such as peptides D2A-Ala and GAAG, which target uPAR–integrin complexes and alter their functions, have already shown a great efficacy in vitro (Degryse et al. 2005).

References

Abu-Ali S., Sugiura T., Takahashi M., et al. (2005). Expression of the urokinase receptor regulates focal adhesion assembly and cell migration in adenoid cystic carcinoma cells. J. Cell. Physiol. 203:410–419.

Adachi Y., Lakka S. S., Chandrasekar N., et al. (2001). Down-regulation of integrin alpha(v)beta (3) expression and integrin-mediated signaling in glioma cells by adenovirus-mediated transfer of antisense urokinase-type plasminogen activator receptor (uPAR) and sense p16 genes. J. Biol. Chem. 276:47171–47177.

Aguirre Ghiso J. A., Kovalski K., and Ossowski L. (1999). Tumor dormancy induced by down-regulation of urokinase receptor in human carcinoma involves integrin and MAPK signaling. J. Cell Biol. 147:89–104.

Aguirre-Ghiso J. A., Liu D., Mignatti A., et al. (2001). Urokinase receptor and fibronectin regulate the ERK(MAPK) to p38(MAPK) activity ratios that determine carcinoma cell proliferation or dormancy in vivo. Mol. Biol. Cell. 12:863–879.

Barinka C., Parry G., Callahan J., et al. (2006). Structural basis of interaction between urokinase-type plasminogen activator and its receptor. J. Mol. Biol. 363:482–495.

Bass R., Werner F., Odintsova E., et al. (2005). Regulation of urokinase receptor proteolytic function by the tetraspanin CD82. J. Biol. Chem. 280:14811–14818.

Besta F., Massberg S., Brand K., et al. (2002). Role of beta(3)-endonexin in the regulation of NF-kappaB-dependent expression of urokinase-type plasminogen activator receptor. J. Cell Sci. 115:3879–3888.

Bianchi E., Ferrero E., Fazioli F., et al. (1996). Integrin-dependent induction of functional urokinase receptors in primary T lymphocytes. J. Clin. Invest. 98:1133–1141.

Blasi F. (1999). The urokinase receptor. A cell surface, regulated chemokine. APMIS 107:96–101.

Blasi F. and Carmeliet P. (2002). uPAR: a versatile signalling orchestrator. Nat. Rev. Mol. Cell. Biol. 3:932–943.

Bohuslav J., Horejsi V., Hansmann C., et al. (1995). Urokinase plasminogen activator receptor, beta 2-integrins, and Src-kinases within a single receptor complex of human monocytes. J. Exp. Med. 181:1381–1390.

Brünner N., Nielsen H. J., Hamers M., et al. (1999). The urokinase plasminogen activator receptor in blood from healthy individuals and patients with cancer. APMIS. 107:160–167.

Cabodi S., Moro L., Bergatto E., et al. (2004). Integrin regulation of epidermal growth factor (EGF) receptor and of EGF-dependent responses. Biochem. Soc. Trans. 32:438–442.

Carriero M. V., Del Vecchio S., Capozzoli M., et al. (1999). Urokinase receptor interacts with alpha(v)beta5 vitronectin receptor, promoting urokinase-dependent cell migration in breast cancer. Cancer Res. 59:5307–5314.

Chapman H. A., Wei Y., Simon D. I., et al. (1999). Role of urokinase receptor and caveolin in regulation of integrin signaling. Thromb. Haemost. 82:291–297.

Chaurasia P., Aguirre-Ghiso J. A., Liang O. D., et al. (2006). A region in urokinase plasminogen receptor domain III controlling a functional association with alpha5beta1 integrin and tumor growth. J. Biol. Chem. 281:14852–14863.

Chavakis T., May A. E., Preissner K. T., et al. (1999). Molecular mechanisms of zinc-dependent leukocyte adhesion involving the urokinase receptor and beta2-integrins. Blood 93:2976–2983.

Chavakis T., Kanse S. M., Lupu F., et al. (2000). Different mechanisms define the antiadhesive function of high molecular weight kininogen in integrin- and urokinase receptor-dependent interactions. Blood 96:514–522.

Chintala S. K., Mohanam S., Go Y., et al. (1997). Altered in vitro spreading and cytoskeletal organization in human glioma cells by downregulation of urokinase receptor. Mol. Carcinog. 20:355–365.

Ciambrone G. J. and McKeown-Longo P. J. (1992). Vitronectin regulates the synthesis and localization of urokinase-type plasminogen activator in HT-1080 cells. J. Biol. Chem. 267:13617–13622.

Colman R. W., Pixley R. A., Najamunnisa S., et al. (1997). Binding of high molecular weight kininogen to human endothelial cells is mediated via a site within domains 2 and 3 of the urokinase receptor. J. Clin. Invest. 100:1481–1487.

Conese M., Nykjaer A., Petersen C. M., et al. (1995). Alpha-2 macroglobulin receptor/Ldl receptor-related protein(Lrp)-dependent internalization of the urokinase receptor. J. Cell Biol. 131:1609–1622.

Cunningham O., Andolfo A., Santovito M. L., et al. (2003). Dimerization controls the lipid raft partitioning of uPAR/CD87 and regulates its biological functions. EMBO J. 22:5994–6003.

Czekay R.-P., Kuemmel T. A., Orlando R. A., et al. (2001). Direct binding of occupied urokinase receptor (uPAR) to LDL receptor-related protein is required for endocytosis of uPAR and regulation of cell surface urokinase activity. Mol. Biol. Cell. 12:1467–1479.

Czekay R.-P., Aertgeerts K., Curriden S. A., et al. (2003). Plasminogen activator inhibitor-1 detaches cells from extracellular matrices by inactivating integrins. J. Cell Biol. 160:781–791.

Degryse B. (2003). Is uPAR the centre of a sensing system involved in the regulation of inflammation? Curr. Med. Chem Anti-Inflammatory & Anti-Allergy Agents 2:237–259.

Degryse B., Resnati M., Rabbani S. A., et al. (1999). Src-dependence and pertussis-toxin sensitivity of urokinase receptor-dependent chemotaxis and cytoskeleton reorganization in rat smooth muscle cells. Blood 94:649–662.

Degryse B., Orlando S., Resnati M., et al. (2001a). Urokinase/urokinase receptor and vitronectin/alpha(v)beta(3) integrin induce chemotaxis and cytoskeleton reorganization through different signaling pathways. Oncogene 20:2032–2043.

Degryse B., Sier C. F., Resnati M., et al. (2001b). PAI-1 inhibits urokinase-induced chemotaxis by internalizing the urokinase receptor. FEBS Lett. 505:249–254.

Degryse B., Resnati M., Czekay R. -P., et al. (2005). Domain 2 of the urokinase receptor contains an integrin-interacting epitope with intrinsic signaling activity: generation of a new integrin inhibitor. J. Biol. Chem. 280:24792–24803.

Demetriou M. C., Pennington M. E., Nagle R. B., et al. (2004). Extracellular alpha 6 integrin cleavage by urokinase-type plasminogen activator in human prostate cancer. Exp. Cell Res. 294:550–558.

Dumler I., Weis A., Mayboroda O. A., et al. (1998). The Jak/Stat pathway and urokinase receptor signaling in human aortic vascular smooth muscle cells. J. Biol. Chem. 273:315–321.

Dumler I., Kopmann A., Weis A., et al. (1999a). Urokinase activates the Jak/Stat signal transduction pathway in human vascular endothelial cells. Arterioscler. Thromb. Vasc. Biol. 19:290–297.

Dumler I., Kopmann A., Wagner K., et al. (1999b). Urokinase induces activation and formation of Stat4 and Stat1-Stat2 complexes in human vascular smooth muscle cells. J. Biol. Chem. 274:24059–24065.

Edwards S., Lalor P. F., Tuncer C., et al. (2006). Vitronectin in human hepatic tumours contributes to the recruitment of lymphocytes in an alpha v beta3-independent manner. Br. J. Cancer 95:1545–1554.

Ellis V., Scully M. F., and Kakkar V. V. (1989). Plasminogen activation initiated by single-chain urokinase-type plasminogen activator. Potentiation by U937 monocytes. J. Biol. Chem. 264:2185–2188.

Fazioli F., Resnati M., Sidenius N., et al. (1997). A urokinase-sensitive region of the human urokinase receptor is responsible for its chemotactic activity. EMBO J. 16:7279–7286.

Ffrench-Constant C. and Colognato H. (2004). Integrins: versatile integrators of extracellular signals. Trends Cell Biol. 14:678–686.

Franco P., Vocca I., Carriero M. V., et al. (2006). Activation of urokinase receptor by a novel interaction between the connecting peptide region of urokinase and alpha v beta 5 integrin. J. Cell Sci. 119:3424–3434.

Furlan F., Orlando S., Laudanna C., et al. (2004). The soluble D2D3(88–274) fragment of the urokinase receptor inhibits monocyte chemotaxis and integrin-dependent cell adhesion. J. Cell Sci. 117:2909–2916.

Galbraith C. G., Yamada K. M., and Galbraith J. A. (2007). Polymerizing actin fibers position integrins primed to probe for adhesion sites. Science 315:992–995.

Gellert G. C., Goldfarb R. H., and Kitson R. P. (2004). Physical association of uPAR with the alphaV integrin on the surface of human NK cells. Biochem. Biophys. Res. Commun. 315:1025–1032.

Ghosh S., Brown R., Jones J. C., et al. (2000). Urinary-type plasminogen activator (uPA) expression and uPA receptor localization are regulated by alpha 3beta 1 integrin in oral keratinocytes. J. Biol. Chem. 275:23869–23876.

Ghosh S., Johnson J. J., Sen R., et al. (2006). Functional relevance of urinary-type plasminogen activator receptor-alpha3beta1 integrin association in proteinase regulatory pathways. J. Biol. Chem. 281:13021–13029.

Giancotti F. G. and Tarone G. (2003). Positional control of cell fate through joint integrin/receptor protein kinase signaling. Annu. Rev. Cell Dev. Biol. 19:173–206.

Ginsberg M. H., Partridge A., and Shattil S. J. (2005). Integrin regulation. Curr. Opin. Cell Biol. 17:509–516.

Gondi C. S., Kandhukuri N., Kondraganti S., et al. (2006). Down-regulation of uPAR and cathepsin B retards cofilin dephosphorylation. Int. J. Oncol. 28:633–639.

Hapke S., Gawaz M., Dehne K., et al. (2001a). beta(3)A-integrin downregulates the urokinase-type plasminogen activator receptor (u-PAR) through a PEA3/ets transcriptional silencing element in the u-PAR promoter. Mol. Cell. Biol. 21:2118–2132.

Hapke S., Kessler H., Arroyo de Prada N., et al. (2001b). Integrin alpha(v)beta(3)/vitronectin interaction affects expression of the urokinase system in human ovarian cancer cells. J. Biol. Chem. 276:26340–26348.

Hehlgans S., Haase M., and Cordes N. (2007). Signalling via integrins: implications for cell survival and anticancer strategies. Biochim. Biophys. Acta 1775:163–180.

Hoyer-Hansen G., Ronne E., Solberg H., et al. (1992). Urokinase plasminogen activator cleaves its cell surface receptor releasing the ligand-binding domain. J. Biol. Chem. 267:18224–18229.

Huai Q., Mazar A. P., Kuo A., et al. (2006). Structure of human urokinase plasminogen activator in complex with its receptor. Science 311:656–659.

Huang M., Mazar A. P., Parry G., et al. (2005). Crystallization of soluble urokinase receptor (suPAR) in complex with urokinase amino-terminal fragment (1–143). Acta Crystallogr. D Biol. Crystallogr. 61:697–700.

Jin H., Song Y. P., Boel G., et al. (2005). Group A streptococcal surface GAPDH, SDH, recognizes uPAR/CD87 as its receptor on the human pharyngeal cell and mediates bacterial adherence to host cells. J. Mol. Biol. 350:27–41.

Khatib A. M., Nip J., Fallavollita L., et al. (2001). Regulation of urokinase plasminogen activator/plasmin-mediated invasion of melanoma cells by the integrin vitronectin receptor alphaV-beta3. Int. J. Cancer 91:300–308.

Kim K. S., Hong Y. K., Joe Y. A., et al. (2003a). Anti-angiogenic activity of the recombinant kringle domain of urokinase and its specific entry into endothelial cells. J. Biol. Chem. 278:11449–11456.

Kim K. S., Hong Y. K., Lee Y., et al. (2003b). Differential inhibition of endothelial cell proliferation and migration by urokinase subdomains: amino-terminal fragment and kringle domain. Exp. Mol. Med. 35:578–585.

Kim C. K., Hong S. H., Joe Y. A., et al. (2007). The recombinant kringle domain of urokinase plasminogen activator inhibits in vivo malignant glioma growth. Cancer Sci. 98:253–258.

Kinashi T. (2005). Intracellular signalling controlling integrin activation in lymphocytes. Nat. Rev. Immunol. 5:546–559.

Kindzelskii A. L., Xue W., Todd R. F. 3rd., et al. (1994). Aberrant capping of membrane proteins on neutrophils from patients with leukocyte adhesion deficiency. Blood 83:1650–1655.

Kindzelskii A. L., Eszes M. M., Todd R. F. 3rd., et al. (1997). Proximity oscillations of complement type 4 (alphaX beta2) and urokinase receptors on migrating neutrophils. Biophys. J. 73:1777–1784.

Kjøller L. and Hall A. (2001). Rac mediates cytoskeletal rearrangements and increased cell motility induced by urokinase-type plasminogen activator receptor binding to vitronectin. J. Cell Biol. 152:1145–1157.

Koshelnick Y., Ehart M., Hufnagl P., et al. (1997). Urokinase receptor is associated with the components of the JAK1/STAT1 signaling pathway and leads to activation of this pathway upon receptor clustering in the human kidney epithelial tumor cell line TCL-598. J. Biol. Chem. 272:28563–28567.

Kwak S. H., Mitra S., Bdeir K., et al. (2005). The kringle domain of urokinase-type plasminogen activator potentiates LPS-induced neutrophil activation through interaction with {alpha}V {beta}3 integrins. J. Leukoc. Biol. 78:937–945.

Leksa V., Godár S., Cebecauer M., et al. (2002). The N terminus of mannose 6-phosphate/insulin-like growth factor 2 receptor in regulation of fibrinolysis and cell migration. J. Biol. Chem. 277:40575–40582.

Llinas P., Le Du M. H., Gardsvoll H., et al. (2005). Crystal structure of the human urokinase plasminogen activator receptor bound to an antagonist peptide. EMBO J. 24:1655–1663.

Liu D., Aguirre Ghiso J., Estrada Y., et al. (2002). EGFR is a transducer of the urokinase receptor initiated signal that is required for in vivo growth of a human carcinoma. Cancer Cell. 1:445–457.

Madsen C. D., Ferraris G. M., Andolfo A., et al. (2007). uPAR-induced cell adhesion and migration: vitronectin provides the key. J. Cell Biol. 177:927–939.

Margheri F., Manetti M., Serrati S., et al. (2006). Domain 1 of the urokinase-type plasminogen activator receptor is required for its morphologic and functional, beta2 integrin-mediated connection with actin cytoskeleton in human microvascular endothelial cells: failure of association in systemic sclerosis endothelial cells. Arthritis Rheum. 54:3926–3938.

May A. E., Kanse S. M., Lund L. R., et al. (1998). Urokinase receptor (CD87) regulates leukocyte recruitment via beta 2 integrins in vivo. J. Exp. Med. 188:1029–1037.

May A. E., Neumann F. J., Schomig A., et al. (2000). VLA-4 (alpha(4)beta(1)) engagement defines a novel activation pathway for beta(2) integrin-dependent leukocyte adhesion involving the urokinase receptor. Blood 96:506–513.

May A. E., Schmidt R., Kanse S. M., et al. (2002). Urokinase receptor surface expression regulates monocyte adhesion in acute myocardial infarction. Blood 100:3611–3617.

Mazar A. P. (2001). The urokinase plasminogen activator receptor (uPAR) as a target for the diagnosis and therapy of cancer. Anticancer Drugs 12:387–400.

Mazzieri R., D'Alessio S., Kenmoe R. K., et al. (2006). An uncleavable uPAR mutant allows dissection of signaling pathways in uPA-dependent cell migration. Mol. Biol. Cell 17:367–378.

Montuori N., Rossi G., and Ragno P. (1999). Cleavage of urokinase receptor regulates its interaction with integrins in thyroid cells. FEBS Lett. 460:32–36.

Montuori N., Carriero M. V., Salzano S., et al. (2002). The cleavage of the urokinase receptor regulates its multiple functions. J. Biol. Chem. 277:46932–46939.

Myohanen H. T., Stephens R. W., Hedman K., et al. (1993). Distribution and lateral mobility of the urokinase-receptor complex at the cell surface. J. Histochem. Cytochem. 41:1291–1301.

Naldini L., Tamagnone L., Vigna E., et al. (1992). Extracellular proteolytic cleavage by urokinase is required for activation of hepatocyte growth factor/scatter factor. EMBO J. 11:4825–4833.

Naldini L., Vigna E., Bardelli A., et al. (1995). Biological activation of pro-HGF (hepatocyte growth factor) by urokinase is controlled by a stoichiometric reaction. J. Biol. Chem. 270:603–611.

Nguyen D. H., Catling A. D., Webb D. J., et al. (1999). Myosin light chain kinase functions downstream of Ras/ERK to promote migration of urokinase-type plasminogen activator-stimulated cells in an integrin-selective manner. J. Cell Biol. 146:149–164.

Nip J., Rabbani S. A., Shibata H. R., et al. (1995). Coordinated expression of the vitronectin receptor and the urokinase-type plasminogen activator receptor in metastatic melanoma cells. J. Clin. Invest. 95:2096–2103.

Nykjaer A., Petersen C. M., Moller B., et al. (1992). Purified alpha 2-macroglobulin receptor/LDL receptor-related protein binds urokinase.plasminogen activator inhibitor type-1 complex. Evidence that the alpha 2-macroglobulin receptor mediates cellular degradation of urokinase receptor-bound complexes. J. Biol. Chem. 267:14543–14546.

Nykjaer A., Kjøller L., Cohen R. L., et al. (1994). Regions involved in binding of urokinase-type-1 inhibitor complex and pro-urokinase to the endocytic alpha 2-macroglobulin receptor/low density lipoprotein receptor-related protein. Evidence that the urokinase receptor protects pro-urokinase against binding to the endocytic receptor. J. Biol. Chem. 269:25668–25676.

Nykjaer A., Conese M., Christensen E. I., et al. (1997). Recycling of the urokinase receptor upon internalization of the uPA:serpin complexes. EMBO J. 16:2610–2620.

Nykjaer A., Christensen E. I., Vorum H., et al. (1998). Mannose 6-phosphate/insulin-like growth factor-II receptor targets the urokinase receptor to lysosomes via a novel binding interaction. J. Cell Biol. 141:815–828.

Odekon L. E., Sato Y., and Rifkin D. B. (1992). Urokinase-type plasminogen activator mediates basic fibroblast growth factor-induced bovine endothelial cell migration independent of its proteolytic activity. J. Cell Physiol. 150:258–263.

Odekon L. E., Blasi F., and Rifkin D. B. (1994). Requirement for receptor-bound urokinase in plasmin-dependent cellular conversion of latent TGF-beta to TGF-beta. J. Cell Physiol. 158:398–407.

Pawar S. C., Demetriou M. C., Nagle R. B., et al. (2007). Integrin alpha6 cleavage: a novel modification to modulate cell migration. Exp. Cell Res. 313:1080–1089.

Ploug M. and Ellis V. (1994). Structure-function relationships in the receptor for urokinase-type plasminogen activator. Comparison to other members of the Ly-6 family and snake venom alpha-neurotoxins. FEBS Lett. 349:163–168.

Pluskota E., Soloviev D. A., and Plow E. F. (2003). Convergence of the adhesive and fibrinolytic systems: recognition of urokinase by integrin alpha Mbeta 2 as well as by the urokinase receptor regulates cell adhesion and migration. Blood 101:1582–1590.

Pluskota E., Soloviev D. A., Bdeir K., et al. (2004). Integrin alphaMbeta2 orchestrates and accelerates plasminogen activation and fibrinolysis by neutrophils. J. Biol. Chem. 279:18063–18072.

Pöllänen J., Hedman K., Nielsen L. S., et al. (1988). Ultrastructural localization of plasma membrane-associated urokinase-type plasminogen activator at focal contacts. J. Cell Biol. 106:87–95.

Porter J. C. and Hogg N. (1998). Integrins take partners: cross-talk between integrins and other membrane receptors. Trends Cell Biol. 8:390–396.

Ragno P. (2006). The urokinase receptor: a ligand or a receptor? Story of a sociable molecule. Cell. Mol. Life Sci. 63:1028–1037.

Ragno P., Montuori N., Covelli B., et al. (1998). Differential expression of a truncated form of the urokinase-type plasminogen-activator receptor in normal and tumor thyroid cells. Cancer Res. 58:1315–1319.

Reinartz J., Schafer B., Batrla R., et al. (1995). Plasmin abrogates alpha v beta 5-mediated adhesion of a human keratinocyte cell line (HaCaT) to vitronectin. Exp. Cell Res. 220:274–282.

Resnati M., Guttinger M., Valcamonica S., et al. (1996). Proteolytic cleavage of the urokinase receptor substitutes for the agonist-induced chemotactic effect. EMBO J. 15:1572–1582.

Resnati M., Pallavicini I., Wang J. M., et al. (2002). The fibrinolytic receptor for urokinase activates the G protein-coupled chemotactic receptor FPRL1/LXA4R. Proc. Natl. Acad. Sci. USA 99:1359–1364.

Roztocil E., Nicholl S. M., and Davies M. G. (2007). Mechanisms of kringle fragment of urokinase-induced vascular smooth muscle cell migration. J. Surg. Res. 141:83–90.

Sawai H., Okada Y., Funahashi H., et al. (2006). Interleukin-1alpha enhances the aggressive behavior of pancreatic cancer cells by regulating the alpha6beta1-integrin and urokinase plasminogen activator receptor expression. BMC Cell Biol. 7:8.

Schwab W., Gavlik J. M., Beichler T., et al. (2001). Expression of the urokinase-type plasminogen activator receptor in human articular chondrocytes: association with caveolin and beta 1-integrin. Histochem. Cell Biol. 115:317–323.

Sidenius N. and Blasi F. (2000). Domain 1 of the urokinase receptor (uPAR) is required for uPAR-mediated cell binding to vitronectin. FEBS Lett. 470:40–46.

Sidenius N., Andolfo A., Fesce R., et al. (2002). Urokinase regulates vitronectin binding by controlling urokinase receptor oligomerization. J. Biol. Chem. 277:27982–27990.

Sier C. F., Sidenius N., Mariani A., et al. (1999). Presence of urokinase-type plasminogen activator receptor in urine of cancer patients and its possible clinical relevance. Lab. Invest. 79:717–722.

Silvestri I., Longanesi Cattani I., Franco P., et al. (2002). Engaged urokinase receptors enhance tumor breast cell migration and invasion by upregulating alpha(v)beta5 vitronectin receptor cell surface expression. Int. J. Cancer 102:562–571.

Simon D. I., Rao N. K., Xu H., et al. (1996). Mac-1 (CD11b/CD18) and the urokinase receptor (CD87) form a functional unit on monocytic cells. Blood 88:3185–3194.

Simon D. I., Wei Y., Zhang L., et al. (2000). Identification of a urokinase receptor–integrin interaction site. Promiscuous regulator of integrin function. J. Biol. Chem. 275:10228–10234.

Simons K. and Toomre D. (2000). Lipid rafts and signal transduction. Nat. Rev. Mol. Cell Biol. 1:31–39.

Sitrin R. G., Todd R. F. 3rd., Albrecht E., et al. (1996). The urokinase receptor (CD87) facilitates CD11b/CD18-mediated adhesion of human monocytes. J. Clin. Invest. 97:1942–1951.

Sitrin R. G., Pan P. M., Harper H. A., et al. (1999). Urokinase receptor (CD87) aggregation triggers phosphoinositide hydrolysis and intracellular calcium mobilization in mononuclear phagocytes. J. Immunol. 163:6193–6200.

Sitrin R. G., Pan P. M., Harper H. A., et al. (2000). Clustering of urokinase receptors (uPAR; CD87) induces proinflammatory signaling in human polymorphonuclear neutrophils. J. Immunol. 165:3341–3349.

Stahl A. and Mueller B. M. (1995). The urokinase-type plasminogen activator receptor, a GPI-linked protein, is localized in caveolae. J. Cell Biol. 129:335–344.

Stahl A. and Mueller B. M. (1997). Melanoma cell migration on vitronectin: regulation by components of the plasminogen activation system. Int. J. Cancer 71:116–122.

Stefansson S. and Lawrence D. A. (1996). The serpin PAI-1 inhibits cell migration by blocking integrin $\alpha v\beta 3$ binding to vitronectin. Nature 383:441–443.

Stepanova V., Jerke U., Sagach V., et al. (2002). Urokinase-dependent human vascular smooth muscle cell adhesion requires selective vitronectin phosphorylation by ectoprotein kinase CK2. J. Biol. Chem. 277:10265–10272.

Stephens R. W., Nielsen H. J., Christensen I. J., et al. (1999). Plasma urokinase receptor levels in patients with colorectal cancer: relationship to prognosis. J. Natl. Cancer Inst. 91:869–874.

Tarui T., Mazar A. P., Cines D. B., et al. (2001a). Urokinase-type plasminogen activator receptor (CD87) is a ligand for integrins and mediates cell-cell interaction. J. Biol. Chem. 276:3983–3990.

Tarui T., Miles L. A., and Takada Y. (2001b). Specific interaction of angiostatin with integrin alpha(v)beta(3) in endothelial cells. J. Biol. Chem. 276:39562–39568.

Tarui T., Akakura N., Majumdar M., et al. (2006). Direct interaction of the kringle domain of urokinase-type plasminogen activator (uPA) and integrin alpha v beta 3 induces signal transduction and enhances plasminogen activation. Thromb. Haemost. 95:524–534.

Wang G. J., Collinge M., Blasi F., et al. (1998). Posttranscriptional regulation of urokinase plasminogen activator receptor messenger RNA levels by leukocyte integrin engagement. Proc. Natl. Acad. Sci. USA. 95:6296–6301.

Weaver A. M., Hussaini I. M., Mazar A., et al. (1997). Embryonic fibroblasts that are genetically deficient in low density lipoprotein receptor-related protein demonstrate increased activity of the urokinase receptor system and accelerated migration on vitronectin. J. Biol. Chem. 272:14372–14379.

Webb D. J., Nguyen D. H., Sankovic M., et al. (1999). The very low density lipoprotein receptor regulates urokinase receptor catabolism and breast cancer cell motility in vitro. J. Biol. Chem. 274:7412–7420.

Webb D. J., Nguyen D. H., and Gonias S. L. (2000). Extracellular signal-regulated kinase functions in the urokinase receptor-dependent pathway by which neutralization of low density lipoprotein receptor-related protein promotes fibrosarcoma cell migration and matrigel invasion. J. Cell Sci. 113:123–134.

Wei Y., Lukashev M., Simon D. I., et al. (1996). Regulation of integrin function by the urokinase receptor. Science 273:1551–1555.

Wei Y., Yang X., Liu Q., et al. (1999). A role for caveolin and the urokinase receptor in integrin-mediated adhesion and signaling. J. Cell Biol. 144:1285–1294.

Wei Y., Eble J. A., Wang Z., et al. (2001). Urokinase receptors promote beta1 integrin function through interactions with integrin alpha3beta1. Mol. Biol. Cell 12:2975–2986.

Wei Y., Czekay R. P., Robillard L., et al. (2005). Regulation of alpha5beta1 integrin conformation and function by urokinase receptor binding. J. Cell Biol. 168:501–511.

Wei Y., Tang C. H., Kim Y., et al. (2007). Urokinase receptors are required for alpha 5 beta 1 integrin-mediated signaling in tumor cells. J. Biol. Chem. 282:3929–3939.

Wong W. S., Simon D. I., Rosoff P. M., et al. (1996). Mechanisms of pertussis toxin-induced myelomonocytic cell adhesion: role of Mac-1(CD11b/CD18) and urokinase receptor (CD87). Immunology. 88:90–97.

Xia Y., Borland G., Huang J., et al. (2002). Function of the lectin domain of Mac-1/complement receptor type 3 (CD11b/CD18) in regulating neutrophil adhesion. J. Immunol. 169:6417–6426.

Xue W., Kindzelskii A. L., Todd R. F. 3rd., et al. (1994). Physical association of complement receptor type 3 and urokinase-type plasminogen activator receptor in neutrophil membranes. J. Immunol. 152:4630–4640.

Xue W., Mizukami I., Todd R. F., et al. (1997). Urokinase-type plasminogen activator receptors associate with beta1 and beta3 integrins of fibrosarcoma cells: dependence on extracellular matrix components. Cancer Res. 57:1682–1689.

Yebra M., Parry G. C., Stromblad S., et al. (1996). Requirement of receptor-bound urokinase-type plasminogen activator for integrin alphavbeta5-directed cell migration. J. Biol. Chem. 271:29393–29399.

Yebra M., Goretzki L., Pfeifer M., et al. (1999). Urokinase-type plasminogen activator binding to its receptor stimulates tumor cell migration by enhancing integrin-mediated signal transduction. Exp. Cell Res. 250:231–240.

Chapter 24
Measuring uPAR Dynamics in Live Cells

Moreno Zamai, Gabriele Malengo, and Valeria R. Caiolfa

Abstract Urokinase-type plasminogen activator receptor (uPAR) is a key component of the urokinase plasminogen activation (uPA) system, which plays important roles in physiological processes as well as in tumor invasion and metastasis. Besides uPAR's well-established role in the regulation of pericellular proteolysis, a large body of evidences suggests that several of uPAR-mediated events do not require the proteolytic activity of uPA. The common accepted notion is that uPAR transduces signals through direct lateral physical interactions in multimolecular complexes involving membrane-spanning proteins and extracellular surface proteins. However, none of these interactions have ever been visualized and confirmed in living cells, at steady state and in the absence of crosslinkers or antibody clustering agent. Thus, the physical engagement of uPAR in its monomeric or oligomeric forms in multimolecular complexes needs to be convincingly verified observing uPAR at work. This chapter discusses how fully functional fluorescent protein-tagged uPAR chimeras can be generated and used for determining distribution, recruitment, mobility, monomer-dimer exchange, and biologically relevant uPAR interactions in living cells, and in unperturbed conditions.

Introduction: The Urokinase-Type Plasminogen
Activator Receptor

The cellular receptor for the urokinase-type plasminogen activator receptor (uPAR) is a single polypeptide chain of 313 amino acid residues, with a signal peptide of 21 residues. The receptor is composed of three domains (as numbered from the N-terminus)–domain DI (amino acid residues 1–77), domain DII (residues 93–177), and domain DIII (residues 193–272)–that are homologous with respect to the

V.R. Caiolfa
Department of Molecular Biology and Functional Genomics, San Raffaele Scientific Institute, DIBIT 4A1 room 16, Via Olgettina 58, 20132 Milano-Italy, e-mail: valeria.caiolfa@hsr.it

D. Edwards et al. (eds.), *The Cancer Degradome.*
© Springer Science + Business Media, LLC 2008

arrangement of disulfide bonds (four to five disulphide bonds), but differ in their amino acid sequence (Casey et al. 1994). The three domains are members of the Ly-6/uPAR, a neurotoxin protein domain family. At the posttranslational level, after cleavage and removal of the last 30 C-terminal residues, a glycosyl-phosphatydil-inositol (GPI) anchor is attached to Gly 283, anchoring the receptor to the cell surface (Ploug et al. 1993).

Soluble forms of uPAR (suPAR), generated by either hydrolytic activity of GPI-specific phospholipases (Wilhelm et al. 1999) or juxtamembrane proteolytic cleavage (Beaufort et al. 2004), have been identified in biological fluids, both *in vitro* and *in vivo*. Purified uPAR shows a single 55–60 kDa band after sodium dodecyl sulfate-polyacrylamide gel electrophoresis and silver staining. It is a heavily glycosylated protein, as the deglycosylated polypeptide chain comprises only 35 kDa (Moller et al. 1993).

uPAR is a key component of the urokinase plasminogen activation (uPA) system, a well-characterized system of serine proteases (Preissner et al. 2000). The uPA–uPAR system plays important roles in physiological processes such as wound healing, inflammation, and stem cell mobilization, as well as in severe pathological conditions, like tumor invasion, metastasis, and HIV-1 infection (Alfano et al. 2002, Gyetko et al. 2004, Lund et al. 2006, Selleri et al. 2005, Sidenius and Blasi, 2003). Besides uPAR's well-established role in the regulation of pericellular proteolysis, an expanding body of evidence suggests that several uPAR-mediated events do not require the proteolytic activity of uPA but involve transmembrane signaling. These include adhesion and chemotactic movement of myeloid cells (Gyetko et al. 1994, Waltz et al. 1993), cell migration in human epithelial cells (Busso et al. 1994) and bovine endothelial cells (Odekon et al. 1992), and cell growth (Aguirre Ghiso et al. 1999, Fischer et al. 1998, Mazzieri et al. 2006, Rabbani et al. 1992). Because uPAR is bound to the plasma membrane by a GPI anchor, it has no direct access to the cytoplasm for engaging intracellular signaling intermediates. The common accepted notion is that uPAR transduces signals through direct lateral physical interactions in multimolecular complexes involving membrane-spanning proteins, extracellular surface proteins, or second messengers at the intracellular site. uPAR-mediated cell signaling has been shown to involve vitronectin (Vn)-, fibronectin (Fn)-, and laminin-binding integrins (Chaurasia et al. 2006, D'Alessio and Blasi, in preparation, Hoyer-Hansen et al. 1997, Sidenius and Blasi 2000, Wei et al. 1994), G-protein coupled chemotactic receptors such as FPRL1/LXA4R and FRP (Resnati et al. 2002, Selleri et al. 2006), and members of the low-density lipoprotein receptor-related protein family (LDLR) (Chazaud et al. 2000, Conese et al. 1995, Li et al. 2002). Attempts have been made to identify the regions in uPAR involved in specific interaction with Vn and integrins (Chaurasia et al. 2006, Degryse et al. 2005, Li et al. 2003, Wei et al. 2007). Furthermore, binding of uPA to uPAR induces additional intracellular signaling events in some cells involving the Jak/Stat signaling pathway (Bohuslav et al. 1995, Degryse et al. 2005, Dumler et al. 1998, Koshelnick et al. 1997).

The uPAR Interactome

The so-called uPAR interactome is the subject of recent reviews, which we invite the reader to refer to (Binder et al. 2007, Ragno 2006). The uPAR interactome has recently found some rationale in the crystal structure of suPAR solved in association with a competitive peptide inhibitor of the uPA–uPAR interaction (Llinas et al. 2005). This initial study and the following ones (Barinka et al. 2006, Huai et al. 2006) provide the first structural basis toward understanding how uPAR may organize its multiple molecular interactions. These studies reveal that uPAR is composed of three consecutive three-finger domains organized in an almost circular manner, generating both a deep internal cavity where the antagonist peptide binds in a helical conformation and a large external surface. The models propose that the receptor-binding module of uPA engages the uPAR central cavity, thus leaving the external receptor surface accessible for other protein interactions (e.g., Vn and integrins).

Finally, it was also suggested that dimerization regulates the biological activity of the receptor by determining differential ligand binding and lipid raft partitioning, as detergent-resistant membrane fractions (DRM) were enriched in uPAR dimers and coincided with higher Vn-binding activity (Cunningham et al. 2003). More recently, the same group has demonstrated that direct uPAR–Vn interaction is required and sufficient to initiate downstream changes leading to cell migration and signal transduction (Madsen et al. 2007).

The entire uPA-uPAR interactome has been derived from coimmunoprecipitation and antibody clustering experiments and from immunofluorescence imaging. None of the direct physical interactions reported to mediate uPAR-signaling has ever been visualized and confirmed in living cells, at steady state and, more importantly, in the absence of any crosslinker or antibody clustering agent. In addition, the existence, distribution, and regulation of uPAR monomers and dimers, as well as the uPAR monomers-dimers involvement in specific multiprotein complexes, have not been documented in living cells yet. Increasing evidences are arising from studies on GPI-anchored proteins that underscore the high dynamic nature of GPI-protein organization in the cell membrane (Simons and Toomre 2000, Suzuki et al. 2007). The dynamic exchange in membrane microdomains (Kusumi and Suzuki 2005, Lenne et al. 2006) may explain the involvement of uPAR in signal transduction processes. As for other GPI-proteins, the lateral association of uPAR with membrane-spanning proteins must be dynamically modulated, and it might be heavily affected by external cross-linking or clustering agents.

Thus, the physical engagement of uPAR in its monomeric or oligomeric forms in multimolecular complexes needs to be convincingly verified observing uPAR at work, as "the most important question becomes which of the many molecular interactions are really essential to mediate uPAR function" (Madsen et al. 2007). The study of the physical interactions of uPAR in multimolecular complexes at real time and in living cells should lead to a detailed characterization of the following: (a) which are the molecules forming physical complexes with uPAR, (b) at which uPAR expression, (c) in which cellular status or condition (e.g., resting or stimulated states), (d) where these interactions occur in the cell, and (e) how they are dynamically regulated. In other words, the spatiotemporal regulation of uPAR interactions in live cells is a key for understanding the specific activation of uPAR-mediated signal transduction cascades.

Rapid advances in live-cell imaging and microspectroscopy technologies, combined with the use of the last generation of genetically encoded fluorescent proteins (FPs), have resulted in a revolution in cell biology, since it is now possible to track the assembly of protein complexes within the organized microenvironment of a living cell.

The next sections discuss the first living cell models for uPAR, expressing fully functional FP-uPAR chimeras and their use in molecular dynamics and dimerization studies in real time, in resting living cells, and in the absence of any crosslinker or antibody-clustering agent. The most promising combinations of imaging and microspectroscopy tools will also be discussed for their potential to study uPAR dynamics in a manner that is both qualitative and quantitative.

The First Live Cell Fluorescent Model for uPAR

Fluorescence imaging and microspectroscopy in live cells is mainly based on the use of FP-tagged chimeras either transiently or stably transfected in cells. The main assumption is that the insertion of an FP in the sequence of the target protein does not alter the correct folding and sorting of the protein under study. This assumption arises from the evidence that FPs are relatively small and compact beta-barrel proteins of 27 kDa, which may form an additional independent domain in the chimeric sequence. The evidence that the chimeric protein translocates correctly in the cell is a generally accepted criterion for assuming retention of function. However, this criterion is not sufficient per se for considering FP-tagged uPAR chimeras suitable tools to investigate the dynamics of the physical interactions of the receptor in living cells, as the wild-type receptor (wt-uPAR) is endowed with several biological functions. These functions are dependent on (1) uPAR binding and activation of pro-uPA that promotes pericellular plasminogen activity and subsequent cleavage of the receptor at the DI domain; (2) uPAR-dependent cell adhesion to Vn; (3) binding, internalization, and recycling induced by the interaction with the uPA–PAI1 complex; and (4) shedding from the GPI-anchor. Furthermore, it is also expected to partially recover the GPI-anchored FP-tagged uPAR in the DRM fractions, similarly to untagged uPAR.

To the best of our knowledge, the first fully functional FP-chimeras of uPAR have only recently been generated and described (Caiolfa et al. 2007, Malengo et al. 2008). The fluorescent uPAR chimeras were constructed by inserting the sequence encoding the monomeric FPs, EGFP (Zacharias et al. 2002), or mRFP1 (Campbell et al. 2002) between the third domain of uPAR and the GPI-anchoring sequence at a position where N. Sidenius and collaborators had previously epitope-tagged uPAR without disrupting receptor function (Cunningham et al. 2003). The FP-tagged uPAR chimeras expressed in HEK293 cells were correctly GPI-anchored, sorted to the cell surface, and partitioned partially to DRM very similarly to untagged uPAR (Caiolfa et al. 2007). The chimeras also retained the normal affinity binding for uPA, promoted pericellular plasminogen activation, supported uPAR-dependent cell adhesion to Vn, and internalized after binding uPA-PAI1 (Caiolfa et al. 2007).

The notions that the expression of uPAR induces profound changes in cell morphology and migration, and it correlates with the malignant phenotype of cancers, are well documented in the literature (de Bock and Wang 2004, Laufs et al. 2006). Therefore, uPAR expression must be well controlled in any novel fluorescent cell model system generated for dynamic interaction studies. The HEK293 cell line is one of the few human cell lines that do not express wt-uPAR and do not secrete pro-uPA. It has been reported that uPAR expression in HEK293 modulates the adhesion and mobility of the cells through interactions with Vn, integrins, and G-protein coupled receptors (Chaurasia et al. 2006, Degryse et al. 2005, Gargiulo et al. 2005, Resnati et al. 2002, Wei et al. 1994, 1996, 1999, 2001). In accordance with these observations, a more recent study has confirmed that expression of human uPAR in HEK293 cells induces changes in cell morphology, migration, and signaling (Madsen et al. 2007). The morphological changes observed include a general flattening of the cells, reduced cell–cell contact, disappearance of membrane ruffles, formation of extensive lamellipodia, and a complete reorganization of the matrix-proximal F-actin cytoskeleton. The changes in cell morphology as well as the threefold increased ERK1/2 activation also reflected the cell mobility induced by uPAR expression in these cells (Madsen et al. 2007). The same morphological changes were observed in HEK293, stably expressing the functional uPAR-EGFP-GPI (Caiolfa et al. 2007) and uPAR-mRFP1-GPI (Caiolfa V.R. unpublished results).

Having a functional fluorescent uPAR, which efficiently reproduces the fundamental functions of the wt-receptor in living cells, the first questions that one tries to answer are those related to uPAR distribution and mobility at the cell surface, potential segregation in membrane microdomains, dimerization, and regulation (if any) of the monomer–dimer dynamics by direct interaction with well-characterized extracellular ligands of uPAR (e.g., Vn, pro-uPA, uPA-PAI1). These studies are the logical prerequisite for exploring the uPAR interactome at work.

The Diffusing Forms of uPAR

How does uPAR diffuse at the cell surface in resting and unstimulated steady-state conditions? Although several good reviews addressed to a general cell biology audience provide an exhaustive overview of the various approaches that can be applied for protein diffusion studies (Bacia et al. 2006, Bates et al. 2006, Vukojevic et al. 2005), we will review briefly the principles of fluorescence recovery after photobleaching (FRAP) (Axelrod et al. 1976), since this is the most popular technique for measuring the mobility of a protein in a minimally perturbed living cell. In FRAP experiments, a single protein is labeled with a fluorescent tag and the fluorophores within a small area are irreversibly photobleached by a short and intense laser pulse. The subsequent movement of surrounding nonbleached fluorescent molecules into the photobleached area is recorded at low laser power. By monitoring the levels and rates of fluorescence recovery with time, one can determine kinetic parameters such as the mobile fraction (Mf) and the diffusion coefficient (D_{diff}). The refilling of the entire bleached spot is necessary to obtain a complete recovery curve. The bleached area can be refilled with fluorophores diffusing from any subcellular pool, from very distant pools, as well as from adjacent ones. As a consequence, various processes, such as membrane flow, molecular interactions, and trafficking, may simultaneously contribute to the overall recovery kinetics, which make data difficult to interpret. Although FRAP is a simple and powerful method, it needs to be combined with appropriate mathematical models in order to interpret the data properly and analyze the complex dynamic processes involved. Recent advances in FP technology and confocal microscopy have made it possible to extend the FRAP approach to intracellular dynamic studies, such as those on trafficking processes (Snapp et al. 2006), or diffusion of membrane proteins (Kenworthy et al. 2004). Nevertheless, FRAP does not really measure the diffusion of the molecules locally illuminated by the light beam, as the recovery curve only informs on the overall diffusion of all mobile fluorophores in the cell, providing only an indirect description of the protein mobility. Truly, the "local" information on the lateral mobility of single protein molecules is missing in FRAP analysis. In addition, FRAP cannot provide any information on the molecular forms of the diffusing species.

These are major limitations for the use of FRAP in uPAR dynamics. In fact, one would like to determine whether other membrane-spanning proteins or extracellular ligands control the recruitment and mobility of uPAR and its dimerization. Thus, we need to follow at the same time the local dynamics and the assembly of uPAR molecules at the cell surface.

Alternatively to FRAP, the combination of fluorescence correlation spectroscopy (FCS) (Berland et al. 1995) and photon counting histogram (PCH) analyses (Chen et al. 1999) can reach this level of molecular details. Latest technological advances have revived FCS as a useful technique for measuring translational mobility in the cytoplasm and nucleus as well as in cellular membranes. FCS analysis has been the subject of several reviews (Elson 2004, Hess et al. 2002, Levin and Carson 2004, Thompson et al. 2002) and a book (Rigler and Elson 2001). The less popular PCH approach is fully complementary to FCS for associating D_{diff} with the state of aggregation of the mobile species (Chen et al. 2000, 2002, 2003b). More in general, FCS and PCH are known as "single molecule" fluorescence fluctuation spectroscopy (FFS) because they allow the determination of the D_{diff} and the aggregation state of single protein molecules or single protein assemblies.

The measurement of the fluctuations of the fluorescence intensity signal is the common base for the two complementary methods (Fig. 24.1). The most stringent requirement for FCS and PCH to work is the possibility to observe the fluorescence signal in a small volume, as that defined by a 2-photon excitation, <0.1 ($<1\ \mu^3$), and at very high sensitivity and dynamic range (Fig. 24.1a). Only if the volume is so

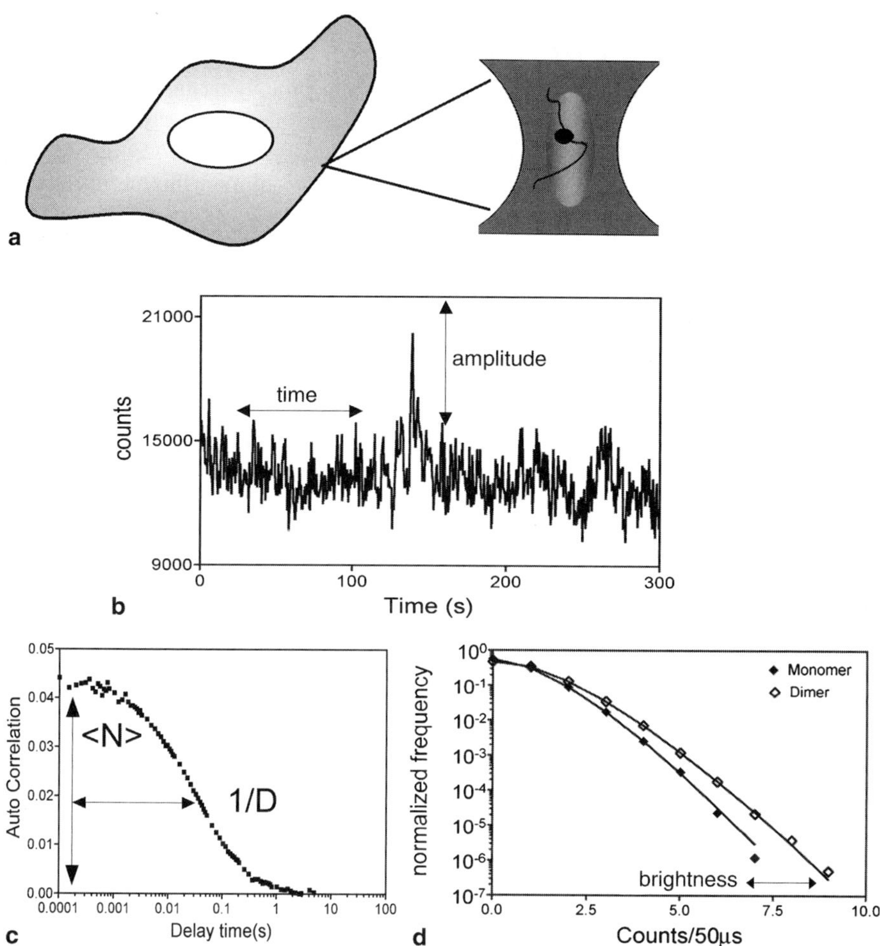

Fig. 24.1 Measuring uPAR dynamics and aggregation by fluorescence correlation spectroscopy (*FCS*) and photon counting histogram (*PCH*). The 2-photon excitation illuminates a microvolume in the cell (**a**). The fluorescence fluctuations due to the diffusion of one (or very few) molecule can be recorded from the same volume. The fluorescence signal has a time and amplitude structure (**b**). The time structure is analyzed by FCS through the autocorrelation function (**c**), providing the average number of molecules that diffuse under the light beam, $<N>$, and the diffusion time (inversely related to the diffusion coefficient, D_{diff}). The amplitude structure is analyzed by PCH (**d**) and it determines the brightness of the molecules diffusing under the light beam

small, it can contain just one or few molecules at any instant of time. If the number of molecules changes, because they diffuse in and out of the illuminated volume, the fluorescence intensity will change with time (Fig. 24.1b). The time of the diffusion process causes characteristic frequencies (fluctuations of the signal) to appear in the fluorescence intensity trace. This is called the "time structure" of the intensity and can be temporally autocorrelated to reveal information about the local concentration (i.e., local average number of molecules, $<N>$) and dynamics (i.e., diffusion mechanisms and D_{diff}) of the fluorescent species in the selected region of the cell (Fig. 24.1c) (Berland et al. 1995).

The intensity trace can also provide information on the aggregation state of the fluorophores. If two identical proteins with one fluorescent probe each (e.g., uPAR-EGFP-GPI:uPARE-GFP-GPI) diffuse synchronously (together) and enter the observation volume, there will be a burst in intensity (higher fluctuation). The diffusing aggregate, in fact, has twice the brightness of a monomer because it carries twice the number of fluorescent moieties. Clearly, the amplitude of the fluctuation provides information on the brightness of the diffusing particles (Chen et al. 2003b). Thus, the time and amplitude structures of the fluorescence intensity trace are affected by the underlying molecular species, and the dynamic processes (translational mobility) that cause the change in the fluorescence intensity. The analysis of the PCH recovers the information contained in the amplitude of the fluorescence intensity trace determining the brightness of the diffusing species, which is the number of photons per second per molecule (Fig. 24.1d). Knowing the brightness of monomeric uPAR-EGFP-GPI, one can determine the oligomerization state of the diffusing uPAR-EGFP-GPI species.

PCH analysis of uPAR-EGFP-GPI stably expressed in HEK293 cells has demonstrated that the receptors diffuse predominantly as monomers. More rarely the synchronous diffusion of two uPAR-EGFP-GPI molecules was observed, while higher order, mobile uPAR-EGFP-GPI oligomers were not found in resting unstimulated cells (Caiolfa et al. 2007, Malengo et al. 2008).

The parallel FCS analysis provided information on the nature of translational mobility and D_{diff} of the uPAR-EGFP-GPI molecules. Monomers and dimers diffused anomalously at the cell membrane. The D_{diff} indicated that diffusion was more restrained and slower at the basal membrane in cells seeded on serum or Vn matrices. The asymmetry in mobility was accompanied by the recruitment of the receptors in the basal membranes. The same did not happen in cells seeded on Fn matrices. On Fn, the distribution of uPAR at the cell membrane was not asymmetric, and the D_{diff} were similar in both basal and apical membranes (Caiolfa et al. 2007). The anomalous diffusion of proteins at the cell surface was documented already several years ago (Feder et al. 1996), and since then it has been the subject of a number of biophysical studies (Banks and Fradin 2005, Marguet et al. 2006, Ritchie et al. 2005, Sheets et al. 1997, Smith et al. 1999, Weiss et al. 2003). The lateral mobility of a protein at the cell membrane cannot be simply defined by models of Brownian motion, as it is "locally" confined by obstacles and barriers (Morone et al. 2006, Ritchie et al. 2003, Simson et al. 1998, Suzuki et al, 2005), molecular crowding (Banks and Fradin 2005), and/or lipid–protein segregation (Jacobson et al. 2007, Kusumi et al. 2004, Kusumi and Suzuki 2005, Lagerholm et al. 2005, Lenne et al. 2006, Marguet et al. 2006, Ritchie and Kusumi 2004, Simons and Vaz 2004). These mechanisms are not static and give rise to the complex and dynamic pattern of protein lateral organization in the cell membrane.

The partition of GPI-anchored uPAR in both detergent-resistant and in detergent-soluble membrane fractions (Cunningham et al. 2003) is commonly interpreted as the partitioning of uPAR in and out of the cholesterol-enriched submembrane domains, say rafts (Rajendran and Simons, 2005, Simons and Toomre 2000, Simons and Vaz 2004). However, FCS analysis does not support the existence of two distinct uPAR populations in the plasma membrane of HEK293 cells, which could be associated with discrete and long-lasting membrane microdomains. This is not surprising, as the "raft" concept of discrete microdomains is being challenged, considering that various conventional imaging approaches (cells fixed on glass) as well as the use of chemical clustering or antibody stimuli introduces major artifacts in the organization and dynamics of GPI-anchored proteins (Morone et al. 2006, Ritchie and Kusumi 2004, Suzuki et al. 2005). On one hand, it seems that none of the available optical, spectroscopic, and biochemical techniques is well suited for capturing and visualizing a raft (Jacobson et al. 2007, Lagerholm et al. 2005). On the other hand, the dynamic nature of membrane microdomains gathers increasing experimental support. Deviations from free diffusion of GPI-anchored proteins (i.e., anomalous diffusion) have recently been explained by a dynamic partition mechanism by which the GPI-proteins diffuse into and out of the permeable microdomains, and are only transiently confined in the absence of molecular cross-linking (Lenne et al. 2006, Marguet et al. 2006).

The fluorescence intensity traces acquired on the basal membrane in green HEK293 cells seeded on serum or Vn matrices also showed irreversible photobleaching. The irreversible photobleaching indicated the presence of an immobile fraction of receptors. The bleached receptors were not replaced by bright uPAR-EGFP-GPI molecules diffusing into the observation volume. Irreversible photobleaching was never observed in other submembrane regions, confirming thatVn, directly engaging uPAR in cell adhesion, recruited and restricted the mobility of the receptors at the basal side. Conversely, in green HEK293 cells seeded on Fn matrices, no recruitment and no irreversible photobleaching or decreased mobility and confinement of uPAR at the basal membrane were observed, strongly suggesting that uPAR did not participate in Fn-mediated HEK293 adhesion (Caiolfa et al. 2007).

Dimerization of uPAR at the Surface of Living Cells

From the above studies, two major questions arise: (a) Do two molecules of uPAR-EGFP-GPI diffusing together (i.e., undergoing correlated diffusion) form a true dimer? Or, in other words, how "physically" interacting uPAR–uPAR complexes can be convincingly detected in live cells? (b) In which molecular form (monomer versus dimer) is the receptor immobilized at the basal membrane of cells adhered to Vn-matrices?

Conditional colocalization fluorescence-based imaging cannot obviously be taken as a proof for direct physical interaction between two or more proteins bearing different fluorescent tags. The optical resolution limit of conventional confocal microscopy is, in fact, by far above interprotein distances in "physically interacting" assemblies. This limit is due to the use of visible light (i.e., fluorescence). It varies within 200–300 nm in the x,y-confocal plane and it is in the 1 μ range in the perpendicular z-plane, hampering any conclusion about the proteins forming a "stable complex" under conditions that are not triggered by chemical cross-linking (cell fixation) or clustering agents. Furthermore, once again it should be remembered that uPAR is a GPI-anchored protein, and as such, it is sensitive to lipid–cholesterol distribution in the cell membrane and to antibody-stimulated clustering.

On one hand, in many protocols for immunofluorescence (or immunoelectron) visualization, the cells are first fixed using paraformaldehyde and/or glutaraldehyde, and then labeled with the specific antibodies or ligands. As discussed by Kusumi and Suzuki (2005), the cell fixation protocols cause artifacts with GPI-anchored proteins (or raftophilic molecules in general). The use of low concentrations of paraformaldehyde, which is generally assumed to "fix" the amino-containing molecules at their intrinsic locations, actually may enhance the clustering of raftophilic molecules as the addition of multivalent antibodies and ligands, rather than blocking the redistribution of these molecules in situ (Kusumi and Suzuki 2005).

On the other hand, coimmunoprecipitation experiments give only a batch view of protein complexes and are equally exposed to biases due to stimulus (antibody)-induced clustering.

For these reasons, efforts should be made to confirm physical interactions in steady-state cells (in the absence of artificial stimulation), taking into account the dynamic nature (variable in time and location) of functionally productive protein assemblies.

In the last few years, Forster resonance energy transfer (FRET) approaches have experienced an extraordinary renaissance in cell biology (Jares-Erijman and Jovin 2006, Wallrabe and Periasamy 2005). FRET is a phenomenon by which, under light excitation, a fluorescent donor molecule transfers part of its excitation to an acceptor molecule that becomes fluorescent without being directly excited by the light. The parameter that mainly determines the efficiency of FRET is the distance between the donor and acceptor molecules, so much that FRET is defined as a molecular ruler with nanometric units (Siegel et al. 2000). The nanometric units vary according to the donor–acceptor pair. For FP pairs, the distance at which FRET efficiency can be as high as 50% is in the range of 3–5 nm (Patterson et al. 2000). Thus, FP-FRET pairs are sensitive reporters of protein assemblies. Even so, measuring FRET in living cells having variable coexpression of the donor and acceptor molecules and relevant autofluorescence is a challenging task. The number of publications on this subject testifies the efforts made during the past years to improve the reliability of FRET approaches in living cells. The discussion of these protocols goes beyond the aims of this chapter, as several good reviews are available in the literature (Jares-Erijman and Jovin 2006, Schmid and Birbach 2007, Soon et al. 2007). Nevertheless, the most sensitive approach for determining FRET in living cells is by far the donor fluorescence lifetime imaging (donor-FLIM) (Bastiaens and Squire 1999, Chen et al. 2003a, van Munster and Gadella 2005, Wallrabe and Periasamy 2005). Conceptually, this is the easiest method: if part of the energy of the donor is transferred to an acceptor molecule, then the donor molecule will spend a shorter time in its excited state and will decay faster (i.e., the donor fluorescence lifetime is shorter in the presence of FRET). Practically, however, the application of donor-FLIM is still restricted to few expert laboratories of fluorescence spectroscopy because FLIM equipments are sophisticated and, more importantly, the analysis of donor-FLIM data is especially difficult. Despite the fact that many papers describe the use of donor-FLIM for detecting protein–protein interactions, the analytical problem finds only approximate (and very time- and computer power-consuming) solutions, when FPs are used as donor–acceptor FRET pairs in living cells. This is because all FPs in cells have complex fluorescence lifetime decays (Suhling et al. 2002), for which no robust mathematical decomposition is possible. Working on HEK293 cells cotransfected with uPAR-EGFP-GPI and uPAR-mRFP1-GPI, we also had to face the issue of convoluted decays in FLIM images, which were made even more complex by the presence of a cell-to-cell variable autofluorescence that affected the donor, uPAR-EGFP-GPI, lifetime as FRET would do (Caiolfa V.R. and Zamai M. unpublished). The consequence was that FRET could be evaluated only in a very approximate and not reproducible manner. This experience has prompted us to change radically the analytical approach to donor-FLIM and has contributed to inspire the revolutionary phasor-FLIM analysis (Digman et al. 2008).

After recording the fluorescence intensity of a HEK293/uPAR-EGFP-GPI cell (Fig. 24.2a, left top panel), we measured pixel-by-pixel the fluorescence lifetime decays in the 2-photon time correlated single photon counting (TCSPC) mode, as it is done in the classical donor-FLIM method. A decay (no matter how complex) can be transformed in a phasor (a vector), using the universal circle of the polar coordinates representation, and the phasor ensembles determined in each image can be shown in contour plot (Fig. 24.2a, left bottom panel). In this representation, only single exponential decays fall on the universal circle (Fig. 24.2a, left bottom panel, phasor 1). If the decay at one pixel is a convolution of lifetimes, due to multiple species (i.e., unquenched donor plus cellular autofluorescence, or multiple decays of the EGFP fluorophore) contributing the fluorescence intensity, the phasor falls inside the universal circle (Fig. 24.2a, left bottom panel, phasor 2) (Clayton et al. 2004, Gratton et al. 1984, Redford and Clegg 2005). In the HEK293/uPAR-EGFP-GPI apical or basal membranes (shown in the example in Fig. 24.2), there are pixels that give rise to complex decays, as most of the phasors are inside the universal circle (Fig. 24.2a, left bottom panel). The cellular autofluorescence was measured in untransfected HEK293 cells, and the position of phasor 3 in Fig. 24.2a (left bottom panel) indicates the mean of the phasor distribution detected as auto-fluorescence. Thus, the experimentally derived green line in the plot represents the mean of the phasor distributions for HEK293/uPAR-EGFP-GPI cells with different contribution of cellular autofluorescence (Fig. 24.2a, left bottom panel, green line). FRET is expected to shift the donor phasors on the right, in the space between the green and the red lines, the latter representing uPAR-EGFP-GPI phasors quenched by 50% FRET by uPAR-mRFP1-GPI and having different cell autofluorescence contributions (Fig. 24.2a, left bottom panel, red line). The phasor representation allows the selection of subsets of phasors (i.e., decays), and their localization in the correspondent FLIM image. In the HEK293/uPAR-EGFP-GPI cell, for example in Fig. 24.2a, the majority of the phasors, >98%, can be selected in a narrow area of the plot (Fig. 24.2a, right bottom panel, black circle). The position of these phasors

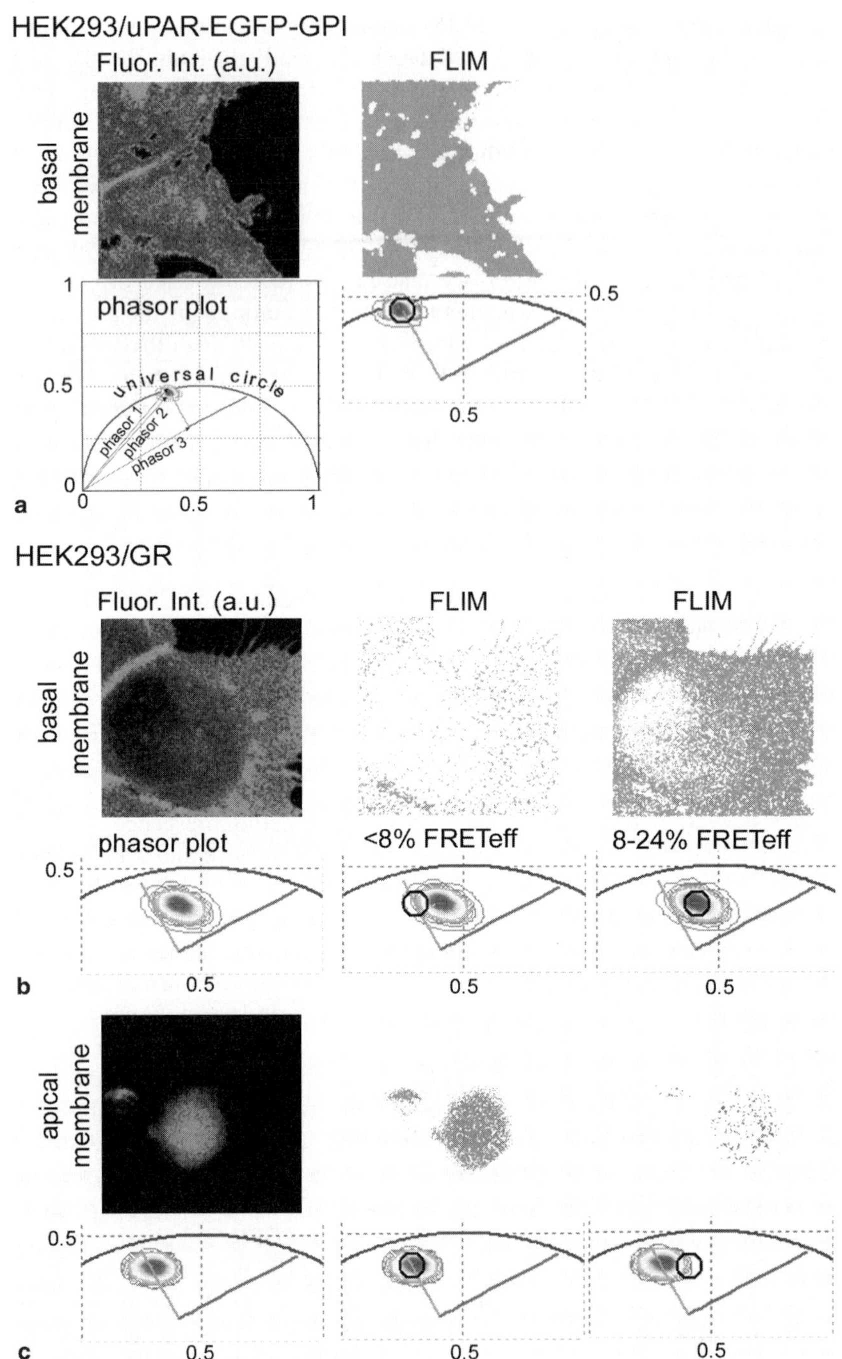

Fig. 24.2 uPAR dimers detected by Forster resonance energy transfer (*FRET*) using the novel phasor-FLIM analysis of donor lifetime decays. (a) A typical donor-FLIM experiment: 2-photon fluorescence intensity (left top panel) and donor fluorescence lifetime decays are acquired pixel-by-pixel in parallel in the time correlated single photon counting (*TCSPC*) mode. Then, the decay

in the image (Fig. 24.2a, right top panel) indicates that the lifetime distribution of uPAR-EGFP-GPI is indeed complex but uniformly distributed at the basal membrane. The same result was obtained in apical membrane sections (Caiolfa et al. 2007). The spread of the phasor distribution (i.e., the size of the black circle) in HEK293/uPAR-EGFP-GPI cells determined the confidence limit for FRET analysis, which was found equal to 8% FRET (i.e., FRET below 8% was not significant). Thus, we used a circle with identical radius for analyzing the phasor distributions in HEK293/uPAR-EGFP-GPI/uPAR-mRFP1-GPI cells. Figure 24.2b illustrates an example of phasor FLIM analysis on the basal membrane of a HEK293 cotransfected with uPAR-EGFP-GPI and uPARmRFP1-GPI cell (termed HEK293/GR). In HEK293/GR cells adhered on Vn matrices we measured FRET efficiencies in the range of 8–24% at the basal side, as the majority of the phasors were found outside the area of unquenched donor and cellular autofluorescence (Fig. 24.2b, right panel). In few regions of the basal membrane, and in particular in the membrane border excluded from the large lamellipodia, FRET was negligible (Fig. 24.2b, mid panel). A remarkably different FRET distribution was observed in apical membranes (Fig. 24.2c). Only few phasors were associated with FRET (Fig. 24.2c, right panel), while the majority were unquenched by FRET (Fig. 24.2c, mid panel).

The FRET analysis clearly demonstrated that uPAR dimers exist in living cells, and that they are mainly recruited at the basal membrane. Likely, they are the main constituents of the immobile fraction observed in diffusion studies.

The morphological changes that were recognized in HEK293 cells coexpressing the green and red uPAR chimeras when seeded in Vn matrices as well as the asymmetric distribution of the receptors at the cell membrane were not reproduced

---

Fig. 24.2 (Continued) in each pixel is transformed in a phasor (left bottom panel). Phasor 1 identifies a pixel in which the fluorescence decay was monoexponential, and therefore it falls on the universal circle of the polar coordinates. Phasor 2 identifies a pixel in which the fluorescence decay was complex (multiexponential) as it falls inside the universal circle. Phasor 3 identifies a typical decay of untransfected HEK293 cells. The ensemble of all phasors from the image is represented in contour plot. The green line depicts the values of the phasor distributions in HEK293/uPAR-EGFP-GPI cells at different contribution of cell autofluorescence (determined in untrasfected HEK293 cells). The red line marks the position of uPAR-EGFP-GPI phasors quenched by 50% FRET and having all possible cell autofluorescence contributions (from 0% to 100%). In the zoomed graph (right bottom panel), the black circle selects 98% of all phasors in the image. The localization of these phasors is shown in the correspondent FLIM image (right top panel). In this example, the contribution of cell autofluorescence was <1% in average. (**b**) FRET efficiencies determined in the basal membrane of an HEK293/GR cell by the phasor-FLIM analysis. The figure shows on the left panel the fluorescence intensity image (top) and the phasor plot (bottom). The phasor distribution is almost completely outside the green trajectory, indicating that donor lifetime is quenched by FRET. Less than 10% of pixels on this membrane section do not show FRET (<8%, mid top and bottom panels). The majority, 89%, of pixels in the image shows FRET efficiencies in the 8–24% range (right top and bottom panels). In this example, the average contribution of cell autofluorescence was 34–36%. (**c**) FRET efficiencies determined in the apical membrane of the HEK293/GR cell shown in (**b**). The fluorescence intensity image and phasor plot are shown in the left top and bottom panels. The same two subsets of FRET efficiencies (<8% and 8–24%) are reproduced in the mid and right panels. In this example, the average contribution of cell autofluorescence was 27–38% (*See also* Color Insert II)

in Fn matrices (Caiolfa et al. 2007, Madsen et al. 2007). Similar FRET studies on HEK293/GR cells seeded on Fn-matrices demonstrated that uPAR-EGFP-GPI: uPAR-mRFP1-GPI complexes were rare in both basal and apical membranes (Caiolfa et al. 2007).

These studies confirmed the previous observations made by N. Sidenius and collaborators (Cunningham et al. 2003), and proved that dimers exist in live, resting cells and that they are recruited and stabilized by the interaction with Vn, participating in Vn-mediated cell adhesion. These results also strongly suggested that uPAR was not engaged in the Fn-mediated cell adhesion, under unstimulated, steady-state conditions, in HEK293 cells.

Future Perspectives

The most recent studies on the dynamics and dimerization of FP-tagged uPAR in living cells have set off a novel scenario for determining which among the many "physical interactions" of uPAR are decisive in mediating relevant mechanisms in cancer: cell adhesion, migration, and proliferation. Very likely, from now on, uPAR interactions in live cells can be locally discriminated and quantitatively determined under a variety of conditions. Furthermore, the role of the physiological uPAR ligands in the distribution and exchange of monomers versus dimers can be defined. Part of this information is already available, as it has been shown that Vn-mediated effects on uPAR in HEK293 cells are reversibly counteracted by the binding of the uPA–PAI1 complex, internalization, and recycling (Caiolfa et al. 2007). Additional studies on the effect of pro-uPA or its N-terminal fragment, ATF (Stoppelli et al. 1985), cholesterol depletion, and uPAR mutations will also contribute to elucidate the role of two molecular forms of uPAR.

Then, further advances in the exploration of functionally relevant uPAR-mediated signaling pathways are expected to arise from the application of increasingly informative microspectroscopic approaches that need to be developed for FP-tagged uPAR live cell models. These approaches should include fluorescence cross-correlation spectroscopy (FCCS) (Bacia et al. 2006, Bacia and Schwille 2003, Schwille et al. 1997) and dual-color PCH (Chen et al. 2005). By FCCS it will be possible to study the synchronous diffusion of two proteins, each tagged with an FP of different color, and, therefore, investigate the codiffusion of uPAR with each of its interatome partners (i.e., integrins, FPRL1, etc.). The parallel analysis by dual PCH might help in defining the stoichiometry of the codiffusing uPAR assemblies, determining how many molecules of uPAR and each of its interatome partners diffuse together. There are already reports in which FCCS has been employed for detecting the codiffusion of two different proteins using variants of the green fluorescent chimeras that do not spectrally overlap (Bacia et al. 2002, Bacia and Schwille 2003, Rosales et al. 2007). In contrast, dual-color PCH analysis has been developed just three years ago (Chen et al. 2005) and attempted in live cell only in 2006 (Hillesheim et al. 2006).

These very innovative tools combined to donor-FLIM experiments, which are now easily and quickly performed using the new phasor-FLIM analysis, will definitely contribute to define the functional map of the uPAR interactome and the molecular forms of uPAR involved in these interactions.

Acknowledgments We are grateful to Enrico Gratton, Francesco Blasi, and Nicolai Sidenius for their support, and for their criticism and efficacious collaboration in every aspect of our work. We thank the Cariplo Foundation, Milano, Italy, for cofunding the uPAR interactome project in our laboratory. We also thank Beniamino Barbieri and ISS Inc. (Champaign, Urbana, IL, USA) for their continual assistance and for providing us the ALBA correlator.

References

Aguirre Ghiso J.A., Kovalski K., and Ossowski L. (1999). Tumor dormancy induced by down-regulation of urokinase receptor in human carcinoma involves integrin and MAPK signaling. J Cell Biol. 147.89–104.

Alfano M., Sidenius N., Panzeri B. et al. (2002). Urokinase-urokinase receptor interaction mediates an inhibitory signal for HIV-1 replication. Proc Natl Acad Sci U S A. 99:8862–7.

Axelrod D., Koppel D.E., Schlessinger J. et al. (1976). Mobility measurement by analysis of fluorescence photobleaching recovery kinetics. Biophys J. 16:1055–69.

Bacia K., Schwille P. (2003). A dynamic view of cellular processes by *in vivo* fluorescence auto-and cross-correlation spectroscopy. Methods. 29:74–85.

Bacia K., Majoul I.V., Schwille P. (2002). Probing the endocytic pathway in live cells using dual-color fluorescence cross-correlation analysis. Biophys J. 83:1184–93.

Bacia K., Kim S.A., Schwille P. (2006). Fluorescence cross-correlation spectroscopy in living cells. Nat Methods. 3:83–9.

Banks D.S., Fradin C. (2005). Anomalous diffusion of proteins due to molecular crowding. Biophys J. 89:2960–71.

Barinka C., Parry G., Callahan J. et al. (2006). Structural basis of interaction between urokinase-type plasminogen activator and its receptor. J Mol Biol. 363:482–95.

Bastiaens P.I., Squire A. (1999). Fluorescence lifetime imaging microscopy: spatial resolution of biochemical processes in the cell. Trends Cell Biol. 9:48–52.

Bates I.R., Wiseman P.W., Hanrahan J.W. (2006). Investigating membrane protein dynamics in living cells. Biochem Cell Biol. 84:825–31.

Beaufort N., Leduc D., Rousselle J.C. et al. (2004). Proteolytic regulation of the urokinase receptor/CD87 on monocytic cells by neutrophil elastase and cathepsin G. J Immunol. 172:540–9.

Berland K.M., So P.T., Gratton E. (1995). Two-photon fluorescence correlation spectroscopy: method and application to the intracellular environment. Biophys J. 68:694–701.

Binder B.R., Mihaly J., Prager G.W. (2007). uPAR-uPA-PAI-1 interactions and signaling: a vascular biologist's view. Thromb Haemost. 97:336–42.

Bohuslav J., Horejsi V., Hansmann C. et al. (1995). Urokinase plasminogen activator receptor, beta 2-integrins, and Src-kinases within a single receptor complex of human monocytes. J Exp Med. 181:1381–90.

Busso N., Masur S.K., Lazega D., et al. (1994). Induction of cell migration by pro-urokinase binding to its receptor: possible mechanism for signal transduction in human epithelial cells. J Cell Biol. 126:259–70.

Caiolfa V.R., Zamai M., Malengo G., et al. (2007). Monomer dimer dynamics and distribution of GPI-anchored uPAR are determined by cell surface protein assemblies. J Cell Biol. 179:1067–82.

Campbell, R.E., Tour O., Palmer A.E., et al. (2002). A monomeric red fluorescent protein. Proc Natl Acad Sci USA. 99:7877–82.

Casey J.R., Petranka J.G., Kottra J. et al. (1994). The structure of the urokinase-type plasminogen activator receptor gene. Blood. 84:1151–6.

Chaurasia P., Aguirre-Ghiso J.A., Liang O.D. et al. (2006). A region in urokinase plasminogen receptor domain III controlling a functional association with alpha5beta1 integrin and tumor growth. J Biol Chem. 281:14852–63.

Chazaud B., Bonavaud S., Plonquet A., et al. (2000). Involvement of the [uPAR:uPA:PAI-1:LRP] complex in human myogenic cell motility. Exp Cell Res. 258:237–44.

Chen Y., Muller J.D., So P.T. et al. (1999). The photon counting histogram in fluorescence fluctuation spectroscopy. Biophys J. 77:553–67.

Chen Y., Muller J.D., Tetin S.Y. et al. (2000). Probing ligand protein binding equilibria with fluorescence fluctuation spectroscopy. Biophys J. 79:1074–84.

Chen Y., Muller J.D., Ruan Q. et al. (2002). Molecular brightness characterization of EGFP *in vivo* by fluorescence fluctuation spectroscopy. Biophys J. 82:133–44.

Chen Y., Mills J.D., Periasamy A. (2003a). Protein localization in living cells and tissues using FRET and FLIM. Differentiation. 71:528–41.

Chen Y., Wei L.N., Muller J.D. (2003b). Probing protein oligomerization in living cells with fluorescence fluctuation spectroscopy. Proc Natl Acad Sci USA. 100:15492–7.

Chen Y., Tekmen M., Hillesheim L. et al. (2005). Dual-color photon-counting histogram. Biophys J. 88:2177–92.

Clayton A.H., Hanley Q.S., Verveer P.J. (2004). Graphical representation and multicomponent analysis of single-frequency fluorescence lifetime imaging microscopy data. J Microsc. 213:1–5.

Conese M., Nykjaer A., Petersen C.M. et al. (1995). alpha-2 Macroglobulin receptor/Ldl receptor-related protein(Lrp)-dependent internalization of the urokinase receptor. J Cell Biol. 131:1609–22.

Cunningham O., Andolfo A., Santovito M.L. et al. (2003). Dimerization controls the lipid raft partitioning of uPAR/CD87 and regulates its biological functions. EMBO J. 22:5994–6003.

de Bock C.E., Wang Y. (2004). Clinical significance of urokinase-type plasminogen activator receptor (uPAR) expression in cancer. Med Res Rev. 24:13–39.

Degryse B., Resnati M., Czekay R.P. et al. (2005). Domain 2 of the urokinase receptor contains an integrin-interacting epitope with intrinsic signaling activity: Generation of a new integrin inhibitor. J Biol Chem.

Digman M.A., Caiolfa V.R., Zamai M., Gratton E. (2008). The phasor approach to fluorescence lifetime imaging analysis. Biophys J. 94:L14–6.

Dumler I., Weis A., Mayboroda O.A. et al. (1998). The Jak/Stat pathway and urokinase receptor signaling in human aortic vascular smooth muscle cells. J Biol Chem. 273:315–21.

Elson E.L. (2004). Quick tour of fluorescence correlation spectroscopy from its inception. J Biomed Opt. 9:857–64.

Feder T.J., Brust-Mascher I., Slattery J.P. et al. (1996). Constrained diffusion or immobile fraction on cell surfaces: a new interpretation. Biophys J. 70:2767–73.

Fischer K., Lutz V., Wilhelm O. et al. (1998). Urokinase induces proliferation of human ovarian cancer cells: characterization of structural elements required for growth factor function. FEBS Lett. 438:101–5.

Gargiulo L., Longanesi-Cattani I., Bifulco K. et al. (2005). Cross-talk between fMLP and vitronectin receptors triggered by urokinase receptor-derived SRSRY peptide. J Biol Chem.

Gratton E., Jameson D.M., Hall R.D. (1984). Multifrequency phase and modulation fluorometry. Annu Rev Biophys Bioeng. 13:105–24.

Gyetko M.R., Todd R.F. 3rd, Wilkinson C.C. et al. (1994). The urokinase receptor is required for human monocyte chemotaxis *in vitro*. J Clin Invest. 93:1380–7.

Gyetko M.R., Aizenberg D., Mayo-Bond L. (2004). Urokinase-deficient and urokinase receptor-deficient mice have impaired neutrophil antimicrobial activation *in vitro*. J Leukoc Biol. 76:648–56.

Hess S.T., Huang S., Heikal A.A. et al. (2002). Biological and chemical applications of fluorescence correlation spectroscopy: a review. Biochemistry. 41:697–705.

Hillesheim L.N., Chen Y., Muller J.D. (2006). Dual-color photon counting histogram analysis of mRFP1 and EGFP in living cells. Biophys J. 91:4273–84.

Hoyer-Hansen G., Behrendt N., Ploug M. et al. (1997). The intact urokinase receptor is required for efficient vitronectin binding: receptor cleavage prevents ligand interaction. FEBS Lett. 420:79–85.

Huai Q., Mazar A.P., Kuo A. et al. (2006). Structure of human urokinase plasminogen activator in complex with its receptor. Science. 311:656–9.

Jacobson K., Mouritsen O.G., Anderson R.G. (2007). Lipid rafts: at a crossroad between cell biology and physics. Nat Cell Biol. 9:7–14.

Jares-Erijman E.A., Jovin T.M. (2006). Imaging molecular interactions in living cells by FRET microscopy. Curr Opin Chem Biol. 10:409–16.

Kenworthy A.K., Nichols B.J., Remmert C.L. et al. (2004). Dynamics of putative raft-associated proteins at the cell surface. J Cell Biol. 165:735–46.

Koshelnick Y., Ehart M., Hufnagl P. et al. (1997). Urokinase receptor is associated with the components of the JAK1/STAT1 signaling pathway and leads to activation of this pathway upon receptor clustering in the human kidney epithelial tumor cell line TCL-598. J Biol Chem. 272:28563–7.

Kusumi A., Suzuki K. (2005). Toward understanding the dynamics of membrane-raft-based molecular interactions. Biochim Biophys Acta. 1746:234–51.

Kusumi A., Koyama-Honda I., Suzuki K. (2004). Molecular dynamics and interactions for creation of stimulation-induced stabilized rafts from small unstable steady-state rafts. Traffic. 5:213–30.

Lagerholm B.C., Weinreb G.E., Jacobson K. et al. (2005). Detecting microdomains in intact cell membranes. Annu Rev Phys Chem. 56:309–36.

Laufs S., Schumacher J., Allgayer H. (2006). Urokinase-receptor (u-PAR): an essential player in multiple games of cancer: a review on its role in tumor progression, invasion, metastasis, proliferation/dormancy, clinical outcome and minimal residual disease. Cell Cycle. 5:1760–71.

Lenne P.F., Wawrezinieck L., Conchonaud F. et al. (2006). Dynamic molecular confinement in the plasma membrane by microdomains and the cytoskeleton meshwork. EMBO J. 25:3245–56.

Levin M.K., Carson J.H. (2004). Fluorescence correlation spectroscopy and quantitative cell biology. Differentiation. 72:1–10.

Li Y., Knisely J.M., Lu W. et al. (2002). Low density lipoprotein (LDL) receptor-related protein 1B impairs urokinase receptor regeneration on the cell surface and inhibits cell migration. J Biol Chem. 277:42366–42371.

Li Y., Lawrence D.A., Zhang L. (2003). Sequences within domain II of the urokinase receptor critical for differential ligand recognition. J Biol Chem. 278:29925–32.

Llinas P., Helene Le Du M., Gardsvoll H. et al. (2005). Crystal structure of the human urokinase plasminogen activator receptor bound to an antagonist peptide. EMBO J. 24:1655–63.

Lund L.R., Green K.A., Stoop A.A. et al. (2006). Plasminogen activation independent of uPA and tPA maintains wound healing in gene-deficient mice. EMBO J. 25:2686–97.

Madsen C.D., Ferraris G.M., Andolfo A. et al. (2007). uPAR-induced cell adhesion and migration: vitronectin provides the key. J Cell Biol. 177:927–39.

Malengo G., Andolfo A., Sidenius N., et al. (2008). Fluorescence Correlation Spectroscopy and Photon Counting Histogram on membrane proteins: Functional dynamics of the GPI-anchored Urokinase Plasminogen Activator Receptor. Journal Biomedical Optics. In press.

Marguet D., Lenne P.F., Rigneault H. et al. (2006). Dynamics in the plasma membrane: how to combine fluidity and order. EMBO J. 25:3446–57.

Mazzieri R., D'Alessio S., Kenmoe R.K. et al. (2006). An uncleavable uPAR mutant allows dissection of signaling pathways in uPA-dependent cell migration. Mol Biol Cell. 17:367–78.

Moller L.B., Pollanen J., Ronne E. et al. (1993). N-linked glycosylation of the ligand-binding domain of the human urokinase receptor contributes to the affinity for its ligand. J Biol Chem. 268:11152–9.

Morone N., Fujiwara T., Murase K. et al. (2006). Three-dimensional reconstruction of the membrane skeleton at the plasma membrane interface by electron tomography. J Cell Biol. 174:851–62.

Odekon L.E., Sato Y., Rifkin D.B. (1992). Urokinase-type plasminogen activator mediates basic fibroblast growth factor-induced bovine endothelial cell migration independent of its proteolytic activity. J Cell Physiol. 150:258–63.

Patterson G.H., Piston D.W., Barisas B.G. (2000). Forster distances between green fluorescent protein pairs. Anal Biochem. 284:438–40.

Ploug M., Kjalke M., Ronne E. et al. (1993). Localization of the disulfide bonds in the NH2-terminal domain of the cellular receptor for human urokinase-type plasminogen activator. A domain structure belonging to a novel superfamily of glycolipid-anchored membrane proteins. J Biol Chem. 268:17539–46.

Preissner K.T., Kanse S.M., May A.E. (2000). Urokinase receptor: a molecular organizer in cellular communication. Curr Opin Cell Biol. 12:621–8.

Rabbani S.A., Mazar A.P., Bernier S.M. et al. (1992). Structural requirements for the growth factor activity of the amino-terminal domain of urokinase. J Biol Chem. 267:14151–6.

Ragno P. (2006). The urokinase receptor: a ligand or a receptor? Story of a sociable molecule. Cell Mol Life Sci. 63:1028–37.

Rajendran L., Simons K. (2005). Lipid rafts and membrane dynamics. J Cell Sci. 118:1099–102.

Redford G.I., Clegg R.M. (2005). Polar plot representation for frequency-domain analysis of fluorescence lifetimes. J Fluoresc. 15:805–15.

Resnati M., Pallavicini I., Wang J.M. (2002). The fibrinolytic receptor for urokinase activates the G protein-coupled chemotactic receptor FPRL1/LXA4R. Proc Natl Acad Sci USA 99: 1359–64.

Rigler R., Elson E. (2001). Fluorescence Correlation Spectroscopy: Theory and Applications. Springer, Berlin.

Ritchie K., Kusumi A. (2004). Role of the membrane skeleton in creation of microdomains. Subcell Biochem. 37:233–45.

Ritchie K., Iino R., Fujiwara T. et al. (2003). The fence and picket structure of the plasma membrane of live cells as revealed by single molecule techniques. Mol Membr Biol. 20:13–8.

Ritchie K., Shan X.Y., Kondo J. et al. (2005). Detection of non-Brownian diffusion in the cell membrane in single molecule tracking. Biophys J. 88:2266–77.

Rosales T., Georget V., Malide D. et al. (2007). Quantitative detection of the ligand-dependent interaction between the androgen receptor and the co-activator, Tif2, in live cells using two color, two photon fluorescence cross-correlation spectroscopy. Eur Biophys J. 36:153–61.

Schmid J.A., Birbach A. (2007). Fluorescent proteins and fluorescence resonance energy transfer (FRET) as tools in signaling research. Thromb Haemost. 97:378–84.

Schwille P., Meyer-Almes F.J., Rigler R. (1997). Dual-color fluorescence cross-correlation spectroscopy for multicomponent diffusional analysis in solution. Biophys J. 72:1878–86.

Selleri C., Montuori N., Ricci P. et al. (2005). Involvement of the urokinase-type plasminogen activator receptor in hematopoietic stem cell mobilization. Blood. 105:2198–205.

Selleri C., Montuori N., Ricci P. et al. (2006). In vivo activity of the cleaved form of soluble urokinase receptor: a new hematopoietic stem/progenitor cell mobilizer. Cancer Res. 66:10885–90.

Sheets E.D., Lee G.M., Simson R. et al. (1997). Transient confinement of a glycosylphosphatidylinositol-anchored protein in the plasma membrane. Biochemistry. 36:12449–58.

Sidenius N., Blasi F. (2000). Domain 1 of the urokinase receptor (uPAR) is required for uPAR-mediated cell binding to vitronectin. FEBS Lett. 470:40–6.

Sidenius N., Blasi F. (2003). The urokinase plasminogen activator system in cancer: recent advances and implication for prognosis and therapy. Cancer Metastasis Rev. 22:205–22.

Siegel R.M., Chan F.K., Zacharias D.A. et al. (2000). Measurement of molecular interactions in living cells by fluorescence resonance energy transfer between variants of the green fluorescent protein. Sci STKE. 2000:PL1.

Simons K., Toomre D. (2000). Lipid rafts and signal transduction. Nat Rev Mol Cell Biol. 1:31–9.

Simons K., Vaz W.L. (2004). Model systems, lipid rafts, and cell membranes. Annu Rev Biophys Biomol Struct. 33:269–95.

Simson R., Yang B., Moore S.E. et al. (1998). Structural mosaicism on the submicron scale in the plasma membrane. Biophys J. 74:297–308.

Smith P.R., Morrison I.E., Wilson K.M. et al. (1999). Anomalous diffusion of major histocompatibility complex class I molecules on HeLa cells determined by single particle tracking. Biophys J. 76:3331–44.

Snapp E.L., Sharma A., Lippincott-Schwartz J. et al. (2006). Monitoring chaperone engagement of substrates in the endoplasmic reticulum of live cells. Proc Natl Acad Sci USA. 103:6536–41.

Soon L., Braet F., Condeelis J. (2007). Moving in the right direction-nanoimaging in cancer cell motility and metastasis. Microsc Res Tech. 70:252–7.

Stoppelli M.P., Corti A., Soffientini A. et al. (1985). Differentiation-enhanced binding of the amino-terminal fragment of human urokinase plasminogen activator to a specific receptor on U937 monocytes. Proc Natl Acad Sci USA. 82:4939–43.

Suhling K., Siegel J., Phillips D. et al. (2002). Imaging the environment of green fluorescent protein. Biophys J. 83:3589–95.

Suzuki K., Ritchie K., Kajikawa E. et al. (2005). Rapid hop diffusion of a G-protein-coupled receptor in the plasma membrane as revealed by single-molecule techniques. Biophys J. 88:3659–80.

Suzuki K.G., Fujiwara T.K., Sanematsu F. et al. (2007). GPI-anchored receptor clusters transiently recruit Lyn and G alpha for temporary cluster immobilization and Lyn activation: single-molecule tracking study 1. J Cell Biol. 177:717–30.

Thompson N.L., Lieto A.M., Allen N.W. (2002). Recent advances in fluorescence correlation spectroscopy. Curr Opin Struct Biol. 12:634–41.

van Munster E.B., Gadella T.W. (2005). Fluorescence lifetime imaging microscopy (FLIM). Adv Biochem Eng Biotechnol. 95:143–75.

Vukojevic V., Pramanik A., Yakovleva T. et al. (2005). Study of molecular events in cells by fluorescence correlation spectroscopy. Cell Mol Life Sci. 62:535–50.

Wallrabe H., Periasamy A. (2005). Imaging protein molecules using FRET and FLIM microscopy. Curr Opin Biotechnol. 16:19–27.

Waltz D.A., Sailor L.Z., Chapman H.A. (1993). Cytokines induce urokinase-dependent adhesion of human myeloid cells. A regulatory role for plasminogen activator inhibitors. J Clin Invest. 91:1541–52.

Wei Y., Eble J.A., Wang Z. et al. (2001). Urokinase receptors promote beta1 integrin function through interactions with integrin alpha3beta1. Mol Biol Cell. 12:2975–86.

Wei Y., Lukashev M., Simon D.I. et al. (1996). Regulation of integrin function by the urokinase receptor. Science. 273:1551–5.

Wei Y., Waltz D.A., Rao N. et al. (1994). Identification of the urokinase receptor as an adhesion receptor for vitronectin. J Biol Chem. 269:32380–8.

Wei Y., Yang X., Liu Q. et al. (1999). A role for caveolin and the urokinase receptor in integrin-mediated adhesion and signaling. J Cell Biol. 144:1285–94.

Wei Y., Tang C.H., Kim Y. et al. (2007). Urokinase receptors are required for alpha 5 beta 1 integrin-mediated signaling in tumor cells. J Biol Chem. 282:3929–39.

Weiss M., Hashimoto H., Nilsson T. (2003). Anomalous protein diffusion in living cells as seen by fluorescence correlation spectroscopy. Biophys J. 84:4043–52.

Wilhelm O.G., Wilhelm S., Escott G.M. et al. (1999). Cellular glycosylphosphatidylinositol-specific phospholipase D regulates urokinase receptor shedding and cell surface expression. J Cell Physiol. 180:225–35.

Zacharias D.A., Violin J. D., Newton A. C., Tsien R.Y. (2002). Partitioning of lipid-modified monomeric GFPs into membrane microdomains of live cells. Science. 296:913–6.

Chapter 25
Janus-Faced Effects of Broad-Spectrum and Specific MMP Inhibition on Metastasis

Charlotte Kopitz and Achim Krüger

Abstract In this chapter, we summarize the current knowledge of the effects of matrix metalloproteinase (MMP) inhibition on metastasis. While it was initially perceived that MMP inhibition should be the most potent therapeutic antimetastatic strategy, it has now been appreciated that more basic research has to be performed before another, more sophisticated, attempt should be made to use MMPs as targets in the clinic. The summary makes it evident that the complexity of preclinical models may obscure conclusive results. This may have led to the overestimated expectations and premature introduction of MMP inhibitors (MMPIs) into clinical trials. In addition to discussion of the options and limitations of different metastasis models in the search for improved strategies to inhibit MMPs, we emphasize the necessity to simplify the readout of models at this stage of research when the biology of interference with the proteolytic network is still not clear. Further, we stress the effects of MMP inhibition on the modulation of the host microenvironment, which critically influences the metastatic potential of tumor cells. This latter aspect is one of the new appreciations explaining the Janus-faced effects of broad-spectrum and specific MMP inhibition on metastasis.

Introduction: MMPs as Targets for Antimetastatic Intervention

Metastasis, the key feature of malignant tumors, comprises the formation of tumor cell colonies in distant organs. Detachment of tumor cells from the primary tumor is a necessary but not sufficient parameter of metastasis. Eventually, the efficacy of metastasis is determined by the ability of tumor cells to invade and colonize distant organs. Proteolytic degradation of the extracellular matrix (ECM) is one necessary prerequisite for the different steps of tumor cell dissemination, which include invasion

A. Krüger

Klinikum rechts der Isar der Technischen Universität München, Institut für Experimentelle Onkologie und Therapieforschung, e-mail: achim.krueger@lrz.tu-muenchen.de

D. Edwards et al. (eds.), *The Cancer Degradome.*   495
© Springer Science + Business Media, LLC 2008

of the surrounding tissue, vessel boundaries (intravasation and extravasation), and distant organ tissue (Stetler-Stevenson 2001). The proteolytic activity is provided by different proteases secreted not only by tumor cells but also by a variety of activated host cells (fibroblasts, immune cells, etc.) within the tumor microenvironment (Mueller and Fusenig 2002). Indeed, members of many protease families including serine- (Reuning et al. 1998), cysteine- (Joyce and Hanahan 2004), aspartyl- (Liaudet-Coopman et al. 2006), and metalloproteases (Egeblad and Werb 2002) were shown to be associated with tumor progression. Among these, the matrix metalloproteinases (MMPs) are unique, as they are able to degrade most of the substrates within the ECM, including fibrillar collagens (Coussens et al. 2002), a major structural molecule in the interstitium. Initially, the assumption prevailed that all MMPs promote the invasion of tumor cells. We meanwhile learned that MMPs substantially contribute to virtually all aspects of tumor progression, including angiogenesis, differentiation, proliferation, and apoptosis (Egeblad and Werb 2002). In addition, at least some MMPs may have features that are counterproductive for tumor progression (Montel et al. 2004, Abraham et al. 2005, López-Otín and Matrisian 2007), and it is not yet clear whether this relies only on their proteolytic activity or whether other, as yet unknown, features of MMPs are involved. The conceptual consequence of this new knowledge, which has unfortunately only been accumulated after the failure of clinical trials with broad-spectrum MMP inhibitors (MMPIs) (Zucker et al. 2000, Pavlaki and Zucker 2003), is that we need to elucidate the beneficial time-points and locations at which specific MMPs should be inhibited in order to be able to prevent or suppress metastasis. This strategy of MMP inhibition is undoubtedly worthy of pursuit, as there are still no efficient antimetastatic strategies available.

In different cancer diseases high expression of MMPs correlates with poor prognosis and shorter survival of patients with colorectal (Zucker and Vacirca 2004), breast (Duffy et al. 2000), brain (Miyata et al. 2006), bladder (Kanayama 2001), and renal cancer (Miyata et al. 2006), and non-Hodgkin's lymphoma (Kossakowska et al. 2000), indicating that MMP activity promotes late steps of metastasis, leading to the devastating colonization of sometimes several organs and the explosion of numbers of scattered metastatic foci toward the end of the disease.

In this chapter, we summarize experimental evidence of the consequences of broad-spectrum or specific MMP inhibition on metastasis. We also discuss the ways in which MMP inhibition can provoke changes in tissue homeostasis via changed gene expression, resulting in alterations in the susceptibility of target organs to metastases.

Reexamination of Expectations: Options and Limitations of Murine Metastasis Models for the Evaluation of MMP-Inhibitory Effects *In Vivo*

The view that MMPIs should not be pursued further for drug development because they have failed in the clinic has an element of schadenfreude about it: it is perhaps a consequence of the elevated expectations for their efficacy in preventing metastatic spread, angiogenesis, and the growth of primary tumors and metastases that were associated with their introduction (Brown 1995) (see also Chap. 36). These expectations had been fed by, what was at that time, an incomplete knowledge of the biology of MMPs and their inhibition. Although MMPs are involved in many steps of the metastatic cascade (Chambers and Matrisian 1997), their respective functions are so diverse that the overall inhibition of this class of molecules would result in a severe disturbance of tissue homeostasis and lead to unpredictable consequences during tumor progression. In addition, the therapy windows of MMPIs have to be clearly defined with the aim to inhibit MMPs as specifically as possible in respect to their spatiotemporal expression or even more importantly, their activity. In order to fill the still existing knowledge gap in these respects, it is necessary to employ *in vivo* models that allow the examination of particular steps of tumor progression as well as underlying cellular and molecular mechanisms in which the effects of MMP inhibition are studied.

Regarding the complexity or clinical relevance of such *in vivo* models we are, at this stage of elucidation of the "Proteolytic Web", reminded of Werner Heisenberg's quote that "any increase in precision of one variable is associated with a decrease in precision of the other variable in the conjugate pair" (Heisenberg 1927), which describes a basic principle in quantum physics. This notion can be translated to experimental approaches of virtually all scientific questions and definitely also to the investigation of complex biological events such as metastasis. Therefore, for an in-depth analysis of the role of MMPs in aspects of metastasis it is certainly preferable to reduce the number of variables. With the knowledge established in simple models – step-by-step – more complex tumor models can be approached, ideally also those that, if at all possible, represent the clinical situation more adequately.

Generation of Metastases in Murine Models

Models that reliably resemble aspects of the metastatic cascade are prerequisites for the analysis of the cellular and molecular effects of MMP-interference on metastasis. The minimum requirement of model systems in biology is that some clearly defined biological processes can be studied. This applied to genetic questions where research on bacteria and yeast greatly promoted our knowledge as well as to questions of developmental biology, where fruit flies and worms had great impact on our knowledge in the field today. The surprises encountered in the context of the development of MMPIs clearly showed that we do not know enough about the biology of MMPs, not to mention of metastasis, to be able to estimate the prospects and consequences of MMP inhibition. Therefore, basic research with simple models is still necessary. For the generation of metastases in animal models, we distinguish "experimental" from "spontaneous" metastasis models and will explain their options and limits below.

Experimental Metastasis Models

The simplest way to mimic parts of the metastatic cascade is the intravasal inoculation of a defined number of tumor cells, leading to the generation of so-called experimental metastases. These metastases are colonies of tumor cells generated after extravasation of inoculated tumor cells from the vasculature, invasion into the subendothelial matrix of and growth in the respective target tissue (Schirrmacher 1985). Subsequently, secondary invasion of tumor cells from established metastases can occur, leading to the destruction of the target organ's integrity and function. Inhibition of the late steps of metastasis is discussed to be the most promising antimetastatic strategy (Radinsky and Ellis 1996, Berman et al. 2001).

Information obtained from this experimental approach is independent from all events at or in the primary tumor site such as angiogenesis and growth rate (Schirrmacher 1985). Therefore, all additional experimental interventions (such as MMP inhibition with synthetic inhibitors or genetic modulation of the target organ of metastasis) will exclusively affect the late phase of metastasis. This dissemination depends on the metastatic potential of the investigated tumor cell line (the "SEED") and/or the susceptibility of the target tissue (the "SOIL") (Fidler 2001). The impact of the experimental intervention is quantifiable by the evaluation of the number and size of metastases in target tissues. Changes in these parameters reflect the impact of the MMP-inhibitory intervention on several parameters (Fig. 25.1):

- survival of the inoculated tumor cells
- efficacy of their extravasation (number)
- proliferative capacity of metastases in the target tissue (size)
- tumor cell invasiveness (secondary invasion from primary metastases, which forms a characteristic metastatic pattern within the target organ)
- organ-specificity (interorgan metastasis pattern)

With experimental metastasis models it is even possible to mimic some clinically relevant situations, such as iatrogenically induced intraoperative dissemination of tumor cells (Atkin et al. 2005) and metastasis of circulating tumor cells, which are often detectable in cancer patients even after removal of the primary tumor (Terstappen et al. 2000). It is important to recognize that experimental metastasis models represent a reductionist approach, which leaves out important steps of the metastatic cascade (unlike spontaneous metastasis models, see below). However, the benefit is that the results obtained are not obscured by events surrounding the establishment of the primary tumor. These events determine the number of disseminating tumor cells and are hardly controllable by the investigator.

Spontaneous Metastasis Models

In spontaneous metastasis models, tumor cells detach from the primary tumor tissue, pass each critical step of the entire metastatic cascade, and establish

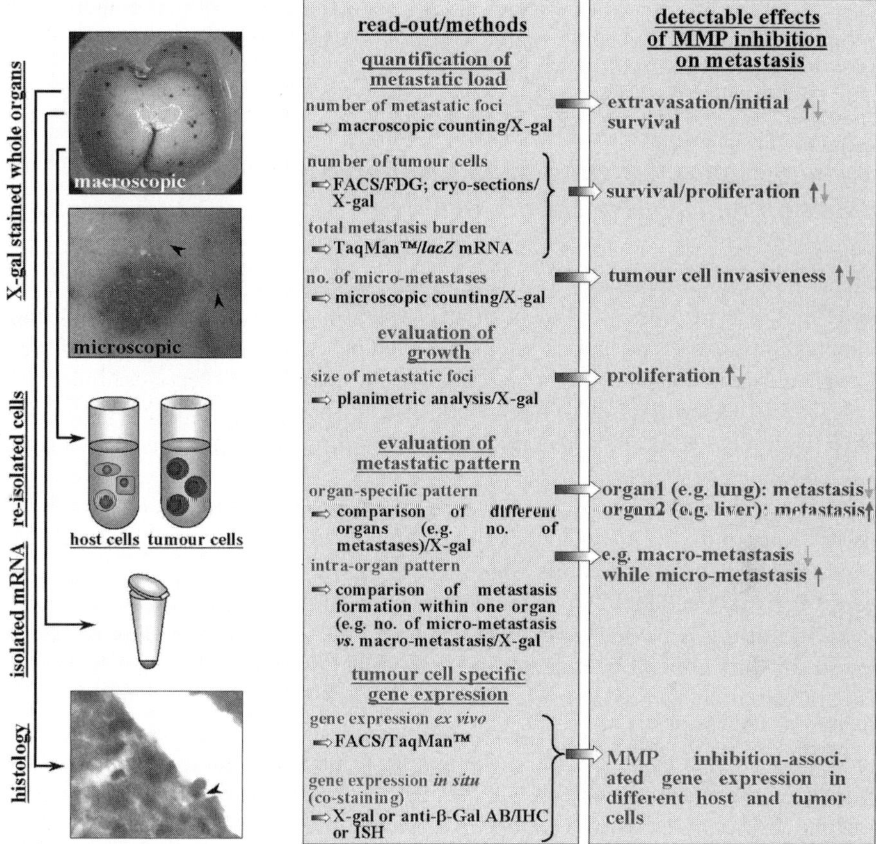

Fig. 25.1 Applications of a genetic tag for evaluation of the effects of matrix metalloproteinase (*MMP*) inhibition on metastasis—*lacZ*-tag as an example. After inoculation of *lacZ*-tagged tumor cells the following methods can be applied to detect these tumor cells: 5-bromo-4-chloro-3-indolyl-β-D-galactopyranoside (*X-gal*) staining of whole organs, re-isolation of tumor cells from target organs by FACS and fluorescein-β-D-galactopyranoside (*FDG*) staining, isolation of total ribonucleic acid (*RNA*) from metastasis-bearing organs and detection of *lacZ* mRNA by quantitative RT-PCR (TAqMan™), and X-gal staining of tissue sections

metastases in distant organs. Generation of primary tumors can be induced by transgenic expression of oncogenes, application of specific carcinogens, or extra-vasal application of tumor cells into the host tissue (Eccles 2001, Guy and Cardoso 2001, Price 2001). Thus, in such models, the efficacy of metastasis depends not only on the metastatic potential of tumor cells but also on features of the primary tumor, such as proliferation and angiogenesis, as well as on its position (Eccles 2001). In such models, the efficacy of organ colonization often depends on the efficacy of the initial steps of metastasis. It became possible only recently, by the use of complex technical approaches such as intravital multiphoton microscopy, to study and dissect at least some aspects of the many biological events occurring during this phase of tumor progression (Wolf et al. 2003, 2007). Only with such technically sophisticated approaches is it possible to reveal at which step of the early phases of tumor cell dissemination inhibition of MMPs is effective.

Advantages of a Genetic Tag to Visualize Effects of MMP Inhibition on Invasive Events In Vivo

Detection of micrometastases in distant organs has always been a challenge in the clinic, and also in most preclinical models used for evaluation of anti-metastatic therapies. Because of the lack of a sensitive read out it is an extremely difficult task to evaluate antimetastatic therapies (such as antiinvasive protease inhibitors) in clinical trials. Determination of therapeutic efficacy basically has to rely on survival data (Bissett et al. 2005, Levin et al. 2006) or systemic tumor markers (Rosenbaum et al. 2005).

Researchers working with *in vivo* models have the opportunity to circumvent this difficulty of the clinic by genetically tagging the tumor cells of interest. Several useful genetic tags are available, which can be stably expressed by the tumor cells and all their descendants. As genetically tagged tumor cells can be injected into mice and this tag is passed on to each daughter cell, all cells, including those that metastasize *in vivo*, can be detected and quantified by the products of the marker gene, for example, β-D-galactosidase (Krüger et al. 1998), luciferase (Jenkins et al. 2003), or enhanced green fluorescent protein (eGFP) (Oyajobi et al. 2007), in metastasis-afflicted organs. One of the simplest options is the marker gene *lacZ*. This bacterial gene codes for β-D-galactosidase, which hydrolyses the chromogenic substrate 5-bromo-4-chloro-3-indolyl-β-D-galactopyranoside (X-gal) leading to the precipitation of indigo blue in each tagged cell (Krüger et al. 1998). While detection of *lacZ* is a tool for analysis of excised tissue, eGFP (Oyajobi et al. 2007) and luciferase (Jenkins et al. 2003) allow detection in the living animal. However, detection of the latter requires more sophisticated equipment, such as a fluorescence illuminator (Oyajobi et al. 2007) or a cooled CCD camera mounted in a light-tight specimen box (Jenkins et al. 2003) and does not easily yield to identification of single metastatic cells due to background fluorescence of many tissues, and if so, only at the cost of employment of high-resolution microscopy. Resolution of the indigo blue precipitate in X-gal-stained whole organ mounts or histological sections containing *lacZ*-expressing cells is extremely high even at the single tumor cell level (Fig. 25.1, upper and lower left pictures) (Krüger et al. 1994). This high resolution and clear boundaries of the colored areas allow the identification of stromal areas within tumors and metastatic foci. The versatility of the different detection methods for *lacZ* expression at the mRNA and protein levels, even of re-isolated living cells, allows the exact quantification of the metastastic load within a target organ, along with evaluation of the growth of metastases and assessment of the patterns of invasiveness in different organs (summarized in Fig. 25.1). The employment of fluorescein-β-D-galactopyranoside (FDG), another substrate of β-D-galactosidase, allows detection of viable *lacZ*-tagged cells. This technique can be applied when dissection and (re-)isolation and characterization of host and tumor cells by flow cytometry is desired (Rocha et al. 2001). Subsequent gene expression analysis of isolated subpopulations allows differentiation of the respective contribution of tumor or host cells to the genetic profile of a metastasis-bearing organ. Also, treatment-provoked changes in gene expression can be assigned to particular cell types (Rocha et al. 2001) (Fig. 25.1). *LacZ*-tagging has been providing a means to detect the effects of MMP interference on particular steps of metastasis at high resolution (Krüger et al. 1998). Indeed, when this technique was applied to mice treated with the broad-spectrum MMPI batimastat, it was evident that liver metastasis was induced while the number of lung metastases was reduced (Della Porta et al. 1999, Krüger et al. 2001). The advantage of simple X-gal staining of several organs of the treated mice quickly revealed quantifiable data on this organ-specific effect, which had not been reported throughout the preclinical testing period of MMPIs in the 1990s. This finding coincided with the halting of most phase III clinical trials with MMPIs due to adverse effects.

Anti- and Pro-metastatic Effects of Broad-Spectrum Inhibition of Metalloproteinases on Metastasis

Natural Broad-Spectrum Metalloproteinase Inhibitors

Under normal physiological conditions, the activity of the 24 known MMPs is tightly regulated by their natural inhibitors (Egeblad and Werb 2002). MMPs can be bound and inhibited by α2-macroglobulin, TFPI-2 (tissue factor pathway inhibitor), PCPE (procollagen C-terminal proteinase enhancer), RECK (reversion-inducing cysteine-rich protein with kazal motifs), and the four known mammalian tissue inhibitors of metalloproteinases (TIMPs), the latter being their major inhibitors (Baker et al. 2002). TIMPs are broad-spectrum inhibitors of MMPs and of some ADAMs (a disintegrin and metalloproteinase) with only slight differences in their specificity profile (Brew et al. 2000). TIMP-1 has the unique feature of binding proMMP-9 (Goldberg et al. 1992), but is a poor inhibitor of MMP-19 and of membrane-bound MT-MMPs (membrane-type MMPs) (Brew et al. 2000). TIMP-1 shares with TIMP-3 the ability to inhibit ADAM-10 activity (Amour et al. 2000).

Since deregulated activity of MMPs can enhance dissemination of tumor cells, it was initially believed that overexpression of TIMPs, either by tumor or by host cells, can prevent metastasis (Brand et al. 2000). Indeed, increased expression of all four known TIMPs by the host or tumor cells significantly reduced metastatic dissemination into different organs in several spontaneous and experimental metastasis models (Brand 2002). However, as the size of the primary tumor was also affected by the overexpression of TIMPs in the spontaneous metastasis models (Krüger et al. 1997, Wang et al. 1997, Li et al. 2001), their postulated anti-invasive action was not conclusive. Use of experimental metastasis models shed light on the anti-invasive activity of all four TIMPs: overexpression of TIMP-1, -2, or -4 leads to significant reduction of metastases in different organs (Yoneda et al. 1997, Brand et al. 2000, Elezkurtaj et al. 2004), and a potential anti-invasive role of host-derived TIMP-3 became evident when knockout of TIMP-3 led to increased experimental metastasis in different organs (Cruz-Munoz et al. 2006, see also Chap. 20). *In vitro* studies showing reduced invasion of tumor cells by overexpression of each particular TIMP (Brand 2002) underlined that TIMPs interfere with metastasis by suppressing the invasive steps. Thus, the concept of "impregnation" (Brand et al. 2000) of a target organ of metastasis by an intrinsic overexpression of natural broad-spectrum MMPIs as well as the "disarmament" of tumor cells by overexpression of TIMPs could be maintained. At least for TIMP-1 it was demonstrated that reduced experimental liver metastasis is most likely due to diminished extravasation of the tumor cells (Elezkurtaj et al. 2004) (Fig. 25.2a, left panel). The reduced extravasation in the T-cell lymphoma model was most likely due to inhibition of MMP-9 (Fig. 25.2a) as its participation in extravasation was previously demonstrated (Arlt et al. 2002). This explanation leads to the assumption that achievement of an effective antimetastatic activity of TIMP-1 relies on an effective TIMP-1 threshold. This assumption is supported by the following facts:

- Increased TIMP-1 expression in the liver of transgenic mice does not reduce experimental liver metastasis of murine T-cell lymphoma cells (Krüger et al. 1997), while the manyfold higher TIMP-1 expression in the liver achieved by adenoviral gene transfer prevented experimental metastasis of this cell line by more than 90% (Elezkurtaj et al. 2004).
- Intraperitoneal application of recombinant TIMP-1 was not able to inhibit spontaneous liver metastasis of orthotopically inoculated human colon carcinoma cells. In contrast, three clones of the human colon cancer cell line transfected to overexpress TIMP-1 displayed an expression level-dependent effect on spontaneous liver metastasis (Yamauchi et al. 2001).

Interestingly, TIMP-1 levels are important not only for the achievement of an antimetastatic effect, but also for the localization of its expression. Local expression of TIMP-1 in the lungs of transgenic mice was not sufficient to inhibit spontaneous metastasis of murine melanoma cells, while exclusive expression of TIMP-1 in the liver with high TIMP-1 levels in the blood significantly reduced metastasis to the lung (de Lorenzo et al. 2003).

Although in animal models a potential antimetastatic activity of all TIMPs has been demonstrated, elevated expression of TIMP-1 and -2 often correlates with poor outcome/shorter survival of patients with different malignancies such as breast (Remacle et al. 2000, Schrohl et al. 2004), ovarian (Davidson et al. 2002, Manenti et al. 2003), gastric (Yoshikawa et al. 2001, Wu et al. 2007), or colorectal cancer (Hammer et al. 2006, Roca et al. 2006). We call this correlation the "TIMP-paradox". One explanation for these observations is that the elevated expression of MMPs in aggressive tumors leads to an upregulation of TIMPs in an ultimately

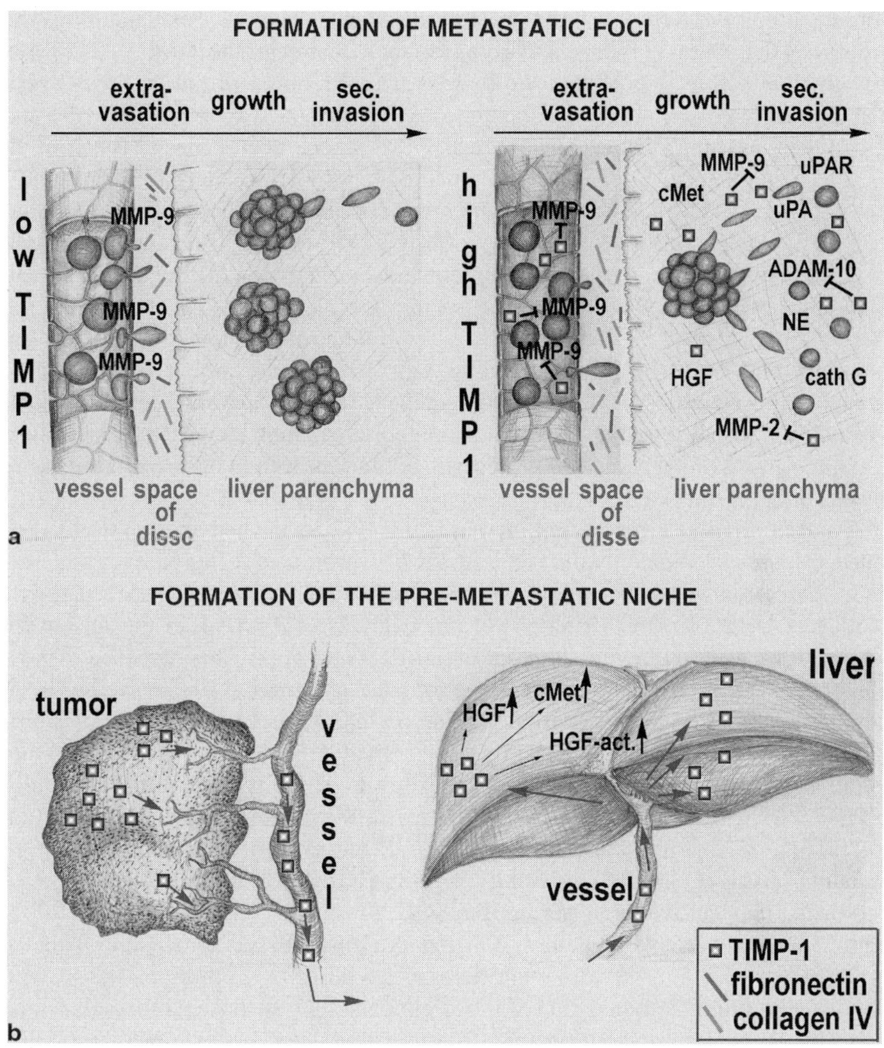

Fig. 25.2 Elucidation of the "TIMP-1-paradox": modulation of a premetastatic niche in the liver by tissue inhibitors of metalloproteinases (TIMP)-1—theoretical development and molecular mechanism. (**a**) In livers with low TIMP-1 expression (left), matrix metalloproteinase (MMP)-9-dependent extravasation, as well as subsequent growth and secondary invasion, occurs as usual. In livers with high TIMP-1 levels, MMP-9-dependent extravasation of tumor cells is reduced by TIMP-1 present in the circulation. High TIMP-1 levels within the liver parenchyma induce expression of MMP-9, MMP-2, ADAM-10, neutrophil elastase (NE), cathepsin *G* (cath *G*), urokinase plasminogen activator (uPA), and urokinase-type plasminogen activator receptor (uPAR). Also, the HGF protein level and the amount of activatable membrane-bound cMet are increased by TIMP-1. Altogether these factors efficiently induce HGF-signaling in the liver, promoting secondary invasion of tumor cells from primary metastases. (**b**) Elevated expression of TIMP-1 in the primary tumor, either by the tumor or by stroma cells (left side), leads to accumulation of TIMP-1 protein in the bloodstream. Circulating blood transports TIMP-1 into the

fruitless attempt to reestablish the proteolytic balance. The other explanation is the hypothesis that there is an intrinsic protumorigenic feature of the TIMPs. All TIMPs are principally able to contribute to increased proliferation of tumor cells (Baker et al. 2002). TIMP-1 and -2 were shown to be able to reduce apoptosis, TIMP-1 and -4 can increase tumorigenesis, and TIMP-1 can induce angiogenesis (Baker et al. 2002). While primary tumor growth and angiogenesis affect metastasis only indirectly, proinvasive activity is one of the main drivers of tumor cell dissemination. However, a direct proinvasive feature of any TIMP has, until recently, not been described. Since the death of cancer patients relies usually on the metastatic spread of tumor cells within the body, the association of increased TIMP-1 levels and shorter survival of the patients indicates a possible causal relationship between metastasis and TIMP-1 expression and activity. Therefore, examination of possible prometastatic activities of TIMP-1 could reveal whether increased expression of this inhibitor in cancer patients with more aggressive malignancies represents a host response toward the deregulated proteolytic activity or whether these inhibitors are causal for aggressiveness.

A causal relationship between increased TIMP-1 levels and shorter survival of cancer patients is evident by the fact that TIMP-1 promotes scattered liver metastasis by induction of the hepatocyte growth factor (HGF)/cMet-signaling pathway (Kopitz et al. 2007). In this study, elevated levels of TIMP-1 in the host were mimicked by adenoviral gene transfer of TIMP-1 into mice. The increased expression of TIMP-1 led to an augmented level of HGF protein in livers, as well as to an increased expression of pro-HGF-activating proteases, such as urokinase plasminogen activator (uPA), tissue plasminogen activator (tPA), and matriptase (Fig. 25.2a, right panel). Also, increased protein levels of the tyrosine-kinase receptor of HGF, cMet, were found in livers with elevated levels of TIMP-1 (Kopitz et al. 2007) (Fig. 25.2a, right panel). Increase in all these factors necessary for HGF-signaling (Jiang et al. 2005) efficiently triggers HGF-signaling even in metastasis-free livers, thus cultivating a susceptible SOIL for tumor cells (Kopitz et al. 2007) (Fig. 25.2a, right panel). This new, TIMP-1-provoked, tissue homeostatic balance led to increased scattering and total metastasis load in livers after challenge with two different tumor cell lines (Fig. 25.2a, right panel). Only those tumor cells which were able to extravasate could benefit from the modulated tissue homeostasis (Fig. 25.2a). This finding offers the explanation that in livers with drastically high overexpression of TIMP-1, no prometastatic effect was observed as extravasation of nearly all inoculated tumor cells was prevented (Elezkurtaj et al. 2004). We also demonstrated an organ specificity of this novel feature of TIMP-1: while elevated host TIMP-1 significantly reduced experimental lung metastasis of human fibrosarcoma cells (HT1080), metastasis of the same tumor cell line to the liver was significantly increased (Kopitz et al. 2007).

Fig 25.2 (Continued) liver (right). In the liver tissue, elevated levels of TIMP-1 increase HGF protein levels, induce expression of HGF activators, and preserve activatable cMet protein on the cell surfaces by inhibition of MMPs and ADAM-10. Together, these effects lead to formation of a TIMP-1-provoked premetastatic niche in the liver

Since HGF-signaling is a highly conserved pathway involved in the repair and regeneration of mammalian livers, we hypothesized that elevated TIMP-1 levels may induce similar effects in livers of human cancer patients. Indeed, the gene expression signature in livers associated with elevated levels of TIMP-1 in patients with colorectal cancer resembled the expression signature that was provoked by elevated TIMP-1 expression in mice (Kopitz et al. 2007). As this gene expression signature is linked to generation of a premetastatic niche in the liver of mice, we suggest that in humans also this gene expression signature creates a particular environment in the liver that favors the aggressive scattering of tumor cells. In fact, the elevated TIMP-1 levels, and, therefore, also the TIMP-1-associated gene expression signature, significantly correlate with an earlier development of liver metastases as compared to patients with significantly lower TIMP-1 expression (Kopitz et al. 2007).

As the MMP-inhibitory N-terminal domain of TIMP-1 (N-TIMP-1) was sufficient to provoke the prometastatic effect in the animal model, MMP inhibition by TIMP-1 is most likely the basis for this activity. Furthermore, increase in cell-associated cMet relies most likely on inhibition of ADAM-10 mediated shedding of cMet. This link between MP-inhibition and enhancement of HGF/cMet-signaling was revealed by an *in vitro* biochemical proof-of-concept study: elevated levels of TIMP-1 and N-TIMP-1 preserved activatable membrane-bound cMet by reduced cMet shedding, demonstrating that the proteolytic release of cMet is inhibited either by direct or by indirect inhibition of a cMet sheddase. Interestingly, suppression of ADAM-10 also prevented shedding of cMet, indicating that ADAM-10 is involved in the regulation of cMet shedding (Kopitz et al. 2007). Sheddases are known for their capacity to regulate cell surface-associated tyrosine kinase receptors (Blobel 2005). These data indicate that preservation of cMet is induced by TIMP-1 or N-TIMP-1, and may be a consequence of direct inhibition of ADAM-10 and thereby loss of its sheddase activity (Amour et al. 2000), or indirectly by inhibition of ADAM-10 activation by a different metalloprotease that is inhibitable by TIMP-1/N-TIMP-1. Also, other studies revealed protumorigenic effects of TIMP-1, which depend on MMP inhibition. It has been shown that MMP inhibition by TIMP-1 alters the gene expression of breast cancer cells (Porter et al. 2005). Some proproliferative activities of TIMP-1 rely on MMP inhibition (Porter et al. 2004), while some are independent of its MMP inhibitory activity (Chesler et al. 1995). These MMP inhibition-independent effects of TIMP-1 and of other TIMPs are described in the section "Metalloproteinase Inhibition-Independent Effects of Natural Metalloproteinase Inhibitors on Metastasis" of this chapter.

Synthetic Broad-Spectrum MMPIs

The initial concept was that broad-spectrum MMP inhibition should block tumor cell spread by decreasing the invasive potential of the tumor cells. Consequently, MMP-inhibitory compounds were designed in order to prevent these death-determining invasive steps of tumor progression. The first generation of synthetic MMP inhibitors (MMPIs) was designed to target effectively all members of the MMPs in order to achieve maximal antimetastatic efficacy. However, these compounds were too broad spectrum, and were unable even to discriminate related zinc-dependent proteases, such as some ADAMs (a disintegrin and metalloproteinase) or bacterial metalloendopeptidases (Wojtowicz-Praga et al. 1997, Brown 1998, see also Chaps. 36 and 38). This first-generation MMPIs were mostly hydroxamic acid derivates, such as batimastat (BB-94), the first MMPI evaluated in cancer patients, and its orally bioavailable successors, marimastat (BB-2516), GM6001 (Pavlaki and Zucker 2003), MMI270 (CGS27023A) (Levitt et al. 2001), and GI129471 (Maquoi et al. 2000, Fisher and Mobashery 2006).

Similar to the results of natural broad-spectrum inhibitors (TIMPs), some synthetic broad-spectrum MMPIs reached significant antimetastatic efficacies in different mouse models. For example, the synthetic broad-spectrum inhibitors batimastat and MMI270 significantly reduced spontaneous metastasis into different organs, including liver and lung (Wang et al. 1994, Prontera et al. 1999, Ogata et al. 2006). But similar to the effects achieved with TIMPs, the concomitant inhibition of primary tumor growth influenced the result (Wang et al. 1994, Prontera et al. 1999). In some studies, the primary tumor was removed before application of the inhibitor, simulating a clinically more relevant situation. In this setting, regrowth of a tumor after its resection could also be affected by the treatment with the used MMPI. Therefore, in this situation, reduced spontaneous metastasis could still reflect the impact of the MMPI on primary tumor growth (Sledge et al. 1995). When the primary tumor did not recur (Ogata et al. 2006), reduced spontaneous metastasis reflected most likely the reduced MMP-dependent invasive steps of the metastatic cascade.

An experimental metastasis model with rat mammary carcinoma cells in syngeneic rats revealed significant reduction of total experimental metastasis burden in lungs of mice treated with batimastat (Eccles et al. 1996). This result alone was not sufficient to state whether batimastat reduced formation (extravasation and early survival of inoculated tumor cells) or growth (survival, proliferation of extravasated tumor cells) of metastases. A follow-up experiment of the same group provides an indication for the step in which batimastat interferes with metastasis: an additional application of batimastat 6 h after tumor cell inoculation increased the antimetastatic effect of batimastat, indicating that an early step of experimental metastasis occurring around 6 h after tumor cell inoculation is affected by batimastat treatment, for example, "a period of active tumor cell invasion or establishment of a favorable microenvironment for growth that is particularly sensitive to MMPIs" (Eccles et al. 1996). Also, other studies demonstrated that application of different broad-spectrum MMPIs, like batimastat and GM6001, reduced the metastatic burden of intravascularly inoculated tumor cells to different organs, including the liver (Jimenez et al. 2000, Winding et al. 2002).

The broad-spectrum MMPI batimastat inhibited the dissemination of the tumor cells to the target organ of metastasis by inhibiting extravasation (Krüger et al.

2001). This direct effect of batimastat on metastasis was evident by quantification of *lacZ*-tagged foci of human breast carcinoma cells in lungs after intravenous inoculation (Krüger et al. 2001). But other features of broad-spectrum MMPIs could also indirectly affect metastasis, such as reduction of proliferation. The reduced primary tumor growth (Wang et al. 1994, Zervos et al. 1997, Prontera et al. 1999) seen in treatments with different synthetic MMPIs has already been interpreted as a measure of the antiproliferative activity of these MMPIs. Indeed, metastasis can also be affected by this antiproliferative activity of synthetic broad-spectrum MMPIs: batimastat treatment reduced the size but not the number of experimental metastases of murine melanoma cells by inhibiting angiogenesis (Wylie et al. 1999). This study indicates that inhibition of MMPs not only blocks the path-clearing function of MMPs (broad-spectrum MMP inhibition) but also interferes with other actions of MMPs, such as activation, degradation, release, or shedding of receptors, growth factors, chemokines, or other receptor ligands (Egeblad and Werb 2002). Therefore, broad-spectrum inhibition of MMPs with synthetic compounds also interferes with gene expression, which consequently modulates tissue homeostasis, resulting in increased susceptibility of the liver to metastasis, as already described for TIMP-1 (see "Natural Broad-Spectrum Metalloproteinase Inhibitors").

Most of the clinical trials with these MMPIs failed because of a lack of therapeutic efficacy and unacceptable side effects (Zucker et al. 2000, Coussens et al. 2002; see also Chap. 36 by Fingleton). This lack of antitumorigenic activity of broad-spectrum MMPIs was surprising, as only a small number of publications had reported on limited effectiveness of these compounds. However, one study, employing intrasplenic inoculation of murine melanoma cells and intravital microscopy, showed that batimastat treatment did not affect extravasation and early survival of extravasated tumor cells (Wylie et al. 1999). But this study alone is not able to explain the massive failure of MMPIs in the clinic. Some light on adverse effects was shed by our studies showing that broad-spectrum inhibition can lead to promotion of liver metastasis (Krüger et al. 2001). Interestingly, this promotion of liver metastasis was associated with an increased expression of Mmp-9, Mmp-2, and Hgf (Krüger et al. 2001, Arlt et al. 2002) and batimastat treatment alone was sufficient to provoke this upregulation in the liver (Krüger et al. 2001).

In addition to the effects of broad-spectrum MMP inhibition on the "Soil", the gene expression of tumor cells can also be changed when they are incubated with synthetic broad-spectrum MMPIs. Expression of MMP-9 by human fibrosarcoma cells was increased by incubation of these tumor cells either with batimastat or with GI129471 (Maquoi et al. 2002, 2004). Since the gelatinases (MMP-2 and -9) seem to be important mediators of liver metastasis (Arlt et al. 2002), the increased expression of MMP-9 by the tumor cells could further promote the metastatic potential of the tumor cells when MMPI-levels are not sufficient to block all MMP-derived proteolytic activity. These results can, at least partially, explain the failure of broad-spectrum MMPIs in the clinical trials.

Specific Inhibition of Metalloproteinases

Specific Suppression of MMP Expression

Natural MMPIs are more or less broad-spectrum inhibitors. However, analysis of the role of particular MMPs during metastasis requires maximum specificity. Also, potential side effects can much better be attributed to individual MMPs when specific inhibition is achieved. The best way to specifically suppress the activity of single MMPs is modulation of their gene expression by knockout or knockdown of gene expression by antisense-, ribozyme-, or RNA-interference technology. Thus, it was demonstrated that specific deficiency in either MMP-2 (Itoh et al. 1998) or -9 (Acuff et al. 2006) in the host leads to reduced experimental metastasis of different tumor cell lines. Transfer of bone marrow cells from wild-type mice into syngeneic MMP-deficient mice or *vice versa* enables the analysis of the contribution of inflammatory cell-derived MMPs to metastasis. Bone marrow transfer showed that MMP-9 expressed by bone marrow-derived cells enhances metastasis by increasing the survival of extravasated tumor cells (Acuff et al. 2006). Also, knockdown of the expression of MMP-1, -2, -9, -11, or -14 in tumor cells reduced the invasive potential of these tumor cells *in vitro* and in spontaneous metastasis models *in vivo* (Yuan et al. 2005, Chetty et al. 2006, Petrella and Brinckerhoff 2006, Jia et al. 2007). Interestingly, knockout of specific MMPs does not always lead to suppressed metastasis. In MMP-8$^{-/-}$ mice, not only was the incidence of chemically induced tumors significantly increased but tumor progression was increased as well, since undifferentiated stage III fibrosarcomas were developed only in MMP-8$^{-/-}$ mice (Balbin et al. 2003). Although metastasis was not tested in this study, the higher stage of tumors indicates a higher rate of spontaneous metastasis. A different study revealed the antimetastatic activity of tumor cell-derived MMP-8: in a pair of breast cancer tumor cell lines it was revealed that the nonmetastatic cell line expressed significantly higher levels of MMP-8 than did the metastatic cell line (Agarwal et al. 2003). A causal relationship between the lower expression of MMP-8 and aggressiveness was evident by the fact that knockdown of MMP-8 increased invasion of the tumor cells *in vitro* (Agarwal et al. 2003). Knockdown of MMP-8 in breast cancer cells by ribozyme technology increased spontaneous metastasis of these cells from orthotopic tumors into lymph nodes and lungs as compared to the parental cell line (Montel et al. 2004). Also, it has been demonstrated that specific inhibition of MMP-3 in knockout mice led to reduced tumor progression and demonstrated a possible tumor-suppressing activity of MMP-3: chemically induced squamous cell carcinomas in MMP-3$^{-/-}$ mice grow faster while the incidence of tumors is equal as compared to wild-type littermates. This increased tumor growth also resulted in a higher rate of spontaneous metastasis to the lung (McCawley et al. 2004). In MMP-9 knockout mice the reduced formation of tumstatin by MMP-9 activity led to increased tumor growth of subcutaneously inoculated murine Lewis lung carcinoma cells (Hamano et al. 2003). Studies with MMP-7$^{-/-}$ mice revealed a possible progression/stage-specific activity of

MMP-7: while the number of experimental lung metastases is increased in MMP-$7^{-/-}$ mice as compared to that in the wild-type control (Acuff et al. 2006), formation and growth of benign intestinal tumors was significantly reduced by MMP-7 deficiency (Wilson et al. 1997). Thus, MMP-7 seems to promote early steps of tumor progression while suppressing the invasive steps of metastasis. These data show strong evidence for a tumor-inhibiting effect of MMPs.

It has to be taken into account that proteases are incorporated in a complex network with overlapping substrate specificities (Overall and Dean 2006). Thus, knockout of one protease in a living system can trigger a cascade of events leading to a reestablishment of a new homeostasis in all organs. We therefore expect a correlation between the importance of a given MMP for maintenance of organ homeostasis and the intensity of alterations provoked by its knockout. Because of the overlapping substrate specificities of MMPs (DeClerck et al. 1997), it is possible that lack of one protease can be counter-regulated by increased expression of other proteases. Thus, especially in experimental metastasis models, the tumor cells are confronted with new metastatic niches due to the adapted tissue environment. This may well favor metastasis.

Specific Synthetic MMPIs

Intensive effort has been put into the development of synthetic MMPIs with increased specificity (Fisher and Mobashery 2006). One of the first MMPIs with a slightly higher specificity toward MMP-2, -3, and -9 was prinomastat (AG3340) (see also Chap. 36). This compound is a nonpeptidomimetic hydroxamate-type MMPI, an advancement of peptidomimetic broad-spectrum inhibitors, such as batimastat. In contrast to the induction of liver metastasis by batimastat, prinomastat significantly reduced metastasis of the T-cell lymphoma cell line to the liver (Arlt et al. 2002). This antimetastatic activity of prinomastat relies most likely on its higher specificity toward MMP-9 (Arlt et al. 2002). Interestingly, the less soluble progenitor compound of prinomastat (AG3319), with a specificity profile similar to that of prinomastat, significantly promoted tumor growth and spontaneous lung metastasis (Santos et al. 1997). We assume that the prometastatic effect of AG3319 is based on a different mechanism than the prometastatic effect of batimastat, since metastasis to different organs was promoted. Most likely, the increase in spontaneous metastases by AG3319 relies on the augmented proliferation of the primary tumor.

A higher specificity toward MMP-2, MMP-9, and MMP-14 was achieved with the pyrimidine-2,4,6-trione-type inhibitors Ro28-2653 and Ro206-0222 (Grams et al. 2001). Employment of these inhibitors into the experimental metastasis model with murine T-cell lymphoma cells revealed a significant correlation between MMP-9-specificity of the MMPI and its antimetastatic efficacy in the liver (Arlt et al. 2002). A single treatment with Ro206-0222 before tumor cell inoculation was sufficient to significantly reduce the number of metastatic foci in the liver as compared to the treatment starting one day after tumor cell inoculation (unpublished data). This finding indicates that Ro206-0222 inhibits liver metastasis, most likely by inhibition of extravasation and early survival of the tumor cells. Another inhibitor, which also efficiently inhibits MMP-2, -9, and -14, but not MMP-1, -3, and -7, is the *N*-sulfonylamino acid derivate MMI-166 (Maekawa et al. 2000). This inhibitor significantly reduced primary tumor growth and subsequent spontaneous metastasis (Fujino et al. 2005) as well as experimental metastasis to liver and lung (Maekawa et al. 2000). Notably, for these inhibitors with increased specificity toward MMP-2, -9, and -14, no prometastatic effects have been described (see also Chap. 36).

Another level of selectivity for MMP-2 and -9 was achieved with the mechanism-based inhibitor SB-3CT. Because of the unique structural features of gelatinases, the reactive species of SB-3CT is formed only in the active site of gelatinases by coordination with residues of the inhibited MMP (Bernardo et al. 2002). This inhibitor achieved the highest antimetastatic efficacy in the T-cell lymphoma liver metastasis model as compared to all other tested MMPIs (Krüger et al. 2005). SB-3CT reduced the number as well as the size of liver metastases, demonstrating the impact of gelatinases on extravasation of tumor cells and outgrowth of metastatic foci (Krüger et al. 2005). Since this mechanism-based inhibitor is a potent inhibitor of both gelatinases, it was so far not possible to ascribe specific roles during metastasis to a particular gelatinase.

Before the translation of results from functional genomic approaches to the design of therapeutic interventions it has to be considered that knockout also prevents possible proteolysis-independent activities of the targeted MMP. In contrast, specific inhibitors interfere only with the proteolytic activity of the MMP. These possible proteolysis-independent activities of MMPs, such as binding to cell surface molecules (Brooks et al. 1998), have to be analyzed before treatment strategies with MMPIs are designed.

Metalloproteinase Inhibition-Independent Effects of Natural Metalloproteinase Inhibitors on Metastasis

Initially, TIMP-1 was identified as erythroid potentiating activity (Gasson et al. 1985). TIMP-1 and TIMP-2 are able to increase proliferation of a variety of cell lines including different tumor cell lines (Baker et al. 2002). Also, TIMP-2 is able to promote cell proliferation (Wingfield et al. 1999). Mutations of the MMP-inhibitory domain of TIMP-1 or TIMP-2 revealed that the proproliferative activity and inhibition of MMPs are distinct features of the full-length TIMP-1 (Chesler et al. 1995) or TIMP-2 molecule (Wingfield et al. 1999), respectively. The growth factor-like activity of TIMP-1 was assumed to be based on binding of TIMP-1 to a cell surface receptor. Recently, CD63, a cell surface molecule of the tetraspanin family, was identified as a receptor for TIMP-1 (Jung et al. 2006). Binding of TIMP-1 to CD63 can induce signaling (Jung et al. 2006). The biological role of TIMP-1/CD63 signaling in physiological and pathophysiological situations, such as cancer, remains to be investigated.

For TIMP-2 a proliferation-independent potentially protumorigenic activity has been described: it is known that TIMP-2 is involved in the cell surface-associated activation of proMMP-2. TIMP-2 binds to proMMP-2 and MT1-MMP (MMP-14), thus recruiting proMMP-2 to the cell surface where it can subsequently be activated by a second active MT1-MMP molecule (Nagase 1998). Since MMP-2 is known to exhibit prometastatic activity (Itoh et al. 1998), increased proMMP-2 activation by elevated TIMP-2 levels could lead to promotion of metastasis. A potentially anti-metastatic MMP-independent activity of TIMP-2 is the inhibition of proliferation *in vitro* and angiogenesis of endothelial cells *in vivo*. The underlying mechanism depends on binding of TIMP-2 to $\alpha3\beta1$ integrin, subsequently leading to reduced activation of tyrosine phosphatase and dissociation of the phosphatase SHP-1 from $\beta1$ (Seo et al. 2003).

Conclusions

In the past few years the concept of MMPs as predominantly path clearing enzymes has undergone radical revision. MMPs participate at many stages during tumor progression and some of their specific actions can lead to inhibition rather than promotion of tumor malignancy (Egeblad and Werb 2002). Thus, inhibition of MMPs can have varying effects on metastasis depending on the pool of inhibited MMPs. Within the complex proteolytic network it is important to know the regulatory effects of MMP inhibition, whether a simple MMP or many MMPs are inhibited. Both modes of inhibition, specific or broad-spectrum, may have varying consequences for the homeostasis of the proteolytic network and, even more importantly, on the homeostasis of metastasis-susceptible tissues. In other words, the interference with the proteolytic network will change the susceptibility of an organ to tumor cells and these changes of the "SOIL" have to be studied in detail in order to be able to predict therapeutic outcome. It seems as if specific inhibition of selected target MMPs is preferable and better manageable, as the molecular side effects may be controllable. Functional genomic studies can identify those MMPs that exclusively promote metastasis—if any exist—as targets for the design of new rationale antiprotolytic treatment strategies.

We have summarized here the effects of broad-spectrum MMPIs on the formation of a premetastatic niche in the liver. This MMP inhibition-dependent activity can in part explain the failure of the clinical trials with synthetic broad-spectrum MMPIs as well as the paradoxical association of increased expression of natural broad-spectrum MMPIs and shorter survival of cancer patients. As illustrated in Fig. 25.2b, we assume that overexpression of TIMP-1 by tumors (Ree et al. 1997, Kallakury et al. 2001, Culhaci et al. 2004) leads, due to efficient secretion of TIMP-1, to increased plasma levels of TIMP-1 (Hammer et al. 2006, Lipton et al. 2007). Therefore, via the bloodstream, TIMP-1 also reaches the liver. The broad-spectrum inhibition of MMPs by TIMP-1 then modulates the homeostasis of the liver, leading to increased HGF-signaling and formation of a premetastatic niche (Fig. 25.2b). The identification of the pathway induced by broad-spectrum MMP inhibition enables the design of novel treatment strategies: (1) It is possible that cancer patients with elevated TIMP-1 levels benefit from antityrosine kinase treatments. (2) A combinatorial approach comprising synthetic MMPIs and tyrosine kinase inhibitors may lead to beneficial results in cancer patients. This combination may lead to inhibited invasion of the tumor cells and prevention of formation of premetastatic niches. Identification of the pool of metalloproteinases whose inhibition is responsible for the creation of the premetastatic tissue environment in the liver will further improve the design of novel antimetastatic treatment strategies.

Acknowledgments Representing all graduate students who have contributed to this work we thank Matthias Arlt, Michael Gerg, and Rita Söltl.

References

Abraham R., Schafer J., Rothe M., et al. (2005). Identification of MMP-15 as an anti-apoptotic factor in cancer cells. J Biol Chem 280:34123–34132.

Acuff H. B., Carter K. J., Fingleton B., et al. (2006). Matrix metalloproteinase-9 from bone marrow-derived cells contributes to survival but not growth of tumor cells in the lung microenvironment. Cancer Res 66:259–266.

Agarwal D., Goodison S., Nicholson B., et al. (2003). Expression of matrix metalloproteinase 8 (MMP-8) and tyrosinase-related protein-1 (TYRP-1) correlates with the absence of metastasis in an isogenic human breast cancer model. Differentiation 71:114–125.

Amour A., Knight C. G., Webster A., et al. (2000). The *in vitro* activity of ADAM-10 is inhibited by TIMP-1 and TIMP-3. FEBS Lett 473:275–279.

Arlt M., Kopitz C., Pennington C., et al. (2002). Increase in gelatinase-specificity of matrix metalloproteinase inhibitors correlates with antimetastatic efficacy in a T-Cell lymphoma model. Cancer Res 62:5543–5550.

Atkin G., Chopada A. and Mitchell I. (2005). Colorectal cancer metastasis: in the surgeon's hands? Int Semin Surg Oncol 2:5.

Baker A. H., Edwards D. R. and Murphy G. (2002). Metalloproteinase inhibitors: biological actions and therapeutic opportunities. J Cell Sci 115:3719–3727.

Balbin M., Fueyo A., Tester A. M., et al. (2003). Loss of collagenase-2 confers increased skin tumor susceptibility to male mice. Nat Genet 35:252–257.

Berman R. S., Portera C. A., Jr. and Ellis L. M. (2001). Biology of liver metastases. Cancer Treat Res 109:183–206.

Bernardo M. M., Brown S., Li Z. H., et al. (2002). Design, synthesis, and characterization of potent, slow-binding inhibitors that are selective for gelatinases. J Biol Chem 277:11201–11207.

Bissett D., O'Byrne K. J., von Pawel J., et al. (2005). Phase III study of matrix metalloproteinase inhibitor prinomastat in non-small-cell lung cancer. J Clin Oncol 23:842–849.

Blobel C. P. (2005). ADAMs: key components in EGFR signalling and development. Nat Rev Mol Cell Biol 6:32–43.

Brand K. (2002). Cancer gene therapy with tissue inhibitors of metalloproteinases (TIMPs). Curr Gene Ther 2:255–271.

Brand K., Baker A. H., Perez-Canto A., et al. (2000). Treatment of colorectal liver metastases by adenoviral transfer of tissue inhibitor of metalloproteinases-2 into the liver tissue. Cancer Res 60:5723–5730.

Brew K., Dinakarpandian D. and Nagase H. (2000). Tissue inhibitors of metalloproteinases: evolution, structure and function. Biochim Biophys Acta 1477:267–283.

Brooks P. C., Silletti S., von Schalscha T. L., et al. (1998). Disruption of angiogenesis by PEX, a noncatalytic metalloproteinase fragment with integrin binding activity. Cell 92:391–400.

Brown P. D. (1995). Matrix metalloproteinase inhibitors: a novel class of anticancer agents. Adv Enzyme Regul 35:293–301.

Brown P. D. (1998). Matrix metalloproteinase inhibitors. Breast Cancer Res Treat 52:125–36.

Chambers A. F. and Matrisian L. M. (1997). Changing views of the role of matrix metalloproteinases in metastasis. J Natl Cancer Inst 89:1260–1270.

Chesler L., Golde D. W., Bersch N., et al. (1995). Metalloproteinase inhibition and erythroid potentiation are independent activities of tissue inhibitor of metalloproteinases-1. Blood 86:4506–4515.

Chetty C., Bhoopathi P., Joseph P., et al. (2006). Adenovirus-mediated small interfering RNA against matrix metalloproteinase-2 suppresses tumor growth and lung metastasis in mice. Mol Cancer Ther 5:2289–2299.

Coussens L. M., Fingleton B. and Matrisian L. M. (2002). Matrix metalloproteinase inhibitors and cancer: trials and tribulations. Science 295:2387–2392.

Cruz-Munoz W., Sanchez O. H., Di Grappa M., et al. (2006). Enhanced metastatic dissemination to multiple organs by melanoma and lymphoma cells in timp-3$^{-/-}$ mice. Oncogene 25:6489–6496.

Culhaci N., Metin K., Copcu E., et al. (2004). Elevated expression of MMP-13 and TIMP-1 in head and neck squamous cell carcinomas may reflect increased tumor invasiveness. BMC Cancer 4:42–49.

Davidson B., Goldberg I., Gotlieb W. H., et al. (2002). The prognostic value of metalloproteinases and angiogenic factors in ovarian carcinoma. Mol Cell Endocrinol 187:39–45.

de Lorenzo M. S., Ripoll G. V., Yoshiji H., et al. (2003). Altered tumor angiogenesis and metastasis of B16 melanoma in transgenic mice overexpressing tissue inhibitor of metalloproteinases-1. In Vivo 17:45–50.

DeClerck Y. A., Imren S., Montgomery A. M., et al. (1997). Proteases and protease inhibitors in tumor progression. Adv Exp Med Biol 425:89–97.

Della Porta P., Soeltl R., Krell H. W., et al. (1999). Combined treatment with serine protease inhibitor aprotinin and matrix metalloproteinase inhibitor Batimastat (BB-94) does not prevent invasion of human esophageal and ovarian carcinoma cells *in vivo*. Anticancer Res 19:3809–3816.

Duffy M. J., Maguire T. M., Hill A., et al. (2000). Metalloproteinases: role in breast carcinogenesis, invasion and metastasis. Breast Cancer Res 2:252–257.

Eccles N. A. (2001). Basic principles for the study of metstasis using animal models. In Metastasis Research Protocols, ed. S. A. Brooks and U. Schumacher, pp. 161–172. Totowa, New Jersey: Humana Press.

Eccles S. A., Box G. M., Court W. J., et al. (1996). Control of lymphatic and hematogenous metastasis of a rat mammary carcinoma by the matrix metalloproteinase inhibitor batimastat (BB-94). Cancer Res 56:2815–2822.

Egeblad M. and Werb Z. (2002). New functions for the matrix metalloproteinases in cancer progression. Nat Rev Cancer 2:161–174.

Elezkurtaj S., Kopitz C., Baker A. H., et al. (2004). Adenovirus-mediated overexpression of tissue inhibitor of metalloproteinases-1 in the liver: efficient protection against T-cell lymphoma and colon carcinoma metastasis. J Gene Med 6:1228–1237.

Fidler I. J. (2001). Seed and soil revisited: contribution of the organ microenvironment to cancer metastasis. Surg Oncol Clin N Am 10:257–269, vii–viiii.

Fisher J. F. and Mobashery S. (2006). Recent advances in MMP inhibitor design. Cancer Metastasis Rev 25:115–136.

Fujino H., Kondo K., Ishikura H., et al. (2005). Matrix metalloproteinase inhibitor MMI-166 inhibits lymphogenous metastasis in an orthotopically implanted model of lung cancer. Mol Cancer Ther 4:1409–1416.

Gasson J. C., Bersch N. and Golde D. W. (1985). Characterization of purified human erythroid-potentiating activity. Prog Clin Biol Res 184:95–104.

Goldberg G. I., Strongin A., Collier I. E., et al. (1992). Interaction of 92-kDa type IV collagenase with the tissue inhibitor of metalloproteinases prevents dimerization, complex formation with interstitial collagenase, and activation of the proenzyme with stromelysin. J Biol Chem 267:4583–4591.

Grams F., Brandstetter H., D'Alo S., et al. (2001). Pyrimidine-2,4,6-Triones: a new effective and selective class of matrix metalloproteinase inhibitors. Biol Chem 382:1277–1285.

Guy C. T. and Cardoso G. (2001). Transgenic animal models. In Metastasis Research Protocols, ed. S. A. Brooks and U. Schumacher, pp. 231–205. Totowa: Humana Press.

Hamano Y., Zeisberg M., Sugimoto H., et al. (2003). Physiological levels of tumstatin, a fragment of collagen IV alpha3 chain, are generated by MMP-9 proteolysis and suppress angiogenesis via alphaV beta3 integrin. Cancer Cell 3:589–601.

Hammer J. H., Basse L., Svendsen M. N., et al. (2006). Impact of elective resection on plasma TIMP-1 levels in patients with colon cancer. Colorectal Dis 8:168–172.

Heisenberg W. (1927). Über die Grundprinzipien der Quantenmechanik. Forschungen und Fortschritte 3:83.

Itoh T., Tanioka M., Yoshida H., et al. (1998). Reduced angiogenesis and tumor progression in gelatinase A-deficient mice. Cancer Res 58:1048–1051.

Jenkins D. E., Oei Y., Hornig Y. S., et al. (2003). Bioluminescent imaging (BLI) to improve and refine traditional murine models of tumor growth and metastasis. Clin Exp Metastasis 20:733–744.

Jia L., Wang S., Cao J., et al. (2007). siRNA targeted against matrix metalloproteinase 11 inhibits the metastatic capability of murine hepatocarcinoma cell Hca-F to lymph nodes. Int J Biochem Cell Biol 39:2135–2142.

Jiang W. G., Martin T. A., Parr C., et al. (2005). Hepatocyte growth factor, its receptor, and their potential value in cancer therapies. Crit Rev Oncol Hematol 53:35–69.

Jimenez R. E., Hartwig W., Antoniu B. A., et al. (2000). Effect of matrix metalloproteinase inhibition on pancreatic cancer invasion and metastasis: an additive strategy for cancer control. Ann Surg 231:644–654.

Joyce J. A. and Hanahan D. (2004). Multiple roles for cysteine cathepsins in cancer. Cell Cycle 3:1516–1619.

Jung K. K., Liu X. W., Chirco R., et al. (2006). Identification of CD63 as a tissue inhibitor of metalloproteinase-1 interacting cell surface protein. Embo J 25:3934–3942.

Kallakury B. V., Karikehalli S., Haholu A., et al. (2001). Increased expression of matrix metalloproteinases 2 and 9 and tissue inhibitors of metalloproteinases 1 and 2 correlate with poor prognostic variables in renal cell carcinoma. Clin Cancer Res 7:3113–3119.

Kanayama H. (2001). Matrix metalloproteinases and bladder cancer. J Med Invest 48:31–43.

Kopitz C., Gerg M., Bandapalli O. R., et al. (2007). Tissue inhibitor of metalloproteinases-1 promotes liver metastasis by induction of hepatocyte growth factor signaling. Cancer Res 67:8615–8623.

Kossakowska A. E., Urbanski S. J. and Janowska-Wieczorek A. (2000). Matrix metalloproteinases and their tissue inhibitors — expression, role and regulation in ruman malignant Non-Hodgkin's lymphomas. Leuk Lymphoma 39:485–493.

Krüger A., Schirrmacher V. and von Hoegen P. (1994). Scattered micrometastases visualized at the single-cell level: detection and re-isolation of lacZ-labeled metastasized lymphoma cells. Int J Cancer 58:275–284.

Krüger A., Fata J. E. and Khokha R. (1997). Altered tumor growth and metastasis of a T-cell lymphoma in Timp-1 transgenic mice. Blood 90:1993–2000.

Krüger A., Schirrmacher V. and Khokha R. (1998). The bacterial lacZ gene: an important tool for metastasis research and evaluation of new cancer therapies. Cancer Metastasis Rev 17:285–294.

Krüger A., Soeltl R., Sopov I., et al. (2001). Hydroxamate-type matrix metalloproteinase inhibitor batimastat promotes liver metastasis. Cancer Res 61:1272–1275.

Krüger A., Arlt M. J., Gerg M., et al. (2005). Antimetastatic activity of a novel mechanism-based gelatinase inhibitor. Cancer Res 65:3523–3526.

Levin V. A., Phuphanich S., Yung W. K., et al. (2006). Randomized, double-blind, placebo-controlled trial of marimastat in glioblastoma multiforme patients following surgery and irradiation. J Neurooncol 78:295–302.

Levitt N. C., Eskens F. A., O'Byrne K. J., et al. (2001). Phase I and pharmacological study of the oral matrix metalloproteinase inhibitor, MMI270 (CGS27023A), in patients with advanced solid cancer. Clin Cancer Res 7:1912–1922.

Li H., Lindenmeyer F., Grenet C., et al. (2001). AdTIMP-2 inhibits tumor growth, angiogenesis, and metastasis, and prolongs survival in mice. Hum Gene Ther 12:515–526.

Liaudet-Coopman E., Beaujouin M., Derocq D., et al. (2006). Cathepsin D: newly discovered functions of a long-standing aspartic protease in cancer and apoptosis. Cancer Lett 237:167–179.

Lipton A., Ali S. M., Leitzel K., et al. (2007). Elevated plasma tissue inhibitor of metalloproteinase-1 level predicts decreased response and survival in metastatic breast cancer. Cancer 109:1933–1939.

López-Otín C. and Matrisian L. M. (2007). Emerging roles of proteases in tumour suppression. Nat Rev Cancer 7:800–808.

Maekawa R., Maki H., Wada T. T. (2000). Anti-metastatic efficacy and safety of MMI-166, a selective matrix metalloproteinase inhibitor. Clin Exp Metastasis 18:61 66.

Manenti L., Paganoni P., Floriani I., et al. (2003). Expression levels of vascular endothelial growth factor, matrix metalloproteinases 2 and 9 and tissue inhibitor of metalloproteinases 1 and 2 in the plasma of patients with ovarian carcinoma. Eur J Cancer 39:1948–1956.

Maquoi E., Frankenne F., Noel A., et al. (2000). Type IV collagen induces matrix metalloproteinase 2 activation in HT1080 fibrosarcoma cells. Exp Cell Res 261:348–359.

Maquoi E., Munaut C., Colige A., et al. (2002). Stimulation of matrix metalloproteinase-9 expression in human fibrosarcoma cells by synthetic matrix metalloproteinase inhibitors. Exp Cell Res 275:110–121.

Maquoi E., Sounni N. E., Devy L., et al. (2004). Anti-invasive, antitumoral, and antiangiogenic efficacy of a pyrimidine-2,4,6-trione derivative, an orally active and selective matrix metalloproteinases inhibitor. Clin Cancer Res 10:4038–4047.

McCawley L. J., Crawford H. C., King L. E., et al. (2004). A protective role for matrix metalloproteinase-3 in squamous cell carcinoma. Cancer Res 64:6965–6972.

Miyata Y., Iwata T., Maruta S., et al. (2006). Expression of matrix metalloproteinase-10 in renal cell carcinoma and its prognostic role. Eur Urol 52:791–797.

Montel V., Kleeman J., Agarwal D., et al. (2004). Altered metastatic behavior of human breast cancer cells after experimental manipulation of matrix metalloproteinase 8 gene expression. Cancer Res 64:1687–1694.

Mueller M. M. and Fusenig N. E. (2002). Tumor-stroma interactions directing phenotype and progression of epithelial skin tumor cells. Differentiation 70:486–497.

Nagase H. (1998). Cell surface activation of progelatinase A (proMMP-2) and cell migration. Cell Res 8:179–186.

Ogata Y., Matono K., Nakajima M., et al. (2006). Efficacy of the MMP inhibitor MMI270 against lung metastasis following removal of orthotopically transplanted human colon cancer in rat. Int J Cancer 118:215–221.

Overall C. M. and Dean R. A. (2006). Degradomics: systems biology of the protease web. Pleiotropic roles of MMPs in cancer. Cancer Metastasis Rev 25:69–75.

Oyajobi B. O., Munoz S., Kakonen R., et al. (2007). Detection of myeloma in skeleton of mice by whole-body optical fluorescence imaging. Mol Cancer Ther 6:1701–1708.

Pavlaki M. and Zucker S. (2003). Matrix metalloproteinase inhibitors (MMPIs): the beginning of phase I or the termination of phase III clinical trials. Cancer Metastasis Rev 22:177–203.

Petrella B. L. and Brinckerhoff C. E. (2006). Tumor cell invasion of von Hippel Lindau renal cell carcinoma cells is mediated by membrane type-1 matrix metalloproteinase. Mol Cancer 5:66–79.

Porter J. F., Shen S. and Denhardt D. T. (2004). Tissue inhibitor of metalloproteinase-1 stimulates proliferation of human cancer cells by inhibiting a metalloproteinase. Br J Cancer 90:463–470.

Porter J. F., Sharma S., Wilson D. L., et al. (2005). Tissue Inhibitor of Metalloproteinases-1 Stimulates Gene Expression in MDA-MB-435 Human Breast Cancer Cells by Means of its Ability to Inhibit Metalloproteinases. Breast Cancer Res Treat 94:185–193.

Price J. E. (2001). Xenograft models in immunodeficient animals, I. Nude mice: spontaneous and experimental metastasis models. In Metastasis Research Protocols, ed. S. A. bBrooks and U. Schumacher, pp. 205–214. Totowa: Humana Press.

Prontera C., Mariani B., Rossi C., et al. (1999). Inhibition of gelatinase A (MMP-2) by batimastat and captopril reduces tumor growth and lung metastases in mice bearing Lewis lung carcinoma. Int J Cancer 81:761–766.

Radinsky R. and Ellis L. M. (1996). Molecular determinants in the biology of liver metastasis. Surg Oncol Clin N Am 5:215–229.

Ree A. H., Florenes V. A., Berg J. P., et al. (1997). High levels of messenger RNAs for tissue inhibitors of metalloproteinases (TIMP-1 and TIMP-2) in primary breast carcinomas are associated with development of distant metastases. Clin Cancer Res 3:1623–1628.

Remacle A., McCarthy K., Noel A., et al. (2000). High levels of TIMP-2 correlate with adverse prognosis in breast cancer. Int J Cancer 89:118–121.

Reuning U., Magdolen V., Wilhelm O., et al. (1998). Multifunctional potential of the plasminogen activation system in tumor invasion and metastasis (review). Int J Oncol 13:893–906.

Roca F., Mauro L. V., Morandi A., et al. (2006). Prognostic value of E-cadherin, beta-catenin, MMPs (7 and 9), and TIMPs (1 and 2) in patients with colorectal carcinoma. J Surg Oncol 93:151–160.

Rocha M., Schirrmacher V. and Umansky V. (2001). Dissection of tumor and host cells from metastasized organs for testing gene expression directly ex vivo. In Metasasis Research Protocols, ed. S. A. Brooks and U. Schumacher, pp. 277–284. Totowa, New Jersey: Humana Press.

Rosenbaum E., Zahurak M., Sinibaldi V., et al. (2005). Marimastat in the treatment of patients with biochemically relapsed prostate cancer: a prospective randomized, double-blind, phase I/II trial. Clin Cancer Res 11:4437–4443.

Santos O., McDermott C. D., Daniels R. G., et al. (1997). Rodent pharmacokinetic and anti-tumor efficacy studies with a series of synthetic inhibitors of matrix metalloproteinases. Clin Exp Metastasis 15:499–508.

Schirrmacher V. (1985). Cancer metastasis: experimental approaches, theoretical concepts, and impacts for treatment strategies. Adv Cancer Res 43:1–73.

Schrohl A. S., Holten-Andersen M. N., Peters H. A., et al. (2004). Tumor tissue levels of tissue inhibitor of metalloproteinase-1 as a prognostic marker in primary breast cancer. Clin Cancer Res 10:2289–2298.

Seo D. W., Li H., Guedez L., et al. (2003). TIMP-2 mediated inhibition of angiogenesis: an MMP-independent mechanism. Cell 114:171–180.

Sledge G. W., Jr., Qulali M., Goulet R., et al. (1995). Effect of matrix metalloproteinase inhibitor batimastat on breast cancer regrowth and metastasis in athymic mice. J Natl Cancer Inst 87:1546–1550.

Stetler-Stevenson W. G. (2001). The role of matrix metalloproteinases in tumor invasion, metastasis, and angiogenesis. Surg Oncol Clin N Am 10:383–392.

Terstappen L. W., Rao C., Gross S., et al. (2000). Peripheral blood tumor cell load reflects the clinical activity of the disease in patients with carcinoma of the breast. Int J Oncol 17:573–578.

Wang M., Liu Y. E., Greene J., et al. (1997). Inhibition of tumor growth and metastasis of human breast cancer cells transfected with tissue inhibitor of metalloproteinase 4. Oncogene 14:2767–2774.

Wang X., Fu X., Brown P. D., et al. (1994). Matrix metalloproteinase inhibitor BB-94 (batimastat) inhibits human colon tumor growth and spread in a patient-like orthotopic model in nude mice. Cancer Res 54:4726–4728.

Wilson C. L., Heppner K. J., Labosky P. A., et al. (1997). Intestinal tumorigenesis is suppressed in mice lacking the metalloproteinase matrilysin. Proc Natl Acad Sci U S A 94:1402–1407.

Winding B., NicAmhlaoibh R., Misander H., et al. (2002). Synthetic matrix metalloproteinase inhibitors inhibit growth of established breast cancer osteolytic lesions and prolong survival in mice. Clin Cancer Res 8:1932–1939.

Wingfield P. T., Sax J. K., Stahl S. J., et al. (1999). Biophysical and functional characterization of full-length, recombinant human tissue inhibitor of metalloproteinases-2 (TIMP-2) produced in Escherichia coli. Comparison of wild type and amino-terminal alanine appended variant with implications for the mechanism of TIMP functions. J Biol Chem 274:21362–21368.

Wojtowicz-Praga S. M., Dickson R. B. and Hawkins M. J. (1997). Matrix metalloproteinase inhibitors. Invest New Drugs 15:61–75.

Wolf K., Mazo I., Leung H., et al. (2003). Compensation mechanism in tumor cell migration: mesenchymal-amoeboid transition after blocking of pericellular proteolysis. J Cell Biol 160:267–277.

Wolf K., Wu Y. I., Liu Y., et al. (2007). Multi-step pericellular proteolysis controls the transition from individual to collective cancer cell invasion. Nat Cell Biol 9:893–904.

Wu C. Y., Wu M. S., Chen Y. J., et al. (2007). Clinicopathological significance of MMP-2 and TIMP-2 genotypes in gastric cancer. Eur J Cancer 43:799–808.

Wylie S., MacDonald I. C., Varghese H. J., et al. (1999). The matrix metalloproteinase inhibitor batimastat inhibits angiogenesis in liver metastases of B16F1 melanoma cells. Clin Exp Metastasis 17:111–117.

Yamauchi K., Ogata Y., Nagase H., et al. (2001). Inhibition of liver metastasis from orthotopically implanted colon cancer in nude mice by transfection of the TIMP-1 gene into KM12SM cells. Surg Today 31:791–798.

Yoneda T., Sasaki A., Dunstan C., et al. (1997). Inhibition of osteolytic bone metastasis of breast cancer by combined treatment with the bisphosphonate ibandronate and tissue inhibitor of the matrix metalloproteinase-2. J Clin Invest 99:2509–2517.

Yoshikawa T., Tsuburaya A., Kobayashi O., et al. (2001). Intratumoral concentrations of tissue inhibitor of matrix metalloproteinase 1 in patients with gastric carcinoma a new biomartker for invasion and its impact on survival. Cancer 91:1739–1744.

Yuan J., Dutton C. M. and Scully S. P. (2005). RNAi mediated MMP-1 silencing inhibits human chondrosarcoma invasion. J Orthop Res 23:1467–1474.

Zervos E. E., Norman J. G., Gower W. R., et al. (1997). Matrix metalloproteinase inhibition attenuates human pancreatic cancer growth *in vitro* and decreases mortality and tumorigenesis *in vivo*. J Surg Res 69:367–371.

Zucker S. and Vacirca J. (2004). Role of matrix metalloproteinases (MMPs) in colorectal cancer. Cancer Metastasis Rev 23:101–117.

Zucker S., Cao J. and Chen W. T. (2000). Critical appraisal of the use of matrix metalloproteinase inhibitors in cancer treatment. Oncogene 19:6642–6650.

Chapter 26
Cytokine Substrates: MMP Regulation of Inflammatory Signaling Molecules

Jennifer H. Cox and Christopher M. Overall

Abstract Inflammation is a highly regulated process involving tissue parenchyma, vascular and connective tissues, as well as host defense and immune cells, and is an essential component of both innate and acquired immunity. MMPs were traditionally considered to assist in the inflammatory response through the breakdown of extracellular matrix molecules, providing an environment condusive to leukocyte infiltration, edema, wound healing, and tissue repair with continued upregulation contributing to the onset of chronic inflammation. It is now evident, however, that the repertoire of MMP substrates, the substrate degradome, extends far beyond the extracellular matrix and includes many key bioactive molecules, the processing of which results in altered actions. In the context of inflammation, several MMPs have been shown biochemically to efficiently and selectively process cytokines and chemokines, as well as associated molecules, thereby changing their bioactive properties. Many of these substrates have been validated physiologically in genetic models of inflammatory disease, where a remarkable array of inflammatory phenotypes has been observed. As a whole, research involving MMPs in inflammation has demonstrated their integral role in orchestrating both the initiation and the termination of inflammatory processes, rather than only tissue degradation.

MMPs and Inflammation

Inflammation is an essential process for the elimination of invading pathogens and in wound healing. Inflamed tissue is a dynamic interactive multicellular and multi-tissue system embedded in normal stroma. A wide variety of signaling networks

Christopher M. Overall
Centre for Blood Research, University of British Columbia, 4.401 Life Sciences Centre, 2350 Health Sciences Mall, Vancouver, BC V6T 1Z3, Canada

D. Edwards et al. (eds.), *The Cancer Degradome.*
© Springer Science + Business Media, LLC 2008

operate in multidirectional communication between the component cells and tissues—the connective tissue cells and epithelium, vascular endothelium, reactive peripheral stroma, and host defense cells—that orchestrate innate immunity, inflammation, resolution, and healing. Cytokine pathways that are perturbed in unresolved acute inflammation drive tissue destruction and inflammatory/immune cell responses in chronic inflammatory disease. Dysregulated inflammation has been linked to numerous pathologies including autoimmune diseases (Wucherpfennig 2001), tumorigenesis (Dvorak 1986), and atherosclerosis (Hansson and Libby 2006). Stringent regulation of the inflammatory process is critical for both resolution and to buffer against pathological destruction. Proteolytic enzymes are now regarded as key effectors and regulators in inflammation and immunity, through not only protein degradation and turnover but also precise and efficient cleavage events leading to altered bioactivity of inflammatory mediators (Overall 2004, Turk 2006).

Matrix metalloproteinases (MMPs) were historically thought to be exclusively involved in turnover of extracellular matrix components. Accordingly, attractive hypotheses evolved that involved matrix degradation in normal development and physiological turnover as well as in many of the steps in inflammation that appeared to be dependent on extracellular matrix turnover. However, with the exception of MMP-14, none of the MMP-deficient mice generated thus far exhibit severe defects in matrix turnover for normal development or homeostasis. In addition, drug trials with broad-spectrum MMP inhibitors have not demonstrated significant abnormalities in extracellular matrix although a small number of patients did develop joint strictures (Bramhall et al. 2002). In contrast, MMP-deficient mice display a remarkable array of inflammatory and immune phenotypes (Table 26.1). Although MMPs have generally been considered proinflammatory molecules, promoting leukocyte migration through the breakdown of basement membrane and extracellular matrix, the observed phenotypes clearly show that MMPs can also function in an anti-inflammatory capacity and the mechanisms underlying these phenotypes are now beginning to emerge. As such, it is now evident that MMPs have a multitude of pleotropic functions in regulating inflammation.

In recent years, substrate discovery efforts have uncovered a plethora of novel bioactive substrates that link MMPs with both the onset and the resolution of inflammation (McCawley and Matrisian 2001, Parks et al. 2004). Notably, MMPs are now known to have pro- and antitarget effects in cancer (Overall and Kleifeld 2006), and analogous opposing roles appear to exist in inflammation as well. As such, several authors have proposed that MMPs, by playing critical roles in regulating the activity of numerous inflammatory mediators, are homeostatic regulators and controllers of inflammation. Here we describe the conclusions from genetic models of inflammation as well as biochemical and proteomic evidence of the direct proteolytic actions of MMPs on cytokines and chemokines, and also indirect effects of MMPs on the inflammatory response.

Table 26.1 Inflammatory and immune phenotypes of MMP-deficient mice

Mouse	Phenotypes relating to inflammatory processes	References
$Mmp2^{-/-}$	More severe antibody-induced arthritis	Itoh et al. 2002
	Diminished egress of lung inflammatory cells	Corry et al. 2002
	Earlier and more severe EAE due to increased MMP-9	Esparza et al. 2004
	Prolonged cardiac allograft survival and lower cellular infiltration	Campbell et al. 2005
	More severe myocardial inflammation and dysfunction	Matsusaka et al. 2005
	Delay in inflammation-associated corneal neovascularization	Samolov et al. 2005
	Exacerbated experimental colitis	Garg et al. 2006
$Mmp3^{-/-}$	Markedly impaired contact hypersensitivity	Wang et al. 1999
	Reduced disc resorption and generation of macrophage chemoattractant	Haro et al. 2000a
	Reduced PMN recruitment in acute lung injury	Warner et al. 2001
	Reduced macrophages in atherosclerotic plaques	Silence et al. 2001
	Impaired clearance of intestinal bacteria	Li et al. 2004
	Delayed macrophage and lymphocyte recruitment in cardiac remodeling	Mukherjee et al. 2005
	Marked resistance to *Salmonella typhimurium* infection	Handley and Miller 2007
$Mmp7^{-/-}$	Increased bacterial survival, lack of intestinal α-defensins	Wilson et al. 1999
	Decreased epithelial cell apoptosis due to reduced FasL shedding	Powell et al. 1999
	Reduced macrophages in herniated discs, loss of TNF-α shedding	Haro et al. 2000b
	Reduced PMNs in lung due to altered chemokine gradients	Li et al. 2002
	Less severe septic arthritis	Gjertsson et al. 2005
$Mmp8^{-/-}$	Altered PMN infiltration in skin carcinogenesis model	Balbin et al. 2003
	Increased PMN infiltration in allergen-induced airway inflammation	Gueders et al. 2005
	Resistant to TNF-induced lethal hepatitis	Van Lint et al. 2005
	Decreased PMN infiltration in LPS-treated air pouch	Tester et al. 2007
	Increased inflammation delays wound healing	Gutierrez et al. 2007
$Mmp9^{-/-}$	More persistant inflammation in contact hypersensitivity	Wang et al. 1999
	Less susceptible to experimental autoimmune encephalomyelitis	Dubois et al. 1999
	Resistant to subepidermal blisters with reduced PMN recruitment	Liu et al. 2000
	Reduced macrophages in experimental cardiac infarction	Ducharme et al. 2000
	Impaired hematopoietic stem cell motility due to reduced sKitL	Heissig et al. 2002
	Less severe antibody-induced arthritis	Itoh et al. 2002
	Impaired cellular infiltration in airway inflammation	Cataldo et al. 2002
	Impaired defense in bacterial meningitis	Bottcher et al. 2003
	Reduced hematopoietic stem cell mobilization	Pelus et al. 2004
	Impaired dendritic cell migration through tracheal epithelium	Ichiyasu et al. 2004

(*continued*)

Table 26.1 (continued)

Mouse	Phenotypes relating to inflammatory processes	References
	Enhanced allergen-induced airway inflammation	McMillan et al. 2004
	Inflammatory cell accumulation in allergic lung model	Corry et al. 2004
	Attenuated dextran sulphate-induced colitis	Santana et al. 2006
	Increased severity of septic arthritis	Calander et al. 2006
	Reduced resistance against *E. coli* peritonitis	Renckens et al. 2006
Mmp10$^{-/-}$	Increased inflammation in models of infection and wound healing	W.C. Parks Unpublished
Mmp12$^{-/-}$	Reduced macrophages in smoke-induced emphysema	Hautamaki et al. 1997
	Less TNF-α shedding in smoke-induced inflammation	Churg et al. 2003
	Improved recovery after spinal cord injury, reduced inflammation	Wells et al. 2003
	Reduction in OVA-induced airway eosinophilia	Pouladi et al. 2004
	Reduction in antigen-induced airway inflammation	Warner et al. 2004
	Reduced infiltrating macrophages in aortic aneurysms	Longo et al. 2005
	Less experimental autoimmune encephalomyelitis, higher Th1/Th2 ratio	Weaver et al. 2005
	Reduced macrophages in atherosclerotic plaques	Johnson et al. 2005
	Decreased macrophages in ligament healing	Wright et al. 2006
	No change in bleomycin-induced lung fibrosis	Manoury et al. 2006
	BALF lacks chemotactic elastin fragments	Houghton et al. 2006
Mmp28$^{-/-}$	Increased inflammatory response	W.C. Parks Unpublished

Genetic Models of Inflammation

Murine models of inflammation and innate immunity have demonstrated diverse and unexpected functions of MMPs in the inflammatory context (Table 26.1). Currently, knockout mice have been generated for at least 15 MMPs: MMP-2, -3, -7, -8, -9, -10, -11, -12, -13, -14, -19, -20, -23, -24, and -28 (Page-McCaw et al. 2007). Of these, MMP-2, -3, -7, -8, -9, -10, -12, and -28 have been reported to have altered inflammatory phenotypes. Because of the significant skeletal defects and short life span of the MMP-14-null mice, inflammatory studies using these mice are not feasible; however, organ transplantation allows for tissue-specific evaluation of MMP-14 deficiency.

Individual MMPs show remarkably divergent physiological functions in inflammatory disease models. For instance, *Mmp2*-null mice have increased inflammation and consequent disease severity in models of arthritis (Itoh et al. 2002), experimental autoimmune encephalomyelitis (Esparza et al. 2004), and experimental colitis (Garg et al. 2006), suggesting a general anti-inflammatory role for the enzyme. This protective role of MMP-2 is consistent with its near-constitutive expression, where a proinflammatory role would be very destructive. In contrast, MMP-7-deficient mice exhibit reduced inflammation in models of septic arthritis (Gjertsson et al. 2005), herniated discs (Haro et al. 2000b), and acute lung injury (Li et al. 2002), demonstrating a proinflammatory function for MMP-7. MMP-3 and MMP-12 can also be generalized as proinflammatory effector molecules as these knockouts are characterized by decreased infiltrating leukocytes in several disease models (summarized in Table 26.1).

The functions of certain MMPs cannot be classified so easily. For example, the neutrophil protease MMP-8 seems to have opposing effects in acute and chronic inflammatory situations. In acute models such as the LPS-treated air pouch (Tester et al. 2007) and tumor necrosis factor (TNF)-induced hepatitis (Van Lint et al. 2005), MMP-8 promotes neutrophil infiltration, at least in part due to ELR$^+$ chemokine processing and activation. In more chronic models of skin carcinogenesis (Balbin et al. 2003) and wound healing (Gutierrez et al. 2007), MMP-8 is involved in terminating and resolving inflammation, potentially through induction of neutrophil apoptosis. MMP-9 also demonstrates complex regulatory properties. For instance, *Mmp9*-null mice have reduced PMN recruitment in subepidermal blisters (Liu et al. 2000) and decreased macrophages in response to cardiac infarction (Ducharme et al. 2000), but conversely show enhanced allergen-induced airway inflammation (Corry et al. 2004) and increased severity in a model of septic arthritis (Calander et al. 2006). Therefore, both MMP-8 and MMP-9 can exert differential regulatory effects depending on the particular stimulus and the temporal status. However, in the neutrophil MMP-8 dominates as its absence leads to decreased neutrophil infiltration despite normal or elevated MMP-9 levels.

MMP Regulation of Cytokines

Cytokines have a central role straddling both innate and adaptive immunity, functioning in both proinflammatory and anti-inflammatory manners. Cytokines are produced by a variety of cell types and signal either *in cis* or *in trans* to nearby cells or throughout the organism. Through the binding of specific cell-surface receptors, cytokines trigger signaling cascades resulting in altered gene and protein expression during an inflammatory reaction. Therefore, precise control of cytokine activity is essential for the accurate control of inflammation. Selective proteolysis of cytokines provides a rapid and effective means to modulate the initiation and termination of an inflammatory response.

Interleukin-1β

Interleukin-1β is an early and potent proinflammatory cytokine that has been linked to fever, leukocytosis, anemia, and elevated acute phase proteins (Dinarello 2005). IL-1β, along with TNF-α, are classical inducers of MMPs in numerous cell types. The production of IL-1β is primarily in macrophages, monocytes, and PMNs, where secretion occurs through nonclassical pathways as IL-1β lacks a secretory signal peptide (Andrei et al. 1999). It is translated as an inactive 33-kDa precursor, some of which seems to be secreted through specialized lysosomes. Caspase-1, also known as IL-1β-converting enzyme/ICE, is an intracellular cysteine protease that cleaves latent IL-1β at Asp^{116}-Ala^{117} releasing the N-terminus to generate the fully active 17-kDa molecule (Kostura et al. 1989). Although caspase-1 is thought to play a dominant role in IL-1β activation, studies of local inflammation with caspase-1-deficient mice suggest that alternate mechanisms exist (Fantuzzi et al. 1997).

IL-1β processing by MMPs has been described *in vitro*, generating biologically active 17-kDa forms (Schonbeck et al. 1998). Activation occurs with the following MMPs in order of increasing efficiency: MMP-2, MMP-3, and MMP-9, whereas MMP-1 lacks the ability to activate latent IL-1β (Table 26.2). With prolonged exposure and high enzyme concentrations, MMP-1, -2, -3, and -9 degrade IL-1β yielding a loss in activity (Ito et al. 1996, Schonbeck et al. 1998). Hence, the physiological role of MMPs in the regulation of IL-1β activity remains to be determined, but biochemical studies appear to show the potential for both proinflammatory and anti-inflammatory effects *in vivo*. The important caveat that biochemical cleavage does not equate to a physiologically relevant substrate must be recognized, and in the case of proIL-1β, which is normally processed intracellularly, substrate–enzyme colocalization must be considered. Simply, high enzyme–substrate ratios reflect extreme parameters that may not be met *in vivo*. Regardless, IL-1β-induced expression of MMPs possibly involves feed-forward or feed-back loops by which MMPs might further activate the cytokine, enhancing inflammation, or degrade IL-1β in an inhibitory loop to dampen or terminate the response.

Tumor Necrosis Factor-α

TNF-α is a multifunctional cytokine with proinflammatory and immunomodulatory roles (Kodama et al. 2005). Overproduction of TNF-α has been implicated in inflammatory conditions such as rheumatoid arthritis, multiple sclerosis, ankylosing spondylitis, and psoriasis (Taylor et al. 2004). TNF-α is produced as a 26-kDa membrane-associated inactive precursor (proTNF-α) that is activated upon proteolytic shedding, thus releasing the soluble form. This cleavage was first attributed to metalloproteinases based on broad-spectrum metalloprotease inhibitors preventing the shedding of TNF-α in cultured leukocytes and endotoxemia models (Gearing et al. 1994, McGeehan et al. 1994, Mohler et al. 1994). However, cloning and purification studies revealed that the major enzyme responsible is the metalloproteinase disintegrin ADAM-17, also called TNF-α converting enzyme (TACE) (Black et al. 1997, Moss et al. 1997).

Although ADAM-17 is considered to be the major generator of soluble TNF-α, studies with MMP-7 and MMP-12 knockout mice in models of herniated disc resorption (Haro et al. 2000b) and cigarette smoke-induced inflammation (Churg et al. 2003), respectively, support a role for MMPs in the physiological release of TNF-α. In the case of MMP-7, intervertebral discs cocultured with peritoneal macrophages from *Mmp7*-null mice had a defect in macrophage influx. This was

Table 26.2 Bioactive MMP substrates: the proinflammatory and anti-inflammatory proteolysis events by MMPs

Substrate	MMPs	Cleavage site	References
Proinflammatory cleavages			
Interleukin-1β	2, 3, 9	ND	Schonbeck et al. 1998
proTNF-α	1, 2, 3, 7, 9	ND	Gearing et al. 1995
	12	[68]SLISPLA↓QA↓VRSS	Chandler et al. 1996
	14, 15	[68]SL↓ISP↓LA↓QA↓VR	d'Ortho et al. 1997; Tam et al. 2004
	17	[68]SLISPLA↓QAVR	English et al. 2000
CXCL5	1	[4]AAAVLR↓ELRC	Van Den Steen et al. 2003b,
	8, 9	[4]AA↓A↓V↓LRELRC	Tester et al. 2007
CXCL8	1, 8, 9, 13, 14	[1]AVLPR↓SAKE	Van den Steen et al. 2000, Tam et al. 2004, Tester et al. 2007
CX₃CL1	2	[68]QAAA↓LTKN	Dean and Overall 2007
LIX (mCXCL5)	1, 2, 8, 9, 13	[1]APSS↓VIAA	Balbin et al. 2003, Van Den Steen et al. 2003b, Tester et al. 2007
Anti-inflammatory cleavages			
Interleukin-1α	1, 2, 3, 9	Degradation	Ito et al. 1996, Schonbeck et al. 1998
proTNF-α	14, 15	Degradation	d'Ortho et al. 1997
Latent TGF-β	2, 3, 9, 13, 14	ND	Yu and Stamenkovic 2000, D'Angelo et al. 2001, Maeda et al. 2001, Karsdal et al. 2002
CCL2	1, 3, 8	[1]QPDA↓INAP	McQuibban et al. 2002
CCL7	1, 2, 3, 13, 14	[1]QPVG↓INTS	McQuibban et al. 2000, 2002
CCL8	3	[1]QPDS↓VSIP	McQuibban et al. 2002
CCL13	1, 3	[1]QPDA↓LNVP	McQuibban et al. 2002
CXCL1	9	Degradation	Van den Steen et al. 2000
CXCL4	9	Degradation	Van den Steen et al. 2000
CXCL7	9	Degradation	Van den Steen et al. 2000
CXCL11	8, 9, 12	[1]FPMF-KRGR	Cox et al. 2008
CXCL12	1, 2, 3, 9, 13, 14	[1]KPVS↓LSYR	McQuibban et al. 2001
CX₃CL1	2	[1]QHLG↓MRKC	Dean and Overall 2007
Unaltered bioactivity cleavages			
CXCL6	8, 9	[1]GPVS↓A↓V↓LTELR	Van Den Steen et al. 2003b
CXCL9	9	[89]VLK↓VRK↓S↓QRSR	Van den Steen et al. 2003a
	7, 12	[89]VLK↓VRKSQRSR	Cox et al. 2008
CXCL10	8, 9	[66]KAV↓SKE↓MS↓KRSP	Van den Steen et al. 2003a
	12	[66]KAVSKE↓MS↓KRSP	Cox et al 2008
LIX (mCXCL5)	8	[76]KKAK↓RNAL	Tester et al. 2007

attributed to a decrease in soluble TNF-α, which was found to be necessary for MMP-3 induction and the consequent generation of a macrophage chemoattractant. In cigarette smoke-induced inflammation, MMP-12 knockout mice have decreased levels of soluble TNF-α and diminished ability to shed TNF-α from cultured alveolar macrophages in response to smoke exposure. Further, MMP-12-mediated shedding of TNF-α was found to promote endothelial activation, PMN infiltration, and the proteolytic tissue damage associated with emphysema.

Biochemically, several MMPs can cleave proTNF-α to generate the active form *in vitro* (Fig. 26.1). For example, purified MMP-1, MMP-2, MMP-3, MMP-7, and MMP-9 can cleave recombinant GST-TNF-α fusion protein (Gearing et al. 1994, 1995). In addition, MMP-8 (Cox et al., unpublished data), MMP-12 (Chandler et al. 1996), MMP-14 and MMP-15 (d'Ortho et al. 1997), and MMP-17 (English et al. 2000) are capable of cleaving proTNF-α. In a proteomic screen evaluating breast cancer cells overexpressing MT1-MMP, TNF-α was increased in the conditioned medium in the MMP-14-transfected cells compared to vector control and biochemically MMP-14 was shown to cleave and generate the active form (Tam et al. 2004). The MMP cleavage sites in proTNF-α are variable, but lie either at or slightly upstream of the ADAM-17-cleavage site of Ala^{76}-Val^{77} (Table 26.2). Notably, ADAM-17 is stimulated by lipopolysaccharide or experimentally by phorbol myristyl acetate, where its role in TNF-α shedding is well established. Conversely, in the absence of LPS stimulation such as noninfective situations, MMP-mediated shedding of TNF-α may be important and as such a variety of experimental approaches at different levels of complexity are needed to determine the relative contributions of various proteases (Overall and Blobel 2007). It is likely that ADAM-17 and several MMPs have complementary roles in the rapid shedding and activation of proTNF-α in response to varying stimuli.

Transforming Growth Factor-β1

Transforming growth factor-β1 (TGF-β1) is generally considered an immunosuppressive cytokine but encompasses a multitude of functions that cannot be defined so simply. TGF-β1 is known for its ability to instigate and maintain immune tolerance (Shull et al. 1992, Wahl et al. 2006). Under normal circumstances, TGF-β1 is secreted in a latent form where its furin-cleaved 80-kDa N-terminus, also called the latency-associated peptide, remains noncovalently associated and requires proteolytic, conformational, or acidic conditions for its removal (Khalil 1999). MMP proteolysis, among other mechanisms, has been proposed to release the active 25-kDa cytokine homodimer from the latency-associated peptide. In chondrocyte cultures, MMP-3 (Maeda et al. 2001) and MMP-13 (D'Angelo et al. 2001) have been found to activate TGF-β1 and likewise MMP-14 expression in osteoblasts causes activation of the latent cytokine and thereby promotes osteoblast survival (Karsdal et al. 2002). In the context of tumor biology, the gelatinases MMP-2 and MMP-9 were found to activate TGF-β, potentially promoting tumor invasion and angiogenesis (Yu and Stamenkovic 2000). Notably, TGF-β1 induces

Neutrophil
CXCR1/2
GAGs
Endothelial layer
MMP-8
MMP-9
Degraded
CXCL1
CXCL4
CXCL7
MMP-7
MMP-12
Macrophage
CCR1
Antagonist
CCL2 CCL7
CCL8 CCL13
Degraded TNFα
ADAM17
CX₃CR
Antagonist
CX₃CL1
MMP-2
MMPs
CXCR4
Lymphocyte
Loss of agonist
CXCL12
MMP-14
Fibroblasts
MMP-2
Active TGFβ
Termination
Neutrophil
CXCR1/2
GAGs
Endothelial layer
CXCL5
CXCL8
MMP-8
MMP-9
Active
CXCL5
CXCL8
MMP-7
MMP-12
Active TNFα
proTNFα
ADAM17
CX₃CR
Macrophage
CCR1
CCL2 CCL7
CCL8 CCL13
MMP-2
CX₃CL1
CXCR4
CXCL12
Lymphocyte
MMP-14
Fibroblasts
Latent TGFβ
Initiation

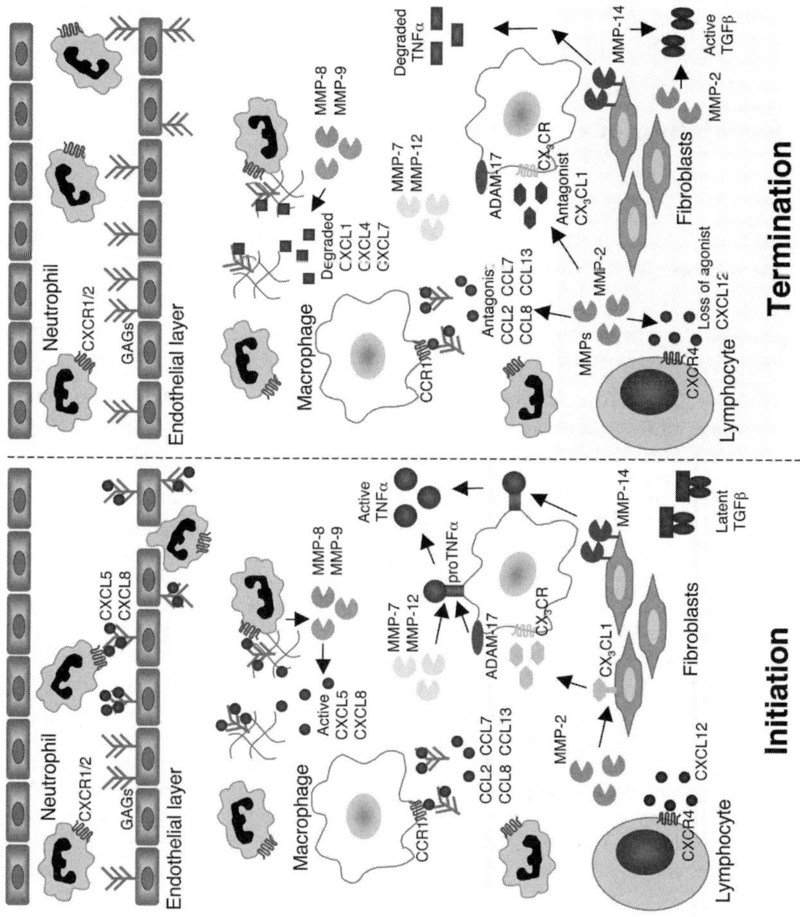

Fig. 26.1

the expression of MMP-2 (Overall et al. 1989, 1991) and MMP-9 (Zhou et al. 1993), perhaps contributing to the potentiation of TGF-β1 activity in a feed-forward manner. Furthermore, TGF-β1 stimulates production of the extracellular matrix, a critical event in wound healing, and through increased collagen deposition leads to induction and activation of MMP-2 and consequent processing of monocyte chemoattractants (CCL2, 7, 8, and 13; see Table 26.2) to generate potent anti-inflammatory molecules, thereby terminating macrophage influx and promoting healing. Although the link between MMPs and TGF-β is yet to be clearly determined in an inflammatory setting, MMP activation of TGF-β could provide a novel anti-inflammatory mechanism by which MMPs function to dampen an immune response.

MMP Regulation of Chemokines

Chemokines are a superfamily of low molecular weight chemotactic cytokines that function in directing the migration of cells in normal leukocyte trafficking and the inflammatory response (Moser et al. 2004). In vivo, chemokines form gradients through interactions with proteoglycan glycosaminoglycans and signal through 7-transmembrane G protein-coupled receptors to induce a chemotactic response. Several chemokines are known to be regulated by proteolysis, many of which are MMP-mediated cleavages, resulting in a multitude of functional consequences including either enhanced or decreased receptor binding, conversion to a receptor

Fig. 26.1 Matrix metalloproteinase (*MMP*) proteolysis of cytokines and chemokines to precisely control the induction (left) and termination (right) of inflammation. Initiation of an inflammatory response (left) is triggered by injury or an infectious agent resulting in the rapid local release of neutrophil chemoattractants and proteases, including MMPs. Circulating neutrophils roll on the endothelial surface through interactions with selectins (not shown). Engagement with chemokines, immobilized by glycosaminoglycan binding, causes firm adhesion to the endothelium through integrins and subsequent extravasation into the underlying tissue. Once in the tissue, neutrophils are activated and degranulate, releasing MMP-8 and MMP-9. MMP-8 processes and activates CXCL5 and CXCL8, resulting in additional PMN recruitment. Monocyte chemoattractant proteins (CCL2, 7, 8, 13) and CXCL12 promote the influx of monocytes and lymphocytes, respectively. The membrane-bound chemokine CX₃CL1 is shed by fibroblast-derived MMP-2 to generate a soluble chemoattractant. Macrophage-derived MMP-7 and MMP-12 are involved in shedding tumor necrosis factor (*TNF*)-α, in cooperation with ADAM-17, yielding a soluble proinflammatory cytokine. TNF-α is also shed by MMP-14, depicted here on fibroblast cells. Once the initiating stimulus is removed, the proinflammatory signals are no longer generated (right). Local chemokine production is decreased and neutrophils no longer enter the tissue. Furthermore, some neutrophil chemokines (CXCL1, CXCL4, CXCL7) are degraded by MMP-9 resulting in inactivation. MMP-2, as well as other MMPs, is involved in dampening the response through the generation of chemokine antagonists. Shown here are CCL2, CCL7, CCL8, CCL13, and CX₃CL1 as substrates of MMPs, forming potent receptor antagonists and inhibiting chemotactic migration. Also, CXCL12 is processed by MMP-2 resulting in a switch in receptor specificity. MMP-14 is potentially involved in downregulating TNF-α through degradation. Furthermore, MMP-2 and MMP-9 (not shown) cleave the latency-associated protein from transforming growth factor (*TGF*)-β to liberate the active TGF-β homodimer with immunosuppressive properties. Active TGF-β induces MMP-2 and MMP-9 expression and extracellular matrix formation in a feedback mechanism

antagonist, changing receptor specificity, and shedding of membrane-anchored chemokines.

Chemokines are characterized by the spacing of conserved cysteine residues that function in the formation of disulphide bonds. Although the sequence similarity among chemokines is quite low, ranging from 20% to 50% homology, the tertiary structure of these proteins is remarkably conserved (Fernandez and Lolis 2002). The general structure consists of an undefined N-terminus, three central antiparallel beta-sheets, and a C-terminal alpha helix. The N-terminal portion of chemokines appears to be most susceptible to proteolysis and cleavage often results in significant changes in activity. In contrast, some C-terminal processing events have been described but no functional changes have been ascribed to these truncations (Van den Steen et al. 2003a, Hensbergen et al. 2004, Tester et al. 2007).

CC Chemokines

The CC family of chemokines is defined by adjacent cysteine residues in the N-terminus and its members typically induce the migration of monocytes, dendritic cells, and natural killer cells. The monocyte chemoattractant proteins (MCPs), also known as CCL2, 7, 8, 13, were among the first chemokines to be identified as MMP substrates. In a yeast genetic screen with the hemopexin domain of MMP-2, CCL7 was found to have a strong interaction with the exosite and was processed efficiently at Gly^4-Ile^5 with recombinant enzyme and by monocytes in culture and *in vivo* (McQuibban et al. 2000, Overall et al. 2002). The truncated chemokine, CCL7 (5–76), still binds its cognate CC receptors but no longer promotes chemotaxis and instead functions as a receptor antagonist with potent anti-inflammatory properties *in vivo*, as determined in mouse models of subcutaneous and peritoneal inflammation. CCL7 is also processed by MMP-1, 3, 13, and 14 (McQuibban et al. 2002) and the cleaved form has been detected in rheumatoid synovial fluid (McQuibban et al. 2000).

All of the MCPs have been identified as MMP substrates (Table 26.2) with characteristic cleavage between residues 4 and 5 in the N-terminus, causing diminished activity and the generation of potent receptor antagonists (McQuibban et al. 2000, 2002). In addition to MMP-2, MMP-1, -3, -8, -13, and -14 cleave MCPs with different profiles of substrate preference, revealing specificity in this interaction (Table 26.2). Notably, this work was the first to demonstrate an unexpected dampening of inflammation by MMPs, traditionally considered to be proinflammatory proteases.

CXC Chemokines

The CXC chemokines predominantly influence the migration of polymorphonuclear neutrophils and T-lymphocytes. A subset of CXC chemokines are characterized by an ELR (glutamic acid-leucine-arginine) sequence proximal to the conserved CXC motif and act exclusively on CXCR1 and CXCR2 receptors, thereby promoting neutrophil chemotaxis. CXCL8, the most potent of the ELR^+ chemokines, is processed by MMP-1, -8, -9, -13, and -14 at Arg^5-Ser^6, causing a tenfold increase in receptor binding affinity and chemotactic activity (Van den Steen et al. 2000, Tam et al. 2004, Tester et al. 2007). In the cellular context, MMP-14-overexpressing cells were discovered to have decreased levels of CXCL8-derived peptides in the conditioned medium compared to vector controls, confirmed biochemically to be due to proteolysis (Tam et al. 2004).

Activation of PMN chemoattractants also applies to the human ELR^+ chemokine CXCL5 as well as the murine chemokine LIX (Balbin et al. 2003, Van Den Steen et al. 2003b, Tester et al. 2007). LIX, also called mCXCL5, is thought to be the mouse ortholog to human CXCL8 in that it is the most potent and abundant of the ELR^+ chemokines. LIX is efficiently cleaved at Ser^4-Val^5 by the neutrophil enzyme MMP-9 *in vitro*, resulting in a significant increase in chemoattractant activity (Van Den Steen et al. 2003b). The generation of the MMP-8-deficient mouse revealed decreased processing of LIX in LPS-stimulated bronchoalveolar lavage fluid, suggesting a specific role for MMP-8 in the activation of LIX (Balbin et al. 2003). This observation is coupled with delayed neutrophil infiltration to the site of challenge in a skin carcinogenesis model in *Mmp8*-null mice, followed by a massive accumulation of neutrophils. Furthermore, in an LPS-treated subcutaneous air pouch model of acute inflammation, *Mmp8*-null mice have a significantly decreased neutrophil influx compared to wild-type controls (Tester et al. 2007). This difference is lost when mice are treated with the truncated chemokines LIX (5–92) or CXCL8 (6–77), demonstrating that LIX and CXCL8 are physiologically relevant substrates of MMP-8 and that the normal cell migration machinery and the ability of neutrophils to migrate through basement membrane and connective tissue is intact in the absence of MMP-8. Interestingly, several MMPs in addition to MMP-8 cleave LIX *in vitro*, yet animal studies indicate a lack of physiological redundancy in this pathway (Tester et al. 2007). This result also questions the importance of MMP-8 in collagen degradation in these matrices, a role traditionally assumed to be essential for neutrophil chemotaxis and extravasation.

Although there is mounting evidence to support the notion that MMPs are involved in promoting PMN infiltration via the proteolytic activation of neutrophil chemoattractants, there is limited data suggesting an opposing role for MMPs in the inactivation of ELR^+ chemokines. This activation mechanism differs from the majority of chemokines, where N-terminal truncation usually results in the loss of agonism coupled with the generation of potent receptor antagonists. However, MMP-9 has been reported to slowly degrade the ELR^+ chemokines CXCL1, CXCL4, and CXCL7 *in vitro*, but the physiological evidence of this interaction is lacking (Van den Steen et al. 2000).

CXCL12, also known as SDF-1, is a multifunctional chemokine with a critical role in development as well as a potent chemoattractant of T lymphocytes, mono-cytes, and $CD34^+$ stem cells. Proteolysis of CXCL12 occurs at Ser^4-Leu^5 by MMP-1, -2, -3, -9, -13, and -14, resulting in a loss of binding to the cognate receptors CXCR4 (McQuibban et al. 2001) and CXCR7 (Balabanian et al. 2005) and, therefore, loss of HIV protection and decreased chemotactic properties. Interestingly, the cleavage product, CXCL12 (5–67), is a potent and specific neurotoxin (Zhang et al. 2003), an effect shown to be mediated through CXCR3 binding and signaling (Vergote et al. 2006), demonstrating for the first time that proteolysis of a chemokine can cause a shift in receptor specificity.

Membrane-Bound Chemokines

CX_3CL1, also known as fractalkine, is unique among the chemokine superfamily in that it can exist in both membrane-anchored and soluble forms, giving it the dual functionality of an adhesion molecule and a chemoattractant (Bazan et al. 1997). When shed from the membrane, CX_3CL1 is a potent chemoattractant for T cells and monocytes. Disintegrin-like metalloproteinases ADAM-10 and ADAM-17 have been reported to shed CX_3CL1 by cleaving adjacent to the transmembrane region, hence regulating CX_3CL1-mediated cell–cell adhesion and chemoattractant activity (Garton et al. 2001, Hundhausen et al. 2003). A proteomic approach recently identified CX_3CL1 as an MMP-2 substrate, where endogenous CX_3CL1 was found to be increased in the conditioned medium of MMP-2 expressing fibroblasts compared to *Mmp2*-null control cells (Dean and Overall 2007). Tandem mass spectrometry coupled with peptide mapping enabled the identification of two cleavage sites: Ala^{71}-Leu^{72} in the C-terminus of the chemokine ectodomain, yielding a soluble chemoattractant, and Gly^4-Met^5, a truncation known to result in potent antagonism of the receptor CX_3CR (Inoue et al. 2005). Hence, it appears that MMP-2 can both activate and inactivate a single chemokine, allowing for precise biphasic regulation of CX_3CL1 activity in the conversion of a cell surface adhesion molecule/agonist to a soluble agonist and antagonist depending on the site of cleavage. The only other membrane-anchored chemokine, CXCL16, is also shed in a metalloproteinase-dependent manner. Solubilization of CXCL16 is inducible by TNF-α and significantly decreased on treatment with GM6001, a broad-spectrum MMP and ADAM inhibitor, but the specific proteases involved are yet to be identified (Hara et al. 2006). Notably, in genetic and inhibitor studies, ADAM-10 and γ-secretases were found to be major components of proteolytic cascades leading to CX_3CL1 and CXCL16 shedding (Schulte et al. 2007).

Proteoglycan-Mediated Chemokine Regulation

Direct processing of chemokines themselves is not the only way by which MMPs regulate chemokine activity. Many chemokines are known to require binding to the glycosaminoglycan side chains of proteoglycans for the formation of gradients and chemotactic activity *in vivo* (Proudfoot et al. 2003, Handel et al. 2005). As such, altering the proteolglycans themselves can affect chemokine gradients and physiological functions. The clearest example of such an interaction came from a study showing that MMP-7-deficient mice have altered neutrophil infiltration in lung inflammation, where neutrophils remain confined to the interstitium without advancing to the alveolar space (Li et al. 2002). This defect was attributed to MMP-7-dependent shedding of the syndecan-1 ectodomain complexed with mCXCL1/KC, a murine ELR$^+$ chemokine, a process required to direct and confine neutrophils to the site of injury. Hence, in the absence of MMP-7, neutrophils are unable to enter the lumen of the lung because the appropriate KC-syndecan-1 gradient has not been generated. Notably, sydecan-1 is also shed by MMP-14 and MMP-16 resulting in enhanced migration *in vitro* (Endo et al. 2003), potentially representing analogous mechanisms involving other MMPs.

Additional MMPs have also been suggested to affect chemokine activity through altering interactions with the extracellular matrix. In a model of TNF-induced acute hepatitis, MMP-8-deficient mice show improved survival coupled with defective neutrophil recruitment at 6 h posttreatment, compared to wild-type controls (Van Lint et al. 2005). This phenotype was proposed to be due to the indirect effect of MMP-8-dependent release of LIX from the extracellular matrix, although direct processing of LIX and modulation of activity has also been observed (Balbin et al. 2003, Van Den Steen et al. 2003b, Tester et al. 2007). Furthermore, MMP-8 processing of LIX does not alter glycosaminoglycan binding, as determined with heparin sulphate affinity assays (Tester et al. 2007). However, C-terminal processing of CXCL11, a Th1 lymphocyte chemoattractant, by MMPs-7, -8, -9, and -12 removes several cationic residues and causes decreased glycosaminoglycan binding (Cox et al. 2008). Therefore, potential regulation of chemokine interactions with proteoglycans in the formation of gradients should be appreciated when considering chemokine bioactivity *in vivo*.

Future Perspectives

Deciphering the individual functions of MMPs in the regulation of inflammatory processes is an overwhelming task but the importance of the field is highlighted by failed clinical trials with MMP inhibitors and the coinciding unpredicted side effects (Zucker et al. 2000, Coussens et al. 2002, Overall and Lopez-Otin 2002, Pavlaki and Zucker 2003, Puente et al. 2003). It is now evident that the physiological roles of MMPs and their respective underlying mechanisms need to be further defined before attempting to inhibit individual MMPs for clinical applications. It is critical that only well-established targets are inhibited and anti-targets are completely avoided (Overall and Kleifeld 2006). With this approach the detrimental actions of MMPs are blocked while still preserving the beneficial anti-target MMP activities. As such, animal model and cell-based approaches complemented by biochemical and proteomic studies will be instrumental in achieving this objective.

Current methods of MMP substrate discovery are predominantly hypothesis driven, which is a very time-consuming approach. However, more systematic approaches to uncover novel MMP substrates and downstream effects of MMP activity are beginning to emerge. For instance, two-dimensional gel electrophoresis was recently used in the analysis of bronchoalveolar lavage fluid of allergen-challenged $Mmp2^{-/-}/Mmp9^{-/-}$ mice, leading to the identification of Ym1, S100A8, and S100A9 as novel MMP substrates (Greenlee et al. 2006). Furthermore, in the cellular context, MMP-14 and MMP-2 transfected cells have been used to identify a variety of new bioactive substrates utilizing mass spectrometry-based proteomic analysis (Tam et al. 2004, Dean and Overall 2007). This work has been extended with the use of MMP inhibitors, thus uncovering additional substrates with the potential to identify adverse inhibitor effects (Butler and Overall 2007).

The remarkable progress in proteomic approaches, such as those listed above, will undoubtedly have far-reaching implications in the field of MMP research, as well as that concerning all proteases (Schilling and Overall 2007). Applying such techniques to *in vivo* models of inflammation will enable rapid and unbiased identification of physiologically relevant substrates and hence a deeper understanding of the numerous inflammation-related phenotypes described here. Regardless of the specific mechanisms, it is undeniable that MMPs are key pleiotropic regulators of inflammation and immunity, demonstrating a wide array of proinflammatory and anti-inflammatory functions, orchestrating both the onset and termination of inflammation. Hence, the former view that the major role of MMPs is to cleave extracellular matrix is no longer tenable. Rather, MMPs are key signaling proteases involved in normal homeostasis and initiating and terminating inflammation through efficient and specific processing of inflammatory mediators.

References

Andrei C., Dazzi C., Lotti L. et al. (1999). The secretory route of the leaderless protein interleukin 1beta involves exocytosis of endolysosome-related vesicles. Mol Biol Cell 10(5): 1463–75.

Balabanian K., Lagane B., Infantino S. et al. (2005). The chemokine SDF-1/CXCL12 binds to and signals through the orphan receptor RDC1 in T lymphocytes. J Biol Chem 280(42): 35760–6.

Balbin M., Fueyo A., Tester A. M. et al. (2003). Loss of collagenase-2 confers increased skin tumor susceptibility to male mice. Nat Genet 35(3): 252–7.

Bazan J. F., Bacon K. B., Hardiman G. et al. (1997). A new class of membrane-bound chemokine with a CX3C motif. Nature 385(6617): 640–4.

Black R. A., Rauch C. T., Kozlosky C. J. et al. (1997). A metalloproteinase disintegrin that releases tumour-necrosis factor-alpha from cells. Nature 385(6618): 729–33.

Bottcher T., Spreer A., Azeh I. et al. (2003). Matrix metalloproteinase-9 deficiency impairs host defense mechanisms against Streptococcus pneumoniae in a mouse model of bacterial meningitis. Neurosci Lett 338(3): 201–4.

Bramhall S. R., Hallissey M. T., Whiting J. et al. (2002). Marimastat as maintenance therapy for patients with advanced gastric cancer: a randomised trial. Br J Cancer 86(12): 1864–70.

Butler G.S., Dean R.A., Tam E.M. et al. (2008). Pharmacoproteomics of a metalloproteinase hydroxamate inhibitor in breast cancer cells: Dynamics of matrix metalloproteinase-14 (MT1-MMP) mediated membrane protein shedding. Mol Cell Biol. In press.

Calander A. M., Starckx S., Opdenakker G. et al. (2006). Matrix metalloproteinase-9 (gelatinase B) deficiency leads to increased severity of Staphylococcus aureus-triggered septic arthritis. Microbes Infect 8(6): 1434–9.

Campbell L. G., Ramachandran S., Liu W. et al. (2005). Different roles for matrix metalloproteinase-2 and matrix metalloproteinase-9 in the pathogenesis of cardiac allograft rejection. Am J Transplant 5(3): 517–28.

Cataldo D. D., Tournoy K. G., Vermaelen K. et al. (2002). Matrix metalloproteinase-9 deficiency impairs cellular infiltration and bronchial hyperresponsiveness during allergen-induced airway inflammation. Am J Pathol 161(2): 491–8.

Chandler S., Cossins J., Lury J. et al. (1996). Macrophage metalloelastase degrades matrix and myelin proteins and processes a tumour necrosis factor-alpha fusion protein. Biochem Biophys Res Commun 228(2): 421–9.

Churg A., Wang R. D., Tai H. et al. (2003). Macrophage metalloelastase mediates acute cigarette smoke-induced inflammation via tumor necrosis factor-alpha release. Am J Respir Crit Care Med 167(8): 1083–9.

Corry D. B., Rishi K., Kanellis J. et al. (2002). Decreased allergic lung inflammatory cell egression and increased susceptibility to asphyxiation in MMP2-deficiency. Nat Immunol 3(4): 347–53.

Corry D. B., Kiss A., Song L. Z. et al. (2004). Overlapping and independent contributions of MMP2 and MMP9 to lung allergic inflammatory cell egression through decreased CC chemokines. Faseb J 18(9): 995–7.

Coussens L. M., Fingleton B., and Matrisian L. M. (2002). Matrix metalloproteinase inhibitors and cancer: trials and tribulations. Science 295(5564): 2387–92.

Cox J.H., Dean R.A., Roberts C.R. et al. (2008). Matrix metalloproteinase processing of CXCL11/ I-TAC results in loss of chemoattractant activity and altered glycosaminoglycan binding. J Biol Chem. In press.

D'Angelo M., Billings P. C., Pacifici M. et al. (2001). Authentic matrix vesicles contain active metalloproteases (MMP). a role for matrix vesicle-associated MMP-13 in activation of transforming growth factor-beta. J Biol Chem 276(14): 11347–53.

d'Ortho M. P., Will H., Atkinson S. et al. (1997). Membrane-type matrix metalloproteinases 1 and 2 exhibit broad-spectrum proteolytic capacities comparable to many matrix metalloproteinases. Eur J Biochem 250(3): 751–7.

Dean R. A. and Overall C. M. (2007). Proteomics discovery of metalloproteinase substrates in the cellular context by iTRAQ labeling reveals a diverse MMP-2 substrate degradome. Mol Cell Proteomics 6(4): 611–23.

Dinarello C. A. (2005). Blocking IL-1 in systemic inflammation. J Exp Med 201(9): 1355–9.

Dubois B., Masure S., Hurtenbach U. et al. (1999). Resistance of young gelatinase B-deficient mice to experimental autoimmune encephalomyelitis and necrotizing tail lesions. J Clin Invest 104(11): 1507–15.

Ducharme A., Frantz S., Aikawa M. et al. (2000). Targeted deletion of matrix metalloproteinase-9 attenuates left ventricular enlargement and collagen accumulation after experimental myocardial infarction. J Clin Invest 106(1): 55–62.

Dvorak H. F. (1986). Tumors: wounds that do not heal. Similarities between tumor stroma generation and wound healing. N Engl J Med 315(26): 1650–9.

Endo K., Takino T., Miyamori H. et al. (2003). Cleavage of syndecan-1 by membrane type matrix metalloproteinase-1 stimulates cell migration. J Biol Chem 278(42): 40764–70.

English W. R., Puente X. S., Freije J. M. et al. (2000). Membrane type 4 matrix metalloproteinase (MMP17) has tumor necrosis factor-alpha convertase activity but does not activate pro-MMP2. J Biol Chem 275(19): 14046–55.

Esparza J., Kruse M., Lee J. et al. (2004). MMP-2 null mice exhibit an early onset and severe experimental autoimmune encephalomyelitis due to an increase in MMP-9 expression and activity. Faseb J 18(14): 1682–91.

Fantuzzi G., Ku G., Harding M. W. et al. (1997). Response to local inflammation of IL-1 beta-converting enzyme- deficient mice. J Immunol 158(4): 1818–24.

Fernandez E. J. and Lolis E. (2002). Structure, function, and inhibition of chemokines. Annu Rev Pharmacol Toxicol 42: 469–99.

Garg P., Rojas M., Ravi A. et al. (2006). Selective ablation of matrix metalloproteinase-2 exacerbates experimental colitis: contrasting role of gelatinases in the pathogenesis of colitis. J Immunol 177(6): 4103–12.

Garton K. J., Gough P. J., Blobel C. P. et al. (2001). Tumor necrosis factor-alpha-converting enzyme (ADAM17) mediates the cleavage and shedding of fractalkine (CX3CL1). J Biol Chem 276(41): 37993–8001.

Gearing A. J., Beckett P., Christodoulou M. et al. (1994). Processing of tumour necrosis factor-alpha precursor by metalloproteinases. Nature 370(6490): 555–7.

Gearing A. J., Beckett P., Christodoulou M. et al. (1995). Matrix metalloproteinases and processing of pro-TNF-alpha. J Leukoc Biol 57(5): 774–7.

Gjertsson I., Innocenti M., Matrisian L. M. et al. (2005). Metalloproteinase-7 contributes to joint destruction in Staphylococcus aureus induced arthritis. Microb Pathog 38(2–3): 97–105.

Greenlee K. J., Corry D. B., Engler D. A. et al. (2006). Proteomic identification of *in vivo* substrates for matrix metalloproteinases 2 and 9 reveals a mechanism for resolution of inflammation. J Immunol 177(10): 7312–21.

Gueders M. M., Balbin M., Rocks N. et al. (2005). Matrix metalloproteinase-8 deficiency promotes granulocytic allergen-induced airway inflammation. J Immunol 175(4): 2589–97.

Gutierrez-Fernandez A., Inada M., Balbin M. et al. (2007). Increased inflammation delays wound healing in mice deficient in collagenase-2 (MMP-8). Faseb J 21(10): 2580–91.

Handel T. M., Johnson Z., Crown S. E. et al. (2005). Regulation of protein function by glycosaminoglycans–as exemplified by chemokines. Annu Rev Biochem 74: 385–410.

Handley S. A. and Miller V. L. (2007). General and specific host responses to bacterial infection in Peyer's patches: a role for stromelysin-1 (matrix metalloproteinase 3) during Salmonella enterica infection. Mol Microbiol 64(1): 94–110.

Hansson G. K. and Libby P. (2006). The immune response in atherosclerosis: a double-edged sword. Nat Rev Immunol 6(7): 508–19.

Hara T., Katakai T., Lee J. H. et al. (2006). A transmembrane chemokine, CXC chemokine ligand 16, expressed by lymph node fibroblastic reticular cells has the potential to regulate T cell migration and adhesion. Int Immunol 18(2): 301–11.

Haro H., Crawford H. C., Fingleton B. et al. (2000a). Matrix metalloproteinase-3-dependent generation of a macrophage chemoattractant in a model of herniated disc resorption. J Clin Invest 105(2): 133–41.

Haro H., Crawford H. C., Fingleton B. et al. (2000b). Matrix metalloproteinase-7-dependent release of tumor necrosis factor-alpha in a model of herniated disc resorption. J Clin Invest 105(2): 143–50.

Hautamaki R. D., Kobayashi D. K., Senior R. M. et al. (1997). Requirement for macrophage elastase for cigarette smoke-induced emphysema in mice. Science 277(5334): 2002–4.

Heissig B., Hattori K., Dias S. et al. (2002). Recruitment of stem and progenitor cells from the bone marrow niche requires MMP-9 mediated release of kit-ligand. Cell 109(5): 625–37.

Hensbergen P. J., Verzijl D., Balog C. I. et al. (2004). Furin is a chemokine-modifying enzyme: *in vitro* and *in vivo* processing of CXCL10 generates a C-terminally truncated chemokine retaining full activity. J Biol Chem 279(14): 13402–11.

Houghton A. M., Quintero P. A., Perkins D. L. et al. (2006). Elastin fragments drive disease progression in a murine model of emphysema. J Clin Invest 116(3): 753–9.

Hundhausen C., Misztela D., Berkhout T. A. et al. (2003). The disintegrin-like metalloproteinase ADAM10 is involved in constitutive cleavage of CX3CL1 (fractalkine) and regulates CX3CL1-mediated cell-cell adhesion. Blood 102(4): 1186–95.

Ichiyasu H., McCormack J. M., McCarthy K. M. et al. (2004). Matrix metalloproteinase-9-deficient dendritic cells have impaired migration through tracheal epithelial tight junctions. Am J Respir Cell Mol Biol 30(6): 761–70.

Inoue A., Hasegawa H., Kohno M. et al. (2005). Antagonist of fractalkine (CX3CL1) delays the initiation and ameliorates the progression of lupus nephritis in MRL/lpr mice. Arthritis Rheum 52(5): 1522–33.

Ito A., Mukaiyama A., Itoh Y. et al. (1996). Degradation of interleukin 1beta by matrix metalloproteinases. J Biol Chem 271(25): 14657–60.

Itoh T., Matsuda H., Tanioka M. et al. (2002). The role of matrix metalloproteinase-2 and matrix metalloproteinase-9 in antibody-induced arthritis. J Immunol 169(5): 2643–7.

Johnson J. L., George S. J., Newby A. C. et al. (2005). Divergent effects of matrix metalloproteinases 3, 7, 9, and 12 on atherosclerotic plaque stability in mouse brachiocephalic arteries. Proc Natl Acad Sci USA 102(43): 15575–80.

Karsdal M. A., Larsen L., Engsig M. T., et al. (2002). Matrix metalloproteinase-dependent activation of latent transforming growth factor-beta controls the conversion of osteoblasts into osteocytes by blocking osteoblast apoptosis. J Biol Chem 277(46): 44061–7.

Khalil N. (1999). TGF-beta: from latent to active. Microbes Infect 1(15): 1255–63.

Kodama S., Davis M., and Faustman D. L. (2005). The therapeutic potential of tumor necrosis factor for autoimmune disease: a mechanistically based hypothesis. Cell Mol Life Sci 62(16): 1850–62.

Kostura M. J., Tocci M. J., Limjuco G. et al. (1989). Identification of a monocyte specific pre-interleukin 1 beta convertase activity. Proc Natl Acad Sci USA 86(14): 5227–31.

Li C. K., Pender S. L., Pickard K. M. et al. (2004). Impaired immunity to intestinal bacterial infection in stromelysin-1 (matrix metalloproteinase-3)-deficient mice. J Immunol 173(8): 5171–9.

Li Q., Park P. W., Wilson C. L. et al. (2002). Matrilysin shedding of syndecan-1 regulates chemokine mobilization and transepithelial efflux of neutrophils in acute lung injury. Cell 111(5): 635–46.

Liu Z., Zhou X., Shapiro S. D. et al. (2000). The serpin alpha1-proteinase inhibitor is a critical substrate for gelatinase B/MMP-9 *in vivo*. Cell 102(5): 647–55.

Longo G. M., Buda S. J., Fiotta N. et al. (2005). MMP-12 has a role in abdominal aortic aneurysms in mice. Surgery 137(4): 457–62.

Maeda S., Dean D. D., Gay I. et al. (2001). Activation of latent transforming growth factor beta1 by stromelysin 1 in extracts of growth plate chondrocyte-derived matrix vesicles. J Bone Miner Res 16(7): 1281–90.

Manoury B., Nenan S., Guenon I. et al. (2006). Macrophage metalloelastase (MMP-12) deficiency does not alter bleomycin-induced pulmonary fibrosis in mice. J Inflamm (Lond) 3: 2.

Matsusaka H., Ikeuchi M., Matsushima S. et al. (2005). Selective disruption of MMP-2 gene exacerbates myocardial inflammation and dysfunction in mice with cytokine-induced cardiomyopathy. Am J Physiol Heart Circ Physiol 289(5): H1858–64.

McCawley L. J. and Matrisian L. M. (2001). Matrix metalloproteinases: they're not just for matrix anymore! Curr Opin Cell Biol 13(5): 534–40.

McGeehan G. M., Becherer J. D., Bast R. C., Jr. et al. (1994). Regulation of tumour necrosis factor-alpha processing by a metalloproteinase inhibitor. Nature 370(6490): 558–61.

McMillan S. J., Kearley J., Campbell J. D. et al. (2004). Matrix metalloproteinase-9 deficiency results in enhanced allergen-induced airway inflammation. J Immunol 172(4): 2586–94.

McQuibban G. A., Gong J. H., Tam E. M. et al. (2000). Inflammation dampened by gelatinase A cleavage of monocyte chemoattractant protein-3. Science 289(5482): 1202–6.

McQuibban G. A., Butler G. S., Gong J. H. et al. (2001). Matrix metalloproteinase activity inactivates the CXC chemokine stromal cell-derived factor-1. J Biol Chem 276(47): 43503–8.

McQuibban G. A., Gong J. H., Wong J. P. et al. (2002). Matrix metalloproteinase processing of monocyte chemoattractant proteins generates CC chemokine receptor antagonists with anti-inflammatory properties *in vivo*. Blood 100(4): 1160–7.

Mohler K. M., Sleath P. R., Fitzner J. N. et al. (1994). Protection against a lethal dose of endotoxin by an inhibitor of tumour necrosis factor processing. Nature 370(6486): 218–20.

Moser B., Wolf M., Walz A. et al. (2004). Chemokines: multiple levels of leukocyte migration control. Trends Immunol 25(2): 75–84.

Moss M. L., Jin S. L., Milla M. E. et al. (1997). Cloning of a disintegrin metalloproteinase that processes precursor tumour-necrosis factor-alpha. Nature 385(6618): 733–6.

Mukherjee R., Bruce J. A., McClister D. M., Jr. et al. (2005). Time-dependent changes in myocardial structure following discrete injury in mice deficient of matrix metalloproteinase-3. J Mol Cell Cardiol 39(2): 259–68.

Overall C. M. (2004). Dilating the degradome: matrix metalloproteinase 2 (MMP-2) cuts to the heart of the matter. Biochem J 383(Pt. 3): e5–7.

Overall C. M. and Blobel C. P. (2007). In search of partners: linking extracellular proteases to substrates. Nat Rev Mol Cell Biol 8(3): 245–57.

Overall C. M. and Kleifeld O. (2006). Tumour microenvironment–opinion: validating matrix metalloproteinases as drug targets and anti-targets for cancer therapy. Nat Rev Cancer 6(3): 227–39.

Overall C. M. and Lopez-Otin C. (2002). Strategies for MMP inhibition in cancer: innovations for the post-trial era. Nat Rev Cancer 2(9): 657–72.

Overall C. M., Wrana J. L. and Sodek J. (1989). Transforming growth factor-beta regulation of collagenase, 72 kDa-progelatinase, TIMP and PAI-1 expression in rat bone cell populations and human fibroblasts. Connect Tissue Res 20(1–4): 289–94.

Overall C. M., Wrana J. L. and Sodek J. (1991). Transcriptional and post-transcriptional regulation of 72-kDa gelatinase/type IV collagenase by transforming growth factor-beta 1 in human fibroblasts. Comparisons with collagenase and tissue inhibitor of matrix metalloproteinase gene expression. J Biol Chem 266(21): 14064–71.

Overall C. M., McQuibban G. A. and Clark-Lewis I. (2002). Discovery of chemokine substrates for matrix metalloproteinases by exosite scanning: a new tool for degradomics. Biol Chem 383 (7–8): 1059–66.

Page-McCaw A., Ewald A. J. and Werb Z. (2007). Matrix metalloproteinases and the regulation of tissue remodelling. Nat Rev Mol Cell Biol 8(3): 221–33.

Parks W. C., Wilson C. L. and Lopez-Boado Y. S. (2004). Matrix metalloproteinases as modulators of inflammation and innate immunity. Nat Rev Immunol 4(8): 617–29.

Pavlaki M. and Zucker S. (2003). Matrix metalloproteinase inhibitors (MMPIs): the beginning of phase I or the termination of phase III clinical trials. Cancer Metastasis Rev 22(2–3): 177–203.

Pelus L. M., Bian H., King A. G. et al. (2004). Neutrophil-derived MMP-9 mediates synergistic mobilization of hematopoietic stem and progenitor cells by the combination of G-CSF and the chemokines GRObeta/CXCL2 and GRObetaT/CXCL2delta4. Blood 103(1): 110–9.

Pouladi M. A., Robbins C. S., Swirski F. K. et al. (2004). Interleukin-13-dependent expression of matrix metalloproteinase-12 is required for the development of airway eosinophilia in mice. Am J Respir Cell Mol Biol 30(1): 84–90.

Powell W. C., Fingleton B., Wilson C. L. et al. (1999). The metalloproteinase matrilysin proteolytically generates active soluble Fas ligand and potentiates epithelial cell apoptosis. Curr Biol 9(24): 1441–7.

Proudfoot A. E., Handel T. M., Johnson Z. et al. (2003). Glycosaminoglycan binding and oligomerization are essential for the *in vivo* activity of certain chemokines. Proc Natl Acad Sci USA 100(4): 1885–90.

Puente X. S., Sanchez L. M., Overall C. M. et al. (2003). Human and mouse proteases: a comparative genomic approach. Nat Rev Genet 4(7): 544–58.

Renckens R., Roelofs J. J., Florquin S. et al. (2006). Matrix metalloproteinase-9 deficiency impairs host defense against abdominal sepsis. J Immunol 176(6): 3735–41.

Samolov B., Steen B., Seregard S. et al. (2005). Delayed inflammation-associated corneal neovascularization in MMP-2-deficient mice. Exp Eye Res 80(2): 159–66.

Santana A., Medina C., Paz-Cabrera M. C. et al. (2006). Attenuation of dextran sodium sulphate induced colitis in matrix metalloproteinase-9 deficient mice. World J Gastroenterol 12(40): 6464–72.

Schilling O. and Overall C. M. (2007). Proteomic discovery of protease substrates. Curr Opin Chem Biol 11(1): 36–45.

Schonbeck U., Mach F., and Libby P. (1998). Generation of biologically active IL-1 beta by matrix metalloproteinases: a novel caspase-1-independent pathway of IL-1 beta processing. J Immunol 161(7): 3340–6.

Schulte A., Schulz B., Andrzejewski M. G. et al. (2007). Sequential processing of the transmembrane chemokines CX3CL1 and CXCL16 by alpha- and gamma-secretases. Biochem Biophys Res Commun 358(1): 233–40.

Shull M. M., Ormsby I., Kier A. B. et al. (1992). Targeted disruption of the mouse transforming growth factor-beta 1 gene results in multifocal inflammatory disease. Nature 359(6397): 693–9.

Silence J., Lupu F., Collen D. et al. (2001). Persistence of atherosclerotic plaque but reduced aneurysm formation in mice with stromelysin-1 (MMP-3) gene inactivation. Arterioscler Thromb Vasc Biol 21(9): 1440–5.

Tam E. M., Morrison C. J., Wu Y. I. et al. (2004). Membrane protease proteomics: isotope-coded affinity tag MS identification of undescribed MT1-matrix metalloproteinase substrates. Proc Natl Acad Sci USA 101(18): 6917–22.

Taylor P. C., Williams R. O. and Feldmann M. (2004). Tumour necrosis factor alpha as a therapeutic target for immune-mediated inflammatory diseases. Curr Opin Biotechnol 15(6): 557–63.

Tester A. M., Cox J. H., Connor A. R. et al. (2007). LPS responsiveness and neutrophil chemotaxis *in vivo* require PMN MMP-8 activity. PLoS ONE 2: e312.

Turk B. (2006). Targeting proteases: successes, failures and future prospects. Nat Rev Drug Discov 5(9): 785–99.

Van den Steen P. E., Proost P., Wuyts A. et al. (2000). Neutrophil gelatinase B potentiates interleukin-8 tenfold by aminoterminal processing, whereas it degrades CTAP-III, PF-4, and GRO-alpha and leaves RANTES and MCP-2 intact. Blood 96(8): 2673–81.

Van den Steen P. E., Husson S. J., Proost P. et al. (2003a). Carboxyterminal cleavage of the chemokines MIG and IP-10 by gelatinase B and neutrophil collagenase. Biochem Biophys Res Commun 310(3): 889–96.

Van Den Steen P. E., Wuyts A., Husson S. J. et al. (2003b). Gelatinase B/MMP-9 and neutrophil collagenase/MMP-8 process the chemokines human GCP-2/CXCL6, ENA-78/CXCL5 and mouse GCP-2/LIX and modulate their physiological activities. Eur J Biochem 270(18): 3739–49.

Van Lint P., Wielockx B., Puimege L. et al. (2005). Resistance of collagenase-2 (matrix metalloproteinase-8)-deficient mice to TNF-induced lethal hepatitis. J Immunol 175(11): 7642–9.

Vergote D., Butler G. S., Ooms M. et al. (2006). Proteolytic processing of SDF-1alpha reveals a change in receptor specificity mediating HIV-associated neurodegeneration. Proc Natl Acad Sci USA 103(50): 19182–7.

Wahl S. M., Wen J. and Moutsopoulos N. (2006). TGF-beta: a mobile purveyor of immune privilege. Immunol Rev 213: 213–27.

Wang M., Qin X., Mudgett J. S. et al. (1999). Matrix metalloproteinase deficiencies affect contact hypersensitivity: stromelysin-1 deficiency prevents the response and gelatinase B deficiency prolongs the response. Proc Natl Acad Sci USA 96(12): 6885–9.

Warner R. L., Beltran L., Younkin E. M. et al. (2001). Role of stromelysin 1 and gelatinase B in experimental acute lung injury. Am J Respir Cell Mol Biol 24(5): 537–44.

Warner R. L., Lukacs N. W., Shapiro S. D. et al. (2004). Role of metalloelastase in a model of allergic lung responses induced by cockroach allergen. Am J Pathol 165(6): 1921–30.

Weaver A., Goncalves da Silva A., Nuttall R. K. et al. (2005). An elevated matrix metalloproteinase (MMP) in an animal model of multiple sclerosis is protective by affecting Th1/Th2 polarization. Faseb J 19(12): 1668–70.

Wells J. E., Rice T. K., Nuttall R. K. et al. (2003). An adverse role for matrix metalloproteinase 12 after spinal cord injury in mice. J Neurosci 23(31): 10107–15.

Wilson C. L., Ouellette A. J., Satchell D. P. et al. (1999). Regulation of intestinal alpha-defensin activation by the metalloproteinase matrilysin in innate host defense. Science 286(5437): 113–7.

Wright R. W., Allen T., El-Zawawy H. B. et al. (2006). Medial collateral ligament healing in macrophage metalloelastase (MMP-12)-deficient mice. J Orthop Res 24(11): 2106–13.

Wucherpfennig K. W. (2001). Mechanisms for the induction of autoimmunity by infectious agents. J Clin Invest 108(8): 1097–104.

Yu Q. and Stamenkovic I. (2000). Cell surface-localized matrix metalloproteinase-9 proteolytically activates TGF-beta and promotes tumor invasion and angiogenesis. Genes Dev 14(2): 163–76.

Zhang K., McQuibban G. A., Silva C. et al. (2003). HIV-induced metalloproteinase processing of the chemokine stromal cell derived factor-1 causes neurodegeneration. Nat Neurosci 6(10): 1064–71.

Zhou H., Bernhard E. J., Fox F. E. et al. (1993). Induction of metalloproteinase activity in human T-lymphocytes. Biochim Biophys Acta 1177(2): 174–8.

Zucker S., Cao J. and Chen W. T. (2000). Critical appraisal of the use of matrix metalloproteinase inhibitors in cancer treatment. Oncogene 19(56): 6642–50.

Chapter 27
Matrix Metalloproteinases as Key Regulators of Tumor–Bone Interaction

Conor C. Lynch and Lynn M. Matrisian

Abstract Bone metastasis is a common occurrence for several tumor types. In the tumor–bone microenvironment, tumor cells manipulate the normal cells of the bone to promote bone resorption and the release of sequestered growth factors from the bone matrix. In turn, these factors support tumor growth, thus completing what has been described as the vicious cycle. Despite medical advances, the treatment options for patients with metastatic bone disease are limited and are often palliative rather than curative. Clearly, new therapies that will prevent the vicious cycle are required. To achieve this, a better understanding of how tumor cells communicate with the normal host cells of the bone is necessary. Given the evidence accumulated over the last decade, it has become clear that matrix metalloproteinases and other proteinase members of the metzincin family are important players in the execution of the vicious cycle. This chapter will focus on how matrix metalloproteinases can control cell–cell communication at the tumor–bone interface via the processing of matrix and nonmatrix substrates.

Introduction: The "Vicious Cycle" of Tumor Progression in the Bone

Upon arrival in the bone microenvironment, tumor cells hijack the normal bone remodeling process and perturb the balance between bone-synthesizing osteoblasts and bone-resorbing osteoclasts to yield areas of extensive bone formation and/or degradation. Our understanding of the molecules that control cell–cell communication at the tumor–bone interface can be summarized as follows. Upon arrival in the bone environment, metastatic tumor cells secrete factors, such as interleukins-1,

C.C. Lynch
Departments of Orthopaedics and Rehabilitation and Cancer Biology, Vanderbilt University Nashville, TN, USA, e-mail: conor.lynch@vanderbilt.edu

D. Edwards et al. (eds.), *The Cancer Degradome.*

-6, -8, and -11 and parathyroid hormone-related peptide (PTHrP), that stimulate the osteoblasts lining the endo-osteal surfaces to produce macrophage colony-stimulating factor (M-CSF) and receptor activator of nuclear kappa B ligand (RANKL) (for review, see Mundy 2002, Bendre et al. 2003). These factors, among others, are essential for the recruitment and differentiation of osteoclast precursor cells. Once mature, osteoclasts form a resorbtive seal on the bone surface and by lowering the pH and secreting cysteine proteinases and matrix metalloproteinases (MMPs), the osteoclasts can resorb the bone matrix (Blair et al. 1989, Delaisse et al. 2000). Degradation of the bone matrix subsequently leads to the generation of collagen products and the activation of latent growth factors sequestered in the bone matrix such as transforming growth factorβ (TGFβ) (Guise and Chirgwin 2003). The release of these factors stimulates tumor growth, thus completing and perpetuating the vicious cycle (Fig. 27.1).

Based on observations in bone and other tissues (Lynch and Matrisian 2002), metalloproteinases can control the communication between the cell types at the tumor–bone interface via (1) classical processing of organic bone matrix components and (2) modifying the activity profile of growth factors and cytokines such as PTHrP, RANKL, and TGFβ.

The Impact of Direct MMP Bone Matrix Degradation on Tumor: Bone Cell Behavior

Over 90% of the organic bone matrix is composed of type-I collagen, making MMPs with collagenolytic activity potentially important mediators of the removal of demineralized bone matrix. Much of our understanding of how MMPs function in the bone has been derived from the study of skeletogenesis in MMP-deficient mice. Surprisingly, many MMPs have transient or no reported effects on skeletogenesis and depending on the bone type (intramembranous vs. endochondral) the MMPs can have increased or decreased roles in bone matrix turnover (Delaisse et al. 2003). Furthermore, in MMP-deficient mice that have transient effects on skeletogenesis such as MMP-9, the effect on bone formation is not due to a lack of bone matrix resorption but is due to an inability of the osteoclasts to migrate into areas requiring bone resorption (Engsig et al. 2000). While developmental studies can provide clues as to the individual contribution of MMPs in bone matrix remodeling, it is important to note that the repertoire of MMPs involved in pathological versus developmental scenarios can be very different and therefore, MMPs that do not function in skele-togenesis should not be discounted as playing a role in tumor-mediated bone matrix degradation. While MMP expression can be induced in the host cells in response to the tumor and vice versa, the impact of MMPs on direct bone degradation and the contribution of the MMP-degraded products on cell behavior will be examined in the context of the major cellular sources of MMPs in the tumor-bone microenvironment, namely, the osteoclasts, osteoblasts, and tumor cells.

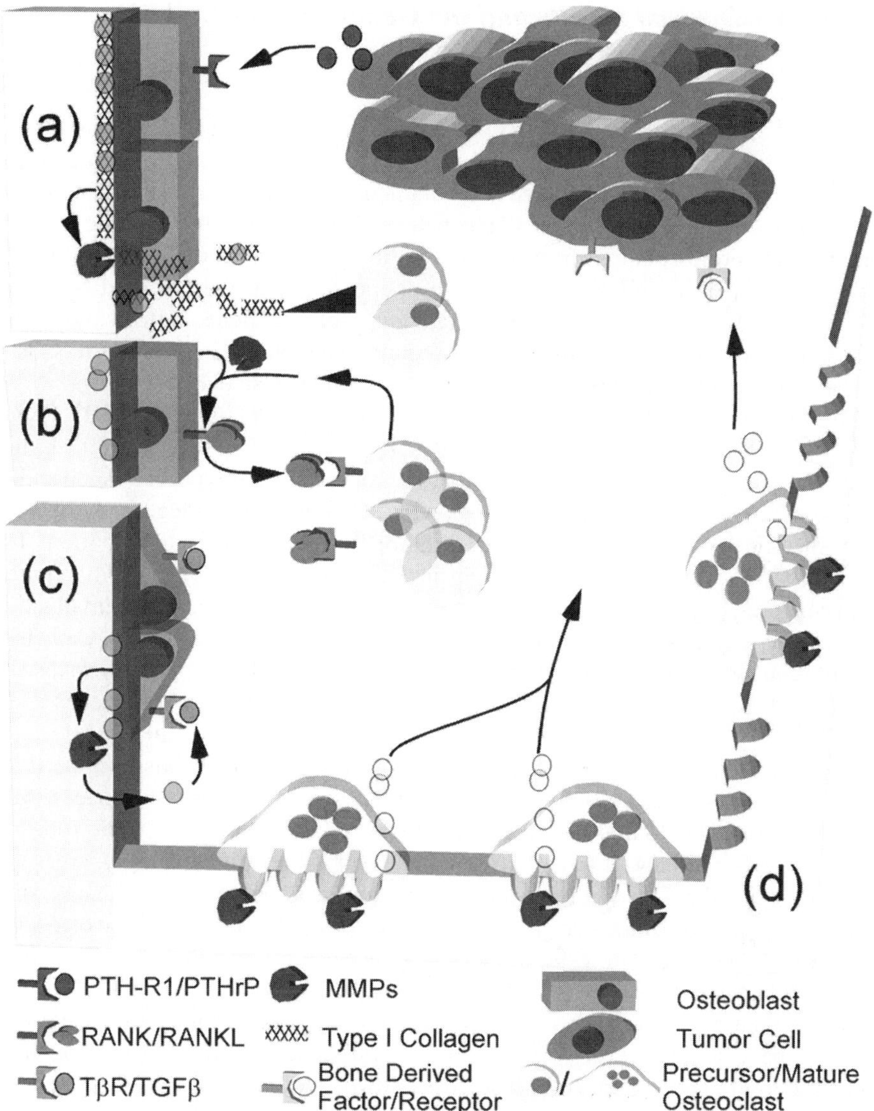

Fig. 27.1 MMP control of cell–cell communication of the tumor–bone interface. (**a**) Metastatic tumor cells induce the expression of MMPs and RANKL in bone lining osteoblasts. The degradation of the organic matrix by osteoblast-derived MMPs generates type-1 collagen fragments (ICTP) that can recruit osteoclast precursor cells to the tumor-bone microenvironment. (**b**) The solubilization of RANKL promotes osteoclast maturation. (**c**) MMP-mediated release of TGFβ from the bone matrix results in osteoblast retraction from the mineralized bone matrix. (**d**) Mature osteoclasts resorb the mineralized matrix via acidification. The expression of cathepsin K and MMPs results in the breakdown of the bone matrix and the release of bone-sequestered growth factors such as TGFβ and IGF-1, which, in turn, can promote tumor growth and the continuation of the vicious cycle

Osteoclast-Derived MMPs and the Generation of Cell Signals from the Bone Matrix

In order to degrade the bone matrix, osteoclast precursors must, under the correct cues from the microenvironment, fuse together to form giant multinucleated cells. Integrins, predominantly $\alpha v \beta 3$, and actin filament fibers allow the osteoclast to form a tight seal with mineralized bone (Teti et al. 1992, Ross et al. 1993). Using chloride channels and proton pumps, the osteoclast acidifies the subosteoclastic zone, thereby lowering the pH and allowing for the demineralization of the calcified bone (for review, see Blair 1998). The demineralization of the bone matrix allows for subsequent enzymatic degradation of the organic bone matrix in the subosteoclastic zone by collagenases that have low pH activity profiles. Osteoclasts express several MMPs including MMP-1, -2, -3, -7, -9, -12, -13, and -14 (Delaisse et al. 1993, Okada et al. 1995, Bord et al. 1996, Sato et al. 1997, Dew et al. 2000, Hou et al. 2004, Lynch et al. 2005), many of which have the ability to process fibrillar type-I collagen and its derivatives. However, based on findings with MMP-deficient mice, many of these osteoclast-derived MMPs are not rate limiting for bone matrix degradation. The phenotype of mice deficient in cathepsin K, which can process fibrillar type-I collagen, suggests that this cysteine protease is the principle protease involved in the turnover of endochondral bone (Saftig et al. 1998, Gowen et al. 1999). Cathepsin K mutations have also been demonstrated to be responsible for pycnodysostosis in humans where carriers display dramatically shortened and petrotic long bones (Gelb et al. 1996). Furthermore, the optimal pH for cathepsin K activity is pH 4 while many of the MMPs have optimal activities at a more neutral pH (Delaisse et al. 1991, Nagase and Woessner 1999). Despite the acidity of the resorption lacunae, MMPs have been localized to this area. MMP-1 is localized to the subosteoclastic zone and has been shown to have activity between pH 6 and 7.4 and, interestingly, areas that had undergone demineralization were still immunoreactive for MMP-1, suggesting that the degradation of the bone matrix continued even after the exit of the osteoclast (Vaes 1972, Delaisse et al. 1993). MMP-9 has also been shown to be localized to the subosteoclastic zone and is up to 85% active at pH 2.3, implying that MMP-9 may retain its gelatinolytic activity in this area (Okada et al. 1995). It may also be possible that as the osteoclast transports the bone degradation products from the bone interface to the apical surface in a process known as transcytosis, MMPs localized to more neutral pH transcytotic vesicles could function in the further degradation of the bone matrix components (Salo et al. 1997).

In serum, the processed collagen products yield clues as to whether they have been cleaved by cathepsin K or by MMPs. Collagenolysis by cathepsin K yields C-terminal (CTX) and N-terminal (NTX) cross-linked telopeptides of type-I collagen while MMP-mediated collagenolysis yields a larger cross-linked C-terminal telopeptide of type-I collagen (ICTP) (Garnero et al. 2003). Interestingly, while CTX and NTX fragments are good serum biomarkers for osteoporosis, serum levels of MMP-generated ICTP correlate with the extent of bone destruction associated with bone metastasis (Blomqvist et al. 1996, Berruti et al. 1999). Furthermore, the generation of ICTP fragments in animal models of bone metastasis can be prevented by treatment with a broad-spectrum metalloproteinase inhibitor (Winding et al. 2002, Garnero et al. 2003). These data suggest that in the pathological scenario of tumor-induced bone degradation, MMPs rather than cathepsin K mediate the degradation of the demineralized bone matrix. The generation of collagen fragments by MMPs provides signaling cues for surrounding cell types. MMP-generated ICTP collagen fragments have been demonstrated as being chemotactic for osteoclast precursor cells (Malone et al. 1982), implying that excessive MMP-mediated degradation of the bone matrix by osteoclasts promotes further osteoclast recruitment to the tumor-bone microenvironment, thus ensuring the continuation of the vicious cycle.

The processing of fibrillar collagen can also lead to the exposure of cell adhesion sites within the collagen structure such as arginine-glycine-aspartic acid (RGD) sequences. These sequences allow for the attachment and migration of $\alpha v\beta 3$ integrin-expressing cells such as osteoclasts and may also allow for the migration of tumor cells to areas of bone remodeling (Davis 1992, Ross et al. 1993). Similarly, the processing of osteopontin (OPN) by MMP-3 and -7 can yield multiple OPN peptides that contain integrin-binding sites. In other model systems, OPN, and presumably MMP-generated OPN products, can affect tumor progression and the host response to the tumor (Crawford et al. 1998, Agnihotri et al. 2001). Although unexplored, it is plausible that the interaction between osteoclast-derived MMP-3 and MMP-7 and OPN could modify tumor-host behavior in the bone microenvironment in a similar manner. In contrast to the attachment and migratory functions of MMP processed type-I collagen and OPN, osteonectin (ON), which has a high affinity for binding to type-I collagen, is thought to have antiadhesive properties (Murphy-Ullrich 2001) and enhances the motility of tumor cell lines such as MDA-MB-231 (Campo McKnight et al. 2006). However, ON can bind to several extracellular growth factors such as platelet-derived growth factor (PDGF), vascular endothelial growth factor (VEGF), and fibroblast growth factor-2 (FGF-2) (for review, see Brekken and Sage 2001). ON is a substrate for MMP-3, -7, and -9 (Sasaki et al. 1997), and the secretion of these MMPs by bone-degrading osteoclasts and subsequent cleavage of ON may result in the release of growth factors that control osteoclastogenesis, angiogenesis, and tumor growth in the bone.

Osteoblast-Derived MMPs: Counterintuitive Roles for MMPs in Bone-Synthesizing Cells

The main function of osteoblasts is bone matrix synthesis. Therefore, the expression of several MMPs such as MMP-1, -2, -3, P-8, -9, -13, and -14 (Bord et al. 1996, 1998, Kusano et al. 1998, Breckon et al. 1999, Holmbeck et al. 1999, Dew et al. 2000, Parikka et al. 2005) by osteoblasts and mesenchymal stem cells is surprising given the traditional role of the MMPs in the degradation of the ECM. However, in order to begin the process of bone remodeling, minimal degradation of the organic bone matrix overlaying the calcified bone by osteoblasts and the generation of signaling cues such as collagen fragments and TGFβ are required for osteoblast retraction from the mineralized matrix and the recruitment of osteoclasts to the mineralized surface (Bonfil et al. 2004, Perez-Amodio et al. 2004). Other functions for osteoblast-derived MMPs in controlling cell function are coming to attention based on the study of MMP-deficient animals. For example, MMP-2-deficient mice have impaired skeletogenesis leading to significantly less bone formation, a phenotype that is mirrored in Al-Aqeel Sewairi syndrome, a human form of osteolysis in which carriers are deficient in MMP-2 (Mosig et al. 2007). During the early stages of skeletal development, the MMP-2-deficient mice exhibit transient but significantly lower numbers of osteoblast and osteoclast populations in comparison to their wild-type litter mate controls (Mosig et al. 2007). These findings can be explained by the necessity of MMP-2 for proper formation of the osteocytic canicular network. Osteocytes are terminally differentiated osteoblasts that reside within the bone matrix and act as a sentinel for the integrity of the bone. This security guard function is achieved via the fine network of canules that the osteocyte weaves into the newly synthesizing bone. In the absence of MMP-2, there is a failure of osteocyte canule formation, presumably due to a lack of MMP-2-mediated bone matrix degradation, which results in improper osteocyte function and apoptosis, hence a failure to communicate with and coordinate subsequent osteoblast-osteoclast responses (Inoue et al. 2006). While the absence of MMP-2 may result in a failure to generate proper canule formation, the possibility that MMP-2 is important for the terminal differentiation of the osteocyte cannot be ruled out since the introduction of MMP-2 siRNA into osteoblast cell lines such as MC3T3 decreased proliferation while enhancing differentiation and bone formation (Mosig et al. 2007). MMP-14 is an important mediator of MMP-2 activation (Sato et al. 1994), and the role of MMP-2 for the proper function of osteocytes is also reflected in MMP-14-deficient animals (Holmbeck et al. 2005). Therefore, in the context of the tumor–bone microenvironment, the expression of MMP-2 may contribute to tumor-induced bone formation by enhancing the differentiation of osteoblasts.

Recent reports have demonstrated that MMP-13 activity is rate limiting for organic bone matrix degradation since MMP-13-deficient animals have abnormalities in growth plate development and exhibit thickened trabecular bone compared to wild-type controls (Inada et al. 2004, Stickens et al. 2004). The increased trabecular bone volume in the MMP-13-deficient animals could potentially be explained by the lack of MMP-13-mediated type-I collagen degradation. Surprisingly, MMP-13 is not expressed by osteoclasts but by osteoblasts, osteocytes, and mononuclear periosteoclast cells (Delaisse et al. 2003, Nakashima and Tamura 2006). While the lack of organic bone matrix degradation by these cell types in the MMP-13-deficient animals cannot be ruled out, new findings may shed light on how MMP-13 contributes to bone formation rather than bone destruction. Krane and colleagues suggest that MMP-13 prevents wingless and integrated (Wnt) signaling by directly binding to the Wnt coreceptor, low-density lipoprotein related receptor related-5/6 (LRP-5/6) (S. Krane personal communication). In osteoblasts, Wnt signals through its coreceptors LRP-5/6 and Frizzled (FRZ), leading to the activation of the canonical Wnt pathway and the transcription of bone-related target genes such as runx2 (for review, see Baron et al. 2006). Unchecked Wnt signaling can lead to osteopetrotic phenotypes. By binding to the LRP-5/6 receptor, MMP-13 prevents Wnt signaling, thereby blocking bone formation by osteoblasts. Therefore, in MMP-13 null mice, the Wnt signaling pathway is unchecked and explains why the mice can have a significantly higher trabecular bone mass and increased bone formation in comparison to wild-type controls. Additional data are required, however, to establish the significance of these in vitro observations. This study is one of the first to show that an MMP, independent of its proteolytic activity, can modulate cell behavior by binding to a cell surface receptor. Again in the context of the tumor–bone microenvironment, enhanced expression of MMP-13 by either the host or tumor cells may prevent proper osteoblast function and tip the balance of bone matrix homeostasis in the favor of bone destruction, thus yielding a more lytic lesion.

Tumor-Derived MMPs in the Metastatic Bone Microenvironment

While tumor-derived MMPs are implicated in the process of bone metastasis by virtue of their ability to collectively degrade extracellular matrix (ECM) barriers, they can also make significant contributions to the vicious cycle of tumor progression once established in the bone. MDA-MB-231 cells, which do not express endogenous MMP-2, form osteolytic lesions upon intracardiac inoculation. The introduction of MMP-2 into the MDA-MB-231 cells and the subsequent intracardiac delivery of the cell lines generated significantly higher numbers of osteolytic lesions compared to the parental cell line (Tester et al. 2004). Conversely, the overexpression of TIMP-2, a natural inhibitor of MMP activation, in the MDA-MB-231 cells prevented tumor growth and tumor-induced osteolysis (Yoneda et al. 1997). Intratibial injection of human prostate cancer cell lines (LnCAP and DU-145) in which MMP-14 was overexpressed or silenced via siRNA treatment demonstrated that MMP-14 significantly contributed to tumor growth and tumor-induced osteolysis (Bonfil et al. 2007). This study also demonstrated that the tumor cells could mediate the degradation of the bone matrix via the collagenolytic MMP-14, although, presumably, this would occur subsequent to osteoclast of the bone matrix. Therefore, tumor-derived MMPs, through direct of the bone matrix, can generate signaling cues such as the ICTP fragment of collagen that promote the recruitment of osteoclast precursors which are for the execution of the vicious cycle.

Other Cellular Sources of MMPs in the Bone–Tumor Microenvironment

MMPs derived from other cell types in the tumor–bone microenvironment may also have an impact on cell-cell communication by directly or indirectly mediating bone matrix degradation. For example, macrophages, which are a rich source of MMPs and are related to osteoclasts, can degrade demineralized bone (Athanasou and Sabokbar 1999). T-cells that express several MMP family members can be recruited to the tumor–bone interface by factors such as PTHrP. The antitumor effect of the T-cells may be inhibited by factors liberated from the bone such as TGFβ, thus making the tumor–bone microenvironment an immune privileged site and permissive to tumor progression (Fournier et al. 2006). As shall be discussed, MMPs can play a pivotal role in the bioavailability of these factors. Antitumor immunoglobulins (IgGs) expressed by B-cells in the tumor microenvironment can be degraded by MMPs and as a consequence may also modulate tumor–bone interactions (Gearing et al. 2002, Li et al. 2007). The use of immunocompromised animals restricts our understanding as to how some of these cell populations contribute to or detract from tumor progression in the bone and reenforces the necessity for syngeneic immunocompetent animal models of primary tumor to bone metastasis that more accurately reflect the human clinical scenario.

Collectively, these studies show that MMPs derived from the major cells involved in the vicious cycle can mediate bone-matrix degradation resulting in the generation of multiple signaling cues that can have profound effects on cell behavior.

MMP Solubilization of Nonmatrix Substrates

MMPs and other metalloproteinases are capable of processing multiple nonmatrix factors such as PTHrP, RANKL, and TGFβ that are critical for the successful completion and perpetuation of the vicious cycle (Table 27.1). In addition, recent studies have also implicated the MMPs in the "homing" of metastatic tumor cells to the bone via the processing of factors such as CXCL12.

MMP Inactivation of CXCL12 Promotes Formation of the Premetastatic Bone Niche

The forces that dictate the predilection of certain cancers for bone metastasis are an active area of research. The expression of chemokine receptor 4 (CXCR4) by tumor cells has been shown to play an integral part in the selective metastasis of breast cancer cells to the bone (Muller et al. 2001, Kang et al. 2003). The cognate ligand for the CXCR4 receptor is CXCL12, which is critical for maintaining hematopoietic precursor cells (HPCs) in the bone stem cell niche (Driessen et al. 2003). CXCL12 has been identified as a substrate for MMP-1, -2, -3, -9, and -14 (McQuibban et al. 2001). Processing by these MMPs results in an inactivated CXCL12 ligand that prompts the expansion and mobilization of the HPC compartment, a phenomenon that is also mediated by MMP-9 shedding of Kit ligand (Bergers et al. 2000, McQuibban et al. 2001, Heissig et al. 2002, Kaplan et al. 2007).

Table 27.1 MMP Nonmatrix substrates and the implications for the tumor–bone microenvironment

Substrate		Substrate function	MMP	Resultant product function	Potential impact of MMP product in the bone-tumor microenvironment
Receptors	FGFR1	Receptor for FGF	MMP-2 (Levi et al. 1996)	A soluble FGFR1 that inhibits FGF signaling by binding soluble FGF	MMP inactivation of FGFR1 may prevent the effects of FGFs on osteoblast differentiation and bone formation (Tang et al. 1996)
	LRP	Endocytosis/Wnt signaling	MMP-14 MMP-15 MMP-16 MMP-17 (Rozanov et al. 2004)	Inactivation of LRP attenuates endocytosis and Wnt signaling	Inhibition of osteoblast differentiation and bone formation (Baron et al. 2006)
	IL-2Rα	Interluekin receptor	MMP-2 MMP-9 (Sheu et al. 2001)	Inactivation and prevention of T-cell proliferation	Possible generation of immune privileged site at the tumor-bone interface (Fournier et al. 2006)
	FAS	Fas ligand receptor (Strand et al. 2004)	MMP-7 (Strand et al. 2004)	Inactivation of the receptor	Inactivation of FAS prevents FasL-mediated osteoblast differentiation and bone formation (Kovacic et al. 2007)
	PAR-1	G-protein-coupled receptor activated by proteases	MMP-1 (Pei et al. 2005)	Activation of the receptor and enhanced cell migration and invasion	Activation of PAR-1 on tumor cells in the bone promotes migration (Pei 2005)
	uPAR	Regulates uPA activity and cell signaling	MMP-2 MMP-3 MMP-8 MMP-9 MMP-12 MMP-13 MMP-14 (Andolfo et al. 2002)	Inactivation and a decrease in cell motility	Decreased uPAR expression inhibits prostate cancer metastasis to the bone by up to 80% (Margheri et al. 2005)

(continued)

Table 27.1 (continued)

Substrate		Substrate function	MMP	Resultant product function	Potential impact of MMP product in the bone-tumor microenvironment
Cytokines/growth factors	HB-EGF	Pleiotropic effects on growth differentiation and apoptosis via ErbB-4 and ErbB-1	MMP-3 (Suzuki et al. 1997) MMP-7 (Yu et al. 2002)	Active soluble form of HB-EGF	HPC expansion and differentiation (Krampera et al. 2005) and myeloma progression (Mahtouk et al. 2006)
	IGF-BPs	Sequesters IGFs in a latent form	MMP-1 (Rajah et al. 1996) MMP-2 (Fowlkes et al. 1994, Thrailkill et al. 1995) MMP-3 (Fowlkes et al. 1994) MMP-11 (Manes et al. 1997)	Degradation of IGF-BPs by MMPs prevents sequestration of active IGFs	Proliferation via activation of the IGF receptors (Giles and Singh 2003)
	FasL	Induces apoptosis by activating FAS	MMP-3 (Vargo-Gogola et al. 2002) MMP-7 (Powell et al. 1999)	Soluble active FasL	Selection of apoptosis-resistant tumor cells. Apoptosis of invading immune cells (Fingleton et al. 2001; Mitsiades et al. 2001)
	TNF-α	Pleiotropic effects on immune cells and tumor cells	MMP-1 MMP-2 MMP-3) MMP-7 MMP-9 MMP-12 (Chandler et al. 1996) MMP-14 MMP-15 MMP-17 (D'Ortho et al. 1997)	Soluble active TNF-α	Osteoclast activation (Hotokezaka et al. 2007). Decrease in tumor cell proliferation (Ritchie et al. 1997)
	RANKL	Essential for osteoclast maturation	MMP-1 MMP-3 MMP-7 MMP-14	Soluble active RANKL	Osteoclastogenesis (Lum et al. 1999, 2005)

Cell adhesion molecules	IL-1β	Immune cell activity and infiltration to sites of infection and tumor growth	MMP-1 (Ito et al. 1996) MMP-2 MMP-3 MMP-9 (Schonbeck et al. 1998)	Activation of the proform but also can inactivate thethe mature cytokine	Control of osteoclast maturation (Kurihara et al. 1990). Modulation of tumor immune evasion and invasion (Song et al. 2003)
	E-Cadherin	Epithelial cell–cell contact	MMP-3 (Lochter et al. 1997) MMP-7 (Noe et al. 2001) MMP-14 (Covington et al. 2006)	Degradation, leading to disassembly of the epithelial adherens junctions	Can promote the migration of the epithelial cells in the bone microenvironment
	VE-Cadherin	Vascular endothelial cell–cell contact	MMP-7 (Ichikawa et al. 2006)	Degradation leading to the loss of endothelial cell–cell junctions and the proliferation of endothelial cells	May potentiate efficient angiogenesis in the tumor microenvironment
	Integrins	Cell contact with the basement membrane	MMP-14 (Deryugina et al. 2000)	Maturation of the integrin and increased cell adhesion migration	Osteoclast migration and resorbtion (Faccio et al. 1998)
	CD-44	Cell anchorage to the basement membrane	MMP-14 (Kajita et al. 2001) MMP-15 MMP-16 MMP-24 (Suenaga et al. 2005)	Degradation of CD-44 results in enhanced cell motility	Potential to enhance tumor and bone cell migration
	ICAM-1	Leukocyte infiltration and adhesion	MMP-9 (Fiore et al. 2002)	Inactivation of ICAM-1 can promote tumor immunoevasion	Generation of immune privileged site at the tumor–bone interface

(continued)

Table 27.1 (continued)

Substrate		Substrate function	MMP	Resultant product function	Potential impact of MMP product in the bone-tumor microenvironment
Cell surface glycoproteins	EMMPRIN	Regulation of MMP activity and promotes cell migration?	MMP-1 MMP-2 (Haug et al. 2004) MMP-14 (Egawa et al. 2006) MMP-15 (Egawa et al. 2006)	Soluble EMMPRIN can enhance MMP expression.	EMMPRIN is located on the surface of 90% of breast to bone micrometastases and may have potential for the induction of MMP expression in the bone cells (Reimers et al. 2004)
	Mucin-1	Protection of epithelial cell surfaces	MMP-14 (Thathiah and Carson 2004)	Solublizatino of Mucin-1 Influence on tumor progression and invasion	Expression of MUC-1 by tumor cells in the bone and subsequent processing by MMPs may allow for immunoevasion (Cloosen et al. 2006)
	Sydeca n-1	Cell adhesion to multiple ECM components	MMP-7 (Li et al. 2002) MMP-14 MMP-16 (Endo et al. 2003)	Soluble syndecans can enhance tumor cell migration.	Expressed in tumor cells in the bone microenvironment and processing by MMPs may promote migration in this context (Chen et al. 2004)

MMPs can be upregulated in the premetastatic organ in response to primary tumors (Hiratsuka et al. 2002). Therefore, primary tumor cells with a predilection for bone metastasis can potentially enhance the expression of MMPs and the turnover of CXCL12 in the premetastatic bone leading to the mobilization of the HPCs from the stem cell niche. This modulation of CXCL12 by MMPs can have a twofold effect, (1) HPCs can be recruited to the primary tumor and assist in tumor vascularization and (2) enhanced CXCL12 levels may promote the recruitment of CXCR4-positive tumor cells to the bone stem cell niche, an area of bone that is actively remodeling and is rich in mitogenic signals, thus providing the metastatic tumor cell with the perfect environment in which to establish a secondary tumor.

Protease Modification of PTHrP in the Tumor–Bone Microenvironment

To survive and progress in the bone microenvironment, the metastatic tumor cells must efficiently communicate with the host cells of the bone. A major factor that facilitates this process is tumor-derived PTHrP, which can inhibit the differentiation of parathyroid hormone-1 receptor (PTH-1R)-positive early osteoblasts while promoting their proliferation via enhanced levels of cyclin D1 (Cornish et al. 1999, Du et al. 2000, Miao et al. 2001, Datta et al. 2007). PTHrP also stimulates osteoblast expression of RANKL, which is essential for the maturation and activation of osteoclasts (Thomas et al. 1999). It is also noteworthy that PTHrP and its receptor PTH-1R are coexpressed in a number of primary cancers such as breast, prostate, lung, and renal cancer (for review, see Liao and McCauley 2006). PTHrP can also affect cell proliferation by increasing levels of cyclin D1 in the tumor and, therefore, the role of PTHrP in autocrine and paracrine growth of the tumor cannot be ignored (Datta et al. 2007).

The PTHrP gene gives rise to three pre-pro form splice variants, which are 139, 141, and 173 amino acids in length (Southby et al. 1995), and each isoform requires proteolytic processing to generate the mature active peptide that contains amino acids 1–36 (PTHrP1-36). Cleavage by furin, a serine protease localized in the Golgi apparatus, is the main mechanism through which the mature form of PTHrP is generated (Liu et al. 1995). While PTHrP can induce the expression of a number of MMPs in the tumor–bone microenvironment (Kawashima-Ohya et al. 1998, Luparello et al. 2003), to date no MMP has been shown to process PTHrP. However, in the extracellular environment, neprilysin (NEP), a membrane-bound metallopro-teinase that is expressed by tumor and bone cells, can attenuate the activity of the mature PTHrP1-36 and the larger C-terminal fragments of PTHrP (Ruchon et al. 2000, Dall'Era et al. 2007, Smollich et al. 2007). NEP has been identified as processing PTHrP1-36 to yield PTHrP1-23 and PTHrP1-26 and is also capable of inactivating the C-terminal PTHrP fragment known as osteostatin (PTHrP107-139) to yield PTHrP134-139 and PTHrP107-133 (Ruchon et al. 2000). In addition, prostate serum antigen (PSA), which is a serine protease and a classical marker of prostate cancer, has also been shown to process PTHrP1-36 to generate PTHrP1-23

(Cramer et al. 1996). These PTHrP fragments can have distinct effects on cell behavior in the bone–tumor microenvironment. PTHrP1-36 binding to PTH-1R can activate both the protein kinase A (PKA) and protein kinase C (PKC) pathways which are important mediators of osteoblast proliferation (Abou-Samra et al. 1992). The PTHrP1-26 fragment does not stimulate the PKA pathway since the first two N-terminal amino acids and amino acids 14–34 are required for receptor binding (Rabbani et al. 1988, Caulfield et al. 1990). However, the PTHrP27-36 fragment can stimulate the PKC pathway (Gagnon et al. 1993). Whether PTHrP27-36 is sufficient for osteoblast proliferation or mediates a different effect altogether remains to be determined. PTHrP107-139 has been shown to be a potent inhibitor of osteoclast formation (Fenton et al. 1994, Kaji et al. 1995) and, therefore, inactivation by NEP or PSA can be permissive for osteoclastogenesis.

MMP Solubilization of RANKL: Roles in Osteoclastogenesis and Tumor Homing to Bone

RANKL, a member of the tumor necrosis factor (TNF) family, is an essential mediator of osteoclast maturation since RANKL-deficient mice have severe osteopetrosis due to a lack of bone resorbing osteoclasts (Kong et al. 1999). The effect of RANKL on osteoclast maturation is attenuated by a soluble decoy receptor known as osteoprotegerin (OPG) that prevents the interaction between RANKL and the receptor RANK on the osteoclast precursor cell surface (Simonet et al. 1997). The direct interaction between RANKL-expressing osteoblasts and RANK-expressing osteoclast precursor cells is thought to be required for osteoclast maturation, but new evidence regarding the solubilization of membrane-bound RANKL by metalloproteinases suggests that osteoclast maturation may occur without the necessity of direct cell–cell contact.

RANKL exists as a trimeric molecule on the cell surface supported by a juxtamembrane stalk region. Recent studies have documented that RANKL is sensitive to cleavage at variable sites within the juxtamembrane region by metalloproteinase family members, a disintegrin and metalloproteinase-17 (ADAM-17), ADAM-19, MMP-1, -3, -7, and -14 (Lum et al. 1999, Schlondorff et al. 2001, Chesneau et al. 2003, Lynch et al. 2005). While the potency of the soluble form of RANKL (sRANKL) may vary depending on the cleavage site, results thus far have suggested that sRANKL is equally as potent as the membrane-bound form in the maturation of osteoclasts or in the survival of dendritic cells (Lum et al. 1999, Schlondorff et al. 2001, Chesneau et al. 2003, Lynch et al. 2005). The metalloproteinases capable of RANKL processing, with the exception of MMP-7, are expressed by osteoblasts. Osteoblast expression of these MMPs in response to the metastatic tumor cells in the bone by factors such as interleukin-1α (IL-1α) (Fujisaki et al. 2006) and PTHrP may not only serve to induce RANKL but also the MMPs that are capable of generating sRANKL. Therefore, sRANKL may circumvent the necessity of direct osteoblast-osteoclast-precursor interaction in the tumor–bone microenvironment, thus accelerating osteoclastogenesis.

MMP-7 is expressed by monocytic osteoclast precursors (Busiek et al. 1992, 1995) and the expression of MMP-7 appears to be constitutive during the maturation of osteoclasts. Although an extensive study in the role of MMP-7 in skeletal development and maturation has not been performed, adult MMP-7-null mice have a normal bone phenotype compared to wild-type animals (Wilson et al. 1997 and unpublished observations). However, under pathological conditions such as tumor progression in the bone, MMP-7 plays a role in the rate of bone degradation. In a rodent model of prostate tumor progression in the bone in which prostate adenocarcinoma tissue was transplanted to the calvaria of wild-type and MMP-7-null mice, a significant reduction in the amount of bone degradation in the MMP-7-deficient animals was observed (Lynch et al. 2005). This decrease was concomitant with a decrease in osteoclast numbers and in the amount of sRANKL present at the tumor–bone interface in the MMP-7-deficient mice and suggested that MMP-7-generated sRANKL was important for osteoclast maturation. An additional contribution of MMP-7 to direct bone degradation by the osteoclast cannot be ruled out.

The contribution of tumor-derived RANKL or sRANKL to osteoclastogenesis in the bone-tumor microenvironment has not been explored but is a distinct possibility. While RANKL expression has been identified in 90% of primary breast cancer cells, metastatic breast to bone cancer cells rarely express the ligand but commonly express the RANK receptor, thus potentially facilitating the direct interaction with RANKL-expressing osteoblasts (Bhatia et al. 2005). Conversely, the majority of metastatic prostate cancer cells in the bone express RANKL, which may contribute to tumor-mediated osteoclastogenesis. Keller and colleagues demonstrated that the human prostate cancer cell line C4-2B, which causes mixed lesions in the bone, expressed and generated sRANKL (Zhang et al. 2001). Histology clearly demonstrated that the prostate tumor cells could promote osteoclastogenesis in the absence of osteoblasts, thus supporting the hypothesis that metastatic RANKL-expressing prostate tumor cells in the bone may be able to act as surrogate osteoblasts and promote osteoclastogenesis (Zhang et al. 2001).

sRANKL generation by metalloproteinases enhances tumor-induced bone destruction but functions for sRANKL outside of the tumor–bone microenvironment may also exist. Several studies have demonstrated that the levels of sRANKL in the serum can be used as an indicator of tumor-induced osteolysis (Terpos et al. 2003, Palma and Body 2005, Chen et al. 2006). However, systemic sRANKL may also play a part in the homing of RANK-expressing tumor cells to the bone. Primary breast, prostate, and skin tissues express the RANK receptor (Fata et al. 2000, Brown et al. 2001, Jones et al. 2006). As previously discussed, the primary tumor can often enhance the expression of MMPs in the premetastatic site. Therefore, enhanced MMP expression in the bone environment may lead to increased levels of sRANKL in the serum, which, in turn, could affect the homing of RANK-positive metastatic cells to the bone. To this end, recent evidence has detailed that the metastasis of RANK-positive melanoma cells to bone can be completely blocked via treatment with an anti-RANKL antibody, but the metastasis of the melanoma cells to other tissues remained unaltered (Jones et al. 2006). These data support the hypothesis that metalloproteinase-generated sRANKL not only assists in the continuation of the vicious cycle by mediating osteoclastogenesis without the requirement of cell–cell contact but also that sRANKL derived from the bone can act as a homing signal for metastatic tumor cells (Jones et al. 2006).

MMP Activation of TGFβ Facilitates Tumor–Bone Interaction

Bone is rich in many growth factors such as TGFβ, insulin-like growth factors (IGFs), and bone morphogenetic proteins (BMPs). These factors are typically integrated into the bone matrix during synthesis by the osteoblasts. In order to progress in the bone, the tumor cells must liberate and gain access to these factors. MMPs play an integral part in this process. While many of the MMPs have been demonstrated to promote tumor progression in the bone (see Table 27.1), TGFβ has been shown to be a key regulator of the behavior and interaction of the major cell types involved in the vicious cycle: the osteoblasts, osteoclasts, and tumor cells.

TGFβ is incorporated into the bone matrix in an inactive state via latency-associated peptide (LAP), latent TGFβ-binding proteins (LTBPs), and the ECM protein decorin. These proteins have been demonstrated to be susceptible to processing by several MMPs (Saharinen et al. 1999). MMP-2, -3, and -9 mediate the release of TGFβ from either the LTBP molecule and/or the LAP molecule (Dallas et al. 1995, Yu and Stamenkovic 2000, Maeda et al. 2001, 2002) while MMP-2, -3, and -7 can process decorin, thus releasing sequestered TGFβ (Imai et al. 1997). The uncontrolled release of TGFβ has been shown to (1) support tumor growth by activating TGFβ receptors (TβR) on the tumor cell surface (Yoneda et al. 1994), (2) mediate osteoblast retraction from the bone surface, thus providing the osteoclasts with access to the mineralized matrix (Perez-Amodio et al. 2004), and (3) mediate osteoclast differentiation and activation (Chambers 2000, Quinn et al. 2001, Fox et al. 2003). MMPs can also modify TGFβ signaling. TGFβ signals via the TGFβ receptor family which is composed of three members (TβRI, II, and III). TβRIII, also referred to as betaglycan, has been shown to enhance the binding of active TGFβ to TβRI but the solubilized form of TβRIII can act as a sink for free TGFβ and inhibit TGFβ signaling (Bierie and Moses 2006). Recently, MMP-14 and -16 have been found to mediate the shedding of TβRIII (Velasco-Loyden et al. 2004).

Interestingly, TGFβ and related TGFβ family members BMPs can have profound effects on cell differentiation, in particular with respect to prostate and breast tumor transdifferentiation into cells with a bone-like phenotype (Barnes et al. 2003, Chung et al. 2005). Heightened TGFβ signaling in the bone can induce the expression of the Runx transcription factors in the tumor cells. Runx factors regulate several bone-related genes such as RANKL, OPG, OPN, and BSP, thereby rendering the tumor cells "osteomimetic" or osteoblast like (Banerjee et al. 2001, Barnes et al. 2003, Enomoto et al. 2003, Inman and Shore 2003). The induction of a genetic program controlling the expression of bone-related genes in the prostate and breast cancer cells in the bone environment in turn can further facilitate the interaction of the metastatic cells with host bone cells.

Hope for Metalloproteinase Inhibitors at the Final Frontier

Human clinical trials did not recapitulate the beneficial effects of the MPIs in preventing tumor progression in preclinical animal models. This was mainly due to the selection of patients for clinical trials (late stage tumor vs. early tumors in preclinical models), dose-limiting side effects, and the inhibition of multiple members of the metzincin family, many of which at the time of the clinical trials had not been discovered [for review, see Coussens et al. (2002) and Chap. 36 by Fingleton].

Given that defined metalloproteinases can control multiple facets of the vicious cycle, the use of selective MPIs may provide a therapeutic benefit for patients with bone metastases. Preclinical experiments have reported the efficacy of broad-spectrum MPIs in preventing the progression of prostate and breast tumors in the bone. The breast cancer cell line, MDA-MB-231, causes extensive osteolytic lesions over time when inoculated either via the intracardiac route or the intratibial route. Two independent studies demonstrated that the treatment of mice bearing MDA-MB-231 bone metastases with the broad-spectrum MPIs BB-94 or GM6001 could prevent tumor growth and tumor-induced osteolysis (Lee et al. 2001, Winding et al. 2002). Efficacy of the MPIs in the treatment of preclinical models of prostate cancer has also been demonstrated. The intratibial injection of PC-3 prostate cancer cells into the tibia of athymic causes extensive osteolysis that can be prevented by the administration of BB-94 (Nemeth et al. 2002). Other reagents with documented metalloproteinase inhibition characteristics, such as the shark cartilage derivative Neovastat, also prevent osteolytic tumor progression (Weber et al. 2002). To translate MPIs for human clinical use, the selective inhibition of metalloproteinases is a prerequisite in order to avoid the described drawbacks of the original MPIs. To this end, the MMP-2 and -9 selective inhibitor SB-3CT has been shown to be effective in preventing PC-3 prostate tumor progression with decreased tumor growth, vascularization, and osteolysis being reported (Bonfil et al. 2006). However, dose-limiting side effects in human clinicial trials with the gelatinase-specific inhibitor prinomastat have been reported (Coussens et al. 2002). But despite this, the identification of MMPs that contribute to tumor progression without affecting joint function should allow for the application of selective MPIs for the treatment of patients with debilitating bone metastases.

Conclusions

In the current chapter we have examined the cellular sources of MMPs in the tumor-bone microenvironment and discussed how the MMPs and other metalloproteinases can serve as potent mediators of cell–cell communication at the tumor–bone interface via the processing of (1) bone matrix components such as type-I collagen to yield signaling peptides and (2) nonmatrix factors such as CXCL12, PTHrP, RANKL, and TGFβ, which play key roles in the homing of metastatic tumor cells to the bone and in the vicious cycle of tumor progression in the bone microenvironment.

Given the increasing list of factors susceptible to MMP processing, it is probable that many more roles for the MMPs in tumor–bone interaction exist. Defining these roles will provide insights into how the vicious cycle is perpetuated and assist in the design of selective MPIs that have the potential to be efficacious in the treatment of bone metastases.

References

Abou-Samra A. B., Juppner H., Force T. et al. (1992). Expression cloning of a common receptor for parathyroid hormone and parathyroid hormone-related peptide from rat osteoblast-like cells: a single receptor stimulates intracellular accumulation of both cAMP and inositol trisphosphates and increases intracellular free calcium. Proc Natl Acad Sci USA 89:2732–2736.

Agnihotri R., Crawford H. C., Haro H. et al. (2001). Osteopontin, a novel substrate for matrix metalloproteinase-3 (stromelysin-1) and matrix metalloproteinase-7 (matrilysin). J Biol Chem 276:28261–28267.

Andolfo A., English W. R., Resnati M. et al. (2002). Metalloproteases cleave the urokinase-type plasminogen activator receptor in the D1-D2 linker region and expose epitopes not present in the intact soluble receptor. Thromb Haemost 88:298–306.

Athanasou N. A., and Sabokbar A. (1999). Human osteoclast ontogeny and pathological bone resorption. Histol Histopathol 14:635–647.

Banerjee C., Javed A., Choi J. Y. et al. (2001). Differential regulation of the two principal Runx2/Cbfa1 n-terminal isoforms in response to bone morphogenetic protein-2 during development of the osteoblast phenotype. Endocrinology 142:4026–4039.

Barnes G. L., Javed A., Waller S. M. et al. (2003). Osteoblast-related transcription factors Runx2 (Cbfa1/AML3) and MSX2 mediate the expression of bone sialoprotein in human metastatic breast cancer cells. Cancer Res 63:2631–2637.

Baron R., Rawadi G., and Roman-Roman S. (2006). Wnt signaling: a key regulator of bone mass. Curr Top Dev Biol 76:103–127.

Bendre M., Gaddy D., Nicholas R. W. et al. (2003). Breast cancer metastasis to bone: it is not all about PTHrP. Clin Orthop Relat Res 415:S39–S45.

Bergers G., Brekken R., McMahon G. et al. (2000). Matrix metalloproteinase-9 triggers the angiogenic switch during carcinogenesis. Nat Cell Biol 2:737–744.

Berruti A., Dogliotti L., Gorzegno G. et al. (1999). Differential patterns of bone turnover in relation to bone pain and disease extent in bone in cancer patients with skeletal metastases. Clin Chem 45:1240–1247.

Bhatia P., Sanders M. M., and Hansen M. F. (2005). Expression of receptor activator of nuclear factor-kappaB is inversely correlated with metastatic phenotype in breast carcinoma. Clin Cancer Res 11:162–165.

Bierie B., and Moses H. L. (2006). Tumour microenvironment: TGFbeta: the molecular Jekyll and Hyde of cancer. Nat Rev Cancer 6:506–520.

Blair H. C. (1998). How the osteoclast degrades bone. Bioessays 20:837–846.

Blair H. C., Teitelbaum S. L., Ghiselli R. et al. (1989). Osteoclastic bone resorption by a polarized vacuolar proton pump. Science 245:855–857.

Blomqvist C., Risteli L., Risteli J. et al. (1996). Markers of type I collagen degradation and synthesis in the monitoring of treatment response in bone metastases from breast carcinoma. Br J Cancer 73:1074–1079.

Bonfil R. D., Osenkowski P., Fridman R. et al. (2004). Matrix metalloproteinaes and bone metastasis. Cancer Treat Res 118:173–195.

Bonfil R. D., Sabbota A., Nabha S. et al. (2006). Inhibition of human prostate cancer growth, osteolysis and angiogenesis in a bone metastasis model by a novel mechanism-based selective gelatinase inhibitor 1. Int J Cancer 118:2721–2726.

Bonfil R. D., Dong Z., Trindade Filho J. C. et al. (2007). Prostate cancer-associated membrane type 1-matrix metalloproteinase: a pivotal role in bone response and intraosseous tumor growth. Am J Pathol 170:2100–2111.

Bord S., Horner A., Hembry R. M. et al. (1996). Production of collagenase by human osteoblasts and osteoclasts in vivo. Bone 19:35–40.

Bord S., Horner A., Hembry R. M. et al. (1998). Stromelysin-1 (MMP-3) and stromelysin-2 (MMP-10) expression in developing human bone: potential roles in skeletal development. Bone 23:7–12.

Breckon J. J., Papaioannou S., Kon L. W. et al. (1999). Stromelysin (MMP-3) synthesis is up-regulated in estrogen-deficient mouse osteoblasts in vivo and in vitro. J Bone Miner Res 14:1880–1890.

Brekken R. A., and Sage E. H. (2001). SPARC, a matricellular protein: at the crossroads of cell-matrix communication. Matrix Biol 19:816–827.

Brown J. M., Corey E., Lee Z. D. et al. (2001). Osteoprotegerin and rank ligand expression in prostate cancer. Urology 57:611–616.

Busiek D. F., Ross F. P., McDonnell S. et al. (1992). The matrix metalloproteinase matrilysin (PUMP) is expressed in developing human mononuclear phagocytes. J Biol Chem 267:9087–9092.

Busiek D. F., Baragi V., Nehring L. C. et al. (1995). Matrilysin expression by human mononuclear phagocytes and its regulation by cytokines and hormones. J Immunol 154:6484–6491.

Campo McKnight D. A., Sosnoski D. M., Koblinski J. E. et al. (2006). Roles of osteonectin in the migration of breast cancer cells into bone. J Cell Biochem 97:288–302.

Caulfield M. P., McKee R. L., Goldman M. E. et al. (1990). The bovine renal parathyroid hormone (PTH) receptor has equal affinity for two different amino acid sequences: the receptor binding domains of PTH and PTH-related protein are located within the 14–34 region. Endocrinology 127:83–87.

Chambers T. J. (2000). Regulation of the differentiation and function of osteoclasts. J Pathol 192:4–13.

Chandler S., Cossins J., Lury J. et al. (1996). Macrophage metalloelastase degrades matrix and myelin proteins and processes a tumor necrosis factor-alpha fusion protein. Biochem Biophys Res Commun 228:421–429.

Chen D., Adenekan B., Chen L. et al. (2004). Syndecan-1 expression in locally invasive and metastatic prostate cancer. Urology 63:402–407.

Chen G., Sircar K., Aprikian A. et al. (2006). Expression of RANKL/RANK/OPG in primary and metastatic human prostate cancer as markers of disease stage and functional regulation. Cancer 107:289–298.

Chesneau V., Becherer J. D., Zheng Y. et al. (2003). Catalytic properties of ADAM19. J Biol Chem 278:22331–22340.

Chung L. W., Baseman A., Assikis V. et al. (2005). Molecular insights into prostate cancer progression: the missing link of tumor microenvironment. J Urol 173:10–20.

Cloosen S., Gratama J., van Leeuwen E. B. et al. (2006). Cancer specific Mucin-1 glycoforms are expressed on multiple myeloma. Br J Haematol 135:513–516.

Cornish J., Callon K. E., Lin C. et al. (1999). Stimulation of osteoblast proliferation by C-terminal fragments of parathyroid hormone-related protein. J Bone Miner Res 14:915–922.

Coussens L. M., Fingleton B., and Matrisian L. M. (2002). Matrix metalloproteinase inhibitors and cancer: trials and tribulations. Science 295:2387–2392.

Covington M. D., Burghardt R. C., and Parrish A. R. (2006). Ischemia-induced cleavage of cadherins in NRK cells requires MT1-MMP (MMP-14). Am J Physiol Renal Physiol 290:F43–F51.

Cramer S. D., Chen Z., and Peehl D. M. (1996). Prostate specific antigen cleaves parathyroid hormone-related protein in the PTH-like domain: inactivation of PTHrP-stimulated cAMP accumulation in mouse osteoblasts. J Urol 156:526–531.

Crawford H. C., Matrisian L. M., and Liaw L. (1998). Distinct roles of osteopontin in host defense activity and tumor survival during squamous cell carcinoma progression in vivo. Cancer Res 58:5206–5215.

D'Ortho M. P., Will H., Atkinson S. et al. (1997). Membrane-type matrix metalloproteinases 1 and 2 exhibit broad-spectrum proteolytic capacities comparable to many matrix metalloproteinases. Eur J Biochem 250:751–757.

Dall'Era M. A., True L. D., Siegel A. F. et al. (2007). Differential expression of CD10 in prostate cancer and its clinical implication. BMC Urol 7:3.

Dallas S. L., Miyazono K., Skerry T. M. et al. (1995). Dual role for the latent transforming growth factor-beta binding protein in storage of latent TGF-beta in the extracellular matrix and as a structural matrix protein 6. J Cell Biol 131:539–549.

Datta N. S., Pettway G. J., Chen C. et al. (2007). Cyclin D1 as a target for the proliferative effects of PTH and PTHrP in early osteoblastic cells. J Bone Miner Res 22:951–964.

Davis G. E. (1992). Affinity of integrins for damaged extracellular matrix: alpha v beta 3 binds to denatured collagen type I through RGD sites. Biochem Biophys Res Commun 182:1025–1031.

Delaisse J. M., Ledent P., and Vaes G. (1991). Collagenolytic cysteine proteinases of bone tissue. Cathepsin B, (pro)cathepsin L and a cathepsin L-like 70 kDa proteinase. Biochem J 279 (Pt 1):167–174.

Delaisse J. M., Eeckhout Y., Neff L. et al. (1993). (Pro)collagenase (matrix metalloproteinase-1) is present in rodent osteoclasts and in the underlying bone-resorbing compartment. J Cell Sci 106 (Pt 4):1071–1082.

Delaisse J. M., Engsig M. T., Everts V. et al. (2000). Proteinases in bone resorption: obvious and less obvious roles. Clin Chim Acta 291:223–234.

Delaisse J. M., Andersen T. L., Engsig M. T. et al. (2003). Matrix metalloproteinases (MMP) and cathepsin K contribute differently to osteoclastic activities. Microsc Res Tech 61:504–513.

Deryugina E. I., Bourdon M. A., Jungwirth K. et al. (2000). Functional activation of integrin alpha V beta 3 in tumor cells expressing membrane-type 1 matrix metalloproteinase. Int J Cancer 86:15–23.

Dew G., Murphy G., Stanton H. et al. (2000). Localisation of matrix metalloproteinases and TIMP-2 in resorbing mouse bone. Cell Tissue Res 299:385–394.

Driessen R. L., Johnston H. M., and Nilsson S. K. (2003). Membrane-bound stem cell factor is a key regulator in the initial lodgment of stem cells within the endosteal marrow region. Exp Hematol 31:1284–1291.

Du P., Ye Y., Seitz P. K. et al. (2000). Endogenous parathyroid hormone-related peptide enhances proliferation and inhibits differentiation in the osteoblast-like cell line ROS 17/2.8. Bone 26:429–436.

Egawa N., Koshikawa N., Tomari T. et al. (2006). Membrane type 1 matrix metalloproteinase (MT1-MMP/MMP-14) cleaves and releases a 22-kDa extracellular matrix metalloproteinase inducer (EMMPRIN) fragment from tumor cells. J Biol Chem 281:37576–37585.

Endo K., Takino T., Miyamori H. et al. (2003). Cleavage of syndecan-1 by membrane type matrix metalloproteinase-1 stimulates cell migration. J Biol Chem 278:40764–40770.

Engsig M. T., Chen Q. J., Vu T. H. et al. (2000). Matrix metalloproteinase 9 and vascular endothelial growth factor are essential for osteoclast recruitment into developing long bones. J Cell Biol 151:879–890.

Enomoto H., Shiojiri S., Hoshi K. et al. (2003). Induction of osteoclast differentiation by Runx2 through receptor activator of nuclear factor-kappa B ligand (RANKL) and osteoprotegerin regulation and partial rescue of osteoclastogenesis in Runx2−/− mice by RANKL transgene. J Biol Chem 278:23971–23977.

Faccio R., Grano M., Colucci S. et al. (1998). Activation of alphav beta3 integrin on human osteoclast-like cells stimulates adhesion and migration in response to osteopontin. Biochem Biophys Res Commun 249:522–525.

Fata J. E., Kong Y. Y., Li J. et al. (2000). The osteoclast differentiation factor osteoprotegerin-ligand is essential for mammary gland development. Cell 103:41–50.

Fenton A. J., Martin T. J., and Nicholson G. C. (1994). Carboxyl-terminal parathyroid hormone-related protein inhibits bone resorption by isolated chicken osteoclasts. J Bone Miner Res 9:515–519.

Fingleton B., Vargo-Gogola T., Crawford H. C. et al. (2001). Matrilysin [MMP-7] expression selects for cells with reduced sensitivity to apoptosis. Neoplasia 3:459–468.

Fiore E., Fusco C., Romero P. et al. (2002). Matrix metalloproteinase 9 (MMP-9/gelatinase B) proteolytically cleaves ICAM-1 and participates in tumor cell resistance to natural killer cell-mediated cytotoxicity. Oncogene 21:5213–5223.

Fournier P. G., Chirgwin J. M., and Guise T. A. (2006). New insights into the role of T cells in the vicious cycle of bone metastases. Curr Opin Rheumatol 18:396–404.

Fowlkes J. L., Enghild J. J., Suzuki K. et al. (1994). Matrix metalloproteinases degrade insulin-like growth factor-binding protein-3 in dermal fibroblast cultures. J Biol Chem 269:25742–25746.

Fox S. W., Haque S. J., Lovibond A. C. et al. (2003). The possible role of TGF-beta-induced suppressors of cytokine signaling expression in osteoclast/macrophage lineage commitment in vitro. J Immunol 170:3679–3687.

Fujisaki K., Tanabe N., Suzuki N. et al. (2006). The effect of IL-1alpha on the expression of matrix metalloproteinases, plasminogen activators, and their inhibitors in osteoblastic ROS 17/2.8 cells. Life Sci 78:1975–1982.

Gagnon L., Jouishomme H., Whitfield J. F. et al. (1993). Protein kinase C-activating domains of parathyroid hormone-related protein. J Bone Miner Res 8:497–503.

Garnero P., Ferreras M., Karsdal M. A. et al. (2003). The type I collagen fragments ICTP and CTX reveal distinct enzymatic pathways of bone collagen degradation. J Bone Miner Res 18:859–867.

Gearing A. J., Thorpe S. J., Miller K. et al. (2002). Selective cleavage of human IgG by the matrix metalloproteinases, matrilysin and stromelysin. Immunol Lett 81:41–48.

Gelb B. D., Shi G. P., Chapman H. A. et al. (1996). Pycnodysostosis, a lysosomal disease caused by cathepsin K deficiency. Science 273:1236–1238.

Giles E. D., and Singh G. (2003). Role of insulin-like growth factor binding proteins (IGFBPs) in breast cancer proliferation and metastasis. Clin Exp Metastasis 20:481–487.

Gowen M., Lazner F., Dodds R. et al. (1999). Cathepsin K knockout mice develop osteopetrosis due to a deficit in matrix degradation but not demineralization. J Bone Miner Res 14:1654–1663.

Guise T. A., and Chirgwin J. M. (2003). Transforming growth factor-beta in osteolytic breast cancer bone metastases. Clin Orthop Relat Res 415:S32–S38.

Haug C., Lenz C., Diaz F. et al. (2004). Oxidized low-density lipoproteins stimulate extracellular matrix metalloproteinase inducer (EMMPRIN) release by coronary smooth muscle cells. Arterioscler Thromb Vasc Biol 24:1823–1829.

Heissig B., Hattori K., Dias S. et al. (2002). Recruitment of stem and progenitor cells from the bone marrow niche requires mmp-9 mediated release of kit-ligand. Cell 109:625–637.

Hiratsuka S., Nakamura K., Iwai S. et al. (2002). MMP9 induction by vascular endothelial growth factor receptor-1 is involved in lung-specific metastasis. Cancer Cell 2:289–300.

Holmbeck K., Bianco P., Caterina J. et al. (1999). MT1-MMP-deficient mice develop dwarfism, osteopenia, arthritis, and connective tissue disease due to inadequate collagen turnover. Cell 99:81–92.

Holmbeck K., Bianco P., Pidoux I. et al. (2005). The metalloproteinase MT1-MMP is required for normal development and maintenance of osteocyte processes in bone. J Cell Sci 118:147–156.

Hotokezaka H., Sakai E., Ohara N. et al. (2007). Molecular analysis of RANKL-independent cell fusion of osteoclast-like cells induced by TNF-alpha, lipopolysaccharide, or peptidoglycan. J Cell Biochem 101:122–134.

Hou P., Troen T., Ovejero M. C. et al. (2004). Matrix metalloproteinase-12 (MMP-12) in osteoclasts: new lesson on the involvement of MMPs in bone resorption. Bone 34:37–47.

Ichikawa Y., Ishikawa T., Momiyama N. et al. (2006). Matrilysin (MMP-7) degrades VE-cadherin and accelerates accumulation of beta-catenin in the nucleus of human umbilical vein endothelial cells. Oncol Rep 15:311–315.

Imai K., Hiramatsu A., Fukushima D. et al. (1997). Degradation of decorin by matrix metalloproteinases: identification of the cleavage sites, kinetic analyses and transforming growth factor-beta 1 release. Biochem J 322:809–814.

Inada M., Wang Y., Byrne M. H. et al. (2004). Critical roles for collagenase-3 (Mmp13) in development of growth plate cartilage and in endochondral ossification. Proc Natl Acad Sci U S A 101:17192–17197.

Inman C. K., and Shore P. (2003). The osteoblast transcription factor Runx2 is expressed in mammary epithelial cells and mediates osteopontin expression. J Biol Chem 278:48684–48689.

Inoue K., Mikuni-Takagaki Y., Oikawa K. et al. (2006). A crucial role for matrix metalloproteinase 2 in osteocytic canalicular formation and bone metabolism. J Biol Chem 281:33814–33824.

Ito A., Mukaiyama A., Itoh Y. et al. (1996). Degradation of interleukin 1-beta by matrix metalloproteinases. J Biol Chem 271:14657–14660.

Jones D. H., Nakashima T., Sanchez O. H. et al. (2006). Regulation of cancer cell migration and bone metastasis by RANKL. Nature 440:692–696.

Kaji H., Sugimoto T., Kanatani M. et al. (1995). Carboxyl-terminal peptides from parathyroid hormone-related protein stimulate osteoclast-like cell formation. Endocrinology 136:842–848.

Kajita M., Itoh Y., Chiba T. et al. (2001). Membrane-type 1 matrix metalloproteinase cleaves CD44 and promotes cell migration. J Cell Biol 153:893–904.

Kang Y., Siegel P. M., Shu W. et al. (2003). A multigenic program mediating breast cancer metastasis to bone. Cancer Cell 3:537–549.

Kaplan R. N., Psaila B., and Lyden D. (2007). Niche-to-niche migration of bone-marrow-derived cells. Trends Mol Med 13:72–81.

Kawashima-Ohya Y., Satakeda H., Kuruta Y. et al. (1998). Effects of parathyroid hormone (PTH) and PTH-related peptide on expressions of matrix metalloproteinase-2, -3, and -9 in growth plate chondrocyte cultures. Endocrinology 139:2120–2127.

Kong Y. Y., Yoshida H., Sarosi I. et al. (1999). OPGL is a key regulator of osteoclastogenesis, lymphocyte development and lymph-node organogenesis. Nature 397:315–323.

Kovacic N., Lukic I. K., Grcevic D. et al. (2007). The Fas/Fas ligand system inhibits differentiation of murine osteoblasts but has a limited role in osteoblast and osteoclast apoptosis. J Immunol 178:3379–3389.

Krampera M., Pasini A., Rigo A. et al. (2005). HB-EGF/HER-1 signaling in bone marrow mesenchymal stem cells: inducing cell expansion and reversibly preventing multilineage differentiation. Blood 106:59–66.

Kurihara N., Bertolini D., Suda T. et al. (1990). IL-6 stimulates osteoclast-like multinucleated cell formation in long term human marrow cultures by inducing IL-1 release. J Immunol 144:4226–4230.

Kusano K., Miyaura C., Inada M. et al. (1998). Regulation of matrix metalloproteinases (MMP-2, -3, -9, and -13) by interleukin-1 and interleukin-6 in mouse calvaria: association of MMP induction with bone resorption. Endocrinology 139:1338–1345.

Lee J., Weber M., Mejia S. et al. (2001). A matrix metalloproteinase inhibitor, batimastat, retards the development of osteolytic bone metastases by MDA-MB-231 human breast cancer cells in Balb C nu/nu mice. Eur J Cancer 37:106–113.

Levi E., Fridman R., Miao H. Q. et al. (1996). Matrix metalloproteinase 2 releases active soluble ectodomain of fibroblast growth factor receptor 1. Proc Natl Acad Sci U S A 93:7069–7074.

Li M., Sasaki T., Ono K. et al. (2007). Distribution of macrophages, osteoclasts and the B-lymphocyte lineage in osteolytic metastasis of mouse mammary carcinoma. Biomed Res 28:127–137.

Li Q., Park P. W., Wilson C. L. et al. (2002). Matrilysin shedding of syndecan-1 regulates chemokine mobilization and transepithelial efflux of neutrophils in acute lung injury. Cell 111:635–646.

Liao J. and McCauley L. K. (2006). Skeletal metastasis: Established and emerging roles of parathyroid hormone related protein (PTHrP). Cancer Metastasis Rev 25:559–571.

Liu B., Goltzman D., and Rabbani S. A. (1995). Processing of pro-PTHRP by the prohormone convertase, furin: effect on biological activity. Am J Physiol 268:E832–838.

Lochter A., Galosy S., Muschler J. et al. (1997). Matrix metalloproteinase stromelysin-1 triggers a cascade of molecular alterations that leads to stable epithelial-to-mesenchymal conversion and a premalignant phenotype in mammary epithelial cells. J Cell Biol 139:1861–1872.

Lum L., Wong B. R., Josien R. et al. (1999). Evidence for a role of a tumor necrosis factor-alpha (TNF-alpha)- converting enzyme-like protease in shedding of TRANCE, a TNF family member involved in osteoclastogenesis and dendritic cell survival. J Biol Chem 274:13613–13618.

Luparello C., Sirchia R., and Pupello D. (2003). PTHrP [67–86] regulates the expression of stress proteins in breast cancer cells inducing modifications in urokinase-plasminogen activator and MMP-1 expression 5. J Cell Sci 116:2421–2430.

Lynch C. C., and Matrisian L. M. (2002). Matrix metalloproteinases in tumor-host cell communication. Differentiation 70:561–573.

Lynch C. C., Hikosaka A., Acuff H. B. et al. (2005). MMP-7 promotes prostate cancer-induced osteolysis via the solubilization of RANKL 1. Cancer Cell 7:485–496.

Maeda S., Dean D. D., Gay I. et al. (2001). Activation of latent transforming growth factor beta1 by stromelysin 1 in extracts of growth plate chondrocyte-derived matrix vesicles. J Bone Miner Res 16:1281–1290.

Maeda S., Dean D. D., Gomez R. et al. (2002). The first stage of transforming growth factor beta1 activation is release of the large latent complex from the extracellular matrix of growth plate chondrocytes by matrix vesicle stromelysin-1 (MMP-3). Calcified Tissue Int 70:54–65.

Mahtouk K., Cremer F. W., Reme T. et al. (2006). Heparan sulphate proteoglycans are essential for the myeloma cell growth activity of EGF-family ligands in multiple myeloma. Oncogene 25:7180–7191.

Malone J. D., Teitelbaum S. L., and Griffin G. L. (1982). Recruitment of osteoclast precursors by purified bone matrix constituents. J Cell Biol 92:227–230.

Manes S., Mira E., Barbacid M. M. et al. (1997). Identification of insulin-like growth factor-binding protein-1 as a potential physiological substrate for human stromelysin-3. J Biol Chem 272:25706–25712.

Margheri F., D'Alessio S., Serrati S. et al. (2005). Effects of blocking urokinase receptor signaling by antisense oligonucleotides in a mouse model of experimental prostate cancer bone metastases. Gene Ther 12:702–714.

McQuibban G. A., Butler G. S., Gong J. H. et al. (2001). Matrix metalloproteinase activity inactivates the CXC chemokine stromal cell-derived factor-1. J Biol Chem 276:43503–43508.

Miao D., Tong X. K., Chan G. K. et al. (2001). Parathyroid hormone-related peptide stimulates osteogenic cell proliferation through protein kinase C activation of the Ras/mitogen-activated protein kinase signaling pathway. J Biol Chem 276:32204–32213.

Mitsiades N., Yu W., Poulaki V. et al. (2001). Matrix metalloproteinase-7-mediated cleavage of Fas ligand protects tumor cells from chemotherapeutic drug cytotoxicity. Cancer Res 61:577–581.

Mosig R. A., Dowling O., DiFeo A. et al. (2007). Loss of MMP-2 disrupts skeletal and craniofacial development and results in decreased bone mineralization, joint erosion and defects in osteoblast and osteoclast growth. Hum Mol Genet 16:1113–1123.

Muller A., Homey B., Soto H. et al. (2001). Involvement of chemokine receptors in breast cancer metastasis. Nature 410:50–56.

Mundy G. R. (2002). Metastasis to bone: causes, consequences and therapeutic opportunities. Nat Rev Cancer 2:584–593.

Murphy-Ullrich J. E. (2001). The de-adhesive activity of matricellular proteins: is intermediate cell adhesion an adaptive state? J Clin Invest 107:785–790.

Nagase H. and Woessner J. F. Jr. (1999). Matrix metalloproteinases. J Biol Chem 274:21491–21494.

Nakashima A. and Tamura M. (2006). Regulation of matrix metalloproteinase-13 and tissue inhibitor of matrix metalloproteinase-1 gene expression by WNT3A and bone morphogenetic protein-2 in osteoblastic differentiation. Front Biosci 11:1667–1678.

Nemeth J. A., Yousif R., Herzog M. et al. (2002). Matrix metalloproteinase activity, bone matrix turnover, and tumor cell proliferation in prostate cancer bone metastasis. J Natl Cancer Inst 94:17–25.

Noe V., Fingleton B., Jacobs K. et al. (2001). Release of an invasion promoter E-cadherin fragment by matrilysin and stromelysin-1. J Cell Sci 114:111–118.

Okada Y., Naka K., Kawamura K. et al. (1995). Localization of matrix metalloproteinase 9 (92-kilodalton gelatinase/type IV collagenase = gelatinase B) in osteoclasts: implications for bone resorption. Lab Invest 72:311–322.

Palma M. A. and Body J. J. (2005). Usefulness of bone formation markers in breast cancer. Int J Biol Markers 20:146–155.

Parikka V., Vaananen A., Risteli J. et al. (2005). Human mesenchymal stem cell derived osteoblasts degrade organic bone matrix in vitro by matrix metalloproteinases. Matrix Biol 24:438–447.

Pei D. (2005). Matrix metalloproteinases target protease-activated receptors on the tumor cell surface. Cancer Cell 7:207–208.

Perez-Amodio S., Beertsen W., and Everts V. (2004). (Pre-)osteoclasts induce retraction of osteoblasts before their fusion to osteoclasts. J Bone Miner Res 19:1722–1731.

Powell W. C., Fingleton B., Wilson C. L. et al. (1999). The metalloproteinase matrilysin (MMP-7) proteolytically generates active soluble Fas ligand and potentiates epithelial cell apoptosis. Curr Biol 9:1441–1447.

Quinn J. M., Itoh K., Udagawa N. et al. (2001). Transforming growth factor beta affects osteoclast differentiation via direct and indirect actions. J Bone Miner Res 16:1787–1794.

Rabbani S. A., Mitchell J., Roy D. R. et al. (1988). Influence of the amino-terminus on in vitro and in vivo biological activity of synthetic parathyroid hormone-like peptides of malignancy. Endocrinology 123:2709–2716.

Rajah R., Nunn S. E., Herrick D. J. et al. (1996). Leukotriene D-4 induces MMP-1, which functions as an igfbp protease in human airway smooth muscle cells. Am J Physiol Lung Cell Mol Physiol 15:L1014–L1022.

Reimers N., Zafrakas K., Assmann V. et al. (2004). Expression of extracellular matrix metalloproteases inducer on micrometastatic and primary mammary carcinoma cells. Clin Cancer Res 10:3422–3428.

Ritchie C. K., Andrews L. R., Thomas K. G. et al. (1997). The effects of growth factors associated with osteoblasts on prostate carcinoma proliferation and chemotaxis: implications for the development of metastatic disease. Endocrinology 138:1145–1150.

Ross F. P., Chappel J., Alvarez J. I. et al. (1993). Interactions between the bone matrix proteins osteopontin and bone sialoprotein and the osteoclast integrin alpha v beta 3 potentiate bone resorption. J Biol Chem 268:9901–9907.

Rozanov D. V., Hahn-Dantona E., Strickland D. K. et al. (2004). The low density lipoprotein receptor-related protein LRP is regulated by membrane type-1 matrix metalloproteinase (MT1-MMP) proteolysis in malignant cells. J Biol Chem 279:4260–4268.

Ruchon A. F., Marcinkiewicz M., Ellefsen K. et al. (2000). Cellular localization of neprilysin in mouse bone tissue and putative role in hydrolysis of osteogenic peptides. J Bone Miner Res 15:1266–1274.

Saftig P., Hunziker E., Wehmeyer O. et al. (1998). Impaired osteoclastic bone resorption leads to osteopetrosis in cathepsin-K-deficient mice. Proc Natl Acad Sci U S A 95:13453–13458.

Saharinen J., Hyytiainen M., Taipale J. et al. (1999). Latent transforming growth factor-beta binding proteins (LTBPs)–structural extracellular matrix proteins for targeting TGF-beta action 2. Cytokine Growth Factor Rev 10:99–117.

Salo J., Lehenkari P., Mulari M. et al. (1997). Removal of osteoclast bone resorption products by transcytosis. Science 276:270–273.

Sasaki T., Gohring W., Mann K. et al. (1997). Limited cleavage of extracellular matrix protein bm-40 by matrix metalloproteinases increases its affinity for collagens. J Biol Chem 272:9237–9243.

Sato H., Takino T., Okada Y. et al. (1994). A matrix metalloproteinase expressed on the surface of invasive tumour cells. Nature 370:61–64.

Sato T., Ovejero M. D., Hou P. et al. (1997). Identification of the membrane-type matrix metalloproteinase MT1-MMP in osteoclasts. J Cell Sci 110(5):589–596.

Schlondorff J., Lum L., and Blobel C. P. (2001). Biochemical and pharmacological criteria define two shedding activities for TRANCE/OPGL that are distinct from the tumor necrosis factor alpha convertase. J Biol Chem 276:14665–14674.

Schonbeck U., Mach F., and Libby P. (1998). Generation of biologically active IL-1 beta by matrix metalloproteinases: a novel caspase-1-independent pathway of IL-1 beta processing. J Immunol 161:3340–3346.

Sheu B. C., Hsu S. M., Ho H. N. et al. (2001). A novel role of metalloproteinase in cancer-mediated immunosuppression. Cancer Res 61:237–242.

Simonet W. S., Lacey D. L., Dunstan C. R. et al. (1997). Osteoprotegerin: a novel secreted protein involved in the regulation of bone density. Cell 89:309–319.

Smollich M., Gotte M., Yip G. W. et al. (2007). On the role of endothelin-converting enzyme-1 (ECE-1) and neprilysin in human breast cancer. Breast Cancer Res Treat 106(3):361–369.

Song X., Voronov E., Dvorkin T. et al. (2003). Differential effects of IL-1 alpha and IL-1 beta on tumorigenicity patterns and invasiveness. J Immunol 171:6448–6456.

Southby J., O'Keeffe L. M., Martin T. J. et al. (1995). Alternative promoter usage and mRNA splicing pathways for parathyroid hormone-related protein in normal tissues and tumours. Br J Cancer 72:702–707.

Stickens D., Behonick D. J., Ortega N. et al. (2004). Altered endochondral bone development in matrix metalloproteinase 13-deficient mice 2. Development 131:5883–5895.

Strand S., Vollmer P., van den A. L. et al. (2004). Cleavage of CD95 by matrix metalloproteinase-7 induces apoptosis resistance in tumour cells. Oncogene 23:3732–3736.

Suenaga N., Mori H., Itoh Y. et al. (2005). CD44 binding through the hemopexin-like domain is critical for its shedding by membrane-type 1 matrix metalloproteinase. Oncogene 24:859–868.

Suzuki M., Raab G., Moses M. A. et al. (1997). Matrix metalloproteinase-3 releases active heparin-binding EGF-like growth factor by cleavage at a specific juxtamembrane site. J Biol Chem 272:31730–31737.

Tang K. T., Capparelli C., Stein J. L. et al. (1996). Acidic fibroblast growth factor inhibits osteoblast differentiation in vitro: altered expression of collagenase, cell growth-related, and mineralization-associated genes. J Cell Biochem 61:152–166.

Terpos E., Szydlo R., Apperley J. F. et al. (2003). Soluble receptor activator of nuclear factor kappaB ligand-osteoprotegerin ratio predicts survival in multiple myeloma: proposal for a novel prognostic index. Blood 102:1064–1069.

Tester A. M., Waltham M., Oh S. J. et al. (2004). Pro-matrix metalloproteinase-2 transfection increases orthotopic primary growth and experimental metastasis of MDA-MB-231 human breast cancer cells in nude mice. Cancer Res 64:652–658.

Teti A., Colucci S., Grano M. et al. (1992). Protein kinase C affects microfilaments, bone resorption, and [Ca2$^+$]o sensing in cultured osteoclasts. Am J Physiol 263:C130–139.

Thathiah A. and Carson D. D. (2004). MT1-MMP mediates MUC1 shedding independent of TACE/ADAM17. Biochem J 382:363–373.

Thomas R. J., Guise T. A., Yin J. J. et al. (1999). Breast cancer cells interact with osteoblasts to support osteoclast formation. Endocrinology 140:4451–4458.

Thrailkill K. M., Quarles L. D., Nagase H. et al. (1995). Characterization of insulin-like growth factor-binding protein 5-degrading proteases produced throughout murine osteoblast differentiation. Endocrinology 136:3527–3533.

Vaes G. (1972). The release of collagenase as an inactive proenzyme by bone explants in culture. Biochem J 126:275–289.

Vargo-Gogola T., Crawford H. C., Fingleton B. et al. (2002). Identification of novel matrix metalloproteinase-7 (matrilysin) cleavage sites in murine and human Fas ligand. Arch Biochem Biophys 408:155–161.

Velasco-Loyden G., Arribas J., and Lopez-Casillas F. (2004). The shedding of betaglycan is regulated by pervanadate and mediated by membrane type matrix metalloprotease-1. J Biol Chem 279:7721–7733.

Weber M. H., Lee J., and Orr F. W. (2002). The effect of Neovastat (AE-941) on an experimental metastatic bone tumor model. Int J Oncol 20:299–303.

Wilson C. L., Heppner K. J., Labosky P. A. et al. (1997). Intestinal tumorigenesis is suppressed in mice lacking the metalloproteinase matrilysin. Proc Natl Acad Sci U S A 94:1402–1407.

Winding B., NicAmhlaoibh R., Misander H. et al. (2002). Synthetic matrix metalloproteinase inhibitors inhibit growth of established breast cancer osteolytic lesions and prolong survival in mice. Clin Cancer Res 8:1932–1939.

Yoneda T., Sasaki A., and Mundy G. R. (1994). Osteolytic bone metastasis in breast cancer. Breast Cancer Res Treat 32:73–84.

Yoneda T., Sasaki A., Dunstan C. et al. (1997). Inhibition of osteolytic bone metastasis of breast cancer by combined treatment with the bisphosphonate ibandronate and tissue inhibitor of the matrix metalloproteinase-2. J Clin Invest 99:2509–2517.

Yu Q., and Stamenkovic I. (2000). Cell surface-localized matrix metalloproteinase-9 proteolytically activates TGF-beta and promotes tumor invasion and angiogenesis. Genes Dev 14:163–176.

Yu W. H., Woessner J. F. Jr., McNeish J. D. et al. (2002). CD44 anchors the assembly of matrilysin/MMP-7 with heparin-binding epidermal growth factor precursor and ErbB4 and regulates female reproductive organ remodeling. Genes Dev 16:307–323.

Zhang J., Dai J., Qi Y. et al. (2001). Osteoprotegerin inhibits prostate cancer-induced osteoclastogenesis and prevents prostate tumor growth in the bone. J Clin Invest 107:1235–1244.

Chapter 28
The Plasminogen Activation System as a Source of Prognostic Markers in Cancer

Ib Jarle Christensen, Helle Pappot, and Gunilla Høyer-Hansen

Abstract The components of the plasminogen activation system including the urokinase plasminogen activator, uPA, its cellular receptor, uPAR, and its inhibitor, PAI-1, are involved in cancer invasion and metastasis and our aim was to evaluate their prognostic utility for patients with solid tumors including breast cancer, colorectal cancer, lung cancer, ovarian cancer, and prostate cancer.

Studies reporting the association between uPA, uPAR, and PAI-1 levels and patient outcome have been identified, where samples have been analyzed using quantitative immunoassay platforms. The studies have measured uPA, uPAR, or PAI-1 in either tumor tissue or blood samples.

The identified studies have analyzed the correlation between the levels of uPA, uPAR, and PAI-1 and patient outcome using regression methods. Many of these conclude that there is a statistically significant association between these components and patient survival. Circulating levels of uPAR have been shown to predict outcome in breast cancer, colorectal cancer, ovarian cancer, and prostate cancer. There is evidence that tumor tissue levels of uPAR are not correlated with circulating levels, although both predict outcome. The uPA and PAI-1 levels in breast cancer predict response to adjuvant therapy. For breast cancer, a pooled analysis and a randomized clinical trial have been identified confirming the association between the marker levels and outcome.

The uPA, uPAR, and PAI-1 levels are associated with patient outcome measured in either tumor tissue or blood samples. There is a need for meta-analyses and prospective trials confirming the results in solid tumors with the possible exception of breast cancer where these studies have been performed.

G. Høyer-Hansen
Finsen Laboratory, Copenhagen Biocenter, Ole Maaløes Vej 5, DK-2200 Copenhagen N, Denmark, e-mail: gunilla@finsenlab.dk

D. Edwards et al. (eds.), *The Cancer Degradome.*
© Springer Science + Business Media, LLC 2008

Introduction

Continuous development of our understanding of the molecular basis of cancer is a prerequisite, not only for the identification and design of specific anti-tumor agents but also for the development of new tools in diagnosis, selection, and prognosis of cancer patients. Such tools are important to help improve cancer survival by allowing earlier diagnosis, selection of patients who will benefit from adjuvant therapy, and assessment of prognosis. Three components of the plasminogen activation system—the urokinase plasminogen activator, uPA, its cellular receptor, uPAR, and its inhibitor, PAI-1—are involved in cancer invasion and metastasis (Dano et al. 2005). Their functional characteristics and cellular localization are well described and several mono- and polyclonal antibodies have been raised against them enabling the design of quantitative immunoassays measuring uPA, uPAR, and PAI-1, complexes of these, and for uPAR, the cleavage products. This gives a unique possibility to use these specific components as clinical markers.

After the design of quantitative enzyme-linked immunosorbent assays (ELISAs) with sufficient sensitivity to measure uPA, PAI-1, or uPAR in clinical samples, several studies on the prognostic significance of these components in various cancers have been performed (for review, *see* Andreasen et al. 1997, 2000; Duffy et al. 1999, Duffy 2002, Hoyer-Hansen and Lund 2007). In fact, uPA and PAI-1 were among the first novel tumor biological factors to be validated at the highest level of evidence (LOE I) (Hayes et al. 1996, 1998), with respect to their clinical utility in breast cancer measuring the antigen in tumor tissue extracts (Janicke et al. 2001, Look et al. 2002). In this chapter, we will review prognostic studies where uPA, uPAR, and PAI-1 have been measured by immunoassays on samples of tumor tissue or blood from patients with different forms of solid tumors.

Quantification of uPA, uPAR, and PAI-1

uPA is present in blood in a proform, in an active form and in complex with PAI-1 (Andreasen et al. 1997), and it has been suggested that the major part of uPA in blood is present as uPA–uPAR complexes (de Witte et al. 1998). The same forms of uPA are found in tumor tissue extracts. Since most assays, quantifying uPA, PAI-1, or uPAR in human samples, measure mixtures of the different molecular forms, it follows that the amounts should be expressed in molar concentrations. A source of variation in the analysis of tissue samples is the extraction procedure (Schmitt et al. 2007). This variation was evident in a study where a pool of breast tumor tissue was extracted with three different buffers and the contents of uPAR, uPA, and PAI-1 were determined. Employing an acidic buffer containing Triton X-100 yielded the highest concentrations of uPA and PAI-1, whereas the highest levels of uPAR were retrieved after extraction at pH 8.1 in a Triton X-114 containing buffer (Ronne et al. 1995). Furthermore, because laboratories often use different immunoassays the absolute analyte values differ considerably. This is mainly dependent on the antibodies and detection systems employed (Hoyer-Hansen et al. 2000, Hoyer-Hansen and Lund 2007). In a quality-control study, the levels of uPA and PAI-1 in a pool of breast cancer tissue were found to vary with the immunoassay format used, and the authors suggested that in multicenter studies the same immunoassay should be used for quantitation of the antigen (Sweep et al. 1998). However, this recommendation has not been followed and to enable multicenter studies, values have been normalized for comparison across datasets (Look et al. 2002).

In order to obtain an estimate of the variation in antigen levels measured by different immunoassays, it is often useful to determine the amounts in blood from healthy donors. The concentration of uPA in plasma from healthy donors was 23.7 ± 6 pmol/l measured with an ELISA, using a monoclonal catching antibody reacting with an epitope in the receptor binding A-chain and a biotinylated polyclonal anti-uPA detecting antibody. This antibody combination measures both uPA and uPA–PAI-1 complexes (Grondahl-Hansen et al. 1988). This value is somewhat higher than recent measurements using a different ELISA (American Diagnostica Inc.) on citrate plasma from healthy controls, where the reported uPA concentration was 4.3 pmol/l (Shariat et al. 2007).

PAI-1 is an efficient plasminogen activator inhibitor in its active form and is present in blood in complex with vitronectin. uPA–PAI-1 complexes are also present in blood, but PAI-1 in complex will not bind vitronectin (Andreasen et al. 1997). The latent form of PAI-1 has 100-fold lower affinity for vitronectin and does not inhibit uPA activity. The function and biochemical properties of PAI-1 have been reviewed (Andreasen et al. 1997, 2000; Stefansson et al. 2003, Lijnen 2005). In tumor tissue extracts, both PAI-1 and uPA–PAI-1 complexes are present. There are several different immunoassay formats available for determination of the total PAI-1 concentration, and in addition ELISAs have been developed to measure the level of uPA–PAI-1 complexes (Grebenschikov et al. 1999, Pedersen et al. 2003). The PAI-1 concentration in plasma from healthy donors is $\sim$444 pmol/l (Lijnen 2005) and the concentration of uPA–PAI-1 complexes is in most healthy donors below the detection limit 0.18 pmol/l (Pedersen et al. 1999).

uPAR is a multifunctional protein attached to the cell surface by a glycolipid anchor. Its structural and functional properties have been extensively reviewed (Ploug 2003). uPA can cleave the glycolipid-anchored intact uPAR, uPAR(I–III), in the linker region between domains I and II liberating domain I, uPAR(I), and leaving the cleaved uPAR(II–III) on the cell surface (Hoyer-Hansen et al. 1992, 1997b). This cleavage inactivates the binding potential of uPAR both for uPA and vitronectin (Hoyer-Hansen et al. 1997a). Soluble forms of uPAR are present in blood and other body fluids. Intact and cleaved uPAR are shed from the cell surface possibly by phospholipases (Wilhelm et al. 1999). Because of a conformational difference in the linker region connecting domain I and II between glycolipid-anchored and soluble uPAR, the latter is not cleaved by physiologically relevant concentrations of uPA (Hoyer-Hansen et al. 2001). Whether the soluble uPAR(II–III) is shed from the cell surface or a result of a proteolytic cleavage of soluble uPAR(I–III) by another protease has not been clarified.

Several groups have designed ELISAs for measurements of the collective amounts of uPAR forms in a sample (Mizukami et al. 1995, Stephens et al. 1999, Kotzsch et al. 2000, Riisbro et al. 2002). The uPAR variants measured in the different immunoassays are dependent on the antibody combinations used and have recently been reviewed (Hoyer-Hansen and Lund 2007). Immunoassays for the specific quantification of intact and cleaved uPAR as well as the liberated uPAR(I) have also been designed, and 42 pmol/l of uPAR(I–III), 26 pmol/l of uPAR(I), and 81 pmol/l of uPAR(I–III) and uPAR(II–III) were detected in a citrate plasma pool from healthy volunteers (Piironen et al. 2004). This compares well with the 75 pmol/l measured with an ELISA employing an anti-uPAR polyclonal catching antibody and a monoclonal detecting antibody with an epitope in the carboxy terminal part of uPAR permitting the detection of all uPAR forms except uPAR(I) (Riisbro et al. 2002).

Breast Cancer

Breast cancer was the first cancer disease where prognosis was shown to be associated with the plasminogen activation system (Janicke et al. 1991, 1993). A significant correlation between the tumor tissue levels of uPA and PAI-1 and relapse-free survival (RFS) and overall survival (OS) was demonstrated. A number of subsequent studies showed the same tendency using ELISAs to determine the level of uPA and PAI-1 in tumor tissue (Grondahl-Hansen et al. 1993, 1997; Duffy, 2002, Hansen et al. 2003). The identifiable studies of uPA and PAI-1 in tumor tissue by members of the European Organization for Research and Treatment Cancer-Receptor and Biomarker Group have been analyzed in a pooled analysis comprising 8,377 breast cancer patients (Look et al. 2002). This report included 18 datasets with clinical data. Rank statistics were used to assess the prognostic significance of uPA and PAI-1 with respect to RFS and OS. The results show that elevated levels of both biomarkers demonstrate a highly significant association with shorter RFS as well as OS and independent of the classical indicators such as nodal status, tumor size, age, and receptor status. In addition, both biomarkers contribute independently to the prediction of outcome. High levels of uPA and PAI-1 were the strongest predictors of poor RFS and OS after nodal status. In the multivariate models for RFS and OS, the hazard ratios for uPA and PAI-1 were higher than 2.5. Figure 28.1a shows the hazard ratio with 95% confidence interval for RFS as a function of uPA levels for each participating study. Similarly, Fig. 28.1b shows the hazard ratio for RFS as a function of PAI-1 levels. The results of this study demonstrate that LOE for the prognostic value of uPA and PAI-1 in tumor tissue can be considered to be at level I.

The potential clinical utility of uPA and PAI-1 determinations in the management of breast cancer disease has been shown in a prospective multicenter therapy trial randomizing node-negative patients, with elevated levels of uPA and PAI-1 in tumor tissue to observation or combination chemotherapy (Janicke et al. 2001). Patients with low levels of uPA and PAI-1 were observed. The results of the trial

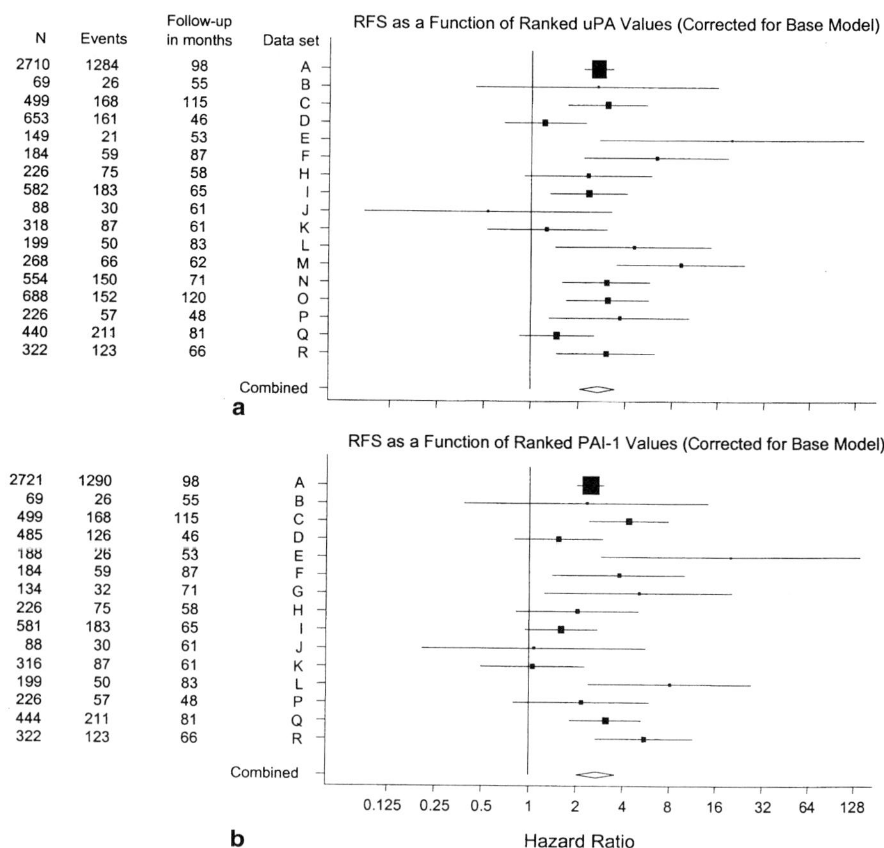

Fig. 28.1 Multivariate analysis for relapse-free survival (*RFS*) as a function of urokinase plasminogen activator (*uPA*) (**a**) and its inhibitor plasminogen activator inhibitor *(PAI-1)* (**b**) per dataset presented as a forest plot. All data are corrected for the base model, including age and menopausal status, tumor size, lymph node status, steroid hormone-receptor status, histologic grade, and adjuvant treatment. For each dataset, the hazard ratio (*HR*) of transformed ranked *uPA* or ranked *PAI-1* values is plotted as a solid square; its area being inversely proportional to the variance of the estimated effect and its 95% confidence interval (*CI*) is plotted as horizontal line. Individual patient data were obtained from both published and unpublished datasets. Diamond represents the combined random effects estimate (middle of the diamond) and its 95% *CI* (extremes of the diamond) of the combined estimates adjusted for the base model. Reproduced from Look et al. (2002) with permission from Oxford University Press

suggest that patients with low levels do not need adjuvant therapy, whereas high-risk patients as defined by their uPA and PAI-1 tumor tissue levels benefit from combination chemotherapy.

In another retrospective study, it was shown that tumor tissue levels of uPA and PAI-1 in high-risk patients show a significant interaction with adjuvant therapy suggesting that these biomarkers could be useful for the identification of patients benefiting from chemotherapy, whereas no interaction was seen for endocrine therapy (Harbeck et al. 2002). Furthermore, high levels of uPA, uPAR, and PAI-1 in primary cytosols from patients with recurrent breast cancer treated with tamoxifen were found to be predictive for poorer response to treatment and for shorter progression-free survival (PFS) (Foekens et al. 1995, Meijer-van Gelder et al. 2004).

The hypothesis that active uPA in complex with PAI-1 could be a stronger prognostic marker than the total amounts of uPA and PAI-1 was investigated in tumor tissue extracts from 342 breast cancer patients. High levels of the complex predicted longer RFS. However, total PAI-1 levels showed superior prognostic power (Pedersen et al. 2000). In another study, where the levels of uPA–PAI-1 complexes were measured in tumor extracts from 1,119 patients with primary breast tumors, high levels were found to correlate with a decrease in RFS time in a multivariate analysis. In this study, the efficacy of adjuvant chemotherapy was found to be associated with uPA–PAI-1 complexes in primary breast cancer and patients with high levels of the complex having longer RFS versus patients having lower levels of the complex, suggesting a predictive role for the complex (Manders et al. 2004).

Measurements of the receptor for uPA (uPAR) in tumor tissue and association with prognosis in primary breast cancer have demonstrated that high levels of uPAR are correlated with poor RFS and OS (Grondahl-Hansen et al. 1995, de Witte et al. 2001). The prognostic impact of uPAR is reduced when PAI-1, in particular, is included in the multivariate model (Foekens et al. 2000). In some studies, subgroup analyses suggest that the prognostic value of uPAR may be related to specific clinical subtypes, for example, postmenopausal node-positive breast cancer patients (Grondahl-Hansen et al. 1995).

In breast tumor tissue, uPA, PAI-1, and uPAR are located in stromal cells. uPA immunreactivity is mainly identified on myofibroblasts and macrophages in ductal breast cancer (Nielsen et al. 2001). The majority of PAI-1-expressing cells in invasive ductal breast carcinoma are myofibroblasts (Offersen et al. 2003), whereas uPAR immunostaining is associated with macrophages immediately surrounding the malignant epithelium (Pyke et al. 1993, Bianchi et al. 1994).

Another approach to the assessment of uPAR's predictive potential is to analyze circulating levels of uPAR (suPAR, soluble urokinase plasminogen activator receptor). Peripheral blood is considerably more homogeneous and easier to collect than tumor tissue. Furthermore, since the pathologist must have priority in selection of material for diagnosis and also because a tumor is rather heterogeneous, the piece that is available for extraction and immunoassay may not be completely representative of the whole tumor. To investigate the prognostic significance of suPAR in blood, the levels were determined in preoperatively collected serum samples from 274 primary breast cancer patients (Riisbro et al. 2002). A significant association, independent of the classical clinical covariates, with RFS as well as OS in these patients was demonstrated. Tumor tissue cytosols were available from a subset of these patients and the content of suPAR in cytosols was predictive for OS, independent of the suPAR level in sera (Riisbro et al. 2002). However, the levels of suPAR in serum and cytosol from the same patient were not correlated. In another group of patients with primary breast cancer ($n = 113$), no significant correlation was found between the levels of uPA, PAI-1, and uPA–PAI-1 complex in EDTA plasma and tumor tissue extracts (Grebenchtchikov et al. 2005).

Colorectal Cancer

The prognostic significance of uPA in colorectal cancer (CRC) was initially investigated analyzing extracts of frozen tumor tissue from histologically verified CRC patients sampled during operation ($n = 92$). The association of uPA levels with OS was statistically significant in univariate analysis (Ganesh et al. 1994a). The multivariate analysis could not show a significant association between the level of uPA and OS; however, the power of this study is rather weak. In another study testing the association of uPA with patient outcome, patients with high levels of uPA had poorer survival independent of stage (Skelly et al. 1997). The concentration of uPAR in extracts from tumor tissue resected from 161 CRC patients identified the uPAR concentration as an independent and significant prognostic factor for 5-year survival (OS) (Ganesh et al. 1994b).

The cellular localization of uPA in colon adenocarcinomas is confined to fibroblast-like cells and endothelial cells in the tumor stroma, while no staining of the malignant epithelial cells was detected (Grondahl-Hansen et al. 1991). A more recent study demonstrated that the PAI-1-positive cells located at the leading edge of the invasive tumor are myofibroblasts, and it is possible that these myofibroblasts also express uPA (Illemann et al. 2004). uPAR is expressed primarily by macrophages and some budding cancer cells located at the leading edge of invasive CRC (Pyke et al. 1994, 1995).

The circulating levels of PAI-1 were measured in EDTA plasma samples and related to OS in a cohort of 609 CRC patients. High levels of PAI-1 were found to be statistically significantly associated with poor survival in a univariate but not in a multivariate analysis (Nielsen et al. 1998). The level of suPAR was analyzed in plasma samples from 591 patients from the same cohort and was demonstrated to significantly predict OS independently of clinical baseline values including stage and age (Stephens et al. 1999). Figure 28.2 shows the Kaplan–Meier estimates of OS for patients with low and high levels of suPAR for each of the Dukes' stage tumors (A, B, C, and D). Since the survival of these patients was recorded as death of all causes, the survival of an age- and gender-matched population was used for comparison (Fig. 28.2). As illustrated in the figure, patients with Dukes' stage B cancer and with suPAR levels below the cutpoint, the survival did not differ from the expected mortality of an age- and gender-matched population (Stephens et al. 1999). A validation study of suPAR's prognostic value confirmed these results in an independent set of rectal cancer patients (Fernebro et al. 2001, Riisbro et al. 2005).

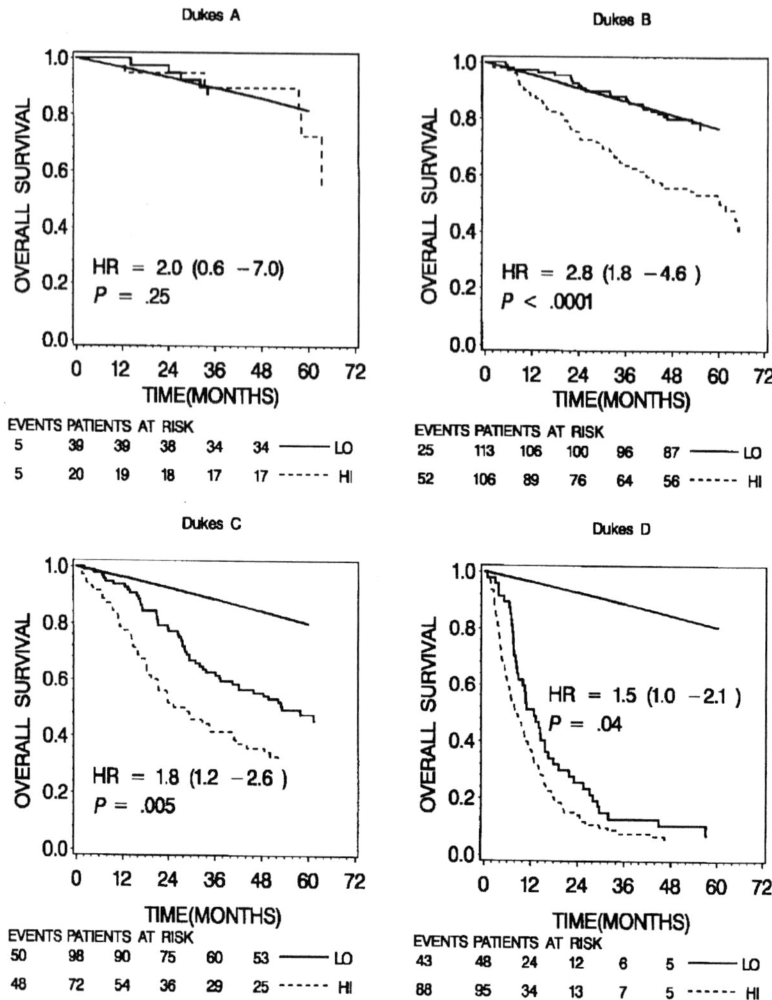

Fig. 28.2 Preoperative plasma soluble urokinase plasminogen activator receptor (*suPAR*) level and overall survival (*OS*) for patients with colorectal cancer (*CRC*) in each of the four Dukes' stages. Patients in each stage were divided into two groups consisting of those above (stepped solid curve) and those below (stepped dashed curve) an arbitrary cut point, which was the median plasma *suPAR* level determined for all patients (44.6 pmol/l). The survival curve for a Dukes' stage-specific, age- and sex-matched cohort drawn from the general Danish population is also plotted (continuous solid line) for each Dukes' stage. The *P* value (two-sided) was calculated by use of the log-rank test, and HRs with 95% confidence intervals (*CIs*) were calculated by use of the Cox regression model. The numbers of patients at risk after each 12-month interval up to 48 months are indicated below each plot. For patients with Dukes' B disease, the survival probabilities at 24 months and 48 months (plus 95% *CIs*) were 92% (87–97) and 80% (72–87), respectively, for the low *suPAR* (LO) group and 74% (66–83) and 55% (46–65), respectively, for the high *suPAR* (HI) group. For patients with Dukes' stage C disease, the survival probabilities at 24 months and 48 months (plus 95% *CIs*) were 77% (69–85) and 54% (44–64), respectively, for the low *suPAR* (LO) group and 51% (40–63) and 34% (23–45), respectively, for the high *suPAR* (HI) group. Reproduced from Stephens et al. (1999) with permission from Oxford University Press

Lung Cancer

Lung cancer can be classified as small-cell lung cancer (SCLC) or non-small-cell lung cancer (NSCLC); the latter can be further classified as squamous cell carcinoma, large-cell carcinoma, or adenocarcinoma according to histological criteria. The prognostic value of uPA, PAI-1, and uPAR has been investigated in the different types of NSCLC, but to our knowledge not yet in SCLC. In patients with pulmonary adenocarcinomas, a high level of PAI-1 but not uPA in tumor tissue extracts ($n = 106$) was found to be a significant predictor of shorter OS (Pedersen et al. 1994b). The same result was obtained when lung tumor tissue homogenates from 147 patients with NSCLC were analyzed for their content of uPA and PAI-1 antigen and correlated with OS (Werle et al. 2004). In another study, tumor tissue extracts from patients with squamous cell lung cancer ($n = 77$) and large-cell lung cancer ($n = 38$) were analyzed for their levels of uPA, uPAR, and PAI-1. None of the components was found to have a statistically significant prognostic impact in the group of large-cell lung cancer, whereas a survival benefit was demonstrated for squamous cell lung cancer patients with low levels of uPAR in tumor extracts (Pedersen et al. 1994a). In two later studies pooling patients with NSCLC with tumors of different histologies ($n = 88$) (Salden et al. 2000) and ($n = 118$) (Offersen et al. 2007), an independent and significant survival difference could not be demonstrated for uPA, PAI-1, and uPAR levels analyzed in tumor extracts. One reason could be that the different histological subgroups have been combined, as they may differ with respect to prognosis (Mountain et al. 1987). Another factor could be that the ELISA formats employed in the studies were different.

It has been hypothesized that cleavage of uPAR is a sign of an active plasminogen activation system, and therefore the cleaved variants should have a stronger prognostic significance than the level of total uPAR (Solberg et al. 1994). This was tested by measuring uPAR(I) in 63 of the 77 tumor extracts from patients with squamous cell lung cancer (Almasi et al. 2005), that were previously analyzed by uPAR ELISA (Pedersen et al. 1994a). The prognostic significance of ELISA-measured uPAR and time-resolved fluorescence immunoassay measured uPAR(I) in the 63 tumor extracts and their impact on survival were analyzed showing that uPAR(I) was statistically significant (Almasi et al. 2005). The prognostic strength of specific uPAR forms compared to the collective amounts of all uPAR forms was also evident in a study of lung tumor tissue homogenates from 95 patients with NSCLC (Werle et al. 2004). The samples were analyzed using three different ELISAs, a commercially available ELISA measuring all forms of uPAR and also the complexed forms (American Diagnostica Inc.), one where the epitope specificity of the detecting antibody is not revealed (HD13), and the third which will only measure uPAR(I–III) and uPAR(I) (IIIF10). In a multivariate analysis, only the levels of the uPAR forms measured with the uPAR(IIIF10) assay were significantly associated with OS (Werle et al. 2004).

Cleavage of uPAR and the levels of liberated uPAR(I) reflect uPA activity. The complex uPA–PAI-1 is another way of monitoring active uPA. The levels of uPA–PAI-1 complexes were determined in tumor tissue extracts from patients with pulmonary adenocarcinoma ($n = 99$), showing that levels of uPA–PAI-1 complexes were not correlated with survival; however, low complex levels in combination with high PAI-1 levels were associated with poor prognosis in lung cancer patients with pulmonary adenocarcinoma (Pappot et al. 2006).

Elevated levels of uPA, PAI-1, and suPAR are present in blood from patients with lung cancer even though their prognostic significance has not been determined. The levels of uPA were significantly higher in EDTA plasma from patients with NSCLC than in plasma from the control subjects and were significantly higher in plasma from patients with squamous cell carcinoma than in plasma from patients with pulmonary adenocarcinoma (Yang et al. 2005). The suPAR level was also significantly increased in citrate plasma from patients with NSCLC compared to that in citrate plasma from healthy donors, whereas there was no increase in the suPAR level in citrate plasma from patients with SCLC (Pappot et al. 1997).

Gynecological Cancer

Ovarian cancer is often characterized by late diagnosis and poor prognosis. Early detection can result in curative intervention and there is a need for biomarkers for improved classification of patient survival. One of the first studies to demonstrate an association between survival and uPA and PAI-1 was restricted to patients ($n = 51$) with advanced ovarian cancer (FIGO IIIc), showing that both markers predict survival with emphasis on patients without residual tumor (Kuhn et al. 1994). These authors confirmed their finding in a later study on 86 patients with ovarian cancer with stage FIGO IIIc, where PAI-1 was shown to be an independent prognostic marker in FIGO IIIc patients with or without residual tumor (Kuhn et al. 1999). To identify patients at risk in the latter study, optimized cutoff values (log-rank statistics) for uPA and PAI-1 were used. In a cohort of primary ovarian cancer of all FIGO stages ($n = 82$), both uPA and PAI-1 levels in tumor tissue extracts were significantly associated with RFS and OS, using log-rank statistics to calculate the optimal cutoff values (Konecny et al. 2001). However, in another study measuring uPA and PAI-1 in cytosols from 90 patients with ovarian cancer and using the median as the cutoff value, no significant association of these biomarkers with RFS or OS could be shown (van der Burg et al. 1996).

Whereas high levels of uPA and PAI-1 in ovarian tumor tissue predict short OS, the opposite was found for uPAR in tumor tissue from patients with ovarian cancer. The levels of uPAR were lower in benign as compared to invasive or borderline tumors. However, among the malignant tumors ($n = 51$), the more advanced and poorly differentiated tumors contained lower levels of uPAR than the well-differentiated, less-advanced tumors (Borgfeldt et al. 2003).

The circulating levels of suPAR have also been measured in preoperatively collected sera from patients with ovarian cancer. When patients ($n = 87$) were ranked according to their FIGO classification, the highest levels were found in stage II patients and then the levels decreased with increasing FIGO stage. Nevertheless, high levels of suPAR in preoperatively collected sera were found to correlate with poor survival (Sier et al. 1998). In another study where citrate plasma samples from 53 patients with ovarian cancer were analyzed, the levels of suPAR were significantly associated with RFS (Riisbro et al. 2001).

Patients with progressive ovarian cancer disease often present with ascites/peritoneal fluid and in a small study ($n = 36$) of primary ascites, secreted uPA and uPAR were not found to be independent prognostic markers (Chambers et al. 1995). Ovarian cysts are detected in some women and can be both benign and malignant. Ascites as well as cystic fluids were collected in a study where the concentrations of suPAR in these body fluids were compared with those in serum from 77 patients admitted for surgery of ovarian tumors (Wahlberg et al. 1998). Elevated suPAR levels were found in serum from patients with more advanced disease. However, the concentrations of suPAR in the body fluids were quite different; in serum the measured concentrations were below 100 pmol/l, in ascites/peritoneal fluid were between 290 and 590 pmol/l, and in cystic fluids were even higher, from 650 to a maximum above 8,000 pmol/l. The concentrations of suPAR in cystic fluids clearly separated benign and malignant cysts with predictive values above 90%. The levels of suPAR in cystic fluids might, therefore, aid the early diagnosis of patients with ovarian cancer. Both intact and cleaved suPAR were present in the cystic fluids (Wahlberg et al. 1998). Different forms of uPAR were also found in tumor tissue, serum, ascites, and urine from patients with ovarian cancer. Whereas all tumor lysates, ascites, and urine contained uPAR(I–III) and uPAR(II–III), uPAR(I) was only detected in urine samples and in serum, only intact suPAR was visualized in Western blots (Sier et al. 2004).

In endometrial cancer, all three biomarkers were measured in tumor tissue extracts and the levels of uPAR did not show any correlation with clinical parameters. However, elevated PAI-1 levels were significantly associated with shorter RFS and OS (Tecimer et al. 2001). The levels of suPAR in endometrial cancer patients ($n = 34$) were elevated compared to those in healthy donors and patients with benign endometrial diseases, but no survival analysis was performed (Riisbro et al. 2001).

Prostate Cancer

In prostate cancer tissue, the levels of uPA, uPAR, and PAI-1 have, to our knowledge, not been determined using quantitative immunoassays, even though the cellular localization of uPAR and PAI-1 has been determined using immunohistochemistry (Usher et al. 2005). uPAR is located on neutrophils and macrophages and PAI-1 in myofibroblasts, macrophages, and endothelial cells. In no cases was staining in cancer cells demonstrated in tissue from 16 patients with histologically confirmed prostate adenocarcinoma (Usher et al. 2005).

The circulating levels of uPA and suPAR in prostate cancer patients have been shown to be associated with OS (Miyake et al. 1999, Shariat et al. 2007). In this study ($n = 72$), a significant association was demonstrated in a univariate analysis.

In a newer study with 429 patients, a significant correlation to PFS was found for uPA and suPAR, and they were also found to be independent of baseline clinical covariates (Shariat et al. 2007). This study also showed that the circulating levels of both uPA and suPAR were higher in patients with prostate cancer than in healthy controls.

In a recent study, the concentrations of uPAR(I) as well as the calculated suPAR(II–III) were found to be significantly elevated in serum samples from patients with prostate cancer compared to those in serum from men with benign prostatic conditions. Specific measurements of uPAR(I) were found to improve specificity of prostate cancer detection (Piironen et al. 2006).

Discussion

An important aspect in the presentation of marker studies is the description of sampling and storage methods as well as the validation of the immunoassay methods (recovery, linearity, specificity) that has been performed. Guidelines for reporting these points have been proposed (McShane et al. 2006). The prognostic impact of uPA, uPAR, and PAI-1 on survival endpoints (RFS, OS, and PFS) presented in this chapter is based on the use of regression analysis to develop prognostic models. These models aim at improving the classification of disease severity and ultimately selection of patients for adjuvant therapies or prediction of patients likely to respond to a particular treatment. The validity of these statistical models should be carefully considered to avoid bias (Simon and Altman 1994). Divergent results cannot be only due to technical aspects of the measurement but also due to statistical methodology. Many of the studies are underpowered and the number of events low, 10 events per covariate is considered to be a minimum (Peduzzi et al. 1996). In many studies, researchers have chosen to dichotomize continuous marker levels for statistical analysis, whereas others use the actual or transformed value, often log transformed making comparisons of results difficult (Altman and Royston 2006). Some analyses are based on optimized cutpoints, which should be considered with some caution (Altman et al. 1994). Proper specification of the statistical model and validation of the analysis is mandatory for correct interpretation of the results (Harrell et al. 1996). These considerations include specification of relevant covariates and interactions taking caution to avoid over-fitting of the model, verification of model assumptions, and checking for lack of fit. Two types of validated models have been proposed, the statistically validated model as described above and the clinically validated model (Altman and Royston 2000). The latter is an assessment of how well a model performs in an independent patient dataset. The published prognostic studies on uPA, PAI-1, and uPAR in cancer vary greatly with respect to sampling techniques, assay methods, sample size, endpoints, and methods of statistical analysis. A more standardized approach to reporting tumor marker studies has been proposed, presenting recommendations for use in publication of these studies (McShane et al. 2006). The development of clinically relevant markers for prognosis and prediction in cancer requires an evidence-based approach, encompassing systematic reviews and meta-analyses as well as prospective, collaborative studies (Altman and Riley 2005).

The present status for the evaluation of the prognostic value of the uPA, PAI-1, and uPAR is that there is a need for properly designed prospective studies to confirm the hypotheses generated from retrospective studies using samples and data accrued for other purposes. The only exception is breast cancer where one large pooled study confirming the hypothesis based on 8,377 breast cancer patients (Look et al. 2002) has been published bringing the LOE to I and a randomized prospective study comparing intervention with adjuvant treatment to low-risk patients with a high-risk profile defined by their uPA/PAI-1 levels (Janicke et al. 2001). The clinical use of uPA and PAI-1 for assessment of node-negative breast cancer patients has been recommended in combination with grading and stage (Thomssen et al. 2003). However, due to decreasing tumor sizes it has been suggested that the sample requirements should be modified, a micro-ELISA should be used, and the assay revalidated before implementation (Hayes 2005).

The levels of circulating uPA and uPAR have been shown to be elevated in several cancers including lung and prostate and to correlate with RFS and OS in patients with breast cancer (Riisbro et al. 2002), to OS in CRC (Stephens et al. 1999), and to PFS in prostate cancer (Shariat et al. 2007). Elevated suPAR levels in blood from patients with infectious diseases have been reported and their prognostic value assessed, thus demonstrating that these markers are not specific for cancer diseases. Although interesting results have been obtained, more studies are needed to confirm these findings, but the present evidence suggests that the blood levels of suPAR could be a marker of severe inflammation and immune activation (Mizukami et al. 1995, Slot et al. 1999, Ostrowski et al. 2005a, b).

Conclusion

A large proportion of the reviewed studies show that high levels of uPA, uPAR, and PAI-1 in solid tumors are associated with poor prognosis defined by endpoints such as overall survival, recurrence-free survival, and progression-free survival, suggesting that these biomarkers have value as prognostic indicators. Although most of these studies are retrospective and meta-analyses as well as prospective studies are required to confirm or reject the posed hypotheses, we believe that the evidence points to a prognostic as well as a predictive role for these biomarkers. The circulating levels of uPA, uPAR, and PAI-1 have also been shown to be associated with prognosis, elevated levels have been found in patients with solid tumors, and the circulating levels are not associated with tumor tissue levels of the same biomarker. Circulating levels of uPA, uPAR, and PAI-1 are not specific for a single cancer or cancer in general and their use should be evaluated in the context of possible co-morbidity.

Acknowledgements The excellent graphic assistance provided by John Post is gratefully acknowledged. This work was supported by EU contract LSHC-CT-2003-503297 as well as the Danish Cancer Society.

References

Almasi C. E., Hoyer-Hansen G., Christensen I. J., et al. (2005). Prognostic impact of liberated domain I of the urokinase plasminogen activator receptor in squamous cell lung cancer tissue. Lung Cancer 48:349–355.

Altman D. G. and Riley R. D. (2005). Primer: an evidence-based approach to prognostic markers. Nat Clin Pract Oncol 2:466–472.

Altman D. G. and Royston P. (2000). What do we mean by validating a prognostic model? Stat Med 19:453–473.

Altman D. G. and Royston P. (2006). The cost of dichotomising continuous variables. BMJ 332:1080.

Altman D. G., Lausen B., Sauerbrei W., et al. (1994). Dangers of using "optimal" cutpoints in the evaluation of prognostic factors. J Natl Cancer Inst 86:829–835.

Andreasen P. A., Kjoller L., Christensen L., et al. (1997). The urokinase-type plasminogen activator system in cancer metastasis: a review. Int J Cancer 72:1–22.

Andreasen P. A., Egelund R., and Petersen H. H. (2000). The plasminogen activation system in tumor growth, invasion, and metastasis. Cell Mol Life Sci 57:25–40.

Bianchi E., Cohen R. L., Thor A. T., et al. (1994). The urokinase receptor is expressed in invasive breast cancer but not in normal breast tissue. Cancer Res 54:861–866.

Borgfeldt C., Bendahl P. O., Gustavsson B., et al. (2003). High tumor tissue concentration of urokinase plasminogen activator receptor is associated with good prognosis in patients with ovarian cancer. Int J Cancer 107:658–665.

Chambers S. K., Gertz R. E., Jr., Ivins C. M. et al. (1995). The significance of urokinase-type plasminogen activator, its inhibitors, and its receptor in ascites of patients with epithelial ovarian cancer. Cancer 75:1627–1633.

Dano K., Behrendt N., Hoyer-Hansen G., et al. (2005). Plasminogen activation and cancer. Thromb Haemost 93:676–681.

de Witte H., Sweep F., Brunner N., et al. (1998). Complexes between urokinase-type plasminogen activator and its receptor in blood as determined by enzyme-linked immunosorbent assay. Int J Cancer 77:236–242.

de Witte J. H., Foekens J. A., Brunner N., et al. (2001). Prognostic impact of urokinase-type plasminogen activator receptor (uPAR) in cytosols and pellet extracts derived from primary breast tumours. Br J Cancer 85:85–92.

Duffy M. J. (2002). Urokinase plasminogen activator and its inhibitor, PAI-1, as prognostic markers in breast cancer: from pilot to level 1 evidence studies. Clin Chem 48:1194–1197.

Duffy M. J., Maguire T. M., McDermott E. W., et al. (1999). Urokinase plasminogen activator: a prognostic marker in multiple types of cancer. J Surg Oncol 71:130–135.

Fernebro E., Madsen R. R., Ferno M., et al. (2001). Prognostic importance of the soluble plasminogen activator receptor, suPAR, in plasma from rectal cancer patients. Eur J Cancer 37:486–491.

Foekens J. A., Look M. P., Peters H. A., et al. (1995). Urokinase-type plasminogen activator and its inhibitor PAI-1: predictors of poor response to tamoxifen therapy in recurrent breast cancer. J Natl Cancer Inst 87:751–756.

Foekens J. A., Peters H. A., Look M. P., et al. (2000). The urokinase system of plasminogen activation and prognosis in 2780 breast cancer patients. Cancer Res 60:636–643.

Ganesh S., Sier C. F., Griffioen G., et al. (1994a). Prognostic relevance of plasminogen activators and their inhibitors in colorectal cancer. Cancer Res 54:4065–4071.

Ganesh S., Sier C. F., Heerding M. M. (1994b). Urokinase receptor and colorectal cancer survival. Lancet 344:401–402.

Grebenchtchikov N., Maguire T. M., Riisbro R., et al. (2005). Measurement of plasminogen activator system components in plasma and tumor tissue extracts obtained from patients with breast cancer: an EORTC Receptor and Biomarker Group collaboration. Oncol Rep 14:235–239.

Grebenschikov N., Sweep F., Geurts A., et al. (1999). ELISA for complexes of urokinase-type and tissue-type plasminogen activators with their type-1 inhibitor (uPA-PAI-1 and tPA-PAI-1). Int J Cancer 81:598–606.

Grondahl-Hansen J., Agerlin N., Munkholm-Larsen P., et al. (1988). Sensitive and specific enzyme-linked immunosorbent assay for urokinase-type plasminogen activator and its application to plasma from patients with breast cancer. J Lab Clin Med 111:42–51.

Grondahl-Hansen J., Ralfkiaer E., Kirkeby L. T., et al. (1991). Localization of urokinase-type plasminogen activator in stromal cells in adenocarcinomas of the colon in humans. Am J Pathol 138:111–117.

Grondahl-Hansen J., Christensen I. J., Rosenquist C., et al. (1993). High levels of urokinase-type plasminogen activator and its inhibitor PAI-1 in cytosolic extracts of breast carcinomas are associated with poor prognosis. Cancer Res 53:2513–2521.

Grondahl-Hansen J., Peters H. A., van Putten W. L., et al. (1995). Prognostic significance of the receptor for urokinase plasminogen activator in breast cancer. Clin Cancer Res 1:1079–1087.

Grondahl-Hansen J., Christensen I. J., Briand P., et al. (1997). Plasminogen activator inhibitor type 1 in cytosolic tumor extracts predicts prognosis in low-risk breast cancer patients. Clin Cancer Res 3:233–239.

Hansen S., Overgaard J., Rose C., et al. (2003). Independent prognostic value of angiogenesis and the level of plasminogen activator inhibitor type 1 in breast cancer patients. Br J Cancer 88:102–108.

Harbeck N., Kates R. E., Look M. P., et al. (2002). Enhanced benefit from adjuvant chemotherapy in breast cancer patients classified high-risk according to urokinase-type plasminogen activator (uPA) and plasminogen activator inhibitor type 1 (n = 3424). Cancer Res 62:4617–4622.

Harrell F. E., Jr., Lee K. L., and Mark D. B. (1996). Multivariable prognostic models: issues in developing models, evaluating assumptions and adequacy, and measuring and reducing errors. Stat Med 15:361–387.

Hayes D. F., Bast R. C., Desch C. E. et al. (1996). Tumor marker utility grading system: A framework to evaluate clinical utility of tumor markers. J Natl Cancer Inst 88:1456–1466.

Hayes D. F. (1998) Determination of clinical utility of tumor markers: a tumor marker utility grading system. Rec Res Cancer Res 152:71–85.

Hayes D. F. (2005). Prognostic and predictive factors revisited. Breast 14:493–499.

Hoyer-Hansen G. and Lund I. K. (2007). Urokinase receptor variants in tissue and body fluids. Adv Clin Chem 44:65–102.

Hoyer-Hansen G., Ronne E., Solberg H., et al. (1992). Urokinase plasminogen activator cleaves its cell surface receptor releasing the ligand-binding domain. J Biol Chem 267:18224–18229.

Hoyer-Hansen G., Behrendt N., Ploug M., et al. (1997a). The intact urokinase receptor is required for efficient vitronectin binding: receptor cleavage prevents ligand interaction. FEBS Lett 420:79–85.

Hoyer-Hansen G., Ploug M., Behrendt N., et al. (1997b). Cell-surface acceleration of urokinase-catalyzed receptor cleavage. Eur J Biochem 243:21–26.

Hoyer-Hansen G., Hamers M. J., Pedersen A. N., et al. (2000). Loss of ELISA specificity due to biotinylation of monoclonal antibodies. J Immunol Methods 235:91–99.

Hoyer-Hansen G., Pessara U., Holm A., et al. (2001). Urokinase-catalysed cleavage of the urokinase receptor requires an intact glycolipid anchor. Biochem J 358:673–679.

Illemann M., Hansen U., Nielsen H., et al. (2004). Leading-edge myofibroblasts in human colon cancer express plasminogen activator inhibitor-1. Am J Clin Pathol 122:256–265.

Janicke F., Schmitt M., and Graeff H. (1991). Clinical relevance of the urokinase-type and tissue-type plasminogen activators and of their type 1 inhibitor in breast cancer. Semin Thromb Hemost 17:303–312.

Janicke F., Schmitt M., Pache L., et al. (1993). Urokinase (uPA) and its inhibitor PAI-1 are strong and independent prognostic factors in node-negative breast cancer. Breast Cancer Res Treat 24:195–208.

Janicke F., Prechtl A., Thomssen C., et al. (2001). Randomized adjuvant chemotherapy trial in high-risk, lymph node-negative breast cancer patients identified by urokinase-type plasminogen activator and plasminogen activator inhibitor type 1. J Natl Cancer Inst 93:913–920.

Konecny G., Untch M., Pihan A., et al. (2001). Association of urokinase-type plasminogen activator and its inhibitor with disease progression and prognosis in ovarian cancer. Clin Cancer Res 7:1743–1749.

Kotzsch M., Luther T., Harbeck N., et al. (2000). New ELISA for quantitation of human urokinase receptor (CD87) in cancer. Int J Oncol 17:827–834.

Kuhn W., Pache L., Schmalfeldt B., et al. (1994). Urokinase (uPA) and PAI-1 predict survival in advanced ovarian cancer patients (FIGO III) after radical surgery and platinum-based chemotherapy. Gynecol Oncol 55:401–409.

Kuhn W., Schmalfeldt B., Reuning U., et al. (1999). Prognostic significance of urokinase (uPA) and its inhibitor PAI-1 for survival in advanced ovarian carcinoma stage FIGO IIIc. Br J Cancer 79:1746–1751.

Lijnen H. R. (2005). Pleiotropic functions of plasminogen activator inhibitor-1. J Thromb Haemost 3:35–45.

Look M. P., van Putten W. L., Duffy M. J., et al. (2002). Pooled analysis of prognostic impact of urokinase-type plasminogen activator and its inhibitor PAI-1 in 8377 breast cancer patients. J Natl Cancer Inst 94:116–128.

Manders P., Tjan-Heijnen V. C., Span P. N., et al. (2004). Predictive impact of urokinase-type plasminogen activator: plasminogen activator inhibitor type-1 complex on the efficacy of adjuvant systemic therapy in primary breast cancer. Cancer Res 64:659–664.

McShane L. M., Altman D. G., Sauerbrei W., et al. (2006). REporting recommendations for tumor MARKer prognostic studies (REMARK). Breast Cancer Res Treat 100:229–235.

Meijer-van Gelder M. E., Look M. P., Peters H. A., et al. (2004). Urokinase-type plasminogen activator system in breast cancer: association with tamoxifen therapy in recurrent disease. Cancer Res 64:4563–4568.

Miyake H., Hara I., Yamanaka K., et al. (1999). Elevation of serum levels of urokinase-type plasminogen activator and its receptor is associated with disease progression and prognosis in patients with prostate cancer. Prostate 39:123–129.

Mizukami I. F., Faulkner N. E., Gyetko M. R., et al. (1995). Enzyme-linked immunoabsorbent assay detection of a soluble form of urokinase plasminogen activator receptor in vivo. Blood 86:203–211.

Mountain C. F., Lukeman J. M., Hammar S. P., et al. (1987). Lung cancer classification: the relationship of disease extent and cell type to survival in a clinical trials population. J Surg Oncol 35:147–156.

Nielsen B. S., Sehested M., Duun S., et al. (2001). Urokinase plasminogen activator is localized in stromal cells in ductal breast cancer. Lab Invest 81:1485–1501.

Nielsen H. J., Pappot H., Christensen I. J., et al. (1998). Association between plasma concentrations of plasminogen activator inhibitor-1 and survival in patients with colorectal cancer. BMJ 316:829–830.

Offersen B. V., Nielsen B. S., Hoyer-Hansen G., et al. (2003). The myofibroblast is the predominant plasminogen activator inhibitor-1-expressing cell type in human breast carcinomas. Am J Pathol 163:1887–1899.

Offersen B. V., Pfeiffer P., Andreasen P., et al. (2007). Urokinase plasminogen activator and plasminogen activator inhibitor type-1 in nonsmall-cell lung cancer: relation to prognosis and angiogenesis. Lung Cancer 56:43–50.

Ostrowski S. R., Piironen T., Hoyer-Hansen G., et al. (2005a). High plasma levels of intact and cleaved soluble urokinase receptor reflect immune activation and are independent predictors of mortality in HIV-1-infected patients. J Acquir Immune Defic Syndr 39:23–31.

Ostrowski S. R., Ullum H., Goka B. Q., et al. (2005b). Plasma concentrations of soluble urokinase-type plasminogen activator receptor are increased in patients with malaria and are associated with a poor clinical or a fatal outcome. J Infect Dis 191:1331–1341.

Pappot H., Hoyer-Hansen G., Ronne E., et al. (1997). Elevated plasma levels of urokinase plasminogen activator receptor in non-small cell lung cancer patients. Eur J Cancer 33:867–872.

Pappot H., Pedersen A. N., Brunner N., et al. (2006). The complex between urokinase (uPA) and its type-1 inhibitor (PAI-1) in pulmonary adenocarcinoma: relation to prognosis. Lung Cancer 51:193–200.

Pedersen A. N., Brunner N., Hoyer-Hansen G., et al. (1999). Determination of the complex between urokinase and its type-1 inhibitor in plasma from healthy donors and breast cancer patients. Clin Chem 45:1206–1213.

Pedersen A. N., Christensen I. J., Stephens R. W., et al. (2000). The complex between urokinase and its type-1 inhibitor in primary breast cancer: relation to survival. Cancer Res 60:6927–6934.

Pedersen A. N., Mouridsen H. T., Tenney D. Y., et al. (2003). Immunoassays of urokinase (uPA) and its type-1 inhibitor (PAI-1) in detergent extracts of breast cancer tissue. Eur J Cancer 39:899–908.

Pedersen H., Brunner N., Francis D., et al. (1994a). Prognostic impact of urokinase, urokinase receptor, and type 1 plasminogen activator inhibitor in squamous and large cell lung cancer tissue. Cancer Res 54.4671–4675.

Pedersen H., Grondahl-Hansen J., Francis D., et al. (1994b). Urokinase and plasminogen activator inhibitor type 1 in pulmonary adenocarcinoma. Cancer Res 54:120–123.

Peduzzi P., Concato J., Kemper E., et al. (1996). A simulation study of the number of events per variable in logistic regression analysis. J Clin Epidemiol 49:1373–1379.

Piironen T., Laursen B., Pass J., et al. (2004). Specific immunoassays for detection of intact and cleaved forms of the urokinase receptor. Clin Chem 50:2059–2068.

Piironen T., Haese A., Huland H., et al. (2006). Enhanced discrimination of benign from malignant prostatic disease by selective measurements of cleaved forms of urokinase receptor in serum. Clin Chem 52:838–844.

Ploug M. (2003). Structure-function relationships in the interaction between the urokinase-type plasminogen activator and its receptor. Curr Pharm Des 9:1499–1528.

Pyke C., Graem N., Ralfkiaer E., et al. (1993). Receptor for urokinase is present in tumor-associated macrophages in ductal breast carcinoma. Cancer Res 53:1911–1915.

Pyke C., Ralfkiaer E., Ronne E., et al. (1994). Immunohistochemical detection of the receptor for urokinase plasminogen activator in human colon cancer. Histopathology 24:131–138.

Pyke C., Salo S., Ralfkiaer E., et al. (1995). Laminin-5 is a marker of invading cancer cells in some human carcinomas and is coexpressed with the receptor for urokinase plasminogen activator in budding cancer cells in colon adenocarcinomas. Cancer Res 55:4132–4139.

Riisbro R., Stephens R. W., Brunner N., et al. (2001). Soluble urokinase plasminogen activator receptor in preoperatively obtained plasma from patients with gynecological cancer or benign gynecological diseases. Gynecol Oncol 82:523–531.

Riisbro R., Christensen I. J., Piironen T., et al. (2002). Prognostic significance of soluble urokinase plasminogen activator receptor in serum and cytosol of tumor tissue from patients with primary breast cancer. Clin Cancer Res 8:1132–1141.

Riisbro R., Christensen I. J., Nielsen H. J., et al. (2005). Preoperative plasma soluble urokinase plasminogen activator receptor as a prognostic marker in rectal cancer patients. An EORTC-Receptor and Biomarker Group collaboration. Int J Biol Markers 20:93–102.

Ronne E., Hoyer-Hansen G., Brunner N., et al. (1995). Urokinase receptor in breast cancer tissue extracts. Enzyme-linked immunosorbent assay with a combination of mono- and polyclonal antibodies. Breast Cancer Res Treat 33:199–207.

Salden M., Splinter T. A., Peters H. A., et al. (2000). The urokinase-type plasminogen activator system in resected non-small-cell lung cancer. Rotterdam Oncology Thoracic Study Group. Ann Oncol 11:327–332.

Schmitt M., Mengele K., Schueren E., et al. (2007). European organisation for research and treatment of cancer (EORTC) pathobiology group standard operating procedure for the preparation of human tumour tissue extracts suited for the quantitative analysis of tissue-associated biomarkers. Eur J Cancer 43:835–844.

Shariat S. F., Roehrborn C. G., McConnell J. D., et al. (2007). Association of the circulating levels of the urokinase system of plasminogen activation with the presence of prostate cancer and invasion, progression, and metastasis. J Clin Oncol 25:349–355.

Sier C. F., Stephens R., Bizik J., et al. (1998). The level of urokinase-type plasminogen activator receptor is increased in serum of ovarian cancer patients. Cancer Res 58:1843–1849.

Sier C. F., Nicoletti I., Santovito M. L., et al. (2004). Metabolism of tumour-derived urokinase receptor and receptor fragments in cancer patients and xenografted mice. Thromb Haemost 91:403–411.

Simon R. and Altman D. G. (1994). Statistical aspects of prognostic factor studies in oncology. Br J Cancer 69:979–985.

Skelly M. M., Troy A., Duffy M. J., et al. (1997). Urokinase-type plasminogen activator in colorectal cancer: relationship with clinicopathological features and patient outcome. Clin Cancer Res 3:1837–1840.

Slot O., Brunner N., Locht H., et al. (1999). Soluble urokinase plasminogen activator receptor in plasma of patients with inflammatory rheumatic disorders: increased concentrations in rheumatoid arthritis. Ann Rheum Dis 58:488–492.

Solberg H., Romer J., Brunner N., et al. (1994). A cleaved form of the receptor for urokinase-type plasminogen activator in invasive transplanted human and murine tumors. Int J Cancer 58:877–881.

Stefansson S., McMahon G. A., Petitclerc E., et al. (2003). Plasminogen activator inhibitor-1 in tumor growth, angiogenesis and vascular remodeling. Curr Pharm Des 9:1545–1564.

Stephens R. W., Nielsen H. J., Christensen I. J., et al. (1999). Plasma urokinase receptor levels in patients with colorectal cancer: relationship to prognosis. J Natl Cancer Inst 91:869–874.

Sweep C. G., Geurts-Moespot J., Grebenschikov N., et al. (1998). External quality assessment of trans-European multicentre antigen determinations (enzyme-linked immunosorbent assay) of urokinase-type plasminogen activator (uPA) and its type 1 inhibitor (PAI-1) in human breast cancer tissue extracts. Br J Cancer 78:1434–1441.

Tecimer C., Doering D. L., Goldsmith L. J., et al. (2001). Clinical relevance of urokinase-type plasminogen activator, its receptor, and its inhibitor type 1 in endometrial cancer. Gynecol Oncol 80:48–55.

Thomssen C., Janicke F., and Harbeck N. (2003). Clinical relevance of prognostic factors in axillary node-negative breast cancer. Onkologie 26:438–445.

Usher P. A., Thomsen O. F., Iversen P., et al. (2005). Expression of urokinase plasminogen activator, its receptor and type-1 inhibitor in malignant and benign prostate tissue. Int J Cancer 113:870–880.

van der Burg M. E., Henzen-Logmans S. C., Berns E. M., et al. (1996). Expression of urokinase-type plasminogen activator (uPA) and its inhibitor PAI-1 in benign, borderline, malignant primary and metastatic ovarian tumors. Int J Cancer 69:475–479.

Wahlberg K., Hoyer-Hansen G., and Casslen B. (1998). Soluble receptor for urokinase plasminogen activator in both full-length and a cleaved form is present in high concentration in cystic fluid from ovarian cancer. Cancer Res 58:3294–3298.

Werle B., Kotzsch M., Lah T. T., et al. (2004). Cathepsin B, plasminogen activator-inhibitor (PAI-1) and plasminogen activator-receptor (uPAR) are prognostic factors for patients with non-small cell lung cancer. Anticancer Res 24:4147–4161.

Wilhelm O. G., Wilhelm S., Escott G. M., et al. (1999). Cellular glycosylphosphatidylinositol-specific phospholipase D regulates urokinase receptor shedding and cell surface expression. J Cell Physiol 180:225–235.

Yang S. F., Hsieh Y. S., Lin C. L., et al. (2005). Increased plasma levels of urokinase plasminogen activator and matrix metalloproteinase-9 in nonsmall cell lung cancer patients. Clin Chim Acta 354:91–99.

Chapter 29
Cysteine Cathepsins and Cystatins
as Cancer Biomarkers

Tamara T. Lah, Nataša Obermajer, María Beatriz Durán Alonso,
and Janko Kos

Abstract Cysteine cathepsins are lysosomal cysteine proteases that are involved
in a number of important biological processes, including intracellular protein
turnover, propeptide and prohormone processing, apoptosis, bone remodelling,
and reproduction. In cancer, the cathepsins have been linked to extracellular matrix
remodelling and to the promotion of tumour cell motility, invasion, angiogenesis
and metastasis, resulting in poor outcome of the disease. The levels of cathepsins as
well as of their endogenous inhibitor cystatins in clinical samples have been
suggested as potential biomarkers, and this chapter is focused on their role to
predict the diagnosis, risk of recurrence and death, and response to therapy in
patients with cancer.

Introduction

In recent years, the discovery of new biomarkers has become an important part in
cancer research. The term "biomarker" refers to any measurable diagnostic indicator that is used to assess the risk or presence of the disease. In a broader sense
(Gutman and Kessler 2006), the term "diagnostic" may also imply the possibility to
determine the prognosis or staging of the disease, monitor disease progression,
predict the risk of recurrence and death and/or optimize treatment outcome by
enabling clinicians to select the most effective therapy for individual patient. The
measurements of biomarkers include various techniques which are applied on
collected patients' samples, namely, whole tumour tissue preparations (homogenates, extracts, embedded sections, etc.), laser-micro-dissected tumour areas,

J. Kos
Faculty of Pharmacy, University of Ljubljana, Aškerčeva 6, 1000 Ljubljana, Slovenia,
e-mail: Janko.kos@ffa.uni-lj.si

D. Edwards et al. (eds.), *The Cancer Degradome.*
© Springer Science + Business Media, LLC 2008

comprising specific cell types and tissue structures, and body fluids. The field of biomarkers is expanding from measuring a single biomarker molecule to an increasing array of cutting-edge techniques, including tests for genetic alterations, gene expression arrays, proteomic profiles and antibody immunoassays. New biomarkers should be independent of the clinical–pathological features, if additional benefit and new information on the disease progression pathways are expected. Significantly, new biomarkers are needed in lower stages of the disease progression, as high stage and high histological grade already provide sufficient prognostic and predictive impact.

Among the various biomolecules involved in cancer development and progression, proteases and their inhibitors may provide new diagnostic and prognostic information for patients with cancer, thereby presenting new biomarkers. Among them, cathepsin D, urokinase-type plasminogen activator (uPA), plasminogen activator inhibitor-1 (PAI-1) and tissue inhibitor of metalloproteinase-1 (TIMP-1) have been the subjects of previous and ongoing multicentral studies, whereas other cancer degradome components still need to be either evaluated or approved for larger confirmatory trials. In this chapter, only selected, most investigated cancers are discussed, and a list of the relevant references is given in Tables 29.1–29.4.

Cysteine Peptidases

Peptidases, more commonly known as proteases, comprise a group of 561 genes reported in the human genome, structurally classified by their gene sequence homology but functionally classified according to their catalytic mechanism as cysteine, serine, threonine, aspartate and metalloproteases, or unknown (Barrett et al. 2004; http://www.merops.sanger.ac.uk). Their activity is ultimately regulated by the activation of (pre)proforms, by the balance between the levels of proteases and their endogenous protein inhibitors and by their concentration and localization in different cellular or tissue compartments. So far, 156 inhibitor genes have been identified (http://www.merops.sanger.ac.uk). It has been shown that alterations in gene expression of proteases and their inhibitors have critical effects on various pathological stages, including cancer progression (reviewed by Turk 2006). Despite a lack of precise knowledge regarding their function, proteases and their antagonists are studied as possible markers for various endpoints in cancer progression, including the relapse and death of patients after the tumour removal or application of other therapeutic regimens. A number of proteases have also been targets for cancer therapy, as described in recent reviews (Overall and López-Otín 2002, Lah et al. 2006, Turk 2006).

There are in total 11 cysteine proteases of the C1A papain family (Barrett et al. 2004) in humans, which are trivially called cathepsins, sharing a general mechanism of nucleophilic attack of sulphur anion on the carbonyl carbon of an amide bond. According to structural and conformational differences, they are

Table 29.1 Cathepsin B as biomarker in selective types of cancer

Cathepsin B	Clinical sample	Application
	Serum	Prognosis[a] and diagnosis[b] in hepatocellular carcinoma (Leto et al. 1997)
	Serum	Diagnosis in ovarian patients (Warwas et al. 1997)
	Serum	Prognosis in colon carcinoma (Kos et al. 1998)
	Serum	Diagnosis in head and neck cancer (Strojan et al. 2001)
	Serum	Diagnosis in tongue cancer (Saleh et al. 2006)
	Serum	Diagnosis of bladder cancer (Eijan et al. 1997)
	Urine	Diagnosis of bladder carcinoma (Eijan et al. 2000)
	Tumour tissue	Prognosis in breast cancer (Lah et al. 1992, 1997, 2000b; Foekens et al. 1998, Levičar et al. 2002)
	Tumour tissue	(Thomssen et al. 1995, 1998; Harbeck et al. 2001, 2002)
	Tumour tissue	Differential diagnosis in breast cancer (Karkola et al. 2003)
	Tissue sections[c]	Diagnosis and prognosis in breast cancer (Lah et al. 2000a)
	Tissue sections	Diagnosis and prognosis in prostate cancer (Sinha et al. 1993, 2001, 2002, 2007)
	Tumour tissue	Diagnosis in prostate cancer (Fernandez et al. 2001)
	Tumour tissue	Diagnosis and relation to metastatic potential in prostate carcinoma (Chu et al. 2006)
	Tumour tissue	Diagnosis in ovarian carcinoma (Warwas et al. 2000)
	Tumour tissue	Diagnosis in ovarian cancer (Vazquez-Otrin et al. 2005)
	Tissue sections	Prognosis in ovarian carcinoma (Scorilas et al. 2002)
	Tumour tissue	Altered expression in lung tumours (Ledakis et al. 1996, Krepela et al. 1998, Kayser et al. 2003, Werle et al. 2006)
	Tumour tissue	Prognosis in lung tumours (Ebert et al. 1994, Werle et al. 1997)
	Tumour tissue	Diagnosis in colorectal carcinoma (Shuja et al. 1991, Hazen et al. 2000)
	Tumour tissue	Prognosis and diagnosis in colorectal carcinoma (Campo et al. 1994, Herszenyl et al. 1999, Troy et al. 2004)
	Tissue sections	Prognosis and diagnosis in colorectal carcinoma (Talieri et al. 2004)
	Tumour tissue	Prognosis in head and neck carcinoma (Russo et al. 1995)
	Tumour tissue	Diagnosis in head and neck carcinoma (Strojan et al. 2000)
	Tumour tissue	Diagnosis in oesophagous carcinoma (Hughes et al. 1998, Luo et al. 2004)
	Tissue sections	Diagnosis in tongue carcinoma (Saleh et al. 2006)
	Tumour tissue	Diagnosis of oral SCC (Vigneswaran et al. 2000, Kawasaki et al. 2002)
	Tumour tissue	Diagnosis in blad2 TCC (Visscher et al. 1994, Eijan et al. 2003, Staak et al. 2004)
	Tissue sections	Prognosis in pancreatic adenocarcinoma (Niedergethmann et al. 2004)
	Tumour tissue	Diagnosis in melanoma (Kageshita et al. 1995, Froelich et al. 2001)
	Tumour tissue	Prognosis in melanoma (Yoshii et al. 1995, Goldman et al. 1999)

(*continued*)

Table 29.1 (continued)

Cathepsin B	Clinical sample	Application
	Tissue sections	Diagnosis of aggressive meningioma (Strojnik et al. 2001, Trinkaus et al. 2003)
	Tumour tissue	Diagnosis of aggressive meningioma (Trinkaus et al. 2005, Lah et al. 2007, submitted)
	Tumour tissue	Altered expression in glioblastoma (Rempel et al. 1994, Mikkelsen et al. 1995)
	Tissue sections	Prognosis of glioma (Strojnik et al. 1999, Strojnik et al. 2000, Wang et al. 2005)
	Tumour tissue	Prognosis of glioma (Nakabayashi et al. 2005)

SCC Squamous cell carcinoma, *TCC* Transitional cell carcinoma.

[a] Prognosis means that the biomarker is related to disease-free and/or overall survival.

[b] Diagnosis means that a biomarker is altered in a neoplasm or is related to tumour progression.

[c] Tissue sections mean that immunohistochemistry and/or in situ hybridization and/or laser-dissected lesions were used to study the biomarker.

endopeptidases (e.g. cathepsins L and S), exopeptidases (aminopeptidase cathepsin H and amino-dipeptidase cathepsin C, carboxypeptidase cathepsin X), or both, such as cathepsin B, which is a carboxydipeptidase and an endopeptidase. They are synthesized as 30–50 kDa precursors, which are phosphorylated and glycosylated in the Golgi apparatus to be bound to mannose phosphate receptors, and thereby directed to endosomal—lysosomal compartments. However, differential splicing of cathepsin B (Muentener et al. 2003, reviewed by Yan and Sloane 2003) and cathepsin L (Arora and Chauhan 2002, Caserman et al. 2006), more abundant in cancer cells, may result in enzyme isoforms that are directed to various subcellular destinations, including the nucleus and/or to an excessive secretion, as first observed by Mason et al. (1987).

In tumours, alterations in expression, processing and localization have been described at various levels when compared to their normal and benign tissue counterparts (Sloane et al. 1994). Cathepsins play specific roles in cancer progression, which depend also on the tumour and tumour-associated microenvironment, as observed, for example, with the ubiquitous cathepsins B, L and H, expressed in most tissues and cell types. The entire clan of human cysteine cathepsins—B, C, H, F, J, K, L, O, S, L2/V, W and X/Z—is now being screened in relation to tumour progression (reviewed by Joyce et al. 2004, Gochareva et al. 2006, Mohamed and Sloane 2006, Gochareva and Joyce 2007), in both in vitro and animal experiments. Universal cellular distribution and a rather broad pH optimum range convey a very versatile and selective protease signalling. This type of signalling may involve one or more proteolytic steps, as proposed for the invasion-related proteolytic cascade in the early 1990s by Schmitt et al. (1992). This cascade, involving the cathepsins as the initial step, may interfere with oncogene and/or tumour suppressor gene signalling, leading to cancer progression. However, deletion of cathepsins B, C, L and S in an animal model of pancreatic islet cell cancer, RIP1-Tag2 (RT2) mice,

Table 29.2 Cathepsins L, H, S, K, F and X as biomarkers in selective types of cancer

Cysteine protease	Clinical sample	Application
Cathepsin L	Serum	Diagnosis in ovarian carcinoma (Nishida et al. 1995)
	Tumour tissue	Prognosis in head and neck carcinoma (Kos et al. 1995)
	Tumour tissue	Prognosis in breast cancer (Lah et al. 1992, 1997, Foekens et al. 1998, Levičar et al. 2001, Thomssen et al. 1995, 1998, Harbeck et al. 2001, 2002)
	Tumour tissue	Diagnosis and prognosis in breast carcinoma (Lah et al. 2000a)
	Tissue sections	Diagnosis and relation to metastatic potential in prostate carcinoma (Chu et al. 2006)
	Tumour tissue	Diagnosis in ovarian cancer (Vazquez-Otrin et al. 2005)
	Tumour tissue	Diagnosis in lung carcinoma (Werle et al. 1995, 2004, Ledakis et al. 1996, Krepela et al. 1998, Kayser et al. 2003)
	Tumour tissue	Diagnosis in colorectal carcinoma (Shuja et al. 1991, Hazens et al. 2000)
	Tumour tissue	Prognosis and diagnosis in colorectal carcinoma (Campo et al. 1994, Herszenyl et al. 1999, Troy et al. 2004)
	Tumour tissue	Diagnosis in head and neck carcinoma (Strojan et al. 2000)
	Tumour tissue	Diagnosis of gingival and tongue cancer (Macabeo-Ong et al. 2003)
	Tumour tissue	Prognosis in pancreatic adenocarcinoma (Niedergethmann et al. 2004)
	Tumour tissue	Diagnosis of aggressive meningioma (Strojnik et al. 2001, Trinkaus et al. 2003)
	Tissue sections	Prognosis in meningioma (Lah et al. 2007, submitted)
	Tumour tissue	Diagnosis in glioma (Sivaparavathi et al. 1996b, Strojnik et al. 2000)
Cathepsin H	Serum	Prognosis in colon carcinoma (Schweiger et al. 2004)
	Serum	Prognostic in melanoma (Kos et al. 1997)
	Tumour tissue	Altered expression in cervical carcinoma (Vazquez-Otrin et al. 2005)
	Tumour tissue	Altered expression in advanced colorectal carcinoma (de Re et al. 2000)
	Tumour tissue	Altered expression in head and neck cancer (Kos et al. 1995)
	Tissue sections	Diagnostic in melanoma (Kageshita et al. 1995) Prognosis in melanoma (Kos et al. 1997)
	Tumour tissue	Diagnosis in bladder transitional cell carcinoma (Staak et al. 2004)
Cathepsin S	Serum	No correlation with diagnosis/prognosis of cathepsin S in lung tumours (Kos et al. 2001)
	Tumour tissue	Prognosis in tumour and in adjacent lung parenchyma
	Tumour tissue	Diagnosis in prostate cancer (Fernandez et al. 2001)
	Tissue sections	Diagnosis in ovarian cancer (Vazquez-Otrin et al. 2005)
	Tumour tissue	Prognosis in glioblastoma (Flannery et al. 2003, 2006)
Cathepsin K	Serum	Diagnosis in prostate cancer (Brubaker et al. 2003)
	Tumour tissue	Diagnosis in prostate cancer (Brubaker et al. 2003)

(*continued*)

Table 29.2 (continued)

Cysteine protease	Clinical sample	Application
	Tumour tissue	Diagnosis of early lung tumour in stromal host tissue (Linnerth et al. 2005, Acuff et al. 2006)
	Tumour tissue	Diagnosis of bladder carcinoma (Blaveri et al. 2005)
Cathepsin X	Tissue sections	Diagnosis of invasive prostate carcinoma (Naegler et al. 2004)
Cathepsin F	Tissue sections	Diagnosis in ovarian cancer (Vazquez-Otrin et al. 2005)

TCC Transitional cell carcinoma.

resulted in unexpected effects on tumour cell proliferation and tumour growth, apoptosis, angiogenic switch and tumour micro-vascular density in addition to those on tumour invasion (Bell-McGuinn et al. 2007). Furthermore, activation of cathepsins and cystatins plays a role in the immune response to cancer, reviewed by Honey and Rudensky (2003) and Obermajer et al. (2006), respectively, leading to tumour regression. Therefore, many of cathepsins' signalling pathways in cancer development are not yet fully understood.

Cystatins

Cystatins are a superfamily of evolutionarily related protein inhibitors of cysteine proteases. Type I cystatins (the stefins), stefins A and B, are cytosolic proteins, in contrast to type II cystatins, which are secreted into the extracellular environment (Abrahamson et al. 1986). The seven members of the latter group are cystatins C, E/M, D, F, S, SA and SN, together with male reproductive tract cystatins 8 (CRES, cystatin-related epididymal and spermatogenic), 9 (testatin), 11 and 12 (cystatin T), bone marrow-derived cystatin-like molecule (CLM, cystatin 13) and secreted phosphoprotein (SPP-24, cystatin 14). Type III cystatins, the kininogens, are large multifunctional plasma proteins, containing three type II cystatin-like domains. Fetuins and latexins are constituted by two tandem cystatin-like domains; however, they do not possess inhibitory activity against cysteine proteases.

Cystatins of different families possess different biochemical properties. In general, cystatins are tight-binding inhibitors of the C1 family of cysteine proteases. Nevertheless, the inhibitory profile of a particular cystatin is rather specific. Type II cystatins possess also a second reactive site for inhibition of the C13 family of cysteine proteases (legumain), thus being involved in antigen presentation (Alvarez-Fernandez et al. 1999), or act independently of their inhibitory properties. Besides selective inhibition of cysteine proteases, the secretion of mostly type II cystatins may also regulate the inhibitory potential against intracellular and extracellular targets. Cystatins have been determined in different tissues and biological

Table 29.3 Inhibitors of cysteine proteases cystatins type I as biomarkers in cancer

Cysteine protease inhibitor	Clinical sample	Application
Stefin A	Serum	Prognosis in head and neck cancer (Budihna et al. 1996)
	Tumour tissue	Prognosis and diagnosis in hepatocellular carcinoma (Leto et al. 1997)
	Tumour tissue	Diagnosis and prognosis in breast cancer (Lah et al. 1992, 1997; Levičar et al. 2001)
	Tissue sections	Diagnosis and prognosis in breast cancer (Kuopio et al. 1998)
	Tissue sections	Diagnosis and prognosis in breast tumours and metastasis (Parker et al. 2007)
	Tissue sections	Diagnosis of prostate tumours (Soderstrom et al. 1995)
	Tissue sections	Diagnosis of high-grade prostatic adenocarcinoma (Mirtti et al. 2003)
	Tissue sections	Prognosis and diagnosis in prostate cancer (Sinha et al. 2001, 2002, 2007)
	Tumour tissue	Prognosis in head and neck cancer (Strojan et al. 2000)
	Tumour tissue	Diagnosis in lung tumours (Ebert et al. 1997, Krepela et al. 1998, Bianchi et al. 2004)
	Tumour tissue	Prognosis in patients with non-small-cell lung carcinoma (Leinonen et al. 2007, Werle et al. 2006)
	Tumour tissue	Prognosis in patients with small-cell lung carcinoma (Werle et al. 2006)
	Tumour tissue	Prognosis in head and neck cancer (Budihna et al. 1996, Strojan et al. 2000, 2007)
	Tumour tissue	Diagnosis and prognosis in oesophagus carcinoma (Luo et al. 2004, Li et al. 2005)
	Tumour tissue	Altered expression in meningioma progression (Lah et al. 2007, submitted)
	Tumour tissue	Prognosis and diagnosis in hepatocellular carcinoma (Leto et al. 1997)
Stefin B	Serum	Diagnosis and prognosis in colorectal cancer (Kos et al. 2000)
	Tumour tissue	Prognosis of primary invasive breast carcinoma (Levičar et al. 2002)
	Tumour tissue	Diagnosis in lung tumours (Ebert et al. 1997, Krepela et al. 1998)
	Tumour tissue	Prognosis in patients with non-small-cell lung carcinoma (Ebert et al. 1997, Werle et al. 2006)
	Tumour tissue	Diagnosis of prostatic adenocarcinoma (Mirtti et al. 2003)
	Tumour tissue	Altered expression of stefin B in oesophagus carcinoma (Shiraishi et al. 1998)
	Tumour tissue	Diagnosis and prognosis in colorectal cancer (Kos et al. 2000)
	Tumour tissue	Diagnosis and prognosis of oesophageal cancer (Shiraishi et al. 1998)
	Tumour tissue	Prognosis of head and neck cancer (Budihna et al. 1996)
	Tumour tissue	Risk of disease recurrence and death in head and neck cancer (Strojan et al. 2000)
	Tumour tissue	Prognosis in meningioma (Lah et al. 2007, submitted)

Table 29.4 Inhibitors of cysteine proteases, cystatins Type II, as biomarkers in cancer

Cysteine protease inhibitor	Clinical sample	Application
Cystatin C	Serum	Diagnosis and monitoring in head and neck cancer (Strojan et al. 2004)
	Serum	Prognosis in head and neck cancer (Strojan et al. 2004)
	Serum	Prognosis in colorectal cancer (Kos et al. 2000)
	Serum	Prognosis of cathepsin B:cystatin ratio (Zore et al. 2001)
	Serum	Diagnosis in tongue cancer (Saleh et al. 2006)
	Serum	Diagnosis of bladder carcinoma (Tokyol et al. 2006)
	Serum	Prognosis in melanoma (Kos et al. 1997)
	Pleural fluid	Diagnosis of lung cancer (Werle et al. 2003)
	Tumour tissue	Diagnosis in breast cancer (Yano et al. 2001)
	Tumour tissue	Diagnosis in breast cancer (Vigneswaran et al. 2000)
	Tumour tissue	Diagnosis in lung tumours (Ebert et al. 1997, Krepela et al. 1998)
	Tissue sections	Prognosis in colorectal cancer (Kos et al. 2000)
	Tumour tissue	Diagnosis and monitoring in head and neck cancer (Strojan et al. 2004)
	Tumour tissue	Prognosis in lung cancer (Werle et al. 2006)
	Tissue sections	Prediction of clinical response in lung cancer (Petty et al. 2006)
	Tumour tissue	Diagnosis in colorectal carcinoma (Hirai et al. 1999)
	Tumour tissue	Prognosis in oesophagus carcinoma (Dreilich et al. 2005)
	Tissue sections	Diagnosis of tongue carcinoma (Saleh et al. 2006)
	Serum	Prognosis in non-Hodgkin B-cell lymphoma (Mulaomerovic et al. 2007)
	Tumour tissue	Diagnosis and prognosis of meningioma (Lah et al. 2007, submitted)
	Tumour tissue	Diagnosis in glioma (Lignelid et al. 1997)
	Tumour tissue	Prognosis in gliomas (Nakabayashi et al. 2005)
	Serum	Prognosis in haemoblastosis (Poteriaeva et al. 2003)
Cystatin E/M	Pleural fluid	Diagnosis and altered expression in lung cancer (Werle et al. 2003)
	Tumour tissue	Diagnosis-altered expression in breast cancer (Sotiropoulou et al. 1997)
	Tissue sections	Inverse correlation with breast cancer progression (Zhang et al. 2004)
	Tumour tissue	Diagnosis-altered expression in breast cancer (Vigneswaran et al. 2000)
	Tumour tissue	Diagnosis-altered expression in lung tumours (Ebert et al. 1997)
	Tumour tissue	Diagnosis-altered expression in colorectal cancer (Saleh et al. 2005)

fluids, such as saliva, tears, seminal fluid, ascites, and bronchoalveolar, cerebrospinal and synovial fluids. Type I cystatins are up-regulated in tumour tissue and, up to a certain level, could counterbalance the over-expressed tumour-associated proteolytic activity. On the contrary, type II cystatins are in most cases down-regulated in tumours. Although their role remains protective, their lower levels give way to a surplus of harmful tumour-associated proteolytic activity, as recently reviewed by Kos and Lah (2006). Outside the cell, high levels of type II cystatins may impair the extracellular activity of cysteine proteases associated with the degradation of extracellular matrix (ECM), resulting in tumour cell invasion and metastasis. However, they can also be involved in processes that lead to tumour regression, such as anti-tumour immune responses and apoptosis, as well as in prevention of cell migration and seeding.

As the main physiological role of cystatins is believed to be the regulation of excessive cysteine proteinase activity, their levels in tissues as well as in body fluids could serve as biological markers for different stages of tumour progression. At the first glance, enhanced expression of cystatins would be expected to diminish the tumour-associated proteolytic activity responsible for spreading tumour. In fact, there is evidence showing higher levels of stefins A and B and cystatin C in tumour tissues, correlating with a favourable prognosis in patients with cancer. However, higher levels of stefins A and B and cystatin C in body fluids have been associated with a poor prognosis for patients with cancer. Hypothetically, alterations in secretion may result in higher extracellular and lower intracellular levels of cystatin and, therefore, a reverse correlation with patients' survival might be expected (Kos and Lah 2006). On the contrary, the correlation of high levels of cystatins in body fluids with poor clinical outcome of patients with cancer may also be due to a role in the regulation of processes involved in cancer regression.

Clinical Research on Cathepsins and Cystatins as Tumour Biomarkers

Several extensive reviews on cathepsins in cancer have been published over the past years, starting with Sloane et al. (1994), Kos and Lah (1998, 2006), Lah and Kos (1998), Yan and Sloane (2003), Berdowska (2004), Joyce and Hanahan (2004), Jedeszko and Sloane (2004) and Lah et al. (2006). Therefore, in this chapter we will only briefly summarize the old data and add some of the recent reports.

Cysteine cathepsins and cystatins have either been used as single biomarkers or as the ratio between specific cathepsin and inhibitor, as this may actually represent a fraction of the "active" enzyme, and proved as a more significant endpoint marker, as, for example, in lung cancer (Ebert et al. 1997, Werle et al. 2006) and in colon carcinomas (Zore et al. 2001). However, because of a variety of cathepsins and cystatins present in tumours homogenates and body fluids, any conclusion on their balance in tumour tissue is difficult. This is also because they bind with different affinities and rates and have different subcellular distributions, as discussed by Calkins et al. (1998) and Zajc et al. (2006).

Carcinomas

Breast carcinoma is the most common malignancy in women in the industrialized world. Routine clinical management of breast cancer relies on traditional classification, including tumour grade, histopathologic tumour type, carcinoma size and lymph node status. Despite the overall association of these variables with prognosis and outcome, this system remains a relatively weak predictor of behaviour in some circumstances. Tumours of apparently homogenous morphology vary in response to therapy and lead to divergent outcomes, which is also the case with many other types of tumours. In female breast, a variety of tumours arise from large, medium or small ducts. The most common histological subtypes are ductal carcinomas, which grow within the lumen of ducts, and lobular carcinomas, which arise from small ends of the ducts. These subtypes differ both in disease progression and treatment. Carcinomas comprise malignancies of cells of epithelial origin, progressing from atypical and dysplastic neoplasms to benign (adenocarcinomas) and later to malignant carcinomas. This transition is on one hand associated with disruption of components that maintain cell–cell and cell–ECM interactions as well as ECM and/or basement membrane components, and on the other hand with the transition from the epithelial cell to the migratory mesenchymal cell morphology, which may also be associated with an altered expression of cysteine cathepsins and cystatins (Fig. 29.1). However, as suggested by Friedl and Wolf (2003), under certain environmental and/or therapy-induced stress conditions, cancer cells can also undergo a "step backwards in evolutionary time", towards a more primitive, ameboid type of cell, which may be associated with decreased, instead of increased, protease activity, presumably a requirement for stroma invasion. With this in mind, we are presenting and discussing sometimes contradictory (pre)clinical data on cathepsins and cystatins

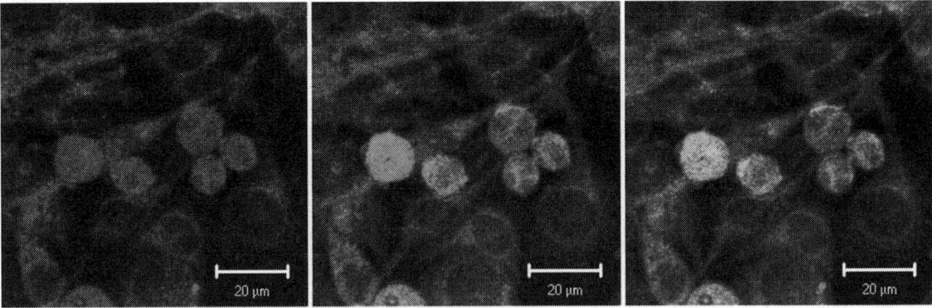

Fig. 29.1 Confocal microscopy of cathepsins X (*red*) and L (*green*) in a coculture of highly invasive MCF10A neoT cells and pro-monocytic U937 cells. Cathepsins L and X are expressed in both cell types, their expression and colocalization is more intense in U937 cells (*See also* Color Insert II)

in different carcinomas. Here, we first review the most common gender-related cancers, breast and prostate carcinoma, followed by ovarian and cervical carcinoma, and then by more frequent carcinomas such as those of the lung and colon.

Breast Carcinoma

Cathepsins

In breast carcinoma, aspartic protease cathepsin D was the first and most thoroughly studied lysosomal enzyme, and was found to be an independent marker for poor prognosis, being correlated with the incidence of clinical metastasis, as reviewed recently by Liaudet-Coopman et al. (2006). The first clinical study on cysteine cathepsins in patients with tumour was published by Lah et al. (1991), relating cathepsin L activity to a potential application in prognosis. This was followed by the determination of protein levels of both cathepsins L and B and their inhibitors in breast tumour cytosols (Lah et al. 1997), which was then expanded to a larger tumour population (Lah et al. 2000b). Besides in cytosols, immunohistochemical analyses were also carried out (Lah et al. 2000a). Taken together, our data showed a diversity of cellular distribution among cathepsins B, L and D, demonstrating high cathepsin B expression in myoepithelial and endothelial cells present in neovascular structures in invasive ductal carcinoma, whereas cathepsin L was mostly associated with the breast tumour cells. Cathepsin L correlated with the high histological grade, whereas cathepsin B antigen in the tumour cells was not related to any other features of tumour progression. The same was observed recently by Vigneswaran et al. (2005) in breast cancer specimens, utilizing laser capture microdissection/RT-PCR (reverse transcriptase-polymerase chain reaction), Western blotting and immunohistochemistry, and, in addition, showing no distinction in the levels of either cathepsin between metastatic and non-metastatic primary tumours. Moreover, Karkola et al. (2003), using gene profiling and real time polymerase chain reaction (RT-PCR), found differential expression of the cathepsin B gene in ductal versus lobular breast carcinoma, the latter achieving its invasive growth through the loss of E-cadherin, whereas the invasiveness showed by the ductal type is presumably accompanied by alterations in other sets of genes, including an over-expression of cathepsin B.

With respect to prognosis, our data showed that total tissue cathepsin B protein was a more powerful prognostic factor for disease-free survival (DFS) and overall survival (OS) than cathepsin L, particularly in the lymph node-negative patients (Lah et al. 2000b, Levičar et al. 2002). This is of significance as staging lymph node-negative patients is difficult and is needed to avoid over-treating the patients. Foekens et al. (1998) also reported high and comparable prognostic significance of both cathepsins B and L in a larger population of 1,500 patients with breast cancer. High impact of cathepsin L on prognosis was simultaneously reported by Thomssen et al. (1998), especially when the authors combined high cathepsin L with high PAI-1 expression (Thomssen et al. 1995). In contrast, Harbeck et al. (2001) stated that cathepsin L, but not cathepsin B, was a good prognostic, actually predictive, factor for both DFS and OS in the treated, lymph node-positive patients, having an application in treatment decision-making. Multivariate comparison of protease profile that includes the uPA/PAI-1 system and cathepsins may prove to be a key issue when making decisions on treatment, as suggested in a report on multicentric breast cancer study by Harbeck et al. (2002). Some of the discrepancies observed in breast and other cancer studies may be mostly due to difference in experimental design with respect to the tumour tissue analysis, often limited size and/or biased selection of the patient population, different therapeutic regimens and also due to inherent biological variability, even within the same tumour histology.

Cystatins

Sotiropoulou et al. (1997) were the first to report that the cystatin M gene was down-regulated in primary breast tumours. Zhang et al. (2004) confirmed the inverse correlation of cathepsin M with breast cancer progression in a study comprising cancer cell lines, animal models and human breast cancer tissues. Using laser capture microdissection and quantitative RT-PCR, consistent expression of cystatin M was found in normal human breast epithelial cells, whereas expression was decreased by 86% in invasive ductal carcinoma cells of stage I–IV tumours. Complete loss of expression of cystatin M was observed in two of three invasive ductal carcinoma cells from stage IV patients. Thus, cystatin M was suggested as a novel candidate tumour suppressor gene for breast cancer (Zhang et al. 2004). This was in agreement with previous studies reporting on decreased cystatin C expression relative to cathepsin B (Yano et al. 2001), and the very first study, measuring lower total cysteine protease inhibitory activity in homogenates of advanced breast cancer correlating to bad prognosis (Lah et al. 1992). However, these findings have been recently challenged by Vigneswaran et al. (2005), who have demonstrated increased mRNA expression levels of both cystatin M and cystatin C in the tumour cells dissected from breast tumour tissue, particularly in larger, higher tumour stage, although both markers were not related to the metastatic dissemination of these tumours.

Stefins, the intracellular cathepsin inhibitors, were also investigated in breast cancer cytosols (Lah et al. 1992) and it was demonstrated that the decrease in stefin A protein was significantly related to poor prognosis in terms of DFS ($p < 0.008$) and OS ($p < 0.02$). Later studies by Levičar et al. (2002) in a population of breast carcinoma of mixed histologies could not confirm the significance of stefin levels in the total population of patients, but both stefin A and B protein levels were of borderline significance in lymph node-negative patients. However, higher ratio between cathepsins (B and L) and the stefins (A and B) was significantly associated with early relapse of the patients. Meanwhile, Kuopio et al. (1998) reported in an immunohistochemical study of infiltrative breast carcinoma a positive staining pattern of stefin A in about 10% of the cases, most of them of the ductal infiltrative type, whereas no lobular carcinomas showed positive staining. This would support the notion (Karkola et al. 2003) of a selective involvement of cysteine cathepsins in the invasion of ductal and not lobular breast carcinoma. Tumours positive for stefin A were of larger size and had higher mitotic activity than were stefin A-negative tumours. Focal stefin positivity was seen in myoepithelial cells of benign ducts, similar to that reported for cathepsin B (Lah et al. 2000a). Surprisingly, in this study, stefin A was prognostic for OS, the risk for breast cancer-related death being significantly higher ($p < 0.01$) in patients with stefin A-positive tumours than in those with negative ones, as shown in total and lymph node-negative populations. The discrepancy among these studies might have been recently resolved by Parker et al. (2007), who reported that high stefin A levels correlated with longer DFS, remaining a significant independent prognostic factor ($p < 0.0014$) also in a multivariate analysis. The authors confirmed this notion using a clinically relevant murine model of spontaneous mammary metastasis to bone, where the stefin A gene was highly expressed in metastatic mammary tumours, along with increased cathepsin B, which was co-expressed in the tumour cells interacting with the stroma. Moreover, enforced expression of stefin A in the highly metastatic 4T1 breast cancer cell line significantly reduced bone metastasis, following an orthotopic injection into mammary gland. These data support and expand our initial suggestions regarding stefin A (Lah et al. 1992, 1997) not only as a tumour but also as a metastasis supressor gene.

Prostate Carcinoma

Cathepsins

A variety of cysteine cathepsins, such as cathepsin B (Sinha et al. 1993), cathepsin H (Waghray et al. 2001), cathepsin S (Fernandez et al. 2001) and cathepsin K (Brubaker et al. 2003), as well as aspartic cathepsin D (Chu et al. 2006), have been found to be associated with the invasive and metastatic potential of prostate carcinoma, as reviewed by Jedeszko and Sloane (2004). Naegler et al. (2004) investigated the carboxypeptidase cathepsin X, using immunolabelling and RT-PCR in patients with prostatic intraepithelial neoplasia and prostate carcinoma of various Gleason stages. Immunohistochemical analysis revealed that cathepsin X was highly increased in prostate intraepithelial neoplasia (PIN) and prostate carcinoma; however, it was not related to Gleason stage, indicating that it may play a role in early tumorigenesis of prostate cancer. In contrast, relatively weak and heterogenous staining was observed for cathepsins F, B and L. As no differences at the genomic and the mRNA level were observed, cathepsin X up-regulation most likely occurs in the absence of genomic amplification. In a series of publications, Sinha and co-workers (Sinha et al. 1993, 2001, 2002, 2007) thoroughly investigated cathepsin B expression in the progression of prostate cancer and also found considerable variability in the individual localization of cathepsin B within and between Gleason histological scores. However, prostate carcinoma within an individual Gleason score is a heterogenous tumour that contains aggressive and less aggressive clones or subpopulations that can be defined by the ratio of cathepsin B:stefin A. The tumour with an aggressive clone may be identified as having a ratio that is higher than 1 and less aggressive clones are defined by a ratio equal or lower than 1. In one of these studies (Sinha et al. 2002), the authors also found a significant positive association ($p = 0.0066$) between the cathepsin B:stefin A ratio and the incidence of pelvic lymph node metastases. Moreover, mortality rates were higher in patients with cathepsin B levels greater than those of stefin A. This ratio can, therefore, be used in the differential diagnosis and treatment of patients. In a very recent publication (Sinha et al. 2007), needle biopsies from Gleason pattern $3 + 3$ (score 6) were characterized, showing a trend towards an inverse relationship between relatively high cathepsin B:stefin A ratio and clinical stage that may be indicative of an early invasive tumour. A larger patient population needs now to be screened for this diagnostic parameter to be validated in needle biopsy samples.

Cystatins

Stefin A was expressed in the basal cells in all types of benign prostatic hyperplasias (Soderstrom et al. 1995, Mirtti et al. 2003). It was demonstrated that the secretory epithelial cells do not express stefin A. In the hyperplastic prostate, the expression of stefin A was decreased and expressed more locally in the basal cells (Soderstrom et al. 1995). In adenocarcinomas, the absence of stefin A expression was obvious and if the sections contained both benign and malignant cells, only the benign glandular structures expressed stefin A, while malignant carcinomas showed no stefin A staining (Soderstrom et al. 1995). Stefin B showed both nuclear and cytoplasmic expression in the columnar epithelial cells, and there was a significant decrease in cytoplasmic stefin B staining in carcinomas, compared to other lesions (Mirtti et al. 2003). Therefore, expression of stefins A and B alone could be used as an aid in the diagnosis of prostatic adenocarcinoma. Moreover, stefin A can discriminate between high-grade prostatic intraepithelial neoplasias and grade I carcinoma (Mirtti et al. 2003). Cystatin C is highly expressed and widely distributed throughout the male genital tract, as shown by immunohistochemical and mRNA analyses (Jiborn et al. 2004). However, to our knowledge, no studies related to diagnosis or prognosis of prostate cancer have been carried out so far.

Ovarian and Uterine Cervix Carcinoma

Cathepsins

Cathepsins B and L both have been found to be increased in ovarian ascites and in the sera of patients with ovarian and cervical carcinoma, as reviewed by Kos and Lah (1998) and Berdowska (2004). Briefly, in ovarian cancer cathepsin B levels were first demonstrated to be enhanced both in human cancer tissues (Scorilas et al. 2002) and in sera from patients (Warwas et al. 1997). Warwas et al. (2000) showed that cathepsin B activity was a useful marker in preoperative differential diagnosis of malignant ovarian and uterine tumours, particularly when used in combination with commonly used serum marker, CA125. Scorilas et al. (2002) proposed that immunohistochemical analysis of cathepsin B could serve as an independent marker for progression-free survival and OS in patients with ovarian cancer with 7-year follow-up, cathepsin B expression being associated with 1.5-fold increased risk of relapse and death. Similarly, serum levels of cathepsin L were related to other serum markers, CA125 and CA72-4 (Nishida et al. 1997). Cathepsin F expression in normal tissues is quite variable, depending on the tissue type, whereas its induction in numerous cancer cells lines suggests a specific role in cancer progression (Vazquez-Otrin et al. 2005). The authors, using cDNA arrays and RT-PCR, reported that in cervical tumour tissue cathepsins F and B were up-regulated, whereas cathepsins L, S, H and C were down-regulated, compared to the normal counterparts. Further investigations on cathepsin profiles in cervical cancer progression will, therefore, be of interest.

Cystatins

Electrophoresis and immunohistochemical and Western blotting analyses have revealed increased cystatin C levels in malignant, but not in benign, ovarian tumours (Nishikawa et al. 2004). Stefin A was distributed throughout the epithelium, except in the basal and parabasal cell layers. In low-grade cervical intraepithelial neoplasia, reduced staining was observed in the lower third of the epithelium, and in high-grade cervical intraepithelial neoplasia, a significant reduction in the staining intensity was seen in the middle and upper thirds. The immunoreactivity in intraepithelial neoplasia was closely related to the degree of morphological maturation. In highly cellular and poorly differentiated cervical intraepithelial neoplasias III, stefin A staining in epithelia and nuclei was negative (Pollanen et al. 1995). In squamous cell carcinoma (SCC), stefin A was often abundant in highly differentiated areas and almost absent in poorly differentiated ones (Eide et al. 1997), which is in agreement with a generally observed down-regulation of stefin A during progression of epithelial tumours. Similarly, Kastelic et al. (1994) showed marginal levels of stefin A and high levels of stefin B in ovarian carcinoma, compared to control epithelial tissue. These inhibitors were also found in ascites from patients with ovarian carcinoma, where they were only poorly associated with cysteine cathepsins (Lah et al. 1991).

Lung Carcinoma

Cathepsins

A large body of literature reports on increased expression of cathepsins B, L and S, but not that of cathepsin C, and lower cathepsin H levels in lung tumour tissues, compared to normal adjacent counterparts (Berdowska 2004). In summary, in patients with non-small cell lung carcinoma (NSCLC), tumours with advanced lymph node metastasis displayed increased cathepsin B levels and their survival was poor (Ebert et al. 1997, Werle et al. 2000, 2003, 2004, 2006; Kayser et al. 2003). The authors showed that cathepsin B activity levels in tumour cells of infiltrated lymph nodes in patients with NSCLC were also a prognostic for survival. In multivariate analysis, the prognostic significance of cathepsin B followed that of lymph node involvement (pN) and tumour site (pT). Also, the studies of other authors (as listed by Jedeszko and Sloane 2004) have described high cathepsin B protein levels in lung tumours in relation to disease progression and to metastasis, although showing a variability in the involvement of this enzyme in the progression of lung tumours of different histologies. Although discrepancies among the studies exist, likely due to the methodology used for protein determination (immunohistochemistry and ELISA, enzyme-linked immunosorbent assay), there are strong indications that the presence of cathepsin B in tumour cells, along with its expression in inflammatory cells (mostly histiocytes type I and II), is a key factor that presumably acts in favour of tumour progression. Interestingly, in a multivariate analysis comparing cathepsin B to the PAI-1/uPAR system in a large cohort of patients with NSCLC, among a total of 10 parameters, cathepsin B activity and concentration were prognostic for OS (Werle et al. 2004). However, only PAI-1 and uPAR added an independent prognostic information with regard to the established clinical and histomorphological factors. Among other cathepsins, cathepsin L, although elevated in lung tumour tissue, was not relevant for prognosis, possibly due to a great variability in its cellular distribution within tumour tissue. Cathepsin S levels were also elevated in tumour tissue versus adjacent tissue and also found highly expressed in non-infiltrated and infiltrated lymph nodes (Kos et al. 2001). Surprisingly, they correlated with good prognosis, which was explained by the presence of cathepsin S in cells of the immune response against the tumour, thereby contributing to a higher survival rate. Cathepsin K elevation in lung tumours was shown by RT-PCR analysis (Buhling et al. 2000) and recently by Acuff et al. (2006) by micro-array analysis. In the orthotopic lung cancer model, where human lung cancer A549 cells were directly injected into mouse lungs, the authors demonstrated that cathepsin K was up-regulated (along with metalloproteinases-12 and -13) in murine tumour tissue compared to normal lung, proving that the up-regulation of host-derived proteinases is an important part of the microenvironmental response to the tumour. Cathepsin H level was increased in lung tumours, though not correlating with DFS (Schweiger et al. 2004). However, in the group of long-term cigarette smokers, cathepsin H had an impact on prognosis and also appeared to be associated with chronic inflammation. It is thus not surprising that its levels were highly elevated both in the sera of patients with lung cancer and in the sera of patients with other lung diseases. Recently, Linnerth et al. (2005), using a transgenic mouse model to identify markers of human lung tumours by cDNA arrays, found cathepsin H to be one of three proteins that consistently showed increased levels in murine tumours compared to normal lung, the enzyme being localized to lamellar bodies of type II pneumocytes and elevated in early stage (I) node-negative tumours, which makes it a useful indicator for early tumour detection and diagnosis.

Cystatins

Significantly, higher cysteine protease inhibitory potential and increased levels of stefin A protein, but not of stefin B or cystatin C, in stage I SCC were reported by Krepela et al. (1998). Expression of stefin A in lung tumours was found to be higher in early detected tumours and was lost during tumour progression (Bianchi et al. 2004), confirming that stefin A may have a tumour-suppressor function (Abrahamson et al. 2003), which is in agreement with the observations made on oesophagus (Palungwachira et al. 2002), prostatic adenocarcinoma (Mirtti et al. 2003), squamous carcinoma of the skin (Luo et al. 2004) and in breast cancer (Lah et al. 1992, Parker et al. 2007). Staining for stefin A has shown significantly different appearance in various stages of bronchial epithelium differentiation and also in various lung tumour histologies (Leinonen et al. 2007). Cystatin C was compared to cathepsin B in patients' sera, where Zore et al. (2001) demonstrated that the cathepsin B–cystatin C complex was significantly lower in sera of patients bearing malignant lung tumours than in those with non-cancerous lung diseases or healthy controls. Expression of cystatin C relative to its target protease cathepsin B, determined by cDNA micro-arrays, was also found independently predictive for the response to platinum-based chemotherapy in non-small-cell lung cancer therapy (Petty et al. 2006). Werle et al. (2003) also reported that pleural effusions of different origin contain high levels of cystatin C, perhaps constituting the major part of an inhibitor reservoir. The levels of cystatin E/M appeared to be significantly related to primary pleural tumours, whereas cystatin F correlated more with inflammatory processes of lung disorder.

In the most recent comparative study on five cystatins, Werle et al. (2006) demonstrated that median levels of stefins A and B in NSCLC tumours were significantly greater in tumour than in lung tissue, whereas those of cystatin C and cystatin E/M were lower. The levels of cystatin F remained below the detection limit. Similarly to cathepsins, cystatins were present in tumour cells, macrophages and neutrophils. In univariate analysis over a 7-year observation period, patients with high levels of stefins A and B and cysteine proteinase inhibitory (CPI) activity exhibited a significantly better survival probability, as suggested earlier (Ebert et al. 1997). In contrast, cystatin C and E/M provided no prognostic information. In multivariate analysis, the stefins A and B prognostic impact followed immediately that of pathological tumour-node-metastasis (pTNM) stage and the authors suggested that these two markers may add independent prognostic information for better assessment of low- and high-risk patients with NSCLC (Werle et al. 2006). In a similar study, Leinonen et al. (2007) found reduced stefin A expression in whole lung tissue samples from patients with NSCLC, associated with lower DFS. In conclusion, these studies indicate that cystatins are involved in the progression of lung cancer: the cystatins type I (stefins A and B) are up-regulated in tumour tissue and may to a certain extent counterbalance cathepsin activity, particularly by binding tightly to cathepsins L and H. However, their function as tumour suppressors is supported by survival analyses, which correlate with their higher levels. On the contrary, cystatins type II (C, E and M) are down-regulated in tumours, thereby possibly contributing to elevated cathepsin activities. Therefore, in individual types of histology, cathepsin B, individual cystatins or their ratios may be used for patient prognosis and/or evaluating response to therapy.

Colon Carcinoma

Cathepsins

Molecular progression of colon carcinoma is best investigated among all cancers and often involves inactivation in Ras oncogene, which is reportedly associated with higher cathepsin L expression and altered cathepsin B trafficking in colorectal carcinoma cells (Cavallo-Medved et al. 2003). There is clear evidence of cathepsin B over-expression in cancerous versus normal colon mucosa. In particular, this is a case in the invasive edges of the tumour, where cathepsin B is expressed in infiltrated macrophages as well as in tumour cells (Campo et al. 1994, Hazen et al. 2000). It appears to be associated with the degradation of basement membranes as adenomas progress to carcinoma (reviewed extensively by Jedeszko and Sloane 2004). Along with cathepsin L, cathepsin B protein was elevated in metastatic colon carcinoma tissue (Herszenyl et al. 1999). Initial studies (Shuja et al. 1991) reported elevated levels of cathepsins B and L in Dukes' stages A and B, decreasing in stage D. This was later on confirmed by Troy et al. (2004), who found significant correlations between tumour and normal cathepsins B and L protein and activity ratio, gradually decreasing with tumour progression. Survival of early-stage patients was inversely related to both cathepsins, but this was not observed in patients in advanced stages. This was confirmed by immunohistochemical studies of paired samples providing correlation of cathepsin B with differentiation grade and lymph node involvement (Talieri et al. 2004, Hirai et al. 1999). Taken together, the data show that during the progression of colon carcinoma transiently elevated tumour cathepsin B levels may not represent an independent prognostic marker.

In contrast, in a cohort of 325 sera from patients with cancer, Kos et al. (1998) found highly significant correlation between secreted cathepsin B and advanced Dukes' stages but no association with other factors, including carcinoembryonic antigen (CEA). In survival analyses, patients with high cathepsin B levels had a significantly lower survival probability, although in multivariate analysis Dukes' stage was a better prognosticator. When CEA and cathepsin B data were combined, and CEA-positive patients were further separated according to cathepsin B levels, those with high cathepsin B expression had a significantly shorter survival (HR, hazard ratio, 2.2, $p < 0.0001$). However, in a study of 300 patients, no correlation was found between cathepsin H levels in serum and Dukes' stage, nor with other disease markers (Schweiger et al. 2004). While carrying out survival analysis, the authors found a significant difference between the groups of patients with low cathepsin H (first tertile) who had a poor prognosis and the remaining group of patients ($p < 0.003$). Moreover, combining cathepsin H values with CEA, patients with high CEA and low cathepsin H had the highest risk of death, with an HR of 2.7 ($p < 0.0001$). This is rather surprising and points to two opposing roles of cathepsin H in these processes, one to facilitate and the other to prevent cancer progression (e.g. invasion opposed to immune response). This question is not solved yet, and it would suggest completely different roles from those played by cathepsins B and L, at least in this type of tumour.

Cystatins

Although the cystatin C mRNA levels were not very different in malignant versus non-malignant colon carcinoma tissues (Hirai et al. 1999), several studies provided opposite data on cystatin C protein levels. For example, in primary colorectal cancer tissues, cystatin C was present in the cytoplasm and on the cell surface, whereas non-malignant tissues were negative (Saleh et al. 2005) The extent of positive staining was by 30% higher in adenocarcinoma than in carcinoma. Anti-papain inhibitory activity was increased in malignant tissues, from well differentiated through moderately to poorly differentiated carcinoma, and invasive adeno-carcinomas had higher inhibitory activity than non-invasive ones (Saleh et al. 2005). In the sera of 345 patients with colorectal cancer, Kos et al. (2000) reported on the levels of stefin A and cystatin C, which were moderately but significantly increased, but did not correlate with advanced tumour stages. In contrast, stefin B correlated significantly with tumour progression, its levels being the highest in stage D ($p < 0.007$). Both stefin B and cystatin C levels, but not stefin A levels, also correlated with patients' survival. These data revealed that increased secretion of cysteine proteinase inhibitors has an impact on shorter patients' survival. The levels of cathepsin B–cystatin C complex, being significantly lower in advanced Dukes' stages (C and D), represent another prognostic parameter in colon cancer (Zore et al. 2001).

Head and Neck Carcinoma

Cathepsins

Squamous cell carcinoma of the head and neck (SCCHN) is among the most prevalent cancers with poor prognosis and patients usually die of persistent or recurrent local-regional disease. In this cancer type significantly higher protein levels of cathepsins B and L are found in solid tumour homogenates versus normal tissue compartments, with laryngeal tissue containing slightly higher levels than normal oral cavity and pharynx (Strojan et al. 2000); however, they were not predictive for the disease outcome. This is in contrast to the previously reported bad prognostic impact of high cathepsin B activity (Russo et al. 1995). The latter may be related to the prognostic relevance of lower stefin A levels, also reported by Strojan et al. (2000), who found that the risk of disease recurrence in SCCHN was significantly higher in patients with low levels of stefins A and B. Also, in a multivariate analysis, standardized values of stefin A remained the strongest independent factor for both disease-free and disease-specific survival. This was recently confirmed by pooled analysis, as described below (Strojan et al. 2007).

In sera of patients with SCCHN, taken at diagnosis and after the therapy, the levels of cathepsins B and L decreased after therapy, but only cathepsin L levels correlated significantly with the risk of relapse and death. Patients who suffered relapse had lower cathepsin L levels at the time of diagnosis than did those without recurrence. In contrast, cathepsin H appeared to be down-regulated in head and neck cancer tissues (Kos et al. 1995), as well as in serum, but in this case no significant correlation with patients' prognosis was observed.

Cystatins

Strojan et al. (2000) reported on slightly but significantly higher median levels of both stefins A and B in tumour tissues and recently (Strojan et al. 2007) demonstrated up-regulation of stefins A and B proteins in some and down-regulation in other subset of tumours, as found in matched pairs of patients with breast cancer (Levičar et al. 2002). The concentrations of stefins A and B correlated with the pN classification and stage of the disease. The group of patients at high risk of disease progression was characterized by significantly lower levels of stefin A in tumour compared with those in the non-tumourous mucosa. In patients' sera, significantly higher stefin A and lower stefin B concentrations were measured before therapy, and the levels of stefin A were found to be significantly higher compared to those after therapy (Strojan et al. 2001), which may classify stefin A as a relevant biomarker for this malignancy, both in tumour tissue and in serum. Recent proteomic profiling in conjunction with protein identification have also confirmed a differential expression of stefin A between healthy mucosa and head and neck SCC (Roesch-Ely et al. 2007).

The median level of cystatin C in tumour tissue of patients with SCCHN was 1.18-fold lower from that in corresponding mucosa, also depending on the site of sampling: it was lower in non-laryngeal tissue (oral cavity, orto- or hypo-pharynx) than in laryngeal tissue. The tumour cystatin C level correlated inversely with pN-stage, whereas a trend towards lower cystatin C levels was observed in the group with extra-nodal tumour extension compared to those with no extra-nodal spread (Strojan et al. 2004b). Tumour cystatin C levels correlated inversely with pN-stage and the patients with low cystatin C levels exhibited poor DFS and disease-specific survival in univariate and multivariate analysis, although this was of a lower significance than that reported for stefin A. In the sera of patients with SCCHN, a significant increase in cystatin C expression was found compared to a control group of healthy donors (Strojan et al. 2004a). A trend towards lower cystatin C levels in the serum of patients with no relapse of the disease was also observed (Strojan et al. 2004a).

Oesophageal Carcinoma

Cathepsins

Oral and oesophageal carcinomas may be considered as specific tumours in the head and neck region. In oesophagus adenocarcinoma, the cathepsin B gene appeared to be amplified as a novel amplicon on chromosome 8p22-23 in 12% of the patients investigated, and elevated mRNA levels were found in all cases (Hughes et al. 1998). This parallelled the observed increased protein levels of cathepsin B in these tumours, thus indicating a role of cathepsin B in oesophagus progression, although because of the limited population and the overall bad outcome of this cancer, its prognostic relevance could not be confirmed. Later on, Luo et al. (2004), using DNA micro arrays, found about threefold increase in cathepsin B levels in oesophageal SCC tissues.

Cystatins

In the same study, Luo et al. (2004) reported on a six- to sevenfold decrease in stefin A levels in oesophagus squamous cell carcinoma. This was followed by a recent demonstration (Li et al. 2005) that over-expression of stefin A delayed the in vitro and in vivo growth of cells and significantly inhibited the incidence of lung metastasis, compared to control (50%) in xenograft mice with the EC9706 cell lines or empty vectors, respectively. Transfection with stefin A resulted in a dramatic reduction in angiogenesis, tumour growth and invasion, associated with metastasis, suggesting a therapeutic application of stefin A. Previous to that, the expression of stefin B in tumour tissue from human oesophageal carcinoma was found to be markedly decreased compared to control tissue (Shiraishi et al. 1998). These authors also suggested that stefin B may be a useful marker for predicting the biologic aggressiveness of human oesophageal carcinoma, as the cases with a tumour:normal ratio of less than 0.5 were of advanced clinical stage and had higher frequency of lymph node metastasis, thereby associating a decrease in stefin B with metastasis. Cystatin C levels were shown by Dreilich et al. (2005) to be significantly correlated with those of the angiogenic cytokine vascular endothelial growth factor (VEGF) and suggested to be used for prognosis in patients with oesophageal carcinoma, as the tumour values correlated with survival in univariate analysis.

Oral and Laryngeal Carcinoma

In oral carcinoma, the major problem is high mortality, because patients are diagnosed at an advanced stage and with lymph node involvement, which reduces the survival rate by at least 50%. Histologically, most of them are SCC with the most invasive basaloid subtype.

Cathepsins and Cystatins

Elevated levels of cathepsin B in oral carcinoma were associated with advanced tumour stage and poor grade, whereas cathepsin L was less abundant, as revealed by immunohistochemistry by Vigneswaran et al. (2000). However, aspartic protease cathepsin D correlated with metastatic spread and the authors suggested cathepsins as additional prognostic tools and potential therapeutic targets. The same conclusions were made by Kawasaki et al. (2002), who also found higher immunohistochemical scores of cathepsin B (and H), as well as D, but not of cathepsin L, in oesophagus tumours at various locations (tongue, oral cavity and gingiva), increasing with higher tumour-node-metastasis (TNM) stage tumours. Cathepsin D, but not B, was particularly associated with node metastasis and survival. In contrast to this, Macabeo-Ong et al. (2003) found significantly elevated levels of cathepsin L transcripts in a cohort of mostly gingival and tongue cancers. Cathepsin L mRNA and protein were progressively elevated in both dysplasias and cancer tissues and cathepsin L over-expression was an independent marker of dysplasias grade. The latest study on cathepsin B and cysteine proteinase inhibitory activities in the sera of human tongue cancer patients showed elevated cathepsin B levels in correlation with progression (stage) and invasion of this cancer (Saleh et al. 2006). In this study also, elevated levels of cystatin C and cathepsin B were found in tongue cancer tissue sections. Russo et al. (1995) reported that cathepsin B, but not cathepsin L, activity ratio in tumours versus normal mucosa of laryngeal cancers was found of prognostic significance along with histopathological and ploidy biomarkers in univariate, but not in multivariate, analysis in a cohort of 71 patients.

Bladder and Pancreatic Carcinomas

Cathepsins

Bladder carcinoma presents itself as either a superficial or a muscle-invasive disease, the latter comprising only 20% of the cases, although these have worse prognosis. Histopathologically, 90% of bladder cancers appear as urothelial carcinoma, for example, transitional cell carcinoma (TCC), and the rest as pure SCC. In a recent study, Blaveri et al. (2005) compared 80 bladder tumours and 9 bladder cancer cell lines with normal bladder tissues, using DNA micro-array analysis, and could clearly differentiate between TCC and SCC and between tumours with good and bad prognosis, as well as tumour stages by distinct gene expression profiles that were further verified on another independent data set. The validation of these gene sets was also done on tissue micro-arrays using immunohistochemistry and revealed the aspartic protease cathepsin E as a marker for the progression of superficial tumours. This was also suggested independently in a cohort of 67 patients with bladder cancer by Wild et al. (2005). Blaveri et al. (2005) also found about fivefold higher cathepsin K gene expression in the muscle-invasive type, compared to superficial tumours and the authors suggested, also based on early reports (reviewed by Berdowska 2004, Jedeszko and Sloane 2004), that cathepsins B, K and possibly L may promote muscle invasion, whereas other cathepsins (E and H) may be active in superficial tumours. Indeed, Eijan et al. (2003) found that cathepsin B was a diagnostic marker correlating with TCC tumours progression (also found by Visscher et al. 1994), which most likely release this enzyme, as cathepsin B activity in plasma (Eijan et al. 1997) and cathepsin B protein in the urine of patients with bladder cancer was measured (Eijan et al. 2000). Furthermore, since a different pattern of cathepsin B expression was observed even in normal peritumoural tissue compared to normal mucosa, they suggested that this protease is involved in the evolution of pre-neoplastic lesions. A similar conclusion was made by Staak et al. (2004), demonstrating elevated cathepsin B (and H), but not cathepsin L, activities in tumour versus normal bladder tissue at an early stage, however, without any further increase during tumour progression. They also suggested an elevated release of cysteine cathepsins from the tissue, as seen in tissue cell cultures.

Cystatins

Staak et al. (2004) measured the total cysteine proteinase inhibitory activity in bladder TCC, but these were not significantly altered. However, recently Tokyol et al. (2006) reported significantly increased levels of cystatin C in the sera of patients with bladder cancer, compared to controls.

Pancreatic Carcinoma

Although excellent animal models exist, such as RIP1-Tag 2(RT2) mouse model of pancreatic islet cell cancer, which are studied regarding the alteration on proteolytic profile during tumour progression (also within the scope of CANCERDEGRA-DOME), not many data relate to cathepsins in clinical progression of pancreatic carcinoma. Niedergethmann et al. (2004) reported on strong and independent prognostic impact of cathepsins B and L in resectable pancreatic adenocarcinoma in tissue sections of a cohort of 70 patients, defining subgroups of patients with early recurrence and poor outcome. A recent systemic review on molecular prognostic markers in pancreatic cancer did not reveal cathepsins as relevant for diagnosis and prognosis of this cancer (Garcea et al. 2005)

Liver Cancer–Hepatocellular Carcinoma

Cathepsins

Cathepsins are highly expressed in liver tissue, but no alteration in their levels was reported on hepatocellular carcinoma. A diffuse pattern of cathepsin B cytoplasmic staining was observed by Terada et al. (1995), and no differences in the level of activities in tissue or serum levels were found in patients with hepatocellular carcinoma (Niewczas et al. 2002).

Cystatins

Cystatin C and stefin A concentrations were significantly higher in the sera of patients with hepatocellular carcinoma than in healthy controls (Takeuchi et al. 2001). Leto et al. (1997) previously reported that stefin A levels were increased in the sera of hepatocellular patients compared to healthy subjects ($p < 0.02$) and a significant relationship was observed between stefin A serum levels and tumour size, the number of neoplastic lesions and a cancer serum marker, alpha fetoprotein. Stefin A was, therefore, suggested as an additional biochemical parameter to monitor the therapeutic response to treatment. In the same study, stefin A was also suggested as a predictive marker for response to therapy in patients with hepatocellular carcinoma and also as a marker to discriminate patients with cirrhosis who were developing precancerous lesions.

Melanoma

Melanoma progresses from normal skin via dysplastic nevi into malignant melanoma, which represents about 1% of all patients with cancer, rapidly increasing in incidence and mortality with a relatively short survival rate compared to carcinomas. Therefore, there is a particular need for markers that indicate early stages of malignant progression in this type of cancer to improve treatment planning.

Cathepsins

Kageshita et al. (1995) observed markedly higher immunohistochemical labelling of cathepsins B, L and H in malignant melanoma tissue sections than in early primary melanoma and dysplastic nevi. This was followed by additional evaluation of activities and transcription rates (Froelich et al. 2001), shown to be elevated for cathepsins B and L in melanoma and associated with its progression. These two studies differ with respect to cathepsin H, as Froelich et al. (2001) described an inverse correlation between this protease and the invasive potential of melanoma. Cathepsin B was found to be elevated not only at the invasive tumour front but also in the surrounding stroma (Froelich et al. 2001). This is in agreement with previous report (Ulmer et al. 1998), demonstrating that cathepsin B activity was fivefold higher in fibroblasts isolated from primary melanoma than in those from normal skin, suggesting a strong tumour–stroma interaction during melanoma progression. Goldman et al. (1999) have also reported that cathepsin B expression at the invasive front was indicative of bad prognosis, whereas Yoshii et al. (1995) demonstrated highly increased cathepsin B versus cystatin activity in most recurrent tumours. With respect to prognosis, Kos et al. (1997) observed significantly shorter OS in metastatic melanoma patients who presented high contents of cathepsins B and H ($p < 0.003$ and $p < 0.006$, respectively) in their sera. No relation of cathepsin L to prognosis was observed.

Cystatins

Kos et al. (1997) also demonstrated that the median levels of cystatin C were significantly higher ($p < 0.02$) within the group of metastatic melanoma patients compared to healthy controls. In contrast, the levels of stefin A were not statistically different among the groups of control, non-metastatic and metastatic melanoma patients, which is strikingly different from breast, head and neck and colon carcinoma. No correlation with survival was found for stefin A and cystatin C in patients with melanoma. Also, there was no difference in cystatin C serum levels between responders and non-responders to immunochemotherapy.

Intracranial Tumours

Intracranial tumours represent a unique and rather heterogenous population of both benign and malignant neoplasms. Recent data showed that the most abundant among the benign tumours are meningiomas, representing a prevalence of about 20% of intracranial malignancies. Among the malignant tumours, however, the most abundant are astrocytic tumours, for example, gliomas, with about 25% prevalence within all brain neoplasms. Meningiomas have a high average survival rate, ranging from 5 to 45 years, whereas gliomas range from 6 to 24 months survival in 90% of cases. Therefore, meningioma patients would be the ones to benefit most from additional biomarkers that could suggest more aggressive treatments in subpopulations of the invasive meningioma, whereas for glioma the emphasis for research and development is more on finding new potential therapeutic regimens. In both cases, these cancers are more difficult to treat compared to other types of tumours, due to their particular anatomical occurrence and the blood–brain barrier, limiting access to the tumour site.

Meningioma

Meningiomas stem from the arachnoidal cells in brain meninges. Tumour invasion into brain tissue is less common, but if it occurs, these tumours are malignant and have a high rate of recurrence. This is a more common feature of atypical (WHO II) and anaplastic (WHO III) meningioma, although also within benign meningiomas

(WHO I) a fraction of these tumours may present more aggressive behaviour and progress faster.

Cathepsins and Cystatins

Higher immunohistochemical staining of cathepsin B in border benign and atypical meningioma was demonstrated in a series of 88 meningiomas (Strojnik et al. 2001), and it was suggested that this marker may be useful to distinguish the clear benign tumours from the histomorphologically benign, but invasive, meningiomas, as confirmed later in total tumour extracts (Trinkaus et al. 2003, 2005)—Fig. 29.2. However, in a recent study in a larger cohort (119) of tumour extracts from benign, atypical and malignant meningioma, cathepsin B levels were not significant bio-markers (Lah et al. 2007, submitted), and it was suggested that meningioma invasiveness may be due to only a small fraction of the tumour cells that are located

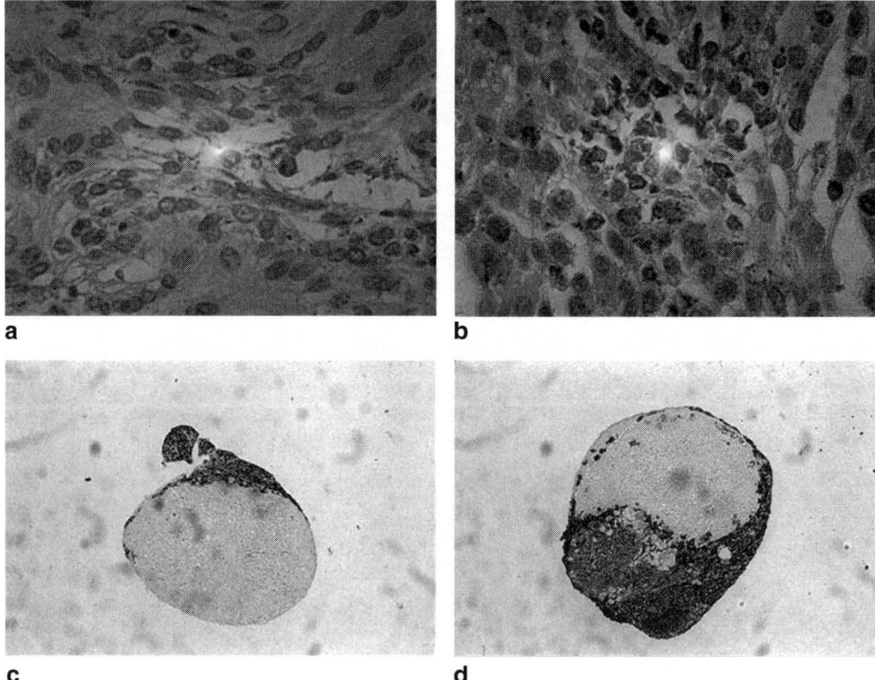

Fig. 29.2 Immunohistological staining of cathepsin B in human meningiomas (**a, b**). Clear benign meningioma (**a**) differ from histologically diagnosed atypical and anaplastic meningioma by more intense cathepsin B staining (**b**) (Strojnik et al. 2001). Chick heart invasion assay (**c, d**). Tumour explants were incubated in a coculture with the chick heart spheroids and labelled for cathepsin B. The meningioma spheroid is either not invasive (**c**) or shows different pattern of invasive behaviour (**d**) in chick heart matrix (*See also* Color Insert II)

at the tumour edge and express higher levels of cathepsin B. In this cohort, higher levels of cathepsin L, stefins A and B and cystatin C were also found in atypical meningiomas, but between the histological tumour grades only stefin A was significant. However, regarding a risk of relapse, cathepsin L, cystatin C and particularly stefin B ($p < 0.008$) levels were significant prognosticators; stefin B remained an independent prognosticator also in multivariate analysis.

Cathepsins B and H activities have been found increased and cystatin C concentration decreased in the cerebrospinal fluid of patients with leptomeningeal metastasis as compared with three control groups (without involvement of the central nervous system, multiple sclerosis cases and healthy volunteers, Nagai et al. 2003). Intense cathepsins B and H immunohistochemical staining was also detected in the metastasizing leptomeningeal tissues, although the presence and the activities in the CSF were not related with patients' survival rates. Collectively, these data indicate that the cysteine cathepsin system is involved in process(es) related to early relapse of meningioma, regardless of the histological classification and invasiveness.

Glioma

Cathepsins

Various cysteine cathepsins have been reported to be up-regulated in gliomas, including cathepsins B, L, H and S along with aspartic cathepsin D. This up-regulation can be observed at both the mRNA and the protein levels and often correlates with alterations in the subcellular localization and secretion of these cathepsins. Moreover, these changes in expression are more often marked at the invading front of the tumours, suggesting a role of these enzymes in glioma invasion. Rempel et al. (1994) observed increased cathepsin B transcript abundance, protein and activity levels in a set of primary brain tumours, when compared to normal brain tissue samples. Cathepsin B levels of expression, as well as changes in the subcellular localization of the protein, were correlated with histomorphological and clinical evidence of invasion. Expression of cathepsin B in tumour vessels was also reported, as confirmed in later studies (Mikkelsen et al. 1995, Sivaparvathi et al. 1995, Strojnik et al. 1999). Strojnik et al. (1999) also confirmed higher expression of cathepsin B protein in glioma cells, correlating with clinical and histopathological parameters in a cohort of 100 glioma patients. Intensity of cathepsin B immunostaining on either the endothelial cells alone or both the endothelial cells and the tumour cells together were highly significant prognostic factors. Additionally, a high cathepsin B total score (i.e. cathepsin B levels in both tumour and endothelial cells) is a significant predictor for shorter OS (Strojnik et al. 1999, 2000). Furthermore, intense cathepsin B staining of endothelial cells in glioblastoma is on its own a significant marker for shorter survival. Correlation of glioma grade with an increase in cathepsin B protein expression was recently reported by Wang et al. (2005), who also established a positive correlation between cathepsin B staining and microvessel density ($p < 0.05$). Nakabayashi and co-workers reported on correlation between high cathepsin B levels/low cystatin C expression and WHO glioma grade (Nakabayashi et al. 2005). This study on 57 glioma samples showed that patients with high cathepsin B levels presented significantly shorter DFS than did those with low cathepsin B ($p < 0.0001$). Also, gliomas with a relatively high ratio of cathepsin B to cystatin C tended to recur, indicating that depressed expression of cystatin C protein and elevated expression of cathepsin B protein are unfavourable prognostic factors in patients with glioma. Cathepsin B, therefore, appears to be an important factor in glioma progression, as the effects observed following its down-regulation, that is, reduced growth, invasiveness and angiogenesis (Gondi et al. 2004) of glioma tumours, were observed not only in in vitro studies but also in animal models (Lakka et al. 2004).

Regarding cathepsins L and H, their protein and activity levels are also significantly higher in glioblastoma and anaplastic astrocytomas than in low-grade gliomas and normal brain tissue samples (Sivaparvathi et al. 1996a, Sivaparvathi et al. 1996b, Strojnik et al. 2005). Differently from cathepsin B, prognostic significance of cathepsin L has not been confirmed, despite a significant correlation between cathepsin L immunohistochemical score and histological grading ($p < 0.005$). Expression of cathepsin S in astrocytoma biopsies was first reported by Flannery et al. (2003), demonstrating that cathepsin S was absent from normal astrocytes, oligodendrocytes, neurons and endothelial cells but expressed in astrocytoma cells, microglia and macrophages, and the highest activity was observed in grade IV tumours. A trend towards higher cathepsin S expression in glioblastoma tumours was observed and a high cathepsin S immunohistochemical score in glioblastoma has been associated with significantly shorter survival of patients. The cathepsin S score in glioblastoma was shown to be an independent predictor of survival ($p = 0.033$) in a multivariate analysis that included patients' age, radiation dose and Karnofsky score (Flannery et al. 2006, 2007).

Cystatins

In brain tumours, stefin A mRNA was detected in benign but not in malignant tissues (Strojnik et al. 2000). In the study of Lignelid et al. (1997), only a minority of cystatin C mRNA-positive brain and pituitary tumours revealed cystatin C-immunoreactive tumour cells and cystatin C did not correlate with any specific brain tumour type. However, the fact that the protein could be demonstrated more frequently in astrocytomas than in their more malignant counterparts suggests that cellular production and secretion of cystatin C changes with malignant progression of these tumours (Lignelid et al. 1997). This is in line with the report of Nakabayashi et al. (2005) showing that the cystatin C mRNA tended to be higher in low-grade astrocytomas than in glioblastomas and low cystatin C expression in glioma correlates with tumour recurrence. Also, multivariate analysis demonstrated prognostic impact of high tumour grade and low cystatin C expression on shorter DFS ($p < 0.0001$) (Nakabayashi et al. 2005).

Rare Cancers

There are some scattered data on a variety of other rare and less rare cancers where the cathepsins and cystatins have not been investigated to a great extent. These are only listed as a reference in Tables 29.1 and 29.2 and not discussed within this chapter.

Conclusions–Summary

This chapter is focused on breast and prostate carcinoma, the main subject of the CANCERDGRADOME integrated project, followed by other gender-related cancers (ovarian and cervical carcinoma) and continues with other common tumours like lung, colon, and head and neck carcinomas, to end with two intracranial malignancies, meningioma and glioma.

In breast cancer, differential expression of cysteine proteases seems to be related to the tumour histology, the most common breast ductal carcinoma gaining the invasiveness, particularly due to the increased expression of cathepsin B. In most studies, cathepsin B was found to be associated with bad prognosis, particularly in lymph node-negative population, whereas in three studies, cathepsin L has been found to be a good predictive factor, particularly in lymph node-positive patients. Invasive ductal carcinoma progression was associated first with high stefin A expression; however, in very recent studies, high stefin A levels in whole tumour extracts were shown to be correlated with a better outcome of the disease. This was confirmed in a clinically relevant murine model of spontaneous mammary metastasis, suggesting stefin A not only as a tumour but also as a metastasis supressor gene.

In prostate cancer, cysteine cathepsins B, L, H, S and K and very recently also cathepsin X have been associated with the invasive and metastatic potential. However, only cathepsin B to stefin A ratio was showing a trend towards an inverse relationship with clinical stage, independently of the Gleason score. This ratio may be indicative of an early invasive tumour and correlates with pelvic node metastasis and mortality rate. Stefin A can also discriminate between high-grade prostatic intraepithelial neoplasia and low-grade (I) carcinoma. A larger patient population needs to be screened to validate this diagnostic parameter in tumours as well as in the needle biopsy samples.

In tumours of the genital tract, such as *ovarian carcinoma*, cathepsin B seems to be most useful marker in pre-operative differential diagnosis of malignant ovarian and uterine tumours, particularly in combination with commonly used serum markers, CA125 and CA72-4. Cathepsin F over-expresion, along with cathepsin B, seems to be relevant in the progression of this type of cancer, whereas cathepsins L, S, H and C were reported as down-regulated. Stefin A was down-regulated in poorly differentiated areas of the advanced tumours and very low amounts were found secreted, compared with high levels of stefin B found in ovarian ascites fluid. In contrast, cystatin C levels increased in malignant tumours, compared to benign ones, although there are no data on the prognostic impact in these types of tumours.

In lung carcinoma, cathepsin B in tumour cells, along with its expression in inflammatory cells, seems to be a key factor associated with tumour progression. In several studies on patients with NSCLC, increased levels of cathepsin B were found in advanced tumours correlating with poor patients' survival. In a multivariate analysis, comparing cathepsin B to the PA1-uPAR system in a large cohort of NSCLC patients, among a total of 10 parameters, cathepsin B activity and concentration were prognostic for OS. However, only PAI-1 and uPAR added an independent prognostic information with regard to established clinical and histomorphological factors. Cathepsins L and H, although both elevated in lung tumour tissue, were not relevant for prognosis. However, in the group of long-term cigarette smokers, cathepsin H had an impact on prognosis and also appeared to be associated with chronic inflammation. In contrast, high cathepsin S levels correlated with good prognosis, due to its increased expression in cells of the immune response, thereby contributing to a higher survival rate. Cathepsin K was shown to be up-regulated in tumour-associated stroma, and may play a role in early stroma response to tumour. The stefins A and B are up-regulated in lung tumour tissue and may to a certain extent counterbalance cathepsin activity. Their function as tumour suppressors is supported by survival analyses. On the contrary, cystatins type II (C, E and M) are down-regulated in lung tumours, thereby possibly contributing to elevated cathepsin activities.

In colon carcinoma, the data of various studies show that elevated cathepsin B and cathepsin L levels may represent a prognostic, although not independent, marker for tumour progression. In patients' sera, cathepsin B and cathepsin H were relevant for prognosis, particularly in combination with CEA. Significantly, increased serum levels of stefin B and cystatin C are also associated with shorter patients' survival. Recently, the levels of cathepsin B–cystatin C complex, being significantly lower in advanced Dukes' stages (C and D), were also proposed as a useful prognostic parameter.

In head and neck cancer, cathepsins B and L were found elevated in tumour tissue of patients with SCC; however, their levels were not predictive for the disease outcome. On the contrary, stefins A and B levels were significantly correlated with the disease-free and disease-specific survival. In multivariate analysis, high stefin A concentration appeared as the strongest independent predictor for favourable DFS when compared to tumour stage, site and extracapsular extension. These results were reconfirmed after pooling the data of various data sets, and stefin A tumour tissue levels can be recognized as promising candidate marker for prognosis in patients with operable carcinoma of head and neck. In oesophagus carcinoma, stefin A has been clearly demonstrated to effect the progression of the disease, suggesting that this inhibitor may have therapeutic applications. Cathepsin B in oral cavity and cathepsin L in the tongue carcinoma were found as significant biomarkers for disease progression.

Bladder carcinoma progression seems to be characterized by over-expression of an aspartic cathepsin E and cathepsin K genes. It was recently suggested that cathepsins B, K and possibly L may promote muscle invasion, whereas other cathepsins (E and H) may be active in superficial tumours. Cathepsin B was also found as a marker for TCC bladder carcinoma progression since these tumours secrete higher amounts of cathepsin B into plasma and urine. Cystatin C levels were also found significantly increased in the serum of patients with bladder cancer.

Pancreatic and liver carcinomas have not been studied to great extent regarding cysteine cathepsins as markers to draw any conclusions. Significantly, stefin A in hepatocellular cancer was increased and significantly related to tumour progression. Stefin A was not only suggested as a predictive marker for response to therapy in patients with hepatocellular carcinoma but also as a marker to discriminate patients with cirrhosis who were developing precancerous lesions.

Melanoma: Cathepsin B was found highly elevated in early and in advanced melanoma, presumably due to intensive tumour–stroma interactions. Cathepsin B alone or in the ratio to cystatin C was prognostic for survival, both in tumours and patients' sera. On the contrary, high serum levels of cystatin C correlate with metastatic melanoma. Also, high cathepsin H levels in tumour tissue patients' sera were of high prognostic significance whereas cathepsin L does not seem to be involved in the progression of melanoma. These data strongly support further clinical research with respect to cathepsins and cystatins, which could prove the prognostic value of these biomarkers in this deadly cancer.

Intracranial tumours: In meningiomas, tumours with the highest incidence among a variety of intracranial tumours, cathepsin L and the cystatins, particularly stefin B, were found as significant prognosticators for tumour relapse. Cathepsin B labelling of the invasive tumour edges, although not elevated in the bulk of tumours, may be used in differential diagnosis of aggressive histologically benign meningioma. Cathepsins B, L, H and S were found increased also in glioma; among them cathepsin B appears to be an important factor in glioma progression, as shown in in vitro studies and animal models. It may be used as a prognostic parameter for patients with glioma, in conjunction with magnetic resonance imaging (MRI) scans and traditional histomorphological grading. Of interest, higher cathepsin B expression is associated not only with tumour cells but also with endothelial cells in new vessels, representing another possibility for prognosis of shorter survival. Another of the cysteine proteases up-regulated in glioma, cathepsin S, has also been implicated in angiogenesis (Shi et al. 2003, Wang et al. 2006) and its immunohistochemical score in glioblastoma, grade IV glioma, was demonstrated as a prognostic factor. In contrast, increased levels of cathepsin L in advanced glioma were not related to patients' survival.

Acknowledgements This chapter included research supported by the Cancerdegradome Integrated Project (Partners#17 and #34) and many of the data listed in this chapter, including those from the authors, were obtained through collaborative efforts within the EORTC PathoBiology Group. The authors also acknowledge the programme support from the Slovene Research Agency: P1-0245 (granted to TL) and P4-0127 (granted to JK).

References

Abrahamson M., Barrett A.J., Salvesen G., et al. (1986). Isolation of cysteine proteinase inhibitors from human urine. Their physiological and enzyme kinetic properties and concentrations in biological fluids. J. Biol. Chem. 261:11282–11289.

Abrahamson M., Alvarez-Fernandez M., Nathanson C.M. (2003). Cystatins. Biochem. Soc. Symp. 70:179–199.

Acuff H.B., Sinnamon M., Fingleton B., et al. (2006). Analysis of host- and tumour-derived proteinases using a custom dual species microarray reveals a protective role for stromal matrix metalloproteinase-12 in non-small cell lung cancer. Cancer Res. 66:7968–7975.

Aho H.J., Ping W., Soderstrom K.O., et al. (1995). Acid cysteine proteinase inhibitor in cutanus lymphocytic infiltrates. Am. J. Dermatopathol. 17:115–125.

Alvarez-Fernandez M., Barrett A.J., Gerhartz B., et al. (1999). Inhibition of mammalian legumain by some cystatins is due to a novel second reactive site. J. Biol. Chem. 274:19195–19203.

Arora S., and Chauhan S.S. (2002). Identification and characterization of a novel human cathepsin L splice variant. Gene 293:123–131.

Barrett A.J., Rawlings N.D., Woessner J.F. (2004). The Handbook of Proteolytic Enzymes, 2nd ed. Academic Press. London.

Bell-McGuinn K.M., Garfall A.L., Bogyo M., et al. (2007). Inhibition of cysteine cathepsin protease activity enhances chemotherapy regimens by decreasing tumour growth and invasiveness in a mouse model of multistage cancer. Cancer Res. 67:7378–7385.

Berdowska I. (2004). Cysteine proteases as disease markers. Clin. Chim. Acta 342:41–69.

Bianchi F., Hu J., Pelos S.G., et al. (2004). Lung cancers detected by screening with spiral computed tomography have a malignant phenotype when analysed by cDNA microarray. Clin. Cancer Res. 10:6023–6028.

Blaveri E., Simko J.P., Korkola J.E., et al. (2005). Bladder cancer outcome and subtype classification. Clin. Cancer Res. 11:4044–4056.

Brubaker K.D., Vessella R.L., True L.D., et al. (2003). Cathepsin K mRNA and protein expression in prostate cancer progression. J. Bone Miner. Res. 18:220–230.

Budihna M., Strojan P., Smid L., et al. (1996). Prognostic value of cathepsins B, H, L, D and their endogenous inhibitors stefins A and B in head and neck carcinoma. Biol. Chem. Hoppe Seyler 377:385–390.

Buhling F., Waldburg N., Gerbe A., et al. (2000). Cathepsin K expression in human lung. Adv. Exp. Med. Biol. 477:281–286.

Calkins C.C., Sameni M., Koblinski J., et al. (1998). Differential localization of cysteine proteinase inhibitors and target cysteine protease, cathepsin B by immuno-confocal microscopy. J. Histochem. Cytochem. 46:745–751.

Campo E., Munoz J., Miguel R., et al. (1994). Cathepsin B expression in colorectal carcinomas correlates with tumour progression and shortened patiet survival. Am. J. Pathol. 145:310–319.

Caserman S., Kenig S., Sloane B.F., et al. (2006). Cathepsin L splice variants in human breast cell lines. Biol. Chem. 387:629–634.

Cavallo-Medved D., Dosescu J., Linebaugh B.E., et al. (2003). Mutant K-ras regulates cathepsin B localization on the surface of human colorectal carcinoma cells. Neoplasia 5:1–12.

Chu J.H., Sun Z.Y., Meng X.L., et al. (2006). Differential metastasis-associated gene analysis of prostate carcinoma cell derived from primary tumour and spontaneous lymphatic metastasis in nude mice with orthotopic implantantion of PC-3M cells. Cancer Lett. 233:70–88.

de Re E.C., Shuja S., Cai J., et al. (2000). Alterations in cathepsin H activity and protein pattern in human colorectal carcinoma. Br. J. Cancer 82:1317–1326.

Dreilich M., Wagenius G., Bergstrom S., et al. (2005). The role of cystatin C and the angiogenic cytokines VEGF and bFGF in patients with esophageal carcinoma. Med. Oncol. 22:29–38.

Ebert W., Knoch H., Werle B., et al. (1994). Prognostic value of increased lung tumor tissue cathepsin B. Anticancer Res. 14:895–900.

Ebert E., Werle B., Julke B., et al. (1997). Expression of cysteine protease inhibitors stefin A, stefin B, and cystatin C in human lung tumour tissue. Adv. Exp. Med. Biol. 421:259–265.

Eide T.J., Jarvinen M., Hopsu-Havu V.K., et al. (1997). Immunolocalization of cystatin A in neoplastic, virus and inflammatory lesions of the uterine cervix. Acta Histochem. 93:241–248.

Eijan A.M., Casabe A.B., Puricelli L., et al. (1997). Levels of plasma cysteine—proteinase activity in bladder cancer patients. Oncol. Rep. 4:447–450.

Eijan A.M., Sandes E.O., Puricelli L., et al. (2000). Cathepsin B levels in urine from bladder cancer patients. Oncol. Rep. 7:1395–1399.

Eijan A.M., Sandes E.O., Riveros M.D., et al. (2003). High expression of cathepsin B in transitional bladder carcinoma correlates with tumour invasion. Cancer 98:262–268.

Fernandez P.L., Farre X., Nadal A., et al. (2001). Expression of cathepsins B and S in the progression of prostate carcinoma. Int. J. Cancer 95:51–55.

Flannery T., Gibson D., Mirakhur M., et al. (2003). The clinical significance of cathepsin S expression in human astrocytomas. Am. J. Pathol. 163:175–182.

Flannery T., McQuaid S., McGoohan C., et al. (2006). Cathepsin S expression: An independent prognostic factor in glioblastoma tumours—A pilot study. Int. J. Cancer 119:854–860.

Flannery T., McConnell R.S., McQuaid S., et al. (2007). Detection of cathepsin S cysteine protease in human brain tumour microdialysates in vivo. Br. J. Neurosurg. 21:204–209.

Foekens J.A., Kos J., Peters H.A., et al. (1998). Prognostic significance of cathepsin B and L in primary human breast cancer. J. Clin. Oncol 16:1013–1021.

Friedl P., and Wolf K. (2003). Tumour-cell invasion and migration: Diversity and escape mechanisms. Nat. Rev. Cancer 3:362–374.

Froelich E., Schlagenhauff B., Mohrle M., et al. (2001). Activity, expression and transcription rate of the cathepsins B, D, H and L in cutaneous malignant melanoma. Cancer 91:972–982.

Garcea G., Neal C.P., Pattenden C.J., et al. (2005). Molecular prognostic markers in pancreatic cancer: A systemic review. Eur. J. Cancer 41:2213–2236.

Gochareva V., and Joyce J.A. (2007). Cysteine cathepsins and the cutting edge of cancer invasion. Cell Cycle 6:60–64.

Gochareva V., Zeng W., Ke D., et al. (2006). Distinct roles for cysteine cathepsin genes in multistage tumorigenesis. Genes Dev. 20:543–556.

Goldman T., Suler L., Ribbert D., et al. (1999). The expression of proteolytic enzymes at the dermal invading front of primary cutaneous melanoma predicts metastasis. Pathol. Res. Pract. 195:171–175.

Gondi C.S., Lakka S.S., Dinh D.H., et al. (2004). RNAi-mediated inhibition of cathepsin B and uPAR leads to decreased cell invasion, angiogenesis and tumor growth in gliomas. Oncogene 23:8486–8496.

Gutman S., and Kessler L.G.F. (2006). The US Food and drug administration perspective on cancer biomarker development. Nat. Rev. Cancer 6:565–571.

Harbeck N., Alt U., Krueger A., et al. (2001). Prognostic impact of proteolytic factors (urokinase-type plasminogen activator, plasminogen activator inhibitor 1 and cathepsins B, D and L) in primary breast cancer reflects effects of adjuvant systemic therapy. Clin. Cancer Res. 7:2757–2764.

Harbeck N., Kates R.E., Look M.P., et al. (2002). Enhanced benefit from adjuvant chemotherapy in breast cancer patients classified high-risk according to urokinase-type plasminogen activator (uPA) and plasminogen activator inhibitor type 1 (n = 3424). Cancer Res. 62:4617–4622.

Hazen L.G., Bleeker F.E., Lauritzen B., et al. (2000). Comparative localisation of cathepsin B protein and activity in colorectal cancer. J. Histochem. Cytochem. 48:1421–1430.

Herszenyl L., Plebani M., Carraro P., et al. (1999). The role of cysteine and serine proteases in colorectal carcinoma. Cancer 86:1135–1142.

Hirai K., Yokoyama M., Asano G., et al. (1999). Expression of cathepsin B and cystatin C in human colorectal cancer. Hum. Pathol. 30:680–688.

Honey K., and Rudensky A.Y. (2003). Lysosomal cysteine proteases regulate antigen presentation. Nat. Rev. Immunol. 3:472–482.

Hughes S.J., Glover T.W., Zhu X.-X. et al. (1998). A novel amplification at 8p22–23 results in overexpression of cathepsin B in esophageal adenocarcinoma. Proc. Nat. Acad. Soc. USA 95:12410–12415.

Jedeszko C., and Sloane B.F. (2004). Cysteine cathepsins in human cancer. Biol. Chem. 385:1017–1027.

Jiborn T., Abrahamson M., Wallin H., et al. (2004). Cystatin C is highly expressed in human male reproductive system. J. Androl. 25:564–572.

Joyce J.A., and Hanahan D. (2004). Multiple roles for cysteine cathepsins in cancer. Cell cycle 3:1516–1519.

Joyce J.A., Baruch A., Chehade Kmeyer-Morse N., et al. (2004). Cathepsin cysteine proteases are effectors of invasive growth and angiogenesis during multistage tumorigenesis. Cancer Cell 5:443–453.

Kageshita T., Yoshii A., Kimura T., et al. (1995). Biochemical and immunohistochemical analysis of cathepsins B, H, L and D in human melanocytic tumours. Arch. Dermatol. Res. 287:266–272.

Karkola J.E., DeVries S., Fridlyan J., et al. (2003). Differentiation of lobular versus ductal breast carcinoma by expression microarray analysis. Cancer Res. 63:7167–7175.

Kastelic L., Turk B., Kopitar-Jerala N., et al. (1994). Stefin B, the major low molecular weight inhibitor in ovarian carcinoma. Cancer Lett. 82:81–88.

Kawasaki G., Kato Y., Mizuno Y.A., et al. (2002). Cathepsins expression in oral squamous cell carcinoma: Relationship with clinicopathologic factors. Oral. Surg. Oral. Med. Oral. Pathol. Oral. Radiol. Endod. 93:446–454.

Kayser K., Richter N., Hufnage P., et al. (2003). Expression, proliferation activity and clinical significance of cathepsin B and cathepsin L in operated lung cancer. Anticancer Res. 23:2767–2772.

Kos J., and Lah T.T. (1998). Cysteine proteinases and their endogenous inhibitors: Target proteins for prognosis, diagnosis and therapy in cancer (review). Oncol. Rep. 5:1349–1361.

Kos J., and Lah T.T. (2006). Cystatins in cancer. In: Human Stefins and Cystatins, eds. E. Zerovnik, N. Kopitar-Jerala. pp. 152–165. New York: Nova Science Publishers, Inc.

Kos J., Šmid A., Krašovec M., et al. (1995). Lysosomal proteases cathepsins D, B, H, L and their inhibitors stefins A and B in head and neck cancer. Biol. Chem. Hoppe-Seyler 376:401–405.

Kos J., Štabuc B., Schweiger A., et al. (1997). Cathepsins B, H and L and their inhibitors in stefin A and cystatin C in sera of melanoma patients. Clin. Cancer Res. 3:1815–1822.

Kos J., Nielsen H.-J., Krašovec M., et al. (1998). Prognostic values of cathepsin B and carcinoembryonic antigen in sera of patients with colorectal cancer. Clin. Cancer Res. 4:1511–1516.

Kos J., Krašovec M., Cimerman N., et al. (2000). Cysteine protease inhibitors stefin A, stefin B and cystatin C in sera from patients with colorectal cancer: Relation to prognosis. Clin. Cancer Res. 6:505–511.

Kos J., Sekirnik A., Kopitar G., et al. (2001). Cathepsin S in tumours, regional lymph nodes and sera of patients with lung cancer: Relation to prognosis. Br. J. Cancer 85:1193–1200.

Krepela E., Prochazka J., Karova B., et al. (1998). Cysteine proteases and cysteine protease inhibitors in non-small cell lung cancer. Neoplasma 45:318–331.

Kuopio T., Kankaanranta A., Jalava P., et al. (1998). Cysteine protease inhibitor cystatin A in breast cancer. Cancer Res. 58:432–436.

Lah T.T., and Kos J. (1998). Cysteine proteinases in cancer progression and their clinical relevance for prognosis. Biol. Chem. 379:125–130.

Lah T.T., Kokalj-Kunovar M., Kastelic L., et al. (1991). Cystatins and stefins in ascites fluid from ovarian carcinoma. Cancer Lett. 61:243–253.

Lah T.T., Kokalj-Kunovar M., Štrukelj B., et al. (1992). Stefins and lysosomal cathepsins B, L and D in human breast carcinoma. Int. J. Cancer 50:36–44.

Lah T.T., Kos J., Blejec A., et al. (1997). The expression of lysosomal proteases and their inhibitors in breast cancer: Possible relationship to prognosis of the disease. Pathol. Oncol. Res. 3(2):89–99.

Lah T.T., Kalman E., Najjar D., et al. (2000a). Cells producing cathepsins D, B and L in human breast carcinoma and their association with prognosis. Hum. Pathol. 31:149–160.

Lah T.T., Čerček M., Blejec A., et al. (2000b). Cathepsin B, a prognostic indicator in lymph node-negative breast carcinoma patients: Comparison with cathepsin D, cathepsin L and other clinical indicators. Clin. Cancer Res. 6:578–584.

Lah T.T., Durán Alonso M.B., and Van Noorden C.J.V. (2006). Antiprotease therapy in cancer: Hot or not? Expert Opin. Biol. Ther. 6:257–279.

Lah T.T., Nanni I., Trinkaus M., et al. (2008). Towards better diagnosis and prognosis of aggressive meningioma: A role of cysteine proteases. (submitted)

Lakka S.S., Gondi C.S., Yanamandra N., et al. (2004). Inhibition of cathepsin B and MMP-9 gene expression in glioblastoma cell line via RNA interference reduces tumor cell invasion, tumor growth and angiogenesis. Oncogene 23:4681–4689.

Ledakis P., Tester W., Rosenberg N., et al. (1996). Cathepsins D, B and L in malignant human lung tissue. Clin. Cancer Res. 5:561–568.

Leinonen T., Pirinen R., Bohm J., et al. (2007). Biological and prognostic role of acid cysteine proteinase inhibitor (ACPI, cystatin A) in non-small-cell lung cancer. J. Clin. Pathol. 60:515–519.

Leto G., Tumminello F.M., Pizzolanti G., et al. (1997). Lysosomal cathepsins B and L and stefin A blood levels in patients with hepatocellular carcinoma and/or liver cirrhosis: Potential clinical implications. Oncology 54:79–83.

Levičar N., Kos J., Blejec A., et al. (2002). Comparison of potential biological markers cathepsin B, cathepsin L, stefin A and stefin B with urokinase and plasminogen-activator inhibitor-1 and clinico-pathological data of breast cancer patients. Cancer Detect. Prev. 26:42–49.

Li W., Ding F., Zhang L., et al. (2005). Overexpression of stefin A in human esophageal squamous cell carcinoma cells inhibit tumour cell growth, angiogenesis, invasion and metastasis. Clin. Cancer Res. 11:8753–8762.

Liaudet-Coopman E., Beaujouin M., Derocq D., et al. (2006). Cathepsin D: A newly discovered functions of a long-standing aspartic protease in cancer and apoptosis. Cancer Lett. 237:167–179.

Lignelid H., Collins V.P., Jacobsson B. (1997). Cystatin C and transthyretin expression in normal and neoplastic tissues of the human brain and pituitary. Acta Neuropathol. (Berl) 93:494–500.

Linnerth N.M., Sitbovan K., Moorehead R.A. (2005). Use of a transgenic model to identify markers of human lung tumours. Int. J. Cancer 114:977–982.

Luo A., Kong J., Hu G., et al. (2004). Discovery of Ca(2)-relevant and differentiation-associated genes downregulated in esophageal squamous cell carcinoma using cDNA microarray. Oncogene 23:1291–1299.

Macabeo-Ong M., Shiboski C.H., Silerman S., et al. (2003). Quantitative analysis of cathepsin L mRNA and protein expression during oral cancer progression. Oral. Oncol. 39:638–647.

Mason R.W., Gal S., Gottesman M.M. (1987). The identification of the major excreted protein (MEP) from a transformed mouse fibroblasts cell line as a catalytically active precursor form of cathepsin L. Biochem. J. 248:449–454.

Mikkelsen T., Yan P.S., Ho K.L., et al. (1995). Immunolocalization of cathepsin B in human glioma: Implications for tumor invasion and angiogenesis. J. Neurosurg. 83:285–290.

Mirtti T., Alanen K., Kallajoki M. et al. (2003). Expression of cystatins, high molecular weight cytokeratin, and proliferation markers in prostatic adenocarcinoma and hyperplasia. Prostate 54:290–298.

Mohamed M.M., and Sloane B.F. (2006). Cysteine cathepsins: Multifunctional enzymes in cancer. Nat. Rev. Cancer 6:764–775.

Muentener K., Zwicky R., Scus G., et al. (2003). The alternative use of exons 2 and 3 in cathepsin B mRNA controls enzyme trafficking and triggers nuclear fragmentation in human cells. Histochem. Cell Biol. 119:93–101.

Mulaomerovic A., Halilbašić A., Čičkušić E., et al. (2007). Cystatin C as a potential marker for relapse in patients with non-Hodgkin B-cell lymphoma. Cancer Lett. 248:192–197.

Naegler D.K., Hrueger S., Kellner A., et al. (2004). Up-regulation of cathepsin X in prostate cancer and prostatic intraepithelial neoplasia. Prostate 60:109–119.

Nagai A., Terashim M., Harada T., et al. (2003). Cathepsin B and H activities and cystatin C concentrations in cerebrospinal fluid from patients with leptomeningeal metastasis. Clin. Chim. Acta 329:53–60.

Nakabayashi H., Hara M., Shimuzu K. (2005). Clinicopathologic significance of cystatin C expression in gliomas. Hum. Pathol. 36:1008–1015.

Niedergethmann M., Wostbrock B., Sturm J.W., et al. (2004). Prognostc impact of cysteine proteases cathepsin B and cathepsin L in pancreatic adenocarcinoma. Pancreas 2:204–211.

Niewczas M., Paczek L., Krawczyk M., et al. (2002). Enzymatic activities of cathepsin B, cathepsin B and L, plasmin, trypisn and colagenase in hepatocellular carcinoma. Pol. Arch. Med. Wewn. 108:653–662.

Nishida Y., Kohno K., Kawamata T., Morimotsu K., Kuwano M., Miyakawa I. (1995). Increased cathepsin L levels in serum in some patients with ovarian cancer: comparison with CA125 and CA72-4. Gynecol. Oncol. 56:357–361.

Nishikawa H., Ozaki Y., Nakanishi T., et al. (2004). The role of cathepsin B and cystatin C in the mechanisms of invasion by ovarian cancer. Gynecol. Oncol. 92:881–886.

Obermajer N., Doljak B., Kos J. (2006). Cysteine cathepsins: Regulators of antitumour immune response. Expert Opin. Biol. Ther. 6:1295–1309.

Overall C.M., and López-Otín C. (2002). Strategies for MMP inhibition in cancer: Innovations for the post-trial era. Nat. Rev. Cancer 2:657–672.

Palungwachira P., Kakuta M., Yamazaki M., et al. (2002). Immunohistochemical localization of cathepsin L and cystatin A in normal skin and skin tumors. J. Dermatol. 29:573–579.

Parker B.S., Bidwell B.N., Slavin J.L., et al. (2007). Enhanced expression of the cathepsin inhibitor stefin A is associated with metastasis in breast cancer. Clin. Exp. Metastasis 24:211–316.

Petty R.D., Kerr K.M., Murray G.I., et al. (2006). Tumour transcriptome reveals the predictive and prognostic impact of lysosomal protease inhibitors in non-small cell lung cancer. J. Clin. Oncol. 24:1729–1744.

Pollanen R., Pyykkonen K., Jarvinen M., et al. (1995). Immunolocalization of cystatin A in condylomatous and dysplastic lesions of the human uterine cervix: Correlation with the presence and type of human papillomavirus infection. Int. J. Ginecol. Pathol. 14:217–222.

Poteriaeva O.N., Usova T.A., Levina O.A., et al. (2003). Immunoenzyme technique for analysis of cystatin C in serum of patients with hemoblastosis. Klin. Lab. Diagn. 7:35–38.

Rempel S.A., Rosenblum M.L., Mikkelsen T., et al. (1994). Cathepsin B expression and localization in glioma progression and invasion. Cancer Res. 54:6027–6031.

Roesch-Ely M., Nees M., Karsai S., et al. (2007). Proteomic analysis reveals successive aberrations in protein expression from healthy mucosa to invasive head and neck cancer. Oncogene 26:54–64.

Russo A., Bazan V., Gebbia N., et al. (1995). Flow cytometry DNA analysis and lysosomal cathepsins B and L in locally advanced laryngeal cancer. Cancer (Phila) 76:1757–1764.

Saleh Y., Sebzda T., Warwas M., et al. (2005). Expression of cystatin M in clinical human colorectal cancer tissues. J. Exp. Theor. Oncol. 5:49–53.

Saleh Y., Wnukiewicz J., Andrzejak R., et al. (2006). Cathepsin B and cysteine protease inhibitors in human tongue cancer: Correlation with tumour staging and in vitro inhibition of cathepsin B by chicken cystatins. J. Cancer Mol. 2:67–72.

Schmitt M., Jaenicke F., Graeff H. (1992). Protease, matrix degradation and tumour-cell spread. Fibrinolysis 6:1–17.

Schweiger A., Christensen I.J., Nielsen H.J., et al. (2004). Serum cathepsin H as a potential prognostic marker in patients with colorectal cancer. Int. J. Biol. Markers 19:254–257.

Scorilas A., Fotiou S., Tsiambas E., et al. (2002). Determination of cathepsin B expression may offer additional prognostic information for ovarian cancer patients. Biol. Chem. 383:1297–1303.

Shi G.P., Sukhova G.K., Kuzuya M., et al. (2003). Deficiency of the cysteine protease cathepsin S impairs microvessel growth. Circ. Res. 92:493–500.

Shiraishi T., Mori M., Tanaka S., et al. (1998). Identification of cystatin B in human esophageal carcinoma, using differential displays in which gene expression is related to lymph node metastasis. Int. J. Cancer 79:175–178.

Shuja S., Sheahan K., Murnane M.J. (1991). Cysteine endopeptidase activity levels in normal human tumours, colorectal adenoma and carcinoma. Int. J. Cancer 49:341–346.

Sinha A.A., Gleason D.F., DeLeon O.F., et al. (1993). Localisation of a biotonylated cathepsin B oligonucleotide probe in human prostate including invasive cells and invasive edges by in situ hybridization. Anat. Rec. 235:233–240.

Sinha A.A., Quast B.J., Wilson M.J., et al. (2001). Ratio of cathepsin B to stefin A identifies heterogeneity within Gleason histologic scores for human prostate cancer. Prostate 48:274–284.

Sinha A.A., Quast B.J., Wilson M.J., et al. (2002). Prediction of pelvic lymph node metastasis by the ratio of cathepsin B to stefin A in patients with prostate carcinoma. Cancer 94:3141–3149.

Sinha A.A., Morgan J.L., Wood N., et al. (2007). Heterogeneity of cathepsin B and stefin A expression in Gleason pattern 3 + 3 (score 6) prostate cancer needle biopsies. Anticancer Res. 27:1407–1413.

Sivaparvathi M., Sawaya R., Wang S.W., et al. (1995). Overexpression and localization of cathepsin B during the progression of human gliomas. Clin. Exp. Metastasis 13:49–56.

Sivaparvathi M., Sawaya R., Gokaslan Z.L., et al. (1996a). Expression and the role of cathepsin H in human glioma progression and invasion. Cancer Lett. 104:121–126.

Sivaparvathi M., Yamamoto M., Nicolson G.L. et al. (1996b). Expression and immunohistochemical localization of cathepsin L during the progression of human gliomas. Clin. Exp. Metastasis 14:27–34.

Sloane B.F., Moin K., Lah T.T. (1994). Lysosomal enzymes and their endogenous inhibitors in neoplasia. In: Biochemical and molecular aspects of selected cancers, eds. T.G. Pretlow, T.P. Pretlow. pp. 411–466. New York: Academic Press.

Soderstrom K.O., Laato M., Wu P., et al. (1995). Expression of acid cysteine protease inhibitor (ACPI) in normal human prostate, benign prostatic hyperplasia and adenocarcinoma. Int. J. Cancer 62:1–4.

Sotiropoulou G., Anisowitz A., Sager R. (1997). Identification, cloning and characterisation of cystatin M, a novel cysteine proteinase inhibitor, down-regulated in breast cancer. Cancer Res. 272:903–910.

Staak A., Tolic D., Kristiansen G., et al. (2004). Expression of cathepsins B, H and L and their inhibitors as markers of transitional cell carcinoma of the bladder. Urology 63:1089–1094.

Strojan P., Budihna M., Šmid L., et al. (2000). Prognostic significance of cysteine proteases cathepsin B and L and their endogenous inhibitor stefins A and B in patients with squamous cell carcinoma with head and neck cancer. Clin. Cancer Res. 6:1052–1062.

Strojan P., Budihna L., Šmid L., et al. (2001). Cathepsin B and L and stefin A and B levels as serum tumor markers in squamous cell cacinoma of head and neck. Neoplasma 48:66–71.

Strojan P., Svetic B., Šmid L., et al. (2004a). Serum cystatin C in patients with head and neck carcinoma. Clin. Chem. Acta 344:155–161.

Strojan P., Oblak I., Svetic B., et al. (2004b). Cystaine proteinase inhibitor cystatin C in squamous cell carcinoma of the head and neck: Relation to prognosis. Br. J. Cancer 90:1961–1968.

Strojan P., Anicin A., Svetic B., et al. (2007). Stefin A and stefin B: Markers for prognosis in operable squamous cell carcinoma of head and neck. Int. J. Radiat. Oncol. Biol. Phys. 68:1335–1341.

Strojnik T., Kos J., Zidanik B., et al. (1999). Cathepsin B immunohistochemical staining in tumor and endothelial cells is a new prognostic factor for survival in patients with brain tumors. Clin. Cancer Res. 5:559–567.

Strojnik T., Zajc I., Bervar A., et al. (2000). Cathepsin B and its inhibitor stefin A in brain tumors. Pflugers Arch. 439:R122–R123.

Strojnik T., Židanik B., Kos J., et al. (2001). Cathepsins B and L are markers for clinically invasive types of meningiomas. Neurosurgery 48:598–605.

Strojnik T., Kavalar R., Trinkaus M., et al. (2005). Cathepsin L in glioma progression: Comparison with cathepsin B. Cancer Detect. Prev. 29:448–455.

Takeuchi M., Fukuda Y., Nakano I., et al. (2001). Elevation of cystatin C concentrations in patients with chronic liver disease. Eur. J. Gastroenterol. Hepatol. 13:951–955.

Talieri M., Papdopoulu S., Scorilas A., et al. (2004). Cathepsin B and cathepsin D expression in the progression of colorectal adenoma to carcinoma. Cancer Lett. 205:97–106.

Terada T., Ohta T., Minato H., Nakanuma Y. (1995). Expression of pancreatic trypsinogen/trypsin and cathepsin B in human cholangiocarcinomas and hepatocellular carcinomas. Hum. Pathol. 26:746–752.

Thomssen C., Schmitt M., Goretzki L., et al. (1995). Prognostic value of the cysteine proteases cathepsins B and L in human breast cancer. Clin. Cancer Res. 1:741–746.

Thomssen C., Oppelt P., Jaenicke F., et al. (1998). Identification of low risk node-negative breast cancer patients by tumour biological factors PAI-1 and cathepsin L. Anticancer Res. 18:2173–2180.

Tokyol C., Koken T., Demirbas M., et al. (2006). Expression of cathepsin D in bladder carcinoma: Correlation with biological features and serum cystatin C levels. Tumori 92:230–235.

Trinkaus M., Vranič A., Dolenc V.V., et al. (2003). Cathepsin L in human meningiomas. Radiol. Oncol. 37:89–99.

Trinkaus M., Vranič A., Dolenc V.V., et al. (2005). Cathepsins B and L and their inhibitors stefin B and cystatin C as markers for malignant progression of benign meningioma. Int. J. Biol. Markers 20:50–59.

Troy A.M., Sheahan K., Mulcahy H.E., et al. (2004). Expression of cathepsins B and L antigen and activity is associated with early colorectal cancer progression. Eur. J. Cancer 40:1610–1616.

Turk B. (2006). Targeting proteases: Successes, failures and future prospects. Nat. Rev. Drug Discov. 5:785–799.

Ulmer A., Korber V., Schmid H., et al. (1998). Increased activity of cathepsin B in fibroblasts isolated from primary melanoma in comparison to fibroblasts from normal skin. Exp. Dermatol. 7:14–17.

Vazquez-Otrin G., Pina-Sanchez P., Vazquez K. et al. (2005). Overexpression of cathepsin F, matrix metalloproteinases 11 and 12 in cervical cancer. BMC Cancer 5:1–11.

Vigneswaran N., Zhao W., Dassanayake A., et al. (2000). Variable expression of cathepsin B and D correlates with highly invasive and metastatic phenotype of oral cancer. Hum. Pathol. 31:931–937.

Vigneswaran N., Wu J., Muller S., et al. (2005). Expression analysis of cystatin C and M in laser-capture microdissectioned human breast cancer cells—A preliminary study. Pathol. Res. Pract. 200:753–762.

Visscher D.W., Sloane B.F., Sameni M., et al. (1994). Clinicopathologic significance of cathepsin B immunostaining in transitional neoplasia. Mod. Pathol. 7:76–81.

Waghray A., Keppler D., Sloane B.F., et al. (2002). Identification of a truncated form of human prostate tumour cells. J. Biol. Chem. 277:11533–11538.

Wang M., Tang J., Liu S., et al. (2005). Expression of cathepsin B and microvascular density increases with higher grade of astrocytomas. J. Neurooncol. 71:3–7.

Wang B., Sun J., Kitamoto S., et al. (2006). Cathepsin S controls angiogenesis and tumor growth via matrix-derived angiogenic factors. J. Biol. Chem. 281:6020–6029.

Warwas M., Haczynska H., Gerber J., et al. (1997). Cathepsin B-like activity as a serum tumour marker in ovarian carcinoma. Eur. J. Clin. Chem. Clin. Biochem. 35:301–304.

Warwas M., Gerber J., Haczynska H., et al. (2000). Comparison of cathepsin B-like activity with CA125 assays use in ovarian cancer diagnostic. Ginekol. Pol. 71:400–405.

Werle B., Ebert W., Klein W., et al. (1995). Assessment of cathepsin L activity by use of the inhibitor Ca-074 compared to cathepsin B activity in human lung tumour tissue. Biol. Chem. Hoppe-Seyler 376:157–164.

Werle B., Juelke B., Lah T., et al. (1997). Cathepsin B fraction active at physiological pH of 7.5 is of prognostic significance in squamous cell carcinoma of human lung. Br. J. Cancer 75:1137–1143.

Werle B., Kraft C., Kos J., et al. (2000). Cathepsin B in infiltrated lymph nodes is of prognostic significance for patients with non-small cell lung carcinoma. Cancer 89:2282–2289.

Werle B., Sauckel K., Nathanson C.M., et al. (2003). Cystatin C, E/M and F in human pleural fluids of patients with neoplastic and inflammatory lung disorders. Biol. Chem. 384:281–287.

Werle B., Kotzsch M., Lah T.T., et al. (2004). Cathepsin B, plasminogen activator—inhibitor (PAI-1) and plasminogen activator–receptor (uPAR) are prognostic factors for patients with non small cell lung cancer. Anticancer Res. 24:4147–4162.

Werle B., Schanzerbacher U., Lah T.T., et al. (2006). Cystatins in non-small cell lung cancer: Tissue levels, localization and relation to prognosis. Oncol. Rep. 16:647–656.

Wild P.J., Herr A., Wissmann C., et al. (2005). Gene expression profiling of progressive papillary noninvasive carcinomas of the urinary bladder. Clin. Cancer Res. 11:4415–4429.

Yan S., and Sloane B.F. (2003). Molecular regulation of human cathepsins B implication in pathologies. Biol. Chem. 384:845–854.

Yano M., Hirai K., Naito Z., et al. (2001). Expression of cathepsin B and cystatin C in human breast cancer. Surg. Today 31:385–391.

Yoshii A., Kageshita T., Tsushima H., et al. (1995). Clinical relevance of cathepsin B-like enzyme activity and cysteine proteinase inhibitor in metastatic tumours. Arch Dermatol. Res. 287:209–213.

Zajc I., Bervar A., Lah T.T. (2006). Cysteine cathepsins, stefins and extracellular matrix degradation during invasion of transformed human breast cancer cell lines. Radiol. Oncol. 40:259–271.

Zhang J., Shridhar J., Dai Q., et al. (2004). Cystatin M: A novel tumour suppressor gene for breast cancer. Cancer Res. 64:6957–6964.

Zore I., Krašovec M., Cimerman N., et al. (2001). Cathepsin B/cystatin C complex levels in sera from patients with lung and colorectal cancer. Biol. Chem. 382:805–810.

Chapter 30
Novel Degradome Markers in Breast Cancer

Caroline J. Pennington, Simon Pilgrim, Paul N. Span, Fred C. Sweep, and Dylan R. Edwards

Abstract Microarray studies have identified several gene signatures that are associated with tumour metastasis or adverse outcome in several cancer types. Among these are the 17-gene solid tumour metastasis signature of Ramaswamy et al. (2003), the 70-gene good versus poor outcome predictor of van't Veer et al. (2002), a wound response signature of Chang et al. (2004), and a 21-gene recurrence-score indicator in tamoxifen-treated node-negative breast cancer. These signatures show only moderate concordance, reflecting the complexity of the genetic pathways that contribute to malignancy. Unsurprisingly, degradome genes are represented in most of these predictors, and the utility of several more protease and related genes as biomarkers has been established by other profiling studies and functional analysis in model systems over the past few years. This chapter draws together information on key metalloproteinases and inhibitors that has emerged from array and quantitative real time polymerase chain reaction (qRT-PCR) analyses in the past few years, thereby highlighting candidates for further investigation. In particular, *MMP1, MMP7, MMP9, MMP11, MMP23* and *PLAUR* (uPAR) have all featured in poor prognosis expression signatures, and *MMP1* and *MMP2* are included in a 4-gene determinant of breast metastasis to the lungs. Moreover, from qRT-PCR studies, *MMP8* and *ADAMTS15* are shown to be novel markers of good prognosis in breast cancer, whereas *MMP11, ADAMTS8, ADAM12* and *ADAM28* are associated with poor outcome.

Introduction

Clinical and histopathological factors such as tumor grade, stage and lymph node involvement can provide general insights that help to identify patients with cancer most at risk from their disease. However, patients with the same clinicopathological

C.J. Pennington
Biomedical Research Centre, School of Biological Sciences, University of East Anglia, Norwich, NR4 7TJ, UK, e-mail: c.j.pennington@uea.ac.uk

D. Edwards et al. (eds.), *The Cancer Degradome.*
© Springer Science + Business Media, LLC 2008

characteristics can differ markedly in the course of their disease and response to therapy. Expression profiling has emerged recently as a powerful analytical tool that can improve the biological classification of cancer subtypes and provide refined prognostic and predictive information (Sotiriou and Piccart 2007). Several gene signatures have been developed and validated in multiple independent patient cohorts and different technology platforms (Fan et al. 2006). But, in addition to the promise of their clinical applications, these signatures can provide powerful insights into the cellular origins and molecular machinery of cancer. Given the central importance of the degradome in cancer biology, it is unsurprising to find proteases and inhibitors among the genes that comprise these expression-based predictors. The identification of a gene as having prognostic significance provides circumstantial evidence that its products may participate in tumour growth and metastasis, and indeed the importance of several genes is already clear from studies with cell- and animal-based models. Expression profiling, involving both high-complexity microarrays and low-complexity-focussed degradome technologies discussed in previous chapters (see Chaps. 2, 3 and 4), is therefore proving to be useful in identifying candidate genes for detailed functional investigation. The counterpoint of this is that knowledge of degradome gene function could prove useful in optimization of molecular profiling for diagnostic purposes.

In this chapter, we will review current information on a group of expression signatures with a focus on breast cancer and will provide a dissection of their relative degradome content. We will also report on related profiling efforts that have identified additional cancer-associated protease and inhibitor genes that have potential as biomarkers.

Expression Signatures in Breast Cancer

Bioinformatic analysis of microarray expression data on RNA isolated from primary breast cancer has yielded expression signatures that are prognostic as regards patient outcome or predictive of response to therapy. Signatures such as the 70-gene poor prognosis predictor (Mamma Print) developed in Amsterdam (van't Veer et al. 2002) and the (Oncotype Dx) recurrence-score profile (Paik et al. 2004) that identify patients in the node-negative, estrogen receptor-positive category who are at higher risk of recurrence are now undergoing evaluation in large-scale clinical trials (Sotiriou and Piccart 2007). Table 30.1 lists several recent gene signatures that have been developed, including the 70-gene Amsterdam good versus poor outcome (van't Veer et al. 2002, van de Vijver et al. 2002); the 21-gene recurrence score predictor in tamoxifen-treated, node-negative breast cancer (Paik et al. 2004); the 76-gene Rotterdam signature for lymph node metastasis (Wang et al. 2005); a 512-gene wound response signature, based on supervised analysis of similarities between serum-stimulated fibroblasts and mammary cancer (Chang et al. 2004, 2005); a 17-gene solid tumor metastasis signature (Ramaswamy et al. 2003); a 95-gene lung metastasis signature (LMS) (Minn et al. 2005); a 186-gene

Table 30.1 Degradome genes present in breast cancer gene expression signatures

Recurrence score (Oncotype DX 21 genes) Paik et al. (2004)	Poor prognosis (van't Veer; MammaPrint 70 genes) van't Veer et al. (2002)	Rotterdam (76 genes) Wang et al. (2005)	Lung metastasis signature (95 genes) Minn et al. (2005)	Solid tumour metastasis (128 genes) Ramaswamy et al. (2003)	Invasiveness signature (186 genes) Liu et al. (2007)	Intrinsic subtype (306 genes) Sorlie et al. (2001; 2003) Hu et al. (2006)	Wound response (Core serum response 512 genes) Chang et al. (2004, 2005)
MMP11	*MMP9*	*MMP23A*	*MMP1*	*CPN1* Carboxypeptidase N	*MMP7*	*MMP7*	CTSF Cathepsin F
CTSL2 cathepsin L2		*CANP* Ca-activated neutral proteinase	*MMP2*	*CPM* Carboxypeptidase M	*CASP8* Caspase-8	*MMP1*	*NLN* Neurolysin
		C3 Complement component C3	*CASP1*	*PSEN2* Presenilin 2	*PSMA5* proteasome subunit, alpha 5	*CASP1* Caspase-1	*DPP7* Dipeptidylpeptidase 7
			QPCT Glutaminyl cyclase	*BIRC5* Survivin	*PLAUR* Urokinase plasminogen activator	*CASP3* Caspase-3	*PGCP* Plasma glutamate carboxypeptidase
			MBTPS2 Membrane-bound transcription factor peptidase, site 2			*CASP7* Caspase-7	*TIMP1*
			CST7 Cystatin F			*UCHL1* Ubiquitin C-terminal hydrolase 1	*ADAMTS1*
			SERPINE2 Serpin E2 *SPINK4* Serine P1 kazal type 4				*SERPING1* Serpin G *SLPI* Secretory leukocyte protease inhibitor (antileukoproteinase)

(continued)

Table 30.1 (continued)

Recurrence score (Oncotype DX 21 genes) Paik et al. (2004)	Poor prognosis (van't Veer; MammaPrint 70 genes) van't Veer et al. (2002)	Rotterdam (76 genes) Wang et al. (2005)	Lung metastasis signature (95 genes) Minn et al. (2005)	Solid tumour metastasis (128 genes) Ramaswamy et al. (2003)	Invasiveness signature (186 genes) Liu et al. (2007)	Intrinsic subtype (306 genes) Sorlie et al. (2001; 2003) Hu et al. (2006)	Wound response (Core serum response 512 genes) Chang et al. (2004, 2005)
			EPHX2 Epoxide hydrolase				*TFPI2* Tissue factor pathway inhibitor 2
							APP Amyloid beta (A4) precursor protein (protease nexin-II)
							USP1 Ubiquitin specific protease 1
							BIRC5 Survivin
							CST3 Cystatin C
							PSMA7 Proteasome (prosome, macropain) subunit, alpha type, 7
							CASP3 Caspase 3
							PCSK7 Proprotein convertase subtilisin/kexin type 7
							PLAUR Urokinase plasminogen activator receptor
							GGH Gamma-glutamyl hydrolase

The table lists degrandome genes contained within each of the indicate prognostic/predictive breast cancer expression signatures.

invasiveness signature, based on comparison of tumorigenic CD44+/CD24−/low breast cancer cells with normal mammary epithelium (Liu et al. 2007) and Intrinsic Subtype profiles that are based on molecular phenotyping of breast cancers into basal-like, ERBB2+, normal breast-like, luminal A and luminal B subtypes (Sorlie et al. 2001, 2003; Hu et al. 2006).

Each of these profiles provides prognostic information that is independent of well-established clinicopathological factors, such as patient age and menopausal status, tumor grade, stage, estrogen receptor (ER)-positivity and lymph node involvement, indicating that they report information on the basic molecular characteristics driving the tumorigenic process. Concerns have been raised about the lack of concordance between these gene signatures, and the potential for multiple groupings of genes to be equally informative in terms of prognostication. However, it is encouraging that many signatures have now been validated across multiple data sets and different microarray platforms, indicating their robustness and clinical utility (Chang et al. 2005, Hu et al. 2006, Fan et al. 2006). Differences between the genes comprising the predictors indicate that individual genes are probably less important than overall functionality, for instance, as reflected in gene ontology (GO) groupings. In fact, as is to be expected, many of the genes represented are broadly associated with cell cycle control, apoptosis, cell–cell signalling and cell–matrix interactions. Given the pervasive involvement of degradome genes in all aspects of cell control and tissue homeostasis, it is also not surprising that most signatures include proteases or associated genes. We have dissected the published gene signatures listed in Table 30.1 to identify these degradome components. The wound response signature is particularly intriguing from the degradome standpoint. The inclusion of 18 degradome genes in the set reflects the overall size of the signature (512 genes), but more significantly, this gene set emerged from analysis of the serum response of human fibroblasts, which originally generated a 677-gene subset, from which 165 genes linked to cell cycle progression were removed, leaving the remaining genes as a core serum response (CSR). The remaining genes in the core serum response are associated primarily with matrix remodelling, cytoskeletal architecture and cell–cell signaling (Chang et al. 2004). This marker set may, therefore, be the best model in the contribution of the stroma in mammary cancer, which we know from many localisation studies to be the primary site of expression of many protease genes (reviewed in Egeblad and Werb 2002, Pedersen et al. 2005).

It is clear that the predictive power of the signatures depends on the collection of genes as a whole, rather than the individual components, which reflects genetic heterogeneity among patients and tumors. Indeed, a two-gene predictor of recurrence in tamoxifen-treated patients, based on the ratio of HOXB13: IL17BR expression (Ma et al. 2004), fares poorly in independent data sets (Fan et al. 2007). However, some of the array studies were able to generate minimal gene sets from larger signatures—for instance, the 17-gene metastasis signature was derived from a larger 128-gene collection (Ramaswamy et al. 2003). It is, therefore, possible that the gene sets that comprise the signatures could be refined further and optimized by selection of members that have definitive prognostic value as individual markers, as is the case for the uPA/uPAR/PAI-1 system discussed in Chap. 28 by Christensen et al.

Focussed Degradome Expression Profiling

In addition to whole genome or broad coverage microarrays, two groups have developed focussed degradome arrays, namely, the HuMu protein oligonucleotide array for the Affymetrix platform developed by Sloane and co-workers (See Chap. 3 by Schwartz et al.), and the CLIP–CHIP glass slide oligonucleotide array of Overall and Lopez-Otin and their collaborators (Chap. 2 by Kappelhopp et al.). These arrays are particularly suited for analysis of degradome genes in experimental systems, and offer the advantage of independent detection of both tumor cell and host contributions in human xenograft studies in mice. Comparison of breast cancer specimens with normal mammary tissue using the CLIP–CHIP has indicated up-regulation of several genes in infiltrating ductal carcinoma, including *ADAMTS17*, carboxypeptidases A5 and M, tryptase-gamma and matriptase-2 (Overall et al. 2004).

Quantitative real-time polymerase chain reaction (qRT-PCR) is being used extensively to provide detailed insights into specific subsets of degradome genes in numerous cancer types (see Chap. 4 by Pennington et al.). We recently developed a TaqMan® low-density array analysis using a panel of 380 genes that encompass the entire human metalloproteinase and serine proteinase families, and their inhibitors, along with selected cathepsins. Figure 30.1 shows details of the 19 degradome genes that showed the highest altered expression in a panel of 42 tumours (regardless of histological grade) compared to 9 normal mammary tissues. Genes that were up-regulated included *ADAMDEC1*, *ADAMTS14*, *CSTL2*, *CST1*, *MMP1*, *3, 10, 11, 12, 13* and *KLK4*. Down-regulated genes were *FREM1* and *RELN* (reelin). Although this analysis focussed on the normal/tumor comparison and may, therefore, be subject to changes in representation of epithelial versus stromal tissue compartments, the identification of many genes that have previously been recognized to have roles in breast cancer (such as cathepsin L, cystatin 1, and the MMPs) is encouraging, and suggests that further investigation is merited for the other members of this set.

Potential Metalloproteinase and Inhibitor Markers in Breast Cancer

MMPs and TIMPs

There is an extensive literature linking matrix metalloproteinases (MMPs) and tissue inhibitor of metalloproteinases (TIMPs) with breast cancer from studies of clinical specimens and mouse models that have been documented thoroughly elsewhere (see supplementary Table 3 in Egeblad and Werb 2002). A few key points will be made here with respect to particular genes:

MMP1 can be expressed by both cancer and stromal cells, and its elevated expression has been shown to be a poor prognostic marker (Cheng et al. 2007)

ADAMDEC1	ADAM-like, decysin 1
ADAMTS14	ADAM metallopeptidase with thrombospondin type 1 motif, 14
CPA6	Carboxypeptidase A6
CST1	Cystatin SN
CST2	Cystatin SA
CTSD	Cathepsin D
CTSL2	Cathepsin L2
DPEP1	Dipeptidase 1 (renal)
FAP	Fibroblast activation protein, alpha
FREM1	Fras1-related extracellular matrix Protein 1
HTRA4	HtrA serine peptidase 4
KLK4	Kallikrein-related peptidase 4
MMP1	Matrix metallopeptidase 1
MMP10	Matrix metallopeptidase 10
MMP11	Matrix metallopeptidase 11
MMP12	Matrix metallopeptidase 12
MMP13	Matrix metallopeptidase 13
MMP3	Matrix metallopeptidase 3
RELN	Reelin

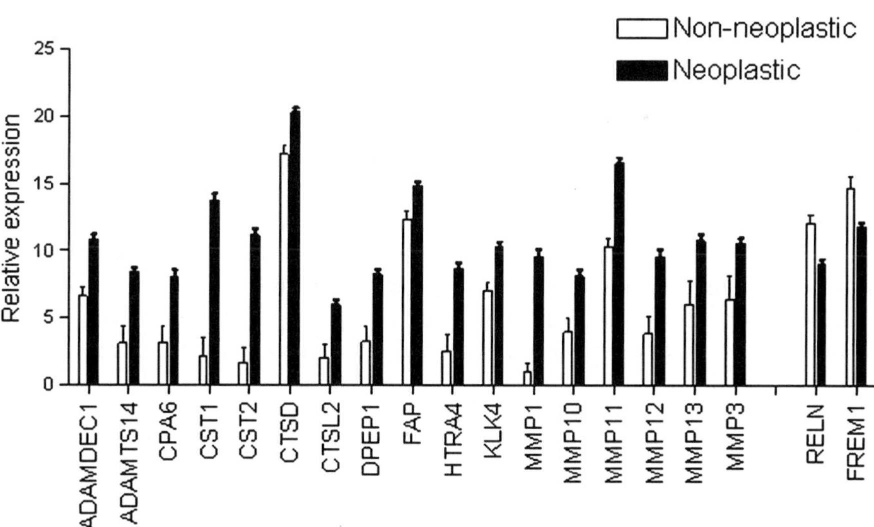

Fig. 30.1 TaqMan® analysis of complete metalloproteinase and serine proteinases degradome genes in normal mammary tissue and breast cancer. The figure shows the top 19 genes (all $p < 0.001$) from a 380-gene screen showing their levels of expression (expressed relative to 18S RNA and displayed as $40-C_t$:larger values, higher expression).

and predictive of development of invasive carcinomas in pre-malignant lesions (Poola et al. 2005). As discussed in Chap. 32 by Laxton and Ye, the 2G allele of the 1G/2G promoter polymorphism that is associated with elevated expression has been linked with local invasion and lymph node metastasis (Przybylowska et al. 2006). *MMP1* appears as a component of the signature of the basal-like (or triple-negative) subtype in the intrinsic subtype classification, which is one of the groupings associated with worse prognosis (Hu et al. 2006). *MMP1* (along with *MMP2*) is also part of a 54-gene signature observed in clonal variants of MDA-MB-231 that show high metastasis to the lungs (Minn et al. 2005). This LMS has been validated in clinical specimens (Minn et al. 2007). Moreover, using gene knock-downs and pharmacological inhibition, *MMP1* and *MMP2* together with cyclooxygenase-2 (COX2) and epiregulin (*EREG*) constitute the four genes within this that have been recently shown to be functionally important for vascularization and growth of primary tumors formed by orthotopic xenografts in the mammary fat pad, and for subsequent metastasis to the lung (Gupta et al. 2007). Thus, *MMP1* is perhaps one of the strongest degradome markers of breast malignancy, and a bona fide target for anti-cancer therapy (Overall and Kleifeld 2006).

MMP2. In addition to being part of the LMS, higher levels of activation of pro-MMP2 have been observed in lymph node-positive primary breast tumors compared to lymph node-negative primary breast tumors (Iwata et al. 1996) and although expressed predominantly by stromal cells (Tetu et al. 2006, Brummer et al. 1999), staining for MMP2 is observed on cell membranes for malignant tumors (Jones et al. 1999). Other data from mouse models are consistent with a role for MMP2 as a cancer target (Overall and Kleifeld 2006); however, since its expression at the RNA level tends to be less strongly regulated than other *MMPs*, it is probably not ideal as a prognostic signature gene.

MMP3 expression is increased in mammary tumors versus normal breast, and has been localised to both tumor and stromal components (Brummer et al. 1999) and to infiltrating T cells (Iwata et al. 1996), with immunohistochemical staining around invasive epithelial cells. Moreover, in mouse models, ectopic expression of MMP3 under a mammary-specific promoter led to spontaneous development of malignant lesions (Sternlicht et al. 2000). MMP3 induces epithelial-mesenchymal transition and genomic instability by an as yet unknown mechanism that involves activation of Rac1b and reactive oxygen species (Radisky et al. 2005). Thus, MMP3 has functional links as a promoter of mammary tumorigenesis, though this may not hold in every tumor setting (Overall and Kleifeld 2006).

MMP7 appears as a marker of the basal-like phenotype in the intrinsic subtype signatures, and also in the invasiveness signature. In MDA-MB-231 xenografts, ribozyme-mediated knock-down of *MMP7* attenuated tumor growth (Jiang et al. 2005). Matrilysin is expressed primarily by tumor cells and its higher expression is linked with poor survival (Jiang et al. 2005), though it is also present in infiltrating mononuclear inflammatory cells, the presence of which is associated with distant metastasis (Gonzalez et al. 2007).

MMP8 (neutrophil collagenase) is an enigmatic protease which we choose to discuss here despite its absence from all of the expression signatures. Expression of

MMP8 inhibits mammary tumor metastasis in xenograft models (Montel et al. 2004), and *Mmp8*-null mice display enhanced tumorigenesis (Balbin et al. 2003, Gutiérriez-Fernández et al. 2008). Elevated MMP8 expression detected by qRT-PCR in a cohort of 229 Dutch patients with breast cancer was associated with improved survival (Gutiérriez-Fernández et al. 2008). Thus, MMP8 is an anti-target that performs a metastasis-suppressive function, though its mechanism of action remains unclear. It is likely that its low level of expression, which is at the detection limit for TaqMan qRT-PCR, makes its quantification problematic on DNA micro-arrays.

MMP9 is the only degradome gene to appear in the van't Veer 70-gene poor prognosis signature. Knock-down of MMP9 in mammary cancer cell lines reduces tumor growth and angiogenesis in xenografts (Kunigal et al. 2006); however, MMP9 appears to be both a tumor target and an anti-target, since in mouse models its absence leads to reduced tumor burden, though tumors that do form tend to be more aggressive (Overall and Kleifeld 2006). MMP9 has been observed to be expressed by malignant epithelial cells, infiltrating mononuclear cells, and stromal fibroblasts and endothelial cells; thus its contributions to tumorigenesis may be multifaceted.

MMP11 is one of the most consistently up-regulated genes in expression studies contrasting normal tissues and malignant tumors (e.g. Richardson et al. 2006), whose stromal expression contributes to malignancy (Masson et al. 1998). Elevated *MMP11* expression was associated with poor overall survival (OS) in our qRT-PCR analysis of a Dutch patient cohort (Fig. 30.2). As discussed in Chap. 19 by Andarawewa and Rio, MMP11 is a negative regulator of mammary adipogenesis, and its expression by peri-tumoral fibroblasts provides a survival cue for malignant epithelial cells (Andarawewa et al. 2005).

MMP12 is expressed primarily in macrophages, and mouse model data suggest that it performs protective functions in lung cancer (Acuff et al. 2006), potentially by mediating generation of angiogenesis inhibitors (Overall and Kleifeld 2006). *MMP12* has been identified in the in silico transcriptomic studies, described in Chap. 31 by Iljin et al., as their top hit with regard to over-expression in multiple cancer types.

MMP13 is expressed by myofibroblasts and its expression increases during the transition from ductal carcinoma in situ to invasive carcinoma (Nielsen et al. 2001, Schuetz et al. 2006). It may thus serve as one of the earliest markers of cancer.

MMP23A is identical in coding sequence to *MMP23B*, reflecting recent gene duplication. Little is known about this gene, though it appears in the Rotterdam poor prognosis signature. It has been linked to cranial suture closure during deve-lopment, whose loss results in craniosynostosis (Gajecka et al. 2005). Oncomine analysis (Fig. 30.3) shows that MMP23 expression decreases with increasing tumor grade, and is lower in primary tumors with metastases compared to those without. This suggests that *MMP23A/B* is an interesting candidate for further functional evaluation.

TIMP1 is part of the wound response signature, and a gene that is frequently up-regulated in many cancer types compared to normal tissues. TIMP1 has an anti-apoptotic effect on breast cancer cells by protecting epithelial cells from TNF-related apoptosis-inducing ligand (TRAIL)-induced cell death (Liu et al.

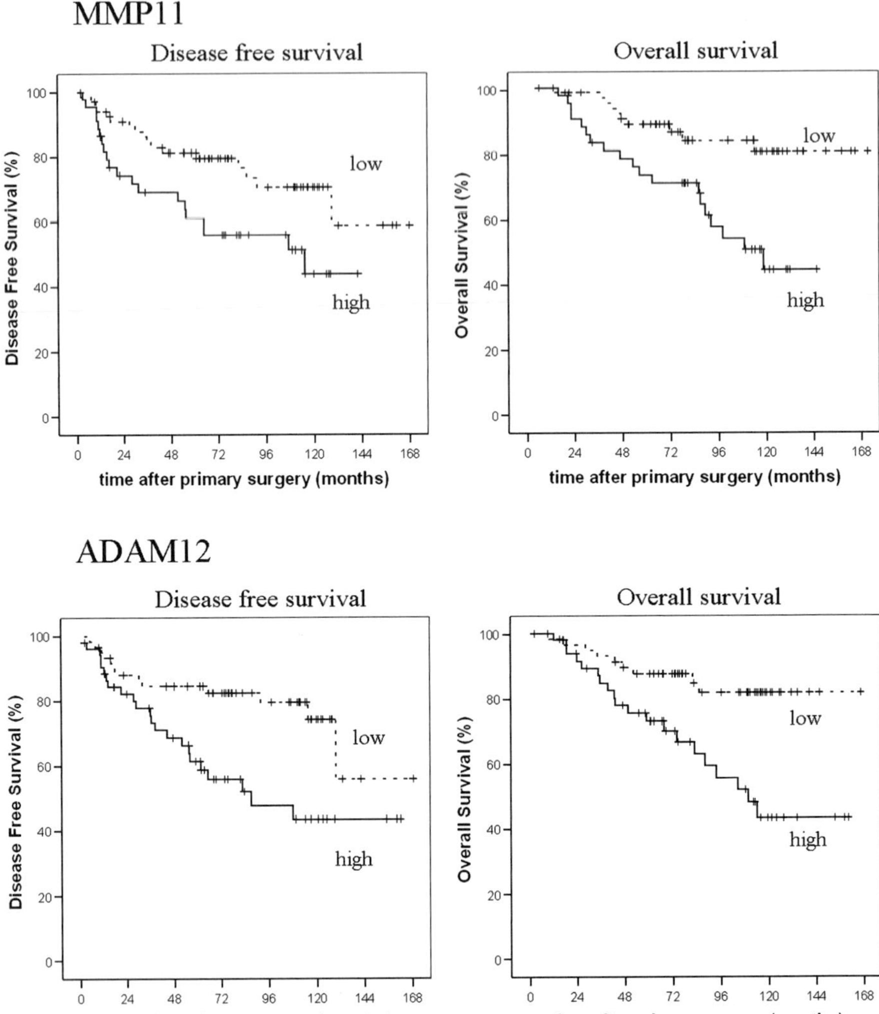

Fig. 30.2 Kaplan—Meier survival analysis showing relationships between expression levels of MMP11 and ADAM12 transcripts and disease-free survival (*DFS*) and overall survival (*OS*) in breast cancer. Data are displayed for all patients who did not receive adjuvant therapy (112 cases). For MMP11, *DFS*: HR = 2.17, 95% CI = 1.14–4.17, *P* =0.019; *OS*: HR = 3.15, 95% CI = 1.46–6.79, *P* = 0.003. For ADAM12, *DFS*: HR = 2.60, 95% CI = 1.31–5.15, *P* = 0.006; *OS*: HR = 3.25, 95% CI = 1.48–7.16, *P* = 0.003

2005). In addition to its role as a MMP inhibitor, this mechanism may potentially involve its association with the tetraspanin CD63 and co-localisation with integrin β1 (Jung et al. 2006). High levels of TIMP1 protein detected by ELISA in extracts from primary tumors are associated with poor survival (Schrohl et al. 2004) and the presence of distant metastases (Vizoso et al. 2007), and predict a poor response to

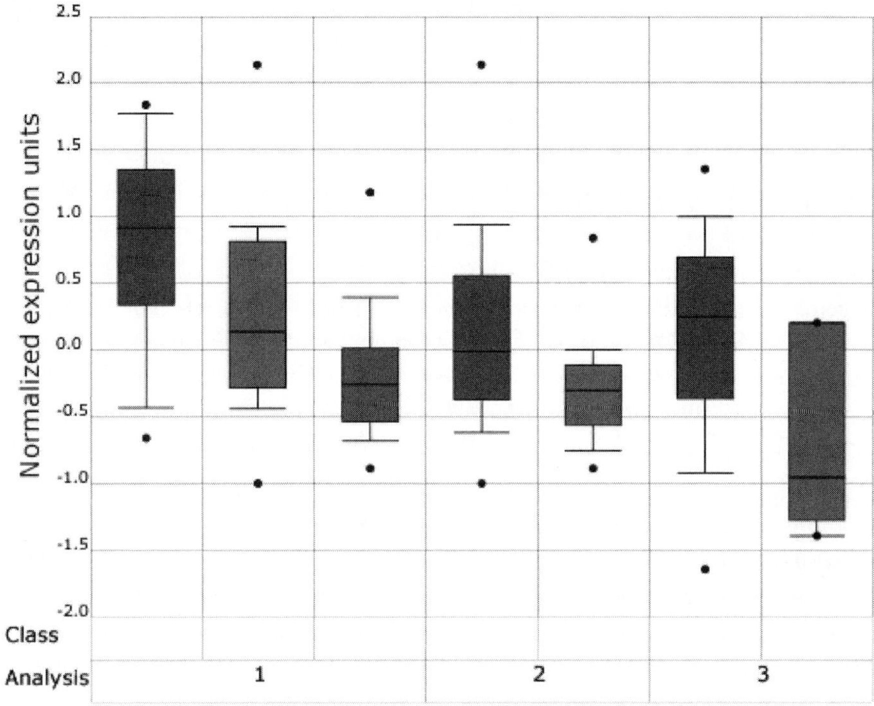

Fig. 30.3 Oncomine analysis of MMP23A/B expression in the public databases. Study 1: Breast carcinoma grade 1 versus grade 2 versus grade 3 (van de Vijver et al. 2002). Study 2: Lymphocytic infiltrate-negative versus lymphocytic infiltrate-positive (van't Veer et al. 2002). Study 3: Breast carcinoma versus metastatic breast carcinoma (Radvanyi et al. 2005)

chemotherapy (Schrohl et al. 2006). However, the situation with *TIMP1* transcripts is somewhat complex, since two variant isoforms have now been identified—a full length form and a truncated form (see Chap. 33 by Frederiksen et al.)—and ELISA-determined protein levels are not concordant with transcript levels, which may indicate translational regulation of TIMP1 expression or a significant contribution from circulating TIMP1 (Sieuwerts et al. 2007).

TIMP2 is primarily stromal in origin in breast cancer and transcript levels generally increase with histological grade (Brummer et al. 1999) but do not show associations with survival (Span et al. 2004). Increased TIMP2 immunostaining has been linked with distant metastases in some studies (Vizoso et al. 2007, Gonzalez et al. 2007), but not in others (Jones et al. 1999, Kuvaja et al. 2005). However, *TIMP2*, like *MMP2*, is unlikely to be a useful tumor marker since it is expressed in a largely constitutive fashion by most cell types.

TIMP3 transcripts are found in tumor-associated fibroblasts in mammary cancer (Byrne et al. 1995) and elevated levels have been correlated with success of adjuvant endocrine therapy (Span et al. 2004) and disease-free survival (DFS) (Kotzsch et al. 2005). This is consistent with a well-established pro-apoptotic role for over-expressed TIMP3 (Ahonen et al. 2003), and the observation that *TIMP3* is frequently subject to epigenetic silencing in several types of cancer including breast (Esteller et al. 2001).

ADAMs and ADAMTSs

The ADAMs (a disintegrin and metalloproteinases) have functions in cell adhesion and ectodomain shedding of cell surface proteins (Blobel 2005), while their relatives, the ADAMTSs (ADAM with thrombospondin motifs), are secreted enzymes involved in processing of fibrillar collagens (ADAMTS2, -3 and -14), cleavage of matrix proteoglycans (aggrecanases, ADAMTS1, -4, -5, -8, -9, -15) and platelet function (ADAMTS13, Porter et al. 2005). In particular, ADAM17 is a principal activator of pro-TNFα (Blobel 2005) and ADAM10 and -17 both participate in shedding and activation of epidermal growth factor receptor (EGFR) ligands (Sahin et al. 2004). Several *ADAM/ADAMTS* genes have been implicated in cancer pathogenesis, though none are represented in the breast expression signatures (Table. 30.1). We comment on a few selected members below:

ADAM12 is expressed in breast cancer cells, and it accelerates tumor development in the MMTV-PyMT mouse mammary cancer model, potentially by inhibiting epithelial cell apoptosis while enhancing stromal cell death (Kveiborg et al. 2005). It has been reported to be a broad-spectrum protease with gelatinase activity, which itself is detected at elevated levels in the urine of patients with breast cancer (Roy et al. 2004). It interacts via its cysteine-rich domain with syndecans, leading to promotion of integrin β1-dependent adhesion and cell spreading. Interestingly, *ADAM12* appears among a list of 122 genes found to be mutated at elevated frequency in comprehensive sequencing of a set of breast cancer genomes (Sjoblom et al. 2006). Our data indicate that *ADAM12* is indeed a robust marker of worse DFS and OS in patients with breast cancer who had not received adjuvant therapy (Fig. 30.2).

ADAM15 has been reported to be over-expressed in breast and prostate cancers (Kuefer et al. 2006). Our data and work from others have indicated that the gene has multiple splice variants, all of which affect the cytoplasmic domain: these splice variants show differential expression in normal and malignant mammary tissue (Ortiz et al. 2004, Zhong et al. 2008).

ADAM17 has been functionally linked to the activation of EGFR ligands in the development of breast cancer, and has been detected at higher levels in tumors compared to normal tissue by Western blot (Borrell-Pages et al. 2003). Knock-down or inhibition of ADAM17 prevented mobilisation of TGF-α and amphiregulin in breast cancer cell lines and 3D-colony growth (Kenny and Bissell 2007).

Moreover, *ADAM17* has been correlated with poor prognosis (Kenny and Bissell 2007) and lymph node metastasis (McGowan et al. 2007).

ADAM28 is expressed at higher levels in breast tumors compared to normal mammary tissue (Mitsui et al. 2006). It has been linked to breast cancer growth by its ability to cleave insulin-like growth factor binding protein-3 (IGFBP-3), thereby enhancing IGF-1-mediated cell proliferation (Mitsui et al. 2006).

ADAMTSs. Eleven out of the nineteen genes in the ADAMTS family are dysregulated in breast cancer, with *ADAMTS1, 3, 5, 8, 9, 10* and *18* being consistently down-regulated compared to normal mammary tissue and *ADAMTS4, 6, 14* and *20* being up-regulated (Porter et al. 2004). However, subsequent studies have indicated that of the genes that encode aggrecanase functions, *ADAMTS8* and *ADAMTS15* (but not *ADAMTS1*) RNA levels are associated with prognosis, with high levels of *ADAMTS8* and low levels of *ADAMTS15* transcripts in combination identifying a group of patients at high risk of recurrence and poor OS (Porter et al. 2006). Both ADAMTS1 and ADAMTS8 are anti-angiogenic, ADAMTS1 by binding and sequestration of vascular endothelial growth factor (VEGF) (Luque et al. 2003). But the roles of ADAMTSs in cancer may be complex, since cleavage of ADAMTS1 in its internal spacer region converts it from an enhancer of tumor metastasis and angiogenesis to an inhibitor, suggesting that the post-translational processing of these proteins is critical in understanding their functions (Liu et al. 2006). This could be a factor in the association of *ADAMTS8* transcript levels with poor outcome (Porter et al. 2006). In non-small-cell lung cancer, the ADAMTS8 locus shows hypermethylation, consistent with a role as a tumor-suppressor function (Dunn et al. 2004).

ADAMTS18 has also been suggested to be a tumor-suppressor function that is epigenetically silenced in diverse cancers (Jin et al. 2007), and likewise *ADAMTS9* in oesophageal cancer (Lo et al. 2007).

Perspectives

Prognostic signatures that have emerged from studies of several tumor types provide a rich seam of information for data mining to select genes for functional evaluation of their roles in carcinogenesis and metastasis. Several degradome genes have emerged in this fashion, including *MMP23A/B, ADAM15, ADAM28, ADAMTS8, ADAMTS14* and *ADAMTS15*, along with others whose roles are already actively being pursued. Taking this avenue further, the development of supervised expression signatures of collections of genes based on prior knowledge of the association of individual genes with aspects of malignancy could be a useful way to develop optimised predictive and prognostic marker sets. The sensitivity, speed and accuracy of TaqMan qRT-PCR (*see* Chap. 4 by Pennington et al.) for small gene sets recommend it for implementation of optimised gene signature analysis in the clinical setting, and so the potential for development of degradome-based signatures looks promising.

References

Acuff H. B., Sinnamon M. et al. (2006). Analysis of host- and tumor-derived proteinases using a custom dual species microarray reveals a protective role for stromal matrix metalloproteinase-12 in non-small cell lung cancer. Cancer Res 66(16): 7968–75.

Ahonen M., Poukkula M. et al. (2003). Tissue inhibitor of metalloproteinases-3 induces apoptosis in melanoma cells by stabilization of death receptors. Oncogene 22(14): 2121–34.

Andarawewa K. L., Motrescu E. R. et al. (2005). Stromelysin-3 is a potent negative regulator of adipogenesis participating to cancer cell-adipocyte interaction/crosstalk at the tumor invasive front. Cancer Res 65(23): 10862–71.

Balbin M., Fueyo A. et al. (2003). Loss of collagenase-2 confers increased skin tumor susceptibility to male mice. Nat Genet 35(3): 252–7.

Blobel C. P. (2005). ADAMs: key components in EGFR signalling and development. Nat Rev Mol Cell Biol 6(1): 32–43.

Borrell-Pages M., Rojo F. et al. (2003). TACE is required for the activation of the EGFR by TGF-alpha in tumors. EMBO J 22(5): 1114–24.

Brummer O., Athar S. et al. (1999). Matrix-metalloproteinases 1, 2, and 3 and their tissue inhibitors 1 and 2 in benign and malignant breast lesions: an in situ hybridization study. Virchows Arch 435(6): 566–73.

Byrne J. A., Tomasetto C. et al. (1995). The tissue inhibitor of metalloproteinases-3 gene in breast carcinoma: identification of multiple polyadenylation sites and a stromal pattern of expression. Mol Med 1(4): 418–27.

Chang H. Y., Sneddon J. B. et al. (2004). Gene expression signature of fibroblast serum response predicts human cancer progression: similarities between tumors and wounds. PLoS Biol 2(2): E7.

Chang H. Y., Nuyten D. S. et al. (2005). Robustness, scalability, and integration of a wound-response gene expression signature in predicting breast cancer survival. Proc Natl Acad Sci U S A 102(10): 3738–43.

Cheng S., Tada M. et al. (2007). High MMP-1 mRNA expression is a risk factor for disease-free and overall survivals in patients with invasive breast carcinoma. J Surg Res. 146(1): 104–9.

Dunn J. R., Panutsopulos D. et al. (2004). METH-2 silencing and promoter hypermethylation in NSCLC. Br J Cancer 91(6): 1149–54.

Egeblad M. and Werb Z. (2002). New functions for the matrix metalloproteinases in cancer progression. Nat Rev Cancer 2(3): 161–74.

Esteller M., Corn P. G. et al. (2001). A gene hypermethylation profile of human cancer. Cancer Res 61(8): 3225–9.

Fan C., Oh D. S. et al. (2006). Concordance among gene-expression-based predictors for breast cancer. N Engl J Med 355(6): 560–9.

Gajecka M., Yu W. et al. (2005). Delineation of mechanisms and regions of dosage imbalance in complex rearrangements of 1p36 leads to a putative gene for regulation of cranial suture closure. Eur J Hum Genet 13(2): 139–49.

Gonzalez L. O., Pidal I. et al. (2007). Overexpression of matrix metalloproteinases and their inhibitors in mononuclear inflammatory cells in breast cancer correlates with metastasis-relapse. Br J Cancer 97(7): 957–63.

Gupta G. P., Nguyen D. X. et al. (2007). Mediators of vascular remodelling co-opted for sequential steps in lung metastasis. Nature 446(7137): 765–70.

Gutiériez-Fernández A., Fueyo A., et al. (2008). Collagenase-2 (MMP-8) functions as a metastasis suppressor through modulation of tumor cell adhesion and invasiveness. Cancer Res. 68(8): 2755–63.

Hu Z., Fan C. et al. (2006). The molecular portraits of breast tumors are conserved across microarray platforms. BMC Genomics 7: 96.

Iwata H., Kobayashi S. et al. (1996). Production of matrix metalloproteinases and tissue inhibitors of metalloproteinases in human breast carcinomas. Jpn J Cancer Res 87(6): 602–11.

Jiang W. G., Davies G. et al. (2005). Targeting matrilysin and its impact on tumor growth in vivo: the potential implications in breast cancer therapy. Clin Cancer Res 11(16): 6012–9.

Jin H., Wang X. et al. (2007). Epigenetic identification of ADAMTS18 as a novel 16q23.1 tumor suppressor frequently silenced in esophageal, nasopharyngeal and multiple other carcinomas. Oncogene 26(53): 7490–8.

Jones J. L., Glynn P. et al. (1999). Expression of MMP-2 and MMP-9, their inhibitors, and the activator MT1-MMP in primary breast carcinomas. J Pathol 189(2): 161–8.

Jung K. K., Liu X. W. et al. (2006). Identification of CD63 as a tissue inhibitor of metalloproteinase-1 interacting cell surface protein. EMBO J 25(17): 3934–42.

Kenny P. A. and Bissell M. J. (2007). Targeting TACE-dependent EGFR ligand shedding in breast cancer. J Clin Invest 117(2): 337–45.

Kotzsch M., Farthmann J. et al. (2005). Prognostic relevance of uPAR-del4/5 and TIMP-3 mRNA expression levels in breast cancer. Eur J Cancer 41(17): 2760–8.

Kuefer R., Day K. C. et al. (2006). ADAM15 disintegrin is associated with aggressive prostate and breast cancer disease. Neoplasia 8(4): 319–29.

Kunigal S., Lakka S. S. et al. (2007). RNAi-mediated downregulation of urokinase plasminogen activator receptor and matrix metalloprotease-9 in human breast cancer cells results in decreased tumor invasion, angiogenesis and growth. Int J Cancer 121(10): 2307–16.

Kuvaja P., Talvensaari-Mattila A. et al. (2005). The absence of immunoreactivity for tissue inhibitor of metalloproteinase-1 (TIMP-1), but not for TIMP-2, protein is associated with a favorable prognosis in aggressive breast carcinoma. Oncology 68(2–3): 196–203.

Kveiborg M., Frohlich C. et al. (2005). A role for ADAM12 in breast tumor progression and stromal cell apoptosis. Cancer Res 65(11): 4754–61.

Liu R., Wang X. et al. (2007). The prognostic role of a gene signature from tumorigenic breast-cancer cells. N Engl J Med 356(3): 217–26.

Liu X.W., Taube M.E. (2005). Tissue inhibitor of metalloproteinase-1 protects human breast epithelial cells form extrinsic cell death: a potential oncogenic activity of tissue inhibitor of metalloproteinase-1. Cancer Res. 65(3): 898–906.

Liu Y. J., Xu Y. et al. (2006). Full-length ADAMTS-1 and the ADAMTS-1 fragments display pro- and antimetastatic activity, respectively. Oncogene 25(17): 2452–67.

Lo P. H., Leung A. C. et al. (2007). Identification of a tumor suppressive critical region mapping to 3p14.2 in esophageal squamous cell carcinoma and studies of a candidate tumor suppressor gene, ADAMTS9. Oncogene 26(1): 148–57.

Luque A., Carpizo D. R. et al. (2003). ADAMTS1/METH1 inhibits endothelial cell proliferation by direct binding and sequestration of VEGF165. J Biol Chem 278(26): 23656–65.

Ma X. J., Wang Z. et al. (2004). A two-gene expression ratio predicts clinical outcome in breast cancer patients treated with tamoxifen. Cancer Cell 5(6): 607–16.

Masson R., Lefebvre O. et al. (1998). In vivo evidence that the stromelysin-3 metalloproteinase contributes in a paracrine manner to epithelial cell malignancy. J Cell Biol 140(6): 1535–41.

McGowan P. M., Ryan B. M. et al. (2007). ADAM-17 expression in breast cancer correlates with variables of tumor progression. Clin Cancer Res 13(8): 2335–43.

Minn A. J., Gupta G. P. et al. (2005). Genes that mediate breast cancer metastasis to lung. Nature 436(7050): 518–24.

Minn A. J., Gupta G. P. et al. (2007). Lung metastasis genes couple breast tumor size and metastatic spread. Proc Natl Acad Sci U S A 104(16): 6740–5.

Mitsui Y., Mochizuki S. et al. (2006). ADAM28 is overexpressed in human breast carcinomas: implications for carcinoma cell proliferation through cleavage of insulin-like growth factor binding protein-3. Cancer Res 66(20): 9913–20.

Montel V., Kleeman J. et al. (2004). Altered metastatic behavior of human breast cancer cells after experimental manipulation of matrix metalloproteinase 8 gene expression. Cancer Res 64(5): 1687–94.

Nielsen B. S., Rank F. et al. (2001). Collagenase-3 expression in breast myofibroblasts as a molecular marker of transition of ductal carcinoma in situ lesions to invasive ductal carcinomas. Cancer Res 61(19): 7091–100.

Ortiz R. M., Karkkainen I. et al. (2004). Aberrant alternative exon use and increased copy number of human metalloprotease-disintegrin ADAM15 gene in breast cancer cells. Genes Chromosomes Cancer 41(4): 366–78.

Overall C. M. and Kleifeld O. (2006). Tumour microenvironment—opinion: validating matrix metalloproteinases as drug targets and anti-targets for cancer therapy. Nat Rev Cancer 6(3): 227–39.

Overall C. M., Tam E. M. et al. (2004). Protease degradomics: mass spectrometry discovery of protease substrates and the CLIP-CHIP, a dedicated DNA microarray of all human proteases and inhibitors. Biol Chem 385(6): 493–504.

Paik S., Shak S. et al. (2004). A multigene assay to predict recurrence of tamoxifen-treated, node-negative breast cancer. N Engl J Med 351(27): 2817–26.

Pedersen T. X., Pennington C. J. et al. (2005). Extracellular protease mRNAs are predominantly expressed in the stromal areas of microdissected mouse breast carcinomas. Carcinogenesis 26(7): 1233–40.

Poola I., DeWitty R. L. et al. (2005). Identification of MMP-1 as a putative breast cancer predictive marker by global gene expression analysis. Nat Med 11(5): 481–3.

Porter S., Scott S. D. et al. (2004). Dysregulated expression of adamalysin-thrombospondin genes in human breast carcinoma. Clin Cancer Res 10(7): 2429–40.

Porter S., Clark I. M. et al. (2005). The ADAMTS metalloproteinases. Biochem J 386(Pt. 1): 15–27.

Porter S., Span P. N. et al. (2006). ADAMTS8 and ADAMTS15 expression predicts survival in human breast carcinoma. Int J Cancer 118(5): 1241–7.

Przybylowska K., Kluczna A. et al. (2006). Polymorphisms of the promoter regions of matrix metalloproteinases genes MMP-1 and MMP-9 in breast cancer. Breast Cancer Res Treat 95(1): 65–72.

Radisky D. C., Levy D. D. et al. (2005). Rac1b and reactive oxygen species mediate MMP-3-induced EMT and genomic instability. Nature 436(7047): 123–7.

Radvanyi L., Singh-Sandhu D. et al. (2005). The gene associated with trichorhinophalangeal syndrome in humans is overexpressed in breast cancer. Proc Natl Acad Sci U S A 102(31): 11005–10.

Ramaswamy S., Ross K. N. et al. (2003). A molecular signature of metastasis in primary solid tumors. Nat Genet 33(1): 49–54.

Richardson A. L., Wang Z. C. et al. (2006). X chromosomal abnormalities in basal-like human breast cancer. Cancer Cell 9(2): 121–32.

Roy R., Wewer U. M. et al. (2004). ADAM 12 cleaves extracellular matrix proteins and correlates with cancer status and stage. J Biol Chem 279(49): 51323–30.

Sahin U., Weskamp G. et al. (2004). Distinct roles for ADAM10 and ADAM17 in ectodomain shedding of six EGFR ligands. J Cell Biol 164(5): 769–79.

Schrohl A. S., Holten-Andersen M. N. et al. (2004). Tumor tissue levels of tissue inhibitor of metalloproteinase-1 as a prognostic marker in primary breast cancer. Clin Cancer Res 10(7): 2289–98.

Schrohl A. S., Meijer-van Gelder M. E. et al. (2006). Primary tumor levels of tissue inhibitor of metalloproteinases-1 are predictive of resistance to chemotherapy in patients with metastatic breast cancer. Clin Cancer Res 12(23): 7054–8.

Schuetz C. S., Bonin M. et al. (2006). Progression-specific genes identified by expression profiling of matched ductal carcinomas in situ and invasive breast tumors, combining laser capture microdissection and oligonucleotide microarray analysis. Cancer Res 66(10): 5278–86.

Sieuwerts A. M., Usher P. A. et al. (2007). Concentrations of TIMP1 mRNA splice variants and TIMP-1 protein are differentially associated with prognosis in primary breast cancer. Clin Chem 53(7): 1280–8.

Sjoblom T., Jones S. et al. (2006). The consensus coding sequences of human breast and colorectal cancers. Science 314(5797): 268–74.

Sorlie T., Perou C. M. et al. (2001). Gene expression patterns of breast carcinomas distinguish tumor subclasses with clinical implications. Proc Natl Acad Sci U S A 98(19): 10869–74.

Sorlie T., Tibshirani R. et al. (2003). Repeated observation of breast tumor subtypes in independent gene expression data sets. Proc Natl Acad Sci U S A 100(14): 8418–23.

Sotiriou C. and Piccart M. J. (2007). Taking gene-expression profiling to the clinic: when will molecular signatures become relevant to patient care? Nat Rev Cancer 7(7): 545–53.

Span P. N., Lindberg R. L. et al. (2004). Tissue inhibitors of metalloproteinase expression in human breast cancer: TIMP-3 is associated with adjuvant endocrine therapy success. J Pathol 202(4): 395–402.

Sternlicht M. D., Bissell M. J. et al. (2000). The matrix metalloproteinase stromelysin-1 acts as a natural mammary tumor promoter. Oncogene 19(8): 1102–13.

Tetu B., Brisson J. et al. (2006). The influence of MMP-14, TIMP-2 and MMP-2 expression on breast cancer prognosis. Breast Cancer Res 8(3): R28.

van de Vijver M. J., He Y. D. et al. (2002). A gene-expression signature as a predictor of survival in breast cancer. N Engl J Med 347(25): 1999–2009.

van't Veer L. J., Dai H. et al. (2002). Gene expression profiling predicts clinical outcome of breast cancer. Nature 415(6871): 530–6.

Vizoso F. J., Gonzalez L. O. et al. (2007). Study of matrix metalloproteinases and their inhibitors in breast cancer. Br J Cancer 96(6): 903–11.

Wang Y., Klijn J. G. et al. (2005). Gene-expression profiles to predict distant metastasis of lymph-node-negative primary breast cancer. Lancet 365(9460): 671–9.

Zhong J.L., Poghosyan Z. (2008). Distinct functions of natural ADAM-15 cytoplasmic domain variants in human mammary carcinoma. Mol Cancer Res. 6(3): 383–394.

Chapter 31
Meta-Analysis of Gene Expression Microarray Data: Degradome Genes in Healthy and Cancer Tissues

Kristiina Iljin, Sami Kilpinen, Johanna Ivaska, and Olli Kallioniemi

Abstract The "degradome" set of genes comprises proteases and their inhibitors. These genes may have distinct roles in normal tissues as well as in tumor development and progression. In order to systematically investigate which proteases are overexpressed in cancers, we studied the mRNA expression patterns of degradome genes across healthy and malignant human tissues using our In Silico Transcriptomics (IST) database which covers gene expression data from over 70 different normal tissue types and 50 tumor types. The analysis of nearly 500 degradome gene expression profiles across all major human tissues and malignancies gave a comprehensive view of the expression of proteases in healthy and diseased tissues. Interestingly, the most distinct clusters enriched in protease inhibitors were detected from normal tissues, such as liver, highlighting the tight control of extracellular matrix remodeling. Furthermore, normal tissues and their corresponding cancer tissues clustered separately in most cases, indicating significant alteration in protease expression during carcinogenesis. More detailed analysis of matrix metalloproteinase (MMP) gene expression patterns validated many previously published findings such as overexpression of MMP13 gene in several cancer types as well as down-regulation of MMP28 in colorectal cancers. In addition, meta-analysis revealed novel aspects on MMPs, such as elevated MMP12 gene expression in several cancer types where it had not been previously studied. Furthermore, we performed in silico coexpression analyses to obtain insights on biological processes which may be associated with MMP12 expression. The most significant associations were found with mitosis and inflammatory response. Taken together, we demonstrate novel, unprecedented possibilities for rapid discovery of biological insights, putative biomarkers, and therapeutic targets using in silico analysis of existing gene expression datasets.

K. Iljin
Medical Biotechnology, VTT Technical Research Centre of Finland and University of Turku, Turku, Finland

D. Edwards et al. (eds.), *The Cancer Degradome.*
© Springer Science + Business Media, LLC 2008

Introduction: Gene Expression Microarray Databases

Genome-wide mRNA expression studies are widely used in all fields of biological and biomedical research. The number of published studies on gene expression microarrays has continued to increase almost exponentially in the past 8 years. However, in many cases the microarray study is performed with a relatively small set of samples representing one cell type, tissue, or tumor, and the data are often interpreted with a significant bias toward the interests and expertise of the investigators performing the study. Since the same data could be useful to a number of researchers in other fields as well, efforts have been made to launch large public microarray data depositories, such as NCBI GEO and ArrayExpress (Barrett et al. 2005, Sarkans et al. 2005), where people are encouraged to submit their data. However, the enormous variability of microarray technologies and platforms as well as lack of standardization has made it difficult to make use of the existing microarray data.

In order to facilitate the application of public gene expression microarray data for biomedical research, we have developed a systematic In Silico Transcriptomics (IST) database and data mining capability covering data from almost 8,000 individual samples. All the samples have been analyzed on the Affymetrix platform (Kilpinen et al. 2008). At present, the IST database (version 1.1) covers gene expression data from 70 different normal tissue types, 50 tumor types, and 176 different functional experiments on cell lines. All samples were manually annotated including anatomical location, cell type, disease, and treatment data as well as patient demographics (e.g., age, sex, and race). This has resulted in a unique capability to explore the role of individual genes, or groups of genes, across all cell and tissue types, both in normal and pathological conditions. This provides rapid possibilities for discovery research, based on the analysis of the vast amounts of data in silico. The IST analysis could be considered a "virtual RT-PCR" experiment across 8,000 samples.

Figure 31.1 illustrates the IST analysis of kallikrein 3 (KLK3) across normal and pathological tissues. KLK3, also known as prostate-specific antigen (PSA), is a serine protease and the commonly used prostate cancer biomarker (Bradford et al. 2006). As expected, the IST analysis indicates that KLK3 mRNA expression is prostate specific, but not cancer specific, as high expression is detected both from normal and malignant prostate. Similar analyses using other known tissue-specific markers have been used to validate the IST approach. Furthermore, sophisticated bioinformatic approaches have been used to validate the normalization procedure and data mining tools. For example, before any normalization, gene expression patterns clustered based on the Affymetrix platform version, whereas after normalization the samples clustered according to anatomical site (Kilpinen et al. 2008). Because of the detailed clinical information available from many samples, IST analysis also makes it possible to correlate the expression profile of a set of genes with clinicopathological features of the disease. For example, one can explore the heterogeneity of a particular cancer by dividing it into molecular subtypes. Gene network analyses can be applied to the expression data to identify clusters of genes with similar expression profiles across tissues, diseases, or experiments.

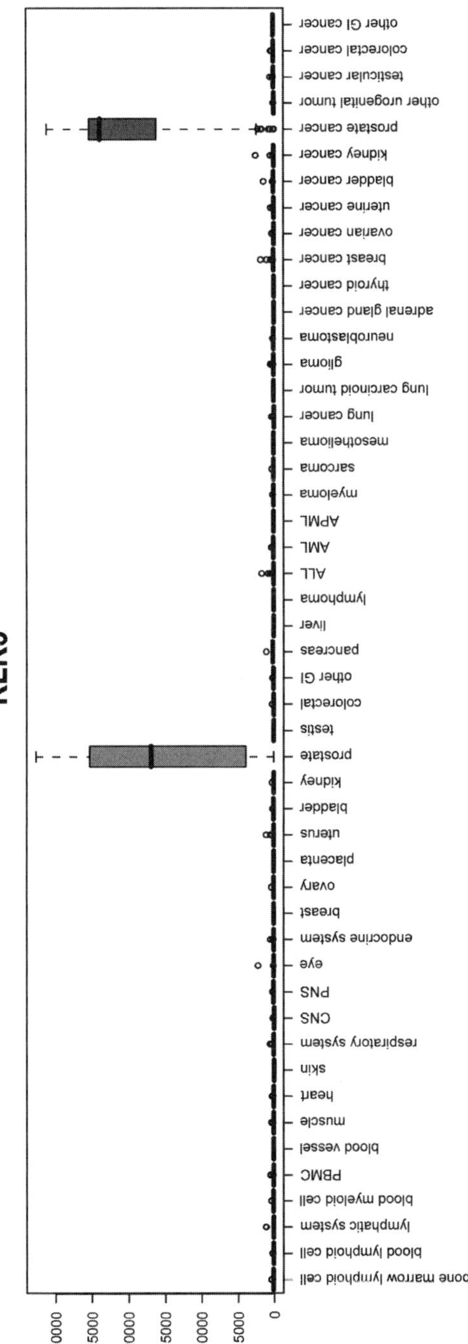

Fig. 31.1 Kallikrein 3 (*KLK3*) mRNA expression across 49 tissues. *KLK3* mRNA expression is prostate specific, but not cancer specific, since high expression is detected both in normal prostate and in prostate cancer. Relative *KLK3* expression level is presented on *y*-axis and tissues on *x*-axis. Abbreviations: *CNS*, central nervous system; *PNS*, peripheral nervous system; other *GI*, other gastrointestinal tissues; *ALL*, acute lymphoblastic leukemia; *AML*, acute myeloid leukemia; *APML*, acute promyelocytic leukemia

Overview of Mean Degradome Gene Expression Profiles in Normal Tissue and Cancers

Here, we performed a systematic, large-scale IST analysis of mRNA expression levels of 493 degradome genes consisting of aspartic proteases, cysteine proteases, metalloproteases, serine proteases, threonine proteases, and protease inhibitors. The mean expression levels of all these genes were first clustered across all the major tissue types present in the database. The results presented in Fig. 31.2 reveal four important general observations. First, clustering separated degradome genes into two main branches, those predominantly expressed in blood-derived tissues and others that were more widely expressed. Second, the same clustering also led to identification of subsets of degradome genes. For example, the hematological cluster was enriched in cysteine and threonine proteases (relative enrichments 1.93 and 3.00 with p values of 0.0006 and 0.003). Third, normal tissues and corresponding cancer tissues clustered separately in most cases. This observation is important as it indicates that the expression pattern of degradome genes is altered significantly in cancers. Fourth, the most distinct clusters of highly expressed degradome genes were detected in normal tissues, indicating that these genes are needed for normal tissue architecture and function. Especially strong cluster consisting mainly of protease inhibitors and serine proteases was observed in liver.

The high expression of protease inhibitors in normal tissues may relate to the turnover and remodeling of extracellular matrix, which must be tightly controlled to prevent excessive degradation. The correct balance between proteases and their inhibitors in the liver is known to be important since the deregulated expression of degradome genes is known to play a significant role in liver fibrosis and cirrhosis (Iredale 2003). Fibrosis may result from sustained wound healing leading to increased collagen and other matrix protein levels disrupting eventually normal liver function (Friedman 2000). Previous studies indicate that matrix degradation occurs in advanced cirrhosis as a result of decreased tissue inhibitor of MMP expression and simultaneous expression of matrix-degrading metalloproteinases (Iredale 2003). Besides liver, several other normal organs such as testis, pancreas, bone marrow, and kidney had distinct clusters of proteases or their inhibitors highly expressed. The expression profiles of kallikrein-1 (KLK1) and renin (REN) precursor are shown in Fig. 31.3 and were picked as examples of degradome genes present in pancreatic and renal clusters. KLK1 is a serine protease, named also as kidney, pancreas, and salivary gland kallikrein, according to its expression profile. High KLK1 expression in pancreas has been previously shown by immunohistochemical (IHC) stainings, supporting our mRNA expression data (Wolf et al. 1998). REN is an aspartyl protease that plays an important role in blood pressure regulation, electrolyte balance, and renal development (Nishimura and Ichikawa 1999). Previous studies have shown high REN expression in kidney as well as in cultured uterine-placental cells, supporting also the mRNA expression data obtained from the database (Baxter et al. 1991). In addition to distinct normal tissue-specific clusters, the average expression levels of degradome genes can distinguish several bigger clusters formed from blood-related, gastrointestinal, and nervous tissues, which are indicated in Fig. 31.2.

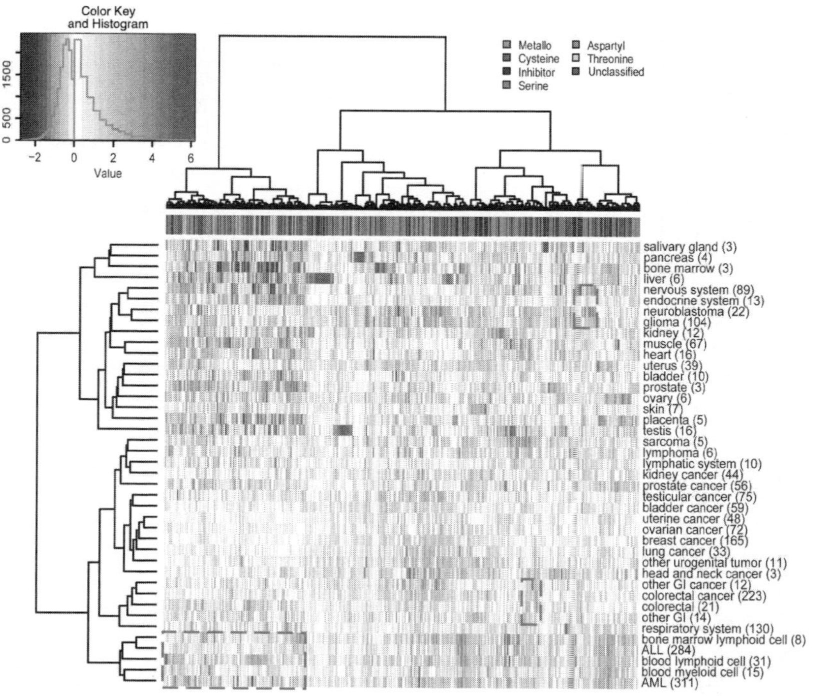

Fig. 31.2 Hierarchical clustering of mean degradome gene expression profiles for 18 malignant and 23 healthy tissues. The number of samples for each tissue is marked in the parentheses. Values of heatmap are scaled genewise to mean 0 and standard deviation 1. Degradome genes ($n = 493$) consist of 145 metalloproteases (presented by the green color above the image), 103 cysteine proteases (red), 101 protease inhibitors (blue), 112 serine proteases (light blue), 18 threonine proteases (yellow), and 13 aspartic proteases (cyan). Few significant degradome gene clusters are marked by the dark green dashed rectangles. The leftmost one indicates blood-related degradome genes, the middle cluster contains mainly gastrointestinal genes, and the rightmost indicates nervous system-specific degradome genes (*See also* Color Insert II)

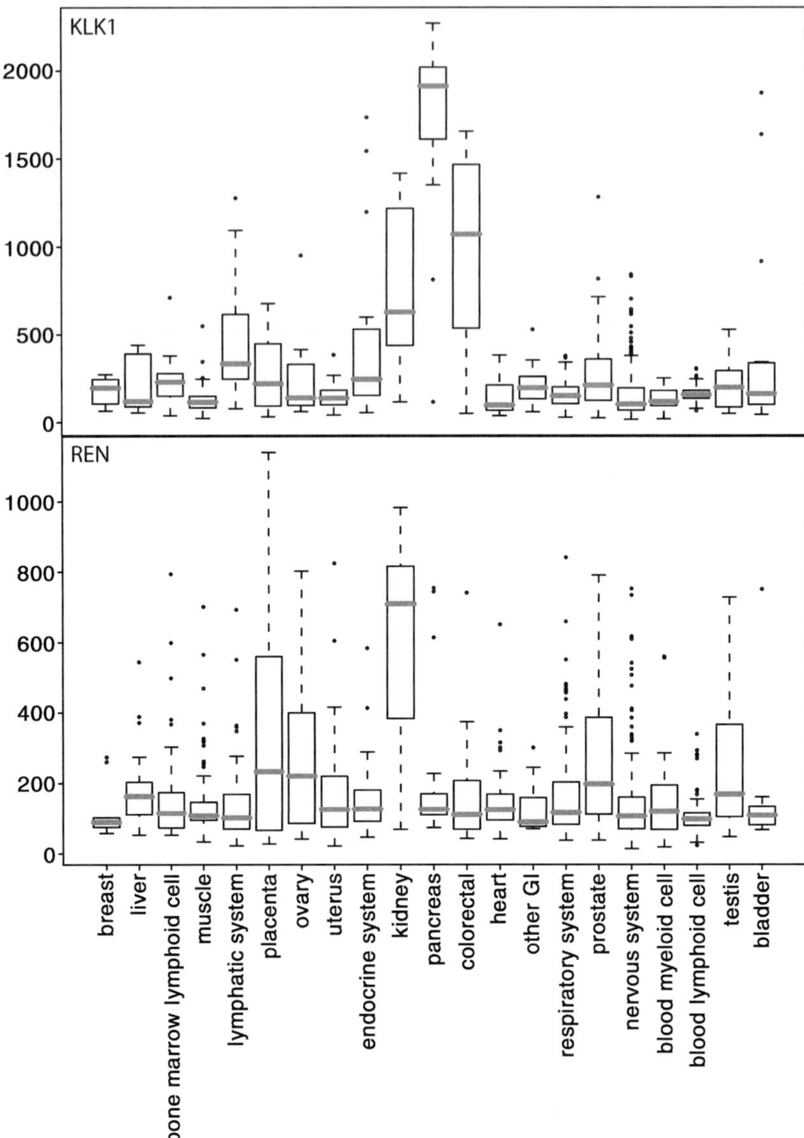

Fig. 31.3 Box-plot presentation of normalized expression values for kallikrein 1 (*KLK1*) and REN mRNA expression across major normal tissues. The box extends from the first to the third quartile of the data and the median is indicated by the line in the middle. The whiskers extend to the extreme values unless there are outliers. The data observations which lie more than 1.5* inter-quartile range (*IQR*) lower than the first quartile, or 1.5**IQR* higher than the third quartile, are considered outliers and are indicated separately. *KLK1* mRNA is predominantly expressed in pancreas, whereas REN in kidney among normal tissues

In Silico Expression Gene Expression Profiling of Matrix Metalloproteinases

Dissemination of tumor cells in the body gives rise to metastases which are the major factors in mortality of cancer. Invasion of tumor cells is dependent on degradation of connective tissue in the underlying basement membrane and in the underlying stroma. Matrix degradation is catalyzed by several proteinases, the most important ones being MMPs. MMPs have been associated with tumor invasion due to their capacity to degrade extracellular matrix and basement membranes. In addition to its scaffold function, extracellular matrix is a reservoir of biologically active molecules, which can be modulated via cleavage by MMPs, resulting in an altered cellular behavior such as induced migration or apoptosis (Visse and Nagase 2003). There are a lot of data on MMPs and cancer, illustrated by the fact that PubMed search with these keywords resulted in 4,500 publications (in July 2007). However, in many studies, the emphasis has been limited to single or a few MMPs in a given tumor type and the number of samples studied has been relatively small. To get a comprehensive view of MMP expression profiles in healthy and malignant tissues, we performed hierarchical clustering with the mean MMP expression values in 18 malignant and 23 normal tissues. The results are presented in Fig. 31.4. As expected, the expression profile of MMPs varies significantly among different tissues and cancers also within a particular organ.

MMP-focussed clustering separated two main branches, one consisting of the majority of normal tissues as well as brain, blood, and most urogenital cancers. The other branch consisted mainly of epithelial tumors as well as a few normal tissues such as colorectum, testis, ovary, uterus, and placenta. Uterine and ovarian cancers have very similar MMP gene expression patterns with high expression of MMP7, MMP10, and MMP11. Increased expression of MMP7 in uterine carcinomas has been described earlier and it has been associated with cancer invasion, metastasis, and poor prognosis (Misugi et al. 2005). MMP10 expression has not been studied in uterine and ovarian cancers although previous results indicate that MMP10 is expressed in endometrium and overexpressed in Ras-transformed ovarian surface epithelial cells (Ulkü et al. 2003, Vassilev et al. 2005). MMP11 has been previously shown to be overexpressed in uterine leiomyomas and ovarian carcinomas, supporting our data on MMP expression profiling (Palmer et al. 1998, Mueller et al. 2000). The cluster consisting of lymphoid cells from bone marrow and blood as well as leukemia (ALL and AML) samples expresses mainly MMP17, MMP8, and MMP20. These MMPs are known to be expressed in blood, highlighted by the aliases of MMP8, which is also known as neutrophil collagenase, and MMP20, which is also known as leukolysin (Puente et al. 1996, Gauthier et al. 2003, Van Wart 1992, Pei 1999).

In comparison to the "lymphoid" cluster, the cluster consisting of most epithelial tumor types as well as testis expresses several of the MMPs, such as MMP28, MMP12, MMP1, MMP3, MMP14, MMP10, MMP13, and MMP11, at high levels. More detailed analysis of the results indicates that the median values of MMP13

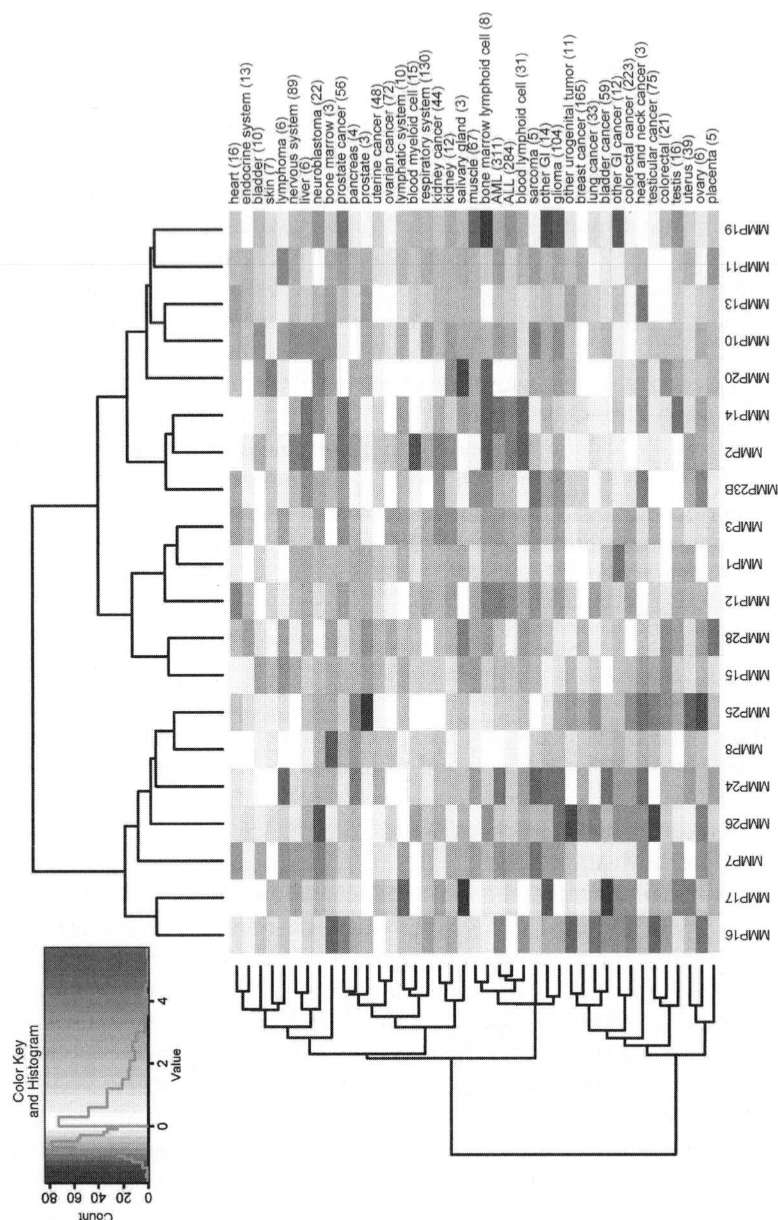

Fig. 31.4 Hierarchical clustering of 20 matrix metalloproteinase genes across 18 malignant and 23 healthy tissues. The number of samples for each tissue is marked in the parentheses. Values of heatmap are scaled genewise to mean 0 and standard deviation 1 (*See also* Color Insert II)

and MMP11 are highest among MMPs in breast cancer samples, whereas lung cancers express MMP12, MMP10, and MMP13 and colorectal cancers MMP12 and MMP3 at highest levels. The elevated expression of MMP13 and MMP11, especially in invasive breast cancer, has been described earlier (Freije et al. 1994, Nielsen et al. 2001, Wolf et al. 1993). MMP11 enhances cancer cell survival and high levels of MMP11 expression have been associated with tumor invasion and poor prognosis (Boulay et al. 2001, Wu et al. 2001). In addition to breast tumors, MMP13 expression has been also associated with tumor cell invasion and metastasis in lung cancers, supporting our clustering data (Bodey et al. 2001, Hsu et al. 2006). Our data favor, therefore, the notion of MMP13 as a cancer-specific protease, even though MMP13 is also expressed by tumor stromal cells and in chronic inflammation (Brinckerhoff et al. 2000). As seen in Fig. 31.4, all normal tissues express low levels of MMP13, whereas 8 out of 15 cancer types show elevated expression of MMP13.

Although the expression of MMPs usually increases during carcinogenesis, this is not always the case. MMP28, also known as epilysin, has been shown to be expressed in normal intestine and IHC analyses indicate that its expression is downregulated in colon cancers (Bister et al. 2004). Reduced MMP28 mRNA expression in colorectal cancers was also detected by IST analysis of the mean MMP28 mRNA expression values in tumors versus normal tissues. This indicates that also downregulation in mRNA expression can be detected by meta-analyses of microarray data.

In our data, the expression profile of MMP2 correlated well with that of MMP14, as shown in Fig. 31.4. Numerous studies have demonstrated an important role for MMP2 in cancer invasion due to its ability to degrade type IV collagen, a component of the basement membrane. As most MMPs, also MMP2 is produced as a latent pro-peptide that has to be activated via proteolytic cleavage. MMP14 (also called MT1-MMP) is a membrane-spanning protease that activates MMP2 on the cell surface (Sato et al. 1994). Similar expression patterns in tissues may, therefore, reflect the functional link between these two MMPs.

MMP12 As An Example of In Silico Discovery—A Potential Biomarker for Several Cancers

One of the most striking examples of an MMP that was specifically upregulated in cancer compared to the normal tissue is MMP12. MMP12 was initially named as macrophage elastase, due to its expression in macrophages and capacity to degrade elastin in addition to other substrates (Shapiro et al. 1993, Chandler et al. 1996). The expression profile of MMP12 is shown in Fig. 31.5. As reported earlier, MMP12 is highly expressed in a subset of lung cancers (marked here as respiratory system) where its expression has been shown to correlate with local recurrence and metastatic disease (Hofmann et al. 2005). MMP12 expression has been associated with

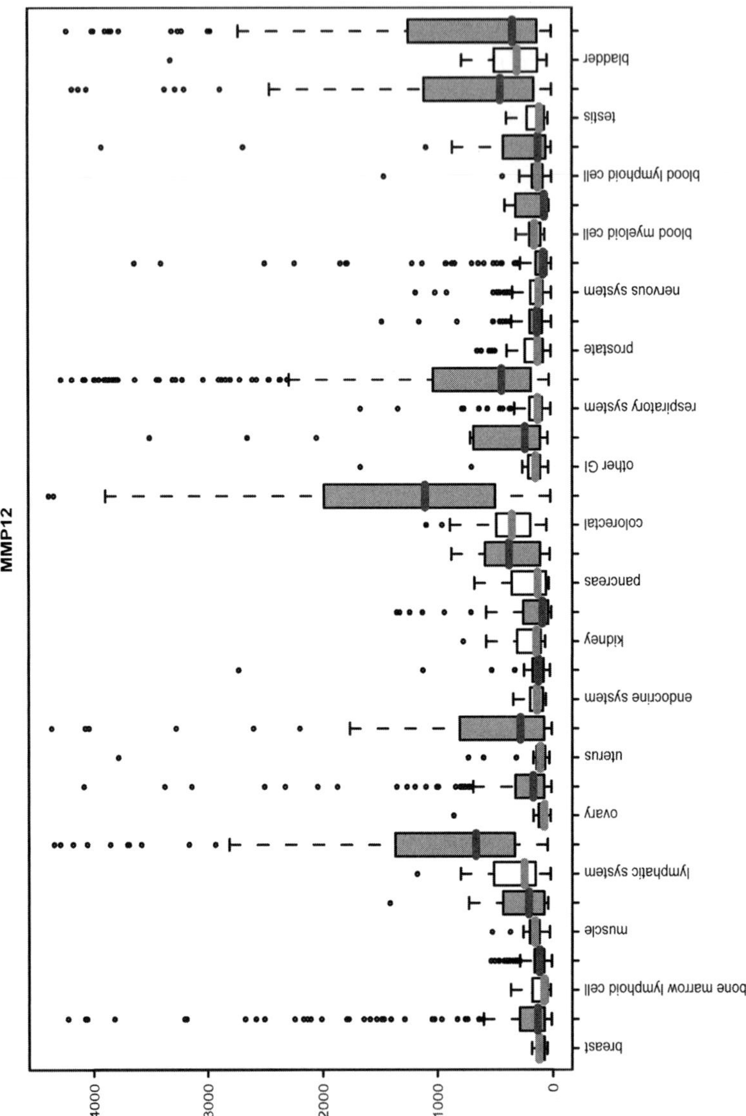

Fig. 31.5 Box-plot presentation of normalized expression values for MMP12 mRNA expression across major normal (white boxes) and cancer (gray boxes) tissues. The box extends from the first to the third quartile of the data and the median is indicated by the line in the middle. The whiskers extend to the extreme values unless there are outliers. The data observations which lie more than 1.5* inter-quartile range (*IQR*) lower than the first quartile, or 1.5**IQR* higher than the third quartile, are considered outliers and are indicated separately

poor prognosis in pancreatic cancers (Balaz et al. 2002). However, the role of MMP-12 has not been studied in many of the other cancer types where it seems to be overexpressed. Our analysis reveals that MMP12 is highly overexpressed also in breast, colorectal, bladder, testicular, ovarian, and uterine cancers as well as in lymphomas. In addition to tissue samples, the IST database contains expression data obtained from primary and established cell lines, some of which have been functionally treated. The analysis of MMP12 mRNA expression throughout this dataset revealed that its expression is very low in both cultured normal and malignant cell lines with the exception of peripheral blood mononuclear cells, macrophages, dendritic cells, and endometrial stromal cells. Interestingly, TNF-alpha was also "re-discovered" as an inducer of MMP12 expression in human microvascular endothelial cells (Viemann et al. 2006). Taken together, the results from these analyses suggest that the high expression of MMP12 in tumor samples results from an inflammatory response with overrepresentation of macrophages present within tumor samples in comparison to the corresponding normal tissues.

Gene Coexpression Analyses to Explore *In Vivo* Functions of MMP12

MMP12 overexpression in many cancer types is an interesting finding although its biological function supporting tumorigenesis is not well known. To find out more about MMP12 biological function and putative interaction partners, we, therefore, first performed a correlation analysis of MMP12 with other key genes in the same pathway. MMP12 has been shown to function as an anti-angiogenic protein due to its ability to covert plasminogen (PLG) to angiostatin and collagen XVIII to endostatin (Cornelius et al. 1998, Wen et al. 1999). Our results from the correlation analysis of MMP12 against PLG or COL18A1 indicate that in general, these genes are not coexpressed in cancers. As an example, correlation plots are presented from lung cancer samples in Fig. 31.6. However, MMP12 and COLXVIII show a weak positive correlation in some cancer types such as in kidney carcinoma, thyroid gland papillary cancer, small-cell lung cancer, and bladder cancer. In addition to regulating the levels of anti-angiogenic molecules, MMP12 has been shown to play a role in vascular mimicry in a transgenic mouse model of ischemic cardiomyopathy (Moldovan et al. 2000). It has been suggested that a similar mechanism may exist in tumors (Maniotis et al. 1999, Zhang et al. 2007). However, it is not known whether MMP12 mRNA upregulation participates in this process or whether it has other roles in addition to the ones that have been previously characterized.

As the correlation analysis with previously associated molecules failed to reveal potential functional clues to MMP12, we performed a more thorough analysis of MMP12 coexpression networks. The transcriptome of the cell, in any given moment, reflects its contextual information, such as its differentiation status and effects of the cellular microenvironment. Therefore, by studying the coexpression

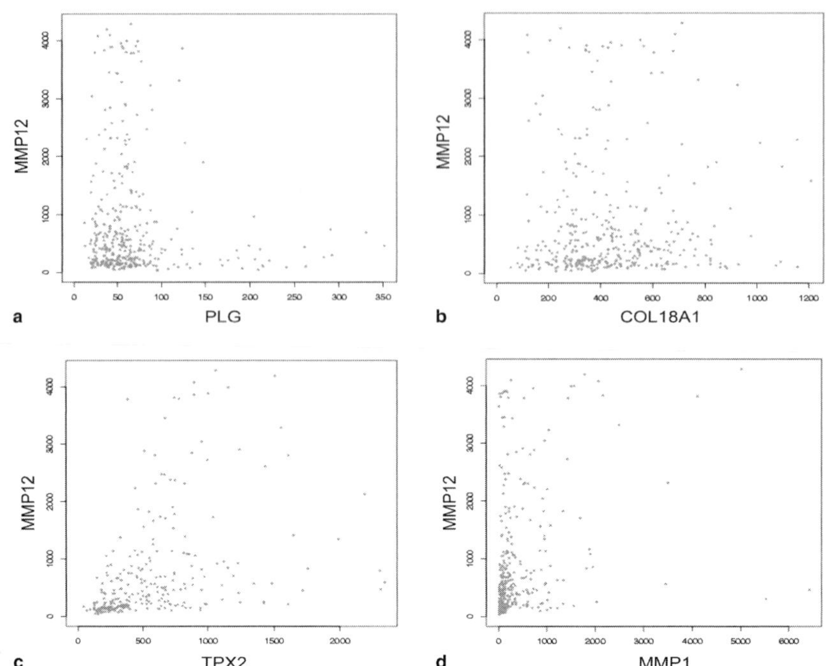

Fig. 31.6 Coexpression plots of MMP12 and plasminogen (*PLG*), collagen 18A1 (*COL18A1*), microtubule-associated protein homologue TPX2, and MMP1 in lung cancer samples

"environment" or coexpression network of genes it is possible to study their putative functions. Here, we demonstrated this approach by uncovering putative functions of MMP12 in colorectal, lung, breast, and bladder cancers. First, we identified genes showing most prominent coexpression with MMP12 in tumor samples. The correlation analysis results obtained from lung cancers ($n = 250$) indicate that the most strongly associated gene is microtubule-associated protein homologue TPX2 ($r = 0.3836, p < 0.001$). Interestingly, TPX2 has been previously identified as a differentially expressed gene between cancerous and noncancerous lung cells (Manda et al. 1999). MMP12 and TPX2 mRNA expression is correlated also in breast cancers ($r = 0.3836, p < 0.001$), whereas in colorectal and bladder cancers these genes are not clearly coexpressed. Interestingly, IST gene expression profiling results indicated that TPX2 mRNA expression was induced in all cancer types, whereas its expression in normal tissues was prominent only in testis, colorectal tissues, liver, and lymphatic system-derived tissues. Previously published data indicate that TPX2 is required for targeting Aurora A kinase to the mitotic spindle apparatus and it has been recently identified as a promising cancer target for multiple human tumor cell lines (Kufer et al. 2002, Morgan-Lappe et al. 2007). In breast cancers ($n = 360$), the gene showing the strongest correlation with MMP12 mRNA expression was interleukin-8 ($r = 0.55, p < 0.001$). MMP12 and IL8 mRNA expressions were associated with all four cancer types analyzed.

Interestingly, in colorectal cancer, MMP1 showed the strongest association with MMP12 ($r = 0.56$, $p < 0.001$). In fact, MMP1 mRNA expression was correlated with MMP12 in all cancer types analyzed, supporting the clustering results obtained with the mean expression values presented in Fig. 31.4, where MMP1 and MMP12 are placed adjacent to each other. In bladder cancers ($n = 59$), osteopontin (SPP1), a secreted phosphoprotein, showed the strongest association with MMP12 expression. IST results indicate that osteopontin 1 and MMP12 gene expression correlated with the other cancer types analyzed. Previous literature clearly associates elevated osteopontin mRNA expression with tumor aggressiveness in several cancer types, and SSP1 silencing has been shown to reduce tumor growth and invasive properties both *in vitro* and *in vivo* (Shevde et al. 2006, Chakraborty et al. 2006).

We created a clustered heatmap of the most strongly MMP12-associated genes ($n = 97$) in colorectal, lung, breast, and bladder cancers (Fig. 31.7). The results indicate that there is distinct group of genes correlating positively with MMP12 and a group with an inverse correlation. Additionally, the expression of these genes reveals several subgroups of MMP12-coexpressed genes. These sets of genes could reveal the functional context of MMP12 and its role in these cancer types. We then linked these coexpressed genes with ki-67 mRNA expression, well-known marker of cell proliferation (Seigneurin and Guillaud 1991). The results indicate strong association between the expression levels of ki-67 and a set of genes correlating positively with MMP12 especially in lung cancers. Therefore, our results indicate that MMP12 mRNA expression is associated with cell proliferation particularly in lung cancers.

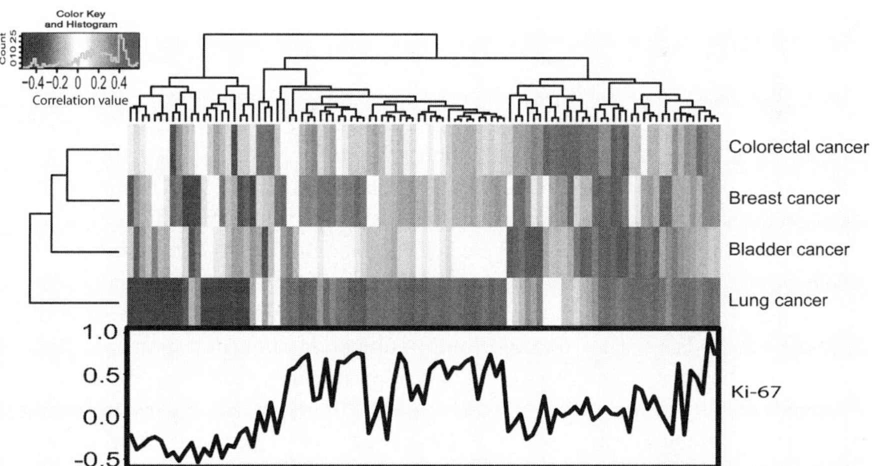

Fig. 31.7 Clustered heatmap of 97 genes having strongest correlations ($-0.4 > r > 0.4$ with Bonferroni corrected p value < 0.01) with MMP12 in colorectal, breast, bladder, and lung cancers. Below the image is a correlation coefficient plot of Ki-67 gene across in vivo-derived 3,928 datapoints of In Silico Transcriptomics (*IST*) database (*See also* Color Insert II)

We also performed gene ontology (GO) analysis of the MMP12 correlating genes in order to further identify biological processes associated with MMP12 expression. Elevated MMP12 mRNA expression has been described in a mouse model for chronic obstructive pulmonary disease (COPD) together with increased immune and inflammatory response as well as mitosis (Meng et al. 2006) Also, the GO analysis of MMP12-associated genes in breast cancer indicates overrepresentation of mitotic genes, such as cyclins, TKK protein kinase, aurora kinase A, BUB1, and centromere protein E. In addition, several GO terms for stress responses (pathogens, wounding, external stimulus, inflammation) were significantly overrepresented. Some of these genes, like IL1B, have been previously associated with MMP12 (Xie et al. 2005). Furthermore, several other GO terms such as chemotaxis were significantly enriched among MMP12-associated genes.

Taken together, IST analysis of MMP expression across different tissues supported previously found gene expression patterns of some MMPs, like MMP13. In addition, it revealed previously unknown cancer-specific expression patterns for less-studied members of the family. In particular, MMP12 was found to be a potentially strong biomarker in several cancers. The MMP12 gene coexpression analysis suggested that MMP12 overexpression is associated with cell proliferation, particularly in lung cancer, as well as inflammatory response. In conclusion, our results suggested several potential functional associations for MMP12 *in vivo* in human cancer that could now be explored in more detail by experimental research.

Conclusions

In silico gene expression analysis ("meta-analysis") provides a rapid, cost-effective method for acquiring valuable biological insights on gene function, gene regulation, and coexpression. In comparison to reverse transcriptase -polymerase chain reaction (RT-PCR) and IHC analyses, which may take from days to months, IST profiling can be performed for individual genes in minutes. The tissue coverage present in the analysis is significantly higher in IST profiling with most normal and tumor tissues, leukemias and established cell lines represented in comparison to the other methods. The results obtained from IST profiling are quantitative, whereas IHC studies are at best semiquantitative, and largely dependent on antibodies, detection methods, tissue fixation, and subjective interpretation of staining intensity. Furthermore, coexpression of two or more genes at the time can be rapidly performed by IST profiling, whereas with other methods, this increases the workload proportionally to the number of genes. However, one should bear in mind that the mRNA gene expression levels do not always correlate with protein expression, as these represent distinct steps in the gene regulatory process.

IST profiling has made it possible to perform genome-wide studies throughout the human body in health and disease which has not been possible before with other methods. Here, we have demonstrated that in addition to the validation of known expression profiles, meta-analysis may give several novel insights into gene function by revealing mRNA expression in tissues which have not been previously described for the gene of interest. Systematic analyses of global gene expression patterns will also facilitate the development of genomics-based diagnostics and therapeutics.

References

Balaz P., Friess H., Kondo Y., et al. (2002). Human macrophage metalloelastase worsens the prognosis of pancreatic cancer. Ann. Surg. 235(4):519–527.

Barrett T., Suzek T., Troup D., et al. (2005). NCBI GEO: mining millions of expression profiles-database and tools. Nucleic Acids Res. 33:D562–6.

Baxter J. D., Duncan K., Chu W., et al. (1991). Molecular biology of human rennin and its gene. Recent Prog. Horm. Res. 47:211–257.

Bister V. O., Salmela M. T., Karjalainen-Lindsberg M. L., et al. (2004). Differential expression of three matrix metalloproteinases, MMP-19, MMP-26, and MMP-28, in normal and inflamed intestine and colon cancer. Dig. Dis. Sci. 49(4):653–661.

Bodey B., Bodey B., Gröger A. M. et al. (2001). Invasion and metastasis: the expression and significance of matrix metalloproteinases in carcinomas of the lung. *In Vivo* 15(2):175–180.

Boulay A., Masson R., Chenard M. P., et al. (2001). High cancer cell death in syngeneic tumors developed in host mice deficient for the stromelysin-3 matrix metalloproteinase. Cancer Res. 61:2189–2193.

Bradford T. J., Tomlins S. A., Wang X., et al. (2006). Molecular markers of prostate cancer. Urol. Oncol. 24(6):538–551.

Brinckerhoff C. E., Rutter J. L., and Benbow U. (2000). Interstitial collagenases as markers of tumor progression. Clin. Cancer Res. 6(12):4823–4830.

Chakraborty G., Jain S., Behera R., et al. (2006). The multifaceted roles of osteopontin in cell signaling, tumor progression and angiogenesis. Curr. Mol. Med. 2006 (8):819–830.

Chandler S., Cossins J., Lury J., et al. (1996). Macrophage metalloelastase degrades matrix and myelin proteins and processes a tumour necrosis factor-alpha fusion protein. Biochem. Biophys. Res. Commun. 228(2):421–429.

Cornelius L. A., Nehring L. C., Harding E., et al. (1998). Matrix metalloproteinases generate angiostatin: effects on neovascularization. J. Immunol. 161(12):6845–6852.

Freije J. M., Díez-Itza I., Balbín M., et al. (1994). Molecular cloning and expression of collagenase-3, a novel human matrix metalloproteinase produced by breast carcinomas. J. Biol. Chem. 269(24):16766–16773.

Friedman S. L. (2000). Molecular regulation of hepatic fibrosis, an integrated cellular response to tissue injury. J. Biol. Chem. 275:2247–2250.

Gauthier M. C., Racine C., Ferland C., et al. (2003). Expression of membrane type-4 matrix metalloproteinase (metalloproteinase-17) by human eosinophils. Int. J. Biochem. Cell. Biol. 35 (12):1667–1673.

Hofmann H. S., Hansen G., Richter G., et al. (2005). Matrix metalloproteinase-12 expression correlates with local recurrence and metastatic disease in non-small cell lung cancer patients. Clin. Cancer Res. 11(3):1086–1092.

Hsu C. P., Shen G. H., and Ko J. L. (2006). Matrix metalloproteinase-13 expression is associated with bone marrow microinvolvement and prognosis in non-small cell lung cancer. Lung Cancer 52(3):349–357.

Iredale J. P. (2003). Cirrhosis: new research provides a basis for rational and targeted treatments. Br.Med. J. 327:143–147.

Kilpinen S., Autio R., Ojala K., et al. (2008). Systematic bioinformatic analysis of expression levels of 17330 human genes across 9783 samples from 175 types of healthy and pathological tissues. Submitted.

Kufer T. A., Silljé H. H., Körner R., et al. (2002). Human TPX2 is required for targeting Aurora-A kinase to the spindle. J. Cell. Biol. 158(4):617–623.

Manda R., Kohno T., Matsuno Y., et al. (1999). Identification of genes (SPON2 and C20orf2) differentially expressed between cancerous and noncancerous lung cells by mRNA differential display. Genomics 61(1):5–14.

Maniotis A. J., Folberg R., Hess A., et al. (1999). Vascular channel formation by human melanoma cells *in vivo* and *in vitro*: vasculogenic mimicry. Am. J. Pathol. 155(3):739–752.

Meng Q. R., Gideon K. M., Harbo S. J., et al. (2006). Gene expression profiling in lung tissues from mice exposed to cigarette smoke, lipopolysaccharide, or smoke plus lipopolysaccharide by inhalation. Inhal. Toxicol. 18(8):555–568.

Misugi F., Sumi T., Okamoto E., et al. (2005). Expression of matrix metalloproteinases and tissue inhibitors of metalloproteinase in uterine endometrial carcinoma and a correlation between expression of matrix metalloproteinase-7 and prognosis. Int. J. Mol. Med. 16(4):541–546.

Moldovan N. I., Goldschmidt-Clermont P. J., Parker-Thornburg J., et al. (2000). Contribution of monocytes/macrophages to compensatory neovascularization: the drilling of metalloelastase-positive tunnels in ischemic myocardium. Circ. Res. 87(5):378–384.

Morgan-Lappe S. E., Tucker L. A., Huang X., et al. (2007). Identification of Ras-related nuclear protein, targeting protein for xenopus kinesin-like protein 2, and stearoyl-CoA desaturase 1 as promising cancer targets from an RNAi-based screen. Cancer Res. 67(9):4390–4398.

Mueller J., Brebeck B., Schmalfeldt B., et al. (2000). Stromelysin-3 expression in invasive ovarian carcinomas and tumours of low malignant potential. Virchows. Arch. 437(6):618–624.

Nielsen B. S., Rank F., López J. M., et al. (2001). Collagenase-3 expression in breast myofibroblasts as a molecular marker of transition of ductal carcinoma in situ lesions to invasive ductal carcinomas. Cancer Res. 61(19):7091–7100.

Nishimura H. and Ichikawa I. (1999). What have we learned from gene targeting studies for the rennin angiotensin system of the kidney? Int. Med. 38:315–323.

Palmer S. S., Haynes-Johnson D., Diehl T., et al. (1998). Increased expression of stromelysin 3 mRNA in leiomyomas (uterine fibroids) compared with myometrium. J. Soc. Gynecol. Investig. 5(4):203–209.

Pei D. (1999). Leukolysin/MMP25/MT6-MMP: a novel matrix metalloproteinase specifically expressed in the leukocyte lineage. Cell Res. 9(4):291–303.

Puente X. S., Pendás A. M., Llano E., et al. (1996). Molecular cloning of a novel membrane-type matrix metalloproteinase from a human breast carcinoma. Cancer Res. 56(5):944–949.

Sarkans U., Parkinson H., Lara G., et al. (2005). The ArrayExpress gene expression database: a software engineering and implementation perspective. Bioinformatics 21:1495–1501.

Sato H., Takino T., Okada Y., et al. (1994). A matrix metalloproteinase expressed on the surface of invasive tumour cells. Nature 370(6484):61–65.

Seigneurin D. and Guillaud P. (1991). Ki-67 antigen, a cell cycle and tumor growth marker. Pathol. Biol. 39(10):1020–1028.

Shapiro S. D., Kobayashi D. K., and Ley T. J. (1993). Cloning and characterization of a unique elastolytic metalloproteinase produced by human alveolar macrophages. J. Biol. Chem. 268 (32):23824–23829.

Shevde L. A., Samant R. S., Paik J. C., et al. (2006). Osteopontin knockdown suppresses tumorigenicity of human metastatic breast carcinoma, MDA-MB-435. Clin. Exp. Metastasis 23(2):123–133.

Ulkü A. S., Schäfer R., and Der C. J. (2003). Essential role of Raf in Ras transformation and deregulation of matrix metalloproteinase expression in ovarian epithelial cells. Mol. Cancer Res. 1(14):1077–1088.

Van Wart H. E. (1992). Human neutrophil collagenase. Matrix Suppl. 1:31–36.

Vassilev V., Pretto C. M., Corne P. B., et al. (2005). Response of matrix metalloproteinases and tissue inhibitors of metalloproteinases messenger ribonucleic acids to ovarian steroids in human endometrial explants mimics their gene- and phase-specific differential control *in vivo*. J. Clin. Endocrinol. Metab. 90(10):5848–5857.

Viemann D., Goebeler M., Schmid S., et al. (2006). TNF induces distinct gene expression programs in microvascular and macrovascular human endothelial cells. J. Leukoc. Biol. 80 (1):174–185.

Visse R. and Nagase H. (2003). Matrix metalloproteinases and tissue inhibitors of metalloproteinases: structure, function, and biochemistry. Circ. Res. 92(8):827–839.

Wen W., Moses M. A., Wiederschain D., et al. (1999). The generation of endostatin is mediated by elastase. Cancer Res. 59(24):6052–6.

Wolf C., Rouyer N., Lutz Y., et al. (1993). Stromelysin 3 belongs to a subgroup of proteinases expressed in breast carcinoma fibroblastic cells and possibly implicated in tumor progression. Proc. Natl. Acad. Sci U S A. 90(5):1843–1847.

Wolf W. C., Harley R. A., Sluce D., et al. (1998). Cellular localization of kallistatin and tissue kallikrein in human pancreas and salivary glands. Histochem. Cell Biol. 110:477–484.

Wu E., Mari B. P., Wang F., et al. (2001). Stromelysin-3 suppresses tumor cell apoptosis in a murine model. J. Cell. Biochem. 82:549–555.

Xie S., Issa R., Sukka M. B., et al. (2005). Induction and regulation of matrix metalloproteinase-12 in human airway smooth muscle cells. Respir. Res. 6:148.

Zhang S., Zhang D., and Sun B. (2007). Vasculogenic mimicry: current status and future prospects. Cancer Lett. 254:157–164.

Chapter 32
Degradome Gene Polymorphisms

Ross Laxton and Shu Ye

Abstract DNA sequence variations contribute to inter-individual variability in susceptibility to and/or severity of common diseases such as cancer, diabetes, and asthma. About 90% of all sequence variations in the human genome are single nucleotide polymorphisms. Each single nucleotide polymorphism arises from a single nucleotide substitution, resulting in the presence of two alleles which differ by one base pair of DNA. The second most common form of DNA sequence variation is microsatellite polymorphism, which is also known as short tandem repeat polymorphism. A microsatellite is a run of tandem repeats of a short DNA sequence, usually 1–4 base pairs. A microsatellite polymorphism can be di-allelic or multi-allelic, with the different alleles differing in the number of repeat units, and hence in length. It is likely that only a very small percentage of DNA sequence variations in the human genome are functionally important, exerting effects on gene expression or function. A DNA sequence variant that is functionally neutral can still be found to be associated with disease susceptibility, arising from linkage disequilibrium with a functional variant located on the same chromosome. Identification of genetic variants that contribute to disease susceptibility and/or severity can provide understanding of the molecular basis of the disease, could have diagnostic and prognostic value, and may provide useful molecular targets for developing novel therapeutics. There is evidence suggesting that polymorphisms in certain matrix metalloproteinase (MMP) genes could influence cancer susceptibility and/or prognosis. This chapter will highlight some of these findings, focusing on MMP1, MMP2, MMP3, MMP9, and MMP12.

S. Ye
Clinical Pharmacology, William Harvey Research Institute, Barts and the London School of Medicine, John Vane Science Centre, Charterhouse Square, London EC1M 6BQ, United Kingdom, e-mail: s.ye@qmul.ac.uk

D. Edwards et al. (eds.), *The Cancer Degradome.*
© Springer Science + Business Media, LLC 2008

Introduction

DNA sequence variations contribute to inter-individual variability in susceptibility to and/or severity of common diseases such as cancer, diabetes, and asthma. About 90% of all sequence variations in the human genome are single nucleotide polymorphisms (SNPs). Each SNP arises from a single nucleotide substitution, resulting in the presence of two alleles which differ by one base pair of DNA. The second most common form of DNA sequence variation is microsatellite polymorphism, which is also known as short tandem repeat polymorphism. A microsatellite is a run of tandem repeats of a short DNA sequence, usually 1–4 base pairs. A microsatellite polymorphism can be di-allelic or multi-allelic, with the different alleles differing in the number of repeat units, and hence in length.

It is likely that only a very small percentage of DNA sequence variations in the human genome are functionally important, exerting effects on gene expression or function. A DNA sequence variant that is functionally neutral can still be found to be associated with disease susceptibility, arising from linkage disequilibrium with a functional variant located on the same chromosome. Identification of genetic variants that contribute to disease susceptibility and/or severity can provide understanding of the molecular basis of the disease, could have diagnostic and prognostic value, and may provide useful molecular targets for developing novel therapeutics.

There is evidence suggesting that polymorphisms in certain matrix metalloproteinase (MMP) genes could influence cancer susceptibility and/or prognosis. This chapter will highlight some of these findings, focusing on MMP1, MMP2, MMP3, MMP9, and MMP12 (Table 32.1).

MMP1

The MMP1 gene contains 10 exons and spans 8,244 base pairs on chromosome 11q22.2. The promoter region of the gene contains a number of cis-elements that play important roles in the regulation of MMP1 expression, including two activator protein-1 (AP-1)-binding sites (from nucleotide position -186 and -73, respectively, relative to the transcriptional start site), a polyoma enhancer A binding protein-3 (PEA3)-binding site (from -89), a transforming growth factor (TGF)- β inhibitory element (from -245), a CCAAT/enhancer binding protein-β site (from -2010), and a TATA box (from -32) (Yan and Boyd 2007).

The first reported polymorphism in the MMP1 gene is the 1G/2G polymorphism arising from the insertion/deletion of a guanosine (G) at nucleotide position–1607 (relative to the transcriptional start site) in the MMP1 gene promoter (Rutter et al. 1998). The polymorphism is assigned the number rs1799750 in the dbSNP database. In addition to the 1G/2G polymorphism, there are a number of other common polymorphisms (with minor allele frequency > 0.05) in the MMP1 gene promoter, including–839G > A (rs473509), -755G > T (rs498186), -519A > G (rs1144393),

Table 32.1 Reported functional polymorphisms in MMP genes

Gene	Polymorphism	Functional studies		Cancers reported to be associated with the polymorphism
		Difference in promoter activity	Transcription factor involved	
MMP1	-1607G/GG	2G allele > 1G allele	Ets	Lung cancer (Zhu et al. 2001, Su et al. 2005, Zhang et al. 2006)
				Head and neck squamous cell carcinoma (Hashimoto et al. 2004, O-charoenrat et al. 2006)
				Oral squamous cell carcinoma (Lin et al. 2004a, Cao and Li 2006)
				Cutaneous malignant melanoma (Ye et al. 2001)
				Breast cancer (Przybylowska et al. 2004, 2006)
				Ovarian cancer (Kanamori et al. 1999, Six et al. 2006)
				Cervical cancer (Nishioka et al. 2003, Lai et al. 2005)
				Endometrial carcinoma (Nishioka et al. 2000)
				Colorectal cancer (Ghilardi et al. 2001, Hinoda et al. 2002, Zinzindohoue et al. 2005, Elander et al. 2006)
				Renal cell carcinoma (Hirata et al. 2003, 2004)
MMP2	-1575G > A	G allele > A allele	Estrogen receptor	Lung cancer (Yu et al. 2002, Zhou et al. 2005, Rollin et al. 2007)
				Breast cancer (Zhou et al. 2004)
	-1306C > T	C allele > T allele	SP1	Head and neck squamous cell carcinoma (O-charoenrat and Khantapura 2006)
	-735C > T	C allele > T allele	SP1	Oral squamous cell carcinoma (Lin et al. 2004b)
				Esophageal squamous cell carcinoma (Yu et al. 2004)
				Gastric cardia adenocarcinoma (Miao et al. 2003)
				Colorectal cancer (Xu et al. 2004)
MMP3	-1612 5A/6A	5A allele > 6A allele	NFkB, ZBP89	Lung cancer (Fang et al. 2005)
				Breast cancer (Ghilardi et al. 2002, Krippl et al. 2004)
				Head and neck squamous cell carcinoma (Blons et al. 2004)
				Esophageal squamous cell carcinoma (Zhang et al. 2004)

(continued)

Table 32.1 (continued)

Gene	Polymorphism	Functional studies		Cancers reported to be associated with the polymorphism
		Difference in promoter activity	Transcription factor involved	
MMP9	-1562C > T	T allele > C allele	Unknown	Gastric cancer (Matsumura et al. 2005)
	(CA)n	(CA)$_{20-23}$ > (CA)$_{14}$	Unknown	Prostate cancer (Sfar et al. 2007)
MMP12	-82A > G	A allele > G allele	AP1	Lung cancer (Heist et al. 2006)
				Bladder cancer (Kader et al. 2006)

-422T > A (rs475007), -340C > T (rs514921), and 320C > T (rs494379) (Pearce et al. 2005). Studies have shown that the 1G/2G polymorphism has a functional effect on MMP1 expression (further discussed below), and a recent study shows that other polymorphisms in the MMP1 gene promoter could also influence its transcriptional activity in tumour cells (Pearce et al. 2007). There are three common SNPs (rs10488, rs470558, and rs1051121) in the coding region of the gene, which are located in exons 2, 5, and 6, respectively (data from the dbSNP database). None of these coding SNPs (which are all synonymous substitutions) results in a change in the amino acid sequence of the MMP1 protein.

Studies have shown that the 1G/2G polymorphism exerts an effect on the transcriptional activity of the MMP1 gene promoter. The insertion of the G nucleotide at position–1607 in the 2G allelic promoter creates an Ets-binding site (with a core sequence -GGA-) (Rutter et al. 1998). In vitro assays have shown that this site interacts with a transcription factor belonging to the Ets family, and that this transcription factor forms a complex with the transcription factor AP-1 which binds to a nearby AP-1-binding site located at position -1602 (Rutter et al. 1998, Tower et al. 2002). It has also been shown that the 2G allele has between 2- and 29-fold higher promoter activity than the 1G allele in various types of cells, including melanoma cell lines (Rutter et al. 1998, Tower et al. 2002, 2003a), breast cancer cell lines (Benbow et al. 1999, Tower et al. 2003b), fibroblasts (Rutter et al. 1998), and amnion cells (Fujimoto et al. 2002). It is likely that the difference in promoter activity between the two alleles is a result of the differential binding of the above-mentioned transcription factors to the two alleles. Studies have also indicated that these transcription factors can be activated through an ERK1/2-dependent signal transduction pathway and a Jun N-terminal Kinase (JNK)-dependent pathway (Ranganathan et al. 2001, Benbow et al. 2002, Tower et al. 2002, Nelson et al. 2003, 2006) and that compared with the 1G allele, the 2G allele has a greater response to stimuli that trigger these pathways (Ranganathan et al. 2001, Fujimoto et al. 2002, Nelson et al. 2003, 2006).

It appears that the allele-specific effect on promoter activity results in a difference in MMP1 expression level in tumour tissues. Examinations of MMP1 mRNA levels in ovarian cancer tissues (Kanamori et al. 1999, Wenham et al. 2003) and in head and neck cancer tissues (O-charoenrat et al. 2006) showed that the levels are highest in tumour tissues from 2G homozygotes, intermediate in heterozygotes, and lowest in 1G homozygotes. MMP1 protein levels in breast cancer tissues have also been shown to be highest in 2G homozygotes, intermediate in heterozygotes, and lowest in 1G homozygotes (Przybylowska et al. 2006).

In addition, a study showed that 83% of metastatic melanomas with loss of heterozygosity at the MMP1 locus retain the 2G allele, whereas only 17% retain the 1G allele (Noll et al. 2001). It is thought that although loss of either the 1G or 2G allele from 1G/2G heterozygotes is random, retention of the transcriptionally more active 2G allele would promote tumour invasion and metastasis, and therefore metastatic tumours are more likely to contain the 2G allele as opposed to the 1G allele (Noll et al. 2001).

There is emerging evidence suggesting a possible association of MMP1 gene variation with susceptibility and/or prognosis of certain cancers. The evidence has derived from studies of the 1G/2G polymorphism. Currently, there is no reported study of any other MMP1 polymorphism in relation to cancer susceptibility or progression.

A study of 456 European American patients with lung cancer and 451 healthy controls showed that the 2G/2G genotype was significantly more prevalent in cases than in controls (46.1% vs 31.9%, $p < 0.001$) (Zhu et al. 2001). The study estimated that overall, the odds ratio of developing lung cancer was 1.82 [95% confidence interval (CI) 1.38–2.39] for individuals of the 2G/2G genotype compared with individuals of the 1G/1G or 1G/2G genotype. In this study, the association of the 2G/2G with increased lung cancer susceptibility was more pronounced in men than in women, odds ratio being 2.15 (95% CI 1.42–3.26) and 1.34 (95% CI 0.84–2.15), respectively. Another study of European Americans also found the relationship between the 1G/2G polymorphism and lung cancer in men (919 cases and 603 controls), although the estimated odds ratio was smaller, being 1.30 (95% CI 1.00–1.75) for the 2G/2G genotype and 1.23 (95% CI 0.88–1.73) for the 1G/2G genotype compared with the 1G/1G genotype (Su et al. 2005). The association between the 2G/2G genotype and increased susceptibility to lung cancer was also observed in a study in a Chinese population (150 cases and 200 controls), with an odds ratio of 1.77 for the 2G/2G genotype compared with the 1G/1G and 1G/2G genotype (Zhang et al. 2006).

The 1G/2G polymorphism was also found to be associated with head and neck squamous cell carcinoma in a study of 300 cases and 300 controls of Thai descent, with an odds ratio of 2.28 (95% CI 1.58–3.27) for the 2G/2G genotype compared with the 1G/1G and 1G/2G genotypes, which remained significant after adjusting for age, gender, smoking, and alcohol drinking habit (O-charoenrat et al. 2006). This association was also detected in an investigation in Japanese, involving 140 head and neck squamous cell carcinoma cases and 223 controls, in which the odds ratio was estimated to be 1.56 (95% CI 1.02–2.38) for 2G/2G compared with 1G/1G and 1G/2G (Hashimoto et al. 2004). The 2G/2G genotype was also found to be associated with oral squamous cell carcinoma in two studies in Chinese, with estimated odds ratios of over 2, compared with the 1G/1G and 1G/2G genotypes (Lin et al. 2004a, Cao and Li 2006).

In addition, there is evidence of association of the 1G/2G polymorphism with susceptibility to colorectal cancer (Ghilardi et al. 2001, Hinoda et al. 2002, Elander et al. 2006), renal cell carcinoma (Hirata et al. 2003, 2004), ovarian cancer (Kanamori et al. 1999), and endometrial carcinoma (Nishioka et al. 2000).

Studies have also suggested that progression and metastasis of certain cancers is influenced by MMP1 genotypes. In a study of European British patients with cutaneous malignant melanoma ($n = 139$), the 2G/2G genotype was found to be associated with deep invasive primary tumours and poorer prognosis, the frequency of the 2G/2G genotype being higher in patients with vertical growth phase tumour than in patients with horizontal growth phase tumour (34% vs 17%), and disease-free survival after surgery being lower in patients of the 2G/2G genotype (Ye et al. 2001). Similarly, in a study of 201 French patients with colorectal cancer who were followed up for a median of 30 months, patients of the 2G/2G genotype had significantly reduced survival than did those of the 1G/1G or 1G/2G genotype, and a multivariate analysis indicated that the 2G/2G genotype was an independent poor prognostic factor with adjustment for age, disease stage, and adjuvant chemotherapy (Zinzindohoue et al. 2005). Other studies have suggested a relationship of the 2G allele with lymph node metastasis in breast cancer (Przybylowska et al. 2004, 2006), shortened survival in patients with ovarian cancer (Six et al. 2006), and metastasis and poorer prognosis of cervical cancer (Nishioka et al. 2003, Lai et al. 2005).

MMP2

The MMP2 gene contains 13 exons and occupies 27.52 kb pairs on chromosome 16q12.2. The promoter of the gene does not have a TATA box, but contains SP-1, AP-2, p53, and oestrogen receptor-binding sites, which participate in regulating MMP2 gene transcription (Harendza et al. 2003, Yan and Boyd 2007).

There are several common SNPs in the promoter region of the MMP2 gene, including -1575G > A (rs243866), -1306C > T (rs243865), -955C > A (rs2285052), -790T > G (rs243864), -735CT (rs2285053), and -168G > T (rs17859829) (Price et al. 2001, Yu et al. 2004). The coding region also has several common SNPs, with two (rs1132896 and rs1053605) in exon 5, one (rs243849) in exon 7, one (rs2287074) in exon 9, and two (rs10775332 and rs11541998) in exon 12 (data from dbSNP database). All these coding SNPs are synonymous substitutions.

The $-1306C > T$ substitution abolishes an SP-1-binding site (CCACC) in this region. Experiments carried out in an epithelial cell line (293), a macrophage cell line (RAW264.7), and a smooth muscle cell line (A10) showed that the $-1306C >$ T SNP had an allele-specific transcriptional effect, with the C allelic promoter having over twofold higher activity than the T allelic promoter (Price et al. 2001). In vitro experiments also showed that transcription factor SP-1 binds to the region around the polymorphic site in the C allele but does not bind to the T allele (Price et al. 2001), which is a likely explanation for the higher promoter activity of the C allele as compared with the T allele.

The -735C > T substitution also abolishes an SP-1-binding site (from CCCTCC on the C allele to CTCTCC on the T allele) (Yu et al. 2004). In vitro assays showed that transcription factor SP-1 binds to this region in the MMP2 promoter of the C allele but does not bind to the T allele (Yu et al. 2004). It was also shown in experiments using an embryonic kidney cell line (HEK293) that the C allele had over 1.6-fold higher promoter activity than the T allele, and that the -1306C > T and -735C > T SNPs had an additive effect, such that the -1306C-735C haplotype had the highest promoter activity, followed by the -1306C and -735T haplotype and then the -1306T and -735C haplotype, with -1306T and -735T haplotype having the lowest promoter activity (Yu et al. 2004).

The -1575G > A SNP is located in a half-palindromic-binding site (TGACC) for the oestrogen receptor, a nuclear receptor which acts as a transcription factor. In vitro experiments showed that the DNA sequence at this promoter region interacted with a nuclear protein from an oestrogen receptor-positive breast carcinoma cell line (MCF-7) and that this interaction was more readily detectable with the G allele than with the A allele (Harendza et al. 2003). In contrast, there was no such interaction in experiments using nuclear proteins from an oestrogen receptor-negative breast carcinoma cell line (MDA-MB-231). Further assays confirmed that the nuclear protein binding to this region of the MMP2 promoter was oestrogen receptor (Harendza et al. 2003). In the oestrogen receptor-positive breast carcinoma cell line (MCF-7), the G allele had approximately twofold higher promoter activity than the A allele (Harendza et al. 2003). In contrast, in the oestrogen receptor-negative breast carcinoma cell line (MDA-MB-231), there was no difference in promoter activity between the two alleles. However, the twofold difference in promoter activity between the G and A alleles could be brought about when the oestrogen receptor-negative breast carcinoma MDA-MB-231 cells were transfected with an oestrogen receptor-α-expressing plasmid and incubated with estradiol (Harendza et al. 2003).

Examinations of MMP2 mRNA levels in oesophageal tissues showed that the levels were higher in individuals of the -1306C/C genotype than in those of the −1306C/T or −1306T/T genotype, and were higher in individuals of the -735C/C genotype than in those of the -735C/T or -735T/T genotype (Yu et al. 2004). When both SNPs were considered, the tissues from individuals of the -1306C and -735C haplotype had higher MMP2 mRNA levels than did tissues from individuals of the other haplotypes (Yu et al. 2004).

In a study of 781 Chinese patients with lung cancer and 852 age- and sex-matched controls, the MMP2 -1306C > T and -735 C > T SNPs were found to be associated with susceptibility to lung cancer (Yu et al. 2002, Zhou et al. 2005). It was estimated that odds ratio for lung cancer was 2.18-fold (95% CI 1.70–2.79) higher in individuals who were homozygous for the -1306C allele compared with individuals of the -1306C/T or -1306T/T genotype, and was 1.57-fold (95% CI 1.27–1.95) higher for individuals of the -735C/C genotype compared with those of the -735C/T or -735T/T genotype. When both SNPs were taken into consideration, the -1306C and -735C haplotype was associated with a fivefold higher risk for lung cancer as compared with the -1306T and -735T haplotype. The study also showed that MMP2 gene variation and smoking had additive effects. In addition, a study in European French patients with non-small-cell lung cancer suggested that the -735C/C genotype was a risk factor for poor survival (Rollin et al. 2007).

The -1306C > T and -735 C > T SNPs were also found to be associated with susceptibility to oesophageal squamous cell carcinoma in a study of 527 Chinese cases and 777 controls (Yu et al. 2004). Odds ratios for the disease were estimated to be 1.52 (95% CI 1.17-1.96) for -1306C homozygotes compared with those of the -1306C/T or -1306T/T genotype, and 1.30 (95% CI 1.04-1.63) for -735C homozygotes compared with the -735C/T or -735T/T genotype. When both SNPs were taken into account, the -1306C-735C haplotype was associated with 6.53-fold (95% CI 2.78-15.33) higher risk of oesophageal squamous cell carcinoma compared with the -1306T-735T haplotype, and was associated with increased risk for distant metastasis (odds ratio 3.34, 95% CI 1.16-9.63). The same research group have also reported an association of the −1306C > T SNP with gastric cardia adenocarcinoma in a study of 356 Chinese cases and 789 controls, with C allele homozygotes having 3.36-fold (95% CI 2.34-4.97) higher risk for developing the disease, compared with individuals of the C/T or T/T genotype (Miao et al. 2003). Other groups have reported an association of the -1306C/C genotype with increased risk of head and neck squamous cell carcinoma (O-charoenrat and Khantapura 2006), oral squamous cell carcinoma (Lin et al. 2004b), and colorectal cancer (Xu et al. 2004).

A relationship of the -1306C > T polymorphism with breast cancer susceptibility was also detected in a study of 462 Chinese cases and 509 control women, with C allele homozygotes having over twofold higher risk for the disease, compared with individuals of the T/T or C/T genotype (Zhou et al. 2004).

MMP3

The MMP3 gene contains 10 exons and spans 7,815 base pairs on chromosome 11q22.2. MMP3 is also subject to tight transcriptional regulation, involving a number of cis-elements in the promoter of the gene, including a platelet-derived growth factor responsive element (from nucleotide position -1659 relative to the transcriptional start site), an interleukin-1 responsive element (from -1614), two PEA3-binding sites (from -216 and -208, respectively), two AP-1-binding sites (from -189 and -70, respectively), and a TATA box (from −30) (Kirstein et al. 1996, Borghaei et al. 1999, Yan et al. 2007).

The first reported polymorphism in the MMP3 gene is the 5A/6A polymorphism (rs3025058), with one allele having a run of five adenosines (5A) and another allele having six adenosines (6A), from nucleotide position -1612 (relative to the transcriptional start site) in the MMP3 gene promoter (Ye et al. 1995). The allele frequencies significantly differ between populations, with 5A allele frequencies being lowest in Africans (e.g. 0.01 in Cameroon and 0.02 in New Guinea), higher in Asians (e.g. 0.16 in India and 0.17 in China) and higher still in Europeans (e.g. 0.42

in Italy), and increasing from the south to the north in European populations (0.42 in Southern Italian, 0.48 in Czech, 0.50 in Northern Italian, 0.51 in German, 0.52 in British, and 0.54 in Swedish populations) (Rockman et al. 2004). There are data suggesting that the higher frequencies of the 5A allele in Europeans, particularly in Northern Europeans, have arisen from a positive natural selection due to an unknown cause (Rockman et al. 2004).

Other common polymorphisms (with a minor allele frequency > 0.05) have subsequently been identified in the promoter, including -1986T $>$ C (rs645419), -1346A $>$ C (rs632478), -709A $>$ G (rs522616), and -376G $>$ C (rs617819) (Beyzade et al. 2003). In the coding region, there are three common SNPs, two (rs679620 and rs602128) located in exon 2 and one (rs520540) in exon 8. One (rs679620) of the exon 2 SNPs is a non-synonymous substitution resulting in a change of lysine to glutamic acid at amino acid residue 45 (data from dbSNP database).

The 5A/6A polymorphism is located within the interleukin-1 responsive element located at nucleotide position from -1614 to -1595 relative to the transcriptional start site. In vitro experiments showed that the 5A/6A polymorphism had an allele-specific effect on MMP3 expression, with the 5A allelic promoter having greater transcriptional activity than the 6A allelic promoter (Ye et al. 1996). This difference was observed in several cell types including macrophages, smooth muscle cells, and fibroblasts (Ye et al. 1996, Beyzade et al. 2003). In agreement, studies of the levels of MMP3 mRNA and protein in ex vivo tissues from individuals of different genotypes for the 5A/6A polymorphism showed that the levels were highest in 5A homozygotes, intermediate in heterozygotes, and lowest in 6A homozygotes (Medley et al. 2003, Lichtinghagen et al. 2003).

In vitro experiments showed that the sequence surrounding the 5A/6A polymorphic site interacts with two nuclear proteins, one of which binds to the 6A allele more effectively than to the 5A allele (Ye et al. 1996). Studies suggest that one of these nuclear proteins is transcription factor ZBP89 (also named ZNF148), which has a similar affinity with the 5A and 6A alleles and acts as a transcriptional enhancer (Ye et al. 1999). It has been suggested that the other nuclear protein is transcription factor NFκB and that it has different affinity for the 5A and 6A alleles (Borghaei et al. 2004), although this needs to be confirmed.

Most studies of MMP3 gene variation in relation to cancer have focused on the 5A/6A polymorphism. A study of this polymorphism in Austrian patients with breast cancer ($n = 500$) and controls ($n = 500$) showed that the genotype frequencies in patients did not significantly differ from those in controls; however, lymph node metastasis was more prevalent in patients of the 5A/5A genotype than in patients of the 5A/6A or 6A/6A genotype (Krippl et al. 2004). The odds ratio for lymph node metastasis was estimated to be 1.78 (95% CI 1.14–2.76) for patients with the 5A/5A genotype compared with those of the 5A/6A or 6A/6A genotype (Krippl et al. 2004). The association between the 5A/5A genotype and increased susceptibility to lymph node metastasis was also observed in a study of Italian patients with breast cancer ($n = 86$) (Ghilardi et al. 2002). In addition, a relationship of the 5A/5A and 5A/6A genotypes with lymphatic metastasis was also observed in Chinese patients with oesophageal squamous cell carcinoma ($n = 234$) (Zhang et al.

2004) and Chinese patients with non-small-cell lung carcinoma (Fang et al. 2005). Moreover, a study of French patients with head and neck squamous cell carcinoma ($n = 148$) showed that patients of the 5A/5A genotype had a poor response to chemotherapy (Blons et al. 2004). Compared with patients of the 5A/5A genotype, those of the 5A/6A genotype and those of the 6A/6A were, respectively, 1.7- and 6.7-fold more likely to respond to chemotherapy (Blons et al. 2004).

MMP9

MMP9 contains 13 exons and spans 7,653 base pairs of DNA on chromosome 20q13.12. Transcriptional regulation of the MMP9 gene involves an NFκB-binding site (from nucleotide position -600 relative to the transcriptional start site), an SP-1-binding site (from -558), a PEA3-binding site (from -540), two AP-1-binding sites (from -533 and -79, respectively), and a TATA box (from -29) (Yan et al. 2007).

Several common polymorphisms (with minor allele frequency > 0.05) have been identified within the proximal 2 kb promoter region of the MMP9 gene, including the -1702T > A (rs3918241) and -1562C > T (rs3918242) SNPs, and the -90(CA)n microsatellite polymorphism (rs5841617). In the coding region, there are several common non-synonymous SNPs, that is, Arg279Gln (rs17576) in exon 6, Pro574Arg (rs2250889) in exon 10, and Arg668Gln (rs2274756) in exon 12, as well as two synonymous SNPs, that is, Gly607Gly (rs13969) in exon 11 and Val694Val (rs13925) in exon 13 (data from dbSNP database).

In vitro studies indicated a functional effect of the -1562C > T SNP, with the T allele having a higher promoter activity than the C allele (Zhang et al. 1999). In agreement, a study using ex vivo aortic tissues showed that MMP9 mRNA levels, MMP9 protein levels, and MMP9 activity were higher in -1562T allele carriers than in non-carriers (Medley et al. 2004). In addition, studies of plasma MMP9 levels showed that the levels were higher in -1562T allele carriers than in non-carriers (Zhang et al. 1999, Blankenberg et al. 2003).

There is evidence suggesting that the (CA)n polymorphism also has an effect on MMP9 transcription (Shimajiri et al. 1999, Peters et al. 1999). In vitro assays showed that the alleles containing 21, 22, or 23 CA repeats have higher promoter activity than those containing 14 or 18 CA repeats (Shimajiri et al. 1999, Peters et al. 1999), and that a nuclear protein binds more effectively to the alleles that have higher promoter activity than to the alleles with low promoter activity (Shimajiri et al. 1999).

A study of Japanese patients with gastric cancer ($n = 177$) suggested an association of the MMP9 -1562C > T SNP with susceptibility to lymphatic invasion, with an odds ratio of 2.27 (95% CI 1.09–4.74) for -1562T carriers, compared with non-carriers (Matsumura et al. 2005). An association of carriage of the -1562T allele with susceptibility and metastasis of prostate cancer was also observed in a study of 101 Tunisian cases and 106 controls (Sfar et al. 2007).

The Arg279Gln and Pro574Arg polymorphisms were found to be associated with susceptibility to lung cancer in a case-control study in Chinese involving 744 cases and 747 controls. The study showed that the odds ratio for lung cancer was 2.16 (95% CI 1.30–3.59) for individuals carrying one or two copies of the 279Arg and/or 574Pro alleles, and 2.44 (95% CI 1.48–4.03) for those carrying more than two copies of these alleles, compared with individuals who carried neither of these alleles (Hu et al. 2005).

MMP12

The MMP12 gene contains 10 exons distributing over 12.25 kb pairs on chromosome 11q22.2. Cis-elements identified in the MMP12 gene promoter include two T-cell factor-4/-catenin-binding sites (from nucleotide position -1658 and -1538, respectively), a leader-binding protein-binding site (from -1576), a PEA3-binding site (from –335), an octamer-binding protein-binding site (from -166), an AP-1-binding site (from -74), and a TATA box (from -29) (Yan et al. 2007).

There are several common polymorphisms in the promoter region of the MMP12 gene, including -82A > G (rs2276109), -1079T ins/del (28360355), and -1839C > G (rs1277718), and there are also two non-synonymous SNPs located in exons 6 and 8, respectively, that is, Gln279Arg (rs17368582) and Asn357Ser (rs652438) (Jormsjo et al. 2000, Joos et al. 2002) (dbSNP database).

A study of European American patients with bladder cancer ($n = 560$) and controls ($n = 560$) indicated an association of the -82G/G genotype with over fourfold higher risk of invasive bladder cancer (Kader et al. 2006). In addition, a study of American patients with non-small-cell lung cancer ($n = 382$) suggested a relationship between the Asn357Ser SNP and prognosis, patients carrying the Ser allele having poorer recurrence-free survival (odds ratio 1.74, 95% CI 1.18–2.58) and overall survival (odds ratio 1.53, 95% CI 1.05–2.23), which remained significant after adjusting for age, sex, smoking, and histological cancer subtype (Heist et al. 2006).

Summary

Studies have indicated associations of the MMP1, MMP2, MMP3, MMP9, and MMP12 gene variations with susceptibility to and/or progression of several types of cancer. The associations could potentially be explained by increased MMP expression in individuals of certain genotype, in particular increased MMP1 in individuals carrying the MMP1 -1607 2G allele, higher MMP2 expression in individuals of the MMP2 -1306C/C and -735C/C genotypes, elevated MMP3 in carriers of the MMP3 -1612 5A allele, and increased MMP9 in MMP9 -1562T carriers.

However, some of the reported associations were not observed in some other studies (Wenham et al. 2003, Shin et al. 2005, Kader et al. 2006). The discrepancies between different studies could be due to small samples size which would increase the chances of false-negative and false-positive findings, or due to differences in genetic background and/or environment factors which can also influence cancer susceptibility and progression. In view of the multifactorial nature of cancers, ascertainment of a moderate effect from a genetic variant would require large samples, and the required sample sizes depend on the effect sizes and the allele frequency of the polymorphism (Zondervan and Cardon 2004). Further studies of the MMP gene variations in larger samples would be warranted to assess their potential influences on cancer susceptibility and/or prognosis.

References

Benbow U., Rutter J. L., Lowrey C. H., et al. (1999). Transcriptional repression of the human collagenase-1 (MMP-1) gene in MDA231 breast cancer cells by all-trans-retinoic acid requires distal regions of the promoter. Br. J. Cancer 79: 221–228.

Benbow U., Tower G. B., Wyat C. A., et al. (2002). High levels of MMP-1 expression in the absence of the 2G single nucleotide polymorphism is mediated by p38 and ERK1/2 mitogen-activated protein kinases in VMM5 melanoma cells. J. Cell Biochem. 86: 307–319.

Beyzade S., Zhang S., Wong Y., et al. (2003). Influences of matrix metalloproteinase-3 gene variation on extent of coronary atherosclerosis and risk of myocardial infarction. J. Am. Coll. Cardiol. 41: 2130–2137.

Blankenberg S., Rupprecht H. J., Poirier O., et al. (2003). Plasma concentrations and genetic variation of matrix metalloproteinase 9 and prognosis of patients with cardiovascular disease. Circulation 107: 1579–1585.

Blons H., Gad S., and Zinzindohoue F. (2004). Matrix metalloproteinase 3 polymorphism: A predictive factor of response to neoadjuvant chemotherapy in head and neck squamous cell carcinoma. Clin. Cancer Res. 10: 2594–2599.

Borghaei R. C., Sullivan C., and Mochan E. (1999). Identification of a cytokine-induced repressor of interleukin-1 stimulated expression of stromelysin 1 (MMP-3). J. Biol. Chem. 274: 2126–2131.

Borghaei R. C., Rawlings P. L., Jr., Javadi M., et al. (2004). NF-kappaB binds to a polymorphic repressor element in the MMP-3 promoter. Biochem. Biophys. Res. Commun. 316: 182–188.

Cao Z. G., and Li C. Z. (2006). A single nucleotide polymorphism in the matrix metalloproteinase-1 promoter enhances oral squamous cell carcinoma susceptibility in a Chinese population. Oral Oncol. 42: 32–38.

Elander N., Soderkvist P., and Fransen K. (2006). Matrix metalloproteinase (MMP) -1, -2, -3 and -9 promoter polymorphisms in colorectal cancer. Anticancer Res. 26: 791–795.

Fang S., Jin X., Wang R., et al. (2005). Polymorphisms in the MMP1 and MMP3 promoter and non-small cell lung carcinoma in North China. Carcinogenesis 26: 481–486.

Fujimoto T., Parry S., Urbanek M., et al. (2002). A single nucleotide polymorphism in the matrix metalloproteinase-1 (MMP-1) promoter influences amnion cell MMP-1 expression and risk for preterm premature rupture of the fetal membranes. J. Biol. Chem. 277: 6296–6302.

Ghilardi G., Biondi M. L., Mangoni J., et al. (2001). Matrix metalloproteinase-1 promoter polymorphism 1G/2G is correlated with colorectal cancer invasiveness. Clin. Cancer Res. 7: 2344–2346.

Ghilardi G., Biondi M. L., Caputo M., et al. (2002). A single nucleotide polymorphism in the matrix metalloproteinase-3 promoter enhances breast cancer susceptibility. Clin. Cancer Res. 8: 3820–3823.

Harendza S., Lovett D. H., Panzer U., et al. (2003). Linked common polymorphisms in the gelatinase a promoter are associated with diminished transcriptional response to estrogen and genetic fitness. J. Biol. Chem. 278: 20490–20499.

Hashimoto T., Uchida K., Okayama N., et al. (2004). Association of matrix metalloproteinase (MMP)-1 promoter polymorphism with head and neck squamous cell carcinoma. Cancer Lett. 211: 19–24.

Heist R. S., Marshall A. L., Liu G., et al. (2006). Matrix metalloproteinase polymorphisms and survival in stage I non-small cell lung cancer. Clin. Cancer Res. 12: 5448–5453.

Hinoda Y., Okayama N., Takano N., et al. (2002). Association of functional polymorphisms of matrix metalloproteinase (MMP)-1 and MMP-3 genes with colorectal cancer. Int. J. Cancer 102: 526–529.

Hirata H., Naito K., Yoshihiro S., et al. (2003). A single nucleotide polymorphism in the matrix metalloproteinase-1 promoter is associated with conventional renal cell carcinoma. Int. J. Cancer 106: 372–374.

Hirata H., Okayama N., Naito K., et al. (2004). Association of a haplotype of matrix metalloproteinase (MMP)-1 and MMP-3 polymorphisms with renal cell carcinoma. Carcinogenesis 25: 2379–2384.

Hu Z., Huo X., Lu D., et al. (2005). Functional polymorphisms of matrix metalloproteinase-9 are associated with risk of occurrence and metastasis of lung cancer. Clin. Cancer Res. 11: 5433–5439.

Joos L., He J. Q., Shepherdson M. B., et al. (2002). The role of matrix metalloproteinase polymorphisms in the rate of decline in lung function. Hum. Mol. Genet. 11: 569–576.

Jormsjo S., Ye S., Moritz J., et al. (2000). Allele-specific regulation of matrix metalloproteinase-12 gene activity is associated with coronary artery luminal dimensions in diabetic patients with manifest coronary artery disease. Circ. Res. 86: 998–1003.

Kader A. K., Shao L., Dinney C. P., et al. (2006). Matrix metalloproteinase polymorphisms and bladder cancer risk. Cancer Res. 66: 11644–11648.

Kanamori Y., Matsushima M., Minaguchi T., et al. (1999). Correlation between expression of the matrix metalloproteinase-1 gene in ovarian cancers and an insertion/deletion polymorphism in its promoter region. Cancer Res. 59: 4225–4227.

Kirstein M., Sanz L., Quinones S., et al. (1996). Cross-talk between different enhancer elements during mitogenic induction of the human stromelysin-1 gene. J. Biol. Chem. 271: 18231–18236.

Krippl P., Langsenlehner U., Renner W., et al. (2004). The 5A/6A polymorphism of the matrix metalloproteinase 3 gene promoter and breast cancer. Clin. Cancer Res. 10: 3518–3520.

Lai H. C., Chu C. M., Lin Y. W., et al. (2005). Matrix metalloproteinase 1 gene polymorphism as a prognostic predictor of invasive cervical cancer. Gynecol. Oncol. 96: 314–319.

Lichtinghagen R., Bahr M. J., Wehmeier M., et al. (2003). Expression and coordinated regulation of matrix metalloproteinases in chronic hepatitis C and hepatitis C virus-induced liver cirrhosis. Clin. Sci. (Lond) 105: 373–382.

Lin S. C., Chung M. Y., Huang J. W., et al. (2004a). Correlation between functional genotypes in the matrix metalloproteinases-1 promoter and risk of oral squamous cell carcinomas. J. Oral Pathol. Med. 33: 323–326.

Lin S. C., Lo S. S., Liu C. J., et al. (2004b). Functional genotype in matrix metalloproteinases-2 promoter is a risk factor for oral carcinogenesis. J. Oral Pathol. Med. 33: 405–409.

Matsumura S., Oue N., Nakayama H., et al. (2005). A single nucleotide polymorphism in the MMP-9 promoter affects tumor progression and invasive phenotype of gastric cancer. J. Cancer Res. Clin. Oncol. 131: 19–25.

Medley T. L., Kingwell B. A., Gatzka C. D., et al. (2003). Matrix metalloproteinase-3 genotype contributes to age-related aortic stiffening through modulation of gene and protein expression. Circ. Res. 92: 1254–1261.

Medley T. L., Cole T. J., Dart A. M., et al. (2004). Matrix metalloproteinase-9 genotype influences large artery stiffness through effects on aortic gene and protein expression. Arterioscler. Thromb. Vasc. Biol. 24: 1479–1484.

Miao X., Yu C., Tan W., et al. (2003). A functional polymorphism in the matrix metalloproteinase-2 gene promoter (–1306C/T) is associated with risk of development but not metastasis of gastric cardia adenocarcinoma. Cancer Res. 63: 3987–3990.

Nelson K. K., Ranganathan A. C., Mansouri J., et al. (2003). Elevated sod2 activity augments matrix metalloproteinase expression: Evidence for the involvement of endogenous hydrogen peroxide in regulating metastasis. Clin. Cancer Res. 9: 424–432.

Nelson K. K., Subbaram S., Connor K. M., et al. (2006). Redox-dependent matrix metalloproteinase-1 expression is regulated by JNK through Ets and AP-1 promoter motifs. J. Biol. Chem. 281: 14100–14110.

Nishioka Y., Kobayashi K., Sagae S., et al. (2000). A single nucleotide polymorphism in the matrix metalloproteinase-1 promoter in endometrial carcinomas. Jpn. J. Cancer Res. 91: 612–615.

Nishioka Y., Sagae S., Nishikawa A., et al. (2003). A relationship between Matrix metalloproteinase-1 (MMP-1) promoter polymorphism and cervical cancer progression. Cancer Lett. 200: 49–55.

Noll W. W., Belloni D. R., Rutter J. L., et al. (2001). Loss of heterozygosity on chromosome 11q22–23 in melanoma is associated with retention of the insertion polymorphism in the matrix metalloproteinase-1 promoter. Am. J. Pathol. 158: 691–697.

O-charoenrat P., and Khantapura P. (2006). The role of genetic polymorphisms in the promoters of the matrix metalloproteinase-2 and tissue inhibitor of metalloproteinase-2 genes in head and neck cancer. Oral Oncol. 42: 257–267.

O-charoenrat P., Leksrisakul P., and Sangruchi S. (2006). A functional polymorphism in the matrix metalloproteinase-1 gene promoter is associated with susceptibility and aggressiveness of head and neck cancer. Int. J. Cancer 118: 2548–2553.

Pearce E., Tregouet D. A., Samnegard A., et al. (2005). Haplotype effect of the matrix metalloproteinase-1 gene on risk of myocardial infarction. Circ. Res. 97: 1070–1076.

Pearce E. G., Laxton R. C., Pereira A. C., et al. (2007). Haplotype effects on matrix metalloproteinase-1 gene promoter activity in cancer cells. Mol. Cancer Res. 5: 221–227.

Peters D. G., Kassam A., St Jean P. L., et al. (1999). Functional polymorphism in the matrix metalloproteinase-9 promoter as a potential risk factor for intracranial aneurysm. Stroke 30: 2612–2616.

Price S. J., Greaves D. R., and Watkins H. (2001). Identification of novel, functional genetic variants in the human matrix metalloproteinase-2 gene: Role of Sp1 in allele-specific transcriptional regulation. J. Biol. Chem. 276: 7549–7558.

Przybylowska K., Zielinska J., Zadrozny M., et al. (2004). An association between the matrix metalloproteinase 1 promoter gene polymorphism and lymphnode metastasis in breast cancer. J. Exp. Clin. Cancer Res. 23: 121–125.

Przybylowska K., Kluczna A., Zadrozny M., et al. (2006). Polymorphisms of the promoter regions of matrix metalloproteinases genes MMP-1 and MMP-9 in breast cancer. Breast Cancer Res. Treat. 95: 65–72.

Ranganathan A. C., Nelson K. K., Rodriguez A. M., et al. (2001). Manganese superoxide dismutase signals matrix metalloproteinase expression via H2O2-dependent ERK1/2 activation. J. Biol. Chem. 276: 14264–14270.

Rockman M. V., Hahn M. W., Soranzo N., et al. (2004). Positive selection on MMP3 regulation has shaped heart disease risk. Curr. Biol. 14: 1531–1539.

Rollin J., Regina S., Vourc'h P., et al. (2007). Influence of MMP-2 and MMP-9 promoter polymorphisms on gene expression and clinical outcome of non-small cell lung cancer. Lung Cancer 56: 273–280.

Rutter J. L., Mitchell T. I., Buttice G., et al. (1998). A single nucleotide polymorphism in the matrix metalloproteinase-1 promoter creates an Ets binding site and augments transcription. Cancer Res. 58: 5321–5325.

Sfar S., Saad H., Mosbah F., et al. (2007). TSP1 and MMP9 genetic variants in sporadic prostate cancer. Cancer Genet. Cytogenet. 172: 38–44.

Shimajiri S., Arima N., Tanimoto A., et al. (1999). Shortened microsatellite d(CA)21 sequence down-regulates promoter activity of matrix metalloproteinase 9 gene. FEBS Lett. 455: 70–74.

Shin A., Cai Q., Shu X. O., et al. (2005). Genetic polymorphisms in the matrix metalloproteinase 12 gene (MMP12) and breast cancer risk and survival: The Shanghai Breast Cancer Study. Breast Cancer Res. 7: R506–R512.

Six L., Grimm C., Leodolter S., et al. (2006). A polymorphism in the matrix metalloproteinase-1 gene promoter is associated with the prognosis of patients with ovarian cancer. Gynecol. Oncol. 100: 506–510.

Su L., Zhou W., Park S., et al. (2005). Matrix metalloproteinase-1 promoter polymorphism and lung cancer risk. Cancer Epidemiol. Biomarkers Prev. 14: 567–570.

Tower G. B., Coon C. C., Benbow U., et al. (2002). Erk 1/2 differentially regulates the expression from the 1G/2G single nucleotide polymorphism in the MMP-1 promoter in melanoma cells. Biochim. Biophys. Acta 1586: 265–274.

Tower G. B., Coon C. I., Belguise K., et al. (2003a). Fra-1 targets the AP-1 site/2G single nucleotide polymorphism (ETS site) in the MMP-1 promoter. Eur. J. Biochem. 270: 4216–4225.

Tower G. B., Coon C. I., and Brinckerhoff C. E. (2003b). The 2G single nucleotide polymorphism (SNP) in the MMP-1 promoter contributes to high levels of MMP-1 transcription in MCF-7/ADR breast cancer cells. Breast Cancer Res. Treat. 82: 75–82.

Wenham R. M., Calingaert B., Ali S., et al. (2003). Matrix metalloproteinase-1 gene promoter polymorphism and risk of ovarian cancer. J. Soc. Gynecol. Investig. 10: 381–387.

Xu E., Lai M., Lv B., et al. (2004). A single nucleotide polymorphism in the matrix metalloproteinase-2 promoter is associated with colorectal cancer. Biochem. Biophys. Res. Commun. 324: 999–1003.

Yan C., and Boyd D. D. (2007). Regulation of matrix metalloproteinase gene expression. J. Cell Physiol. 211: 19–26

Ye S., Watts G. F., Mandalia S., et al. (1995). Genetic variation in the human stromelysin promoter is associated with progression of coronary atherosclerosis. Br. Heart J. 73: 209–215.

Ye S., Eriksson P., Hamsten A., et al. (1996). Progression of coronary atherosclerosis is associated with a common genetic variant of the human stromelysin-1 promoter which results in reduced gene expression. J. Biol. Chem. 271: 13055–13060.

Ye S., Whatling C., Watkins H., et al. (1999). Human stromelysin gene promoter activity is modulated by transcription factor ZBP-89. FEBS Lett. 450: 268–272.

Ye S., Dhillon S., Turner S. J., et al. (2001). Invasiveness of cutaneous malignant melanoma is influenced by matrix metalloproteinase 1 gene polymorphism. Cancer Res. 61: 1296–1298.

Yu C., Pan K., Xing D., et al. (2002). Correlation between a single nucleotide polymorphism in the matrix metalloproteinase-2 promoter and risk of lung cancer. Cancer Res. 62: 6430–6433.

Yu C., Zhou Y., Miao X., et al. (2004). Functional haplotypes in the promoter of matrix metalloproteinase-2 predict risk of the occurrence and metastasis of esophageal cancer. Cancer Res. 64: 7622–7628.

Zhang B., Ye S., Herrmann S. M., et al. (1999). Functional polymorphism in the regulatory region of gelatinase B gene in relation to severity of coronary atherosclerosis. Circulation 99: 1788–1794.

Zhang J., Jin X., Fang S., et al. (2004). The functional SNP in the matrix metalloproteinase-3 promoter modifies susceptibility and lymphatic metastasis in esophageal squamous cell carcinoma but not in gastric cardiac adenocarcinoma. Carcinogenesis 25: 2519–2524.

Zhang W. Q., Lin H., Zhou Y. A., et al. (2006). [Association of MMP1 -1607(1G – 2G) single nucleotide polymorphism with susceptibility to lung cancer in Northwestern Chinese population of Han nationality]. Zhonghua Yi. Xue. Yi. Chuan Xue. Za Zhi. 23: 313–315.

Zhou Y., Yu C., Miao X., et al. (2004). Substantial reduction in risk of breast cancer associated with genetic polymorphisms in the promoters of the matrix metalloproteinase-2 and tissue inhibitor of metalloproteinase-2 genes. Carcinogenesis 25: 399–404.

Zhou Y., Yu C., Miao X., et al. (2005). Functional haplotypes in the promoter of matrix metalloproteinase-2 and lung cancer susceptibility. Carcinogenesis 26: 1117–1121.

Zhu Y., Spitz M. R., Lei L., et al. (2001). A single nucleotide polymorphism in the matrix metalloproteinase-1 promoter enhances lung cancer susceptibility. Cancer Res. 61: 7825–7829.

Zinzindohoue F., Lecomte T., Ferraz J. M., et al. (2005). Prognostic significance of MMP-1 and MMP-3 functional promoter polymorphisms in colorectal cancer. Clin. Cancer Res. 11: 594–599.

Zondervan K. T., and Cardon L. R. (2004). The complex interplay among factors that influence allelic association. Nat. Rev. Genet. 5: 89–100.

Chapter 33
TIMP-1 as a Prognostic Marker in Colorectal Cancer

Camilla Frederiksen, Anne Fog Lomholt, Hans Jørgen Nielsen, and Nils Brünner

Abstract Colorectal cancer (CRC) accounts for about 1 million new cases per year worldwide and is the third most prevalent cancer in Europe, and the second most important in relation to cancer-specific death. Outcome in patients with CRC relates to the stage of disease at diagnosis. The staging is based on histopathological characteristics and has limitations especially concerning the prediction of prognosis for patients presenting with the intermediate tumour stages (II and III). A number of publications have suggested tissue inhibitor of metalloproteinases 1 (TIMP-1) as a new and valid prognostic marker in CRC. This chapter reviews the literature on TIMP-1 as a prognostic marker in CRC.

Introduction

Colorectal cancer (CRC) accounts for about 1 million new cases per year worldwide (Parkin et al. 2005) and is the third most prevalent cancer in Europe, and the second most important in relation to cancer-specific death (Ferlay et al. 2007). CRC is most prevalent in developed countries with an individual lifetime risk of 5% and is, therefore, considered related to the lifestyle of the individual (Levin and Dozois 1991, Weitz et al. 2005).

As in most cancers, clinical outcome in patients with CRC relates to the stage of disease at diagnosis. Staging in CRC is accomplished using the TNM system (TNM Classification of Malignant Tumors NCI 2002). Presently, the TNM staging, based on the histopathological characteristics of the tumour, is the strongest predictor of patient outcome, and thereby an important tool in patient management following

N. Brünner
Section of Biomedicine, Department of Veterinary Pathobiology, Faculty of Life Sciences, University of Copenhagen, DK-1870. Frederiksberg C, Denmark, e-mail: nbr@life.ku.dk

D. Edwards et al. (eds.), *The Cancer Degradome.*
© Springer Science + Business Media, LLC 2008

Table 33.1 TNM-Staging and Estimated 5-Year Survival for Patients with CRC

TNM	Tumour	Regional lymph node	Distant metastasis	5-year survival
Stage 0	Tis	N0	M0	>90%
Stage I	T1	N0	M0	>90%
	T2	N0	M0	80–85%
Stage IIA	T3	N0	M0	70–75%
Stage IIB	T4	N0	M0	70–75%
Stage IIIA	T1, T2	N1	M0	60%
Stage IIIB	T3, T4	N1	M0	40%
Stage IIIC	Any T	N2	M0	25%
Stage IV	Any T	Any N	M1	<5%

T = primary tumour, N = regional lymph nodes, M = distant metastasis.
CRC colorectal cancer.

primary surgical tumour resection. However, as it appears from Table 33.1, which shows the different tumour stages according to the TNM system, including the expected 5-year survival for each stage (Weitz et al. 2005, Compton and Greene 2004), the TNM staging system does not offer a precise prediction of prognosis for the individual patient. At the time of diagnosis, 70–80% of the patients do not show signs of distant metastases (Guillou et al. 2005) and are consequently offered surgical resection with curative intent. Patients presenting with low rectal cancers are offered preoperative radio- and/or chemotherapy (Rodel and Sauer 2005), aiming to improve resectability of the tumour. Postoperatively, all patients with CRC diagnosed with stage III disease are offered adjuvant systemic chemotherapy, while patients with stage I or stage II CRC most often do not receive adjuvant treatment (Gill et al. 2004). Overall, survival of patients in stages I–III has been reported to range from 35% to 90% in different studies (Obrand and Gordon 1997, Eisenberg et al. 1982, George et al. 2006, Staib et al. 2002).

The remaining 20–30% of patients who are presenting with disseminated disease (stage IV) at diagnosis are offered various types of treatment, for example, surgery of the primary tumour and—when possible—of distant metastases, and chemotherapy and/or irradiation, and different types of biological treatment such as antibodies against the epidermal growth factor (EGF) receptor or vascular epidermal growth factor (VEGF). Despite this variety of treatment modalities, the 5-year survival in stage IV patients is still disappointing (Table 33.1) (NCI 2007; www.seer.cancer.gov).

Approximately 35% of patients with stage III CRC will be cured by the primary surgery, but they are still offered adjuvant treatment; thus, these patients will be over-treated. At the same time, in patients with stage II CRC the 5-year survival rate is ~75%. Since stage II patients rarely receive adjuvant systemic therapy, ~25% of these patients must be considered as being under-treated.

To overcome this dilemma, improved and differentiated staging of patients with CRC is needed. One approach is to identify biomarkers (genes and proteins) in tumour tissue and different body fluids (blood, sputum, urine) that may improve staging of the patients. Several tumour markers have been proposed as potential candidates for prognostic evaluation in CRC, but the number of markers that actually have proven clinically useful is small (Hayes et al. 1996, Bast et al. 2001, Schilsky and Taube 2002).

With the many reports on new tumour markers, the need for international guidelines for the establishment and validation of new tumour markers is evident. The American Society of Clinical Oncology (ASCO) (Locker et al. 2006) and the European Group of Tumour Makers (EGTM) (Duffy et al. 2003) regularly issue evidence-based clinical practice guidelines for the use of tumour markers in cancer by analyzing systematic review and meta-analyses of published tumour marker studies. Hayes and co-workers (Hayes et al. 1996) introduced a Tumour Marker Utility Grading System (TMUGS) for the clinical utility of tumour markers. Through the TMUGS, a potential tumour marker can be evaluated and assigned a level of evidence (LOE) utility score ranging from V to I. Level V evidence is considered weak and is derived from case reports and small pilot studies. Levels IV and III are retrospective studies of smaller (IV) or larger size (III). Level II can be obtained from prospective studies not specifically designed to test marker utility. Finally, level I evidence can be obtained from a single, high powered, prospective, randomized trial or a meta-analysis/overview of multiple well-designed studies.

Several soluble markers have been proposed as potential candidates for prognostic evaluation in patients with CRC. Carcino embryonic antigen (CEA), which is used in the postoperative monitoring of patients with CRC, has been proposed as a potential prognostic marker but has not yet met the criteria (LOE I) needed for clinical implementation (Goldstein and Mitchell 2005, Wang et al. 2007, Locker et al. 2006). Since the specificity and sensitivity of CEA is relative, alternative markers are still vigorously searched for, aiming at improving the prognostic differentiation of stage II patients in particular and thereby increasing CRC survival.

Over the last 7 years, a number of publications have suggested tissue inhibitor of metalloproteinases 1 (TIMP-1) as a new and valid prognostic marker in CRC. Of particular interest is that TIMP-1 in many of these studies has demonstrated independency from the TNM classification, which means that TIMP-1 measurements may be used to further estimate prognosis in each of the patients with stages I–III CRC.

In the following, a review on the literature on TIMP-1 as a prognostic marker in CRC will be given. In addition, the LOE for each of the studies has been evaluated.

Based on a systematic search in the Medline database (http://www.ncbi.nlm.nih.gov/entrez) from January 1, 1990, through May 2007, the terms: "Tissue Inhibitor of Metalloproteinases 1 or TIMP-1", "prognosis" and "patient outcome" and "colorectal or colon or rectal" and "cancer or malignancy or neoplasm" were combined with the operator "AND", and yielded a total of 22 original articles. Eleven of these met the criteria of evaluating or proposing TIMP-1 as a prognostic marker in patients with CRC. The articles are listed in Table 33.2.

Table 33.2 The Eleven Articles Reviewed Evaluating or Proposing TIMP-1 as a Prognostic Marker in Patients with CRC

Author (Year)	Study type	Patients	Specimen	Follow-up (range)	Analysis (Assay Kit/Brand)	TIMP-1 correlated with tumour stage	TIMP-1 as prognostic marker	Level of evidence (LOE)
Zeng et al. (1995)	Descriptive	56	Tissue	–	Northern blotting and IHC	No	–	V
Oberg et al. (2000)	Case-control	158	Serum	46 months (17–68)	ELISA (in-house)	No	No	IV
Holten-Andersen et al. (2000)	Retrospective	588	Plasma	82 months (68–95)	ELISA (in-house)	No[a]	Yes ($p < 0.0001$)	III
Pellegrini et al. (2000)	Descriptive	41	Serum	–	ELISA (Amersham)	Yes ($p = 0.02$)	–	V
Simpson et al. (2000)	Prospective case-control	41	Plasma	–	ELISA (Amersham)	No	–	IV
Yukawa et al. (2001)	Descriptive	44	Plasma	–	ELISA (FUJI)	Yes (see text)	–	V
Ishida et al. (2003)	Prospective	123	Serum	Curative 48 months (36–67) non curative 16 months (2–36)	ELISA (FUJI)	Yes ($p< 0.03$)	No ($p = 0.62$)	III

Holten-Andersen et al. (2004)	Retrospective	352	Plasma	43 months (20–95)	ELISA (in-house)	No [b]	Yes ($p = 0.01$)	III
Yukawa et al. (2004)	Prospective	87	Plasma	–	ELISA (FUJI)	No	Yes[a] ($p = 0.03$)	III
Waas et al. (2005)	Prospective case-control	94 + 51	Plasma	27.5 months (0–56)	ELISA (Amersham)	Yes ($p < 0.0001$)	Yes ($p = 0.0003$)	III
Roca et al. (2006)	Descriptive	84	Tissue	60 months (survivors)	IHC	–	No	V
Holten-Andersen et al. (2006)	Retrospective	280	Plasma	94 months (82–109)	ELISA (in-house)	No	Yes ($p = 0.002$)	III

List of publications on TIMP-1 and patient outcome in colorectal cancer. (*ELISA* enzyme-linked immunosorbent assay, *IHC* immunohistochemistry, *ISH* in situ hybridization, *p* p value, *TIMP*-1 tissue inhibitor of metalloproteinases 1, – = no data.

[a]Test for significant differences between all stages.

[b]These studies (Holten-Andersen et al. 2004, 2000) demonstrate that patients with Duke's stages A, B and C disease have similar plasma TIMP-1 values while patients with Duke's stage D disease have significantly higher plasma TIMP-1 values.

Function of TIMP-1 in Cancer

TIMP-1 in General

The TIMP family consists presently of four different members (TIMP-1–4), which share many structural features. The TIMP-1 gene is located on chromosome Xp11.1–p11.4 and it codes for a soluble 28.5 kDa glycoprotein that forms non-covalent 1:1 stoichiometric complexes with matrix metalloproteinases (MMPs), thereby inhibiting the proteolytic activity of these molecules (Brew et al. 2000, Egeblad and Werb 2002).

Accumulating evidence indicates that TIMP-1 can act on other molecular targets in addition to MMPs, and thereby promotes tumour progression. TIMP-1 has been shown to inhibit apoptosis, to stimulate cell growth and to regulate angiogenesis, as will be described in the following sections. The promoting role of TIMP-1 in cancer growth is supported by the many studies that demonstrate an association between increased levels of TIMP-1 in cancer tissue or in blood from patients with cancer and a poor clinical outcome (Holten-Andersen et al. 2000, Yukawa et al. 2001, Schrohl et al. 2004).

TIMP-1 and Regulation of Cell Growth

TIMP-1 was originally discovered as an eryhtroid-potentiating activity protein, a humeral protein that enhances the proliferation of human erythroid progenitors and certain cancer cells (Niskanen et al. 1988, Docherty et al. 1985, Murate et al. 1993, Golde et al. 1980). Since then the role of TIMP-1 in cell survival and proliferation has been extensively studied and stimulation of growth has been observed in many different cell types including osteosarcoma cells, lymphoma cells, keratinocytes, fibroblasts and breast cancer cells (Hayakawa et al. 1992, Bertaux et al. 1991, Chesler et al. 1995, Avalos et al. 1988). However, an anti-proliferative activity (through cell cycle arrest in G_1 phase) of TIMP-1 has been demonstrated in non-malignant human mammary epithelial cells (Taube et al. 2006, Fata et al. 1999), suggesting a cell type-related specific effect of TIMP-1 on cell proliferation.

A few studies have demonstrated that the ability of TIMP-1 to stimulate cell growth is dependent on its ability to inhibit MMPs (Fata et al. 1999, Porter et al. 2004). However, some studies have reported TIMP-1 to stimulate cell proliferation through a cell surface-binding protein independently of its MMP inhibitory effect (Docherty et al. 1985, Golde et al. 1980, Chesler et al. 1995, Avalos et al. 1988). Indication that TIMP-1 may be signalling through a cell surface-binding protein was first reported by Avalos and colleagues (Avalos et al. 1988, Bertaux et al. 1991). Luparello and colleagues (Luparello et al. 1999) subsequently showed that added human TIMP-1 could elicit a significant proliferative response of human breast carcinoma cells in a dose-dependent manner. Findings by Ritter and colleagues (Ritter et al. 1999) demonstrated that TIMP-1 can bind to human breast carcinoma cells and translocate to the nucleus. Several other studies have subsequently investigated how TIMP-1 by way of the cell membrane initiates an intracellular signalling cascade, resulting in cell proliferation. Binding of TIMP-1 to the cell surface has been shown to stimulate cell growth through a signal transduction cascade mediated by Ras (Wang et al. 2002). It remains to be elucidated how TIMP-1 interacts with its cell surface-binding protein and how TIMP-1 initiates growth signal transduction pathways.

TIMP-1 and Regulation of Apoptosis

Accumulating evidence has established that TIMP-1 is able to inhibit apoptosis induced by various apoptotic stimuli. TIMP-1 may regulate cell survival by at least two pathways. One pathway correlates with its ability to inhibit MMPs (MMP-dependent) and the other by a pathway independent of MMP inhibition (MMP-independent).

Murphy and colleagues (Murphy et al. 2002) demonstrated that TIMP-1 had a direct, consistent and significant anti-apoptotic effect on human and rat stellate cells mediated via a MMP-dependent mechanism. It appeared that TIMP-1 was able to preserve the integrity of the ECM and stabilize the cell matrix interactions through inhibition of MMPs (Jiang et al. 2002), and thereby inhibit apoptosis.

Several other studies have reported the anti-apoptotic effect of TIMP-1 to be independent of its MMP inhibitory activity. Reduced-alkylated TIMP-1, completely devoid of MMP inhibitory activity, has been shown to suppress apoptosis of Burkitts lymphoma cell lines (Guedez et al. 1998) and human breast epithelial cells (Li et al. 1999, Liu et al. 2005) in vitro. It was recently shown that TIMP-1 binds to the cell surface protein CD63, thereby regulating cell survival and polarization (Jung et al. 2006).

TIMP-1 may regulate apoptosis by the MMP-dependent and MMP-independent pathways, both of which may coexist in the cell. Thus, TIMP-1 is a significant determinant for cell death/cell survival, and thereby regulation of the microenvironment.

TIMP-1 and Regulation of Angiogenesis

The pathological formation of new blood vessels from existing vessels (angiogenesis) is essential for tumours to survive and metastasize. Angiogenesis can develop by several different cellular mechanisms, including sprouting, intussusceptions or by incorporation of bone marrow-derived endothelial precursors (Carmeliet and Jain 2000).

Endothelial cell invasion into the surrounding stroma/tissue is an essential step during angiogenesis and requires proteases such as MMPs (Chun et al. 2004). The ability of TIMP-1 to inhibit MMPs has been reported to induce both pro-angiogenic and anti-angiogenic effects.

Preclinical in vitro and in vivo (Fisher et al. 1994, Johnson et al. 1994, Reed et al. 2003, Fernandez et al. 1999) studies have shown that TIMP-1 is capable of suppressing angiogenesis through inhibition of MMP. Akahane and colleagues (Akahane et al. 2004) demonstrated that recombinant human TIMP-1 suppressed migration of endothelial cells in a dose-dependent manner, indicating that inhibition of angiogenesis is related to MMP inhibition. However, Cornelius and colleagues (Cornelius et al. 1998) showed that MMPs act on plasminogen to generate angiostatin, and thereby limiting tumour vascularization. Therefore, TIMPs that inhibit MMPs and thus the production of angiostatin may promote angiogenesis.

In addition, several studies have shown that TIMP-1 also has a direct effect on endothelial cell proliferation, which is separate from MMP inhibition (Hayakawa et al. 1992, Akahane et al. 2004). Yoshiji and colleagues (Yoshiji et al. 1998) reported that over-expression of TIMP-1 enhances VEGF expression in mammary carcinoma and promote neo-vascularization of the retina by VEGF induction (Yamada et al. 2001).

Further research on the pleiotopic actions of TIMP-1 in the cellular microenvironment is warranted, especially in the light of the many clinical studies showing that elevated levels of TIMP-1 are associated with poor prognosis of patients with cancer.

TIMP-1 and Prognosis in CRC

Descriptive Studies, LOE V

Zeng and colleagues (Zeng et al. 1995) reported on the level of expression of TIMP-1 mRNA in tumour tissue from 56 patients with CRC. TIMP-1 mRNA was significantly elevated in the cancer tissue when compared with the corresponding adjacent normal tissue ($p = 0.0001$). In addition, TIMP-1 mRNA levels were correlated with presence or absence of metastatic disease, for example, higher TIMP-1 mRNA levels were observed in tissue from patients with advanced disease. This descriptive study was not aiming to prove a prognostic value of TIMP-1 mRNA, but it was the first study showing a correlation between TIMP-1 level and tumour stage in CRC, hence suggesting an association between patient prognosis and TIMP-1 mRNA levels. Pellegrini and colleagues (Pellegrini et al. 2000) performed a study on 41 patients with CRC undergoing surgical resection. By principal components analysis, serum TIMP-1 levels were found to be associated with progression of disease from stage III to IV ($p = 0.02$). Roca and colleagues (Roca et al. 2006) did immunohistochemical (IHC) staining of 84 paraffin-embedded CRC tumours.

High TIMP-1 immunoreactivity (more than 25% of cells stained) was shown to correlate insignificantly with poor patient outcome (p value not given), and TIMP-1 immunoreactivity was not retained as a prognostic marker in the final multivariate Cox analysis.

Yukawa and colleagues (Yukawa et al. 2001) performed a study on plasma TIMP-1 in 44 patients undergoing surgery for CRC and found that TIMP-1 values correlated with serosal invasion ($p = 0.011$), Dukes' stage ($p = 0.036$) and liver metastasis ($p = 0.025$). The prognostic value of TIMP-1 was not studied, but the results suggested a possible prognostic association since serosal invasion, liver metastasis and higher Dukes' stage are known to correlate with overall survival.

Case/Monitoring Series, LOE IV

Oberg and colleagues (Oberg et al. 2000) performed a case-control study including 158 patients with CRC to evaluate the prognostic impact of serum TIMP-1 (among other potential markers). A correlation between tumour stage and serum TIMP-1 level ($p < 0.001$) was described while no significant correlation with patient outcome was observed for Dukes' stages A–C patients ($n = 133$) in this study.

Simpson and colleagues (Simpson et al. 2000) performed a study on plasma TIMP-1 from 41 patients with CRC. Plasma TIMP-1 was significantly higher in those patients with metastatic disease compared with those with localized disease ($p = 0.02$). No significant difference in plasma TIMP-1 levels was observed among patients with Dukes' stages A–C disease, but no data were presented on a possible prognostic impact of plasma TIMP-1 in these patients.

Retrospective Studies, LOE III

Waas and colleagues (Waas et al. 2005) evaluated the prognostic value of TIMP-1 in 94 patients with CRC. Patients presenting with a high plasma TIMP-1 value had significantly shorter survival ($p = 0.0003$) than did patients with low plasma TIMP-1 values. Like in other studies, TIMP-1 was shown to correlate with age, tumour stage and distant metastasis.

Yukawa and colleagues (Yukawa et al. 2004) performed a study on plasma TIMP-1 in 82 patients undergoing elective surgery for CRC. Patients were grouped according to a plasma TIMP-1 cut-off value of 170.0 ng/ml calculated based on TIMP-1 values in the group of patients diagnosed with early-stage CRC. Based on this grouping, TIMP-1 level was shown to predict patient outcome when analyzed univariately ($p = 0.03$), that is, high plasma TIMP-1 levels correlated with shorter patient survival ($p = 0.02$). However, plasma TIMP-1 was not retained in the final multivariate model in which patients with distant metastases and peritoneal metastases were excluded.

Holten-Andersen and co-workers (Holten-Andersen et al. 2000) conducted a retrospective study of 588 patients with CRC undergoing elective surgery with an average follow-up of 89 months. Plasma TIMP-1 was measured in samples obtained preoperatively. Treated as a continuous variable, plasma TIMP-1 level ($\log_e$ TIMP-1) was found to be significantly associated with survival [(HR 3.3 (95% CI: 2.6–4.2); $p < 0.0001$)]. The correlation between plasma TIMP-1 and survival was found for both rectal cancer and for colancancer. Plasma TIMP-1 was not significantly different among Dukes' stages A, B and C patients, whereas patients with Dukes' stage D disease had significantly higher plasma TIMP-1 values ($p < 0.0001$). They also found a correlation between age and plasma TIMP-1 levels ($p < 0.001$). Multivariate analysis showed that plasma TIMP-1 was retained in the final model [(HR 2.5 (95% CI: 1.7–3.7); $p < 0.0001$)]. In a subsequent study, Holten-Andersen and co-workers measured plasma TIMP-1 in samples obtained $\sim$6 months following the primary surgery (Holten-Andersen et al. 2006). This study confirmed the association between plasma TIMP-1 levels and patient survival, and the authors suggest that plasma TIMP-1 could be a marker of minimal residual disease. Holten-Andersen and colleagues (Holten-Andersen et al. 2004) also performed a retrospective cohort study of 352 patients with rectal cancer (RC) undergoing elective surgery. The follow-up period was 43 months (range, 20–95 months) for survivors. Plasma TIMP-1 was measured in samples obtained preoperatively. Plasma TIMP-1 levels were found to correlate with patient outcome in a univariate analysis ($p < 0.0001$). Performing a multivariate analysis, plasma TIMP-1 levels were shown to be independently associated with patient survival [(HR 2.2 (95% CI: 1.2–4.1); $p = 0.01$)]. This study thus confirmed the first study showing an independent association between plasma TIMP-1 levels and survival of patients with RC.

Ishida and co-workers (2003) performed a cohort study of 123 patients undergoing elective resection for CRC. TIMP-1 was analyzed in serum samples. The 112 patients undergoing curatively intended resection were followed for 48 months (range, 36–67 months) and the remaining 11 patients undergoing palliative resection were followed for 16 months (range, 2–36 months). Serum TIMP-1 levels were associated with Dukes' stage ($p = 0.03$), lymph node metastasis ($p = 0.04$) and liver metastasis ($p < 0.001$). A cut-off value was calculated based on TIMP-1 values from 20 healthy individuals. This cut-off value was used to group patients into high-/low-level groups, and based on this grouping TIMP-1 was not found to predict outcome in the study population.

Discussion

This chapter has evaluated TIMP-1 as a prognostic marker in CRC. Eleven publications met the criteria for inclusion in the evaluation of TIMP-1 as a prognostic marker in patients with CRC. The studies included quantification of TIMP-1 mRNA or TIMP-1 IHC in tumour tissue or TIMP-1 measurements in serum and in plasma. Most of the studies included only a limited number of patients and must, therefore, be categorized as exploratory in nature. In many of the studies, no uni-variate survival analyses had been performed, but the authors proposed TIMP-1 as a prognostic variable based on associations between TIMP-1 and already established prognostic parameters such as tumour stage. However, most of the studies found no significant differences in TIMP-1 levels among stages I–III patients (Dukes' stages A–C), while a significant increase was observed in the subgroup of stage IV (Dukes' stage D) patients. These results strongly suggest that TIMP-1 levels can be used to further stratify the prognosis of patients in each of stages I–III, which is what is presently needed. The only IHC study included (Roca et al. 2004) reported on TIMP-1 immunoreactivity in the cancer cells as well as in the tumour stroma. It has since unequivocally been demonstrated that TIMP-1 immunoreactivity and mRNA expression is located in the infiltrating stromal cells in the tumour tissue and no mRNA expression or immunoreactivity is present in the tumour cells. Less importance should, therefore, be given to the study by Roca et al. (2006). Three studies measured TIMP-1 in serum (Oberg et al. 2000, Ishida et al. 2003, Pellegrini et al. 2000). It has previously been shown that TIMP-1 levels in serum are signifi-cantly higher and more variable than in plasma. It was proposed that this difference may be caused by release of TIMP-1 from platelets during blood coagulation, and thus not related to the presence of the cancer disease. It is noteworthy that two of these three serum studies failed to demonstrate any association between TIMP-1 levels and patient survival. In contrast, the five studies on plasma TIMP-1 in which an association between TIMP-1 levels and patient survival was presented, all came to the same conclusion that plasma TIMP-1 levels are significantly associated with patient outcome, for example, high plasma TIMP-1 levels being associated with poor patient survival and low levels with long survival. These results support further studies on plasma TIMP-1 as a prognostic variable in CRC. The final aim is to reach LOE 1, which will allow plasma TIMP-1 to be included in the daily management of patients with CRC.

By reviewing the identified papers, it became obvious that various clinical and methodological aspects have to be discussed before continuing studies on plasma TIMP-1 as a prognostic marker in CRC. In studies evaluating prognosis, the study population has to be well characterized aiming to prevent misinterpretation caused by bias or confounders. A satisfactory description of the study population includ-ing description of sample selection, inclusion criteria, diagnostic criteria, clinical and demographic characteristics and length of follow-up should be presented at publication (Altman 2001).

Plasma TIMP-1 level is known to increase with age—as was shown in some of the presented studies (Holten-Andersen et al. 2000, Waas et al. 2005); consequently, it is of importance to stratify for age when evaluating TIMP-1. Additionally, increased plasma TIMP-1 levels have been described in various cancers and non-malignant diseases (Hofmann et al. 2000, Schrohl et al. 2004, Gouyer et al. 2005, Gorogh et al. 2006, Klimiuk et al. 2002, Zureik et al. 2005, Arthur et al. 1998); consequently, this should be considered when performing an evaluation of TIMP-1 levels in plasma. In most of the reviewed plasma studies age and sex is accounted for, while the issue of co-morbidity has not been included yet.

Information concerning pre-analytical factors, such as specimen collection, handling and storage, is another important issue in the evaluation of potential biomarkers. Plasma TIMP-1 measurements have been shown to be influenced by several pre-analytical variables (Jung et al. 1996, Lomholt et al. 2007), but in most of the present studies this information is not available.

Also, analytical differences between studies may influence the overall conclusions. The assays employed for plasma TIMP-1 measurement are not directly comparable; hence, some kind of objective measures should be available allowing for comparison of results from different arrays. For example, inclusion of a common external reference plasma sample in all assays would allow for comparison. In addition, such a reference sample could also be used to estimate intra- and inter-assay variations within and among laboratories. Finally, it should be obvious that a thorough validation (including tests for specificity, sensitivity, linearity and recovery) of the assay in question is an absolute necessity.

Various statistical methods were applied, which makes comparison between the studies even more difficult. Only four papers presented a numerical summary of the prognostic strength of TIMP-1, such as hazard ratio (Holten-Andersen et al. 2004, 2006, 2000, Roca et al. 2006), although the latter presents hazard ratios only for significant markers and thus not for TIMP-1. Length of follow-up of patients were recorded only in five studies (Roca et al. 2006, Holten-Andersen et al. 2000, Oberg et al. 2000, Curran et al. 2004, Waas et al. 2005). Furthermore, in several publications the follow-up period was described inadequately in respect to the median follow-up for both survivors and dead individuals.

Analysis of statistical data can also be biased when cut-off points are handled in various ways, for example, Yukawa et al. (2001) presented a cut-off value of 170 ng/ml for plasma TIMP-1 and defined patients with positive or negative plasma TIMP-1 according to this cut-off value. Other studies had a more conservative approach using either the median TIMP-1 value or divided TIMP-1 values into tertiles or quartiles (Holten-Andersen et al. 2006, Ishida et al. 2003, Oberg et al. 2000). Holten-Andersen et al. (2006) used two approaches to scoring TIMP-1 for analysis of prognostic value: (1) a cut-point defined by the 95th percentile of healthy donors adjusted for age and gender; (2) TIMP-1 on a continuous scale (log transformed). The more basic approach with treating TIMP-1 values as continuous variables and then comparing to patient outcome was presented only in three of the studies (Holten-Andersen et al. 2000, 2002, 2006).

The completeness of the required information will reflect the LOE of the individual study. For example, it is obviously easy to describe the methodological variables when a study is performed on material collected for the actual study, and a study performed in this manner will consequently reach a higher LOE, than a study based on archive biological material where access to this information may be difficult or even impossible. In this review, only two (Waas et al. 2005, Yukawa et al. 2004) of the presented studies used biological material aimed for the actual study.

Future Perspectives – Bringing TIMP-1 to LOE I?

Keeping all of these variables in mind, it is obviously difficult to draw a final conclusion on the utility of TIMP-1 as a prognostic marker in CRC. In order to test if plasma TIMP-1 can be brought to LOE I, a large prospectively controlled study that is designed to test the association between plasma TIMP-1 and patient survival should consider all imaginable confounders and bias possible. Thus, standard operating procedures (SOPs) for sample collection, handling and storage should be in place before the study is initiated. All relevant patient characteristics should be collected prospectively and inclusion and exclusion criteria for the study should be defined. Statistical strength calculations should be performed when designing such studies, to allow valid statistical calculations of the end-points. Also, SOPs for

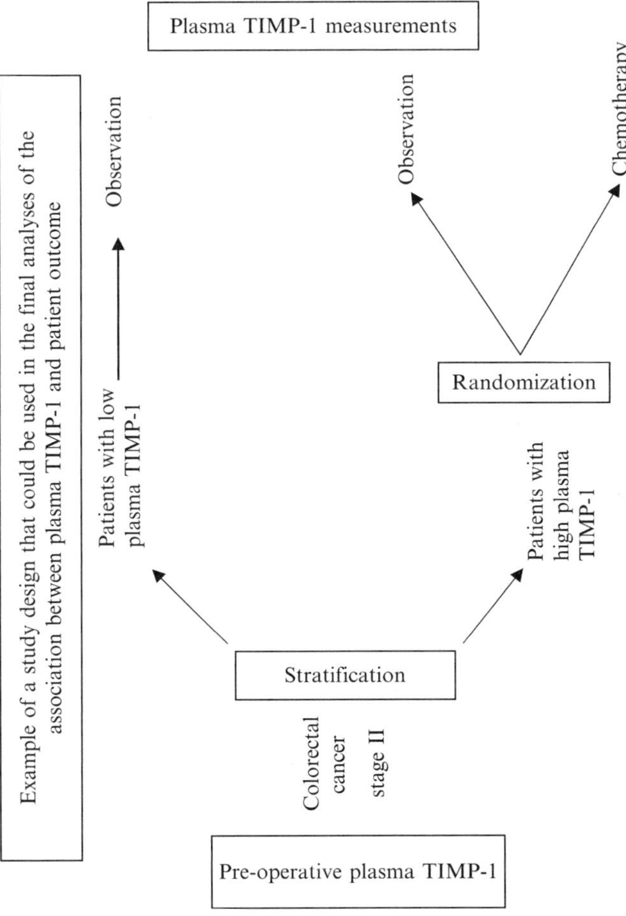

Fig. 33.1 Example of a study design that could be used in the final analyses of the association between plasma TIMP-1 and patient outcome

the plasma TIMP-1 analyses should be worked out before the initiation of the study and as suggested above, internal and external reference material should be available and included on each assay run. For the post-analytical variables, a statistical analysis procedure (SAP) should clearly describe which statistical methods, including plasma TIMP-1 values (cut-off points or continuous variable), are to be used in the final analyses of the association between plasma TIMP-1 and patient outcome. Figure 33.1 shows an example of a study design that could be used for such a clinical study. In this study, the primary aim will be to validate the prognostic impact of plasma TIMP-1 in patients with stage II RC. The secondary aim will be to test if "high TIMP-1" patients will benefit from adjuvant therapy. Patients with low TIMP-1 plasma levels will be observed (current practice) while patients with high plasma TIMP-1 levels will be randomized to either observation (current practice) or adjuvant chemotherapy. End-points will be disease-free survival and overall survival. The study could include CC as well as RC stage II patients. A second plasma sample should be collected at the end of chemotherapy and subsequently being analyzed for TIMP-1, the hypothesis being that patients in whom plasma TIMP-1 levels do not return to a level comparable to the levels found in healthy individuals might be those having minimal residual disease (Holten-Andersen et al. 2006) despite adjuvant treatment.

Preferably such a study should compare the results of the plasma TIMP-1 measurements with currently available predictors of prognosis, for example, histological grading and serum CEA levels, to evaluate its future potential. Actually, some of the studies included in the present review have included such comparisons and the data suggest plasma TIMP-1 to be an independent predictor of survival in patients with CRC, that is, plasma TIMP-1 can add to the prognostic information that can be obtained by these two other factors.

References

Akahane T, Akahane M, Shah A et al. (2004). TIMP-1 inhibits microvascular endothelial cell migration by MMP-dependent and MMP-independent mechanisms. Exp Cell Res; (301): 158–167.

Altman D G. (2001). Systematic reviews of evaluations of prognostic variables. BMJ; (323): 224–228.

Arthur M J, Mann D A, Iredale J P. (1998). Tissue inhibitors of metalloproteinases, hepatic stellate cells and liver fibrosis. J Gastroenterol Hepatol; (13 Suppl): S33–S38.

Avalos B R, Kaufman S E, Tomonaga M et al. (1988). K562 cells produce and respond to human erythroid-potentiating activity. Blood; (71): 1720–1725.

Bast R C, Jr., Ravdin P, Hayes D F et al. (2001). 2000 update of recommendations for the use of tumor markers in breast and colorectal cancer: Clinical practice guidelines of the American Society of Clinical Oncology. J Clin Oncol; (19): 1865–1878.

Bertaux B, Hornebeck W, Eisen A Z et al. (1991). Growth stimulation of human keratinocytes by tissue inhibitor of metalloproteinases. J Invest Dermatol; (97): 679–685.

Brew K, Dinakarpandian D, Nagase H. (2000). Tissue inhibitors of metalloproteinases: Evolution, structure and function. Biochim Biophys Acta; (1477): 267–283.

Carmeliet P, Jain R K. (2000). Angiogenesis in cancer and other diseases. Nature; (407): 249–257.

Chesler L, Golde D W, Bersch N et al. (1995). Metalloproteinase inhibition and erythroid potentiation are independent activities of tissue inhibitor of metalloproteinases-1. Blood; (86): 4506–4515.

Chun T H, Sabeh F, Ota I et al. (2004). MT1-MMP-dependent neovessel formation within the confines of the three-dimensional extracellular matrix. J Cell Biol; (167): 757–767.

Compton C C, Greene F L. (2004). The staging of colorectal cancer: 2004 and beyond. CA Cancer J Clin; (54): 295–308.

Cornelius L A, Nehring L C, Harding E et al. (1998). Matrix metalloproteinases generate angiostatin: Effects on neovascularization. J Immunol; (161): 6845–6852.

Curran S, Dundas S R, Buxton J et al. (2004). Matrix metalloproteinase/tissue inhibitors of matrix metalloproteinase phenotype identifies poor prognosis colorectal cancers. Clin Cancer Res; (10): 8229–8234.

Docherty A J, Lyons A, Smith B J et al. (1985). Sequence of human tissue inhibitor of metallo-proteinases and its identity to erythroid-potentiating activity. Nature; (318): 66–69.

Duffy M J, van D A, Haglund C et al. (2003). Clinical utility of biochemical markers in colorectal cancer: European Group on Tumour Markers (EGTM) guidelines. Eur J Cancer; (39): 718–727.

Egeblad M, Werb Z.(2002). New functions for the matrix metalloproteinases in cancer progression. Nat Rev Cancer; (2): 161–174.

Eisenberg B, Decosse J J, Harford F et al. (1982). Carcinoma of the colon and rectum: the natural history reviewed in 1704 patients. American Cancer Society; 49(6): 1131–1134.

Fata J E, Leco K J, Moorehead R A et al. (1999). Timp-1 is important for epithelial proliferation and branching morphogenesis during mouse mammary development. Dev Biol; (211): 238–254.

Ferlay J, Autier P, Boniol M et al. (2007). Estimates of the cancer incidence and mortality in Europe in 2006. Ann Oncol; (18): 581–592.

Fernandez H A, Kallenbach K, Seghezzi G et al. (1999). Inhibition of endothelial cell migration by gene transfer of tissue inhibitor of metalloproteinases-1. J Surg Res; (82): 156–162.

Fisher C, Gilbertson-Beadling S, Powers E A et al. (1994). Interstitial collagenase is required for angiogenesis in vitro. Dev Biol; (162): 499–510.

George S, Primrose J, Talbot R et al. (2006). Will Rogers revisited: Prospective observational study of survival of 3592 patients with colorectal cancer according to number of nodes examined by pathologists. Br J Cancer; (95): 841–847.

Gill S, Loprinzi C L, Sargent D J et al. (2004). Pooled analysis of fluorouracil-based adjuvant therapy for stage II and III colon cancer: Who benefits and by how much? J Clin Oncol; (22): 1797–1806.

Golde D W, Bersch N, Quan S G et al. (1980). Production of erythroid-potentiating activity by a human T-lymphoblast cell line. Proc Natl Acad Sci U S A; (77): 593–596.

Goldstein M J, Mitchell E P.(2005). Carcinoembryonic antigen in the staging and follow-up of patients with colorectal cancer. Cancer Invest; (23): 338–351.

Gorogh T, Beier U H, Baumken J et al. (2006). Metalloproteinases and their inhibitors: Influence on tumor invasiveness and metastasis formation in head and neck squamous cell carcinomas. Head Neck; (28): 31–39.

Gouyer V, Conti M, Devos P et al. (2005). Tissue inhibitor of metalloproteinase 1 is an independent predictor of prognosis in patients with nonsmall cell lung carcinoma who undergo resection with curative intent. Cancer; (103): 1676–1684.

Guedez L, Stetler-Stevenson W G, Wolff L et al. (1998). In vitro suppression of programmed cell death of B cells by tissue inhibitor of metalloproteinases-1. J Clin Invest; (102): 2002–2010.

Guillou P J, Quirke P, Thorpe H et al. (2005). Short-term endpoints of conventional versus laparoscopic-assisted surgery in patients with colorectal cancer (MRC CLASICC trial): Multicentre, randomised controlled trial. Lancet; (365): 1718–1726.

Hayakawa T, Yamashita K, Tanzawa K et al. (1992). Growth-promoting activity of tissue inhibitor of metalloproteinases-1 (TIMP-1) for a wide range of cells. A possible new growth factor in serum. FEBS Lett; (298): 29–32.

Hayes D F, Bast R C, Desch C E et al. (1996). Tumor marker utility grading system: A framework to evaluate clinical utility of tumor markers. J Natl Cancer Inst; (88): 1456–1466.

Hofmann U B, Westphal J R, van Muijen G N et al. (2000). Matrix metalloproteinases in human melanoma. J Invest Dermatol; (115): 337–344.

Holten-Andersen M, Christensen I J, Nilbert M et al. (2004). Association between preoperative plasma levels of tissue inhibitor of metalloproteinases 1 and rectal cancer patient survival. A validation study. Eur J Cancer; (40): 64–72.

Holten-Andersen M N, Stephens R W, Nielsen H J et al. (2000). High preoperative plasma tissue inhibitor of metalloproteinase-1 levels are associated with short survival of patients with colorectal cancer. Clin Cancer Res; (6): 4292–4299.

Holten-Andersen M N, Christensen I J, Nielsen H J et al. (2002). Total levels of tissue inhibitor of metalloproteinases 1 in plasma yield high diagnostic sensitivity and specificity in patients with colon cancer. Clin Cancer Res; (8): 156–164.

Holten-Andersen M N, Nielsen H J, Sorensen S et al. (2006). Tissue inhibitor of metalloproteinases-1 in the postoperative monitoring of colorectal cancer. Eur J Cancer; (42): 1889–1896.

Ishida H, Murata N, Hayashi Y et al. (2003). Serum levels of tissue inhibitor of metalloproteinases-1 (TIMP-1) in colorectal cancer patients. Surg Today; (33): 885–892.

Jiang Y, Goldberg I D, Shi Y E. (2002). Complex roles of tissue inhibitors of metalloproteinases in cancer. Oncogene; (21): 2245–2252.

Johnson M D, Kim H R, Chesler L et al. (1994). Inhibition of angiogenesis by tissue inhibitor of metalloproteinase. J Cell Physiol; (160): 194–202.

Jung K, Nowark L, Lein M et al. (1996). Role of specimen collection in preanalytical variation of metalloproteinases and their inhibitors in blood. Clin Chem; (42): 2043–2045.

Jung K K, Liu X W, Chirco R et al. (2006). Identification of CD63 as a tissue inhibitor of metalloproteinase-1 interacting cell surface protein. EMBO J; 25(17): 3934–3942.

Klimiuk P A, Sierakowski S, Latosiewicz R et al. (2002). Serum matrix metalloproteinases and tissue inhibitors of metalloproteinases in different histological variants of rheumatoid synovitis. Rheumatology (Oxford); (41): 78–87.

Levin K E, Dozois R R. (1991). Epidemiology of large bowel cancer. World J Surg; (15): 562–567.

Li G, Fridman R, Kim H R. (1999). Tissue inhibitor of metalloproteinase-1 inhibits apoptosis of human breast epithelial cells. Cancer Res; (59): 6267–6275.

Liu X W, Taube M E, Jung K K et al. (2005). Tissue inhibitor of metalloproteinase-1 protects human breast epithelial cells from extrinsic cell death: A potential oncogenic activity of tissue inhibitor of metalloproteinase-1. Cancer Res; (65): 898–906.

Locker G Y, Hamilton S, Harris J et al. (2006). ASCO 2006 update of recommendations for the use of tumor markers in gastrointestinal cancer. J Clin Oncol; (24): 5313–5327.

Lomholt A F, Frederiksen C B, Christensen I J et al. (2007). Plasma tissue inhibitor of metalloproteinases-1 as a biological marker? Pre-analytical considerations. Clin Chim Acta; (380): 128–132.

Luparello C, Avanzato G, Carella C et al. (1999). Tissue inhibitor of metalloprotease (TIMP)-1 and proliferative behaviour of clonal breast cancer cells. Breast Cancer Res Treat; (54): 235–244.

Murate T, Yamashita K, Ohashi H et al. (1993). Erythroid potentiating activity of tissue inhibitor of metalloproteinases on the differentiation of erythropoietin-responsive mouse erythroleukemia cell line, ELM-I-1-3, is closely related to its cell growth potentiating activity. Exp Hematol; (21): 169–176.

Murphy F R, Issa R, Zhou X et al. (2002). Inhibition of apoptosis of activated hepatic stellate cells by tissue inhibitor of metalloproteinase-1 is mediated via effects on matrix metalloproteinase inhibition: Implications for reversibility of liver fibrosis. J Biol Chem; (277): 11069–11076.

NCI. (2007). Surveillamce, Epidemiology, and End Results (SEER) Program. Available from URL: www.seer.cancer.gov.

Niskanen E, Gasson J C, Teates C D et al. (1988). In vivo effect of human erythroid-potentiating activity on hematopoiesis in mice. Blood; (72): 806–810.

Oberg A, Hoyhtya M, Tavelin B et al. (2000). Limited value of preoperative serum analyses of matrix metalloproteinases (MMP-2, MMP-9) and tissue inhibitors of matrix metalloproteinases (TIMP-1, TIMP-2) in colorectal cancer. Anticancer Res; (20): 1085–1091.

Obrand D I, Gordon P H. (1997). Incidence and patterns of recurrence following curative resection for colorectal carcinoma. Dis Colon Rectum; (40): 15–24.

Parkin D M, Bray F, Ferlay J et al. (2005). Global cancer statistics, 2002. CA Cancer J Clin; (55): 74–108.

Pellegrini P, Contasta I, Berghella A M et al. (2000). Simultaneous measurement of soluble carcinoembryonic antigen and the tissue inhibitor of metalloproteinase TIMP1 serum levels for use as markers of pre-invasive to invasive colorectal cancer. Cancer Immunol Immunother; (49): 388–394.

Porter J F, Shen S, Denhardt D T. (2004). Tissue inhibitor of metalloproteinase-1 stimulates proliferation of human cancer cells by inhibiting a metalloproteinase. Br J Cancer; (90): 463–470.

Reed M J, Koike T, Sadoun E et al. (2003). Inhibition of TIMP1 enhances angiogenesis in vivo and cell migration in vitro. Microvasc Res; (65): 9–17.

Ritter L M, Garfield S H, Thorgeirsson U P. (1999). Tissue inhibitor of metalloproteinases-1 (TIMP-1) binds to the cell surface and translocates to the nucleus of human MCF-7 breast carcinoma cells. Biochem Biophys Res Commun; (257): 494–499.

Roca F, Mauro L V, Morandi A et al. (2006). Prognostic value of E-cadherin, beta-catenin, MMPs (7 and 9), and TIMPs (1 and 2) in patients with colorectal carcinoma. J Surg Oncol; (93): 151–160.

Rodel C, Sauer R. (2005). Neoadjuvant radiotherapy and radiochemotherapy for rectal cancer. Recent Results Cancer Res; (165): 221–230.

Schilsky R L, Taube S E. (2002). Tumor markers as clinical cancer tests—Are we there yet? Semin Oncol; (29): 211–212.

Schrohl A S, Holten-Andersen M N, Peters H A et al. (2004). Tumor tissue levels of tissue inhibitor of metalloproteinase-1 as a prognostic marker in primary breast cancer. Clin Cancer Res; (10): 2289–2298.

Simpson R A, Hemingway DM, Thompson M M. (2000). Plasma TIMP-1—A marker of metastasis in colorectal cancer. Colorectal Dis; (2): 100–105.

Staib L, Link K H, Blatz A et al. (2002). Surgery of colorectal cancer: Surgical morbidity and five- and ten-year results in 2400 patients—Monoinstitutional experience. World J Surg; (26): 59–66.

Taube M E, Liu X W, Fridman R et al. (2006). TIMP-1 regulation of cell cycle in human breast epithelial cells via stabilization of p27(KIP1) protein. Oncogene; (25): 3041–3048.

TNM Classification of Malignant Tumours. (2002). John Wiley & Sons, Hoboken, New Jersey.

Waas E T, Hendriks T, Lomme R M et al. (2005). Plasma levels of matrix metalloproteinase-2 and tissue inhibitor of metalloproteinase-1 correlate with disease stage and survival in colorectal cancer patients. Dis Colon Rectum; (48): 700–710.

Wang J Y, Lu C Y, Chu K S et al. (2007). Prognostic significance of pre- and postoperative serum carcinoembryonic antigen levels in patients with colorectal cancer. Eur Surg Res; (39): 245–250.

Wang T, Yamashita K, Iwata K et al. (2002). Both tissue inhibitors of metalloproteinases-1 (TIMP-1) and TIMP-2 activate Ras but through different pathways. Biochem Biophys Res Commun; (296): 201–205.

Weitz J, Koch M, Debus J et al. (2005). Colorectal cancer. Lancet; (365): 153–165.

Yamada E, Tobe T, Yamada H et al. (2001). TIMP-1 promotes VEGF-induced neovascularization in the retina. Histol Histopathol; (16): 87–97.

Yoshiji H, Harris S R, Raso E et al. (1998). Mammary carcinoma cells over-expressing tissue inhibitor of metalloproteinases-1 show enhanced vascular endothelial growth factor expression. Int J Cancer; (75): 81–87.
Yukawa N, Yoshikawa T, Akaike M et al. (2001). Plasma concentration of tissue inhibitor of matrix metalloproteinase 1 in patients with colorectal carcinoma. Br J Surg; (88): 1596–1601.
Yukawa N, Yoshikawa T, Akaike M et al. (2004). Prognostic impact of tissue inhibitor of matrix metalloproteinase-1 in plasma of patients with colorectal cancer. Anticancer Res; (24): 2101–2105.
Zeng Z S, Cohen A M, Zhang Z F et al. (1995). Elevated tissue inhibitor of metalloproteinase 1 RNA in colorectal cancer stroma correlates with lymph node and distant metastases. Clin Cancer Res; (1): 899–906.
Zureik M, Beaudeux J L, Courbon D et al. (2005). Serum tissue inhibitors of metalloproteinases 1 (TIMP-1) and carotid atherosclerosis and aortic arterial stiffness. J Hypertens; (23): 2263–2268.

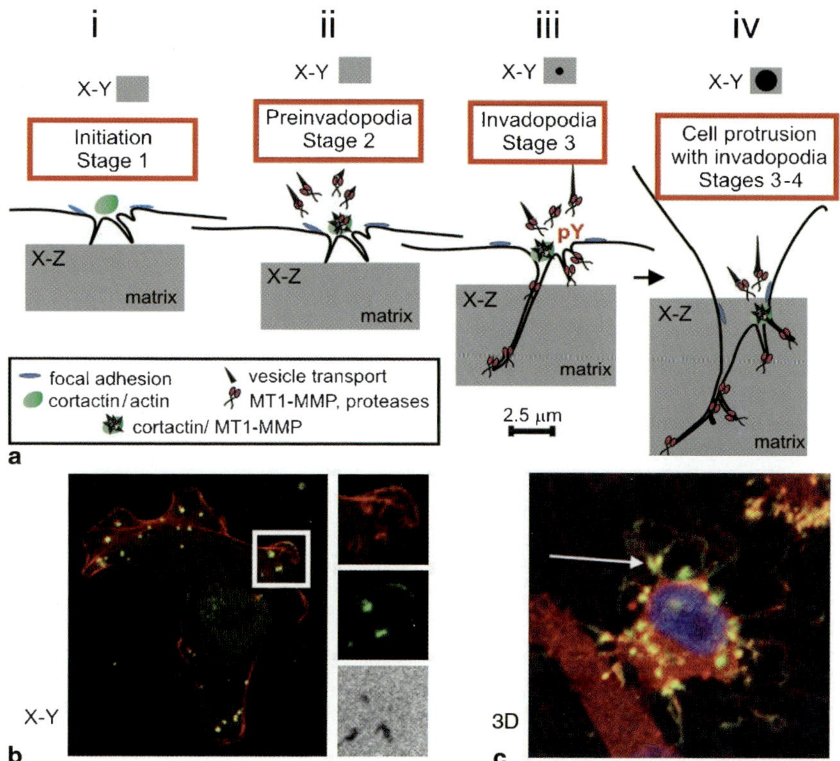

Fig. 21.1 Stages of invadopodia in 2D and 3D culture settings. **a** *i–iv* illustrate the initial formation of invadopodia beginning with cortactin recruitment (**i**), followed by recruitment of proteases such as MT1-MMP (**ii**), extension of invadopodia membrane into matrix facilitated by membrane-associated proteases and (**iii**), the enlargement of the site of degradation and invasion as the cell extends a pseudopodia/protrusion containing invadopodia and adhesion sites. **b** MDA-MB-231/c-Src(Y527F) cells breast cancer cells were immunostained using anticortactin 4F11 mAb (*green*) and counterstained with phalloidin (*red*) and imaged using confocal microscopy. The gelatin matrix (*grey*) contained sites of degradation and colocalizing F-actin and cortactin (*contained in insets*). **c** MDA-MB-231/c-Src(Y527F) cells were cultured on a thicker gelatin matrix whose autofluorescence is visualized together with cortactin in the red channel, DAPI in the blue channel, and antiphosphotyrosine 4G10 mAb staining in the green channel of this confocal image. Invadopodia are seen projecting into the matrix around the cell and aggregates containing colocalized cortactin and phosphotyrosine are visualized as yellow dots (*arrow*)

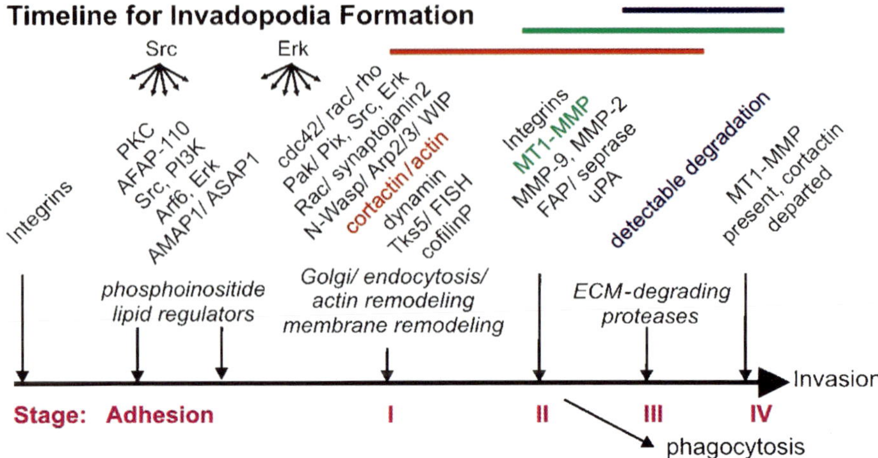

Fig. 21.2 Time line for invadopodia formation. The stages of invadopodia are presented along a time line and prominent invadopodia-associated proteins are hypothetically assigned to positions along the time line according to studies colocalizing them with cortactin/actin, MT1-MMP, or gelatin degradation. In color, the template for the time line is illustrated by cortactin/actin (*red*), MT1-MMP (*green*), and detectable degradation (*blue*)

Fig. 24.2 uPAR dimers detected by Forster resonance energy transfer (*FRET*) using the novel phasor-FLIM analysis of donor lifetime decays. (**a**) A typical donor-FLIM experiment: 2-photon fluorescence intensity (left top panel) and donor fluorescence lifetime decays are acquired pixel-by-pixel in parallel in the time correlated single photon counting (*TCSPC*) mode. Then, the decay (Continued) in each pixel is transformed in a phasor (left bottom panel). Phasor 1 identifies a pixel in which the fluorescence decay was monoexponential, and therefore it falls on the universal circle of the polar coordinates. Phasor 2 identifies a pixel in which the fluorescence decay was complex (multiexponential) as it falls inside the universal circle. Phasor 3 identifies a typical decay of untransfected HEK293 cells. The ensemble of all phasor from the image is represented in contour plot. The green line depicts the values of the phasor distributions in the HEK293/uPAR-EGFP-GPI cells at different contribution of cell autofluorescence (determined in untrasfected HEK293 cells). The red line marks the position of uPAR-EGFP-GPI phasors quenched by 50% FRET and having all possible cell autofluorescence contributions (from 0% to 100%). In the zoomed graph (right bottom panel), the black circle selects 98% of all phasors in the image. The localization of these phasors is shown in the correspondent FLIM image (right top panel). In this example, the contribution of cell autofluorescence was <1% in average. (**b**) FRET efficiencies determined in the basal membrane of an HEK293/GR cell by the phasor-FLIM analysis. The figure shows on the left panel the fluorescence intensity image (top) and the phasor plot (bottom). The phasor distribution is almost completely outside the green trajectory, indicating that donor lifetime is quenched by FRET. Less than 10% of pixels on this membrane section do not show FRET (<8%, mid top and bottom panels). The majority, 89%, of pixels in the image shows FRET efficiencies in the 8–24% range (right top and bottom panels). In this example, the average contribution of cell autofluorescence was 34–36%. (**c**) FRET efficiencies determined in the apical membrane of the HEK293/GR cell shown in (**b**). The fluorescence intensity image and phasor plot are shown in the left top and bottom panels. The same two subsets of FRET efficiencies (<8% and 8–24%) are reproduced in the mid and right panels. In this example, the average contribution of cell autofluorescence was 27–38%

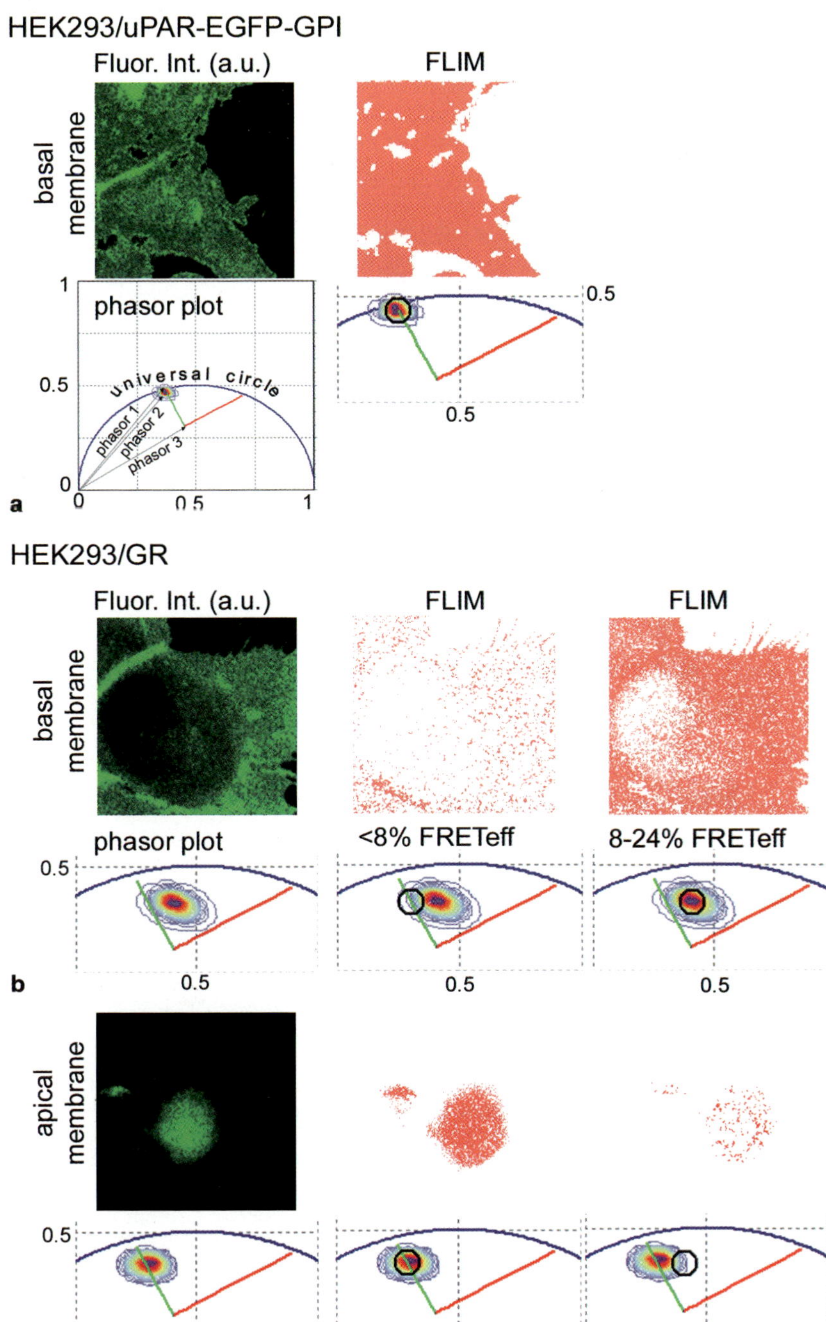
HEK293/uPAR-EGFP-GPI
Fluor. Int. (a.u.)
FLIM
basal membrane
phasor plot
1
0.5
universal circle
phasor 1
phasor 2
phasor 3
0
0
0.5
1
0.5
0.5
a
HEK293/GR
Fluor. Int. (a.u.)
FLIM
FLIM
basal membrane
phasor plot
<8% FRETeff
8-24% FRETeff
0.5
0.5
0.5
0.5
0.5
b
apical membrane
0.5
0.5
0.5
0.5
0.5
c

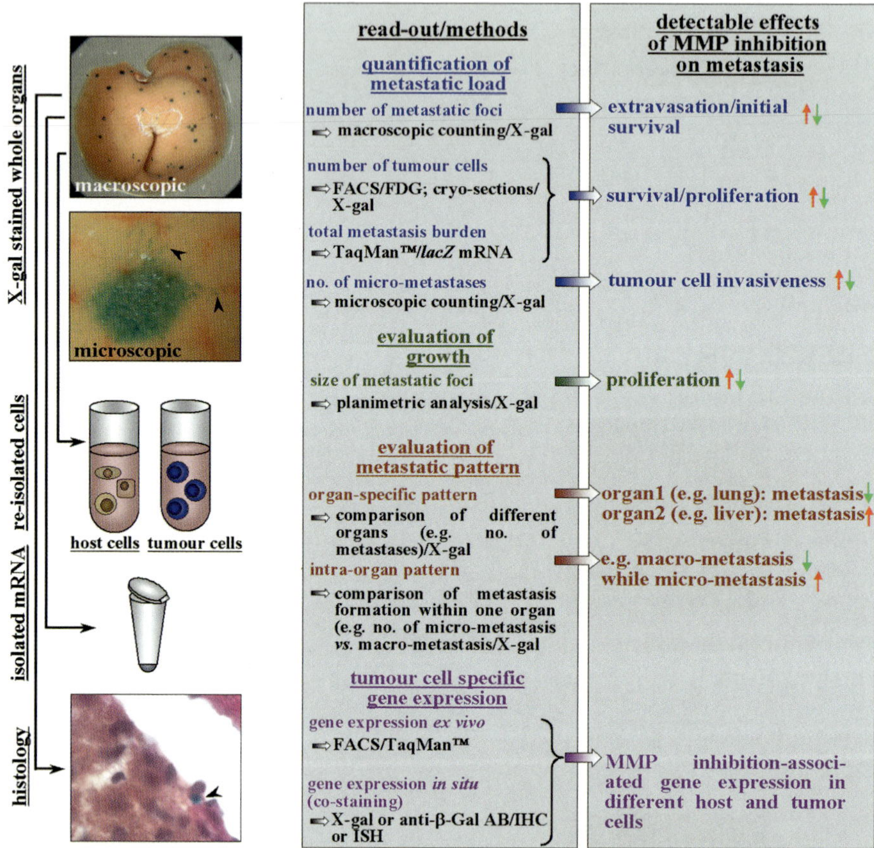

Fig. 25.1 Applications of a genetic tag for evaluation of the effects of matrix metalloproteinase (*MMP*) inhibition on metastasis—*lacZ*-tag as an example. After inoculation of *lacZ*-tagged tumor cells the following methods can be applied to detect these tumor cells: 5-bromo-4-chloro-3-indolyl-β-ᴅ-galactopyranoside (*X-gal*) staining of whole organs, re-isolation of tumor cells from target organs by FACS and fluorescein-β-ᴅ-galactopyranoside (*FDG*) staining, isolation of total ribonucleic acid (*RNA*) from metastasis-bearing organs and detection of *lacZ* mRNA by quantitative RT-PCR (TAqMan™), and X-gal staining of tissue sections

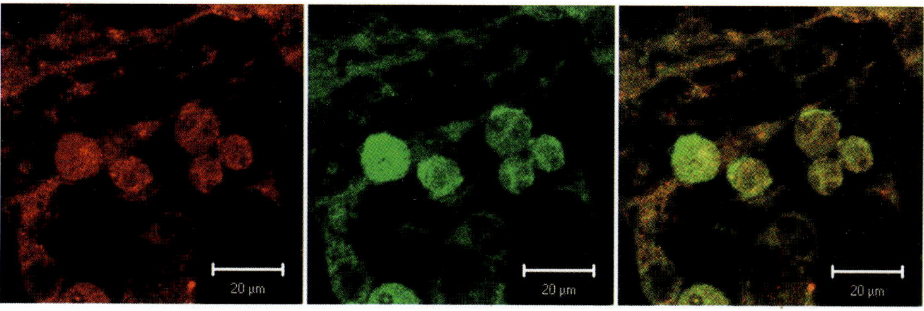

Fig. 29.1 Confocal microscopy of cathepsins X (*red*) and L (*green*) in a coculture of highly invasive MCF10A neoT cells and pro-monocytic U937 cells. Cathepsins L and X are expressed in both cell types, their expression and colocalization is more intense in U937 cells

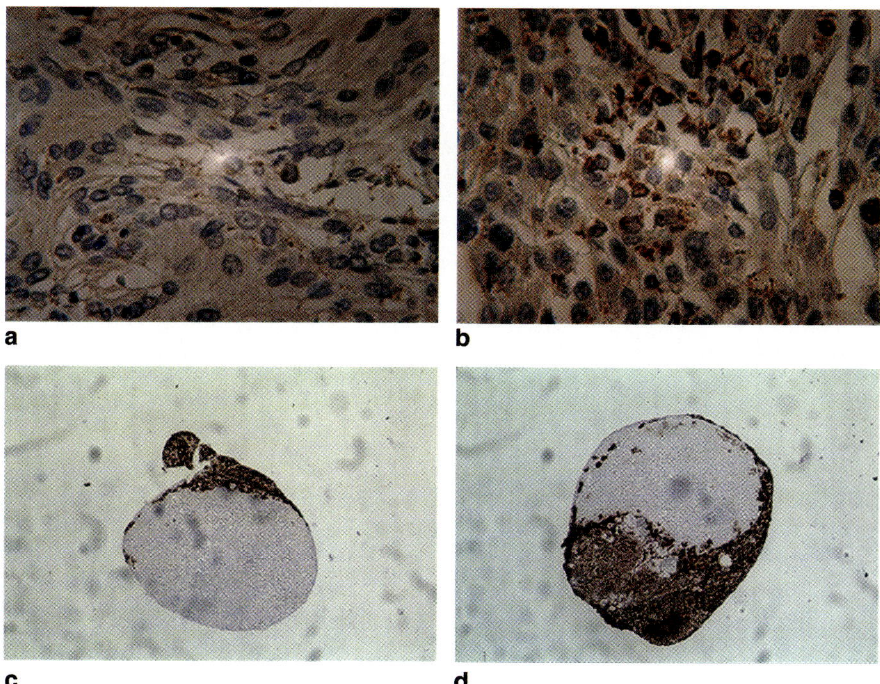

Fig. 29.2 Immunohistological staining of cathepsin B in human meningiomas (**a**, **b**). Clear benign meningioma (**a**) differ from histologically diagnosed atypical and anaplastic meningioma by more intense cathepsin B staining (**b**) (Strojnik et al. 2001). Chick heart invasion assay (**c**, **d**). Tumour explants were incubated in a coculture with the chick heart spheroids and labelled for cathepsin B. The meningioma spheroid is either not invasive (**c**) or shows different pattern of invasive behaviour (**d**) in chick heart matrix

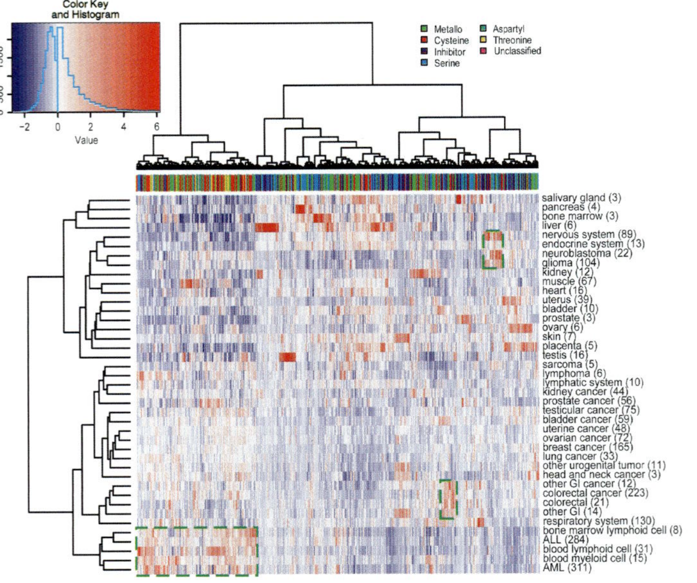

Fig. 31.2 Hierarchical clustering of mean degradome gene expression profiles for18 malignant and 23 healthy tissues. The number of samples for each tissue is marked in the parentheses. Values of heatmap are scaled genewise to mean 0 and standard deviation 1. Degradome genes ($n = 493$, listed below the heatmap) consist of 145 metalloproteases (presented by the green color above the image), 103 cysteine proteases (red), 101 protease inhibitors (blue), 112 serine proteases (light blue), 18 threonine proteases (yellow), and 13 aspartic proteases (cyan). Few significant degradome gene clusters are marked by the dark green dashed rectangles. The leftmost one indicates blood-related degradome genes, the middle cluster contains mainly gastrointestinal genes, and the rightmost indicates nervous system-specific degradome genes

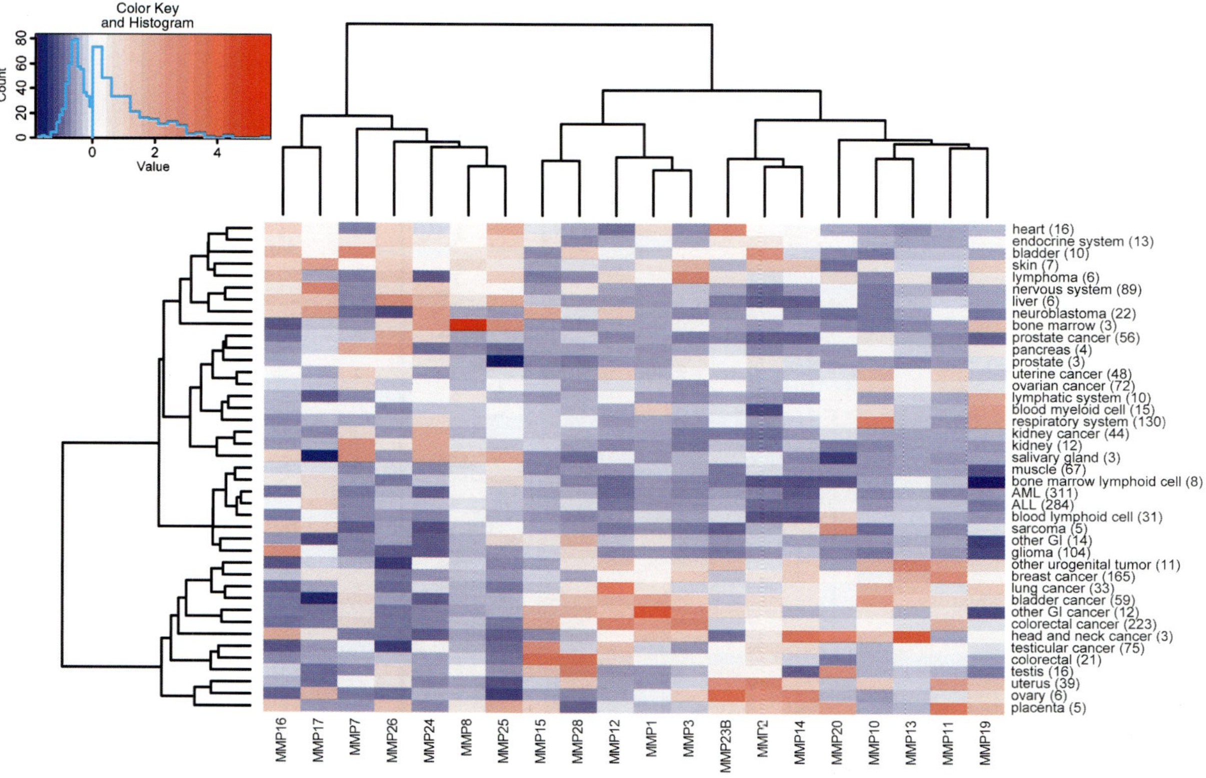

Fig. 31.4 Hierarchical clustering of 20 matrix metalloproteinase genes across 18 malignant and 23 healthy tissues. The number of samples for each tissue is marked in the parentheses. Values of heatmap are scaled genewise to mean 0 and standard deviation 1

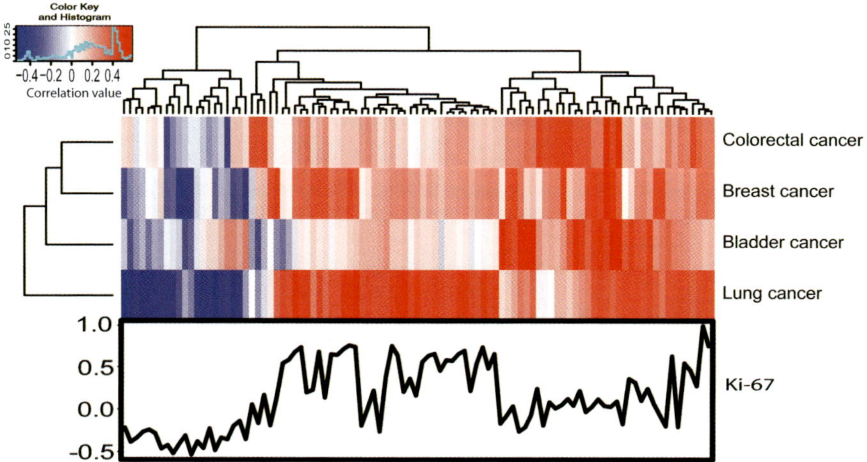

Fig. 31.7 Clustered heatmap of 97 genes having strongest correlations ($-0.4 > r > 0.4$ with Bonferroni corrected p value < 0.01) with MMP12 in colorectal, breast, bladder, and lung cancers. Below the image is a correlation coefficient plot of Ki-67 gene across in vivo-derived 3,928 datapoints of In Silico Transcriptomics (*IST*) database

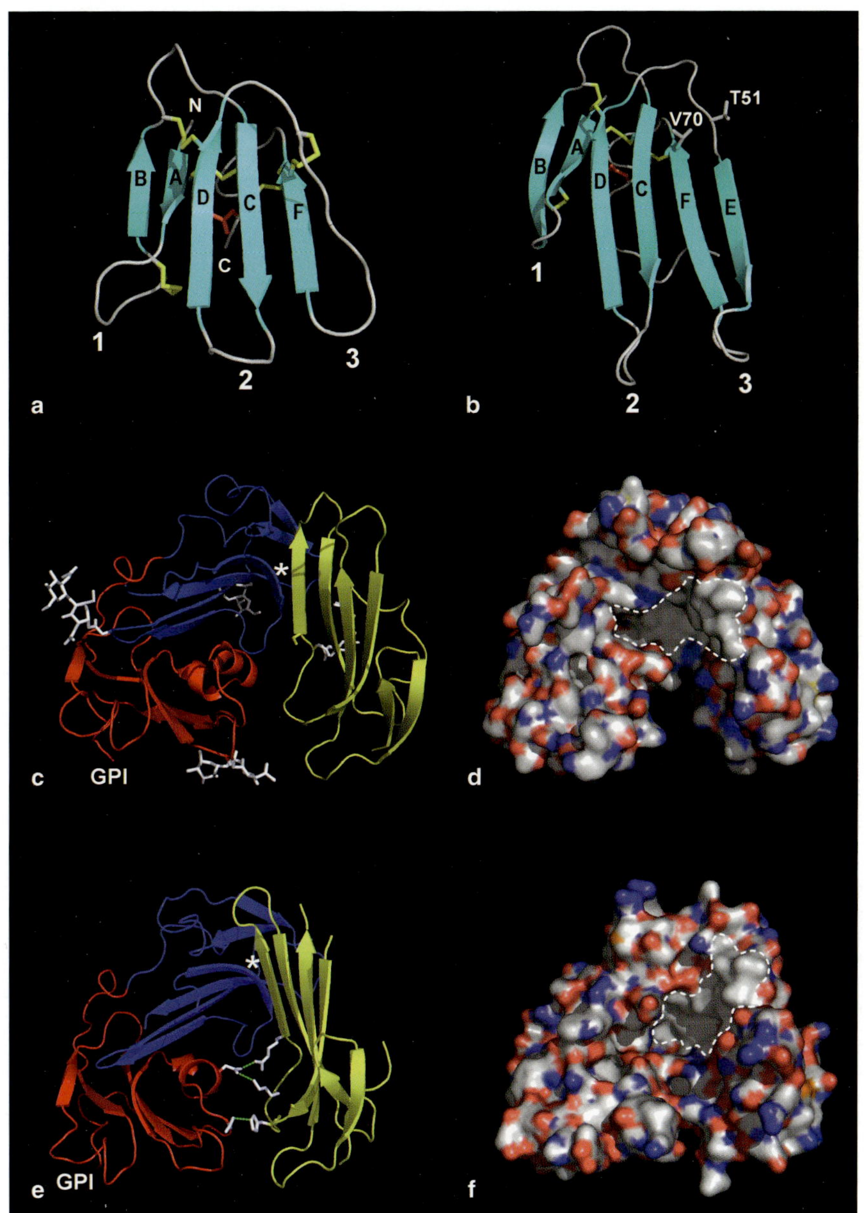

Fig. 34.1 Crystal structure of human urokinase-type plasminogen activator receptor (*uPAR*). The prototype structure for the three-finger fold is shown in panel **a** by a cartoon representation of the nuclear magnetic resonance (*NMR*) structure solved for the weak α-neurotoxin bucandin isolated from the Malayan krait (1IJC). The disulfide bonds are shown as sticks in yellow, the invariant asparagine in red, and the β-sheets in cyan with their identification code in capital letters. For comparison of the folding topology, the structure of *uPAR* DI is depicted in panel **b** in a similar orientation and color coding (1YWH). The position of the missing consensus disulfide in *uPAR* DI between βIE and βIF is highlighted by the side chains of Thr^{51} and Val^{70} occupying this position. Cartoon representations of the crystal structures solved for *uPAR*–peptide (1YWH) and *uPAR*–*ATF* (amino-terminal fragment) (2FD6) complexes are shown in panels **c** and **e**, respectively, after removal of the bound ligands. The individual Ly-6/uPAR (*LU*) domains in *uPAR* are colored yellow (DI), blue (DII), and red (DIII) and the positions of the *GPI* (glycosylphosphatidylinositol)-anchor are indicated. The positions of the carbohydrates are indicated in panel **c** by the white sticks of the innermost two *N*-acetyl-glucosamines, whereas the bent β-strand βIID is highlighted by an asterisk in both panels. The interdomain hydrogen bonds established between DI and DIII in the *uPAR*–*ATF* structure (His^{47}-Asn^{259}, Arg^{53}-Asp^{254}, and Lys^{50}-Asp^{254}) are highlighted in green in panel **e**. The corresponding molecular surface representations of the *uPAR*–peptide and *uPAR*–*ATF* complexes are shown in panels **d** and **f** colored by atom: carbon (white), nitrogen (blue), oxygen (red), and sulfur (yellow). The conspicuous hydrophobic walls of the binding cavities in the two structures are delimited by white hatched lines. These molecular representations were created by PyMOL (DeLano Scientific)

Fig. 34.2 Structures of human urokinase-type plasminogen activator receptor (*uPAR*) in complex with amino-terminal fragment (*ATF*), somatomedin B (*SMB*), and a synthetic peptide antagonist. The crystal structure of *uPAR*–*ATF* (2FD6) complexes are shown as a cartoon representation in panel **a** and as a combined surface and cartoon representation in panel **b**. Surface-exposed residues in *uPAR* causing a $\Delta\Delta G > 0.5$ kcal/mol on urokinase-type plasminogen activator (*uPA*) binding upon alanine substitution (Gårdsvoll et al. 2006) are highlighted in panel **b** by red. The solution structure of *SMB* (2JQ8) is shown in panel **c** with the residues important for *uPAR* binding shown as blue sticks. A model for the ternary *SMB*–*uPAR*–*ATF* complex with residues important for this interaction is shown as white sticks in panel **d** (Gårdsvoll and Ploug 2007). The crystal structure of *uPAR*–AE147 (1YWH) is shown as a cartoon representation in panel **e** and in a close up as a combined cartoon and surface representation in panel **f**. The four hydrophobic hot spot residues of AE147 (K-S-D-Cha-F-s-k-Y-L-W-S-K) are shown as white sticks and the target sites in *uPAR* for photoactivatable analogues of AE147 are shown as yellow surfaces (i.e., His^{251} being target for a photoprobe replacing Trp^{10}, whereas Leu^{66} and Arg^{53} are target sites for insertions from photoprobes replacing Phe^{5}). Capital letters denote amino acids in L-configuration, whereas lower case letters signify a D-configuration. Cha is cyclo-*L*-hexylalanine. These molecular representations were created by PyMOL (DeLano Scientific)

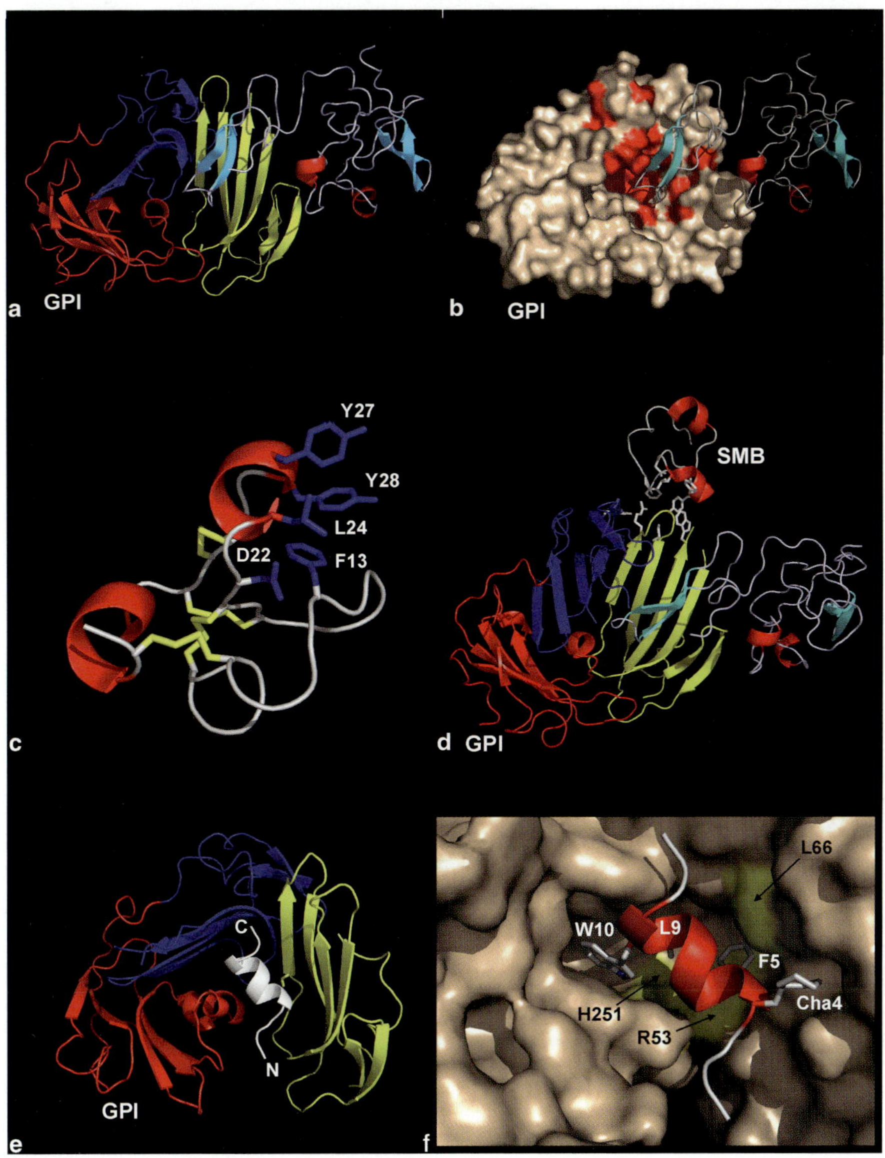

a GPI
b GPI
Y27
Y28
L24
D22
F13
c
d GPI
SMB
e GPI
C
N
f
L66
W10 L9
F5
H251
R53
Cha4

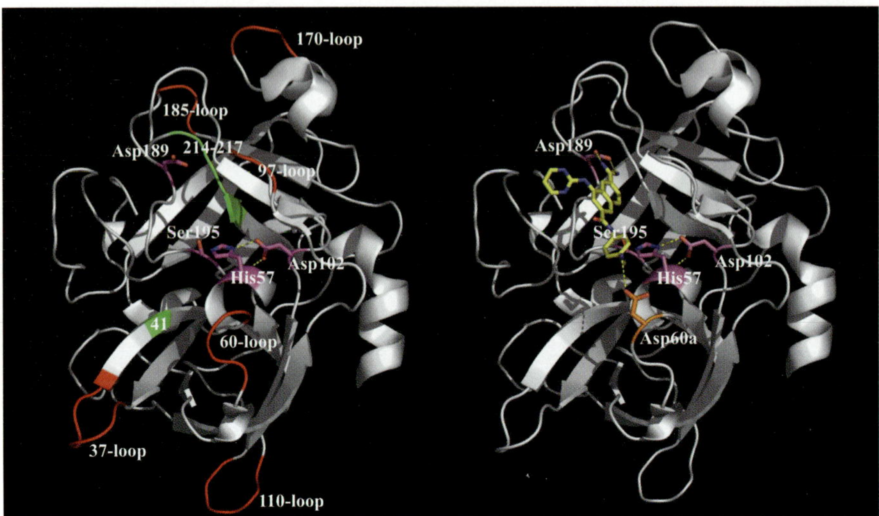

Fig. 35.2 The three-dimensional structure of human urokinase-type plasminogen activator (uPA). The left figure shows the catalytic triad Ser195, His57, and Asp102, and Asp189 in the S1 pocket (pink), the main chain binding area around residues 214–217 and 41 (green) and *uPA's*-specific surface loops (red). The right figure shows the structure of human uPA in complex with *N*-[4-(aminomethyl)phenyl]-6-carbamimidoyl-4-(pyrimidin-2-ylamino)naphthalene-2-carboxamide. The inhibitor is shown in CPK, the catalytic triad in pink, and Asp 60a in orange. The structures are displayed with PyMOL Viewer v0.99 on the basis of the PDB file 1sqa (Wendt et al. 2004). Note that the amino group of the inhibitor hydrogen bonds to the side chain of Asp60a

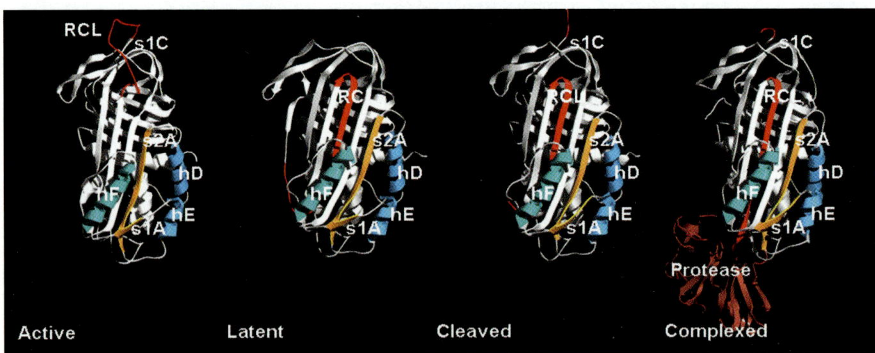

Fig. 35.3 Different conformations of plasminogen activator inhibitor-1 (PAI-1). The three-dimensional structures of active PAI-1 [(Sharp et al. 1999) pdb file 1B3K)], latent PAI-1 [(Stein et al., unpublished) pdb file 1LJ5], P_1-P_1' cleaved PAI-1 [(Aertgeerts et al. 1995) pdb file 9PAI], and the trypsin-α_1-proteinase inhibitor complex [(Huntington et al. 2000) pdb file 1EZX] are shown. The indicated structural elements of the serpins are the surface exposed or inserted RCL (red); α-helices D and E (blue); α-helix hF (blue-green); β-strands 1A and 2A (orange). The serine protease moiety of the complex is coloured dark red. The structures are displayed with Swiss-PBD Viewer v3.7b2. No structure is available of a plasminogen activator–PAI-1 complex. Therefore, the structure of a protease–serpin complex is illustrated by the trypsin–α_1-proteinase inhibitor complex. Please note that part of the structure of the protease moiety of the complex was not solved because of crystallographic disordering

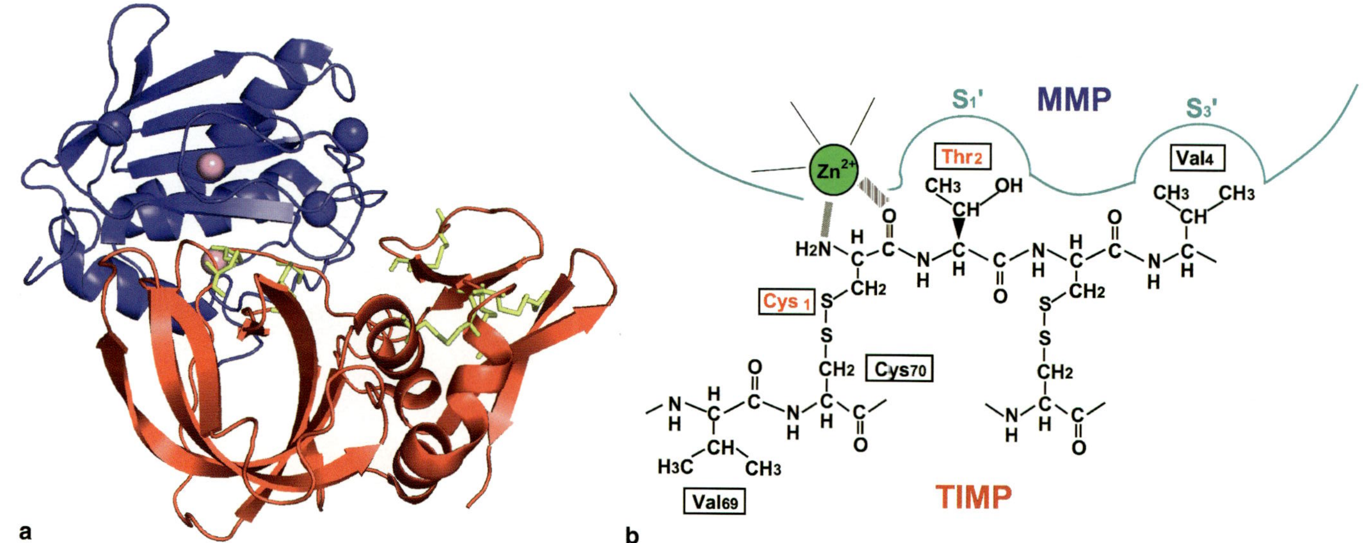

Fig. 37.1 The mode of interaction of tissue inhibitor of metalloproteinase (*TIMP*)-1 and the catalytic domain of matrix metalloproteinase (*MMP*)-3. **a** Ribbon diagram of the complex of TIMP-1 (red) and the catalytic domain of MMP-3 (blue). Zn^{2+} and Ca^{2+} are shown as pink and blue spheres, and the disulphide bonds in TIMP-1 are shown in yellow. The image was prepared from the Brookhaven Protein Data Bank (*PDB*) entry (1UEA) using the PyMOL program by W. L. DeLano (http://www.pymol.org). **b** Schematic representation of N-terminal region of TIMP-1 that interacts with the active site of MMP

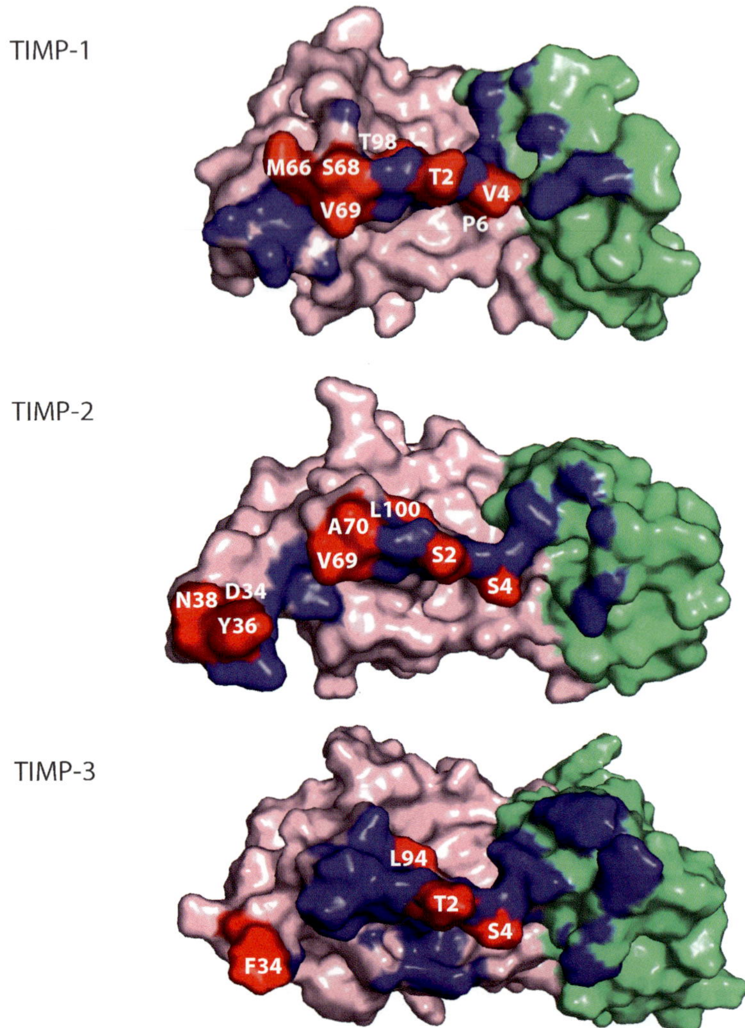

Fig. 37.2 Reactive sites of tissue inhibitor of metalloproteinase (*TIMP*)-1, TIMP-2 and TIMP-3. The N-terminal domain and the C-terminal domain of each TIMP are shown in pink and green, respectively. The residues in reactive ridge that are within 4 Å contact with the catalytic domain of a matrix metalloproteinase (*MMP*) are shown in blue and the residues that were subjected to mutagenesis studies are shown in red. TIMP-1 and TIMP-2 models were created from the Brookhaven Protein Data Bank (*PDB*) entry 1UEA and 1BUV, respectively, as in Fig. 37.1. The TIMP-3 structure was modelled based on crystal structures of TIMP-1 and TIMP-2

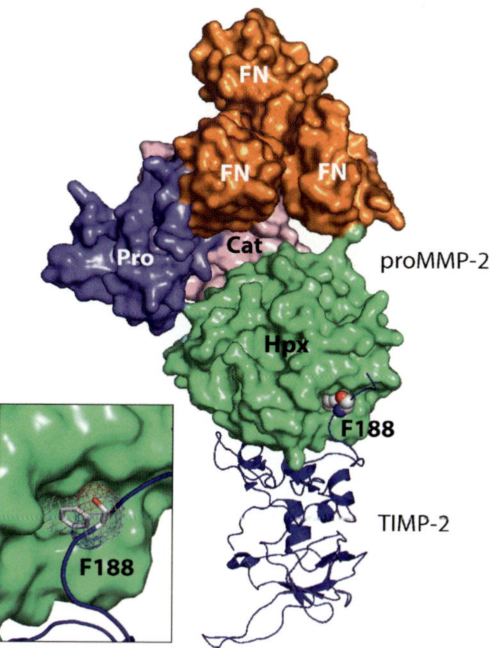

Fig. 37.3 Three-dimensional structure of the pro-MMP-2–TIMP-2 complex. Pro-MMP-2 is shown as surface representation and tissue inhibitor of metalloproteinase (*TIMP*)-2 as a ribbon diagram. Phe 188 of TIMP-2 is shown in sphere. Inset, the pocket in the Hpx domain with which Phe 188 of TIMP-2 interacts. This figure was prepared from the Brookhaven Protein Data Bank (*PDB*) entry 1GXD as in Fig. 37.1. Pro, pro-domain; Cat, catalytic domain; FN, fibronectin type II motif; Hpx, haemopexin domain

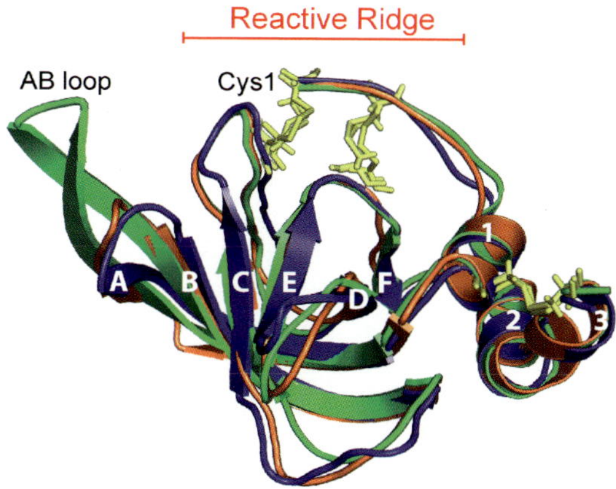

Fig. 37.4 Comparison of N-TIMP-1, N-TIMP-2 and N-TIMP-3 structures. Image of the N-terminal domain of tissue inhibitor of metalloproteinase (*TIMP*)-1 (blue) and that of TIMP-2 (green) were created from the Brookhaven Protein Data Bank (*PDB*) entry 1UEA and 1BUV, respectively, as in Fig. 37.1. N-TIMP-3 (orange) structure was modelled based on crystal structures of TIMP-1 and TIMP-2. Disulphide bonds are shown in yellow. TIMP-2 has a longer AB β strands than TIMP-1 and TIMP-3

Section V
Novel Therapeutic Strategies

Chapter 34
Structure and Inhibition of the Urokinase-Type Plasminogen Activator Receptor

Benedikte Jacobsen, Magnus Kjaergaard, Henrik Gårdsvoll, and Michael Ploug

Abstract The urokinase-type plasminogen activator receptor (uPAR/CD87) is a glycolipid-anchored receptor that is involved in focalizing plasminogen activation to the cell surface due to its high-affinity binding to the urokinase-type plasminogen activator (uPA). This chapter describes recent accomplishments in the molecular understanding of the structural biology of uPAR and its interactions with the cognate ligands, uPA and vitronectin. Furthermore, the structural basis for the pharmacological inhibition of uPAR by monoclonal antibodies, recombinant fusion proteins, and synthetic peptide antagonists are discussed. These compounds may prove valuable as drug candidates in combined intervention strategies targeting tumor invasion and metastasis.

Abbreviations

ANS	8-anilino-1-naphthalene sulfonate;
ATF	Amino-terminal fragment;
GFD	Growth factor-like domain;
GPI	Glycosylphosphatidylinositol;
LU	Ly-6/uPAR;
mAb	Monoclonal antibody;
PNH	Paroxysmal nocturnal hemoglobinuria;
SMB	Somatomedin B;
tPA	Tissue-type plasminogen activator;
uPA	Urokinase-type plasminogen activator;
uPAR	uPA receptor.

M. Ploug
Finsen Laboratory Section 3735, Copenhagen Biocenter, Rigshospitalet, Ole Maaløes Vej 5, Room 3.3.31, DK-2200 Copenhagen N, Denmark, e-mail: m-ploug@finsenlab.dk

D. Edwards et al. (eds.), *The Cancer Degradome.*
© Springer Science + Business Media, LLC 2008

Introduction

The first indications for the existence of a cellular binding site for the serine protease urokinase-type plasminogen activator (uPA) on cultured cells were reported more than three decades ago (Quigley 1976). The direct demonstration of a membrane protein responsible for this high-affinity uPA binding (later denoted the uPA receptor or uPAR) remained nonetheless elusive for almost another decade (Stoppelli et al. 1985, Vassalli et al. 1985, Nielsen et al. 1988). Despite a huge research effort, another two decades passed before the first three-dimensional structure of uPAR finally was solved by X-ray crystallography (Llinas et al. 2005). During this period, the involvement of uPAR in facilitating plasminogen activation and focusing the proteolytic activity to the cell surface through its high-affinity binding of pro-uPA ($K_D < 1$ nM) was being unraveled mostly by traditional biochemical and enzyme kinetic studies. In brief, these functional studies revealed that cells expressing uPAR are endowed with a highly efficient potential for cell surface-associated plasminogen activation due to the inherent amplification of the proteolytic activity by reciprocal zymogen activation, that is, pro-uPA activation by plasmin and plasminogen activation by uPA (Ellis et al. 1989). The primary role of uPAR in this reaction is to focus pro-uPA to the cell surface, where an optimal alignment of the two zymogens, that is, uPAR-bound pro-uPA and plasminogen bound to the cell surface via its lysine-binding kringle domains, lowers the apparent K_m for plasminogen activation in this particular microenvironment well below the plasma concentration of 2 μM plasminogen (Ellis and Danø 1993). This arrangement facilitates the initiation and amplification of this enzyme cascade. Inhibitor repression exerted by the cognate serpins plasminogen activator inhibitor type-1 (PAI-1) and α_2-antiplasmin further adds to the confinement of plasminogen activation to the cell surface, as α_2-antiplasmin selectively inhibits solution-phase plasmin due to its requirement for unoccupied lysine-binding kringles in the target protease (Ellis et al. 1991). Elegant genetic studies in mouse models clearly demonstrate that the above-mentioned biochemical pathway for uPAR–pro-uPA-mediated plasminogen activation is indeed operational and significant *in vivo* (Zhou et al. 2000, Liu et al. 2003, Bolon et al. 2004). It should, however, be emphasized that mice with targeted ablation of the uPAR gene nevertheless display no overt phenotypic abnormalities (Bugge et al. 1995b, Dewerchin et al. 1996), in contrast to mice genetically deficient in plasminogen (Bugge et al. 1995a). This observation illustrates the remarkable functional redundancy that exists in plasminogen activation during normal homeostasis and under pathological conditions, where plasminogen can be activated by other proteases besides the bona fide activators uPA and tissue-type plasminogen activator (tPA) (Selvarajan et al. 2001, Lund et al. 2006).

A number of so-called nonproteolytic functions of uPAR have also been reported, which implicate uPAR in cell adhesion and migration via a direct interaction with the matrix protein vitronectin (Waltz and Chapman 1994, Wei et al. 1994, Madsen et al. 2007), by modulation of the activity of certain integrins, for example,

$\alpha_5\beta_1$, $\alpha_M\beta_2$, and $\alpha_3\beta_1$ (Wei et al. 1996, 2001, 2007; Chaurasia et al. 2006), or by activation of a G-coupled receptor (FPRL1, formyl peptide receptor like 1) by defined proteolytic uPAR fragments (Resnati et al. 2002). As these interactions will be covered separately by other chapters in this volume, they will be discussed only briefly in the following sections when they have a direct bearing on established structure—function relationships in uPAR.

Structure of Human uPAR

Since the initial identification, purification, and sequencing of human uPAR were accomplished around 1990 (Nielsen et al. 1988, Behrendt et al. 1990, Roldan et al. 1990), a large literature has accumulated providing more than 1,000 entries in PubMed, when this database is searched by the keyword "uPAR." This chapter will, however, mainly focus on recent developments in our understanding of the structure, function, and inhibition of human uPAR, with a view to the two crystal structures recently solved for two different uPAR complexes (Llinas et al. 2005, Huai et al. 2006). Early data on the uPAR biochemistry will be treated only superficially in the section "Posttranslational Modification," and the reader is referred to a more comprehensive review on this subject (Ploug 2003).

Posttranslational Modification

The gene for human uPAR is located on chromosome 19q13 (Børglum et al. 1992), and its seven exons spanning 23 kb encode a 335 residues long precursor polypeptide, which is further processed by removal of the traditional N-terminal signal peptide dictating secretion (Roldan et al. 1990, Casey et al. 1994). As described in the following section, the mature uPAR comprises a single polypeptide of only 283 amino acids, as an additional signal peptide, which is required for the tethering of uPAR to the plasma membrane by a glycosyl-phosphatidylinositol (GPI) anchor, is removed from the C-terminus during processing (Ploug et al. 1991).

Membrane Attachment by a GPI-Anchor

Initially, uPAR was thought to span the plasma membrane with a conventional hydrophobic transmembrane type I domain (Roldan et al. 1990). Subsequent biochemical studies on purified uPAR demonstrated that this was clearly not the case, as uPAR is covalently attached directly to the outer leaflet of the lipid bilayer of the cell membrane by the addition of a C-terminal GPI-anchor (Ploug et al. 1991). This late posttranslational modification is accomplished at the luminal face of the endoplasmatic reticulum, where the translocated protein is processed by a 5-subunit transamidase complex (GPI8). This enzyme initially cleaves the C-terminal signal sequence, and subsequently catalyzes the amide bond formation between the newly formed carboxyl of the cleaved protein and a terminal ethanolamine phosphate of a preformed GPI-moiety (Orlean and Menon 2007). This mode of glycolipid anchoring has a number of important implications for the function of uPAR. First, the saturated fatty acyl chains of the GPI-anchor promote the temporal sequestering of uPAR in detergent-resistant membrane microdomains that are enriched with cholesterol and sphingolipids (i.e., lipid rafts). The local high density of uPAR in such relatively immobile microenvironments may provide uPAR with an apparent "functional multivalency," which could be significant in, for example, boosting the weak monovalent interaction of uPAR with the somatomedin B (SMB) domain of vitronectin (Gårdsvoll and Ploug 2007). With a view to this proposition, it is noteworthy that vitronectin predominantly binds to lipid raft-associated uPAR in HEK293 cells induced to overexpress this protein (Cunningham et al. 2003). Second, being deprived of a transmembrane domain, uPAR cannot per se transduce signals directly to the intracellular compartment and any signaling events involving uPAR must, therefore, rely on secondary interactions with other signaling competent proteins. Third, uPAR can be shed from the cell surface by GPI-specific phospholipases. Fourth, the physical property of the glycan moiety of the GPI-anchor allows uPAR a considerable spatial freedom relative to the plasma membrane, thus enabling it to adopt an optimal orientation relative to its interaction partners (Chevalier et al. 2006).

A pathological implication inherent to utilizing this type of membrane anchoring is the defective expression of uPAR on the surface of peripheral blood cells isolated from patients with the hematological stem cell disorder paroxysmal nocturnal hemoglobinuria (PNH) (Ploug et al. 1992b). The predominant genetic defects underlying this disease are single mutations affecting the X-linked phosphatidylinositol glycan-class A (PIG-A) gene, which encodes a subunit of the *N*-acetylglucosamine phosphatidylinositol transferase that catalyzes the first step in the biosynthesis of GPI-anchors (Orlean and Menon 2007). In the absence of suitable preformed GPI-substrates, the GPI8 transamidase hydrolyzes the peptide bond flanking the C-terminal signal sequence, and leukocytes affected by PNH consequently secrete a soluble form of uPAR lacking the GPI-anchor as well as the C-terminal signal sequence, but maintaining its high-affinity for uPA binding (Ploug et al. 1992a). Accordingly, patients with PNH have elevated plasma levels of soluble uPAR (Rønne et al. 1995).

N-Linked Glycosylation

Human uPAR contains five potential N-linked glycosylation sites and studies using recombinant uPAR expressed in Chinese hamster ovary cells or Drosophila S2 cells clearly show that only four of these sites are utilized, that is, Asn^{52}, Asn^{162}, Asn^{172}, and Asn^{200}, whereas the last position at Asn^{233} generally is unmodified (Ploug et al. 1998a, Gårdsvoll et al. 2004). Glycosylation at position Asn^{52} moderately influences the affinity of the uPA–uPAR complex (Møller et al. 1993, Ploug et al. 1998a, Gårdsvoll et al. 1999).

Ly-6/uPAR/α-neurotoxin Protein Domain Family

The first hint to the overall folding topology of uPAR came from primary structure considerations, demonstrating that this protein is composed of three internally repeated sequence motifs that are homologous to a large group of secreted snake venom α-neurotoxins (Ploug and Ellis 1994). The consensus sequence defining these Ly-6/uPAR/α-neurotoxin (LU) domains covers ~90 residues and is primarily based on conservation of eight cysteine residues and a single asparagine residue terminating the motif. Disulfide assignments and studies by limited proteolysis further consolidated the notion that these sequence repeats in uPAR really represent autonomous domain structures (Ploug et al. 1993). Finally, the organization of the uPAR gene, where each sequence repeat is encoded by separate exon-sets flanked by symmetrical phase-1 introns (Casey et al. 1994), implies that these LU domains represent evolutionary mobile protein modules prone to exon-shuffling by intronic recombination. The validity of this proposition is further advanced by the fact that this domain type is also found as the extracellular ligand-binding domain in the otherwise nonhomologous transforming growth factor-β receptor family (Table 34.1). Intriguingly, a small uPAR-like gene cluster covering 600 kb is found on chromosome 19q13, which has probably arisen by gene duplication, as it comprises all the hitherto known GPI-anchored proteins with more than one LU domain, that is, uPAR, C4.4A, PRV-1, PRO4353, and TX101 (Table 34.1). The prototype folding topology for the LU protein domain family is represented by the so-called three-finger fold found in numerous snake venom α-neurotoxins. As shown in Fig. 34.1a, this structure is dominated by a relatively large, antiparallel β-sheet projecting three loops ("fingers") from a globular core that contains the four consensus disulfide bonds and the invariant asparagine. Accordingly, it was proposed that

Table 34.1 Proteins belonging to the Ly-6/uPAR/α-Neurotoxin protein domain family[a]

Secreted single domains

    Snake venom α-neurotoxins:

    α-bungarotoxin (1KC4), κ-bungarotoxin (1KAB), erabutoxin a (1QXD), bucandin (1IJC), cardiotoxin (2BHI), α-cobratoxin (1YI5), fasciculin (1MAH), denmotoxin A (2H5F)

    Human proteins:

    SLURP-1 and -2, SP-10

GPI-anchored human proteins

    Single-domain proteins:

    MIRL/CD59 (1CDQ), SAMP14, LY6D/E48, LY6E/RIG-E, LY6H, PSCA

    Multidomain proteins:

    uPAR (1YWH, 2FD6, 2I9B), C4.4A, PRV-1/CD177, TX101, PRO4356

Type I transmembrane human proteins:

    TGF-β receptor I (1VJY) and II (1KTZ), activin receptor IIB (1LX5), bone morphogenetic protein receptor IA (1REW) and II (2HLR)

[a] This table provides a nonexhaustive list of members of the Ly-6/uPAR/α-neurotoxin protein domain family, showing the functional and structural diversity of the modular proteins encompassing this domain type. The protein database entries are provided in parentheses where the three-dimensional structures have been determined

uPAR was composed of three homologous LU domains, each adopting the three-finger fold found in the snake venom α-neurotoxins (Ploug and Ellis 1994)—a proposition that was corroborated a decade later, when the crystal structure of uPAR was solved, as shown in Fig. 34.1b, c (Llinas et al. 2005).

Assembly of a Functional Multidomain uPAR

Several lines of biochemical evidence have established a close correlation between maintenance of the intact multidomain assembly of uPAR and preservation of high-affinity ligand binding. Disruption of interdomain interactions by specific cleavage of the linker region between uPAR domain I (DI) and domain II (DII) by limited proteolysis thus impairs high-affinity uPA, as well as vitronectin binding (Ploug et al. 1994, Høyer-Hansen et al. 1997). Along the same lines, the very high sensitivity of uPA binding to low concentrations of guanidine hydrochloride and moderately lowered pH probably reflects subtle rearrangements in the interdomain assemblies of the individual LU domains in uPAR (Ploug et al. 1994, Ploug 1998, Jacobsen et al. 2007). Finally, the intact, unoccupied uPAR possesses a solvent-exposed hydrophobic patch that can be probed by the extrinsic fluorophore 8-anilino-1-naphthalene sulfonate (ANS). Importantly, this exposed hydrophobic site is lost after cleavage of the linker region between uPAR DI–DII and by complex formation with pro-uPA or low-molecular weight peptide antagonists (Ploug et al. 1994, 1998b, 2001). The ANS fluorescence hence reports on the availability of an unoccupied and functional high-affinity binding site on uPAR for uPA.

The crystal structures recently solved for uPAR nicely illustrate how the three-finger fold can be assembled into a modular protein, creating a complex, high-affinity binding site for the serine protease uPA (Llinas et al. 2005, Barinka et al. 2006, Huai et al. 2006). As shown in Fig. 34.1c, e, the three individual LU domains in uPAR are intimately assembled in a right-handed orientation, creating a large 13-stranded antiparallel β-sheet. A pronounced domain interface is established between the β-sheets of DI and DII, where the unusual bending of the βIID strand (centered on Gly^{146}) enables its "bifurcate" interaction with β-strands in both DI (βIE) and DII (βIIC). Also contributing to the total DI–DII interface, which covers more than 1,000 Å^2, is the interaction governed by residues 94–115 located in the small detached β-sheet formed by βIIA and βIIB. An equivalently organized interface is created between uPAR DII and DIII by the engagement of the analogous β-strands, that is, a bent βIIE and βIIID. This domain assembly creates a unique and dynamic topology for uPAR with a 19 Å deep hydrophobic ligand-binding cavity on the "front" side with the long flexible linker regions and glycosylation sites residing on the "back" of the molecule (Fig. 34.1c–f). Depending on the ligand used for cocrystallization, uPAR can adopt either an "open" croissant-like topology with a breach between DI and DIII (Fig. 34.1c, d), as observed for the uPAR–peptide antagonist complex (Llinas et al. 2005), or a more closed conformation with a hydrogen-bonding network between DI and DIII (Fig. 34.1e, f), as found in the uPAR–ATF complex (Huai et al. 2006). A

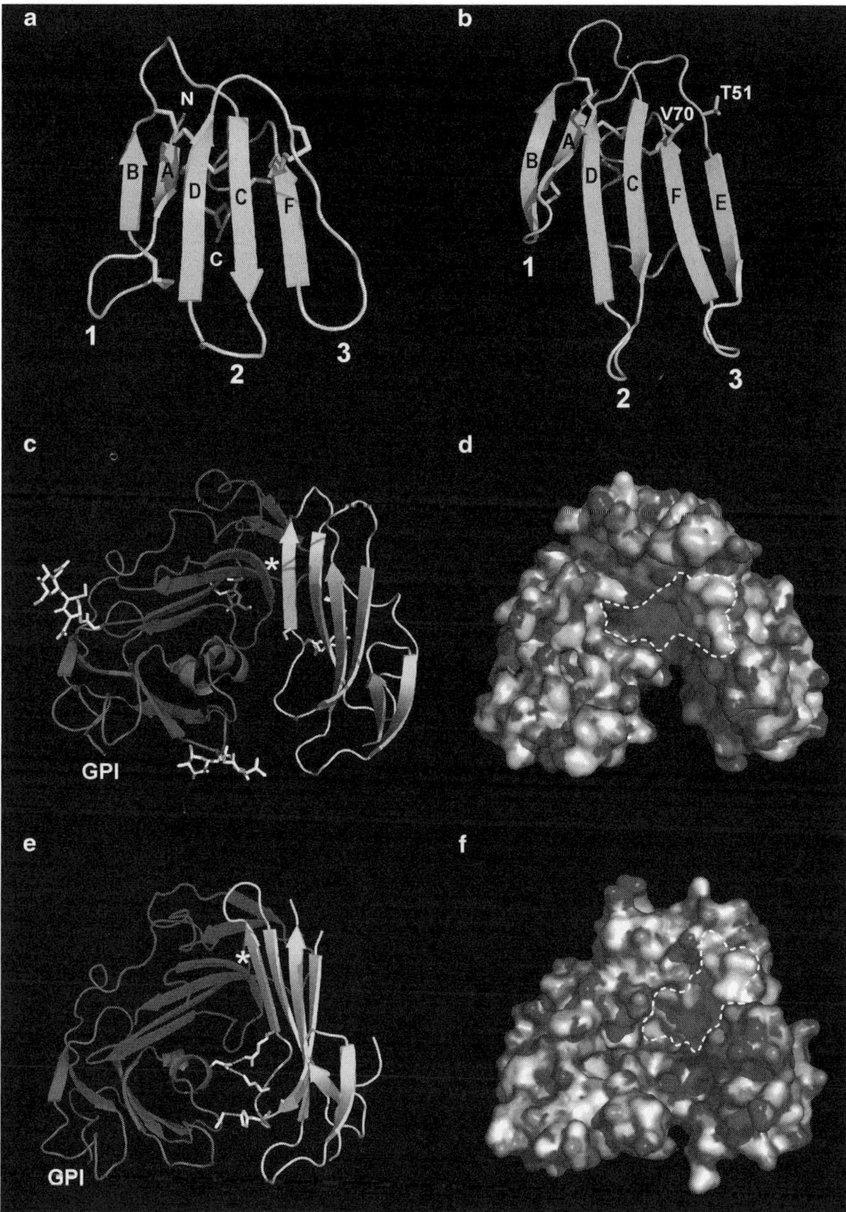

Fig. 34.1 Crystal structure of human urokinase-type plasminogen activator receptor (*uPAR*). The prototype structure for the three-finger fold is shown in panel **a** by a cartoon representation of the nuclear magnetic resonance (*NMR*) structure solved for the weak α-neurotoxin bucandin isolated from the Malayan krait (1IJC). The disulfide bonds are shown as sticks in yellow, the invariant asparagine in red, and the β-sheets in cyan with their identification code in capital letters. For comparison of the folding topology, the structure of *uPAR* DI is depicted in panel **b** in a similar orientation and color coding (1YWH). The position of the missing consensus disulfide in *uPAR* DI

detailed comparison of these two structures reveals the dynamic flexibility of the overall folding topology of uPAR, where DI can undergo a 20° rotation relative to DII. This rotation is centered on the aforementioned hinge between the bent βIID and βIE and enforces a significant movement of various loop regions. Those connecting βIB-βIC (residues 16–23) and βID-βIE (residues 46–53) are, for example, displaced by 13 and 9.5 Å, respectively, to accommodate the two different ligands.

As illustrated in Fig. 34.1d, f, both conformations of uPAR harbor a deep and very hydrophobic ligand-binding cavity, which is occupied by the bound ligands. The pronounced hydrophobicity of this cavity is mainly governed by aliphatic side chains residing in DI (Val^{29}, Leu^{31}, Leu^{38}, Leu^{40}, Leu^{55}, Tyr^{57}, and Leu^{66}) and DII (Leu^{123}, Val^{125}, Leu^{144}, Leu^{150}, Pro^{151}, and Leu^{168}), but in the complex, this patch is completely shielded from solvent exposure by the bound ligand. With a view to this, it has been speculated that the entropic costs of maintaining this architecture of the hydrophobic-binding cavity in the unoccupied state is so unfavorable that uPAR is likely to undergo some kind of rearrangement to shield the vacant ligand-binding cavity (Yuan and Huang 2007). Whether such a "latent" uPAR conformation actually exists remains to be established experimentally, but it should be emphasized that the ligand-free uPAR does bind ANS, demonstrating that at least some residual hydrophobic surface is still solvent-exposed in this state (Ploug et al. 1994).

Another important feature of the structures solved for uPAR is the exposed and flexible nature of the linker peptide between DI and DII, where residues 84–90 were defined in neither of the complexes due to lack of sufficient electron densities (Llinas et al. 2005, Huai et al. 2006). This is in excellent agreement with biochemical findings showing that this particular region is extremely sensitive to proteolysis by a wide variety of proteases, for example, chymotrypsin (Behrendt et al. 1991), uPA (Høyer-Hansen et al. 1992), MMP-12 (Koolwijk et al. 2001), and the type-II transmembrane human airway trypsin-like protease (Beaufort et al. 2007). Cleavage of this linker peptide not only renders uPAR deficient in uPA and vitronectin binding but also seems to uncover an alleged chemotatic neo-epitope on uPAR DII–

Figure 34.1 (Continued) between βIE and βIF is highlighted by the side chains of Thr^{51} and Val^{70} occupying this position. Cartoon representations of the crystal structures solved for *uPAR*–peptide (1YWH) and *uPAR–ATF* (amino-terminal fragment) (2FD6) complexes are shown in panels **c** and **e**, respectively, after removal of the bound ligands. The individual Ly-6/uPAR (*LU*) domains in *uPAR* are colored yellow (DI), blue (DII), and red (DIII) and the positions of the *GPI* (glycosylphosphatidylinositol)-anchor are indicated. The positions of the carbohydrates are indicated in panel **c** by the white sticks of the innermost two *N*-acetyl-glucosamines, whereas the bent β-strand βIID is highlighted by an asterisk in both panels. The interdomain hydrogen bonds established between DI and DIII in the *uPAR–ATF* structure (His^{47}-Asn^{259}, Arg^{53}-Asp^{254}, and Lys^{50}-Asp^{254}) are highlighted in green in panel **e**. The corresponding molecular surface representations of the *uPAR*–peptide and *uPAR–ATF* complexes are shown in panels **d** and **f** colored by atom: carbon (white), nitrogen (blue), oxygen (red), and sulfur (yellow). The conspicuous hydrophobic walls of the binding cavities in the two structures are delimited by white hatched lines. These molecular representations were created by PyMOL (DeLano Scientific) (*See also* Color Insert II)

DIII that engages and activates the G-coupled FPRL1 receptor (Resnati et al. 1996, 2002).

Structural Aspects of Ligand Binding to uPAR

Through the combination of biochemical binding analyses using a vast number of single-site uPAR mutants and structure elucidation by crystallography, a detailed knowledge on the structure–function relationships in the uPA–uPAR interaction has now been accomplished. Data on structure–function aspects of uPAR–vitronectin binding have also emerged recently using an equivalent mutagenesis approach.

Structure of uPA–uPAR Complexes

The primary ligand for uPAR is the serine protease uPA, which is a modular enzyme composed of the N-terminal growth factor-like domain (GFD, residues 1–46), followed by a kringle domain (47–135), and finally the serine protease domain. Structures for neither the zymogen pro-uPA nor the corresponding two-chain uPA with a cleaved activation site at Lys^{158}-Ile^{159} have been determined. In contrast, several structures have been solved for both the amino-terminal fragment (ATF, residues 1–135) and the serine protease domain as separate entities (Hansen et al. 1994, Barinka et al. 2006). As opposed to uPAR, all determinants required for the high-affinity of the pro-uPA–uPAR interaction ($K_D < 1$ nM) are retained within a single module, that is, GFD^{1-48} of uPA (Ploug et al. 1998b). Studies by site-directed mutagenesis implicate the major β-hairpin (Ω-loop) of GFD in uPAR binding, where Tyr^{24}, Phe^{25}, Ile^{28}, and Trp^{30} are particularly important (Magdolen et al. 1996).

Recently, the crystal structure was solved for human uPAR in a ternary ATF–uPAR–Fab complex (Huai et al. 2006), and this structure proved identical to the one subsequently solved for the bimolecular ATF–uPAR complex (Barinka et al. 2006). As illustrated in Fig. 34.2a, b, the β-hairpin of GFD efficiently inserts into the deep hydrophobic-binding cavity in uPAR, resulting in the burial of $\sim$2,300 Å^2 solvent-exposed molecular surface at the interface of the ATF–uPAR complex. Accordingly, the previously mentioned hot spot residues in GFD for the high-affinity binding to uPAR are all intimately involved in van der Waals contacts with the hydrophobic residues from DI and DII lining the walls of the binding cavity. The energetic contributions of each individual amino acid side chain in uPAR to the uPA interaction were mapped by a comprehensive mutagenesis analysis (Gårdsvoll et al. 2006). From this study, it is evident that no single residue in uPAR provides the major contribution to the free energy of binding as is the case for uPA, but the high-affinity binding is rather accomplished by the added contributions from the previously mentioned hydrophobic residues in DI and DII that line the binding cavity (Fig. 34.2b). Interestingly, this study also identified Asp^{140} as a key residue for the ATF–uPAR complex, and being situated in the loop connecting βIIC and βIID

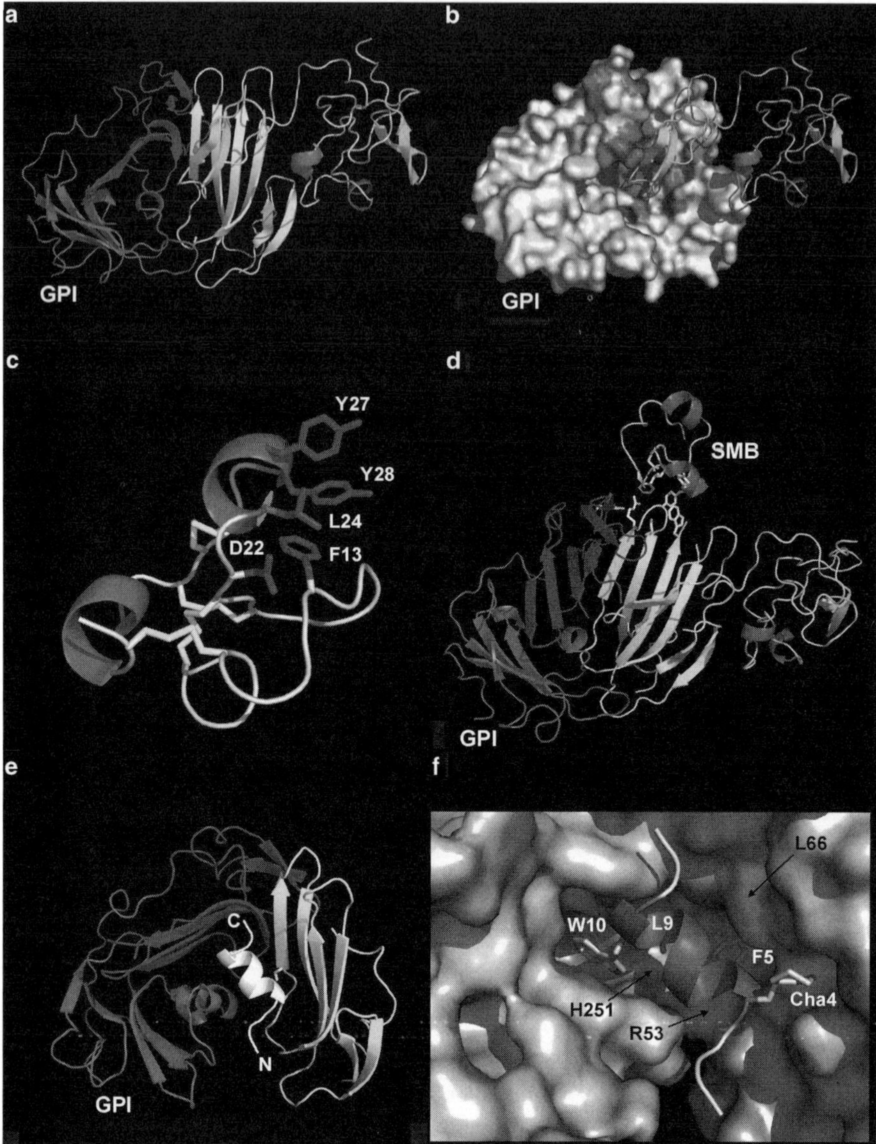

Fig. 34.2 Structures of human urokinase-type plasminogen activator receptor (*uPAR*) in complex with amino-terminal fragment (*ATF*), somatomedin B (*SMB*), and a synthetic peptide antagonist. The crystal structure of *uPAR–ATF* (2FD6) complexes are shown as a cartoon representation in panel **a** and as a combined surface and cartoon representation in panel **b**. Surface-exposed residues in *uPAR* causing a $\Delta\Delta G > 0.5$ kcal/mol on urokinase-type plasminogen activator (*uPA*) binding upon alanine substitution (Gårdsvoll et al. 2006) are highlighted in panel **b** by red. The solution structure of *SMB* (2JQ8) is shown in panel **c** with the residues important for *uPAR* binding shown as blue sticks. A model for the ternary *SMB–uPAR–ATF* complex with residues important for this interaction is shown as white sticks in panel **d** (Gårdsvoll and Ploug 2007). The crystal structure of

at the rim of the binding cavity, it undergoes a major repositioning upon ATF binding. The unique "knot and hole" topology of the ATF–uPAR interface may provide an optimal geometry for the targeting of uPAR by low-molecular weight antagonists with a view to therapeutic intervention in cancer (as discussed in the section "Peptide Antagonists").

Both structures of the ATF–uPAR complex position the kringle of uPA close to uPAR DI outside the central binding cavity, where van der Waals contact is established between Asp[11] in DI and His[87] of the kringle (Fig. 34.2a, b). Although some undefined role of the kringle in "stabilizing" the uPA–uPAR complex has been reported (Bdeir et al. 2003), this interaction is probably too weak to translate into a measurable change in affinity upon mutation of any residues in uPAR DI including Asp[11], and accordingly, GFD[1–48] and ATF[1–135] display comparable affinities for uPAR (Ploug et al. 1998b, Gårdsvoll et al. 2006). Nonetheless, the orientation of the kringle in the ATF–uPAR complex provides a molecular basis for the observation that human but not murine uPA–uPAR complexes can be stabilized covalently by amine-reactive homobifunctional cross-linkers (Estreicher et al. 1989). The target sites for this conjugation are Lys[43] in uPAR DI and Lys[98] in the kringle of uPA (Gårdsvoll et al. 2006). In the crystal structure, these are positioned 10 Å apart within the effective range of the employed cross-linkers. As the target site in human uPAR (Lys[43]) is replaced by a nonreactive Arg[43] in the mouse uPAR, the equivalent covalent complex cannot be established in the latter species, despite the formation of a stable noncovalent ATF–uPAR complex.

Structure of uPAR–Vitronectin Complexes

Another important function of uPAR is its involvement in cell adhesion and migration on vitronectin-rich matrices, as first demonstrated for cytokine-stimulated human myeloid cells (Waltz and Chapman 1994). Interestingly, the efficiency of uPAR-mediated cell adhesion to vitronectin is controlled by the level of uPA saturation, setting the stage for its possible regulation. In cultured primary human microvascular endothelial cells, uPAR and uPA colocalize to focal adhesion sites in a vitronectin-dependent process (Salasznyk et al. 2007). Among other factors, this subcellular redistribution is controlled by the N-terminal SMB domain of vitronectin, which is also the major binding determinant for the vitronectin–uPAR complex (Deng et al. 1996). In transfected cell lines having high expression levels of uPAR,

---

Figure 34.2 (Continued) *uPAR*–AE147 (1YWH) is shown as a cartoon representation in panel **e** and in a close up as a combined cartoon and surface representation in panel **f**. The four hydrophobic hot spot residues of AE147 (K-S-D-Cha-F-s-k-Y-L-W-S-K) are shown as white sticks and the target sites in *uPAR* for photoactivatable analogues of AE147 are shown as yellow surfaces (i.e., His[251] being target for a photoprobe replacing Trp[10], whereas Leu[66] and Arg[53] are target sites for insertions from photoprobes replacing Phe[5]). Capital letters denote amino acids in L-configuration, whereas lower case letters signify a D-configuration. Cha is cyclo-*L*-hexylalanine. These molecular representations were created by PyMOL (DeLano Scientific) (*See also* Color Insert II)

this uPA dependence on vitronectin adhesion is uncoupled (Kjøller and Hall 2001, Madsen et al. 2007). Notwithstanding this limitation, such transfected cell lines were instrumental for deciphering the pronounced changes in cell morphology and activation of signaling pathways that accompany the increased cell adhesion and motility induced by the vitronectin–uPAR interaction (Kjøller and Hall 2001, Madsen et al. 2007).

No structures have as yet been solved for intact vitronectin, but the structure of its SMB domain has been solved both in the free form (Kjaergaard et al. 2007) and in complex with PAI-1 (Zhou et al. 2003). The functional binding sites on SMB for the serpin PAI-1 and uPAR are overlapping and encompass Phe^{13}, Asp^{22}, Leu^{24}, Tyr^{27}, and Tyr^{28} (Deng et al. 1996, Gårdsvoll and Ploug 2007). The complementary binding interface on uPAR was mapped by an exhaustive alanine-scanning mutagenesis interrogating 244 individual positions in human uPAR (Gårdsvoll and Ploug 2007). Intriguingly, only five positions in uPAR proved important for vitronectin binding, and these occupy contiguous positions on uPAR DI (Trp^{30}, Arg^{58}, and Ile^{63}) and the flanking linker region between DI and DII (Arg^{91} and Tyr^{92}). This epitope is also critically involved in uPAR-induced cell adhesion and migration on vitronectin-coated matrices (Madsen et al. 2007). Docking the SMB domain manually on this epitope indicates that Arg^{91} would interact with Asp^{22} in SMB (Fig. 34.2d), thus bearing a striking resemblance to the organization of the interface in the SMB–PAI-1 interface, where Arg^{101} in PAI-1 forms an ionic bond with Asp^{22} (Zhou et al. 2003). In our model, no interaction seems possible between SMB and ATF in the ternary complex SMB–uPAR–ATF (Fig. 34.2d). In concordance with this proposition, the affinities of SMB for preformed complexes between uPAR and GFD, ATF, or pro-uPA are indistinguishable (Gårdsvoll and Ploug 2007). This model is now verified experimentally by the crystal structure solved for SMB-uPAR-ATF (Huai et al. 2008).

In contrast to the high-affinity of the GFD–uPAR interaction, SMB exhibits a much weaker binding for unoccupied uPAR with a K_D of ~2 μM. Importantly, this affinity is increased fourfold to 0.4 μM if the hydrophobic binding cavity in uPAR is preoccupied with either GFD, ATF, or pro-uPA, signifying a possible allosteric regulation of the vitronectin-binding site by uPA (Gårdsvoll and Ploug 2007).

Targeting the uPA–uPAR Interaction

Studies on some human malignant conditions, such as colon and breast carcinomas, have implicated uPAR as an adverse factor for disease progression, where high uPAR levels in plasma or resected tumor tissue lysates are correlated to poor prognosis (Stephens et al. 1999, Foekens et al. 2000). Histological studies have shown that uPAR is predominantly expressed by the activated tumor-associated stroma and in a few detached solitary tumor cells, as typified by the expression profiles in the invasive areas of colon carcinomas and in ductal carcinoma in situ (DCIS) with microinvasion of the breast (Rømer et al. 2004, Nielsen et al. 2007). Consequently, uPAR is considered a possible therapeutic target for the development of drugs with implications for treatment of invasion and metastasis in certain human cancers. A large literature exists on downregulation of uPAR expression by gene therapy, but this is considered outside the primary scope of this chapter and the reader is referred to a recent review on this interesting topic (Pillay et al. 2007). The following section will briefly describe some of the avenues exploited to develop specific antagonists targeting the uPA–uPAR interaction (Rømer et al. 2004).

Monoclonal Antibodies

During the last decade, the use of monoclonal antibodies (mAbs) as treatment modalities in cancer therapy has become widely accepted (Reichert and Valge-Archer 2007). In line with this, a number of monoclonal anti-uPAR antibodies have been developed that inhibit the uPA–uPAR interaction. Interestingly, inhibitory mAbs that are reactive against uPAR DI seem to bind one of the two nonoverlapping antigenic sites on this domain. While the one group constitutes conventional competitive inhibitors of the uPA–uPAR interaction, the other group acts as allosteric inhibitors forming a transient ternary uPA–uPAR–mAb complex (List et al. 1999, Pass et al. 2007). The molecular basis for the latter mode of inhibition is still unknown, but it probably relates to the above-mentioned dynamic nature of the three-domain assembly in uPAR that directly modulates the architecture of the hydrophobic-binding cavity.

While a plethora of studies have demonstrated the inhibitory effects of anti-uPAR mAbs in various *in vitro* settings, the use of such mAbs in animal models is still in its infancy. A recent study has nevertheless reported a remarkable efficacy of systemic administration of an anti-uPAR mAb on primary tumor growth, peritoneal invasion, and liver metastasis of an orthotopically transplanted pancreatic carcinoma cell line (Bauer et al. 2005). Notwithstanding this impressive effect, it is important to emphasize that in human cancer uPAR is predominantly expressed by the tumor-associated stroma, and this is not targeted in the above kind of xenotransplanted tumors. It is becoming increasingly evident that the tumor microenvironment is actively involved in the malignant progression of dysfunctional epithelial cells and, therefore, represents an obvious target during therapeutic intervention (Albini and Sporn 2007). To enable targeting of uPAR in tumor as well as stroma in genetic mouse cancer models, a number of inhibitory mouse mAbs against murine uPAR were developed in uPAR-deficient mice (Pass et al. 2007). Although the *in vivo* efficacy of these antibodies after systemic administration is well documented by their inhibition of receptor-bound uPA-mediated plasminogen activation (Pass et al. 2007) and fibrin surveillance in the liver (Jögi et al. 2007), no data are yet available in tumor models.

One unintended corollary of therapeutically targeting a GPI-anchored molecule like uPAR present on peripheral blood cells is the risk of developing a transient PNH phenotype during sustained mAb administration. This phenomenon has been observed in several patients treated for chronic lymphocytic leukaemia with

Alemtuzumab (Campath-1H), an immunosuppressive mAb reactive with the GPI-anchored CD52 (Taylor et al. 1997).

Recombinant Fusion Proteins

Taking advantage of the distinct topography of the receptor-binding module in uPA, several hybrid proteins have been designed exploiting GFD or ATF for the specific targeting of uPAR. In some cases, these receptor-binding modules were fused to larger proteins to prolong their half-lives and were used as competitive inhibitors of the uPA–uPAR interaction (Min et al. 1996), whereas in other cases specific protease inhibitor domains were attached, such as bovine pancreatic trypsin inhibitor or urinary trypsin inhibitor, to obtain a dual inhibitory effect (Kobayashi et al. 1998, Quax et al. 2001). In one case, an exposed loop region of cystatin was even replaced by just the minimal receptor-binding β-hairpin of GFD and this construct bound human uPAR with high affinity (Muehlenweg et al. 2000). These examples clearly illustrate the accessibility and versatility of the ligand-binding cavity in uPAR for targeting by such hybrid proteins.

Peptide Antagonists

Although protein–protein interfaces in general are challenging targets for small molecule inhibitor design, the unique topology of the deep and hydrophobic-binding cavity in uPAR may render this particular receptor more amenable for such approaches. Several attempts at developing small molecule inhibitors of the uPA–uPAR interaction by traditional screening methods in chemical libraries have accordingly resulted in the disclosure of a few hydrophobic compound families having IC_{50} values in the lower nanomolar range (Rosenberg 2001). In this section, we will, however, primarily focus on the exploration of two principally different vistas toward the development of specific peptide-based inhibitors of the uPA–uPAR interaction.

Cyclic Peptides Derived from the β-Hairpin of GFD

The intimate engagement of the β-hairpin of GFD (residues 18–32) in the interaction with uPAR led Magdolen and co-workers to explore this motif as a potential lead for the development of a small peptide-based inhibitor of the uPA–uPAR interaction (Bürgle et al. 1997, Magdolen et al. 2001, Schmiedeberg et al. 2002). In particular, the optimal presentation of the residues in the Ω-loop connecting the two major β-strands in GFD (i.e., [22]NKYFSNI[28]) seemed to be crucial for achieving high affinity and efficacy in competing the uPA–uPAR interaction. After a comprehensive synthesis program, they arrived at the cyclic decapeptide <u>c</u>NKYFSNI<u>C</u>W, where the establishment of a disulfide bond between the two underlined cysteines was decisive for its activity (c denotes D-cysteine). This compound exhibited an

IC_{50} of 40 nM for the inhibition of uPA binding to uPAR. Residues in bold are essential for the antagonistic properties of this peptide and accordingly the NMR-derived solution structure of the peptide showed that these side chains tend to form a hydrophobic cluster on one side of the molecule, thus mimicking the organization in the Ω-loop of ATF (Schmiedeberg et al. 2002). Most likely, this modified cyclic peptide still retains it original binding specificity, and thus targets the hydrophobic-binding cavity of uPAR (Fig. 34.1d, f). This assumption is substantiated by the observation that a corresponding cyclic peptide representing uPA[19–31] inhibits the uPAR-induced ANS fluorescence with a 1:1 stoichiometry, as do the genuine protein ligands (Ploug et al. 1998b). To improve the protease stability of the peptide and increase its half-life in biological solutions, the single lysine present in the peptide needed to be replaced by norleucine. Daily intraperitoneal administration of this cyclic peptide was capable of reducing the overall tumor burden of an orthotopic xenograft of human ovarian cystadenoma carcinoma cell line in nude mice (Sato et al. 2002).

Linear Peptide Antagonists Derived by Combinatorial Chemistry

In contrast to the above-mentioned rational design of uPA-derived cyclic peptides, Goodson and coworkers chose an unbiased approach using selection in a naive phage-display library by cell lines expressing high levels of uPAR (Goodson et al. 1994). This resulted in a large number of 15-mer inhibitory peptides displaying IC_{50} values ranging from 10 nM to 10 μM for the uPA–uPAR interaction. Further studies on structure-activity relationships and affinity maturation by combinatorial chemistry identified a novel 9-mer inhibitory peptide denoted AE105[1] that forms a very tight 1:1 complex with uPAR displaying a K_D of ~0.4 nM (Ploug et al. 2001, Jørgensen et al. 2004). This peptide competes ATF binding to cells with an IC_{50} of 10 nM and as it also inhibits the uPAR-mediated ANS fluorescence with a 1:1 stoichiometry, the most likely target is the hydrophobic ligand-binding cavity in uPAR. Early studies by photoaffinity labeling further substantiated this proposition, as illustrated in Fig. 34.2f, where His[251], Arg[53], and Leu[66] located at the floor of the cavity served as targets for the specific photoinsertions derived from peptides having photoprobes replacing Trp[10] and Phe[5] (Ploug 1998). Derivatives of AE105 were instrumental for solving the first crystal structure of uPAR (Llinas et al. 2005). In the uPAR–AE147 complex, the peptide antagonist adopts a right-handed 3.6 α-helix which is tightly embedded in the central binding cavity where it buries ~2,000 Å^2 of the solvent-accessible surface (Fig. 34.2e, f). In particular, the hot spot residues (Phe[5], Leu[9], and Trp[10]) are firmly "anchored" to the hydrophobic floor of the cavity, thus providing a structural argument for their large contribution

---

[1] AE105 has the sequence D-Cha-F-s-r-Y-L-W-S, where capital letters denote amino acid in L-configuration and lower case D-configuration. Cha is cyclo-LG for uPAR binding. AE120 is a pseudosymmetrical dimer of two AE105 peptides linked at their carboxytermi: $[AE105]_2$-βA-K, where βA is β-L-alanine (Ploug et al. 2001)

to the high-affinity interaction of uPAR–AE105. Kinetic studies by amide ^{1}H/^{2}H exchange on uPAR–AE105 complexes in solution demonstrate the long-lasting protection of seven amide backbone hydrogen atoms in the peptide by uPAR, in accordance with the α-helical structure of the bound peptide (Jørgensen et al. 2004). In future studies, the noncorrelated ^{1}H/^{2}H exchange observed for the bound peptide may furthermore provide valuable information on changes in the dynamic properties of the hydrophobic cavity upon binding of the aforementioned mAbs that induce allosteric inhibition of uPA binding.

AE105 and its pseudosymmetrical dimer AE120[1] in particular are relatively resilient to proteolytic degradation due to their unique composition of both natural and unnatural amino acids (Ploug et al. 2001). This has enabled their use in various experimental settings using biological materials. First, AE120 was instrumental for the development of a time-resolved immunofluorescence assay specific for the quantitative detection of various cleaved forms of uPAR in plasma from patients with cancer (Piironen ct al. 2004). Second, immobilized AE120 serves as an excellent affinity matrix for the efficient one-step purification of uPAR (Jacobsen et al. 2007). Third, intravasation of human HEp-3 carcinoma cells into the circulation of the chicken embryo by invading the chorioallantoic membrane is specifically inhibited by AE120 (Ploug et al. 2001). The strict species specificity of this class of peptide antagonist has unfortunately limited their application to cancer studies in xenograft mouse models, as they do not inhibit the corresponding murine uPA–uPAR interaction.

The topology of the uPAR–AE147 complex, where the N-terminal end of the peptide is positioned in the breach between DI and DIII (Fig. 34.2e, f), suggests that this region can be modified without devastating losses in affinity or specificity. Supporting this notion are the experimental observations that N-terminal extensions of AE105 affect neither the binding kinetics with uPAR (Ploug et al. 2001) nor the protection of the core α-helix of the bound peptide from ^{1}H/^{2}H exchange (Jørgensen et al. 2004). More importantly, the added amide-hydrogen atoms in the extension of AE105 exchange freely with the solvent (Jørgensen et al. 2005). Combined, these properties make AE105 and its derivatives obvious candidates as reporter substances for imaging of uPAR expression *in vivo*, using positron emission tomography after introduction of a chelated radionucleotide at the N-terminus of the peptide (Li et al. 2008). Ultimately, such peptide conjugates could also find a possible application in uPAR-targeted radiotherapy using a chelated α-emitter such as ^{213}Bi (Knör et al. 2007).

Conclusions

The emergence of the crystal structures of uPAR in complex with a peptide antagonist and the receptor-binding module of its cognate ligand uPA has revealed how a small three-finger fold, primarily known for its toxic properties in the snake venom α-neurotoxins, can be assembled into a complex multidomain receptor, creating a surprisingly large and hydrophobic ligand-binding cavity. It will be interesting to unravel how this dynamic domain assembly can be modulated by physiological ligands as well as therapeutic substances.

Acknowledgments We thank Dr. Mingdong Huang (Fujian Institute of Research on the Structure of Matter, Chinese Academy of Sciences) for many helpful discussions. Photographer John Post is thanked for excellent assistance in creating the photographic artwork. This work was supported by the European Union Contract LSHC-CT2003-503297.

References

Albini A, Sporn M B. (2007). The tumour microenvironment as a target for chemoprevention. Nat Rev Cancer 7:139–147.

Barinka C, Parry G, Callahan J et al. (2006). Structural basis of interaction between urokinase-type plasminogen activator and its receptor. J Mol Biol 363:482–495.

Bauer T W, Liu W, Fan F et al. (2005). Targeting of urokinase plasminogen activator receptor in human pancreatic carcinoma cells inhibits c-Met- and Insulin-like growth factor-1 receptor-mediated migration and invasion and orthotopic tumor growth in mice. Cancer Res 65:7775–7781.

Bdeir K, Kuo A, Sachais B S et al. (2003). The kringle stabilizes urokinase binding to the urokinase receptor. Blood 102:3600–3608.

Beaufort N, Leduc D, Eguchi H et al. (2007). The human airway trypsin-like protease modulates the urokinase receptor (uPAR, CD87) structure and functions. Am J Physiol Lung Cell Mol Physiol 292:L1263–1272.

Behrendt N, Rønne E, Ploug M et al. (1990). The human receptor for urokinase plasminogen activator. NH_2-terminal amino acid sequence and glycosylation variants. J Biol Chem 265:6453–6460.

Behrendt N, Ploug M, Patthy L et al. (1991). The ligand-binding domain of the cell surface receptor for urokinase-type plasminogen activator. J Biol Chem 266:7842–7847.

Bolon I, Zhou H M, Charron Y et al. (2004). Plasminogen mediates the pathological effects of urokinase-type plasminogen activator overexpression. Am J Pathol 164:2299–2304.

Børglum A D, Byskov A, Ragno P et al. (1992). Assignment of the urokinase-type plasminogen activator receptor gene (PLAUR) to chromosome 19q13.1–q13.2. Am J Hum Genet 50:492–497.

Bugge T H, Flick M J, Daugherty C C et al. (1995a). Plasminogen deficiency causes severe thrombosis but is compatible with development and reproduction. Genes Dev 9:794–807.

Bugge T H, Suh T T, Flick M J et al. (1995b). The receptor for urokinase-type plasminogen activator is not essential for mouse development or fertility. J Biol Chem 270:16886–16894.

Bürgle M, Koppitz M, Riemer C et al. (1997). Inhibition of the interaction of urokinase-type plasminogen activator (uPA) with its receptor (uPAR) by synthetic peptides. Biol Chem 378:231–237.

Casey J R, Petranka J G, Kottra J et al. (1994). The structure of the urokinase-type plasminogen activator receptor gene. Blood 84:1151–1156.

Chaurasia P, Aguirre-Ghiso J A, Liang O D et al. (2006). A region in urokinase plasminogen receptor domain III controlling a functional association with $\alpha_5\beta_1$ integrin and tumor growth. J Biol Chem 281:14852–14863.

Chevalier F, Lopez-Prados J, Groves P et al. (2006). Structure and dynamics of the conserved protein GPI anchor core inserted into detergent micelles. Glycobiology 16:969–980.

Cunningham O, Andolfo A, Santovito M L et al. (2003). Dimerization controls the lipid raft partitioning of uPAR/CD87 and regulates its biological functions. Embo J 22:5994–6003.

Deng G, Curriden S A, Wang S et al. (1996). Is plasminogen activator inhibitor-1 the molecular switch that governs urokinase receptor-mediated cell adhesion and release? J Cell Biol 134:1563–1571.

Dewerchin M, Nuffelen A V, Wallays G et al. (1996). Generation and characterization of urokinase receptor-deficient mice. J Clin Invest 97:870–878.

Ellis V, Danø K. (1993). Potentiation of plasminogen activation by an anti-urokinase monoclonal antibody due to ternary complex formation. A mechanistic model for receptor-mediated plasminogen activation. J Biol Chem 268:4806–4813.

Ellis V, Scully M F, Kakkar V V. (1989). Plasminogen activation initiated by single-chain urokinase-type plasminogen activator. Potentiation by U937 monocytes. J Biol Chem 264:2185–2188.

Ellis V, Behrendt N, Danø K. (1991). Plasminogen activation by receptor-bound urokinase. A kinetic study with both cell-associated and isolated receptor. J Biol Chem 266:12752–12758.

Estreicher A, Wohlwend A, Belin D et al. (1989). Characterization of the cellular binding site for the urokinase-type plasminogen activator. J Biol Chem 264:1180–1189.

Foekens J A, Peters H A, Look M P et al. (2000). The urokinase system of plasminogen activation and prognosis in 2780 breast cancer patients. Cancer Res 60:636–643.

Gårdsvoll H, Ploug M. (2007). Mapping of the vitronectin binding site on the urokinase receptor: involvement of a coherent receptor interface consisting of residues from both domain I and the flanking interdomain linker region. J Biol Chem 282:13561–13572.

Gårdsvoll H, DanøK, Ploug M. (1999). Mapping part of the functional epitope for ligand binding on the receptor for urokinase-type plasminogen activator by site-directed mutagenesis. J Biol Chem 274:37995–38003.

Gårdsvoll H, Werner F, Søndergaard L et al. (2004). Characterization of low-glycosylated forms of soluble human urokinase receptor expressed in Drosophila Schneider 2 cells after deletion of glycosylation-sites. Protein Expr Purif 34:284–295.

Gårdsvoll H, Gilquin B, Le Du M H et al. (2006). Characterization of the functional epitope on the urokinase receptor. Complete alanine scanning mutagenesis supplemented by chemical cross-linking. J Biol Chem 281:19260–19272.

Goodson R J, Doyle M V, Kaufman S E et al. (1994). High-affinity urokinase receptor antagonists identified with bacteriophage peptide display. Proc Natl Acad Sci USA 91:7129–7133.

Hansen A P, Petros A M, Meadows R P et al. (1994). Solution structure of the amino-terminal fragment of urokinase-type plasminogen activator. Biochemistry 33:4847–4864.

Høyer-Hansen G, Rønne E, Solberg H et al. (1992). Urokinase plasminogen activator cleaves its cell surface receptor releasing the ligand-binding domain. J Biol Chem 267:18224–18229.

Høyer-Hansen G, Behrendt N, Ploug M et al. (1997). The intact urokinase receptor is required for efficient vitronectin binding: receptor cleavage prevents ligand interaction. FEBS Lett 420:79–85.

Huai Q, Mazar A P, Kuo A et al. (2006). Structure of human urokinase plasminogen activator in complex with its receptor. Science 311:656–659.

Huai Q, Zhou A, Lin L et al. (2008). Crystal structures of two human vitronectin, urokinase and urokinase receptor complexes. Nat Struct Mol Biol 15: 422–423.

Jacobsen B, Gårdsvoll H, Juhl Funch G et al. (2007). One-step affinity purification of recombinant urokinase-type plasminogen activator receptor using a synthetic peptide developed by combinatorial chemistry. Protein Expr Purif 52:286–296.

Jögi A, Pass J, Høyer-Hansen G et al. (2007). Systemic administration of anti-urokinase plasminogen activator receptor monoclonal antibodies induce hepatic fibrin deposition in tissue-type plasminogen activator deficient mice. J Thromb Haemost 9:1936–1944.

Jørgensen T J, Gårdsvoll H, Danø K et al. (2004). Dynamics of urokinase receptor interaction with peptide antagonists studied by amide hydrogen exchange and mass spectrometry. Biochemistry 43:15044–15057.

Jørgensen T J, Gårdsvoll H, Ploug M et al. (2005). Intramolecular migration of amide hydrogens in protonated peptides upon collisional activation. J Am Chem Soc 127:2785–2793.

Kjaergaard M, Gårdsvoll H, Hirschberg D et al. (2007). Solution structure of recombinant somatomedin B domain from vitronectin produced in Pichia pastoris. Protein Sci 19:1934–1945.

Kjøller L, Hall A. (2001). Rac mediates cytoskeletal rearrangements and increased cell motility induced by urokinase-type plasminogen activator receptor binding to vitronectin. J Cell Biol 152:1145–1157.

Knör S, Sato S, Huber T et al. (2007). Development and evaluation of peptidic ligands targeting tumour-associated urokinase-type plasminogen activator receptor (uPAR) for use in a-emitter therapy of disseminated ovarian cancer. Eur J Nuclear Med Mol Imaging 35: 53–64.

Kobayashi H, Sugino D, She M Y et al. (1998). A bifunctional hybrid molecule of the aminoterminal fragment of urokinase and domain II of bikunin efficiently inhibits tumor cell invasion and metastasis. Eur J Biochem 253:817–826.

Koolwijk P, Sidenius N, Peters E et al. (2001). Proteolysis of the urokinase-type plasminogen activator receptor by metalloproteinase-12: implication for angiogenesis in fibrin matrices. Blood 97:3123–3131.

Li Z-B, Niu G, Wang H et al. (2008). Imaging of urokinase-type plasminogen activator receptor expression using a 64Cu-labeled linear peptide antagonist by microPET. Clin Cancer Res DOI: 10.1158/1078-0432.CCR-07-4434.

List K, Høyer-Hansen G, Rønne E et al. (1999). Different mechanisms are involved in the antibody mediated inhibition of ligand binding to the urokinase receptor: a study based on biosensor technology. J Immunol Methods 222:125–133.

Liu S, Aaronson H, Mitola D J et al. (2003). Potent antitumor activity of a urokinase-activated engineered anthrax toxin. Proc Natl Acad Sci USA 100:657–662.

Llinas P, Le Du M H, Gårdsvoll H et al. (2005). Crystal structure of the human urokinase plasminogen activator receptor bound to an antagonist peptide. Embo J 24:1655–1663.

Lund L R, Green K A, Stoop A A et al. (2006). Plasminogen activation independent of uPA and tPA maintains wound healing in gene-deficient mice. Embo J 25:2686–2697.

Madsen C D, Ferraris G M, Andolfo A et al. (2007). uPAR-induced cell adhesion and migration: vitronectin provides the key. J Cell Biol 177:927–939.

Magdolen V, Rettenberger P, Koppitz M et al. (1996). Systematic mutational analysis of the receptor-binding region of the human urokinase-type plasminogen activator. Eur J Biochem 237:743–751.

Magdolen V, Bürgle M, de Prada N A et al. (2001). Cyclo^{19,31}[D-Cys^{19}]-uPA$_{19-31}$ is a potent competitive antagonist of the interaction of urokinase-type plasminogen activator with its receptor (CD87). Biol Chem 382:1197–1205.

Min H Y, Doyle L V, Vitt C R et al. (1996). Urokinase receptor antagonists inhibit angiogenesis and primary tumor growth in syngeneic mice. Cancer Res 56:2428–2433.

Møller L B, Pöllanen J, Rønne E et al. (1993). N-linked glycosylation of the ligand-binding domain of the human urokinase receptor contributes to the affinity for its ligand. J Biol Chem 268:11152–11159.

Muehlenweg B, Assfalg-Machleidt I, Parrado S G et al. (2000). A novel type of bifunctional inhibitor directed against proteolytic activity and receptor/ligand interaction. Cystatin with a urokinase receptor binding site. J Biol Chem 275:33562–33566.

Nielsen B S, Rank F, Illemann M et al. (2007). Stromal cells associated with early invasive foci in human mammary ductal carcinoma in situ coexpress urokinase and urokinase receptor. Int J Cancer 120:2086–2095.

Nielsen L S, Kellerman G M, Behrendt N et al. (1988). A 55,000–60,000 Mr receptor protein for urokinase-type plasminogen activator. Identification in human tumor cell lines and partial purification. J Biol Chem 263:2358–2363.

Orlean P, Menon A K. (2007). Thematic review series: lipid posttranslational modifications. GPI anchoring of protein in yeast and mammalian cells, or: how we learned to stop worrying and love glycophospholipids. J Lipid Res 48:993–1011.

Pass J, Jögi A, Lund I K et al. (2007). Murine monoclonal antibodies against murine uPA receptor produced in gene-deficient mice: inhibitory effects on receptor mediated uPA activity *in vitro* and *in vivo*. Thromb Haemost 97:1013–1022.

Piironen T, Laursen B, Pass J et al. (2004). Specific immunoassays for detection of intact and cleaved forms of the urokinase receptor. Clin Chem 50:2059–2068.

Pillay V, Dass C R, Choong P F. (2007). The urokinase plasminogen activator receptor as a gene therapy target for cancer. Trends Biotechnol 25:33–39.

Ploug M. (1998). Identification of specific sites involved in ligand binding by photoaffinity labeling of the receptor for the urokinase-type plasminogen activator. Residues located at equivalent positions in uPAR domains I and III participate in the assembly of a composite ligand-binding site. Biochemistry 37:16494–16505.

Ploug M. (2003). Structure-function relationships in the interaction between the urokinase-type plasminogen activator and its receptor. Curr Pharm Des 9:1499–1528.

Ploug M, Ellis V. (1994). Structure-function relationships in the receptor for urokinase-type plasminogen activator. Comparison to other members of the Ly-6 family and snake venom α-neurotoxins. FEBS Lett 349:163–168.

Ploug M, Ellis V, Danø K. (1994). Ligand interaction between urokinase-type plasminogen activator and its receptor probed with 8-anilino-1-naphthalenesulfonate. Evidence for a hydrophobic binding site exposed only on the intact receptor. Biochemistry 33:8991–8997.

Ploug M, Rønne E, Behrendt N et al. (1991). Cellular receptor for urokinase plasminogen activator. Carboxyl-terminal processing and membrane anchoring by glycosyl-phosphatidylinositol. J Biol Chem 266:1926–1933.

Ploug M, Eriksen J, Plesner T et al. (1992a). A soluble form of the glycolipid-anchored receptor for urokinase-type plasminogen activator is secreted from peripheral blood leukocytes from patients with paroxysmal nocturnal hemoglobinuria. Eur J Biochem 208:397–404.

Ploug M, Plesner T, Rønne E et al. (1992b). The receptor for urokinase-type plasminogen activator is deficient on peripheral blood leukocytes in patients with paroxysmal nocturnal hemoglobinuria. Blood 79:1447–1455.

Ploug M, Kjalke M, Rønne E et al. (1993). Localization of the disulfide bonds in the NH_2-terminal domain of the cellular receptor for human urokinase-type plasminogen activator. A domain structure belonging to a novel superfamily of glycolipid-anchored membrane proteins. J Biol Chem 268:17539–17546.

Ploug M, Rahbek-Nielsen H, Nielsen P F et al. (1998a). Glycosylation profile of a recombinant urokinase-type plasminogen activator receptor expressed in Chinese hamster ovary cells. J Biol Chem 273:13933–13943.

Ploug M, Østergaard S, Hansen L B et al. (1998b). Photoaffinity labeling of the human receptor for urokinase-type plasminogen activator using a decapeptide antagonist. Evidence for a composite ligand-binding site and a short interdomain separation. Biochemistry 37:3612–3622.

Ploug M, Østergaard S, Gårdsvoll H et al. (2001). Peptide-derived antagonists of the urokinase receptor. Affinity maturation by combinatorial chemistry, identification of functional epitopes, and inhibitory effect on cancer cell intravasation. Biochemistry 40:12157–12168.

Quax P H, Lamfers M L, Lardenoye J H et al. (2001). Adenoviral expression of a urokinase receptor-targeted protease inhibitor inhibits neointima formation in murine and human blood vessels. Circulation 103:562–569.

Quigley J P. (1976). Association of a protease (plasminogen activator) with a specific membrane fraction isolated from transformed cells. J Cell Biol 71:472–486.

Reichert J M, Valge-Archer V E. (2007). Development trends for monoclonal antibody cancer therapeutics. Nat Rev Drug Discov 6:349–356.

Resnati M, Guttinger M, Valcamonica S et al. (1996). Proteolytic cleavage of the urokinase receptor substitutes for the agonist-induced chemotactic effect. Embo J 15:1572–1582.

Resnati M, Pallavicini I, Wang J M et al. (2002). The fibrinolytic receptor for urokinase activates the G protein-coupled chemotactic receptor FPRL1/LXA4R. Proc Natl Acad Sci USA 99:1359–1364.

Roldan A L, Cubellis M V, Masucci M T et al. (1990). Cloning and expression of the receptor for human urokinase plasminogen activator, a central molecule in cell surface, plasmin dependent proteolysis. Embo J 9:467–474.

Rømer J, Nielsen B S, Ploug M. (2004). The urokinase receptor as a potential target in cancer therapy. Curr Pharm Des 10:2359–2376.

Rønne E, Pappot H, Grøndahl-Hansen J et al. (1995). The receptor for urokinase plasminogen activator is present in plasma from healthy donors and elevated in patients with paroxysmal nocturnal haemoglobinuria. Br J Haematol 89:576–581.

Rosenberg S. (2001). New developments in the urokinase-type plasminogen activator system. Expert Opin Ther Targets 5:711–722.

Salasznyk R M, Zappala M, Zheng M et al. (2007). The uPA receptor and the somatomedin B region of vitronectin direct the localization of uPA to focal adhesions in microvessel endothelial cells. Matrix Biol 26:359–370.

Sato S, Kopitz C, Schmalix W A et al. (2002). High-affinity urokinase-derived cyclic peptides inhibiting urokinase/urokinase receptor-interaction: effects on tumor growth and spread. FEBS Lett 528:212–216.

Schmiedeberg N, Schmitt M, Rolz C et al. (2002). Synthesis, solution structure, and biological evaluation of urokinase type plasminogen activator (uPA)-derived receptor binding domain mimetics. J Med Chem 45:4984–4994.

Selvarajan S, Lund L R, Takeuchi T et al. (2001). A plasma kallikrein-dependent plasminogen cascade required for adipocyte differentiation. Nat Cell Biol 3:267–275.

Stephens R W, Nielsen H J, Christensen I J et al. (1999). Plasma urokinase receptor levels in patients with colorectal cancer: relationship to prognosis. J Natl Cancer Inst 91:869–874.

Stoppelli M P, Corti A, Soffientini A et al. (1985). Differentiation-enhanced binding of the amino-terminal fragment of human urokinase plasminogen activator to a specific receptor on U937 monocytes. Proc Natl Acad Sci USA 82:4939–4943.

Taylor V C, Sims M, Brett S et al. (1997). Antibody selection against CD52 produces a paroxysmal nocturnal haemoglobinuria phenotype in human lymphocytes by a novel mechanism. Biochem J 322 (Pt. 3):919–925.

Vassalli J D, Baccino D, Belin D. (1985). A cellular binding site for the Mr 55,000 form of the human plasminogen activator, urokinase. J Cell Biol 100:86–92.

Waltz D A, Chapman H A. (1994). Reversible cellular adhesion to vitronectin linked to urokinase receptor occupancy. J Biol Chem 269:14746–14750.

Wei Y, Waltz D A, Rao N et al. (1994). Identification of the urokinase receptor as an adhesion receptor for vitronectin. J Biol Chem 269:32380–32388.

Wei Y, Lukashev M, Simon D I et al. (1996). Regulation of integrin function by the urokinase receptor. Science 273:1551–1555.

Wei Y, Eble J A, Wang Z et al. (2001). Urokinase receptors promote b1 integrin function through interactions with integrin $\alpha_3\beta_1$. Mol Biol Cell 12:2975–2986.

Wei Y, Tang C H, Kim Y et al. (2007). Urokinase receptors are required for a_5b_1 integrin-mediated signaling in tumor cells. J Biol Chem 282:3929–3939.

Yuan C, Huang M. (2007). Does the urokinase receptor exist in a latent form? Cell Mol Life Sci 64:1033–1037.

Zhou A, Huntington J A, Pannu N S et al. (2003). How vitronectin binds PAI-1 to modulate fibrinolysis and cell migration. Nat Struct Biol 10:541–544.

Zhou H M, Nichols A, Meda P et al. (2000). Urokinase-type plasminogen activator and its receptor synergize to promote pathogenic proteolysis. Embo J 19:4817–4826.

Chapter 35
Engineered Antagonists of uPA and PAI-1

M. Patrizia Stoppelli, Lisbeth M. Andersen, Giuseppina Votta, and Peter A. Andreasen

Abstract It is now beyond reasonable doubt that plasminogen activation, catalyzed by urokinase-type plasminogen activator (uPA), plays an important role in the growth and dissemination of malignant tumours. The plasmin generated facilitates spread of tumour cells by catalyzing degradation of basement membranes and the extracellular matrix (ECM). In addition, uPA participates in cancer cell-directed tissue remodelling of the surrounding stroma. The function of uPA relies not only on plasmin generation but also on a complex set of pericellular, molecular, and functional interactions with cell surface receptors, adhesion molecules, and ECM proteins. In particular, a delicate balance between uPA and its fast and specific inhibitor, plasminogen activator inhibitor-1 (PAI-1), appears to contribute strongly to tumour dissemination. Here, we review recent advances in engineering compounds inhibiting each of the molecular interactions of uPA and PAI-1. Such compounds include organochemicals, peptides, and monoclonal antibodies, derived by structure-based rational design or directed evolution. Such compounds will help to decipher the tumour biological functions of each molecular interaction of uPA and PAI-1 and provide leads for the eventual use of uPA and PAI-1 as therapeutic targets.

Abbreviations

ATF,	N-terminal uPA fragment;
CTRs,	Complement-type repeats;
ECM,	Extracellular matrix;
ELISA,	Enzyme-linked immunosorbent assay;
FPR,	Formyl peptide receptor;
GFD,	Growth factor domain;
GPI,	Glycosyl phosphatidyl inositol;

P.A. Andreasen
Department of Molecular Biology, University of Aarhus, 10C Gustav Wied's Vej, 8000 Aarhus C, Denmark. e-mail: pa@mb.au.dk

D. Edwards et al. (eds.), *The Cancer Degradome.*
© Springer Science | Business Media, LLC

LDLR,   Low-density lipoprotein receptor;
MMP,   Matrix metalloproteinase;
PAI-1,   Plasminogen activator inhibitor-1;
PI3K,   Phosphoinositide 3-kinase;
RCL,   Reactive centre loop;
RNAi,   RNA interference;
TIMP,   Tissue inhibitor of metalloproteases;
TTSP,   Type-two transmembrane serine protease;
tPA,   Tissue-type plasminogen activator;
uPA,   Urokinase-type plasminogen activator;
uPAR,   Urokinase-type plasminogen activator receptor.

Introduction

The urokinase-type plasminogen activator (uPA) system is a serine protease system with a complex pericellular organization regulating cell adhesion and initiating intracellular signalling. The proteolytic conversion of plasminogen to the active protease plasmin is catalyzed by soluble or receptor (uPAR)-bound uPA. uPA is generated by proteolytic conversion of the initially secreted pro-enzyme (pro-uPA). uPA is inhibited by its primary serpin inhibitor plasminogen activator inhibitor-1 (PAI-1). PAI-1 also exhibits a high affinity for the extracellular matrix (ECM) protein vitronectin. The uPA–uPAR complex and integrins bind to vitronectin in competition with PAI-1. uPAR-bound uPA–serpin complexes are endocytosed by receptors related to the low-density lipoprotein receptor (LDLR). It is now beyond reasonable doubt that uPA plays a causal role in tumour growth, invasion, and metastasis. There is also good evidence that the other molecular interactions among the various members of the uPA system are of importance for the malignant phenotype (for reviews, see Dano et al. 1985, Andreasen et al. 1997, 2000; Blasi and Carmeliet 2002, Durand et al. 2004, Andreasen 2007). There is, therefore, great interest in generating specific inhibitors of the components of the system, both with a view to their use for elucidating the tumour biological functions of the various molecular interactions and to generate leads for drug development. We here review the biochemical and cell biological bases, possibilities, and perspectives of targeting uPA and PAI-1 in cancer.

Plasminogen Activation and Cancer Dissemination

Since the 1970s, experiments with cell cultures and animal models have strongly indicated a decisive tumour biological role of uPA-catalyzed plasminogen activation (for reviews, see Dano et al. 1985, Andreasen et al. 1997, 2000). Most recently, several studies involving mice with specific disruption of the genes for plasminogen (plasminogen$^{-/-}$ mice) and uPA (uPA$^{-/-}$ mice) were all in agreement with the idea that uPA-catalyzed plasmin generation is rate limiting for tumour growth, local invasion, and/or metastasis (Shapiro et al. 1996, Bugge et al. 1997, 1998, Sabapathy et al. 1997, Gutierrez et al. 2000, Bajou et al. 2001, Frandsen et al. 2001, Almholt et al. 2005). In 1990, a high level of uPA in extracts of primary breast carcinomas was reported to predict an early relapse (Duffy et al. 1990, Schmitt et al. 1990). uPAR was found to occur in higher concentrations in breast carcinomas than in benign breast lesions (Del Vecchio et al. 1993) and later also to be a prognostic marker (for a review, see Andreasen et al. 2000). These relationships were later confirmed with other cancer types (for a review, see Duffy and Duggan 2004).

The role of the uPA system in cancer should also be seen in relation to the fact that the growth and spread of solid tumours not only depend on the proliferative and invasive properties of the cancer cells themselves but also require the formation of a vascularized supporting tumour stroma by processes like angiogenesis and desmoplasia (for a review, see Mueller and Fusenig 2004). Since these processes require an active communication across the tumour–host interface, they may be referred to as cancer cell-directed tissue remodelling. The hypothesis of the importance of stromal uPA is consistent with the findings that different types of tumours transplanted onto uPA$^{-/-}$ or plasminogen$^{-/-}$ hosts grew slower, disseminated slower, and/or became less vascularized than when transplanted onto wild-type hosts (Bugge et al. 1997, Bajou et al. 2001, Frandsen et al. 2001), and that a genetically mammary gland tumour, in which uPA is expressed mainly by stromal cells, induced disseminated slower in uPA$^{-/-}$ mice than in wild-type mice (Almholt et al. 2005). Moreover, uPA is expressed by myofibroblasts in human breast carcinomas (Nielsen et al. 1996, 2001). Thus, uPA may also support tumour growth, invasion, and metastasis by being involved in angiogenesis and in shaping a desmoplastic stroma favourable for cancer cell migration.

One would, therefore, expect that protease inhibitors would tend to limit tumour growth, invasion, and metastasis. It therefore initially came as a surprise that measurements of the uPA and PAI-1 levels in breast tumour extracts by enzyme-linked immunosorbent assay (ELISA) showed the two to be correlated (Reilly et al. 1992) and that a high PAI-1 level in breast tumour extracts was found to be associated with a poor prognosis (Janicke et al. 1993, Grondahl-Hansen et al. 1993). The PAI-1 level in primary tumours is now known to be one of the most informative biochemical prognostic markers in several cancers (for reviews, see Andreasen et al. 1997, 2000; Duffy and Duggan 2004, Harbeck et al. 2004). The prognostic value of combined measurements of uPA and PAI-1 levels in breast cancer has been supported by so-called level I studies (Janicke et al. 2001, Look et al. 2002). Results from a multicentre prospective randomized therapy trial showed that patients with node-negative breast cancer with low levels of uPA and PAI-1 in their primary tumour have a very good prognosis, and may thus be candidates for being spared the burden of adjuvant chemotherapy. In contrast, node-negative patients with high uPA and PAI-1 levels are at substantially increased risk of disease recurrence, comparable to that of patients with three or more tumour cell positive axillary lymph nodes (for a review, see Harbeck et al. 2004).

But what is exactly the role of PAI-1 in tumours? Does it play a causal role in tumour growth and spread? A high level of PAI-1 in tumours could help to protect the tumour tissue itself from the destruction associated with a high uPA production and direct the tissue destruction towards the surrounding ECM. In this scenario, PAI-1 would be a target for anti-cancer therapy. Or is PAI-1 expression a defence mechanism from the normal tissue against tumour-induced proteolysis? In this scenario, PAI-1 would be an anti-target. Is the PAI-1 level merely a coincidental reflection of tumour aggressiveness caused by other factors? The combined use of information obtained from the use of biochemical studies, cell culture model systems, histological techniques, and mouse model systems has in the latest few years made it possible to draw some new conclusions concerning the tumour biological functions of PAI-1 or at least to pinpoint the problems. Some recent results do point to PAI-1 being causally involved in tumour growth and spread, since PAI-1 expressed by myofibroblasts is likely to participate in shaping a tumour stroma optimal for cancer cell migration and invasion, and PAI-1 expressed by endothelial cells is likely to participate in angiogenesis. But other recent results have suggested that PAI-1 expressed by other cell types, like malignant epithelial cells and myoepithelial cells, may have other tumour biological functions. Such other functions may or may not promote tumour metastasis. It cannot be excluded that PAI-1 expressed by some cell types may restrict tumour growth and spread (for a review, see Andreasen 2007).

It has been known for a long time that hyperactivation of the coagulation system, resulting in deep vein thrombosis or pulmonary embolism, can be the first manifestation of a tumour (Trousseau 1865). Recently, information has been accumulating about the possible mechanistic link between cancer and haemostasis (for reviews, see Boccaccio and Medico 2006, Rak et al. 2006). In particular, also PAI-1 has been implicated in Trousseau's syndrome. Being a fast and specific inhibitor also of tissue-type plasminogen activator (tPA), PAI-1 has an anti-fibrinolytic effect, and a high blood plasma level of PAI-1 is a risk factor for thrombotic diseases (for a review, see Vaughan 2002). PAI-1 leaking from tumours into the bloodstream may contribute to thrombotic complications in cancer. Thus, transduction of the onco-gene MET into mouse liver was shown to induce deep vein thrombosis, even before the appearance of the first preneoplastic lesions. PAI-1 was one of the two genes increased most in the transformed hepatocytes, and the blood plasma level of PAI-1 was increased threefold (Boccaccio et al. 2005). Moreover, PAI-1-stabilized micro-thrombi may facilitate arrest of cancer cells in the circulation, and thus their invasion into organs distant from the primary tumour. Although lung colonization by intravenously injected melanoma cells was reported to be unaffected by the PAI-1 level of the host (Eitzman et al. 1996), other results were in agreement with this hypothesis. Thus, transfection of intravenously injected human fibrosarcoma cells with PAI-1 cDNA was reported to increase their lung colonization (Tsuchiya et al. 1997). Also, studies with fibrinogen$^{-/-}$ mice suggested that arrest of circulating cancer cells by microthrombi facilitates establishment of metastasis in target organs (Palumbo et al. 2000).

Biochemical Properties of uPA

Domain Structure of uPA

uPA is secreted from cells as a 411 residues long, inactive, single-chain pro-enzyme (pro-uPA), consisting of a growth factor domain (GFD) (residues 1–49), a kringle (residues 50–131), an interdomain linker or connecting peptide (residues 132–158), and a serine protease domain (residues 16/159–250/411, using a dual numbering system of the residues in the serine protease domain, first the number according to the chymotrypsin template numbering, then the numbering from the N-terminus of uPA). Pro-uPA becomes activated by proteolytic cleavage at Lys158-Ile16/159. The N-terminal A-chain and the C-terminal B-chain of the mature enzyme are linked by disulphide bond between Cys148 and Cys122/279 (Fig. 35.1). Cleavage of the linker between the kringle and Cys148 results in an N-terminal fragment (ATF, residues 1–135) and a C-terminal, so-called low molecular weight uPA (for a review, see Andreasen et al. 1997).

The Catalytic Properties of uPA

The serine protease domain of uPA catalyzes the proteolytic conversion of the inactive zymogen plasminogen to the active serine protease plasmin, which is able to degrade many extracellular proteins, including fibrin and laminin, and catalyze proteolytic activation of zymogen forms of some metalloproteases (for a review, see Andreasen et al. 2000). Plasmin generation, therefore, regulates turnover of the ECM. The serine protease domain of uPA has the same overall fold as other serine proteases of the trypsin clan, consisting of two antiparallel β-barrels. The catalytic triad, His57/204, Asp102/255, and Ser195/356, is localized on loops between the two barrels. uPA's serine protease domain has specific, surface-exposed loops around residues 37/180, 60/209, 97/248, 110/263, 170/327, and 185/344 (Fig. 35.2) (Spraggon et al. 1995).

Serine protease-catalyzed peptide bond hydrolysis is initiated by the P1 residue of the substrate docking into the specificity or S1 pocket of the protease. Next, there is a nucleophilic attack of the carbon atom of the carbonyl group of the substrate P1 amino acid by the hydroxyl group of Ser195, with His57 and Asp102 acting as a charge relay system. This step leads to the formation of a tetrahedral transition state, which is stabilized by the interaction of the now charged oxygen atom of the P1 carbonyl group with the oxyanion hole composed of the amide NH groups of Ser195 and Gly193 and by main chain β-strand-type hydrogen bonds between the P1-P3 and P2′ amino acids of the substrate and residues 214–217 and residue 41 of the enzyme, respectively, in the polypeptide-binding cleft on either side of S195. Subsequently, there is an aqueous deacylation step involving a second tetrahedral transition state (for a review, see Hedstrom 2002).

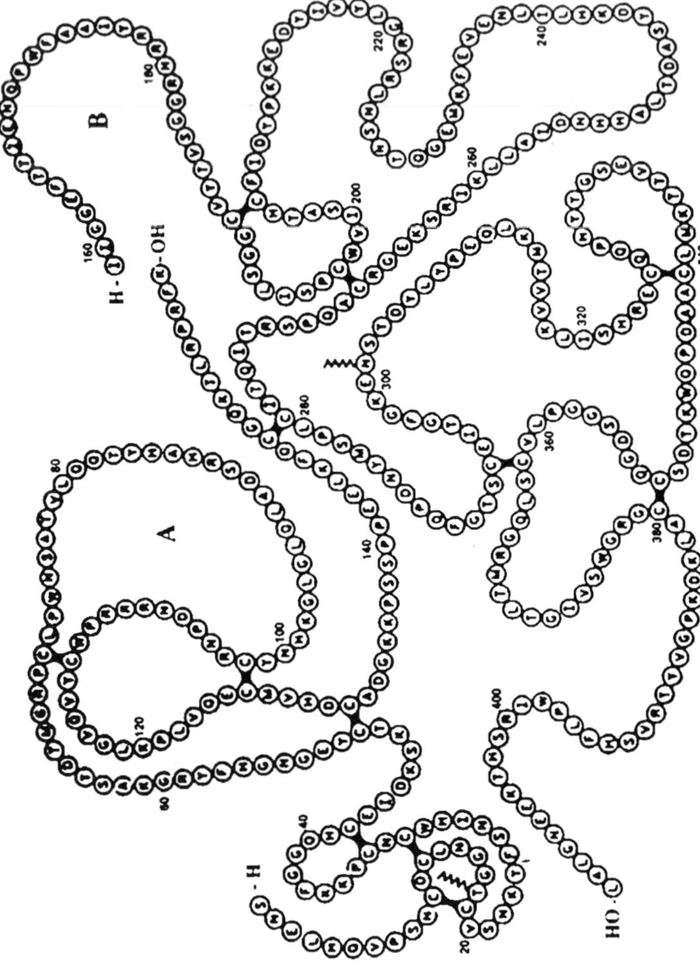

Fig. 35.1 The domain structure of urokinase-type plasminogen activator (uPA). The primary structure of human *uPA* in the two-chain form is depicted. The amino acid sequence is indicated in the one-letter code. Also indicated are glycosylation sites, by a zigzag line, phosphorylation sites (P), and disulphide bonds. The N-terminal A chain consists of the growth factor domain (GFD) and the kringle domain, each with three intradomain disulphide bonds. The C-terminal B chain contains five intradomain disulphide bonds. The intradomain linker is connected to the C-terminal B chain by a disulphide bond

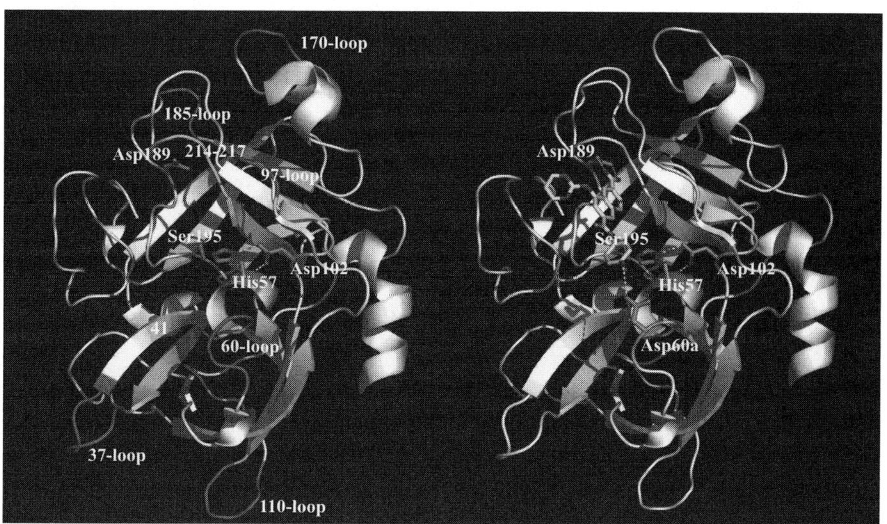

Fig. 35.2 The three-dimensional structure of human urokinase-type plasminogen activator (uPA). The left figure shows the catalytic triad Ser195, His57, and Asp102, and Asp189 in the S1 pocket (pink), the main chain binding area around residues 214–217 and 41 (green) and *uPA's*-specific surface loops (red). The right figure shows the structure of human uPA in complex with *N*-[4-(aminomethyl)phenyl]-6-carbamimidoyl-4-(pyrimidin-2-ylamino)naphthalene-2-carboxamide. The inhibitor is shown in CPK, the catalytic triad in pink, and Asp 60a in orange. The structures are displayed with PyMOL Viewer v0.99 on the basis of the PDB file 1sqa (Wendt et al. 2004). Note that the amino group of the inhibitor hydrogen bonds to the side chain of Asp60a (*See also* Color Insert II)

uPA has a restricted substrate specificity, cleaving only specific peptide bonds in a few specific proteins, always on the C-terminal side of Arg residues (for a review, see Dano et al. 1985). In contrast to other serine proteases, the activation sequence of plasminogen is present in a disulphide bridge-constrained loop: ... Cys-Pro-Gly-Arg-Val-Val-Gly-Gly-Cys ... (for a review, see Parry et al. 2000). uPA has also been reported to catalyze cleavage of the sequence ... Asn-Ser-Gly-Arg-Ala-Val ... in the loop connecting the C-terminal and the middle domain of the three domains of uPAR (Hoyer-Hansen et al. 1992). Screening a phage-displayed peptide library, Ke et al. (1997) found that the optimal uPA substrate is the sequence Ser-Gly-Arg-Ser-Ala. However, that library did not contain disulphide bridge-constrained peptides, which may be cleaved with a different efficiency. Like with other serine proteases, the substrate specificity of uPA is strongly influenced by the subsites S3, S2, S1, S1′, S2′, and S3′, binding the P3, P2, P1, P1′, P2′, and P3′ residues of the substrate. The primary P1 residue specificity of uPA is determined by the properties of the S1 pocket, in particular the localization of an Asp residue in position 189/351 at the bottom of the pocket. In this respect, uPA has an S1 pocket similar to that of trypsin. However, the S1 pockets of different Arg-specific proteases have some differences that provide each enzyme with distinct S1 subsites (for a review, see Hedstrom 2002). In addition, uPA has a restricted, less accessible, hydrophobic S2 pocket and a solvent-accessible S3 pocket, which can accommodate a wide range of residues (Spraggon et al. 1995). The protease-specific surface loops surrounding the active site are also involved in exosite interactions of uPA with its proteinaceous substrates. Thus, uPA's 37-loop, 60-loop, 97-loop, and autolysis loop (residues 142–152) have been proposed to be involved in determining its specificity for plasminogen (Ke et al. 1997, Zhang et al. 1997, Parry et al. 2000, Wang et al. 2000).

Single-chain pro-uPA has an activity that is at least 250-fold lower than that of mature two-chain uPA. The three-dimensional structure of pro-uPA has not been determined. However, studies of other serine proteases show that the amino group of the new N-terminal amino acid resulting from the activating cleavage forms a salt bridge to the side chain of Asp194 which then holds the oxyanion hole in an active conformation (for a review, see Hedstrom 2002). In the case of pro-uPA, the cleavage can be catalyzed by plasmin, in a positive feedback loop (for a review, see Dano et al. 1985). However, pro-uPA activation occurs also in plasminogen-deficient mice. In their urine, pro-uPA activation is catalyzed by glandular kallikrein (List et al. 2000). A number of other proteases have been reported to catalyze activation of pro-uPA in vitro (Ichinose et al. 1986, Koivunen et al. 1989, Brunner et al. 1990, Kobayashi et al. 1991, Goretzki et al. 1992, Wolf et al. 1993, Stack and Johnson 1994, Yoshida et al. 1995, Lee et al. 2000, Kannemeier et al. 2001, Roemisch et al. 2002, Suzuki et al. 2004, Yasuda et al. 2005, Kilpatrick et al. 2006, Moran et al. 2006). The efficiency of the conversion varies strongly between these proteases. The best candidates for physiological pro-uPA activators are plasmin and the type-two transmembrane serine proteases (TTSPs) matriptase, hepsin, and tryptase ε/PRSS22, as these proteases can activate uPAR-bound pro-uPA efficiently at cell surfaces (Ellis et al. 1989, Lee et al. 2000, Suzuki et al. 2004, Yasuda et al. 2005, Kilpatrick et al. 2006, Moran et al. 2006). Plasmin, which is not a transmembrane protease, accumulates at cell surfaces by binding to membrane proteins with C-terminal lysines via the kringle domains (Ellis et al. 1989).

GFD and uPAR

The GFD of uPA binds to uPAR (Appella et al. 1987). uPAR was discovered by Vassalli et al. (1985) and Stoppelli et al. (1985) as a cell surface-associated binding activity with a K_D in the range 100–500 pM. uPAR is attached to the cell membrane by a glycosyl phosphatidyl inositol (GPI) anchor (Ploug et al. 1991, Moller et al. 1992). Pro-uPA and two-chain active uPA bind with the same affinity (Cubellis et al. 1986). Detailed information about the structural basis for binding of uPA to uPAR has been provided by recent X-ray crystal structure analyses (Llinas et al. 2005, Barinka et al. 2006, Huai et al. 2006). uPAR also binds to the N-terminal somatomedin B domain of vitronectin. The binding is stimulated by uPA but competed by PAI-1 (Wei et al. 1994, Deng et al. 1996, Kanse et al. 1996). uPAR's vitronectin interaction surface has been mapped by site-directed mutagenesis (Gardsvoll and Ploug 2007, Madsen et al. 2007). Also, uPAR colocalizes with integrins in focal adhesion sites and may bind directly to integrins (for reviews, see Andreasen et al. 1997, 2000; Chapman 1997).

A main function of uPAR is the acceleration of both plasminogen activation and pro-uPA activation by co-accumulation of uPA, pro-uPA, plasminogen, and plasmin at cell surfaces (Ellis et al. 1989, 1991). Its interaction with other cell surface molecules endows uPAR with a ligand-dependent ability to stimulate intracellular signalling, resulting in regulation of proliferation, apoptosis, adhesion, migration, and invasion. The affinity of uPAR to vitronectin enables it to function as an adhesion receptor and its ability to interact with integrins enables it to regulate integrin-dependent adhesion (for reviews, see Ossowski and Aguirre-Ghiso 2000, Blasi and Carmeliet 2002, Sidenius and Degryse, this volume).

The Kringle Domain of uPA

In contrast to the kringles of plasmin, uPA's kringle does not contain a lysine binding site. Instead, it has some affinity for polyanions like heparin (Stephens et al. 1992), mediated by a basic cluster of Arg108, Arg109, and Arg110 (Petersen et al. 2001). uPA's kringle has also been reported to bind, with a relative low affinity, to $\alpha_v\beta_3$ integrin (Tarui et al. 2006). There are also a few reports of effects of isolated uPA kringle on various cellular parameters (Mukhina et al. 2000, Kim et al. 2003a, 2007) and of intra-molecular interactions between the kringle and the GFD, stabilizing uPA binding to uPAR (Bdeir et al. 2003). Largely, however, no specific biochemical functions have as yet been established for the uPA kringle.

The Linker or Connecting Peptide of uPA

Evidence for a functional role of the connecting peptide was originally provided by the analysis of uPA serine phosphorylation. In A431 human carcinoma cells, uPA may be phosphorylated on Ser138 and/or Ser303 (Franco et al. 1997). This post-translational modification occurs before secretion and is inhibited by protein kinase C inhibitors (Franco et al. 1998). Phosphorylation of uPA at these residues affects neither the catalytic efficiency nor its binding to uPAR, but phosphorylated uPA lacks the ability to stimulate cell adhesion and migration. An S138E uPA variant also lacks these activities, therefore mimicking the phosphorylated molecule (Franco et al. 1997). Interestingly, Ser138 is conserved among different mammalian species, like mouse, rat, yellow baboon, bovine, and orangutan (although it is replaced by a Phe in pig, which on the other hand also has a nine-residues insertion between positions 134 and 135). The conservation suggests a relevant functional role (Votta and Stoppelli, unpublished data). Further evidence indicating the relevance of the connecting peptide region is based on the effects of Å6, a peptide corresponding to uPA residues 136–143 which inhibits tumour progression and angiogenesis (Guo et al. 2000).

Recently, an isolated peptide corresponding to uPA residues 135–158 was reported to bind to $\alpha_v\beta_5$ integrin with high affinity, to be chemotactic at picomolar concentrations, and to stimulate the colocalization of uPAR and $\alpha_v\beta_5$ integrin (Franco et al. 2006). The most ready interpretation of these observations is that the linker of uPA induces a change in $\alpha_v\beta_5$ integrin, which promotes its association with uPAR and favours the functional interaction between the two molecules, but that this activity is prevented by phosphorylation of Ser138.

Biochemical Properties of PAI-1

Interaction of PAI-1 with Its Target Proteases

PAI-1 belongs to the serpin class of serine protease inhibitors and the primary physiological inhibitor of both uPA and tPA. The serpin inhibitory mechanism begins with the protease attacking the P_1-$P_1{'}$ bond in the serpin's reactive centre loop (RCL), but at the enzyme-acyl intermediate stage, the N-terminal part of the RCL inserts into a central β-sheet A, pulls the protease to the serpin's opposite pole, distorts it, and halts its completion of the catalytic cycle. At some reaction conditions, RCL insertion is delayed, resulting in abortive complex formation and production of an uncomplexed P_1-$P_1{'}$ bond cleaved serpin. This type of reaction is referred to as serpin substrate behaviour. Premature loop insertion, in the absence of protease, results in conversion of serpins to an inactive, so-called latent form (Fig. 35.3) (Ye and Goldsmith 2001, Wind et al. 2002, Huntington 2006). RCL insertion is associated with conformational changes in PAI-1's flexible joint region, including α-helices D and E and β-strands 1A and 2A (Mottonen et al. 1992, Aertgeerts et al. 1995, Sharp et al. 1999, Nar et al. 2000, Stout et al. 2000).

Interaction of PAI-1 with Vitronectin

Active PAI-1 binds to vitronectin with a K_D around 1 nM, while protease-complexed, protease-cleaved, and latent PAI-1 bind with an about 100-fold lower affinity (Lawrence et al. 1997). PAI-1's vitronectin-binding site is localized in the flexible joint region (Fig. 35.3) (Lawrence et al. 1994, van Meijer et al. 1994, Padmanabhan and Sane 1995, Arroyo De Prada et al. 2002, Jensen et al. 2002, Zhou et al. 2003). This localization is in good agreement with the different conformations of the flexible joint region in active PAI-1 and PAI-1 forms with an inserted RCL and their differential vitronectin affinity. On the contrary, the latency transition of PAI-1 is delayed approximately twofold by its binding to vitronectin (Declerck et al. 1988), probably as a result of vitronectin-induced conformational changes of PAI-1 (Fa et al. 1995, Gibson et al. 1997, Hansen et al. 2001).

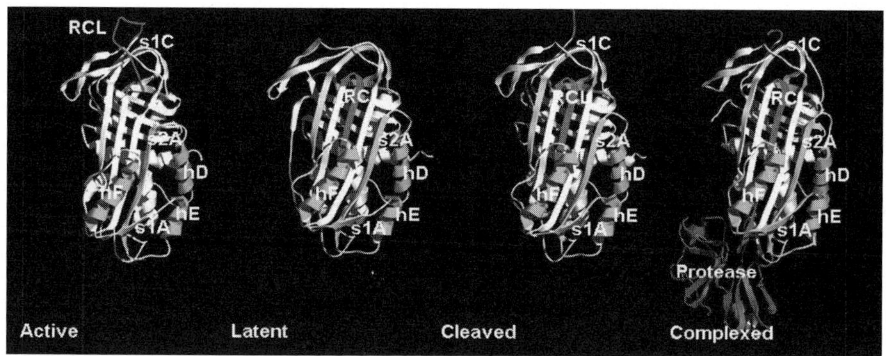

Fig. 35.3 Different conformations of plasminogen activator inhibitor-1 (PAI-1). The three-dimensional structures of active PAI-1 [(Sharp et al. 1999) pdb file 1B3K)], latent PAI-1 [(Stein et al., unpublished) pdb file 1LJ5], P_1-P_1' cleaved PAI-1 [(Aertgeerts et al. 1995) pdb file 9PAI], and the trypsin-α_1-proteinase inhibitor complex [(Huntington et al. 2000) pdb file 1EZX] are shown. The indicated structural elements of the serpins are the surface exposed or inserted RCL (red); α-helices D and E (blue); α-helix hF (blue-green); β strands 1A and 2A (orange). The serine protease moiety of the complex is coloured dark red. The structures are displayed with Swiss-PBD Viewer v3.7b2. No structure is available of a plasminogen activator–PAI-1 complex. Therefore, the structure of a protease–serpin complex is illustrated by the trypsin–α_1-proteinase inhibitor complex. Please note that part of the structure of the protease moiety of the complex was not solved because of crystallographic disordering (*See also* Color Insert II)

Interaction of PAI-1 with Endocytosis Receptors

uPA–PAI-1 complex can bind to certain clathrin-coated pit-localized receptors of the LDLR family, including LRP-1A, LRP-1B, LRP-2, VLDLR, and sorLA, with K_D values in the nanomolar range (for reviews, see Andreasen et al. 1994, Nykjaer and Willnow 2002, Durand et al. 2004). Each of the receptors of the LDLR family mediates endocytosis of many structurally unrelated ligands. Moreover, some of them have a signalling function (for reviews, see Nykjaer and Willnow 2002, Gonias et al. 2004, Lillis et al. 2005). The binding of the uPA–PAI-1 complex to these receptors may function in the clearance of the complex both locally in the tissues and systemically in the liver. uPAR-bound uPA–PAI-1 complex is endocytosed and degraded much faster than uPA–PAI-1 complex free in the fluid phase, presumably because the relatively low-affinity binding to the endocytosis receptors is facilitated by accumulation of the complex on the cell surface by the relatively high-affinity binding to uPAR (Nykjaer et al. 1992, 1994; Zhang et al. 1998). uPAR is, as a consequence of its binding to the uPA moiety of the uPA–PAI-1-endocytosis receptor complex, co-endocytosed (Conese et al. 1995) but later recirculated to the cell surface (Nykjaer et al. 1997).

In general, ligand binding by receptors of the LDLR family occurs to clusters of seven or more mutually highly homologous complement-type repeats (CTRs) in the receptors (for a review, see Gliemann 1998). The ligand-receptor binding specificity seems to depend on each individual ligand making contacts to a specific set of

CTRs in each receptor. Expressing overlapping two CTR fragments of one of the ligand-binding CTR clusters of LRP, the uPA–PAI-1 complex was found to bind specifically to two specific, adjacent CTRs. By site-directed mutagenesis, the binding was found to be dependent on conserved acidic and hydrophobic residues placed centrally in the CTRs (Andersen et al. 2001). On the ligand side, site-directed mutagenesis showed that the uPA–PAI-1 complex has an extended receptor interaction surface with a majority of basic and hydrophobic residues in α-helix D of PAI-1 and the 37-loop of uPA (Horn et al. 1998, Rodenburg et al. 1998, Stefansson et al. 1998, Skeldal et al. 2006) (Fig. 35.1, Fig. 35.2 and Fig. 35.3). uPA–PAI-1 complex binds to LRP with an affinity which is ~100-fold higher than that of free uPA, active, latent, and cleaved PAI-1 (Nykjaer et al. 1994), in good agreement with the fact that both uPA and PAI-1 are implicated in the interaction surface. There is as yet no structural information about any endocytosis receptor-bound serine protease or serpin, but structural analysis of other ligands in complex with two or more CTRs from such receptors has shown that the ligand–receptor interfaces are composed of basic and hydrophobic residues in the ligand and in the receptors (Verdaguer et al. 2004, Fisher et al. 2006, Jensen et al. 2006a, Lazic et al. 2006).

How Do the Different Molecular Interactions Integrate to Support the Malignant Phenotype?

Cell adhesion, migration, invasion, proliferation, and survival are basic cellular properties of strong relevance to tumour progression. A large number of reports on effects of uPA and PAI-1 on these parameters define a complex network of pericellular and membrane interactions, which should be taken into consideration in novel targeting strategies. Overall, these effects, observed at cellular level, may contribute to the link between the overproduction of uPA and PAI-1 and tumour metastatic ability.

Cell Adhesion

Cell adhesion may be counteracted by uPA because of its capability to generate plasmin for proteolytic degradation of adhesion receptors and ECM proteins. Therefore, PAI-1 would be expected to promote cell adhesion because of its anti-proteolytic effect (for a review, see Andreasen et al. 2000). In particular, the binding of PAI-1 to vitronectin can result in protection of vitronectin against proteolysis and, because vitronectin is a ligand for integrins and for uPAR, confer a pro-adhesive effect to PAI-1 (Ciambrone and Mckeown-Longo 1990, Huang et al. 2003). Moreover, PAI-1 has been demonstrated to cause oligomerization and ECM deposition of vitronectin (Seiffert and Loskutoff 1996) and would also by that mechanism be expected to promote cell adhesion.

On the contrary, early reports assigned to uPA and its amino-terminal region a strong pro-adhesive ability on myelomonocytic cell lines pre-exposed to differentiating agents (for a review, see Chapman 1997). Later, it became clear that uPA stimulates uPAR-dependent adhesion of cells to vitronectin, and thereby acts pro-adhesively. uPAR may cis-activate integrins and modify their specificity. Thus, interaction between uPAR and α_1 integrin leads to a decreased adhesion to fibronectin and an increased adhesion to vitronectin. Cell adhesion is further modulated by the uPA–uPAR interaction through the control of focal adhesion turnover and actin polymerization. In this scenario, an anti-adhesive effect may result from the competition between PAI-1 and membrane receptors (integrins and uPAR) for binding to vitronectin (for reviews, see Andreasen et al. 2000, Blasi and Carmeliet 2002, Kugler et al. 2003, Durand et al. 2004, Alfano et al. 2005).

Cell Migration

Cell locomotion has consistently been reported to be stimulated by uPA in a proteolysis-independent manner, through interaction of its GFD with uPAR and activation of intracellular signalling cascades (for reviews, see Blasi and Carmeliet 2002, Kugler et al. 2003). It has been observed that cells exposed to proteolytically inactivated uPA, to its N-terminal fragment (ATF), or to a growth factor-derived peptide retaining uPAR-binding activity (residues 12–32 of the human sequence), undergo clearcut cytoskeletal rearrangements and extension of protrusions (Carriero et al. 1999, Degryse et al. 2001). Regarding the mechanistic aspects of the uPA-dependent migration, proximal signalling involves other membrane receptors, like epidermal growth factor receptor, formyl peptide receptor (FPR), and many integrins ($\alpha_v\beta_5$, $\alpha_v\beta_3$, $\alpha_5\beta_1$, and others) (Ossowski and Aguirre-Ghiso 2000). Intracellular mediators include proteins associated with focal contacts, like the focal adhesion kinase (FAK), Src family kinases as well as actin-associated proteins like paxillin, talin, and Wiskott-Aldrich syndrome protein (WASP) (Chiaradonna et al. 1999, Sturge et al. 2002). In addition, the connecting peptide (residues 135–158) was reported to possess a uPAR-dependent motogen activity (Franco et al. 2006), dependent on its specific binding to $\alpha_v\beta_5$ integrin (see section The Linker or Connecting Peptide of uPA). Furthermore, a remarkable chemotactic activity was assigned to an isolated uPA kringle (Mukhina et al. 2000; see also section The Kringle Domain of uPA). Moreover, pro-migratory effects of uPA may result from plasmin generation, which could lower the resistance of the ECM (for a review, see Andreasen et al. 2000).

The conclusions concerning the role of PAI-1 in cell migration in cell cultures are less consistent: in some reports, PAI-1 was found to be pro-migratory, while in others it was found to be anti-migratory. PAI-1 would be expected to inhibit plasminogen activation-dependent cell migration and the pro-migratory effect of uPA–uPAR binding to vitronectin at the leading cell edge, and indeed, both anti-proteolytic and non-anti-proteolytic anti-migratory effects of PAI-1 have been reported. But the ability of PAI-1 to compete for uPAR–vitronectin and integrin–vitronectin binding has also been reported to stimulate cell migration, perhaps because PAI-1 in these reports was presented to the migrating cell at its trailing edge. Third, based on their ability to mediate endocytosis of uPAR-bound uPA–PAI-1 complex, endocytosis receptors could be hypothesized to be important for uPA and PAI-1 regulation of cell migration. During the process of endocytosis, not only the uPA–PAI-1 complex but also uPAR is endocytosed, but while uPA and PAI-1 are degraded in endosomes or lysosomes, uPAR is recycled to the cell surface. In cell cultures with PAI-1 production, the continued rapid endocytosis receptor-mediated recycling of uPAR has been reported to be required for maintenance of a steady-state level of cell surface uPAR, and thus for maintenance of a steady-state level of cell surface-associated plasminogen activation activity. Assuming that uPAR could recycle in a polarized endocytic cycle, the endocytosis receptors could be hypothesized to regulate the pericellular distribution of uPA-mediated proteolysis during migration. In addition, the pro-migratory intracellular signalling cascades stimulating by binding of uPA to uPAR and the lateral interaction between uPAR and integrins would be terminated by removal of uPA–uPAR complex from the cell surface by endocytosis after binding of PAI-1. Moreover, the binding of uPA–PAI-1 complex to endocytosis receptors has been reported to initiate signalling in some cell types. Considering this complexity, it is not surprising that apparently mutually conflicting results were reported concerning the possible role of endocytosis of the uPA–PAI-1 complex in cell migration (for reviews, see Andreasen et al. 1997, 2000; Durand et al. 2004, Andreasen 2007).

Cell Invasion

Cell invasion in cell culture model systems was consistently stimulated by uPA by both proteolytic and non-proteolytic mechanisms (for reviews, see Andreasen et al. 1997, 2000). Consequently, one would expect PAI-1 to inhibit uPA-dependent invasion, but again, variable results were reported, PAI-1 inhibiting invasion in some reports but stimulating it in others (for a review, see Durand et al. 2004).

Cell Proliferation, Survival, and Apoptosis

Many reports describe a uPAR-dependent pro-proliferative effect of uPA (for a review, see Alfano et al. 2005). In contrast, the isolated uPA kringle was reported to possess growth-inhibitory properties (Kim et al. 2003b, 2007).

There is also good evidence that binding of uPA to uPAR may act anti-apoptotically in a proteolysis-independent, signalling-dependent manner, thus contributing to the establishment of the malignant phenotype. One example is provided by

T241 fibrosarcoma cells implanted in uPA$^{-/-}$ mice, which exhibit decreased proliferative and increased apoptotic indices, suggesting that alterations in host expression of uPA may affect the balance between tumour cell death and proliferation (Gutierrez et al. 2000). Also, reducing uPAR expression by SNB19 glioblastoma cells by antisense uPAR constructs resulted in in vivo inhibition of SNB19 tumour formation associated with loss of mitochondrial transmembrane potential, release of cytochrome C from mitochondria, subsequent activation of caspase-9, and apoptosis (Yanamandra et al. 2000). The anti-apoptotic ability of uPA–uPAR binding may be due, at least in part, to activation of the Ras–ERK signalling pathway. Thus, MDA-MB-231 breast cancer cells cultured in the presence of anti-uPA antibodies, which block the binding of uPA to uPAR, led to a decrease in the level of phosphorylated ERK and promotion of apoptosis (Ma et al. 2001). U87MG glioblastoma cells bearing an antisense to uPA exhibit a reduced level of phosphorylated phosphoinositide 3-kinase (PI3K) and the serine/threonine kinase Akt as well as impaired migration and survival (Chandrasekar et al. 2003). uPA has been reported to protect retinal pigment epithelial cells from UV-, cis-platin, and detachment-induced apoptosis. This effect is due to the interaction of GFD with uPAR, which results in the activation of the PI3K–Akt pathway and leads to the transcriptional activation of BcL-XL (Alfano et al. 2006).

In a number of articles, PAI-1 was reported to be anti-apoptotic and pro-proliferative (Kwaan et al. 2000, Soeda et al. 2001, 2006; Chen et al. 2004, 2006; Lademann et al. 2005, Romer et al. 2005). However, in other articles, PAI-1 was reported to be pro-apoptotic and anti-proliferative (Ploplis et al. 2004, Balsara et al. 2006, Kortlever et al. 2006). The apparent contradiction between the two sets of observations remains to be resolved. The protection by PAI-1 against apoptosis may stem from its anti-proteolytic effect, which could promote cell adhesion and survival: uPA, by virtue of its proteolytic activity, may generate active plasmin, which has been shown to be pro-apoptotic (Rossignol et al. 2004). But interference with uPAR- and endocytosis-receptor-initiated signalling may also confer PAI-1 with pro-apoptotic activity.

Antagonists of the Enzyme Activity of uPA

Antibody Inhibitors of uPA

Polyclonal antibodies against murine uPA were shown to inhibit its plasminogen activation activity (Dano et al. 1980). Inhibition of plasminogen activation was used as a screening principle for identifying hybridoma clones with anti-human uPA antibodies, resulting in derivation of the first anti-uPA monoclonal antibody (Kaltoft et al. 1982). Since then, many more inhibitory anti-uPA monoclonal antibodies have been raised. Some monoclonal anti-catalytic anti-human uPA antibodies were reported to cross-react with tPA; on the basis of proteolytic fragmentation of uPA and uPA–tPA sequence comparison, they were reported to have an epitope spanning the sequence around the active site Ser (Takahashi and Naora 1985). By site-directed mutagenesis, several anti-human uPA monoclonal antibodies, which inhibit uPA's ability to activate plasminogen, were found to have overlapping epitopes composed of variable combinations of Arg35/178, Arg36/179, His37/180, Arg37a/181, Tyr60b/209, Lys61/211, and Asp63/214 in the 37- and 60-loops near the active site (Petersen et al. 2001). The latter antibodies do not inhibit uPA-catalyzed hydrolysis of small peptidylic substrates, consistent with the notion that they inhibit plasminogen activation by sterically hindering uPA–plasminogen exosite interactions. The information about the inhibitory mechanism of antibodies targeting structures outside the highly conserved active site may form the basis for future design of more specific low molecular mass uPA inhibitors interfering with uPA–plasminogen exosite interactions.

Protein Inhibitors of uPA

The dimeric, non-specific serine protease inhibitor ecotin binds two protease molecules per ecotin dimer through interactions with both the protease active site and an exosite. With some success, ecotin has been converted into a high-affinity uPA inhibitor by engineering each of the two interaction sites (Wang et al. 1995, Yang and Craik 1998, Laboissiere et al. 2002, Stoop and Craik 2003).

Small Organochemical Inhibitors of uPA

uPA can be inhibited by classical serine protease inhibitors like diisopropylphosphofluoridate (for a review, see Walker and Lynas 2001). It is also inhibited by p-aminobenzamidine (compound 1, Table 35.1, Fig. 35.4), which inhibits, with variable efficiency, all serine proteases which prefer Arg as a P1 residue (for a review, see Magdolen et al. 2000). In fact, most attempts to generate specific organochemical uPA inhibitors were based on the ability of p-aminobenzamidine or closely related aryl amidine, aryl guanidine, and acyl guanidine Arg homologues to insert into the P1-binding pocket and addition of various additional chemical structures to achieve specificity. In many cases, this goal has been pursued by rational design based on X-ray crystal structure analysis of uPA–inhibitor complexes. The additional chemical structures have been targeting the S2 and S3 pockets and subsites within the P1-binding pocket, and, in a few cases, also surface loops outside the active site. Hundreds of different compounds have been investigated. An excellent and comprehensive review on this type of uPA inhibitors was given by Rockway et al. (2002). Here, we will focus on recent and representative examples of such inhibitors.

Table 35.1 Low molecular mass inhibitors of the enzyme activity of uPA

Compound number	Compound name	Systematic name	K_i for uPA	Specificity[a]	References
1	PAB	*p*-aminobenzamidine	82 μM	–	Magdolen et al. 2000
2	Amiloride	3,5-diamino-6-chloro-*N*-(diaminomethylidene) pyrazine-2-carboxamide	7 μM	2.4-fold over trypsin	Vassalli and Belin 1987
3	WX-293T	*N*-(1-adamantyl)-*N'*-(4-guanidinobenzyl) urea	2.4 μM	250-fold over thrombin	Sperl et al. 2000
4	CA-11	2-[5-(amino-azaniumylidene-methyl)-6-fluoro-1H-benzoimidazol-2-yl]-6-[(1S,2S)-2-methylcyclohexyl]oxy-phenolate	11 nM	100-fold over plasmin	Katz et al. 2004
5	23b	*N*-[4-(aminomethyl)phenyl]-6-carbamimidoyl-4-(pyrimidin-2-ylamino)naphthalene-2-carboxamide	0.64 nM	32-fold over trypsin	Wendt et al. 2004
6	WX-UK1	(*N*-α-2,4,6-triisopropyl-phenylsulfonyl-3-amidino-(L)-phenylalanine-4-ethoxycarbonyl-piperazide	0.41 μM	0.09-fold over trypsin	Stürzebecher et al. 1999
7	UK371,804	2-{[(4-chloro-1-guanidino-7-isoquinolinyl) sulfonyl)]amino}-isobutyric acid	10 nM	2,700-fold over tPA	Fish et al. 2007
8	5j	di-4-methylsulfonophenyl *N*-(benzyloxycarbonylamino)-4-guanidinophenyl-methane phosphonate trifuoroacetate	53 nM	434-fold over trypsin	Sienczyk and Oleksyszyn 2006
9	UK122	4-[(5-Oxo-2-phenyl-4(5H)-Oxazolylidene) methyl]-benzenecarboximidamide	0.2 μM	500-fold over trypsin	Zhu et al. 2007
10	Cyclic hexapeptide		41 nM	1:1 for trypsin	Wakselman et al. 1993
11		phenylethylsulfonyl-D-ser-ala-D-aminomethyl benzamidine	3.1 nM	100-fold over plasmin	Tamura et al. 2000
12	Cyclic peptide, upain-1	CSWRGLENHMRC	2.1 μM	50-fold over trypsin	Hansen et al. 2005

[a] Specificity defined as K_i for the other serine protease with the lowest K_i among those tested, divided by the K_i for uPA.

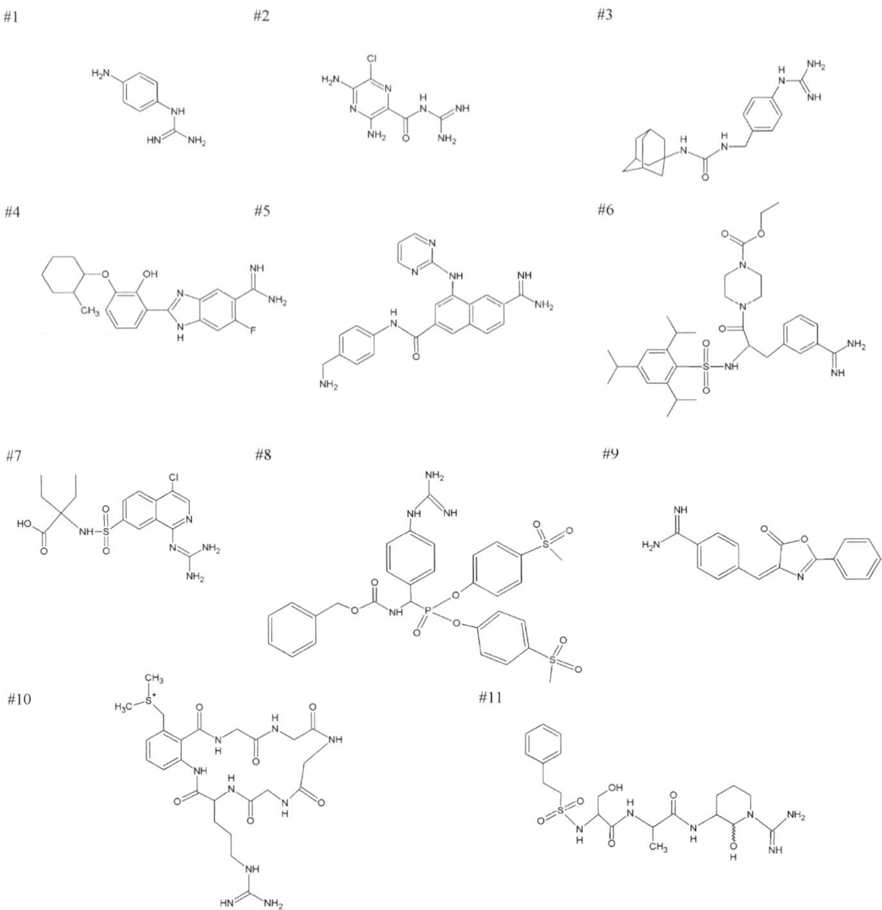

Fig. 35.4 Chemical structures of urokinase-type plasminogen activator (uPA) inhibitors. The numbers of the compounds refer to Table 35.1

Amiloride (compound 2, Table 35.1, Fig. 35.4) inhibits uPA with a K_i of 7 μM (Vassalli and Belin 1987). Since the K_i for inhibition of tPA and plasmin is more than 100 μM, it has been a useful agent for discriminating between uPA and tPA in plasminogen activation assays. However, since it inhibits trypsin with a K_i value of 17 μM and is a diuretic drug, its use for pharmacological purposes is excluded. X-ray crystal structure analysis of the uPA–amiloride complex (Nienaber et al. 2000b, Zeslawska et al. 2000) showed the expected insertion of the guanidine group into the S1 pocket and, interestingly, of the chloro-atom into a small hydrophobic pocket referred to as the S1β subsite, a feature also observed with other uPA inhibitors containing halogen atoms (see below).

A 4-substituted phenylguanidine analogue (compound 3, Table 35.1, Fig. 35.4) inhibits uPA with a K_i of 2.4 μM. The phenylguanidine group binds in the S1 pocket, while the amadamantyl group binds in a shallow hydrophobic pocket close to the Cys42-Cys58 disulphide bond. This inhibitor is thus unusual by achieving specificity by targeting an area of uPA closer to the S' pockets than to the S pockets (Sperl et al. 2000).

Researchers at the Celera company developed a series of uPA inhibitors based on 4-aminoarylguanidine, p-aminobenzamidine, 2-(2-phenol)-indole, or benzimidazole, culminating with a derivative of 6-fluoro-5-amidinobenzimidazole (compound 4, Table 35.1, Fig. 35.4). This compound inhibits uPA with a K_i of 11 nM and exhibits strong preference for uPA over plasmin, trypsin, tryptase, tPA, thrombin, factor Xa, and hepsin. The specificity of this inhibitor rests on halogen atoms inserting into uPA's S1β pocket (Katz et al. 2001, 2003, 2004; Mackman et al. 2001, 2002; Spencer et al. 2002).

A series of naphtamidine-based inhibitors was developed by researchers at the Abbott company (Nienaber et al. 2000a, Klinghofer et al. 2001, Wendt et al. 2004, Bruncko et al. 2005). The most promising compound developed (compound 5, Table 35.1, Fig. 35.4) has a high affinity for human uPA (K_i 0.64 nM) and a good selectivity, with a 32-fold higher K_i for trypsin inhibition. X-ray crystal structure analysis showed that an important feature of this compound is a salt bridge between the Asp60a/208 carboxylate and the charged amino group of the inhibitor (Fig. 35.3) (Wendt et al. 2004). This interaction adds both affinity and selectivity to the uPA inhibitor binding, as an Asp residue in this position is an unusual feature in the trypsin family. Inhibitors targeting this position will even show selectivity for human uPA over murine uPA, as the latter has a Gln residue in this position. In the rodent tumour models tested, Abbott's uPA inhibitors showed unsatisfactory results (Giranda, V., personal communication).

WX-UK1, a derivative of 3-aminophenylalanine (compound 6, Table 35.1, Fig. 35.4), inhibits human uPA with a K_i of 0.41 μM (Sturzebecher et al. 1999, Zeslawska et al. 2000, Setyono-Han et al. 2005). It has been reported to inhibit the metastatic spread of cancer cells in rodents (Ertongur et al. 2004, Setyono-Han et al. 2005, Killeen et al. 2007). However, it has a poor selectivity for uPA, in fact inhibiting several other human serine proteases better than uPA (Setyono-Han et al. 2005). It is unknown whether the reported anti-metastatic effect was in fact due to inhibition of uPA. This compound was found to be well tolerated by humans in early clinical trials.

After a series of studies (Barber and Dickinson 2002, Barber et al. 2002, 2004), researchers at the Pfizer company derived a compound (compound 7, Table 35.1, Fig. 35.4) which inhibits uPA with a K_i of 10 nM and exhibited a 4,000-fold selectivity over tPA (Fish et al. 2007).

Sienczyk and Oleksyszyn (2006) synthesized a carboxybenzyl-amino-(4-guanidino-phenyl)methanephosphonate aromatic ester derivative, an irreversible serine protease inhibitor which shows a good selectivity for uPA over trypsin (compound 8, Table 35.1, Fig. 35.4).

Zhu et al. (2007) reported a 4-oxazolidinone analogue which inhibits human uPA with a K_i of 0.2 μM and an at least 500-fold selectivity over tPA, plasmin, thrombin, and trypsin (compound 9, Table 35.1, Fig. 35.4).

Peptides Inhibitors of uPA

Chloromethylketone derivatives of amino acids are classical protease inhibitors (Baker 1967, Shaw 1970). Attaching tripeptides to the chloromethylketone group provides some specificity (Lijnen et al. 1984), and Glu-Gly-Arg-chloromethylketone has been a valuable tool for biophysical studies of uPA (see, for instance, Spraggon et al. 1995). However, for in vivo use, such inhibitors do not seem to be of interest.

Wakselman et al. (1993) reported a cyclic hexapeptide, which inhibits uPA with a K_i of 41 nM, which did not observably inhibit plasmin, tPA, or thrombin, but which was unable to discriminate against trypsin and uPA (compound 10, Table 35.1, Fig. 35.4). Its Arg residue was assumed to bind in the S1 pocket of uPA. It is not known why this peptide is an inhibitor and not a substrate.

On the basis of identification of a peptide sequence which is an optimal uPA substrate (Ke et al. 1997) (see section 1.2), Tamura et al. (2000) designed a series of peptide-based inhibitors of uPA, phenethylsulfonyl-D-Ser-Ala-D-aminomethyl benzamidine being the most potent (compound 11, Table 35.1, Fig. 35.4). It inhibited uPA with a K_i value of 3.1 nM and exhibited 100- and 800-fold selectivity for uPA over plasmin and tPA, respectively, but showed a poor selectivity for uPA over trypsin (Tamura et al. 2000). An inhibitor of this series was shown to inhibit lung colonization by human HT-1080 fibrosarcoma cells injected into the tail vein of mice (Schweinitz et al. 2004). Screening phage-displayed random, disulfide bridge-constrained peptide repertoires with human uPA as a bait, a uPA-binding sequence, Cys-Ser-Trp-Arg-Gly-Leu-Glu-Asn-His-Arg-Met-Cys (upain-1), was isolated (compound 12, Table 35.1).

Upain-1 inhibits human uPA competitively, with a K_i value of around 1 μM. The cyclical structure of upain-1 is indispensable for activity. Among several other human serine proteases, including trypsin, upain-1 is highly selective for human uPA and does not even measurably inhibit murine uPA (Hansen et al. 2005). An X-ray crystal structure analysis of the three-dimensional structure of upain-1–uPA complex (Zhao et al. 2007) revealed that the bound peptide adopts a rigid conformation stabilized by two tight β-turns formed by the Leu6-His9 and His9-Cys12 segments, respectively. The active conformation of upain-1 is maintained by intramolecular hydrogen bonds between the Glu7 residue and the main chain nitrogen atoms of residues 4, 5, and 6. Together with site-directed mutagenesis of the peptides as well as of the enzyme (Hansen et al. 2005), the structure analysis (Zhao et al. 2007) indicates that binding specificity depends on extended binding interactions involving specific surface loops of the enzymes and several residues of the peptides, including the insertion of the Arg4 of upain-1 into the uPA's S1 pocket; the localization of upain-1's Ser2 residue close to the S1β pocket of uPA; the occupancy of the S2 pocket and the oxyanion hole of uPA by the Gly5 and Glu7 residues of upain-1, respectively; and the binding of the Asn8 residue of upain-1 to the 37-loop and 60-loop of uPA, explaining the strong specificity of upain-1 for uPA. The steric hindrance of the side chain of Glu7 and the indole ring of Trp3 residue of upain-1 prevents the carboxyl group of Arg4 residue at the cleavage site from being accessible by the catalytic Ser195 residue, explaining why upain-1 is an inhibitor rather than a substrate. Based on the structure, a new pharmacophore for the design of highly specific uPA inhibitors can be proposed (Zhao et al. 2007). Upain-1 was found to inhibit cell surface-associated plasminogen activation (Hansen et al. 2005), but its effect in xenotransplant cancer models remains to be investigated.

Antagonists of the uPA Growth Factor Domain

Much effort has been put into developing antagonists of the binding of uPA's GFD to uPAR. The first antagonist to be developed was a peptide corresponding to residues 18–32 of uPA, the minimal sequence allowing uPAR binding (Appella et al. 1987). Since then, a variety of antagonists have been developed, including antibodies against uPAR or the GFD of uPA and peptides binding to uPAR and competitively inhibiting its binding to uPA. These antagonists will be covered by another chapter in this volume (Ploug, M., this volume).

It should be emphasized that while peptides derived from the uPA GFD may antagonize the binding of uPA or ATF to uPAR, they may be agonistic in other respects, for instance, stimulating uPAR-dependent cell migration (Franco et al. 2006).

Antagonists of the uPA Kringle

The epitopes for two monoclonal anti-uPA antibodies included Arg108, Arg109, and Arg110 in the kringle (Petersen et al. 2001). The ability of these antibodies to block the binding of uPA to polyanions (Stephens et al. 1992) correlated with a reduced uPA-polyanion affinity after substitution of the three Arg residues (Petersen et al. 2001). However, antibodies are not well suited for searching for specific effects of the kringle, as they may inhibit the uPA–uPAR interaction by steric hindrance (Corti et al. 1989).

Antagonists of the uPA Linker Region

The functional effects of the linker region of uPA are inhibited by phosphorylation of Ser138 or a S138E substitution (see section The Linker or Connecting Peptide of uPA). Interestingly, the phosphorylated or Glu-substituted variants are dominant negative (Franco et al. 2006) and may, therefore, be used as antagonists when searching for molecular and functional interactions of the linker region.

At micromolar concentrations, Å6, a capped eight amino acid peptide (residues 136–143), inhibits angiogenesis, tumour cell migration, and invasion in vitro and in vivo (Guo et al. 2000). Å6 was also reported to inhibit the growth and migration of estrogen-receptor-positive Mat B-III rat breast cancer cells in vitro and in vivo.

Histological analysis of primary tumours showed a decrease in new blood vessel density and increased tumour cell death in Å6- and tamoxifen-treated animals, and these effects were greater in experimental animals receiving Å6 and tamoxifen in combination (Guo et al. 2002). Similarly, Å6 has been reported to inhibit the growth and neovascularization of human U87MG glioblastoma cells growing on nude mice, particularly so when combined with cis-platin (Mishima et al. 2000). The mechanism of action of Å6 remains unknown. Surface plasmon resonance analyses showed that Å6 acts at micromolar concentrations to inhibit uPA–uPAR binding, suggesting that its cellular effects are caused by interference of the uPA–uPAR interaction (Guo et al. 2000). This observation is unexpected since the uPA–uPAR binding can be structurally accounted for by uPA's GFD (Barinka et al. 2006, Huai et al. 2006), but a secondary, low affinity, interaction between the CP and uPAR does not seem to be excluded. Alternatively, Å6 may be an antagonist of the functional effects of the full-length uPA connecting peptide (residues 135–158) (see above). Recently, a phase I clinical trial of Å6 in gynaecologic malignancies has been completed and indicated no toxicity (Berkenblit et al. 2005).

Antagonists of the Anti-Proteolytic Activity of PAI-1

Antibody Antagonists of PAI-1

The first reported antagonist of the anti-proteolytic effect of PAI-1 was a monoclonal antibody (Nielsen et al. 1986). Since then, many monoclonal anti-PAI-1 antibodies, antagonising PAI-1's anti-proteolytic activity by several mechanisms, have been described (for a review, see Gils and Declerck 2004). First, some monoclonal antibodies have epitopes close to the RCL and sterically hinder the access of target proteases (Fig. 35.3). Second, some monoclonal antibodies, with epitopes in α-helix F (Fig. 35.3), inhibit the conformational change of PAI-1 associated with protease complex formation and induce serpin substrate behaviour. Third, some monoclonal antibodies convert PAI-1 to the inactive latent form (for a review, see Gils and Declerck 2004). The latter type of antibodies have epitopes either composed of residues from the α-helix D-β-strand 2A loop and β-strand 3B or located in the area exposed upon extraction of β-strand 1C during latency transition. Such antibodies are believed to act by stabilization of a "prelatent" form, a form postulated to resemble the latent form in some respects, but in contrast to the latter to exist in equilibrium with the active form (Verhamme et al. 1999, Gorlatova et al. 2003, Naessens et al. 2003, Dupont et al. 2006).

Small Organochemical Inhibitors of PAI-1

A variety of low molecular mass organochemicals have been found to inhibit PAI-1's anti-proteolytic activity. The first organochemical PAI-1 inhibitor to be developed was a diketopiperazine derivative (Bryans et al. 1996). A variety of related compounds have been developed since then (Charlton et al. 1996, 1997; Friederich et al. 1997, Folkes et al. 2001, Wang et al. 2002, De Nanteuil et al. 2003, Ye et al., 2004). Various biochemical analyses showed that the diketopiperazine derivatives induce a stable, inactive PAI-1 conformation, which is different from other known PAI-1 conformations (Einholm et al. 2003). A number of other, mostly negative charged, amphipathic organochemical compounds of diverse chemical structures were reported to inactivate the anti-proteolytic effects of PAI-1 (Urano et al. 1992, Munch et al. 1993, Bjorquist et al. 1998, Chikanishi et al. 1999, Neve et al. 1999, Egelund et al. 2001, Gils et al. 2002, Pedersen et al. 2003, Crandall et al. 2004, Elokdah et al. 2004, Gopalsamy et al. 2004, Hu et al. 2005, Liang et al. 2005, Gardell et al. 2007). Available evidence, based on competition with monoclonal antibodies with known epitopes, site-directed mutagenesis, and molecular modelling, suggests that these compounds bind in a hydrophobic pocket beneath α-helix D (Bjorquist et al. 1998, Egelund et al. 2001, Gorlatova et al. 2007). Such compounds were found to induce PAI-1 substrate behaviour within seconds, followed by PAI-1 polymerization (Egelund et al. 2001, Pedersen et al. 2003, Gardell et al. 2007, Gorlatova et al. 2007). High concentrations of non-ionic detergents were found to induce substrate behaviour and latency transition (Ehnebom et al. 1997, Gils and Declerck 1998, Andreasen et al. 1999, Gils et al. 2000, 2003). Some other compounds inactivated PAI-1 by less characterized mechanisms (Gardsvoll et al. 1998, Chikanishi et al. 1999, Neve et al. 1999).

Insertion Peptides Inactivating PAI-1

Peptides with a sequence identical to that of the RCL can insert into PAI-1's β-sheet A in the place of the RCL and induce serpin substrate behaviour (Kvassman et al. 1995). A single report seems to suggest that such peptides may also be effective in vivo (Eitzman et al. 1995).

In Vivo Use of Antagonists of PAI-1's Anti-Proteolytic Effect

A couple of points should be noted concerning the in vivo use of the antagonists of PAI-1's anti-proteolytic effect described above. First, the reported organochemical antagonists have low affinity to PAI-1, with K_i values in the micromolar range, may not be specific for PAI-1, and bind tightly to serum albumin. Second, when tested in vitro, most, probably all, organochemical or peptidylic PAI-1 inactivating compounds have a much higher IC_{50} in the presence than in the absence of vitronectin (Eitzman et al. 1995, Egelund et al. 2001, Leik et al. 2006, Gorlatova et al. 2007). Considering that the PAI-1–vitronectin binding has a K_D around 1 nM and that vitronectin is very abundant, having a blood plasma concentration of around 1 μM (for a review, see Andreasen et al. 1997), one would expect that such antagonists would work quite inefficiently in vivo. Of course, it is not known how fast PAI-1, after having been released from cells, reacts with vitronectin relative to its reaction with its target proteases. But in contrast, the IC_{50} values for monoclonal antibodies antagonizing the anti-proteolytic effect of PAI-1 are either decreased (Schousboe et al. 2000, Wind et al. 2001, Komissarov et al. 2005) or unaffected (Andreasen, P.A., unpublished data) by vitronectin. Moreover, monoclonal antibodies bind with undoubted specificity and high affinity. Therefore, at the moment, monoclonal antibodies seem more promising than organochemicals and insertion peptides.

In in vivo studies with organochemical antagonists of uncertain specificity, special measures must be undertaken to control that the observed physiological effects are really caused by inhibition of PAI-1, and not of other proteins. PAI-1$^{-/-}$ mice are suitable controls in that connection.

An effect on tumour growth, invasion, and/or metastasis of targeting PAI-1 in an animal tumour model remains to be reported. But a PAI-1 inactivating agent has in a single case been demonstrated to inhibit a cellular process relevant to cancer spread. In an in vivo Matrigel implantation angiogenesis assay, Leik et al. (2006) demonstrated that oral administration of the PAI-1 antagonist tiplaxtinin inhibits angiogenesis. Importantly, in the study of Leik et al. (2006), angiogenesis in PAI-1$^{-/-}$ mice was significantly lower than in wild-type mice and not affected by tiplaxtinin.

Antagonists of PAI-1 Vitronectin Binding

The anti-proteolytically inactive PAI-1 forms induced by organochemicals or monoclonal antibodies have a strongly reduced affinity for vitronectin (Verhamme et al. 1999, Egelund et al. 2001, Einholm et al. 2003, Pedersen et al. 2003, Gorlatova et al. 2007). But there are, to date, no reports of compounds antagonizing exclusively the binding of PAI-1 to vitronectin. Only a PAI-1 variant mutated in the RCL and without anti-proteolytic activity, but still with intact vitronectin affinity, was reported to enhance plasmin-mediated matrix degradation in experimental nephritis, presumably by competing endogenous wild-type PAI-1 off vitronectin (Huang et al. 2003).

Antagonists of uPA–PAI-1 Binding to Endocytosis Receptor

A PAI-1-binding peptide derived from a phage-displayed peptide library was found to antagonize the binding of the uPA–PAI-1 complex to VLDLR and LRP-1A, presumably by steric hindrance of the access of the receptors to their binding site in PAI-1's flexible joint region. Unfortunately, since the peptide inhibits PAI-1 with a μM K_i and is quite hydrophobic, the window between its specific, PAI-1-dependent effects and its non-specific, detergent-like effects on cultured cells is quite small (Jensen et al. 2006b).

Novel Inhibitory Strategies

An alternative to antagonizing the various pericellular activities of uPA and PAI-1 is inhibiting their expression. Antisense technologies have been employed since the early 1990s to study the tumour biological functions of the components of the uPA system in cellular and animal model systems. More recently, RNA interference (RNAi) has been recognized as an effective methodology to silence specific genes (for reviews, see Wang 2001, Nozaki et al. 2006, Pillay et al. 2007). Furthermore, nucleic acid molecules with high affinity for a target transcription factor can be introduced into cells as decoy cis-elements to bind these factors and alter gene expression. According to a novel strategy, a peptide nucleic acid chimera containing a Sp1-binding sequence downmodulates the expression of Sp1 target genes, thus preventing uPAR-dependent breast tumour invasion (Zannetti et al. 2005). Although encouraging, these studies are still preliminary and will have to be extended to animal models, which mimic tumour progression using clinically acceptable modes of nucleic acid delivery. A drawback of these strategies is that they do not allow targeting exclusively specific molecular and functional interaction of a given gene product—they target them all concomitantly.

Conclusions and Perspectives

The potential use of uPA and PAI-1 as therapeutic targets in cancer should be viewed in light of the latest new knowledge about proteolytic enzyme systems and cancer. The classical hypothesis about the role of proteases in cancer is considered too simple today. Proteases participate not only in invasion through basement membranes but also in all aspects of the abnormal tissue development at which tumour growth, invasion, and metastasis must be considered. Moreover, in many cases, protease inhibitors, like PAI-1, do not have the expected anti-invasive effects but probably promote tumour invasion. The tumour biological functions of proteolytic enzyme systems seem to depend on a delicate balance between the activities of proteases and inhibitors. The idea of proteases and inhibitors having multiple functions is based on the finding that each of them has multiple molecular interactions and is expressed by multiple cell types in human tumours.

The following are important questions to pursue. Do uPA and PAI-1 expressed by different cell types have the same or different tumour biological functions? Do different molecular interactions of uPA and PAI-1 have different tumour biological functions? Do molecular interactions of uPA and PAI-1 besides the classical ones have tumour biological functions? How do the confusingly complex network of molecular interactions of uPA and PAI-1 integrate to support specific cell biological functions? Is it possible to target different tumour biological functions of the same biochemical activity of uPA and PAI-1 expressed by different cell types? With both uPA and PAI-1 being potential therapeutic targets, should each of them be inhibited at different stages of tumour progression, or should they be inhibited simultaneously? Does PAI-1 leaking from tumours contribute to thrombotic complications in cancer?

To answer these questions, it will be relevant to develop new agents specifically inactivating each of the multiple molecular interactions of uPA and PAI-1, either by rational design based on thorough biophysical studies of the interactions of uPA and PAI-1 with their inhibitors, including X-ray crystal structure analysis of uPA and PAI-1 in complex with inhibitors, and/or directed evolution. Although the affinities of the developed organochemical uPA's proteolytic activity antagonists were often in the low nanomolar range, the documentation for their specificity was far from satisfactory. Often, only a few additional serine proteases other than uPA were studied with respect to inhibition. Even among the serine proteases investigated, the specificity was often not satisfactory. The organochemical inhibitors of PAI-1 all had a low affinity and an undocumented specificity. In most cases, the ability of the inhibitors to specifically inhibit uPA or PAI-1 in mouse cancer models was not investigated. Inhibitory monoclonal antibodies and inhibitory recombinant proteins have the required high affinity and high specificity and seem more promising tools at the moment. The use of murine monoclonal antibodies, raised in knockout animals, seems promising tools for investigating the functions of uPA and PAI-1 in mouse cancer models. On a longer perspective, humanization of murine monoclonal antibodies may allow their use in humans. Peptides have a specificity comparable to that of monoclonal antibodies and a size allowing for chemical modification and may become valuable tools after proper affinity maturation.

When testing the effects of organochemical inhibitors in in vivo cancer models, it is important to document that any observed effect on tumours is really due to inhibition of uPA or PAI-1 and not other proteins. This requirement can be met by studying the effect of the inhibitors on tumours growing in $uPA^{-/-}$ or $PAI-1^{-/-}$ mice, in which no effect of the inhibitors should be expected. This requirement has not been met with any uPA inhibitor as yet and only in a single case with a PAI-1 inhibitor. The situation with uPA and PAI-1 may be very similar to that observed with other proteases and protease inhibitors. Studies of uPA and PAI-1 may contribute to forming new general paradigms concerning tumour biological functions of specific proteins.

A clear understanding of their molecular interactions, their cell-specific and tumour stage-specific expression, and availability of highly specific inhibitors should be established. Then, but not before, proteases and their inhibitors may be inhibited in clinical trials and eventually become versatile targets for anti-cancer therapy.

References

Aertgeerts K., De Bondt H. L., De Ranter C. J. et al (1995). Mechanisms contributing to the conformational and functional flexibility of plasminogen activator inhibitor-1. Nat Struct Biol 2: 891–7.

Alfano D., Franco P., Vocca I. et al (2005). The urokinase plasminogen activator and its receptor: role in cell growth and apoptosis. Thromb Haemost 93(2): 205–11.

Alfano D., Iaccarino I. and Stoppelli M. P. (2006). Urokinase signaling through its receptor protects against anoikis by increasing BCL-xL expression levels. J Biol Chem 281: 17758–67.

Almholt K., Lund L. R., Rygaard J. et al (2005). Reduced metastasis of transgenic mammary cancer in urokinase-deficient mice. Int J Cancer 113: 525–32.

Andersen O. M., Petersen H. H., Jacobsen C. et al (2001). Analysis of a two-domain binding site for the urokinase-type plasminogen activator-plasminogen activator inhibitor-1 complex in low-density-lipoprotein-receptor-related protein. Biochem J 357: 289–96.

Andreasen P. A. (2007). PAI-1 – a potential therapeutic target in cancer. Curr Drug Targets 8: 1030–41.

Andreasen P. A., Sottrup-Jensen L., Kjoller L. et al (1994). Receptor-mediated endocytosis of plasminogen activators and activator/inhibitor complexes. FEBS Lett 338: 239–45.

Andreasen P. A., Kjoller L., Christensen L. et al (1997). The urokinase-type plasminogen activator system in cancer metastasis: a review. Int J Cancer 72: 1–22.

Andreasen P. A., Egelund R., Jensen S. et al (1999). Solvent effects on activity and conformation of plasminogen activator inhibitor-1. Thromb Haemost 81: 407–14.

Andreasen P. A., Egelund R. and Petersen H. H. (2000). The plasminogen activation system in tumor growth, invasion, and metastasis. Cell Mol Life Sci 57: 25–40.

Appella E., Robinson E. A., Ullrich S. J. et al (1987). The receptor-binding sequence of urokinase. A biological function for the growth-factor module of proteases. J Biol Chem 262: 4437–40.

Arroyo De Prada N., Schroeck F., Sinner E. K. et al (2002). Interaction of plasminogen activator inhibitor type-1 (PAI-1) with vitronectin. Eur J Biochem 269: 184–92.

Bajou K., Noël A., Gerard R. D. et al (1998). Absence of host plasminogen activator inhibitor 1 prevents cancer invasion and vascularization. Nat Med 4: 923–8.

Baker B. R. (1967). Design of active-site-directed irreversible enzyme inhibitors; Wiley, New York.

Balsara R. D., Castellino F. J. and Ploplis V. A. (2006). A novel function of plasminogen activator inhibitor-1 in modulation of the AKT pathway in wild-type and plasminogen activator inhibitor-1-deficient endothelial cells. J Biol Chem 281: 22527–36.

Barber C. G. and Dickinson R. P. (2002). Selective urokinase-type plasminogen activator (uPA) inhibitors. Part 2: (3-substituted-5-halo-2-pyridinyl) guanidines. Bioorg Med Chem Lett 12: 185–7.

Barber C. G., Dickinson R. P. and Horne V. A. (2002). Selective urokinase-type plasminogen activator (uPA) inhibitors. Part 1: 2-pyridinylguanidines. Bioorg Med Chem Lett 12: 181–4.

Barber C. G., Dickinson R. P. and Fish P. V. (2004). Selective urokinase-type plasminogen activator (uPA) inhibitors. Part 3: 1-isoquinolinylguanidines. Bioorg Med Chem Lett 14: 3227–30.

Barinka C., Parry G., Callahan J. et al (2006). Structural basis of interaction between urokinase-type plasminogen activator and its receptor. J Mol Biol 363: 482–95.

Bdeir K., Kuo A., Sachais B. S. et al (2003). The kringle stabilizes urokinase binding to the urokinase receptor. Blood 102: 3600–8.

Berkenblit A., Matulonis U. A., Kroener J. F. et al (2005). A6, a urokinase plasminogen activator (uPA)-derived peptide in patients with advanced gynecologic cancer: a phase I trial. Gynecol Oncol 99: 50–7.

Bjorquist P., Ehnebom J., Inghardt T. et al (1998). Identification of the binding site for a low-molecular-weight inhibitor of plasminogen activator inhibitor type 1 by site-directed mutagenesis. Biochemistry 37: 1227–34.

Blasi F. and Carmeliet P. (2002). uPAR: a versatile signalling orchestrator. Nat Rev Mol Cell Biol 3: 932–43.

Boccaccio C. and Medico E. (2006). Cancer and blood coagulation. Cell Mol Life Sci 63: 1024–7.

Boccaccio C., Sabatino G., Medico E. et al (2005). The MET oncogene drives a genetic programme linking cancer to haemostasis. Nature 434: 396–400.

Bruncko M., McClellan W. J., Wendt M. D. et al (2005). Naphthamidine urokinase plasminogen activator inhibitors with improved pharmacokinetic properties. Bioorg Med Chem Lett 15: 93–8.

Brunner G., Simon M. M. and Kramer M. D. (1990). Activation of pro-urokinase by the human T cell-associated serine proteinase HuTSP-1. FEBS Lett 260: 141–4.

Bryans J., Charlton P., Chicarelli-Robinson I. et al (1996). Inhibition of plasminogen activator inhibitor-1 activity by two diketopiperazines, XR330 and XR334 produced by Streptomyces sp. J Antibiot (Tokyo) 49: 1014–21.

Bugge T. H., Kombrinck K. W., Xiao Q. et al (1997). Growth and dissemination of Lewis lung carcinoma in plasminogen-deficient mice. Blood 90: 4522–31.

Bugge T. H., Lund L. R., Kombrinck K. K. et al (1998). Reduced metastasis of Polyoma virus middle T antigen-induced mammary cancer in plasminogen-deficient mice. Oncogene 16: 3097–104.

Carriero M. V., Del Vecchio S., Capozzoli M. et al (1999). Urokinase receptor interacts with alpha (v)beta5 vitronectin receptor, promoting urokinase-dependent cell migration in breast cancer. Cancer Res 59: 5307–14.

Chandrasekar N., Mohanam S., Gujrati M. et al (2003). Downregulation of uPA inhibits migration and PI3k/Akt signaling in glioblastoma cells. Oncogene 22: 392–400.

Chapman H. A. (1997). Plasminogen activators, integrins, and the coordinated regulation of cell adhesion and migration. Curr Opin Cell Biol 9: 714–24.

Charlton P. A., Faint R. W., Bent F. et al (1996). Evaluation of a low molecular weight modulator of human plasminogen activator inhibitor-1 activity. Thromb Haemost 75: 808–15.

Charlton P., Faint R., Bent F. et al (1997). Evaluation of a low molecular weight modulator of human plasminogen activator inhibitor-1 activity. Fibrinolysis & Proteolysis 11: 51–6.

Chen Y., Kelm R. J. Jr., Budd R. C. et al (2004). Inhibition of apoptosis and caspase-3 in vascular smooth muscle cells by plasminogen activator inhibitor type-1. J Cell Biochem 92: 178–88.

Chen Y., Budd R. C., Kelm R. J. et al (2006). Augmentation of proliferation of vascular smooth muscle cells by plasminogen activator inhibitor type 1. Arterioscler Thromb Vasc Biol 26: 1777–83.

Chiaradonna F., Fontana L., Iavarone C. et al (1999). Urokinase receptor-dependent and -independent p56/59(hck) activation state is a molecular switch between myelomonocytic cell motility and adherence. EMBO J 18: 3013–23.

Chikanishi T., Shinohara C., Kikuchi T. et al (1999). Inhibition of plasminogen activator inhibitor-1 by 11-keto-9(E),12(E)-octadecadienoic acid, a novel fatty acid produced by Trichoderma sp. J Antibiot (Tokyo) 52: 797–802.

Ciambrone G. J. and McKeown-Longo P. J. (1990). Plasminogen activator inhibitor type I stabilizes vitronectin-dependent adhesions in HT-1080 cells. J Cell Biol 111: 2183–95.

Conese M., Nykjaer A., Petersen C. M. et al (1995). α-2 Macroglobulin receptor/Ldl receptor-related protein(Lrp)-dependent internalization of the urokinase receptor. J Cell Biol 131: 1609–22.

Corti A., Sarubbi E., Soffientini A. et al (1989). Epitope mapping of the anti-urokinase monoclonal antibody 5B4 by isolated domains of urokinase. Thromb Haemost 62: 934–9.

Crandall D. L., Elokdah H., Di L. et al (2004). Characterization and comparative evaluation of a structurally unique PAI-1 inhibitor exhibiting oral in-vivo efficacy. J Thromb Haemost 2: 1422–8.

Cubellis M. V., Nolli M. L., Cassani G. et al (1986). Binding of single-chain prourokinase to the urokinase receptor of human U937 cells. J Biol Chem 261: 15819–22.

Dano K., Nielsen L. S., Moller V. et al (1980). Inhibition of a plasminogen activator from oncogenic virus-transformed mouse cells by rabbit antibodies against the enzyme. Biochim Biophys Acta 630: 146–51.

Dano K., Andreasen P. A., Grondahl-Hansen J. et al (1985). Plasminogen activators, tissue degradation, and cancer. Adv Cancer Res 44: 139–266.

De Nanteuil G., Lila-Ambroise C., Rupin A. et al (2003). New fibrinolytic agents: benzothiophene derivatives as inhibitors of the t-PA-PAI-1 complex formation. Bioorg Med Chem Lett 13: 1705–8.

Declerck P. J., De Mol M., Alessi M. C. et al (1988). Purification and characterization of a plasminogen activator inhibitor 1 binding protein from human plasma. Identification as a multimeric form of S protein (vitronectin). J Biol Chem 263: 15454–61.

Degryse B., Orlando S., Resnati M. et al (2001). Urokinase/urokinase receptor and vitronectin/alpha(v)beta(3) integrin induce chemotaxis and cytoskeleton reorganization through different signaling pathways. Oncogene 20: 2032–43.

Del Vecchio S., Stoppelli M. P., Carriero M. V. et al (1993). Human urokinase receptor concentration in malignant and benign breast tumors by in vitro quantitative autoradiography: comparison with urokinase levels. Cancer Res 53: 3198–206.

Deng G., Curriden S. A., Wang S. et al (1996). Is plasminogen activator inhibitor-1 the molecular switch that governs urokinase receptor-mediated cell adhesion and release? J Cell Biol 134: 1563–71.

Duffy M. J. and Duggan C. (2004). The urokinase plasminogen activator system: a rich source of tumour markers for the individualised management of patients with cancer. Clin Biochem 37: 541–8.

Duffy M. J., Reilly D., O'Sullivan C. et al (1990). Urokinase-plasminogen activator, a new and independent prognostic marker in breast cancer. Cancer Res 50: 6827–9.

Dupont D. M., Blouse G. E., Hansen M. et al (2006). Evidence for a pre-latent form of the serpin plasminogen activator inhibitor-1 with a detached beta-strand 1C. J Biol Chem 281: 36071–81.

Durand M. K., Bodker J. S., Christensen A. et al (2004). Plasminogen activator inhibitor-I and tumour growth, invasion, and metastasis. Thromb Haemost 91: 438–49.

Egelund R., Einholm A. P., Pedersen K. E. et al (2001). A regulatory hydrophobic area in the flexible joint region of plasminogen activator inhibitor-1, defined with fluorescent activity-neutralizing ligands. Ligand-induced serpin polymerization. J Biol Chem 276: 13077–86.

Ehnebom J., Bjorquist P., Anderson J.-O. et al (1997). Detergent tween-80 modifies the specific activity of PAI-1. Fibrinolysis & Proteolysis 11: 165–70.

Einholm A. P., Pedersen K. E., Wind T. et al (2003). Biochemical mechanism of action of a diketopiperazine inactivator of plasminogen activator inhibitor-1. Biochem J 373: 723–32.

Eitzman D. T., Fay W. P., Lawrence D. A. et al (1995). Peptide-mediated inactivation of recombinant and platelet plasminogen activator inhibitor-1 in vitro. J Clin Invest 95: 2416–20.

Eitzman D. T., Krauss J. C., Shen T. et al (1996). Lack of plasminogen activator inhibitor-1 effect in a transgenic mouse model of metastatic melanoma. Blood 87: 4718–22.

Ellis V., Scully M. F. and Kakkar V. V. (1989). Plasminogen activation initiated by single-chain urokinase-type plasminogen activator. Potentiation by U937 monocytes. J Biol Chem 264: 2185–8.

Ellis V., Behrendt N. and Dano K. (1991). Plasminogen activation by receptor-bound urokinase. A kinetic study with both cell-associated and isolated receptor. J Biol Chem 266: 12752–8.

Elokdah H., Abou-Gharbia M., Hennan J. K. et al (2004). Tiplaxtinin, a novel, orally efficacious inhibitor of plasminogen activator inhibitor-1: design, synthesis, and preclinical characterization. J Med Chem 47: 3491–4.

Ertongur S., Lang S., Mack B. et al (2004). Inhibition of the invasion capacity of carcinoma cells by WX-UK1, a novel synthetic inhibitor of the urokinase-type plasminogen activator system. Int J Cancer 110: 815–24.

Fa M., Karolin J., Aleshkov S. et al (1995). Time-resolved polarized fluorescence spectroscopy studies of plasminogen activator inhibitor type 1: conformational changes of the reactive center upon interactions with target proteases, vitronectin and heparin. Biochemistry 34: 13833–40.

Fish P. V., Barber C. G., Brown D. G. et al (2007). Selective urokinase-type plasminogen activator inhibitors. 4. 1-(7-sulfonamidoisoquinolinyl)guanidines. J Med Chem 50: 2341–51.

Fisher C., Beglova N. and Blacklow S. C. (2006). Structure of an LDLR-RAP complex reveals a general mode for ligand recognition by lipoprotein receptors. Mol Cell 22: 277–83.

Folkes A., Roe M. B., Sohal S. et al (2001). Synthesis and in vitro evaluation of a series of diketopiperazine inhibitors of plasminogen activator inhibitor-1. Bioorg Med Chem Lett 11: 2589–92.

Franco P., Iaccarino C., Chiaradonna F. et al (1997). Phosphorylation of human pro-urokinase on Ser138/303 impairs its receptor-dependent ability to promote myelomonocytic adherence and motility. J Cell Biol 137: 779–91.

Franco P., Massa O., Garcia-Rocha M. et al (1998). Protein kinase C-dependent in vivo phosphorylation of prourokinase leads to the formation of a receptor competitive antagonist. J Biol Chem 273: 27734–40.

Franco P., Vocca I., Carriero M. V. et al (2006). Activation of urokinase receptor by a novel interaction between the connecting peptide region of urokinase and alpha v beta 5 integrin. J Cell Sci 119: 3424–34.

Frandsen T. L., Holst-Hansen C., Nielsen B. S. et al (2001). Direct evidence of the importance of stromal urokinase plasminogen activator (uPA) in the growth of an experimental human breast cancer using a combined uPA gene-disrupted and immunodeficient xenograft model. Cancer Res 61: 532–7.

Friederich P. W., Levi M., Biemond B. J. et al (1997). Novel low-molecular-weight inhibitor of PAI-1 (XR5118) promotes endogenous fibrinolysis and reduces postthrombolysis thrombus growth in rabbits. Circulation 96: 916–21.

Gardell, S.J., Krueger, J.A., Antrilli, T.A., Elokdah, H., Mayer, S., Orcutt, S.J., Crandall, D.L., and Vlasuk, G.P. (2007). Neutralisation of plasminogen activator inhibitor I (PAI-1) by the synthetic antagonist PAI-749 via a dual mechanism of action. Mol. Pharmacol. 72: 897–906.

Gardsvoll H. and Ploug M. (2007). Mapping of the vitronectin-binding site on the urokinase receptor: involvement of a coherent receptor interface consisting of residues from both domain I and the flanking interdomain linker region. J Biol Chem 282: 13561–72.

Gardsvoll H., van Zonneveld A. J., Holm A. et al (1998). Selection of peptides that bind to plasminogen activator inhibitor 1 (PAI-1) using random peptide phage-display libraries. FEBS Lett 431: 170–4.

Gibson A., Baburaj K., Day D. E. et al (1997). The use of fluorescent probes to characterize conformational changes in the interaction between vitronectin and plasminogen activator inhibitor-1. J Biol Chem 272: 5112–21.

Gils A. and Declerck P. J. (1998). Modulation of plasminogen activator inhibitor 1 by Triton X-100— identification of two consecutive conformational transitions. Thromb Haemost 80: 286–91.

Gils A. and Declerck P. J. (2004). The structural basis for the pathophysiological relevance of PAI-I in cardiovascular diseases and the development of potential PAI-I inhibitors. Thromb Haemost 91: 425–37.

Gils A., Knockaert I., Brouwers E. et al (2000). Glycosylation-dependent conformational transitions in plasminogen activator inhibitor-1: evidence for the presence of two active conformations. Fibrinolysis & Proteolysis 14: 58–64.

Gils A., Stassen J. M., Nar H. et al (2002). Characterization and comparative evaluation of a novel PAI-1 inhibitor. Thromb Haemost 88: 137–43.

Gils A., Pedersen K. E., Skottrup P. et al (2003). Biochemical importance of glycosylation of plasminogen activator inhibitor-1. Thromb Haemost 90: 206–17.

Gliemann J. (1998). Receptors of the low density lipoprotein (LDL) receptor family in man. Multiple functions of the large family members via interaction with complex ligands. Biol Chem 379: 951–64.

Gonias S. L., Wu L. and Salicioni A. M. (2004). Low density lipoprotein receptor-related protein: regulation of the plasma membrane proteome. Thromb Haemost 91: 1056–64.

Gopalsamy A., Kincaid S. L., Ellingboe J. W. et al (2004). Design and synthesis of oxadiazolidinediones as inhibitors of plasminogen activator inhibitor-1. Bioorg Med Chem Lett 14: 3477–80.

Goretzki L., Schmitt M., Mann K. et al (1992). Effective activation of the proenzyme form of the urokinase-type plasminogen activator (pro-uPA) by the cysteine protease cathepsin L. FEBS Lett 297: 112–8.

Gorlatova N. V., Elokdah H., Fan K. et al (2003). Mapping of a conformational epitope on plasminogen activator inhibitor-1 by random mutagenesis. Implications for serpin function. J Biol Chem 278: 16329–35.

Gorlatova N. V., Cale J. M., Elokdah H. et al (2007). Mechanism of inactivation of plasminogen activator inhibitor-1 by a small molecule inhibitor. J Biol Chem 282: 9288–96.

Grondahl-Hansen J., Christensen I. J., Rosenquist C. et al (1993). High levels of urokinase-type plasminogen activator and its inhibitor PAI-1 in cytosolic extracts of breast carcinomas are associated with poor prognosis. Cancer Res 53: 2513–21.

Guo Y., Higazi A. A., Arakelian A. et al (2000). A peptide derived from the nonreceptor binding region of urokinase plasminogen activator (uPA) inhibits tumor progression and angiogenesis and induces tumor cell death in vivo. FASEB J 14: 1400–10.

Guo Y., Mazar A. P., Lebrun J. J. et al (2002). An antiangiogenic urokinase-derived peptide combined with tamoxifen decreases tumor growth and metastasis in a syngeneic model of breast cancer. Cancer Res 62: 4678–84.

Gutierrez L. S., Schulman A., Brito-Robinson T. et al (2000). Tumor development is retarded in mice lacking the gene for urokinase-type plasminogen activator or its inhibitor, plasminogen activator inhibitor-1. Cancer Res 60: 5839–47.

Hansen M., Busse M. N. and Andreasen P. A. (2001). Importance of the amino-acid composition of the shutter region of plasminogen activator inhibitor-1 for its transitions to latent and substrate forms. Eur J Biochem 268: 6274–83.

Hansen M., Wind T., Blouse G. E. et al (2005). A urokinase-type plasminogen activator-inhibiting cyclic peptide with an unusual P2 residue and an extended protease binding surface demonstrates new modalities for enzyme inhibition. J Biol Chem 280: 38424–37.

Harbeck N., Kates R. E., Gauger K. et al (2004). Urokinase-type plasminogen activator (uPA) and its inhibitor PAI-I: novel tumor-derived factors with a high prognostic and predictive impact in breast cancer. Thromb Haemost 91: 450–6.

Hedstrom L. (2002). Serine protease mechanism and specificity. Chem Rev 102: 4501–24.

Horn I. R., van den Berg B. M., Moestrup S. K. et al (1998). Plasminogen activator inhibitor 1 contains a cryptic high affinity receptor binding site that is exposed upon complex formation with tissue-type plasminogen activator. Thromb Haemost 80: 822–8.

Hoyer-Hansen G., Ronne E., Solberg H. et al (1992). Urokinase plasminogen activator cleaves its cell surface receptor releasing the ligand-binding domain. J Biol Chem 267: 18224–9.

Hu B., Jetter J. W., Wrobel J. E. et al (2005). Synthesis and SAR of 2-carboxylic acid indoles as inhibitors of plasminogen activator inhibitor-1. Bioorg Med Chem Lett 15: 3514–8.

Huai Q., Mazar A. P., Kuo A. et al (2006). Structure of human urokinase plasminogen activator in complex with its receptor. Science 311: 656–9.

Huang Y., Haraguchi M., Lawrence D. A. et al (2003). A mutant, noninhibitory plasminogen activator inhibitor type 1 decreases matrix accumulation in experimental glomerulonephritis. J Clin Invest 112: 379–88.

Huntington J. A. (2006). Shape-shifting serpins—advantages of a mobile mechanism. Trends Biochem Sci 31: 427–35

Huntington J. A., Read R. J. and Carrell R. W. (2000). Structure of a serpin-protease complex shows inhibition by deformation. Nature 407: 923–6.

Ichinose A., Fujikawa K. and Suyama T. (1986). The activation of pro-urokinase by plasma kallikrein and its inactivation by thrombin. J Biol Chem 261: 3486–9.

Janicke F., Schmitt M., Pache L. et al (1993). Urokinase (uPA) and its inhibitor PAI-1 are strong and independent prognostic factors in node-negative breast cancer. Breast Cancer Res Treat 24: 195–208.

Janicke F., Prechtl A., Thomssen C. et al (2001). Randomized adjuvant chemotherapy trial in high-risk, lymph node-negative breast cancer patients identified by urokinase-type plasminogen activator and plasminogen activator inhibitor type 1. J Natl Cancer Inst 93: 913–20.

Jensen J. K., Wind T. and Andreasen P. A. (2002). The vitronectin binding area of plasminogen activator inhibitor-1, mapped by mutagenesis and protection against an inactivating organo-chemical ligand. FEBS Lett 521: 91–4.

Jensen G. A., Andersen O. M., Bonvin A. M., Bjerrum-Bohr I. et al (2006a). Binding site structure of one LRP-RAP complex: implications for a common ligand-receptor binding motif. J Mol Biol 362: 700–16.

Jensen J. K., Malmendal A., Schiott B., Skeldal S. et al (2006b). Inhibition of plasminogen activator inhibitor-1 binding to endocytosis receptors of the low-density-lipoprotein receptor family by a peptide isolated from a phage display library. Biochem J 399: 387–96.

Kaltoft K., Nielsen L. S., Zeuthen J. et al (1982). Monoclonal antibody that specifically inhibits a human Mr 52,000 plasminogen-activating enzyme. Proc Natl Acad Sci USA 79: 3720–3.

Kannemeier C., Feussner A., Stohr H. A. et al (2001). Factor VII and single-chain plasminogen activator-activating protease: activation and autoactivation of the proenzyme. Eur J Biochem 268: 3789–96.

Kanse S. M., Kost C., Wilhelm O. G. et al (1996). The urokinase receptor is a major vitronectin-binding protein on endothelial cells. Exp Cell Res 224: 344–53.

Katz B. A., Sprengeler P. A., Luong C. et al (2001). Engineering inhibitors highly selective for the S1 sites of Ser190 trypsin-like serine protease drug targets. Chem Biol 8: 1107–21.

Katz B. A., Elrod K., Verner E. et al (2003). Elaborate manifold of short hydrogen bond arrays mediating binding of active site-directed serine protease inhibitors. J Mol Biol 329: 93–120.

Katz B. A., Luong C., Ho J. D. et al (2004). Dissecting and designing inhibitor selectivity determinants at the S1 site using an artificial Ala190 protease (Ala190 uPA). J Mol Biol 344: 527–47.

Ke S. H., Coombs G. S., Tachias K. et al (1997). Optimal subsite occupancy and design of a selective inhibitor of urokinase. J Biol Chem 272: 20456–62.

Killeen S. D., Andrews E. J., Wang J. H. et al (2007). Inhibition of urokinase plasminogen activator with a novel enzyme inhibitor, WXC-340, ameliorates endotoxin and surgery-accelerated growth of murine metastases. Br J Cancer 96: 262–8.

Kilpatrick L. M., Harris R. L., Owen K. A. et al (2006). Initiation of plasminogen activation on the surface of monocytes expressing the type II transmembrane serine protease matriptase. Blood 108: 2616–23.

Kim K. S., Hong Y. K., Lee Y., Shin J. Y. et al (2003a). Differential inhibition of endothelial cell proliferation and migration by urokinase subdomains: amino-terminal fragment and kringle domain. Exp Mol Med 35: 578–85.

Kim H. K., Lee S. Y., Oh H. K., Kang B. H. et al (2003b). Inhibition of endothelial cell proliferation by the recombinant kringle domain of tissue-type plasminogen activator. Biochem Biophys Res Commun 304: 740–6.

Kim C. K., Hong S. H., Joe Y. A. et al (2007). The recombinant kringle domain of urokinase plasminogen activator inhibits in vivo malignant glioma growth. Cancer Sci 98: 253–8.

Klinghofer V., Stewart K., McGonigal T. et al (2001). Species specificity of amidine-based urokinase inhibitors. Biochemistry 40: 9125–31.

Kobayashi H., Schmitt M., Goretzki L. et al (1991). Cathepsin B efficiently activates the soluble and the tumor cell receptor-bound form of the proenzyme urokinase-type plasminogen activator (Pro-uPA). J Biol Chem 266: 5147–52.

Koivunen E., Huhtala M. L. and Stenman U. H. (1989). Human ovarian tumor-associated trypsin. Its purification and characterization from mucinous cyst fluid and identification as an activator of pro-urokinase. J Biol Chem 264: 14095–9.

Komissarov A. A., Andreasen P. A., Bodker J. S. et al (2005). Additivity in effects of vitronectin and monoclonal antibodies against alpha-helix F of plasminogen activator inhibitor-1 on its reactions with target proteinases. J Biol Chem 280: 1482–9.

Kortlever R. M., Higgins P. J. and Bernards R. (2006). Plasminogen activator inhibitor-1 is a critical downstream target of p53 in the induction of replicative senescence. Nat Cell Biol 8: 877–84.

Kugler M. C., Wei Y. and Chapman H. A. (2003). Urokinase receptor and integrin interactions. Curr Pharm Des 9: 1565–74.

Kvassman J. O., Lawrence D. A. and Shore J. D. (1995). The acid stabilization of plasminogen activator inhibitor-1 depends on protonation of a single group that affects loop insertion into beta-sheet A. J Biol Chem 270: 27942–7.

Kwaan H. C., Wang J., Svoboda K. et al (2000). Plasminogen activator inhibitor 1 may promote tumour growth through inhibition of apoptosis. Br J Cancer 82: 1702–8.

Laboissiere M. C., Young M. M., Pinho R. G. et al (2002). Computer-assisted mutagenesis of ecotin to engineer its secondary binding site for urokinase inhibition. J Biol Chem 277: 26623–31.

Lademann U., Romer M. U., Jensen P. B. et al (2005). Malignant transformation of wild-type but not plasminogen activator inhibitor-1 gene-deficient fibroblasts decreases cellular sensitivity to chemotherapy-mediated apoptosis. Eur J Cancer 41: 1095–100.

Lawrence D. A., Berkenpas M. B., Palaniappan S. et al (1994). Localization of vitronectin binding domain in plasminogen activator inhibitor-1. J Biol Chem 269: 15223–8.

Lawrence D. A., Palaniappan S., Stefansson S. et al (1997). Characterization of the binding of different conformational forms of plasminogen activator inhibitor-1 to vitronectin. Implications for the regulation of pericellular proteolysis. J Biol Chem 272: 7676–80.

Lazic A., Dolmer K., Strickland D. K. et al (2006). Dissection of RAP-LRP interactions: binding of RAP and RAP fragments to complement-like repeats 7 and 8 from ligand binding cluster II of LRP. Arch Biochem Biophys 450: 167–75.

Lee S. L., Dickson R. B. and Lin C. Y. (2000). Activation of hepatocyte growth factor and urokinase/plasminogen activator by matriptase, an epithelial membrane serine protease. J Biol Chem 275: 36720–5.

Leik C. E., Su E. J., Nambi P. et al (2006). Effect of pharmacologic plasminogen activator inhibitor-1 inhibition on cell motility and tumor angiogenesis. J Thromb Haemost 4: 2710–5.

Liang A., Wu F., Tran K. et al (2005). Characterization of a small molecule PAI-1 inhibitor, ZK4044. Thromb Res 115: 341–50.

Lijnen H. R., Uytterhoeven M. and Collen D. (1984). Inhibition of trypsin-like serine proteinases by tripeptide arginyl and lysyl chloromethylketones. Thromb Res 34: 431–7.

Lillis A. P., Mikhailenko I. and Strickland D. K. (2005). Beyond endocytosis: LRP function in cell migration, proliferation and vascular permeability. J Thromb Haemost 3: 1884–93.

List K., Jensen O. N., Bugge T. H. et al (2000). Plasminogen-independent initiation of the pro-urokinase activation cascade in vivo. Activation of pro-urokinase by glandular kallikrein (mGK-6) in plasminogen-deficient mice. Biochemistry 39: 508–15.

Llinas P., Le Du M. H., Gardsvoll H. et al (2005). Crystal structure of the human urokinase plasminogen activator receptor bound to an antagonist peptide. EMBO J 24: 1655–63.

Look M. P., van Putten W. L., Duffy M. J. et al (2002). Pooled analysis of prognostic impact of urokinase-type plasminogen activator and its inhibitor PAI-1 in 8377 breast cancer patients. J Natl Cancer Inst 94: 116–28.

Ma Z., Webb D. J., Jo M. et al (2001). Endogenously produced urokinase-type plasminogen activator is a major determinant of the basal level of activated ERK/MAP kinase and prevents apoptosis in MDA-MB-231 breast cancer cells. J Cell Sci 114: 3387–96.

Mackman R. L., Katz B. A., Breitenbucher J. G. et al (2001). Exploiting subsite S1 of trypsin-like serine proteases for selectivity: potent and selective inhibitors of urokinase-type plasminogen activator. J Med Chem 44: 3856–71.

Mackman R. L., Hui H. C., Breitenbucher J. G. et al (2002). 2-(2-Hydroxy-3-alkoxyphenyl)-1H-benzimidazole-5-carboxamidine derivatives as potent and selective urokinase-type plasminogen activator inhibitors. Bioorg Med Chem Lett 12: 2019–22.

Madsen C. D., Ferraris G. M., Andolfo A. et al (2007). uPAR-induced cell adhesion and migration: vitronectin provides the key. J Cell Biol 177: 927–39.

Magdolen V., Arroyo de Prada N., Sperl S. et al (2000). Natural and synthetic inhibitors of the tumor-associated serine protease urokinase-type plasminogen activator. Adv Exp Med Biol 477: 331–41.

Mishima K., Mazar A. P., Gown A. et al (2000). A peptide derived from the non-receptor-binding region of urokinase plasminogen activator inhibits glioblastoma growth and angiogenesis in vivo in combination with cisplatin. Proc Natl Acad Sci USA 97: 8484–9.

Moller L. B., Ploug M. and Blasi F. (1992). Structural requirements for glycosyl-phosphatidylinositol-anchor attachment in the cellular receptor for urokinase plasminogen activator. Eur J Biochem 208: 493–500.

Moran P., Li W., Fan B. et al (2006). Pro-urokinase-type plasminogen activator is a substrate for hepsin. J Biol Chem 281: 30439–46.

Mottonen J., Strand A., Symersky J. et al (1992). Structural basis of latency in plasminogen activator inhibitor-1. Nature 355: 270–3.

Mueller M. M. and Fusenig N. E. (2004). Friends or foes—bipolar effects of the tumour stroma in cancer. Nat Rev Cancer 4: 839–49.

Mukhina S., Stepanova V., Traktouev D. et al (2000). The chemotactic action of urokinase on smooth muscle cells is dependent on its kringle domain. Characterization of interactions and contribution to chemotaxis. J Biol Chem 275: 16450–8.

Munch M., Heegaard C. W. and Andreasen P. A. (1993). Interconversions between active, inert and substrate forms of denatured/refolded type-1 plasminogen activator inhibitor. Biochim Biophys Acta 1202: 29–37.

Naessens D., Gils A., Compernolle G. et al (2003). Elucidation of the epitope of a latency-inducing antibody: identification of a new molecular target for PAI-1 inhibition. Thromb Haemost 90: 52–8.

Nar H., Bauer M., Stassen J. M. et al (2000). Plasminogen activator inhibitor 1. Structure of the native serpin, comparison to its other conformers and implications for serpin inactivation. J Mol Biol 297: 683–95.

Neve J., Leone P. A., Carroll A. R. et al (1999). Sideroxylonal C, a new inhibitor of human plasminogen activator inhibitor type-1, from the flowers of Eucalyptus albens. J Nat Prod 62: 324–6.

Nielsen B. S., Sehested M., Timshel S. et al (1996). Messenger RNA for urokinase plasminogen activator is expressed in myofibroblasts adjacent to cancer cells in human breast cancer. Lab Invest 74: 168–77.

Nielsen B. S., Sehested M., Duun S. et al (2001). Urokinase plasminogen activator is localized in stromal cells in ductal breast cancer. Lab Invest 81: 1485–501.

Nielsen L. S., Andreasen P. A., Grondahl-Hansen J. et al (1986). Monoclonal antibodies to human 54,000 molecular weight plasminogen activator inhibitor from fibrosarcoma cells—inhibitor neutralization and one-step affinity purification. Thromb Haemost 55: 206–12.

Nienaber V. L., Davidson D., Edalji R. et al (2000a). Structure-directed discovery of potent non-peptidic inhibitors of human urokinase that access a novel binding subsite. Structure 8: 553–63.

Nienaber V. L., Wang J., Davidson D. et al (2000b). Re-engineering of human urokinase provides a system for structure-based drug design at high resolution and reveals a novel structural subsite. J Biol Chem 275: 7239–48.

Nozaki S., Endo Y., Nakahara H. et al (2006). Targeting urokinase-type plasminogen activator and its receptor for cancer therapy. Anticancer Drugs 17: 1109–17.

Nykjaer A. and Willnow T. E. (2002). The low-density lipoprotein receptor gene family: a cellular Swiss army knife? Trends Cell Biol 12: 273–80.

Nykjaer A., Petersen C. M., Moller B. et al (1992). Purified alpha 2-macroglobulin receptor/LDL receptor-related protein binds urokinase plasminogen activator inhibitor type-1 complex. Evidence that the alpha 2-macroglobulin receptor mediates cellular degradation of urokinase receptor-bound complexes. J Biol Chem 267: 14543–6.

Nykjaer A., Kjoller L., Cohen R. L. et al (1994). Regions involved in binding of urokinase-type-1 inhibitor complex and pro-urokinase to the endocytic alpha 2-macroglobulin receptor/low density lipoprotein receptor-related protein. Evidence that the urokinase receptor protects pro-urokinase against binding to the endocytic receptor. J Biol Chem 269: 25668–76.

Nykjaer A., Conese M., Christensen E. I. et al (1997). Recycling of the urokinase receptor upon internalization of the uPA: serpin complexes. EMBO J 16: 2610–20.

Ossowski L. and Aguirre-Ghiso J. A. (2000). Urokinase receptor and integrin partnership: coordination of signaling for cell adhesion, migration and growth. Curr Opin Cell Biol 12: 613–20.

Padmanabhan J. and Sane D. C. (1995). Localization of a vitronectin binding region of plasminogen activator inhibitor-1. Thromb Haemost 73: 829–34.

Palumbo J. S., Kombrinck K. W., Drew A. F. et al (2000). Fibrinogen is an important determinant of the metastatic potential of circulating tumor cells. Blood 96: 3302–9.

Parry M. A., Zhang X. C. and Bode I. (2000). Molecular mechanisms of plasminogen activation: bacterial cofactors provide clues. Trends Biochem Sci 25: 53–9.

Pedersen K. E., Einholm A. P., Christensen A. et al (2003). Plasminogen activator inhibitor-1 polymers, induced by inactivating amphipathic organochemical ligands. Biochem J 372: 747–55.

Petersen H. H., Hansen M., Schousboe S. L. et al (2001). Localization of epitopes for monoclonal antibodies to urokinase-type plasminogen activator: relationship between epitope localization and effects of antibodies on molecular interactions of the enzyme. Eur J Biochem 268: 4430–9.

Pillay V., Dass C. R. and Choong P. F. (2007). The urokinase plasminogen activator receptor as a gene therapy target for cancer. Trends Biotechnol 25: 33–9.

Ploplis V. A., Balsara R., Sandoval-Cooper M. J. et al (2004). Enhanced in vitro proliferation of aortic endothelial cells from plasminogen activator inhibitor-1-deficient mice. J Biol Chem 279: 6143–51.

Ploug M., Ronne E., Behrendt N. et al (1991). Cellular receptor for urokinase plasminogen activator. Carboxyl-terminal processing and membrane anchoring by glycosyl-phosphatidylinositol. J Biol Chem 266: 1926–33.

Rak J., Yu J. L., Luyendyk J. et al (2006). Oncogenes, trousseau syndrome, and cancer-related changes in the coagulome of mice and humans. Cancer Res 66: 10643–6.

Reilly D., Christensen L., Duch M. et al (1992). Type-1 plasminogen activator inhibitor in human breast carcinomas. Int J Cancer 50: 208–14.

Rockway T. W., Nienaber V. and Giranda V. L. (2002). Inhibitors of the protease domain of urokinase-type plasminogen activator. Curr Pharm Des 8: 2541–58.

Rodenburg K. W., Kjoller L., Petersen H. H. et al (1998). Binding of urokinase-type plasminogen activator-plasminogen activator inhibitor-1 complex to the endocytosis receptors alpha2-macroglobulin receptor/low-density lipoprotein receptor-related protein and very-low-density lipoprotein receptor involves basic residues in the inhibitor. Biochem J 329: 55–63.

Roemisch J., Feussner A., Nerlich C. et al (2002). The frequent Marburg I polymorphism impairs the pro-urokinase activating potency of the factor VII activating protease (FSAP). Blood Coagul Fibrinolysis 13: 433–41.

Romer M. U., Kirkebjerg Due A., Knud Larsen J. et al (2005). Indication of a role of plasminogen activator inhibitor type I in protecting murine fibrosarcoma cells against apoptosis. Thromb Haemost 94: 859–66.

Rossignol P., Ho-Tin-Noe B., Vranckx R. et al (2004). Protease nexin-1 inhibits plasminogen activation-induced apoptosis of adherent cells. J Biol Chem 279: 10346–56.

Sabapathy K. T., Pepper M. S., Kiefer F. et al (1997). Polyoma middle T-induced vascular tumor formation: the role of the plasminogen activator/plasmin system. J Cell Biol 137: 953–63.

Schmitt M., Janicke F. and Graeff H. (1990). Tumour-associated fibrinolysis: the prognostic relevance of plasminogen activators uPA and tPA in human breast cancer. Blood Coagul Fibrinolysis 1: 695–702.

Schousboe S. L., Egelund R., Kirkegaard T. et al (2000). Vitronectin and substitution of a beta-strand 5A lysine residue potentiate activity-neutralization of PA inhibitor-1 by monoclonal antibodies against alpha-helix F. Thromb Haemost 83: 742–51.

Schweinitz A., Steinmetzer T., Banke I. J. et al (2004). Design of novel and selective inhibitors of urokinase-type plasminogen activator with improved pharmacokinetic properties for use as antimetastatic agents. J Biol Chem 279: 33613–22.

Seiffert D. and Loskutoff D. J. (1996). Type 1 plasminogen activator inhibitor induces multimerization of plasma vitronectin. A suggested mechanism for the generation of the tissue form of vitronectin in vivo. J Biol Chem 271: 29644–51.

Setyono-Han B., Sturzebecher J., Schmalix W. A. et al (2005). Suppression of rat breast cancer metastasis and reduction of primary tumour growth by the small synthetic urokinase inhibitor WX-UK1. Thromb Haemost 93: 779–86.

Shapiro R. L., Duquette J. G., Roses D. F. et al (1996). Induction of primary cutaneous melanocytic neoplasms in urokinase-type plasminogen activator (uPA)-deficient and wild-type mice: cellular blue nevi invade but do not progress to malignant melanoma in uPA-deficient animals. Cancer Res 56: 3597–604.

Sharp A. M., Stein P. E., Pannu N. S. et al (1999). The active conformation of plasminogen activator inhibitor 1, a target for drugs to control fibrinolysis and cell adhesion. Structure 7: 111–8.

Shaw E. (1970). Selective chemical modification of proteins. Physiol Rev 50: 244–96.

Sienczyk M. and Oleksyszyn J. (2006). Inhibition of trypsin and urokinase by Cbz-amino(4-guanidinophenyl)methanephosphonate aromatic ester derivatives: the influence of the ester group on their biological activity. Bioorg Med Chem Lett 16: 2886–90.

Skeldal S., Larsen J. V., Pedersen K. E. et al (2006). Binding areas of urokinase-type plasminogen activator-plasminogen activator inhibitor-1 complex for endocytosis receptors of the low-density lipoprotein receptor family, determined by site-directed mutagenesis. FEBS J 273: 5143–59.

Soeda S., Oda M., Ochiai T. et al (2001). Deficient release of plasminogen activator inhibitor-1 from astrocytes triggers apoptosis in neuronal cells. Brain Res Mol Brain Res 91: 96–103.

Soeda S., Shinomiya K., Ochiai T. et al (2006). Plasminogen activator inhibitor-1 aids nerve growth factor-induced differentiation and survival of pheochromocytoma cells by activating both the extracellular signal-regulated kinase and c-Jun pathways. Neuroscience 141: 101–8.

Spencer J. R., McGee D., Allen D. et al (2002). 4-Aminoarylguanidine and 4-aminobenzamidine derivatives as potent and selective urokinase-type plasminogen activator inhibitors. Bioorg Med Chem Lett 12: 2023–6.

Sperl S., Jacob U., Arroyo de Prada N. et al (2000). (4-Aminomethyl)phenylguanidine derivatives as nonpeptidic highly selective inhibitors of human urokinase. Proc Natl Acad Sci USA 97: 5113–8.

Spraggon G., Phillips C., Nowak U. K. et al (1995). The crystal structure of the catalytic domain of human urokinase-type plasminogen activator. Structure 3: 681–91.

Stack M. S. and Johnson D. A. (1994). Human mast cell tryptase activates single-chain urinary-type plasminogen activator (pro-urokinase). J Biol Chem 269: 9416–9.

Stefansson S., Muhammad S., Cheng X. F. et al (1998). Plasminogen activator inhibitor-1 contains a cryptic high affinity binding site for the low density lipoprotein receptor-related protein. J Biol Chem 273: 6358–66.

Stephens R. W., Bokman A. M., Myohanen H. T. et al (1992). Heparin binding to the urokinase kringle domain. Biochemistry 31: 7572–9.

Stoop A. A. and Craik C. S. (2003). Engineering of a macromolecular scaffold to develop specific protease inhibitors. Nat Biotechnol 21: 1063–8.

Stoppelli M. P., Corti A., Soffientini A. et al (1985). Differentiation-enhanced binding of the amino-terminal fragment of human urokinase plasminogen activator to a specific receptor on U937 monocytes. Proc Natl Acad Sci USA 82: 4939–43.

Stout T. J., Graham H., Buckley D. I. et al (2000). Structures of active and latent PAI-1: a possible stabilizing role for chloride ions. Biochemistry 39: 8460–9.

Sturge J., Hamelin J. and Jones G. E. (2002). N-WASP activation by a beta1-integrin-dependent mechanism supports PI3K-independent chemotaxis stimulated by urokinase-type plasminogen activator. J Cell Sci 115: 699–711.

Sturzebecher J., Vieweg H., Steinmetzer T. et al (1999). 3-Amidinophenylalanine-based inhibitors of urokinase. Bioorg Med Chem Lett 9: 3147–52.

Suzuki M., Kobayashi H., Kanayama N. et al (2004). Inhibition of tumor invasion by genomic down-regulation of matriptase through suppression of activation of receptor-bound pro-urokinase. J Biol Chem 279: 14899–908.

Tamura S. Y., Weinhouse M. I., Roberts C. A. et al (2000). Synthesis and biological activity of peptidyl aldehyde urokinase inhibitors. Bioorg Med Chem Lett 10: 983–7.

Takahashi K. and Naora H. (1985). A monoclonal antibody against human urokinase: the epitope structure and sequence homology with a human tissue-type plasminogen activator. Cell Struct Funct 10: 195–208.

Tarui T., Akakura N., Majumdar M. et al (2006). Direct interaction of the kringle domain of urokinase-type plasminogen activator (uPA) and integrin alpha v beta 3 induces signal transduction and enhances plasminogen activation. Thromb Haemost 95: 524–34.

Trousseau A. (1865). Phlegmasia Alba Dolens. Clinique Medicale de l'Hotel-Dieu de Paris, France: 654–712.

Tsuchiya H., Sunayama C., Okada G. et al (1997). Plasminogen activator inhibitor-1 accelerates lung metastasis formation of human fibrosarcoma cells. Anticancer Res 17: 313–6.

Urano T., Strandberg L., Johansson L. B. et al (1992). A substrate-like form of plasminogen-activator-inhibitor type 1. Conversions between different forms by sodium dodecyl sulphate. Eur J Biochem 209: 985–92.

van Meijer M., Gebbink R. K., Preissner K. T. et al (1994). Determination of the vitronectin binding site on plasminogen activator inhibitor 1 (PAI-1). FEBS Lett 352: 342–6.

Vassalli J. D. and Belin D. (1987). Amiloride selectively inhibits the urokinase-type plasminogen activator. FEBS Lett 214: 187–91.

Vassalli J. D., Baccino D. and Belin D. (1985). A cellular binding site for the Mr 55,000 form of the human plasminogen activator, urokinase. J Cell Biol 100: 86–92.

Vaughan D. E. (2002). Angiotensin and vascular fibrinolytic balance. Am J Hypertens 15: 3S–8S.

Verdaguer N., Fita I., Reithmayer M. et al (2004). X-ray structure of a minor group human rhinovirus bound to a fragment of its cellular receptor protein. Nat Struct Mol Biol 11: 429–34.

Verhamme I., Kvassman J. O., Day D. et al (1999). Accelerated conversion of human plasminogen activator inhibitor-1 to its latent form by antibody binding. J Biol Chem 274: 17511–7.

Wakselman M., Xie J., Mazaleyrat J. P. et al (1993). New mechanism-based inactivators of trypsin-like proteinases. Selective inactivation of urokinase by functionalized cyclopeptides incorporating a sulfoniomethyl-substituted m-aminobenzoic acid residue. J Med Chem 36: 1539–47.

Walker B. and Lynas J. F. (2001). Strategies for the inhibition of serine proteases. Cell Mol Life Sci 58: 596–624.

Wang C. I., Yang Q. and Craik C. S. (1995). Isolation of a high affinity inhibitor of urokinase-type plasminogen activator by phage display of ecotin. J Biol Chem 270: 12250–6.

Wang S., Golec J., Miller W. et al (2002). Novel inhibitors of plasminogen activator inhibitor-1: development of new templates from diketopiperazines. Bioorg Med Chem Lett 12: 2367–70.

Wang X., Terzyan S., Tang J. et al (2000). Human plasminogen catalytic domain undergoes an unusual conformational change upon activation. J Mol Biol 295: 903–14.

Wang Y. (2001). The role and regulation of urokinase-type plasminogen activator receptor gene expression in cancer invasion and metastasis. Med Res Rev 21: 146–70.

Wei Y., Waltz D. A., Rao N. et al (1994). Identification of the urokinase receptor as an adhesion receptor for vitronectin. J Biol Chem 269: 32380–8.

Wendt M. D., Rockway T. W., Geyer A. et al (2004). Identification of novel binding interactions in the development of potent, selective 2-naphthamidine inhibitors of urokinase. Synthesis, structural analysis, and SAR of N-phenyl amide 6-substitution. J Med Chem 47: 303–24.

Wind T., Jensen M. A. and Andreasen P. A. (2001). Epitope mapping for four monoclonal antibodies against human plasminogen activator inhibitor type-1: implications for antibody-mediated PAI-1-neutralization and vitronectin-binding. Eur J Biochem 268: 1095–106.

Wind T., Hansen M., Jensen J. K. et al (2002). The molecular basis for anti-proteolytic and non-proteolytic functions of plasminogen activator inhibitor type-1: roles of the reactive centre loop, the shutter region, the flexible joint region and the small serpin fragment. Biol Chem 383: 21–36.

Wolf B. B., Vasudevan J., Henkin J. et al (1993). Nerve growth factor-gamma activates soluble and receptor-bound single chain urokinase-type plasminogen activator. J Biol Chem 268: 16327–31.

Yanamandra N., Konduri S. D., Mohanam S. et al (2000). Downregulation of urokinase-type plasminogen activator receptor (uPAR) induces caspase-mediated cell death in human glioblastoma cells. Clin Exp Metastasis 18: 611–5.

Yang S. Q. and Craik C. S. (1998). Engineering bidentate macromolecular inhibitors for trypsin and urokinase-type plasminogen activator. J Mol Biol 279: 1001–11.

Yasuda S., Morokawa N., Wong G. W. et al (2005). Urokinase-type plasminogen activator is a preferred substrate of the human epithelium serine protease tryptase epsilon/PRSS22. Blood 105: 3893–901.

Ye B., Chou Y. L., Karanjawala R. et al (2004). Synthesis and biological evaluation of piperazine-based derivatives as inhibitors of plasminogen activator inhibitor-1 (PAI-1). Bioorg Med Chem Lett 14: 761–5.

Ye S. and Goldsmith E. J. (2001). Serpins and other covalent protease inhibitors. Curr Opin Struct Biol 11: 740–5.

Yoshida E., Ohmura S., Sugiki M. et al (1995). Prostate-specific antigen activates single-chain urokinase-type plasminogen activator. Int J Cancer 63: 863–5.

Zannetti A., Del Vecchio S., Romanelli A. et al (2005). Inhibition of Sp1 activity by a decoy PNA-DNA chimera prevents urokinase receptor expression and migration of breast cancer cells. Biochem Pharmacol 70: 1277–87.

Zeslawska E., Schweinitz A., Karcher A. et al (2000). Crystals of the urokinase type plasminogen activator variant beta(c)-uPAin complex with small molecule inhibitors open the way towards structure-based drug design. J Mol Biol 301: 465–75.

Zhang J. C., Sakthivel R., Kniss D. et al (1998). The low density lipoprotein receptor-related protein/alpha2-macroglobulin receptor regulates cell surface plasminogen activator activity on human trophoblast cells. J Biol Chem 273: 32273–80.

Zhang Y., Wisner A., Maroun R. C. et al (1997). Trimeresurus stejnegeri snake venom plasminogen activator. Site-directed mutagenesis and molecular modeling. J Biol Chem 272: 20531–7.

Zhao G., Yuan C., Wind T. et al (2007). Structural basis of specificity of a peptidyl urokinase inhibitor, upain-1. J Struct Biol 160: 1–10.

Zhou A., Huntington J. A., Pannu N. S. et al (2003). How vitronectin binds PAI-1 to modulate fibrinolysis and cell migration. Nat Struct Biol 10: 541–4.

Zhu M., Gokhale V. M., Szabo L. et al (2007). Identification of a novel inhibitor of urokinase-type plasminogen activator. Mol Cancer Ther 6: 1348–56.

Chapter 36
MMP Inhibitor Clinical Trials – The Past, Present, and Future

Barbara Fingleton

Abstract Various pharmacological inhibitors of matrix metalloproteinases (MMPs) have been tested in phase I, II, and III trials of multiple cancer types and no significant evidence of efficacy has emerged. This overwhelming failure has understandably led to questions regarding MMPs as suitable drug targets in the oncology setting. In this chapter, a synopsis of the various trials and their results is presented along with a discussion of dose-limiting toxicities and the contrasting successful preclinical testing of these agents. Thoughtful application of the lessons learnt from the MMP inhibitor clinical experience should improve the likelihood of successful proteinase inhibitors in the future.

Introduction

Small-molecule inhibitors of matrix metalloproteinases (MMPs), also called MMP inhibitors or MMPIs, have been in development since the 1980s. In various iterations, they have gone through extensive clinical testing only to fail at the final stage, that is, in large-scale phase III clinical trials. All but one of these clinical trials has been in cancer settings. The resounding failure of these trials significantly dampened enthusiasm for continued development of MMP-targeting agents with perhaps, overlapping negativity for other protease targets in cancer. In this chapter, we will see what can be learned from the foregoing failures and if there is any future for MMPs as a drug target in cancer. We will examine the rationale behind the drugs used as well as the disease settings, the trouble with interpreting the negative results, as well as the questions that remain to be answered. The goal, of course, is to avoid repeating mistakes of the past, in the future. Other chapters in this section discuss structure and inhibitor design (Maskos and Bode) as well as novel synthetic

B. Fingleton
Dept. of Cancer Biology, Vanderbilt University School of Medicine, 734 PRB 2220 Pierce Ave, Nashville, TN 37232-6840, USA, e-mail: barbara.fingleton@vanderbilt.edu

D. Edwards et al. (eds.), *The Cancer Degradome.*
© Springer Science + Business Media, LLC

inhibitors (Yiotakis and Dive) and the reader is referred to them for additional information.

Targeting MMPs

As has been obvious, no doubt, from other chapters in this volume as well as a vast amount of accumulated literature over the years, there is significant evidence that members of the MMP family play contributing roles in many physiological and pathological processes (Parks et al. 2004, Page-McCaw et al. 2007, Cauwe et al. 2007). MMP expression is increased in multiple tumor types (Egeblad and Werb 2002) and, in general, these increases correlate with decreased survival (Fingleton 2003a). The roles of MMPs in various cancers are now understood to be myriad and complex. However, for a long time, MMPs were thought to play quite simple roles – they were the enzymes that degraded the basement membrane; thus their activity defined malignancy (Liotta et al. 1980), as only when a solid neoplasm invades through the basement membrane is it deemed truly cancerous. Since an invasive cancer then has the capacity to spread throughout the body, the MMPs were identified with the twinned processes of invasion and metastasis and were thus viewed as very attractive candidates to inhibit (Brown 1999). The lethal aspect of most cancers is metastasis; hence, there was an assumption that by targeting MMPs, metastasis could be prevented and cancer would become a readily treatable disease.

The Clinical Trial Process

Before a new drug can be approved for widespread use, it must go through a clinical trials process with the goals of demonstrating both safety and efficacy. In the United States, the Food and Drug Administration (FDA) has oversight for the trial process and is ultimately responsible for approving a drug for sale. In Europe, the European Medicines Evaluation Agency (EMEA) performs a similar function although individual countries also have their own regulatory authorities; while in Japan, the Pharmaceuticals and Medical Devices Agency is the relevant authority. The process usually involves at least three stages or phases (Kummar et al. 2006). In phase I trials, the goal is to determine the safety of the drug and to find the best dose to be used in subsequent phases. Phase I trials are typically small, involving 20–60 people, and of limited duration. The next phase, phase II, is for establishing effectiveness of the treatment as well as further analysis of safety and toxicity. These trials involve a greater number of subjects than in phase I, with participation typically in the low hundreds. The main test of drug efficacy is in phase III where the new drug is tested usually in large numbers of patients often in a blinded, randomized fashion against either a placebo or the currently used treatment. The agent being tested may be given in combination with other treatments such as conventional chemotherapy or radiation therapy. Since so many people are involved in these trials, this is also where low-frequency side effects as well as long-term toxicities can first be noted. In oncology, the gold standard determinant of efficacy is whether a drug increases survival of treated patients (Rothenberg et al. 2003). However, other end points can be used including improved quality of life, time to progression, and disease-free survival for drugs in the adjuvant setting (Johnson et al. 2003). These last two are surrogates for a better life and possibly for survival, although that can be dependent on the cancer type. They can also be difficult-to-interpret end points with bias of different types a frequent problem.

MMPIs were among the first group of anticancer drugs that were "molecularly-targeted agents," rather than classical cytotoxics, where traditional measurements of tumor shrinkage may not be appropriate (Rasmussen and McCann 1997, Hidalgo and Eckhardt 2001, Kummar et al. 2006). Moreover, in phase I trials, the classically defined "maximum tolerated dose (MTD)," which is the maximum dose that can be safely administered before evident adverse events outweigh possible benefits, was not considered a realistic end point (Rasmussen and McCann 1997). Since the goal was to inhibit detrimental enzyme activity maximally, it was felt that dosing in excess of this was unnecessary as well as potentially linked to nonspecific toxicities. Therefore, the ideal end point was the "optimal biological dose," that is, the dose at which greatest biological effect could be seen. Unfortunately for MMPI development, an accurate and generally accepted method for assessing "optimal biological dose" has not yet been established. Realistically, before any new drugs go into clinical testing for oncology indications, a reliable method for pharmaco-dynamic assessment will be required. Currently, perhaps the most likely source of such assays is from the realm of imaging, and readers are referred to Sect. 3 of this volume for a thorough discussion of the state of the art in imaging protease activity.

The Past – Early Trials of MMPIs

Table 36.1 lists clinical trials that have been conducted with MMPIs in patients with cancer. In the following paragraphs, some of the important aspects of these drugs and the trials will be discussed.

Batimastat (BB94)

The first drug tested in cancer clinical trials because of its ability to inhibit MMPs was the British Biotech plc compound batimastat, also known as BB-94. This compound had a very broad inhibition profile with IC$_{50}$s against most MMPs in the low nanomolar range. Batimastat suffered from poor solubility and was not readily orally bioavailable (Rasmussen and McCann 1997). To circumvent this problem, phase I trials used batimastat delivered intraperitoneally in patients with

Table 36.1 Cancer Clinical Trials with (MMP) Inhibitors

Drug	Company	Type	Trial phase	Dose	Patient group/ indication	Principal adverse events	Results	References
Batimastat	British Biotech plc	Hydroxamate	I	600–1800 mg/m^2	Advanced malignancy	Abdominal pain	Some stable disease	(Wojtowicz-Praga et al., 1996)
Batimastat	British Biotech plc	Hydroxamate	I/II	600–1350 mg/m^2	Malignant ascites	Abdominal discomfort, nausea, vomiting, fever	Some reduction in ascites accumulation	(Parsons Watson and Steele 1997)
Batimastat	British Biotech plc	Hydroxamate	I	15–300 mg/m^2	Malignant pleural effusions	Fever, abnormal liver enzymes	Reduction in pleural aspiration/ drainage	(Macaulay et al., 1999)
Batimastat	British Biotech plc	Hydroxamate	I	150–1350 mg/m^2	Malignant ascites	Nausea, vomiting, fatigue, abdominal pain	Reduced peritoneal drainage	(Beattie and Smyth 1998)
Marimastat	British Biotech plc	Hydroxamate	I	25–800 mg (single dose) 50–200 mg bid[*]	Healthy volunteers	Elevated liver transaminases	Marimastat well-tolerated	(Millar et al., 1998)
Marimastat (+ captopril/fragmin)	British Biotech plc	Hydroxamate	I	10 mg bid	Advanced cancer	Myalgia	Combination has biological activity	(Jones et al., 2004)
Marimastat (+ paclitaxel/ carboplatin)	British Biotech plc	Hydroxamate	I	10 & 20 mg bid	NSCLC	MSS; myelosuppression	No PK interaction with chemotherapy	(Goffin et al., 2005)
Marimastat (+ paclitaxel)	British Biotech plc	Hydroxamate	I	10 mg	Advanced malignancy	Neutropenia	Marimastat can be administered with paclitaxel	(Toppmeyer et al., 2003)

Marimastat	British Biotech plc	Hydroxamate	I	50 mg bid & 25 mg qd	Gastric cancer	MSS	Suggestion of biological effect on tumor	(Tierney et al., 1999)
Marimastat	British Biotech plc	Hydroxamate	I	5–75 mg bid	Pancreatic cancer	MSS	Dose-associated changes in CA19/9 suggest biological activity	(Rosemurgy et al., 1999)
Marimastat	British Biotech plc	Hydroxamate	I	5–50 mg	Recurrent colorectal cancer	MSS	Dose dependent changes in CEA levels	(Primrose et al., 1999)
Marimastat	British Biotech plc	Hydroxamate	I	25, 50, & 100 mg bid	Lung cancer	MSS	50 mg bid is MTD	(Wojtowicz-Praga et al., 1998)
Marimastat	British Biotech plc	Hydroxamate	I/II	5, 20, or 40 mg	Prostate cancer	MSS	Changes in PSA suggest biological activity	(Rosenbaum et al., 2005)
Marimastat vs gemcitabine	British Biotech plc	Hydroxamate	III	5, 10, or 25 mg bid	Pancreatic cancer	MSS	No OS difference between gemcitabine and 25 mg dose of marimastat	(Bramhall et al., 2001)
Marimastat (+TMZ)	British Biotech plc	Hydroxamate	II	25 mg bid	Anaplastic glioma	MSS	Similar to single-agent TMZ, better PFS associated with MSS	(Groves et al., 2006)
Marimastat (+TMZ)	British Biotech plc	Hydroxamate	II	50 mg daily	Glioblastoma multiforme	MSS	PFS for combination better than historical value	(Groves et al., 2002)
Marimastat (+gamma knife)	British Biotech plc	Hydroxamate	II	10 mg bid	glioma	MSS	No advantage in grade 4, slight survival improvement in grade 3	(Larson et al., 2002)
Marimastat	British Biotech plc	Hydroxamate	II	10 & 100 mg bid	melanoma	MSS	10 mg bid is MTD	(Quirt et al., 2002)
Marimastat		Hydroxamate	II			MSS		

(continued)

Table 36.1 (continued)

Drug	Company	Type	Trial phase	Dose	Patient group/ indication	Principal adverse events	Results	References
	British Biotech plc			5–10 mg bid for 12 months	Early-stage breast cancer		Trough plasma levels too low for biological activity. No adjuvant trial warranted	(Miller et al., 2002)
Marimastat	British Biotech plc	Hydroxamate	II	10, 25, or 100 mg bid	pancreatic	MSS	Some responses including changes in CA19–9	(Evans et al., 2001)
Marimastat	British Biotech plc	Hydroxamate	III	10 mg bid	GBM	MSS	No change in OS; no difference in quality of life	(Levin et al., 2006)
Marimastat	British Biotech plc	Hydroxamate	III	10 mg bid	Colorectal liver metastases	MSS	No increase in OS; better survival in patients with MSS	(King et al., 2003)
Marimastat	British Biotech plc	Hydroxamate	III	10 mg bid	Gastric cancer	MSS	Difference in OS not significant, improved survival with marimastat evident over longer term and in prior-treated subgroup	(Bramhall et al., 2002a)
Marimastat (+gemcitabine)	British Biotech plc	Hydroxamate	III	10 mg bid	Pancreatic cancer	MSS	No improvement in OS	(Bramhall et al., 2002b)
Marimastat	British Biotech plc	Hydroxamate	III	10 mg bid	Breast cancer	MSS	Inferior survival associated with MSS	(Sparano et al., 2004)
Marimastat	British Biotech plc	Hydroxamate	III	10 mg bid	SCLC [responsive to chemo]	MSS, Lethargy, Anorexia, Nausea	No difference in OS, PFS or RR	(Shepherd et al., 2002)
Prinomastat	Agouron/ Pfizer Global	Hydroxamate	I	1–100 mg bid	Solid tumors	Fatigue, MSS	Doses of 5–10 mg bid identified for future trials	(Hande et al., 2004)

Prinomastat (+paclitaxel)	Agouron	Hydroxamate	II	15 mg bid	Metastatic melanoma	MSS	No benefit in PFS or OS	(Collier et al., 2002)
Prinomastat (+cis/5-FU/pac/RT)	Agouron	Hydroxamate	II	15 mg bid	Esophageal adenocarcinoma	MSS	Closed early	(Heath et al., 2006)
Prinomastat	Agouron	Hydroxamate	II	5 or 25 mg bid	Progressive breast cancer	MSS	No objective disease responses	(Rugo et al., 2001)
Prinomastat (+TMZ)	Agouron	Hydroxamate	II	25 mg bid	Glioblastoma multiforme	MSS	No difference in OS or PFS	(Levin et al., 2002)
Prinomastat (+Gem/Cis)	Agouron	Hydroxamate	III	15 mg bid	NSCLC	MSS	No difference in OS, RR, PFS	(Bissett et al., 2005)
Prinomastat (+Pac/Carb)	Agouron	Hydroxamate	III	5–15 mg bid	NSCLC	MSS	No difference in OS, PFS, RR	(Smylie et al., 2001)
Prinomastat (+mito/pred)	Agouron	Hydroxamate	III	5–10 mg bid	Hormone-refractory prostate cancer	MSS	No difference in PSA RR, RPFS, SPFS, OS	(Ahmann et al., 2001)
Tanomastat	Bayer	Biphenyl	I	100–1600 mg/day	Advanced malignancy	Liver enzyme alterations, thrombo-cytopenia	PK used to select 800 mg bid dose, plasma TIMP-2 levels increase with drug dose	(Rowinsky et al., 2000)
Tanomastat	Bayer	Biphenyl	I	100–1600 mg/day	Advanced malignancy	Liver enzyme alterations, thrombo-cytopenia	Selected 800 mg bid dose for phase III	(Heath et al., 2001)
Tanomastat	Bayer	Biphenyl	III	600–2400 mg/day	Ovarian cancer	Fatigue, nausea, thrombo-cytopenia, anemia	1200 mg qd dose identified	(Hirte et al., 2006)
Tanomastat	Bayer	Biphenyl	III	800 mg bid	NSCLC	None reported	At unplanned interim analysis, PFS better in treatment arm	(Rigas et al., 2003)
Tanomastat	Bayer	Biphenyl	III	800 mg bid	SCLC	None reported	At interim analysis, Sig. shorter time to progression in treatment arm. Trial halted	(Rigas et al., 2003)

(continued)

Table 36.1 (continued)

Drug	Company	Type	Trial phase	Dose	Patient group/ indication	Principal adverse events	Results	References
Tanomastat	Bayer	Biphenyl	III	800 mg bid	Pancreatic cancer	None reported	Gemcitabine significantly better	(Moore et al., 2000)
BMS275291	Bristol Myers Squibb	Mercaptoacyl	I	600–2400 mg/day	Advanced metastatic cancer	Transaminitis, rash, low-grade MSS	Suggestion of disease stabilization	(Rizvi et al., 2004)
BMS275291	Bristol Myers Squibb	Mercaptoacyl	II	1200 or 2400 mg/day	Prostate cancer	No DLT; rare thrombosis, fatigue, neuropathy	No responses	(Lara et al., 2006)
BMS275291	Bristol Myers Squibb	Mercaptoacyl	II	1200 mg qd	Breast cancer	MSS; hypersensitivity, rash	Discontinuation rate too high due to toxicity	(Miller et al., 2004)
BMS275291 (+carboplatin/taxol)	Bristol Myers Squibb	Mercaptoacyl	II/III	1200 mg qd	NSCLC	Flu-like; rash, hypersensitivity, febrile neutropenia	No difference in RR, PFS, or OS	(Douillard et al., 2004, Leighl et al., 2005)
CGS27023A/ MMI270	Ciba-Geigy/ Novartis	Hydroxamate	I	50 mg qd–600 mg tid	Advanced solid cancers	Rash, MSS;	Identified 300 mg bid (MTD) as dose for phase II	(Levitt et al., 2001)
CGS27023A/ MMI270 (+5FU/ FA)	Ciba-Geigy/ Novarits	Hydroxamate	I	50 mg qd–300 mg bid	Advanced colorectal cancer	MSS; rash	PR or SD in 23/30 patients	(Eatock et al., 2005)
S-3304	Shionogi	Sulfonamide	I	10–800 mg	Healthy volunteers	No DLT; mild-moderate headache, myalgia	Drug well tolerated	(van Marle et al., 2005)
S-3304	Shionogi	Sulfonamide	I	800–3200 mg bid	Advanced solid cancers	No DLT, some GI effects	Inhibition in local gelatinolytic activity with drug	(Chiappori et al., 2007)

Name	Company	Compound	Phase	Dose	Indication	Toxicity	Result	Reference
Metastat/Col-3/CMT-3	Collagenex	Chemically modified tetracycline	I	36–98 mg/m²/day	Advanced cancers	Photosensitivity; infrequent lupus; anemia	Disease stabilization in several patients	(Rudek et al., 2001)
Metastat	Collagenex	Chemically modified tetracycline	I	25–70 mg/m²/day	AIDS-associated Kaposi's sarcoma	Photosensitivity; rash, headache	44% response rate, serum MMP2 levels decreased in responders	(Cianfrocca et al., 2002)
Metastat	Collagenex	Chemically modified tetracycline	I	36–98 mg/m²/day	Advanced cancers	Photosensitivity	Some disease stabilization, MTD is 50 mg/m²/day	(Syed et al., 2004)
Metastat	Collagenex	Chemically modified tetracycline	II	50 or 100 mg/day	AIDS-associated Kaposi's sarcoma	Photosensitivity	Response rate greater than prespecified target, reduced MMP2, MMP9 levels in responder plasma	(Dezube et al., 2006)
Metastat	Collagenex	Chemically modified tetracycline	II	50 mg/day	Advanced soft tissue sarcoma	Photosensitivity	No relevant clinical activity	(Chu et al., 2007)
Neovastat	Aeterna	Shark cartilage extract	I/II	30–240 ml/day	Advanced cancers	No DLT; hypoglycemia in one patient	Increased survival in NSCLC subgroup	(Berger et al., 2001)
Neovastat	Aeterna	Shark cartilage extract	I/II	30–240 ml/day	NSCLC	No DLT	Increased survival with higher doses	(Latreille et al. 2003)
Neovastat	Aeterna	Shark cartilage extract	II	60 or 240 ml/day	Renal cell carcinoma	No DLT; taste alteration	Better survival with higher dose	(Batist et al. 2002)

(continued)

Table 36.1 (continued)

Drug	Company	Type	Trial phase	Dose	Patient group/ indication	Principal adverse events	Results	References
Neovastat	Aeterna	Shark cartilage extract	III	240 ml/day	Renal cell carcinoma	None reported	No improvement in OS, better median survival in single met subgroup	(Aeterna Zentaris Inc. 2003, Escudier et al. 2003)
Neovastat (+IC/ CRT)	Aeterna	Shark cartilage extract	III	120 ml bid	NSCLC	None specific to neovastat	Not reported	(Lu et al. 2005)

*Abbreviations: *bid* twice daily, *CA19/9* cancer antigen 19/9, *carb* carboplatin, *CEA* carcinoembryonic antigen, *cis* cisplatin, *CRT* chemoradiation therapy, *DLT* dose-limiting toxicity, *5FU* 5-fluorouracil, *FA* folinic acid, *Gem* gemcitabine, *GI* gastrointestinal, *IC* induction chemotherapy, *met* metastasis, *Mito* mitoxantrone, *MSS* musculoskeletal syndrome, *MTD* maximum tolerated dose, *NSCLC* non-small-cell lung cancer; *OS* overall survival, *Pac* paclitaxel, *PFS* progression-free survival, *PK* pharmacokinetic, *Pred* prednisone, *PSA* prostate-specific antigen, *qd* once daily, *RPFS* radiographic progression-free survival, *RR* response rate, *SCLC* small-cell lung cancer; *SD* stable disease, *SPFS* symptomatic progression-free survival, *tid* three times daily, *TMZ* temozolomide.

malignant ascites (Wojtowicz-Praga et al. 1996, Parsons et al. 1997) and malignant pleural effusion (Macaulay et al. 1999). Since these were phase I trials designed with the intentions of dose-finding and identifying toxicity, there was no real evaluation of efficacy; nevertheless, some patients appeared to have a reduction in the requirement for drainage of ascites or pleural effusion (Parsons et al. 1997, Macaulay et al. 1999). Batimastat was not used further in oncology and was replaced with the orally bioavailable drug marimastat (BB-2516, British Biotech plc). Another early MMPI that had similar problems with oral bioavailability was galardin, also known as ilomastat or GM6001. This was initially tested in an eyedrop format as a treatment of corneal ulcers (Galardy et al. 1994).

Marimastat (BB2516)

As was the case with batimastat and galardin, marimastat was designed as a broad-spectrum MMPI, capable of targeting multiple members of the MMP family including collagenases, gelatinases, and stromelysins (Whittaker et al. 1999). In addition, marimastat showed inhibitory activity against related enzyme families such as the adamalysin or ADAM (a disintegrin and metalloprotease) family of transmembrane proteases. All of these original drugs were designed as peptidomimetics, that is, their structure incorporated features similar to the protease cleavage site of the MMP substrate, collagen. This substrate mimetic was intended to target the drugs to MMPs. In addition to the targeting moiety, a zinc-chelating group was also an essential part of the drug. For these early drugs as well as some later nonpeptidomimetics that were considered "next-generation MMPIs," the chelating group was based on hydroxamic acid, a potent binder of zinc (Whittaker et al. 1999, Sang et al. 2006).

For phase I testing, marimastat was tested as a single agent in healthy volunteers (Millar et al. 1998), in patients with advanced cancers (Nemunaitis et al. 1998), solid tumors, and specifically in patients with gastric (Tierney et al. 1999), pancreatic (Rosemurgy et al. 1999), lung (Wojtowicz-Praga et al. 1998), prostate (Rosenbaum et al. 2005), and colorectal cancer (Primrose et al. 1999). There were also phase I trials in which marimastat was combined with paclitaxel (Toppmeyer et al. 2003) plus carboplatin (Goffin et al. 2005) or fragmin/captropril (Jones et al. 2004). An interesting feature of marimastat was that trials of this drug were among the first to explore novel biological markers as indicators of drug efficacy (Nemunaitis et al. 1998, 2000, Primrose et al. 1999, North et al. 2000). Most controversially, British Biotech used changes in the rate of rise of serum tumor markers as a method to monitor drug effects (Rasmussen and McCann 1997, Steward and Thomas 2000). The idea was that an optimal biological dose could be determined from the degree of impact on tumor markers such as CA-125 in ovarian cancer and carcinoembryonic antigen (CEA) in colon cancer. However, there were significant limitations to the method, the most important being that the natural fluctuations in serum markers were not established, and so changes in these proteins may be completely unrelated to drug administration (Gore et al. 1996). Because of these limitations, it is difficult to draw reliable conclusions from the trials incorporating tumor markers. There is some indication, however, that when all the trials are grouped, reductions in the rate of rise of various markers correlate with drug exposure (Nemunaitis et al. 1998).

Marimastat has been the most widely tested MMPI. Phase II and III trials in many different cancer types have been reported. While indications of efficacy were seen in two of these larger trials, no trial could be regarded as truly successful and primary end points were never met. The first indication of efficacy was a trial in pancreatic cancer where patients were treated either with gemcitabine, the current standard treatment for pancreatic carcinoma, or with marimastat (Bramhall et al. 2001). Patients on the highest dose of marimastat had similar, but not better, survival times as the gemcitabine-treated patients. However, it should be noted that the marimastat-treated patients had inferior quality of life. Moreover, while gemcitabine is the current standard, it is still a relatively ineffective drug prolonging life by an average of only 5–6 weeks (Hochster et al. 2006). Another trial in which patients with pancreatic cancer were treated with gemcitabine plus marimastat or placebo showed that marimastat added no significant benefit to the gemcitabine treatment although there was a trend toward improved survival (Bramhall et al. 2002b). A more hopeful result was obtained in a trial of inoperable patients with gastric cancer. Participants, including a proportion who had had chemotherapy before the trial, were randomized to receive placebo or marimastat (Bramhall et al. 2002a). Once again, the trial failed as the primary end point was not met. Nevertheless, secondary analyses, either concentrating on the previously treated patient group or focusing on a longer timepoint than was used for the primary end point, did show a survival enhancement associated with marimastat treatment. While this is still a failed trial, it does indicate that MMPI therapy can have an effect in a human patient population and suggests that future MMPIs should be tested in patients with gastric cancer who have had debulking chemotherapy. In case this result is interpreted as indicating debulked, chemotherapy-responsive tumors of multiple types are an optimal setting for MMPIs, a result from a different setting should be considered. In a trial of patients with small-cell lung cancer (SCLC) who had a response (either complete or partial) to first-line chemotherapy, marimastat treatment failed to affect progression-free survival or overall survival (Shepherd et al. 2002). Further, quality of life was significantly adversely affected in the marimastat-treated population. One potentially important difference between the SCLC and gastric cancer trials was the possibility that marimastat concentrated in the stomach, so that localized higher levels were achieved at the relevant tumor site (Bramhall et al. 2002a).

Prinomastat (AG3340)

Even in phase I trials of marimastat, side effects were evident (discussed in a later section). This prompted other companies to look at new ways of designing MMPIs to potentially avoid such side effects. Agouron, now part of Pfizer, used information gleaned from crystal structures to rationally design the compound prinomastat (Brown 2000). Unlike the previous broad-spectrum peptidomimetics, prinomastat showed selectivity for gelatinases (MMP2 and MMP9) and against collagenase (MMP1). Since collagenase inhibition was a suspected culprit for MMPI side effects, the prinomastat inhibitory profile was considered a potentially safer MMPI with a larger therapeutic window. Phase I trials of prinomastat, however, revealed joint and muscle pain that was exposure-associated (Hande et al. 2004). A dosage of 5–10 mg twice a day was identified from a combination of pharmacokinetic data and toxicity assessment as a useful dose for future larger studies. Only four patients treated at this level in the phase I trial showed a dose-limiting toxicity (DLT) with two of these becoming evident within the first month of treatment. There appeared a clear link between development of DLTs and dose level as well as dose duration. Prinomastat is described as a selective MMPI; however, it should be noted that while the IC_{50} for MMP1 is 166-fold higher than for MMP2 and 32-fold higher than for MMP9, it is still in the low nanomolar range (Brown 2000).

Prinomastat has been tested in a number of phase II settings and in phase III trials of patients with non-small-cell lung cancer (NSCLC), in combination either with gemcitabine and cisplatin or with paclitaxel and carboplatin (Smylie et al. 2001, Bissett et al. 2005). In addition, prinomastat was assessed in a phase III trial in hormone-refractory patients with prostate cancer who were also treated with mitoxantrone (Ahmann et al. 2001). No efficacy was demonstrated in any of these trials and significant toxicities, principally musculoskeletal in nature, were reported.

Tanomastat (BAY12-9566)

As a biphenyl-type inhibitor, tanomastat is quite different structurally from other MMPIs. This drug was developed by Bayer Corporation and it shows a selective inhibitory profile with little activity against the so-called deep pocket proteases MMP1 and MMP7. Another feature of this drug is its high degree of protein binding with over 99% bound after systemic administration (Heath et al. 2001). Phase I studies in patients with malignant disease found no DLT and the dose of 800 mg twice daily recommended for further studies was identified based on pharmacokinetic properties (Rowinsky et al. 2000, Heath et al. 2001). Although no DLTs were determined, there were some side effects identified. In particular, changes in liver enzymes and some cases of thrombocytopenia were noted (Rowinsky et al. 2000, Heath et al. 2001). Hepatic enzyme changes had also been noted in animal toxicology studies (Rowinsky et al. 2000). Attempts were made with clinical trials of tanomastat to assess biological activity. Interestingly, levels of plasma TIMP-2 appeared to increase with dose (Rowinsky et al. 2000), thus potentially reinforcing drug activity with the endogenous inhibitor. Other investigators using a fluorometric assay for gelatinolytic and type IV collagenolytic activity showed tanomastat dose-dependent reduction in enzyme activity in plasma levels from patients on phase I trials treated for 15 days (Duivenvoorden et al. 2001). As might be anticipated, these changes could not be detected by standard zymography analysis.

Tanomastat has become infamous because of events in a phase III clinical trial of SCLC where patients appeared to have faster disease progression on the tanomastat arm in comparison to placebo (Rigas et al. 2003). These negative results led to early termination of this trial and of all other trials of tanomastat. There have been suggestions that patients in a phase III trial of pancreatic cancer were also adversely affected by tanomastat treatment, but this is not clear. In that trial, patients were treated either with gemcitabine, the standard for pancreatic cancer, or with tanomastat. Tanomastat had inferior efficacy compared to gemcitabine, leading to a significant survival difference and early closure of the trial (Moore et al. 2003). However, this result does not mean that tanomastat caused harm, rather that gemcitabine is a better agent in the pancreatic cancer setting. In a phase III trial in ovarian cancer, tanomastat was well tolerated but had no impact on progression-free or overall survival (Hirte et al. 2006). While there is a possibility that tanomastat is an inherently unsafe drug, there is some indication from a small phase III trial of tanomastat in NSCLC that the important factor is matching the drug with the appropriate disease setting. In patients with NSCLC treated with tanomastat, unscheduled interim analysis suggested a significant improvement in time to disease progression compared to placebo in direct contrast to the SCLC result (Rigas et al. 2003). However, this did not appear to translate to enhanced overall survival.

BMS275291

In another attempt to avoid side effects, the biotechnology company Celltech designed a broad-spectrum MMPI that had no activity against members of the related protease family, the ADAMs. This drug, D-2163, was then renamed as BMS275291 and was put into clinical development by Bristol Myers Squibb. Although the name is not widely used, this drug is also known as rebimastat.

A phase I dose-escalation study of BMS275291 identified 1200 mg/day as the dosage for further trials (Rizvi et al. 2004). This was based on pharmacokinetic data since dose-limiting effects were not evident. At the 1200 mg/day dosage, the steady-state plasma concentration exceeded the calculated IC_{90} for MMP2 and MMP9 and it was considered that higher doses would offer no advantage. Adverse events were reported, largely myalgia and arthralgia, but these were generally low grade and did not appear to be dose related.

As a response to indications from preclinical models that earlier stage disease appeared to be more responsive to MMPI treatment than late-stage aggressive established tumors, a group based at Indiana University designed a trial of BMS275291 in early-stage breast cancers (Miller et al. 2004). The goal of this trial was to determine the discontinuation rate among participants, as a full adjuvant trial was feasible only if patients were willing to remain on long-term treatment. Hence, no relapse-free or overall survival data were analyzed from this trial. Unfortunately, a significant number of BMS275291-treated patients developed toxicities, either musculoskeletal or a hypersensitivity response. While the level of musculoskeletal side effects was lower than seen with other MMPIs such as marimastat, it was still high enough to warrant early termination of the trial and the foregoing of a full adjuvant trial.

A full phase III trial of BMS275291 in NSCLC was conducted by the National Cancer Institute of Canada Clinical Trials Group (Leighl et al. 2005). This was a large study with 774 patients with stage IIB/IV NSCLC randomly assigned to receive standard chemotherapy (paclitaxel and carboplatin) with or without the addition of 1200 mg BMS275291 daily. The trial design was similar to phase III trials in NSCLC of both marimastat and prinomastat. Although these trials failed, the rationale for a further MMPI trial with BMS275291 was the high discontinuation rate of MMPI-treated patients in the previous trials due to toxicity. BMS275291 was considered to be a much less toxic drug, and so it was thought that improved compliance could potentially reveal clinical activity of the drug. However, an interim safety analysis indicated that BMS275291 was causing significant toxicity while having no benefit in terms of survival, progression-free survival, or relapse rate; thus study treatment was halted.

Metastat

Unlike the previously described small-molecule inhibitors of MMPs, metastat (Col-3/CMT-3) is a multifunctional drug that was developed as a nonantibiotic derivative of tetracycline (Fingleton 2003b). It is one of a series of agents designated as "IMPACs" or "inhibitors of multiple proteases and cytokines" by their manufacturer Collagenex Pharmaceuticals Inc. Collagenex is known for marketing the gum disease treatment Periostat, a low-dose doxycycline, which is the first FDA drug approved for its MMP-inhibitory activity. Metastat was the tetracycline derivative thought likely to be the most effective in cancer settings based on its inhibitory profile, which favors the gelatinases (Fingleton 2003b). An advantage of this agent is that it inhibits both the activity and the expression of targeted MMPs, meaning that targeting efficacy can be monitored relatively easily by measuring levels of MMPs.

In phase I trials of metastat in patients with advanced malignancy or with AIDS-related Kaposi's sarcoma, a dose-limiting cutaneous phototoxicity was identified despite mandatory use of sunblock (Rudek et al. 2001, Cianfrocca et al. 2002, Syed et al. 2004). Other rare but serious side effects that appeared to be drug-related were the development of lupus or of reversible sideroblastic anemia (Rudek et al. 2001). There was some indication that in patients who responded to treatment (including disease stabilization), MMP2 levels decreased (Rudek et al. 2001). This effect was most significant in the Kaposi's sarcoma trial (Cianfrocca et al. 2002).

The results of phase II trials of metastat in advanced soft tissue sarcomas (Chu et al. 2007) or in AIDS-related Kaposis's sarcoma (Dezube et al. 2006) have recently been reported. There were no objective responses seen in the soft tissue sarcoma trial. AIDS-related Kaposi's sarcoma, however, appears to be an appropriate setting for the use of metastat. In the phase II trial, which was not placebo-controlled, the number of responses seen at the lower dose (50 mg) was significantly higher than the prespecified target (Dezube et al. 2006). In addition, plasma levels of MMP2 and MMP9 significantly decreased with treatment in those patients who responded and the reduction in MMP levels appeared related to metastat dose.

Neovastat

Nevoastat (AE-941), from Aeterna Laboratories in Canada, is a natural compound extract from shark cartilage. It has been reported to have multiple antiangiogenic activities including inhibition of MMPs (Gingras et al. 2001, Dupont et al. 2002). A phase I/II trial in patients with advanced cancer reported no DLT and relatively minor side effects except for one case of hypoglycemia (Berger et al. 2001). There were also some hints of efficacy in a retrospective subgroup analysis of patients with NSCLC where an increase in overall survival was reported (Berger et al. 2001). Another phase I/II trial that concentrated specifically on patients with NSCLC indicated a significant survival advantage in patients receiving higher doses although no tumor responses were observed (Latreille et al. 2003). A phase II trial in renal cell carcinoma also provided evidence of a dose response with respect to overall survival (Batist et al. 2002). Since renal cell carcinoma is a disease with very poor prognosis and limited treatment options, there was considerable interest in demonstrating that neovastat was a viable therapeutic candidate (Bukowski 2003). In a phase III trial of patients with metastatic renal cell carcinoma that is refractory to immunotherapy, 302 patients were treated in a randomized, double-blind fashion with either placebo or neovastat (Escudier et al. 2003). Unfortunately, the trial failed to meet the primary end point of improved overall survival. Similar to the marimastat gastric cancer trial, subgroup analyses did reveal survival benefit of neovastat in a specific patient group, those with minimal metastasis and clear cell histology (Aeterna Zentaris Inc. 2003). The results of another phase III trial in patients with stage III NSCLC who also receive induction chemotherapy and concomitant radiotherapy has not yet been reported (Lu et al. 2005).

Dose-Limiting Toxicities

One of the most significant problems that emerged during MMPI clinical trials was the development of DLT, usually musculoskeletal in nature. Often, the musculo-skeletal syndrome (MSS) developed only with chronic dosing, and hence was not always apparent from phase I/II trials (Nemunaitis et al. 1998, Peterson 2006). The MSS effects have been described in various ways including tendonitis, arthralgia, myalgia, rheumatoid arthritis-like, resembling Dupuytren's contracture, and so forth (Nemunaitis et al. 1998, Peterson 2006). These effects are very serious and resulted in patients on multiple trials either reducing doses or withdrawing from treatment altogether. In addition, in the two phase III trials of marimastat where suggestions of antitumor efficacy were reported, patients in the marimastat arm tended to report a reduced quality of life because of the side effects of the treatment (Bramhall et al. 2001, Bramhall et al. 2002a). As yet, the cause of these effects is unknown.

There have been various theories to explain the association between MMPI exposure and MSS effects, although some of these are becoming less likely as more evidence accumulates. The first hypothesis was that a basal level of collagenase activity is required for connective tissue turnover and, if this is inhibited, then contracture and other joint problems could develop. Second-generation MMPIs, including the Ciba-Geigy/Novartis compound CGS27023A/MMI270 and the Pfizer/Agouron compound prinomastat, were therefore developed to be collagenase-sparing. However, MSS-type effects were reported with these drugs also. In contrast, the compound known as trocade, developed by Roche specifically for the treatment of rheumatoid arthritis, was designed to target collagenase yet it did not appear to be associated with MSS effects (Hemmings et al. 2001).

A second theory was that MSS effects were a result of inhibition of proteases from families other than MMPs. Specifically, the ADAM family of proteases, members of which act as sheddases for various cell surface proteases, were considered possible culprits. The Celltech/Bristol Myers Squibb compound BMS275291 was developed as a broad-spectrum MMPI with no activity against sheddases. Early trials of this drug looked promising as MSS effects were not seen. However, some, but not all, longer-term trials did report development of MSS specifically in BMS275291-exposed patients (Miller et al. 2004, Leighl et al. 2005). A similar possibility is unintended inhibition of the related ADAM-TS (a disintegrin and metalloprotease with thrombospondin motifs) family of proteases, since members of this family contribute to collagen processing and connective tissue disorders. A form of the hereditary disease Ehlers–Danlos syndrome (type VII C), which is characterized by, among other problems, skin fragility, is due to a mutation in the ADAM-TS2 gene (Le Goff et al. 2006). However, while severely debilitating, the MSS effects were limited in comparison with the spectrum of problems associated with Ehlers–Danlos and other similar diseases.

Examination of mouse models generated to be genetically deficient in specific MMPs identified another possible cause of the MSS effects. Mice deficient in MMP14, unlike other MMP-null mice which were generally healthy, showed severe skeletal abnormalities and died at a young age (Holmbeck et al. 1999, Zhou et al. 2000). Although the phenotype of the MMP14-null animal is much more severe than anything seen in drug-treated patients, it should be remembered that the MMP14-null mouse was deficient from conception, and hence had developmental defects which were unlikely to be seen in adults treated with an MMPI. Mice with inactivating mutations in the MMP2 gene show subtle phenotypes that are also related to skeletal remodeling defects, specifically alterations in number and function of osteoblasts and osteoclasts, the cells responsible for bone generation and turnover (Mosig et al. 2007). Intriguingly, a human kindred has been identified with inactivating mutations within MMP2 (Martignetti et al. 2001). The phenotype observed in carriers of the genetic defects has been described as multicentric osteolyses with arthropathy or "vanishing bone" syndrome. This is characterized by distinctive facial features as well as early onset arthritis and joint contractures. A rodent model of marimastat-induced MSS indicated that the predominant result of the drug treatment is a fibroplasia within the joints (Renkiewicz et al. 2003). This is quite different to the osteolysis seen with both MMP14 and MMP2 deficiencies, and hence it appears unlikely that inhibition of either MMP14 or MMP2 is a significant factor in the generation of MSS. Further studies in rodent models are required, however, to fully explore the potential involvement of these MMPs.

An alternative possibility is that the effects are due to properties of the zinc-binding groups used in the drugs rather than the fact that these drugs target MMPs. The most frequently used zinc-binding group for inhibitors of MMPs has been hydroxamic acid and it is well established that this moiety can chelate other metals besides zinc (Marmion et al. 2004). Indeed, the MMPIs with the lowest reported incidences of MSS are the nonhydroxamates tanomastat (a biphenyl, carboxylate MMPI) and BMS275291, a mercaptoacyl-type inhibitor. There were no reports of MSS with tanomastat, although trials of this compound were discontinued after trial participants on the drug treatment arms were found to have reduced survival in comparison to control-treated arms. As noted above, there have been reports of MSS in a trial of BMS275291 (Miller et al. 2004) but not in others (Leighl et al. 2005). Phase I studies of a novel nonhydroxamate MMPI, called S-3304 from Shionogi & Co. Ltd., were recently reported (Chiappori et al. 2007). This drug is a tryptophan derivative and is selective against MMP1, MMP3, and MMP7. No MSS effects were seen either in healthy volunteers (van Marle et al. 2005) or in patients with cancer (Chiappori et al. 2007) although this could be due to the selectivity profile rather than its nonhydroxamate nature. Finally, the two compounds neovastat and metastat, which are described as having MMP-inhibitory activity in addition to other activities, are also not hydroxamates. Both metastat, a tetracycline derivative, and neovastat, a shark cartilage extract, have been used in multiple clinical trials and neither has been associated with MSS effects.

Two animal models have been shown to be useful in identifying MSS effects. These are marmoset monkeys (Drummond et al. 1999), a model usually confined to later stages of preclinical testing because of expense, and rats (Renkiewicz et al. 2003), which are more amenable to standard preclinical analysis. In the marmoset model, marimastat reportedly caused tendonitis in 7 out of 8 animals dosed at 100 mg/kg/day and in 4 out of 8 animals treated with dosage of 10 or 30 mg/kg/day (Wojtowicz-Praga et al. 1998). Interestingly, an inactive enantiomer of marimastat was reported not to cause MSS in marmosets, suggesting that the effects are related to mechanism of action (Drummond et al. 1999). There is some question, however, as to the overall sensitivity of these models. Indeed BMS275291, which did show MSS effects in some clinical trials albeit at lower levels than did other MMPIs, did not appear to adversely affect marmosets (Rizvi et al. 2004).

Not all reported toxicities were MSS related. In the phase III trial of BMS275291 conducted in patients with NSCLC, the principal drug-specific toxicities were hypersensitivity reactions, febrile neutropenia, flu-like symptoms, and dermatological symptoms including rash (Leighl et al. 2005). Macropapular rash was also a significant finding in patients treated with the Novartis compound CGS27023A (Levitt et al. 2001, Eatock et al. 2005). In tanomastat trials, although not dose limiting, thrombocytopenia was evident in a number of patients (Rowinsky et al. 2000, Heath et al. 2001). Metastat, the tetracycline derivative, was principally associated with cutaneous photosensitivity although other potentially serious effects include lupus and sideroblastic anemia (Rudek et al. 2001).

Lack of Efficacy

One of the most perplexing aspects of the MMPI story has been the disconnection between very promising preclinical studies with these drugs and the complete failure of the same drugs when tested clinically. An excellent overview of many of the reported preclinical studies has been written by Pavlaki and Zucker (2003). In that article, the authors point out the vast difference in tumor biology between mouse models and human cancer. In particular, they allude to the limited stromal involvement in rodent tumors, whether syngenic or xenogenic, and the fact that rodents can carry large tumor burdens without suffering from the systemic effects that are significant sources of morbidity to human patients with much smaller (as a percentage of body weight) tumor loads. In addition, in cancer models, drug treatment is most often initiated in animals with minimal metastatic disease unlike human patients with extensive metastatic burden (Kerbel 1999). A further complication for clinical trials is that patients have often been heavily pretreated and the remaining tumor is refractory to standard therapies, making the bar for demonstrating efficacy very high. In the case of MMPs, a seminal study from the Hanahan Laboratory demonstrated stage-specific efficacy of the drug batimastat in the rat insulin promoter T antigen (RIP-Tag) mouse model of pancreatic malignancy (Bergers et al. 1999). The data show efficacy of the drug at early stages of tumor progression but absolutely no effect at the invasive malignant stage. Since the majority of patients in all of the MMPI phase III trials corresponded to this later, unresponsive stage, it is perhaps not surprising that clinical testing of MMPIs failed to replicate the positive results from preclinical studies. Exceptions are the subgroups of patients from the marimastat gastric cancer trial with low disease burden who apparently showed increased survival due to marimastat treatment (Bramhall et al. 2002a), and the subgroup of neovastat-treated renal carcinoma patients with minimal metastasis (Aeterna Zentaris Inc. 2003). Since subgroup analysis is not a valid method for determining success of a trial, overall, these were still failed efforts.

One argument that has arisen against the use of protease inhibitors is that the process of metastasis is not a realistic target. Conceptually, the idea of MMPIs preventing metastasis would suggest that these drugs would have to be administered before initiation of a metastatic cascade. Since patients often present with malignant cancers, the likelihood is that metastasis has already occurred; thus antimetastasis agents might not be considered useful. However, it is important to recognize that the process of metastasis includes more than basement membrane degradation, migration through tissue vasculature, and extravasation (Eccles and Welch 2007). For a metastatic cell to be clinically relevant, it must grow at the secondary site. Much of the recent work on specific MMPs has identified contributions of MMPs to tumor growth (McCawley and Matrisian 2000). Even basement membrane degradation can be linked to tumor growth as elegant experiments from the lab of Steve Weiss have shown (Hotary et al. 2003). In these studies, MT1-MMP activity is necessary to create space within the matrix to accommodate the increasing number of cells from a growing tumor. Conversely, other investigators have demonstrated that specific MMPs can have protective, antitumorigenic effects (Martin and Matrisian 2007). Mice deficient in MMP3 (McCawley et al. 2004), MMP8 (Balbin et al. 2003), MMP12 (Houghton et al. 2006, Acuff et al. 2006), or MMP19 (Pendás et al. 2004) have all shown enhanced tumor development and/or progression in various models. In addition, MMP9-null mice had fewer but more aggressive tumors in a skin cancer model (Coussens et al. 2000). Together, these data indicate that all MMPs are not equal and broad-spectrum inhibition may be unwise. Hence, there is good reason to think that MMPIs did not fail merely because "the horse was out of the barn" before they were administered, but more likely because of a lack of appreciation for the complex biology in which MMPs are involved, as well as problems with the drugs themselves.

The Present – The Current Status of MMPI Development

In the last several years, development of MMPIs has considerably diminished. There have been some new drugs and associated trials in the cardiovascular arena; however, efficacy was not demonstrated, possibly related to side effects and inadequate dosing (Peterson 2006). A new MMPI with broad-spectrum activity against MMPs and ADAMs is currently being tested clinically in the setting of diabetic nephropathy (ClinicalTrials.gov 2006). This drug, XL784, from Exelixis is a hydroxamate designed to be MMP1 sparing (Exelixis 2007), and it will be interesting to see if any MSS side effects are associated with its use. Thus far, phase I trials have demonstrated safety (Exelixis 2007).

Despite the lack of success of MMPIs, or perhaps in part because of it, research into MMP roles in cancer has flourished. We now have a much better picture of the multiple complex roles these enzymes can play. Most important, we recognize that not all MMP activities are protumorigenic, and therefore the blunt weapon of broad-spectrum inhibition may not be a good strategy (Overall and Kleifeld 2006a). This knowledge has begun to impact new drugs in development. Unfortunately, designing inhibitors of MMPs that are specific to one member of the family is extremely difficult due to the very similar structures of the enzymes (Overall and Kleifeld 2006b). Two family members that appear to allow specific inhibition are MMP12 and MMP13, and a number of agents described as inhibitors of these enzymes have been described in the chemical literature (Chen et al. 2000, Dublanchet et al. 2005). Alantos Pharmaceuticals, a small biotechnology company recently acquired by Amgen Inc., completed preclinical testing of a specific MMP13 inhibitor which they were pursuing as a therapeutic for osteoarthritis. This inhibitor has been described as binding within the active site of MMP13 but without chelating the zinc. Such a novel strategy may be the reason for the specificity although since the structure has not yet been released, this is difficult to determine. A different approach to designing inhibitors with better selectivity profiles is to use a "suicide inhibitor" where the target is covalently modified by the inhibitor binding. Mobashery and colleagues have designed a series of thiirane-based gelatinase selective inhibitors using this approach (Ikejiri et al. 2005), which have shown efficacy in a number of preclinical models (Kruger et al. 2005, Bonfil et al. 2006). Another alternative strategy for obtaining specific inhibitors is to use an antibody approach rather than small molecules. Dyax Inc. has recently reported on its MMP14 neutralizing antibody which has given excellent results in preclinical breast cancer and prostate cancer models (Devy et al. 2007). If this agent can be demonstrated not to cause MSS effects in rodent or marmoset models, then it has a good chance of having successful results in the clinic given the multiple functions of MMP14 in tumor progression.

The Future – Is There One?

MMP inhibition as a therapeutic concept for cancer is severely handicapped by the well-publicized clinical failure of early MMPIs. For there to be a future in this setting, pharmaceutical companies and their investors, clinicians, and regulatory agencies will all have to be convinced that it is worthwhile committing limited resources to a new effort. Inspiring confidence will require demonstration that we have learned the lessons of the past and have in place all the factors necessary for successful trials. Such factors include a thorough understanding of the nature of the target – which MMPs are truly contributory to the disease?; a method for assessing pharmacodynamic activity of the drugs, which ideally would also be a surrogate marker for antitumor efficacy; and drugs that are tolerable for protracted dosing as all indications are that MMPIs will be of most use in the minimal disease setting.

Previous chapters in this volume dealing with imaging offer updates on some of the approaches being taken to develop methods for assessing pharmacodynamic efficacy of inhibitors and the reader is referred to them for a discussion of this idea. Not knowing the causes of the various toxicities that have been observed with previous MMPIs as well as taking into account the various pro- and antitumor activities of different MMP family members, selective or specific inhibition seems a better idea than the broad-spectrum approaches of the past.

Oncology is a difficult setting for new drugs. Generally, new agents are first tested on patients who have been heavily pretreated with cytotoxics and having a beneficial effect on such a refractory background is difficult. However, there have been several recent articles discussing how best to proceed with testing molecularly targeted agents so that efficacy can be demonstrated (Rothenberg et al. 2003, Michaelis and Ratain 2006, Kummar et al. 2006). Overall, the combination of new MMPIs having appropriate selectivity along with better designed trials is likely to result in the successful addition of drugs that inhibit MMPs to the array of therapeutic weapons at the disposal of patients with cancer. When this happens, which will hopefully be in the not too distant future, a large factor in the success will have been the previous failures and all we have learnt from them.

References

Acuff H. B. et al. (2006). Analysis of host- and tumor-derived proteinases using a custom dual species microarray reveals a protective role for stromal matrix metalloproteinase-12 in non-small cell lung cancer. Cancer Res 66: 7968–75.

Aeterna Zentaris Inc. (2003). Aeterna Laboratories reports phase III trial results in renal cell carcinoma with neovastat. http://www.aeterna.com/en/page.php?p=60&q=46. Accessed 22 July 2008.

Ahmann F. R. et al. (2001). Interim results of a phase III study of the matrix metalloprotease inhibitor prinomastat in patients having metastatic, hormone refractory prostate cancer (HRPC). Proc Am Soc Clin Oncol: Abstr 692.

Balbin M. et al. (2003). Loss of collagenase-2 confers increased skin tumor susceptibility to male mice. Nat Genet 35: 252–7.

Batist G. et al. (2002). Neovastat (AE-941) in refractory renal cell carcinoma patients: report of a phase II trial with two dose levels. Ann Oncol 13: 1259–63.

Beattie G. J. and Smyth J. F. (1998). Phase I study of intraperitoneal metalloproteinase inhibitor BB94 in patients with malignant ascites. Clin Cancer Res 4: 1899–902.

Berger F. et al. (2001). Phase I/II trials on the safety, tolerability and efficacy of AE-941 (Neovastat) in patients with solid tumors. Proc Am Soc Clin Oncol: Abstr 2861.

Bergers G. et al. (1999). Effects of angiogenesis inhibitors on multistage carcinogenesis in mice. Science 284: 808–12.

Bissett D. et al. (2005). Phase III study of matrix metalloproteinase inhibitor prinomastat in non-small-cell lung cancer. J Clin Oncol 23: 842–9.

Bonfil R. D. et al. (2006). Inhibition of human prostate cancer growth, osteolysis and angiogenesis in a bone metastasis model by a novel mechanism-based selective gelatinase inhibitor. Int J Cancer 118: 2721–6.

Bramhall S. R. et al. (2001). Marimastat as first-line therapy for patients with unresectable pancreatic cancer: a randomized trial. J Clin Oncol 19: 3447–55.

Bramhall S. R. et al. (2002a). Marimastat as maintenance therapy for patients with advanced gastric cancer: a randomized trial. Br J Cancer 86: 1864–70.

Bramhall S. R. et al. (2002b). A double-blind placebo-controlled, randomised study comparing gemcitabine and marimastat with gemcitabine and placebo as first line therapy in patients with advanced pancreatic cancer. Br J Cancer 87: 161–7.

Brown P. D. (1999). Clinical studies with matrix metalloproteinase inhibitors. APMIS 107: 174–80.

Brown P. D. (2000). Ongoing trials with matrix metalloproteinase inhibitors. Expert Opin Investig Drugs 9: 2167–77.

Bukowski R. M. (2003). AE-941, a multifunctional antiangiogenic compound: trials in renal cell carcinoma. Expert Opin Investig Drugs 12: 1403–11.

Cauwe B., Van den Steen P. E. and Opdenakker G. (2007). The biochemical, biological, and pathological kaleidoscope of cell surface substrates processed by matrix metalloproteinases. Crit Rev Biochem Mol Biol 42: 113–85.

Chen J. M. et al. (2000). Structure-based design of a novel, potent, and selective inhibitor for MMP-13 utilizing NMR spectroscopy and computer-aided molecular design. J Am Chem Soc 122: 9648–54.

Chiappori A. A. et al. (2007). A phase I pharmacokinetic and pharmacodynamic study of s-3304, a novel matrix metalloproteinase inhibitor, in patients with advanced and refractory solid tumors. Clin Cancer Res 13: 2091–9.

Chu Q. S. et al. (2007). A phase II and pharmacological study of the matrix metalloproteinase inhibitor (MMPI) COL-3 in patients with advanced soft tissue sarcomas. Investig New Drugs 25: 359–67.

Cianfrocca M. et al. (2002). Matrix metalloproteinase inhibitor Col-3 in the treatment of AIDS-related Kaposi's sarcoma: a phase I AIDS malignancy consortium study. J Clin Oncol 20: 153–9.

Clinical Trials.gov. (2006). Study of XL784 in Patients with Albuminuria due to Diabetic Nephropathy. http://clinicaltrials.gov/ct/show/NCT00312780. Accessed 17 July 2008.

Collier M. A. et al. (2002) Phase II parallel-design study of the matrix metalloprotease inhibitor prinomastat in combination with weekly paclitaxel in patients with metastatic melanoma. Proc Am Soc Clin Oncol: Abstr 1828.

Coussens L. M. et al. (2000). MMP-9 supplied by bone marrow-derived cells contributes to skin carcinogenesis. Cell 103: 481–90.

Devy L. et al. (2007). Antitumor efficacy of DX-2400, a potent and selective human antibody MMP-14 inhibitor discovered using phage display technology. Proc Am Assoc Cancer Res: Abstr 5618.

Dezube B. J. et al. (2006). Randomized phase II trial of matrix metalloproteinase inhibitor COL-3 in AIDS-related Kaposi's sarcoma: an AIDS malignancy consortium study. J Clin Oncol 24: 1389–94.

Douillard J. Y. et al. (2004). Randomized phase II feasibility study of combining the matrix metalloproteinase inhibitor BMS-275291 with paclitaxel plus carboplatin in advanced non-small cell lung cancer. Lung Cancer 46: 361–8.

Drummond A. H. et al. (1999). Preclinical and clinical studies of MMP inhibitors in cancer. Ann NY Acad Sci 878: 228–35.

Dublanchet A. C. et al. (2005). Structure-based design and synthesis of novel non-zinc chelating MMP-12 inhibitors. Bioorg Med Chem Lett 15: 3787–90.

Duivenvoorden W. C., Hirte H. W. and Singh G. (2001). Quantification of matrix metalloproteinase activity in plasma of patients enrolled in a BAY 12–9566 phase I study. Int J Cancer 91: 857–62.

Dupont E. et al. (2002). Antiangiogenic and antimetastatic properties of Neovastat (AE-941), an orally active extract derived from cartilage tissue. Clin Exp Metastas 19: 145–53.

Eatock M. et al. (2005). A dose-finding and pharmacokinetic study of the matrix metalloproteinase inhibitor MMI270 (previously termed CGS27023A) with 5-FU and folinic acid. Cancer Chemother Pharmacol 55: 39–46.

Eccles S. A. and Welch D. R. (2007). Metastasis: recent discoveries and novel treatment strategies. Lancet 369: 1742–57.

Egeblad M. and Werb Z. (2002). New functions for the matrix metalloproteinases in cancer progression. Nat Rev Cancer 2: 161–174.

Escudier B. et al. (2003). Phase III trial of neovastat in metastatic renal cell carcinoma patients refractory to immunotherapy. Proc Am Soc Clin Oncol: Abstr 844.

Evans J. D. et al. (2001). A phase II trial of marimastat in advanced pancreatic cancer. Br J Cancer 85: 1865–70.

Exelixis. (2007). Pipeline – XL784. http://www.exelixis.com/pipeline_xl784.shtml. Accessed 20 July 2008.

Fingleton B. (2003a). Matrix metalloproteinase inhibitors for cancer therapy: the current situation and future prospects. Expert Opin Ther Targets 7: 385–97.

Fingleton B. (2003b). CMT-3. Collagenex. Curr Opin Investig Drugs 4(12): 1460–7.

Galardy R. E. et al. (1994). Low molecular weight inhibitors in corneal ulceration. Ann NY Acad Sci 6: 315–23.

Gingras D., Batist G. and Beliveau R. (2001). AE-941 (Neovastat): a novel multifunctional anti-angiogenic compound. Expert Rev Anticancer Ther 1: 341–7.

Goffin J. R. et al. (2005). Phase I trial of the matrix metalloproteinase inhibitor marimastat combined with carboplatin and paclitaxel in patients with advanced non-small cell lung cancer. Clin Cancer Res 11: 3417–24.

Gore M. et al. (1996). Tumour marker levels during marimastat therapy. Lancet 348: 263–4.

Groves M. D. et al. (2002). Phase II trial of temozolomide plus the matrix metalloproteinase inhibitor, marimastat, in recurrent and progressive glioblastoma multiforme. J Clin Oncol 20: 1383–8.

Groves M. D. et al. (2006). Phase II trial of temozolomide plus marimastat for recurrent anaplastic gliomas: a relationship among efficacy, joint toxicity and anticonvulsant status. J Neuro-Oncol 80: 83–90.

Hande K. R. et al. (2004). Phase I and pharmacokinetic study of prinomastat, a matrix metalloprotease inhibitor. Clin Cancer Res 10: 909–15.

Heath E. I. et al. (2001). Phase I trial of the matrix metalloproteinase inhibitor BAY12–9566 in patients with advanced solid tumors. Cancer Chemother Pharmacol 48: 269–74.

Heath E. I. et al. (2006). Phase II, parallel-design study of preoperative combined modality therapy and the matrix metalloprotease (mmp) inhibitor prinomastat in patients with esophageal adenocarcinoma. Investig New Drugs 24: 135–40.

Hemmings F. J. et al. (2001). Tolerability and pharmacokinetics of the collagenase-selective inhibitor Trocade in patients with rheumatoid arthritis. Rheumatol 40: 537–43.

Hidalgo M. and Eckhardt S. G. (2001). Development of matrix metalloproteinase inhibitors in cancer therapy. J Natl Cancer Inst 93: 178–93.

Hirte H. et al. (2006). A phase III randomized trial of BAY 12–9566 (tanomastat) as maintenance therapy in patients with advanced ovarian cancer responsive to primary surgery and paclitaxel/platinum containing chemotherapy: a National Cancer Institute of Canada Clinical Trials Group Study. Gynecol Oncol 102: 300–8.

Hochster H. S. et al. (2006). Consensus report of the international society of gastrointestinal oncology on therapeutic progress in advanced pancreatic cancer. Cancer 107: 676–85.

Holmbeck K. et al. (1999). MT1-MMP-deficient mice develop dwarfism, osteopenia, arthritis, and connective tissue disease due to inadequate collagen turnover. Cell 99: 81–92.

Hotary K. B. et al. (2003). Membrane type I matrix metalloproteinase usurps tumor growth control imposed by the three-dimensional extracellular matrix. Cell 114: 33–45.

Houghton A. M. et al. (2006). Macrophage elastase (matrix metalloproteinase-12) suppresses growth of lung metastases. Cancer Res 66: 6149–55.

Ikejiri M. et al. (2005). Potent mechanism-based inhibitors for matrix metalloproteinases. J Biol Chem 280: 33992–4002.

Johnson J. R., Williams G. and Pazdur R. (2003). End points and United States Food and Drug Administration approval of oncology drugs. J Clin Oncol 21: 1404–11.

Jones P. H. et al. (2004). Combination antiangiogenesis therapy with marimastat, captopril and fragmin in patients with advanced cancer. Br J Cancer 91: 30–6.

Kerbel R. S. (1999). What is the optimal rodent model for anti-tumor drug-testing? Cancer Metastas Rev 17: 301–4.

King J. et al. (2003). Randomised double blind placebo control study of adjuvant treatment with the metalloproteinase inhibitor, Marimastat in patients with inoperable colorectal hepatic metastases: significant survival advantage in patients with musculoskeletal side-effects. Anticancer Res 23: 639–45.

Kruger A. et al. (2005). Antimetastatic activity of a novel mechanism-based gelatinase inhibitor. Cancer Res 65: 3523–6.

Kummar S. et al. (2006). Drug development in oncology: classical cytotoxics and molecularly targeted agents. Br J Clin Pharmacol 62: 15–26.

Lara P. N. Jr et al. (2006). A randomized phase II trial of the matrix metalloproteinase inhibitor BMS-275291 in hormone-refractory prostate cancer patients with bone metastases. Clin Cancer Res 12: 1556–63.

Larson D. A. et al. (2002). Phase II study of high central dose Gamma Knife radiosurgery and marimastat in patients with recurrent malignant glioma. Int J Rad Oncol Biol Phys 54: 1397–404.

Latreille J. et al. (2003). Phase I/II trial of the safety and efficacy of AE-941 (Neovastat) in the treatment of non-small-cell lung cancer. Clin Lung Cancer 4: 231–6.

Le Goff C. et al. (2006). Regulation of procollagen amino-propeptide processing during mouse embryogenesis by specialization of homologous ADAMTS proteases: insights on collagen biosynthesis and dermatosparaxis. Development 133: 1587–96.

Leighl N. B. et al. (2005). Randomized phase III study of matrix metalloproteinase inhibitor BMS-275291 in combination with paclitaxel and carboplatin in advanced non-small-cell lung cancer: National Cancer Institute of Canada-Clinical Trials Group Study BR.18. J Clin Oncol 23: 2831–9.

Levin V. A. et al. (2002). Randomized phase II study of temozolomide (TMZ) with and without the matrix metalloprotease (MMP) inhibitor prinomastat in patients (pts) with glioblastoma multiforme (GBM) following best surgery and radiation therapy. Proc Am Soc Clin Oncol: Abstr 100.

Levin V. A et al. (2006). Randomized, double-blind, placebo-controlled trial of marimastat in glioblastoma multiforme patients following surgery and irradiation. J Neuro-Oncol 78: 295–302.

Levitt N. C. et al. (2001). Phase I and pharmacological study of the oral matrix metalloproteinase inhibitor, MMI270 (CGS27023A), in patients with advanced solid cancer. Clin Cancer Res 7: 1912–22.

Liotta L. A et al. (1980). Metastatic potential correlates with enzymatic degradation of basement membrane. Nature 284: 67–8.

C. Lu et al. (2005). A phase III study of AE-941 with induction chemotherapy (IC) and concomitant chemoradiotherapy (CRT) for stage III non-small cell lung cancer (NSCLC) (NCI T99-0046, RTOG 02-70, MDA 99-303): an interim report of toxicity and response. Proc Am Soc Clin Oncol: Abstr 7144.

Macaulay V. M. et al. (1999). Phase I study of intrapleural batimastat (BB-94), a matrix metalloproteinase inhibitor, in the treatment of malignant pleural effusions. Clin Cancer Res 5: 513–20.

Marmion C. J., Griffith D. and Nolan K. B. (2004). Hydroxamic acids – an intriguing family of enzyme inhibitors and biomedical ligands. Eur J Inorg Chem 2004: 3003–16.

Martignetti J. A. et al. (2001). Mutation of the matrix metalloproteinase 2 gene (MMP2) causes a multicentric osteolysis and arthritis syndrome. Nat Genet 28: 261–5.

Martin M. D. and Matrisian L. M. (2007). The other side of MMPs: protective roles in tumor progression. Cancer Metastas Rev 26: 717–24.

McCawley L. J. and Matrisian L. M. (2000). Matrix metalloproteinases: multifunctional contributors to tumor progression. Mol Med Today 6: 149–56.

McCawley L. J. et al. (2004). A protective role for matrix metalloproteinase-3 in squamous cell carcinoma. Cancer Res 64: 6965–72.

Michaelis L. C. and Ratain M. J. (2006). Measuring response in a post-RECIST world: from black and white to shades of grey. Nat Rev Cancer 6: 409–14.

Millar A. W. et al. (1998). Results of single and repeat dose studies of the oral matrix metalloproteinase inhibitor marimastat in healthy male volunteers. Br J Clin Pharmacol 45: 21–6.

Miller K. D. et al. (2002). A randomized phase II pilot trial of adjuvant marimastat in patients with early-stage breast cancer. Ann Oncol 13: 1220–4.

Miller K. D. et al. (2004). A randomized phase II feasibility trial of BMS-275291 in patients with early stage breast cancer. Clin Cancer Res 10: 1971–5.

Moore M. et al. (2000). A comparison between gemcitabine (GEM) and the matrix metalloproteinase (MMP) inhibitor BAY12-9566 (9566) in patients (PTS) with advanced pancreatic cancer. Proc Am Soc Clin Oncol: Abstr 930.

Moore M. J. et al. (2003). Comparison of gemcitabine versus the matrix metalloproteinase inhibitor BAY 12-9566 in patients with advanced or metastatic adenocarcinoma of the pancreas: a phase III trial of the National Cancer Institute of Canada Clinical Trials Group. J Clin Oncol 21: 3296–302.

Mosig R. A. et al. (2007). Loss of MMP-2 disrupts skeletal and craniofacial development and results in decreased bone mineralization, joint erosion and defects in osteoblast and osteoclast growth. Hum Mol Genet 16: 1113–23.

Nemunaitis J. et al. (1998). Combined analysis of studies of the effects of the matrix metalloproteinase inhibitor marimastat on serum tumor markers in advanced cancer: selection of a biologically active and tolerable dose for longer-term studies. Clin Cancer Res 4: 1101–9.

Nemunaitis J. et al. (2000). Prognostic role of K-ras in patients with progressive colon cancer who received treatment with Marimastat (BB2516). Cancer Investig 18: 185–90.

North H., King J. and Morris D. L. (2000). Effect of marimastat on serum tumour markers in patients with colorectal cancer. Int J Surg Investig 2: 213–7.

Overall C. M. and Kleifeld O. (2006a). Tumour microenvironment – opinion: validating matrix metalloproteinases as drug targets and anti-targets for cancer therapy. Nat Rev Cancer 6: 227–39.

Overall C. M. and Kleifeld O. (2006b). Towards third generation matrix metalloproteinase inhibitors for cancer therapy. Br J Cancer 94: 941–6.

Page-McCaw A., Ewald A. J. and Werb Z. (2007). Matrix metalloproteinases and the regulation of tissue remodelling. Nat Rev Mol Cell Biol 8: 221–33.

Parks W. C., Wilson C. L. and López-Boado Y. S. (2004). Matrix metalloproteinases as modulators of inflammation and innate immunity. Nat Rev Immunol 4: 617–29.

Parsons S. L., Watson S. A. and Steele R. J. (1997). Phase I/II trial of batimastat, a matrix metalloproteinase inhibitor, in patients with malignant ascites. Eur J Surg Oncol 23: 526–31.

Pavlaki M. and Zucker S. (2003). Matrix metalloproteinase inhibitors (MMPIs): the beginning of Phase I or the termination of Phase III clinical trials. Cancer Metastasis Rev 22: 177–203.

Pendás A. M. et al. (2004). Diet-induced obesity and reduced skin cancer susceptibility in matrix metalloproteinase 19-deficient mice. Mol Cell Biol 24: 5304–13.

Peterson J. T. (2006). The importance of estimating the therapeutic index in the development of matrix metalloproteinase inhibitors. Cardiovasc Res 69: 677–87.

Primrose J. N. et al. (1999). Marimastat in recurrent colorectal cancer: exploratory evaluation of biological activity by measurement of carcinoembryonic antigen. Br J Cancer 79: 509–14.

Quirt I. et al. (2002). Phase II study of marimastat (BB-2516) in malignant melanoma: a clinical and tumor biopsy study of the National Cancer Institute of Canada Clinical Trials Group. Investig New Drugs 20: 431–7.

Rasmussen H. S. and McCann P. P. (1997). Matrix metalloproteinase inhibition as a novel anticancer strategy: a review with special focus on batimastat and marimastat. Pharmacol Ther 75: 69–75.

Renkiewicz R. et al. (2003). Broad-spectrum matrix metalloproteinase inhibitor marimastat-induced musculoskeletal side effects in rats. Arthr Rheum 48: 1742–9.

Rigas J. R. et al. (2003). Randomized placebo-controlled trials for the matrix metalloproteinase inhibitor (MMPI), BAY12-9566 as adjuvant therapy for patients with small cell and non-small cell lung cancer. Proc Am Soc Clin Oncol: Abstr 2525.

Rizvi N. A. et al. (2004). A phase I study of oral BMS-275291, a novel nonhydroxamate sheddase-sparing matrix metalloproteinase inhibitor, in patients with advanced or metastatic cancer. Clin Cancer Res 10: 1963–70.

Rosemurgy A. et al. (1999). Marimastat in patients with advanced pancreatic cancer: a dose-finding study. Am J Clin Oncol 22: 247–52.

Rosenbaum E. et al. (2005). Marimastat in the treatment of patients with biochemically relapsed prostate cancer: a prospective randomized, double-blind, phase I/II trial. Clin Cancer Res 11: 4437–43.

Rothenberg M. L., Carbone D. P. and Johnson D. H. (2003). Improving the evaluation of new cancer treatments: challenges and opportunities. Nat Rev Cancer 3: 303–9.

Rowinsky E. K. et al. (2000). Phase I and pharmacologic study of the specific matrix metalloproteinase inhibitor BAY 12-9566 on a protracted oral daily dosing schedule in patients with solid malignancies. J Clin Oncol 18: 178–86.

Rudek M. A. et al. (2001). Phase I clinical trial of oral COL-3, a matrix metalloproteinase inhibitor, in patients with refractory metastatic cancer. J Clin Oncol 19: 584–92.

Rugo H. S. et al. (2001). Phase II study of the matrix metalloprotease inhibitor prinomastat in patients with progressive breast cancer. Proc Am Soc Clin Oncol: Abstr 187.

Sang Q. X. et al. (2006). Matrix metalloproteinase inhibitors as prospective agents for the prevention and treatment of cardiovascular and neoplastic diseases. Curr Top Med Chem 6: 289–316.

Shepherd F. A. et al. (2002). Prospective, randomized, double-blind, placebo-controlled trial of marimastat after response to first-line chemotherapy in patients with small-cell lung cancer: a trial of the National Cancer Institute of Canada-Clinical Trials Group and the European Organization for Research and Treatment of Cancer. J Clin Oncol 20: 4434–9.

Smylie M. et al. (2001). Phase III study of the matrix metalloprotease (MMP) inhibitor prinomastat in patients having advanced non-small cell lung cancer (NSCLC). Proc Am Soc Clin Oncol: Abstr 1226.

Sparano J. A. et al. (2004). Randomized phase III trial of marimastat versus placebo in patients with metastatic breast cancer who have responding or stable disease after first-line chemotherapy: Eastern Cooperative Oncology Group trial E2196. J Clin Oncol 22: 4683–90.

Steward W. P. and Thomas A. L. (2000). Marimastat: the clinical development of a matrix metalloproteinase inhibitor. Expert Opin Investig Drugs 9: 2913–22.

Syed S. et al. (2004). A phase I and pharmacokinetic study of Col-3 (Metastat), an oral tetracycline derivative with potent matrix metalloproteinase and antitumor properties. Clin Cancer Res 10: 6512–21.

Tierney G. M. et al. (1999). A pilot study of the safety and effects of the matrix metalloproteinase inhibitor marimastat in gastric cancer. Eur J Cancer 35: 563–8.

Toppmeyer D. L. et al. (2003). A phase I and pharmacologic study of the combination of marimastat and paclitaxel in patients with advanced malignancy. Med Sci Monit 9: PI99–104.

van Marle S. et al. (2005). Safety, tolerability and pharmacokinetics of oral S-3304, a novel matrix metalloproteinase inhibitor, in single and multiple dose escalation studies in healthy volunteers. Int J Clin Pharmacol Ther 43: 282–93.

Whittaker M. et al. (1999). Design and therapeutic application of matrix metalloproteinase inhibitors. Chem Rev 99: 2735–76.

Wojtowicz-Praga S. et al. (1996). Phase I trial of a novel matrix metalloproteinase inhibitor batimastat (BB-94) in patients with advanced cancer. Investig New Drugs 14: 193–202.

Wojtowicz-Praga S. et al. (1998). Phase I trial of Marimastat, a novel matrix metalloproteinase inhibitor, administered orally to patients with advanced lung cancer. J Clin Oncol 16: 2150–6.

Zhou Z. et al. (2000). Impaired endochondral ossification and angiogenesis in mice deficient in membrane type matrix metalloproteinase I. Proc Natl Acad Sci USA 97: 4052–7.

Chapter 37
Tailoring TIMPs for Selective Metalloproteinase Inhibition

Hideaki Nagase and Gillian Murphy

Abstract Members of the metzincin clan of zinc endopeptidases (metalloproteinases, MPs) are key players in the proteolytic modification of cell surface and extracellular matrix proteins which are the bases of many cellular responses. They are regulated by a small family of endogenous inhibitors, the tissue inhibitors of metalloproteinases (TIMPs). TIMPs were originally identified as natural regulators of the matrix metalloproteinases (MMPs), but they exhibit distinctive structural features and biochemical properties, suggesting that each has specific roles *in vivo*. Interactions with the disintegrin metalloproteinases, ADAMs and ADAMTSs, appear to be even more selective. The ability of these proteins to inhibit the MPs is largely due to the interaction of a wedge-shaped ridge on the N-domain which binds within the active-site cleft of the target proteinase. Apart from structural studies, features of the specific interactions between individual TIMPs and MPs have been ascertained by a combination of engineering and kinetic studies. This has also led to the concept of redesigning or 'tailoring' TIMPs for more specific anti-MP functions, as potential tools for interrogation of biological systems, and as therapeutic agents in diseases such as cancer with a strong MP involvement. This chapter will review current progress towards this goal.

Introduction

Cell–matrix and cell–cell interactions influence a diverse range of cellular functions, including proliferation, differentiation, migration and survival. Proteolytic degradation or activation of cell surface and extracellular matrix proteins can mediate rapid cellular responses to their microenvironment, and hence modulate

G. Murphy
Department of Oncology, Cambridge University, Cancer Research UK Cambridge Research Institute, Li ka Shing Centre, Cambridge CB2 0RE UK, e-mail: gm290@cam.ac.uk

D. Edwards et al. (eds.), *The Cancer Degradome.*
© Springer Science + Business Media, LLC 2008

cell behaviour. Members of the metzincin clan of zinc endopeptidases (metalloproteinases, MPs) are key players in such activities and the endogenous inhibitors, the tissue inhibitors of metalloproteinases (TIMPs), are significant regulators of the matrix metalloproteinases (MMPs) and some members of the disintegrin metalloproteinases (ADAMs) and ADAMTS (ADAM with *t*hrombo*s*pondin motifs; Baker et al. 2002, Nagase et al. 2006).

There are four TIMPs (TIMP-1 to TIMP-4) in mammals that show about 40% identity in their amino acid sequences. They were originally identified as natural regulators of the MMPs and all four TIMPs have many basic similarities, but they exhibit distinctive structural features and biochemical properties, suggesting that each has specific roles *in vivo* (Brew et al. 2000, Baker et al. 2002). Their inhibitory activities towards different MMPs are partially selective (Table 37.1), but interactions with the ADAMs and ADAMTSs appear to be more selective (Table 37.2). The structures of TIMP-1 and TIMP-2 have been solved and it is likely that the other TIMPs have a homologous protein fold and topology, which consists of an N-terminal domain (N-domain) with three disulphide bonds and a C-terminal domain (C-domain) with three disulphide bonds. The ability of these proteins to inhibit the MMPs is largely due to the interaction of a wedge-shaped ridge on the N-domain which binds within the active-site cleft of the target MMP. Apart from structural studies, the nature of the specific interactions between individual TIMPs and other family of metalloproteinases (ADAMs, ADAMTSs) has been ascertained by a combination of engineering and kinetic studies. This has also led to the concept of redesigning or 'tailoring' TIMPs for more specific anti-MP functions as potential tools for interrogation of biological systems and as therapeutic agents in diseases such as cancer with a strong MP involvement.

General Biochemical Properties of TIMPs

The TIMPs are composed of identifiable N-terminal and C-terminal subdomains, each stabilized by three disulphide bonds. Expression of a recombinant form of the N-subdomain of TIMP-1 and kinetic analyses showed that it could form complexes with active forms of the MMPs, indicating that the major structural features for specific interaction with these enzymes reside in these three loops (Murphy et al. 1991). Subsequent structural studies showed that this domain consisted of a five-strand β sheet rolled into a β barrel with a oligosaccharide/oligonucleotide binding (OB) fold in the case of TIMP-2 (Williamson et al. 1994), and the identification of features of the MMP interaction site were subsequently mapped by chemical shift perturbation. The site around the Val69-Cys70 of the N-subdomain of the TIMPs was also predicted to interact with MMPs based on proteinase protection studies (Nagase et al. 1997). The mechanism of MMP inhibition by TIMP was clarified by the crystal structure of the complex of TIMP-1 and the catalytic domain of MMP-3 (Gomis-Rüth et al. 1997, Fig. 37.1). The structure showed that TIMP-1 is a 'wedge-shaped' molecule. In this complex, about 75% of the protein–protein

Table 37.1 Apparent Dissociation Constants for the Inhibition of MMPs by TIMPs

MMP	TIMP-1	N-TIMP-1	TIMP-2	N-TIMP-2 (K_i^{app})	TIMP-3 (nM)	N-TIMP-3	TIMP-4	N-TIMP-4
MMP-1	0.38						6.7	
MMP-1 (cat.)	2.1	3.0				1.2	2.6	
MMP-2	<0.009	0.25	0.000006	0.04		4.3	<0.009	
MMP-3	1.5	5.3					0.63	
MMP-3 (cat.)	1.9	1.2				67	1.1	
MMP-7	0.37	1.3		2.0	Yes		Yes	
MMP-8	Yes	Yes	9.7	19				
MMP-8 (cat.)			37	81				
MMP-9	8.5		43		Yes		Yes	
MMP-10	Yes		Yes					
MMP-11	Yes							
MMP-12	Yes							
MMP-13		0.081		0.004				
MT1-MMP (cat.)	147		0.07	3		0.057	0.73	0.83
MMP-15 (cat.)	partial		Yes		Yes			
MMP-16 (cat.)			0.17			0.008	0.34	
MMP-19	52		0.68		0.53		0.20	
MMP-24 (cat.)	65		3.6		0.25		0.53	
MMP-25 (cat.)	0.2		2.0					
MMP-26		0.54						

MMP matrix metalloproteinase, *TIMP* tissue inhibitor of metalloproteinase.

Table 37.2 TIMP Inhibition Profiles of ADAMs and ADAMTSs

Enzyme	TIMP-1	TIMP-2	TIMP-3 (K_i^{app}) (nM)	N-TIMP-3	TIMP-4	References
ADAM						
ADAM-10	~0.1	No	~0.9		No	Amour et al. 2000
ADAM-12	No	No	Inhibits at 500, none at 100		No	Roy et al. 2004
ADAM-12	No	No	50% inhibition at 10			Loechel et al. 2000
ADAM-17 (cat.)	760	11	1.1		92	Zou et al. 2004
ADAM-17 (cat.)	No	25% inhibition at 130	0.182		25% inhibition at 100	Amour et al. 1998
ADAM-17 (cat.)			0.2	0.22		Lee et al. 2002a
ADAM-17 (full ectodomain)			0.74	1.75		Lee et al. 2002b
ADAM-28	No	No	Yes		Yes	Mochizuki et al. 2004
ADAM-33	No	1,400	60		320	Zou et al. 2004
ADAMTS						
ADAMTS-1	No (at 500)	~50% inhibition at 500	Yes (at 500), No (at 100)		No (at 500)	Rodriguez-Manzaneque et al. 2002
ADAMTS-2	No	No	Yes	602 (−heparin) 160 (+heparin)	No	Wang et al. 2006
ADAMTS-4	No	No		3.3		Kashiwagi et al. 2001
ADAMTS-4	350	420	7.9		35% inhibition at 250	Hashimoto et al. 2001
ADAMTS-5	No	No		0.66		Kashiwagi et al. 2001

MMP matrix metalloproteinase, *TIMP* tissue inhibitor of metalloproteinase, *ADAM* a disintegrin and metalloproteinase, *ADAMTS* a disintegrin and metalloproteinase with thrombospondin motifs.

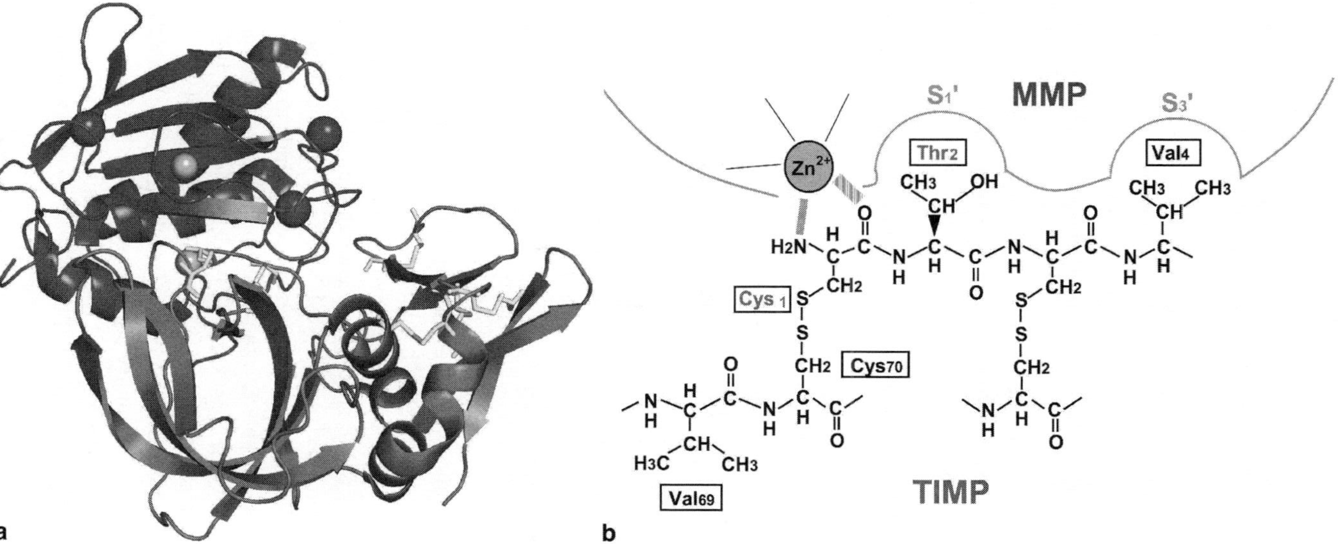

Fig. 37.1 The mode of interaction of tissue inhibitor of metalloproteinase (*TIMP*)-1 and the catalytic domain of matrix metalloproteinase (*MMP*)-3. **a** Ribbon diagram of the complex of TIMP-1 (red) and the catalytic domain of MMP-3 (blue). Zn^{2+} and Ca^{2+} are shown as pink and blue spheres, and the disulphide bonds in TIMP-1 are shown in yellow. The image was prepared from the Brookhaven Protein Data Bank (*PDB*) entry (1UEA) using the PyMOL program by W.L. DeLano (http://www.pymol.org). **b** Schematic representation of N-terminal region of TIMP-1 that interacts with the active site of MMP (*See also* Color Insert II)

contacts in TIMP-1 are from the N-terminal stretch of Cys1 to Val4 and the loop region of Met66 to Val69 that are linked by the Cys1-Cys70 disulphide bond forming a ridge-like structure which slots into the active site of MMPs. The residues 1–4 of TIMP-1 bind to the enzyme in an analogous way as the P1-P1'-P2'-P3' residues of a peptide substrate (proteolytic cleavage occurs at the peptide bond between the P1 and P1'residues), where the side chain of Thr2 extends into the large S1' specificity pocket (substrate-binding pocket that accepts the P1'residue of the substrate) of the enzyme. Ser68 and Val69 occupy the part of the active site (S2 and S3 subsites, respectively), but they are oriented in an opposite direction from that of a bound peptide substrate. The α-amino and the carbonyl groups of the N-terminal Cys1 chelate the Zn^{2+} of the enzyme's active site, and the OH group of Thr2 interacts with Glu202 of the enzyme. This interaction displaces a water molecule essential for peptide hydrolysis from the active site. The crystal structure of TIMP-2-MT1-MMP catalytic domain complex (Fernandez-Catalan et al. 1998) demonstrated that the main feature of the inhibition mechanism of TIMPs is essentially the same, and it is likely that other TIMPs interact with MMPs in a similar manner.

Detailed kinetic studies have been carried out for a number of MMP–TIMP combinations (Table 37.1); TIMP-1 inhibits most MMPs, but is a poor inhibitor of MT1-MMP. TIMP-2, TIMP-3 and TIMP-4 inhibit all MMPs tested so far. It is notable that MMP-2 is inhibited very tightly by TIMP-1, TIMP-2 and TIMP-4 (Hutton et al. 1998, Butler et al. 1999, Bigg et al. 2001, Troeberg et al. 2002). TIMP-3 also inhibits several members of the ADAM and ADAMTS families (Table 37.2). Target proteinases include ADAM-10 (Amour et al. 2000), ADAM-12 (Loechel et al. 2000), ADAM-17/tumour necrosis factor α (TNFα)-converting enzyme (TACE) (Amour et al. 1998), ADAMTS-1 (Hashimoto et al. 2001, Rodriguez-Manzaneque et al. 2002), ADAMTS-4 (Kashiwagi et al. 2001) and ADAMTS-5 (Kashiwagi et al. 2001), whereas ADAM-8 and ADAM-9 are not inhibited by TIMP-1, TIMP-2 or TIMP-3 (Amour et al. 2002). The structural bases of these differential inhibitory features among TIMPs are not well understood. However, mutational studies of the binding ridge within the N-subdomains of TIMP-1, TIMP-2 and TIMP-3 have identified the key interacting residues which can be mutated to modify MMP specificity, as discussed below.

In addition to interacting with active MMPs, some TIMPs bind to the zymogens of MMPs (pro-MMPs). Pro-MMP-2 (progelatinase A) binds to TIMP-2, TIMP-3 or TIMP-4 through interactions between the C-terminal haemopexin domain and the C-domain of the TIMP (Willenbrock et al. 1993, Butler et al. 1999, Morgunova et al. 2002, Troeberg et al. 2002). Also, pro-MMP-9 (progelatinase B) binds to TIMP-1 or TIMP-3 through their C-domains (Goldberg et al. 1992, O'Connell et al. 1994, Butler et al. 1999). These binary complexes inhibit MMPs as the N-terminal inhibitory domain of the TIMP is exposed to the solvent. The formation of a ternary complex of pro-MMP-2, TIMP-2 and MT1-MMP is essential for the activation of pro-MMP-2 by MT1-MMP on the cell surface. The TIMP-2-bound MT1-MMP acts as a 'receptor' of pro-MMP-2 and the adjacent MT1-MMP, free of TIMP-2, acts as an 'activator' (see below).

Mutational Studies to Investigate Inhibition Mechanism of MMPs by TIMPs

Following on from the observation that the N-terminal domain of TIMP-1 could fold independently and inhibit the MMPs (Murphy et al. 1991), this became a target for more detailed studies of the structure–function relationships. The major sites subjected to mutagenesis studies in four TIMPs are shown in Fig. 37.2.

An extensive mutational study of N-TIMP-1 by Huang et al. (1997) suggested that the reactive site of TIMPs located in or close to the reactive ridge region plays important roles in inhibition of MPs and mutation of these residues alter the specificity of TIMPs. Among them, one N-TIMP-1 mutant with a substitution of Ala for Thr2 (Thr2Ala) had different activities against three MMPs: compared with the wild type; the affinity for MMP-1 decreased about 1600-fold, 30-fold for MMP-2 and 250-fold for MMP-3. Mutations at other sites resulted in small changes in TIMP-1 activity, but these were similar for all three MMPs. Meng et al. (1999) investigated the effect of position 2 variants in further detail by generating 13 mutants (Table 37.3), and they were tested against 3 prototypes of collagenase (MMP-1), gelatinase A (MMP-2) and stromelysin (MMP-3). Among them the Gly mutant was the weakest to inhibit these MMPs, and the affinity was reduced 3 to 5 orders of magnitude. This represents a loss of 33–55% of the free energy of interaction. The structure of the MMP-3–TIMP-1 complex indicated that $\sim$1,300 Å^2 of the accessible surface of each molecule is buried upon formation of the complex (Gomis-Rüth et al. 1997), but Thr2 occupies only 8% (108/1,300 Å^2) of it (Fig. 37.1). This disproportionately small area for its effect on MMP inhibition let them conclude that Thr2 in TIMP-1 is a 'hot spot' in the MMP–TIMP interaction interface (Meng et al. 1999).

The S1' specificity pockets of MMPs differ in size, and therefore influence their individual substrate specificity. TIMPs have Thr or Ser in position 2; hence, it was postulated that side chain sizes at this position may discriminate the ability to inhibit MMPs (Meng et al. 1999). As listed in Table 37.3, mutation of residue 2 of N-TIMP-1 significantly influences the affinity for three MMPs tested. When these results were compared with an amino acid in the P1' position of a peptide substrate, there were very poor correlations. For example, the best N-TIMP-1 mutant for inhibiting MMP-1 is the Val2 variant, but a peptide with Val2 at P1' position is a poor substrate. Similarly, Ser and Arg at P1' position in the substrate are not favourable for MMP-2 or MMP-3 (Nagase and Fields 1996), but N-TIMP-1 mutants with those residues were well tolerated for these two enzymes. It was postulated that these discrepancies are due to a greater loss of conformational entropy associated with peptide substrate–MMP interaction compared with the TIMP–MMP interaction. The reactive ridges of TIMPs are rigid due to two disulphide bonds Cys1-Cys70 and Cys3-Cys99 and the conserved proline at position 5. This rigid structure and the interaction of Cys1 with the active site Zn^{2+} orient residue 2 of TIMP-1 in a strict manner, which thus influences the specificity in a different manner from a flexible oligopeptide. Recently, Hamze et al. (2007) showed that substitutions for Thr2 of N-TIMP-1 strongly influenced MMP selectivity;

TIMP-1

TIMP-2

TIMP-3

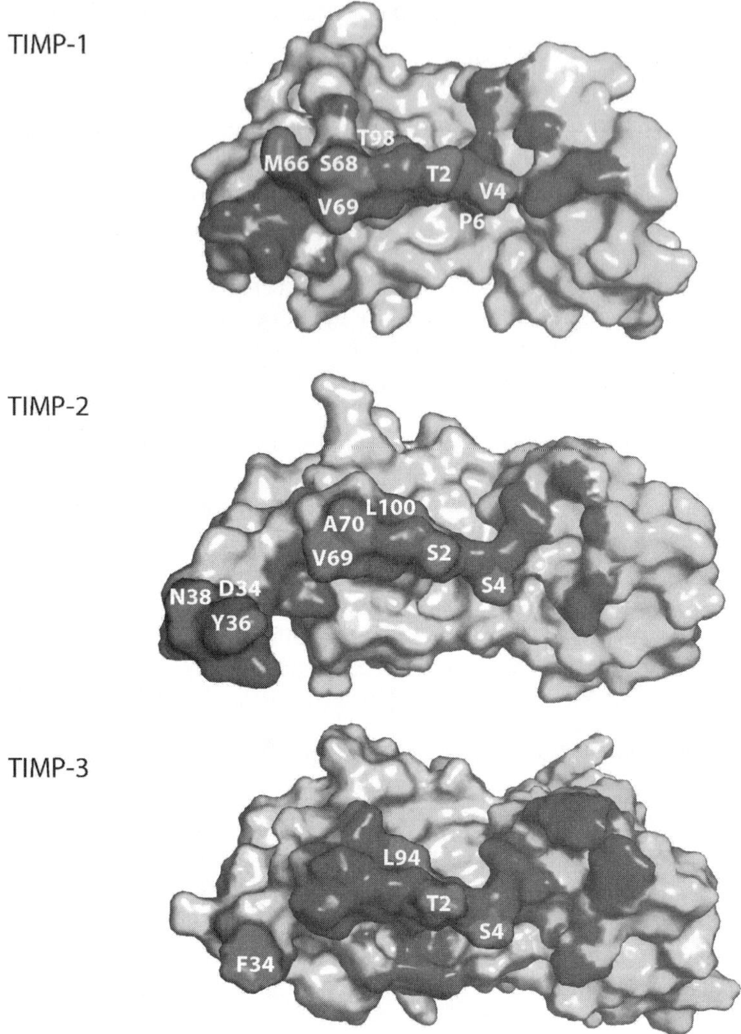

Fig. 37.2 Reactive sites of tissue inhibitor of metalloproteinase (*TIMP*)-1, TIMP-2 and TIMP-3. The N-terminal domain and the C-terminal domain of each TIMP are shown in pink and green, respectively. The residues in reactive ridge that are within 4 Å contact with the catalytic domain of a matrix metalloproteinase (*MMP*) are shown in blue and the residues that were subjected to mutagenesis studies are shown in red. TIMP-1 and TIMP-2 models were created from the Brookhaven Protein Data Bank (*PDB*) entry 1UEA and 1BUV, respectively, as in Fig. 37.1. The TIMP-3 structure was modelled based on crystal structures of TIMP-1 and TIMP-2 (*See also* Color Insert II)

Arg and Gly, which generally reduce MMP affinity, have less effect on binding to MMP-9. When the Arg mutation is added to an N-TIMP-1(AB2) mutant (harbour-ing the AB loop features of TIMP-2; see below), it produces a gelatinase-specific inhibitor with K_i values of 2.8 and 0.4 nM for MMP-2 and -9, respectively.

Table 37.3 K_i Values(nM) of TIMPs and TIMP-1 Variants AB2, TIMP-2 AB loop (nM)

Inhibitor	MMP-1	MMP-2	MMP-3	MMP-9
N-TIMP-1				
Wild type	0.4	0.4	0.2	0.2
T2S	10	0.8	0.2	
T2G	7,000	>40,000	560	2.1
T2A	836	15	50	
T2L	37	0.4	1.3	
T2V	0.6	108	1.2	
T2N	788	6.4	18	
T2Q	348	4.8	12	
T2D	3,250	500	440	
T2E	2,290	173	187	
T2K	668	12	28	
T2R	~2,000	4.8	11	9
V4S	2.1	0.4	1.6	
V4K	8.8	1.1	2.6	
S68Y	61	2.7	0.3	
S68A	14	5.2	32	
S68E	16	3.2	14	
S68R	48	18	49	
T2L/V4S	140	0.3	6	
T2L/V4S/S68A	~3000	4.0	14	
T2S/V4A/S68Y	15	0.1	0.05	
T2R/V4I	Inactive	10	4	
N-TIMP-2	0.6	0.5	9.3	
N-TIMP-3	1.2	4.3	67	
AB2	0.2	1.4	12	0.6
T2G/AB2	124	67	33	5.7
T2R/AB2	~2,000	4.8	11	0.4

The data are from Meng et al. (1999), Wei et al. (2003) and Hamze et al. (2007) *MMP* matrix metalloproteinase, *TIMP* tissue inhibitor of metalloproteinase.

Interestingly, the Gly mutant has a K_i of 2.1 nM for MMP-9 and >40 μM for MMP-2, which is the first indication that engineered TIMPs can discriminate between the two gelatinases (Table 37.3).

Since the TIMP-1–MMP-3 crystal structure showed interaction of the Val4 of TIMP-1 with the S3' subsite of the enzyme and of the Ser68 with the S2, mutations at these residues were prepared to test their role in specificity, but the changes in binding kinetics observed were moderate (Wei et al. 2003, Table 37.3). Val4Ser and Val4Lys mutants are more selective for MMP-2 than the wild type. The Ser68Tyr mutant inhibits MMP-3 much more strongly than MMP-1 or -2. However, double and triple mutants in combination showed further selectivity. The double mutant Thr2Leu/Val4Ser maintained the excellent inhibitory activity for MMP-2, but it decreased the affinity for MMP-1 and MMP-3 by 350- and 30-fold, respectively. The triple mutant Thr2Ser/Val4Ala/Ser68Ala has a high affinity for MMP-3 with a

K_i value of 50 pM and an improved affinity for MMP-2, but the binding to MMP-1 is reduced (Fig. 37.2). Those results suggest that structures of the local substrate-binding site among MMPs are significantly different, so that it may be possible to design further selective TIMP variants to specific MMPs.

The structural basis as to why TIMP-1 displays negligible inhibitory activity against some MMPs was investigated by Lee et al. (2003, 2003a) (Table 37.4). Molecular modelling was used to define the epitopes of the binding ridge that potentially make TIMP-1 less active against MT1-MMP. Individual mutations were carried out and it was shown that TIMP-1 could be transformed into an active inhibitor against MT1-MMP by the mutation of a single residue, Thr98 to Leu (Fig. 37.2). This mutant displayed the inhibitory characteristics of a slow, tight-binding inhibitor. The potency of the mutant could be further enhanced by mutating Val4 to Ala and Pro6 to Val mutations. The inhibitory profile of the triple mutant (Val4Ala/Pro6Val/Thr98Leu) of N-TIMP-1 is indistinguishable from those of other TIMPs against MT1-MMP (Table 37.4, Fig. 37.2). In terms of a potential mechanistic explanation for this, modelling of Thr98Leu mutant into the TIMP-2-MT1-MMP structure suggests that Thr98Leu would not enhance contacts between the enzyme and the inhibitor in the final complex but is critical for the initiation of binding (Lee et al. 2003, 2003a).

Table 37.4 K_i Values(nM) of N-TIMP-1 Variants for MT1-MMP

N-TIMP-1 Variant	MMP-2	MMP-3	MMP-13	MT1-MMP (cat)
Wild-type	0.54	5.3	0.08	178
T2S				220
V4A				66
V4S				81
P6V				78
P6S				95
P6A				165
TIMP2-AB loop				77
TIMP3-AB loop				117
M66K				462
M66D				>500
M66A				199
M66G				254
T98L	0.29	4.6	0.13	11
T98G				>1,000
T98A				>1,000
T98F				>1,000
T98D				>1,000
T98K				>1,000
T98L+V4A/P6L				1.6
T98L+V4A/P6V/TIMP2-AB loop				5.2

The data are from Lee et al. (2003). *MMP* matrix metalloproteinase, *TIMP* tissue inhibitor of metalloproteinase.

Activation of Pro-MMP-2 and the Role of TIMP-2

The activation of MMPs by sequential proteolysis of the propeptide blocking the active-site cleft is regarded as one of the key levels of regulation of these proteinases (Nagase 1997). Pro-MMP-2 activation by cells expressing membrane-bound MT1-MMP involves a unique two-step activation mechanism, with an MT1-MMP 'initiation' cleavage followed by MMP-2 self-cleavage (Atkinson et al. 1995). The process appears to involve binding of the pro-MMP-2 to an MT1-MMP–TIMP-2 complex which forms a 'receptor' at the surface of the cell through interaction of the haemopexin domain of pro-MMP-2 with the C-terminal domain of TIMP-2. By establishing a trimolecular complex consisting of MT1-MMP, TIMP-2 and pro-MMP-2, the components of the 'activation cascade' are concentrated on the cell surface. Processing of pro-MMP-2 within this complex to an intermediate is effectively carried out by an adjacent functional MT1-MMP molecule. This initial cleavage event destabilizes the structure of the propeptide and autolytic cleavage to generate the fully mature enzyme proceeds in an MMP-2 concentration-dependent manner.

Formation of the Pro-MMP-2–TIMP-2 Complex

The interaction between TIMP-2 and pro-MMP-2 through the haemopexin domain was defined in the pro-MMP-2–TIMP-2 crystal structure (Morgunova et al. 2002, Fig. 37.3). This showed that there are at least two sites of interaction: the 'tail' sequence of TIMP-2 establishes five salt bridges with residues within blades III and IV of the haemopexin domain of pro-MMP-2-flanking hydrophobic interactions established by Phe188 with a hydrophobic groove within the enzyme (Fig. 37.3). The importance of the C-terminal 'tail' of TIMP-2 for this complex formation was also established by kinetic analysis. Studies by Willenbrock et al. (1993) showed the effect of salt concentration on the rate of association, k_{on}, indicating that ionic interactions are predominant in the association of the TIMP-2 tail with the haemopexin domain of MMP-2. This was further supported by the mutation of Glu192 Asp193 to Ile192, Asn193 (the corresponding residues found in TIMP-4), which reduced the k_{on} for pro-MMP-2 by 20-fold (Rapti et al. 2006).

Formation of the TIMP-2–MT1-MMP Complex

The interaction of TIMP-2 with MT1-MMP involves mainly the N-terminal domains of the inhibitor and the enzyme. Kinetic studies on the interaction of $\Delta_{128-194}$ TIMP-2 (N-TIMP-2) mutants and MT1-MMP showed that residues that had been identified at the enzyme–inhibitor interface in the MMP-3–TIMP-1 complex (Gomis-Rüth et al. 1997) were also important in this case. Mutation of

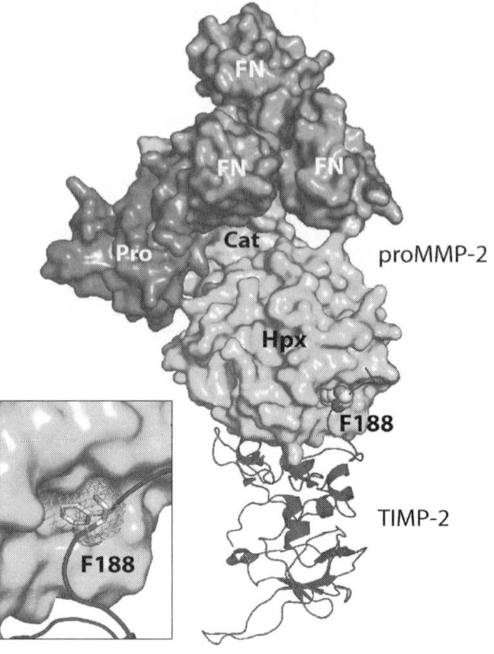

Fig. 37.3 Three-dimensional structure of the pro-MMP-2–TIMP-2 complex. Pro-MMP-2 is shown as surface representation and tissue inhibitor of metalloproteinase (*TIMP*)-2 as a ribbon diagram. Phe 188 of TIMP-2 is shown in sphere. Inset, the pocket in the Hpx domain with which Phe 188 of TIMP-2 interacts. This figure was prepared from the Brookhaven Protein Data Bank (*PDB*) entry 1GXD as in Fig. 37.1. Pro, pro-domain; Cat, catalytic domain; FN, fibronectin type II motif; Hpx, haemopexin domain (*See also* Color Insert II)

residues Ser2, Ala70, Val71 and Gly73, which are located on the reactive ridge surface of N-TIMP-2 (Fig. 37.2), increased the association rate constants and apparent K_i^{app} values for MT1-MMP as well as for MMP-2, MMP-7 and MMP-13 (Butler et al. 1998). The mutational study demonstrated a specific interaction of Tyr36 located in the hairpin turn of the A and B β strands of TIMP-2 and MT1-MMP is important as shown in the crystal structure of the complex where Tyr36 of TIMP-2 fits into a cavity on the surface of MT1-MMP bordered by the so-called 'MT loop' and the side chains of Asp212, Ser189 and Phe180 of the enzyme (Fernandez-Catalan et al. 1998). The crystal structure indicated that there are several other significant interactions between the MT1-MMP surface and residues of the AB loop in TIMP-2 (Fig. 37.4). Alignments of the hairpin loop between the A and B β strands of free TIMP-1 to -4 and enzyme-bound TIMPs were made using nuclear magnetic resonance and X-ray diffraction data for TIMP-1 and -2 and modelling based on the known properties of the OB protein fold (Williamson et al. 2001). In free TIMP-2, the A and B strands are very well defined in the nuclear magnetic resonance structure and the loop end forms a type I β-turn with a G1-β bulge. In the crystal structure of the TIMP-2–MT1-MMP complex, the strand alignment is less clear, and there is some evidence of strand realignment in the middle section (Glu-26 to Asp-30) that may occur as part of the large conformation change upon binding to MT1-MMP. The structural data for TIMP-1 show no significant difference in strand alignment between the free and MMP-3-bound inhibitor, and the predicted hydrogen bonding pattern suggests that this hairpin

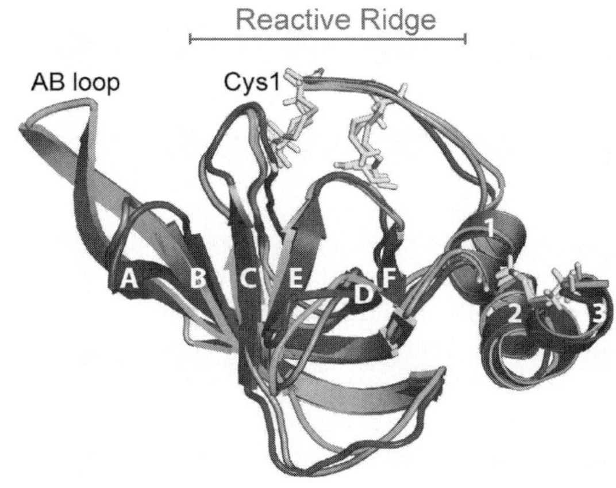

Fig. 37.4 Comparison of N-TIMP-1, N-TIMP-2 and N-TIMP-3 structures. Image of the N-terminal domain of tissue inhibitor of metalloproteinase (*TIMP*)-1 (blue) and that of TIMP-2 (green) were created from the Brookhaven Protein Data Bank (*PDB*) entry 1UEA and 1BUV, respectively, as in Fig. 37.1. N-TIMP-3 (orange) structure was modelled based on crystal structures of TIMP-1 and TIMP-2. Disulphide bonds are shown in yellow. TIMP-2 has a longer AB β strands than TIMP-1 and TIMP-3 (*See also* Color Insert II)

results in a 4-residue turn (i.e. four residues in the loop positions). No structural data are currently available for either TIMP-3 or -4, but their predicted strand alignments and hydrogen-bonding pattern suggest that both these hairpins may end in type I β-turns (i.e. two residues in the loop positions) (Williamson et al. 2001).

While the side chain of Tyr36 is the most important feature of the AB hairpin loop in terms of both initial association and final binding to this MMP (Butler et al. 1998), Tyr36 is not important for the binding to MMP-2. Although the position of the extended AB hairpin will necessitate its close contact with a proteinase bound at the inhibitory site of TIMP-2, this interaction need not contribute to the overall binding affinity in all cases and could, in some cases, weaken the overall binding interaction by making unfavourable contacts with the proteinase (Lee et al. 2003, Table 37.4). It is clear, however, that TIMP-2 may be engineered to abrogate MTI-MMP binding, whereas its binding properties for other MMPs, including MMP-2, are maintained.

Investigation of the N-TIMP-4–MT1-MMP complex formation indicated that N-TIMP-4 reacts with MT1-MMP at a 20-fold slower rate than do N-TIMP-2, probably due to the lack of a residue comparable to Tyr36 in the AB β-turn (Williamson et al. 2001). However, N-TIMP-4 has been shown to have a similar K_i^{app} value for MT1-MMP binding as N-TIMP-2, suggesting that the lack of the binding contribution from Tyr36 is compensated for by other interactions elsewhere. In the final complex, the charged and polar residues Asp34 and Asn38 were considered as potentially important residues for TIMP-2–MT1-MMP association, because they occupy the same positions as Asp34 and Asp37 in TIMP-4. Mutation of both residues to Ala did markedly reduce the rate of initial binding to MT1-MMP (K_{on} value was decreased by 180-fold) but had little effect on the final apparent K_i^{app} value (Rapti et al. 2006). Interestingly, the modification of Asn38 to Asp in

TIMP-2 (making it comparable to TIMP-4) resulted in a 50-fold decrease in the value of K_i^{app} for binding to MT1-MMP. The side chain of Asn38 is involved in making a hydrogen bond to its own backbone (amide O to HN), which makes weaker electrostatic interactions across the hairpin via the side chain of Asp34 to the proteinase with the side chain of Asn208 and the backbone HN of Ile209 (Fernandez-Catalan et al. 1998). Substitution of Asn38 with the negatively charged Asp may cause some structural rearrangement at this site, allowing stronger interactions to form between the tip of the AB hairpin and the catalytic domain of MT1-MMP. It is interesting to speculate that Asp37 in N-TIMP-4 may help compensate for the lack of a residue equivalent to Tyr36, allowing the overall binding constant for MT1-MMP to be similar to that measured for N-TIMP-2. The precise biological significance of the unique interactions between the AB loop of TIMP-2 and MT1-MMP can only be speculated upon, but they may play an important role in the stabilization of the TIMP-2–MT1-MMP complex in which the C-terminal region of TIMP-2 is free to bind to the haemopexin-like domain of pro-MMP-2. This would represent the basis for the formation of a cell surface MT1-MMP-TIMP-2-pro-MMP-2 complex, leading to the activation of pro-MMP-2.

Formation of the Trimeric Pro-MMP-2-TIMP-2-MT1-MMP Complex

It is well documented that pro-MMP-2, TIMP-2 and MT1-MMP form a complex and this trimeric complex formation is crucial to activate pro-MMP-2 on the cell surface (Butler et al. 1998). However, TIMP-4 is unable to promote pro-MMP-2 activation by MT1-MMP, even though it can form a trimolecular complex of pro-MMP-2-TIMP-4-MT1-MMP (Bigg et al. 1997, Hernandez-Barrantes et al. 2001, Worley et al. 2003, Rapti et al. 2006). Mutagenesis studies were conducted to investigate the structural basis to explain why TIMP-4 cannot participate in pro-MMP-2 activation by MT1-MMP. The unique features of the TIMP-2 structure relative to TIMP-4 were used to solve this problem. TIMP-4 is known to bind to pro-MMP-2 via the C-terminal domain of the enzyme with the binding site being complementary to that in TIMP-2. (Bigg et al. 1997, 2001; Worley et al. 2003). However, from the crystal structure data of the TIMP-2–pro-MMP-2 complex, it can be deduced that the C-terminal interactions in a TIMP-4–pro-MMP-2 complex are likely to be considerably weaker, for example, only three salt bridges could be formed between the C-terminal tail of TIMP-4 and the haemopexin domain of pro-MMP-2. In fact, the K_{on} value is somewhat smaller than that of TIMP-2 and the K_i^{app} value slightly larger (Rapti et al. 2006). A TIMP-2 mutant with the replaced C-terminal domain of TIMP-4 showed loss of pro-MMP-2 activation, indicating that the C-terminal domain of TIMP-2 is important in establishing the trimolecular complex of MT1-MMP, TIMP-2 and pro-MMP-2. This was confirmed by analysis of a TIMP-4 mutant containing the replaced C-terminal domain of TIMP-2, which formed a trimolecular complex and promoted pro-MMP-2 processing by

MT1-MMP. Mutants encoding TIMP-4 from Cys1 to Leu185 and a partial tail sequence of TIMP-2 showed some gain of activating capability relative to TIMP-4. The identified residues were subsequently mutated in TIMP-2 (Glu192Asp193 to Ileu192Gln193), and this inhibitor showed a significantly reduced ability to facilitate pro-MMP-2 processing by MT1-MMP. Furthermore, the tail-deletion mutant $\Delta_{186-194}$TIMP-2 was completely incapable of promoting pro-MMP-2 activation by MT1-MMP. Thus, the C-terminal tail residues of TIMP-2 are important determinants for the formation of a stable trimolecular complex of TIMP-2, pro-MMP-2 and MT1-MMP and plays an important role in MT1-MMP-mediated processing of the intermediate and final active forms of MMP-2 at the cell surface (Worley et al. 2003). Although the differences in AB loop between the two inhibitors were also considered, the transfer of the TIMP-2 AB loop to TIMP-4 alone did not endow the inhibitor with pro-MMP-2-activating activity; transfer of both the AB loop and C-terminal domain of TIMP-2 to TIMP-4 generates a mutant that can activate pro-MMP-2 and so confirmed that the C-terminal tail of TIMP-2 is important for the activation process (Rapti et al. 2006).

The Regulation of ADAM and ADAMTS Activity by TIMPs

The potential role of TIMPs in the regulation of ectodomain-shedding events mediated by MPs has been a source of much interest to cell biologists. Of the human ADAMs, 13 are predicted to be active MPs based on the presence of the HEXXHXXGXXH zinc-binding motif. Of these, ADAM-17, or TACE, is the most thoroughly characterized member. In addition to processing membrane-bound TNFα precursor to its soluble form, ADAM-17 also cleaves other membrane proteins (Black 2002). Some of the ADAMs predicted to be active MPs have also been shown to participate in similar proteolytic activities as ADAM-17 in cell-based systems, so the knowledge of TIMP and other inhibitor profiles is of significance. The catalytic activities of purified recombinant ADAMs have been studied using α2-macroglobulin and myelin basic protein (MBP), as well as various peptides (Roghani et al. 1999, Amour et al. 2000). These assays have allowed their susceptibility to TIMPs, potential physiological regulators of ADAM proteolytic activity *in vivo*, to be evaluated. Of the four TIMPs, only TIMP-3 was found to inhibit ADAM-17 and ADAM-12 (Amour et al. 1998, Loechel et al. 2000), whilst both TIMP-1 and TIMP-3 could inhibit ADAM-10 (Amour et al. 2000) (Table 37.2). Furthermore, TIMP-3 also inhibited the aggrecanases ADAMTS-4 and ADAMTS-5 as well as ADAMTS-1, which are members of the related family of disintegrin MPs with thrombospondin domains (Kashiwagi et al. 2001, Rodriguez-Manzaneque et al. 2002). In contrast, ADAM-8 or ADAM-9 were not inhibited by any of the TIMPs (Amour et al. 2002). In the few cases where TIMP-2 has been found to be an inhibitor of a proteolytic shedding event, it seems likely that an MMP is involved (Schlöndorff et al. 2001). In cell-based studies of ectodomain shedding, TIMP-3 has frequently been shown to be an effective inhibitor of the processing of

TNFα, L-selectin, IL6 receptor, CD30 and the p55 TNF receptor 1. In many cases, this may be due to the inhibition of ADAM-17, but substantial further study of this is necessary. Since TIMP-3 is primarily associated with the extracellular matrix, it is effectively localized to the pericellular environment of cells and may represent a significant physiological regulator of membrane MPs, including those involved in ectodomain shedding. This has not been definitively established, however. Studies of the TIMP-3$^{-/-}$ mouse have indicated the importance of TIMP-3 in the regulation of extracellular matrix turnover in the lung and the involuting mammary gland (Leco et al. 2001, Fata et al. 2001). An abnormal inflammatory response with an increase in TNF α after partial hepatectomy of TIMP-3 null mice indicated that ADAM-17 is one of the key enzymes regulated by TIMP-3 *in vivo* (Mohammed et al. 2004).

To investigate the structural basis of TIMP-3 to inhibit aggrecanase, a series of chimeric N-TIMPs were generated by introducing different portions of TIMP-3 into TIMP-1 from the N-terminus (M Kashiwagi and H Nagase, unpublished results). The chimera consisting of N-TIMP-3 (1–68) and N-TIMP-1 (71–124) contains all the residues present in the reactive-site ridge of N-TIMP-3, but it did not inhibit ADAMTS-4. Nevertheless, this protein inhibits MMP-1, MMP-2 and MMP-3 with inhibition constants similar to those of wild-type N-TIMP-3 and it is, therefore, essentially a TIMP-1 rather than a TIMP-3. The replacement of the stretch of the sequence EASESL (62–67) in TIMP-3 by PAMESV (64–69) derived from TIMP-1 did not alter the activity of N-TIMP-3 significantly as an inhibitor of ADAMTS-4. These results suggest that the MMP-reactive ridge alone of TIMP-3 is not sufficient to inhibit ADAMTS-4, but that additional elements are required.

Structural Features of TIMP-3 that Modulate Its Affinity for ADAM-17

Although the overall three-dimensional structures of ADAM and MMP catalytic domains show some similarity, their levels of sequence identity are low and the crystallographic structure of the ADAM-17 catalytic domain indicates that ADAMs have some unique structural features including an additional α-helix and a multiple-turn loop, but lack the structural zinc and calcium ions shared by the MMPs (Maskos et al. 1998). ADAM-17 differs from the MMPs in having a deep S3' pocket merging with the hydrophobic S1' specificity pocket. In the absence of a structure of a TIMP-3–ADAM-17 complex, Lee et al. (2002a) modelled the structure of TIMP-3 using the known structures of TIMP-1 and TIMP-2 and were able to dock this with the catalytic domain of ADAM-17 in a manner similar to that in the two known TIMP–MMP complexes. This suggests that the mechanism of TIMP-3 inhibition of ADAM-17 could be similar to that for MMPs and it has been shown that full-length TIMP-3 and a $\Delta_{122-188}$ TIMP-3 (N-TIMP-3) had similar K_{on} and K_i^{app} values for interaction with a soluble form of the ADAM-17 catalytic domain ($\Delta_{474-824}$ ADAM-17, ADAM17–473) or the whole ectodomain ($\Delta_{652-824}$

ADAM-17, ADAM-17–651; Lee et al. 2002b, 2003b), although the whole ectodomain showed weaker interaction than the catalytic domain. This is in comparison with its interaction with MMP-2, where the loss of its C-terminal three loops markedly abrogates binding (Butler et al. 1999). TIMP-3 binding to the active site of ADAM-17 can be enhanced by modifying Ser4 to Met, Tyr, Lys or Arg, leading to a greater than threefold drop in K_i^{app} value (Table 37.5; Lee et al. 2002a). The affinity to MMP-2 was concomitantly decreased about tenfold; hence an element of selectivity can be introduced into TIMP-3 by the modification of a single residue. This suggests that further mutagenesis studies should allow further specificity to be engineered. Brew and co-workers (Wei et al. 2005) made mutations in N-TIMP-3 designed to disrupt inhibitory activity towards MMPs based on the known structures of the TIMP-1–MMP-3 complex and the TIMP-2-MT1-MMP complex and previous mutational studies with TIMPs (Wingfield et al. 1999). The specific mutations are (i) The addition of an N-terminal alanine extension (-1Ala) to perturb the interaction of Cys1 with the active site Zn^{2+}; this mutation in N-TIMP-1 and TIMP-2 drastically curtailed the inhibitory activity for MMPs; (ii) A Thr2 to Gly mutation which removes the side chain of residue 2; this residue interacts with the S1' specificity pocket of MMPs and this mutation in N-TIMP-1 reduces the affinity for MMP-1, MMP-2, and MMP-3 about 1,000-fold (Meng et al. 1999, Table 37.5). The study identified significant differences between the inhibition of the soluble form of the complete ectodomain of ADAM-17 and MMPs by TIMP-3. Thr2Gly and -1Ala mutants of N-TIMP-3 were potent inhibitors of ADAM-17, but extremely weak inhibitors of the four representative MMPs (MMP-1, MMP-2 and MMP-3 and MT1-MMP) and are also likely to be weak inhibitors of other MMPs. The presence of any extension N-terminal to the α-amino group in TIMPs has been shown to drastically reduce inhibitory activity for MMPs (Wingfield et al. 1999, Wei et al. 2005), presumably because such extensions prevent the interaction of

Table 37.5 TIMP Mutants that Inhibit ADAM-17 (K_i^{app} nM)

TIMP	ADAM-17	MMP-2	References
N-TIMP-1	356	0.54	Lee et al. 2004a
N-TIMP-1 (V4S/TIMP3-AB loop/V69L/T98L)	0.14		Lee et al. 2004a
N-TIMP-3	0.22	0.25	Lee et al. 2003
N-TIMP-3 (S4M)	<0.06	3.2	Lee et al. 2002b
N-TIMP-3 (S4Y)	0.06	3.0	Lee et al. 2002b
N-TIMP-3 (S4K)	<0.06	2.8	Lee et al. 2002b
N-TIMP-3 (S4R)	<0.06	4.9	Lee et al. 2002b
N-TIMP-2	893	0.04	Lee et al. 2004b
N-TIMP-2 (S2T/TIMP-3 AB loop/A70S/V71L)	1.49		Lee et al. 2004b
N-TIMP-3	13	4.3	Wei et al. 2005
N-TIMP-3 (T2G)	36	4,000	Wei et al. 2005
(-1A)N-TIMP-3	34	614	Wei et al. 2005
TIMP-4	8.7	0.014	Lee et al. 2005
TIMP-4 (S2T/T38P/E39F/K40G)	0.73		Lee et al. 2005

MMP matrix metalloproteinase, *TIMP* tissue inhibitor of metalloproteinase, *ADAM* a disintegrin and metalloproteinase.

Cys1 with the catalytic Zn^{2+}. The fact that the -1Ala mutant of N-TIMP-3 is an effective inhibitor of ADAM-17, but not MMPs, suggests that the interaction of the inhibitor with the active site Zn^{2+} may be relatively unimportant for the strength of binding to ADAM-17. As compared with most MMPs, the S1' pocket of ADAM-17 is deep and very hydrophobic. However, substitution of Thr2 of N-TIMP-3 by residues with larger hydrophobic side chains that should fit better into the S1' site of ADAM-17 failed to improve the binding of the inhibitor to this enzyme (Lee et al. 2002a). Mutation of this residue into glycine, which lacks a side chain for potential interaction with the S1' pocket of the proteinase, results in a major reduction in the affinity for MMPs but has little effect on the inhibition of ADAM-17 (Wei et al. 2005). This suggests that this site of interaction also contributes little to the free energy of binding. Using stopped-flow X-ray spectroscopy methods together with transient kinetic analyses, Solomon et al. (2007) demonstrated that the catalytic zinc ion of the catalytic domain of ADAM-17 undergoes dynamic charge transitions before substrate binding to the metal ion. This indicates that distal protein sites influence the enzyme catalytic cleft. The observed charge transitions are synchronized with distinct phases in the reaction kinetics and changes in metal coordination chemistry mediated by the binding of the peptide substrate to the catalytic metal ion and product release.

It has been noted that there is a significant difference in susceptibility to TIMP-3 inhibition between the truncated catalytic domain of ADAM-17 and the full-length ectodomain (Lee et al. 2002a). The disintegrin, cysteine-rich and the crambin-like non-catalytic domains have been shown to influence substrate specificity in ADAM-17 and other ADAMs. Wei et al. (2005) found that the inhibition of the whole ectodomain of ADAM-17 by wild-type N-TIMP-3 and two mutants displays positive cooperativity with Hill coefficients of 1.9–3.5. Positive cooperativity arises from the presence of multiple interacting binding sites and alternative conformational states, but its structural basis in ADAM-17 is currently unknown. However, positive cooperativity has been previously described for the hydrolysis of a synthetic peptide substrate by a similar form of ADAM-17 (Jin et al. 2002). Cooperativity was only observed with a peptide substrate derivatized at the N-and C-termini, whereas unblocked peptides exhibited normal hyperbolic saturation curves (Jin et al. 2002). This apparent allosteric behaviour could have important implications for the regulation of ADAM-17 activity. The extra-catalytic domains of ADAM-17 weaken the binding of TIMP-3 to the catalytic domain. Lee et al. (2003) engineered a new generation of N-TIMP-3 mutants that displayed markedly improved binding affinity for the whole ectodomain of ADAM-17 by the formation of stronger electrostatic bonds with the catalytic domain of the enzyme. With K_i^{app} values of <0.1 nM, these mutants were dramatically better than the wild-type N-TIMP-3 [K_i^{app} 1.7 nM]. It was proposed that Glu31, an acidic residue situated at the base of the AB loop of N-TIMP-3, is drawn into contact with Lys315, a prominent basic residue adjacent to the ADAM-17 catalytic site. The mutagenesis strategy involved reorientation of the edge of N-TIMP-3, in particular, the β-strand A where Glu31 was located. These results suggest that the non-catalytic domains modulate the properties of the catalytic domain and emphasize the importance of considering the inhibitory properties of the longer enzyme forms in developing specific inhibitors for possible use *in vivo*.

Conversion of TIMPs into ADAM-17 Inhibitors

A series of mutagenesis studies by Lee et al. (2002a, 2004, 2005) have elucidated the critical residues of TIMP-3 involved in ADAM-17 inhibition. They were carried out using gain of function studies using TIMP-1, TIMP-2 and TIMP-4 which have little or weaker inhibitory activity for ADAM-17.

TIMP-1 Mutants

Replacement of Thr98 residue with Leu in TIMP-1 transformed it into a versatile inhibitor against an array of MPs otherwise insensitive to wild-type TIMP-1; examples include ADAM 17, MMP 19 and MT5-MMP (Lee 2004a). Using Thr98Leu as the scaffold, the N-TIMP-1 mutant (V4S/TIMP-3-AB loop/V69L/T98L) had K_i^{app} values of 0.14 nM for ADAM-17 catalytic domain, which is equivalent to that of the wild-type N-TIMP-3 (K_i^{app} 0.22 nM) (Table 37.5, Fig. 37.2). The K_i^{app} values of this mutant for MMP-19 and MT5-MMP were 1.2 nM (43-fold improvement) and 15.3 nM (fourfold improvement), respectively. The requirement for leucine at position 98 is absolute for the transformation in inhibitory pattern (Fig. 37.2). On the contrary, the mutation has minimal impact on the MMPs already well inhibited by wild-type TIMP-1, such as gelatinase A and stromelysin 1.

TIMP-2 Mutants

By systematic replacement of the surface epitopes of TIMP-2 based on those of TIMP-3 and a TIMP-1 variant V4S/TIMP-3 AB loop/V69L/T98L, Lee et al. (2004b) also created TIMP-2 mutants that exhibit inhibitory potency almost equal to that of the TIMP-3. The best N-TIMP-2 mutant reported was S2T/TIMP-3 AB loop. A70S/V71L which gave a K_i^{app} of 1.49 nM for ADAM-17, a marked improvement in comparison to that of the wild-type N-TIMP-2 (K_i 893 nM) (Table 37.5). The inhibitory pattern of the mutant was typical of that of a slow, tight-binding inhibitor. They also found that Phe34 within the AB loop, a residue unique to TIMP-3, was shown to be a vital element in ADAM-17 association. A series of mutagenesis studies carried out on Leu100 (equivalent to Thr98 in TIMP-1 and Leu94 in TIMP-3) (see Fig. 37.2) also indicated that the previous finding of a leucine on the EF loop is critical for ADAM-17 recognition. Replacement of the residue by other amino acids resulted in a dramatic decrease in binding affinity, although isoleucine and methionine are still capable of producing the slow, tight-binding effect (Lee 2004b).

TIMP-4 Mutants

Although full-length TIMP-4 displayed negligible activity against ADAM-17, N-TIMP-4 is a slow, tight-binding inhibitor with low nanomolar-binding affinity, suggesting that the C-terminal domains of the TIMPs have a significant negative effect on their activities with the ADAMs. This contrasts with TIMP-3 of which both N-TIMP-3 and the full-length inhibitor equally inhibit ADAM-17. To elucidate further the molecular basis that underpins TIMP–ADAM-17 interactions, Lee et al. (2005) subjected N-TIMP-4 to mutagenesis studies. Transplantation of only three residues, Pro-Phe-Gly, onto the AB loop of N-TIMP-4 resulted in a tenfold enhancement in binding affinity and the K_i values of the resultant mutant were almost comparable with that of TIMP-3. When those residues were included as part of AB loop transplantation into TIMP-1 and TIMP-2, those two mutants exhibited low nanomolar inhibition constants (Lee et al. 2004a, b). Thus, these residues play important role in ADAM-17 inhibition. N-TIMP-4 harbours Leu101 at the equivalent position of Leu94 in TIMP-3 and Thr98 in TIMP-1. Leu at this position was concluded to be crucial for ADAM-17 inhibition, but Pro, Trp and Asp at this position are detrimental (Lee 2004b, 2005). Other residues important to inhibit ADAM-17 are Leu67 in the CD loop that interacts with the S2 subsite of the enzyme and Thr2 at position 2. TIMP-3 has Leu67, but TIMP-1 and TIMP-2 have Val69 and Val71 at the former site. While those are important findings, it is yet to be investigated why full-length TIMP-4 is a weak inhibitor of ADAM-17 (K_i^{app} 180 nM).

Conclusions and Future Prospects

The TIMPs are important regulators of ECM catabolism and cell–cell interactions through their abilities to control the activities of MMPs and some ADAM and ADAMTS MPs. A series of mutational studies have provided us with TIMP variants that have selective inhibitory activity for particular MMPs. Although inhibitors that are specific for a single MMP have not yet been designed, the results obtained so far encourage future studies in this direction. The studies have also highlighted the unique nature of the reactive-site ridge in TIMPs. The requirement for a hydrophobic side chain in the P1' position of peptide substrates of MMPs is well established, but this requirement does not correlate with that of the inhibitory TIMP molecule. P1' side chains that produce a poor substrate are often found to be excellent for inhibitory activity when substituted at the corresponding site (residue 2) in TIMP-1, and in this location significantly influence the inhibitory specificity. This feature of residue 2 and the influence of other residues in the reactive ridge on the affinity of TIMP for different MPs are worthy of investigation as a route to generating more selective inhibitors.

TIMPs were originally found as MMP inhibitors and they have been used as reagents to discriminate MMPs from MPs of other families such as thermolysin, neprilysin and astacins. However, recent studies have shown that TIMP-3 can inhibit a number of ADAM and ADAMTS MPs. This finding is of considerable interest to many researchers as these MPs play important roles in inflammatory processes, shedding of cell surface molecules and degradation and processing of ECM molecules. Mutagenesis studies has provided the information about the residues involved in ADAM-17 inhibition, but those in TIMP-3 involved in ADAMTS-4 and ADAMTS-5 (aggrecanases) inhibition are not clear. While the search for such features is underway, the structural basis for this inhibitory specificity requires the determination of the structure of TIMP-3.

Because MMPs have been implicated in the progression of many diseases associated with aberrant ECM turnover, numerous synthetic MMP inhibitors have been designed and some were clinically tested, but with little success. The reasons for the failure of low molecular weight inhibitors are not clear, but it could result from the inhibition of non-targeted MPs or the fact that the disease had progressed to a point where it could no longer be reversed by the inhibition of MMPs. It is also possible that the inhibitor concentration in the target tissue did not reach an effective level. Animal model studies with selective TIMP variants that exploit tissue-specific gene transfer technology may be useful to investigate which MMPs, ADAMs and ADAMTSs are involved in disease progression and to further develop therapeutic interventions for diseases linked with enhanced ECM degradation. To this end, and to design highly selective inhibitors, we need detailed structural information about the modes of interactions between various target MPs and TIMP variants.

Acknowledgements We thank our colleagues who carried out all the studies described: Keith Brew, Meng Huee Lee, Shuo Wei, Masahide Kashiwagi, Magdalini Rapti and others, as cited. We also express our special gratitude to Ngee Han Lim for assembling Tables 37.1 and 37.2 and Rob Visse for preparation of Figs. 37.1–37.4. The work was supported by grants from the European Union, MRC UK, BBSRC UK, Cancer Research UK, Arthritis Research Campaign and NIH.

References

Amour A, Slocombe P M, Webster A et al. (1998) TNF-alpha converting enzyme (TACE) is inhibited by TIMP-3. FEBS Lett 435: 39–44.

Amour A, Knight C G, Webster A et al. (2000) The *in vitro* activity of ADAM-10 is inhibited by TIMP-1 and TIMP-3. FEBS Lett 473: 275–279.

Amour A, Knight C G, English W R et al. (2002) The enzymatic activity of ADAM8 and ADAM9 is not regulated by TIMPs. FEBS Lett 524: 154–158.

Atkinson S J, Crabbe T, Cowell S, Ward R V, Butler M J, Sato H, Seiki M, Reynolds J J, Murphy G. (1995) Intermolecular autolytic cleavage can contribute to the activation of progelatinase A by cell membranes. J Biol Chem 270: 30479–30485.

Baker A H, Edwards D R and Murphy G. (2002) Metalloproteinase inhibitors: biological actions and therapeutic opportunities. J Cell Sci 115: 3719–3727.

Bigg H F, Shi Y E, Liu Y L E et al. (1997) Specific, high affinity binding of tissue inhibitor of metalloproteinases-4 (TIMP-4) to the COOH-terminal hemopexin-like domain of human gelatinase A. TIMP-4 binds progelatinase A and the COOH-terminal domain in a similar manner to TIMP-2. J Biol Chem 272: 15496–15500.

Bigg H F, Morrison C J, Butler G S et al. (2001) Tissue inhibitor of metalloproteinases-4 inhibits but does not support the activation of gelatinase A via efficient inhibition of membrane type 1-matrix metalloproteinase. Cancer Res 61: 3610–3618.

Black R A. (2002) Tumor necrosis factor-alpha converting enzyme. Int J Biochem Cell Biol 34: 1–5.

Brew K, Dinakarpandian D and Nagase H. (2000) Tissue inhibitors of metalloproteinases: evolution, structure and function. Biochim Biophys Acta 1477: 267–283.

Butler G S, Butler M J, Atkinson S J et al. (1998) The TIMP2 membrane type 1 metalloproteinase "receptor" regulates the concentration and efficient activation of progelatinase A. A kinetic study. J Biol Chem 273: 871–880.

Butler G S, Apte S S, Willenbrock F et al. (1999) Human tissue inhibitor of metalloproteinases 3 interacts with both the N- and C-terminal domains of gelatinases A and B. Regulation by polyanions. J Biol Chem 274: 10846–10851.

Fata J E, Leco K J, Voura E B et al. (2001) Accelerated apoptosis in the Timp-3-deficient mammary gland. J Clin Invest 108: 831–841.

Fernandez-Catalan C, Bode W, Huber R, et al. (1998) Crystal structure of the complex formed by the membrane type 1-matrix metalloproteinase with the tissue inhibitor of metalloproteinases-2, the soluble progelatinase A receptor. EMBO J 17: 5238–5248.

Goldberg G I, Strongin A, Collier I E et al. (1992) Interaction of 92-kDa type IV collagenase with the tissue inhibitor of metalloproteinases prevents dimerization, complex formation with interstitial collagenase, and activation of the proenzyme with stromelysin. J Biol Chem 267: 4583–4591.

Gomis-Rüth F-X, Maskos K, Betz M et al. (1997) Mechanism of inhibition of the human matrix metalloproteinase stromelysin-1 by TIMP-1. Nature 389: 77–79.

Hamze A B, Wei S, Bahudhanapati H et al. (2007) Constraining specificity in the N-domain of tissue inhibitor of metalloproteinases-1; gelatinase-selective inhibitors. Protein Sci 16:1905–1913.

Hashimoto G, Aoki T, Nakamura H et al. (2001) Inhibition of ADAMTS4 (aggrecanase-1) by tissue inhibitors of metalloproteinases (TIMP-1, -2, -3 and -4). FEBS Lett 494: 192–195.

Hernandez-Barrantes S, Shimura Y, Soloway P D et al. (2001) Differential roles of TIMP-4 and TIMP-2 in pro-MMP-2 activation by MT1-MMP. Biochem Biophys Res Commun 281(1): 126–130.

Huang W, Meng Q, Suzuki K et al. (1997) Mutational study of the amino-terminal domain of human tissue inhibitor of metalloproteinases 1 (TIMP-1) locates an inhibitory region for matrix metalloproteinases. J Biol Chem 272: 22086–22091.

Hutton M, Willenbrock F, Brocklehurst K et al. (1998) Kinetic analysis of the mechanism of interaction of full-length TIMP-2 and gelatinase A: evidence for the existence of a low-affinity intermediate. Biochemistry 37: 10094–10098.

Jin G, Huang X, Black R et al. (2002) A continuous fluorimetric assay for tumor necrosis factor-alpha converting enzyme. Anal Biochem 302: 269–275.

Kashiwagi M, Tortorella M, Nagase H et al. (2001) TIMP-3 is a potent inhibitor of aggrecanase 1 (ADAM-TS4) and aggrecanase 2 (ADAM-TS5). J Biol Chem 276: 12501–12504.

Leco K J, Waterhouse P, Sanchez O H et al. (2001) Spontaneous air space enlargement in the lungs of mice lacking tissue inhibitor of metalloproteinases-3 (TIMP-3). J Clin Invest 108(6): 817–829.

Loechel F, Fox J W, Murphy G et al. (2000) ADAM 12-S cleaves IGFBP-3 and IGFBP-5 and is inhibited by TIMP-3. Biochem Biophys Res Commun 278: 511–515.

Lee M H, Verma V, Maskos K et al. (2002a) The C-terminal domains of TACE weaken the inhibitory action of N-TIMP-3. FEBS Lett 520(1–3): 102–106.

Lee M H, Verma V, Maskos K et al. (2002b) Engineering N-terminal domain of tissue inhibitor of metalloproteinase (TIMP)-3 to be a better inhibitor against tumour necrosis factor-alpha-converting enzyme. Biochem J 364: 227–234.

Lee M H, Rapti M and Murphy G. (2003a) Unveiling the surface epitopes that render tissue inhibitor of metalloproteinase-1 inactive against membrane type 1-matrix metalloproteinase. J Biol Chem 278: 40224–40230.

Lee M H, Dodds P, Verma V et al. (2003b) Tailoring tissue inhibitor of metalloproteinases-3 to overcome the weakening effects of the cysteine-rich domains of tumour necrosis factor-alpha converting enzyme. Biochem J 371: 369–376.

Lee M H, Rapti M, Knäuper V., Murphy G. (2004a) Threonine 98, the pivotal residue of tissue inhibitor of metalloproteinases (TIMP)-1 in metalloproteinase recognition. J Biol Chem 279, 17562-17569.

Lee M H, Rapti M, Murphy G. (2004b) Delineating the molecular basis of the inactivity of tissue inhibitor of metalloproteinase-2 against tumor necrosis factor-alpha-converting enzyme. J Biol Chem 279: 45121–45129.

Lee M H, Rapti M, Murphy G. (2005) Total conversion of tissue inhibitor of metalloproteinase (TIMP) for specific metalloproteinase targeting: fine-tuning TIMP-4 for optimal inhibition of tumor necrosis factor-{alpha}-converting enzyme. J Biol Chem 280: 15967–15975.

Maskos K, Fernandez-Catalan C, Huber R et al. (1998) Crystal structure of the catalytic domain of human tumor necrosis factor-alpha-converting enzyme. Proc Natl Acad Sci U S A 95: 3408–3412.

Meng Q, Malinovskii V, Huang W et al. (1999) Residue 2 of TIMP-1 is a major determinant of affinity and specificity for matrix metalloproteinases but effects of substitutions do not correlate with those of the corresponding P1' residue of substrate. J Biol Chem 274: 10184–10189.

Meng Q, Malinovskii V, Huang W, Hu Y J, Chung L, Nagase H, Bode W, Maskos K, Brew K, (1999) Residue 2 of TIMP-1 is a major determinant of afiinity and specificity for matrix metalloproteinases but effects of substitutions do not correlate with those of the corresponding P1' residue of substrate. J Biol Chem 274: 10184–10189.

Mohammed F F, Smookler D S, Taylor S E et al. (2004) Abnormal TNF activity in Timp3-/- mice leads to chronic hepatic inflammation and failure of liver regeneration. Nat Genet 36: 969–977.

Morgunova E, Tuuttila A, Bergmann U et al. (2002) Structural insight into the complex formation of latent matrix metalloproteinase 2 with tissue inhibitor of metalloproteinase 2. Proc Natl Acad Sci U S A 99: 7414–7419.

Murphy G, Houbrechts A, Cockett M I et al. (1991) The N-terminal domain of tissue inhibitor of metalloproteinases retains metalloproteinase inhibitory activity. Biochemistry 30: 8097–8102.

Nagase H. (1997) Activation mechanisms of matrix metalloproteinases. Biol Chem 378(3–4): 151–160.

Nagase H and Fields G B. (1996) Human matrix metalloproteinase specificity studies using collagen sequence-based synthetic peptides. Biopolymers 40: 399–416.

Nagase H, Suzuki K, Cawston T E et al. (1997) Involvement of a region near valine-69 of tissue inhibitor of metalloproteinases (TIMP)-1 in the interaction with matrix metalloproteinase 3 (stromelysin 1). Biochem J 325: 163–167.

Nagase H, Visse R and Murphy G. (2006) Structure and function of matrix metalloproteinases and TIMPs. Cardiovasc Res 69: 562–573.

O'Connell J P, Willenbrock F, Docherty A J et al. (1994) Analysis of the role of the COOH-terminal domain in the activation, proteolytic activity, and tissue inhibitor of metalloproteinase interactions of gelatinase B. J Biol Chem 269: 14967–14973.

Rapti M, Knäuper V, Murphy G et al. (2006) Characterization of the AB loop region of TIMP-2. Involvement in pro-MMP-2 activation. J Biol Chem 28: 23386–23394.

Rodriguez-Manzaneque J C, Westling J, Thai S N et al. (2002) ADAMTS1 cleaves aggrecan at multiple sites and is differentially inhibited by metalloproteinase inhibitors. Biochem Biophys Res Commun 293: 501–508.

Roghani M, Becherer J D, Moss M L, Atherton R E, Erdjument-Bromage H, Arribas J, Blackburn R K, Weskamp G, Tempst P, Blobel C P. (1999) Metalloprotease-disintegrin MDC9: intracellular maturation and catalytic activity. J Biol Chem 274: 3531–3540.

Schlöndorff J, Lum L and Blobel C P. (2001) Biochemical and pharmacological criteria define two shedding activities for TRANCE/OPGL that are distinct from the tumor necrosis factor alpha convertase. J Biol Chem 276: 14665–14674.

Solomon A, Akabayov B, Frenkel A et al. (2007) Key feature of the catalytic cycle of TNF-alpha converting enzyme involves communication between distal protein sites and the enzyme catalytic core. Proc Natl Acad Sci U S A 104(12): 4931–4936.

Troeberg L, Tanaka M, Wait R et al. (2002) E. coli expression of TIMP-4 and comparative kinetic studies with TIMP-1 and TIMP-2: insights into the interactions of TIMPs and matrix metallo-proteinase 2 (gelatinase A). Biochemistry 41: 15025–15035.

Wei S, Chen Y, Chung L et al. (2003) Protein engineering of the tissue inhibitor of metallopro-teinase 1 (TIMP-1) inhibitory domain. In search of selective matrix metalloproteinase inhibi-tors. J Biol Chem 278: 9831–9834.

Wei S, Kashiwagi M, Kota S et al. (2005) Reactive site mutations in tissue inhibitor of metallo-proteinase-3 disrupt inhibition of matrix metalloproteinases but not tumor necrosis factor-alpha-converting enzyme. J Biol Chem 280: 32877–32882.

Willenbrock F, Crabbe T, Slocombe P M et al. (1993) The activity of the tissue inhibitors of metalloproteinases is regulated by C-terminal domain interactions: a kinetic analysis of the inhibition of gelatinase A. Biochemistry 32: 4330–4337.

Williamson R A, Martorell G, Carr M D, Murphy G, Docherty A J P, Freedman R B, Feeney J. (1994) Solution structure of the active domain of tissue inhibitor of metalloproteinases-2. A new member of the OB fold protein family. Biochemistry 33: 11745–11759.

Williamson R A, Hutton M, Vogt G et al. (2001) Tyrosine 36 plays a critical role in the interaction of the AB loop of tissue inhibitor of metalloproteinases-2 with matrix metalloproteinase-14. J Biol Chem 276: 32966–32970.

Williamson R A, Hutton M, Vogt G et al. (2001) Tyrosine 36 plays a critical role in the interaction of the AB loop of tissue inhibitor of metalloproteinases-2 with matrix metalloproteinase-14. J Biol Chem 276: 32966–32970.

Wingfield P T, Sax J K, Stahl S J et al. (1999) Biophysical and functional characterization of full-length, recombinant human tissue inhibitor of metalloproteinases-2 (TIMP-2) produced in Escherichia coli. Comparison of wild type and amino-terminal alanine appended variant with implications for the mechanism of TIMP functions. J Biol Chem 274: 21362–21368.

Worley J R, Thompkins P B, Lee M H et al. (2003) Sequence motifs of tissue inhibitor of metalloproteinases 2 (TIMP-2) determining progelatinase A (proMMP-2) binding and activa-tion by membrane-type metalloproteinase 1 (MT1-MMP). Biochem J 372: 799–809.

Chapter 38
Third-Generation MMP Inhibitors: Recent Advances in the Development of Highly Selective Inhibitors

Athanasios Yiotakis and Vincent Dive

Abstract The association of matrix metalloproteinases (MMPs) with a variety of pathological states has stimulated impressive efforts over the past 20 years to develop synthetic compounds able to block potently and selectively the uncontrolled activity of these enzymes. Extremely potent inhibitors of MMPs have been developed, but in most cases these compounds act as broad-spectrum inhibitors of MMPs. Retrospective analysis suggests that the use of strong zinc-binding groups, like the hydroxamate function, to achieve potent MMP inhibition is responsible not only for the development of inhibitors displaying poor selectivity towards MMP members but also in their ability to potently block other unrelated zinc proteinases. The use of less avid zinc-binding group, like the phosphoryl group present in phosphinic peptide transition-state analogues, has led to a second generation of highly selective MMP inhibitors (MMP-12 selective inhibitors). The third generation of highly selective MMP inhibitors (MMP-13 selective inhibitors) possess no zinc-binding group and exploit the deep S_1' cavity present in some MMPs. Past research on the development of MMP inhibitors has probably underestimated the role of flexibility in the MMP active site and its impact in accommodating different inhibitor structures. Combined use of several biophysical techniques, like nuclear magnetic resonance, X-ray crystallography and isothermal titration experiments, should greatly improve our understanding of the specific structural and dynamic features that can be exploited to obtain series of inhibitors able to specifically block each MMP validated as a therapeutic target.

V. Dive
CEA, iBiTecS, Service d'Ingénierie Moleculaires des Protéines, Gif Sur Yvette, F-91191, France,
e-mail: vincent.dive@cea.fr

D. Edwards et al. (eds.), *The Cancer Degradome.*
© Springer Science + Business Media, LLC 2008

Introduction

The association of matrix metalloproteinases (MMPs) with a variety of pathological states has stimulated impressive efforts over the past 20 years to develop synthetic compounds (Babine and Bender 1997, Whittaker et al. 1999) able to block potently and selectively the uncontrolled activity of these enzymes (Fingleton 2007). MMPs form a group of 23 proteins in humans all of which contain a catalytic domain belonging to the zinc metalloproteinase family (Bode and Maskos 2003). Extremely potent inhibitors of MMPs have been developed, but in most cases these compounds act as broad-spectrum inhibitors of MMPs (Brown et al. 2004). Marked sequence similarity between the catalytic domain of MMPs, well-conserved enzyme active-site topology and high flexibility of a loop segment that plays a key role in MMP specificity are the main factors that may explain difficulties in identifying inhibitors able to fully differentiate one MMP from the others. Recently, extremely selective inhibitors of MMP-12 and -13 have been developed. In parallel, new experimental approaches are being explored that should help the definition of the specific structural and dynamic features of each MMP, so as to tailor an inhibitor with the desired selectivity. As several review articles covering the field of MMP inhibitors have been recently published (Cuniasse et al. 2005, Matter and Schudok 2004, Skiles et al. 2004), the aim of this chapter is to focus on recent advances in the development of highly selective MMP inhibitors, the third generation of MMP inhibitors (Overall and Kleifeld 2006).

MMP Active-Site Topology

MMPs are secreted as latent pro-enzymes, whose activation involves the loss of a pro-sequence of about 80 amino acids. In their active forms, MMPs have in common a catalytic domain (160–170 amino acids) which is usually connected through a hinge region of variable size (2–72 amino acids) to a carboxy-terminal haemopexin domain (250 amino acids). The catalytic domain in isolation is sufficient for peptide substrate hydrolysis (Turk et al. 2001) and is thought to have the same sequence specificity towards such small substrates as the full-form MMPs. For protein substrates, other domains of MMPs play a key role in substrate recognition and cleavage, as in collagenases (MMP-1, -8 and -13) in which the assistance of the haemopexin domain is critical for collagen cleavage (Clark and Cawston 1989, Visse and Nagase 2003). Efficient hydrolysis of peptide substrates by the catalytic domain of MMPs is observed only for peptides having a minimum size of six residues (Netzel-Arnett et al. 1993, Seltzer et al. 1990, Stack and Gray 1989), a property which seems to be related to the occurrence in the MMP active site of six subsites (from S_3 to S_3'), with the catalytic zinc atom occupying a central position, as shown in Fig. 38.1a. The presence of these subsites in MMPs can theoretically be exploited to design MMP inhibitors mimicking the binding of short

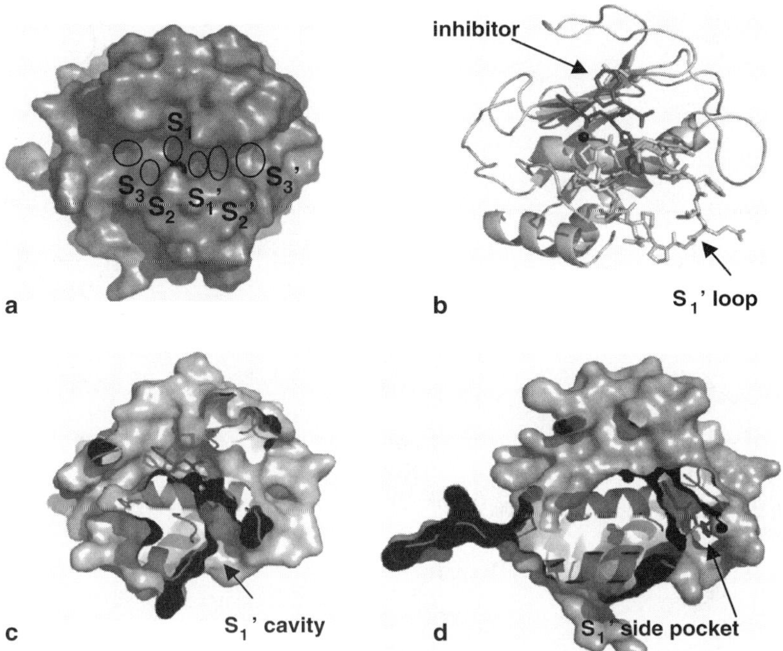

Fig. 38.1 a Molecular surface representation of the X-ray structure of the mini-MMP-9 catalytic domain (PDB code: 2OVZ, comprising Phe^{107} to Gly^{215} and Gln^{391} to Tyr^{443}). The N-terminal segment Phe^{110}-Asp^{113} has been omitted for the surface calculation. The locations of the six subsites in the active-site cleft are indicated by closed circles and the catalytic zinc atom appears as a black sphere. **b** Ribbon representation of mini-MMP-9 in complex with compound 5. Compound 5 (dark grey) and the S$_1$' loop are represented as a stick model. The catalytic zinc atom appears as a black sphere. **c** Molecular cut surface representation of the X-ray structure of the mini-MMP-9 catalytic domain in complex with compound 5. This figure shows that the P$_1$' residue of compound 5 only partially fills the deep S$_1$'cavity. **d** Molecular cut surface representation of the X-ray structure of MMP-13 catalytic domain in complex with compound 8 (PDB code: 1XUR). No closed contact is observed between the catalytic zinc atom (black sphere) and compound 8, represented as a stick model. The distal part of compound 8 points into the S$_1$' side pocket

peptide substrates. However, it turns out that extremely potent MMP inhibitors can be developed without the need to exploit six MMP subsites (Babine and Bender 1997, Whittaker et al. 1999). This has made it possible to keep the molecular weight of many MMP inhibitors below 300 kDa, a critical issue for the potential drugability of such inhibitors.

First-Generation Non-Selective Inhibitors

The presence of a zinc atom in the catalytic domain of MMPs and the particular topology of the MMP active site have greatly influenced the design of synthetic inhibitors. Thus, the development of the first MMP inhibitor generation has relied

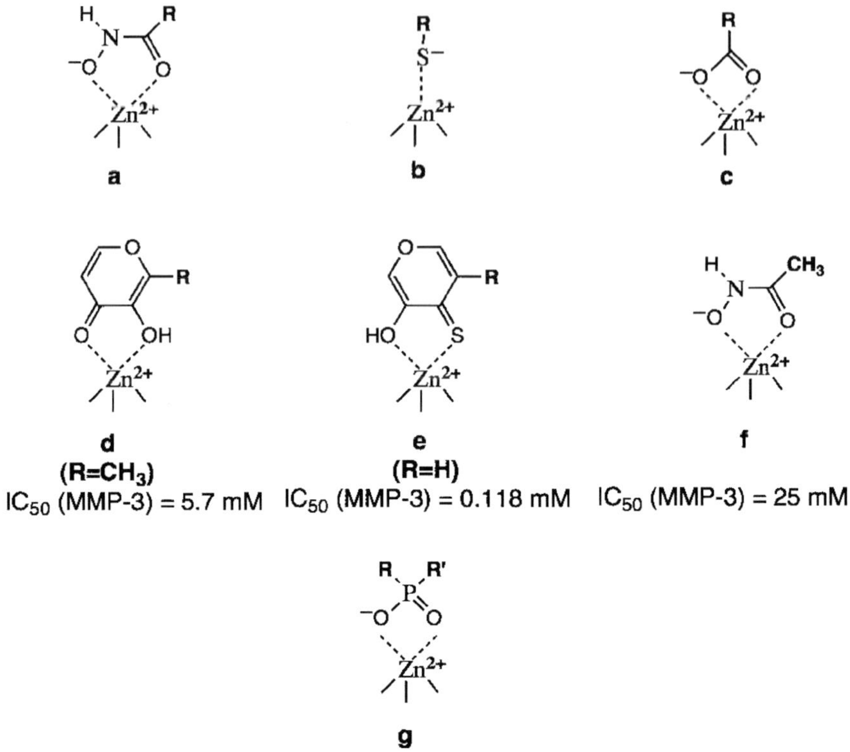

Scheme 38.1 Schematic representation of seven chemical groups (a: hydroxamate, b: thiolate, c: carboxylate, d: hydroxypyrone, e: hydroxythiopyrone, f: acetohydroxamate and g: phosphinate) in complex with zinc atom

on the use of a peptide sequence, recognised by MMPs, to which was grafted a chemical function able to interact potently with the zinc ion located in the MMP active site (Babine and Bender 1997, Whittaker et al. 1999). In this class of compounds, the vast majority of MMP inhibitors developed contain a hydroxamic acid moiety (compound a in Scheme 38.1) as zinc-binding group. This situation is explained by the fact that the use of the hydroxamate function results in the development of extremely potent MMP inhibitors (compound 1, Scheme 38.2). Another good reason for pharmaceutical companies to invest in this class of inhibitors was that no hydroxamate compound was in clinical use at that time and few patents were filled, so the situation was full of opportunities. The first major challenge that chemists had to face in developing hydroxamate inhibitors was the poor in vivo stability of these compounds. Rapid hydrolysis of the hydroxamate function can be observed in vivo, producing hydroxylamine and carboxylate derivatives as metabolic products (Peng et al. 1999). Another major hurdle was the generally poor selectivity of hydroxamate inhibitors. Thus, these inhibitors are unable to discriminate between the different MMPs, with the exception of MMP-1 and -7, which possess a particular S_1' specificity subsite. But more confounding was

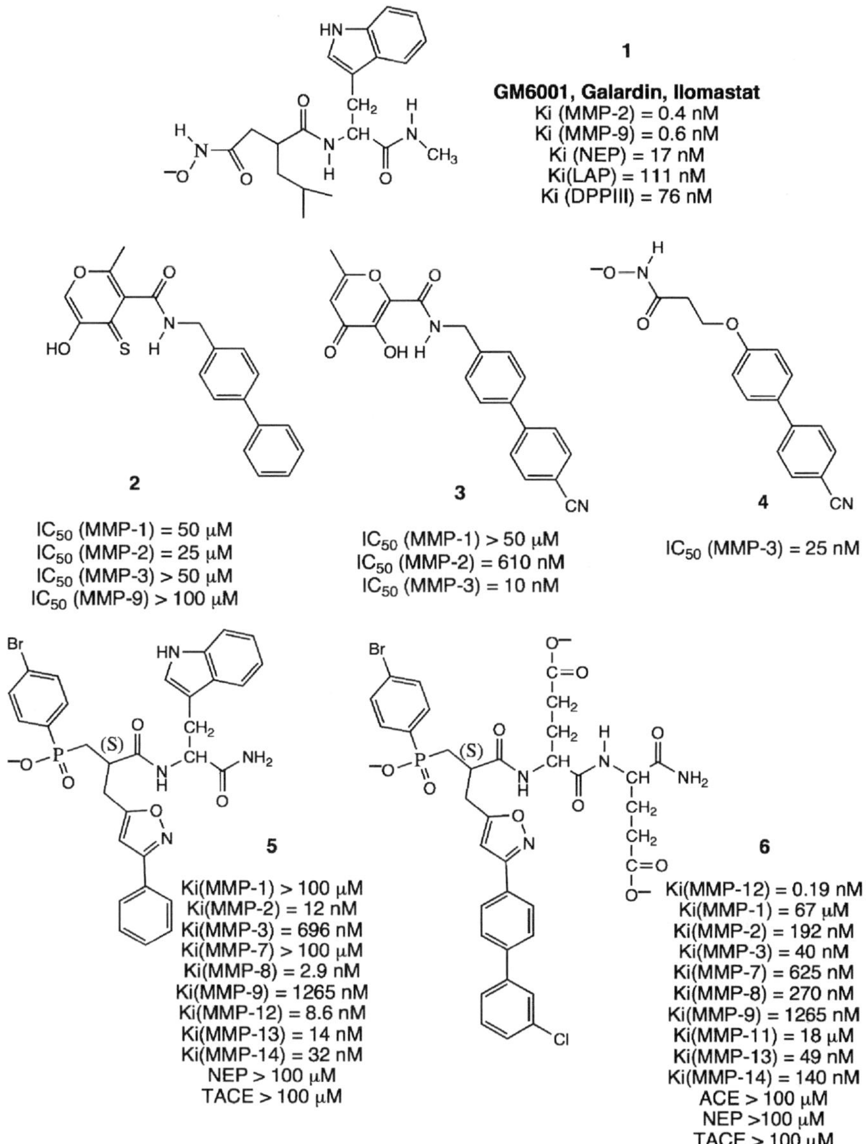

Scheme 38.2 Chemical structures of matrix metalloproteinase (*MMP*) inhibitors containing a zinc-binding group and their potency towards *MMPs*

the fact that these MMP hydroxamate inhibitors also turned out to be highly and sometimes more potent inhibitors of other zinc metalloproteinases, like TACE (*t*umour necrosis factor-*a*lpha-*c*onverting *e*nzyme) (Amour et al. 1998), a protease of the ADAMs family (membrane-bound zinc metalloproteinases, containing *a* *d*isintegrin and *m*etalloprotease domain). Given the similarity between the topology of the active site of TACE and MMPs, this result could be anticipated (Maskos et al. 1998). However, more unexpected was the fact that compound 1 (GM6001 or ilomastat) was proven to potently block zinc metallopeptidases, which display no sequence homology with MMPs and possess a very different active-site topology (Saghatelian et al. 2004). The use of inhibitors displaying broad-spectrum activities towards MMPs and able to block many other zinc proteases may in part explain the failure of clinical trials in patients with advanced-stage cancer (Coussens et al. 2002, Egeblad and Werb 2002). The difficulty in controlling the selectivity of hydroxamate compounds is probably related to the strong interaction of the hydroxamate function with the zinc atom. This suggests that the use of less avid zinc-binding groups could be a reasonable strategy in developing more selective MMP inhibitors.

New Zinc-Binding Groups

Several functional groups have been considered as replacement for the hydroxamate function in MMP inhibitors, like thiol and carboxylate groups (compounds b and c, Scheme 38.1). Recently, many substituted cyclic compounds (compounds d and e, Scheme 38.1) were shown to be more effective MMP-3 binders than the simple acetohydroxamic acid (compound f, Scheme 38.1) (Puerta et al. 2004). Thus, on MMP-3, compound e with an IC_{50} of 120 µM is approximately two orders of magnitude more potent than compound f (25 mM towards MMP-3). Linking a biphenyl group to compound e has yielded compound 2 (Scheme 38.2), which displays micromolar potency towards MMP-1, -2, -3 and -9 (Yan and Cohen 2007). Following the same strategy, but using the pyrone d, which is 4.4 times more potent towards MMP-3 than compound f, yields a potent MMP-3 inhibitor [compound 3, IC_{50} of 10 nM (Puerta et al. 2005)]. When tested on MMP-2, this compound displays an IC_{50} value of 610 nM, a result supporting the proposal that pyrone-based inhibitors could be more selective than hydroxamate compounds. However, the selectivity observed for compound 3 may arise from the biphenyl substituent, which is known to be well tolerated by the large and deep MMP-3 S_1' cavity (Hajduk et al. 1997), but should be more solvent exposed in complex with MMP-2, due to the shorter tunnel-like S_1' subsite of this MMP. In addition, on MMP-3, compound 3 has a potency similar to that of the hydroxamate compound 4, which also harbours the same substituted biphenyl group. This is in contrast with the 4.4-fold difference in potency observed between e and f. This observation suggests that the presence of a P_1' group, like the biphenyl, may subtly change the binding affinity of pyrone-based chelators, or that inhibitor potency and selectivity is critically dependent on the position of the pyrone cycle when the P_1' group is introduced. More systematic evaluation of these new zinc-binding groups is needed to determine whether they may provide very selective MMP inhibitors. As stated for the hydroxamate group, the use of very efficient zinc-binding groups (d and e in Scheme 38.1) could greatly influence the binding mode of the P_1' group in a way that prevents particular interactions critical for selective inhibitor binding. All types of inhibitors discussed above can be viewed as mimics of the products generated by substrate cleavage, to which have been grafted a zinc-chelating group. A different strategy for developing potent enzyme inhibitors is to mimic the structure taken by the substrate in the so-called "transition-state" along the hydrolysis pathway.

Phosphinic Transition-State Analogues

In the case of the zinc metalloproteinase family, phosphinic peptides have been proposed as good mimics of the transition state and turned out in fact to behave as extremely potent inhibitors of this protease family (Dive et al. 2004), provided that the structures of these peptides are appropriately optimised for each target. X-ray structures of several phosphinic peptides in interaction with different zinc proteinases support the view that phosphinic peptides are transition-state analogues (Corradi et al. 2007, Gall et al. 2001, Grams et al. 1996, Tochowicz et al. 2007). As compared with hydroxamate moiety, in the context of the MMP active site, the phosphoryl group (PO_2^-) (g in Scheme 38.1) is expected to act as a weaker zinc-binding group, thus interfering less with the role of the P_1' residue in inhibitor selectivity. Quantum chemistry calculations performed on simple models of inter-action between zinc ions and various zinc-binding groups support this view (Cheng et al. 2002). The above remarks apply only to MMPs and not to families of zinc proteases containing in their active sites additional residues stabilising the transi-tion-state structure. In this case, PO_2^- through its oxygen atoms interacts with these residues, like the tyrosine residue in the active site of astacin (Grams et al. 1996) and ACE (Corradi et al. 2007), increasing the contribution of PO_2^- to inhibitor–enzyme stability.

The X-ray structure of compound 5 in complex with the catalytic domain of MMP-9 has recently been determined (Fig. 38.1b) (Tochowicz et al. 2007). This compound exhibits nanomolar potency towards different MMPs, but at 100 μM does not block MMP-1, -7, TACE and NEP (Neutral EndoPeptidase 24-11). Analysis of this complex indicates that PO_2^- of this compound interacts through its two oxygen atoms with the zinc atom (Scheme 38.2). This structure also shows how the isoxazole side chain of this inhibitor fills part of the deep S_1' cavity that is present in MMP-9 (Fig. 38.1c). The general structure of this inhibitor can be described as follows: $\Phi\text{-}(PO_2\text{-}CH_2)\text{-}P_1'\text{-}P_2'\text{-}NH_2$. Based on the structure of com-pound 5 in this complex, libraries of inhibitors of general formula $Br\text{-}\Phi\text{-}(PO_2\text{-}CH_2)\text{-}P_1'\text{-}P_2'\text{-}P_3'\text{-}NH_2$ have been designed and synthesised by combinatorial chemistry. In these libraries, chemical diversity has been generated by using different iso-xazole side chains in the P_1' position and a combination of natural amino acids in the P_2' and P_2' positions. Screening and deconvolution of these libraries against a panel of ten MMPs led to the identification of a phosphinic peptide (compound 6,

Scheme 38.2) exhibiting a Ki value of 0.2 nM towards MMP-12 (macrophage elastase) and which is more than 2–4 orders of magnitude less potent towards other MMPs (MMP-1, -2, -3, -7, -8, -9, -11, -13 and -14) (Devel et al. 2006). Interestingly, the structure of compound 6 contains a Glu-Glu motif in the P_2' and P_3' positions. However, the selectivity of compound 6 does not arise from a particular preference of MMP-12 to interact with such an acidic motif, as other di-peptide sequences in the P_2' and P_3' inhibitor positions provided more potent MMP-12 inhibitors than compound 6. Thus, it seems more likely that MMP-12 tolerates this motif, while most other MMPs cannot accommodate this motif. Based on a model of interaction of this compound with MMP-12, it has been proposed that the unique tolerance of MMP-12 for this acidic motif might be due to the presence of two unique polar residues, not observed in other MMPs. Even if this explanation seems reasonable, a full understanding of compound 6 selectivity towards MMP-12 in molecular terms will require determination of the X-ray structure of this compound in interaction with MMP-12 and mutagenesis of the MMP-12 residues observed in proximity to the Glu-Glu motif, and thus possibly involved in compound 6 selectivity.

According to the results of the above library screening, with the exception of MMP-12, it appears that all of the MMPs tested do not exhibit a preference for a particular sequence in the P_2' and P_3' positions. Thus, other positions of the inhibitor should be considered in developing selective inhibitors. Full examination of the different inhibitor structures developed in last 15 years indicates that, despite the fact that the MMP active site has long been known to have an extremely deep S_1' cavity, no pseudo-peptide inhibitor in which the P_1' side chain completely fills this cavity has been developed so far. Such an inhibitor may prove to have an interesting selectivity profile. Indeed, when the different MMPs' three-dimensional structures are superimposed, the most important variability between MMPs is observed at the bottom part of the S_1' cavity (Cuniasse et al. 2005). This situation arises from the fact that part of the S_1' cavity is defined by a loop, the S_1' loop (Fig. 38.1b), whose size and sequence varies between the different MMPs. To probe the whole S_1' cavity, new phosphinic inhibitors have been developed by systematically increasing the size of the P_1' side chain. Although this strategy has not yet provided highly selective inhibitors, the observed trends are that inhibitors with the longest P_1' side chain are the most selective, confirming the proposal that probing the bottom part of the S_1' cavity may lead to the identification of the desired selectivity. Obviously, the inherent flexibility of the S_1' loop makes it difficult to predict with accuracy all of the interactions that may take place in solution between the distal part of the P_1' side chain and the different residues lying in the bottom part of the S_1' cavity. Thus, efficient chemistry will be needed to develop P_1' side chains with great chemical diversity in their distal part to identify more selective inhibitors. As discussed in the next section, X-ray structures of enzyme–inhibitor complexes at high resolution provide interesting, but not unambiguous, information from which critical enzyme–inhibitor interactions might be identified and exploited to enhance inhibitor affinity.

New MMP Inhibitors with No Zinc-Binding Groups

High-throughput screening (HTS) of non-peptide libraries has led to the discovery of unusual MMP inhibitors, especially for MMP-3 and -13. These inhibitors represent a new class of MMP inhibitors, as these compounds do not possess a zinc-binding group and thus provide no direct interaction with the zinc active-site atom. They are expected to act as non-competitive inhibitors and thus do not prevent substrate binding. Compound 7 is an example of an "unusual MMP inhibitor" identified by HTS, exhibiting micromolar potency towards MMP-13 and no activity against MMP-1 and MMP-9 (Scheme 38.3) (Chen et al. 2000). The structure of this inhibitor in complex with MMP-13 was solved in solution by nuclear magnetic resonance (NMR) studies (Chen et al. 2000). This inhibitor binds deep in the S_1' cavity of MMP-13, with no apparent interaction between the inhibitor atoms and the catalytic zinc atom. Interestingly, grafting hydroxamate function to such unusual inhibitors increases their potency towards MMP-13 (nanomolar potency), but reduces their MMP selectivity (Chen et al. 2000). This again illustrates how the strong binding of the hydroxamate group to the zinc atom may worsen the contribution to the selectivity of inhibitor-binding group sitting in the S_1' cavity. This effect has stimulated researchers to seek potent MMP-13 inhibitors that do not incorporate a hydroxamate group. From X-ray structure analyses of a lead structure identified by HTS and successive steps of potency optimisation, a highly potent and selective inhibitor of MMP-13 was identified (compound 8, 8 nM towards MMP-13 and no activity detected against MMP-1, -2, -3, -7, -8, -9, -10, -11, -12, -14 and -16 when tested at 100 μM) (Engel et al. 2005). Compound 8 binds in the lower part of the S_1' cavity of MMP-13 and extends into an additional cavity termed the "S_1' side pocket" (Fig. 38.1d). This side pocket results from a particular conformation taken by the S_1' loop of MMP-13. The long loop size of MMP-13 has been suggested to be a determinant factor for the selective binding of compound 8 to this MMP. Thus, MMPs with shorter loop size were suggested to possess an S_1' cavity too shallow and narrow to bind compound 8 (Engel et al. 2005). However, many MMPs (-7, -8, -14, -20 and -24) possess an S_1' loop of size similar to that of MMP-13. For these MMPs, it has been argued that inhibitor recognition depends also on the unique presence of a glycine residue in the MMP-13 S_1' loop sequence and on the particular conformation adopted by this glycine (Gly^{227}) when MMP-13 binds compound 8. Indeed, the local conformation taken by glycine in this complex is forbidden to other amino acids containing a side chain. This proposal probably also explains why other MMPs possessing a longer S_1' loop, like MMP-3, -10 and -12, do not potently interact with compound 8. MMP-25 and -17, which have the longest S_1' loop in the MMP family, have not yet been tested with compound 8. This should be borne in mind, as these two MMPs also possess two glycine residues in their loop, albeit in the same position as in MMP-13, and so are potential candidates in which a side S_1' pocket can be unmasked by inhibitor binding.

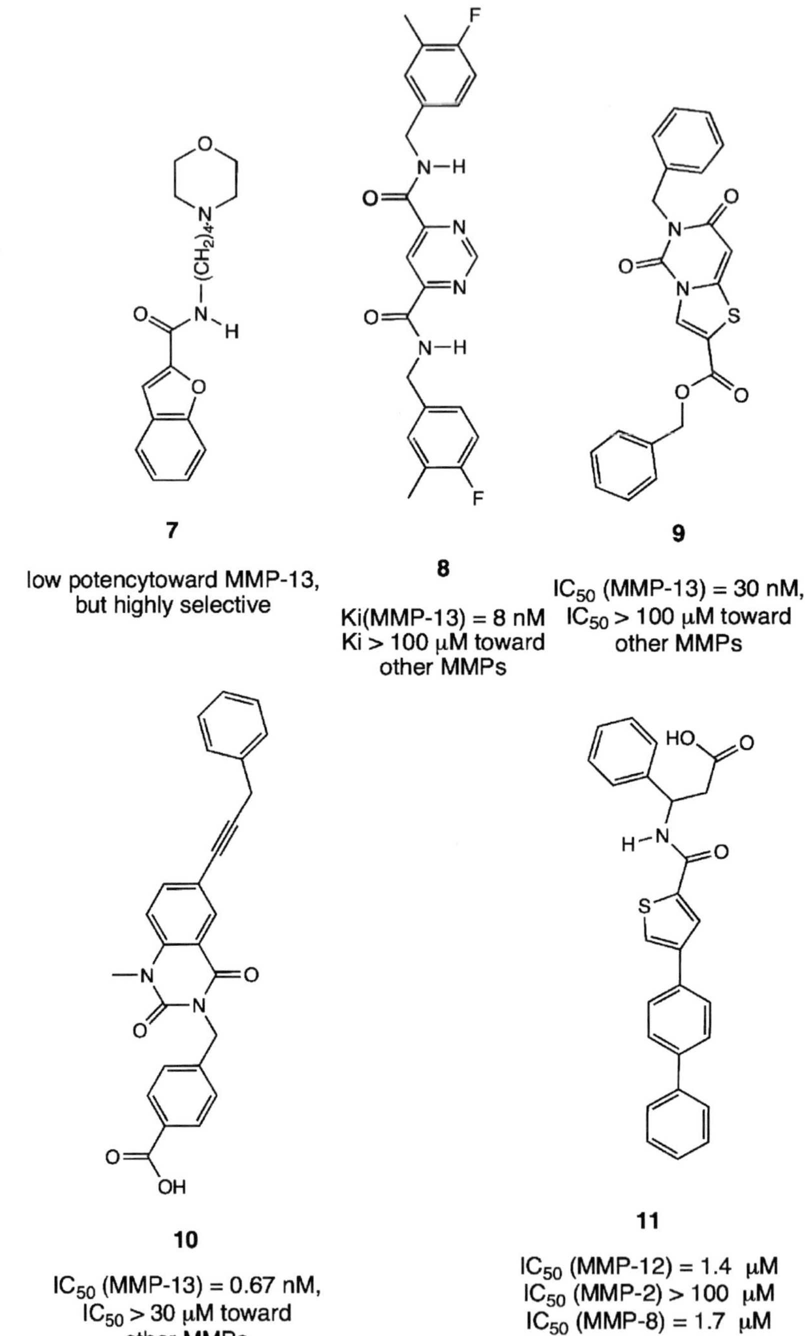

Scheme 38.3 Chemical structures of matrix metalloproteinase (*MMP*) inhibitors with no zinc-binding group and their potency towards *MMPs*

Recently, new highly potent and selective MMP-13 inhibitors (9 and 10, Scheme 38.3) that act like compound 8 have been reported (Johnson et al. 2007). As compared with compound 8, these MMP-13 selective inhibitors possess a very different central core structure, which suggests that many possibilities exist to develop other selective MMP-13 inhibitors. In complex with MMP-13, the central core of these inhibitors is in contact with the wall of the S_1' cavity on one face, but the other face of this core points towards the solvent. This binding mode offers a possible explanation for why different types of core structures can be used to develop selective MMP-13 inhibitors. For MMPs with a longer S_1' loop than MMP-13 (-3, -10, -12, -25 and -17), it cannot be excluded that a similar "side S_1' pocket" can be exploited to develop extremely selective non-competitive inhibitors. Alternatively, structures able to tightly fit in the deep S_1' cavity of these MMPs may yield interesting inhibitors. Along this line, the X-ray structure of a potent MMP-13 inhibitor (compound 11) in interaction with MMP-12 has been reported (Morales et al. 2004). This structure could be the starting point for the development of more potent and extremely selective inhibitors of MMP-12.

Ongoing Research

Despite the impressive efforts that have been devoted to the synthesis of several hundred synthetic inhibitors of MMPs, and the availability of more than 100 X-ray-structures of MMPs in complex with some of these inhibitors, the development of inhibitors able to differentiate different MMPs remains a major challenge. We do not yet have a clear view of all the structural and dynamic parameters that can be exploited to develop highly selective MMP inhibitors. Understanding the basis for protein–inhibitor interactions requires not only extensive structural studies but also the full characterisation of the thermodynamics of their binding to determine the exact participation of the enthalpic (ΔH) and entropic contributions (ΔS) to the free energy (ΔG) of association. In cases where the binding of a series of inhibitors is enthalpy driven, variations in binding affinities can be correlated with specific interactions between the inhibitor and its target, when the X-ray structure of the complex is available. This information can be used to increase the inhibitor's affinity for its target. However, in general, the binding of a compound is a combination of many specific interactions that may vary subtly from one compound to another. Thus, it has been suggested that in silico approaches are not accurate enough to provide a meaningful level of description of the interactions defining the inhibitor affinity (Bertini et al. 2007). A better level of description may rely on the determination of high-resolution X-ray structures of complexes between a particular MMP and a series of inhibitors. Applied to MMP-12, this approach has indicated possible ways of increasing inhibitor potency towards this MMP (Bertini et al. 2007). However, to be helpful in terms of inhibitor selectivity improvement, the above approach may have to characterise the inhibitor series not only for a single MMP but also for the whole family, and this would mean solving at high resolution the structures of tremendous numbers of inhibitor–MMP complexes, which is a challenging task.

The interaction of small inhibitors with MMPs is not always dominated by enthalpy contributions (Bertini et al. 2007, Parker et al. 1999, 2000). For example, it has been reported that the interaction of compound 1 is enthalpy driven when it binds to MMP-3 (K_i = 5.5 nM, ΔH = −8.82 kcal/mol and $-T\Delta S$ =2.34 kcal/mol, ΔS = 7.93 cal/mol) (Parker et al. 1999), but entropically driven when it binds to MMP-12 (K_i = 7.9 nM, ΔH = −4.07 kcal/mol and $-T\Delta S$ = −7 kcal/mol, ΔS = 23.5 cal/mol) (Bertini et al. 2007). Part of the enthalpic contribution determined for the binding of compound 1 to MMP-3 has been proposed to result from hydrogen bond formation between the inhibitor backbone and the enzyme active site, and favourable van der Waals interactions between the Trp of compound 1 and the S_2' subsite (Parker et al. 1999). If this is true, it is hard to explain how the binding enthalpy of compound 1 can vary by a factor of 2 from MMP-3 to -12, as compound 1 is expected to adopt a similar binding mode and exploit similar interactions with two MMPs sharing great similarities in their active-site topology. Part of the entropy changes observed for the binding of compound 1 to MMP-3 has been attributed to some loss of flexibility of the S_1' loop. For inhibitors having a rather short side chain in the P_1' position, a conformational shift of the MMP-3 S_1' loop has been observed, resulting in constriction of loop residues about the inhibitor's P_1' residue (Li et al. 1998). This effect might be specific to MMP-3, as several X-ray structures of MMP-12 with inhibitors possessing P_1' residues of various sizes do not reveal a similar loop conformational shift (Bertini et al. 2006, Morales et al. 2004, Nar et al. 2001). Thus, the more favourable entropic term observed for the binding of compound 1 to MMP-12 may suggest that the S_2' loop of MMP-12 retains more flexibility in the bound state than in MMP-3. Recent NMR studies of MMP-12 in complex with a hydroxamate inhibitor (containing in P_1' position a residue whose size is similar to that in compound 1) support this view (Bertini et al. 2005). More systematic studies of MMPs, in the free and bound state, by NMR spectroscopy in solution are required to gain a better description of the binding-induced conformational shift occurring in the different S_1' loops of these enzymes and of the extent of mobility existing both at the enzyme level and at the inhibitor level (Moy et al. 2002). This is a great challenge, given the number of MMP members and the various inhibitor structures to consider. Each inhibitor, depending on the size of the residue in the P_1' position, is expected to have specific structural and dynamic effects on the S_1' loop, in the particular context of each MMP. Thus, to fully characterise a single MMP, several X-ray and NMR structures have to be solved. Furthermore, these dynamic and structural data should be coupled with thermodynamic studies that may provide insights into the magnitude of the conformational entropy upon inhibitor binding.

Past research on the development of MMP inhibitors has probably overestimated the importance of a strong zinc-binding group in developing inhibitors and underestimated the role of flexibility in the MMP active site and its impact in accommodating different inhibitor structures. These remarks may partly explain the difficulties encountered in the development of highly selective MMP inhibitors.

The optimisation of inhibitor affinity by a structure-based design approach relies on the manipulation of specific interactions between the inhibitor and the enzyme active site (enthalpic contributions). In the context of protein mobility, which is not a specific trait of MMPs, rational approaches to improve inhibitor affinity are less straightforward, as a complex interplay of different contributions is affected by protein mobility. However, progress in the characterisation of MMP dynamics using several biophysical techniques (NMR, X-ray crystallography and isothermal titration calorimetry) should greatly assist the design of more selective inhibitors. On the contrary, the plasticity of the MMP active sites has led to the development of original chemical structures, the most promising being the selective MMP-13 inhibitors. The possibility of exploiting only the S_1' cavity in developing highly selective inhibitors for other MMPs, particularly those containing a similar long size S_1' loop, is an open and interesting issue. This perspective is extremely attractive not only in terms of chemistry but also in developing novel screening methods to identify the third generation of highly selective inhibitors for MMPs identified as valuable therapeutic targets.

References

Amour A, Slocombe P M, Webster A et al. (1998) TNF-alpha converting enzyme (TACE) is inhibited by TIMP-3. FEBS Letters 435: 39–44.

Babine R E, Bender S L (1997) Molecular recognition of proteinminus signLigand complexes: applications to drug design. Chemical Reviews 97: 1359–1472.

Bertini I, Calderone V, Cosenza M et al. (2005) Conformational variability of matrix metalloproteinases: beyond a single 3D structure. Proceedings of the National Academy of Sciences of the United States of America 102: 5334–5339.

Bertini I, Calderone V, Fragai M et al. (2006) Snapshots of the reaction mechanism of matrix metalloproteinases. Angewandte Chemie (International Edition in English) 45: 7952–7955.

Bertini I, Calderone V, Fragai M et al. (2007) Exploring the subtleties of drug-receptor interactions: the case of matrix metalloproteinases. Journal of the American Chemical Society 129: 2466–2475.

Bode W, Maskos K (2003) Structural basis of the matrix metalloproteinases and their physiological inhibitors, the tissue inhibitors of metalloproteinases. Biological Chemistry 384: 863–872.

Brown S, Meroueh S O, Fridman R et al. (2004) Quest for selectivity in inhibition of matrix metalloproteinases. Current Topics in Medicinal Chemistry 4: 1227–1238.

Chen J M, Nelson F C, Levin J I et al. (2000) Structure-based design of a novel, potent and selective inhibitor for MMP-13 utilizing NMR spectroscopy and computer-aided molecular design. Journal of the American Chemical Society 122: 9648–9654.

Cheng F, Zhang R, Luo X et al. (2002) Quantum chemistry study on the interaction of the exogenous ligands and the catalytic zinc ion in matrix metalloproteinases. The Journal of Physical Chemistry B 106: 4552–4559.

Clark I M, Cawston T E (1989) Fragments of human fibroblast collagenase. Purification and characterization. The Biochemical Journal 263: 201–206.

Corradi H R, Chitapi I, Sewell B T et al. (2007) The structure of testis angiotensin-converting enzyme in complex with the C domain-specific inhibitor RXPA380. Biochemistry 46: 5473–5478.

Coussens L M, Fingleton B, Matrisian L M (2002) Matrix metalloproteinase inhibitors and cancer: trials and tribulations. Science (New York, NY) 295: 2387–2392.

Cuniasse P, Devel L, Makaritis A et al. (2005) Future challenges facing the development of specific active-site-directed synthetic inhibitors of MMPs. Biochimie 87: 393–402.

Devel L, Rogakos V, David A et al. (2006) Development of selective inhibitors and substrate of matrix metalloproteinase-12. The Journal of Biological Chemistry 281: 11152–11160.

Dive V, Georgiadis D, Matziari M et al. (2004) Phosphinic peptides as zinc metalloproteinase inhibitors. Cellular and Molecular Life Sciences 61: 2010–2019.

Egeblad M, Werb Z (2002) New functions for the matrix metalloproteinases in cancer progression. Nature Reviews 2: 161–174.

Engel C K, Pirard B, Schimanski S et al. (2005) Structural basis for the highly selective inhibition of MMP-13. Chemistry & Biology 12: 181–189.

Fingleton B (2007) Matrix metalloproteinases as valid clinical targets. Current Pharmaceutical Design 13: 333–346.

Gall A L, Ruff M, Kannan R et al. (2001) Crystal structure of the stromelysin-3 (MMP-11) catalytic domain complexed with a phosphinic inhibitor mimicking the transition-state. Journal of Molecular Biology 307: 577–586.

Grams F, Dive V, Yiotakis A et al. (1996) Structure of astacin with a transition-state analogue inhibitor. Nature Structural Biology 3: 671–675.

Hajduk P J, Sheppard G, Nettesheim D G et al. (1997) Discovery of potent nonpeptide inhibitors of stromelysin using SAR by NMR. Journal of the American Chemical Society 119: 5818–5827.

Johnson A R, Pavlovsky A G, Ortwine D F et al. (2007) Discovery and characterization of a novel inhibitor of matrix metalloprotease-13 (MMP13) that reduces cartilage damage in vivo without joint fibroplasia side effects. The Journal of Biological Chemistry 282: 27781–27791.

Li Y C, Zhang X, Melton R et al. (1998) Solution structure of the catalytic domain of human stromelysin-1 complexed to a potent, nonpeptidic inhibitor. Biochemistry 37: 14048–14056.

Maskos K, Fernandez-Catalan C, Huber R et al. (1998) Crystal structure of the catalytic domain of human tumor necrosis factor-alpha-converting enzyme. Proceedings of the National Academy of Sciences of the United States of America 95: 3408–3412.

Matter H, Schudok M (2004) Recent advances in the design of matrix metalloprotease inhibitors. Current Opinion in Drug Discovery & Development 7: 513–535.

Morales R, Perrier S, Florent J M et al. (2004) Crystal structures of novel non-peptidic, non-zinc chelating inhibitors bound to MMP-12. Journal of Molecular Biology 341: 1063–1076.

Moy F J, Chanda P K, Chen J et al. (2002) Impact of mobility on structure-based drug design for the MMPs. Journal of the American Chemical Society 124: 12658–12659.

Nar H, Werle K, Bauer M M et al. (2001) Crystal structure of human macrophage elastase (MMP-12) in complex with a hydroxamic acid inhibitor. Journal of Molecular Biology 312: 743–751.

Netzel-Arnett S, Sang Q X, Moore W G et al. (1993) Comparative sequence specificities of human 72- and 92-kDa gelatinases (type IV collagenases) and PUMP (matrilysin). Biochemistry 32: 6427–6432.

Overall C M, Kleifeld O (2006) Towards third generation matrix metalloproteinase inhibitors for cancer therapy. British Journal of Cancer 94: 941–946.

Parker M H, Lunney E A, Ortwine D F et al. (1999) Analysis of the binding of hydroxamic acid and carboxylic acid inhibitors to the stromelysin-1 (matrix metalloproteinase-3) catalytic domain by isothermal titration calorimetry. Biochemistry 38: 13592–13601.

Parker M H, Ortwine D F, O'Brien P M et al. (2000) Stereoselective binding of an enantiomeric pair of stromelysin-1 inhibitors caused by conformational entropy factors. Bioorganic & Medicinal Chemistry Letters 10: 2427–2430.

Peng S X, Strojnowski M J, Hu J K et al. (1999) Gas chromatographic-mass spectrometric analysis of hydroxylamine for monitoring the metabolic hydrolysis of metalloprotease inhibitors in rat and human liver microsomes. Journal of Chromatography 724: 181–187.

Puerta D T, Lewis J A, Cohen S M (2004) New beginnings for matrix metalloproteinase inhibitors: identification of high-affinity zinc-binding groups. Journal of the American Chemical Society 126: 8388–8389.

Puerta D T, Mongan J, Tran B L et al. (2005) Potent, selective pyrone-based inhibitors of stromelysin-1. Journal of the American Chemical Society 127: 14148–14149.

Saghatelian A, Jessani N, Joseph A et al. (2004) Activity-based probes for the proteomic profiling of metalloproteases. Proceedings of the National Academy of Sciences of the United States of America 101: 10000–10005.

Seltzer J L, Akers K T, Weingarten H et al. (1990) Cleavage specificity of human skin type IV collagenase (gelatinase). Identification of cleavage sites in type I gelatin, with confirmation using synthetic peptides. The Journal of Biological Chemistry 265: 20409–20413.

Skiles J W, Gonnella N C, Jeng A Y (2004) The design, structure, and clinical update of small molecular weight matrix metalloproteinase inhibitors. Current Medicinal Chemistry 11: 2911–2977.

Stack M S, Gray R D (1989) Comparison of vertebrate collagenase and gelatinase using a new fluorogenic substrate peptide. The Journal of Biological Chemistry 264: 4277–4281.

Tochowicz A, Maskos K, Huber R et al. (2007) Crystal structures of MMP-9 complexes with five inhibitors: contribution of the flexible Arg424 side-chain to selectivity. Journal of Molecular Biology 371: 989–1006.

Turk B E, Huang L L, Piro E T et al (2001) Determination of protease cleavage site motifs using mixture-based oriented peptide libraries. Nature Biotechnology 19: 661–667.

Visse R, Nagase H (2003) Matrix metalloproteinases and tissue inhibitors of metalloproteinases: structure, function, and biochemistry. Circulation Research 92: 827–839.

Whittaker M, Floyd C D, Brown P, Gearing A J (1999) Design and therapeutic application of matrix metalloproteinase inhibitors. Chemical Reviews 99: 2735–2776.

Yan Y L, Cohen S M (2007) Efficient synthesis of 5-amido-3-hydroxy-4-pyrones as inhibitors of matrix metalloproteinases. Organic Letters 9: 2517–2520.

Chapter 39
Protease-Activated Delivery and Imaging Systems

Gregg B. Fields

Abstract Proteolysis has been cited as an important contributor to cancer initiation and progression. But advantage can be taken of tumor-associated proteases to selectively deliver therapeutic or imaging agents. Protease-activated prodrugs, nanotechnology-based drug delivery systems, hydrogels, gene delivery systems, and imaging systems have been described for cancer applications. Activation is modulated by substrates designed for hydrolysis by members of the matrix metalloproteinase family, prostate-specific antigen, hK2, plasmin, urokinase plasminogen activator, legumain, neprilysin/CD10, or cathepsins B, D, or L. The first generation of protease-activated agents has demonstrated proof of principle as well as provided impetus for in vivo applications. One common problem has been a lack of agent stability at nontargeted tissues and organs due to activation by multiple proteases. Second-generation agents may need to incorporate more selective substrates, which can be achieved by consideration of both the sequence specificity and the topological preferences of proteases.

Introduction

A major goal in drug delivery is to effectively direct therapeutic agents to their intended biological target without deleterious side effects. In principle, targeted drug delivery would minimize toxicities while delivering an effective dose of the therapeutic agent where desired. Targeted delivery relies upon the identification of biomolecules that are disease associated. The initiation and progression of cancer has often been linked to proteolytic activity, as discussed elsewhere in this volume for the matrix metalloproteinases (MMPs), Ser proteases such as prostate-specific antigen (PSA), hK2, plasmin, and urokinase plasminogen activator (uPA), the cell surface metalloprotease

G. Fields
Department of Chemistry & Biochemistry, Florida Atlantic University, 777 Glades Road, Boca Raton, FL 33431 USA, e-mail: fieldsg@fau.edu

D. Edwards et al. (eds.), *The Cancer Degradome.*   827
© Springer Science + Business Media, LLC 2008.

neprilysin/CD10/common acute lymphoblastic leukemia antigen, and lysosomal proteases legumain and cathepsins B, D, and L (see Chaps. 1 and 8–11).

One potential delivery strategy is to create inactive agents that, upon proteolytic processing, are converted into active forms. In theory, this could allow for drug delivery to tumors based on the association of specific proteases with cancer progression. Several variations of this approach have been described (Meers 2001, Vartak and Gemeinhart 2007). Most simply, a cytotoxic agent is attached to a protease-sensitive peptide sequence (see also Chap. 40). The cytotoxic agent has greatly reduced activity in such a construct. Upon proteolysis, cytotoxic activity is restored. Numerous protease-activated prodrugs have been created in this fashion. Alternatively, cytotoxic agents may be stored within a carrier and are physically unavailable until the carrier opens. Protease-activated nanotechnology-based drug delivery systems (nano-DDSs) and hydrogels operate on this principle. Prodrug, nano-DDS, and hydrogel technologies may also be utilized for visualization of tumor-associated protease activity by replacing cytotoxic agents with optical or other imaging agents (see also Chap. 7). This chapter examines protease-activated drug delivery and imaging systems that have been applied for the study or eradication of cancer, and discusses some of the shortfalls of these approaches and potential strategies to overcome them.

Protease-Activated Prodrugs

The concept of an anticancer prodrug is straightforward (Fig. 39.1a). A cytotoxic agent is attached to a protease labile sequence, altering the activity of the cytotoxic agent. Upon proteolysis, the initial cytotoxic activity is restored. This approach allows for selective delivery of cytotoxic agents to tumors with discrete protease profiles. To improve circulation times, a carrier may be added to the prodrug (Fig. 39.1b). Sequences utilized for protease-activated prodrugs are compiled in Table 39.1, and an extensive discussion of protease-activated prodrugs is given in Chap. 40. A few additional protease-activated prodrugs are discussed below.

A combinatorial peptide approach was utilized to develop an hK2-activated prodrug (Janssen et al. 2004). The sequence deduced from library screening, Gly-Lys-Ala-Phe-Arg-Arg, was C-terminally linked to 12-aminododecanoyl thapsigargin (12ADT) to create the Gly-Lys-Ala-Phe-Arg~Arg~Leu-12ADT prodrug. A fourfold difference in cytotoxic activity was observed for hK2-producing and -nonproducing tumor cell lines (C4-2B human prostate cancer and TSU human bladder cancer, respectively). The prodrug was selective for hK2 as compared to cathepsins B and D and urokinase, but was hydrolyzed 6 times faster by plasmin than by hK2 (Janssen et al. 2004). In vivo analysis in an LNCaP human prostate cancer mouse xenograft model indicated that Gly-Lys-Ala-Phe-Arg~Arg~Leu-12ADT was stable (<0.5% free Leu-12ADT was observed in plasma after 24 h) and reduced tumor size significantly following four daily doses (Janssen et al. 2006). However, the prodrug was also rapidly cleared and was toxic following prolonged intravenous administration (Janssen et al. 2006).

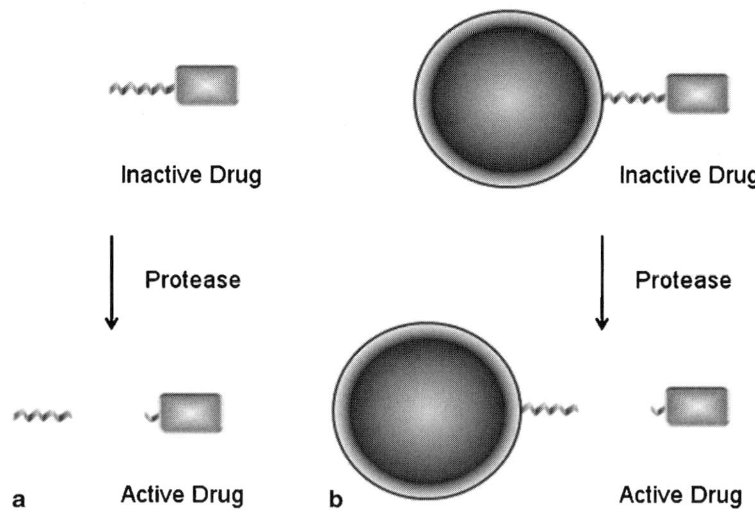

Fig. 39.1 Protease-activated prodrug strategies. The drug is inactivated by attachment of the protease-labile sequence. Protease-catalyzed hydrolysis releases an active form of the drug. The prodrug can simply be the substrate and drug (**a**) or the substrate and drug with an additional component (HSA, human serum albumin, *PEG*, polyethylene glycol, or dextran) (**b**) to improve circulation times

The prodrugs β-Ala-Leu-Ala-Leu-Dox and *N*-succinyl-β-Ala-Leu-Ala-Leu-Dox [CPI-0004Na, where Dox = doxorubicin] were developed for activation by proteases in the tumor microenvironment (Fernandez et al. 2001, Trouet et al. 2001). Succinylation did not change the susceptibility of the sequence to cleavage by cancer-related peptidases or its in vitro blood stability (Fernandez et al. 2001). β-Ala-Leu-Ala-Leu-Dox treatment of MCF-7/6 human breast carcinoma cells and MRC-5 fibroblasts resulted in 14 times more accumulation of Dox in the tumor cells. This contrasted greatly with free Dox, where uptake was slightly higher for the fibroblasts. β-Ala-Leu-Ala-Leu-Dox was 9 times less toxic than Dox and 4 times less toxic than Leu-Dox in mice (Trouet et al. 2001). CPI-0004Na was devoid of intravenous acute toxicity, and both β-Ala-Leu-Ala-Leu-Dox and CPI-0004Na were active via intraperitoneal administration in inhibiting growth of MCF-7/6 human breast tumor xenografts (Fernandez et al. 2001, Trouet et al. 2001). Follow-up studies demonstrated that CPI-0004Na was more active and less toxic in vivo than Dox–HCl (Dubois et al. 2002). This stems from the fact that administration of CPI-0004Na resulted in lower spleen, kidney, and lung exposure to Dox compared with Dox–HCl administration (Dubois et al. 2002). CPI-0004Na was activated by the cell surface protease CD10 (Pan et al. 2003a). Increased potency was observed in CD10$^+$ cell lines, such as B-cell lymphoma, leukemia, and prostate, breast, colorectal, and lung carcinomas (Pan et al. 2003a).

Intracellularly activated prodrugs have been developed to take advantage of lysosomal proteolysis. Cathepsin B, found in all mammalian lysosomes, has been targeted for this purpose. The initial cathepsin B-activated prodrug series was based on X-Lys-p-aminobenzyloxycarbonyl (PABC)-Dox, where Dox release rates and stability to human plasma were examined (Dubowchik and Firestone 1998). Based on these results, benzyloxycarbonyl (Z)-Phe-Lys-PABC-Dox, tert-butyloxycarbonyl (Boc)-Phe-Lys-PABC-mitomycin C, Z-Phe-Lys-PABC-2'-paclitaxel, and Boc-Phe-Lys-PABC-7-paclitaxel were specifically examined for lysosomal activation (Dubowchik et al. 1998). The paclitaxel-containing prodrugs were activated more slowly by cathepsin B, but all four had half-lives of similar magnitudes (19–66 min) with rat liver lysosomes (Dubowchik et al. 1998). A Val-Cit linker was incorporated between an anti-CD30 mAb and either monomethyl auristatin E (MMAE) or monomethyl auristatin F (MMAF) to create a targeted, cathepsin B-activated prodrug (Kung Sutherland et al. 2006). The prodrugs cAC10vc-MMAE and cAC10vc-MMAF interacted with the CD30 cell surface antigen, and were internalized into lysosomes. Both conjugates were cytotoxic for the CD30$^+$ lymphoma Karpas-299 and L540cy cell lines, with subnanomolar IC$_{50}$ values, while exhibiting no activity toward CD30$^-$ Ramos cells (Kung Sutherland et al. 2006). Cys protease inhibitors blocked the cytotoxic effects of cAC10vc-MMAE and cAC10vc-MMAF.

In an oncolytic virotherapy approach related to toxin prodrugs, an MMP-cleavable sequence was engineered within the Sendai virus (SeV) to create a recombinant virus that was highly fusogenic and spread from cell to cell following MMP activation (Kinoh et al. 2004). The matrix protein (M) gene was deleted from the virus, which effectively prevented virus maturation into particles. The viral fusion (F) cleavage site was altered to render it susceptible to MMPs or uPA. MMP-subII SeV/ΔM (which contained the MMP-9 cleavage site Pro-Leu-Gly~Met-Thr-Ser) was found to spread extensively in the MMP-expressing HT1080 human fibrosarcoma cell line, but not in human stomach cancer MKN28 cells, which did not express MMPs. HT1080 tumor growth in nude mice was significantly inhibited by direct tumor injection of MMP-subII SeV/ΔM (Kinoh et al. 2004).

All prodrugs require the chemical conjugation of drugs or drug carriers to the targeting moiety. The conjugation of drugs directly to the targeting ligand, however, can negatively affect the targeting molecule in a manner that disrupts receptor/ligand recognition (Backer et al. 2004) and may alter the cytotoxicity of the drug (Chau et al. 2004, Kline et al. 2004). While it is anticipated that such prodrugs can be obtained, their pharmacokinetics may be unfavorable. Thus, the development of drug carriers that are activated by the targeted proteases and whose pharmacokinetic properties are well understood represents an alternative approach for the creation of protease-activated nanomedicines.

Protease-Activated Nanotechnology-Based Drug Delivery Systems

Drug delivery systems (DDSs) can improve the pharmacological properties of conventional drugs by altering drug pharmacokinetics and biodistribution, as well as functioning as drug reservoirs (Allen and Cullis 2004). Nanotechnology-based

DDSs (nano-DDSs), in which the drug carriers have diameters of $\sim$100 nm or less, have seen recent popularity due to the favorable physical, chemical, and biological properties of biomolecules of that size (Allen and Cullis 2004, Willis 2004). Nano-DDSs include liposomes, dendrimers, micelles, and polymeric and ceramic nano-particles (Sahoo and Labhasetwar 2003, Vine et al. 2006). These nano-DDSs have been widely studied for delivery of various drugs to cellular targets, but each does not possess inherent targeting capabilities. Micelles, liposomes, and nanoparticles can be easily modified to incorporate targeting moieties that allow for more specific or guided delivery of the drug. This includes the concept of enzyme-activated targeting of liposomes (Meers 2001).

Because of a variety of innovations, liposomes have recently begun to realize their potential as drug delivery vehicles. Modification with polyethylene glycol (PEG) or N-(2-hydroxypropyl)methacrylamide (HPMA) has improved liposome circulation times, achieved via decreased interaction with the reticuloendothelial system (RES) (Klibanov et al. 1990, Allen and Hansen 1991, Allen et al. 1991, Duzgunes and Nir 1999, Maruyama et al. 1999, Oku 1999, Whiteman et al. 2001, Jamil et al. 2004, Torchilin 2005, Gabizon et al. 2006). Although not targeted, PEG-stabilized liposomes are in clinical use for doxorubicin delivery to patients with Kaposi's sarcoma (DaunoXome) and ovarian carcinoma (Doxil) (Allen and Cullis 2004, Jamil et al. 2004, Torchilin 2005). The potential of targeted nano-DDSs could further extend the applicability of liposomes (Drummond et al. 1999, Sahoo and Labhasetwar 2003, Allen and Cullis 2004, Torchilin 2005, Gabizon et al. 2006).

The surface of liposomes may be modified to incorporate a sequence hydrolyzed by tumor-associated proteases. Protease-activated liposomes have been designed to either (a) enhance liposomal fusion with tumor or tumor microenvironment cells, facilitating targeted drug delivery (Fig. 39.2) (Pak et al. 1998, 1999; Kondo et al. 2004, Terada et al. 2006), or (b) become substantially destabilized upon proteolysis, resulting in drug release extracellularly (Fig. 39.3) (Hu et al. 1986, Sarkar et al. 2005). Examples of both mechanisms are described below.

Human leukocyte elastase-targeted liposomes have been designed to allow the conversion of liposomes to a more cationic state upon proteolysis, promoting cell fusion (Meers 2001). Because conversion to a more cationic liposome occurs only in the tumor microenvironment, general problems with cationic liposomes (such as short-circulation lifetime and nonspecific toxicity) are avoided. Human leukocyte elastase prefers uncharged amino acid side chains, especially short sequences of Ala or Val. Initially, the peptide-lipid acetyl-Ala-Ala-[1,2-dioleoyl-sn-glycero-3-phosphatidylethanolamine] (N-Ac-AA-DOPE) was constructed for creating elastase-targeted liposomes (Pak et al. 1998). Human leukocyte elastase and proteinase K both hydrolyzed N-Ac-AA-DOPE when the peptide-lipid was incorporated into dioleoyl trimethylammonium propane (DOTAP)/phosphatidylethanolamine liposomes. Liposomal fusion with red blood cells was promoted by proteolysis (Pak et al. 1998). Subsequently, the peptide-lipid N-methoxy-succinyl-Ala-Ala-Pro-Val-DOPE (MeO-suc-AAPV-DOPE) was applied for liposomal targeting, as it was more readily cleaved by human leukocyte elastase than N-Ac-AA-DOPE yet was less sensitive to proteinase K (Pak et al. 1999). Human leukocyte elastase treatment

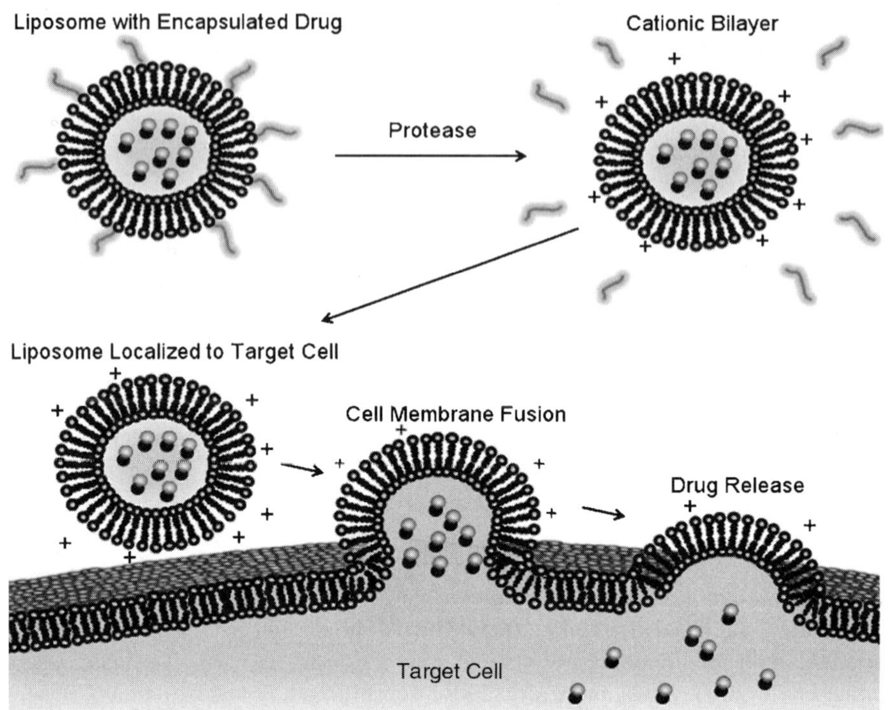

Fig. 39.2 Protease-activated liposomal drug delivery to cells. The liposome incorporates a protease-susceptible sequence on the exterior. Upon hydrolysis, the surface of the liposome becomes more cationic, promoting fusion with nearby cells. Fusion results in intracellular drug delivery

of MeO-suc-AAPV-DOPE/dioleoyl dimethylammonium propane (DODAP) liposomes triggered fusion between these liposomes and HL60 human leukemia cells and cellular delivery of dextrans (Pak et al. 1999).

MT1-MMP has been used to promote cellular delivery of 5'-O-dipalmitoylphosphatidyl 2'-C-cyano-2'-deoxy-1-β-D-arabino-pentofuranosylcytosine (DPP-CNDAC) by incorporating the MT1-MMP ligand stearoyl-Gly-Pro-Leu-Pro-Leu-Arg into 1,2-distearoyl-sn-glycero-3-phosphocholine (DSPC)/cholesterol liposomes (Kondo et al. 2004). Gly-Pro-Leu-Pro-Leu-Arg liposomes bound to HUVECs and accumulated at tumor sites in a mouse model of Colon 26 NL-17 carcinoma. Interestingly, the MT1-MMP substrate Gly-Pro-Leu-Gly-Leu-Arg did not promote specific liposome accumulation at angiogenic sites. A similar observation was made for liposomes possessing an anti-MT1-MMP antibody (Hatakeyama et al. 2007). Gly-Pro-Leu-Pro-Leu-Arg liposomes containing DPP-CNDAC suppressed tumor growth more efficiently than did the control (non-MT1-MMP targeting) liposomes (Kondo et al. 2004). This effect may have been due to accelerated tumor uptake (Hatakeyama et al. 2007).

MMP-2-activated liposomes have been developed to target hepatocellular carcinoma (HCC). In this construct, PEG-Gly-Pro-Leu-Gly~Ile-Ala-Gly-Gln-DOPE

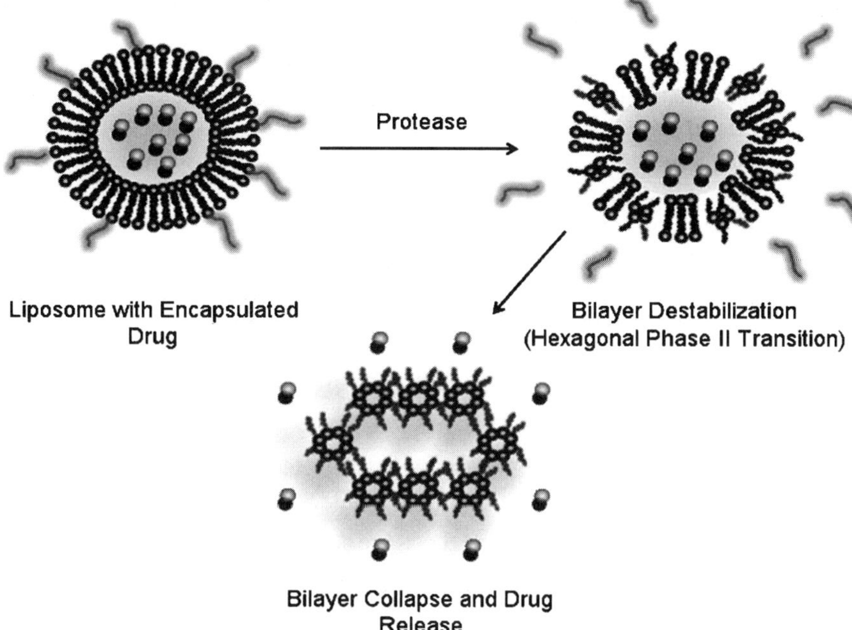

Fig. 39.3 Protease-activated liposomal extracellular drug delivery. The liposome incorporates a protease-susceptible sequence on the exterior. Upon hydrolysis, the liposomal bilayer is destabilized and undergoes a hexagonal phase II transition. The collapse of the liposomal bilayer results in extracellular drug delivery

was incorporated into galactosylated liposomes. The surface coating of PEG was postulated to sterically interfere with liposome uptake by normal hepatocytes. However, the production of MMP-2 by HCC would result in hydrolysis of PEG-Gly-Pro-Leu-Gly~Ile-Ala-Gly-Gln-DOPE and removal of the PEG from the liposome surface. Upon PEG liberation, the galactosylated liposomes would be taken up by HCC. MMP-2 pretreatment was found to enhance uptake of the liposomes by HepG2 cells, in a concentration-dependent manner. The cytotoxicity activity of PEG-Gly-Pro-Leu-Gly~Ile-Ala-Gly-Gln-DOPE liposomes containing N^4-octade-cyl-1-β-D-arabinofuranosylcytosine (NOAC) toward HepG2 cells was enhanced upon treatment with MMP-2 as compared to no treatment (Terada et al. 2006).

Protease-mediated extracellular destabilization of liposomes and drug release was initially described for liposomes containing glycophorin and DOPE (Hu et al. 1986). Trypsin cleaved the hydrophilic region of glycophorin, resulting in liposome conversion to inverted micellar structures (Fig. 39.3) and release of calcein (Hu et al. 1986). This approach was more recently examined for an MMP-9-mediated liposomal system (Sarkar et al. 2005). An MMP cleavage site sequence from type I collagen was prepared as a triple-helical "peptide-amphiphile" (Yu et al. 1996, 1998, 1999), where stearic acid (octadecanoic acid, designated C_{18}) was acylated to the N-terminus of Gly-Pro-Gln-Gly~Ile-Ala-Gly-Gln-Arg-(Gly-Pro-Hyp)$_4$-Gly-Gly

(LP1). LP1 was then used to form liposomes with 1,2-distearoyl-sn-glycero-3-phosphocholine (DSPC) at a ratio of 1:9 (mol%) (Sarkar et al. 2005). This liposome construct was "uncorked" by MMP-9, resulting in release of the liposomal contents. The nonlipidated sequence was cleaved by trypsin, but LP1 was not, most likely due to the triple-helix structure of LP1. Unfortunately, this liposome was not particularly stable, and would likely need the addition of cholesterol and/or PEG lipids to achieve useful stability.

In addition to liposomes, MMP-activated drug release has been described for micelles. PEG-Gly-Pro-Leu-Gly~Val-Dox and PEG-Gly-Pro-Leu-Gly~Val-Arg-Gly-Dox were used to form MMP-2-activated conjugate micelles (Lee et al. 2007a). The conjugate micelles could also be loaded with Dox. Dox-loaded micelles inhibited tumor growth to a greater degree than did the conjugate micelles alone, and also maintained higher concentrations of Dox in plasma.

Protease-Activated Hydrogel Delivery Systems

Complementary to circulating nano-DDSs are those that are applied locally. Protease-activated local drug delivery has been achieved via hydrogels (Fig. 39.4). Protease-activated hydrogels were initially developed for susceptibility to collagenase (MMP-1) or plasmin (West and Hubbell 1999). The protease substrates were Acr-Ala-Pro-Gly-Leu-PEG-Ala-Pro-Gly-Leu-Acr for collagenase and Acr-Val-Arg-Asn-PEG-Val-Arg-Asn-Acr for plasmin. Hydrogels generated by photopolymerization containing either substrate appeared to have similar networks. Acr-Ala-Pro-Gly-Leu-PEG-Ala-Pro-Gly-Leu-Acr hydrogels were hydrolyzed by 2.0 and 0.2 mg/ml collagenase [which may have been *Clostridium histolyticum* (bacterial) collagenase] but not by 0.2 mg/ml plasmin. Acr-Val-Arg-Asn-PEG-Val-Arg-Asn-Acr

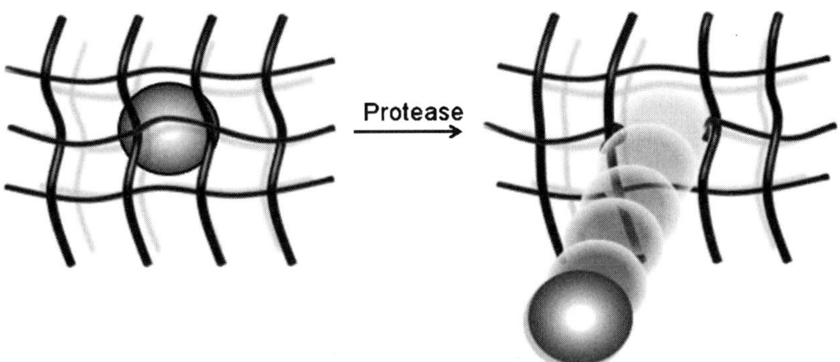

Fig. 39.4 Protease-activated hydrogel drug delivery. The hydrogel incorporates a protease-susceptible sequence that further cross-links and contracts the hydrogel, trapping the drug inside. Upon hydrolysis, the expanded hydrogel releases the drug

hydrogels were hydrolyzed by 2.0 and 0.2 mg/ml plasmin but not 0.2 mg/ml collagenase (West and Hubbell 1999).

PEG-based hydrogels were subsequently developed for MMP-mediated drug delivery (Lutolf et al. 2003a). MMP-2, MMP-9, and MT1-MMP were potential activating agents based on the substrate sequence of acetyl-Gly-Cys-Arg-Asp-Gly-Pro-Gln-Gly~Ile-Trp-Gly-Gln-Asp-Arg-Cys-Gly-NH$_2$. The hydrogel was formed by cross-linking vinyl sulfone-functionalized multiarm PEG with the substrate in triethanolamine-buffered saline, and used to deliver recombinant human bone morphogenetic protein-2 for tissue remodeling (Lutolf et al. 2003a, b). Comparison to control hydrogels demonstrated that bone regeneration was dependent upon MMP activity (Lutolf et al. 2003a).

PEG-based hydrogels have subsequently been used for MMP-mediated delivery of chemotherapeutic agents (cisplatin) to malignant glioma cell lines (Tauro and Gemeinhart 2005a, b). Cisplatin (CDDP) was complexed with acetyl-Cys-Gly~Leu-Asp-Asp, an MMP-sensitive peptide meant to target MMP-2 and MMP-9. Optimized hydrogel wafers were generated using a PEG$_{4000}$ diacrylate. MMP-2 or MMP-9 caused an increase in the release rate of CDDP from the hydrogel compared to no MMP controls and a corresponding enhanced level of cytotoxicity for U-87 malignant glioma cells (Tauro and Gemeinhart 2005b). Comparison of hydrogels formed with PEG$_{574}$ diacrylate versus PEG$_{4000}$ diacrylate found mesh sizes of 21 ± 4 nm and 79 ± 7 nm, respectively (Tauro and Gemeinhart 2005a). It was hypothesized that the larger mesh size allowed MMPs to diffuse into the hydrogel and liberate cisplatin efficiently (Tauro and Gemeinhart 2005a).

Protease-Activated Gene Delivery

Two MMP-activated potential gene delivery systems have been described for cancer applications (Fig. 39.5). In the first example, the MMP-labile Pro-Leu-Gly~Leu-Trp-Ala sequence was incorporated into a chimeric envelope between an epidermal growth factor (EGF) domain and the 4070A murine leukemia virus (MLV) envelope glycoprotein (Peng et al. 1997). The chimeric envelope (E.MMP.A) was expressed and incorporated into viral particles, and the EGF domain could be cleaved by MMP-2 from the viral particle surface. E.MMP.A bound to EGF receptors on human epithelial carcinoma A431 cells but could not infect the cells unless exogenous MMP-2 was added. In contrast, E.MMP.A could infect HT1080 cells, with endogenous MT1-MMP contributing to the infection efficiency (Peng et al. 1997). E.MMP.A showed selective transduction for HT1080 tumor xenografts compared with A431 tumor xenografts (Peng et al. 1999).

The second MMP-activated potential gene delivery system (Fig. 39.5) used a slightly longer MMP-labile sequence (Gly-Gly-Pro-Leu-Gly~Leu-Trp-Ala-Gly-Gly) incorporated between the C-terminal extracellular domain of CD40 ligand (CD40L) and the Gibbon Ape Leukemia Virus (GALV) envelope glycoprotein (Johnson et al. 2003). Transfection of glioma U87, U118, and U251 cell lines with the chimeric envelope (GALV M40) induced fusion to an extent that correlated

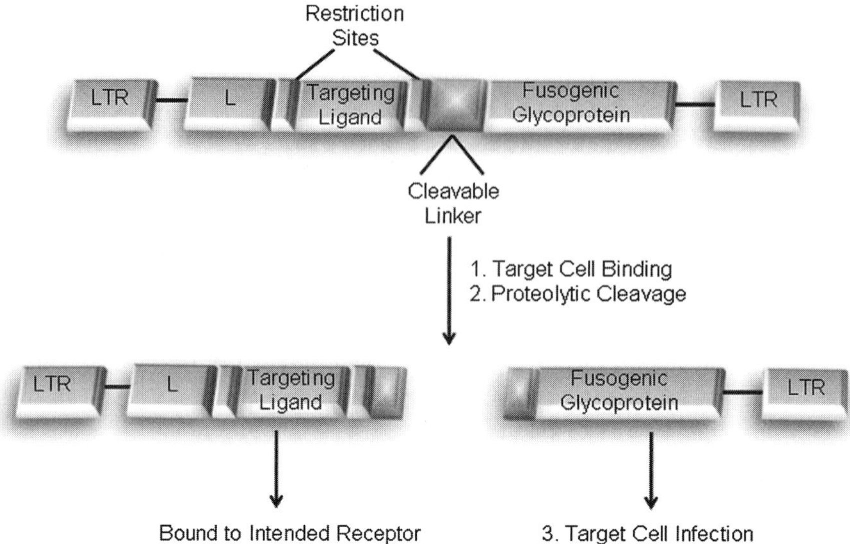

Fig. 39.5 Protease-activated gene delivery. The construct incorporates a substrate between a cell surface receptor targeting sequence and a fusogenic glycoprotein sequence. Binding to the cell surface allows for protease hydrolysis and separation of the targeting protein from the fusogenic glycoprotein. The target cell is then infected by the glycoprotein. Without protease activity, the construct does not infect the target cell

with the relative MMP-2 activity levels for the cell lines. Both soluble MMPs and membrane-bound MMPs were found capable of activating GALV M40. GALV M40 exhibited no fusogenic activity with normal human astrocytes. GALV M40 retained cytotoxic activity in vivo against U87 cells (where transfected cells were administered subcutaneously in nude mice), although activity was lower than for GALV (Johnson et al. 2003). An adenoviral variation of GALV M40 (designated AdM40), where the positions of the GALV and CD40L were reversed and CD40L was linked to the green fluorescent protein (GFP) gene and AD5 DNA via an internal ribosomal entry site (IRES), was also examined for activity against gliomas (Allen et al. 2004). The AdM40 adenovirus infected MMP-2-producing glioma cell lines as described above, and had no activity toward normal human astrocytes. Tumor regression was seen when AdM40 was administered to U87 xenografts in mice, as was a statistically significant improvement in animal survival compared to no treatment. AdM40 was also more effective than AdN40, where the MMP-cleavable sequence was replaced by Gly-Gly-Gly-Gly-Ser (Allen et al. 2004).

Protease-Activated Imaging Agents

Many of the principles utilized to develop fluorescence resonance energy transfer (FRET) substrates have been applied to generate protease-activated imaging agents (see Chap. 7). Modification of activity-based protein profiling (ABPP) strategies to create quenched activity-based probes (qABPs) is described in detail in Chap. 7 and thus is not repeated here. Our discussion focuses on agents developed primarily for optical-based molecular imaging with potential medical applications at the forefront (Weissleder and Mahmood 2001, Bremer et al. 2003, Tung 2004).

Near-infrared fluorescence (NIRF) optical imaging of lysosomal (Cys/Ser) proteases in tumor cells was achieved using Cy5.5 bound to a synthetic graft copolymer composed of poly-L-Lys (PL) and methoxyPEG (MPEG) (Mahmood et al. 1999, Weissleder et al. 1999). The $(Cy5.5)_{11}$-PL-MPEG$_{92}$ construct exhibited reasonable self-quenching, with a 12-fold increase in NIRF signal between trypsin-treated and -untreated constructs (Weissleder et al. 1999). Imaging of BT-20 breast adenocarcinoma and LX-1 small-cell carcinoma in mice was achieved within 24 h and up to 96 h, with good biodistribution (Mahmood et al. 1999, Weissleder et al. 1999).

An NIRF optical imaging agent for cathepsin D was developed by incorporating [Cy5.5]-Gly-Pro-Ile-Cys(ethyl)-Phe~Phe-Arg-Leu-Gly-Lys(FITC)-Cys-NH$_2$ onto the PL-MPEG$_{92}$ copolymer described above (Tung et al. 1999). The imaging agent was not activated by cathepsin D in vitro at low pH (3.5), but was activated by murine 3Y1-AD12 embryonic tumor cells that had been stably transfected with human cathepsin D. Signal was not present using a control probe (Tung et al. 1999). Intravenous administration resulted in imaging of cathepsin D$^+$ but not cathepsin D$^-$ rat embryonic 3Y1 tumors (Tung et al. 2000).

A similar strategy to that discussed above for cathepsin D has been followed to create an MMP-2 imaging agent (Bremer et al. 2001a, b). In this case, [Cy5.5]-Gly-Pro-Leu-Gly~Val-Arg-Gly-Lys(FITC)-Cys-NH$_2$ was attached to the PL-MPEG$_{92}$ copolymer. A control probe was generated by attaching [Cy5.5]-Gly-Val-Arg-Leu-Gly-Pro-Gly-Lys(FITC)-Cys-NH$_2$. The control probe was not activated by MMP-2. MMP-2 and several other MMPs activated the designed probe, but the relative signal generated by other MMPs at equimolar conditions was below 30% of that for MMP-2 (Bremer et al. 2001b). The MMP-2-targeted probe was able to image HT1080 human fibrosarcoma tumors in mice, while much lower signal was obtained from low MMP-2-producing BT20 breast adenocarcinoma tumors (Bremer et al. 2001a, b).

A modified PL-MPEG$_{92}$ copolymer has been utilized to create a probe for optical zymography of uPA activity (Law et al. 2004, 2005). uPA selectivity was achieved via the sequence Gly-Gly-Ser-Gly-Arg~Ser-Ala-Asn-Ala-Lys(FITC)-Cys-NH$_2$, and Cy5.5 or Cy7 was applied for optical imaging. Ultimately, Cy5.5 was preferred as Cy7 provided a lower increase in fluorescence upon protease activation. The more hydrophobic nature of Cy7 was deemed to be problematic, contributing to fluorophore self-aggregation or intramolecular interactions that inhibited enzyme accessibility (Law et al. 2004). The uPA-targeted probe provided tenfold greater fluorescence for A2058 human melanoma and HT1080 human fibrosarcoma cell media compared with HT-29 human colon carcinoma cell media, consistent with the levels of uPA produced by these cells lines (Law et al. 2005). The probe was also selective for uPA activity in the urine of wild-type versus uPA-knockout mice (Law et al. 2004).

In addition to the NIRF probe described above, targeting of MMPs has been used to create cell-imaging quantum dots (QDs) (Zhang et al. 2006). The QDs are coated with streptavidin and conjugated with biotinylated peptide ligands. The peptide ligand contained an MMP-susceptible sequence sandwiched between a transporting sequence (to allow intracellular delivery of the QD) and a blocking group (that inhibited cellular uptake of the QD). Initial studies showed that $(Arg)_n$, where $n = 4$, 5, 7, or 9, allowed for delivery of QDs into HT1080 cells. For MMP-2-activated delivery, the sequence biotin-$(Arg)_4$-Ahx-Pro-Leu-Gly$\sim$Val-Arg-Gly-$(Glu)_4$ (where Ahx = 6-aminohexanoic acid) was designed. This sequence was conjugated to QDs, and the resulting probe was found to exhibit MMP-2-dependent imaging of HT1080 cells. MMP-7-dependent imaging was achieved with a probe that conjugated QDs to biotin-$(Arg)_3$-Gly-Arg-Pro-Leu-Ala$\sim$Leu-Trp-Arg-Ser-Gly-$(Glu)_5$ (Zhang et al. 2006).

QDs have also been modified with gold nanoparticles (AuNPs), where the AuNPs suppress QD luminescence (Chang et al. 2005). A collagenase-susceptible sequence (Gly-Gly-Leu-Gly-Pro-Ala-Gly-Gly-Cys-Gly) was incorporated between the QD and the AuNP, allowing for protease-activated imaging, and a PEG_{750} coating was applied to the QD to enhance enzyme activity. The probe was activated using *C. histolyticum* collagenase type XI, with a 52% increase in luminescence (Chang et al. 2005). However, it has been noted that this change in signal may not be sufficient for practical applications (Zhou and Ghosh 2006). In an analogous application, QD fluorescence was modulated by the attachment of rhodamine Red-X via a collagenase-susceptible Arg-Gly-Asp-Cys sequence (Shi et al. 2006). This QD probe was utilized to monitor proteolytic activity in the extracelluar matrix of HTB 125 normal breast cells and HTB 126 breast carcinoma cells. After 12 min, the fluorescence from the HTB 126 cells was $\sim$1.7 times that of the fluorescence of the HTB 125 cells, indicating greater proteolytic activity in the carcinoma extracellular matrix (Shi et al. 2006). Control MMP experiments (such as determining MMP levels in the respective cell lines or utilizing an MMP inhibitor) were not performed.

Protease-activated imaging has been achieved by the use of two molecular beacons. The first is a fluorogenic proteolytic beacon for in vivo detection and imaging of tumor-associated MMP-7 activity (McIntyre et al. 2004). The MMP-7 substrate is fluorescein-Ahx-Arg-Pro-Leu-Ala$\sim$Leu-Trp-Arg-Ser-Ahx-Cys (Fl-M7). Fl-M7 is then attached to a polyamido amino (PAMAM) dendrimer core, followed by tetramethylrhodamine (TMR), to generate $(Fl-M7)_m$-PAMAM-$(TMR)_n$. Fl-M7 serves as the optical sensor of protease activity, with TMR effectively quenching fluorescein (Fl) fluorescence in the intact conjugate. In addition, TMR provides an internal reference for the intact conjugate, as TMR can be monitored via its own fluorescent properties. Thus, the cleaved and uncleaved conjugates can be discriminated based on the ratio of green/red (Fl/TMR) fluorescence. The M7 sequence was hydrolyzed much more rapidly by MMP-7 compared with MMP-2, MMP-3, trypsin, and proteinase K. Imaging via $(Fl-M7)_m$-PAMAM-$(TMR)_n$ was utilized to discriminate between MMP-7-positive and MMP-7-negative mouse xenograft tumors originating from SW480 human colon cancer cells (McIntyre et al. 2004).

A second MMP-7-activated molecular beacon was created by incorporating the sequence Gly-Pro-Leu-Gly~Leu-Ala-Arg-Lys between pyropheophorbide (Pyro), which served as a photosensitizer to produce $^1O_2$ and an NIRF probe, and black hole quencher 3 (BHQ3), which served as both a fluorescence and an $^1O_2$ quencher (Zheng et al. 2007). The conjugate, referred to as $PP_{MMP7}B$, was activated upon MMP-7 treatment, but not MMP-2 treatment. The hydrolyzed, Pyro-containing portion of the beacon was internalized by human nasopharyngeal epidermoid carcinoma KB (MMP-7$^+$) cells. Internalization allowed for fluorescent imaging and induction of apoptosis via photodynamic generation of $^1O_2$. Neither effect was observed following $PP_{MMP7}B$ treatment of human breast cancer BT20 (MMP-7$^-$) cells. In vivo application of $PP_{MMP7}B$ in a KB mouse model resulted in imaging of tumors and reduction in tumor size (Zheng et al. 2007).

The Pro-Leu-Gly~Leu-Ala-Gly sequence has been used for MMP-2 activation of cell-penetrating peptides (Jiang et al. 2004). A series of peptides was created where the MMP-2-susceptible sequence was inserted between a cell-penetrating peptide (poly-Arg) and a polyanionic sequence that effectively masked the action of the cell-penetrating peptide by associating with it. A key aspect of the construct is that the MMP-2 cleavage site is located within a hairpin structure that allowed for favorable interaction between the poly-Arg and polyanionic sequences. The best activation and uptake of the cell-penetrating peptide by Jurkat and HT1080 cells was observed for succinoyl-(D-Glu)$_8$-Ahx-Pro-Leu-Gly~Leu-Ala-Gly-(D-Arg)$_9$-Ahx-Lys(Fl) and PEG$_{11000}$-Ahx-(D-Glu)$_9$-Ahx-Pro-Leu-Gly~Leu-Ala-Gly-(D-Arg)$_9$-Ahx-Lys(Cy5). The latter construct was effective for imaging HT1080 tumor xenografts and, with replacement of the PEG$_{11000}$ by PEG$_{5000}$, human squamous cell carcinoma tissue sections (Jiang et al. 2004).

Initial studies have been presented for MMP-activated nanoparticle self-assembly imaging of tumor cells (Harris et al. 2006). Superparamagnetic Fe_3O_4 nanoparticles were coated with neutravidin, followed by attachment of an MMP-2-susceptible sequence (Lys-Gly-Pro-Leu-Gly~Val-Arg-Gly-Cys) to Lys ε-amino groups on the neutravidine. Finally, MPEG-succimidyl α-methylbutanoate (where PEG$_{10000}$ was ultimately shown to be optimal) was linked to the peptide Lys ε-amino groups. Treatment with MMP-2 resulted in nanoparticle self-assembly, which caused shortening of T2 relaxation times in magnetic resonance imaging (MRI). Nanoparticles were incubated with HT1080 cell cultures, and self-assembly led to a substantial shortening of T2 relaxation times, maps of which were generated with MRI (Harris et al. 2006).

Specificity of Activation Sequences

The problem with most protease-activated prodrugs and delivery and imaging systems is a lack of stability at nontargeted tissues and organs due to nonspecific activation (Chau et al. 2006a, b). This is well exemplified by many of the MMP-activated systems described herein (Table 39.1). The Gly-Pro-Gln-Gly~Ile-Ala-Gly-Gln-Arg and Gly-Pro-Gln-Gly~Ile-Trp-Gly-Gln protease linker sequences

Table 39.1 Sequences Used for Protease-Activated Drug Delivery and Imaging Agents

Protease	Sequence	Utility
MMP	Pro-Leu-Gly~Hof-Orn-Leu	Dox prodrug
MMP	Glu-Pro-Cit-Phe~Hof-Tyr-Leu	Dox prodrug
MMP	Gly-Pro-Leu-Gly~Val	Dox prodrug
MMP	Gly-Pro-Leu-Gly~Leu-Trp-Ala-Gln	Anthrax toxin PA prodrug
MMP-2	Pro-Gln-Gly~Ile-Ala-Gly-Gln	Melittin prodrug
MMP-2	Pro-Gln-Gly~Ile-Mel-Gly	Mel prodrug
MMP-2	Gly-Pro-Leu-Gly~Val-Arg-Gly-Lys	TNF prodrug
MMP-2	His-Pro-Val-Gly~Leu-Leu-Ala-Arg	TNF prodrug
MMP-2	Gly-Pro-Leu-Gly~Ile-Ala-Gly-Gln	Fusogenic liposomes
MMP-2	Gly-Pro-Leu-Gly~Val-Arg-Gly	Dox micelles
MMP-2	Pro-Leu-Gly~Leu-Trp-Ala	MLV envelope glycoprotein delivery
MMP-2	Gly-Gly-Pro-Leu-Gly~Leu-Trp-Ala-Gly-Gly	GALV envelope glycoprotein delivery
MMP-2	Gly-Pro-Leu-Gly~Val-Arg-Gly	Cy5.5 imaging agent
MMP-2	Pro-Leu-Gly~Val-Arg-Gly	QD imaging agent
MMP-2	Pro-Leu-Gly~Leu-Ala-Gly	Cy5, fluorescein imaging agents
MMP-2	Lys-Gly-Pro-Leu-Gly~Val-Arg-Gly	Nanoparticle imaging agent
MMP-9	Ala-Ala-Leu-Gly~Nva-Pro	FITC prodrug
MMP-9	Pro-Leu-Gly~Met-Thr-Ser	SeV prodrug
MMP-9	Triple-helical Gly-Pro-Gln-Gly~Ile-Ala-Gly-Gln-Arg	Liposomes
MMP-2, MMP-9	Pro-Leu-Gly~Leu	Auristan, Dox prodrugs
MMP-2, MMP-9	Gly-Pro-Leu-Gly~Ile-Ala-Gly-Gln	Dox prodrug
MMP-2, MMP-9	Pro-Val-Gly~Leu-Ile-Gly	MTX prodrug
MMP-2, MMP-9	Gly-Pro-Leu-Gly~Met-Leu-Ser-Gln	Anthrax toxin PA prodrug
MMP-2, MMP-9	Cys-Gly~Leu-Asp-Asp	Hydrogel
MMP-2,MMP-9, MT1-MMP	Gly-Pro-Gln-Gly~Ile-Trp-Gly-Gln	Hydrogel
MMP-7	Arg-Pro-Leu-Ala~Leu-Trp-Arg-Ser	Molecular beacon
MMP-7	Gly-Pro-Leu-Gly~Leu-Ala-Arg-Lys	Molecular beacon
MMP-7, cathepsin B	Asn-Lys-Ser-Arg-Leu-Gly-Leu-Gly	EqTII prodrug
MT1-MMP	Gly-Pro-Leu-Pro-Leu-Arg	Fusogenic liposomes
PSA	Hyp-Ala-Ser-Chg-Gln-Ser-Leu	Dox, vinblastine prodrugs
PSA	His-Ser-Ser-Lys-Leu-Gln-Leu	Dox, 12ADT prodrugs
hK2	Gly-Lys-Ala-Phe-Arg~Arg-Leu	12ADT prodrug
Plasmin	Ala-Phe-Lys	Dox prodrug
uPA	Ser-Gly-Arg-Ser-Ala	Anthrax toxin PA prodrug
uPA	Gly-Gly-Ser-Gly-Arg-Ser-Ala-Asn-Ala	Cy5.5, Cy7 imaging agent
CD10	β-Ala-Leu-Ala-Leu	Dox prodrug
Legumain	Ala-Ala-Asn~Leu	Dox prodrug
Cathepsin B	Phe-Lys	Dox, paclitaxel, mitomycin prodrugs
Cathepsin B	Val-Cit	Auristatin prodrug

Table 39.1 (continued)

Protease	Sequence	Utility
Cathepsin D	Gly-Pro-Ile-Cys(ethyl)-Phe∼Phe-Arg-Leu-Gly	Cy5.5 imaging agent
Elastase	Ala-Ala-Pro-Val	Fusogenic liposomes

Hof homoPhe, *Cit* citrulline, *Mel* melphalan (Phe mustard), *Chg* cyclohexylglycyl, *PA* protective antigen, *MLV* murine leukemia virus, *GALV* gibbon ape leukemia virus, *QD*, quantum dot, *FITC* fluorescein isothiocyanate, *MTX* methotrexate, *TNF* tumor necrosis factor.

used for liposomes and hydrogels are readily cleaved by numerous MMP family members (Nagase and Fields 1996, Lauer-Fields et al. 2000b), and thus do not represent a truly specific target. In similar fashion, the Pro-Leu-Gly∼Val-Arg-Gly sequence, used for NIRF imaging of MMP-2-positive tumors and MMP-2-mediated cellular uptake of QDs, is additionally hydrolyzed by MMP-1, MMP-7, MMP-8, and MMP-9 (Bremer et al. 2001b, Zhang et al. 2006). The NIRF MMP-2 probe was not selective when tested in MMP-2-deficient mice (Sloane et al. 2006). The Pro-Leu-Gly∼Leu-Ala-Gly sequence, used for MMP-2 activation of cell-penetrating peptides (Jiang et al. 2004), is hydrolyzed by numerous MMPs (Nagase and Fields 1996). Gly-Pro-Leu-Gly∼Ile-Ala-Gly-Gln, Pro-Val-Gly∼Leu-Ile-Gly, Gly-Pro-Leu-Gly∼Val, Gly-Pro-Leu-Gly∼Met-Leu-Ser-Gln, and Gly-Pro-Leu-Gly∼Leu-Trp-Ala-Gln, used to deliver N^4-octadecyl-1-β-D-arabinofuranosylcytosine, methotrexate, Dox, or anthrax toxin to MMP-2/MMP-9-positive tumors (Liu et al. 2000, Bae et al. 2003, Chau et al. 2004, Terada et al. 2006), are susceptible to other MMPs (Nagase and Fields 1996). A few "selective" single-stranded sequences have been developed, such as Pro-Cit-Gly∼Hof-Tyr-Leu, which is proposed to be favored by MMP-2, MMP-9, and MT1-MMP over neprilysin (Albright et al. 2005), and Arg-Pro-Leu-Ala∼Leu-Trp-Arg, which is favored by MMP-7 (McIntyre et al. 2004).

The selectivity of sequences designed for cathepsin activation is also of concern. Poly-L-Lys sequences, used for NIRF imaging of broad cathepsin activity (Tung 2004), are certainly susceptible to hydrolysis by trypsin-like proteases. The cathepsin D selective sequence Gly-Pro-Ile-Cys(ethyl)-Phe∼Phe-Arg-Leu-Gly (Tung et al. 1999, Tung 2004) has also been described as cathepsin B specific (Funovics et al. 2004). As discussed previously (Sloane et al. 2006), the selectivity of cathepsin-activated probes needs to be explored in wild-type versus cathepsin-deficient mice.

In some cases, a lack of selectivity may be based on improper sequence design. Some sequences that were meant for application of collagenolytic MMPs were designed based on bacterial collagenase sequence specificity. For example, the sequence PEG-Gly-Gly-Cys(Bodipy)-Leu∼Gly-Pro-Ala-Cys(Bodipy)-Gly-Lys-PEG was utilized for developing collagenase-sensitive PEG hydrogels that would allow for fluorescent imaging of mammalian collagenase activity (Lee et al. 2007b). Proteolysis of the substrate was examined using *C. histolyticum* (bacterial) collagenase, which cleaves Y∼Gly-X bonds, where Leu and Pro can be favored in the Y and X positions, respectively (Van Wart 2004). Subsequently, it was shown that fibroblasts caused degradation of the substrate. The conclusion that collagenases secreted from the fibroblasts (most likely MMP-1 and/or MMP-2) were responsible for this activity seems unlikely, as Leu-Gly-Pro-Ala offers no favorable bonds for MMP-1- or MMP-2-mediated hydrolysis (Nagase and Fields 1996). The PEG-Gly-Gly-Cys(Bodipy)-Leu-Gly-Pro-Ala-Cys(Bodipy)-Gly-Lys-PEG substrate was cleaved rapidly by proteinase K, indicating its susceptibility to other proteases. In similar fashion, FRET substrates were designed with a "collagenase"-susceptible sequence, Ala-Leu-Aib-Ala-Ala~Gly-Gly-Pro-Ala-Cys, between an N-terminal His$_6$ and a QXL-520 dark quenching acceptor attached to the C-terminal Cys (Medintz et al. 2006). The peptide self-associated via the His$_6$ region to dihydrolipoic acid-capped QDs. Although a collagenase from the MMP family was the ultimate intended target, the substrate was tested only with *C. histolyticum* collagenase (Medintz et al. 2006).

For enzyme-activated prodrug and delivery and imaging systems, specificity could be significantly improved if a substrate was introduced that was stable to the in vivo environment during delivery while maintaining a high degree of selectivity for a specific enzyme. In general, there are several problems associated with simple peptides serving as selective substrates. First, the specificity and affinity of such ligands is usually not high. This problem is obviated by induction of well-defined secondary and/or tertiary structures within peptides. A collagen-like triple-helical conformation has been shown to enhance the activity of some MMP family members (Lauer-Fields et al. 2001, 2003a). Second, peptides are typically extremely susceptible to general proteolysis. The conformational restriction or stereochemical manipulation (Pierschbacher and Ruoslahti 1987, Li et al. 1997) of such sequences often reduces general proteolytic activity. Thus, one approach for improving peptide in vivo activity is to create biomolecules with distinct structural elements,that is, that are conformationally constrained. The triple-helical peptide (THP) represents such a construct. The development of THP prodrugs, nano-DDSs, hydrogels, and imaging agents that are activated by only selected MMPs could represent a second generation in protease-targeted medicines.

Our laboratory has described a selective MMP-2/MMP-9/MMP-12 FRET THP substrate, (1(V)436–447 fTHP [(Gly-Pro-Hyp)$_5$-Gly-Pro-Lys(Mca)-Gly-Pro-Pro-Gly~Val-Val-Gly-Glu-Lys(Dnp)-Gly-Glu-Gln-(Gly-Pro-Hyp)$_5$-NH$_2$] (Lauer-Fields et al. 2003a). This MMP-2/MMP-9/MMP-12 triple-helical substrate has been utilized to develop a potentially selective imaging agent by replacing the Mca and Dnp with NIR dyes, where self-quenching occurs in the intact peptide (Edwards et al. 2007). Initial studies found a 1.6-fold increase in 5-carboxy-fluorescein fluorescence upon treatment of the substrate with MMP-2 (Edwards et al. 2007).

We have also previously described a number of FRET THP substrates that are either suitable for most collagenolytic MMPs or selective for different collagenolytic MMPs (Table 39.2) (Lauer-Fields et al. 2001, 2002, 2003a, b, 2004; Baronas-Lowell et al. 2004, Hurst et al. 2004, Minond et al. 2004, 2006, 2007). Selectivity can be modulated by adding hexanoic acid (C$_6$) or decanoic (C$_{10}$) to the *N*-terminus of the fTHP (Table 39.2); addition of these "pseudo-lipids" increases the thermal stability of the triple-helix (Yu et al. 1996, 1998). In addition, THP studies have led to the assignment of unique triple-helical peptidase behaviors for most of the

Table 39.2 Sequences of (Gly-Pro-Hyp)$_5$-Gly-Pro-Lys(Mca)-Gly-Pro-P$_2$-P$_1$~P$_1$'-P$_2$'-P$_3$'-P$_4$'-Lys(Dnp)-Gly-Val-Arg-(Gly-Pro-Hyp)$_5$-NH$_2$ fTHPs and Their MMP Preferences

fTHP	P$_2$-P$_1$~P$_1$'-P$_2$'-P$_3$'-P$_4$ sequence	Peptide MMP preference	C$_6$-peptide MMP preference	C$_{10}$-peptide MMP preference
fTHP-4	Gln-Gly~Leu-Arg-Gly-Gln	MT1-MMP > > MMP-2~MMP-1~MMP-8~MMP-13~MMP-9~MT2-MMP	ND	MMP-8 > MMP-13 > MT1-MMP
fTHP-9	Gln-Gly~Cys(Mob)-Arg-Gly-Gln	MMP-8 > MT1-MMP > > MT2-MMP > MMP-13 > > MMP-1~MMP-2	ND	MMP-8 > > MT1-MMP~MT2-MMP > MMP-13
fTHP-10	Orn-Gly~Leu-Arg-Gly-Gln	MT1-MMP > MMP-13 > MT2-MMP~MMP-8~MMP-2 > MMP-1~MMP-9	ND	MT1-MMP > MMP-13~MT2-MMP
fTHP-11	Orn-Gly~Cys(Mob)-Arg-Gly-Gln	MT1-MMP > > MT2-MMP~MMP-8 > MMP-13	ND	MT1-MMP > > MMP-13 > MMP-8~MT2-MMP
fTHP-12	Leu-Gly~Met-Arg-Gly-Gln	MMP-2 > > MMP-13 > MMP-8~MT1-MMP~MT2-MMP~MT5-MMP	MT2-MMP > MMP-13 > MT1-MMP	
fTHP-13	Val-Asn~Phe-Arg-Gly-Gln	MMP-2 > > MMP-13 > MMP-8	MMP-2~MMP-13 > MMP-8~MT1-MMP	ND
fTHP-14	Val-Asn~Phe-Arg-Gly-Pro	ND	MMP-13 > MMP-8 > MT2-MMP > MT1-MMP	ND

MMP matrix metalloproteinase, *ND* not determined, *fTHP* fluorogenic triple-helical peptide.

collagenolytic MMPs (Lauer-Fields et al. 2000a, b, 2001, 2003a; Minond et al. 2004, 2006, 2007). The information on triple-helical substrate selectivity for MMP family members can be utilized to design specific MMP-activated agents. For example, prior MMP-activated hydrogels that relied on single-stranded substrates (Lutolf et al. 2003a,b, Tauro and Gemeinhart 2005a,b) can use the same Michael-type addition reaction to incorporate selective THP substrates.

The effects of substrate conformation on protease activity have been recently examined for one member of the a disintegrin and metalloproteinase with thrombospondin motifs (ADAMTS) family, ADAMTS-4 (Lauer-Fields et al. 2007). Interest in the ADAMTS family role in cancer stems from reports that highly invasive CNS-1 glioma cells utilize ADAMTS-4 to cleave brevican (Matthews et al. 2000), while high expression levels of ADAMTS-8 in combination with low expression levels of ADAMTS-15 are prognostic of poor clinical outcome in patients with breast carcinoma (Porter et al. 2006). Substrate topology modulated the affinity and sequence specificity of ADAMTS-4, with K_M values indicating a preference for triple-helical structure (Lauer-Fields et al. 2007). These results suggested that ADAMTS-4 substrates could be designed based on a combination of sequence and conformation (topology).

Topological specificity may well be an overall guiding principle for proteolytic behavior (Lauer-Fields et al. 2007). There are numerous cases, beyond the previously mentioned triple-helical structure, where substrate topology may serve as a determinant for protease specificity. Serum proteinase inhibitors, such as the human α2-macroglobulin and its homologues, entrap a variety of proteinases with a similar "bait" region via unique three-dimensional structures (Ruben et al. 1988, Enghild et al. 1989, Sottrup-Jensen 1989). Linear peptide models of α2-macroglobulin cleavage site sequences are hydrolyzed much less efficiently by MMP-1 than are linear models of collagen cleavage sites, whereas the opposite is true for the native proteins (Enghild et al. 1989, Upadhye and Ananthanarayanan 1995). In some cases, α2-macroglobulin cleavage sites are not even predicted based on phage display peptide library experiments (Pan et al. 2003b), suggesting protease recognition of this serum proteinase inhibitor is directed by substrate topology. In a similar fashion, the α-secretase activity against amyloid precursor protein (APP) has little sequence specificity, and thus specificity has been attributed to the distance between the membrane-bound protease and substrate and the α-helical conformation of the substrate (Maruyama et al. 1991, Sisodia 1992, Lammich et al. 1999). The enzyme that cleaves thyrotropin hormone receptors (TSHRs) also lacks sequence specificity, and its action has been attributed to a "molecular ruler" mechanism based on the distance between where the enzyme binds and cleaves (Tanaka et al. 2000). In the case of TSHRs, the initial binding event could well be dictated by the three-dimensional structure of the substrate, in combination with proximity effects based on the membrane-binding sites of the enzyme and substrate. This same scenario may direct ADAM17/tumor-necrosis factor-α-converting enzyme (TACE) shedding of a great variety of cell surface proteins that maintain α-helical structure in their juxtamembrane regions (Arribas and Merlos-Suarez 2003,

Tsakadze et al. 2006). The sheddase activity of MMPs may also be directed by cell surface protein secondary structures (Cauwe et al. 2007).

The development of conformationally constrained substrates has often involved an iterative process that combines knowledge of protein substrate sequences and secondary structures (Lauer-Fields et al. 2007). One could also use computer simulations of peptides to design conformationally constrained substrates. A relatively new physical-principles-based computer simulation method, called ZAM (Zipping & Assembly Method), has been used to determine the structures and properties of peptides and proteins and to design peptides (Ozkan et al. 2007). The ZAM-based strategy is unique among computational methods, in two respects. First, unlike bioinformatics methods, it is based on a first-principles all-atom physics-based potential function [AMBER96 plus the implicit GB/SA solvation model of Onufriev, Bashford, and Case (OBC)] (Onufriev et al. 2002). This means ZAM can handle amino acid sequences not found in the protein data bank, and noncanonical conformations or monomers, such as D-amino acids, or experimental laboratory conditions, or binding or partitioning situations, conformational transitions, unfolded states, and other physical properties.

Second, unlike traditional physics-based modeling, which requires supercomputing-level computational resources even for short peptides, the ZAM approach is effective at sampling the important regions of conformational space, even up to 110-mer peptides and proteins. ZAM does this by following putative physical folding routes of the peptide from unfolded to folded states. ZAM first explores local conformations of peptide sub-pieces of the chain, then growing and assembling those pieces into increasingly larger structures. Computationally, this involves sequential recognition of metastable protein substructures found by the forcefield, then an imposition of restraints to enforce those partial structures, then further searching to grow the structure. Thus, ZAM can sample the essential conformational space of large peptides or small proteins. This sampling efficiency is crucial for modeling peptide structures and energetics, because peptides typically have a great deal of conformational flexibility.

Protease topological preferences could work in concert with secondary-binding sites (exosites) to allow for the selective catabolism of complex substrates. Such a precedent has been established previously for γ-secretase, where hydrolysis of transmembrane domains is regulated by exosite binding to an α-helical structure (Das et al. 2003, Vooijs et al. 2004, Kornilova et al. 2005). It should be possible to identify proteases that recognize specific topologies, and design substrates based on these topologies. Such substrates would enhance the specificity of protease-activated agents.

Acknowledgments I gratefully acknowledge the National Institutes of Health (CA98799, EB000289, and MH078948) for support of my laboratory's research on metalloproteases and drug delivery systems. I am most grateful to Allison Price for compiling the database of references and David Khan for constructing the figures.

References

Albright C.F., Graciani N., Han W. et al. (2005). Matrix metalloproteinase-activated doxorubicin prodrugs inhibit HT1080 xenograft growth better than doxorubicin with less toxicity. Mol. Cancer Ther. 4: 751–760.

Allen C., McDonald C., Giannini C. et al. (2004). Adenoviral vectors expressing fusogenic membrane glycoproteins activated via matrix metalloproteinase cleavable linkers have significant antitumor potential in the gene therapy of gliomas. J. Gene Med. 6: 1216–1227.

Allen T.M. and Cullis P.R. (2004). Drug delivery systems: Entering the mainstream. Science 303: 1818–1822.

Allen T.M. and Hansen C. (1991). Pharmacokinetics of stealth versus conventional liposomes: Effect of dose. Biochim. Biophys. Acta 1068: 133–141.

Allen T.M., Hansen C., Martin F. et al. (1991). Liposomes containing synthetic lipid derivatives of poly(ethylene glycol) show prolonged circulation half-lives in vivo. Biochim. Biophys. Acta 1066: 29–36.

Arribas J. and Merlos-Suarez A. (2003). Shedding of plasma membrane proteins. Cell Surface Proteases: Current Topics in Developmental Biology, Vol. 54. S. Zucker and W.-T. Chen, 125–144, San Diego: Academic Press.

Backer M.V., Gaynutdinov T.I., Patel V. et al. (2004). Adapter protein for site-specific conjugation of payloads for targeted drug delivery. Bioconjugate Chem. 15: 1021–1029.

Bae M., Cho S., Song J. et al. (2003). Metalloprotease-specific poly(ethylene glycol) methyl ether-peptide-doxorubicin conjugate for targeting anticancer drug delivery based on angiogenesis. Drugs Exp. Clin. Res. 29: 15–23.

Baronas-Lowell D., Lauer-Fields J.L., Borgia J.A. et al. (2004). Differential modulation of human melanoma cell metalloproteinase expression by a2b1 integrin and CD44 triple-helical ligands derived from type IV collagen. J. Biol. Chem. 279: 43503–43513.

Bremer C., Bredow S., Mahmood U. et al. (2001a). Optical imaging of matrix metalloproteinase-2 activity in tumors: Feasibility study in a mouse model. Radiology 221: 523–529.

Bremer C., Tung C.-H. and Weissleder R. (2001b). In vivo molecular target assessment of matrix metalloproteinase activity. Nature Med. 7: 743–748.

Bremer C., Ntzachristos V. and Weisslender R. (2003). Optical-based molecular imaging: Contrast agents and potential medical applications. Eur. Radiol. 13: 231–243.

Cauwe B., Van den Steen P.E. and Opdenakker G. (2007). The biochemical, biological, and pathological kaleidoscope of cell surface substrates processed by matrix metalloproteinases. Crit. Rev. Biochem. Mol. Biol. 42: 113–185.

Chang E., Miller J.S., Sun J. et al. (2005). Protease-activated quantum dot probes. Biochem. Biophys. Res. Commun. 334: 1317–1321.

Chau Y., Tan F.E. and Langer R. (2004). Synthesis and characterization of dextran-peptide-methotrexate conjugates for tumor targeting via mediation by matrix metalloproteinase II and matrix metalloproteinase IX. Bioconjugate Chem. 15: 931–941.

Chau Y., Dang N.M., Tan F.E. et al. (2006a). Investigation of targeting mechanism of new dextran-peptide-methotrexate conjugates using biodistribution study in matrix-metalloproteinase-overexpressing tumor xenograft model. J. Pharm. Sci. 95: 542–551.

Chau Y., Padera R.F., Dang N.M. et al. (2006b). Antitumor efficacy of a novel polymer-peptide-drug conjugate in human tumor xenograft models. Int. J. Cancer 118: 1519–1526.

Das C., Berezovska O., Diehl T.S. et al. (2003). Designed helical peptides inhibit an intramembrane protease. J. Am. Chem. Soc. 125: 11794–11795.

Drummond D.C., Meyer O., Hong K. et al. (1999). Optimizing liposomes for delivery of chemotherapeutic agents to solid tumors. Pharmacol. Rev. 51: 691–744.

Dubois V., Dasnois L., Lebtahi K. et al. (2002). CPI-0004Na, a new extracellularly tumor-activated prodrug of doxorubicin: In vivio toxicity, activity and tissue distribution confirm tumor cell selectivity. Cancer Res. 62: 2327–2331.

Dubowchik G.M. and Firestone R.A. (1998). Cathepsin B-sensitive dipeptide prodrugs. 1. A model study of structural requirements for efficient release of doxorubicin. Bioorg. Med. Chem. Lett. 8: 3341–3346.

Dubowchik G.M., Mosure K., Knipe J.O. et al. (1998). Cathepsin B-sensitive dipeptide prodrugs. 2. Models of anticancer drugs paclitaxel (Taxol®), mitomycin C and doxorubicin. Bioorg. Med. Chem. Lett. 8: 3347–3352.

Duzgunes N. and Nir S. (1999). Mechanisms and kinetics of liposome-cell interactions. Adv. Drug Deliv. Rev. 40: 3–18.

Edwards W.B., Cheney P.P., Fields G.B. et al. (2007). Synthesis and evaluation of triple helical peptides with quenched probes for MMP-2 and MMP-9 optical imaging. Joint Molecular Imaging Conference. Providence, Rhode Island.

Enghild J.J., Salvesen G., Brew K. et al. (1989). Interaction of human rheumatoid synovial collagenase (matrix metalloproteinase 1) and stromelysin (matrix metalloproteinase 3) with human a2-macroglobulin and chicken ovastatin. J. Biol. Chem. 264: 8779–8785.

Fernandez A.-M., Van derpoorten K., Dasnois L. et al. (2001). N-succinyl-(β-alanyl-L-leucyl-L-alanyl-L-leucyl) doxorubicin: An intracellularly tumor-activated prodrug devoid of intravenous acute toxicity. J. Med. Chem. 44: 3750–3753.

Funovics M.A., Weissleder R. and Mahmood U. (2004). Catheter-based in vivo imaging of enzyme activity and gene expression: Feasibility study in mice. Radiology 231: 659–666.

Gabizon A.A., Shmeeda H. and Zalipsky S. (2006). Pros and cons of the liposome platform in cancer drug targeting. J. Liposome Res. 16: 175–183.

Harris T.J., von Maltzahn G., Derfus A.M. et al. (2006). Proteolytic actuation of nanoparticle self-assembly. Angew. Chem. Int. Ed. Eng. 45: 3161–3165.

Hatakeyama H., Akita H., Ishida E. et al. (2007). Tumor targeting of doxorubicin by anti-MT1-MMP antibody-modified PEG liposomes. Int. J. Pharm. 342: 194–200.

Hu L., Ho R.J.Y. and Huang L. (1986). Trypsin induced destabilization of liposomes composed of dioleoylphosphatidylethanolamine and glycorphorin. Biochem. Biophys. Res. Commun. 141: 973–978.

Hurst D.R., Schwartz M.A., Ghaffari M.A. et al. (2004). Catalytic- and ecto-domains of membrane type 1-matrix metalloproteinase have similar inhibition profiles but distinct endopeptidase activities. Biochem. J. 377: 775–779.

Jamil J., Sheikh S. and Ahmad I. (2004). Liposomes: The next generation. Mod. Drug Discovery 7 (1): 36–39.

Janssen S., Jakobsen C.M., Rosen D.M. et al. (2004). Screening a combinatorial peptide library to develop a human glandular kallikrein 2-activated prodrug as targeted therapy for prostate cancer. Mol. Cancer Ther. 3: 1439–1450.

Janssen S., Rosen D.M., Ricklis R.M. et al. (2006). Pharmacokinetics, biodistribution, and antitumor efficacy of a human glandular kallikrein 2 (hK2)-activated thapsigargin prodrug. Prostate 66: 358–368.

Jiang T., Olson J.E.S., Nguyen Q.T. et al. (2004). Tumor imaging by means of proteolytic activation of cell penetrating peptides. Proc. Natl. Acad. Sci. USA 101: 17867–17872.

Johnson K.J., Peng K.-W., Allen C. et al. (2003). Targeting the cytotoxicity of fusogenic membrane glycoproteins in gliomas through protease-substrate interaction. Gene Ther. 10: 725–732.

Kinoh H., Inoue M., Washizawa K. et al. (2004). Generation of a recombinant Sendai virus that is selectively activated and lyses human tumor cells expressing matrix metalloproteinases. Gene Ther. 11: 1137–1145.

Klibanov A.L., Maruyama K., Torchilin V.P. et al. (1990). Amphipathic polyethyleneglycols effectively prolong the circulation time of liposomes. FEBS Lett. 268: 235–237.

Kline T., Torgov M.Y., Mendelsohn B.A. et al. (2004). Novel antitumor prodrugs designed for activation by matrix metalloproteinases-2 and -9. Mol. Pharmaceutics 1: 9–22.

Kondo M., Asai T., Katanasaka Y. et al. (2004). Anti-neovascular therapy by liposomal drug targeted to membrane type-1 matrix metalloprotease. Int. J. Cancer 108: 301–306.

Kornilova A.Y., Bihel F., Das C. et al. (2005). The initial substrate-binding site of g-secretase is located on presenilin near the active site. Proc. Natl. Acad. Sci. USA 102: 3230–3235.

Kung Sutherland M.S., Sanderson R.J., Gordon K.A. et al. (2006). Lysosomal trafficking and cysteine protease metabolism confer target-specific cytotoxicity by peptide-linked anti-CD30-auristatin conjugates. J. Biol. Chem. 281: 10540–10547.

Lammich S., Kojro E., Postina R. et al. (1999). Constitutive and regulated a-secretase cleavage of Alzheimer's amyloid precursor protein by a disintegrin metalloprotease. Proc. Natl. Acad. Sci. USA 96: 3922–3927.

Lauer-Fields J.L. and Fields G.B. (2002). Triple-helical peptide analysis of collagenolytic protease activity. Biol. Chem. 383: 1095–1105.

Lauer-Fields J.L., Nagase H. and Fields G.B. (2000a). Use of Edman degradation sequence analysis and matrix-assisted laser desorption/ionization mass spectrometry in designing substrates for matrix metalloproteinases. J. Chromatogr. A. 890: 117–125.

Lauer-Fields J.L., Tuzinski K.A., Shimokawa K. et al. (2000b). Hydrolysis of triple-helical collagen peptide models by matrix metalloproteinases. J. Biol. Chem. 275: 13282–13290.

Lauer-Fields J.L., Broder T., Sritharan T. et al. (2001). Kinetic analysis of matrix metalloproteinase triple-helicase activity using fluorogenic substrates. Biochemistry 40: 5795–5803.

Lauer-Fields J.L., Sritharan T., Stack M.S. et al. (2003a). Selective hydrolysis of triple-helical substrates by matrix metalloproteinase-2 and -9. J. Biol. Chem. 278: 18140–18145.

Lauer-Fields J.L., Kele P., Sui G. et al. (2003b). Analysis of matrix metalloproteinase activity using triple-helical substrates incorporating fluorogenic L- or D-amino acids. Anal. Biochem. 321: 105–115.

Lauer-Fields J.L., Nagase H. and Fields G.B. (2004). Development of a solid-phase assay for analysis of matrix metalloproteinase activity. J. Biomolecular Techniques 15: 305–316.

Lauer-Fields J.L., Sritharan T., Kashiwagi M. et al. (2007). Substrate conformation modulates aggrecanase (ADAMTS-4) affinity and sequence specificity: Suggestion of a common topological specificity of functionally diverse proteases. J. Biol. Chem. 282: 142–150.

Law B., Curino A., Bugge T.H. et al. (2004). Design, synthesis, and characterization of urokinase plasminogen-activator sensitive near-infrared reporter. Chem. Biol. 11: 99–106.

Law B., Hsiao J.-K., Bugge T.H. et al. (2005). Optical zymography for specific detection of urokinase plasminogen activator activity in biological samples. Anal. Biochem. 338: 151–158.

Lee G.Y., Park K., Kim S.Y. et al. (2007a). MMPs-specific PEGylated peptide-DOX conjugate micelles that can contain free doxorubicin. Eur. J. Pharm. Biopharm. 67: 646–654.

Lee S.-H., Moon J.J., Miller J.S. et al. (2007b). Poly (ethylene glycol) hydrogels conjugated with a collagenase-sensitive fluorogenic substrate to visualize collagenase activity during three-dimensional cell migration. Biomaterials 28: 3163–3170.

Li C., McCarthy J.B., Furcht L.T. et al. (1997). An all-D amino acid peptide model of a1(IV)531–543 from type IV collagen binds the a3b1 integrin and mediates tumor cell adhesion, spreading, and motility. Biochemistry 36: 15404–15410.

Liu S., Netzel-Arnett S., Birkedal-Hansen H. et al. (2000). Tumor cell-selective cytotoxicity of matrix metalloproteinase-activated anthrax toxin. Cancer Res. 60: 6061–6067.

Lutolf M.P., Lauer-Fields J.J., Schmoekel H.G. et al. (2003a). Synthetic matrix metalloproteinase-sensitive hydrogels for the conduction of tissue regeneration: Engineering cell-invasion characteristics. Proc. Natl. Acad. Sci. USA 100: 5413–5418.

Lutolf M.P., Weber F.E., Schmoekel H.G. et al. (2003b). Repair of bone defects using synthetic mimetics of collagenous extracellular matrices. Nat. Biotechnol. 21: 513–518.

Mahmood U., Tung C.-H., Bogdanov J. et al. (1999). Near-infrared optical imaging of protease activity for tumor detection. Radiology 213: 866–870.

Maruyama K., Kametani F., Usami M. et al. (1991). "Secretase," Alzheimer amyloid protein precursor secreting enzyme is not sequence-specific. Biochem. Biophys. Res. Commun. 179: 1670–1676.

Maruyama K., Ishida O., Takizawa T. et al. (1999). Possibility of active targeting to tumor tissues with liposomes. Adv. Drug Deliv. Rev. 40: 89–102.

Matthews R.T., Gary S.C., Zerillo C. et al. (2000). Brain-enriched hyaluronan binding (BEHAB)/ brevican cleavage in a glioma cell line is mediated by a disintegrin and metalloproteinase with thrombospondin motifs (ADAMTS) family member. J. Biol. Chem. 275: 22695–22703.

McIntyre J.O., Fingleton B., Wells K.S. et al. (2004). Development of a novel fluorogenic proteolytic beacon for in vivo detection and imaging of tumor-associated matrix metalloproteinase-7 activity. Biochem. J. 377: 617–628.

Medintz I.L., Clapp A.R., Brunel F.M. et al. (2006). Proteolytic activity monitored by fluorescence resonance energy transfer through quantum-dot-peptide conjugates. Nat. Mater. 5: 581–589.

Meers P. (2001). Enzyme-activated targeting of liposomes. Adv. Drug Deliv. Rev. 53: 265–272.

Minond D., Lauer-Fields J.L., Nagase H. et al. (2004). Matrix metalloproteinase triple-helical peptidase activities are differentially regulated by substrate stability. Biochemistry 43: 11474–11481.

Minond D., Lauer-Fields J.L., Cudic M. et al. (2006). The roles of substrate thermal stability and P2 and P1' subsite identity on matrix metalloproteinase triple-helical peptidase activity and collagen specificity. J. Biol. Chem. 281: 38302–38313.

Minond D., Lauer-Fields J.L., Cudic M. et al. (2007). Differentiation of secreted and membrane-type matrix metalloproteinase activities based on substitutions and interruptions of triple-helical sequences. Biochemistry 46: 3724–3733.

Nagase H. and Fields G.B. (1996). Human matrix metalloproteinase specificity studies using collagen sequence-based synthetic peptides. Biopolymers 40: 399–416.

Oku N. (1999). Anticancer therapy using glucuronate modified long-circulating liposomes. Adv. Drug Deliv. Rev. 40: 63–73.

Onufriev A., Case D.A. and Bashford D. (2002). Effective Born radii in the generalized Born approximation: The importance of being perfect. J. Comput. Chem. 23: 1297–1304.

Ozkan S.B., Wu A.G.H., Chodera J. et al. (2007). Protein folding by zipping and assembly. Proc. Natl. Acad. Sci. USA 104: 11987–11992.

Pak C.C., Ali S., Janoff A.S. et al. (1998). Triggerable lipsomal fusion by enzyme cleavage of a novel peptide-lipid conjugate. Biochim. Biophys. Acta 1372: 13–17.

Pak C.C., Erukulla R.K., Ahl P.L. et al. (1999). Elastase activated liposomal delivery to nucleated cells. Biochim. Biophys. Acta 1419: 111–126.

Pan C., Cardarelli P.M., Nieder M.H. et al. (2003a). CD10 is a key enzyme involved in the activation of tumor-activated peptide prodrug CPI-0004Na and novel analogues: Implications for the design of novel peptide prodrugs for the therapy of CD10+ tumors. Cancer Res. 63: 5526–5531.

Pan W., Arnone M., Kendall M. et al. (2003b). Identification of peptide substrates for human MMP-11 (stromelysin-3) using phage display. J. Biol. Chem. 278: 27820–27827.

Peng K.W., Morling F.J., Cosset F.L. et al. (1997). A gene delivery system activatable by disease-associated matrix metalloptoteinases. Hum. Gen. Ther. 8: 729–738.

Peng K.-W., Vile R.G., Cosset F.-L. et al. (1999). Selective transduction of protease-rich tumors by matrix-metalloproteinase-targeted retroviral vectors. Gene Ther. 6: 1552–1557.

Pierschbacher M.D. and Ruoslahti E. (1987). Influence of stereochemistry of the sequence Arg-Gly-Asp-Xaa on binding specificity in cell adhesion. J. Biol. Chem. 262: 17297–17298.

Porter S., Span P.N., Sweep F.C.G. et al. (2006). ADAMTS8 and ADAMTS15 expression predicts survival in human breast carcinoma. Int. J. Cancer 118: 1241–1247.

Ruben G.C., Harris E.D., Jr. and Nagase H. (1988). Electron microscopic studies of free and proteinase-bound duck ovostatins (ovomacroglobulins). J. Biol. Chem. 263: 2861–2869.

Sahoo S.K. and Labhasetwar V. (2003). Nanotech approaches to drug delivery and imaging. Drug Discov. Today 8: 1112–1120.

Sarkar N.R., Rosendahl T., Krueger A.B. et al. (2005). "Uncorking" of liposomes by matrix metalloproteinase-9. Chem. Commun.: 999–1001.

Shi L., De Paoli V., Rosenzweig N. et al. (2006). Synthesis and application of quantum dots FRET-based protease sensors. J. Am. Chem. Soc. 128: 10378–10379.

Sisodia S.S. (1992). β-Amyloid precursor protein cleavage by a membrane-bound protease. Proc. Natl. Acad. Sci. USA 89: 6075–6079.

Sloane B.F., Sameni M., Podgorski I. et al. (2006). Functional imaging of tumor proteolysis. Annu. Rev. Pharmacol. Toxicol. 46: 301–315.

Sottrup-Jensen L. (1989). Alpha-macroglobulins: Structure, shape, and mechanism of proteinase complex formation. J. Biol. Chem. 264: 11539–11542.

Tanaka K., Chazenbalk G.D., McLachlan S.M. et al. (2000). Evidence that cleavage of the thyrotropin receptor involves a "molecular ruler" mechanism: Deletion of amino acid residues 305–320 causes a spatial shift in cleavage site 1 independent of amino acid motif. Endocrinology 141: 3573–3577.

Tauro J.R. and Gemeinhart R.A. (2005a). Extracellular protease activation of chemotherapeutics from hydrogel matrices: A new paradigm for local chemotherapy. Mol. Pharm. 2: 435–438.

Tauro J.R. and Gemeinhart R.A. (2005b). Matrix metalloprotease triggered delivery of cancer chemotherapeutics from hydrogel matrixes. Bioconjugate Chem. 16: 1133–1139.

Terada T., Iwai M., Kawakami S. et al. (2006). Novel PEG-matrix metalloproteinase-2 cleavable peptide-lipid containing galactosylated liposomes for hepatocellular carcinoma-selective targeting. J. Controlled Release 111: 333–342.

Torchilin V.P. (2005). Recent advances with liposomes as pharmaceutical carriers. Nature Rev. Drug Disc. 4: 145–160.

Trouet A., Passioukov A., Van derpoorten K. et al. (2001). Extracellularly tumor-activated prodrugs for the selective chemotherapy of cancer: Application to doxorubicin and preliminary in vitro and in vivo studies. Cancer Res. 61: 2843–2846.

Tsakadze N.L., Sithu S.D., Sen U. et al. (2006). Tumor necrosis factor-a converting enzyme (TACE/ADAM-17) mediates the ectodomain cleavage of intercellular adhesion molecule-1 (ICAM-1). J. Biol. Chem. 281: 3157–3164.

Tung C.-H. (2004). Fluorescent peptide probes for in vivo diagnostic imaging. Biopolymers (Pept. Sci.) 76: 391–403.

Tung C.-H., Bredow S., Mahmood U. et al. (1999). Preparation of a cathepsin D sensitive near-infrared fluorescence probe for imaging. Bioconj. Chem. 10: 892–896.

Tung C.-H., Mahmood U., Bredow S. et al. (2000). In vivo imaging of proteolytic enzyme activity using a novel molecular reporter. Cancer Res. 60: 4953–4958.

Upadhye S. and Ananthanarayanan V.S. (1995). Interaction of peptide substrates of fibroblast collagenase with divalent cations: Ca^{++} binding by substrate as a suggested recognition signal for collagenase action. Biochem. Biophys. Res. Commun. 215: 474–482.

Van Wart H.E. (2004). Clostridium collagenases. Handbook of Proteolytic Enzymes, Second Edition. A. J. Barrett, N. D. Rawlings and J. F. Woessner, 416–419, London: Elsevier Academic Press.

Vartak D.G. and Gemeinhart R.A. (2007). Matrix metalloproteinases: Underutilized targets for drug delivery. J. Drug. Targeting 15: 1–20.

Vine W., Gao K., Zegelman J.L. et al. (2006). Nanodrugs: Fact, fiction & fantasy. Drug Delivery Technol. 6(5): 34–39.

Vooijs M., Schroeter E.H., Pan Y. et al. (2004). Ectodomain shedding and intramembrane cleavage of mammalian Notch proteins are not regulated through oligomerization. J. Biol. Chem. 279: 50864–50873.

Weissleder R. and Mahmood U. (2001). Molecular imaging. Radiology 219: 316–333.

Weissleder R., Tung C.-H., Mahmood U. et al. (1999). In vivo imaging of tumors with protease-activated near-infrared fluorescent probes. Nat. Biotech. 17: 375–378.

West J.L. and Hubbell J.A. (1999). Polymeric biomaterials with degradation sites for proteases involved in cell migration. Macromolecules 32: 241–244.

Whiteman K.R., Subr V., Ulbrich K. et al. (2001). Poly(HPMA)-coated liposomes demonstrate prolonged circulation in mice. J. Liposome Res. 11: 153–164.

Willis R.C. (2004). Good things in small packages: Nanotech advances are producing mega-results in drug delivery. Mod. Drug Discovery 7: 30–36.

Yu Y.-C., Berndt P., Tirrell M. et al. (1996). Self-assembling amphiphiles for construction of protein molecular architecture. J. Am. Chem. Soc. 118: 12515–12520.

Yu Y.-C., Tirrell M. and Fields G.B. (1998). Minimal lipidation stabilizes protein-like molecular architecture. J. Am. Chem. Soc. 120: 9979–9987.

Yu Y.-C., Roontga V., Daragan V.A. et al. (1999). Structure and dynamics of peptide-amphiphiles incorporating triple-helical proteinlike molecular architecture. Biochemistry 38: 1659–1668.

Zhang Y., So M.K. and Rao J. (2006). Protease-modulated cellular uptake of quantum dots. Nano Lett. 6: 1988–1992.

Zheng G., Chen J.M., Stefflova K. et al. (2007). Photodynamic molecular beacon as an activatable photosensitizer based on protease-controlled singlet oxygen quenching and activation. Proc. Natl. Acad. Sci. USA 104: 8989–8994.

Zhou M. and Ghosh I. (2006). Quantum dots and peptides: A bright future together. Biopolymers (Pept. Sci.) 88: 325–339.

Chapter 40
Development of Tumour-Selective and Endoprotease-Activated Anticancer Therapeutics

Jason H. Gill and Paul M. Loadman

Abstract The therapeutic index for current chemotherapeutic agents is low based on the high frequency of systemic toxicities and the lack of selectivity between tumour and normal tissue. Our greater understanding both of the basis of cancer and the mechanisms that drive cancer growth has led to the design of therapeutics which exploit defined abnormalities responsible for the causation, maintenance, expansion or metastatic potential of the disease. Targeting enzymes centrally involved in the characteristic features of cancer is an attractive strategy for tumour-selective prodrug development since these will exploit the known phenotypic differences between 'normal' and tumour cells. One class of enzymes which satisfy all criteria for tumour-selective prodrug development and have been heavily implicated in tumour development and progression is the extracellular endopeptidases of the degradome. Tumour survival and expansion relies heavily upon the increased expression and activity of diverse extracellular endoproteases from multiple enzymatic classes, particularly the metallo-, serine, threonine, cysteine and aspartic proteases. In this chapter, studies utilising the increased endoprotease activity of tumours for selective drug delivery will be described. When considering this prodrug strategy for improvement of cancer treatment, it is important to remember that in addition to improving tumour-selective delivery of therapeutics, this whole strategy functions in parallel to reduce the restrictive toxicity of these agents against normal tissues, including liver, heart and bone marrow. In this sense, these prodrugs can be seen to have enormous clinical potential in any disease state involving increased activity of these endoproteases, including rheumatoid arthritis and other inflammatory diseases.

J.H. Gill
Institute Cancer Therapeutics, University of Bradford, West Yorkshire, BD7 1DP, UK,
e-mail: j.gill1@bradford.ac.uk

D. Edwards et al. (eds.), *The Cancer Degradome.*

Introduction

Cancer is a broad and complex family of diseases that remains one of the most frequent causes of death worldwide (Varmus 2006). The disease exhibits a wide range of genetic, physiological and histological features, and is characterised by upregulated cell growth leading to invasion of surrounding tissues and metastasis to different parts of the body. Surgical removal of cancer remains the first priority for treatment of solid tumours, although for the majority cure by surgery is not realistic due to tumour accessibility, tumour pathology or the presence of tumour spread and metastasis. Therefore, cancer chemotherapy remains the main treatment arm in the majority of cases, either as a single-approach option or as an adjuvant to localised surgery or radiotherapy. For the majority of these chemotherapeutic agents their discovery was empirical with no knowledge of their mechanism of action or often their actual target within the tumour cell (Newell 2005). Whilst only limited success is achieved using these 'classical' antiproliferative cytotoxic agents, many still form the basis of many current treatment regimens today (Collins and Workman 2006). It is without question that these drugs are highly potent and have the ability to kill large numbers of tumour cells, but their clinical value is severely limited by their unavoidable and significant toxicity to 'normal' cells and hence their lack of tumour selective targeting. The therapeutic index for such agents is low based on the high frequency of systemic toxicities and the differential selectivity between tumour versus normal tissue is modest due to the necessity for large concentrations of active drug at the tumour site (Verweij and de Jonge 2000). In addition, in many situations it is often the case that tumours develop resistance to these drugs, presumably as a result of such limitations and the need for prolonged treatment (Rooseboom et al. 2004).

Consequently, it is felt that a plateau of effectiveness has been reached with current agents and there is a very strong need to develop new anticancer therapeutics with a much improved therapeutic selectivity and potential.

Molecular Targeted Anticancer Therapy

Over recent years, numerous advances have been made in the area of anticancer drug development (Newell 2005, Collins and Workman 2006). Improvements in our understanding of the molecular pathology of cancer have provided a mechanistic structure to tumourigenesis, cancer progression and cancer biology (Newell 2005). Our greater understanding of both the basis of cancer and the mechanisms that drive cancer growth has led to the design compounds which exploit defined abnormalities responsible for the causation, maintenance, expansion or metastatic potential of the disease (Newell 2005, Collins and Workman 2006). This has facilitated the evolution of cancer chemotherapy from the development of classical cytotoxic and genotoxic agents to target-led, rationally designed molecules. For the reasons outlined above, the development of new therapeutics offering improved tumour selectivity, lower systemic toxicity and thereby a larger therapeutic index and anticancer efficacy is a major aim for improvement and advancement of cancer treatment.

Despite several successes in this arena, such as Gleevec (Imatinib) and Zolinza (Vorinostat), the interrogation of intracellular signalling pathways to identify such critical targets for drug development is complicated because of both the characteristic genetic instability of cancer and the diverse nature of genetic changes involved in the tumourigenic process (Huang and Oliff 2001, Collins and Workman 2006). An alternative approach to develop more effective and targeted cancer therapeutics is to take advantage of the tumour phenotype, such as differential cell–matrix interactions, altered cell surface receptor expression or increased capacity for proteolytic degradation (Huang and Oliff 2001). One example of such a clinically viable drug target is the epidermal growth factor (EGFR), demonstrated by the success of Iressa (Gefitinib) and Tarceva (Erlotinib) (Collins and Workman 2006). Taken together, these approaches have indeed resulted in lower toxicities against normal tissues and a subsequent improvement in cancer treatment. However, because of the relatively recent introduction of these agents into the clinic, no information is as yet available to judge their impact on cancer treatment and the overall global burden of cancer deaths.

Tumour-Selective Anticancer Drug Delivery

An alternative approach to overcome the systemic toxicities of chemotherapeutic agents has been to develop strategies to focus the delivery of agents specifically to the tumour. This approach is designed to deliver a potentially potent and proven cytotoxic to the tumour or tumour microenvironment as a non-toxic prodrug (Denny 2001). These prodrugs are non-toxic compounds administered systemically which are activated selectively in the tumour environment by exploiting a unique physiological, metabolic or genetic difference enabling discrimination between tumour and normal tissue (Denny 2001), the objective being to achieve a high local concentration of antitumour drugs and to decrease unwanted side effects (Denny 2001, Rooseboom et al. 2004). These therapeutics may be given in large doses as the potent cytotoxic is inactive until its cytotoxicity is triggered by unique phenotypic differences present in the tumour architecture or environment (Denny 2001). To specifically activate prodrugs in a tumour, either the enzyme involved in activation must be selectively present in the tumour or the target tumour should selectively take up the prodrug (Denny 2001, Rooseboom et al. 2004). One potential limitation to tumour-specific targeting of prodrugs is that unlike bacteria and viruses, cancer cells do not contain molecular targets completely foreign to the host (Dubowchik and Walker 1999, Rooseboom et al. 2004).

In general, these tumour-selective prodrugs are composed of a minimum of three domains: a trigger, a linker and an effector, where the 'trigger' (controlling selectivity) is joined to the 'effector' (responsible for therapeutic effect) by a 'linker', the prodrug remaining non-toxic until activation of the trigger (depicted in Fig. 40.1).

The principal step during the development of tumour-selective prodrugs is the identification of a tumour-specific target for activation. This is often an enzyme which is overexpressed or preferably specifically expressed within the tumour environment. This enzyme must be capable of metabolising these agents selectively within tumour tissue. There is considerable evidence demonstrating differential expression of an impressive array of enzymes between normal and tumour tissue, although consistent patterns of enzyme upregulation unique to tumours so far remain elusive (Denny 2001). The ideal characteristics of tumour enzymes suitable for targeting using this approach are as follows:

(i)   Good characterisation of the enzyme or family of enzymes possessing a known role in tumour development and/or progression
(ii) Demonstration of a high affinity of the enzyme for the prodrug

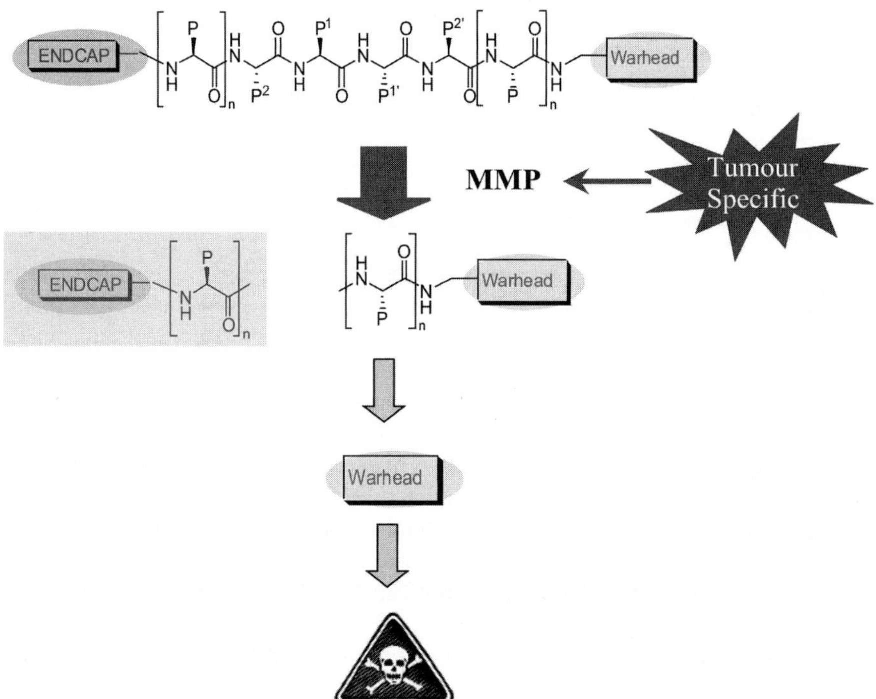

Fig. 40.1 Schematic representation of tumour endoprotease-activated prodrugs (*TAPs*). The prodrug is composed of three domains: a potent therapeutic agent (effector), an endoprotease-cleavable peptide sequence (trigger) and a linker to join the trigger to the effector. This TAP remains inactive until activation of the trigger. The presence of the endoprotease in the tumour but not 'normal' tissues results in activation of the TAP trigger selectively in the tumour, causing the release of the potent effector and a subsequent therapeutic effect

(iii) Significantly elevated expression in the disease state and activity in the tumour environment

(iv) Low or negative expression and lack of enzymatic activity in non-diseased tissue

(v) No presence of the enzyme in a prodrug-activating form in patient serum

(vi) Capability by the enzyme for selective and rapid activation of the prodrug

In the context of tumour-selective anticancer agents, several phenotypic characteristics of the tumour have been suggested as selective prodrug targets involving a range of endogenous enzyme classes, including oxidoreductases, transferases, hydrolases and lyases (Rooseboom et al. 2004). Targeting enzymes centrally involved in the characteristic features of cancer is an attractive strategy for tumour-selective prodrug development since these will provide optimal targets by virtue of the known phenotypic differences between 'normal' and tumour cells. One class of enzymes which satisfy all criteria for tumour-selective prodrug development and have been heavily implicated in tumour development and progression is the proteolytic endoproteases.

Tumour Endoproteases as Targets for Tumour-Selective Anticancer Drug Delivery

The majority of features defining malignancy are dependent upon proteolytic degradation, including acquisition of an improved vasculature system, penetration into surrounding normal tissues and dissemination to distant organ sites. Each of these processes relies heavily upon the increased expression and activity of diverse extracellular endoproteases from multiple enzymatic classes, particularly the metallo-, serine, threonine, cysteine and aspartic proteases (Egeblad and Werb 2002, Borgono et al. 2004, Lee et al. 2004, Mohamed and Sloane 2006, Overall and Kleifeld 2006). Together these proteases constitute the major proportion of the cancer degradome, introduced in Chap. 1 by Puente, Ordóñez and López-Otín.

As proteases represent ~2% of mammalian genes and 5% of all drug targets, there is a major opportunity to exploit their tumour-related association in the development of novel drugs, not least tumour-selective prodrugs (Lee et al. 2004, Overall and Dean 2006, Overall and Kleifeld 2006). The increased enzymatic activity within the tumour environment relative to non-diseased tissue and the capacity of these enzymes to selectively activate specific peptide sequences provides them with the central attributes for development of tumour-selective prodrugs. Accordingly, this increased endoprotease activity within tumours can be harnessed to selectively cleave peptide sequences thus activating non-toxic peptide-conjugated prodrugs of potent anticancer therapeutics, resulting in high levels of the active agent at the tumour and low or negative drug levels in 'normal' tissues. Attempts to exploit the elevated activity of tumour endopeptidases for tumour-selective drug targeting were until very recently largely overlooked, with very little reported in the literature until the late 1990s. This approach has since been steadily gaining momentum and is now proving to be a potentially fruitful direction for improvement of cancer drug development. In this chapter, studies utilising the increased endoprotease activity of tumours for selective drug delivery have been grouped and will be briefly described.

Serine Protease Targeted Tumour-Selective Prodrugs

The serine endoproteases constitute one of the larger families within the degradome. Many members of the serine endoprotease family have been shown to demonstrate increased expression and activity in tumours, supporting their role as major players in both the development and progression of cancer (Del Rosso et al. 2002, Borgono et al. 2004). Within tumours the serine endoproteases have been shown to act in a concerted manner with the matrix metalloproteinases (MMPs) to break down the extracellular matrix (ECM) and basement membrane, facilitating tumour invasion and progression. The increased level of these endoproteases in tumours has resulted in a considerable effort towards the development of prodrugs exploiting these enzymes for tumour-selective drug delivery. In particular, major efforts have focused on prostate-specific antigen (PSA) and the urokinase plasminogen activator (uPA) pathway.

Prostate-Specific Antigen-Activated Prodrugs

PSA is a serine protease, a member of the kallikrein gene family, expressed selectively in prostatic tissue (Borgono et al. 2004). Levels of PSA are significantly elevated in cancers of the prostate relative to non-diseased tissue (Denmeade et al. 2001, Wong et al. 2001, Rao et al. 2007). Consequently, elevated levels of PSA are now utilised routinely in the clinic as a serological diagnostic marker of prostate cancer, with higher levels being indicative of a larger tumour burden or metastatic disease (Rao et al. 2007, Reynolds et al. 2007).

In terms of the suitability of PSA as a target for tumour-selective drug delivery, levels are elevated and the enzyme is proteolytically active in prostate cancer, whereas all normal tissues lack detectable PSA activity. Furthermore, PSA is capable of selective and rapid activation of peptide-conjugated prodrugs (Denmeade et al. 1998, 2003, DeFeo-Jones et al. 2000, Wong et al. 2001, Reynolds et al. 2007). Additionally, though the use of PSA as a diagnostic marker acknowledges its presence in patient serum, this form of PSA is proteolytically inactive and unable to activate targeted prodrugs due to its formation into a complex with the plasma protease inhibitors α_1-antichymotrypisn and α_2-microglobulin (Otto et al. 1998, Reynolds et al. 2007). Therefore, PSA fulfils the defining criteria outlined above and is thereby a potential and strong candidate for the development of tumour-activated prodrugs.

There are currently several examples in the literature of PSA being used to activate prodrug species. Drug conjugates, including doxorubicin (Denmeade et al. 1998, Wong et al. 2001, DeFeo-Jones et al. 2002, DiPaola et al. 2002), vinblastine (DeFeo-Jones et al. 2002), 5-fluorodeoxyuridine (Mhaka et al. 2002), thapsigargin (Denmeade et al. 2003) and paclitaxel (Kumar et al. 2007), have been investigated and resulted in promising preclinical data. The creation of a proteolytic cleavage map of PSA against its physiological substrate semenogelin I led to the identification of a heptapeptide sequence (Hyp-Ala-Ser-Chg-Gln-Ser-Leu) rapidly hydrolysed by PSA (DeFeo-Jones et al. 2000). Covalent linkage of this peptide sequence to the aminoglycoside portion of doxorubicin (a topoisomerase II poison) led to the creation of L-377202, a prodrug hydrolysed by PSA resulting in the release of leucine-doxorubicin and subsequently doxorubicin (DeFeo-Jones et al. 2000). When evaluated in preclinical studies, a substantial therapeutic index was demonstrated for L-377202 and molar equivalent doses of the prodrug eight- to ninefold higher than doxorubicin alone could be administered in vivo without additional toxicity (DeFeo-Jones et al. 2000, DiPaola et al. 2002). Moreover, unlike doxorubicin, treatment with L-377202 did not result in cardiac problems or potential cardiotoxicity. In clinical trials, only a 1.3-fold molar increase in prodrug relative to doxorubicin alone was administered due to dose-limiting neutropenia (DiPaola et al. 2002). Encouragingly, at this dose of L-377202, a substantial reduction in PSA levels was observed, indicating a reduction in tumour volume and consequent therapeutic effect (DiPaola et al. 2002). Further clinical trials are yet to be reported for this agent.

The relative success of L-377202 and corresponding demonstration of proof of concept for this approach led to the development of several other PSA-directed chemotherapeutic prodrugs with mixed success (DeFeo-Jones et al. 2002, Mhaka et al. 2002, Denmeade et al. 2003, Kumar et al. 2007). The synthetic strategy outlined by DeFeo-Jones and coworkers for development of the doxorubicin prodrug, L-377202, was subsequently utilised by the same group to develop PSA-activated prodrugs of the microtubule-targeted agent vinblastine (Fig. 40.2), incorporating an octapeptide sequence (Hyp-Ser-Ser-Chg-Gln-Ser-Ser-Pro) (DeFeo-Jones et al. 2002). In the laboratory, this prodrug once again demonstrated PSA-dependent activation, significantly lower toxicity in vivo compared to desacetyl-vinblastine (non-conjugated drug) alone and a significant inhibition of PSA-producing tumour growth in vivo (DeFeo-Jones et al. 2002). In addition, the therapeutic index and antitumour efficacy of the vinblastine prodrug was significantly greater than that of the original doxorubicin prodrug, L-377202 (DeFeo-Jones et al. 2002). No clinical results of this PSA-activated vinblastine prodrug have been reported to date.

In another study, PSA-activated prodrugs have been developed incorporating the chemotherapeutic agent thapsigargin, a potent inhibitor of an endoplasmic reticulum Ca^{++} pump called SERCA. Unlike 'conventional' chemotherapeutic entities, such as doxorubicin and vinblastine, cytotoxicity of thapsigargin is independent of cellular proliferative status, providing it with the added benefit of potent activity against characteristically slow growing prostate tumours (Denmeade et al. 2003).

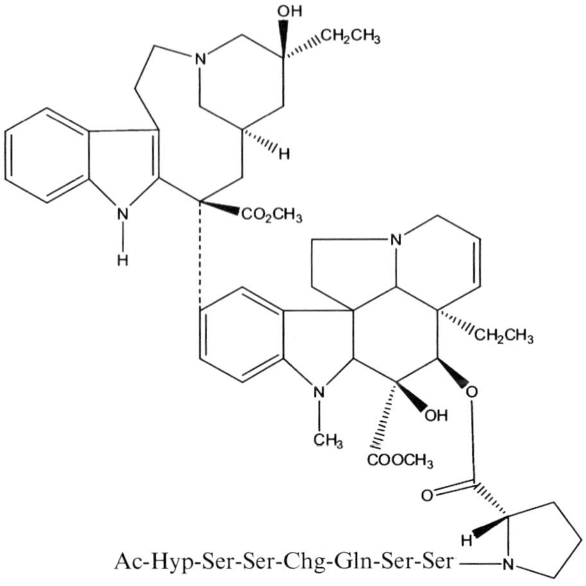

Fig. 40.2 a Structure of tumour-activated prodrug incorporating the microtubule-targeted agent vinblastine targeted at the serine protease, prostate-specific antigen (*PSA*) (DeFeo-Jones et al. 2002). Vinblastine is attached to the fourth position to an octapeptide incorporating a PSA-selective cleavage site. The prodrug is endcapped by acetylation to prevent non-specific exopeptidase activation. **b** Schematic representation of prodrug activation by PSA. The cleavage site for PSA within the peptide is indicated by the arrow. Following the initial cleavage by PSA, the remaining amino acid residues are removed rapidly by exopeptidases

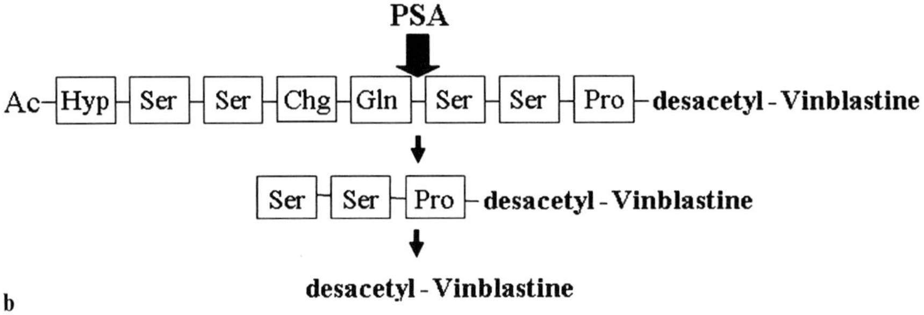

This prodrug comprised thapsigargin coupled via a leucine linker to a PSA-specific hexapeptide (His-Ser-Ser-Lys-Leu-Gln). As with related studies, these thapsigargin prodrugs demonstrated preclinical activity against PSA-producing prostate cells, but not PSA-negative cells (Denmeade et al. 2003). These thapsigargin prodrugs also demonstrated extremely good antitumour activity and PSA-selectivity in vivo, resulting in complete growth inhibition of PSA-producing prostate cancer xenografts but not PSA-negative renal carcinoma xenografts (Denmeade et al. 2003). Following demonstration of both PSA-selectivity and antitumour activity, these PSA-activated prodrugs have reportedly been entered into clinical trial for metastatic prostate cancer, although no results have yet been reported (Denmeade et al. 2003). In terms of PSA-activated chemotherapeutic agents, apart from the doxorubicin, vinblastine and thapsigargin prodrugs outlined above, in vitro studies have also been reported for PSA-activated prodrugs of both paclitaxel and 5-fluorodeoxyuridine (Mhaka et al. 2002, Kumar et al. 2007).

The proven potential for exploitation of PSA as a target for tumour-selective delivery of anticancer therapeutics has resulted in expansion of this strategy from small molecule cytotoxics to biological toxins (Williams et al. 2007). In one such approach, a PSA-targeted protoxin incorporating the bacterial cytolytic protein aerolysin was created (Williams et al. 2007). This was achieved by conversion of the furin recognition sequence within proaerolysin to a PSA-selective sequence, His-Ser-Ser-Lys-Leu-Gln, to produce the protoxin PRX302. Intratumoural administration of PRX302 caused minimal effects against PSA-null tumours, but complete remission of PSA-secreting tumour models. Moreover, no gross toxicity or effects upon tissues directly adjacent to prostate were observed following this injection, supporting the PSA-specificity of this therapy. These encouraging results facilitated the progression of PRX302 towards clinical evaluation as an intraprostatic treatment for recurrent prostate cancer, a study which is currently underway at the time of writing this chapter (Williams et al. 2007).

Tumour-Selective Prodrugs Targeted to the Urokinase Plasminogen Activator Pathway

uPA is a serine protease which in conjunction with its cellular receptor urokinase-type plasminogen activator receptor (uPAR) is a central mediator of ECM remodelling in cancer. The enzyme is secreted as an inactive proenzyme (pro-uPA) which upon binding to uPAR is converted to the active uPA form, localizing to the extracellular cell surface. uPA in turn activates plasminogen to plasmin. Both uPA and uPAR are overexpressed in many types of cancer, including breast (Stephens et al. 1998), colorectal (Baker and Leaper 2003), head and neck (Schmidt and Hoppe 1999), oesophageal (Torzewski et al. 1997, Nekarda et al. 1998), lung (Salden et al. 2000) and haematological malignancies (Scherrer et al. 1999, Abi-Habib et al. 2004). The elevated levels of this proteolytic system in a wide range of tumours coupled to the proteolytic nature of its activity strongly supported its potential for development of tumour-activated anticancer therapeutics (Romer et al. 2004). Through the identification of uPA-cleavable sequences from phage display studies, several such therapeutics have been developed. Furthermore, an increased understanding of the mechanism responsible for bacterial toxin activation has led to the successful development of uPA-targeted anticancer therapies, particularly anthrax and diphtheria protoxins (Liu et al. 2003b, Romer et al. 2004, Abi-Habib et al. 2006, Rono et al. 2006, Su et al. 2007)

Urokinase-Activated Anthrax Protoxin

Anthrax toxin, a three-component toxin, secreted from the bacterium *Bacillus anthracis*, is composed of three individually non-toxic protein components: protective antigen (PrAg), lethal factor and edema factor (Liu et al. 2003c). The cytotoxicity of anthrax arises following binding of PrAg to the cell surface anthrax receptors, tumour endothelial marker 8 (TEM8) or capillary morphogenesis gene 2 (CMG2), and its subsequent cleavage by the cell-surface furin protease resulting in formation of a PrAg heptamer (Liu et al. 2003c, Chen et al. 2007). This heptamer structure enables binding of lethal factor and/or edema factor into the complex. This complex then internalises from the cell surface into acidic endosomes, facilitating translocation of lethal factor and/or edema factor into the cytosol, resulting in cell death (Singh et al. 1999, Liu et al. 2003c, Chen et al. 2007). The absolute dependency of anthrax toxin activation upon proteolytic cleavage of PrAg on the cell surface offers the potential to develop toxins dependent upon endoproteases enriched in the tumour microenvironment, such as the uPA/uPAR system (Liu et al. 2000, 2001, 2003b).

The development of uPA/uPAR-targeted tumour-selective anthrax biotoxins was undertaken by substitution of the furin cleavage fragment (Arg-Lys-Lys-Arg) of native PrAg with a urokinase cleavage sequence (Ser-Gly-Arg-Ser-Ala), termed PrAg-U2 (Liu et al. 2001, 2003b). In order to further increase the cytotoxicity of these biotoxins, lethal factor was also modified by creation of a recombinant cytotoxin incorporating the catalytic domain of Pseudomonas exotoxin A, a fusion protein termed FP59 (Liu et al. 2001). The combined administration of PrAg-U2 together with FP59 was demonstrated to constitute a potent protoxin, with cytotoxicity being strictly dependent upon the integrity of the tumour cell surface-associated plasminogen activation system (Liu et al. 2001, 2003b). The protoxin resulted in potent inhibition of protein synthesis and subsequent cytotoxicity in cells expressing an active cell surface uPA/uPAR system and the cellular anthrax receptor (Liu et al. 2001, 2003b, Rono et al. 2006, Su et al. 2007). The dependency of this protoxin upon uPA/uPAR was further demonstrated in vitro utilising inhibitors of the interaction between uPA and uPAR (Liu et al. 2001), and in vivo by the lack of toxicity in mice null for plasminogen, uPA or uPAR (Liu et al. 2001). Using in vivo murine tumour models, local intradermal administration of the PrAg-U2/FP59 combination adjacent to the tumour resulted in a significant antitumour effect (Liu et al. 2001). In addition, the systemic intraperitoneal administration of PrAg-U2/FP59 was also shown to induce a significant antitumour effect against murine tumour models (Rono et al. 2006). Following demonstration of proof of concept and antitumour activity in murine tumours, the efficacy of this approach against human tumours was subsequently evaluated (Su et al. 2007). Intraperitoneal administration of PrAg-U2/FP59 to mice bearing subcutaneous human non-small cell lung xenografts resulted in tumour regressions in all mice and complete remission for greater than 60 days in 30% of mice, further supporting the potential for this strategy (Su et al. 2007). However, systemic administration of PrAg-U2/FP59 in both the murine and human tumour models elicited off-target toxicity at high-protoxin doses, an observation completely eliminated by concurrent treatment with dexamethasone, suggesting the toxicity to be inflammation related (Rono et al. 2006, Su et al. 2007). The mechanism for antitumour activity, in vivo protoxin pharmacokinetics and efficacy against further clinically relevant human tumour models are reportedly planned, but not yet reported.

Urokinase-Activated Diphtheria Toxin

Diphtheria toxin is a very potent biological agent with significant potential for cancer therapy (Kreitman 2006, Wong et al. 2007). Several fusion proteins incorporating diphtheria toxin have been developed as tumour-selective therapeutics, targeted to specific tumour cell markers or receptors (Kreitman 2006, Wong et al. 2007). These immunotoxins have shown great promise as targeted therapeutics for the treatment of haematological malignancies, including lymphoma and leukaemia, both preclinically and clinically (Kreitman 2006, Wong et al. 2007). One such fusion protein, Denileukin difitox (Ontak®), a recombinant diphtheria toxin targeting interleukin-2 (IL-2) receptor positive cells, is clinically approved for treatment of lymphoma (Wong et al. 2007). Despite the success of this approach, several of these targeted diphtheria toxins have demonstrated normal tissue toxicities when administered in vivo (Abi-Habib et al. 2004, Kreitman 2006, Wong et al. 2007). Therefore, in order to increase specificity, several strategies have been employed to further increase targeting or selectivity towards malignant cells (Abi-Habib et al. 2004).

Taken together the facts that (1) the uPA/uPAR system is elevated in haematological malignancy (Scherrer et al. 1999, Abi-Habib et al. 2004), (2) activation of diphtheria toxin is endoprotease mediated and (3) toxin activation can be directed towards the uPA/uPAR system, the development of protoxins analogous to that of anthrax was suggested (Liu et al. 2001, 2003b, Abi-Habib et al. 2004). This was achieved through the modification of a previously developed diphtheria therapeutic, $DT_{388}GMCSF$, a fusion between the catalytic and translocation domains of diphtheria toxin (DT_{388}) with granulocyte-macrophage colony-stimulating factor (GM-CSF) (Abi-Habib et al. 2004). $DT_{388}GMCSF$ was toxic to leukaemia both preclinically and clinically (Frankel et al. 2002), but unfortunately demonstrated hepatotoxicity (Frankel et al. 2002, Abi-Habib et al. 2004). To improve its specificity towards malignant cells, the furin cleavage site (Arg-Val-Arg-Arg-Ser) within the diphtheria toxin was modified to one cleavable by the uPA/uPAR system (Gly-Ser-Gly-Arg-Ser-Ala), yielding DTU2GMCSF (Abi-Habib et al. 2004). Preclinically, this uPA/uPAR-targeted protoxin was highly toxic to leukaemic cell lines, and this toxicity was greatly inhibited following pre-treatment with anti-uPA and anti-GMCSF antibodies (Abi-Habib et al. 2004). Sensitivity to DTU2GMCSF correlated with expression of uPAR and GMCSF-receptors. Furthermore, DTU2GMCSF was less toxic to normal cells expressing uPAR or GMCSF-R alone. These data supported the improved targeting and proof of principle for this approach. The evaluation of this agent in vivo and its potential for clinical progression have yet to be reported.

Plasmin-Activated Anticancer Chemotherapeutics

The activation of uPA and its localisation to the extracellular surface results in activation of plasminogen to its proteolytically active form, plasmin. The increased levels of uPA/uPAR and, subsequently, plasmin in the tumour microenvironment, coupled to the fact that any plasmin in the systemic circulation is inhibited through binding to α2-antiplasmin and α2-macroglobulin, resulted in the development of anticancer prodrugs activated by plasmin (Chakravarty et al. 1983, de Groot et al. 1999, Devy et al. 2004). In the first reported plasmin prodrug, the anthracycline antitumour drug doxorubicin was conjugated to a plasmin-cleavable tripeptide sequence (Val-Leu-Lys) (Chakravarty et al. 1983). Although demonstrating the potential of this strategy, this prodrug was found to be a poor substrate for plasmin, potentially as a result of chemical steric hindrance (Chakravarty et al. 1983). In order to address these problems, plasmin-targeted prodrugs were designed incorporating a spacer moiety between the tripeptide and doxorubicin (de Groot et al. 1999, Devy et al. 2004). These prodrugs incorporating a linker group demonstrated in vitro cytotoxicity against uPA-expressing tumour cells relative to those expressing low levels (de Groot et al. 1999, Devy et al. 2004). In addition, cytotoxicity was markedly decreased in the presence of the plasmin inhibitor aprotinin (de Groot et al. 1999, Devy et al. 2004). At equimolar concentrations with respect to doxorubicin, the prodrug displayed potent anticancer activity and an antiangiogenic response against uPA-expressing tumour models in vivo (Devy et al. 2004). In contrast to doxorubicin, mice treated with the prodrug demonstrated minimal body weight loss, indicative of a reduction in normal tissue toxicity (Devy et al. 2004).

In a derivation of this approach, the same tripeptide moiety has been attached to paclitaxel through a carbamate or carbonate linker (de Groot et al. 2000). Incubation of this conjugate with plasmin converted this prodrug to paclitaxel, resulting in cytotoxicity. When incubated with non-tumour cells, this prodrug demonstrated ~1000-fold lower toxicity than paclitaxel alone, supporting the chemical stability and plasmin selectivity of this prodrug (de Groot et al. 2000). Of the compounds synthesised and evaluated, the paclitaxel-2'-carbonate-linked prodrug was deemed to be the most promising, leading to further preclinical evaluation, which is as yet unpublished.

It is interesting to note that despite demonstrating potent activity and potential, the direct involvement of the uPA/uPAR system as opposed to plasmin in activation of the majority of these prodrugs has yet to be extensively reported. Despite this, these plasmin-activated anticancer therapeutics are to be progressed into clinical trial (Devy et al. 2004).

Cysteine Protease-Activated Anticancer Prodrugs

In contrast to serine- and metallo- endoproteases, very few studies have been performed to address the extracellular members of the cysteine endoprotease family as anticancer drug targets. One member of this family which has shown promise in this area is legumain, the only asparaginyl endopeptidase in the mammalian genome (Liu et al. 2003a, Wu et al. 2006). The other group of endoproteases resident within this family which have been proposed as targets for tumour-selective drug activation are the cysteine cathepsins, specifically cathepsin B (Potrich et al. 2005, Mohamed and Sloane 2006).

Legumain-Activated Anticancer Therapeutics

Legumain is a novel acidic cysteine endoprotease with restricted substrate specificity, requiring an asparagine at the P1 site within the substrate sequence (Ishii 1994). Expression of legumain is elevated in several tumour types, including breast, colon, prostate and several nervous system tumours, but negative or extremely low in the corresponding normal tissues (Liu et al. 2003a). Legumain, in addition to its 'normal' intracellular expression in the endosomal/lysomal compartment, is present at high levels extracellularly in the tumour microenvironment (Liu et al. 2003a, Wu et al. 2006). Furthermore, legumain is also expressed in intratumoural blood vessels and macrophages, which supported its attractiveness as a target for tumour-selective drug delivery (Wu et al. 2006). As such, several peptide-conjugates of doxorubicin were synthesised by covalent linkage of legumain-cleavable succinyl-blocked peptides to the aminoglycoside of doxorubicin (Wu et al. 2006). In conjunction with the improvement of tumour-selective drug delivery, the rationale behind these legumain-activated agents was the development of prodrugs which were also cell impermeable (Wu et al. 2006). The incorporation of cell membrane permeability inhibiting groups prevented the active drug (doxorubicin) from entering the cell until it had been activated in the tumour microenvironment by legumain. In this way, the activated drug was suggested to have much greater antitumour efficacy by functioning through a 'bystander effect' upon both tumour and stromal cells, rather than just selectively deleting the target-producing cells from the tumour (Wu et al. 2006).

The most effective of the legumain-activated doxorubicin prodrugs, LEG-3, comprised doxorubicin bound to a tetrapeptide (Leu-Asn-Ala-Ala), endcapped with the succinyl group for prevention of cell membrane permeability (Wu et al. 2006). This prodrug was rendered cell permeable in cells overexpressing legumain in vitro, causing the release of leucine-doxorubicin (Leu-Dox), and the resultant liberation of doxorubicin and induction of cytotoxicity (Wu et al. 2006). From in vivo studies, LEG-3 possessed enhanced efficacy compared with doxorubicin alone, in murine syngeneic tumour models and human tumour xenografts, including a doxorubicin-resistant prostate cancer model (Wu et al. 2006). No significant levels of LEG-3 were observed in any normal tissues and much reduced cardiotoxicity and myelosuppression were reported, relative to doxorubicin alone (Wu et al. 2006). These data supported the rationale for this approach and the requirement to undertake further studies.

Extracellular Cathepsin-Activated Anticancer Therapeutics

Cysteine cathepsins are now known to be highly upregulated in a wide range of cancers (Mohamed and Sloane 2006). The majority of the cathepsin family of endoproteases are located intracellularly in endolysosomal vesicles rather than being secreted into the tumour microenvironment (Mohamed and Sloane 2006). It is now known that active forms of cathepsin B are secreted from a wide range of tumour cells (Roshy et al. 2003), and cathepsin H is secreted from prostate tumour cells (Waghray et al. 2002). There are currently only a few studies reported describing the development of anticancer agents designed for activation at the tumour cell surface by cathepsin B (Boyer and Tannock 1993, Panchal et al. 1996, Potrich et al. 2005, Schmid et al. 2007). Two strategies involved conjugation of cathepsin B-cleavable linkers to pore-forming toxins from staphylococcal α-haemolysin and sea anemone (Panchal et al. 1996, Potrich et al. 2005). In both these strategies, cells expressing cathepsin B could activate the protoxins, resulting in cell death (Panchal et al. 1996, Potrich et al. 2005). In another approach, prodrugs of doxorubicin were developed and shown to demonstrate extracellular cleavage by cathepsin B and improved efficacy over doxorubicin alone (Boyer and Tannock 1993). In all of these studies, it remains to be determined whether the prodrugs were actually activated by cathepsins on the cell surface or lysomal cathepsins rather than the suggested secreted cathepsin B. Studies to address the potential for these agents in vivo have yet to be described.

An alternative strategy for cathepsin B-activated prodrugs involved development of macromolecular conjugated peptide prodrugs of the anticancer agents doxorubicin and camptothecin (Loadman et al. 1999, Schmid et al. 2007). In the first of these studies, a prodrug (PK1) comprising doxorubicin conjugated to a macromolecular polymer via a cathepsin-sensitive tetrapeptidyl (Gly-Phe-Leu-Gly) spacer was evaluated by the author of this chapter (Loadman et al. 1999). The high molecular weight of this prodrug restricted its uptake into the tumour cell, resulting in either activation in the extracellular space or uptake into cellular lysosomes and subsequent cathepsin activation. PK1 demonstrated improved efficacy over doxorubicin alone, and dependency upon the vascular properties of the specific tumour. The involvement of extracellular versus lysosomal cathepsin B was not addressed (Loadman et al. 1999). In a variation of this prodrug concept, rather than synthesis of the prodrug in a large macromolecular complex, prodrugs were developed to bind to albumin in the systemic circulation as a macromolecular carrier system (Schmid et al. 2007). Through passive accumulation the levels of these albumin-bound prodrugs is increased within the tumour environment, relative to the non-bound forms. In these prodrugs, the active agent was bound to an alternative cathepsin-sensitive tetrapeptide (Ala-Leu-Ala-Leu), which, in turn, was terminated by an albumin-binding maleimide group (Schmid et al. 2007). Efficient activation of these prodrugs in their albumin-bound form was demonstrated for cathepsin B, but not for cathepsin D in vitro (Schmid et al. 2007). In addition, both prodrugs were also activated to release their respective active agents in human tumour xenograft homogenates ex vivo, a cleavage prevented via incubation in the presence of a cathepsin B selective inhibitor (Schmid et al. 2007). Furthermore, antitumour activity equivalent to administration of the active agent alone was observed following in vivo administration (Schmid et al. 2007). Further preclinical evaluation of these prodrugs is currently underway at the time of writing this chapter.

Matrix Metalloproteinases-Activated Anticancer Therapeutics

The matrix metalloproteinases (MMPs), a family of 24 zinc-dependent endoproteases, are central mediators of tumour development as discussed elsewhere in this book. Historically, the role of MMPs within tumours was believed to be purely degradation of the ECM and subsequent facilitation of tumour cell invasion. It is now known that the MMPs have a considerable number of other non-ECM target substrates, including proteins involved in apoptosis, cell dissociation, cellular communication and cell division (McCawley and Matrisian 2000, Egeblad and Werb 2002, Deryugina and Quigley 2006). Extensive studies have demonstrated overexpression of many of the MMPs in a wide range of human tumour types (Brinckerhoff et al. 2000, Hoekstra et al. 2001, Zucker and Vacirca 2004, Overall and Kleifeld 2006, Atkinson et al. 2007b, Vizoso et al. 2007). In contrast, basal MMP production in normal tissues is generally very low and quite often absent, requiring a major stimulus from situations such as tissue injury or tissue remodelling to elicit MMP gene transcription. At the protein level, there is an additional level of control for regulation of MMP activity provided by the endogenous tissue inhibitors of metalloproteinases (TIMPs). In normal tissues, any active MMP is kept in check by these TIMPs. Conversely in cancer, not only are the MMP levels increased but TIMP levels are generally reduced, resulting in increased levels of proteolytically active MMPs selectively within the tumour microenvironment. Accordingly, the MMPs can be classified as the major family within the degradome as targets for therapeutic exploitation in cancer.

The increased activity of the MMPs within the tumours environment relative to non-diseased tissue coupled to the ability of these enzymes to selectively and specifically cleave short peptide sequences makes this family of enzymes ideal candidates for tumour-selective prodrug development. Over the last 10 years or so, several studies have reported the development of agents which exploit the activity of the MMPs for improved tumour targeting of anticancer therapeutics. In line with an increased understanding of tumourigenesis, MMP biology and MMP-substrate selectivity, these approaches have progressively improved and resulted in the development of agents demonstrating both an improved therapeutic index and anticancer efficacy relative to their therapeutic cargo alone.

MMP-Activated Chemotherapeutics

The majority of MMP-prodrugs developed have been directed towards the secreted MMP family members, particularly MMP-2 and -9 (Timar et al. 1998, Kratz et al.

2001, Mincher et al. 2002, Mansour et al. 2003, Young et al. 2003, Kline et al. 2004, Albright et al. 2005, Van Valckenborgh et al. 2005, Atkinson et al. 2007a). In one of the first reported studies of an MMP-targeted prodrug, the alkylating agent melphalan was incorporated into an MMP-2-cleaveable pentapeptide to create the prodrug MHP (Pro-Gln-Gly-Ile-Melphalan-Gly) (Timar et al. 1998). Although in vitro cytotoxicity of MHP proved disappointing, the melphalan effector was successfully liberated in the presence of MMP-2/-9 and conditioned media from MMP-positive tumour cells, thereby supporting the potential for MMP-targeted prodrugs (Timar et al. 1998).

Subsequent approaches to develop MMP-targeted prodrugs were based upon the structure outlined in Fig. 40.1, in which the effector therapeutic was conjugated to the terminus of the peptide sequence rather than being incorporated within it (as for MHP). One such prodrug was a water-soluble maleimide derivative of doxorubicin, incorporating an octapeptide sequence (Gly-Pro-Leu-Gly-Ile-Ala-Gly-Gln) suggested to be selectively cleaved by MMP-2 (Kratz et al. 2001, Mansour et al. 2003). In addition to having doxorubicin attached at one end of the conjugate, this prodrug was also developed to bind to serum albumin, with the intention of using albumin as a macromolecular carrier to increase tumour accumulation of the prodrug (Kratz et al. 2001, Mansour et al. 2003). Efficient activation of this prodrug to release doxorubicin was demonstrated using both purified MMP-2 and MMP-2-positive tissue homogenates. In vivo, the maximum-tolerated dose (MTD) for albumin-bound doxorubicin was substantially higher than for doxorubicin alone and subsequent studies showed superior activity against A375 melanoma at equitoxic doses (Mansour et al. 2003). The concept of macromolecular delivery of MMP-activated prodrugs has also been investigated by several other groups with differing success (Chau et al. 2004, 2006a, 2006b; Tauro and Gemeinhart 2005).

In contrast to the use of macromolecules as delivery vehicles, MMP-activated prodrugs conforming to the simple 'trigger-linker-effector' structure, shown in Fig. 40.1, have shown significant potential as anticancer therapeutics. In one approach an anthraquinone topoisomerase inhibitor (NU:UB31) was linked to the C-terminus of an MMP-9-cleavable heptapeptide ([D-Ala]-Ala-Ala-Leu-Gly-Nva-Pro), and the N-terminus of the peptide was 'capped' with fluorescein isothiocyanate (FITC) to produce a prodrug, EV1-FITC (Mincher et al. 2002, Young et al. 2003, Van Valckenborgh et al. 2005). The FITC in this prodrug is quenched on conjugation with the anthraquinone and so allows for prodrug cleavage to be observed by fluorescence detection. Using a murine myeloma model, prodrug metabolism in tissues ex vivo was higher in MMP-9 expressing tumour-bearing organs (bone marrow, spleen) relative to other tissues (heart, lung, kidney), as determined by fluorescence from FITC release (Van Valckenborgh et al. 2005). There was, however, a significant level of prodrug activation in non-diseased tissue homogenates, suggesting a lack of specificity towards MMP-9 (Van Valckenborgh et al. 2005). An attempt to further increase MMP-selectivity was addressed using peptidomimetic analogues, incorporating doxorubicin, auristatins and the duocarmycins (Kline et al. 2004). However, these peptidomimetic prodrugs were either not activated by the MMPs or demonstrated no selective activity against MMP-positive versus MMP-negative cells. These studies suggest careful consideration of the active 'warhead' and the peptide sequence is essential for the selective cleavage of these prodrugs.

In one of the most detailed studies reported to date, MMP-targeted prodrugs of doxorubicin were demonstrated to show a much higher therapeutic index than doxorubicin alone, using the MMP-expressing HT1080 preclinical model (Albright et al. 2005). In this strategy, the peptide length was shown to be central to both compound stability and MMP-cleavage efficiency, with a heptapeptide containing three or four amino acids to the carboxy side of the scissile bond being optimal (Albright et al. 2005). Similar to EV1-FITC, these prodrugs were capped on the free peptide end to prevent unrequired exopeptidase degradation. The optimised prodrug (Fig. 40.3) (Doxorubicin-Leu-Tyr-Hof-Gly-Cit-Pro-Glu-Ac) was activated by the secreted MMPs, MMP-2 and -9, and the membrane-tethered MMP, MMP-14 (MT1-MMP), but not the endoprotease neprilysin, reinforcing their selectivity towards the MMP family. In an evolution of previous approaches, this prodrug was shown to be pharmacologically stable in vivo and preferentially metabolised in MMP-expressing tumours relative to heart and plasma, resulting in a tenfold increase in the tumour/heart ratio relative to administration of doxorubicin alone (Albright et al. 2005). In addition, this optimised prodrug resulted in an 80% cure rate against the HT1080 xenograft model, compared to only 10% with doxorubicin alone. One potential downside of this approach was that a significant fraction of Leu-Dox formed in the tumour, as a consequence of prodrug-activation, was not rapidly metabolised to release doxorubicin (Fig. 40.3), allowing for diffusion of Leu-Dox away from the tumour to other tissues before conversion to doxorubicin, with potential consequences for induction of normal tissue toxicity (Albright et al. 2005). Further evaluation of these prodrugs and their progression towards clinical trials has yet to be reported.

Tumour-Selective MMP-Activated Biotoxins

In addition to exploiting MMPs to activate small molecule cytotoxic agents linked to peptides, preliminary work has been undertaken to develop MMP-activated biotoxins, incorporating anthrax toxin, measles toxin, cytolysin, CD95-L or tumour-necrosis factor alpha (TNF-α) (Liu et al. 2000, 2007; Potrich et al. 2005, Gerspach et al. 2006, Springfeld et al. 2006, Watermann et al. 2007). The strategy for these agents is similar to that for MMP-activated chemotherapeutics in that the therapeutic moiety is bound to a MMP-cleavable sequence which acts to inactivate the agent until released selectively in the tumour. In all of these protoxin approaches, MMP activation and the potential for the approach was successfully demonstrated in vitro. In the case of both the measles virus and anthrax protoxins, the success and potency of the approach was also demonstrated in vivo (Springfeld et al. 2006, Liu et al. 2007).

Measles virus is a nonintegrating RNA virus which has shown clinical potential for the treatment of both haematological malignancies and cutaneous lymphoma

a   Ac-Glu-Pro-Cit-Gly-Hof-Tyr-Leu

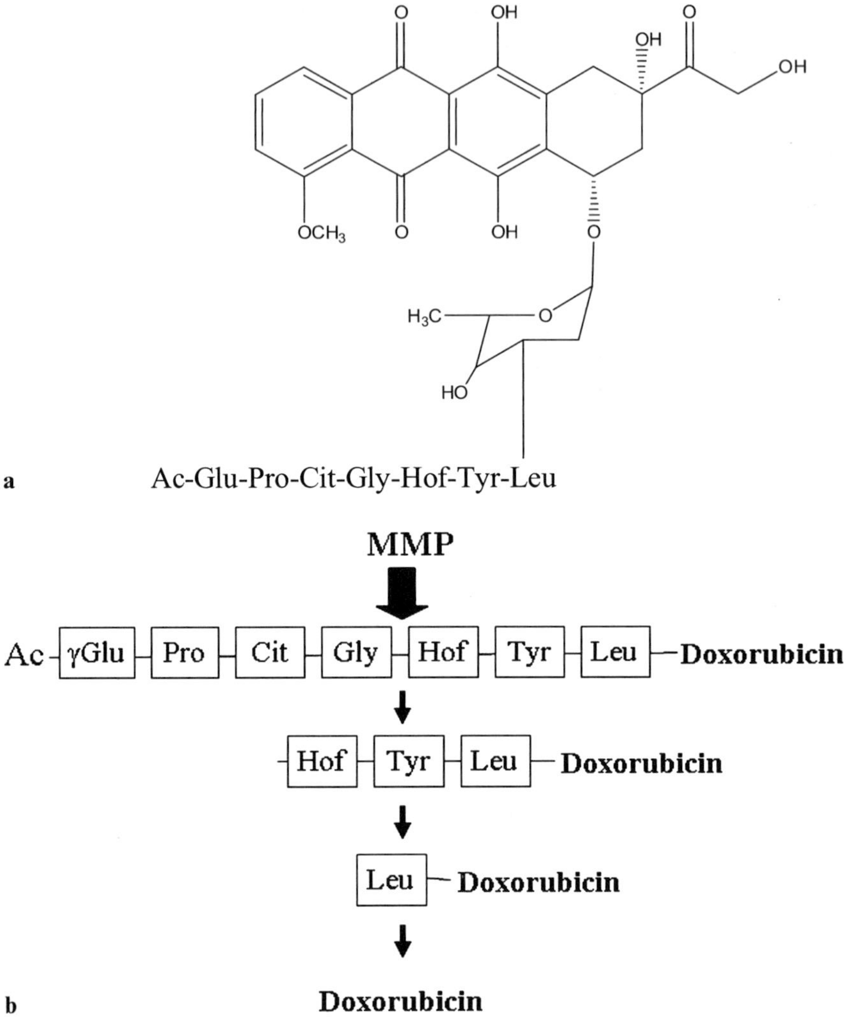

MMP

Ac —[γGlu]—[Pro]—[Cit]—[Gly]—[Hof]—[Tyr]—[Leu]— **Doxorubicin**

—[Hof]—[Tyr]—[Leu]— **Doxorubicin**

[Leu]— **Doxorubicin**

b   **Doxorubicin**

Fig. 40.3 a Structure of matrix metalloproteinase (*MMP*)-targeted tumour-activated prodrug incorporating the cytotoxic agent doxorubicin (Albright et al. 2005). Doxorubicin is attached to a heptapeptide incorporating a MMP-selective cleavage site. The prodrug is endcapped by acetylation to prevent non-specific exopeptidase activation and promote in vivo drug stability. **b** Schematic representation of prodrug activation by MMP with expected cleavage products. The cleavage site for MMP within the peptide is indicated by the arrow. Following the initial cleavage by MMP, the remaining amino acid residues are removed rapidly by exopeptidases

(Springfeld et al. 2006). As with many current anticancer therapies, there is a strong need to increase tumour targeting of virotherapy, to improve both efficacy and reduce off-target effects. Since activity of measles virus is dependent upon the intracellular endoprotease furin to first process its envelope fusion protein component, recombinant variants of the virus targeted at MMP-selective activation were created, with the aim of confining virus activation strictly to tumour tissue (Springfeld et al. 2006). In contrast to the development of other endoprotease protoxins (Liu et al. 2000, 2003b; Abi-Habib et al. 2004), a hexapeptide MMP-cleavage site (Pro-Gln-Gly-Leu-Tyr-Ala) was inserted into the virus without dele-tion of the furin cleavage site (Springfeld et al. 2006). This MMP-activated measles protoxin was shown to be as efficient as the wild-type strain in retarding tumour growth in vivo (Springfeld et al. 2006). In addition, intracerebral injection of the wild-type virus into susceptible mice led to fatal encephalitis, whereas the MMP-activated variant did not, indicating that the protoxin is not pathogenic (Springfeld et al. 2006). The success of this approach has opened up the possibility to improve the therapeutic index of measles virus and other viruses in ongoing clinical trials of oncolysis.

In a similar strategy to that described previously for uPA/uPAR (Liu et al. 2001, 2003b), MMP-targeted anthrax protoxins have been developed (Liu et al. 2007). This was achieved by creation of an attenuated version of the toxin which cannot be cleaved by furin, but is instead activated by MMPs, including MMP-2, -9 and -14 (MT1-MMP) (Liu et al. 2007). The MMP-targeted protoxin was shown to be less toxic than the wild-type toxin to mice in vivo, presumably due to the limited expression of MMPs by non-tumour cells (Liu et al. 2007). Furthermore, following intraperitoneal administration, the MMP-protoxin demonstrated greater antitumour efficacy than the wild-type toxin towards human tumour xenografts (Liu et al. 2007), the antitumour activity being due largely to the effects of the activated toxin upon the tumour vasculature. Furthermore, the MMP-protoxin had a longer plasma half-life than wild-type, a consequence of reduced uptake and clearance of the protoxin by normal cells (Liu et al. 2007). Together these results supported the potential use of this protoxin in cancer therapy, culminating in a proposal for its evaluation in clinical trial.

Conclusions

The increased expression and activity of extracellular endoproteases in cancer makes them attractive targets to be exploited for development of tumour-selective drug delivery. Several strategies and approaches have been evaluated in this arena with varying results, including targeting different endoproteases, variation of the selective peptide 'trigger' and alternative toxic 'warheads', all of which have led to a significant increase in our knowledge and understanding of this area. Determina-tion of the optimal peptide sequence to facilitate both pharmacological stability in vivo and selective cleavage by specific endoproteases is now known to be the most crucial and difficult step in design of these prodrugs. It is only once the require-ments of the specific enzymes are identified that truly proficient and selective prodrugs may be designed. Through the evolution of this area and this approach, many of the more recent prodrugs are demonstrating the necessary requirements and selectivity for the creation of viable and potent clinical candidate prodrugs.

When considering this prodrug strategy for improvement of cancer treatment, it is important to remember that in addition to improving tumour-selective delivery of therapeutics, this whole strategy functions in parallel to reduce the restrictive toxicity of these agents against normal tissues, including liver, heart and bone marrow. In this sense, these prodrugs can be seen to have enomorous clinical potential in any disease state involving increased activity of these endoproteases, including rheumatoid arthritis and other inflammatory diseases. Based on this, efforts to understand the biology of the endoproteases, their substrate selectivity and their expression and localisation within diseased tissues need to be accelerated to realise the clinical potential of these prodrugs.

References

Abi-Habib R. J., Liu S., Bugge T. H. et al. (2004). A urokinase-activated recombinant diphtheria toxin targeting the granulocyte-macrophage colony-stimulating factor receptor is selectively cytotoxic to human acute myeloid leukemia blasts. Blood 104:2143–2148.

Abi-Habib R. J., Singh R., Liu S. et al. (2006). A urokinase-activated recombinant anthrax toxin is selectively cytotoxic to many human tumor cell types. Mol Cancer Ther 5:2556–2562.

Albright C. F., Graciani N., Han W. et al. (2005). Matrix metalloproteinase-activated doxorubicin prodrugs inhibit HT1080 xenograft growth better than doxorubicin with less toxicity. Mol Cancer Ther 4:751–760.

Atkinson J. M., Falconer R. A., Pennington C. J. et al. (2007a). Membrane type 1-matrix metalloproteinase (MT1-MMP) targeted antitumour agents. American Association for Cancer Research Annual Proceedings: Abstract: 2453.

Atkinson J. M., Pennington C. J., Martin S. W. et al. (2007b). Membrane type matrix metalloproteinases (MMPs) show differential expression in non-small cell lung cancer (NSCLC) compared to normal lung: Correlation of MMP-14 mRNA expression and proteolytic activity. Eur J Cancer 43:1764–1771.

Baker E. A., Leaper D. J. (2003). The plasminogen activator and matrix metalloproteinase systems in colorectal cancer: Relationship to tumour pathology. Eur J Cancer 39:981–988.

Borgono C. A., Michael I. P., Diamandis E. P. (2004). Human tissue kallikreins: Physiologic roles and applications in cancer. Mol Cancer Res 2:257–280.

Boyer M. J., Tannock I. F. (1993). Lysosomes, lysosomal enzymes, and cancer. Adv Cancer Res 60:269–291.

Brinckerhoff C. E., Rutter J. L., Benbow U. (2000). Interstitial collagenases as markers of tumor progression. Clin Cancer Res 6:4823–4830.

Chakravarty P. K., Carl P. L., Weber M. J. et al. (1983). Plasmin-activated prodrugs for cancer chemotherapy. 2. Synthesis and biological activity of peptidyl derivatives of doxorubicin. J Med Chem 26:638–644.

Chau Y., Tan F. E., Langer R. (2004). Synthesis and characterization of dextran-peptide-methotrexate conjugates for tumor targeting via mediation by matrix metalloproteinase II and matrix metalloproteinase IX. Bioconjug Chem 15:931–941.

Chau Y., Dang N. M., Tan F. E. et al. (2006a). Investigation of targeting mechanism of new dextran-peptide-methotrexate conjugates using biodistribution study in matrix-metalloproteinase-overexpressing tumor xenograft model. J Pharm Sci 95:542–551.

Chau Y., Padera R. F., Dang N. M. et al. (2006b). Antitumor efficacy of a novel polymer-peptide-drug conjugate in human tumor xenograft models. Int J Cancer 118:1519–1526.

Chen K. H., Liu S., Bankston L. A. et al. (2007). Selection of anthrax toxin protective antigen variants that discriminate between the cellular receptors TEM8 and CMG2 and achieve targeting of tumor cells. J Biol Chem 282:9834–9845.

Collins I., Workman P. (2006). New approaches to molecular cancer therapeutics. Nat Chem Biol 2:689–700.

de Groot F. M., de Bart A. C., Verheijen J. H. et al. (1999). Synthesis and biological evaluation of novel prodrugs of anthracyclines for selective activation by the tumor-associated protease plasmin. J Med Chem 42:5277–5283.

de Groot F. M., van Berkom L. W., Scheeren H. W. (2000). Synthesis and biological evaluation of 2′-carbamate-linked and 2′-carbonate-linked prodrugs of paclitaxel: Selective activation by the tumor-associated protease plasmin. J Med Chem 43:3093–3102.

DeFeo-Jones D., Garsky V. M., Wong B. K. et al. (2000). A peptide-doxorubicin 'prodrug' activated by prostate-specific antigen selectively kills prostate tumor cells positive for prostate-specific antigen in vivo. Nat Med 6:1248–1252.

DeFeo-Jones D., Brady S. F., Feng D. M. et al. (2002). A prostate-specific antigen (PSA)-activated vinblastine prodrug selectively kills PSA-secreting cells in vivo. Mol Cancer Ther 1:451–459.

Del Rosso M., Fibbi G., Pucci M. et al. (2002). Multiple pathways of cell invasion are regulated by multiple families of serine proteases. Clin Exp Metastasis 19:193–207.

Denmeade S. R., Nagy A., Gao J. et al. (1998). Enzymatic activation of a doxorubicin-peptide prodrug by prostate-specific antigen. Cancer Res 58:2537–2540.

Denmeade S. R., Sokoll L. J., Chan D. W. et al. (2001). Concentration of enzymatically active prostate-specific antigen (PSA) in the extracellular fluid of primary human prostate cancers and human prostate cancer xenograft models. Prostate 48:1–6.

Denmeade S. R., Jakobsen C. M., Janssen S. et al. (2003). Prostate-specific antigen-activated thapsigargin prodrug as targeted therapy for prostate cancer. J Natl Cancer Inst 95:990–1000.

Denny W. A. (2001). Prodrug strategies in cancer therapy. Eur J Med Chem 36:577–595.

Deryugina E. I., Quigley J. P. (2006). Matrix metalloproteinases and tumor metastasis. Cancer Metastasis Rev 25:9–34.

Devy L., de Groot F. M., Blacher S. et al. (2004). Plasmin-activated doxorubicin prodrugs containing a spacer reduce tumor growth and angiogenesis without systemic toxicity. Faseb J 18:565–567.

DiPaola R. S., Rinehart J., Nemunaitis J. et al. (2002). Characterization of a novel prostate-specific antigen-activated peptide-doxorubicin conjugate in patients with prostate cancer. J Clin Oncol 20:1874–1879.

Dubowchik G. M., Walker M. A. (1999). Receptor-mediated and enzyme-dependent targeting of cytotoxic anticancer drugs. Pharmacol Ther 83:67–123.

Egeblad M., Werb Z. (2002). New functions for the matrix metalloproteinases in cancer progression. Nat Rev Cancer 2:161–174.

Frankel A. E., Powell B. L., Hall P. D. et al. (2002). Phase I trial of a novel diphtheria toxin/granulocyte macrophage colony-stimulating factor fusion protein (DT388GMCSF) for refractory or relapsed acute myeloid leukemia. Clin Cancer Res 8:1004–1013.

Gerspach J., Muller D., Munkel S. et al. (2006). Restoration of membrane TNF-like activity by cell surface targeting and matrix metalloproteinase-mediated processing of a TNF prodrug. Cell Death Differ 13:273–284.

Hoekstra R., Eskens F. A., Verweij J. (2001). Matrix metalloproteinase inhibitors: Current developments and future perspectives. Oncologist 6:415–427.

Huang P. S., Oliff A. (2001). Drug-targeting strategies in cancer therapy. Curr Opin Genet Dev 11:104–110.

Ishii S. (1994). Legumain: Asparaginyl endopeptidase. Methods Enzymol 244:604–615.

Kline T., Torgov M. Y., Mendelsohn B. A. et al. (2004). Novel antitumor prodrugs designed for activation by matrix metalloproteinases-2 and -9. Mol Pharm 1:9–22.

Kratz F., Drevs J., Bing G. et al. (2001). Development and in vitro efficacy of novel MMP2 and MMP9 specific doxorubicin albumin conjugates. Bioorg Med Chem Lett 11:2001–2006.

Kreitman R. J. (2006). Immunotoxins for targeted cancer therapy. AAPS J 8:E532–551.

Kumar S. K., Williams S. A., Isaacs J. T. et al. (2007). Modulating paclitaxel bioavailability for targeting prostate cancer. Bioorg Med Chem 15:4973–4984.

Lee M., Fridman R., Mobashery S. (2004). Extracellular proteases as targets for treatment of cancer metastases. Chem Soc Rev 33:401–409.

Liu C., Sun C., Huang H. et al. (2003a). Overexpression of legumain in tumors is significant for invasion/metastasis and a candidate enzymatic target for prodrug therapy. Cancer Res 63:2957–2964.

Liu S., Netzel-Arnett S., Birkedal-Hansen H. et al. (2000). Tumor cell-selective cytotoxicity of matrix metalloproteinase-activated anthrax toxin. Cancer Res 60:6061–6067.

Liu S., Bugge T. H., Leppla S. H. (2001). Targeting of tumor cells by cell surface urokinase plasminogen activator-dependent anthrax toxin. J Biol Chem 276:17976–17984.

Liu S., Aaronson H., Mitola D. J. et al. (2003b). Potent antitumor activity of a urokinase-activated engineered anthrax toxin. Proc Natl Acad Sci USA 100:657–662.

Liu S., Schubert R. L., Bugge T. H. et al. (2003c). Anthrax toxin: Structures, functions and tumour targeting. Expert Opin Biol Ther 3:843–853.

Liu S., Wang H., Currie BM, et al. (2008). Matrix metalloproteinase-activated anthrax lethal toxin demonstrates high potency in targeting tumor vasculature. J Biol Chem 283(1):529–540.

Loadman P. M., Bibby M. C., Double J. A. et al. (1999). Pharmacokinetics of PK1 and doxorubicin in experimental colon tumor models with differing responses to PK1. Clin Cancer Res 5:3682–3688.

Mansour A. M., Drevs J., Esser N. et al. (2003). A new approach for the treatment of malignant melanoma: Enhanced antitumor efficacy of an albumin-binding doxorubicin prodrug that is cleaved by matrix metalloproteinase 2. Cancer Res 63:4062–4066.

McCawley L. J., Matrisian L. M. (2000). Matrix metalloproteinases: Multifunctional contributors to tumor progression. Mol Med Today 6:149–156.

Mhaka A., Denmeade S. R., Yao W. et al. (2002). A 5-fluorodeoxyuridine prodrug as targeted therapy for prostate cancer. Bioorg Med Chem Lett 12:2459–2461.

Mincher D. J., Loadman P. M., Lyle J. et al. (2002). Design of tumour-activated oligopeptide prodrugs that exploit the proteolytic activity of matrix metalloproteinases. Eur J Cancer 38:S121–S121.

Mohamed M. M., Sloane B. F. (2006). Cysteine cathepsins: Multifunctional enzymes in cancer. Nat Rev Cancer 6:764–775.

Nekarda H., Schlegel P., Schmitt M. et al. (1998). Strong prognostic impact of tumor-associated urokinase-type plasminogen activator in completely resected adenocarcinoma of the esophagus. Clin Cancer Res 4:1755–1763.

Newell D. R. (2005). How to develop a successful cancer drug—molecules to medicines or targets to treatments? Eur J Cancer 41:676–682.

Otto A., Bar J., Birkenmeier G. (1998). Prostate-specific antigen forms complexes with human alpha 2-macroglobulin and binds to the alpha 2-macroglobulin receptor/LDL receptor-related protein. J Urol 159:297–303.

Overall C. M., Dean R. A. (2006). Degradomics: Systems biology of the protease web. Pleiotropic roles of MMPs in cancer. Cancer Metastasis Rev 25:69–75.

Overall C. M., Kleifeld O. (2006). Tumour microenvironment—Opinion: Validating matrix metalloproteinases as drug targets and anti-targets for cancer therapy. Nat Rev Cancer 6:227–239.

Panchal R. G., Cusack E., Cheley S. et al. (1996). Tumor protease-activated, pore-forming toxins from a combinatorial library. Nat Biotechnol 14:852–856.

Potrich C., Tomazzolli R., Dalla Serra M. et al. (2005). Cytotoxic activity of a tumor protease-activated pore-forming toxin. Bioconjug Chem 16:369–376.

Rao A. R., Motiwala H. G., Karim O. M. (2008) The discovery of prostate-specific antigen. BJU Int. 101(1):5–10.

Reynolds M. A., Kastury K., Groskopf J. et al. (2007). Molecular markers for prostate cancer. Cancer Lett 249:5–13.

Romer J., Nielsen B. S., Ploug M. (2004). The urokinase receptor as a potential target in cancer therapy. Curr Pharm Des 10:2359–2376.

Rono B., Romer J., Liu S. et al. (2006). Antitumor efficacy of a urokinase activation-dependent anthrax toxin. Mol Cancer Ther 5:89–96.

Rooseboom M., Commandeur J. N., Vermeulen N. P. (2004). Enzyme-catalyzed activation of anticancer prodrugs. Pharmacol Rev 56:53–102.

Roshy S., Sloane B. F., Moin K. (2003). Pericellular cathepsin B and malignant progression. Cancer Metastasis Rev 22:271–286.

Salden M., Splinter T. A., Peters H. A. et al. (2000). The urokinase-type plasminogen activator system in resected non-small-cell lung cancer. Rotterdam Oncology Thoracic Study Group. Ann Oncol 11:327–332.

Scherrer A., Wohlwend A., Kruithof E. K. et al. (1999). Plasminogen activation in human acute leukaemias. Br J Haematol 105:920–927.

Schmid B., Chung D. E., Warnecke A. et al. (2007). Albumin-binding prodrugs of camptothecin and doxorubicin with an Ala-Leu-Ala-Leu-linker that are cleaved by cathepsin B: Synthesis and antitumor efficacy. Bioconjug Chem 18:702–716.

Schmidt M., Hoppe F. (1999). Increased levels of urokinase receptor in plasma of head and neck squamous cell carcinoma patients. Acta Otolaryngol 119:949–953.

Singh Y., Klimpel K. R., Goel S. et al. (1999). Oligomerization of anthrax toxin protective antigen and binding of lethal factor during endocytic uptake into mammalian cells. Infect Immun 67:1853–1859.

Springfeld C., von Messling V., Frenzke M. et al. (2006). Oncolytic efficacy and enhanced safety of measles virus activated by tumor-secreted matrix metalloproteinases. Cancer Res 66:7694–7700.

Stephens R. W., Brunner N., Janicke F. et al. (1998). The urokinase plasminogen activator system as a target for prognostic studies in breast cancer. Breast Cancer Res Treat 52:99–111.

Su Y., Ortiz J., Liu S. et al. (2007). Systematic urokinase-activated anthrax toxin therapy produces regressions of subcutaneous human non-small cell lung tumor in athymic nude mice. Cancer Res 67:3329–3336.

Tauro J. R., Gemeinhart R. A. (2005). Matrix metalloprotease triggered delivery of cancer chemotherapeutics from hydrogel matrixes. Bioconjug Chem 16:1133–1139.

Timar F., Botyanszki J., Suli-Vargha H. et al. (1998). The antiproliferative action of a melphalan hexapeptide with collagenase-cleavable site. Cancer Chemother Pharmacol 41:292–298.

Torzewski M., Sarbia M., Verreet P. et al. (1997). Prognostic significance of urokinase-type plasminogen activator expression in squamous cell carcinomas of the esophagus. Clin Cancer Res 3:2263–2268.

Van Valckenborgh E., Mincher D., Di Salvo A. et al. (2005). Targeting an MMP-9-activated prodrug to multiple myeloma-diseased bone marrow: A proof of principle in the 5T33MM mouse model. Leukemia 19:1628–1633.

Varmus H. (2006). The new era in cancer research. Science 312:1162–1165.

Verweij J., de Jonge M. J. (2000). Achievements and future of chemotherapy. Eur J Cancer 36:1479–1487.

Vizoso F. J., Gonzalez L. O., Corte M. D. et al. (2007). Study of matrix metalloproteinases and their inhibitors in breast cancer. Br J Cancer 96:903–911.

Waghray A., Keppler D., Sloane B. F. et al. (2002). Analysis of a truncated form of cathepsin H in human prostate tumor cells. J Biol Chem 277:11533–11538.

Watermann I., Gerspach J., Lehne M. et al. (2007). Activation of CD95L fusion protein prodrugs by tumor-associated proteases. Cell Death Differ 14:765–774.

Williams S. A., Merchant R. F., Garrett-Mayer E. et al. (2007). A prostate-specific antigen-activated channel-forming toxin as therapy for prostatic disease. J Natl Cancer Inst 99:376–385.

Wong B. K., DeFeo-Jones D., Jones R. E. et al. (2001). PSA-specific and non-PSA-specific conversion of a PSA-targeted peptide conjugate of doxorubicin to its active metabolites. Drug Metab Dispos 29:313–318.

Wong B. Y., Gregory S. A., Dang N. H. (2007). Denileukin diftitox as novel targeted therapy for lymphoid malignancies. Cancer Invest 25:495–501.

Wu W., Luo Y., Sun C. et al. (2006). Targeting cell-impermeable prodrug activation to tumor microenvironment eradicates multiple drug-resistant neoplasms. Cancer Res 66:970–980.

Young L., Di Salvo A., Turnbull A. et al. (2003). Design of tumour-activated prodrugs that harness the 'dark side' of MMP-9. Brit J Cancer 88:S27–S27.

Zucker S., Vacirca J. (2004). Role of matrix metalloproteinases (MMPs) in colorectal cancer. Cancer Metastasis Rev 23:101–117.

Chapter 41
Targeting Degradome Genes via Engineered Viral Vectors

Risto Ala-aho, Andrew H. Baker, and Veli-Matti Kähäri

Abstract Members of the degradome play an important role at different levels in carcinogenesis, tumor progression, invasion, and metastasis. Therefore, they are attractive targets for therapeutic intervention in cancer. Accordingly, wild-type or modified genes coding for different inhibitory molecules targeted at different levels of protease expression and activity have been extensively studied in experimental models of cancer using engineered viral vectors for gene delivery. These preclinical studies have served as an important and efficient way of target validation and proved the feasibility of the concept of protease targeting in cancer therapy. In this chapter, we discuss the application of different viral gene delivery vectors for targeting expression and function of the extracellular matrix degrading proteases.

Introduction

Therapeutic applications of gene-based technologies require efficient gene delivery systems for sufficient transduction of genetic material into target cells. An ideal vector for gene transfer is cell specific, efficient, and safe. Viruses are the most common vectors used to introduce new genetic material into cells as they are naturally equipped to infiltrate cells and allow effective intracellular delivery of genetic material in many types of cells. Other methods for delivering new genetic material into cells using nonviral vectors include liposomes and other packaging methods, and physical methods, such as direct injection or "shooting" of genetic material into cells. However, these nonviral methods have limitations in terms of efficiency as compared to viral vectors. In this chapter, we discuss the applicability of different viral gene delivery methods for introducing genetic material to cells for targeting expression and function of the extracellular matrix-degrading proteases in cancer.

A.H. Baker
British Heart Foundation Glasgow, Cardiovascular Resarch Centre, University of Glasgow, Glasow G12 8TA, UK, e-mail: ab11f@clinmed.gla.ac.uk

D. Edwards et al. (eds.), *The Cancer Degradome.*

Viral Vectors for Gene Transfer

In gene-based therapies, genetic material is transferred into cells to treat diseases. Viruses are commonly vectors that are engineered to carry genetic material of interest into target cells or tissues. Viral vectors contain deletions of some or all viral coding sequences making them replication-deficient in the target cells. Viruses from different families have been modified to generate efficient vectors for gene delivery into various types of cells. Most commonly used viral vectors in preclinical studies are retroviruses, adenoviruses, adeno-associated viruses, herpes simplex viruses, and baculoviruses (Table 41.1).

Retrovirus Vectors

Retroviruses are enveloped RNA viruses that use the reverse transcriptase enzyme to copy their genetic material into the DNA of a host cell. The retrovirus structure consists of a dense protein capsid, surrounded by a lipid envelope. The capsid contains two copies of about 10 kb long single-stranded RNA genome and the enzymes reverse transcriptase to convert the single-stranded RNAs into double-stranded DNA copies, integrase to insert the DNA into the host cell genome, and protease to cleave the long-polypeptide chain produced by viral RNA into individual proteins. The simple retroviruses contain three genes, of which group-specific antigen (gag) encodes the structural proteins of the virus, polymerase (pol) encodes reverse transcriptase, protease and integrase, and envelope (env) encodes the coat proteins of retrovirus envelope (Kay et al. 2001). At the each end of the viral genome are cis-acting sequences required for gene expression, replication and viral DNA integration into the host cell genome. The complex retroviruses also have some additional genes.

Attachment and entry of retrovirus to the cell are mediated by a viral envelope glycoprotein complex, which includes an external glycosylated hydrophilic polypeptide (SU) and a transmembrane protein (Hunter 1997). These form together a knob on the surface of the virus particle. The SU domain binds to a specific receptor on the target cell, activating membrane fusion. The interaction between the SU domain and the cellular receptor defines the host range and tissue tropism of a retrovirus. The receptors for retroviral entry appear to be distinct for the different major viral subgroups (Hunter 1997). When a retrovirus infects a cell, it injects its RNA and the reverse transcriptase into the cytoplasm of a host cell (Fig. 41.1). The cDNA produced from the viral RNA genome is inserted directly into a chromosome of the host cell at multiple sites, particularly those that are transcriptionally active. The viral genome is expressed by the host cell machinery for making new copies of viral RNA and structural proteins. New viral particles are assembled, bud from the plasma membrane, and released. In general, retroviruses do not infect quiescent cells, which may be an advantage in cancer therapy although efficiency issues remain.

Table 41.1 Comparison of Different Viral Vector Systems

Feature	Adenoviruses	Retroviruses	Adeno-associated viruses	Herpes simplex viruses	Baculoviruses
Genome	dsDNA	ssRNA (+strand)	ssDNA	dsDNA	dsDNA
Genome size	30–40 kb	8.8–12.3 kb	4.7 kb	152 kb	80–200 kb
Structure	Capsid	Envelope	Capsid	Envelope	Envelope
Maximum insert size	30 kb	7.5 kb	4.7 kb	50 kb	38 kb
Chromosomal integration	No	Yes	Yes	No	No
Duration of expression in vivo	Short	Long	Long	Long	Long
Safety	Extensive inflammatory reactions	Risk of insertional mutagenesis	Immunogenic	Immunogenic, toxic	Immunogenic
Transmitted to quiescent cells	Yes	No Lentiviruses:Yes	Yes	Yes	Yes

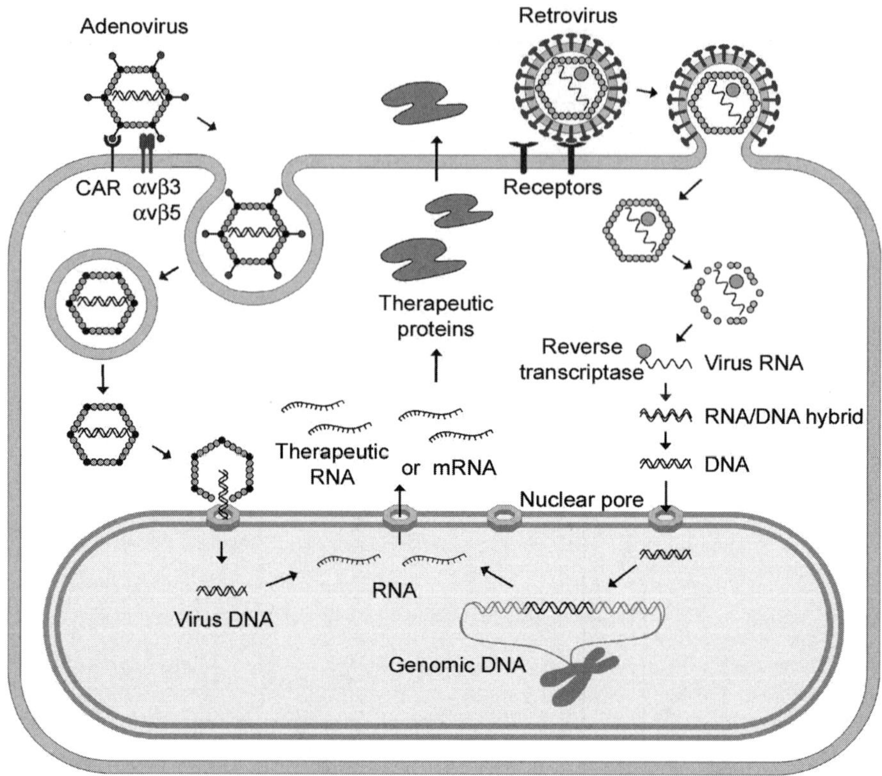

Fig. 41.1 Transduction mechanisms of recombinant adenovirus and retrovirus. The vector particle containing the therapeutic sequences binds to the target cell, enters the cell through receptor-mediated processes. The adenoviral double-stranded genomic DNA enters the nucleus and remains episomal whereas the retroviral RNA sequences are reverse transcribed into DNA, which enters the nucleus and integrates at random positions into host cell genome. Therapeutic RNA molecules or polypeptides are expressed by the host cell machinery

Several viral vectors derived from different retroviruses have been developed for gene delivery. Most retroviral vectors are based on Moloney murine leukemia virus (MMLV), Rous sarcoma virus (RSV), avian leukosis virus (ALV), different lentiviruses, and spumaviruses (Hu and Pathak 2000). Retroviral vectors contain the cis-acting elements needed for gene expression and replication. The retrovirus cis-acting elements contain viral promoters, transcriptional enhancers, and the essential regions for reverse transcription and virus replication. Retroviral vectors are maintained as bacterial plasmids to facilitate the manipulation and production of the vector DNA. Helper contructs are used in packaging cell lines to express required viral proteins and support replication of the genetically modified viruses.

Lentiviruses are members of the retrovirus family, but are more complex, having a more developed genetic structure than do other retroviruses. In addition to the gag, pol, and env gene products, their genome encodes a series of regulatory proteins. In contrast to other retroviruses, lentiviruses are also capable of integrating into the genome of quiescent cells. Many lentiviral vectors used for gene delivery into the cells are based on the human immunodeficiency viruses (HIV) types 1 and 2.

As gene delivery vectors, retroviruses have several advantages over other vectors, especially when permanent gene transfer is the preferred outcome. An important advantage of retroviral vectors is their ability to transform their genetic material stably into the target cell genome. Thus, retroviral vectors can be used to modify permanently the host cell nuclear genome. However, the major risk of retroviral vectors in gene delivery is that they incorporate into the host cell genome in a nonspecific manner, thus causing possible disruption of a host gene at the site of insertion or resulting in an abnormal expression of the nearby host cell gene driven by the viral promoter and transcriptional enhancers. Retroviral delivery of interfering RNA molecules can efficiently induce stable RNA interference, allowing dissection of specific gene function.

Adenovirus Vectors

Adenoviruses are common human pathogens associated with respiratory, gastrointestinal, and ocular infections. There are more than 50 different serotypes of human adenoviruses, which are classified into six major subgroups (A to F) based on their hemagglutination properties and DNA homologies. Adenoviral vectors based on serotype 5 virus from subgroup C have been most commonly used as gene delivery vectors. Adenoviruses are icosahedral nonenveloped viruses, which contain linear, 30–40 kb long double-stranded DNA genome. When they infect a host cell in culture, adenoviruses attach to coxsackie and adenovirus receptor (CAR) with the knob domain of the fiber protein (Fig. 41.1) (Bergelson et al. 1997). However, group B adenoviruses use CD46, which is involved with regulation of complement activation, as their receptor (Gaggar et al. 2003). In addition, adenoviruses can use host-derived blood coagulation factors and potentially other serum proteins for hepatocyte transduction (Parker et al. 2006). The adenovirus penton base protein binds to αV integrins followed by clathrin-dependent endocytosis, and finally, partially disrupted adenovirus capsid is transported to the nucleus and the adenovirus DNA enters the nucleus through the nuclear pore (Stewart et al. 2003).

As gene transfer vectors, recombinant replication-deficient adenoviruses have several advantages, which are beneficial in terms of gene therapy, as well as use in biological research as gene transfer vectors for expression of desired protein or RNA molecule. The adenovirus genome is relatively easy to manipulate and stable as it does not rearrange at a high rate. In addition, adenovirus particles are stable and they replicate efficiently in permissive cells producing high-titer viral stocks. Adenoviruses infect both dividing and quiescent cells, but the adenoviral DNA does not integrate into the genome of the host cell, thus minimizing the risk of insertional mutagenesis. As the adenoviral DNA remains episomal in the cells, adenovirus delivered genetic material is lost on cell division, and treatment with the adenoviruses requires readministration of the therapeutic genetic material in a proliferating cell population. Thus, adenoviral vectors are generally used for therapeutic strategies that require only a short-term expression of the therapeutic transgene. Another disadvantage of using adenoviruses is the host immune response. The humoral immune response generates antibodies against adenovirus proteins and precludes the repeated administration of the adenoviral vector. In addition, T cells can eliminate adenovirus-transduced cells (Dai et al. 1995).

Recombinant adenoviruses can carry relatively large therapeutic genetic material, depending on to which extent adenoviral genes are deleted. In the first generation of adenovirus vectors used for gene delivery, the early gene cassettes E1 and/or E3 were deleted (Russell 2000). The E1 region is essential for adenovirus transformation and the deletion of that region makes adenovirus replication deficient, and allows introducing 4.7 kb of foreign DNA. The E3 gene products subvert the host defense mechanism but are not essential for viral replication. Deletion of the E3 region increases the cloning capacity for inserts up to 8 kb (Bett et al. 1994). The next generation of recombinant adenoviruses was constructed by deleting additional E2 or E4 adenoviral genes (Lusky et al. 1998, Amalfitano et al. 1998). Latest adenoviral vectors lack all of the coding regions of the adenovirus genome, containing only the inverted terminal repeat (ITR) and the packaging sequences around the transgene, but they need a helper virus for provision of the necessary viral genes for virus propagation. These helper-dependent adenoviral vectors have up to 37 kb insert capacity (Parks and Graham 1997). However, helper-dependent adenoviral vectors are easily contaminated with the helper virus, but use of a Cre/loxP helper-dependent system prevented the packaging of the helper virus (Parks et al. 1996). In this system, infection of a Cre recombinase expressing packaging cell line with the helper virus results in excision of the viral packaging signal flanked by loxP recognition sites on the helper virus DNA (Parks et al. 1996).

Adenoviral vectors activate innate and adaptive immune responses *in vivo*. Following entry into cells, the adenoviral capsid activates a number of signaling pathways, including p38 mitogen-activated protein kinase and extracellular signal-regulated kinase1, 2 (ERK1, 2), that lead to expression of proinflammatory cytokines and chemokines (Liu and Muruve 2003). Recent generations of helper-dependent adenoviral vectors have reduced host adaptive immune responses *in vivo* providing an immunologic advantage over the early generations of adenoviral vectors (Muruve et al. 2004). Furthermore, adenoviral vectors based on serotypes other than 5, as well as chimeric serotype 5/35 adenoviruses have been studied to circumvent the host immune response to gene delivery vector (Marsman et al. 2007, Suominen et al. 2006)

Adeno-Associated Virus Vectors

Adeno-associated viruses (AAV) of the parvovirus family are nonpathogenic and the smallest known human viruses. They have a linear, single-stranded DNA genome about 5 kb long, which encodes two genes, rep and cap. As a result of alternative splicing and use of an alternative translation initiation site, these genes encode seven differentproteins needed for viral replication, DNA packaging, and capsid construction.

AAVs have certain advantages as gene delivery vectors over other viruses and are therefore extensively studied. They lack pathogenicity and most people treated with AAV do not develop an immune response to virus or the cells treated with it. In AAV-based vectors, viral genes are replaced by a transgene and its associated regulatory sequences. The low capacity to carry foreign DNA due to small genome is the major limitation of AAV vectors. The propagation of the recombinant AAV requires products from rep and cap genes along with the helper virus gene products (Gonçves 2005). In the absence of helper virus, the wild-type AAV has the ability to stably integrate into the host cell genome, most preferably at a specific site in human chromosome 19. This site-specific integration of AAV requires expression of the rep gene, and the proviral AAV sequences are rescued by the subsequent introduction of a helper virus (Kotin et al. 1992, Surosky et al. 1997). However, the recombinant AAV vectors have lost their site-specific integration. The AAV vectors transduce cells by multiple pathways, and they can integrate randomly into the host cell genome at the site of existing chromosome breaks (Yang et al. 1997, Miller et al. 2004). Because of random integration of the AAV vector into the host cell genome, they may produce distinct mutations, as the new inserted transgene sequence can influence expression of neighboring genes, including possible proto-oncogenes.

AAV-based gene transfer vectors used in preclinical studies are based on the AAV serotype 2 (AAV-2). Their primary receptors are heparin sulphate proteoglycans, but the coreceptors, $\alpha V\beta 5$ integrin and fibroblast growth factor I receptor, give access to wide range of tissue types (Young et al. 2006). However, AAV-2 is rather inefficient at transducing some cell types and its function as gene transfer vector can be hampered by neutralizing antibodies against AAV-2 (Grimm and Kay 2003). Other naturally occurring serotypes have been studied to resolve these limitations, and this has led to generation of new hybrid AAV vectors (Rabinowitz et al. 2004) as well as a host of alternate serotypes of AAV (Gao et al. 2002).

Baculovirus Vectors

Baculoviruses are a group of over 600 species with a double-stranded 80–200 kb long DNA genome (Szewczyk et al. 2006). Baculoviruses contain either single or multiple nucleocapsids embedded in the membrane envelope. Baculoviruses primarily infect insects with a narrow host range but they are not known to replicate in mammalian or other vertebrate animal cells, although they can infect mammalian cells. The most studied baculovirus is Autographa californica multicapsid nucleopolyhedrovirus (AcMNPV). Baculoviruses are generally used as tools for the efficient recombinant protein production in insect cells (O'Reilly et al. 1994) as well as biological pesticides (Cory and Bishop 1997), but their use as viral vectors for targeted gene-based therapies is in continuous expansion, since the baculovirus AcMNPV can infect a variety of mammalian cell lines and express foreign genes under the control of appropriate eukaryotic promoters (Kost et al. 2005). The baculovirus envelope glycoprotein gp64 is responsible for the baculovirus intake into cells. It mediates receptor-binding and pH-dependent clathrin-mediated endocytosis of the virus (van Loo et al. 2001).

Major advantages of recombinant baculoviral vectors include good biosafety profile due to their nonpathogenic nature for humans, and their ability to carry large DNA inserts. Baculoviruses are also relatively easy to manipulate and produce. Proteins produced by baculoviral systems have correct posttranslational modifications.

Since baculoviruses are inactivated in serum by the human complement system, *in vivo* delivery has important limitations. However, several methods to overcome this limitation have been demonstrated. One application is to insert decay-accelerating factor (DAF), which is an important regulator of the complement cascade, into the baculoviral envelope protein gp64 to protect baculovirus against complement attack (Hüser et al. 2001). In addition, expression of avidin or truncated vesicular stomatitis virus G protein has been shown to improve *in vivo* transduction efficiency of baculoviral vectors (Räty et al. 2004, 2006, Kaikkonen et al. 2006).

Herpes Simplex Virus Vectors

Herpes simplex viruses (HSV) are common human pathogens causing infections in many parts of the body, including the mouth, skin, eye, brain, and genitals. There are two serotypes of HSV, of which HSV type 1 (HSV-1) is the most engineered herpes virus for gene transfer (Kay et al. 2001). HSV-1 has a large, 152 kb long linear double-stranded DNA genome, containing over 80 genes, of which about half are nonessential for virus replication. Deletion of these nonessential genes creates capacity for 40–50 kb of foreign DNA.

The genome is packed into an icosahedral capsid coated with a layer of a protein called tegument. At the outermost is a membrane layer containing numerous viral proteins and glycoproteins (Spear 2004). HSV enters the cell by contacting glycosaminoglycan, like heparan sulfate side chains of cell surface proteoglycans. HSV glycoproteins B and C bind to heparan sulfate followed by the binding of viral glycoprotein D into its receptor, which triggers the fusion of HSV envelope with a cell membrane. Cellular receptors for the HSV entry include a member of the TNF receptor family (herpes virus entry mediator), members of the immunoglobulin superfamily (nectin-1 and nectin-2), and a specific site in heparan sulfate (Spear 2004).

Two approaches to produce HSV vectors include HSV amplicon vectors and recombinant genomic vectors. The amplicon vectors are bacterially produced plasmids about 15 kb in size containing viral origin of replication and packaging sequences, and they require a helper virus which provides all the missing structural and regulatory genes. The recombinant HSV vectors are made replication-deficient by deletion of one or more viral genes whose expression is essential for viral replication. The viruses are produced in a complementing cell line that supplies the deleted proteins to the virus in trans for replication. HSV vectors are efficient in transduction of a large number of different cell types resulting in efficient transgene expression. They have successfully been studied in animal models of several diseases, including cancer (Shen and Nemunaitis 2006). The cytotoxic properties of the HSV are a major limitation in practical applications, but deletion of multiple immediate-early genes has been shown to substantially reduce the cytotoxicity of HSV vectors (Krisky et al. 1998). Such mutants may provide efficient gene transfer and long-term transgene expression.

Viral Vectors in Degradome-Targeted Cancer Therapy

There are several strategies to interfere with degradome gene expression and protein function by viral vector systems. Targeted therapy can be directed against extracellular factors or signal transduction pathway molecules mediating degradome gene expression, or it is possible to inhibit transcription by targeted gene silencing of the mRNA of the gene of interest by specific antisense RNA, ribozymes, or small interfering RNA molecules (Fig. 41.2). Furthermore, secretion and function of the

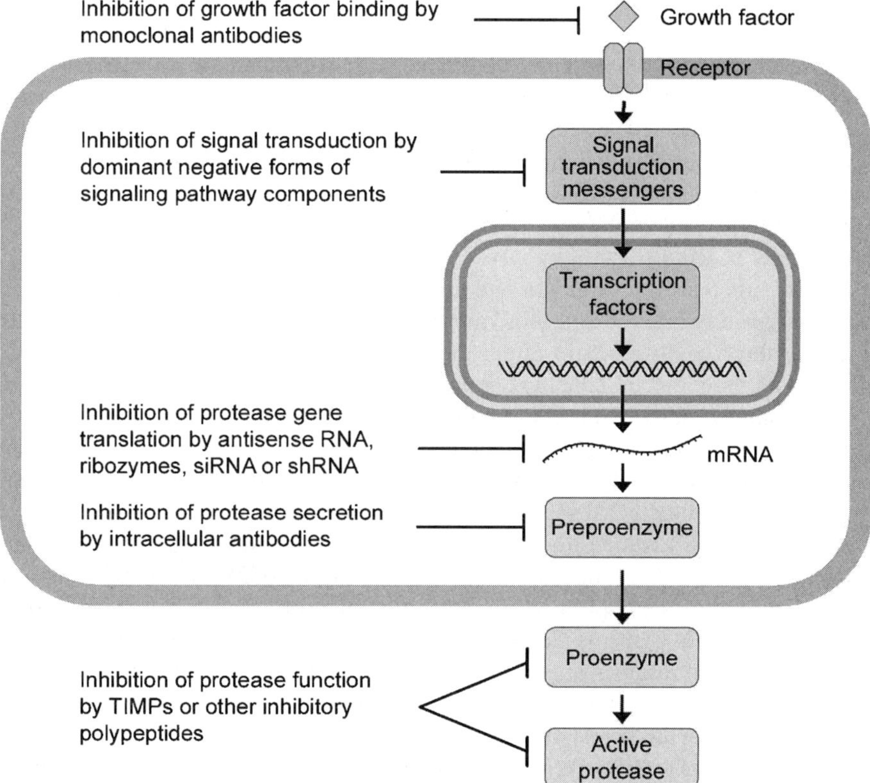

Fig. 41.2 Levels of inhibition of degradome gene expression and activity targeted by viral vector based gene delivery and possible inhibition mechanisms

degradome proteins can be inhibited by intracellular antibodies, and the activation or function of proteases can be blocked by overexpressing the tissue inhibitors of metalloproteinases (TIMPs) or other protease-specific inhibitory proteins.

Inhibition of Degradome Gene Expression by Modulating Gene Regulators

The role of different regulatory pathways of MMP expression has been studied in head and neck squamous cell carcinoma (HNSCC) cell lines. Modulation of the activity of distinct mitogen-activated protein kinases (MAPKs) by dominant negative forms of MAPK pathway mediators has shown potency in the inhibition of MMP expression by cancer cells. Selective inhibition of p38 MAPK signaling by specific adenovirally produced dominant negative forms of p38α and p38δ results in inhibition of MMP-1 and -13 expression by invasive HNSCC cells and suppresses growth of corresponding tumors *in vivo* in SCID mice (Junttila et al. 2007). However, as an untargeted therapy, inhibition of p38 MAPK throughout the body would induce immunosupressive side effects as seen with the chemical p38 inhibitors (Kumar et al. 2003). In addition, p38 has a critical role in inducing apoptosis in normal cells and inhibiting p38 MAPK activity could participate in transformation of normal cells raising extreme concerns for such therapy. TGF-β promotes the invasion of HNSCC cells by upregulating MMP-13 via the Smad pathway. Adenoviral overexpression of an inhibitory Smad, Smad7, in SCC cells inhibited MMP-13 expression and tumor implantation in SCID mice suggesting the Smad-signaling pathway as a novel and specific target for anti-invasive therapy of HNSCCs (Leivonen et al. 2006).

Tumor suppressor p53 has an inhibitory effect on cell growth and induces apoptosis when overexpressed in a variety of tumor cells by adenovirus-mediated gene transfer. The therapeutic effect of p53 overexpression in cancers has been related to its functions in inducing apoptosis, and inhibitory effects on cell proliferation and angiogenesis (Roth 2006). Wild-type (wt) p53 has been shown to inhibit the activity of the human MMP-1 gene promoter (Sun et al. 1999). In addition, the promoter of human MMP-13 contains a putative p53-binding element suggesting that p53 may regulate MMP-13 transcription. Adenovirus-mediated expression of wt p53 in HNSCCs resulted in significant reduction in MMP-13 expression and inhibition of HNSCC cell lines *in vitro* (Ala-aho et al. 2002). The wt p53 can also bind to the MMP-2 gene promoter and adenoviral expression of wt p53 in a p53-mutated melanoma cell line decreased MMP-2 levels and cell invasion (Toschi et al. 2000). These results suggest novel regulatory mechanisms in which adenovirally redelivered wt p53 into p53-mutated tumor cell lines reduces their invasiveness by decreasing the levels of different MMPs.

Nuclear factor-κB (NF-κB) has been shown to regulate the expression of several destructive MMPs in smooth human macrophages and in osteoarthritis (OA) synovial fibroblasts (Chase et al. 2002, Bondeson et al. 2007). Since overexpression of several MMPs, including MMP-1, -3, and -9, is demonstrated in human atherosclerotic plaques, adenovirus-mediated overexpression of NF-κB inhibitor IκBα was used to inhibit MMP expression in human macrophages and rabbit foam cells (Chase et al. 2002). Also, adenoviral expression of IκBα in osteoarthritis synovial fibroblasts resulted in inhibition of MMP-1, -3, -13, and ADAMTS-4 expression (Bondeson et al. 2007). As the expression of IκBα reduces proteolytic activity, it appears to be an attractive therapeutic target in unstable atheromas and OA. In addition, adenovirus-mediated inhibition of some MMP-regulating growth factors, like interleukin-8 (IL-8) or basic fibroblast growth factor (bFGF), has shown potency on possible targets for therapy of cancers (Inoue et al. 2000, 2001).

Degradome Gene Silencing by Antisense Oligonucleotides, siRNA, and Ribozymes

Antisense oligonucleotides and catalytic RNAs such as hammerhead ribozymes are capable of modulating specific gene expression, and they have demonstrated utility in attenuating gene expression (James and Gibson 1998). Ribozymes are site specific and their catalytic potential makes them more efficient in suppressing the specific gene expression than traditional antisense techniques. Several types of viral vectors have been used to suppress degradome gene expression by ribozyme or antisense RNA in variety of cell types.

Retroviral expression of an MMP-7 antisense ribozyme has shown potency to inhibit MMP-7 expression by a breast carcinoma cell line and reduce tumor cell invasiveness and tumor growth in nude mice (Jiang et al. 2005). A retroviral vector has also been used for long-term production of therapeutic RNA molecules in rheumatoid arthritis (RA) synovial fibroblasts (RASFs) in a SCID mouse model (Rutkauskaite et al. 2004, 2005). Elevated levels of collagen-degrading MMPs are associated with progressive matrix degradation in RA and, thus, targeted inhibition of MT1-MMP or MMP-1 production in RASFs could constitute a promising approach to inhibit joint destruction in RA. Inhibition of MMP-1 production by a retrovirally mediated hammerhead MMP-1 antisense ribozyme was maintained for at least 2°months and was accompanied by a significant reduction in the invasiveness of RASFs in the SCID mouse model of RA (Rutkauskaite et al. 2004). Retroviral expression of an MT1-MMP antisense RNA reduced cartilage degradation by RASFs in RA during the 60-day observation period (Rutkauskaite et al. 2005). Adenovirally delivered MMP-13 antisense ribozyme into cutaneous SCC cells inhibited tumor implantation in SCID mice (Ala-aho et al. 2004). Repeated injection of MMP-13 antisense ribozyme encoding adenovirus into SCC xenografts resulted in suppression of tumor growth. However, the increase in the adenoviral dose in intratumoral injections from twice a week to three times a week did not increase the inhibitory effect of MMP-13 antisense ribozyme. This may be due to the short duration of adenoviral gene expression in the cells and the limited transduction efficiency.

Silencing of specific gene expression by small interfering RNA (siRNA) has rapidly become a powerful tool to study gene function and it represents a potential strategy for therapeutic product development (Takeshita and Ochiya 2006). RNA interference induced by siRNA is a transient phenomenon, but high-efficiency transfection of interfering RNA and stable suppression of the specific gene can be achieved with viral vectors expressing short hairpin RNA (shRNA) which is cleaved to active siRNA by cellular ribonuclease Dicer (Bernstein et al. 2001). Each siRNA is then incorporated into RNAi-induced silencing complex that can cleave the target mRNA.

The urokinase plasminogen activator (uPA) system plays a marked role in matrix degradation by converting plasminogen to plasmin and triggers an activation cascade of several MMPs, including MMP-9. The serine protease uPA and its receptor uPAR are produced by several tumor cells and are strongly implicated in tumor progression. Thus, simultaneous inhibition of uPA and MMP-9 expression could have a synergistic effect in inhibiting tumor growth. The adenovirus-mediated simultaneous expression of antisense uPAR and MMP-9 has been used to downregulate uPAR and MMP-9 in glioma cells (Lakka et al. 2003). However, the therapeutic effect of adenovirus-mediated antisense expression was shown to be transient and require relatively high adenovirus doses. Unlike the traditional antisense approach, siRNAs result in the destruction of the target RNA molecule and are more potent in reducing target gene expression. Since siRNA-mediated silencing of gene expression may be superior to the traditional antisense techniques, siRNA-mediated uPA and MMP-9 mRNA degradation were also studied in glioma cells (Lakka et al. 2005). Simultaneous targeted inhibition of the two proteases by RNAi resulted in complete regression of preestablished intracerebral tumor growth in mice.

Since high levels of MMP-2 are associated with the increased propensity for nodal and distant metastases in non-small lung cell carcinoma, it was suggested that MMP-2 inhibition has therapeutic potential for lung cancer. An MMP-2 shRNA adenovirus showed potency in inhibiting tumor growth and formation of lung nodules in a metastatic lung cancer model in mice (Chetty et al. 2006). MMP-9 is strongly expressed in medulloblastoma tumors and inhibition of its expression by adenovirally delivered MMP-9 shRNA has shown efficacy to inhibit medulloblastoma tumor growth and cell invasion in an intracranial model (Rao et al. 2007).

Inhibition of Degradome Protein Activation and Function

In general, gene delivery-based inhibition of proteinase activity can be targeted against activation of the latent form of proteinase or activity of the proteinase itself (Fig. 41.2). Both approaches have been utilized in experimental studies.

The activity of MMPs is specifically inhibited by TIMPs-1–4, which bind to active MMPs in 1:1 molar stoichiometry (Baker et al. 2002). TIMP-1, -2, and -4 are secreted by cells in soluble form, whereas TIMP-3 is associated with ECM (Pavloff et al. 1992). TIMPs also inhibit the activity of metalloproteinases with a disintegrin and metalloproteinase domain (ADAMs), and ADAM-TSs with thrombospondin-like (TS) domains (Baker et al. 2002). TIMP-3 inhibits the activity of tumor necrosis factor-α (TNF-α)-converting enzyme (TACE, ADAM-17), ADAM-10, aggrecanase-1 (ADAM-TS4), and aggrecanase-2 (ADAM-TS5) (Baker et al. 2002). Therefore, TIMP-3 inhibits shedding of several cell surface proteins, including TNF-α, TNF-RI, syndecan-1 and -4, interleukin-6 receptor, and L-selectin. TIMP-3 also functions as a tumor suppressor, as its expression is silenced by hypermethylation of the gene in several cancers (Baker et al. 2002). In addition, TIMP-3 has been shown to promote apoptosis in normal and malignant human cells in culture and *in vivo* (Ahonen et al. 1998, 2002, Baker et al. 1998, 1999). The proapoptotic activity of TIMP-3 resides in the N-terminus of the molecule, which possesses the MMP inhibitory activity (Bond et al. 2000). The proapoptotic effect of TIMP-3 appears to be mediated through stabilization of cell surface death receptors, especially TNF-R1 and Fas (Bond et al. 2002, Ahonen et al. 2003). Lack of TIMP-3 in mice results in increased metastasis of melanoma and lymphoma cells (Cruz-Munoz et al. 2006). In general, because of their wide inhibitory efficacy toward metalloproteinases, TIMPs have served as attractive candidates in gene transfer studies in cancer.

Adenoviral delivery of TIMP-1, -2, and -3 by local intratumoral injection has been shown to inhibit tumor growth in several different animal models of cancer. Adenoviral delivery of TIMP-2 inhibits tumor growth in mouse models (Li et al. 2001) and inhibits mammary tumor growth and angiogenesis (Hajitou et al. 2001) In addition, gene transfer of TIMP-1 using AAV vector inhibits tumor growth in different animal models for cancer (Zacchigna et al. 2004).

Adenoviral delivery of TIMP-3 has been shown to suppress the growth of melanoma and cutaneous SCCs in mouse model more potently than do TIMP-1 (Ahonen et al. 2002, 2003). Similarly, adenovirally delivered TIMP-3 inhibits lung cancer progression in an animal model more potently than do TIMP-1 and -2 (Finan et al. 2006). Retroviral vector-based cell-mediated delivery of TIMP-3 has also been shown to inhibit glioblastoma growth in an animal model (Spurbeck et al. 2003).

In addition to local intratumoral injection of viral vectors coding for TIMPs, attempts to use systemic delivery have also been taken. Adenovirus-mediated expression of TIMP-1 in liver has been shown to protect against lymphoma and colon carcinoma metastasis (Elezkurtaj et al. 2004). In addition, systemically delivered adenoviral expression of TIMP-1 and -2 has been shown to promote the antitumor effect of tumor necrosis factor and interferon-γ providing evidence for feasibility of combining degradome-targeted gene therapy with other cancer therapies (Van Roy et al. 2007). In this context, it is interesting to note that a recent study with an oncolytic adenovirus expressing TIMP-3 showed that including TIMP-3 did not improve the efficacy of the oncolytic adenovirus in a glioma model (Lamfers et al. 2005). However, further preclinical studies on the efficacy of degradome-targeted gene delivery in combination with other antitumor therapies including oncolytic viruses are clearly needed to assess the feasibility of protease-targeted cancer gene therapy.

Furthermore, viral expression of other natural inhibitors of MMPs also may have a role in targeted cancer therapy. Lentiviral vectors generated to deliver expression of angiostatin and endostatin inhibited endothelial cell growth (Shichinohe et al. 2001). In addition, noncatalytic fragments of MMP-2 are shown to inhibit tumor angiogenesis. Since the PEX functions as an endogenous inhibitor of MMP-2 activation, its inhibitory effect could be potentially utilized in cancer therapy. Lentiviral vectors were used to deliver PEX expression in different angiogenesis models to inhibit MMP-2 activity (Pfeifer et al. 2000). As a result, lentiviral expression of the MMP-2 hemopexin domain suppressed angiogenesis in different angiogenesis models, demonstrating that lentiviral vectors are an efficient vehicle for delivery of antiangiogenic factors *in vivo* (Pfeifer et al. 2000).

Concluding Remarks

Several different viral vector systems have been utilized to target degradome genes in experimental cancer models *in vivo*. The feasibility of these gene delivery vectors has been extensively studied in preclinical models of cancer, and they have served as important tools for target validation of proteases in cancer therapy. It is expected that development of vector technology and new ways to target the vectors to specific cells and tissues will result in improved safety and efficacy in gene therapy of cancer. In addition, further studies are clearly required to test the combination of protease-targeted gene therapy with currently available cancer therapies, for example, radiation therapy and chemotherapy. It is expected that targeting the degradome genes via viral gene transfer will improve the efficacy of these conventional treatments for cancer.

Acknowledgments The work of authors has been supported by grants from the Academy of Finland, the Sigrid Jusélius Foundation, the Cancer Research Foundation of Finland, Turku University Central Hospital (project 13336), and European Union Framework Programme 6 (LSHC-CT-2003-503297).

References

Ahonen M, Baker A H, Kähäri V-M (1998) Adenovirus-mediated gene delivery of tissue inhibitor of metalloproteinases-3 inhibits invasion and induces apoptosis in melanoma cells. Cancer Res 58:2310–2315.

Ahonen M, Ala-aho R, Baker A H et al. (2002) Antitumor activity and bystander effect of adenovirally delivered tissue inhibitor of metalloproteinases-3. Mol Ther 6:705–715.

Ahonen M, Poukkula M, Baker A H et al. (2003) Tissue inhibitor of metalloproteinases-3 induces apoptosis in melanoma cells by stabilization of death receptors. Oncogene 22:2121–2134.

Ala-aho R, Grénman R, Seth P, Kähäri V-M (2002) Adenoviral delivery of p53 gene suppresses expression of collagenase-3 (MMP-13) in squamous carcinoma cells. Oncogene 21:1187–1195.

Ala-aho R, Ahonen M, George S J et al. (2004) Targeted inhibition of human collagenase-3 (MMP-13) expression inhibits squamous cell carcinoma growth *in vivo*. Oncogene 23:5111–5523.

Amalfitano A, Hauser M A, Hu H et al. (1998) Production and characterization of improved adenovirus vectors with the E1, E2b, and E3 genes deleted. J Virol 72:926–933.

Baker A H, Zaltsman A B, George S J et al. (1998) Divergent effects of tissue inhibitor of metalloproteinase-1, -2, or -3 overexpression on rat vascular smooth muscle cell invasion, proliferation, and death *in vitro*. TIMP-3 promotes apoptosis. J Clin Invest 101:1478–1487.

Baker A H, George S J, Zaltsman A B et al. (1999) Inhibition of invasion and induction of apoptotic cell death of cancer cell lines by overexpression of TIMP-3. Br J Cancer 79:1347–1355.

Baker A H, Edwards D R, Murphy G (2002) Metalloproteinase inhibitors: biological actions and therapeutic opportunities. J Cell Sci 115:3719–3727.

Bergelson J M, Cunningham J A, Droguett G et al. (1997) Isolation of a common receptor for Coxsackie B viruses and adenoviruses 2 and 5. Science 275:1320–1323.

Bernstein E, Caudy A A, Hammond S M et al. (2001) Role for a bidentate ribonuclease in the initiation step of RNA interference. Nature 409:363–366.

Bett A J, Haddara W, Prevec L et al. (1994) An efficient and flexible system for construction of adenovirus vectors with insertions or deletions in early regions 1 and 3. Proc Natl Acad Sci USA 91:8802–8806.

Bond M, Murphy G, Bennett M R et al. (2000) Localization of the death domain of tissue inhibitor of metalloproteinase-3 to the N terminus. Metalloproteinase inhibition is associated with proapoptotic activity. J Biol Chem 275:41358–41363.

Bond M, Murphy G, Bennett M R et al. (2002) Tissue inhibitor of metalloproteinase-3 induces a Fas-associated death domain-dependent type II apoptotic pathway. J Biol Chem 277:13787–13795.

Bondeson J, Lauder S, Wainwright S et al. (2007) Adenoviral gene transfer of the endogenous inhibitor IkappaBα into human osteoarthritis synovial fibroblasts demonstrates that several matrix metalloproteinases and aggrecanases are nuclear factor-kappaB-dependent. J Rheumatol 34:523–533.

Chase A J, Bond M, Crook M F et al. (2002) Role of nuclear factor-kappa B activation in metalloproteinase-1, -3, and -9 secretion by human macrophages *in vitro* and rabbit foam cells produced *in vivo*. Arterioscler Thromb Vasc Biol 22:765–771.

Chetty C, Bhoopathi P, Joseph P et al. (2006) Adenovirus-mediated small interfering RNA against matrix metalloproteinase-2 suppresses tumor growth and lung metastasis in mice. Mol Cancer Ther 5:2289–2299.

Cory J S, Bishop D H (1997) Use of baculoviruses as biological insecticides. Mol Biotechnol 7:303–313.

Cruz-Munoz W, Sanchez O H, Di Grappa M et al. (2006) Enhanced metastatic dissemination to multiple organs by melanoma and lymphoma cells in timp-3$^{-/-}$ mice. Oncogene 25:6489–6496.

Dai Y, Schwarz E M, Gu D et al. (1995) Cellular and humoral immune responses to adenoviral vectors containing factor IX gene: tolerization of factor IX and vector antigens allows for long-term expression. Proc Natl Acad Sci USA 92:1401–1405.

Elezkurtaj S, Kopitz C, Baker A H et al. (2004) Adenovirus-mediated overexpression of tissue inhibitor of metalloproteinases-1 in the liver: efficient protection against T-cell lymphoma and colon carcinoma metastasis. J Gene Med 6:1228–1237.

Finan K M, Hodge G, Reynolds A M et al. (2006) In vitro susceptibility to the pro-apoptotic effects of TIMP-3 gene delivery translates to greater *in vivo* efficacy versus gene delivery for TIMPs-1 or -2. Lung Cancer 53:273–284.

Gaggar A, Shayakhmetov D, Lieber A (2003) CD46 is a cellular receptor for group B adenoviruses. Nat Med 9:1408–1412.

Gao G P, Alvira M R, Wang L et al. (2002) Novel adeno-associated viruses from rhesus monkeys as vectors for human gene therapy. Proc Natl Acad Sci USA 99:11854–11859.

Gonçalves M A (2005) Adeno-associated virus: from defective virus to effective vector. Virol J 2:43.

Grimm D, Kay M A (2003) From virus evolution to vector revolution: use of naturally occurring serotypes of adeno-associated virus (AAV) as novel vectors for human gene therapy. Curr Gene Ther 3:281–304.

Hajitou A, Sounni N E, Devy L et al. (2001) Down-regulation of vascular endothelial growth factor by tissue inhibitor of metalloproteinase-2: effect on *in vivo* mammary tumor growth and angiogenesis. Cancer Res 61:3450–3457.

Hu W S, Pathak V K (2000) Design of retroviral vectors and helper cells for gene therapy. Pharmacol Rev 52:493–511.

Hunter E (1997) Viral entry and receptors. In: Coffin JM, Hughes SH, Varmus HE (eds) Retroviruses. Cold Spring Harbor Laboratory Press, New York.

Hüser A, Rudolph M, Hofmann C (2001) Incorporation of decay-accelerating factor into the baculovirus envelope generates complement-resistant gene transfer vectors. Nat Biotechnol 19:451–455.

Inoue K, Perrotte P, Wood C G et al. (2000) Gene therapy of human bladder cancer with adenovirus-mediated antisense basic fibroblast growth factor. Clin Cancer Res 6:4422–4431.

Inoue K, Wood C G, Slaton J W et al. (2001) Adenoviral-mediated gene therapy of human bladder cancer with antisense interleukin-8. Oncol Rep 8:955–964.

James H A, Gibson I (1998) The therapeutic potential of ribozymes. Blood 91:371–382.

Jiang W G, Davies G, Martin T A et al. (2005) Targeting matrilysin and its impact on tumor growth *in vivo*: the potential implications in breast cancer therapy. Clin Cancer Res 11:6012–6019.

Junttila M R, Ala-aho R, Jokilehto T et al. (2007) p38α and p38δ mitogen-activated protein kinase isoforms regulate invasion and growth of head and neck squamous carcinoma cells. Oncogene 26:5267–5279.

Kaikkonen M U, Räty J K, Airenne K J et al. (2006) Truncated vesicular stomatitis virus G protein improves baculovirus transduction efficiency *in vitro* and *in vivo*. Gene Ther 13:304–312.

Kay M A, Glorioso J C, Naldini L (2001) Viral vectors for gene therapy: the art of turning infectious agents into vehicles of therapeutics. Nat Med 7:33–40.

Kost T A, Condreay J P, Jarvis D L (2005) Baculovirus as versatile vectors for protein expression in insect and mammalian cells. Nat Biotechnol 23:567–575.

Kotin R M, Linden R M, Berns K I (1992) Characterization of a preferred site on human chromosome 19q for integration of adeno-associated virus DNA by non-homologous recombination. EMBO J 11:5071–5078.

Krisky D M, Wolfe D, Goins W F et al. (1998) Deletion of multiple immediate-early genes from herpes simplex virus reduces cytotoxicity and permits long-term gene expression in neurons. Gene Ther 5:1593–1603.

Kumar S, Boehm J, Lee J C (2003) p38 MAP kinases: key signalling molecules as therapeutic targets for inflammatory diseases. Nat Rev Drug Discov 2:717–726.

Lakka S S, Gondi C S, Yanamandra N et al. (2003) Synergistic down-regulation of urokinase plasminogen activator receptor and matrix metalloproteinase-9 in SNB19 glioblastoma cells efficiently inhibits glioma cell invasion, angiogenesis, and tumor growth. Cancer Res 63:2454–2461.

Lakka S S, Gondi C S, Dinh D H et al. (2005) Specific interference of urokinase-type plasminogen activator receptor and matrix metalloproteinase-9 gene expression induced by double-stranded RNA results in decreased invasion, tumor growth, and angiogenesis in gliomas. J Biol Chem 280:21882–21892.

Lamfers M L, Gianni D, Tung C H et al. (2005) Tissue inhibitor of metalloproteinase-3 expression from an oncolytic adenovirus inhibits matrix metalloproteinase activity *in vivo* without affecting antitumor efficacy in malignant glioma. Cancer Res 65:9398–9405.

Leivonen S K, Ala-aho R, Koli K et al. (2006) Activation of Smad signaling enhances collagenase-3 (MMP-13) expression and invasion of head and neck squamous carcinoma cells. Oncogene 25:2588–2600.

Li H, Lindenmeyer F, Grenet C et al. (2001) AdTIMP-2 inhibits tumor growth, angiogenesis, and metastasis, and prolongs survival in mice. Hum Gene Ther 12:515–526.

Liu Q, Muruve D A (2003) Molecular basis of the inflammatory response to adenovirus vectors. Gene Ther 10:935–940.

Lusky M, Christ M, Rittner K, et al. (1998) In vitro and *in vivo* biology of recombinant adenovirus vectors with E1, E1/E2A, or E1/E4 deleted. J Virol 72:2022–2032.

Marsman W A, Wesseling J G, El Bouch A et al. (2007) Adenoviral serotypes in gene therapy for esophageal carcinoma. J Surg Res 140(1):50–54.

Miller D G, Petek L M, Russell D W (2004) Adeno-associated virus vectors integrate at chromosome breakage sites. Nat Genet 36:767–773.

Muruve D A, Cotter M J, Zaiss A K et al. (2004) Helper-dependent adenovirus vectors elicit intact innate but attenuated adaptive host immune responses *in vivo*. J Virol 78:5966–5972.

O'Reilly D R, Miller L K, Luckov V A (1994) Baculovirus Expression Vectors. A Laboratory Manual, 2nd edn. Oxford University Press, New York.

Parker A L, Waddington S N, Nicol C G et al. (2006) Multiple vitamin K-dependent coagulation zymogens promote adenovirus-mediated gene delivery to hepatocytes. Blood 108:2554–2561.

Parks R J, Graham F L (1997) A helper-dependent system for adenovirus vector production helps define a lower limit for efficient DNA packaging. J Virol 71:3293–3298.

Parks R J, Chen L, Anton M et al. (1996) A helper-dependent adenovirus vector system: removal of helper virus by Cre-mediated excision of the viral packaging signal. Proc Natl Acad Sci USA 93:13565–13570.

Pavloff N, Staskus P W, Kishnani N S et al. (1992) A new inhibitor of metalloproteinases from chicken: ChIMP-3. A third member of the TIMP family. J Biol Chem 267:17321–17326.

Pfeifer A, Kessler T, Silletti S et al. (2000) Suppression of angiogenesis by lentiviral delivery of PEX, a noncatalytic fragment of matrix metalloproteinase. Proc Natl Acad Sci USA 97:12227–12232.

Rabinowitz J E, Bowles D E, Faust S M et al. (2004) Cross-dressing the virion: the transcapsidation of adeno-associated virus serotypes functionally defines subgroups. J Virol 78:4421–4432.

Rao J S, Bhoopathi P, Chetty C et al. (2007) MMP-9 short interfering RNA induced senescence resulting in inhibition of medulloblastoma growth via p16(INK4a) and mitogen-activated protein kinase pathway. Cancer Res 67:4956–4964.

Räty J K, Airenne K J, Marttila A T et al. (2004) Enhanced gene delivery by avidin-displaying baculovirus. Mol Ther 9:282–291.

Räty J K, Liimatainen T, Wirth T et al. (2006) Magnetic resonance imaging of viral particle biodistribution *in vivo*. Gene Ther 13:1440–1446.

Roth J A (2006) Adenovirus p53 gene therapy. Expert Opin Biol Ther 6:55–61.

Russell W C (2000) Update on adenovirus and its vectors. J Gen Virol 81:2573–2604.

Rutkauskaite E, Zacharias W, Schedel J et al. (2004) Ribozymes that inhibit the production of matrix metalloproteinase 1 reduce the invasiveness of rheumatoid arthritis synovial fibroblasts. Arthritis Rheum 50:1448–1456.

Rutkauskaite E, Volkmer D, Shigeyama Y et al. (2005) Retroviral gene transfer of an antisense construct against membrane type 1 matrix metalloproteinase reduces the invasiveness of rheumatoid arthritis synovial fibroblasts. Arthritis Rheum 52:2010–2014.

Shen Y, Nemunaitis J (2006) Herpes simplex virus 1 (HSV-1) for cancer treatment. Cancer Gene Ther 13:975–992.

Shichinohe T, Bochner B H, Mizutani K et al. (2001) Development of lentiviral vectors for antiangiogenic gene delivery. Cancer Gene Ther 8:879–889.

Spear P G (2004) Herpes simplex virus: receptors and ligands for cell entry. Cell Microbiol 6:401–410.

Spurbeck W W, Ng C Y, Vanin E F et al. (2003) Retroviral vector-producer cell-mediated *in vivo* gene transfer of TIMP-3 restricts angiogenesis and neuroblastoma growth in mice. Cancer Gene Ther 10:161–167.

Stewart P L, Dermody T S, Nemerow G R (2003) Structural basis of nonenveloped virus cell entry. Adv Protein Chem 64:455–491.

Sun Y, Sun Y, Wenger L et al. (1999) p53 down-regulates human matrix metalloproteinase-1 (Collagenase-1) gene expression. J Biol Chem 274:11535–11540.

Suominen E, Toivonen R, Grénman R et al. (2006) Head and neck cancer cells are efficiently infected by Ad5/35 hybrid virus. J Gene Med 8:1223–1231.

Surosky R T, Urabe M, Godwin S G et al. (1997) Adeno-associated virus Rep proteins target DNA sequences to a unique locus in the human genome. J Virol 71:7951–7959.

Szewczyk B, Hoyos-Carvajal L, Paluszek M et al. (2006) Baculoviruses - re-emerging biopesticides. Biotechnol Adv 24:143–160.

Takeshita F, Ochiya T (2006) Therapeutic potential of RNA interference against cancer. Cancer Sci 97:689–696.

Toschi E, Rota R, Antonini A et al. (2000) Wild-type p53 gene transfer inhibits invasion and reduces matrix metalloproteinase-2 levels in p53-mutated human melanoma cells. J Invest Dermatol 114:1188–1194.

van Loo N D, Fortunati E, Ehlert E et al. (2001) Baculovirus infection of nondividing mammalian cells: mechanisms of entry and nuclear transport of capsids. J Virol 75:961–970.

Van Roy M, Wielockx B, Baker A et al. (2007) The use of tissue inhibitors of matrix metalloproteinases to increase the efficacy of a tumor necrosis factor/interferon-? antitumor therapy. Cancer Gene Ther 14:372–379.

Yang C C, Xiao X, Zhu X et al. (1997) Cellular recombination pathways and viral terminal repeat hairpin structures are sufficient for adeno-associated virus integration *in vivo* and *in vitro*. J Virol 71:9231–9247.

Young L S, Searle P F, Onion D et al. (2006) Viral gene therapy strategies: from basic science to clinical application. J Pathol 208:299–318.

Zacchigna S, Zentilin L, Morini M et al. (2004) AAV-mediated gene transfer of tissue inhibitor of metalloproteinases-1 inhibits vascular tumor growth and angiogenesis *in vivo*. Cancer Gene Ther 11:73–80.

Appendix

Appendix Table: Inhibitory Activities of Small-Molecule MMPIs

Drug	MMP1	MMP2	MMP3	MMP7	MMP8	MMP9	MMP12	MMP13	MMP14	MMP15	MMP16	MMP24	ADAM17	References
Batimastat/ BB94 (IC_{50})	10	4	20		10	1			3					Whittaker et al. 1999
Batimastat/ BB94 (IC_{50})	3	4	20	6		4								Nelson et al. 2000
Batimastat/ BB94 (IC_{50})	25	32				23		28	19					Arlt et al. 2002
BMS275291 (IC_{50})	25	41	157		10	25		4						Whittaker et al. 1999
BMS275291 (IC_{50})	9	39	157	23		27			40				>100,000	Naglich et al. 2001
CGS27023A/ MMI270 (K_i)	33	20	43			8								Whittaker et al. 1999
CGS27023A/ MMI270 (K_i)	33	11	13	680		8		6	23					Nelson et al. 2000
Galardin/ GM6001 (K_i)	0.4	0.39	26		0.18	0.57								Whittaker et al. 1999
Marimastat/ BB2516 (IC_{50})	5	6	200	20	2	3			1.8				3,800	Whittaker et al. 1999
Marimastat/ BB2516 (IC_{50})	0.15	0.25	20	1.19	0.14	0.68	0.26	0.7	1.5	1.01	0.5	0.8		Peterson 2006
Marimastat/ BB2516 (IC50)	5	6	230	16		3	5							Nelson et al. 2000
Prinomastat/ AG3340 (K_i)	8.2	0.083	0.27	54				0.038						Whittaker et al. 1999
Prinomastat/ AG3340 (K_i)	8.3	0.05	0.3	54		0.26			0.33					Nelson et al. 2000
Prinomostat/ AG3340 (IC_{50})	26	2.4				1		24	12					Arlt et al. 2002

Tanomastat/Bay12-9566 (K_i)	>5,000	11	134		301	1,470		Whittaker et al. 1999
Tanomastat/Bay12-9566 (IC$_{50}$)	>5,000	11	134	51	301	1,470		Sternlicht and Bergers 2000
Trocade/Ro32-3555 (K_i)	3	154	527	4	59	3		Whittaker et al. 1999
Ro 28-2653 (IC$_{50}$)	16,000	12			16	556	10	Arlt et al. 2002
Ro 206-0222 (IC$_{50}$)	4,310	5			2	245	9	Arlt et al. 2002

Notes: (1). All values in nM

(2). Since assay conditions vary, all values should be taken as approximate and can be compared directly only with values in same row

MMP matrix metalloproteinase, *MMPI* matrix metalloproteinase inhibitor, *ADAM* a disintegrin and metalloprotease

References

Arlt M. et al (2002). Increase in gelatinase-specificity of matrix metalloproteinase inhibitors correlates with antimetastatic efficacy in a T-cell lymphoma model. Cancer Res 62: 5543–5550.

Naglich J. G. et al (2001). Inhibition of angiogenesis and metastasis in two murine models by the matrix metalloproteinase inhibitor, BMS-275291. Cancer Res 61: 8480–5.

Nelson A. R. et al (2000). Matrix metalloproteinases: biologic activity and clinical implications. J Clin Oncol 18: 1135–49.

Peterson J. T. (2006). The importance of estimating the therapeutic index in the development of matrix metalloproteinase inhibitors. Cardiovasc Res 69: 677–87.

Sternlicht M. D. and Bergers G. (2000). Matrix metalloproteinases as emerging targets in anticancer therapy: status and prospects. Emerg Ther Targets 4: 609–633.

Whittaker M. et al (1999). Design and therapeutic application of matrix metalloproteinase inhibitors. Chem Rev 99: 2735–2776.

Additional references with useful comparisons of metalloproteinase inhibitors

Hu J. et al (2007) Matrix metalloproteinase inhibitors as therapy for inflammatory and vascular diseases. Nat Rev Drug Disc 6: 480–498.

Pirard B. (2007) Insights into the structural determinants for selective inhibition of matrix metalloproteinases. Drug Disc Today 12: 640–646.

Sang Q.-X. A. et al. (2006) Matrix metalloproteinase inhibitors as prospective agents for the prevention and treatment of cardiovascular and neoplastic diseases. Curr Top Med Chem 6: 289–316.

Index

5A/6A polymorphism and MMP-3 gene
  cancer effects, 671–672
  interleukin-1, 671
  nucleotide position, 670
Active forms, metalloproteases, 113–115
Activity-based probes (ABPs), 144
  applications of
    enzyme inhibitor discovery,
      122–124
    protease activity imaging, 125–128
    tumor biomarkers discovery,
      117–121
  covalent modification, 103–104
  cysteine proteases for, 109
  functional elements of, 103
  linker region and tags, 104
  metalloproteases role in, 106–107
  profiling, 104–106
  serine proteases for, 109–110
  zinc metalloprotease activity in,
    110–115
ADAM-17 enzyme. *See* TNF-α converting
    enzyme (TACE)
  TIMP-1 and TIMP-2 mutants, 805
  TIMP-3 inhibition
    kinetic value, 802–803
    N-TIMP-3 mutation, 803
    zinc ion catalytic domain, 804
  TIMP-4 mutants, 806
ADAMs (a disintegrin and
    metalloproteinases)
  inhibitor, 501
  as markers, 638–639
  TIMPs
    ectodomain shedding, 801–802
    inhibitory activity, 788, 790
ADAMTSs (ADAM with thrombospondin
    motifs)
  as markers, 639
  TIMPs
    ectodomain shedding, 801–802
    inhibitory activity, 788, 790
Adeno-associated virus (AAV) vectors
  advantages, 883
  definition, 882
Adenovirus vectors
  advantages, 881
  innate and adaptive immune responses,
    882
Adipocyte-cancer cell interaction/cross talk,
    366
Adipocyte hypertrophy, 362
Adipocyte protein (aP2), 362
Adipogenesis
  and angiogenesis, 368
  matrix metalloproteinase (MMP)
    adipose homeostasis, 364–365
    proteolytic systems, 362
    in vitro adipocyte differentiation,
      363–364
Adipokines, 367–368
Adiponectin, 367–368
Adipose tissue
  angiogenesis, 368
  homeostasis, 364–365
  organ
    adipocyte-differentiated cells, 362–363
    basement membrane components, 363

formation of adipose tissue, 362
precursor cells, 362
A disintegrin and metalloproteinases
      (ADAMs)
   cellular processes, 378
   Drosophila melanogaster inhibition, 376
   TIMP-1, 378–380
   TIMP-3, 381
Affymetrix dual-species microarray chip
   and Hu/Mu ProtIn chip
      in murine system, 43–44
      probe sets and reference RNAs, 40
      uses of, 42
      xenograft models, 40–41
Agilent, 55–56
Alveolar epithelial cell layer, 352
Amiloride, uPA antagonists, 738
Amine-reactive reagents, 72
   ICPL and dimethyl labeling, 74
   iTRAQ labeling, 73–74
A377MM melanoma cells, 406
Ampli-Chip, 22–23
AnalySIS software, 332
Angiogenesis
   animal model, 315–316
   cathepsin B and S, 293
   cathepsin L, 293–294
   cytokine production, 345
   definition, 305
   differential dynamics of induction, 336
   implication of MMP, 308–309
   malignant tumor formation, 328
   proangiogenic effect, 333–334
   proteolytic systems, 307
   quantification method, 332
   relevance of, 309–310
   role of proteases, in regulation, 329
   stromal MMP contribution, 336–337
   TIMP-1, 685–686
   tumor invasion and, 337
   in vitro assays
      aortic ring assay, 311–313
      cell invasion assay, 310
      tube formation assay, 310–311
   in vivo assays
      CAM Assay, 313–314
      Matrigel plug assay, 314–315
      zebrafish, 315–316

Antagonists
   antibody, 742
   PAI-1
      anti-proteolytic activity, 742–744
      vitronectin binding, 744
   uPA
      enzyme activity, 735–741
      growth factor domain and kringle
         domain, 741
      linker region, 741–742
   uPA–PAI-1 binding
      endocytosis receptor, 744
      inhibitory strategy, 745
Antibody
   antagonists, 742
   inhibitors, 735–736
Anti-cancer strategy, uPAR–integrin
      interaction, 467–468
Anti-Fas antibody treatment. *See* Jurkat T
      lymphocytes
α_2–antiplasmin cleaving enzyme (APCE),
      424
α_2–antiplasmin deficiency, 191–192
Anti-proteolytic activity, PAI-1
   antibody antagonists, 742
   insertion peptides, 743
   small organochemical inhibitors,
      742–743
   in vivo use, 743–744
Antisense oligodeoxynucleotide (asODN)
      therapy, 231–232
Antisense RNA technology, uPAR
      downregulation
   adenovirus constructs, 232–233
   plasmid expression vectors, 232
   shRNA-based RNAi plasmid system, 233
Anti-VEGF agents, 328
Aortic ring assay
   advantages, 312
   ex vivo vascular tissue culture method,
      311
   MMP expression, 312–313
Apoptosis
   induced proteolytic events, 93
   TIMP-1 influence, 685
   uPAR expression, 236
Arg281Gln polymorphism, 672
Array-Ready Oligo Sets^{TM}, 22

Atherosclerotic lesions, 345
Atrial natriuretic peptide (ANP) activation, corin, 270
Autosomal recessive ichthyosis with hypotrichosis (ARIH), 269

Baculovirus vectors
  advantages, 884
  definition, 883
Basement membrane components, 352, 363
Batimastat (BB94)
  clinical trial process, 761, 769
  inhibitor, 506–507
Bidirectional signaling, integrins, 457
Biochemical properties
  with PAI-1 interaction
    endocytosis receptors, 731–732
    target proteases, 730–731
    vitronectin, 730
  uPA
    catalytic properties, 725, 727–728
    connecting peptide, 729–730
    domain structure, 725–726
    GFD and uPAR, 728–729
    kringle domain, 729
Biomarkers
  definition, 587
  measurements of, 587–588
Biotinylated ABP DCG-04, cysteine proteases, 121
Bladder carcinoma
  cathepsins, 608–609
  cystatins, 609
BMS275293 inhibitor, clinical trial process, 772–773
Bone matrix
  cell signal generation, 544–545
  MMPs on remodeling, 542–543
  osteoclast precursor, on degradation, 544
  predilection of metastasis, 553
Bone–tumor microenvironment
  metastatic tumor cells, 541–542
  MMPs
    cell–cell communication, 542–543
    cellular sources, 547–548

TGFβ, 556
  use of MPIs, 557
  osteoclastogenesis and RANKL, 555
  PTHrP protease modification, 553–554
Bortezomib, proteosome inhibitor, 123–124
Bovine aortic endothelial cells (BAEC), 310
Breast cancer
  ADAMs and ADAMTSs as markers, 638–639
  bioinformatic analysis, 628
  degradome genes, 629–630, 632–633
  gene expression profiling, 628–631
  matriptase activity, 269
  MMPs and TIMPs as markers, 632–638
  MMTV-PymT model
    membrane-type MMPs expression, 214
    stromal expression patterns, 213–214
    TIMP-1overexpression in, 215
    tumor growth, 214
  plasminogen activation system
    management of, 572–573
    multivariate analysis, 573–574
    PAI-1 complex and uPA
      progression-free survival (PFS), 574 relapse-free (RFS) and overall survival (OS), 572–574
    uPA receptor (uPAR), 574–575
  TIMP–1 and cell growth, 684–685
Breast carcinoma, cancer biomarkers, 596–597
  cathepsins, 597–598
  cystatins, 598–599
BT20 human breast carcinoma cells, 146

CAM assay
  advantages and limitations, 313
  metastatic process, 314
  tumor-derived MMPs, 313–314
Cancer
  cysteine cathepsins
    angiogenesis, 293–294
    mammary epithelial carcinogenesis, 286

pancreatic islet cell carcinogenesis, 286
skin and cervical cancer, 288
stromal cells and tumor development, 298
tumor cell proliferation and apoptosis, 295–296
tumor invasion, 294–295
uPAR-integrin interactions
antibodies downregulation, 461
anti-cancer strategy, 467–468
α6 integrin cleavage, 459
soluble marker, 454
Cancer-associated dysregulation, 267
Cancer degradome, proteases
activity-based probes (ABP)
covalent modification, 103–104
cysteine proteases, 109
enzyme inhibitor discovery, 122–124
functional elements, 103
linker region and tags, 104
metalloproteases role in, 106–107
profiling, 104–106
protease activity imaging, 125–128
serine proteases, 109–110
tumor biomarkers discovery, 117–121
zinc metalloprotease activity, 110–115
definition, 5
Forster resonance energy transfer (FRET)
fluorescence homotransfer, 108
green fluorescent protein (GFP) role in, 108–109
methodology of, 107–108
Cancer development, proteolytic pathways
advantages of, 173–174
cysteine cathepsin proteases
activation cascades of, 170–171
cathepsin C role, 170
classess of, 169
de novo carcinogenesis models, 171
profiling of, 172
ECM-remodeling proteases, 158–160
matrix metalloproteinase (MMP) activation
anticancer therapeutic targets, 164
cell-mediated mechanism, 161

cofactors, 163
functional roles for, 161–162
metastatis formation, 162–163
MMP-9 proangiogenic roles, 162
protease amplification, 172–173
proteolytic activities of, 157–158
serine proteases regulation
mast cell serine proteases, 165–167
neutrophil elastase (NE), 167–169
plasminogen activators, 164–165
Cancer invasion
Plg activation
mammary gland involution, 211
squamous cell skin cancer, 208
uPARAP/Endo182
invasive tumor growth, 254
tumor-stroma interface, 254–255
Cancer-preventive effect, 365
Cancer prognostic markers. See
Plasminogen activation system, prognostic marker
Cancer therapy, uPAR
antisense RNA technology
adenovirus constructs, 232–233
plasmid expression vectors, 232
asODN therapy, 231–232
RNA interference (RNAi), 233
(CA)n polymorphism
MMP-9 gene, 672
Carcino embryonic antigen (CEA), 699
Catalytic properties, uPA
pro-uPA activators, 728
serine protease domain, 725
substrate specificity, 727–728
Cathepsin B
as biomarker, 589–590
breast carcinoma, 597–599
colon carcinoma, 604–605
deficiency compensation, 121–122
glioma, 613–614
hepatocellular carcinoma, 609–610
increased expression effect, 615–617
melanoma, 610–611
meningioma, 612–613
non-small-cell lung cancer, 602–604
oesophageal carcinoma, 607
oral carcinoma, 608
ovarian cancer, 600–601

prostate carcinoma, 599–600
SCCHN, 605–606
TCC, 609
Cathepsin C, 170–171
Cathepsin F, ovarian carcinoma, 601, 615
Cathepsin H, 616–617
  colon carcinoma, 604–605
  lung carcinoma, 601–602
  melanoma, 610
  SCCHN, 606
Cathepsin K
  bladder carcinoma, 608, 616
  non-small-cell lung cancer, 602
Cathepsin L, 615–617
  bladder carcinoma, 609
  breast carcinoma, 597–598
  colon carcinoma, 604
  meningioma, 613
  oral carcinoma, 608
  ovarian and cervical carcinoma, 601
Cathepsin S, 616–617
  glioma, 614
  prostate carcinoma, 599
Cathepsin V protein, 43–44
Cathepsin X, prostate carcinoma, 599, 615
CC chemokines, 529
CEA. See Carcino embryonic antigen
Cell adhesion, 732–733
Cell growth, TIMP-1, 684–685
Cell invasion, 310, 734
Cell migration, 733–734
  leukocyte
    2D and 3D matrix, 348–352
    endothelial cell interactions, 344–345
    roles of proteinases, 347–348
    transendothelial cell migration,
      integrins, 346
    vs. amoeba, 344
  matricryptic sites
    laminin 5 and proteoglycans, 355
    type-I collagen, 356
    type-IV collagen, 354–355
  metalloproteinase, 343
  in vivo studies
    intravital miscroscopy, 352–353
    roles of proteinases and MMPs, 352
    transendothelial migration of
      lymphocytes, 353

xenopus laevis, 354
zebrafish, 353–354
Cell-permeable filter, 310
Cell proliferation, 734–735
  TIMP, 382–383
Cellular properties, malignant phenotype
  cell adhesion, 732–733
  cell invasion, 734
  cell migration, 733–734
  cell proliferation, survival, and apoptosis,
    734–735
Chemokines, 345
  characterization of, 529
  MMPs regulation
    CC chemokine, 529
    CXC chemokine, 529–531
    CX3CL1, 531
    proteoglycan-mediated process,
      531–532
Chemotactic assay, 347
Chick chorioallantoic membrane (CAM),
    313–314, 349
Chimpanzee degradome, 11
Chondroitinase ABC (ChABC), 355
Choroidal neoangiogenesis, 337
Chronic inflammation, 354
Clinical trials, MMPIs
  batimastat, 761, 769
  BMS275293, 772–773
  development of inhibitors, 778–779
  dose-limiting toxicity, 774–777
  efficacy of drug, 777–778
  marimastat, 769–770
  metastat, 773–774
  neovastat, 774
  prinomastat, 770–771
  tanomastat, 771–772
CLIPCHIP™
  applications and directions of, 32–33
  data analysis
    bioinformatics analysis of, 31, 33
    systematic variation in, 31
  human vs. mouse degradome
    classes of, 20–21
    pseudogenes and nonproteolytic
      homologues, 22
  image analysis
    gridding process, 29

segmentation and data extraction, 30
oligonucleotide array, 632
scanning of, 28–29
Clustered heatmap
  degradome genes, 648–649
  matrix metalloproteinases, 651–652
  MMP-12-associated genes, 657
COFRADIC. *See* Combined fractional
  diagonal chromatography
Collagen gel invasion, 349–350
Collagen internalization
  uPARAP/Endo182
    gelatin, 250
    interdomain organization analysis,
      249
    lysosomal degradation, 250–251
    macrophage mannose receptor, 252
Collagenolysis and MMPs, 544
Colon carcinoma
  cathepsins, 604–605
  cystatins, 605
Colon carcinoma cell, 355
Colorectal cancer (CRC)
  adjuvant systemic chemotherapy, 680
  plasminogen activation system,
    575–576
  TIMP-1, CRC prognosis
    case-control study, 687
    plasma TIMP-1 and cut-off values,
      687–688
    prognostic value analysis, 690
    serum and plasma TIMP-1 levels in,
      686–687
    standard operating procedures (SOPs),
      691–692
    statistical analysis procedure (SAP),
      692
    TIMP-1 mRNA level in, 686
  TNM staging, 679–680
  tumour markers
    carcino embryonic antigen (CEA),
      681
    tissue inhibitor of metalloproteinases 1
      (TIMP–1), 681–683
Combined fractional diagonal
  chromatography (COFRADIC),
    75–76
  apoptosis-induced proteolytic events, 93

methodology, 89, 92
NHS-biotin labelled method, 95
posttranslational proteolytic
  modifications, 92
protease substrate discovery, 90–91
sample simplification, 93–94
strategies of, 95–96
terminal amine isotope labeling, 97–98
time-efficient protocol, 94–95
trypsin digestion method, 94
Connective tissue growth factor
  (CTGF), 18
Corin TTSP family, 270
C-type carbohydrate recognition domains
  (CRDs)
  uPARAP/Endo182
    carbohydrate binding, 248
    cell migration, 252–253
    collagen internalization, 250–251
    interdomain organization analysis,
      249
CXC chemokine, 529–531
CX$_3$CL1 chemokine, 531
Cyclic adenosine monophosphate (cAMP),
  344
Cyclic hexapeptide, uPA antagonists, 740
Cystatin C, 595
  breast cancer, 598
  prostate carcinoma, 600
Cystatins, cancer biomarkers
  bladder carcinoma, 609
  breast carcinoma, 598–599
  colon carcinoma, 605
  glioma, 614
  hepatocellular carcinoma, 610
  lung carcinoma, 603–604
  melanoma, 611
  meningioma, 612–613
  oesophageal carcinoma, 607
  oral and laryngeal carcinoma, 607–608
  ovarian and uterine cervix carcinoma,
    601
  prostate carcinoma, 600
  squamous cell carcinoma of the head
    and neck (SCCHN), 606–607
  type I, II and III, 592–595
Cysteine cathepsins
  angiogenesis, 293–294

cathepsin B and S, 293
cathepsin L, 293–294
cancer biomarkers
    ABP use, 117–120
    bladder carcinoma, 608–609
    breast carcinoma, 597–598
    colon carcinoma, 604–605
    glioma, 613–614
    hepatocellular carcinoma, 609–610
    lung carcinoma, 601–603
    melanoma, 610–611
    meningioma, 612–613
    oesophageal carcinoma, 607
    oral and laryngeal carcinoma,
        607–608
    ovarian and uterine cervix carcinoma,
        600–601
    prostate carcinoma, 599–600
    squamous cell carcinoma of the head
        and neck (SCCHN), 605–606
cancer development
    activation cascades of, 170–171
    cathepsin C role, 170
    classes of, 169
    de novo carcinogenesis models, 171
    profiling of, 172
in cancer, mouse models
    K14-HPV16 model, 288
    MMTV-PyMT model, 286
    RIP1-tag2 model, 286
cathepsin null cancer-prone mice model,
    297–298
characteristics, 282
glycosaminoglycans (GAGs), 282–283
phenotypes, 284–285
stromal cells and tumor development,
    298
subcellular localization, 283–284
synthesis and activation, 282–283
tissue distribution, 284
tumor cell proliferation and apoptosis,
    295–296
tumor invasion
    E-cadherin, 294–295
    ECM degradation and MMP activation,
    295
Cysteine proteases. See also Activity-based
    probes (ABP)

activated anticancer prodrugs
    extracellular cathepsin-activated
        anticancer therapeutics, 866–867
    legumain-activated anticancer
        therapeutics, 865
activity-based probes (ABP) for,
    105–106
breast cancer, 598, 615
cathepsin activity evaluation, 121
DCG-04 ABP in, 103
inhibitors, 592–595 (see also Cystatins,
    cancer biomarkers)
intracranial tumours, 617
lung carcinoma, 603
structure and functions, 588, 590
Cytokines
    MMPs regulation and inflammation
        interleukin-1β, 522–523
        TGF-β1, 526–528
        TNF-α, 524–526
    production impact on angiogenesis, 345
Cytoplasmic domain, 346
    uPARAP/Endo182, 248–249

3D collagen gel invasion, 388
2D-DIGE (Two-dimensional
        differential gel electrophoresis),
        70–71
Decidualization, 210
Degradome genes
    expression profiling, 629–630,
        632–633
    IST analysis and mRNA expression
        heatmap values, 648–649
        kallikrein-1 and renin, 648, 650
Degradome-targeted cancer therapy
    degradome gene silencing, 887–888
    inhibition of proteinase activity, 888–890
    modulating gene regulators, 886–887
DESC1 (Differentially expressed in
        squamous cell carcinoma 1) family,
        266
Detergent-resistant membrane fractions
        (DRM), 477
Diagnostic biomarker, 587
Dictyostelium discoideum, 344
Dipeptidyl peptidase I (DPPI). See
        Cathepsin C

906   Index

Dipeptidyl peptidase IV (DPPIV),
    422–424
Donor fluorescence lifetime imaging
    (donor-FLIM), 484–485
Dose-limiting toxicity (DLT)
  MMPI and MSS effect
    collagenase activity and protease
      inhibition, 775
    genetically deficient model, 775–776
    zinc-binding groups, 776
Doxorubicin (Dox), prodrug, 174
3D reconstituted basement membrane
    (rBM), 141–142
Drosophila melanogaster, 376, 382
DualChip®, 22
Ductal carcinoma in situ (DCIS),
    286–287
Dukes' stage tumors, CRC, 575–576
Dunn chemotaxis chamber, 348
Dye-quenched (DQ) substrates, 141–142.
    See also Tumor proteases, imaging
    activity

ECM-remodeling proteases, cancer
    development, 158–160
Embryogenesis, 316
Encyclopedia of DNA Elements
    (ENCODE) analysis, 45
Endocytic collagen receptor
  biological role of, 255–256
  in cancer
    invasive tumor growth, 254
    tumor–stroma interface, 254–255
  cell migration
    directional and indirectional effect,
      252–253
    orientation effect, 253
  collagen internalization
    gelatin, 250
    lysosomal degradation, 250–251
    macrophage mannose receptor, 252
  C-type carbohydrate recognition
    domains, 248
  Cys-rich domain, 247
  cytoplasmic domain, 248–249
  FN-II domain structure, 247–248
  in healthy tissue, 253

identification, 246
  interdomain organization, 249
  protien structure, 246
  uPA-system interaction, 252
Endocytosis receptor, 731–732, 744–745
Endothelial cell (EC)
  barriers, 347
  cell invasion assay, 310
  lymphatic, 317
  tube formation assay, 310–311
  tube morphogenesis, 387
Endothelial (E-) selectin, 344
Endothelial sprouts, 307
Enteropeptidase
  complex domain structure, 260
  deficiency, 260–262
Enzyme activity, uPA
  antibody inhibitors, 735–736
  peptides inhibitors
    cyclic hexapeptide, 740
    upain-1, 740–741
  protein inhibitors, 736
  small organochemical inhibitors
    amiloride, 738
    serine protease inhibitors, 736
    WX-UK1, 739
Enzyme inhibitor discovery
  cathepsin activity evaluation, 124
  cathepsin B detection, 122
  proteasome-directed ABPs in, 122–123
Enzyme-linked immunosorbent assays
    (ELISAs), 570, 723
  breast cancer and healthy volunteers, 572
  squamous cell lung cancer, 577
  uPA and uPA-PAI-1 complex detection,
    571
Epidemiological data, adiposis and cancer,
    368
Epidermal growth factor receptor (EGFR),
    383
Erythroid potentiating activity (EPA), 378
E-selectin molecule, 353
European Medicines Evaluation Agency
    (EMEA), 760
Even-Ram, S., 348
Experimental metastasis model
  anti-invasive activity of TIMPs, 502
  metastatic cascade mimic, 498

Extracellular matrix (ECM)
    binding, TIMP3, 376
    bound ligands, 382
    cardiac ECM breakdown, 378
    composition, 328
    degradation by MMPs, 545, 547
    driven migration, 343
    homeostasis, 381
    protein vitronectin, 722
    proteolytic degradation, 495
    remodeling, 362, 373
Extracellular matrix-degrading structure
        (EDS), 406, 408
Extracellular-signal-regulated kinase
        (ERK), 383
    uPAR signalling, 228

Fibroblast activation protein-α (FAP),
        420–421, 424
Fibroblast growth factors (FGF), 383
Fibroblastic cells, 328
Fibronectin type-II (FN-II) domain
    in uPARAP/Endo182
        deficiency, 249–250
        structure, 247–248
FIGO classification, ovarian cancer,
        578–579
Fluorescein isothiocyanate (FITC), protein
        substrates, 139–141
Fluorescence correlation spectroscopy
        (FCS), uPAR dynamics
    fluorescence intensity traces, 482–483
    time structure, 481
    translational mobility of cell, 480
Fluorescence fluctuation spectroscopy
        (FFS), 480
Fluorescence recovery after photobleaching
        (FRAP)
    kinetic parameters determination, 480
    single protein label, 479
Fluorescent proteins (FPs) chimers
    HEK295 cell line, 478–479
    wt-uPAR functions, 478
Fluorogenic proteolytic beacon, 838
Fluorophores, 107–108
FMLP receptor-like protein 1 receptor
        (FPRL1), 436
Focal adhesion kinase (FAK), 408, 733

Food and Drug Administration (FDA), 760
Formyl-Met-Leu-Phe (fMLP), 345
Formyl peptide receptor (FPR), 733
Forster resonance energy transfer (FRET)
    fluorescence homotransfer, 108
    green fluorescent protein (GFP) role in,
        108–109
    methodology of, 107–108
    proteases, 484
Fractalkine. *See* CX$_3$CL1 chemokine

Gel-based proteomic methods, 70–71
Gene delivery systems and proteolysis
    chimeric envelope, 835
    GALV envelope, 835–836
Gene expression profiling, 628–631
    collagen XVIII, 655–656
    kallikrein-1 biomarker, 646–647
    matrix metalloproteinases
        clustering and expressions,
            651, 653
        expression values, 651–652
    microarray database, 646
    MMP-12
        coexpression analysis, 655–658
        normalized expression, 653–655
    plasminogen, 655–656
    renin precursor, 648, 650
    TPX2 gene, 656
Gene ontology (GO) analysis, MMP12
        gene, 658
Gene-specific priming (GSP), 58
GeNorm^{TM}, 60
1G/2G polymorphism and MMP1 gene
    allele-specific effect, 666–667
    cancer effects, 667–668
    guanosine change, 664
    transcriptional effect, 666
Glioma
    cathepsins, 613–614
    cystatins, 614
Glycosyl-phosphatydil-inositol (GPI)
        anchor
    uPAR, 476
    wt-uPAR, 478
GPI-anchored membrane attachment
    paroxysmal nocturnal hemoglobinuria
        (PNH), 702

5-subunit transamidase complex (GPI8)
    role, 701
G protein-coupled receptor
    (GPCR), 345
Growth factor domain (GFD)
    antagonists of uPA, 741
    kringle domain and, 729
    and uPAR, 728–729

Hematopoietic stem cell (HSC)
        mobilization
    bone marrow repopulation, 439
    CD34$^+$ adhesion molecules and c-Kit
        expression, 440
    hematopoietic stem cells in,
        438–439
    HSCs differentiation, 439–440
    proteases
        CD26/dipeptidylpeptidase IV (DPPIV),
            441–442
        MMP-9 and serine proteases, 441
        serpin-1and, 442
    uPAR
        5-fluorouracil myeloablation, 444
        plasmin and, 444–445
        protease cleavage and desensitization,
            443
        upregulation, 442
Hepatocellular carcinoma
    cathepsins, 609–610
    cystatins, 610
Hepatocyte growth factor (HGF), 380, 383
Hep3 epidermoid carcinoma cells, uPAR
        expression, 437–438
Hepsin/TMPRSS family
    enteropeptidase
        complex domain structure, 260
        deficiency, 260–262
    hepsin/TMPRSS1
        domain structure and functions, 262
        tumorigenesis and SNPs, 264
    MSP and epitheliasin/TMPRSS2, 264–265
    spinesin/TMPRSS5, 265–266
    TMPRSS3 and TMPRSS4, 265
Herpes simplex virus vectors, 884–885
    amplicon vectors, 884
    cytotoxic properties of, 885
High-fat diet (HFD) treatment, 364

Histone deacetylases (HDACs), 120–121
Human airway tryptase (HAT), 266
Human brain microvascular endothelial
        cells (HBMEC), 310
Human degradome, 20–21
    classification, 5
    features, 4–5
    genomic view of, 6
    protease inhibitor genes in, 9–10
    vs. rodent degradomes complexity
        applications of, 10–11
        protease subfamilies, 9
        reproduction and host
            defense genes, 8
Human invasive ductal carcinomas
        (IDC), 43
Human leukocyte elastase treatment, 831
Human microvascular endothelial cells,
        368
Human papillomaviruses (HPV), 288
Human renal cell carcinomas, 387
Human umbilical vascular endothelial cells.
        See Macrovascular endothelial cells
Hu/Mu ProtIn Chip
    and Affymetrix platform
        in murine system, 43–44
        probe sets and reference RNAs, 40
        uses of, 42
        xenograft models in, 40–41
    CLIP-CHIP array and, 44–45
    cross-hybridization and, 44
    expression profiling, 42–43
    microarray chip, 26–27
HUVEC cells, 368
Hydrogel drug delivery and proteolysis
    PEG-based hydrogels, 835
    photopolymerization, 834
    sequences used, 841–842
Hydroxamate inhibitors, MMPI, 814–815
Hydroxamate mp inhibitor, 353

ICTP collagen, 544–545
Ilomastat, MMP inhibitor, 113
Image analysis algorithm, 332
Imaging agents and proteolysis
    fluorogenic proteolytic beacon, 838
    MMP-7-activated beacon, 839
    NIRF optical imaging, 837

QD imaging, 838
Immunolocalization, 353
Inflammation process, MMP
  CC chemokine, 529
  CXC chemokine, 529–531
  CX_3CL1 chemokine, 531
  genetic models, 522–523
  immune phenotypes, 520–522
  interleukin–1β, 522–523
  proteoglycan-mediated chemokine,
    531–532
  TGF-β1, 526–528
  TNF-α, 524–526
Inflammatory cells, 328
Inhibitors of multiple proteases and
    cytokines (IMPACs), 773
Inhibitory proteoglycan NG2, 355
Insertion peptides, 743
In Silico Transcriptomics (IST) database
  degradome genes, mRNA expression
    heatmap values, 648–649
    renin and kallikrein-1, 648, 650
  kallikrein 3 biomarker, 646–647
  MMP-12 expression, 657–658
Insulin growth factor-binding proteins
    (IGFBPs), 389–390
Intact peritoneal basement membrane
    barriers, 349, 355
Integrins
  $β_1$ integrin, 380
  $α_1$ integrin, 730, 733
  Timp mutant phenocopies, 382
  uPAR, 435–436
    anti-cancer strategy, 467–468
    bidirectional signaling, 457
    intracellular domain, 456
    kringle domain and integrin cleavage,
      459
    spatial distribution regulation
      of, 458
  $α_vβ_3$ integrin, 729
  $α_vβ_5$ integrin, 730, 733
  vitronectin binding, 734
Interactome, 226
Interleukin-1β cytokine
  MMPs and inflammation, 522–523
International Agency for Research on
    Cancer (IARC), 365

Intracellular focal adhesion kinase (FAK),
    386
Intracranial tumours, cathepsins and
    cystatins
  glioma, 613–614
  meningioma, 612–613
Intravital microscopy, 352–354
Intrinsic protease specificity
  active-site cleft and motifs, 68–69
  methods for, 69
Introduction of isotope-coded affinity tags
    (ICATs), 72
Invadopodia
  actin/cortactin cores, 406, 417–418
  characteristics of, 403–404
  3D cultures, 421–422
  2D substrates studies, 408
  EDS in, 406
  filopodia formation, 406–408
  formation process, 409–416
    dynamics, 419–420
    searching and maturation phase, 421
    time lapse imaging approach, 419
  invadopodial complex, 417
  molecular components and functions of,
    405, 409–416
  MT1-MMP in, 417, 420–421, 423
  proteases and tumor microenvironment
    FAP–DPPIV complexes, 423–424
    proangiogenic functions and MMP-9,
      424
  proteolytic activity of
    adhesive and lytic property, 424–425
    ECM degradation and integrins,
      422–423
    microscopic imaging analysis,
      416–417
    MMP-mediated proteolysis, 423
  vinculin rings, 418
  vs. podosomes
    actin core assembly and ECM,
      404–405
    role of, 405
Invasive cancer cell, 366–367
Invasive ductal carcinomas (IDC), 286–287
In vitro adipocyte differentiation
  Mouse embryonic fibroblasts (MEFs),
    363–364

3T3L1 adipocyte precursors, 363
In vitro assays
  aortic ring assay, 311–313
  cell invasion assay, 310
  overexpression model, 386
  tube formation assay, 310–311
  vs. in vivo assays, 309–310
In vivo assays
  CAM Assay, 313–314
  lymphangiogenesis models and
    limitations, 317
  Matrigel plug assay, 314–315
  vs. in vitro assays, 309–310
  zebrafish, 315–316
In vivo models
  anti-proteolytic effect, 743–744
  TIMP function analysis
    cancer development, 382
    cancer phenotypes of, 389
    hepatocytes, 388–389
    IGFBPs, 389–390
    TIMP3 deficiency, 390
    transgenic Timp expression, 388
  tumor–stroma interaction analysis, 329
Isobaric tagging for relative and absolute
    quantitation (iTRAQ)
  advantages and disadvantages, 74
  limitations of, 97
  methodology, 87
  protein guanidination and mas tags, 73
  ratio in, 88
Isotope-coded affinity tags (ICATs)
    proteomics
  advantages, 89
  functional structure, 86
  methodology, 86–87
  vs. iTRAQ, 88
Isotope-coded protein label (ICPL), 74

Jurkat T lymphocytes, 93

Kallikrein–1 (KLK1) biomarker
  IST analysis and gene expression
    mRNA expression of tissues,
      646–647
    vs. renin precursor, 648, 650
Kaplan-Meier survival analysis, 575–576,
    636

Key molecular determinants
  host compartment
    classes of proteases, 332–333
    MMP inhibitors and proteases,
      334–335
    PAI-1 and PAI-2, 333–334
  tumor compartment
    HaCaT cells, 335
    MMP proteases, 335–336
    VEGFR downregulation, 336
Kringle and proteolytic domain, 459

LacZ gene tag, 501
Laminin-5, 355
Legumain, 592
Lentivirus vectors, 880. *See also*
    Retrovirus vectors
Leptin, 367–368
Leukocyte adhesion molecules, 345
Leukocyte cell migration
  2D and 3D studies
    collagen cell invasion, 349–350
    collective-cell migration, 349
    quenched-fluorescent collagen,
      348–349
    short-term and long-term invasion,
      352
    visualisation of cell migration, 348
  endothelial cell interactions
    inflammation, 345
    recruitment of monocytes, 344
    selectin, 344–345
  roles of proteinases, 347–348
  transendothelial cell migration,
    integrins, 346
  vs. amoeba, 344
Leukocyte extravasation, 347
Leukocyte (L-) selectin, 344
Ligneous lesions, plasminogen deficiency,
    187–188
Linear peptide antagonists, uPA-uPAR
    interactions, 713–714
Liposomal drug delivery
  extracellular membrane, 833
  intracellular membrane, 832
  surface modification, 831
Liquid chromatography (LC)
  and ICAT labeling, 72

iTRAQ labeling, 74
  tryptic peptides separation, 75
Low-density lipoprotein receptor (LDLR)
  complement-type repeats (CTRs),
    731–732
  endocytosis receptors, 722, 744
L-selectin shedding, 353
Lung cancer
  corin, 270
  plasminogen activation system, 577–578
Lung carcinoma, cancer biomarkers
  cathepsins, 601–603
  cystatins, 603–604
Lung metastasis signature (LMS), 628
Lymphangiogenesis models
  limitations of, 316
  in vivo models, 317
Lymphatic endothelial cells, 317
Lymphatic ring assay, 316
Lymphoid cell migration, 344
Ly-6/uPAR/a-neurotoxin (LU) protein
  domains, 703–704

Macrophage mannose receptor (MMR)
  collagen clearance, 252
  C-type carbohydrate recognition
    domains, 248
  Cys-rich domain, 247
  recycling of, 248–249
Macrophage migration, 353
Macrovascular endothelial cells, 310
Major excreted protein (MEP), 283
MammaPrint®, 23
Mammary epithelial carcinogenesis
  cathepsin B deficiency, 291
  MMTV-PyMT model of, 286
Mammary gland morphogenesis, 387
Mammary polyoma middle T-induced
  tumorigenesis, 390
MAPKs. See Mitogen-activated protein
  kinases
Marimastat (BB-2518)
  clinical trial process
    phase II and III tests, 770
    phase I test, 769
  inhibitor, 506
Mass spectrometry (MS), protease
  substrates
  characteristics of, 79
  gel-based proteomic methods, 70–71
  solution-based proteomic methods
    amine-reactive reagents, 72–74
    ICAT labeling, 72
    labeled peptides, positive selection of,
      77
    label-free analysis, 77–78
    tryptic peptides, negative selection of,
      74–76
Mast cell serine proteases, cancer
    development
  cell-derived chymases role, 166
  human tryptases role in, 165–166
  mast cell tryptases role in, 166–167
Matrigel implantation angiogenesis assay,
  744
Matrigel plug assay
  drawbacks and solutions, 314–315
  MMP-9, 315
  neovascularization zone delineation, 314
Matriptase
  in cancer, 267
  epidermal differentiation and point
    mutation, 268–269
  epithelial morphogenesis, 268
  matriptase-2 and matriptase-3, 269
  polyserase-1, 269–270
Matrixins. See Matrix metalloproteinases
  (MMPs)
Matrix-inserted surface transplantation
  assay, 331
Matrix metalloprotease-9, HSC
  mobilization, 441
Matrix metalloproteinase inhibitors
  (MMPIs), 496
  active-site topology, 812–813
  chemical structures
    without zinc-binding group, 819–821
    with zinc-binding group, 815–816
  clinical trials on
    batimastat, 761, 769
    BMS275293, 772–773
    development of inhibitors, 778–779
    dose-limiting toxicities, 774–777
    efficacy of drug, 777–778
    marimastat, 769–770
    metastat, 773–774

neovastat, 774
prinomastat, 770–771
tanomastat, 771–772
first-generation inhibitors
hydroxamate compound, 814–815
vs. TACE, 815–816
phosphinic peptides and transition-state
compound 5 and MMP-9, 817
compound 6 and MMP-12, 818
research development
structure-based design approach, 823
thermodynamics and enthalpic
contribution, 821–822
roles of, 37–38
unusual MMP inhibitors
compound 8 and MMP-13, 819, 821
compound 7 structure, 819
Matrix metalloproteinases (MMPs), 61
activated anticancer therapeutics
MMP-activated chemotherapeutics,
867–869
tumour-selective MMP-activated
biotoxins, 869–871
activation and TIMP-2, 797
and ADAMs, 353
adipocytes, in carcinomas
adipocyte–cancer cell interaction/cross
talk, 366
adiposis and cancer, epidemiological
data, 365
invasive cancer cells, 366–367
adipogenesis
adipose homeostasis, 364–365
proteolytic systems, 362
in vitro adipocyte differentiation,
363–364
anticancer therapeutic targets, 164
aortic ring assay, 312–313
bone matrix remodeling, 542–543
CAM assay, 313–314
cell invasion assay, 310
cell-mediated activation mechanism, 161
chemokine regulation and inflammation
CC chemokine, 529
CXC chemokine, 529–531
CX$_3$CL1 chemokine, 531
proteoglycan-mediated process,
531–532
cofactors in, 163
CXCL12 receptor modulation, 548, 553
cytokines and inflammation
interleukin-1β, 522–523
TGF-β1, 526–528
TNF-α, 524–526
2D and 3D studies
collective cell migration, 349
localisation of, 352
vascular remodeling, 348
extracellular matrix degradation, 496
fibrillar type-I collagen, 544–545
functional roles for, 161–162
gene suppression of, 508–509
genetic tags, 500–501
head and neck squamous cell carcinoma
(HNSCC), 886
host compartment, 334–335
and inflammation
genetic models, 522–523
immune phenotypes, 520–522
inhibition mechanism and TIMPs
interaction of MMPs-TIMPs,
793–794
kinetic values of TIMPs, 795–796
inhibitory function
apoptosis, 383–386
TIMP1, 380
IST analysis, 658
laminin 5 and proteoglycans, 355
leukocytes and tumour cell migration,
347
mammary gland involution, 212
as markers, 632, 634–635
Matrigel plug assay, 315
mechanisms, 307–309
membrane-bound enzymes, 206
metastasis bone environment, 547
metastatis formation, 162–163
MMP-2, 18–19
MMP-3
unusual MMP inhibitor, 819
zinc-binding groups and inhibitors, 816
MMP-9, 813, 821
MMP-11, 43
MMP-7 degradation and cleavage of
syndecan-1, 352
MMP-9 proangiogenic roles, 162

Index 913

MMP-12 protein, 42
MMTV-PymT tumors, 214–215
murine metastasis models
  experimental form, 498
  generation, 497
  spontaneous form, 498–500
nonmatrix factors solubilization, 548
nuclear factor-B inhibitor, 886–887
osteoclast-derivative
  bone matrix degradation, 544
  cell function control, 546
  cell signal generation, 544–545
  ECM degradation, 545
  Wnt signaling, 546–547
potential mechanism
  adipose tissue angiogenesis, 368
  regulation by adipokines, 367–368
Pro-MMP-2, 797
proteolytic systems, 362
RANKL solubilization, 554
role of proteases, 329
roles of, 37–38
in silico gene expression profiling
  clustering and expressions, 651, 653
  expression values, 651–652
skin wound healing, 209–210
soluble enzymes, 205
sRANKL protein, 555
stromal contribution, in tumor
  progression, 336–338
synthetic broad-spectrum inhibitors,
  506–508
synthetic MMPI inhibitors
  prinomastat and pyrimidine-2,4,6-
    trione-type, 509
  SB-3CT and MMI-168, 510
TIMPs, natural broad-spectrum inhibitor
  anti-invasive activity, 502
  antimetastatic activity, 502–503
  inhibitory activity, 788–789
  kinetic properties, 792
  liver metastasis, 504–505
  N-TIMP-1 and prometastatic effect,
    505
  proMMP-9 binding, 501
  proproliferative activity, 510–511
  protumorigenic feature, 504
trimeric complex formation, 800–801

tube formation assay, 311
tumor–bone microenvironment
  cell-cell communication, 542–543
  cellular sources, 547–548
  TGFβ, 556
tumor compartment, 335–336
type-I collagen, 356
type-IV collagen, 354–355
uPAR cleavage, 226
zebrafish, 316
Mature lipid-filled adipocytes, 362–363
Maximum tolerated dose (MTD), 761
MCPs. See Monocyte chemoattractant
  proteins
MDA-437 carcinoma cells, 353–354
Melanoma
  cathepsins, 610–611
  cystatins, 611
Membrane-type matrix metalloproteinases,
  378–380, 388
Meningioma, cathepsins and cystatins,
  612–613
Mesodermal stem cells, 362
Metalloproteases activity. See Zinc
  metalloprotease activity
Metastastic colonization, 390
Metastatic tumor cells, 541–542
Metastat inhibitor, 773–774
Microarrays and transcriptomes
  development of, 23
  limitations of, 22
  probe fixation, 23–24
  techniques used, 24
  two-color CLIP-CHIP, 25–26
Microsatellite polymorphism, 664
Microvascular endothelial cells, 310
Minor groove binding (MGB) Taqman
  probes, 52
mirVana™, 54
Mitochondrial serine protease activity.
  See Jurkat T lymphocytes
Mitogen-activated protein kinases
  (MAPKs), 886
MMP-1
  allele-specific effect, 666–667
  cancer effects, 667–668
  guanosine change, 664
  transcriptional effect, 666

MMP-2
  737C/T polymorphism
    lung cancer susceptibility, 669
    squamous cell carcinoma
      susceptibility, 670
  -1308C/T polymorphism
    allele-specific transcriptional effect,
      668
    susceptibility, 669–670
  -1577G/A polymorphism, 669
MMP-3
  5A/6A polymorphism
    cancer effects, 671–672
    interleukin-1, 671
    nucleotide position, 670
  unusual MMP inhibitor, 819
  zinc-binding groups and inhibitors, 816
MMP-9
  polymorphisms, 672–673
  proangiogenic roles, 162
  secondary metastasis formation,
    162–163
MMP-12
  coexpression analysis
    clustered heatmap, 657
    collagen XVIII and plasminogen,
      655–656
    gene ontology analysis, 658
    MMP-1, 656–657
    TPX2 gene, 656
    normalized gene expression, 653–655
    polymorphisms, 673
MMPIs. *See* Matrix metalloproteinases
    inhibitors
Molecular Beacons and Scorpions, 50
Molecularly-targeted agents, 761
Molecular targeted anticancer therapy,
    854–855
Monocyte chemoattractant proteins (MCPs),
    345. *See also* CC chemokines
Monocyte transendothelial migration, 346
Morphometric analysis, 332
Mouse degradome, 20–21
Mouse embryonic fibroblasts (MEFs),
    363–364
Mouse mammary tumor virus (MMTV),
    286
MT loop, 798

Multidomain, uPAR
  crystal structures of, 704–705
  hydrophobic ligand-binding cavity in,
    706
  linker peptide in, 706–707
Multivariate analysis
  breast cancer, 573–574
  CRC, 575
  lung cancer, 577
Murine metastasis models
  experimental form, 498
  generation, 497
  spontaneous form, 498–500
Musculoskeletal syndrome (MSS)
  collagenase activity and protease
    inhibition, 775
  genetically deficient model, 775–776
  zinc-binding groups, 776

NanoDrop®, 54
Nanotechnology-based DDSs (nano-DDSs)
    and proteolysis
  liposomes, drug delivery vehicle
    extracellular delivery, 833
    intracellular delivery, 832
    surface modification, 831
  MMP-activated drug, 834
Naphtamidine-based inhibitors, 739
Near-infrared fluorescence (NIRF) optical
    imaging
  cathepsin D, 837
  sequence used, 841
Neck cancer, DESC1, 266
Neoplastic cells
  production of proteases, 333
  tumor, 327
Neovascularization, 294, 328, 368
Neprilysin (NEP), 553
Neutrophil cell migration, 352
Neutrophil elastase (NE), cancer
    development
  inflammatory responses in, 169
  neutrophil transmigration in, 167–168
  regulation and expression of, 167
  tumorigenesis in, 168
Neutrophil extracellular traps (NET), 167
Neutrophil migration, 353
Neovastat inhibitor

clinical trial process, 774
NHS-biotin labeled method (COFRADIC)
    approach, 95
N-linked glycosylation, uPAR, 702
Non-small cell lung carcinoma (NSCLC)
    cathepsin B activity, 602–603, 616
    plasminogen activation system,
        577–578
Normal-fat diet (NFD), 364
N-terminal domain of TIMP-1 (N-TIMP-1),
    505
N-terminal enrichment methods,
        COFRADIC approach
    apoptosis-induced proteolytic events, 93
    methodology, 89, 92
    NHS-biotin labeled method, 95
    posttranslational proteolytic
        modifications in, 92
    protease substrate discovery, 90–91
    sample simplification, 93–94
    strategies of, 95–96
    terminal amine isotope labeling,
        97–98
    time-efficient protocol, 94–95
    trypsin digestion method, 94
N-terminal fragment (ATF), 733
Nuclear factor-kappa B (NFκB), 383

Oesophageal carcinoma, 607
^{18}O-labeling, COFRADIC method, 76
Oligo GEArray®, 22
Oligo(dT) primers, 58
Oncogenic cell transformation, 236
Oncolytic virotherapy approach, 830
Oncomine analysis, 635, 637
Operon®, 22
Optimal biological dose, 761
Oral and laryngeal carcinoma, 607–608
Organogenesis, 315
Osteoclastogenesis, RANKL
    cleavage of protein, 554
    expression of protein, 555
Osteoclast precursor cells
    bone matrix
        arrival on, 542
        degradation, 544
    cell signals generation, 544–545

MMPs
    cell function control, 546
    ECM degradation, 545
    Wnt signaling, 546–547
Osteonectin (ON), 545
Osteopontin (OPN) processing, 545
Ovarian and uterine cervix carcinoma,
        cancer biomarkers
    cathepsins, 600–601
    cystatins, 601
Ovarian cancer, plasminogen activation
        system, 578–579
P21-activated kinase 1 (PAK1), 149
Paget, S., 328
PAI-1. See Plasminogen activator
        inhibitor-1
Pancreatic cancer
    TMPRSS3 and TMPRSS4 in, 265
    uPAR downregulation, 461
Pancreatic carcinoma, 609
Pancreatic islet cell carcinogenesis
    cathepsin deficiency, 289–291
    RIP1-Tag2 model of, 286
Pathological vascular smooth muscle cell
        migration, 356
p53 downregulation, uPAR, 236
Peptidases, 588, 590, 592. See also
        Cysteine proteases
Peptide antagonists, uPA-uPAR interaction,
        712
Peptide inhibitors
    cyclic hexapeptide, 740
    upain-1, 740–741
Peripheral blood mononuclear cells
        (PBMNCs), 442
Peritumoral fibroblast-like cells, 366
Perivascular cells
    aortic ring assay, 311–312
    vessel stabilization, 307
Peroxisome proliferator-activated receptor
        (PPAR), 362
Phage-displayed peptide library, 727
Phosphatidylinositol-specific phospholipase
        D (PIPL-D), 225
Phosphorylated phosphoinositide 3-kinase
        (PI3K), 735
Photon counting histogram (PCH), uPAR
        dynamics

complementary to FCS, 480
fluorescence intensity traces, 482–483
time structure, 481
Photophores, 110–111
Placental-like growth factor (PlGF), 307
Plasminogen (Plg) activation
activation pathways
tPA role, 185–186
uPA and tPA deficients in,
186–187
uPA role, 186
and cancer dissemination
breast tumour, 723
role of PAI-1, 724
role of uPA system, 722–723
deficiency
congenital effects, 190
epithelial lesions and multiorgan
pathology, 189
fibrinogen expression, 191
ligneous lesions etiology, 187–188
phenotypic abnormalities, 188–189
extracellular protease in, 203–204
inhibitors
α_2-antiplasmin deficiency, 191–192
plasminogen activator inhibitor
(PAI-1), 192–194
matrix metalloproteinase (MMP)
membrane-bound enzymes, 206
soluble enzymes, 205
MMTV-PymT model, 214
prognostic marker
breast cancer, 572–575
colorectal cancer, 575–576
components of, 570
lung cancer, 577–578
ovarian cancer, 578–579
prostate cancer, 579–580
quantification of, 570–572
synthesis and inhibition of, 184
tissue remodeling
embryo implantation, 210–211
postlactational mammary gland
involution, 211–213
skin wound healing, 206–210
uPA/uPAR system, 467
Plasminogen activator inhibitor-1 (PAI-1),
192–194

antagonists
anti-proteolytic activity,
742–744
vitronectin binding, 744
biochemical properties
endocytosis receptors, 731–732
target proteases, 730–731
vitronectin, 730
hyperfibrinolytic state, 193–194
physiological effects of, 192–193
plasminogen activation and cancer
dissemination, 724
prognostic marker
breast cancer, 572–575
colon adenocarcinomas, 575
impact on survival endpoints,
580–581
ovarian cancer, 578–579
prostate cancer, 579
pulmonary adenocarcinoma,
577–578
quantification of, 570–571
Plasminogen activator (PA)–plasmin
system, 332–333
Platelet-derived growth factor receptor
(PDGFR), uPUR, 437
Platelet (P-) selectin, 345
Plg activator inhibitor, 333–334
Podosomes, 403–404. *See also*
Invadopodia
Polyoma virus middle-T-oncogene
(PyMT), 286
Porcine aortic endothelial cells (PAEC),
310
Posttranslational modification, uPAR
GPI-anchored membrane attachment,
701–702
Ly-6/uPAR/a-neurotoxin (LU) protein
domains in, 703–704
N-linked glycosylation, 702
Primary/congenital enteropeptidase
deficiency, 260–262
Prinomastat
clinical trial process, 770–771
inhibitor, 509
Proangiogenic growth factors, 293
Pro576Arg polymorphism, MMP-9 gene,
672

Prodrugs and proteolysis
   cathepsin B-activation, 829–830
   chemical conjugation, 830
   hydrolysis, 828–829
   oncolytic virotherapy, 830
   toxicity, 829
Prognostic markers, cancer. *See*
      Plasminogen activation system,
      prognostic marker
pro-MMP-2–TIMP-2 complex, 797
pro-MMP-2–TIMP-2–MT1-MMP complex,
   800–801
Prostate cancer
   α_6 integrins cleavage, 459
   epitheliasin/TMPRSS2
      autocatalytic cleavage, 264
      proteolysis, 265
      TMPRSS2-ETS gene fusion, 264–265
   hepsin expression in, 264
   plasminogen activation system, 579–580
Prostate carcinoma, cancer biomarkers
   cathepsins, 599–600
   cystatins, 600
Prostate-specific antigen (PSA)
   activated prodrugs, 859, 861
   chemotherapeutic agents, 859–860
   as diagnostic marker, 858
Protease-activated drug delivery
   activation sequences and
      hydrogels, 841–842
      MMP activated system, 839–840
      NIRF imaging, 841
      substrate conformation, 844
      topological specificity, 844
      triple-helical peptide and MMPs,
      842–843
      ZAM approach, 845
   gene delivery systems
      chimeric envelope, 835
      GALV envelope, 835–836
   hydrogel drug delivery
      PEG-based hydrogels, 835
      photopolymerization, 834
   imaging agents
      fluorogenic proteolytic beacon, 838
      MMP-7-activated beacon, 839
      NIRF optical imaging, 837
      QD imaging, 838

   nano-DDSs
      liposomal delivery, 830–834
      MMP-activated delivery, 834
   prodrugs
      cathepsin B-activation, 829–830
      chemical conjugation, 830
      hydrolysis, 828–829
      oncolytic virotherapy, 830
      toxicity, 829
Protease activity imaging
   cysteine cathepsin activity evaluation,
      125–127
   optical imaging techniques in, 125
   proteolytic beacons (PBs) role in,
      127–128
Proteases. *See* Peptidases
   activity-based probes (ABP) for
      covalent modification, 103–104
      cysteine proteases for, 109
      enzyme inhibitor discovery, 122–124
      functional elements of, 103
      linker region and tags, 104
      metalloproteases role in, 106–107
      profiling, 104–106
      protease activity imaging, 125–128
      serine proteases for, 109–110
      tumor biomarkers discovery, 117–121
      zinc metalloprotease activity in,
      110–115
   activity of, 102
   in cancer degradome research
      genomic and functional analysis, 13
      signalling molecules, 12
   features of, 3–4
   fluorescence/Forster resonance energy
      transfer (FRET)
      fluorescence homotransfer, 108
      green fluorescent protein (GFP) role in,
      108–109
      methodology of, 107–108
   hematopoietic stem cell mobilization
      CD26/dipeptidylpeptidase IV (DPPIV),
      441–442
      MMP-9 and serine proteases in, 441
      serpin-1 inhibitor in, 442
   human degradome
      classification, 5
      complexity and tools, 7

genomic view of, 6
importance of, 4–5
vs. rodent degradomes complexity,
7–9
inhibitor genes, 9–10
interaction of PAI-1, 730–731
in physiological processes, 11
protease-associated diseases, 4
targets and anti-targets, 19
Protease substrates
characteristics, 79
gel-based proteomic methods, 70–71
intrinsic specificity
active-site cleft and motifs, 68–69
methods for, 69
solution-based proteomic methods, in
mass spectroscopy
amine-reactive reagents, 72–74
ICAT labeling, 72
labeled peptides, positive selection
of, 77
label-free analysis, 77–78
negative selection of tryptic peptides,
74–76
Protease web, 20
Proteinase inhibitors, 348–349
Protein identification
isotope-coded affinity tags (ICATs)
proteomics, 86–87
MS spectroscopy, 84
shotgun proteomics, 85–86
two-dimensional polyacrylamide gel
electrophoresis (2D-PAGE), 84–85
Protein inhibitors, 736
Proteoglycans, 355
Proteolytic systems, 307
Proteomic profiling, 118
PTHrP protein, 553–554
Pulmonary adenocarcinomas. *See* Lung
cancer, plasminogen activation
system
Pyrimidine-2,4,6-trione-type
inhibitors, 509

qRT-PCR assay, 27–28
Quantitative real-time polymerase chain
reaction (qRT-PCR). *See also*
Taqman qRT-PCR

probe-based detection systems
Molecular Beacons and Scorpion
probes, 52
Taqman probes, 51–52
quantification strategies, 50
and SYBR dye, 50–51
Quantum dots (QDs) imaging, 838
Quenched-fluorescent collagen,
348–349

RANKL protein, osteoclastogenesis
expression of protein, 555
solubilization by MMPs, 554
Reactive centre loop (RCL), 730
Recombinant viral vectors
adenovirus vectors, 881–882
baculoviral vectors, 883–884
Herpes simplex virus vectors,
884–885
Red blood cell colony, 378
Retrovirus vectors
advantages, 881
RNA genome, 878
transduction mechanism, 880
RNA interference (RNAi), cancer therapy,
233
RNAlater®, 54

Secondary/acquired enteropeptidase
deficiency, 262
Secretome
breast cancer metastases, 18–19
extracellular and intracellular
pathways, 18
Serine hydrolase superfamily, 120
Serine protease-catalyzed peptide bond
hydrolysis, 725
Serine proteases regulation, cancer
development
mast cell serine proteases, 165–167
neutrophil elastase (NE), 167–169
plasminogen activators, 164–165
Serine protease targeted tumour-selective
prodrugs
prostate-specific antigen (PSA)
activated prodrugs, 859, 861
chemotherapeutic agents, 859–860
as diagnostic markers, 858

urokinase-type plasminogen activator
    receptor (uPAR), 861
    anthrax protoxin, 862–863
    diphtheria toxin, 863
    plasmin-activated anticancer
        chemotherapeutics, 864
Serine proteinase inhibitor B (SERPINB)
    family, 10
Serine proteinase matriptas, 354
Serine/threonine kinase Akt, 735
Shewanella oneidensis, 78
Short tandem repeat polymorphism. *See*
    Microsatellite polymorphism
Shotgun proteomics, 85–86
SILAC. *See* Stable isotope labeling by
    amino acids in cell culture
Single nucleotide polymorphism
    MMP-1 gene, 664, 666–668
    MMP-2 gene
        C / T polymorphism, 668–670
        G / A polymorphism, 669
    MMP-3 gene, 670–672
    MMP-9 gene, 672–673
    MMP-12 gene, 672–673
Skin and cervical cancer
    cathepsin L deficiency, 291–292
    K14-HPV16 model of, 288
Skin squamous cell carcinoma (SCC)
    MMP expression, 335
    proangiogenic effect of inhibitor,
        333
    surface transplantation model
        keratinocytes, type-I collagen gel,
            329–330
        stages of tumor invasion, 330
        stromal tissue activation, 331
        tumor-stroma interaction analysis,
            329, 336
Small-cell lung cancer (SCLC),
    plasminogen activation system,
    577–578
Small organochemical inhibitors
    PAI-1, 742–743
    uPA
        amiloride, 738
        serine protease inhibitors, 736
    WX-UK1, 739

Snake venom α neurotoxins. *See*
    Ly-6/uPAR/a-neurotoxin (LU)
    protein domains
SNB19 glioblastoma cells, antisense
    uPAR, 236
Solid tumor prognostic markers. *See*
    Plasminogen (Plg) activation
Soluble urokinase plasminogen activator
    receptor (suPAR)
    breast cancer, 574
    CRC patients, 575–576
    generation of, 476
    infectious diseases, 581
    lung cancer, 578
    ovarian cancer, 578–579
    prostate cancer, 579–580
Solution-based quantitative proteomics,
    protein identification
    isotope-coded affinity tags (ICATs)
        proteomics, 86–87
    MS spectroscopy, 84
    shotgun proteomics, 85–86
    two-dimensional polyacrylamide gel
        electrophoresis (2D-PAGE),
        84–85
Sorsby's fundus dystrophy (SFD), 380
Spatial distribution regulation, uPAR, 458
Spontaneous metastasis model, 498–500
Squamous cell carcinoma of the head and
    neck (SCCHN)
    cathepsins, 605–606
    cystatins, 606–607
sRANKL protein (soluble form)
    metalloproteinases generation, 555
    potency of, 554
Stable isotope labeling by amino acids in
    cell culture (SILAC), 76–77
Stefins
    atypical meningioma, 613
    benign prostatic hyperplasia, 600
    breast cancer, 598–599
    lung carcinoma, 616
    NSCLC tumours, 603
    SCCHN, 606
    type I cystatins, 592, 595
Stromal–cellular interface, TIMPs, 373,
    377

Stromal compartment, malignant tumors, 331
Stromal-derived factor 1 (SDF1), HSC mobilization
  dipeptidylpeptidase IV clevage, 441–442
  HSC trafficking, 440
Substrate-based imaging agents, FRET
  fluorescence homotransfer, 108
  green fluorescent protein (GFP) role in, 108–109
  methodology of, 107–108
Surface transplantation model
  proangiogenic effect of PAI-1, 333
  skin squamous cell carcinoma
    epithelial and stromal cells interaction, 329
    keratinocytes, type-I collagen gel, 329–330
    stages of tumor invasion, 330
    stromal tissue activation, 331
    tumor progression, stromal MMP contribution, 336–338
SYBR®, 50
SYPRO®, 70

TACE. *See* Tumour necrosis factor-alpha-converting enzyme
TAg-induced hepatocellular carcinoma, 389–390
Tanomastat inhibitor, clinical trial process, 771–772
TAPs. *See* Tumour endoprotease-activated prodrugs
Taqman®, 50
TaqMan low-density array (TLDA), 61
TaqMan®, 632–633
Taqman qRT-PCR
  in human cancer cell lines and gliomas, 61–62
  primer design, 52–53
  probes, 51–52
  reverse transcription
    oligo(dT) primers and GSP, 58
    reverse transcriptase enzymes used in, 59
  RNA quality and integrity
    alkali degradation, 57

    degradation of RNA, 55–56
    isolation procedure, 54–55
    and PCR efficiency, 55
  standard curve *vs.* D/D Ct post-run analysis, 60
  validation and normalization
    GeNorm analyzes, 60
    using 18S rRNA assessment, 59–60
Terminal amine isotope label method, 97–98
Time correlated single photon counting (TCSPC) mode, 485–486
Time-lapse videomicroscopy, 348, 356
TIMP-1
  cell proliferation, 379–380
  erythroid potentiating activity (EPA), 378
  MMP-inhibitory function, 380
  mutants, 805
TIMP-3
  ADAM-17
    kinetic values, 802–803
    N-TIMP–3 mutation, 803
    zinc ion catalytic domain, 804
  physiological role of, 381
  Sorsby's fundus dystrophy (SFD), 380
TIMP-2–MT1-MMP complex
  kinetic study, 797
  structure analysis
    MT loop formation, 798
    residue comparison, 799–800
TIMP-paradox, 502
Tissue factor pathway inhibitor 2 (TFPI2), 265
Tissue homeostasis
  cardiac, 381
  regulatation, 373, 376
Tissue inhibitor of metalloproteinases-1 (TIMP-1), 38
  in cancer
    anti-apoptotic effect, 685
    cell growth regulation, 684–685
    pro and anti-angiogenic effects, 685–686
  CRC prognosis
    case-control study in, 687

plasma TIMP-1 and cut-off values,
  687–688
prognostic value analysis, 690
serum and plasma TIMP-1 levels in,
  686–687
standard operating procedures (SOPs),
  691–692
statistical analysis procedure (SAP),
  692
TIMP-1 mRNA level in, 686
Tissue inhibitors of metalloproteinase-3
  (TIMP-3)
adenoviral delivery of, 889
extracellular matrix (ECM), 888
Tissue inhibitors of metalloproteinases
  (TIMPs), 61
ADAMSs
  ectodomain shedding, 801–802
  inhibitory activities, 788, 790
ADAMTSs
  ectodomain shedding, 801–802
  inhibitory activity, 788, 790
anti-invasive activity, 502
antimetastatic activity, 502–503
biochemical properties
  disulphide bonds and β sheets, 788
  kinetic studies on MMP, 792
in cancer
  angiogenesis, 385–387
  apoptosis, 383–384, 386
  cell contact and motility, 387–388
  cell proliferation, 382–383, 385
  in vitro studies, 382
  in vivo models, 388–390
collagenase inhibitor, 374
EC tubulogenesis, 311
enzymatic activity regulation, 307
evolution and structure of, 374–376
inhibition mechanism and MMPs
  interaction complex, 793–794
  kinetic values, 795–796
  reactive sites, 794
inhibitory activities
  MMPs, 788–789
liver metastasis, 504–505
as markers, 635–638
N-TIMP-1 and prometastatic effect, 505
N-TIMP-3 mutation, 803

proMMP-9 binding, 501
proproliferative activity, 510–511
protumorigenic feature, 504
targets and phenotypes of
  murine phenotypes, 377
  process of liver regeneration, 378
  TIMP1, 378–380
  TIMP2, 380
  TIMP3, 380–381
  TIMP4, 381–382
  in vivo analysis, 382
TIMP-2
  MT1-MMP complex, 797–800
  mutants, 805
  Pro-MMP-2 activation and complex,
    797
TIMP-3 and ADAM-17
  kinetic values, 802–803
  N-TIMP-3 mutation, 803
  zinc ion catalytic domain, 804
TIMP-1 mutants, 805
TIMP-4 mutants, 806
trimeric complex formation, 800–801
types of, 788
zebrafish genes, 316
Tissue remodeling, Plg activation
embryo implantation
  decidualization, 210
  galardin treatment, 211
  protease inhibitors, 210–211
postlactational mammary gland
  involution
  galardin inhibition and fibrinolysis,
    212
  involution phases and Plg-deficiency,
    211–212
  MMP-3 overexpression, 213
  proteolytic components in, 211
skin wound healing
  MMP inhibitor galardin, 209
  MMPs deficiencies, 209–210
  Plg deficient and healing time,
    208–209
  protease expression, 208
  uPA mRNA expression, 206–208
Tissue-type plasminogen activator (tPA),
  724
tPA, plasminogen activation, 185–186

Transarterial chemoembolization (TACE),
        381
Transcriptome, 22
    Hu/Mu ProtIn chip and CLIP-CHIP,
        26–27
    microarrays and
        development of, 23
        limitations of, 22
        probe fixation, 23–24
    qRT-PCR assay, 27–28
    two-color CLIP-CHIP, 25–26
Transcytosis, 544
Transforming growth factor-β1 (TGF-β1)
    MMPs and inflammation, 526–528
    MMPs and tumor–bone interaction, 556
Transitional cell carcinoma (TCC),
        608–609
Transmembrane domain, 346
Transmembrane glycoproteins, 344
Transplantation chamber assay, 331
2,4,6-Trinitrobenzenesulfonic acid (TNBS),
        75–76
Triple-helical peptide (THP), 842
Trousseau's syndrome, 724
Trypsin-like serine proteases, 8–9. *See also*
        Plasminogen
Tryptic peptides, negative selection of
    biotinylation of, 75
    COFRADIC and TNBS methods, 75–76
Tube formation assay
    advantages of vasculogenic assays, 311
    endothelial cells (EC) morphogenesis,
        310
Tumor angiogenesis, 230–231
Tumor-associated macrophages (TAMs),
        298
Tumor biomarkers discovery
    metalloproteases and histone deacetylases
        (HDACs), 120–121
    proteomic profiling in, 117–119
    serine hydrolase superfamily, 120
    ubiquitin-specific proteases (USPs), 121
Tumor–bone microenvironment. *See*
        Bone–tumor microenvironment
Tumor cell
    extravasation, 345–346
    host stroma interaction, 390
    invasion

    membrane-type MMP contribution,
        335
    MMP deficiency, 337
    quantification methods, 332
    stages of, 330
    stromal activation and, 331
    TIMP, 387
    migration
        collagen cell invasion, 349–350
        collective-cell migration, 349
        quenched-fluorescent collagen,
            348–349
        role of proteinases, 347–348
        short-term and long-term invasion,
            352
        visualisation of cell migration, 348
    proliferation and apoptosis, 295–296
Tumor endoprotease-activated prodrugs
        (TAPs), 856
Tumorigenesis, 366
Tumorigenic cell proliferation, 386
Tumor invasion, cysteine cathepsins, 345
    E-cadherin, 294–295
    ECM degradation and MMP activation,
        295
Tumor Marker Utility Grading System
        (TMUGS), 681
Tumor necrosis factor-α (TNF-α), 524–526
TNF-α-converting enzyme (TACE),
        815–816
Tumor progression
    genetic tumor cells alteration, 327
    inflammation, 345
    role of proteases, 329
    stromal MMP contribution, 336–338
    tumor–stroma interface, 328
Tumor proteases, imaging activity
    data processing and quantification
        methods, 149–151
    live cells and
        cationic cell penetrating peptide (CPP),
            144
        DQsubstrate degradation, 141–142
        interactions in, 144–145
        uPARAP and endocytosis, 142–143
    MMPIs and Hu/Mu ProtIn chip, 138
    modulation of
        cocktails, 148–149

inhibitors in, 147–148
protease-binding partners and β1
    integrin, 148
tumor cell motility and PAK1, 149
synthetic substrates
    FITC-labeled protein substrtes,
        139–141
    histochemical analyses, 139
tumor microenvironment
    DQ-collagen IV proteolysis, 145–146
    spheroids cohesiveness, 146
Tumor-selective anticancer drug delivery
    characteristics of tumour enzymes,
        856–857
    tumour-selective prodrugs, 855–856
Tumor–stroma interactions
    analysis of, 329, 331
    importance of, 328
Tumor therapy, 328, 335
Tumor vascularization, 337
Two-dimensional polyacrylamide gel
        electrophoresis (2D-PAGE),
        70–71
    limitations of, 85
    matrix metalloproteinase (MMP)-14
        substrate discovery, 84–85
Type-I collagen, 356
Type-I collagen gels, 348–349
Type-I IGF receptor, 389
Type II transmembrane serine proteases
        (TTSPs)
    corin, 270
    HAT/DESC subfamily, 266
    hepsin/TMPRSS
        enteropeptidase, 260, 262
        epitheliasin/TMPRSS2, 264–265
        hepsin/TMPRSS1, 262–263
        mosaic serine protease (MSP), 264
        spinesin/TMPRSS5, 265–266
        TMPRSS3 and TMPRSS4, 265
    matriptase
        in cancer, 267
        epidermal differentiation and point
            mutation, 268–269
        epithelial morphogenesis, 268
        matriptase-2 and matriptase-3, 269
        polyserase-1, 269–270
Type-IV collagen, 354–355

Type-two transmembrane serine proteases
        (TTSPs), 728
Tyrosine kinase receptors, uPAR, 437

Ubiquitin-specific proteases (USPs), 121
uPA-induced cell migration, 466
Upain-1, uPA antagonists, 740–741
uPA–PAI-1 complexes, prognostic values
    breast cancer, 574–575
    ELISA, 571
    pulmonary adenocarcinoma, 577–578
uPA, plasminogen activation, 186
uPARAP/Endo182. See Endocytic collagen
        receptor
uPAR–integrin interactions. See also
        Urokinase-type plasminogen
        activator receptor (uPAR)
    anti-cancer strategy, 467–468
    α and β integrins in, 462
    in cancer
        anti-cancer strategy, 467–468
        α6 integrin cleavage, 459
        uPAR antibodies downregulation, 461
        uPAR soluble marker, 454
    crossregulation of, 460
    kringle domain and uPA in, 459
    models of, 463–464
    uPAR domain II and domain III in,
        461–462
uPAR interactome
    physical interactions, 477–478
    Vn interactions, 477
uPAR–vitronectin complex, 709–710
uPA-system interaction, uPARAP/Endo182,
        252
uPA–uPAR complexes
    amine-reactive homobifunctional
        cross-linkers, 709
    ATF–uPAR–Fab complex structure,
        707–708
    cellular responses, 227
    components, 224
    prognostic values, 570, 577
uPA–uPAR interactions
    colon and breast carcinomas, 710–711
    growth factor-like domain (GFD) cyclic
        peptide, 712–713
    linear peptide antagonists, 713–714

monoclonal antibodies (mAbs), 711–712
peptide antagonists, 712
recombinant fusion proteins, 712
Urokinase-type plasminogen activator
        (uPA)
    antagonists of
        enzyme activity, 735–741
        growth factor and kringle domain, 741
        linker region, 741–742
    biochemical properties
        catalytic properties, 725, 727–728
        connecting peptide, 729–730
        domain structure, 725–726
        GFD and uPAR, 728–729
        kringle domain, 729
    breast cancer
        multivariate analysis, 573–574
        relapse-free (RFS) and overall survival
            (OS), 572–574
        uPA-PAI-1 complex, 574–575
    colorectal cancer, 575–576
    impact on survival endpoints, 580
    level of evidence (LOE), 570, 572, 581
    ovarian cancer, 578–579
    plasminogen activation and cancer
        dissemination, 722–723
    prognostic marker
    prognostic studies on, 580–581
    prostate cancer, 579–580
    pulmonary adenocarcinomas, 577–578
    quantification of, 570–571
Urokinase-type plasminogen activator
        receptor (uPAR), 422–423
    amino acid polypeptide and, 225
    apoptosis and, 236
    breast tumour invasion, 745
    cancer and, 230–231
    cell adhesion, 733
    cell migration, 228–229, 733–734
    cell proliferation, 229–230
    cleavage and shedding, 225–226
    colocalization of, 730
    dimers detection
        FRET and donor-FLIM approach,
            484–485
        HEK295/uPAR-EGFP-GPI cell,
            485–488
    domains and funtions, 452

domains of, 475
downregulation, cancer therapy
    antisense oligodeoxynucleotide
        (asODN) therapy, 231–232
    antisense RNA technology in, 232–233
    cellular function and signal
        transduction, 233–234
    and gene modulation, 235–236
    shRNA-based RNAi plasmid, 233
    uPAR/integrin interaction, 234–235
    uPAR/Vn interaction, 234
FCS and PCH
    diffusion of identical proteins, 482
    as fluorescence fluctuation
        spectroscopy (FFS), 480
    irreversible photobleaching, 483
    time structure, 481
    translational mobility of molecules,
        482
fluorescent proteins (FPs) chimers
    HEK295 cell line, 478–479
    wt-uPAR functions, 478
FRAP, 479–480
function of, 729
and GFD, 728–729, 741
G protein-coupled receptors, 436
human HSCs mobilization
    protease cleavage and desensitization,
        443
    uPAR upregulation, 442
integrins
    bidirectional signaling, 457
    intracellular domain, 456
    kringle domain and integrin cleavage,
        459
    spatial distribution regulation, 458
interactome
    physical interactions, 477–478
    Vn interaction, 477
internalization
    CIMPR and LRP-dependent
        internalization, 456
    integrin regulation, 460
    PAI-1-uPA complex formation,
        454–455
membrane-bound ligands, 453–454
mouse HSCs mobilization
    5-fluorouracil myeloablation, 444

plasmin in, 444–445
multidomain assembly of
  crystal structures, 704–705
  hydrophobic ligand-binding cavity, 706
  linker peptide, 706–707
nonproteolytic functions of, 700–701
pericellular proteolysis, 454
plasminogen activation, 700
posttranslational modification
  GPI-anchored membrane attachment,
    701–702
  Ly-6/uPAR/a-neurotoxin (LU)
    protein domains in, 703–704
  N-linked glycosylation, 702
prognostic marker
  breast cancer, 573–574
  CRC patients, 575
  lung cancer, 577
  NSCLC patients, 577–578
  ovarian cancer, 578–579
  prognostic studies on, 580–581
  properties, 571
  prostate cancer, 579–580
  quantification of, 572
pro-proliferative effect, 734
serine protease targeted
    tumour-selective prodrugs, 861
  anthrax protoxin, 862–863
  diphtheria toxin, 863
  plasmin-activated anticancer
    chemotherapeutics, 864
signaling pathway
  in cancer cells, 437–438
  tyrosine kinase receptor, 437
signal transduction and interaction
  extracellular-signal-regulated kinase
    (ERK), 227–228
  integrins, 227
  Vn glycoprotein, 226–227
soluble and purified forms of, 476
soluble ligands and signaling pathway
  of, 453
structural aspects of
  uPAR–vitronectin complex,
    709–710
  uPA–uPAR complex, 707–709
structure, 434
uPA and VN ligands, 434–435

uPA–PAI-1-endocytosis receptor, 731
uPAR–integrins coimmunoprecipitation,
  435–436
uPA system, 476
uPA–uPAR interaction, 741–742
Vascular endothelial cells, 328
Vascular endothelial growth factor (VEGF),
  307, 368, 383
Vascular inflammation, 381
Vascularized granulation tissue, 331
Vascular lumen formation, 307
Vascular smooth muscle cells (VSMC),
  405
Vasculogenesis
  angiogenesis in tumors, 305
  mechanisms of action,
    307–309
  sprouting process, 311
  tube formation assays, 311
Vasculogenic assays, 311
Vessel normalization, 336
Vessel regression, 336
Vessel stabilization, 307
Viral gene delivery vectors
  adeno-associated virus (AAV) vectors
    advantages, 883
    definition, 882
  adenovirus vectors, 881–882
    advantages, 881
    innate and adaptive immune responses,
      882
  baculovirus vectors
    advantages, 884
    definition, 883
  in degradome-targeted cancer therapy
    degradome gene silencing,
      887–888
    inhibition of proteinase activity,
      888–890
    modulating gene regulators,
      886–887
  Herpes simplex virus vectors,
    884–885
    amplicon vectors, 884
    cytotoxic properties of, 885
  retrovirus vectors
    advantages, 881
    RNA genome, 878

transduction mechanism, 880
Vitronectin, 730
von Willebrand A (VWA) domain, 346
v-Src transformed fibroblasts
    immunoelectron microscopy, 408
    podosomes of monocytic cells, 418
Wingless and integrated (Wnt) signals,
    546–547
Wiskott-Aldrich syndrome protein
    (WASP), 733
wt-uPAR (wild-type urokinase-type
    plasminogen activator receptor),
    478–479
WX-UK1, uPA antagonists, 739

Xenografts, Hu/Mu ProtIn array, 41–43.
    See also Hu/Mu ProtIn Chip
Xenopus laevis, 354

ZAM. *See* Zipping & assembly method
Zebrafish

embryonic and organogenic
    angiogenesis, 315
metzincin genes, 315–316
MMP–2, 316
Zinc metalloprotease activity
    active forms of,
        113–115
    activity-based probes (ABP) for
        covalent modification and selectivity,
        115
    cross-linking, 111–112
    hydroxamate groups, 112–113
    photophores, 110–111
    substrate-based imaging agents for,
        115–117
Zinc metalloproteinase family,
    812–817.
    See also Matrix metalloproteinases
        inhibitors (MMPIs)
Zipping & Assembly Method
    (ZAM), 845